REFRIGERATION AND AIR CONDITIONING TECHNOLOGY

REFRIGERATION AND AIR CONDITIONING TECHNOLOGY

Concepts, Procedures, and Troubleshooting Techniques

WILLIAM C. WHITMAN
WILLIAM M. JOHNSON

Central Piedmont Community College
Charlotte, North Carolina

 delmar publishers inc. ®

COVER PHOTOS:

Upper left-hand photo: *Courtesy of Simpson Electric Company, Elgin, Illinois*

Upper right-hand photo: *Courtesy of Bohn Heat Transfer, Division of Wickes Manufacturing Company, Danville, Illinois*

Lower right-hand photo: *Courtesy of Honeywell Inc., Golden Valley, Minnesota*

Lower left-hand photo: *Courtesy of Lennox Industries Inc., Dallas, Texas*

ILLUSTRATOR: Bob Rome

Delmar Staff

Executive Editor: Mark W. Huth
Associate Editor: Jonathan Plant
Managing Editor: Barbara Christie
Production Editor: Eleanor Isenhart
Design Coordinator: Susan Mathews
Production Coordinator: Karen Seebald

For information, address Delmar Publishers Inc.,
2 Computer Drive West, Box 15-015,
Albany, New York 12212

Printed in the United States of America
Published simultaneously in Canada
by Nelson Canada
A Division of International Thomson Limited

10 9 8 7 6 5 4 3 2 1

Library of Congress Cataloging-in-Publication Data
Whitman, William C.
    Refrigeration and air conditioning technology.
    Includes index.
    1. Refrigeration and refrigerating machinery.
2. Air conditioning. I. Johnson, William M. II. Title.
TP492.W6 1987   621.5′6   86-16506
ISBN 0-8273-2416-2
ISBN 0-8273-2417-0 (instructor's guide)
ISBN 0-8273-2418-9 (student guide)

BRIEF CONTENTS

CONTENTS

Refrigeration and Air Conditioning Technology was written because we felt that a text was needed to better suit the students and instructional staff in the Air Conditioning and Refrigeration Technology program at our institution.

The text is flexible enough to meet the needs of most readers. After completing the first three sections, you may concentrate on courses in refrigeration or air conditioning (heating and/or cooling). If your objective is to complete a whole curriculum, you may proceed until you have finished the sequence scheduled by your school's curriculum.

We have tried to make the text easy to read and to present the material in a practical way, using everyday language and occasionally using terms more commonly used by mechanics and technicians. Our approach to electrical application, for instance, differs from standard treatments—we classify components as "power passing" or "power consuming."

Objectives are listed at the beginning of each unit. A summary and review questions are provided at the end of each unit. Students should answer the questions while reviewing what they have read; instructors can use the questions to stimulate class discussion and for classroom unit review. The questions are not necessarily designed for testing.

Practical troubleshooting procedures are a main feature of this text. There are practical component and system troubleshooting suggestions and techniques. In many units practical examples of service technician calls are presented in a down-to-earth situational format.

One unit is dedicated to general safety precautions. This includes general safety procedures and describes tools and equipment that can be used to prevent injury. Specific safety techniques and tips are highlighted throughout the text.

A significant feature of this text is the section on heating. We tried to include heating systems found throughout the country. Obviously some of these heating systems are more appropriate for some geographical areas than others. These heating systems are presented separately by unit so that a unit can easily be skipped if the instructional program does not include it. At the suggestion of several people, we have included a unit on wood heat.

Another section is devoted to all-weather systems and heat pumps. The number of heat pump installations in temperate climates have risen dramatically. Some increase has also occurred in the colder areas.

This text covers some older systems of heating and cooling as well as newer designs since there are thousands of older systems still existing that need servicing.

Refrigeration and Air Conditioning Technology can be used by students as a general reference after completing their program of study. Salespeople, suppliers, contractors, installers, and service technicians can also benefit from the use of this text.

We would like to thank the following reviewers for their help in developing this text. Bill Abernathy, formerly of Orange Coast College; Herschael Blitz, Wayne County Community College; Harvey Castelaz, Milwaukee Area Technical College; Walter Hilmes, Indiana Vocational Technical College; and Henry Puzio, Lincoln Technical Institute.

Section One
Theory of Heat

THEORY

1.1 TEMPERATURE

The word "temperature" is used in everyday discussions or descriptions about comfort, weather, and food preparation. It is used in many conversations and decision-making processes by people who still do not know exactly how far-reaching the word is or what it really means.

Temperature can be thought of as a description of the level of heat. For now, heat can be thought of as energy in the form of molecules in motion. The starting point of temperature is, therefore, the starting point of molecular motion. To describe this in more usable terms, we will describe some more familiar points of reference.

Most people know that the freezing point of water is 32 degrees Fahrenheit (32°F) and that the boiling point is 212 degrees Fahrenheit (212°F). These points are commonly indicated on a thermometer.

Early thermometers were of glass-stem types operating on the theory that when the substance in the bulb was heated it would expand and rise up in the tube, Figure 1–1. Mercury and alcohol are still commonly used today for this application. More information

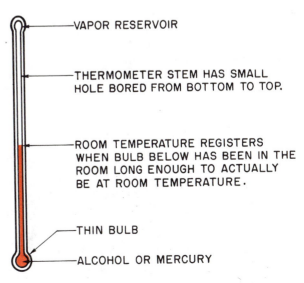

VAPOR RESERVOIR

THERMOMETER STEM HAS SMALL HOLE BORED FROM BOTTOM TO TOP.

ROOM TEMPERATURE REGISTERS WHEN BULB BELOW HAS BEEN IN THE ROOM LONG ENOUGH TO ACTUALLY BE AT ROOM TEMPERATURE.

THIN BULB

ALCOHOL OR MERCURY

Figure 1–1. Thermometer.

about temperature measurement is found in the section on automatic controls.

We must qualify the statement that water boils at 212°F. Pure water boils at precisely 212°F at sea level when the atmosphere is 68°F. This qualification concerns the relationship of the earth's atmosphere to the boiling point and will be covered in detail later in this section in the discussion on pressure. The statement that water boils at 212°F at sea level when the atmosphere is 68°F is important because these are standard conditions that will be applied to actual practice in later units.

Pure water has a freezing point of 32°F. Obviously the temperature can go lower than 32°F, but the question is, how much lower?

The theory is that molecular motion stops at −460°F. See Figure 1–2 for an illustration of the levels of heat or molecular motion shown on a thermometer scale. This is theoretical because molecular motion has never been totally stopped. The complete stopping of molecular motion will be recorded as absolute zero. This has been calculated to be −460°F. Scientists have actually come within a few degrees of reaching absolute zero.

Temperature can also be expressed in degrees Centigrade or, more commonly, Celsius. The weather forecaster often uses the term "Celsius." Celsius and Fahrenheit both express the level of heat, but they do it in different terms, Figure 1–2.

Temperature has been expressed in everyday terms up to this point. It is equally important in the air conditioning, heating, and refrigeration industry to describe temperature in terms engineers and scientists use. Performance ratings of equipment are established

in terms of *absolute* temperature. Equipment is rated to establish criteria for comparing equipment performance. In other words, different manufacturers make similar claims about their products. We can use the equipment rating to evaluate these claims. The Fahrenheit absolute scale is called the *Rankine* scale (named for its inventor W. J. M. Rankine), and the Celsius absolute scale is known as the *Kelvin* scale (named for the scientist, Lord Kelvin). Absolute temperature scales begin where molecular motion starts; they use 0 as the starting point. For instance, 0 on the Fahrenheit absolute scale is called absolute zero or 0° Rankine (0°R). Similarly, 0 on the Celsius absolute scale is called absolute zero or 0° Kelvin (0 K). See Figure 1–3.

The Fahrenheit, Celsius and the Rankine, Kelvin scales are used interchangeably to describe equipment and fundamentals of this industry. Memorization is not very important. To be able to work back and forth from degrees Fahrenheit to degrees Rankin, with Celsius and Kelvin surfacing from time to time, is too much to ask. A working knowledge of these scales and a ready reference table are more practical. Figure 1–3 shows how these four scales are related. The world that we live in accounts for only a small portion of the total temperature spectrum. The thermometer scale illustrated in Figure 1–4 shows some examples of how typical temperatures compare.

Our earlier statement that temperature describes the level of heat or molecular motion can now be explained. As a substance becomes warmer, the molecular motion, and therefore the temperature, increases, Figure 1–5.

1.2   INTRODUCTION TO HEAT

The laws of thermodynamics can help us to understand what heat is all about. One of these laws states that heat can neither be created nor destroyed. This means that all of the heat that the world experiences is not created but is merely converted to usable heat from something that is already here. This heat can also be accounted for when it is transferred from one substance to another.

Heat can now be more fully explored by using temperature as one of the describing factors. Remember, temperature describes the level of heat with reference to no heat. The term used to describe the quantity of heat is known as the *British thermal unit* (Btu). This term explains how much heat is contained in a substance. The rate of heat consumption can be determined by adding time to the picture, but more on this later.

The Btu is defined as the amount of heat required to raise the temperature of 1 lb of water 1°F. For example, when 1 lb of water (about 1 pint) is heated from 68° to 69°F, 1 Btu of heat energy is absorbed into the water, Figure 1–6. To actually measure how much heat is absorbed in a process like this, we need

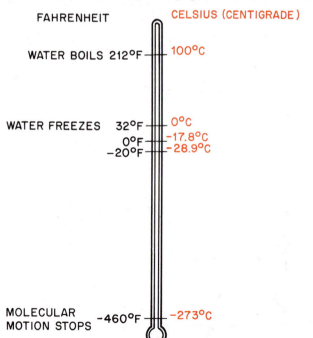

Figure 1–2. Fahrenheit scale compared to Celsius scale.

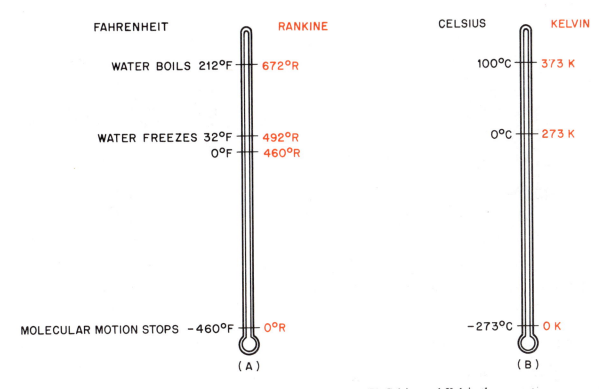

FAHRENHEIT   RANKINE   CELSIUS   KELVIN

WATER BOILS 212°F — 672°R   100°C — 373 K

WATER FREEZES 32°F — 492°R   0°C — 273 K
               0°F — 460°R

MOLECULAR MOTION STOPS −460°F — 0°R   −273°C — 0 K

(A)   (B)

Figure 1-3. (A) Fahrenheit and Rankine thermometer. (B) Celsius and Kelvin thermometer.

an instrument of laboratory quality. This instrument is called a *calorimeter*. Notice the similarity to the word "calorie," the food word for energy.

Heat flows naturally from a warmer substance to a cooler substance. Rapidly moving molecules in the warmer substance give up some of their energy to the slower-moving molecules in the cooler substance. The warmer substance cools because the molecules have slowed. The cooler substance becomes warmer because the molecules are moving somewhat faster.

TEMPERATURE OF STARS	54,000°F	30 000°C
TUNGSTEN LAMP TEMPERATURE	5,000°F	2 760°C
BONFIRE TEMPERATURE	2,500°F	1 370°C
MERCURY BOILS	674°F	357°C
AVERAGE OVEN TEMPERATURE	350°F	177°C
PURE WATER BOILS AT SEA LEVEL	212°F	100°C
BODY TEMPERATURE	98.6°F	37°C
PURE WATER FREEZES	32°F	0°C
	−460°F	−273°C

Figure 1-4. Civilization is generally exposed to a comparatively small range of temperatures.

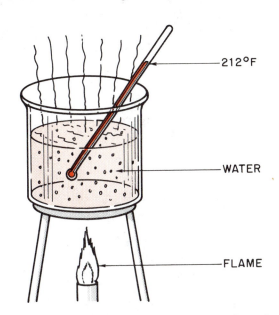

212°F

WATER

FLAME

Figure 1-5. The water in the pot boils because the molecules move faster when heat is applied.

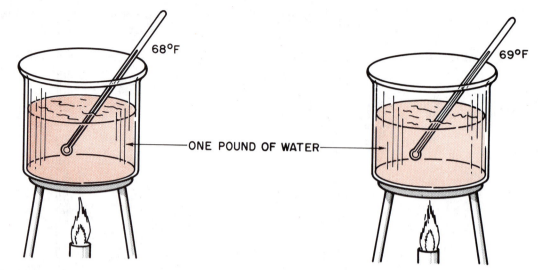

Figure 1-6. 1 British thermal unit (Btu) of heat energy is required to heat 1 lb of water from 68°F to 69°F.

The following example will illustrate the difference in the amount or quantity of heat compared to the level of heat. One tank of water weighing 10 lb (slightly more than 1 gallon) is heated to a temperature level of 200°F. A second tank of water weighing 100,000 lb (slightly more than 12,000 gallons) is heated to 175°F. It is easy to imagine that the 10-lb tank will cool to room temperature much faster than the 100,000-lb tank. The temperature difference of 25°F is not very much, but the cool-down time is much longer for the 100,000-lb tank, Figure 1–7.

A comparison using water is always helpful in showing the level verses the quantity of heat. A well with 200 ft of water would not have nearly as much water as a large lake with 25 ft of water. The depth of water (in feet) tells us the level of water, but it in no way expresses the quantity (gallons) of water.

In practical terms, each piece of heating equipment is rated according to the amount of heat it will produce. If the equipment had no such rating, it would be difficult for a buyer to intelligently choose the correct appliance.

A gas or oil furnace used to heat a home has the rating permanently printed on a nameplate. Either furnace would be rated in Btu per hour, which is a *rate* of energy consumption. Later, this rate will be used to calculate the amount of fuel required to heat a house or a structure. For now, it is sufficient to say that if one needs a 75,000-Btu/h furnace to heat a house on the coldest day, a furnace rated at 75,000 Btu/h should be chosen. If not, the house will begin to get cold on any day the temperature falls below the capacity of the furnace.

1.3 CONDUCTION

Conduction heat transfer can be explained as the energy actually traveling from one molecule to another.

ROOM TEMPERATURE (70°F)

10-POUND TANK OF WATER
IS HEATED TO 200°F
(ABOUT 1 GALLON)

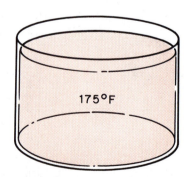

100,000-POUND TANK OF WATER
IS HEATED TO 175°F
(ABOUT 12,000 GALLONS)

Figure 1-7. The smaller tank will cool to room temperature first because there is a smaller quantity of heat.

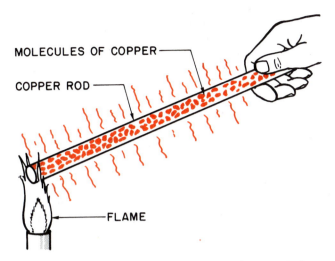

Figure 1-8. The copper rod is held in the flame only for a short time before heat is felt at the far end.

Figure 1-9. The car fender and the fence post are actually the same temperature, but the fender feels colder because the metal conducts heat away from the hand faster than the wooden fence post does.

As a molecule moves faster, it causes others to do the same. For example, if one end of a copper rod is in a fire, the other end gets too hot to handle. The heat travels up the rod from molecule to molecule, Figure 1-8.

Conduction heat transfer is used in many heat transfer applications that are experienced regularly. Heat is transferred by conduction from the hot electric eye on the cookstove to the pan of water. It is then transferred by conduction into the water. Note that there is an orderly explanation for each step.

Heat does not conduct at the same rate in all materials. Copper, for instance, conducts at a different rate than iron does. Glass is a very poor conductor of heat. A glassblower can blow a piece of glass that is red hot on one end while the other end is in the glassblower's mouth.

Touching a wooden fence post or another piece of wood on a cold morning does not give the same sensation as touching a car fender or another piece of steel does. The piece of steel feels much colder. Actually the steel is not colder; it just conducts heat out of the hand faster, Figure 1-9.

The different rates at which various materials conduct heat have an interesting similarity to the conduction of electricity. As a rule, substances that are poor conductors of heat are also poor conductors of electricity. For instance, copper is one of the best conductors of electricity and heat, and glass is one of

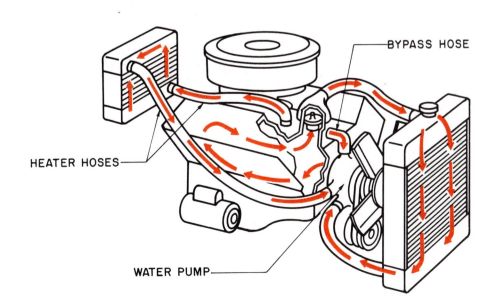

Figure 1-10. The engine generates heat that is transferred into the circulating water by conduction. The circulating water is pumped through the heater coil inside the car (forced convection). Air is passed over the coil where it absorbs heat from the coil by conduction. Warm air is then conveyed to the interior by a fan.

the poorest conductors of both. Glass is actually used as an insulator of electrical current flow.

1.4 CONVECTION

Convection heat transfer is used to move heat from one location to another. When heat is moved, it is normally transferred into some substance that is readily movable, such as air or water. Many large buildings have a central heating plant where water is heated and pumped throughout the building to the final heated space. Notice the similarity of the words "convection" and "convey" (to carry from one place to another).

The automobile heater is a good example of convection heat. Heat from the engine's combustion process is passed by conduction to the water. Hot water from the engine is then passed through a heater coil. The heat in the water is transferred by convection from the water in the engine to the heater coil. The heat is transferred through the coil from the water to the air and conveyed to the car's interior by the heater fan. When a fan or pump is used to convey the heat, the process is called *forced* convection, Figure 1–10.

Another example of heat transfer by convection is that when air is heated, it rises; this is called *natural* convection. When air is heated, it expands, and the warmer air becomes lighter than the surrounding unheated air. This principle is applied in many ways in the air conditioning industry. A familiar example is baseboard heating units, which are normally installed on the outside walls of buildings and use electricity or hot water as the heat source. When the air near the floor is heated, it expands and rises. This draws cooler air into the heater and sets up a natural convection current in the room, Figure 1–11.

1.5 RADIATION

Radiation heat transfer can best be explained by using the sun as an example of the source. The sun is approximately 93,000,000 miles from the earth's surface, yet we can feel its intensity. The sun's surface temperature is extremely hot compared to anything on earth. Heat transferred by radiation travels through space without heating the space and is absorbed by the first solid object that it encounters. The earth does not experience the total heat of the sun because heat transferred by radiation diminishes by the square of the distance travelled. In practical terms this means that every time the distance is doubled, the heat intensity decreases by one-fourth. If you hold your hand close to a light bulb, for example, you feel the heat's intensity, but if you move your hand twice the distance away, you feel only one-fourth of the heat intensity. Keep in mind that, because of the square-of-the-distance explanation, radiant heat does not transfer the actual temperature value. If it did, the earth would be as hot as the sun, Figure 1–12.

Electric heaters that glow red hot, are good practical examples of radiant heat. The electric heater coil glows red hot and radiates heat into the room. It does not

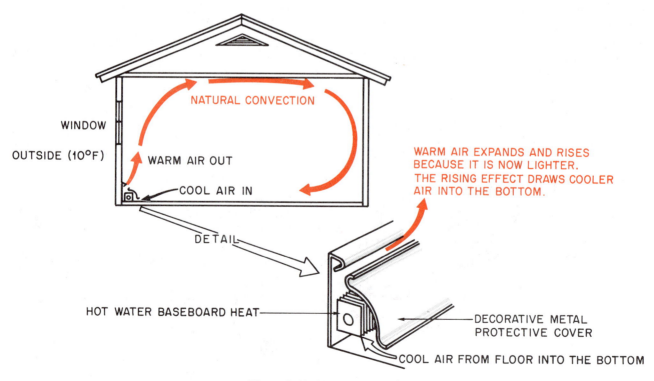

Figure 1–11. Natural convection.

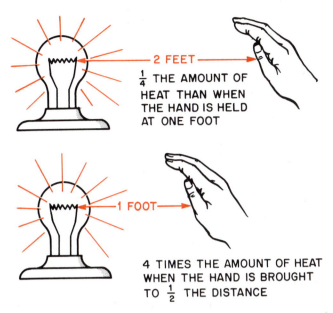

Figure 1–12. The intensity of the heat diminishes by the square of the distance.

heat the air, but it warms the solid objects that the heat rays encounter. Any heater that glows has the same effect.

1.6 SENSIBLE HEAT

The discussion of what heat is and how it is transferred from one substance to another does not complete our usable knowledge about heat. Heat level can readily be measured when it changes the temperature of a substance (remember the example of changing 1 lb of water from 68° to 69°F). This process can be measured with a thermometer and can easily be seen.

When a change of temperature can be registered, we know that the level of heat has changed and is called *sensible heat.*

1.7 LATENT HEAT

Another type of heat is called *latent* or *hidden* heat. In this process heat is known to be added but no temperature rise is noticed. A good example is heat added to water while it is boiling in an open container. Once water is brought to the boiling point, adding more heat only makes it boil faster; it does not raise the temperature.

The following example will give an overall picture of sensible heat and latent heat. The characteristics of water are going to be explored from 0°F through the temperature range to above the boiling point. Examine the chart in Figure 1–13 and notice that temperature is plotted on the left margin, and heat content is plotted along the bottom of the chart. We see that as heat is added the temperature will rise except during the hidden-heat process. This chart is interesting because heat can be added without causing a rise in temperature.

The following statements will help you to understand the chart.

1. Water is in the form of ice at point 1 where the example starts. Point 1 is *not* absolute 0: It is 0°F and is used as a point of departure.
2. Heat added from point 1 to point 2 is sensible heat. This is a registered rise in temperature. Note, it only takes 0.5 Btu of heat to raise 1 lb of ice 1°F; more on this later.
3. When point 2 is reached, the ice is thought of as being saturated with heat. This means that if more

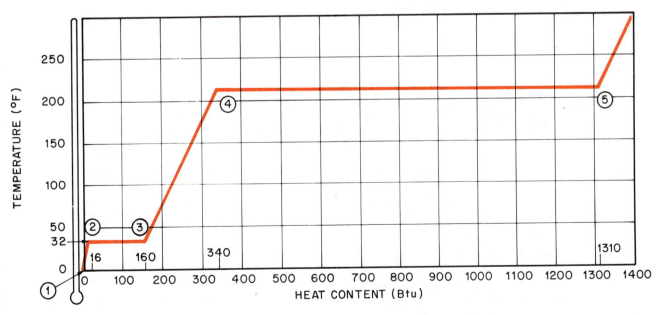

Figure 1–13. Heat/temperature graph.

heat is added, it will be known as latent heat and will serve to melt the ice but not to raise the temperature. Adding 144 Btu of heat will change the 1 lb of ice to 1 lb of water. Removing any heat will cool the ice below 32°F.

4. When point 3 is reached, the substance is now water and is known as a saturated liquid. Adding more heat causes a rise in temperature (this is sensible heat). Removal of any heat at point 3 results in some of the water changing back to ice. This is known as removing latent heat because there is no change in temperature.

5. Heat added from point 3 to point 4 is sensible heat; when point 4 is reached, 180 Btu of heat will have been added: 1 Btu/lb/°F temperature change.

6. Point 4 represents another saturated point. The water is saturated with heat to the point that the removal of any heat causes the liquid to cool off below the boiling point. Heat added is identified as latent heat and causes the water to boil and to start changing to a vapor (steam). Adding 970 Btu makes the 1 lb of liquid boil to point 5 and become a vapor.

7. Point 5 represents another saturated point. The water is now in the vapor state. Heat removed would be latent heat and would change some of the vapor back to a liquid. This is called *condensing the vapor*. Any heat added at point 5 is sensible heat; it raises the vapor temperature above the boiling point. Heating the vapor above the boiling point is called *superheat*. Superheat will be important in future studies. Note that in the vapor state it only takes 0.5 Btu to heat the water vapor (steam) 1°F. Remember the same held true while water was in the ice (solid) state.

SAFETY PRECAUTION: *When examining these principles in practice, be careful because the water and steam are well above body temperature and you could be seriously burned.*

1.8 SPECIFIC HEAT

We now realize that different substances respond differently to heat. When 1 Btu of heat energy is added to 1 lb of water, it changes the temperature 1°F. This only holds true for water. When other substances are heated, different values occur. For instance, we noted that adding 0.5 Btu of heat energy to either ice or steam (water vapor) caused a 1°F rise per pound while in these states. They heated at twice the rate. Adding 1 Btu would cause a 2°F rise. This difference in heat rise is known as *specific heat*.

Specific heat is the amount of heat necessary to raise the temperature of 1 lb of a substance 1°F. Every substance has a different specific heat. Note that the specific heat of water is 1 Btu/lb/°F. See Figure 1–14 for the specific heat of some other substances.

SUBSTANCE	SPECIFIC HEAT Btu/lb/°F
ALUMINUM	0.224
BRICK	0.22
CONCRETE	0.156
COPPER	0.092
ICE	0.504
IRON	0.129
MARBLE	0.21
STEEL	0.116
WATER	1.00
SEA WATER	0.94
AIR	0.24 (AVERAGE)

Figure 1–14. Specific heat table.

1.9 SIZING HEATING EQUIPMENT

Specific heat is so significant because the amount of heat required to change the temperatures of different substances is used to size equipment. Recall the example of the house and furnace earlier in this unit.

The following example shows how this would be applied in practice. A manufacturing company may need to buy a piece of heating equipment to heat steel before it can be machined. The steel may be stored outside in the cold at 0°F and need preheating before machining. The temperature desired for the machining is 70°F. How much heat must be added to the steel if the plant wants to machine 1000 lb/h?

Remember that the steel is coming into the plant at a fixed rate of 1000 lb/h and that heat has to be added at a steady rate to stay ahead of production. Figure 1–14 gives a specific heat of 0.129 Btu/lb/°F for steel. This means that 0.129 Btu of heat energy must be added to 1 lb of steel to raise its temperature 1°F.

Q = Weight × Specific Heat × Temperature Difference

where Q = quantity of heat needed. Substituting in the formula, we get

Q = 1000 lb/hr × 0.129 Btu/lb/°F × 70°F
Q = 9030 Btu/hr required to heat the steel for machining.

The previous example has some known values and a value to be found. The known information is used to find the unknown value with the help of the formula. The formula can be used when adding heat or removing heat and is used often in heat load calculations for sizing both heating and cooling equipment.

1.10 PRESSURE

Pressure is defined as force per unit of area. This is normally expressed in pounds per square inch (psi). Simply stated, when a 1-lb weight rests on an area of

1 square inch (1 in.^{2}), the pressure exerted downward is 1 pound per square inch (psi). Similarly, when a 100-lb weight rests on a 1-in.^{2} area, 100 psi of pressure are exerted, Figure 1–15.

When you swim below the surface of the water, you feel a pressure pushing inward on your body. This pressure is caused by the weight of the water and is very real. A different sensation is felt when flying in an airplane without a pressurized cabin. Your body is subjected to less pressure instead of more, yet you still feel uncomfortable.

It is easy to understand why the discomfort under water. The weight of the water pushes in. In the airplane, the reason is just the reverse. There is less pressure high in the air than down on the ground. The pressure is greater inside your body and is pushing out.

Water weighs 62.4 pounds per cubic foot (lb/ft^3). A cubic foot (7.48 gal) exerts a downward pressure of 62.4 lb/ft^2 when it is in its actual cube shape, Figure 1–16. How much weight is then resting on one square inch? The answer is very simply calculated. The bottom of the cube has an area of 144 in.^{2} (12 in. × 12 in.) sharing the weight. Each square inch has a total pressure of 0.433 lb (62.4 ÷ 144) resting on it. Thus, the pressure at the bottom of the cube is 0.433 psi, Figure 1–17.

1.11 ATMOSPHERIC PRESSURE

The sensation of being underwater and feeling the pressure of the water is familiar to most people. The earth's atmosphere is like an ocean of air that has weight and exerts pressure. The earth's surface can be thought of as being at the bottom of this ocean of air. Different locations are at different depths. For instance, there are sea level locations such as Miami, Florida or mountaintop locations such as Denver,

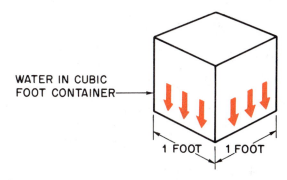

Figure 1–16.. One cubic foot (1 ft^3) of water (7.48 gal) exerts all of its pressure downward. 1 ft^3 of water weighs 62.4 lb spread over 1 ft^2.

Colorado. The atmospheric pressures at these two locations are different and will be considered later. For now, we will consider the fact that we live at the bottom of this ocean of air.

The atmosphere that we live in has weight just as water does, but not as much. Actually the earth's atmosphere exerts a weight or pressure of 14.696 psi at sea level when the surrounding temperature is 68° F. These are standard conditions.

Atmospheric pressure can be measured with an instrument called a *barometer*. The barometer is a glass tube about 36 in. long that is closed on one end and filled with mercury. It is then inserted open-side down into a puddle of mercury and held upright. The mercury will try to run down into the puddle, but it will not all run out. The atmosphere is pushing down on the puddle, and a vacuum is formed in the top of the tube. The mercury in the tube will fall to 29.92 in. at sea level when the surrounding atmospheric tempera-

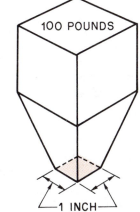

Figure 1–15. Both weights are resting on a 1-square inch (in^2) surface. One weight exerts a pressure of 1 psi, the other 100 psi.

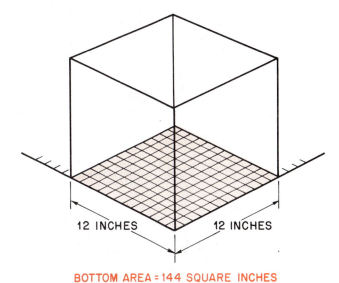

BOTTOM AREA = 144 SQUARE INCHES

Figure 1–17. One cubic foot (1 ft^3) of water exerts a pressure downward of 62.4 lb/ft^2 on bottom surface area.

ture is 68°F, Figure 1–18. This is a standard that is used for comparison in engineering and scientific work. If the barometer is taken up higher, such as on a mountain, the mercury column will start to fall. It will fall about 1 in./1000 ft of altitude. When the barometer is at standard conditions and the mercury drops, it is called a low-pressure system; this means the weather is going to change. Listen closely to the weather report, and the weather forecaster will make these terms more meaningful.

If the barometer is placed inside a closed jar and the atmosphere evacuated, the mercury column falls to a level with the puddle in the bottom, Figure 1–19. When the atmosphere is pushed back into the jar, the mercury again rises, because a vacuum exists above the mercury column in the tube.

The mercury in the column has weight and counteracts the atmospheric pressure of 14.696 psi at standard conditions. A pressure of 14.696 psi then is equal to the weight of a column of mercury (Hg) 29.92 in. high. The expression "inches of mercury" thus becomes an expression of pressure and can be converted to pounds per square inch. The conversion factor is 1 psi = 2.036 in. Hg (29.92 ÷ 14.696); 2.036 is often rounded off to 2 (30 in. Hg ÷ 15 psi).

Another type of barometer is the *aneroid* barometer. This is a more practical instrument to transport. Atmospheric pressure has to be measured in many places, so instruments other than the mercury barometer had to be developed for field use, Figure 1–20.

1.12 PRESSURE GAGES

Measuring pressures in a closed system requires a different method—the Bourdon tube, Figure 1–21. The

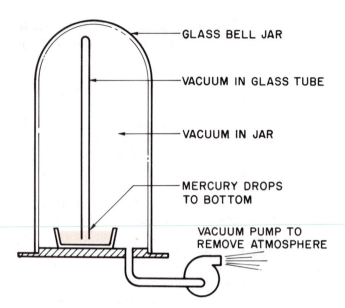

Figure 1–19. When the mercury barometer is placed in a closed glass jar (bell jar) and the atmosphere is removed, the pressure at the top of the column is the same as the pressure in the jar; the column of mercury drops to the level of the puddle.

Bourdon tube is linked to a needle and can measure pressures above and below atmosphere. A common tool used in the refrigeration industry to take readings in the field or shop is a combination of a low-pressure gage (called the *low-side gage*) and a high-pressure gage (called the *high-side gage*), Figure 1–22. Notice that the gage on the left reads pressures above and below atmospheric pressure. It is called a *compound gage*. The gage on the right will read up to 500 psi and is called the *high-pressure gage*. More coverage of these gages is given in the unit on refrigeration.

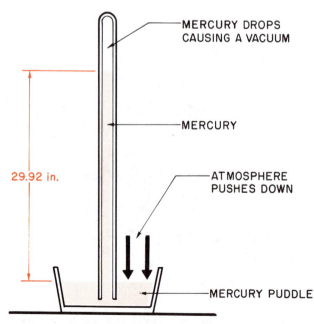

Figure 1–18. Mercury (Hg) barometer.

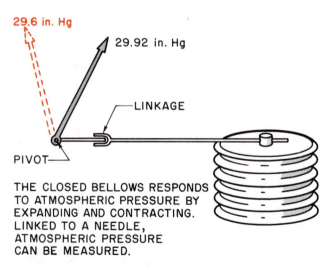

THE CLOSED BELLOWS RESPONDS TO ATMOSPHERIC PRESSURE BY EXPANDING AND CONTRACTING. LINKED TO A NEEDLE, ATMOSPHERIC PRESSURE CAN BE MEASURED.

Figure 1–20. The aneroid barometer uses a closed bellows that expands and contracts with atmospheric pressure changes. -

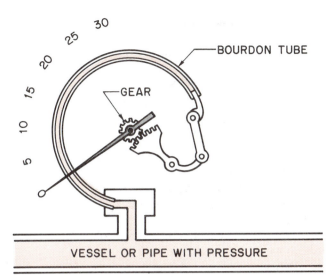

Figure 1-21. The Bourdon tube is made of a thin substance such as brass. It is closed on one end and the other end is fastened to the pressure being checked. When an increase in pressure is experienced, the tube tends to straighten out. When attached to a needle linkage, pressure changes are indicated.

Figure 1-23. This gage reads 50 psig. To convert this gage reading to psia, add the atmospheric pressure, 50 psig + 15 atmosphere = 65 psia. *Photo by Bill Johnson*

Note that these gages read 0 psi when opened to the atmosphere. If they do not, then they should be calibrated to 0 psi. These gages are designed to read pounds-per-square-inch gage pressure (psig). Atmospheric pressure is used as the starting or reference point. If you want to know what the absolute pressure is, you must add the atmospheric pressure to the gage reading. For example, to convert a gage reading of 50 psig to absolute pressure, you must add the atmosphere's pressure of 14.696 psi to the gage reading. Let's round off 14.696 to 15 for this example. Then 50 psig + 15 = 65 psia (pounds per square inch absolute), Figure 1-23.

SAFETY PRECAUTION: *Working with temperatures that are above or below body temperature can cause skin and flesh damage. Proper protection, such as gloves and safety glasses, *must be used.* Pressures that are above or below the atmosphere's can cause bodily injury. A vacuum can cause a blood blister on the skin. Pressures above atmospheric can pierce the skin or inflict damage by flying objects.*

SUMMARY

Figure 1-22. The gage on the left is called a compound gage because it reads below atmospheric pressure in in. Hg and above atmospheric pressure in psi. The right-hand gage reads high pressure up to 500 psi. *Photo by Bill Johnson*

- Thermometers measure temperature. Four temperature scales are Fahrenheit, Celsius, Fahrenheit absolute (Rankine), Celsius absolute (Kelvin).
- Molecules in matter are constantly moving. The higher the temperature, the faster they move.
- The British thermal unit (Btu) describes the quantity of heat in a substance. One Btu is the amount of

heat necessary to raise the temperature of 1 lb of water 1°F.

● The transfer of heat by conduction is the transfer of heat from molecule to molecule. As molecules in a substance move faster and with more energy, they cause others near them to do so.

● The transfer of heat by convection is the actual moving of heat in a fluid (vapor state or liquid state) from one place to another. This can be by natural convection, where heated liquid or air rises naturally, or by forced convection, where liquid or air is moved with a pump or fan.

● Radiant heat is a form of energy transmitted through a medium, such as air, without heating it. Solid objects absorb the energy, become heated, and transfer the heat to the air.

● Difference in the temperature of sensible heat can be measured with a thermometer.

● Latent (or hidden) heat is that heat added to a substance without registering on a thermometer. For example, heat added to melting ice causes ice to melt but does not increase the temperature.

● Specific heat is the amount of heat (measured in Btu) required to raise the temperature of one pound of a substance one degree Fahrenheit. Substances have different specific heats.

● Pressure is the force applied to a specific unit of area. The atmosphere around the earth has weight and therefore exerts pressure. The weight or pressure is greater at sea level (14.696 psi or 29.92 in. Hg at 68°F) than at higher elevations.

● Barometers measure atmospheric pressure in inches of mercury. Two barometers in use are the mercury and the aneroid.

● Gages have been developed to measure pressures in enclosed systems. Two common gages used in the air conditioning, heating, and refrigeration industry are the compound gage and the high-pressure gage. The compound gage reads pressures both above and below atmospheric pressure.

REVIEW QUESTIONS

1. Define temperature.
2. Describe early thermometers and indicate how they work.
3. List the standard conditions necessary for water to boil at 212°F.
4. List four types of temperature scales.
5. Under standard conditions at what point on the Celsius scale will water freeze?
6. At what Fahrenheit temperature do scientists think all molecular motion will stop?
7. Define the British thermal unit.
8. Describe the direction of heat flow from one substance to a substance of a different temperature.
9. Describe heat transfer by conduction.
10. Describe heat transfer by convection.
11. Describe heat transfer by radiation.
12. State the difference between sensible heat and latent heat.
13. What is another term for latent heat?
14. Define specific heat.
15. What is the atmospheric pressure at sea level under standard conditions?
16. Why is the atmospheric pressure less when measured on a mountain top?
17. Describe the differences between a mercury barometer and an aneroid barometer.
18. Approximately how many inches of mercury are equal to atmospheric pressure at sea level under standard conditions?
19. Explain the operation of a pressure gage using a Bourdon tube.
20. Explain the difference between psig and psia.

MATTER AND ENERGY

OBJECTIVES

After studying this unit, you should be able to

- **define matter.**
- **list the three states in which matter is commonly found.**
- **define density.**
- **define specific gravity and specific volume.**
- **state two forms of energy very important to the air conditioning (heating and cooling), and refrigeration industry.**
- **describe work and state the formula used to determine the amount of work in a given task.**
- **define horsepower.**
- **convert horsepower to watts.**
- **convert watts to British thermal units.**

2.1 MATTER

Matter is commonly explained as a substance that takes up space and has weight. The weight comes from the earth's gravitational pull. Matter also exists in three states: *solids, liquids,* and *gases.* The heat content and pressure determine the state of matter. An explanation of each follows. Water will again be used as the example.

Water in the solid state is known as ice. It exerts all of its force downward—it has weight. The molecules of the water are highly attracted to each other, Figure 2–1.

When the water is heated above the freezing point, it begins to change to a liquid state. The molecular

activity is higher, and the water molecules have less attraction for each other. Water in the liquid state exerts a pressure outward and downward. It now seeks its own level by pushing outward and downward, Figure 2–2.

Water heated above the liquid state, 212°F at standard conditions, becomes a vapor. In the vapor state the molecules have even less attraction for each other and are said to travel at random. The vapor exerts pressure more or less in all directions, Figure 2–3.

The study of matter leads to the study of other terms that help to understand how different substances compare to each other. For instance, the following terms will help you understand matter.

2.2 MASS

Mass is a term that is used along with weight to describe matter. The universe is made up of matter that has weight or mass. The air in the atmosphere has weight or mass. When the atmosphere is evacuated out of a jar all of the mass is removed, and a vacuum is created.

2.3 DENSITY

The *density* of a substance describes its mass to volume relationship. The mass contained in a particular volume is the density of that substance. In the British system of units volume is measured in cubic

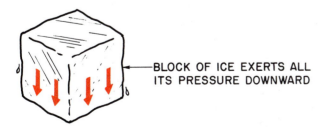

Figure 2–1. Solids exert all their pressure downward. The molecules of solid water have a great attraction for each other and hold together.

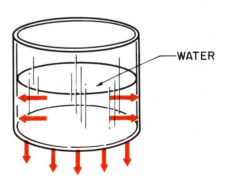

Figure 2–2. The water in the container exerts pressure outward and downward. The outward pressure is what makes water seek its own level. The water molecules still have a small amount of adhesion to each other.

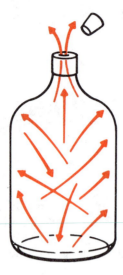

Figure 2–3. Gas molecules travel at random. When a container with a small amount of gas pressure is opened, the gas molecules seem to repel each other and fly out.

feet. Sometimes it is advantageous to compare different substances according to weight per unit volume. Water, for example, has a density of 62.4 lb/ft^3. Wood floats on water because the density (weight per volume) of wood is less than the density of water. In other words, it weighs less per cubic foot. Iron, on the other hand, sinks because it is denser than water. See Figure 2–4 for some typical densities.

2.4 SPECIFIC GRAVITY

Specific gravity compares the densities of various substances. The specific gravity of water is 1. The specific gravity of iron is 7.86. This means that a volume of iron is 7.86 times heavier than an equal volume of water. Since water weighs 62.4 lb/ft^3, a cubic foot of iron would weigh 62.4 $\times$ 7.86 or 490 lb. Aluminum has a specific gravity of 2.7, so it has a density or weight per cubic foot of 2.7 $\times$ 62.4 = 168 lb.

2.5 SPECIFIC VOLUME

Specific volume compares the densities of gases. It indicates the space (volume) a weight of gas will occupy. One pound of clean dry air has a volume of 13.33 ft^3 at standard conditions. Hydrogen has a density of 179 ft^3/lb under the same conditions. Because there are more cubic feet of hydrogen per pound, it is lighter than air. Although both are gases, the hydrogen has a tendency to rise when mixed with air.

Natural gas is explosive when mixed with air, but it is lighter than air and has a tendency to rise like hydrogen. Propane gas is another often-used heating gas and has to be treated differently from natural gas because it is heavier than air. Propane has a tendency to fall and collect in low places and to cause potential danger from ignition.

The specific volumes of various gases that are pumped are invaluable information. They enable the engineer to choose the size of the compressor or vapor pump to do a particular job. The specific volumes for vapors vary according to the pressure the vapor is under. An example is refrigerant-22, which is a common refrigerant used in residential air conditioning units. At 3 psig about 2.5 ft^3 of gas must be pumped to move 1 lb of gas. At the standard design condition of 70 psig, only 0.48 ft^3 of gas needs to be pumped to move a pound of the same gas. A complete breakdown of specific volume can be found in the properties of liquid and saturated vapor tables in engineering manuals for any refrigerant.

2.6 ENERGY

Energy is important because using energy properly to operate equipment is a major goal of the air conditioning and refrigeration industry. Energy in the form of electricity drives the motors; heat energy from the fossil fuels of natural gas, oil, and coal heats homes and industry. What is this energy and how is it used?

First realize that the only new energy we get is from the sun heating the earth. Most of the energy we use is converted to usable heat from something already here (for example fossil fuels). This conversion from fuel to heat can be direct or indirect. An example of direct conversion is a gas furnace, which converts the gas flame to usable heat by combustion. The gas is burned in a combustion chamber, and the heat from combustion is transferred to circulated air by conduction through the heat exchanger wall of thin steel. The heated air is then distributed throughout the heated space, Figure 2–5.

An example of indirect conversion is a fossil-fuel power plant. Gas may be used in the power plant to generate the steam that turns a steam turbine generator to produce electricity. The electricity is then distributed by the local power company and consumed locally as electric heat, Figure 2–6.

SUBSTANCE	DENSITY lb/ft^3	SPECIFIC GRAVITY
ALUMINUM	168	2.7
BRASS	536	8.7
COPPER	555	8.92
GOLD	1204	19.3
ICE	57	0.92
IRON	490	7.86
LITHIUM	33	0.53
TUNGSTEN	1186	19.0
MERCURY	845	13.54
WATER	62.4	1

Figure 2–4. Table of density and specific gravity.

We have already said that most of the energy we use is a result of the sun's shining supporting plant growth for thousands of years. Fossil fuels come from decayed vegetable and animal matter covered by earth and rock during changes in the earth's surface. This decayed matter is in various states, such as gas, oil, or coal, depending on the conditions it was subjected to in the past, Figure 2–7.

2.8 ENERGY CONTAINED IN HEAT

In our discussion on temperature we said that temperature is a measure of the level of heat, and that heat is a form of energy because of the motion of molecules. Since molecular motion does not stop until −460° F, energy is still available in a substance even at very low temperatures. This energy is in relationship to other substances that are at lower temperatures. For example, if two substances at very low temperatures are moved close together, heat will transfer from the warmer substance to the colder one. In Figure 2–8 a substance at −200° F is placed next to a substance at −350° F. As we discussed earlier, the warmer substance gives up heat (energy) to the cooler substance. Thus, heat is said to flow "downhill," that is, the temperature hill. The energy used by home and industry is not at these low levels.

Electric heat is different from gas heat. Electrons flow through a special wire in the heater section that converts electrical energy to heat. A moving air stream is then passed over the heated wire, allowing heat to be transferred to the air by conduction and moved to the heated space by forced convection (the fan).

2.9 ENERGY IN MAGNETISM

Magnetism is another method of converting electron flow to usable energy. Electron flow is used to develop magnetism to turn motors. The motors turn the prime movers—fans, pumps, compressors—of air, water, and refrigerant. In Figure 2–9 an electric motor turns a water pump to boost the water pressure from 20 to 60

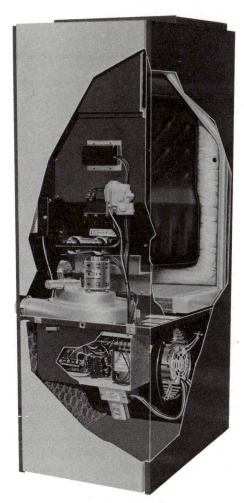

Figure 2–5. Cutaway of a high-efficiency condensing gas furnace. *Courtesy Heil-Quaker Corporation*

2.7 CONSERVATION OF ENERGY

The preceding discussion leads to the law of conservation of energy. The law states that **energy is neither created nor destroyed.** It can now be said that energy can be accounted for.

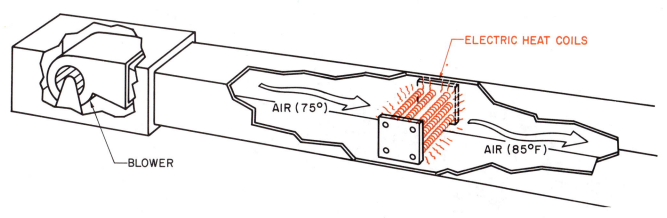

Figure 2–6. An electric heat air stream.

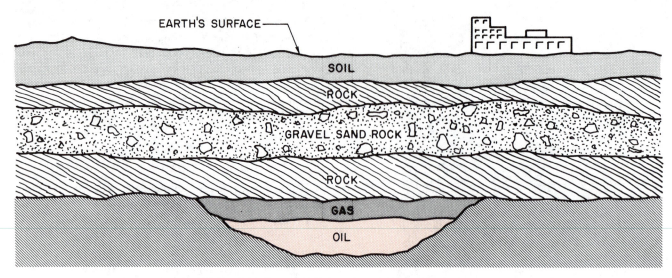

Figure 2–7. Gas and oil deposits settle into depressions.

psig. This takes energy. The energy in this example is purchased from the power company.

The preceding examples serve only as an introduction to the concepts of gas heat, electric heat, and magnetism. Each subject will later be covered in detail.

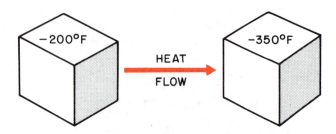

Figure 2–8. Heat energy is still available at these low temperatures and will still transfer from the warmer to the colder substance.

For now it is important to realize that any system furnishing heating or cooling uses energy.

2.10   PURCHASE OF ENERGY

Energy must be transferred from one owner to another and accounted for. This energy is purchased as a fossil fuel or as electric power. Energy purchased as a fossil fuel is normally purchased by the unit. Natural gas is an example. Natural gas flows through a meter that measures how many cubic feet have passed during some time span, such as a month. Fuel oil is normally sold by the gallon, coal by the ton. The amount of heat each of these units contains, is known, so a known amount of heat is purchased: Natural gas, for instance, has a heat content of about 1000 Btu/ft^3, whereas the heat content of coal varies from one type of coal to another.

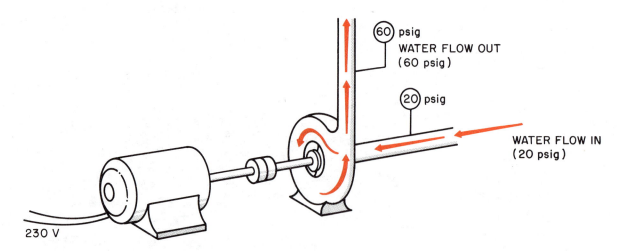

Figure 2–9. Magnetism used in an electric motor is converted to work to boost water pressure to force circulation.

2.11   ENERGY USED AS WORK

Energy purchased from electrical utilities opens up a whole new subject. This energy is known as electric power. *Power* is the rate of doing work. *Work* can be explained as a force moving an object in the direction of the force. It is expressed by this formula.

$$\text{Work} = \text{Force} \times \text{Distance}$$

For instance, when a 150-lb man climbs a flight of stairs 100 ft high (about the height of a 10-story building), he performs work. But how much? The amount of work in this example is equivalent to the amount of work necessary to lift this man the same height. We can calculate the work by using the preceding formula.

$$\text{Work} = 150 \text{ lb} \times 100 \text{ ft}$$
$$= 15,000 \text{ ft-lb}$$

Notice that no time limit has been added. This example can be accomplished by a healthy man in a few minutes. But if the task were to be accomplished by a machine such as an elevator, more information is necessary. Do we want to take seconds, minutes, or hours to do the job? The faster the job is accomplished, the more power is required.

2.12   HORSEPOWER

Power is the rate of doing work. An early expression of power was the *horsepower*. Many years ago it was determined that an average good horse could lift the equivalent of 33,000 lb a height of 1 ft in 1 min, which is the same as 33,000 ft-lb/min or 1 horsepower (hp). This describes a rate of doing work because time has been added. Keep in mind that lifting 330 lb a height of 100 ft in 1 min or 660 lb 50 ft in 1 min is the same amount of work. This seems like a lot of work, and it is, but a workhorse can weigh 1500 to 1800 lb. As a point of reference, the fan motor in the average furnace can be rated at 1/2 hp. See Figure 2–10 for an illustration of the horse lifting 1 hp.

When the horsepower is compared to the man climbing the stairs, the man would have to climb the 100 ft in less than one-half minute to equal 1 hp. That makes the task seem even harder. A 1/2-hp motor could lift the man 100 ft in 1 min if only the man were lifted. The reason is that 15,000 ft-lb of work are required. (Remember that 33,000 ft-lb of work in 1 min equals 1 hp.)

Our purpose in discussing these things is to help you understand how to use power effectively and to understand how power companies determine their methods of charging for power.

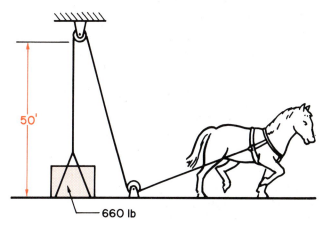

Figure 2–10. When a horse can lift the 660 lb a height of 50 ft in 1 min, it has done the equivalent of 33,000 ft-lb of work in 1 min, or 1 hp.

2.13   ELECTRICAL POWER—THE WATT

The unit of measurement for electrical power is the *watt* (W). This is the unit used by the power company. When converted to electrical energy, 1 hp = 746 W; that is, when 746 W of electrical energy are properly used, the equivalent of 1 hp of work has been accomplished.

Fossil-fuel energy can be compared to electrical energy and one form of energy can be converted to the other. There must be some basis, however, for conversion so that one fuel can be compared to another. The examples we use to illustrate this comparison will not take efficiencies into account. Efficiencies for the various fuels will be covered in the section on applications for each fuel. Some examples of conversions follow.

2.14   THE BRITISH THERMAL UNIT

1. **Converting electric heat rated in kilowatts (kW) to the equivalent gas or oil heat rated in Btu.** Suppose we want to know the capacity in Btu for a 20-kW electric heater (a kilowatt is 1000 watts). 1 kW = 3413 Btu.

   20 kW × 3413 Btu/kW = 68,260 Btu of heat energy

2. **Converting Btu to kW.** Suppose a gas or oil furnace has an output capacity of 100,000 Btu/hr. Since 3413 Btu = 1 kW, we have

   100,000 Btu ÷ 3413 = 29.3 kW

In other words, a 29.3-kW electric heat system would be required to replace the 100,000-Btu/hr furnace.

Contact the local utility company for rate comparisons between different fuels.

SAFETY PRECAUTION: *Any device that consumes power, such as an electric motor or gas furnace, is potentially dangerous. These devices should only be handled or adjusted by experienced people.*

SUMMARY

- Matter takes up space, has weight, and can be in the form of a solid, a liquid, or a gas.
- The mass of a substance is its weight.
- In the British system of units, density is the weight of a substance per cubic foot.
- Specific gravity is the term used to compare the density of various substances.
- Specific volume is the amount of space a pound of a vapor or a gas will occupy.
- Electrical energy and heat energy are two forms of energy used in this industry.
- Fossil fuels are purchased by the unit. Natural gas is metered by the cubic foot, oil is purchased by the gallon, and coal is purchased by the ton. Electricity is purchased from the electric utility company by the kilowatt-hour (kWh).
- Work is the amount of force necessary to move an object: Work = Force × Distance.
- Horsepower is the equivalent of lifting 33,000 lb a height of 1 ft in 1 min, or some combination totaling the same.
- Watts are a measurement of electrical power. One horsepower equals 746 W.
- 3.413 Btu = 1 W. 1 kW (1000 W) = 3413 Btu.

REVIEW QUESTIONS

1. Define matter.
2. What are the three states in which matter is commonly found?
3. What is the term used for water when it is in the solid state?
4. In what direction does a solid exert force?
5. In what direction does a liquid exert force?
6. Describe how vapor exerts pressure.
7. Define density.
8. Define specific gravity.
9. Describe specific volume.
10. Why is information regarding the specific volume of gases important to the designer of air conditioning, heating, and refrigeration equipment?
11. What are the two types of energy most frequently used or considered in this industry?
12. How were fossil fuels formed?
13. What is work?
14. State the formula for determining the amount of work accomplished in a particular task.
15. If an air conditioning compressor weighing 300 lb had to be lifted 4 ft to be mounted on a base, how many foot-pounds of work must be accomplished?
16. Describe horsepower and list the three quantities needed to determine horsepower.
17. How many watts of electrical energy are equal to 1 horsepower?
18. How many Btu are there in 4000 W (4 kW)?
19. How many Btu would be produced in a 12-kW electric heater?
20. What unit of energy does the power company charge the consumer for?

REFRIGERATION AND REFRIGERANTS

OBJECTIVES

After studying this unit, you should be able to

- state three reasons why ice melts in ice boxes.
- discuss applications for high-, medium-, and low-temperature refrigeration.
- describe the term "ton of refrigeration."
- describe the basic refrigeration cycle.
- explain the relationship between pressure and the boiling point of water or other liquids.
- describe the function of the evaporator or cooling coil.
- explain the purpose of the compressor.
- discuss the function of the condensing coil.
- state the purpose of the metering device.
- list the three refrigerants commonly used in residential and light commercial refrigeration and air conditioning systems.
- list four characteristics to consider when choosing a refrigerant for a system.

3.1 HISTORY OF REFRIGERATION

Preserving food is one of the most valuable uses of refrigeration. It was discovered as far back as Roman times that food lasted longer when it was kept cold, above freezing. The reason is that food spoilage slows down as molecular motion slows down. This slows the growth of bacteria that causes food to spoil. Below the frozen hard point, food-spoiling bacteria stop growing. The frozen hard point for most foods is considered to be 0°F. The food temperature range between 35° and 45°F is known in the industry as medium temperature; below 0°F is considered low temperature. These ranges are used to describe many types of refrigeration equipment and applications.

For many years dairy products and other perishables were stored in the coldest room in the house, the basement, the well, or a spring. In the South, temperatures as low as 55°F could be reached in the summer with underground water. This would add to the time that some foods could be kept.

In the North ice was cut from lakes in the winter and stored in insulated buildings (called "coolers") or in sawdust piles. The ice was saved until spring and summer when it was sold to people who had insulated ice boxes. Some of these ice boxes were in high-rise apartment houses, and the ice had to be carried to the top floors. The ice was then loaded in the ice boxes. The ice melted when it absorbed heat from the food stuff in the box, Figure 3–1. Some of the northern ice was transported by barge to the South and sold. Ice was very expensive, and only the wealthy could afford it. It may be hard to imagine a summer in the South without a cold drink or ice cream, but that is the way it used to be.

In the late 1800s a mild winter in the North caused a shortage of ice, which hastened the development of mechanical refrigeration invented earlier in the century. Ice was then manufactured by mechanical refrigeration and sold to people with ice boxes, but still only the wealthy could afford it.

In the early 1900s some companies manufactured the household refrigerator. Like all new items, it took a while to catch on. Now, of course, most houses have a refrigerator with a freezing compartment.

Frozen food was just beginning to become popular about the time World War II began. Because most people did not have a freezer at this time, central frozen food locker plants were established so that a family could have its own locker. Food that is frozen fresh is very appealing because it stays fresh. Refrigerated foods, both medium temperature and low temperature, are so common now that most people take them for granted.

The refrigeration process is now used in the comfort cooling of the home and business and in the air conditioning of automobiles. The air conditioning application of refrigeration is known in this industry as high temperature.

Now that uses for high-, medium-, and low-temperature refrigeration have been established, and some of the applications have been covered, it is time to learn *how* refrigeration works.

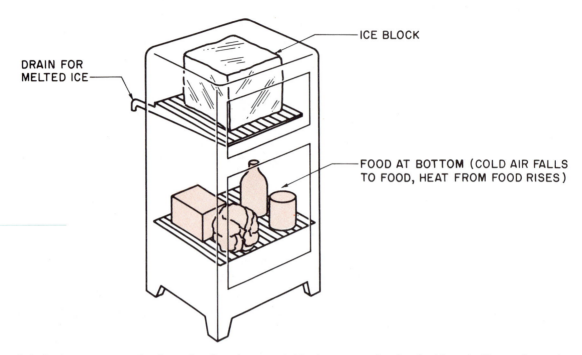

DRAIN FOR
MELTED ICE

ICE BLOCK

FOOD AT BOTTOM (COLD AIR FALLS
TO FOOD, HEAT FROM FOOD RISES)

Figure 3-1. Ice boxes were made of wood at first then metal. The boxes were insulated with cork. If a cooling unit were placed where the ice is, this would be a refrigerator.

3.2 REFRIGERATION

Refrigeration **is the process of removing heat from a place where it is not wanted and transferring that heat to a place where it makes no difference.** Think about the household refrigerator. In any American household, the room temperature from summer to winter is normally between 70° and 90°F. The temperature inside the refrigerator fresh food section should be about 35°F. Remember that heat flows naturally from a warm level to a cold level. Therefore, heat is continuously trying to leak into the refrigerator. The heat on the outside of the refrigerator does no harm. Heat in the room is trying to flow into the refrigerator and it does through the insulated walls, through the door when it is opened, and through warm food placed in the refrigerator, Figures 3-2, 3-3, and 3-4.

3.3 RATING REFRIGERATION EQUIPMENT

Refrigeration equipment must also have a rating system so that equipment can be compared. The method for rating refrigeration equipment goes back to the days of using ice as the source for removing heat. Remember from the heat–temperature graph that it takes 144 Btu of heat energy to melt a pound of ice at 32°F. This same figure will be used again to rate refrigeration equipment.

The term for this rating is the *ton. One ton of refrigeration* is the amount of heat required to melt 1 ton of ice in a 24-hr period. Previously, we saw that it

takes 144 Btu of heat to melt a pound of ice. It would then take 2000 times that much heat to melt a ton of ice (2000 lb/ton):

$$144 \text{ Btu/lb} \times 2000 \text{ lb} = 288,000 \text{ Btu}$$

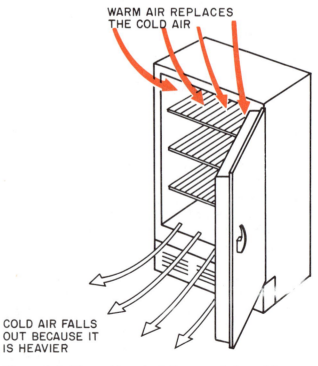

WARM AIR REPLACES
THE COLD AIR

COLD AIR FALLS
OUT BECAUSE IT
IS HEAVIER

Figure 3-2. The colder air falls out of the refrigerator because it is heavier. It is replaced with warmer air from the top. This warm air is a heat leakage.

Figure 3–3. Warm food that is moved from the room or stove adds heat to the refrigerator and is considered heat leakage. This added heat has to be removed, or the inside temperature will rise.

to melt a ton of ice. When accomplished in a 24-hr period, it is known as 1 ton of refrigeration. The same rules apply when removing heat from a substance. For example, an air conditioner, having a 1-ton capacity, will remove 288,000 Btu/24 h or 12,000 Btu/h (288,000 ÷ 24 = 12,000), or 200 Btu/min (12,000 ÷ 60 = 200), Figure 3–5.

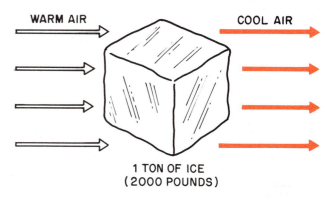

Figure 3–5. 2000 lb of ice requires 144 Btu/lb to melt. 2000 pounds × 144 Btu/lb = 288,000 Btu per 2000 lb. When this is accomplished in 24 hr, it is known as a work rate of 1 ton of refrigeration. This is the same as 12,000 Btu/hr or 200 Btu/min.

3.4 THE REFRIGERATION PROCESS

How does heat flow up the temperature hill from a cold place to a warmer place? The heat leaking into the refrigerator does not necessarily raise the temperature of the contents or food stuff an appreciable amount. If it did, the food would spoil. The heat inside the refrigerator rises to a predetermined level; the refrigeration machine then comes on and pumps the heat out. The refrigerator pump has to pump the heat up the temperature hill from the 35°F or 0°F compartment to the room temperature of 70° to 90°F. The components of the refrigerator are used to accomplish this task, Figure 3–6.

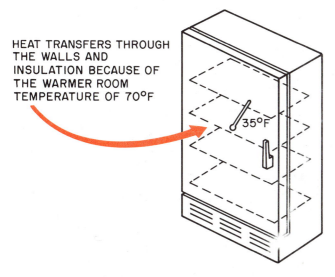

Figure 3–4. Heat transfers through the walls into the box by conduction. The walls have insulation, but this does not stop the leakage completely.

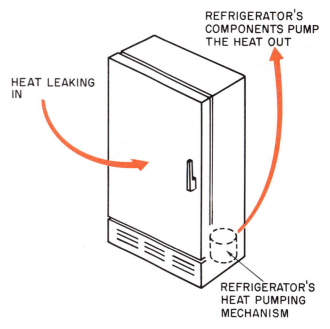

Figure 3–6.. Heat that leaks into the refrigerator from any source has to be removed by the refrigerator's heat pumping mechanism. The heat has to be pumped from the 35°F box to the 70°F room.

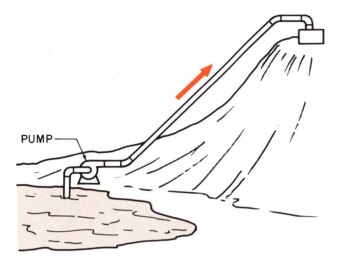

Figure 3-7. Power is required to pump water uphill; the same is true for pumping heat up the temperature hill from the 35°F box to the 70°F room.

The process of pumping heat out of the refrigerator could be compared to pumping water from a valley to the top of a hill. It takes just as much energy to pump water up the hill as it does to carry it. Although a water pump with a motor seems to perform this task without effort, it really does work. If a gasoline engine, for instance, were driving the pump, the gasoline would be burned and converted to work energy. An electric motor uses electric power as work energy, Figure 3-7. Refrigeration is the process of moving of heat. This takes energy that has to be purchased.

We will use the residential air conditioning system to explain the basics of refrigeration. Residential air conditioning, whether window unit or central system, is considered to be high-temperature refrigeration, and is used for comfort cooling. The residential system can be seen from the outside, touched and listened to for the examples that will be given.

The refrigeration concept in the residential air conditioner is the same as in the household refrigerator. It pumps the heat from inside the house to the outside of the house. You may wonder why the air conditioner blows cold air if it is really pumping heat. The answer is simple. When cold air is coming into the house, then heat must be exhausted somewhere. Another part of the system is exhausting this heat. This cold air is only recirculated air. Room air at approximately 75°F from the room goes into the unit, and air at approximately 55°F comes out. This is the same air with some of the heat removed, Figure 3-8.

The following example of a window air conditioner will illustrate this concept. The statements also serve as guidelines to some of the design data used throughout the air conditioning field.

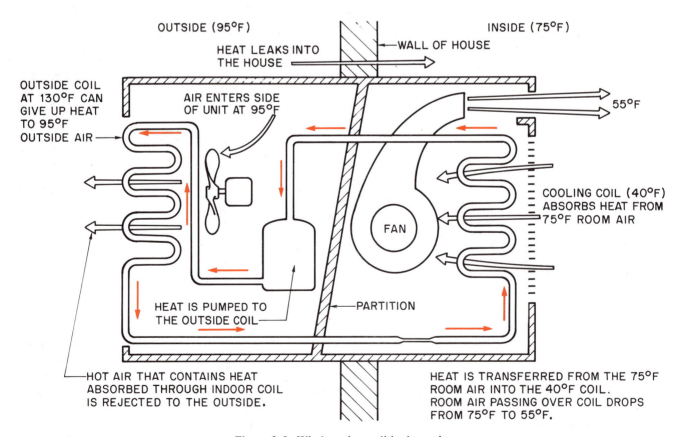

Figure 3-8. Window air conditioning unit.

1. The outside design temperature is 95°F.
2. The inside desired temperature is 75°F.
3. The cooling coil temperature is 40°F. This coil transfers heat from the room into the refrigeration machine. Notice that with a 75°F room temperature and a 40°F cooling coil temperature, heat will transfer from the room air into the coil.
4. This heat transfer makes the air leaving the coil about 55°F.
5. The outside coil temperature is 130°F. This coil transfers heat from the system to the outside air. Notice that when the outside air temperature is 95°F and the coil temperature is 130°F, heat will be transferred from the system to the outside air.

Careful examination of Figure 3–8 shows that heat ... he refrigeration sys- ... sferred to the out- ... tem. The air condi- ... the heat out of the

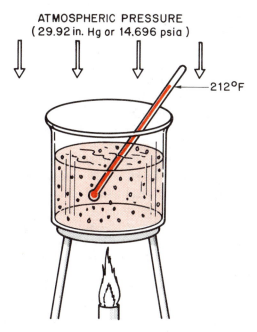

Figure 3–10. Water boils at 212°F when atmospheric pressure is 29.92 in. Hg.

house. The system capacity must be large enough to keep the heat pumped out of the house so that the occupants will not become uncomfortable.

3.5   PRESSURE AND TEMPERATURE RELATIONSHIP

We have been calling the refrigeration system a machine, but we have not yet given any explanation of how the process actually works. To understand the refrigeration process, we must go back to Figure 1–13 (heat/temperature graph), where water was changed to steam. Water boils at 212°F at 29.92 in. Hg pressure. This suggests that water has other boiling points. The next statement is one of the most important in the book. Memorize it. **The boiling point of water can be changed and controlled by controlling the vapor pressure above the water.** Understanding this concept is necessary, because water will be used as the heat transfer medium in this example. The next few paragraphs are critically important for understanding refrigeration.

The *pressure-temperature relationship* correlates the vapor pressure and the boiling point of water and is the basis for controlling the system's temperatures.

We have already said that pure water boils at 212°F at standard conditions of sea level and 68°F because this condition exerts a pressure on the water's surface of 29.92 in. Hg (or 14.696 psi). Find this reference point on the table in Figure 3–9. Also see Figure 3–10, the container of water that boils at sea level. When this same pan of water is taken to a mountaintop, the boiling

		LUTE PRESSURE
	2	in. Hg
		0.063
		0.103
		0.165
		0.180
		0.195
		0.212
		0.229
		0.248
		0.268
		0.289
		0.312
		0.336
		0.362
		0.522
		0.739
		1.032
		1.422
		1.933
		2.597
		3.448
		4.527
		5.881
150	3.719	7.573
160	4.742	9.656
170	5.994	12.203
180	7.512	15.295
190	9.340	19.017
200	11.526	23.468
210	14.123	28.754
212	14.696	29.921

Figure 3–9. Boiling point of water—pressure and temperature relationship table.

point changes, Figure 3–11, because the thinner atmosphere causes a reduction in pressure (about 1 in. Hg/1000 ft). In Denver, Colorado, for example, which is about 5000 ft above sea level, the atmospheric pressure is approximately 25 in. Hg. Water boils at 203.4°F at that pressure. This makes cooking foods such as potatoes and dried beans more difficult because they need a higher temperature, but by placing the food in a closed container that can be pressurized (a pressure cooker) and allowing the pressure to go up to about 15 psi above atmosphere (or 30 psia), the boiling point can be raised to 250°F, Figure 3–12.

Studying the water pressure and temperature table reveals that whenever the pressure is increased, the boiling point increases, and that whenever the pressure is reduced, the boiling point is reduced. If water were boiled at a temperature low enough to absorb heat out of a room, we could have comfort cooling (air conditioning).

Let's place a thermometer in the pan of pure water, put the pan inside a bell jar with a barometer, and start the vacuum pump. Suppose the water is at room temperature (70°F). When the pressure in the jar reaches the pressure that corresponds to the boiling point of water at 70°F, the water will start to boil and vaporize. This point is 0.739 in. Hg (0.363 psia). See Figure 3–13 and the illustration of the container in the jar.

Now let's lower the pressure in the jar to correspond to a temperature of 40°F. This new pressure of 0.248 in. Hg (0.122 psia) will cause the water to boil at 40°F. The water is not hot even though it is boiling. The thermometer in the pan indicates this. If the jar were opened to the atmosphere, the water would be nice and cold to drink.

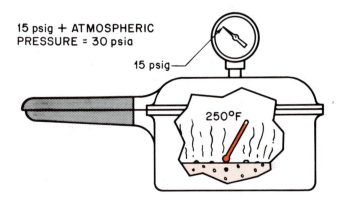

Figure 3–12. The water in the pressure cooker boils at 250°F. As heat is added the water boils to make vapor. The vapor cannot escape, and the vapor pressure rises to 15 psig. The water boils at 250°F because the pressure is 15 psig.

Now let's circulate this water boiling at 40°F through a cooling coil. If room air were passed over it, it would absorb heat from the room air. Since the room air is giving up heat to the coil, the air leaving the coil is cold. See Figure 3–14 for the cooling coil illustration.

When water is used in this way, it is called a *refrigerant*. **A *refrigerant* is a substance that can be changed readily to a vapor by boiling it and then to a liquid by condensing it.** The refrigerant must be able to make this change repeatedly without altering its characteristics. Water is not normally used as a refrigerant in small applications for reasons that will be discussed later. We used it in this example because everyone is familiar with its characteristics.

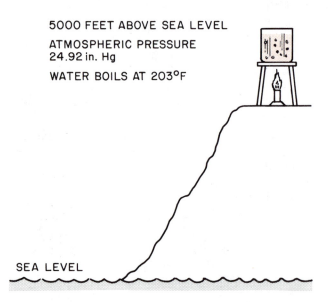

Figure 3–11. Water boils at 203°F when atmospheric pressure is 24.92 in. Hg.

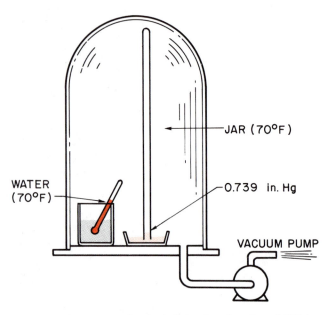

Figure 3–13. Pressure in the bell jar is reduced to 0.739 in. Hg, and the boiling temperature of the water is reduced to 70°F because the pressure is 0.739 in. Hg. (0.363 psia).

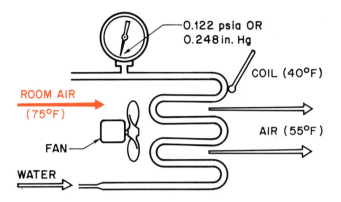

Figure 3–14. The water is boiling at 40°F because the pressure is 0.122 psia or 0.248 in. Hg. The room air is 75°F and gives up heat to the 40°F coil.

An actual refrigeration system should now be easier to understand. To explore how a real system works, we will use a real refrigerant. Refrigerant-22 (R-22) is commonly used in residential air conditioning, so it will be used in the next examples. See Figure 3–15 for the pressure and temperature relationship chart for R-22. This chart is just like that for water but at different temperature and pressure levels. Take a moment to become familiar with this chart; observe that temperature is in the left column expressed in °F, and pressure is in the right column expressed in psig. Now find 40°F in the left column. Read to the right and notice that the gage reading is 68.5 psig. What does this mean in usable terms?

The pressure and temperature of a refrigerant will correspond when both liquid and vapor are present under two conditions:

1. When the change of state (boiling or condensing) is taking place
2. When the refrigerant is at equilibrium (i.e., no heat is added or removed)

Suppose that a drum of R-22 is allowed to set in a room until it reaches the room temperature of 75°F. It

TEMPERATURE °F	REFRIGERANT			TEMPERATURE °F	REFRIGERANT			TEMPERATURE °F	REFRIGERANT		
	12	22	502		12	22	502		12	22	502
−60	19.0	12.0	7.2	12	15.8	34.7	43.2	42	38.8	71.4	83.8
−55	17.3	9.2	3.8	13	16.4	35.7	44.3	43	39.8	73.0	85.4
−50	15.4	6.2	0.2	14	17.1	36.7	45.4	44	40.7	74.5	87.0
−45	13.3	2.7	1.9	15	17.7	37.7	46.5	45	41.7	76.0	88.7
−40	11.0	0.5	4.1	16	18.4	38.7	47.7	46	42.6	77.6	90.4
−35	8.4	2.6	6.5	17	19.0	39.8	48.8	47	43.6	79.2	92.1
−30	5.5	4.9	9.2	18	19.7	40.8	50.0	48	44.6	80.8	93.9
−25	2.3	7.4	12.1	19	20.4	41.9	51.2	49	45.7	82.4	95.6
−20	0.6	10.1	15.3	20	21.0	43.0	52.4	50	46.7	84.0	97.4
−18	1.3	11.3	16.7	21	21.7	44.1	53.7	55	52.0	92.6	106.6
−16	2.0	12.5	18.1	22	22.4	45.3	54.9	60	57.7	101.6	116.4
−14	2.8	13.8	19.5	23	23.2	46.4	56.2	65	63.8	111.2	126.7
−12	3.6	15.1	21.0	24	23.9	47.6	57.5	70	70.2	121.4	137.6
−10	4.5	16.5	22.6	25	24.6	48.8	58.8	75	77.0	132.2	149.1
−8	5.4	17.9	24.2	26	25.4	49.9	60.1	80	84.2	143.6	161.2
−6	6.3	19.3	25.8	27	26.1	51.2	61.5	85	91.8	155.7	174.0
−4	7.2	20.8	27.5	28	26.9	52.4	62.8	90	99.8	168.4	187.4
−2	8.2	22.4	29.3	29	27.7	53.6	64.2	95	108.2	181.8	201.4
0	9.2	24.0	31.1	30	28.4	54.9	65.6	100	117.2	195.9	216.2
1	9.7	24.8	32.0	31	29.2	56.2	67.0	105	126.6	210.8	231.7
2	10.2	25.6	32.9	32	30.1	57.5	68.4	110	136.4	226.4	247.9
3	10.7	26.4	33.9	33	30.9	58.8	69.9	115	146.8	242.7	264.9
4	11.2	27.3	34.9	34	31.7	60.1	71.3	120	157.6	259.9	282.7
5	11.8	28.2	35.8	35	32.6	61.5	72.8	125	169.1	277.9	301.4
6	12.3	29.1	36.8	36	33.4	62.8	74.3	130	181.0	296.8	320.8
7	12.9	30.0	37.9	37	34.3	64.2	75.8	135	193.5	316.6	341.2
8	13.5	30.9	38.9	38	35.2	65.6	77.4	140	206.6	337.2	362.6
9	14.0	31.8	39.9	39	36.1	67.1	79.0	145	220.3	358.9	385.0
10	14.6	32.8	41.0	40	37.0	68.5	80.5	150	234.6	381.5	408.4
11	15.2	33.7	42.1	41	37.9	70.0	82.1	155	249.5	405.1	432.9

Vacuum — Red Figures
Gage Pressure — Bold Figures

Figure 3–15. Pressure and temperature relationship chart in in. Hg vacuum or psig.

will then be in equilibrium—no outside forces are acting on it. The drum and its partial liquid partial vapor contents will now be at the room temperature of 75°F. The pressure and temperature chart indicates a pressure of 132 psig, Figure 3–15.

Now suppose that the same drum of R-22 is moved into a walk-in cooler and allowed to reach the room temperature of 35°F and attain equilibrium. The drum will then reach a new pressure of 61.5 psig, because while the drum is cooling off to 35°F the vapor inside the drum is reacting to the cooling effect by partially condensing; therefore the pressure drops.

If we move the drum (now at 35°F) back into the warmer room (75°F) and allow it to warm up, the liquid inside the drum reacts to the warming effect by boiling slightly and creating vapor. Thus the pressure gradually increases to 132 psig, which corresponds to 75°F.

If we move the drum (now at 75°F) into a room at 100°F, the liquid again responds to the temperature change by slightly boiling and creating more vapor. As the liquid boils and makes vapor, the pressure steadily increases (according to the pressure and temperature chart) until it corresponds to the liquid temperature. This continues until the contents of the drum reach the pressure, 196 psig, corresponding to 100°F, Figures 3–16, 3–17, and 3–18.

Further study of the temperature/pressure chart shows that when the pressure is lowered to atmospheric pressure, R-22 boils at about −41°F. If the valve on the drum of R-22 were opened slowly and the vapor

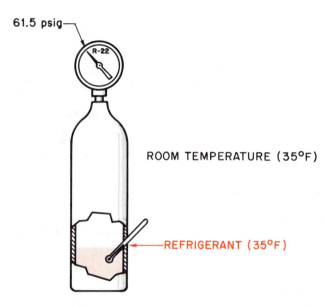

Figure 3–17. The drum is moved into a walk-in cooler and left until the drum and its contents become the same as the inside of the cooler, 35°F. Until the drum and its contents get to 35°F some of the vapor in the drum will be changing to a liquid and reducing the pressure. Soon the drum pressure will correspond to room temperature of 35°F, 61.5 psig.

allowed to escape to the atmosphere, the pressure loss of the vapor would cause the liquid remaining in the drum to drop in temperature. Soon the pressure in the drum would be down to atmospheric pressure, and the drum would frost over and become −41°F. If we connected a gage line to the liquid port on a drum and opened the valve very slowly, holding the end of the gage line in a thick coffee cup, the escaping liquid will

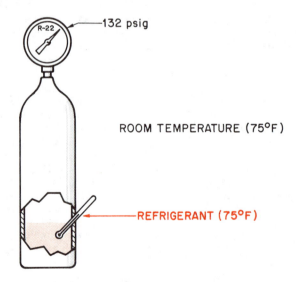

Figure 3–16. The drum of R-22 is left in a 75°F room until the drum and its contents are at room temperature. The drum contains a partial liquid, partial vapor mixture, which, when both become room temperature, will be said to be in equilibrium; no more temperature changes will be taking place. At this time the drum pressure, 132 psig, will correspond to the drum temperature of 75°F.

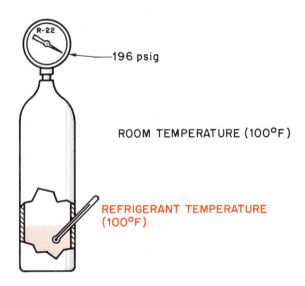

Figure 3–18. The drum is moved into a 100°F room and allowed to reach the point of equilibrium at 100°F, 196 psig. The pressure rise is due to some of the liquid refrigerant boiling to a vapor and increasing the total drum pressure.

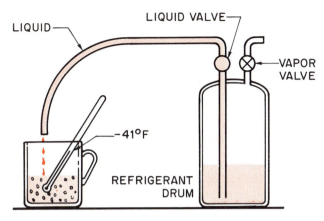

Figure 3–19. When a gage line is attached to the liquid line valve on an R-22 refrigeration drum and liquid is allowed to trickle out of the drum into the cup, the liquid will collect in the cup. It will boil at a temperature of −41°F at atmospheric pressure.

cool the cup to −41°F, Figure 3–19. The liquid will accumulate in the bottom of the cup and will boil.

SAFETY PRECAUTION: ✳ Do not perform this experiment without supervision from your instructor. Wear goggles and do not put your finger in the −41°F liquid. It will burn just like boiling water because the temperature difference between your hand and the R-22 is about the same as the difference between your hand and boiling water. ✳

A crude but effective demonstration can be used to show how air can be cooled. Fasten a long piece of

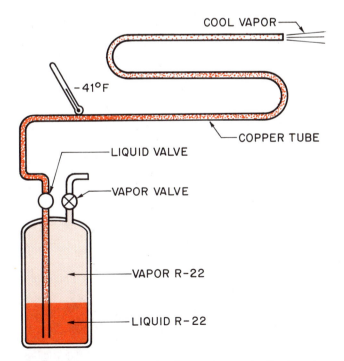

Figure 3–20. When the tubing is attached to the liquid valve on an R-22 drum and liquid is allowed to trickle into the tubing, the liquid will boil at −41°F at atmospheric pressure.

copper tubing to the liquid tap on the refrigerant drum and allow the liquid to trickle into the tube while air is passing over it. The tube has a temperature of −41°F, corresponding to atmospheric pressure, because it is escaping out of the end of the tube at atmospheric pressure. If the tube were coiled up and placed in an airstream, it would cool the air. This system has two big drawbacks:

1. The coil would freeze up and slow down the air.
2. The expensive refrigerant would be wasted and the atmosphere would be polluted, Figure 3–20.

3.6 REFRIGERATION COMPONENTS

By adding some components to the system, these problems can be eliminated. There are four major components to the mechanical refrigeration systems covered in this book:

1. The evaporator
2. The compressor
3. The condenser
4. The refrigerant metering device

3.7 THE EVAPORATOR

The *evaporator* absorbs heat into the system. When the refrigerant is boiled at a lower temperature than that of the substance to be cooled, it absorbs heat from the substance. The boiling temperature of 40°F was chosen in the previous air conditioning examples because it is the design temperature normally used for air conditioning systems. The reason is that room temperature is close to 75°F, which readily gives up heat to a 40°F coil. The 40°F temperature is also well above the freezing point of the coil. See Figure 3–21 for the coil-to-air relationships.

Let's see what happens as the R-22 refrigerant passes through the evaporator coil. The refrigerant enters the coil as a mixture of about 75% liquid and 25% vapor. This liquid–vapor mixture will be discussed in Section 3.9. The mixture is tumbling and boiling down the tube, with the liquid being turned to vapor all along

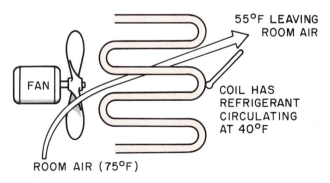

Figure 3–21. The evaporator is operated at 40°F to be able to absorb heat from the 75°F air.

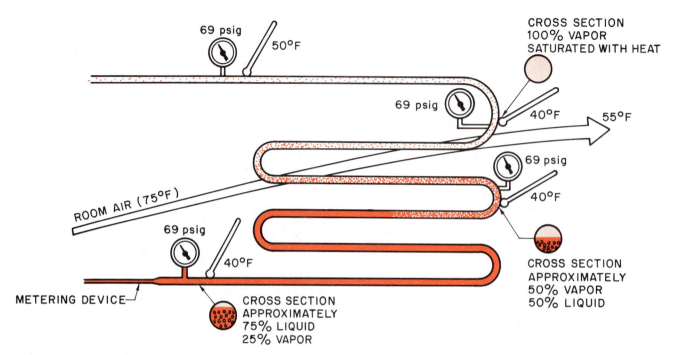

Figure 3–22. The evaporator absorbs heat into the refrigeration system by boiling the refrigerant at a temperature lower than the room air passing over it. The 75°F room air readily gives up heat to the 40°F evaporator by conduction.

the coil because heat is being added to the coil from the air, Figure 3–22. About halfway down the coil, the mixture becomes more vapor than liquid. The purpose of the evaporator is to boil all of the liquid into a vapor just before the end of the coil. This occurs approximately 90% of the way through the coil, when all of the liquid is gone, leaving pure vapor. At this precise point we have a saturated vapor. This is the point where the vapor would start to condense if heat were removed, or become superheated if any heat were added. When a vapor is *superheated,* it no longer corresponds to the pressure and temperature relationship; it will take on sensible heat and its temperature will rise. When metering devices are covered, it will become evident that superheat is refrigeration insurance. Superheat insures that no liquid gets past the evaporator, because when there is superheat, there is no liquid.

Evaporators have many design configurations. But for now just remember that they absorb the heat into the system from the substance to be cooled. The substance may be solid, liquid, or gas, and the evaporator has to be designed to fit the condition. See Figure 3–23 for a typical evaporator. Once absorbed into the system, the heat is now in the refrigerant gas and is drawn into the compressor.

3.8   THE COMPRESSOR

The *compressor* is the heart of the system. It pumps heat through the system in the form of heat-laden vapor. A compressor can be considered a vapor pump.

Two types of compressing mechanisms are covered in this text: (1) the reciprocating compressor (2) the rotary compressor. Both of these compressors are known as *positive displacement pumps.* This basically means that when a cylinder is full of gas, it either will discharge the gas or it will stall.

Because the *reciprocating compressor* is by far the most used, it will be covered in more detail. It is called "reciprocating" because the piston goes back and forth or up and down, not in a circle. The crankshaft converts the circular motion of the electric motor to reciprocating motion. This principle is the same as the crankshaft in an automobile engine. Figure 3–24 shows a reciprocating crankshaft, piston, and cylinder.

The compressor, or vapor pump, boosts the pressure in the system. On a hot day in the South it is not unusual for the compressor to raise the gas pressure in the system from 70 psig on the low-pressure side of the

Figure 3–23. Typical evaporator. *Courtesy Larkin Coils, Inc.*

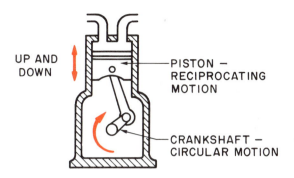

Figure 3-24. The crankshaft converts the circular motion of the motor to the reciprocating or back and forth motion of the piston.

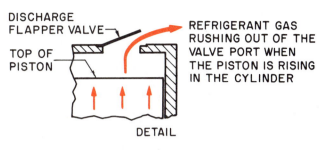

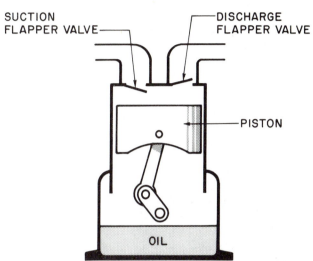

Figure 3-25. Flapper valves and compressor components.

system to 300 psig on the high-pressure side of the system. This requires a lot of work energy and is accomplished through a set of valves. The valves in reciprocating compressors are called *flapper valves* or *reed valves*. These valves use no driven force from the compressor; their operation is based on pressure difference, Figure 3-25.

From the illustrations notice that when the piston comes up it compresses or pressurizes and traps high-pressure gas at the top of the cylinder. This forces the flapper valve to open and let the gas from the low-pressure side of the system into the high-pressure side of the system. The efficiency of the compressor is hurt at this point because of the trapped clearance volume. This describes a simple one-cylinder compressor. The compressor has numerous design configurations, but all operate on the same principle.

The reciprocating compressor is turned by a motor on the outside (an *open-drive* system) or on the inside (a *hermetic* compressor), Figure 3-26. The compressor needs a method to pump positive lubrication to all moving parts and some way to cool the motor if it is inside the compressor. These details will be covered more fully in Sections IV and VII.

The rotary compressor, on the other hand, has never been extensively used, at least in the medium and large sizes. It does the same job in the compression cycle as the reciprocating compressor does: It boosts the pressure from the low-pressure side of the system to the high-pressure side of the system. The rotary compressor has fewer moving parts than the reciprocating compressor and is more efficient because it has no backstroke. Remember that the reciprocating compressor has some volume left at the top of the stroke that reexpands on the downstroke, Figure 3-27. The rotary compressor has no downstroke; it goes around and

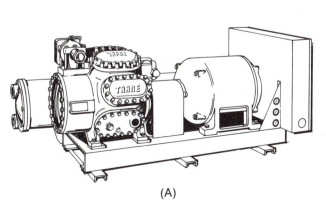

Figure 3-26. (A) Open-drive compressor. (B) Hermetic compressor. *Courtesy (A) Trane Company (B) Copeland Corporation*

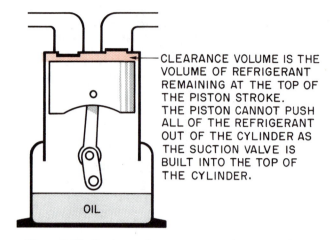

CLEARANCE VOLUME IS THE VOLUME OF REFRIGERANT REMAINING AT THE TOP OF THE PISTON STROKE. THE PISTON CANNOT PUSH ALL OF THE REFRIGERANT OUT OF THE CYLINDER AS THE SUCTION VALVE IS BUILT INTO THE TOP OF THE CYLINDER.

OIL

Figure 3–27. Reciprocating compressor clearance volume.

never backs up. The rotary compressor has no off-center crankshaft; the shaft is straight. The off-center motion is due to the off-center position of the circular piston in the chamber. When the piston makes a complete circle in the chamber, it is equivalent to a complete revolution of the reciprocating compressor. Figure 3–28 shows that all of the gas is squeezed out of the cylinder. No reexpansion of gas occurs.

In Figure 3–29 gas that is cold to the touch is entering the compressor through the inlet (suction) side. Although the gas is cold, it is known to be laden with heat from the boiling process in the evaporator. Remember that heat from the room air entered the evaporator at 75°F and added heat to the boiling refrigerant. This same heat is in this cold vapor entering the compressor.

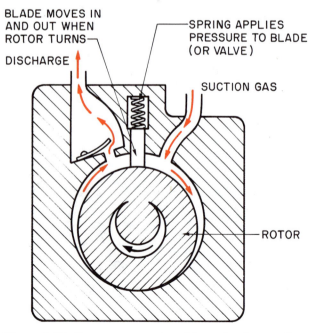

BLADE MOVES IN AND OUT WHEN ROTOR TURNS

DISCHARGE

SPRING APPLIES PRESSURE TO BLADE (OR VALVE)

SUCTION GAS

ROTOR

Figure 3–28. Rotary compressor with motion in one direction and no back stroke.

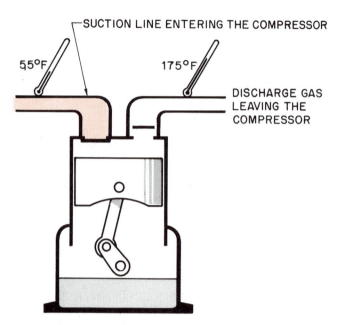

SUCTION LINE ENTERING THE COMPRESSOR

55°F   175°F

DISCHARGE GAS LEAVING THE COMPRESSOR

Figure 3–29. Compressor suction line is cold to the touch, yet it contains the heat absorbed in the house.

When this cold gas enters the compressor housing, it still has to be pulled into the cylinder through the suction valve. The suction valve is a thin piece of spring steel that covers the suction port. The valves in a compressor are different from the valves in an automotive engine. The valves in an automotive engine are physically driven or moved by the camshaft. The valves in a refrigeration compressor are opened and closed by a difference in pressure from one side of the valve to the other, Figure 3–30. The boiling refrigerant causes a pressure at the evaporator side of the valve. This valve is known as a *flapper* valve because it flaps up and down with pressure difference. When the pressure on the cylinder side of the valve is less than the evaporator side, the valve flaps open and lets the cylinder fill with vapor, Figure 3–31.

The refrigerant is drawn into the cylinder by the suction of the piston on the downstroke. When the piston gets to the bottom of the stroke, it reverses direction and starts on the upstroke. As the piston begins to get to the top of the cylinder, the cold gas is packed into the top of the cylinder. This causes the heat in the gas to concentrate, and the gas becomes warm. Some of the work energy that the compressor is exerting is transferred to the compressed refrigerant. This work energy is known as *heat of compression*. The result is that a hot, highly compressed gas is on the cylinder side of the discharge flapper valve trying to push it open. When the pressure inside the cylinder becomes greater than the pressure on the valve outlet side, the valve flaps open and the gas rushes out of the cylinder into the high-pressure side of the system.

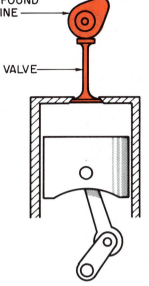

A MECHANICAL FORCE MUST TURN CAMSHAFT
IN TIME WITH THE CRANKSHAFT

CAMSHAFT SUCH AS FOUND
IN AUTOMOBILE ENGINE

VALVE

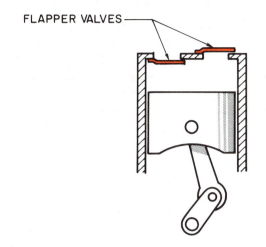

FLAPPER VALVES ARE OPERATED
BY DIFFERENTIAL IN PRESSURE

FLAPPER VALVES

Figure 3–30. Camshaft valves vs. pressure-operated valves.

The gas leaving the compressor is highly superheated; that is, it is much hotter than the condensing temperature. The condensing temperature is 130°F, whereas the temperature of the hot gas leaving the compressor is about 200°F. This gas contains the heat absorbed by the evaporator plus the heat picked up in the compressor (heat of compression). The gas line leaving the compressor is the *hot gas line.* It carries the hot gas to the next component in the system, the condenser.

3.9 THE CONDENSER

The *condenser* rejects heat from the refrigeration system that the evaporator absorbed and the compressor pumped. The condenser receives the hot gas after it leaves the compressor through the short pipe between the compressor and the condenser called the hot gas line, Figure 3–32. The hot gas is forced into the top of the condenser coil by the compressor. The gas is being pushed along at high speed and at high temperature, (about 200°F). The gas is not corresponding to the pressure and temperature relationship because the head pressure is 300 psig. The head pressure for 200°F would be off the pressure and temperature chart. Remember that the temperature at which the change of state would take place is 130°F. This temperature establishes the head pressure of 300 psig.

The gas entering the condenser is so hot compared to the surrounding air that a heat exchange begins to take place immediately in the air. The surrounding air that is being passed over the condenser is 95°F as compared to the near 200°F of the gas entering the condenser. As the gas moves through the condenser, it begins to give up heat to the surrounding air. This causes a drop in gas temperature. The gas keeps cooling off until it reaches the condensing temperature of 130°F, and the change of state begins to take place. The change of state begins slowly at first with small amounts of liquid and gets faster as the combination gas–liquid mixture moves toward the end of the condenser.

When the condensing refrigerant gets about 90% of the way through the condenser, the refrigerant in the pipe becomes almost, or all, pure liquid. Now more heat can be taken from the liquid. The liquid at the end of the condenser is at the condensing temperature of 130°F and can still give up some heat to the surrounding 95°F air. When the liquid at the end of the condenser goes below 130°F (about 110°F), it is called *subcooled,* Figure 3–33.

There are as many types of condensing devices as there are evaporating devices. Remember that the condenser is the component that rejects the heat out of the refrigeration system. The heat may have to be rejected into a solid, liquid, or a gas substance, and a condenser can be designed to do the job. Figure 3–34 shows some typical condensing units.

3.10 THE REFRIGERANT METERING DEVICE

The warm liquid is now moving down the liquid line in the direction of the *metering device.* The liquid temperature is about 115°F and may still give up some

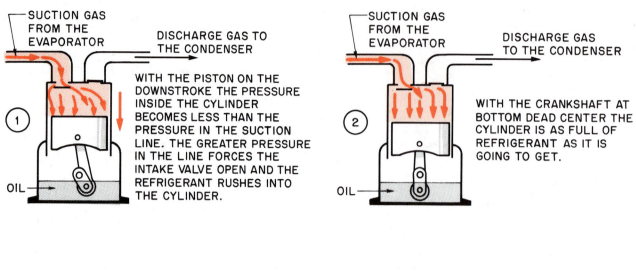

SUCTION GAS FROM THE EVAPORATOR

DISCHARGE GAS TO THE CONDENSER

① WITH THE PISTON ON THE DOWNSTROKE THE PRESSURE INSIDE THE CYLINDER BECOMES LESS THAN THE PRESSURE IN THE SUCTION LINE. THE GREATER PRESSURE IN THE LINE FORCES THE INTAKE VALVE OPEN AND THE REFRIGERANT RUSHES INTO THE CYLINDER.

OIL

SUCTION GAS FROM THE EVAPORATOR

DISCHARGE GAS TO THE CONDENSER

② WITH THE CRANKSHAFT AT BOTTOM DEAD CENTER THE CYLINDER IS AS FULL OF REFRIGERANT AS IT IS GOING TO GET.

OIL

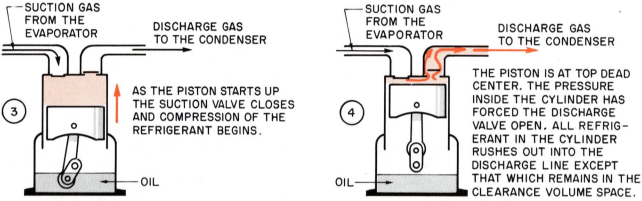

SUCTION GAS FROM THE EVAPORATOR

DISCHARGE GAS TO THE CONDENSER

③ AS THE PISTON STARTS UP THE SUCTION VALVE CLOSES AND COMPRESSION OF THE REFRIGERANT BEGINS.

OIL

SUCTION GAS FROM THE EVAPORATOR

DISCHARGE GAS TO THE CONDENSER

④ THE PISTON IS AT TOP DEAD CENTER. THE PRESSURE INSIDE THE CYLINDER HAS FORCED THE DISCHARGE VALVE OPEN. ALL REFRIGERANT IN THE CYLINDER RUSHES OUT INTO THE DISCHARGE LINE EXCEPT THAT WHICH REMAINS IN THE CLEARANCE VOLUME SPACE.

OIL

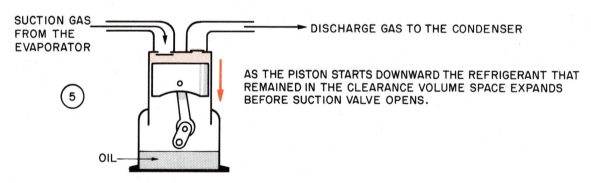

SUCTION GAS FROM THE EVAPORATOR

DISCHARGE GAS TO THE CONDENSER

⑤ AS THE PISTON STARTS DOWNWARD THE REFRIGERANT THAT REMAINED IN THE CLEARANCE VOLUME SPACE EXPANDS BEFORE SUCTION VALVE OPENS.

OIL

Figure 3–31. Compressor sequence of operation.

heat to the surroundings before reaching the metering device. This line may be routed under a house or through a wall where it may easily reach a new temperature of about 110°F. Any heat given off to the surroundings is helpful because it came from within the system and will help the system capacity.

The metering device used in this discussion will be a simple fixed-size type known as an *orifice*. It is a small restriction of a fixed size in the line, Figure 3–35.

This device holds back the full flow of refrigerant and is the dividing point between the high-pressure and the low-pressure sides of the system. Only pure liquid must enter it. The pipe leading up the orifice may be the size of a pencil, and the precision-drilled hole in the orifice may be the size of a very fine sewing needle. As you can see from the figure, the gas flow is greatly restricted here. The liquid refrigerant entering the orifice is at a pressure of 300 psig; the refrigerant leaving the orifice

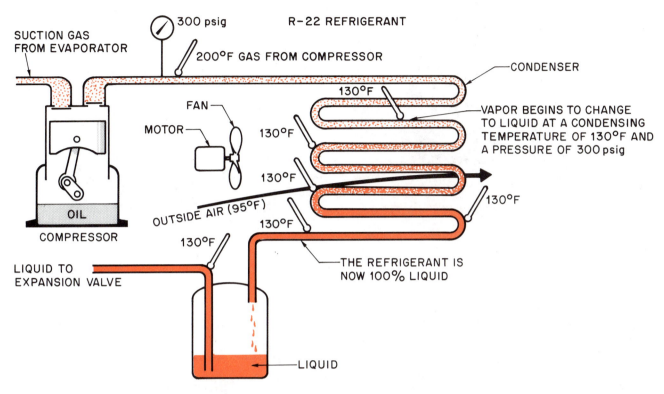

Figure 3–32. Vapor inside the condenser changes to liquid refrigerant.

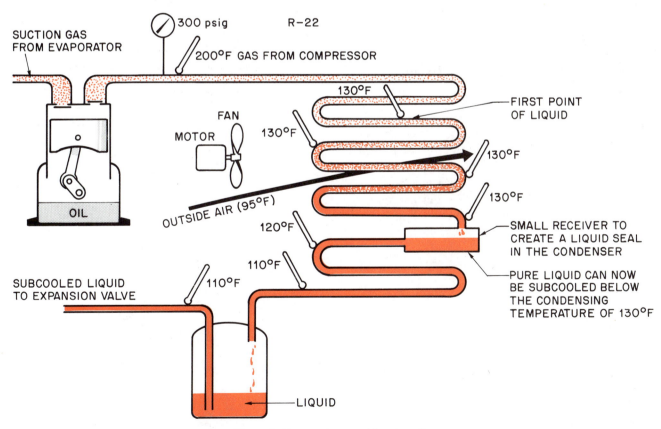

Figure 3–33. Condenser with subcooling.

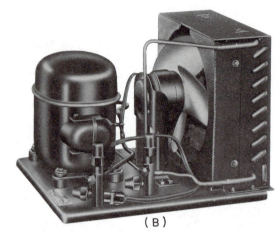

Figure 3-34. Typical condensing units. (A) Air-cooled semihermetic condensing unit. (B) Air-cooled hermetic condensing unit. *Courtesy (A) Copeland Corporation (B) Tecumseh Products Company*

is a *mixture* of about 75% liquid and 25% vapor at a new pressure of 70 psig and a new temperature of 40°F, Figure 3–34.

Two questions usually arise at this time.

1. Why did 25% of the liquid change to a gas?
2. How did the mixture of 100% pure liquid go from 110° to 40°F in such a short space?

These questions can be answered by using a garden hose as the example. If you pinch the outlet end of a water hose under pressure, the water coming out feels cooler, Figure 3–36. It actually is because some of it evaporates and turns to a mist. This evaporation takes heat out of the rest of the water and cools it off. Now when the high-pressure refrigerant passes through the orifice, it does the same thing as the water in the hose does; namely, it changes pressure (300 psig to 70 psig), and some of the refrigerant flashes to a vapor, (called *flash* gas) which cools the remaining gas to the pressure and temperature relationship of 70 psig, or 40°F. The liquid entering the metering device would heat the liquid in the evaporator if it were warmer when it actually reached the evaporator. This quick drop in pressure in the metering device lowers the boiling point of the liquid leaving the metering device.

Several types of metering devices are available for many applications. They will be covered in detail in later units. See Figure 3–37 for some examples of the various types of metering devices.

3.11 REFRIGERATION SYSTEM AND COMPONENTS

The basic components of the mechanical compression system have been described according to function. These components must be properly matched for each specific application. For instance, a low-temperature compressor cannot be applied to a high-temperature application because of the pumping characteristics. Some equipment can be mismatched successfully by using the manufacturer's data, but only someone with

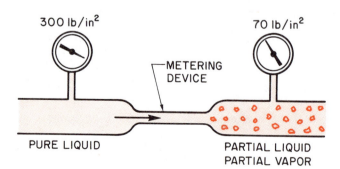

Figure 3-35. An orifice metering device.

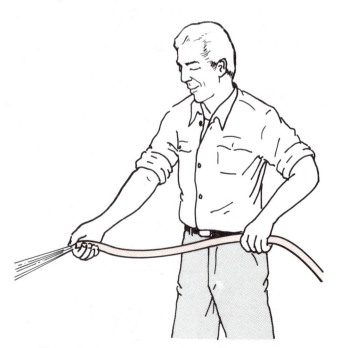

Figure 3-36. Person squeezing end of garden hose.

Figure 3-37. Metering devices. (A) Capillary tube. (B) Automatic. (C) Thermostatic. *Courtesy (A) Parker Hannifin Corporation (B and C) Singer Controls Division, Schiller Park, Illinois*

considerable knowledge and experience should do so.

We now describe a matched system correctly working at design conditions. Later we will explain malfunctions and adverse operating conditions.

Our example is a typical air conditioning system operating at a design temperature of 75°F inside temperature with a humidity (moisture content of the conditioned room air) of 50%. These conditions are to be maintained inside the house. The air in the house gives up heat to the refrigerant. The humidity factor has been brought up at this time because the indoor coil is also responsible for removing some of the moisture from the air to keep the humidity at an acceptable level. This is known as *dehumidifying*.

Moisture removal requires considerable energy. Approximately the same amount (970 Btu) of latent heat removal is required to condense a pound of water from the air as to condense a pound of steam. All air conditioning systems must have a method for dealing with this moisture after it has turned to a liquid. Some units drip, some drain the liquid into plumbing waste drains, and some use the liquid at the outdoor coil to help the system capacity by evaporating at the condenser.

Remember that part of the system is inside the house and part of the system is outside the house. We now describe the complete refrigeration cycle. The numbers in the description correspond to those in Figure 3-38.

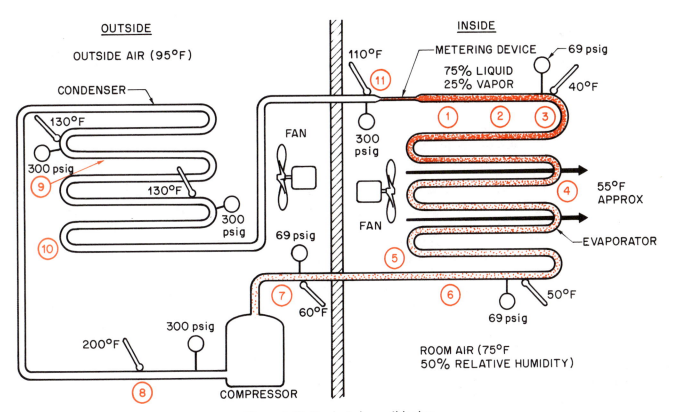

Figure 3-38. Typical air conditioning.

1. A mixture of 75% liquid and 25% vapor leaves the metering device and enters the evaporator. More on this mixture later.

2. The mixture is R-22 at a pressure of 69 psig, which corresponds to a 40°F boiling point. It is important to remember that *the pressure is 69 psig because the evaporating refrigerant is boiling at 40°F.*

3. The mixture tumbles down the tube in the evaporator with the liquid evaporating as it moves along.

4. When the mixture is about halfway through the coil, it is composed of 50% liquid and 50% vapor and still at the same temperature/pressure relationship because the change of state is taking place.

5. The refrigerant is now 100% vapor. In other words, it has reached the *saturation point* of the vapor. Recall the example in the saturated water table; the water reached various points where it was saturated with heat. We say it is saturated with heat because if any heat is removed at this point, some of the vapor changes back to a liquid and if any heat is added, the vapor rises in temperature. This is called *superheat.* (Superheat is sensible heat.) The saturated vapor is still at 40°F and still able to absorb more heat from the 75°F room air.

6. Pure vapor now exists that is normally superheated about 10°F above the saturation point. Examine the line in Figure 3–37 at this point and you will see that the temperature is about 50°F.

Note: To arrive at the correct superheat reading at this point, take the following steps.

a. Note the suction pressure reading from the suction gage: 69 psig.

b. Convert the suction pressure reading to suction temperature using the P & T chart for R-22: 40°F.

c. Use a suitable thermometer to record the actual temperature of the suction line: 50°F.

d. Subtract the saturated suction temperature from the actual suction line temperature: 50°F − 40°F = 10°F of superheat.

This vapor is said to be heat laden because it contains the heat removed from the room air. Remember that the heat was absorbed into the vaporizing refrigerant that boiled off to this vapor in the end of the suction line.

7. The vapor is drawn into the compressor by its pumping action, which creates a low-pressure suction. When the vapor left the evaporator its temperature was about 50°F with 10°F of superheat above the saturated boiling temperature. As the vapor moves along towards the compressor, it is contained in the suction line. This line is usually copper and insulated to keep it from drawing heat into the system from the surroundings; however,

it still picks up some heat. Because the suction line carries vapor, any heat that it picks up will quickly raise the temperature. Remember that it does not take much sensible heat to raise the temperature of a vapor. Depending on the length of the line and the quality of the insulation, the suction line temperature may be 60°F at the compressor end.

8. Highly superheated gas leaves the compressor through the *hot gas line* on the high-pressure side of the system. This line normally is very short because the condenser is usually very close to the compressor. On a hot day the hot gas line may be close to 200°F with a pressure of 300 psig. Since the saturated temperature corresponding to 300 psig is 130°F, the hot gas line has about 70° of superheat that must be removed before condensing can take place. Because the line is so hot and a vapor is present, the line will give up heat readily to the surroundings. The surrounding air temperature is 95°F.

9. The superheat has been removed down to the 130°F condensing temperature, and liquid refrigerant is beginning to form. Now notice that the coil temperature is corresponding to the high-side pressure of 300 psig and 130°F. The high-pressure reading of 300 psig is due to the refrigerant condensing at 130°F. The condensing conditions are arrived at by knowing the efficiency of the condenser. In this example we use a standard condenser which has a condensing temperature about 35°F higher than the surrounding air used to absorb heat from the condenser. In this example 95°F outside air is used to absorb the heat, so 95°F + 35°F = 130°F condensing temperature. Some condensers will condense at 25°F above the surrounding air; these are high-efficiency condensers, and the high-pressure side of the system will be operating under less pressure.

10. The refrigerant is now 100% liquid at the saturated temperature of 130°F. As the liquid continues along the coil, the air continues to cool the liquid to below the actual condensing temperature. The liquid may go as much as 20°F below the condensing temperature of 130°F before it reaches the metering device.

11. The liquid refrigerant reaches the metering device through a pipe, usually copper, from the condenser. This liquid line is often field installed and not insulated. The distance between the two may be long, and the line may give up heat along the way. Heat given up here is leaving the system, and that is good. The refrigerant entering the metering device may be as much as 20°F cooler than the condensing temperature of 130°F, so the liquid line entering the metering device may be 110°F. The process the refrigerant goes through in the metering device

is highly important. This was covered previously when the system was discussed as individual components. Now let's add it to the system's sequence of events.

12. When the refrigerant enters the metering device the liquid is about 110°F. This is 20°F cooler than the condensing temperature and is 100% pure liquid. In the short distance of the metering device's orifice (a pinhole about the size of a small sewing needle) the above liquid is changed to a mixture of about 75% liquid and 25% vapor. The 25% vapor is known as *flash gas* and is used to cool the remaining 75% of the liquid down to 40°F, the boiling temperature of the evaporator.

The refrigerant has now completed the refrigeration cycle and is ready to go around again. It should be evident now that a refrigerant does the same thing over and over, changing from a liquid to a vapor in the evaporator and back to a liquid form in the condenser. The expansion device meters the flow to the evaporator, and the compressor pumps the refrigerant out of the evaporator.

The following statements briefly summarize the refrigeration cycle

1. **The evaporator absorbs heat into the system.**
2. **The condenser rejects heat from the system.**
3. **The compressor pumps the heat-laden vapor through the system.**
4. **The expansion device meters the flow of refrigerant.**

3.12 REFRIGERANTS

Previously we have used water and R-22 as examples of refrigerants. Although many products have the characteristics of a refrigerant, only a few will be covered here. Residential and light commercial air conditioning and refrigeration systems commonly use three refrigerants (yet another is still encountered in existing equipment): R-22, used primarily in air conditioning; R-12, used primarily in medium- and high-temperature refrigeration; R-502, used primarily in low-temperature refrigeration. R-500 can still be found in older equipment.

3.13 REFRIGERANT MUST BE SAFE

A refrigerant must be safe to protect people from sickness or injury, even death, if the refrigerant should escape from its system. For instance, it could be a disaster to use ammonia for the air conditioning system in a public place.

Modern refrigerants are nontoxic, and equipment is designed to use a minimum amount of refrigerant to accomplish its job. A household refrigerator or window air conditioner, for example, normally uses less than 2 lb of refrigerant, yet for years almost a pound of refrigerant was used as the propellent in a 16-oz. aerosol can of hair spray.

SAFETY PRECAUTION: *Because refrigerants are heavier than air, proper ventilation is important. For example, if a leak in a large tank of refrigerant should occur in a basement, the oxygen could be displaced by the refrigerant and a person could be overcome and pass out. Avoid open flame when a refrigerant is present. When refrigeration equipment or drums are located in a room with an open gas flame, such as a pilot light on a gas water heater or furnace, keep the equipment leak free. If the refrigerant escapes and gets to the flame, the flame will sometimes burn an off-blue or blue-green. This means the flame is giving off a toxic and corrosive gas that will deteriorate any steel in the vicinity and burn the eyes and nose and severely hamper the breathing of anyone in the room. The refrigerants themselves will not burn.*

Refrigerants should not be allowed to escape to the atmosphere on purpose. In the 1970s an issue became very prominant that involved some of the common refrigerants that were used as propellants in the aerosol industry. It is believed that the release of these refrigerants to the atmosphere causes a build up of the ozone layer in the upper atmosphere and can cut some of the sun's rays off from the earth. A big drive to eliminate these problems led to different propellants being used. The refrigeration technician should make every effort to keep refrigerants that are being handled from escaping.

The refrigerants mentioned here are not corrosive under normal conditions, so they do not deteriorate equipment or surroundings when they escape. When the refrigerant is escaping, the refrigeration technician can use various methods to detect leaks. This brings up the next quality of a good refrigerant.

3.14 REFRIGERANT MUST BE DETECTABLE

A good refrigerant must be readily detectable. The first leak detection device that can be used for some large leaks is listening for the hiss of the escaping refrigerant, Figure 3–39(A). This is not the best way in all cases as some leaks may be so small they may not be heard by the human ear. However, many leaks can be found in this way.

Soap bubbles are a practical and yet simple leak detector. Commercially prepared products that blow large elastic types of bubbles are used by many service technicians. These are very valuable when it is known that a leak is in a certain area. Soap bubble solution can be applied with a brush to the tubing joint to see exactly where the leak is. Leaking refrigerant will cause bubbles, Figure 3–39(B). There are times when a piece

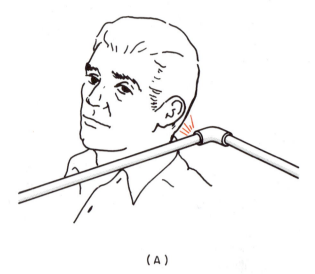

(A)

(B)

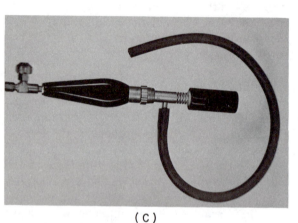

(C)

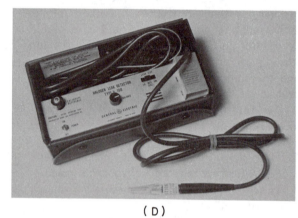

(D)

Figure 3–39. Methods and equipment for determining refrigerant leaks. *Courtesy (D) Yokogawa. Photos (B, C) by Bill Johnson*

of equipment can be submerged in water to watch for bubbles. This is very effective when it can be used.

The halide leak detector, Figure 3–39(C), is available for use with acetylene and propane gas. It operates on the principle that when the refrigerant is run through an open flame in the presence of glowing copper the flame will change color.

The leak detection devices we have mentioned are capable of finding fairly large leaks. More accurate methods are required on some recent equipment because refrigerant charges have been reduced to the bare working minimum. Some residential air conditioning equipment has refrigerant charge specifications that call for half-ounce accuracy. The electronic leak detectors are capable of detecting leak rates down to one-half ounce per year, Figure 3–39(D).

3.15 THE BOILING POINT OF THE REFRIGERANT

The boiling point of the refrigerant should be low at atmospheric pressure so that low temperatures may

be obtained without going into a vacuum. For example, R-502 can be boiled as low as −50° F before it goes into a vacuum, whereas R-12 can only be boiled down to −21° F before it goes into a vacuum. Water would have to be boiled at 29.67 in. Hg vacuum just to boil at 40° F.

Note: When using the compound gage below atmospheric pressure, the scale reads in reverse of the inches of mercury absolute scale. It starts at atmospheric pressure and counts down to a perfect vacuum, called inches of mercury vacuum. When possible, avoid using refrigerants that boil below 0 psig. This is one reason why R-502 is a good choice for a low-temperature system. When a system operates in a vacuum and a leak occurs, the atmosphere is pulled inside the system instead of the refrigerant leaking out of the system.

3.16 PUMPING CHARACTERISTICS

The pumping characteristics have to do with how much refrigerant vapor is pumped per amount of work accomplished. Water was disqualified as a practical refrigerant for small equipment partly for this reason. One pound of water at 40°F has a vapor volume of 2445 ft³ compared to about 0.6 ft³ for R-22. Thus the compressor would have to be very large for a water system.

Modern refrigerants meet all of these requirements better than any of the older types. See Figure 3–40 for pressure and temperature chart for the refrigerants we have discussed.

3.17 REFRIGERANT CHEMICAL MAKEUP

Each refrigerant has a chemical formula and chemical name. Sometimes this formula or name best indicates the refrigerant to use in a particular application. The formulas and names are as follows:

- The chemical formula for R-12 is CCl_2F_2; the chemical name is dichlorodifluoromethane.

- The chemical formula for R-22 is $CHClF_2$; the chemical name is monochlorodifluoromethane.
- There is no chemical formula or name for R-502. It is an azeotropic mixture of 48.8% R-22 and 51.2% R-115. The chemical formula for R-115 is $CClF_2CF_3$; its chemical name is chloropentafluoroethane. This mixture occurs not in the field but while the refrigerant is being manufactured.
- R-500 is an azeotropic mixture of 73.8% R-12 and 26.2% R-152a. The chemical formula for R-152a is CH_3CH_2Cl; its chemical name is difluoroethane. R-500 is also mixed during the manufacturing.

SAFETY PRECAUTION: *All refrigerants that have been discussed in this text are stored in pressurized containers (drums) and should be handled with care. Consult your instructor or supervisor for use and handling of these refrigerants. Goggles should be worn while transferring the refrigerants from the container to the system.*

TEMPERATURE °F	REFRIGERANT			TEMPERATURE °F	REFRIGERANT			TEMPERATURE °F	REFRIGERANT		
	12	22	502		12	22	502		12	22	502
−60	19.0	12.0	7.2	12	15.8	34.7	43.2	42	38.8	71.4	83.8
−55	17.3	9.2	3.8	13	16.4	35.7	44.3	43	39.8	73.0	85.4
−50	15.4	6.2	0.2	14	17.1	36.7	45.4	44	40.7	74.5	87.0
−45	13.3	2.7	1.9	15	17.7	37.7	46.5	45	41.7	76.0	88.7
−40	11.0	0.5	4.1	16	18.4	38.7	47.7	46	42.6	77.6	90.4
−35	8.4	2.6	6.5	17	19.0	39.8	48.8	47	43.6	79.2	92.1
−30	5.5	4.9	9.2	18	19.7	40.8	50.0	48	44.6	80.8	93.9
−25	2.3	7.4	12.1	19	20.4	41.9	51.2	49	45.7	82.4	95.6
−20	0.6	10.1	15.3	20	21.0	43.0	52.4	50	46.7	84.0	97.4
−18	1.3	11.3	16.7	21	21.7	44.1	53.7	55	52.0	92.6	106.6
−16	2.0	12.5	18.1	22	22.4	45.3	54.9	60	57.7	101.6	116.4
−14	2.8	13.8	19.5	23	23.2	46.4	56.2	65	63.8	111.2	126.7
−12	3.6	15.1	21.0	24	23.9	47.6	57.5	70	70.2	121.4	137.6
−10	4.5	16.5	22.6	25	24.6	48.8	58.8	75	77.0	132.2	149.1
−8	5.4	17.9	24.2	26	25.4	49.9	60.1	80	84.2	143.6	161.2
−6	6.3	19.3	25.8	27	26.1	51.2	61.5	85	91.8	155.7	174.0
−4	7.2	20.8	27.5	28	26.9	52.4	62.8	90	99.8	168.4	187.4
−2	8.2	22.4	29.3	29	27.7	53.6	64.2	95	108.2	181.8	201.4
0	9.2	24.0	31.1	30	28.4	54.9	65.6	100	117.2	195.9	216.2
1	9.7	24.8	32.0	31	29.2	56.2	67.0	105	126.6	210.8	231.7
2	10.2	25.6	32.9	32	30.1	57.5	68.4	110	136.4	226.4	247.9
3	10.7	26.4	33.9	33	30.9	58.8	69.9	115	146.8	242.7	264.9
4	11.2	27.3	34.9	34	31.7	60.1	71.3	120	157.6	259.9	282.7
5	11.8	28.2	35.8	35	32.6	61.5	72.8	125	169.1	277.9	301.4
6	12.3	29.1	36.8	36	33.4	62.8	74.3	130	181.0	296.8	320.8
7	12.9	30.0	37.9	37	34.3	64.2	75.8	135	193.5	316.6	341.2
8	13.5	30.9	38.9	38	35.2	65.6	77.4	140	206.6	337.2	362.6
9	14.0	31.8	39.9	39	36.1	67.1	79.0	145	220.3	358.9	385.0
10	14.6	32.8	41.0	40	37.0	68.5	80.5	150	234.6	381.5	408.4
11	15.2	33.7	42.1	41	37.9	70.0	82.1	155	249.5	405.1	432.9

Vacuum — Red Figures
Gage Pressure — Bold Figures

Figure 3–40. Pressure and temperature relationship chart in in. Hg vacuum or psig.

SUMMARY

- Bacterial growth that causes food spoilage slows at lower temperatures.
- The frozen hard point for most foods is considered to be 0°F. Below this point food-spoiling bacteria stops growing.
- Product temperatures above 45°F and below room temperature are considered high-temperature refrigeration.
- Product temperatures between 35° and 45°F are considered to be medium-temperature refrigeration.
- Product temperatures from 0° and −10°F are considered low-temperature refrigeration.
- Refrigeration is the process of removing heat from a place where it is not wanted and transferring it to a place where it makes no difference.
- One ton of refrigeration is the amount of heat necessary to melt one ton of ice in a 24-hr period. It takes 288,000 Btu to melt a ton of ice in a 24-hr period or 12,000 Btu in one hour or 200 Btu in one minute.
- The boiling point of liquids can be changed and controlled by controlling the pressure on the refrigerant. The relationship of the vapor pressure and the boiling point is called the pressure–temperature relationship. When the pressure is increased, the boiling point increases. When the pressure is decreased, the boiling point decreases.
- A compressor can be considered a vapor pump. It lowers the pressure in the evaporator to the desired temperature and increases the pressure in the condenser to a level where the vapor may be condensed to a liquid.
- The liquid refrigerant moves from the condenser to the metering device where it again enters the evaporator. The metering device causes part of the liquid to vaporize, and the pressure and temperature are greatly reduced as the refrigerant enters the evaporator and starts the cycle over again.
- Refrigerants have a very definite chemical makeup and are usually designated with an "R" and a number for field identification. R-22 is primarily used in air conditioning systems, R-12 in high-, medium-, and some low-temperature applications, and R-502 in low- and extra low-temperature applications.
- A refrigerant must be safe, must be detectable, must have a low boiling point, and must have good pumping characteristics.

REVIEW QUESTIONS

1. Name three reasons why ice melts in an icebox.
2. What are the approximate temperature ranges for low-, medium-, and high-temperature refrigeration applications?
3. What is a ton of refrigeration?
4. Describe briefly the basic refrigeration cycle.
5. What is the relationship between pressure and the boiling point of liquids?
6. What is the function of the evaporator in a refrigeration or air conditioning system?
7. What does the compressor do in the refrigeration system?
8. What happens to the refrigerant in the condenser?
9. What happens to heat in the refrigerant while in the condenser?
10. What does the metering device do?
11. Describe the difference between a reciprocating compressor and a rotary compressor.
12. Give a simple definition of a compressor.
13. Name two common refrigerants used in refrigeration.
14. What refrigerant is commonly used in air conditioning systems?
15. List four characteristics a manufacturer must consider when choosing a refrigerant.

Section Two

Shop Practices and Tools

TOOLS AND EQUIPMENT

OBJECTIVES

After studying this unit, you should be able to

- **describe most hand tools used by the air conditioning, heating, and refrigeration technician.**
- **describe most of the equipment used to install and service air conditioning, heating, and refrigeration systems.**

Air conditioning, heating, and refrigeration technicians must be able to properly use hand tools and specialized equipment relating to this field. Use tools and equipment correctly for the job for which they were intended. This unit contains a brief description of most of the general and specialized tools and equipment that technicians use in this field. Some of these are described in more detail in other units as they apply to specific tasks.

4.1 GENERAL HAND TOOLS

Figures 4–1 through 4–6 illustrate many general hand tools used by service technicians.

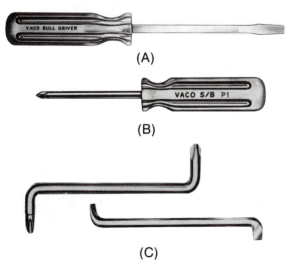

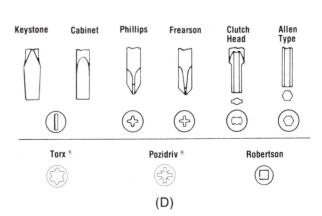

Figure 4-1. Screwdrivers. (A) Straight or slot blade. (B) Phillips tip. (C) Offset. (D) Other standard screwdriver bit types. *Courtesy Vaco Products Company*

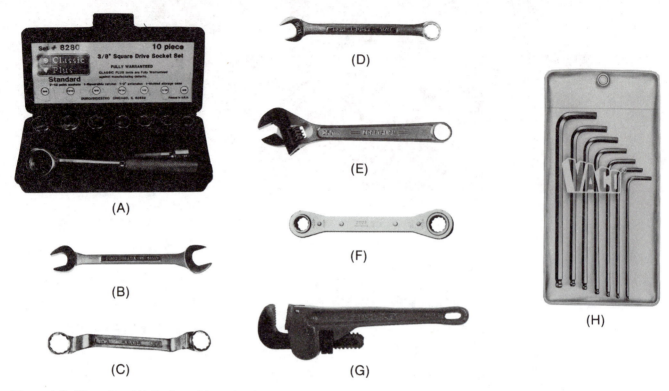

Figure 4–2. Wrenches. (A) Socket with ratchet handle. (B) Open end. (C) Box end. (D) Combination. (E) Adjustable open end. (F) Ratchet box. (G) Pipe. (H) Hex key. *Courtesy (F, H) Vaco Products Company. Photos (A–E, G) by Bill Johnson*

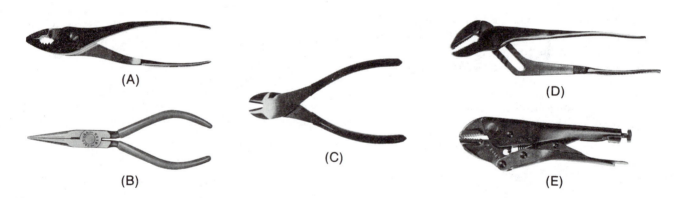

Figure 4–3. Pliers. (A) General purpose. (B) Needle nose. (C) Side cutting. (D) Slip joint. (E) Vise grip. *Photos by Bill Johnson*

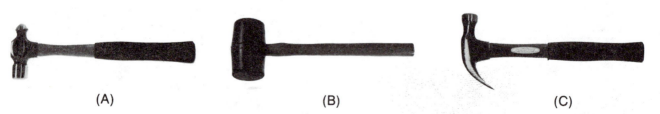

Figure 4–4. Hammers. (A) Ball peen. (B) Soft head. (C) Carpenter's claw. *Photos by Bill Johnson*

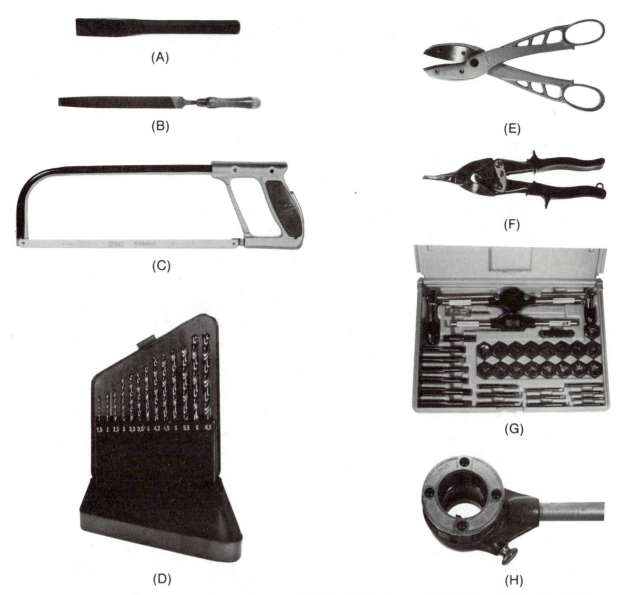

Figure 4-5. General Metal-Cutting Tools. (A) Cold chisel. (B) File. (C) Hacksaw. (D) Drill bits. (E) Straight Metal snips. (F) Aviation metal snips. (G) Tap and die. A tap is used to cut an internal thread. A die is used to cut an external thread. (H) Pipe-threading die. *Photos by Bill Johnson*

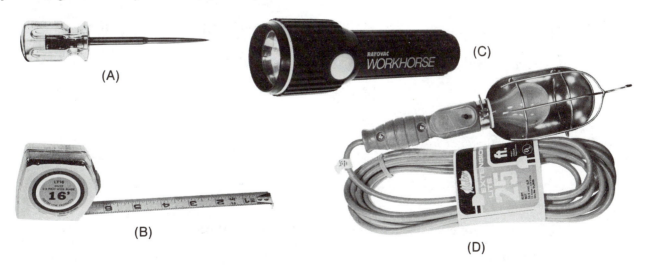

Figure 4-6.. Other general-purpose tools. (A) Awl. (B) Rule. (C) Flashlight. (D) Extension cord/light. *Courtesy (A) Vaco Products Company. Photos (B–D) by Bill Johnson*

4.2 SPECIALIZED HAND TOOLS

The following tools are regularly used by technicians in the air conditioning, heating, and refrigeration field.

Nut Drivers. Nut drivers have a socket head and are used primarily to remove hex head screws from panels on air conditioning, heating, and refrigeration cabinets. They are available with hollow shaft, solid shaft, and extra long or stubby shafts, Figure 4–7. The hollow shaft allows the screw to protrude into the shaft when it extends beyond the nut.

Air Conditioning and Refrigeration Reversible Ratchet Box Wrenches. Ratchet box wrenches, where the ratchet direction can be changed by pushing the lever-like button near the end of the wrench, are used with air conditioning and refrigeration valves and fittings. Two openings on each end of the wrench allow it to be used on four sizes of valve stems or fittings, Figure 4–8.

Flare Nut Wrench. The flare nut wrench is used like a box end wrench. The opening at the end of the wrench allows it to be slipped over the tubing. The wrench can then be placed over the flare nut to tighten or loosen it, Figure 4–9. Ratchet flare nut wrenches for use in tight places are available.

Wrenches should not be used if they are worn because they will round off the corners of fittings made of soft brass. Do not use a pipe to extend the handle on a wrench for more leverage. Fittings can be loosened by heating or by using a penetrating oil.

Wiring and Crimping Tools. Wiring and crimping tools are available in many designs. Figure 4–10 illustrates a combination tool for crimping solderless

Figure 4–9. Flare nut wrench. *Photo by Bill Johnson*

connectors, stripping wire, cutting wire, and cutting small bolts. This figure also illustrates an *automatic wire stripper.* To use this tool, insert the wire into the proper strip-die hole. The length of the strip is determined by the amount of wire extending beyond the die away from the tool. Hold the wire in one hand and squeeze the handles with the other. Release the handles and remove the stripped wire.

Inspection Mirrors. Inspection mirrors are usually available in rectangular or round shapes, with fixed or telescoping handles, some over 30 in. long. The mirrors are used to inspect areas or parts in components that are behind or underneath other parts, Figure 4–11.

Stapling Tackers. Stapling tackers are used to fasten insulation and other soft materials to wood; some types may be used to install low-voltage wiring, Figure 4–12.

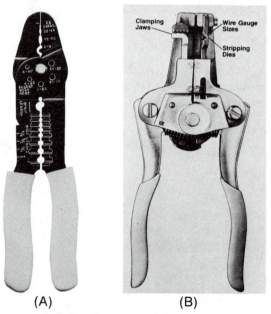

(A)   (B)

Figure 4–10. Wiring and crimping tools. (A) A combination crimping and stripping tool for crimping solderless connectors, stripping wire, cutting wire, and and cutting small bolts. (B) Automatic wire stripper *Courtesy Vaco Products Company*

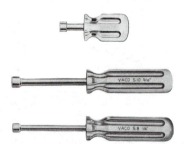

Figure 4–7. Assorted nut drivers. *Courtesy Vaco Products Company*

Figure 4–8. Air conditioning and refrigeration reversible ratchet box wrench. *Courtesy Vaco Products Company*

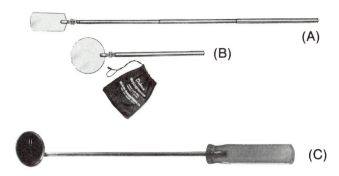

Figure 4–11. Inspection mirrors. *Courtesy (A, B) Delavan Corporation (C) Robinair Division-Sealed Power Corporation*

The following tools are used to install tubing. Most will be described more fully in other units in the book.

Tube Cutter. Tube cutters are available in different sizes and styles. The standard tube cutter is shown in Figure 4–13. They are also available with a ratchet feed mechanism. The cutter opens quickly to insert the tubing and slides to the cutting position. Some models have a flare cutoff groove that reduces the tube loss when removing a cracked flare. Many models also include a retractable reamer to remove inside burrs, and a filing surface to remove outside burrs. Figure 4–14 illustrates a cutter for use in tight spaces where standard cutters do not fit.

Inner-outer Reamers. Inner-outer reamers use three cutters to ream both the inside and trim the outside edges of tubing, Figure 4–15.

Figure 4–12. Stapling tacker. *Courtesy Bostitch Division of Textron Inc.*

Figure 4–13. Tube cutter. *Courtesy Robinair Division-Sealed Power Corporation*

Figure 4–14. Small tube cutter used in tight places. *Courtesy Robinair Division-Sealed Power Corporation*

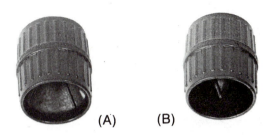

Figure 4–15. Inner-outer reamers. *Photos by Bill Johnson*

Flaring Tools. The flaring tool has a flaring bar to hold the tubing, a slip-on yoke, and a feed screw with flaring cone and handle. Several sizes of tubing can be flared with this tool, Figure 4–16.

Swaging Tools. Swaging tools are available in punch type and lever type, Figure 4–17A, B.

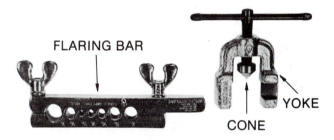

FLARING BAR

YOKE

CONE

Figure 4–16. Flaring tool. This tool has a flaring bar, a yoke and a feed screw with flaring cone. *Photo by Bill Johnson*

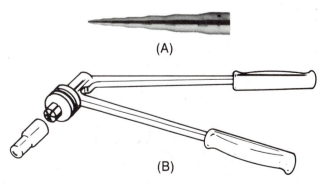

(A)

(B)

Figure 4–17. Swaging tool. (A) Punch type. (B) Lever Type. *Courtesy (A) Robinair Division-Sealed Power Corporation*

Tube Benders. Three types of tube benders may be used: spring type, lever type, and, to a lesser extent, gear type. These tools are used for bending soft copper and aluminum, Figure 4–18.

Tube Brushes. Tube brushes clean the inside and outside of tubing and the inside of fittings. There are types that can be turned by hand or by an electric drill, Figure 4–19.

Plastic Tubing Shear. A plastic tubing shear cuts plastic tubing and non-wire-reinforced plastic or synthetic hose, Figure 4–20.

Tubing Pinch-off Tool. A tubing pinch-off tool is used to pinch shut the short stub of tubing often provided for service, such as the service stub on a

Figure 4–20. Plastic tubing shear. *Courtesy Bramec Inc.*

compressor. This tool is used to pinch shut this stub prior to sealing it by soldering it, Figure 4–21.

Metalworkers Hammer. A metalworkers hammer straightens and forms sheet metal for duct work, Figure 4–22.

Reciprocating Portable Power Saw. The reciprocating portable power saw is used to make rough cuts in plywood and to cut into duct work when necessary to install heating or air conditioning systems.

4.3   SPECIALIZED SERVICE AND INSTALLATION EQUIPMENT

One of the most important of all pieces of refrigeration and air conditioning service equipment is the *gage*

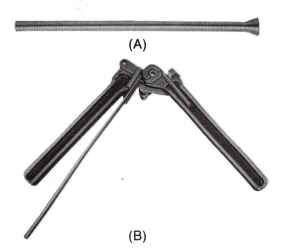

(A)

(B)

Figure 4–18. Tube benders. (A) Spring type. (B) Lever type. *Courtesy Robinair Division-Sealed Power Corporation*

Figure 4–19. Tubing brushes. *Courtesy Shaefer Brushes*

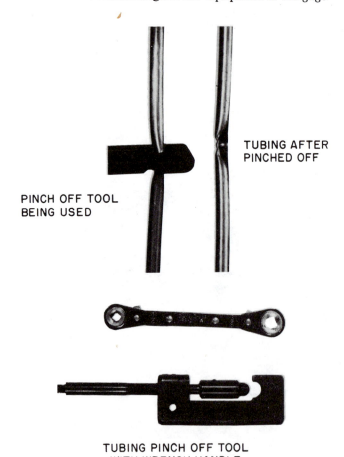

TUBING AFTER PINCHED OFF

PINCH OFF TOOL BEING USED

TUBING PINCH OFF TOOL WITH WRENCH HANDLE

Figure 4–21. Tubing pinch-off tool. *Photos by Bill Johnson*

Figure 4-22. Metalworker's hammer. *Photo by Bill Johnson*

manifold. This equipment normally includes the *compound gage* (low-pressure and vacuum), *high-pressure gage,* the *manifold,* valves, and hoses. The gage manifold may be two-valve or four-valve. The four-valve design has separate valves for the vacuum, low-pressure, high-pressure, and refrigerant drum connections, Figure 4-23.

Charging Cylinder. The charging cylinder allows the service technician to more accurately charge a system with refrigerant. It may have heating elements that help

Figure 4-24. Charging cylinder. *Courtesy Robinair Division-Sealed Power Corporation*

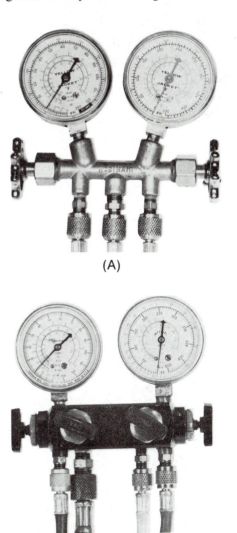

(A)

(B)

Figure 4-23. Gage manifolds. (A) Two-valve gage manifold. (B) Four-valve gage manifold *Photos by Bill Johnson*

to compensate for temperature fluctuations. Instructions for using this cylinder are given in other units, Figure 4-24.

Vacuum Pump. Vacuum pumps designed specifically for servicing air conditioning and refrigeration systems remove the air and noncondensible gases from the system. This is called *evacuating* the system and is necessary because they take up space, contain moisture, and cause excessive pressures, Figure 4-25.

Halide Leak Detector. A halide leak detector, Figure 4-26, detects refrigerant leaks. They are used with acetylene gas or propane gas. When the detector is ignited, the flame heats a copper disc. Air for the combustion is drawn through the attached hose. The end of the hose is passed over or near fittings or other areas where a leak may be suspected. If there is a leak the refrigerant will be drawn into the hose and contact the copper disc. This breaks down the halogen refrigerants into other compounds and changes the color of the flame. The colors range from green to purple, depending on the size of the leak.

Figure 4-25. Vacuum pump. *Photo by Bill Johnson*

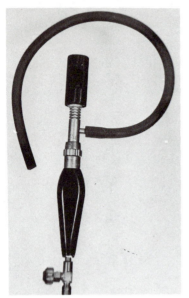

Figure 4-26. Halide leak detector used with acetylene. *Photo by Bill Johnson*

Electronic Leak Detectors. Electronic leak detectors contain an element sensitive to halogen gases. The device may be battery or AC powered and often has a pump to suck in the gas and air mixture. A ticking signal that increases in frequency and intensity as the probe "homes in" on the leak is used to alert the operator. Many also have varying sensitivity ranges that can be adjusted, Figure 4-27.

Thermometers. Thermometers range from simple pocket styles to electronic and recording types. Pocket-style mercury column and dial-indicator thermometers, Figure 4-28, and remote bulb thermometers are frequently used. For more sophisticated temperature measurements an electronic or recording-type thermometer may be used, Figures 4-29, 4-30.

Figure 4-28. Thermometers. (A) Mercury column. (B) Dial indicator. (C) Remote bulb. *Courtesy (B) Robinair Division-Sealed Power Corporation. Photos (A, C) by Bill Johnson*

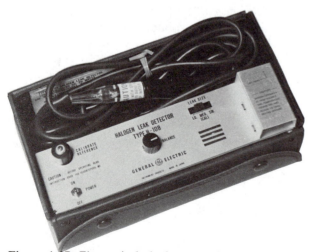

Figure 4-27. Electronic leak detector. *Courtesy Yokogawa Corporation of America*

Figure 4-29. Electronic thermometer. *Courtesy Thermal Engineering*

Figure 4–30. Recording thermometer. *Courtesy United Technologies Bacharach*

Fin Straighteners. Fin straighteners are available in different styles. Figure 4–31 is an example of one type capable of straightening condensor and evaporator coil fins having spacing of 8, 9, 10, 12, 14, and 15 fins per inch.

Heat Guns. Heat guns are needed in many situations to warm refrigerant, melt ice, and do other tasks. Figure 4–32 is a heat gun carried by many technicians.

Hermetic Tubing Piercing Valves. Piercing valves are an economical way to tap a line for charging, testing, or

Figure 4–31. Condenser or evaporator coil fin straightener. *Courtesy Robinair Division-Sealed Power Corporation*

Figure 4–32. Heat gun. *Courtesy Robinair Division-Sealed Power Corporation*

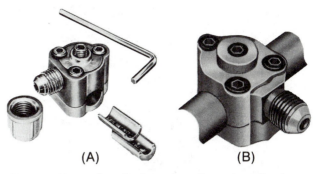

Figure 4–33. Tubing piercing valve. *Photos by Bill Johnson*

purging hermetically sealed units. A valve such as that in Figure 4–33 is clamped to the line. Often these are designed so that the valve stem is turned until a sharp needle pierces the tubing. Then the stem is backed off so that the service operations can be accomplished. The valve can then be left on the tubing for future servicing.

Compressor Oil Charging Pump. Figure 4–34 is an example of a compressor oil charging pump used for charging refrigeration compressors with oil without pumping the compressor down.

Charging Stations. Charging stations contain equipment used to charge a system, an appliance, or an automobile air conditioning system. The setup in Figure 4–35 is often used in a shop. A typical station would include a vacuum pump, a charging cylinder, manifold, gages, lines, and possibly other useful items.

Soldering and Welding Equipment. Soldering and welding equipment is available in many styles, sizes, and qualities.

Soldering Gun. A soldering gun is used primarily to solder electrical connections. It does not produce enough heat for soldering tubing, Figure 4–36.

Propane Gas Torch. Figure 4–37 shows a propane gas torch with a disposable propane gas tank. This is an easy torch to use. The flame adjusts easily and can be used for many soldering operations.

Air Acetylene Unit. Air acetylene units provide sufficient heat for soldering and brazing. They consist of a

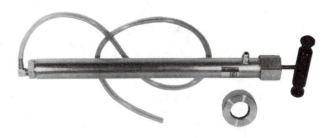

Figure 4–34. Compressor oil charging pump. *Courtesy Robinair Division-Sealed Power Corporation*

Figure 4-35. Charging station. *Courtesy Robinair Division-Sealed Power Corporation*

Figure 4-36. Soldering gun. *Photo by Bill Johnson*

Figure 4-37. Propane gas torch with a disposable propane gas tank. *Photo by Bill Johnson*

Figure 4-38. Air acetylene unit. *Courtesy Wingaersheek*

torch, which can be fitted with several sizes of tips, a regulator, hoses, and an acetylene tank, Figure 4-38.

Oxyacetylene Welding Units. Oxyacetylene welding units are used for light-duty welding applications and heavy-duty brazing. They consist of a torch, regulators, hoses, oxygen tanks, and acetylene gas tanks. When the oxygen and acetylene gases are mixed in the proper proportion, a very hot flame is produced, Figure 4-39.

Sling Psychrometer. The sling psychrometer uses the wet-bulb–dry-bulb principle to obtain relative humidity readings quickly. Two thermometers, a dry bulb and a wet bulb, are whirled together in the air. Evaporation will occur at the wick of the wet-bulb

Figure 4-39. Oxyacetylene welding unit. *Courtesy Wingaersheek*

thermometer, giving it a lower temperature reading. The difference in temperature depends on the humidity in the air. The drier the air, the greater the difference in the temperature readings because the air absorbs more moisture, Figure 4–40. Most manufacturers provide a scale so that the relative humidity can easily be determined. Before you use it, be sure that the wick is clean and wet with pure water if possible.

Motorized Psychrometer. A motorized psychrometer is often used when many readings are taken over a large area.

Nylon Strap Fastener. Figure 4–41 shows the installation tool used to install nylon strap clamps around flexible duct. This tool automatically cuts the clamping strap off flush when a preset tension is reached.

Figure 4–40. Sling psychrometer. *Photo by Bill Johnson*

Figure 4–41. Tool used to install nylon strap clamps around flexible duct. *Courtesy Panduit Corporation*

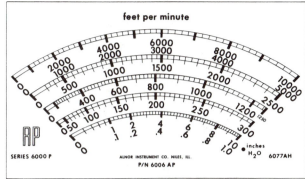

Figure 4–43. Dial face on an air velocity measuring instrument. *Courtesy Alnor Instrument Company*

Air Velocity Measuring Instruments. Air velocity measuring instruments are necessary to balance duct systems, check fan and blower characteristics, and make static pressure measurements. They measure air velocity in feet per minute. Figure 4–42 shows an air velocity measuring kit. This type of instrument makes air velocity measurements from 50 to 10,000 ft/min. Figure 4–43 shows a dial face on one of these instruments. Figure 4–44 shows an air velocity instrument incorporating a microprocessor that will take up to 250 readings across hood openings or in a duct and will display the average air velocity and temperature readings when needed.

Air Balancing Meter. The air balancing meter shown in Figure 4–45 eliminates the need to take multiple readings. The meter will make readings directly from exhaust or supply grilles in ceilings, floors, or walls, and read out in cubic feet per minute (cfm).

Carbon Dioxide (CO_2) and Oxygen (O_2) Indicators. CO_2 and O_2 indicators are used to make flue-gas analy-

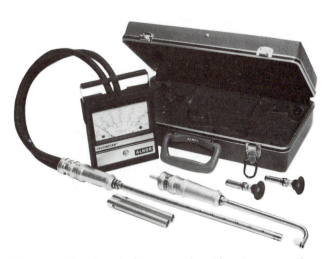

Figure 4–42. Air velocity measuring kit. *Courtesy Alnor Instrument Company*

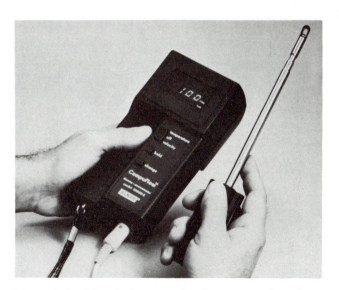

Figure 4–44. Air velocity measuring instrument with microprocessor. *Courtesy Alnor Instrument Company*

Figure 4–45. Air balancing meter. *Courtesy Alnor Instrument Company*

Figure 4–47. Carbon monoxide indicator. *Courtesy United Technologies Bacharach*

ses to determine combustion efficiency in gas or oil furnaces, Figure 4–46.

Carbon Monoxide (CO) Indicator. A CO indicator is used to take flue-gas samples in natural gas furnaces to determine the percentage of carbon monoxide present, Figure 4–47.

Electronic Combustion Analyzer. The electronic combustion analyzer is used to measure oxygen (O_2) concentrations within flue gases. It indicates flue-gas temperature, tests smoke, and tests for carbon monoxide, Figure 4–48. Other models designed to be used with larger heating equipment have built-in microcomputers that automatically compute the percentage of excess air, the percentage of carbon dioxide, and the combustion efficiency.

Draft Gage. A draft gage is used to check the pressure of the flue gas in gas and oil furnaces to insure that the flue gases are moving up the flue at a satisfac-

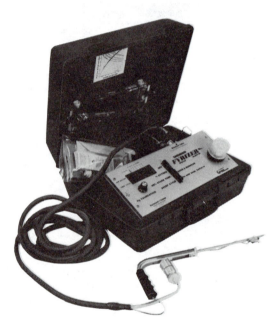

Figure 4–48. Electronic combustion analyzer. *Courtesy United Technologies Bacharach*

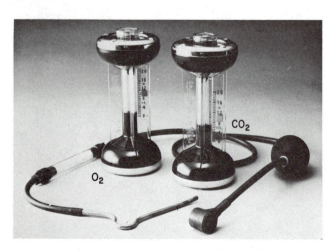

Figure 4–46. Carbon dioxide and oxygen indicators. *Courtesy United Technologies Bacharach*

tory speed, Figure 4–49. The flue gas is normally at a slightly negative pressure.

Volt-Ohm-Milliammeter (VOM). A VOM, often referred to as a multimeter, is an electrical instrument that measures voltage (volts), resistance (ohms), and current (milliamperes). These instruments have several ranges in each mode. They are available in many types, ranges, and quality, either with a regular dial (analog) readout or a digital readout. Figure 4–50 illustrates both an analog and digital type. If purchasing one, be sure to select one with the features and ranges used by technicians in this field.

Figure 4-49. Draft gage. *Courtesy United Technologies Bacharach*

AC Clamp-on Ammeter. An AC clamp-on ammeter is a valuable instrument. It is also called clip-on, tang-type, snap-on, or other names. Some can also measure voltage and/or resistance. Unless you have an ammeter like this, you must interrupt the circuit to place the ammeter in the circuit. With this instrument you simply clamp the jaws around a single conductor, Figure 4-51.

U-Tube Mercury Manometer. A U-tube mercury manometer determines the level of vacuum while evacuating a refrigeration system. It is accurate to approximately 0.5 mm Hg, Figure 4-52A.

A U-Tube Water Manometer. A U-tube water manometer determines natural gas and propane gas pressures during servicing or installation of gas furnaces and gas-burning equipment, Figure 4-52B.

Inclined Water Manometer. An inclined water manometer determines air pressures in very-low-pressure systems, such as to 0.1″ of water column. These are used to analyze air flow in air conditioning and heating air distribution systems, Figure 4-52C.

Electronic Vacuum Gage. An electronic vacuum gage checks the vacuum when evacuating a refrigeration, air conditioning, or heat pump system. It measures a vacuum down to about 1 micron or 0.001 millimeter (mm) (1000 microns = 1 mm), Figure 4-53.

A technician will use many other tools and equipment, but the ones covered in this unit are most commonly used.

(A)

(B)

Figure 4-50. Volt-ohm-milliammeter (VOM). (A) Analog. (B) Digital. *Courtesy (A) Amprobe Instrument, Div. of Core Industries Inc. (B) Simpson Electric Co. Elgin, Illinois*

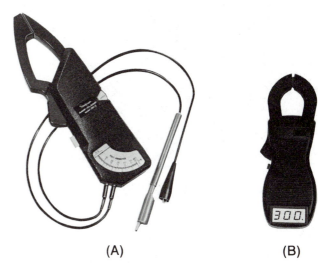

(A)   (B)

Figure 4-51. AC Clamp-on volt-ammeter. *Courtesy (A) Simpson Electric Co. Elgin, Illinois. (B) Amprobe Instrument, Div. of Core Industries Inc.*

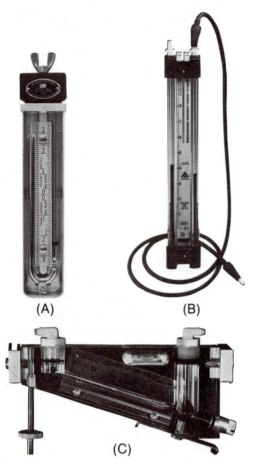

(A)   (B)

(C)

Figure 4-52. Manometers. (A) Mercury U tube. (B) Water U tube. (C) Water inclined. *Courtesy (A, B) Robinair Division-Sealed Power Corporation. Photo (C) by Bill Johnson*

Figure 4-53. Electronic Vacuum Gage. *Courtesy Robinair Division-Sealed Power Corporation*

SUMMARY

- Air conditioning, heating, and refrigeration technicians should be familiar with available hand tools and equipment.
- Technicians should properly use hand tools and specialized equipment.
- ✳These tools and this equipment should be used only for the job for which they were designed. Other use may damage the tool or equipment and may be unsafe for the technician.✳

REVIEW QUESTIONS

1. List two types of screwdriver tips.
2. Give an example of a use for a nut driver.
3. List five different types of wrenches.
4. List four different types of pliers.
5. What is the name of a hammer used primarily for metal work?
6. What is a hacksaw used for?
7. Describe uses for a wiring and crimping tool.
8. What is a tube cutter used for?
9. What are flaring tools used for?
10. Describe two types of benders for soft tubing.
11. List the components usually associated with a gage manifold unit.
12. Why is a charging cylinder used?
13. What is the purpose of a vacuum pump?
14. What are two types of refrigerant leak detectors?
15. List three types of thermometers.
16. What is a soldering gun normally used for?
17. What is an air acetylene torch used for?
18. What is a sling psychrometer used for?
19. When are air velocity measuring instruments used?
20. Describe the purpose of CO_2 and O_2 indicators.
21. What is the purpose of a draft gage?
22. What does a VOM measure?
23. What does an AC clamp-on ammeter measure?

FASTENERS

OBJECTIVES

After studying this unit, you should be able to

- identify common fasteners used with wood.
- identify a common fastener used with sheet metal.
- describe a fastener used in masonry.
- describe hanging devices for piping, tubing, and duct.

As an air conditioning (heating and cooling) and refrigeration technician, you need to know about different types of fasteners and the various fastening systems so that you will use the right fastener or system for the job and securely install all equipment and materials.

5.1 NAILS

Probably the most common fastener used in wood is the nail. Nails are available in many styles and sizes. *Common nails* are large, flat-headed, wire nails with a specific diameter for each length. A *finishing nail* has a very small head so that it can be driven below the surface of the wood; it is used where a good finish is desired.

These nails are sized by using the term penny. This term is abbreviated by the letter *d*. For instance, an 8d common nail will describe the shape and size. These nails can be purchased plain or with a coating, usually zinc or resin. The coating protects against corrosion and helps to prevent the nail from working out of the wood. Figure 5–1 illustrates a few types and sizes of nails.

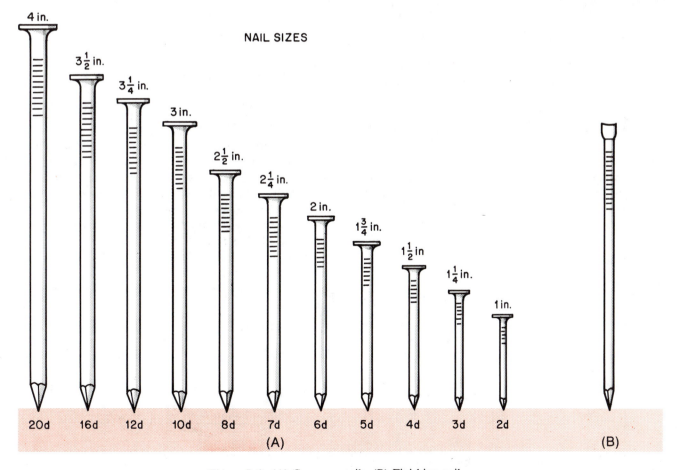

Figure 5–1. (A) Common nails. (B) Finishing nail.

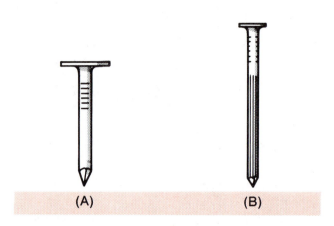

Figure 5–2. (A) Roofing nail. (B) Masonary nail.

A *roofing nail* is used to fasten shingles and other roofing materials to the roof of a building. It has a large head to keep the shingles from tearing and pulling off. A roofing nail can sometimes be used on strapping to hang duct work or tubing, Figure 5–2. *Masonry nails* are made of hardened steel and can be driven into masonry, Figure 5–2.

5.2 STAPLES AND RIVETS

Staples. A *staple* is a fastener made somewhat like a nail but shaped in a U with a point on each end. Staples are available in different sizes for different uses. One use is to fasten wire in place. The staples are simply hammered in over the wire. *Never hammer them too tight because they can damage the wire, Figure 5–3.*

Other types of staples may be fastened in place with a *stapling tacker,* Figure 5–4. By depressing the handle a staple can be driven through paper, fabric, or insulation into the wood. *Outward clinch tackers* will anchor staples inside soft materials. The staple legs spread outward to form a strong tight clinch. They can be used to fasten insulation around heating or cooling pipes and ducts and to install ductboard, Figure 5–5. Other models of tackers are used to staple low-voltage wiring to wood.

Rivets. *Pin rivets* or *blind rivets* are used to join two pieces of sheet metal. They are actually hollow rivets

Figure 5–4. Stapling tacker. *Courtesy Bostitch Division of Textron, Inc.*

assembled on a pin, often called a *mandrel.* Figure 5–6 illustrates a pin rivet assembly, and how it is used to join sheet metal, and a riveting tool or gun. The rivets are inserted and set from only one side of the metal. Thus, they are particularly useful when there is no access to the back side of the metal being fastened.

To use a pin rivet, drill a hole in the metal the size of the rivet diameter. Insert the pin rivet in the hole. Then place the nozzle of the riveting tool over the pin and squeeze the handles. Jaws grab the pin and pull the head of the pin into the rivet, expanding it and forming a head on the other side. Continue squeezing the handles until the head on the reverse side forms a tight joint. When this happens, the pin breaks off and is ejected from the riveting tool.

5.3 THREADED FASTENERS

We will describe only a few of the most common threaded fasteners.

Wood Screws. *Wood screws* are used to fasten many types of materials to wood. They generally have a flat, round, or oval head, Figure 5–7. Flat and oval heads have an angle beneath the head. Holes for these screws should be countersunk so that their angular surface will be recessed. Wood screws can have a straight slotted head or a Phillips recessed head.

Wood screw sizes are specified by length (in inches) and shank diameter (a number from 0 to 24). The larger the number, the larger the diameter or gauge of the shank.

Tapping Screws. *Tapping* or *sheet metal* screws are used extensively by service technicians. These screws

Figure 5–3. Staples used to fasten wire. *Photo by Bill Johnson*

Figure 5–5. Staple clinched outward.

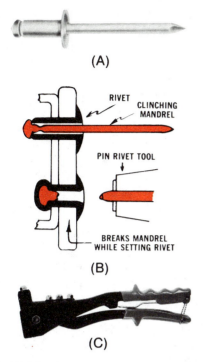

(A)

(B)

(C)

Figure 5-6. (A) Pin rivet assembly. (B) Using pin rivets to fasten sheet metal. (C) Riveting gun. *Courtesy (A, B) Duro Dyne Corp. Photo (C) by Bill Johnson*

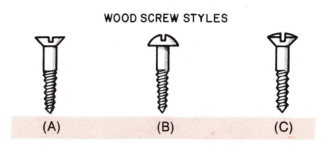

WOOD SCREW STYLES

(A)   (B)   (C)

Figure 5-7. (A) Flat head wood screw. (B) Round head wood screw. (C) Oval head wood screw.

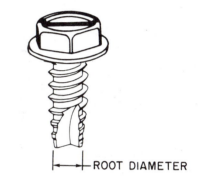

ROOT DIAMETER

Figure 5-8. Tapping screw.

may have a straight slot, Phillips head, or a straight slot and hex head, Figure 5–8. To use a tapping screw, drill a hole into the sheet metal the approximate size of the root diameter of the thread. Then turn the screw into the hole with a conventional screwdriver.

Figure 5–9 shows a tapping screw, often called a self-drilling screw, that can be turned into sheet metal with an electric drill using a chuck similar to the one shown. This screw has a special point used to start the hole. It can be used with light to medium gauges of sheet metal.

Set Screws. *Set screws* have points as illustrated in Figure 5–10. They have square heads, hex heads, or no heads. The headless type may be slotted for a

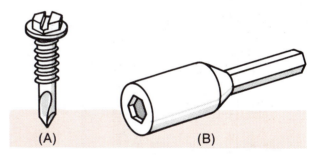

(A)   (B)

Figure 5-9. (A) Self-drilling screw. Note the special point used to start the hole. (B) Drill chuck.

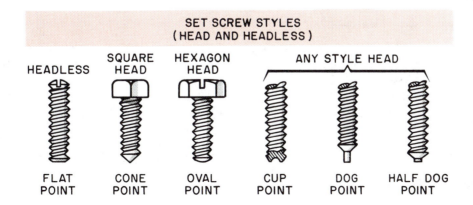

SET SCREW STYLES
(HEAD AND HEADLESS)

HEADLESS   SQUARE HEAD   HEXAGON HEAD   ANY STYLE HEAD

FLAT POINT   CONE POINT   OVAL POINT   CUP POINT   DOG POINT   HALF DOG POINT

Figure 5-10. Styles of set screws.

Figure 5-11. Anchor shield with screw. *Courtesy Rawlplug Company, Inc.*

Figure 5-12. Hollow wall anchor. *Courtesy Rawlplug Company, Inc.*

Figure 5-13. Toggle bolt. *Courtesy Rawlplug Company, Inc.*

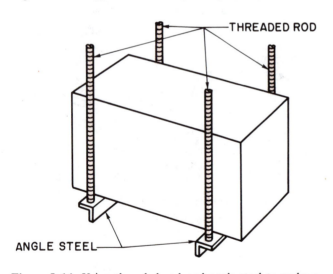

Figure 5-14. Using threaded rod and angle steel to make a hanger.

Figure 5-15. A cotter pin.

Figure 5-16. A wire pipe hook.

screwdriver, or it may have a hexagonal or fluted socket. These screws are used to keep a pulley from turning on a shaft, and other similar applications.

Anchor Shields. *Anchor shields* with bolts or screws are used to fasten objects to masonry or, in some instances, hollow walls. Figure 5-11 illustrates a one-piece multipurpose steel anchor bolt used in masonry material. Drill a hole in the masonry the size of the sleeve. Tap the sleeve and bolt into the hole with a hammer. Turn the bolt head. This expands the sleeve and secures the bolt in the masonry. A variety of head styles are available.

Wall Anchor. A *hollow wall anchor,* Figure 5-12, can be used in plaster, wallboard, gypsum board, and similar materials. Once the anchor has been set, the screw may be removed as often as necessary without affecting the anchor.

Toggle Bolts. *Toggle bolts* provide a secure anchoring in hollow tile, building block, plaster over lath, and gypsum board. Drill a hole in the wall large enough for the toggle in its folded position to go through. Push the toggle through the hole and use a screwdriver to turn the bolt head. You must maintain tension on the toggle or it will not tighten, Figure 5-13.

Threaded Rod and Angle Steel. *Threaded rod* and *angle steel* can be used to custom-make hangers for pipes or components such as an air handler, Figure 5-14. Be sure that all materials are strong enough to adequately support the equipment.

5.4 OTHER FASTENERS

Cotter Pins. *Cotter pins,* Figure 5-15, are used to secure pins. The cotter pin is inserted through the hole in the pin, and the ends spread to retain it.

Pipe Hook. A *wire pipe hook* is a wire bent into a U with a point at an angle on both ends. The pipe or tubing rests in the bottom of the U, and the pointed ends are driven into wooden joists or other wood supports, Figure 5-16.

Pipe Strap. A *pipe strap* is used for fastening pipe and tubing to joists, ceilings, or walls, Figure 5-17(A). The arc on the strap should be approximately the same as the outside diameter of the pipe or tubing. These straps are normally fastened with round head screws. Notice that the underside of the head of a round head screw is flat and provides good contact with the strap.

Perforated Strap. *Perforated strap* may also be used to support pipe and tubing. Round head stove bolts with nuts fasten the strap to itself, and round head wood screws fasten the strap to a wood support, Figure 5-17(B).

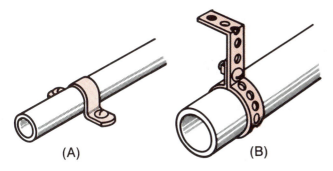

Figure 5-17. (A) A pipe strap. (B) A perforated strap.

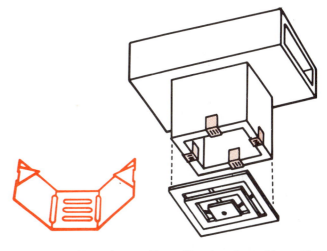

Figure 5-19. Fastening a grille to fiberglass duct with a grille clip.

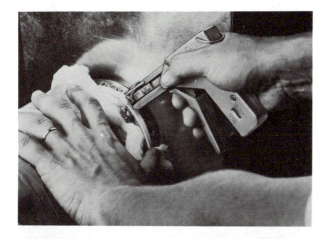

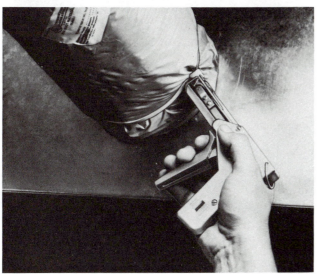

Figure 5-18. A system for fastening round flexible duct to a sheet metal collar. *Courtesy Panduit Corporation*

Nylon Strap. To fasten round flexible duct to a sheet metal collar, apply the inner liner of the duct with a sealer and clamp it with a *nylon strap*. A special tool is manufactured to install this strap, which applies the correct tension and cuts the strap off. Now position the insulation and vapor barrier over the collar, which can then be secured with another nylon strap. Apply duct tape over this strap and duct end to further seal the system. Figure 5-18 illustrates this procedure.

Grille Clip. A *grille clip* fastens grilles to the ends of fiberglass ducts. The clips are bent, Figure 5-19, around the end of the duct with the points pushed into the sides of the duct. Screws fasten the grilles to these clips.

Figure 5-20 illustrates a system to fasten damper regulators, controls, and other components to fiberglass duct. It consists of a drill screw, a head plate, and a backup plate.

Many fasteners not described here are highly specialized. More are developed each year. The intent of this unit is to depict the more common ones and to encourage you to keep up to date by discussing new fastening techniques with your supplier.

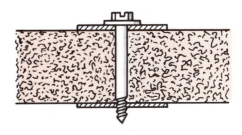

Figure 5-20. A system for fastening fiberglass duct.

SUMMARY

- Technicians need to use a broad variety of fasteners.

- Some of these fasteners are common nails, wood screws, masonry nails, staples, tapping screws, and set screws.

- Anchor shields and other devices are often used with these fasteners to secure them in masonry walls or hollow walls.

- Typical hangers used with pipe, tubing, and duct work are wire pipe hooks, pipe straps, perforated steel straps, and custom hangers made from threaded rod and angle steel.

- There are other specialty fasteners for flexible duct and fiberglass duct.

REVIEW QUESTIONS

1. Name three types of nails.
2. What term is used to describe the size of nails?
3. What is the abbreviation for this term?
4. What are masonry nails made from?
5. Describe two types of staples.
6. Describe the procedure for fastening two pieces of sheet metal together with a pin rivet.
7. Sketch three common types of screw heads.
8. Describe the procedure used for fastening two pieces of sheet metal together with a tapping screw.
9. Describe two types of tapping screws.
10. Describe two types of set screws.
11. How are anchor shields used?
12. What is a pipe strap? What is it used for?
13. What is a grille clip used for?

TUBING AND PIPING

OBJECTIVES

After studying this unit, you should be able to

- **list the different types of tubing used in heating, air conditioning, and refrigeration applications.**
- **describe two common ways of cutting copper tubing.**
- **list procedures used for bending tubing.**
- **discuss procedures used for soldering and brazing tubing.**
- **describe two methods for making flared joints.**
- **state procedures for making swaged joints.**
- **describe procedures for preparing and threading steel pipe ends.**
- **list four types of plastic pipe and describe uses for each.**

6.1 PURPOSE OF TUBING AND PIPING

The correct size, layout, and installation of tubing, piping, and fittings keeps a refrigeration or air conditioning system operating properly and prevents refrigerant loss. The piping system provides passage for the refrigerant to the evaporator, the compressor, the condenser, and the expansion valve. It also provides the way for oil to drain back to the compressor. Tubing, piping, and fittings are used in numerous applications, such as fuel lines for oil and gas burners and water lines for hot-water heating. The tubing, piping and fittings used must be of the proper size, the system must be laid out properly and it must be installed correctly.

SAFETY PRECAUTION: *Careless handling of the tubing and poor soldering or brazing techniques may cause serious damage to system components. You must keep contaminants, including moisture, from air conditioning and refrigeration systems.*

6.2 TYPES AND SIZES OF TUBING

Copper tubing is generally used for plumbing, heating, and refrigerant piping. Aluminum, steel, and stainless steel may occasionally be used for refrigerant piping. Steel and wrought iron pipe are used for gas piping

and, sometimes, hot water heating. Plastic pipe is used for waste drains, condensate drains, water supplies, water-source heat pumps, and venting high-efficiency gas furnaces.

Copper tubing is available as soft copper or hard-drawn copper tubing. Soft copper may be bent or used with elbows, tees, and other fittings. Hard-drawn tubing is not intended to be bent; use it only with fittings to obtain the necessary configurations. Copper tubing used for refrigeration or air conditioning is called *ACR* (air conditioning and refrigeration) tubing. Copper tubing used in plumbing and heating is available in four standard weights: Type K is heavy duty; type L is the standard size and used most frequently; types M and DWV are not used extensively in this industry. The outside diameter of these four types is approximately $\frac{1}{8}$ in. larger than the size indicated; that is, $\frac{1}{2}$-in. tubing has an outside diameter of approximately $\frac{5}{8}$ in. ACR tubing is sized by its outside diameter, $\frac{1}{2}$-in. ACR tubing has an outside diameter of $\frac{1}{2}$ in. For plumbing and heating applications use the inside diameter (ID), and for air conditioning and refrigeration applications use the outside diameter (OD), Figure 6–1. Copper tubing is normally available in diameters from $\frac{3}{16}$ in. to greater than 6 in.

Soft copper tubing is normally available in 25-ft or 50-ft rolls and in diameters from $\frac{3}{16}$ to $\frac{3}{4}$ in. It can be special ordered in 100-ft lengths. ACR tubing is capped on each end to keep it dry and clean inside and often has a charge of nitrogen to keep it free of contaminants. Never pull out the tubing sideways from the roll. Place it on a flat surface and unroll it, Figure 6–2. Cut only what you need and recap the ends. Do not bend or straighten the tubing more than necessary because it will harden. This is called *work hardening*. Work-hardened tubing can be softened by heating and allowing it to cool slowly. This is called *annealing*. Don't use a high concentrated heat in one area, but use a flared flame over one foot at a time. Heat to a cherry red and allow to cool slowly.

Hard-drawn copper tubing is available in 20-ft lengths and in larger diameters than soft copper tubing. Be as careful with hard-drawn copper as with soft copper, and recap the ends when the tubing is not used.

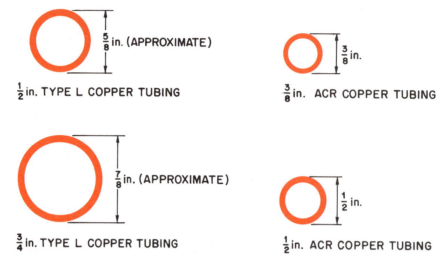

$\frac{1}{2}$ in. TYPE L COPPER TUBING — $\frac{5}{8}$ in. (APPROXIMATE)

$\frac{3}{8}$ in. ACR COPPER TUBING — $\frac{3}{8}$ in.

$\frac{3}{4}$ in. TYPE L COPPER TUBING — $\frac{7}{8}$ in. (APPROXIMATE)

$\frac{1}{2}$ in. ACR COPPER TUBING — $\frac{1}{2}$ in.

Figure 6-1. Tubing used for plumbing and heating is sized by its inside diameter. ACR tubing is sized by its outside diameter.

6.3 TUBING INSULATION

ACR tubing is often insulated on the low-pressure side of an air conditioning or refrigeration system between the evaporator and compressor to keep the refrigerant from absorbing heat. Insulation also prevents condensation from forming on the lines, Figure 6-3. The closed-cell structure of this insulation usually eliminates the need for a vapor barrier. The insulation may be purchased separately from the tubing, or it may be factory installed. If you install the insulation, it is easier, where practical, to apply it to the tubing before assembling the line. The inside diameter of the insulation is usually powdered to allow easy slippage even around most bends. You can buy adhesive to seal the ends of the insulation together, Figure 6-4.

For existing lines, or when it is impractical to insulate before installing the tubing, the insulation can be slit with a sharp utility knife and snapped over the tubing. All seams must be sealed with an adhesive. Do not use tape.

Do not stretch tubing insulation because the wall thickness of the insulation will be reduced and the adhesive may then fail to hold, and the effectiveness of the insulation reduced.

6.4 LINE SETS

Tubing can be purchased as line sets. These sets are charged with refrigerant, sealed on both ends, and may be obtained with the insulation installed. These line sets normally will have fittings on each end for quicker and

Figure 6-3. ACR tubing with insulation. *Photo by Bill Johnson*

Figure 6-2. Roll of soft tubing. Place on a flat surface and unroll. *Photo by Bill Johnson*

Figure 6-4. When joining two ends of tubing insulation, use an adhesive made specifically for this purpose. *Photo by Bill Johnson*

Figure 6–5. A typical line set. *Photo by Bill Johnson*

Figure 6–7. Removing a burr. *Photo by Bill Johnson*

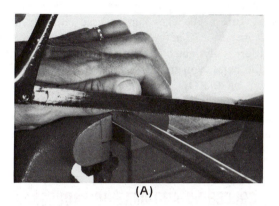

(A)

cleaner field installation, Figure 6–5. Precharging helps to eliminate improper field charging. It also reduces the possibility of contamination in the system, eliminating clogged systems and compressor damage.

6.5   CUTTING TUBING

Tubing is normally cut with a tube cutter or a hacksaw. The tube cutter is most often used with soft tubing and smaller-diameter hard-drawn tubing. A hacksaw is used with larger-diameter hard-drawn tubing. To cut the tubing with a tube cutter, see Figure 6–6:

(A) Place the tubing in the cutter and align the cutting wheel with the cutting mark on the tube. Tighten the adjusting screw until a moderate pressure is applied to the tubing.

(B) Revolve the cutter around the tubing, keeping a moderate pressure applied to the tubing by gradually turning the adjusting screw.

(C) Continue until the tubing is cut. Do not apply excessive pressure because it may break the cutter wheel and constrict the opening in the tubing.

When the cut is finished, the excess material (called a *burr*) pushed into the pipe by the cutter wheel must be removed, Figure 6–7. Burrs cause turbulence and restrict the fluid or vapor passing through the pipe.

To cut the tubing with a hacksaw, make the cut at a 90° angle to the tubing. A fixture may be used to insure an accurate cut, Figure 6–8. After cutting, ream the tubing

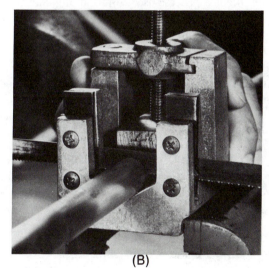

(B)

Figure 6–8. Proper procedure for using a hacksaw. *Photos by Bill Johnson*

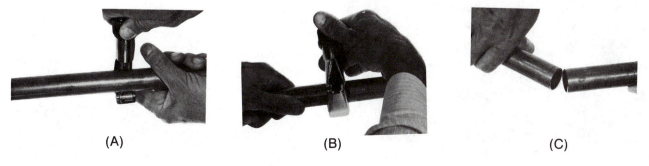

(A)   (B)   (C)

Figure 6–6. Proper procedure for using a tube cutter. *Photos by Bill Johnson*

and file the end. Remove all chips and filings, making sure that no debris or metal particles get into the tubing.

6.6 BENDING TUBING

Only soft tubing should be bent. Use as large a radius bend as possible, Figure 6-9A. All areas of the tubing must remain round. Do not allow it to flatten or kink, Figure 6-9B. Carefully bend the tubing, gradually working around the radius.

Tube bending springs may be used to help make the bend, Figure 6-10. They can be used either inside or outside the tube. They are available in different sizes for different diameter tubing. To remove the spring after the bend, you might have to twist it. If you use a spring on the OD, bend the tube before flaring so that the spring may be removed.

Lever-type tube benders, Figure 6-11, which are available in different sizes, are used to bend soft copper and thin walled steel tubing.

6.7 SOLDERING AND BRAZING PROCESSES

Soldering is a process used to join piping and tubing to fittings. It is used primarily in plumbing and heating systems utilizing copper and brass piping and fittings.

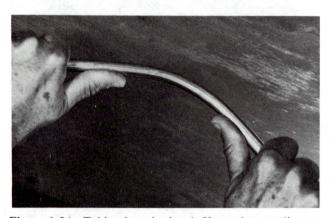

Figure 6-9A. Tubing bent by hand. Use as large radius as possible. *Photo by Bill Johnson*

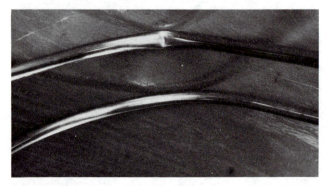

Figure 6-9B. Do not allow the tubing to flatten or kink when bending. *Photo by Bill Johnson*

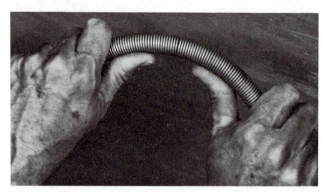

Figure 6-10. Tube bending springs used inside or outside of the tubing. Be sure to use the proper size. *Photo by Bill Johnson*

Large refrigeration systems also use hard tubing and fittings. Soldering, often called soft soldering, is done at temperatures under 800°F, usually in the 375° to 500°F range, Figure 6-12.

The 50/50 tin-lead solder is suitable for moderate pressures and temperatures. For higher pressures, or where greater joint strength is required, 95/5 tin-antimony solder can be used.

Brazing, requiring higher temperatures, is often called silver brazing, and is similar to soldering. It is used to join tubing and piping in air conditioning and refrigeration systems. Don't confuse this with welding brazing. In brazing processes temperatures over 800°F are used. The differences in temperature are necessary due to the different combinations of alloys used in the filler metals.

Brazing filler metals suitable for joining copper tubing are alloys containing 15%-60% silver (BAg), or copper alloys containing phosphorus (BCuP). Brazing filler metals are sometimes referred to as *hard solders* or *silver solders*. These are confusing terms often used by technicians, and it's better to avoid using them.

In soldering and brazing, the base metal (the piping or tubing) is heated to the melting point of the filler

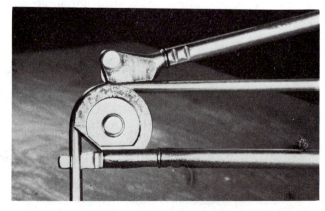

Figure 6-11. Use of a lever-type tube bender. *Photo by Bill Johnson*

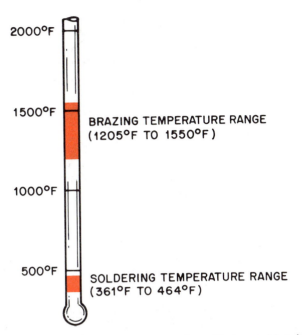

Figure 6-12. Temperature ranges for soldering and brazing.

material. *The piping and tubing must not melt.* When two close-fitting, clean, smooth metals are heated to the point where the filler metal melts, this molten metal is drawn into the close-fitting space by *capillary attraction* (see Figure 6-13 for an explanation of this). If the soldering is properly done, the molten solder will be absorbed into the pores of the base metal, adhere to all surfaces, and form a bond, Figure 6-14.

6.8   HEAT SOURCES FOR SOLDERING AND BRAZING

Propane, butane, or air-acetylene torches are the most commonly used sources of heat when soldering or

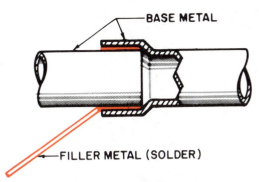

Figure 6-14. The molten solder in a soldered joint will be absorbed into the surface pores of base metal.

brazing. A *propane* or *butane* torch can be easily ignited and adjusted to the type and size of joint being soldered. Various tips are available, Figure 6-15.

An *air-acetylene* unit is the type of heat source used most often by air conditioning and refrigeration technicians. It usually consists of a B tank of acetylene gas, a regulator, a hose, and a torch, Figure 6-16. Various sizes of standard tips are available for a unit like this. The smaller tips are used for small-diameter tubing; larger tips are used for large-diameter tubing and for high-temperature applications. Figure 6-17 illustrates a popular high-velocity tip.

Follow these procedures when setting up, igniting, and using an air-acetylene unit, Figure 6-18.

(A) Before connecting the regulator to the tank, open the tank valve slightly to blow out any dirt that may be lodged at the valve.
(B) Connect the regulator with hose and torch to the tank. Insure that all connections are tight.
(C) Open the tank valve one-half turn.
(D) Adjust the regulator valve to about midrange.

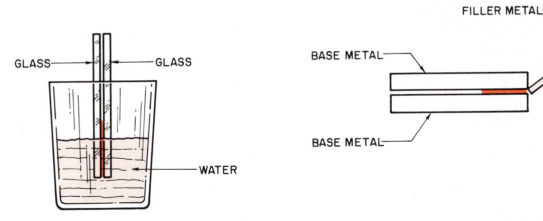

Figure 6-13. Two examples of capillary attraction. On the left are two pieces of glass spaced close together. When inserted in water, capillary attraction draws the water into the space between the two pieces of glass. The water molecules have a greater attraction for the glass than for each other. Therefore, they work their way up between the two pieces of glass. On the right is an illustration showing melted filler metal being drawn into the space between the two pieces of base metal. The molecules in the filler metal have a greater attraction to the base metal than they have for each other. These molecules work their way along the joint, first "wetting" the base metal and then filling the joint.

Figure 6–15. A propane torch with typical tip for soldering. *Photo by Bill Johnson*

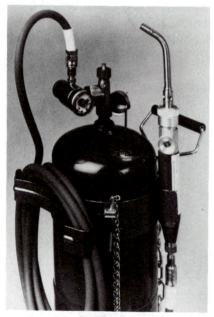

Figure 6–16. A typical air–acetylene setup. *Courtesy Winaersheek*

(E) Open the needle valve on the torch slightly, and ignite the gas with the spark lighter. *Do not use matches or cigarette lighters.* Adjust the flame using the needle valve at the handle so that there will be a sharp inner flame and a blue outer flame. After each use, shut off the valve on the tank and bleed off the acetylene in the hose by opening the

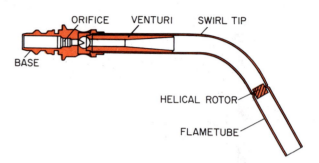

Figure 6–17. A high-velocity tip popular with many technicians. *Courtesy Winaersheek*

valve on the torch handle. Bleeding the acetylene from the hoses relieves the pressure when not in use.

Oxyacetylene torches may be preferred when brazing large-diameter tubing. *This equipment can be very dangerous when not used properly.* Because it is seldom used by the residential and light commercial air conditioning, heating, and refrigeration technician and because the instructions are so extensive, we will not describe how to use it in this text. *We recommend that you thoroughly understand proper instructions before using this equipment. We also highly recommend that when you first begin to use the equipment, use it only under the close supervision of a qualified person.*

6.9   SOLDERING TECHNIQUES

The mating diameters of tubing and fittings are designed or sized to fit together properly. For good capillary attraction there should be a space between the metals of approximately 0.003 in. After the tubing has been cut to size and deburred, you must do the following for good soldered joints:

1. Clean mating parts of the joint.
2. Apply a flux to the male connection.
3. Assemble the tubing and fitting.
4. Heat the joint and apply the solder.
5. Wipe the joint clean.

Cleaning.   The end of the copper tubing and the inside of the fitting must be absolutely clean. Even though these surfaces may look clean, they contain fingerprints, dust, or oxidation. A fine sand cloth, a cleaning pad, or a special wire brush may be used. When the piping system is for a hermetic compressor, the sand cloth should be a nonconducting approved type, Figure 6–19A.

Fluxing.   Apply flux soon after the surfaces are cleaned. For soft soldering flux may be a paste, jelly, or liquid. Apply the flux with a *clean* brush or applicator. Do *not* use a brush that has been used for any

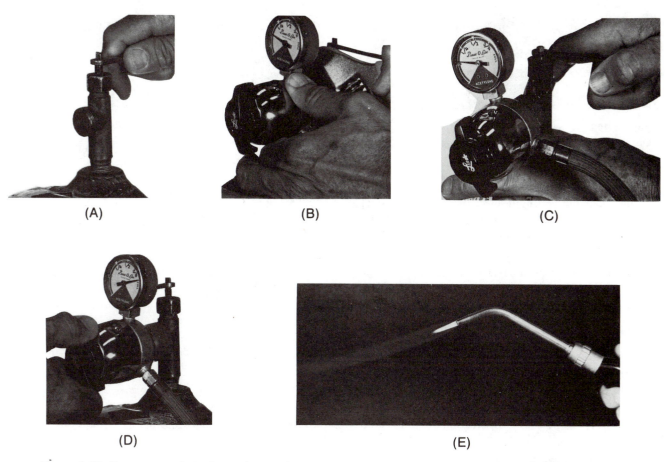

(A) (B) (C)

(D) (E)

Figure 6–18. Proper procedures for setting up, igniting, and using an air–acetylene unit. *Photos by Bill Johnson*

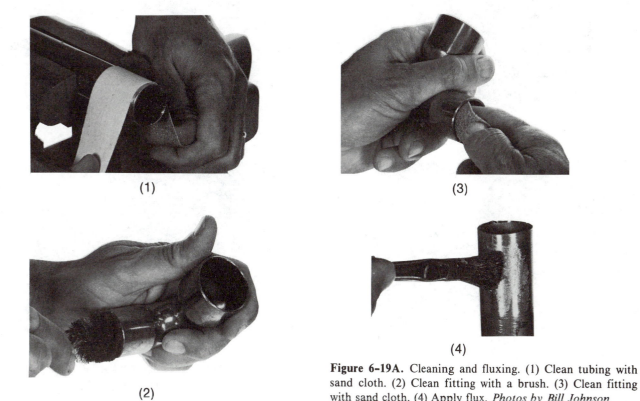

(1) (3)

(2) (4)

Figure 6-19A. Cleaning and fluxing. (1) Clean tubing with sand cloth. (2) Clean fitting with a brush. (3) Clean fitting with sand cloth. (4) Apply flux. *Photos by Bill Johnson*

Figure 6-19B. A properly assembled and supported joint ready to be soldered. *Photo by Bill Johnson*

other purpose. Apply the flux only to the area to be joined, and avoid getting it into the piping system. The flux minimizes oxidation while the joint is being heated. It also helps to float dirt or dust out of the joint.

Assembly. Soon after the flux is applied assemble and support the joint so that it is straight and will not move while being soldered, Figure 6-19B.

Heating and Applying Solder. When soldering, heat the tubing near the fitting first for a short time. Then move the torch from the tubing to the fitting. Keep moving the torch to spread the heat evenly and do not overheat any area. Do not point the flame into the fitting socket, but hold the torch so that the inner cone of the flame just touches the metal. After briefly heating the joint, touch the solder to the joint. If it does not readily melt, remove it and continue heating the joint. Continue to test the heat of the metal with the solder. Do *not* melt the solder with the flame; use the heat in the metal. When the solder flows freely from the heat of the metal, feed enough solder in to fill the joint. Do not use excessive solder. Figure 6-20 is a step-by-step procedure for heating the joint and applying solder.

Wiping. While the joint is still hot, wipe it with a rag. This is not necessary for producing a good bond, but it improves the appearance of the joint.

6.10 BRAZING TECHNIQUES

Cleaning. The cleaning procedures for brazing are very similar to those for soldering. The brazing flux is applied with a brush to the cleaned area of the tube end. Avoid getting flux inside the piping system. Some brands of a silver-copper-phosphorus alloy do not require extensive cleaning when brazing copper to copper. Follow the instructions from the filler-material manufacturer.

Applying Heat for Brazing. ✱Before you heat the joint, it is good practice to force nitrogen or carbon dioxide into the system to purge the air and reduce the

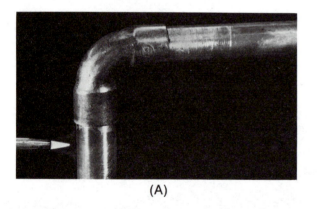

(A)

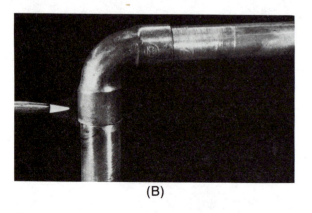

(B)

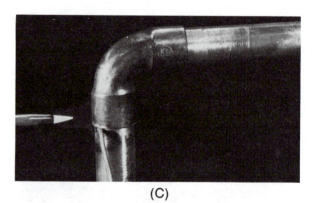

(C)

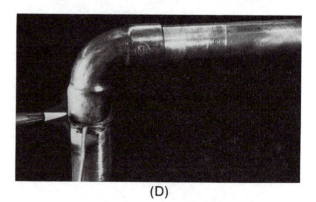

(D)

Figure 6-20. Proper procedures for heating a joint and applying solder. (A) Start by heating the tubing. (B) Keep moving the flame. Do not point flame into edge of fitting. (C) Touch solder to joint to check for proper heat. Do not melt solder with flame. (D) When joint is hot enough, solder will flow. *Photos by Bill Johnson*

possibility of oxidation.* Apply heat to the parts to be joined with an air-acetylene torch. Heat the tube first, beginning about one inch from the edge of the fitting, sweeping the flame around the tube. It is very important to keep the flame in motion and to not overheat any one area. Then switch the flame to the fitting at the base of the cup. Heat uniformly, sweeping the flame from the fitting to the tube. Apply the filler rod or wire at a point where the tube enters the socket. When the proper temperature is reached, the filler metal will flow readily, by capillary attraction, into the space between the tube and the fitting. As in soldering, do not heat the rod or wire itself. The temperature of the metal at the joint should be hot enough to melt the filler metal. When the joint is at the correct temperature, it will be cherry red in color. The procedures are the same as with soldering except for the materials used and the higher heat applied.

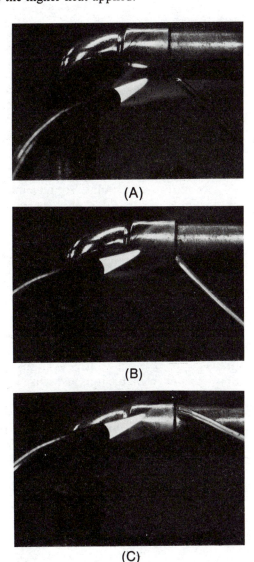

(A)

(B)

(C)

Figure 6-21. When making soldered or brazed joints in the horizontal position, (A) apply the filler metal at the bottom, (B) then to the two sides, and (C) finally to the top, making sure the operations overlap. *Photos by Bill Johnson*

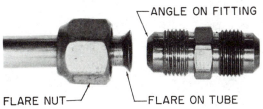

Figure 6-22. Components for a flare joint. *Photo by Bill Johnson*

For horizontal joints it is preferable to apply the filler metal first at the bottom, then to the sides, and finally to the top, making sure the operations overlap, Figure 6-21. On vertical joints it does not matter where the filler is first applied.

The flux used in the brazing process will cause oxidation. When the brazing is done, wash these joints with soap and water if possible.

6.11   MAKING FLARE JOINTS

Another method of joining tubing and fittings is the *flare joint*. This joint uses a flare on the end of the tubing against an angle on a fitting and is secured with a flare nut behind the flare on the tubing, Figure 6-22.

The flare on the tubing can be made with a screw type flaring tool. To make the flare on the end of the tube use the following procedure and see Figure 6-23.

1. Cut the tube to the right length.
2. Ream to remove all burrs and clean all residue from the tubing.
3. Slip the flare nut or coupling nut over the tubing with the threaded end facing the end of the tubing.
4. Clamp the tube in the flaring block, Figure 6-23(A). Adjust it so that the tube is slightly above the block (about one-third of the total height of the flare).
5. Place the yoke on the block with the tapered cone over the end of the tube. Many technicians use a drop or two of refrigerant oil to lubricate the inside of the flare while it is being made, Figure 6-23(B).
6. Turn the screw down firmly, Figure 6-23(C). Continue to turn the screw until the flare is completed.
7. Remove the tubing from the block, Figure 6-23(D). Inspect for defects. If you find any, cut off the flare and start over.
8. Assemble the joint.

6.12   MAKING A DOUBLE-THICKNESS FLARE

A double-thickness flare provides more strength at the flare end of the tube. To make the flare is a two-step operation. Either a punch and block or combination flaring tool is used. Figure 6-24 illustrates the procedure for making double-thickness flares with the combination flaring tool.

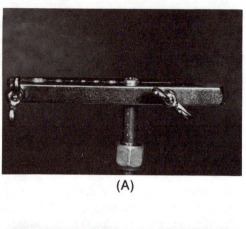
(A)

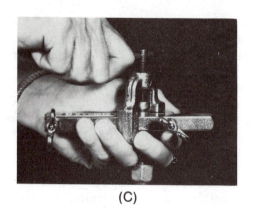
(C)

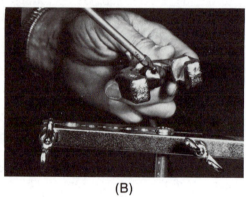
(B)

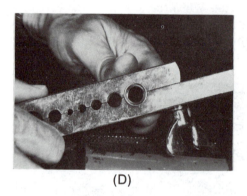
(D)

Figure 6–23. Proper procedure for making a flare joint using a screw-type flaring tool. *Photos by Bill Johnson*

Many fittings are available to use with a flare joint. Each of the fittings has a 45° angle on the end that fits against the flare on the end of the tube, Figure 6–25.

6.13 SWAGING TECHNIQUES

Although swaging is not as common as flaring, you should be aware of it.

Swaging is the joining of two pieces of copper tubing of the same diameter by expanding or stretching the end of one piece to fit over the other so the joint may be soldered or brazed, Figure 6–26. As a general rule, the length of the joint that fits over the other is equal to the approximate outside diameter of the tubing.

You can make a swaged joint by using a punch or a lever-type tool to expand the end of the tubing, Figure 6–27.

Place the tubing in a flare block or an anvil block that has a hole equal to the size of the outside diameter of the tubing. The tube should extend above the block by an amount equal to the outside diameter of the tube plus approximately $\frac{1}{8}$ in., Figure 6–28(A). Place the correct size swaging punch in the tube and strike it with a hammer until the proper shape and length of the joint has been obtained, Figure 6–28(B). Follow the same procedure with screw-type or lever-type tools. A drop or two of refrigerant oil on the swaging tool will help. Assemble the joint. The tubing should fit together easily.

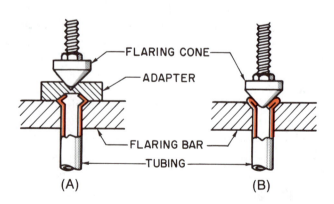

Figure 6–24. Procedure for making a double-thickness flare. (A) Place adapter of combination flaring tool over the tube extended above flaring block. Screw down to bell out tubing. (B) Remove adapter, place cone over tube and screw down to form double flare.

Figure 6–25. Examples of flare fittings. *Photo by Bill Johnson*

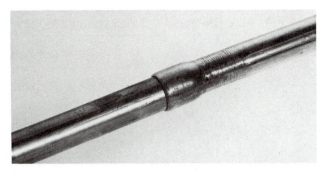

Figure 6-26. Joining tubing by a swaged joint. *Photo by Bill Johnson*

Always inspect the tubing after swaging to see if there are cracks or other defects. If any are seen or suspected, cut off the swage and start over.

SAFETY PRECAUTION: *Field fabrication of tubing is not done under factory-clean conditions, so you need to be observant and careful that *no* foreign materials enter the piping. When the piping is applied to air conditioning or refrigeration, it should be remembered that any foreign matter *will* cause problems. Utmost care *must* be taken.*

6.14 STEEL AND WROUGHT IRON PIPE

The terms "steel pipe," "wrought steel," and "wrought iron" pipe are often used interchangeably and

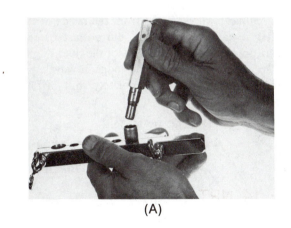

(A)

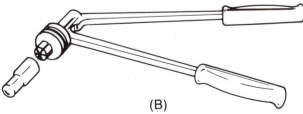

(B)

Figure 6-27. (A) Swaging punch. (B) Lever-type swaging tool. *Photo (A) by Bill Johnson* (B) From Lang, *Air Conditioning: Procedures and Installation,* © 1982 by Delmar Publishers, Inc.

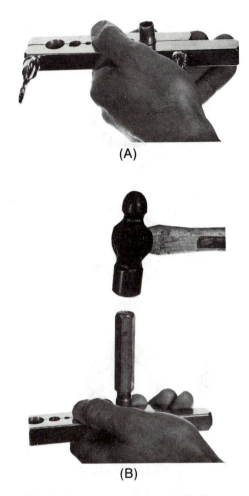

(A)

(B)

Figure 6-28. Making a swaged joint. (A) Tube placed in block for swaging. (B) Swaging punch expanding metal. *Photos by Bill Johnson*

incorrectly. When you want wrought iron pipe, specify "genuine wrought iron" to avoid confusion.

When manufactured, the steel pipe is either welded or produced without a seam by drawing hot steel through a forming machine. This pipe may be painted, left black, or coated with zinc (i.e., *galvanized*) to help resist rusting.

Steel pipe is often used in plumbing, hydronic (hot water) heating, and gas heating applications. The size of the pipe is referred to as the *nominal* size. For pipe sizes 12 in. or less in diameter the nominal size is approximately the size of the inside diameter of the pipe. For sizes larger than 12 in. in diameter the outside diameter is considered the nominal size. The pipe comes in many wall thicknesses but is normally furnished in standard, extra strong, and double extra strong sizes. Figure 6-29 is a cross section showing the different wall thicknesses of a 2-in. pipe. Tables are published indicating this information for each standard pipe diameter. Steel pipe is normally available in 21-ft lengths.

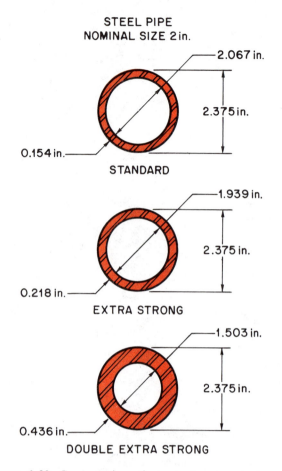

Figure 6-29. Cross section of standard, extra strong, and double extra strong steel pipe.

6.15   JOINING STEEL PIPE

Steel pipe is joined (or joined with fittings) either by welding or by threading the end of the pipe and using threaded fittings. There are two types of American National Standard Pipe Threads: tapered pipe and straight pipe. In this industry only the tapered threads are used because they produce a tight joint and help prevent the pressurized gas or liquid in the pipe from leaking.

Pipe threads have been standardized. Each thread is V-shaped with an angle of 60°. The diameter of the thread has a taper of $\frac{3}{4}$ in./ft or $\frac{1}{16}$ in./in. There should be approximately seven perfect threads and two

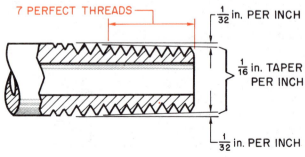

Figure 6-30. Cross section of a pipe thread.

PIPE SIZE (INCHES)	THREADS PER INCH
$\frac{1}{8}$	27
$\frac{1}{4}, \frac{3}{8}$	18
$\frac{1}{2}, \frac{3}{4}$	14
1 to 2	$11\frac{1}{2}$
$2\frac{1}{2}$ to 12	8

Figure 6-31. Threads per inch for some pipe sizes.

or three imperfect threads for each joint, Figure 6-30. Perfect threads must not be nicked or broken, or leaks may occur.

Thread diameters refer to the approximate inside diameter of the steel pipe. The nominal size then will be smaller than the actual diameter of the thread. Figure 6-31 shows the number of threads per inch for some pipe sizes. A thread dimension is written as follows: first the diameter, then the number of threads per inch, then the letters *NPT*, Figure 6-32.

You also need to be familiar with various fittings. Some common fittings are illustrated in Figure 6-33.

Four tools are needed to cut and thread pipe.

- A *hacksaw* (use one with 18 to 24 teeth per inch) or *pipe cutter* is generally used to cut the pipe. A pipe cutter is best because it makes a square cut, but there must be room to swing the cutter around the pipe, Figure 6-34.
- A *reamer* removes burrs from the inside of the pipe after it has been cut. The burrs must be removed because they restrict the flow of the fluid or gas; they also may become loose and lodge in valves, preventing them from being completely shut off, Figures 6-35A and 6-35B.

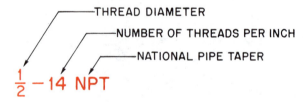

Figure 6-32. Thread specification.

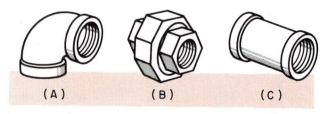

Figure 6-33. Steel pipe fittings. (A) 90° elbow. (B) Union. (C) Coupling.

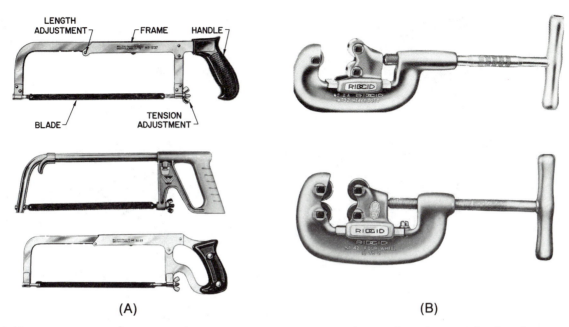

Figure 6–34. (A) Standard hacksaws. (B) Pipe cutters. *Courtesy (A) From Slater and Smith, Basic Plumbing* © 1979 by Delmar Publishers, Inc. *(B) Ridge Tool Company, Elyria, OH*

- A *threader,* also known as a *die.* Most threading devices used in this field are fixed-die threaders, Figure 6–36.
- Holding tools such as the *chain vise, yoke vise,* and *pipe wrench,* Figures 6–37.

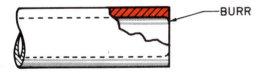

Figure 6–35A. Burr inside a pipe.

Figure 6–35B. Using a reamer to remove a burr. *Photo by Bill Johnson*

When large quantities of pipe are cut and threaded regularly, special machines can be used. These machines are not covered in this text.

Cutting. The pipe must be cut square to be threaded properly. If there is room to revolve a pipe cutter around the pipe, you can use a one wheel cutter. Otherwise, use one with more than one cutting wheel. Hold the pipe in the chain vise or yoke vise if it has not yet been installed. Place the cutting wheel directly over the place where the pipe is to be cut. Adjust the cutter with the T handle until all the rollers or cutters contact the pipe. Apply moderate pressure with the T handle and rotate the cutter around the pipe. Turn the handle about one-quarter turn for each revolution around the pipe. ✱Do not apply too much pressure because it will cause a large burr inside the pipe and excessive wear of the cutting wheel, Figure 6–38.✱

Figure 6–36. Fixed-die-type pipe threader. *Courtesy Ridge Tool Company, Elyria, OH*

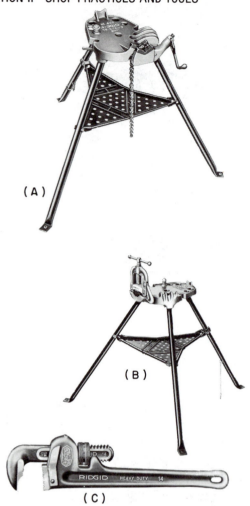

Figure 6–37. (A) Chain vise. (B) Yoke vise. (C) Pipe wrench. *Courtesy Ridge Tool Company, Elyria, OH*

To use a hacksaw, start the cut gently, using your thumb the guide the blade. *Keep your thumb away from the teeth.* A hacksaw will only cut on the forward stroke. Do not apply pressure on the backstroke. Do not force the hacksaw or apply excessive pressure. Let the saw do the work, Figure 6–39.

Reaming. After the pipe is cut, put the reamer in the end of the pipe. Apply pressure against the reamer and turn clockwise. Ream only until the burr is removed Figure 6–35B.

Threading. To thread the pipe, place the die over the end and make sure it lines up square with the pipe Apply cutting oil on the pipe and turn the die once or twice. Then reverse the die approximately one-quarter turn. Then rotate the die one or two turns, and reverse again. Continue this procedure and apply cutting oil liberally until the end of the pipe is flush with the far side of the die, Figure 6–40.

Figure 6–38. Cutting steel pipe with a pipe cutter. *Courtesy Ridge Tool Company, Elyria, OH*

6.16 INSTALLING STEEL PIPE

When installing steel pipe, hold and/or turn the fittings and pipe with pipe wrenches. These wrenches have teeth set at an angle so that the fitting or pipe will be held securely when pressure is applied. Position the wrenches in opposite directions on the pipe and fitting, as shown in Figure 6–41.

When assembling the pipe, use the *correct* pipe thread dope on the male threads. Do not apply the dope closer than two threads from the end of the pipe, Figure 6–42, otherwise it might get into the piping system.

All state and local codes must be followed. You should continually familiarize yourself with all applicable codes.

SAFETY PRECAUTION: *The technician should also realize the importance of installing the size pipe specified. A designer has carefully studied the entire system and has indicated the size that will deliver the correct amount of gas or fluid. Pipe sizes other than those specified should never be substituted.*

Figure 6–39. Cutting steel pipe with a hacksaw. *Photo by Bill Johnson*

Figure 6–40. Threading pipe. *Photos by Bill Johnson*

6.17 PLASTIC PIPE

Plastic pipe is used for many plumbing, venting, and condensate applications. You should be familiar with the following types.

ABS (Acrylonitrilebutadiene Styrene). *ABS* is used for water, drains, waste, and venting. It can withstand heat to 180°F without pressure. Use a solvent cement to join ABS with ABS; use a transition fitting to join ABS to a metal pipe. ABS is rigid and has good impact strength at low temperatures.

Figure 6–41. Holding pipe and turning fitting with pipe wrenches. *Photo by Bill Johnson*

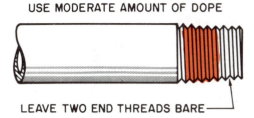

Figure 6–42. Applying pipe dope.

PE (Polyethylene). *PE* is used for water, gas, and irrigation systems. It can be used for water-supply and sprinkler systems and water-source heat pumps. PE is not used with a hot water supply, although it can stand heat with no pressure. It is flexible and has good impact strength at low temperatures. It is normally attached to fittings with two hose clamps. Place the screws of the clamps on opposite sides of the pipe. Figures 6–43.

PVC (Polyvinyl Chloride). *PVC* can be used in high-pressure applications at low temperatures. It can be used for water, gas, sewage, certain industrial processes, and in irrigation systems. It is a rigid pipe with a high impact strength. PVC can be joined to PVC fittings with a solvent cement, or it can be threaded and used with a transition fitting for joining to metal pipe.

CPVC (Chlorinated Polyvinyl Chloride). *CPVC* is very similar to PVC except that it can be used with temperatures up to 180°F at 100 psig. It is used for both hot- and cold-water supplies and is joined to fittings in the same manner as PVC.

To prepare PVC or CPVC for joining, Figure 6–44:

(A) Cut the end square with a plastic tubing shear, a hacksaw, or tube cutter. The tube cutter should have a special wheel for plastic pipe.
(B) Deburr the pipe inside and out with a knife and/or a half-round file.
(C) Clean the pipe end. Apply primer and solvent to both the outside of the pipe and the inside of the fitting. (Follow instructions on primer and cement containers.)

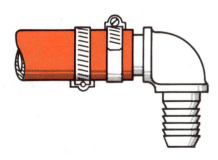

Figure 6–43. Position of clamps on PE pipe.

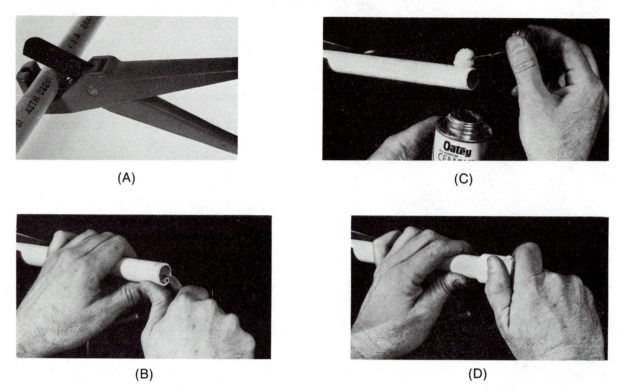

(A)

(C)

(B)

(D)

Figure 6–44. Cutting and joining PVC or CPVC pipe. *Photos by Bill Johnson*

(D) Insert the pipe all the way into the fitting. Turn approximately one-quarter turn to spread the cement.

PVC and CPVC are prepared for joining similar to ABS except that a primer must be applied before applying the solvent cement. The same cement cannot be used for ABS and PVC or CPVC. Schedule #80 PVC and CPVC can be threaded. A regular pipe thread die can be used. *Do not use the same die for metal and plastic pipe. The die used for metal will become too dull to be used for plastic. The plastic pipe die must be kept very sharp. Always follow manufacturer's directions when using any plastic pipe and cement.*

SUMMARY

- *The use of correct tubing, piping, and fittings along with the proper installation is necessary for a refrigeration or air conditioning system to operate properly. Careless handling of the tubing and poor soldering or brazing techniques may cause serious damage to the components of the system.*
- Copper tubing is generally used for plumbing, heating, and refrigerant piping.
- Copper tubing is available in soft- or hard-drawn copper. Type L is the standard size used most frequently in plumbing and heating. ACR tubing is used in air conditioning and refrigeration.
- The size of heating and plumbing tubing refers to the inside diameter of the tubing. The size of ACR tubing refers to the outside diameter.
- A proper adhesive should be used to fasten tubing insulation together.
- ACR tubing can be purchased as line sets charged with refrigerant and sealed on both ends.
- Tubing may be cut with a hacksaw or tube cutter.
- Soft tubing may be bent. Tube bending springs or lever-type benders may be used, or the bend can be made by hand.
- Soldering and brazing fasten tubing and fittings together. Temperatures below 800°F are used for soldering; temperatures over 800°F, for brazing.
- Air acetylene units are used most often for soldering and brazing.
- The flare joint is another method of joining tubing and fittings.
- Soldered swaged joints is a method used to fasten two pieces of copper tubing together but it is not used as much as soldering or brazing fittings to tubing or using flared joints.
- Steel pipe is used in plumbing, hydronic heating, and gas heating applications.
- Steel pipe is joined with fittings by welding or by threaded joints.
- ABS, PE, PVC, and CPVC are four types of plastic pipe; each has a different use.
- All but the PE type are joined to fittings with a solvent cement.

REVIEW QUESTIONS

1. Which type of copper tubing may be bent?
2. What standard type of copper tubing is used most frequently in plumbing and heating?
3. What would the type of $\frac{1}{2}$ in. refer to with regard to copper tubing used in plumbing and heating?
4. What would the size $\frac{1}{2}$ in. refer to with regard to ACR tubing?
5. In what size rolls is soft copper tubing normally available?
6. In what lengths is hard-drawn copper normally available?
7. Why are some ACR tubing lines insulated?
8. What sealant is best for all tubing insulation seams?
9. Describe a line set.
10. Describe the procedure for cutting tubing with a tube cutter.
11. Describe procedures for bending soft copper tubing.
12. What are the approximate temperature ranges used when soldering?
13. What is the minimum temperature for brazing?
14. What type of solder is suitable for moderate pressures and temperatures?
15. What are elements that make up brazing filler metal alloys?
16. Describe how to make a good soldered joint.
17. What type of equipment is used most frequently in soldering and brazing applications?
18. List the procedures used when setting up, igniting, and using an air acetylene unit.
19. Describe the procedures used to make a good brazed joint.
20. Describe a flared joint.
21. Describe the proper procedures for making a flared joint.
22. Describe a double-thickness flare.
23. What is swaging?
24. What should you do if you see or suspect a crack in a flared joint?
25. What are two methods of manufacturing steel pipe?
26. What are some uses of steel pipe?
27. What does the nominal pipe size 12 in. or less in diameter refer to?
28. Describe the procedure for preparing and threading the end of steel pipe.
29. Describe the thread dimension 1/4-18 NPT.
30. List four types of plastic pipe.
31. State a use for each type of plastic pipe.
32. Describe the procedure for joining each type to fittings.

SYSTEM EVACUATION

OBJECTIVES

After studying this unit, you should be able to

- describe a deep vacuum.
- describe two different types of evacuation.
- describe two different types of vacuum measuring instruments.
- choose a proper high-vacuum pump.
- list some of the proper evacuation practices.
- describe a high-vacuum single evacuation.
- describe a triple evacuation.

7.1 PURPOSE OF SYSTEM EVACUATION

Refrigeration systems are designed to operate with only refrigerant and oil circulating inside them. When systems are assembled or serviced, air enters the system. Air contains oxygen, nitrogen, hydrogen and water vapor, all of which are detrimental to the system. These gases cause two problems. The nitrogen is called a noncondensible gas. It will not condense in the condenser and will occupy condenser space that would normally be used for condensing. This will cause a rise in head pressure. See Figure 7–1 for an illustration of a condenser with noncondensible vapors inside. The other gases are noncondensible and also cause chemical reactions that produce acids in the system. Acids in the system cause deterioration of the system's parts, copper plating of the running gear and motor insulation to break down.

These noncondensible gases must be removed from the system if it is to have a normal life expectancy. Many systems have been operated for years with small amounts of these products inside them, but they will not last and give the reliability the customer pays for.

Noncondensible gases are removed by vacuum pumps after the system is leak checked. The pressure inside the system is reduced to an almost perfect vacuum.

7.2 THEORY INVOLVED WITH EVACUATION

To *pull a vacuum* means to lower the pressure in a system below the atmosphere's pressure. The atmosphere exerts a pressure of 14.696 psia (29.92 in. Hg) at sea level at 68°F. Vacuum is commonly expressed in millimeters of mercury (mm Hg). The atmosphere will support a column of mercury 760 mm (29.92 in.) high. To pull a vacuum, in a refrigeration system, for example, the pressure inside the system must be reduced to 0 psia (29.92 in. Hg vacuum) to remove all of the atmosphere.

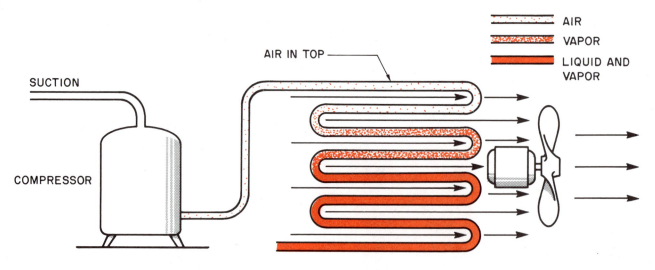

Figure 7–1. Condenser containing noncondensible gases.

A compound gage is often used to indicate the vacuum level. A compound gage starts at 0 in. Hg vacuum and reduces to 30 in. Hg vacuum. When the term "vacuum" is used it is applied to the compound gage. When the term "in. Hg" is used, it is applied to a manometer or barometer. There are methods other than evacuation to remove noncondensibles; these methods will be discussed later.

We can use the bell jar in Unit 1 to describe a typical system evacuation. A refrigeration system contains a volume of gases like the bell jar. The only difference is that a refrigeration system is composed of many small chambers connected by piping. These chambers include the cylinders of the compressor, which may have a reed valve partially sealing it from the system. When the atmosphere is removed from the bell jar, it is called *pulling a vacuum*. When the noncondensibles are removed from the refrigeration equipment, the process is called *pulling a vacuum*. As the atmosphere is pulled out of the bell jar, the barometer inside the jar changes, Figure 7–2. The standing column of mercury begins to drop. When the column drops down to 1 mm, only a small amount of the atmosphere is still in the jar (1/760 of the original volume since 760 mm = 29.92 in.).

7.3   MEASURING THE VACUUM

When the pressure in the bell jar is reduced to 1 mm Hg, the mercury column is very hard to see, so another pressure measurement called the micron is used. (1000 microns = 1 mm Hg). Microns are measured with electronic instruments, Figure 7–3.

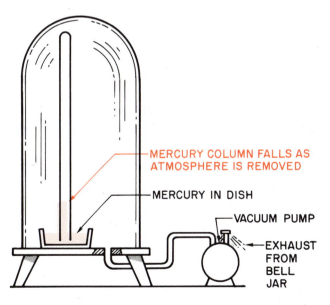

Figure 7–2. The mercury barometer in this bell jar illustrates how the atmosphere will support a column of mercury (Hg). As the atmosphere is removed from the jar, the column of Hg will begin to fall. When all of the atmosphere is removed, the Hg will be in the bottom of the column.

MERCURY COLUMN FALLS AS ATMOSPHERE IS REMOVED

MERCURY IN DISH

VACUUM PUMP

EXHAUST FROM BELL JAR

Figure 7–3. An electronic vacuum (micron) gage used to measure vacuums in the very low range. *Courtesy Robinair Division-Sealed Power Corporation*

Another vacuum gage often used in refrigeration work is the U tube manometer, Figure 7–4, a glass gage closed on one side, which uses mercury as an indicator. The two columns of mercury balance each other. The atmosphere has been removed from one side of the mercury column so that the instrument has a standing column of about 5 in. of mercury. This device can be used for fairly accurate readings down to about 1 mm. Since the columns of mercury are only about 5 in. different in height, this gage starts indicating at about 25 in. Hg vacuum below the atmosphere's pressure. When the gage is attached to the system and the vacuum pump is started, the gage will not read until the vacuum reaches about 25 in. Hg vacuum. Then the gage will gradually fall until the two columns of mercury are equal. At this time the vacuum in the system is between 1 millimeter Hg and a perfect vacuum. The instrument cannot be read much closer than that because the eye cannot see any better, Figure 7–4.

7.4   THE VACUUM PUMP

A pump capable of removing the atmosphere down to a very low vacuum is necessary. The vacuum pumps usually used in the refrigeration field are manufactured with rotary compressors. The pumps that pull the lowest vacuums are two-stage rotary vacuum pumps, Figure 7–5. These vacuum pumps are capable of reducing the pressure in a leak-free vessel down to 0.1 micron. It is not practical to pull a vacuum this low in a field-installed system because the refrigerant oil in the system will boil very slightly and create a vapor. The usual vacuum required by manufacturers is in the neighborhood of 200 microns, although some may require a vacuum as low as 50 microns.

When moisture is in a system, a very low vacuum

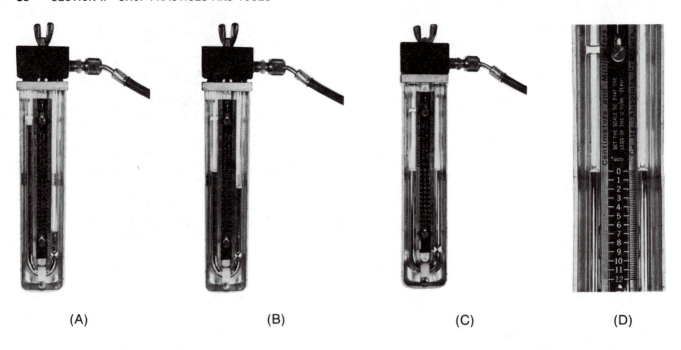

(A)   (B)   (C)   (D)

Figure 7-4. The mercury U tube manometer is at various stages of evacuation. The column on the left is a closed column with no atmosphere above it. When the column on the right is connected to a system that is below atmosphere at a very low vacuum, the column on the left will fall and the column on the right will rise. This will not start until the vacuum on the right is about 25 inches of Hg vacuum or 5 inches of Hg. See the text for an explanation of the difference. As the atmosphere is pulled out of the right hand column, the column rises. It will rise to be exactly parallel to the left hand column at a perfect vacuum. This is very hard to see. *Photos by Bill Johnson*

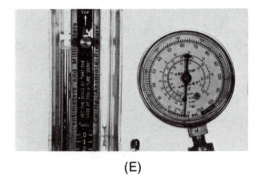

(E)

will cause the moisture to boil to a vapor. This vapor will be removed by the vacuum pump and exhausted to the atmosphere. Small amounts of moisture can be removed this way, but it is not practical to remove large amounts with a vacuum pump because of the large amount of vapor produced by the boiling water. For example, 1 lb of water (about a pint) in a system will turn to 1022 ft^3 of vapor if boiled at 65°F. See Figure 3-9 for a water pressure table. Great quantities

Figure 7-5. Two-stage rotary vacuum pump. *Photo by Bill Johnson*

of moisture should be drained from the system when possible before attaching the vacuum pump.

7.5 DEEP VACUUM

The *deep vacuum* method involves reducing the pressure in the system to about 50 to 200 microns. When the vacuum reaches the desired level, the vacuum pump is valved off and the system is allowed to stand for some time period to see if the pressure rises. If the pressure rises and stops at some point, a material such as water is boiling in the system. If this occurs, continue evacuating. If the pressure continues to rise, there is a leak, and the atmosphere is seeping in the system. In this case the system should be pressured and leak-checked again.

When a system's pressure is reduced to 50–200 microns and the pressure remains constant, no noncondensible gases or moisture is left in the system. This is a slow process. The reason is that when the vacuum pump pulls the system pressure to below about 5 mm, the pumping process slows down. It seems to take forever to pull the last portion of the vacuum. The technician should have other work planned and let the vacuum pump run. Most technicians plan to start the vacuum pump as early as possible and finish other work while the vacuum pump does it's work.

Some technicians leave the vacuum pump running all night. Then the vacuum should be at the desired level the next morning. This is a good practice if some precautions are taken. When the vacuum pump pulls a vacuum, the system becomes a large volume of low pressure with the vacuum pump between this volume and the atmosphere, Figure 7–6. If the vacuum pump shuts off during the night from a power failure, it may lose its lubricating oil to the system by the vacuum in the system. If the power is restored and it starts back up, it will be without adequate lubrication and could be damaged. The oil is pulled out of the vacuum pump by the large vacuum volume, Figure 7–7. This can be prevented by installing a large solenoid valve in the vacuum line entering the vacuum pump and wiring the solenoid valve coil in parallel with the vacuum pump motor. The solenoid valve should have a large port to keep from restricting the flow. This will be discussed in more detail later in this unit, Figure 7–8. Now, if the power fails, or if someone disconnects the vacuum pump (a good possibility at a construction site), the vacuum will not be lost, and the vacuum pump will not lose its lubrication.

7.6 MULTIPLE EVACUATION

Multiple evacuation is used by many technicians for removing the atmosphere to the lowest level of contamination. Multiple evacuation is normally accomplished by evacuating a system to a low vacuum, about 1 or 2 mm, and then allowing a small amount of refrigerant to bleed into the system. The system is then evacuated

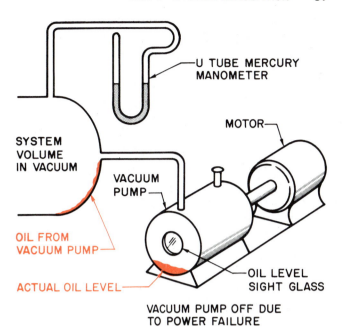

Figure 7–7. This system of stored vacuum has pulled the oil out of the vacuum pump.

until the vacuum is again reduced to 1 mm Hg. The following is a detailed description of a multiple evacuation. This one is performed three times and called a *triple evacuation*. Figure 7–9 is a diagram of the valve arrangements.

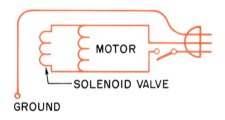

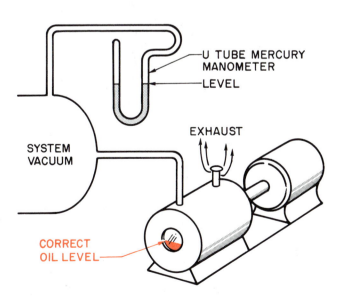

Figure 7–6. The vacuum pump in this system has pulled a vacuum on a large system.

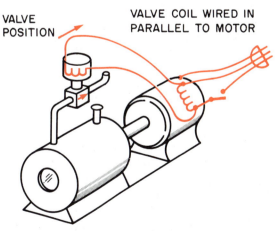

Figure 7–8. Vacuum pump with solenoid valve in the inlet line. Note the direction of the arrow on the solenoid valve. It is installed to prevent flow from the pump. It must be installed in this direction.

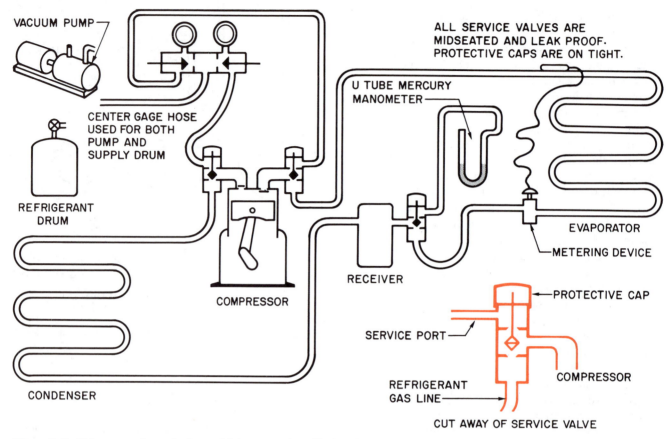

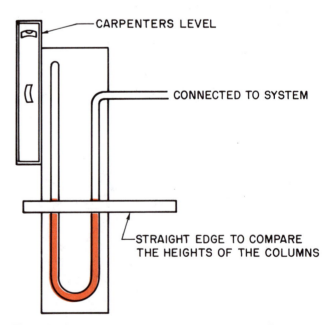

Figure 7–9. This system is ready for multiple evacuation. Notice the valve arrangements and where the U tube manometer is installed.

1. Attach a U tube mercury manometer to the system. The best place is as far from the vacuum port as possible. For example, on a refrigeration system the pump may be attached to the suction and discharge service valves, and the U tube manometer to the liquid receiver valve port. Then start the vacuum pump.

2. Let the vacuum pump run until the manometer reaches 1 mm. The mercury manometer should be positioned straight up and down to take accurate readings. Lay a straightedge across the manometer to help determine the column heights, Figure 7–10.

3. Allow a small amount of refrigerant to enter the system until the vacuum is about 20 in. Hg. This must be indicated on the manifold gage because the mercury in the mercury gage will rise to the top and give no indication. Figure 7–11 shows the manifold gage reading. This small amount of refrigerant vapor will fill the system and absorb and mix with other vapors.

4. Open the vacuum pump valve and start the vapor moving from the system again. Let the vacuum pump run until the vacuum is again reduced to 1 mm Hg. Then repeat step 3.

5. When the refrigerant has been added to the system the second time, open the vacuum pump valve and again remove the vapor. Operate the vacuum pump

for a long time during this third pull down. It is best to operate the vacuum pump until the manometer columns are equal. Some technicians call this *flat out.*

Figure 7–10. Mercury U tube manometer being read as close as possible. The manometer is straight up and down and a straightedge is used to compare the two columns of mercury.

Figure 7–11. Manifold gage reading 20 in. Hg vacuum. *Photo by Bill Johnson*

6. When the vacuum has been pulled the third time, allow refrigerant to enter the system until the system is about 5 psig above the atmosphere. Now remove the mercury manometer (it cannot stand system pressure) and charge the system.

7.7 LEAK DETECTION WHILE IN A VACUUM

We mentioned that if a leak is present in a system, the vacuum gage will start to rise, if the system is still in a vacuum, indicating a pressure rise in the system. The vacuum gage will rise very fast, probably faster than any other method. *Many technicians use this as an indicator that a leak is still in the system, but this is not a recommended leak test procedure.* It allows air to enter the system, and the technician cannot determine from the vacuum where the leak is. Also,

when a vacuum is used for leak checking it is only proving that the system will not leak under a pressure difference of 14.696 psi. If all of the atmosphere is removed from a system, there is only the atmosphere's pressure trying to get back into the system.

When checking for a leak using a vacuum, the technician is using a reverse pressure (the atmosphere trying to get into the system) of only 14.696 psi. The system may be operating under an operating pressure of

$$350 \text{ psig} + 14.696 \text{ psi} = 364.696 \text{ psig}$$

for an R-22 air-cooled condenser on a very hot day when fully loaded, Figure 7–12.

Using a vacuum for leak checking also may hide a leak. For example, if a pin-sized hole is in a solder connection that has a flux buildup on it, the vacuum will tend to pull the flux into the pinhole and may even hide it to the point that a deep vacuum can be achieved. Then when pressure is applied to the system, the flux will blow out of the pinhole, and a leak will show up, Figure 7–13.

7.8 LEAK DETECTION—STANDING PRESSURE TEST

The best leak-checking procedure is a *standing pressure test* using a pressure source that will not change any appreciable amount with temperature changes. Nitrogen is a good gas to use for this. *Never use air.* To perform this test, put a small amount of refrigerant in the system for leak-checking purposes up to about 10 psig of pressure. Then push the pressure up to about 150 psig with the nitrogen. The small amount of

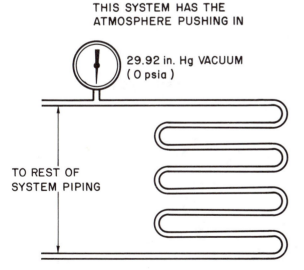

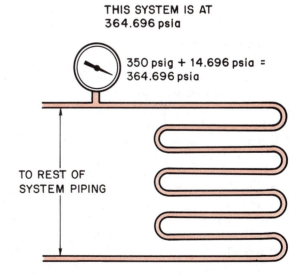

Figure 7–12. These two systems are being compared to each other under different pressure situations. One is evacuated and has the atmosphere's pressure trying to get into the system. The other has 350 psig + 14.696 = 364.696 psia. The one under the most pressure is under the most stress. This shows that using a vacuum as a leak test does not give the system a proper leak test.

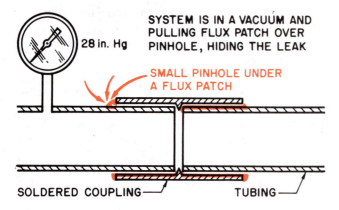

SYSTEM IS IN A VACUUM AND PULLING FLUX PATCH OVER PINHOLE, HIDING THE LEAK

28 in. Hg

SMALL PINHOLE UNDER A FLUX PATCH

SOLDERED COUPLING

TUBING

Figure 7–13. This system was leak checked under vacuum.

refrigerant can be detected with any common leak detector. *Do not pressurize any system above the system's working pressure written on the label of the equipment.* No manufacturer uses a working pressure lower than 150 psig with refrigerants discussed in this book (R-12, R-22, and R-502). Therefore we can assume that 150 psig is safe. When the system pressure is up to 150 psig, tap the gage slightly to make sure the needle is free, and make a mark, Figure 7–14. Let the system stand at this pressure while leak checking. When the leak check is complete, observe the gage reading again. If it has fallen, there is a leak. Do not forget that the gage manifold and connections may leak. When the system is leak checked and no drop is found in the gage, let the system stand for a while. The smaller the system, the shorter the standing time needed. For example, a small beverage cooler may need to stand for only 1 hr to be sure that the system is leak free, whereas a 20-ton system may need to stand under pressure for 12 hr. If the standing time is long, you will have more assurance that there is no leak.

7.9 REMOVING MOISTURE WITH A VACUUM

Removing moisture with a vacuum is the process of using the vacuum pump to remove moisture from a refrigeration system. There are two kinds of moisture in the system, vapor and liquid. When the moisture is in the vapor state, it is easy to remove. When it is in the liquid state, it is much harder to remove. The example earlier in this unit shows that 1022 ft^3 of vapor at 65°F must be pumped to remove 1 lb of water. This is not a complete explanation, because as the vacuum pump begins to remove the moisture, the water will boil and the temperature of the trapped water will drop. For example, if the water temperature drops to 50°F, 1 lb of water will then boil to 1704 ft^3 of vapor that must be removed. This is a pressure level in the system of 0.362 in. Hg or 9.2 mm Hg (0.362 × 25.4 mm/in.). The vacuum level is just reaching the low ranges. As the vacuum pump pulls lower, the water will boil more (if the vacuum pump has the capacity to

pump this much vapor), and the temperature will decrease to 36°F. The water will now create a vapor volume of 2948 ft^3. This is a vapor pressure in the system of 0.212 in. Hg or 5.2 mm Hg (0.212 × 25.4 mm/in.). This illustrates that lowering the pressure level creates more vapor. It takes a large vacuum pump to pull moisture out of a system.

If the system pressure is reduced further, the water will turn to ice and be even more difficult to remove. If large amounts of moisture need to be removed from a system with a vacuum pump, the following procedure will help.

1. Use a large vacuum pump. If the system is flooded, for example if a water-cooled condenser pipe ruptures from freezing, a 5-cfm vacuum pump is recommended for systems up to 10 ton. If the system is larger, a larger pump or a second pump should be used.

2. Drain the system in as many low places as possible. Remove the compressor and pour the water and oil from the system. **Do not add the oil back until the system is ready to be started, after evacuation. If you add it earlier, the oil may become wet and hard to evacuate.**

3. Apply as much heat as possible without damage to the system. If the system is in a heated room, the room may be heated to 90°F without fear of damaging the room and its furnishings, or the system, Figure 7–15. If part of the system is outside, use a heat lamp, Figure 7–16. The entire system, including the interconnecting piping, must be heated to a warm temperature, or the water will boil to a vapor where the heat is applied and condense where the system is cool. For example, if you know water is in the evaporator inside the structure and you apply heat to the evaporator, the water will boil to a vapor. If it is cool outside, the water vapor may condense outside in the condenser piping. The water is only being moved around.

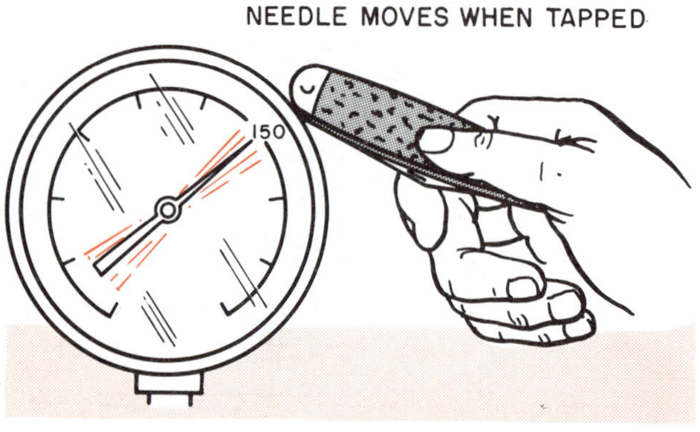

NEEDLE MOVES WHEN TAPPED

150

Figure 7–14. When using a pressure gage for a standing leak test, tap the gage lightly to make sure the needle is free; then mark the gage.

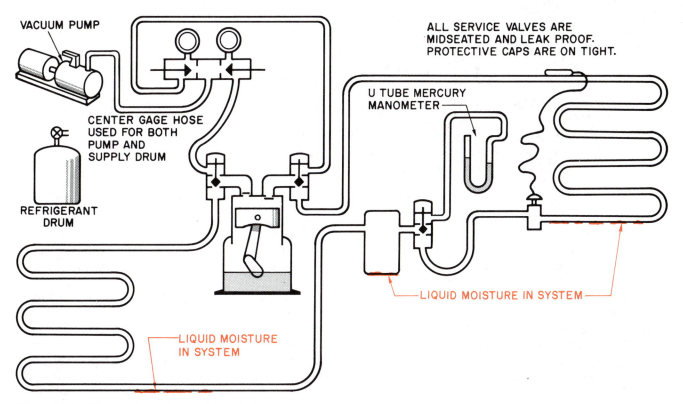

Figure 7-15. When a system has moisture in it and is being evacuated, heat may be applied to the system. This will cause the water to turn to vapor, and the vacuum pump will remove it.

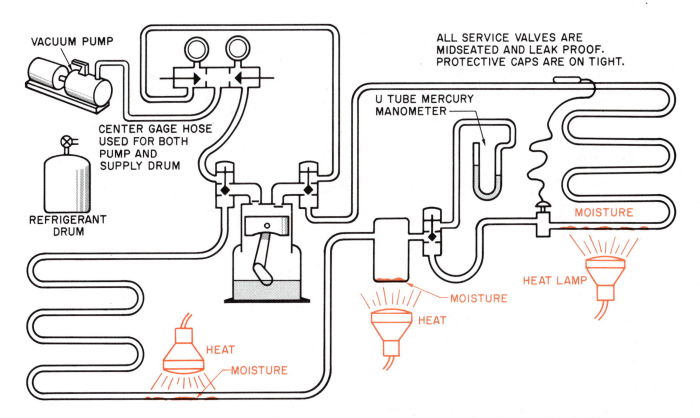

Figure 7-16. When heat is supplied to a large system with components inside and outside, the entire system must be heated. If not, the moisture will condense where the system is cool.

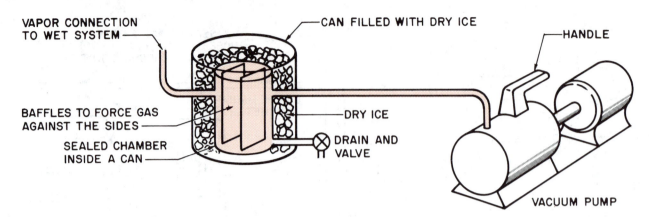

Figure 7-17. A cold trap.

4. Start the vacuum pump and observe the oil level in it. As moisture is removed, some of it will condense in the vacuum pump's crankcase. Some vacuum pumps have a feature called *gas ballast* that introduces some atmosphere between the first and second stages of the two-stage pump. This prevents some of the moisture from condensing in the crankcase. Regardless of the vacuum pump, watch the oil level. ✳The water will displace the oil and raise the oil out of the pump. Soon, water may be the only lubricant in the vacuum pump crankcase and damage may occur to the vacuum pump. They are *very* expensive and should be protected.✳

7.10 GENERAL EVACUATION PROCEDURES

Some general rules apply to deep vacuum and multiple evacuation procedures. If the system is large enough or if you must evacuate the moisture from several systems, you can construct a cold trap to use in the field. The *cold trap* is a refrigerated volume in the vacuum line between the wet system and the vacuum pump. When the water vapor passes through the cold trap, the moisture freezes to the walls of the trap, which is normally refrigerated with dry ice (CO_2), a commercially available product. The trap is heated, pressurized and drained periodically to remove the

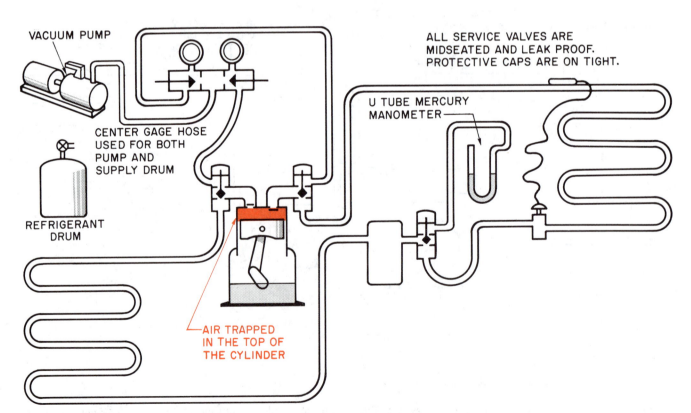

Figure 7-18. Vapor trapped in the cylinder of the compressor.

moisture, Figure 7–17. The cold trap can save a vacuum pump.

Noncondensible gases and moisture may be trapped in a compressor and difficult to release as a vapor that can be pumped out of the system. We said that a compressor may have small chambers, such as cylinders, that may contain air or moisture. Only the flapper valves are setting on top of these chambers, but there is no reason for the air or water to move out of the cylinder while it is under a vacuum. There are times when it is advisable to start the compressor after a vacuum has been tried. This is easy to do with the triple evacuation method. When the first vacuum has been reached, refrigerant can be charged into the system until it reaches atmospheric pressure. The compressor can then be started for a few seconds. All. chambers should be flushed at this time. *Do not start a hermetic compressor while it is in a deep vacuum. Motor damage may occur.* Figure 7–18 is an example of vapor trapped in the cylinder of a compressor.

Water can be trapped in a compressor under the oil. The oil has surface tension, and the moisture may stay under it even under a deep vacuum. During a deep vacuum, the oil surface tension can be broken with vibration, such as striking the compressor housing with a soft face hammer. Any kind of movement that causes the oil's surface to shake will work, see Figure 7–19. Applying heat to the compressor crankcase will also release the water, Figure 7–20.

The technician that evacuates many systems must use time-saving procedures. For example, a typical gage manifold may not be the best choice because it has very small valve ports that slow the evacuation process, Figure 7–21. However, some gage manifolds are manufactured with large valve parts and a special large hose for the vacuum pump connection, Figure 7–22. The gage manifold in Figure 7–23 has four valves and four hoses. The extra two valves are used to control the refrigerant and the vacuum pump lines. When using this manifold, you need not disconnect the vacuum pump and switch the hose line to the refrigerant drum to charge refrigerant into the system. When the time comes to stop the evacuation and charge refrigerant into the system, close one valve and open the other, Figure 7–23. This is a much easier and cleaner method of changing from the vacuum line to the refrigerant line.

When a gage line is disconnected from the vacuum pump, air is drawn into the gage hose. This air must be purged from the gage hose at the top, near the manifold. It is impossible to get all of the air out of the manifold because some will be trapped and pushed into the system, Figure 7–24.

Most gage manifolds have valve stem depressors in the ends of the gage hoses. These depressors are a restriction to the evacuation process. When a vacuum pump pulls down to the very low ranges (1 mm), these valve depressors slow the vacuum process considerably.

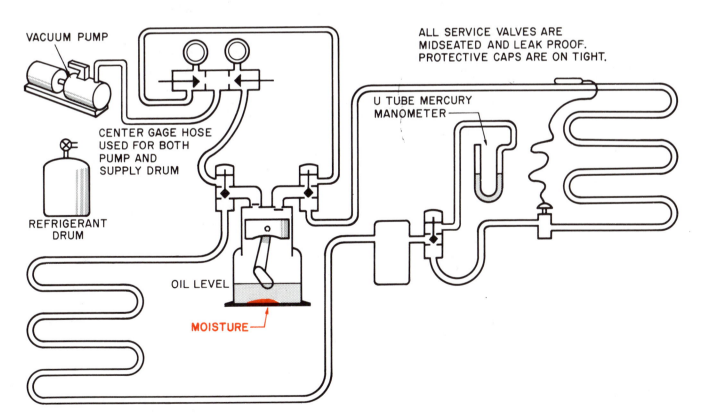

Figure 7–19. Compressor with water under the oil in the crankcase.

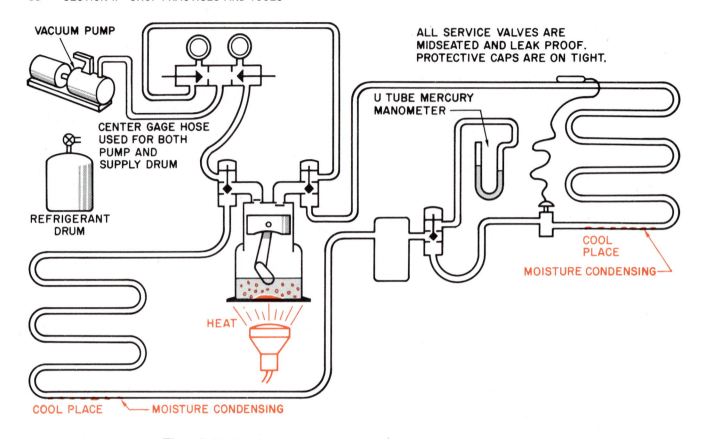

Figure 7-20. Heat applied to compressor to boil the water under the oil.

Many technicians erroneously use oversized vacuum pumps and undersized connectors because they do not realize that the vacuum can be pulled much faster with large connectors. The valve depressors can be removed from the ends of the gage hoses, and adapters can be used when valve depression is needed. Figure 7-25 shows one of these adapters. Figure 7-26 is a small valve that can be used on the end of a gage hose; it will even give the technician the choice as to when the valve stem is depressed.

7.11 SYSTEMS WITH SCHRADER VALVES

A system with Schrader valves for gage ports, will take much longer to evacuate than a system with service valves. The reason is that the valve stems and the depressors act as very small restrictions. An alternative is to remove the valve stems during evacuation and replace them when evacuation is finished. A system with water to be removed will take much time to evacuate if there are Schrader valve stems in the service

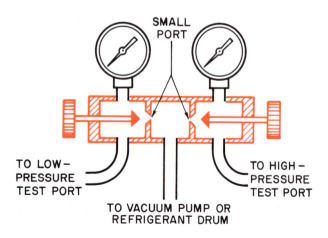

Figure 7-21. Gage manifold with small ports.

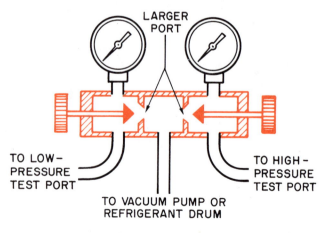

Figure 7-22. Gage manifold with large ports.

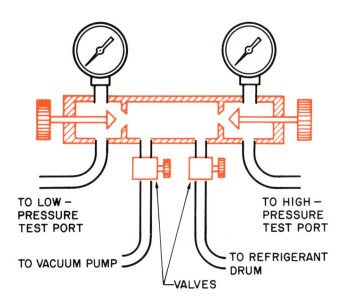

Figure 7-25. This gage adapter can be used instead of the gage depressors that are normally in the end of gage lines. The adapters may be used for gage readings. *Photo by Bill Johnson*

Figure 7-26. This small valve can also be used for controlled gage readings. When the technician wants to read a pressure in a Schrader port, this adapter valve may be used by turning the valve handle down. *Photo by Bill Johnson*

Figure 7-23. Manifold with four valves and four gage hoses.

ports. These valve stems are designed to be removed for replacement, so they can also be removed for evacuation, Figure 7-27.

A special tool, called a *field service valve,* can be used to replace Schrader valve stems under pressure, or it can be used as a control valve during evacuation. The tool has a valve arrangement that allows the technician to evacuate a system through it with the stem backed out of the Schrader valve. The stem is replaced when evacuation is completed.

Schrader valves are shipped with a special cap, which is used to cover the valve when it is not in use.

This cover has a soft gasket and should be the only cover used for Schrader valves. If a standard flare cap is used and overtightened, the Schrader valve top will be distorted and valve stem service will be very difficult, if it can be done at all.

7.12 GAGE MANIFOLD HOSES

The standard gage manifold uses flexible hoses with connectors on the ends. These hoses sometimes get

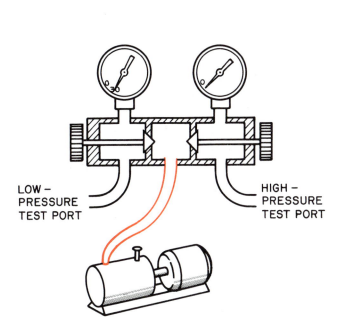

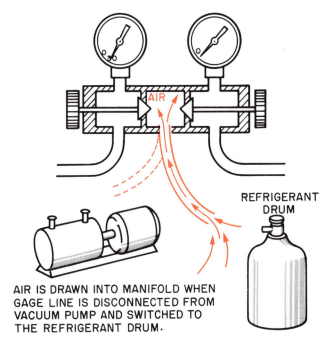

AIR IS DRAWN INTO MANIFOLD WHEN GAGE LINE IS DISCONNECTED FROM VACUUM PUMP AND SWITCHED TO THE REFRIGERANT DRUM.

Figure 7-24. Piping diagram showing how air is trapped in the gage manifold.

Style ATS1

Figure 7-27. Schrader valve assembly. *Courtesy J/B Industries*

Figure 7-28. Gage manifold with copper gage lines used for evacuation. *Photo by Bill Johnson*

pinhole leaks, usually around the connectors, that may leak while under a vacuum but not be evident when the hose has pressure inside it. The reason is that the hose swells when pressurized. If you have trouble while pulling a vacuum and you can't find a leak, substitute soft copper tubing for the gage lines, Figure 7-28.

7.13 SYSTEM VALVES

For a system with many valves and piping runs, perhaps even multiple evaporators, check the system's valves to see if they are open. A system may have a closed solenoid valve. The valve traps air in the liquid line

between the expansion valve and the solenoid valve, Figure 7-29. This valve must be opened for complete evacuation. It may even need a temporary power supply to operate its magnetic coil. Some solenoid valves have a screw on the bottom to manually jack the valve open, Figure 7-30.

7.14 USING DRY NITROGEN

Good workmanship practices while assembling or installing a system can make system evacuation a more

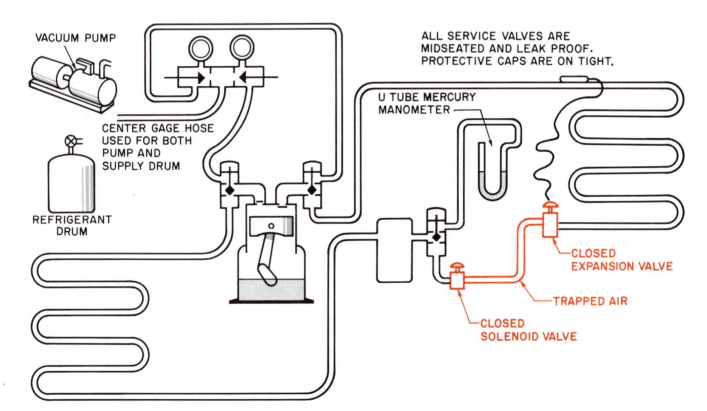

Figure 7-29. Closed solenoid valve trapping air in the system liquid line.

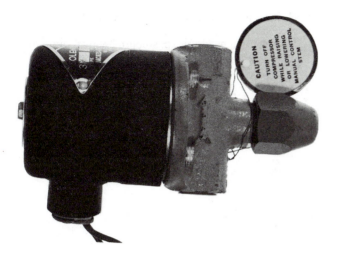

Figure 7-30. Manual opening stem for a solenoid valve. *Photo by Bill Johnson*

pleasant task. When piping is field installed, sweeping dry nitrogen through the refrigerant lines can keep the atmosphere pushed out and clean the pipe. It is relatively inexpensive to use a dry-nitrogen set up, and using it saves time and money.

When a system has been open to the atmosphere for some period of time, it needs evacuation. The task can be quickened by sweeping the system with dry nitrogen or refrigerant before evacuation. See Figure 7–31 for an example of how this is done.

SUMMARY

- Only two products should be circulating in a refrigeration system: refrigerant and oil.
- *Noncondensible gases and moisture are common foreign matter that get into systems during assembly and repair. They must be removed.*
- Evacuation using low vacuum levels involves pumping the system to below atmospheric pressure to remove foreign materials from the system.
- Vapors will be pumped out by the vacuum pump. Liquids must be boiled to be removed with a vacuum.
- Water makes a large volume of vapor when boiled at low pressure levels. It should be drained from a system if possible.
- The two common vacuum gages are the U tube manometer and the electronic micron gage.
- Pumping a vacuum may be quickened with large unrestricted lines.
- Good workmanship and piping practice along with a dry-nitrogen setup will lessen the evacuation time.

REVIEW QUESTIONS

1. What are foreign matters in a refrigeration system that must be removed?
2. What is evacuation?

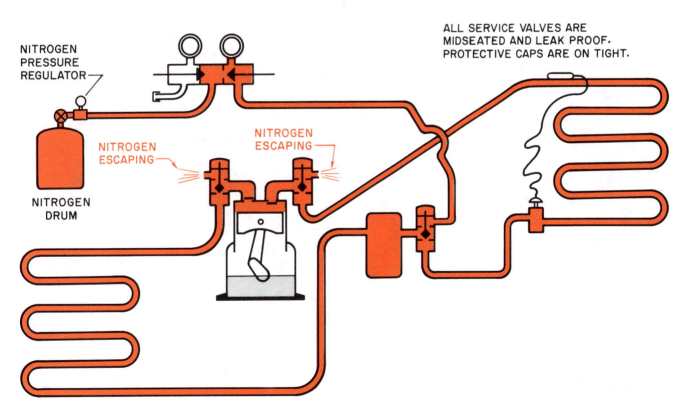

Figure 7-31. The technician is using dry nitrogen to sweep this system before evacuation.

3. What are the only two products that should be circulating in a refrigeration system?

4. Why is it hard to remove water from a refrigeration system with a vacuum pump?

5. Name some things that can be done to help remove water from a refrigeration system.

6. What can be done to get water out from under oil in a compressor crankcase?

7. Name the two evacuation processes.

8. What must be done to solenoid valves and other valves in a system to assure proper evacuation?

9. Name two common vacuum test gages.

10. How many microns are there in a millimeter of mercury?

SYSTEM CHARGING

OBJECTIVES

After studying this unit, you should be able to

- describe how refrigerant is charged into systems in the vapor and the liquid states.
- describe system charging using two different weighing methods.
- state the advantage of using electronic scales for weighing refrigerant into a system.
- describe the graduated cylinder.

8.1 CHARGING A REFRIGERATION SYSTEM

Charging a system refers to the adding of refrigerant to a refrigeration system. The correct charge must be added for a refrigeration system to operate as it was designed to, and this is not always easy to do. Each component in the system must have the correct amount of refrigerant. The refrigerant may be added to the system in the vapor or liquid states by weighing, by measuring, or by using operating pressure charts.

This unit describes how to add the refrigerant correctly and safely in the vapor state and the liquid state. The correct charge for a particular system will be discussed in the unit where that system is covered. Heat pump charging, for example, is covered in the unit on heat pumps.

8.2 VAPOR REFRIGERANT CHARGING

Vapor refrigerant charging of a system is accomplished by allowing vapor to move out of the vapor space of a refrigerant drum and into the low-pressure side of the refrigerant system. When the system is not operating—for example, when a vacuum has just been pulled, or when the system is out of refrigerant—you can add vapor to the low- and high-pressure sides of the system. When the system is running, refrigerant may normally be added only to the low side of the system, because the high side is under more pressure than the refrigerant in the drum. For example, an R-12 system may have a head pressure of 181 psig on a 95°F day (this is determined by taking the outside temperature of 95°F and adding 35°F; this gives a condensing temperature of 130°F, or 181 psig for R-12). The drum

is setting in the same ambient temperature of 95°F but only has a pressure of 108 psig, according to the pressure–temperature chart, Figure 8–1.

The low-side pressure in an operating system is much lower than the drum pressure if the drum is warm. For example, on a 95°F day, the drum will have a pressure of 108 psig, but the evaporator pressure may be only 20 psig. Refrigerant will easily move into the system from the drum, Figure 8–2. In the winter, however, the drum may have been in the back of the truck all night and its pressure then may be lower than the low side of the system, Figure 8–3. In this case the

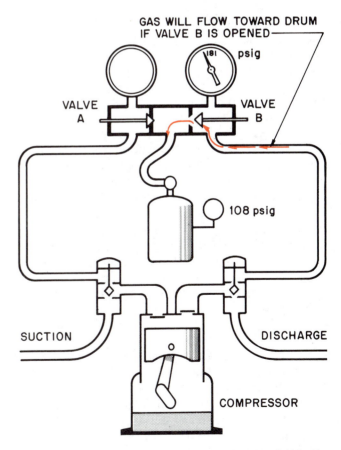

Figure 8-1. This refrigerant drum has a pressure of 108 psig. The high side of the system has a pressure of 181 psig. Pressure in the system will prevent the refrigerant in the drum from moving into the system.

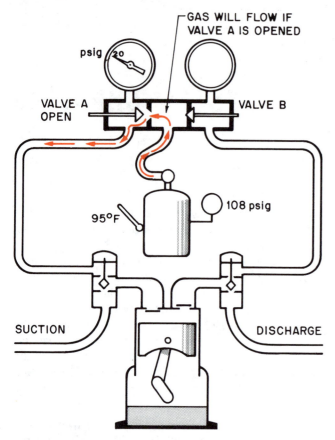

Figure 8–2. Temperature of the drum is 95°F. The pressure inside the drum is 108 psig. The low-side pressure is 20 psig.

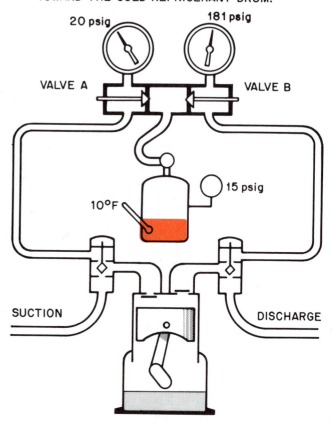

Figure 8–3. The refrigerant in this drum is at a low temperature and pressure because it has been in the back of a truck all night in cold weather. The drum pressure is 15 psig which corresponds to 10°F. The pressure in the system is 20 psig.

drum will have to be heated to get refrigerant to move from the drum to the system. It is a good idea to have a drum of refrigerant stored in the equipment room of large installations. The drum will always be there in case you have no refrigerant in the truck, and the drum will be up to room temperature even in cold weather.

When vapor refrigerant is pulled out of a refrigerant drum, the liquid boils to replace the vapor that is leaving. As more and more vapor is released from the drum, the liquid in the bottom of the drum continues to boil, and its temperature decreases. If enough refrigerant is released, the drum pressure will decrease to the low-side pressure of the system. Heat will have to be added to the liquid refrigerant to keep the pressure up. **✳ But *never* use concentrated heat from a torch. Gentle heat, such as from a warm tub of water, is safer. The water temperature should not exceed 90°F. ✳** This will maintain a drum pressure of 117 psig for R-12 if the refrigerant temperature is kept the same as the water in the tub. Move the refrigerant drum around to keep the liquid in the center of the drum in touch with the warm outside of the drum, Figure 8–4.

The larger the volume of liquid refrigerant in the bottom of the drum, the longer the drum will maintain the pressure. When large amounts of refrigerant must be charged into a system, use the largest drum avail-

able. For example, don't use a 25-lb drum to charge 20 lb of refrigerant into a system if a 125-lb drum is available.

8.3   LIQUID REFRIGERANT CHARGING

Liquid refrigerant charging of a system is normally accomplished in the liquid line. For example, when a system is out of refrigerant, liquid refrigerant can be

Figure 8–4. Refrigerant drum in warm water to keep the pressure up.

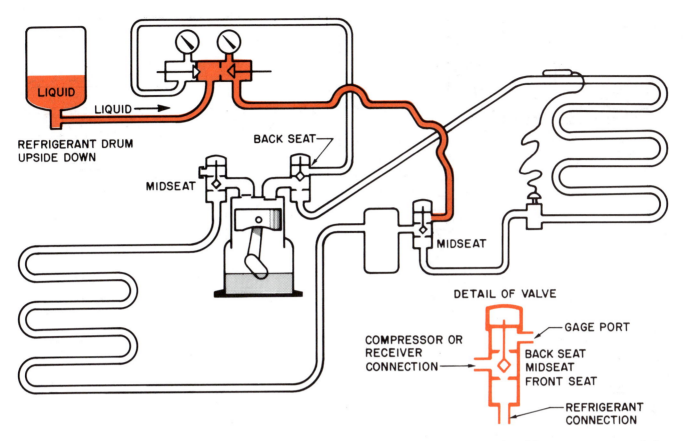

Figure 8–5. This system is being charged while it has no refrigerant in the system. The liquid refrigerant moves toward the evaporator and the condenser when doing this. No refrigerant will enter the compressor.

charged into the *king* valve on the liquid line. If the system is in a vacuum, you can connect the liquid connection of a drum of refrigerant and liquid refrigerant may be allowed to enter the system until it has nearly stopped. The liquid will enter the system and move towards the evaporator and the condenser. When the system is started, the refrigerant is about equally divided between the evaporator and the condenser, and there is no danger of liquid flooding into the compressor, Figure 8–5. When charging with liquid refrigerant, the drum pressure is not reduced. When large amounts of refrigerant are needed, the liquid method is preferable to other methods because it saves time.

When a system has a king valve, it may be *front seated* while the system is operating, and the pressure on the low side of the system will drop. Liquid from the drum may be charged into the system at this time through an extra charging port. The liquid from the drum is actually feeding the expansion device. **Be careful not to overcharge the system,** Figure 8–6. The low-pressure control may have to be bypassed during charging to keep it from shutting the system off. Be sure to remove the electrical bypass when charging is completed, Figure 8–7. *Every manufacturer cautions against charging liquid refrigerant into the suction line of a compressor. To repeat what we stated in the unit of refrigeration, liquid refrigerant must not enter the compressor.*

Some charging devices are commercially available that allow the liquid line to be attached to the suction line for charging a system that is running. They are orifice metering devices that are actually a restriction between the gage manifold and the system's suction line, Figure 8–8. They meter liquid refrigerant into the suction line where it flashes into a vapor. The same thing may be accomplished using the gage manifold valve, Figure 8–9. The pressure in the suction line is maintained at a pressure of not more than 10 lb higher than the system suction pressure. This will meter the liquid refrigerant into the suction line as a vapor. *It should only be used on compressors where the suction gas passes over the motor windings. This will boil any liquid refrigerant that may reach the compressor. If the lower compressor housing becomes cold, stop adding liquid. This method should only be performed under the supervision of an experienced person.*

When a measured amount of refrigerant must be charged into a system, it may be weighed into the system or measured in by using a graduated charging cylinder. Package systems, such as air conditioners and refrigerated cases will have the recommended charge printed on the nameplate. This charge must be added

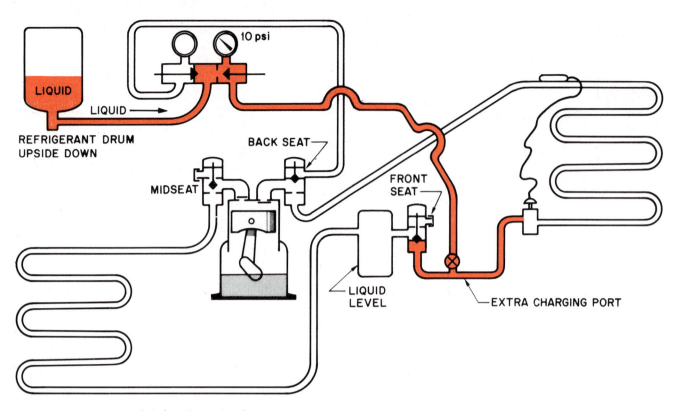

Figure 8–6. This system is being charged by front seating the "king" valve and allowing liquid refrigerant to enter the liquid line.

Figure 8–7. Bypassing the low-pressure control. *Photo by Bill Johnson*

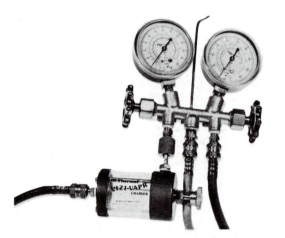

Figure 8–8. Charging device in the gage line between the liquid refrigerant in the drum and the suction line of the system. *Photo by Bill Johnson*

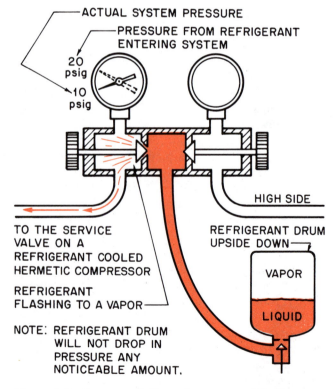

Figure 8–9. Gage manifold used to accomplish the same thing as in Figure 8–8.

Figure 8-10. Scale for measuring refrigerant. *Photo by Bill Johnson*

to the system from a deep vacuum or it will not be correct.

8.4 WEIGHING REFRIGERANT

Weighing refrigerant may be accomplished with various scales. Bathroom and other inaccurate scales should *not* be used. See Figure 8-10 for an accurate dial scale graduated in pounds and ounces. Secure the scales in the truck (make sure they are the portable type) to keep the mechanism from shaking and changing the calibration. Dial scales can be difficult to use, as the next example shows.

Suppose 28 oz of refrigerant is needed. Put the drum of refrigerant on a dial scale and it is found to weigh 24 lb 4 oz. As refrigerant runs into the system, drum weight decreases. Determining the final drum weight is not easy for some technicians.

The calculated final drum weight is

$$24 \text{ lb } 4 \text{ oz} - 28 \text{ oz} = 22 \text{ lb } 8 \text{ oz}$$

To determine this, 24 lb 4 oz was converted to ounces:

$$24 \text{ lb} \times 16 \text{ oz/lb} = 384 \text{ oz} + 4 \text{ oz} = 388 \text{ oz}$$

Now subtract 28 oz from the 388 oz:

$$388 \text{ oz} - 28 \text{ oz} = 360 \text{ oz}$$

is the final drum weight. Because the scales do not read in ounces, you must convert to pounds and ounces:

$$360 \text{ oz} \div 16 \text{ oz/lb} = 22.5 \text{ lb}$$
$$= 22 \text{ lb } 8 \text{ oz}$$

It's easy to make a mistake in this calculation, Figure 8-11.

Electronic scales are often used. These are very accurate but more expensive than dial-type scales. These scales can be adjusted to zero with a full drum, so as refrigerant is added to the system the scales read a positive value. For example: if 28 oz of refrigerant is needed in a system, put the refrigerant drum on the scale and set the scale at 0. As the refrigerant leaves the drum, the scale counts upward. When 28 oz is reached, the refrigerant flow can be stopped, Figure 8-12. This is a time-saving feature that avoids the cumbersome calculations involved with the dial scale.

8.5 USING GRADUATED CYLINDERS

Graduated cylinders are often used to add refrigerant to systems. These cylinders have a visible column of liquid refrigerant, so you can observe the liquid level in the cylinder. Use the pressure gage at the top of the drum to determine the temperature of the refrigerant. The liquid refrigerant has a different volume at differ-

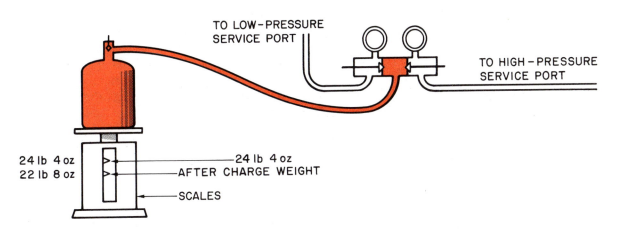

Figure 8-11. Using a set of scales to measure refrigerant into a system.

Figure 8–12. Electronic scales with adjustable zero feature. *Photo by Bill Johnson*

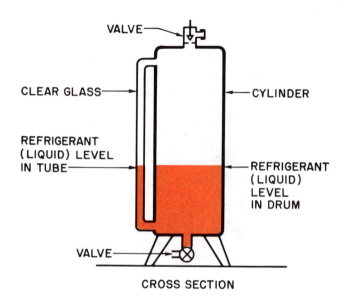

Figure 8–14. Cross section of a graduated cylinder. The refrigerant may be seen in the tube as it moves into the system.

ent temperatures, so the temperature of the refrigerant must be known. This temperature is dialed on the graduated cylinder, Figure 8–13. The final liquid level inside the drum must be calculated much like the previous example, but it is not quite as complicated.

Suppose a graduated cylinder has 4 lb 4 oz of R-12 in the cylinder at 100 psig. Turn the dial to 100 psig and record the level of 4 lb 4 oz. The system charge of 28 oz is subtracted from the 4 lb 4 oz as follows:

$$4 \text{ lb} \times 16 \text{ oz/lb} = 64 \text{ oz} + \text{the remaining } 4 \text{ oz} = 68 \text{ oz}$$

Then

$$68 \text{ oz} - 28 \text{ oz} = 40 \text{ oz}$$

the final drum level.

$$40 \text{ oz} \div 16 \text{ oz/lb} = 2.5 \text{ lb} = 2 \text{ lb } 8 \text{ oz}$$

The advantage of the graduated cylinder is the refrigerant can be seen as the level drops. See Figure 8–14 for an example of this set up.

Some graduated cylinders have heaters in the bottom to keep the refrigerant temperature from dropping when vapor is pulled from the cylinder.

When selecting a graduated cylinder for charging purposes, be sure you select one that is large enough for the systems that you will be working with. It is very hard to use a cylinder twice for one accurate charge. When charging systems with more than one type of refrigerant, you need a charging cylinder for each type of refrigerant. You will also not overcharge the customer or use the wrong amount of refrigerant if you closely follow the above mentioned methods.

Some system manuals give typical operating pressures that may be compared to the gage readings for determining the correct charge. These are called *charging charts.* Figure 8–15 can be used while using the manufacturer's directions. This is discussed in various units in this text that apply to the particular system.

Figure 8–13. Graduated cylinder used to measure refrigerant into a system using volume. *Courtesy Robinair Division-Sealed Power Corporation*

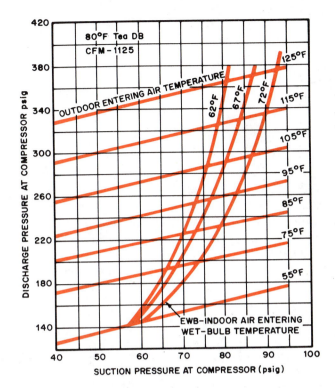

Figure 8–15. Typical system charging chart. Pressures and temperatures are plotted to arrive at the correct charge. *Reproduced courtesy of Carrier Corporation*

SUMMARY

- Refrigerant may be added to the refrigeration system in the vapor state or the liquid state under the proper conditions.
- When refrigerant is added in the vapor state, the refrigerant drum will lose pressure as the vapor is pulled out of the drum.
- *Liquid refrigerant is normally added in the liquid line and only under the proper conditions.*

- *Liquid refrigerant must never be allowed to enter the compressor.*
- Refrigerant is measured into systems using weight and volume.
- It can be difficult to add refrigerant using dial scales because the final drum weight must be calculated. The scales are graduated in pounds and ounces.
- Electronic scales may have a drum-emptying feature that allows the scales to be adjusted to zero with a full drum of refrigerant on the platform.
- Graduated cylinders use the volume of the liquid refrigerant. This volume varies at different temperatures. These may be dialed onto the cylinder for accuracy.

REVIEW QUESTIONS

1. How is liquid refrigerant added to the refrigeration system when the system is out of refrigerant?
2. How is the refrigerant drum pressure kept above the system pressure when charging with vapor from a drum?
3. Why does the refrigerant pressure decrease in a refrigerant drum while charging with vapor?
4. What is the main disadvantage of dial scales?
5. What type of equipment normally has the refrigerant charge printed on the nameplate?
6. What feature of digital electronic scales makes them useful for refrigerant charging?
7. How is refrigerant pressure maintained in a graduated cylinder?
8. How does a graduated cylinder account for the volume change due to temperature changes?
9. What must you remember when purchasing a charging cylinder?
10. What methods besides weighing and measuring are used for charging systems?

CALIBRATING INSTRUMENTS

OBJECTIVES

After studying this unit, you should be able to

- describe instruments used in heating, air conditioning, and refrigeration.
- test and calibrate a basic thermometer at the low- and high-temperature ranges.
- check an ohmmeter for accuracy.
- describe the comparison test for an ammeter and a voltmeter.
- describe procedures for checking pressure instruments above and below atmospheric pressure.
- check flue-gas analysis instruments.

9.1 THE NEED FOR CALIBRATION

The service technician cannot always see or hear what is taking place within a machine or piece of equipment. Instruments such as voltmeters and temperature testers are used to help complete the picture; therefore these instruments must be reliable. Although the instruments are supposed to be calibrated when manufactured, this is not always the case. They may need to be checked before you use them. Even if they are perfectly calibrated the instruments may not remain so due to use and existing field conditions such as moisture and vibration. Instruments may be transported in a dirty truck over rough roads to the job site. The instrument stays in the truck through extremes of hot and cold weather. The instrument compartment may sweat (moisture due to condensation of humidity from the air) and cause the instruments to become damp. All of these things cause stress to the instrument. They will not stay in calibration for ever.

9.2 CALIBRATION

Some instruments cannot be calibrated. *Calibration* means to change the instrument's output or reading to correspond to a standard or correct reading. For example, if a speedometer shows 55 mph for an automobile actually traveling at 60 mph, the speedometer is out of calibration. If the speedometer can be changed to compare to the correct speed, it can be calibrated. If the speedometer cannot be changed to read the actual speed, it cannot be calibrated.

Some instruments are designed for field use and will stay calibrated longer. The new electronic instruments with digital readout features may not be as sensitive to field use as the analog (needle-type) instruments, Figure 9–1.

This unit deals with the most common instruments used for troubleshooting. These instruments measure

(A)

(B)

Figure 9–1. (A) Analog meter. (B) Digital meter. *Courtesy (A) Amprobe Instrument Div. of Core Industries Inc. (B) Simpson Electric Co. Elgin, Illinois*

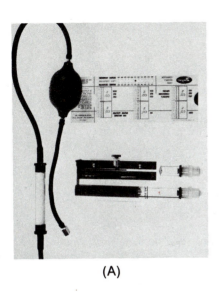

(B)

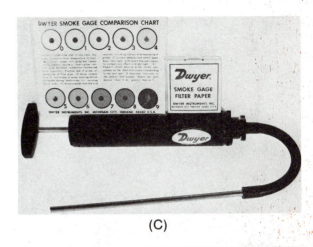

(C)

(A)

Figure 9–2. Flue-gas analysis kit. Note the high-temperature range of the thermometer. (A) CO_2 tester. (B) Thermometer. (C) Smoke tester. (D) Draft gage. *Photos by Bill Johnson*

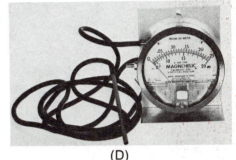

(D)

temperature, pressure, voltage, amperage, resistance, refrigerant leaks, and make flue-gas analyses. To check and calibrate instruments, you must have reference points. Some instruments can be readily calibrated, some must be returned to the manufacturer for calibration, and some cannot be calibrated. We recommend that whenever you buy an instrument, you check some readings against known values. If the instrument is not within the standards the manufacturer states, return it to the supplier or to the manufacturer. Save the box the instrument came in as well as the directions and warranty. They can save you much time.

9.3 TEMPERATURE-MEASURING INSTRUMENTS

Temperature-measuring instruments measure the temperature of vapors, liquids, and solids. Air, water, and refrigerant in copper lines are the common substances measured for temperature level. Regardless of the medium to be measured, the methods for checking the accuracy of the instruments are similar.

Refrigeration technicians must have thermometers that are accurate from −50° to 50°F to measure the refrigerant lines and the inside of coolers. Higher temperatures are experienced when measuring ambient temperatures, such as when the operating pressures for the condenser are being examined. Heating and air conditioning technicians must measure air temperatures from 40°F to 150°F, and water temperatures as high as 220°F for normal service. This can require a wide range of instruments. For temperatures above 250°F, for

example, flue-gas analysis in gas- and oil-burning equipment, special thermometers are used. The thermometer is included in the flue-gas analysis kit, Figure 9–2.

In the past most technicians relied on glass-stem mercury or alcohol thermometers. These are easy to use to measure fluid temperature when the thermometer can be inserted into the fluid, but they are very difficult to use to measure temperature of solids. They are being replaced by the electronic thermometer, which is very popular. Electronic-thermometers are simple, economical, and accurate, Figures 9–3, 9–4. Both

Figure 9–3. Needle-type electronic thermometer. *Courtesy Thermal Engineering*

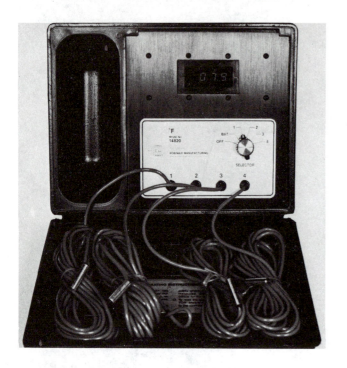

Figure 9-5. Pocket dial-type thermometer. *Photo by Bill Johnson*

Figure 9-4. Digital-type electronic thermometer. *Photo by Bill Johnson*

the analog and digital versions are adequate. Although the digital instrument costs more, it retains accuracy for a longer time under rough conditions.

The pocket-type dial thermometer is often used for field readings, Figure 9–5. It is not intended to be a laboratory-grade accurate instrument. The scale on this unit goes from 0° to 220°F in a very short distance. The distance the needle must move to travel from the bottom of the scale to the top of the scale is only about 2.5 in. (the circumference of the dial).

Three reference points are easily obtainable for checking these instruments: 32°F (ice and water), 98.6°F (body temperature), 212°F (boiling point of water), Figure 9–6. The reference points should be close to the temperature range you are working in. When using any of these as a reference for checking the accuracy of a temperature-measuring device, remember that a thermometer indicates the temperature of the sensing element. The reason for mentioning this is that

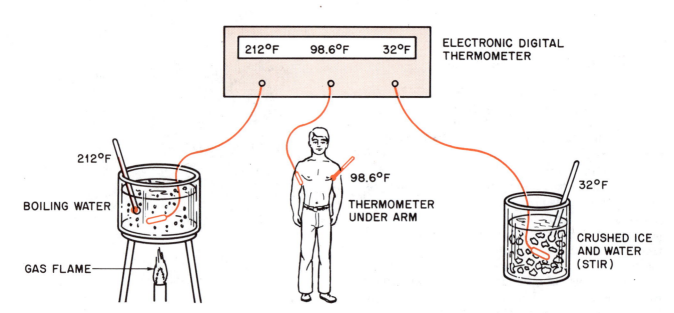

Figure 9-6. Three reference points that a service technician may use.

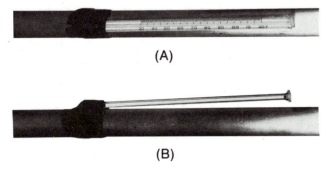

(A)

(B)

Figure 9-7. The technician must remember that a temperature sensing element indicates the temperature of the sensing element. *Photo by Bill Johnson*

many technicians make the mistake of thinking that the sensing element indicates the temperature of the medium being checked. *It does not.* Many inexperienced technicians merely set a thermometer lead on a copper line and read the temperature, but the thermometer sensing element has more contact with the surrounding air than with the copper line, Figure 9–7.

One method of temperature-instrument checking is to submerge the instrument-sensing element into a known temperature condition (such as ice and water while the change of state is taking place) and allow the sensing element to reach the known temperature. The following method checks an electronic thermometer with four plug-in leads that can be moved from socket to socket.

1. Fasten the four leads together as shown in Figures 9–8 and 9–9. Something solid can be fastened with them so that they can be stirred in ice and water.
2. For a low-temperature check, crush about a quart of ice, preferably made from pure water. If pure water is not available, make sure the water has no salt or sugar because they change the freezing point. You must crush the ice very fine (wrap it in a towel and pound it with a hammer), or there may be warm spots in the mixture.
3. Pour enough water, pure if possible, over the ice to almost cover the ice. **Do not cover the ice with the water,** or it will float and may be warmer on the bottom of the mixture. The ice must reach to the bottom of the vessel.

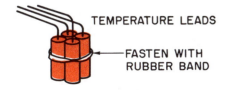

TEMPERATURE LEADS

FASTEN WITH RUBBER BAND

Figure 9-8. The four leads to the temperature tester are fastened at the ends so that they can all be submerged in water at the same time.

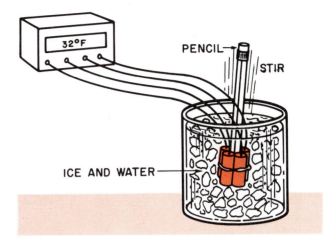

32°F

PENCIL

STIR

ICE AND WATER

Figure 9-9. A pencil is fastened to the group to give the leads some rigidity so that they can be stirred in the ice and water. Note that the ice must reach to the bottom of the pan.

4. Stir the temperature leads in the mixture of ice and water, where the change of state is taking place, for at least 5 min. The leads must have enough time to reach the temperature of the mixture.
5. If the leads vary, note which leads are in error and by how much. The leads must be numbered, and the temperature differences must be marked on the instrument case, or else mark the leads with their error.
6. For a high-temperature check, put a pan of water on a stove-top eye and bring the water to a boil. Make sure the thermometers you are checking indicate up to 212°F. If they do, immerse them in the boiling water. *Do not let them touch the bottom of the pan or they may conduct heat directly from the heat to the lead, Figure 9–10.* Stir the thermometers for at least 5 min and check the readings. It is not critical that the thermometers be accurate to a perfect 212°F, because with products at these temperature levels, a degree or two one way or the

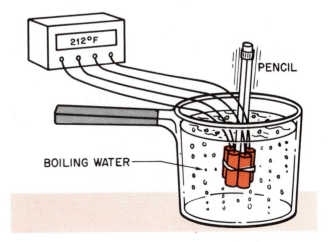

212°F

PENCIL

BOILING WATER

Figure 9-10. High-temperature test for accuracy of four temperature leads.

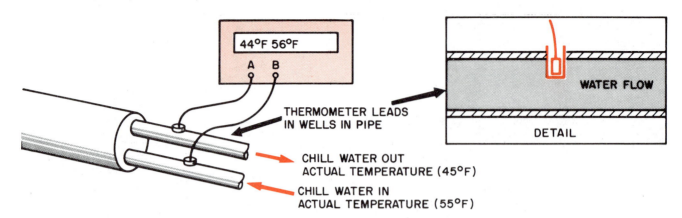

Figure 9–11. Thermometer with two leads accurate to within 1°F.

other does not make a big difference. If any lead reads more than 4°F from 212°F, mark it defective. Remember that water boils at 212°F at sea level at 68°F only. Any altitude above sea level will make a slight difference. If you are more than 1000 ft above sea level, we highly recommend that you use a laboratory glass thermometer as a standard and that you do not rely on the boiling water temperature being correct.

Accuracy is more important in the lower temperature ranges where small temperature differences are measured. A 1°F error does not sound like much until you have one lead that is off +1°F and another off −1°F and try to take an accurate temperature rise across a water heat exchanger that only has a 10°F rise. You have a built-in 20% error, Figure 9–11.

If a digital thermometer with leads that cannot be moved from socket to socket is used, there may be an adjustment in the back of the instrument for each lead. Figure 9–12 shows how this thermometer can be calibrated.

Glass thermometers often cannot be calibrated because the graduations are etched on the stem. If the graduations are printed on the back of the instrument, the back may be adjustable. A laboratory-grade glass thermometer is certified as to its accuracy, and it may be used as a standard for field instruments. It is a good investment for calibrating electronic thermometers, Figure 9–13.

Many dial-type thermometers have built-in means for making adjustments. These instruments may be tested for accuracy as we have described, and calibrated if possible. If not, the dial may need to be marked, Figure 9–14.

Body temperature may also be used as a standard when needed. Remember that the outer extremities, such as the hands, are not at body temperature. The body is 98.6°F in the main blood flow, next to the trunk of the body, Figure 9–15.

9.4 PRESSURE TEST INSTRUMENTS

Pressure test instruments register pressures above and below atmospheric pressure. The gage manifold and its construction were discussed earlier. The technician must be able to rely on these gages and have some reference points with which to check them periodically. This is particularly necessary when there is reason to doubt gage accuracy. This instrument is used frequently and is subject to considerable abuse, Figure 9–16.

Gage readings can be taken from a drum of refrigerant and the pressure–temperature relationship can be compared in the following manner. The gage manifold should be opened to the atmosphere, and both gages should be checked to see that they read 0 psig. If they do not, adjust them. **It is impossible to determine a correct gage reading if the gage is not set at 0 at the start of the test.** Connect the gage manifold

Figure 9–12. Thermometer with leads that cannot be moved or relocated within the instrument. *Photo by Bill Johnson*

Figure 9–13. Laboratory-grade glass thermometer. *Photo by Bill Johnson*

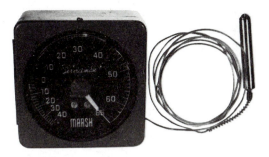

Figure 9–14. Large dial-type thermometers may be calibrated with the calibration screw. *Photo by Bill Johnson*

to a drum of fresh new refrigerant that has been in a room with a fixed temperature for a long time. Purge the gage manifold of air. Using an old drum may lead to errors due to drum pollution. If the drum pressure is not correct due to pollution, it *cannot* be used to check the gages. The drum pressure is the standard and must be reliable. A 1-lb drum may be purchased and kept at a fixed temperature just for the purpose of checking gages, Figure 9–17.

The refrigerant should have a known pressure if the drum temperature is known. Typically, the drum is left in a temperature-controlled office all day, and the readings are taken late in the afternoon. Keep the drum out of direct sunlight. If the refrigerant is R-12 and the office is 75°F, the drum and refrigerant should be 75°F if they have been left in the office for a long enough time. When the gages are connected to the drum and purged of any air, the gage reading should compare to 75°F and read 77 psig. (See the pressure–temperature chart for R-12.) The drum can be connected in such a manner that both gages (the low and high sides) may be checked at the same time, Figure 9–18. The same test can be performed with R-22, and the reading will be higher. The reading for R-22 at 75°F should be 132 psig. Performing the test with both refrigerants checks the gages at two different pressure ranges.

Checking the low-side gage in a vacuum is not as

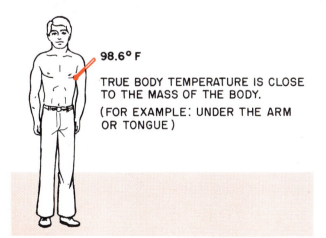

Figure 9–15. Using the human body as a standard.

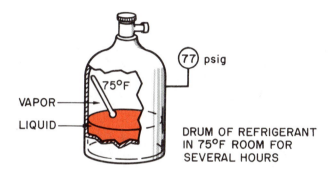

Figure 9–16. Gage manifold. *Photo by Bill Johnson*

Figure 9–17. This refrigerant drum has been left at a known temperature for long enough that the temperature of the refrigerant is the same as the known room temperature. When the gage manifold is attached to this drum, the pressure inside the drum can be obtained from a pressure–temperature chart.

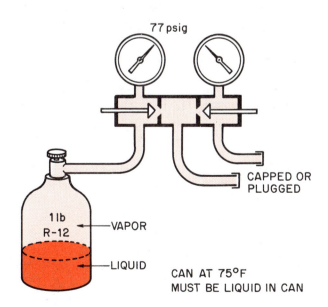

Figure 9–18. Both gages are connected to the drum so that they can be checked at the same time.

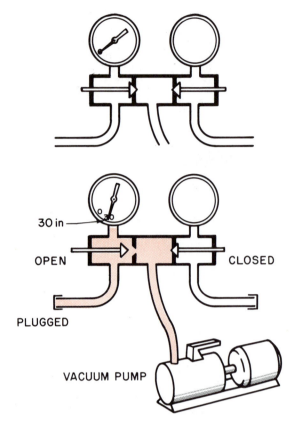

Figure 9-19. When evacuated, the gage should read 30 in. Hg vacuum. If this is the case, all points in between 0 psig and 30 in. Hg should be correct.

easy as checking the gages above atmosphere because we have no readily available known vacuum. One method is to open the gage to atmosphere and make sure that it reads 0 psig. Then connect the gage to a two-stage vacuum pump and start the pump. When the pump has reached its lowest vacuum, the gage should read 30 in. Hg (29.92 in. Hg vacuum), Figure 9-19. *Note:* The vacuum pump will not make as much noise at a low vacuum as it will at a pressure close to the atmosphere's. If the gage is correct at atmospheric pressure and at the bottom end of the scale, you can assume that it is correct in the middle of the scale. If vacuum readings closer than this are needed for monitoring a system that runs in a vacuum, you should buy a larger more accurate vacuum gage, Figure 9-20.

The mercury manometer and the electronic micron gage may be checked in the following manner. This test is a field test and not 100% accurate, but it is sufficient to tell the technician whether the instruments are within a working tolerance or not.

1. Prepare a two-stage vacuum pump for the lowest vacuum that it will pull. Change the oil, if it has been used, to improve the pumping capacity. Connect a gage manifold to the vacuum pump with the mercury manometer and the micron gage as shown in Figure 9-21. The low-pressure gage, the

micron gage, and the mercury manometer may be compared at the same time. Start the vacuum pump. If the micron gage has readings in the 5000-micron range, the mercury manometer and the micron gage may be compared at this point. Remember, 1 mm Hg = 1000 microns so 5 mm Hg = 5000 microns. The low-side manifold gage will read 30 in. Hg, and you cannot easily distinguish movement below 1 mm on the mercury manometer.

2. When the vacuum pump has evacuated the manifold and gages, observe the readings. If the mercury manometer is reading *flat out*—both columns of mercury are at the same level when compared (be sure the instrument is perfectly vertical)—the micron gage should read between 0 and 1000 microns.

Note: if any atmosphere has seeped into the left column of the mercury manometer, then when the vacuum is pulled the right column will rise higher than the left column. This indicates more than a perfect vacuum, which is not possible. Whenever the right column rises higher than the left column, the reading is *wrong*. Check the manometer for a bubble on top of the left column.

It is difficult to compare the mercury manometer closer than this because it is hard to compare the columns. If the mercury columns are flat out and the micron gage is still reading high, you should send the micron gage to the manufacturer for calibration.

Note: If the vacuum pump will not pull the mercury manometer and the micron gage down to a very low level—flat out on the mercury manometer and 50

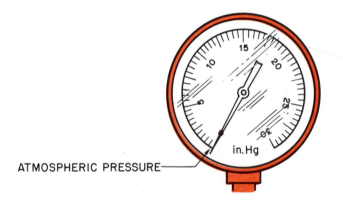

Figure 9-20. A large gage is more accurate for monitoring systems that operate in a vacuum because the needle moves farther from 0 psig to 30 in. Hg.

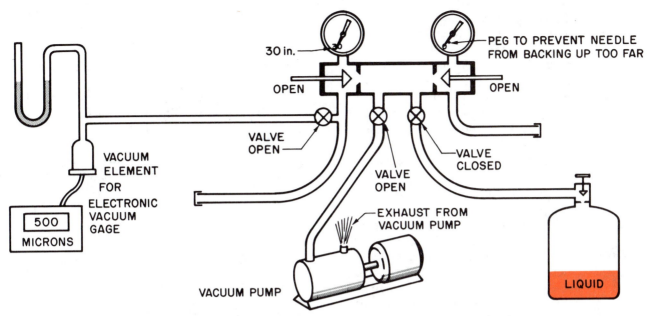

Figure 9–21. Setup to check mercury U tube manometer and an electronic micron gage.

microns on the micron gage—either the vacuum pump is not pumping or the connections are leaking. The connections can be replaced with copper lines. If the vacuum still will not pull down, connect the micron gage directly to the vacuum pump with the shortest possible connection and see if the vacuum pump will pull the gage down. If it will not, check the gauge on another pump to see which is not performing, the gage or the pump.

It is very hard to get instruments to correlate exactly in a vacuum. It is also difficult to determine which

instrument is correct and whether the vacuum pump is evacuating the system. If the evacuation happens too quickly to be observed, a volume, such as an empty refrigerant drum, can be used to retard the pull-down time, Figure 9–22.

9.5   ELECTRICAL TEST INSTRUMENTS

Electrical test instruments are not as easy to calibrate; however, they may be checked for accuracy. The technician must know that the ohm scale, the volt scale, and the ammeter scale are correct. The milliamp

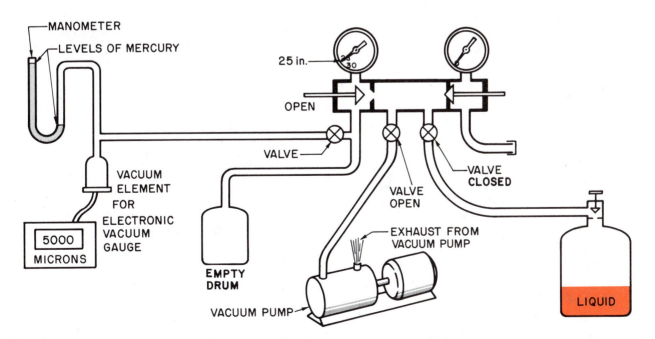

Figure 9–22. If the evacuation in Figure 9–21 happens too fast, a volume, such as an empty refrigerant drum, may be connected.

scale on the meter is seldom used and must be checked by the manufacturer or compared to another meter.

When electrical testing procedures must be relied on for accuracy, it pays to buy a good quality instrument. Buying electrical test instruments is like buying a camera—there are many grades available. If you don't need accuracy from an electrical test instrument, a cheap one may do the job. Good instruments can be depended upon when close measurements are needed.

You should periodically (once a year) check electrical test instruments for accuracy. One way to check these instruments is to compare the instrument readings against known values.

The ohmmeter feature of a volt-ohm-milliammeter (VOM) can be checked by obtaining several high-quality resistors of a known resistance at an electronic supply house. Get different values of resistors, so you can test the ommeter at each end and the midpoint of every scale on the meter, Figure 9–23. Always start each test by a zero adjustment check of the ohmmeter. *If the meter is out of calibration, send it to the experts. Do not try to repair it yourself. Be sure to start each test by a zero adjustment check of the ohmmeter.*

The volt scale is not as easy to check as the ohm scale. A friend at the local power company or technical school may allow you to compare your meter to a high-quality bench meter. This is recommended at least once a year or whenever you suspect your meter is incorrect, Figure 9–24. It is very satisfying to know your meter is correct when you call the local power

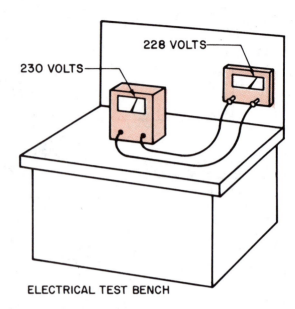

Figure 9–24. VOM being compared to a quality bench meter in the voltage mode.

company and report to them that they have low voltage on a particular job.

The clamp-on ammeter is used most frequently. This instrument clamps around one conductor in an electrical circuit. Like the voltmeter, it can be compared to a high-quality bench meter at the local power company or technical school, Figure 9–25. Some amount of checking can be done by using Ohm's law and comparing the ampere reading to a known resistance heater circuit. For example, Ohm's law states that current (A) is equal to voltage (V) divided by resistance (Ω):

$$I = E \, / \, R.$$

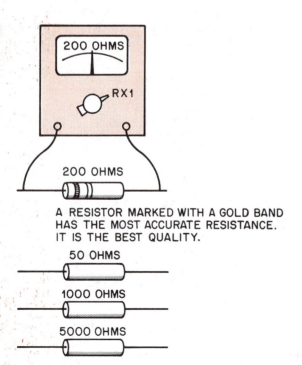

Figure 9–23. Checking an ohmmeter at various resistance ranges with known resistances.

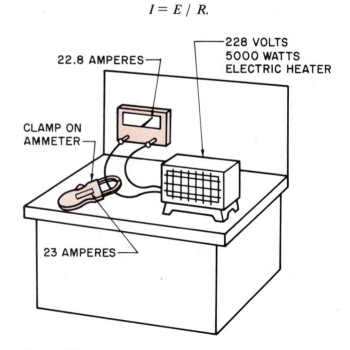

Figure 9–25. Clamp-on ammeter being compared to a quality bench ammeter.

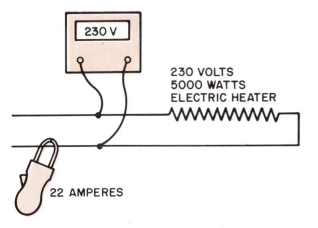

Figure 9–26. Using an electric heater to check the calibration of an ammeter.

If a heater has a resistance of 10 Ω and an applied line voltage of 228 V, the amperage (current) on this circuit should be

$$I = 228 \text{ V} / 10 \text{ }\Omega = 22.8 \text{ A}$$

Remember to read the voltage at the same time as the amperage. You will notice small errors because the resistance of the electric heaters will change when they get hot. The resistance will be greater, and the exact ampere reading will not compare precisely to the calculated one. But the purpose is to check the ammeter, and if it is off by more than 10% send it to the repair shop, Figure 9–26.

9.6 REFRIGERATION LEAK DETECTION DEVICES

Two refrigerant detection devices are commonly used: the halide torch and the electronic leak detector.

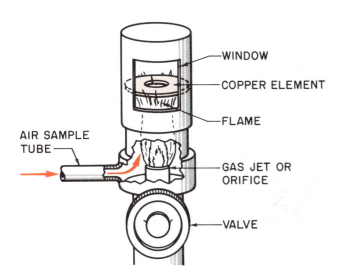

Figure 9–27. Halide torch for detecting refrigerant leaks.

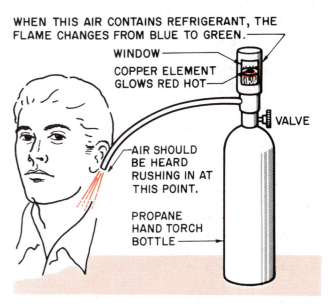

Figure 9–28. A rushing sound may be heard at the end of the sampling tube if the halide torch detector is pulling in air and working correctly.

Halide Torch

The halide torch cannot be calibrated, but it can be checked to make sure that it will detect leaks. It must be maintained for it to be reliable. It will detect a leak rate of about 6 oz per year. The halide torch uses the primary air port to draw air into the burner through a flexible tube. If there is any refrigerant in this air sample, it passes over a copper element and the color of the flame changes from the typical blue of a gas flame to a green color, Figure 9–27. A large leak will extinguish the flame of the halide torch.

The maintenance on this torch consists of keeping the tube clear of debris and keeping a copper element in the burner head. If the sample tube becomes restricted, the flame may burn yellow. You can place the end of the sample tube close to your ear and hear the rushing sound of air being pulled in the tube. If you can't hear it, or if it burns yellow, clean the tube, Figure 9–28.

The copper element is replaceable. If you cannot find an element, you can make a temporary one out of a piece of copper tubing, Figure 9–29.

Electronic Leak Detectors

Electronic leak detectors are much more sensitive than the halide torch (they can detect leak rates of about 1/2 oz per year) and are widely used. The electronic leak detector samples air; if the air contains refrigerant, the detector either sounds an alarm or lights the probe end. These devices are manufactured in both battery powered and 120-VAC powered units.

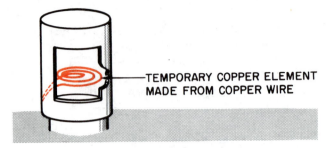

Figure 9-29. Temporary copper element.

Some units may have a pump to pull the sample across the sensing element, and some have the sensing element located in the head of the probe, Figure 9-30.

Some electronic leak detectors have an adjustment that will compensate for background refrigerant. There are some equipment rooms that have so many small leaks that there is a small amount of refrigerant in the air all the time. The electronic leak detector will indi-

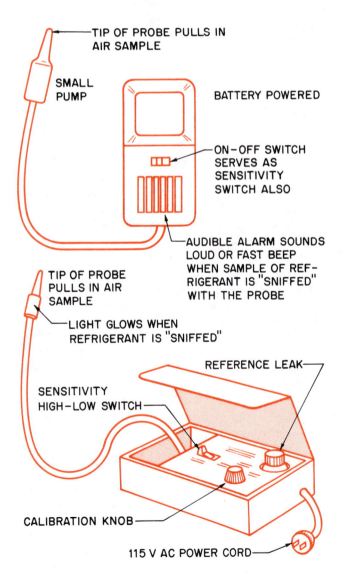

Figure 9-30. Electronic leak detectors.

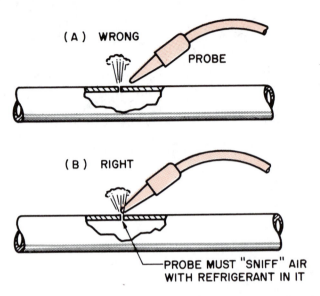

Figure 9-31. (A) Electronic leak detector is not sensing the small pinhole leak because it is spraying past the detector's sensor. (B) Sensor will detect refrigerant leak.

cate all the time unless it has a feature to account for this background refrigerant.

No matter what style of leak detector you use, you must be confident that the detector will actually detect a leak. Remember that the detector only indicates what it samples. If the detector is in the middle of a refrigerant cloud and the sensor is sensing air, it will not sound an alarm or light up. For example, a pinhole leak in a pipe can be passed over with the probe of a leak detector. The sensor is sensing air next to the leak, not the leak itself, Figure 9-31.

Some manufacturers furnish a sample refrigerant container of R-11. The container has a pinhole in the top with a calibrated leak, Figure 9-32. The refrigerant will remain in the container for a long time if the lid is replaced after each use.

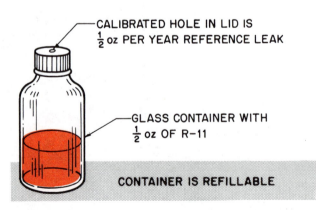

Figure 9-32. This is a small refrigerant vial that serves as a reference leak.

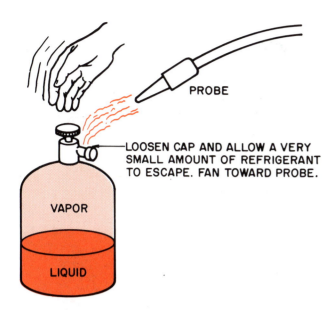

PROBE

LOOSEN CAP AND ALLOW A VERY SMALL AMOUNT OF REFRIGERANT TO ESCAPE. FAN TOWARD PROBE.

VAPOR

LIQUID

Figure 9-33. If a reference leak vial is not available, a small amount of refrigerant may be allowed to escape and to mix with air. This sample may be fanned towards the probe.

SAFETY PRECAUTION: ∗Never spray pure refrigerant into the sensing element. Damage will occur.∗ If you do not have a reference leak canister, a gage line under pressure from a refrigerant drum can be loosened slightly, and the refrigerant can be fanned to the electronic leak detector sensing element. In doing it this way, air is mixed with the refrigerant to dilute it, Figure 9-33.

9.7 FLUE-GAS ANALYSIS INSTRUMENTS

Flue-gas analysis instruments analyze fossil-fuel-burning equipment such as oil and gas furnaces. These instruments are normally sold in kit form with a carrying case. ∗There are chemicals in the flue-gas kit that must not be allowed to contact any tools or other instruments. This chemical is intended to stay in the container or the instrument that uses it. The instrument has a valve at the top that is a potential leak source. It should be checked periodically. It is best to store and transport the kit in the upright position so that if the valve does develop a leak it will not run out of the instrument, Figure 9-34.∗

These are precision instruments that cost money and deserve care and attention. The draft gage is sensitive to a very fine degree. This kit should not be hauled in the truck except when you intend to use it.

A calibration check of these instruments is not necessary. The chemicals in the analyzer should be changed according to the manufacturer's suggestions. These instruments are direct-reading instruments and cannot be calibrated. The only adjustment is the zero adjustment on the sliding scale, Figure 9-35, which is adjusted at the beginning of the test. This is done by venting the sample chamber to the atmosphere. The fluid should fall. If it gets dirty, change it.

The thermometer in the kit is used to register very high temperatures, up to 1000°F. There is no easy reference point except the boiling point of water, 212°F. This may be used as the reference, even though it is very near the bottom of the scale, Figure 9-36.

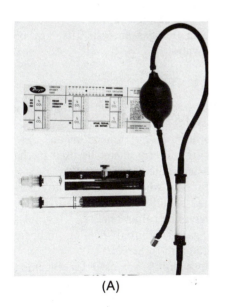

(A)

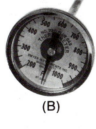

(B)

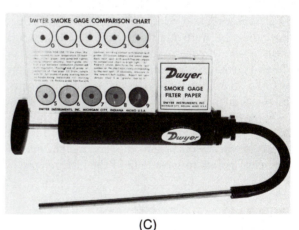

(C)

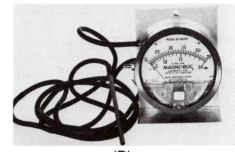

(D)

Figure 9-34. Flue-gas analysis kit. (A) CO_2 tester. (B) Thermometer. (C) Smoke tester. (D) Draft gage. *Photos by Bill Johnson*

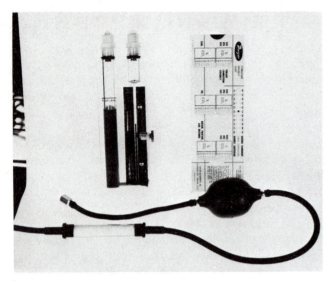

Figure 9–35. Zero adjustment on the sliding scale of flue-gas analysis kit. *Photo by Bill Johnson*

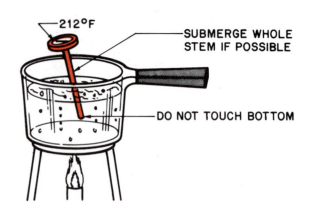

Figure 9–36. Flue-gas kit thermometer being checked in boiling water.

9.8 GENERAL MAINTENANCE

Any instrument with a digital readout will have batteries, usually standard flashlight batteries. They must be maintained. *Buy the best batteries available; inexpensive batteries may cause problems. Good batteries will not leak acid on the instrument components if left unattended and the battery goes dead.*

The instruments you use extend your senses; the instruments therefore must be maintained so that they can be believed. Airplane pilots sometimes have a malady called vertigo. They become dizzy and lose their sensation and relationship with the horizon. Suppose the pilot were in a storm and being tossed around, even upside down at times. The pilot can be upside down and have a sensation that the plane is right side up and diving. Suppose this sensation overcame the pilot's tendency to believe his instruments. The pilot may try to climb. Of course, the plane is diving towards earth while the pilot thinks it is climbing. Instruments must be believable, and the technician must have reference points to have faith in the instruments.

SUMMARY

- The instruments used by technicians must be reliable.
- Reference points for all instruments should be established to give the technician confidence.
- The three easily obtainable reference points for temperature-measuring instruments are ice and water 32°F, body temperature 98.6°F, and boiling temperature 212°F.

- ∗Make sure that the temperature-sensing element reflects the actual temperature of the medium used as the standard.∗
- Pressure-measuring instruments must be checked above and below atmospheric pressure. There are no good reference points below the atmosphere, so a vacuum pump pulling a deep vacuum is used as the reference.
- Some electronic leak detectors have reference leak canisters furnished.
- Flue-gas analysis kits need no calibration except for the sliding scale on the sample chamber.
- The thermometers in flue-gas analysis kits may be checked in boiling water.

REVIEW QUESTIONS

1. Name the three reference points for checking temperature-measuring instruments.
2. What does "calibrating an instrument" mean?
3. How can a glass-stem thermometer be used in the calibration of an electronic thermometer?
4. What should you do when an electronic thermometer's leads are slightly out of calibration?
5. Can all thermometers be calibrated?
6. What should a gage manifold reading indicate when opened to the atmosphere?
7. What two reference points are used for checking vacuum for a gage manifold low-side gage?
8. What reference point may be used for checking a pressure gage?
9. How is a flue-gas instrument calibrated?
10. What types of batteries are suggested for instruments?

10

SAFETY

OBJECTIVES

After studying this unit, you should be able to

- **describe proper procedures for working with pressurized systems and vessels, electrical energy, heat, cold, rotating machinery, chemicals, and moving heavy objects.**
- **avoid various safety hazards.**

The heating, air conditioning, and refrigeration technician works very close to many potentially dangerous situations: liquids and gases under pressure, electrical energy, heat, cold, chemicals, rotating machinery, moving heavy objects, and so on. The job must be completed in a manner that is safe for the technician and the public. A working knowledge of pressure, temperature, electricity, and some simple chemicals forms the basis for safety in this field.

10.1 PRESSURE VESSELS AND PIPING

Pressure vessels and piping are part of most systems that are serviced. For example, a drum of R-22 sitting in the back of an open truck with the sun shining on it may have a drum temperature of 110°F on a summer day. From the temperature/pressure chart the pressure inside the drum is 226 psig. This pressure reading means that the drum has a pressure of 226 lb for each square inch of surface area. A large drum may have a total area of 1500 in.² This gives a total inside pressure (pushing outward) of

$$1500 \text{ in.}^2 \times 226 \text{ psi} = 339,000 \text{ lb}$$

This is equal to

$$339,000 \text{ lb} \div 2000 \text{ lb/ton} = 169.5 \text{ ton}$$

Figure 10–1. This pressure is well contained and will be safe if the drum is protected. Do not drop it. *Move the drum only while a protective cap is on it, if it is designed for one, Figure 10–2.*

The pressure in this drum can be thought of as potential danger. It will not become dangerous unless it is allowed to escape in an uncontrolled fashion. The

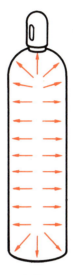

Figure 10–1. Pressure exerted across the entire surface area of refrigerant bottle.

drum has a relief valve in the top in the vapor space. If the pressure builds up to the relief valve setting, approximately 250 psig, it will start relieving vapor. As the vapor pressure is relieved, the liquid in the drum will begin to cool and reduce the pressure, Figure 10–3.

The refrigerant drum has a fusible plug made of a material with a low melting temperature. The plug will melt and blow out if the drum gets too hot. This

Figure 10–2. Refrigerant bottle with protective cap. *Photo by Bill Johnson*

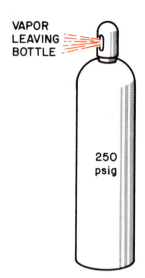

Figure 10-3. Pressure relief valve will reduce pressure in the drum.

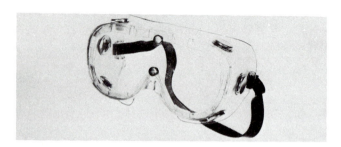

Figure 10-5. Protective eye goggles. *Photo by Bill Johnson*

prevents the drum from bursting and injuring personnel and property around it. *Many technicians apply heat to refrigerant drums while charging a system to keep the pressure from dropping in the drum. This is a very dangerous practice. It is recommended that the drum of refrigerant be set in a container of warm water with a temperature no higher than 90°F, Figure 10-4.*

You will be taking pressure readings on refrigeration and air conditioning systems. This can freeze your hands. Liquid R-22 boils at −41°F when released to the atmosphere. If you are careless and get this refrigerant on your skin, it will quickly cause frostbite. Keep your skin and eyes away from any liquid refrigerant. *If a leak develops and refrigerant is escaping, the best thing to do is stand back and look for a valve with which to shut if off. Do not try to stop it with your hands.*

Always wear protective eye goggles when transferring refrigerant. Blowing refrigerant, and any particles that may become airborne with the blast of vapor, can harm the eyes. These goggles are vented to keep them

from fogging over with condensation and to keep the operator cool, Figure 10-5.

10.2 ELECTRICAL HAZARDS

Electrical shocks and burns are ever-present hazards to deal with. It is impossible to troubleshoot all circuits with the power off, so you must learn safe methods for troubleshooting "live" circuits. As long as electricity is contained in the conductors and the devices where it is supposed to function, there is nothing to fear. When uncontrolled electrical flow takes place (e.g., if you touch two live wires), you are likely to get hurt.

SAFETY PRECAUTION: *Electrical power should always be shut off at the distribution or entrance panel when installing equipment. Whenever possible the power should be shut off when servicing the equipment. Don't ever think that you are good enough or smart enough to work with live electrical power when it is not necessary.*

However, there are times when tests have to be made with the power on. Extreme care should be taken when making these tests. Insure that you know the voltage in the circuit you are checking. Also, make sure that the range selector on the test instrument is set properly before using it. *Don't stand in a wet or damp area when making these checks. Use only proper test equipment, and make sure it is in good condition.* Intelligent and competent technicians take all precautions. You will learn more about using test instruments in other units in this book.

Electrical Shock

Electrical shock occurs when you become part of the circuit. Electricity flows through your body and can damage your heart—stop it from pumping, resulting in death if it is not restarted very quickly. It is a good idea to take a first aid course that includes cardiopulminary methods for life saving.

To prevent electrical shock, don't become a conductor between two live wires or a live (hot) wire and ground. The electricity must have a path to flow through. Don't let your body be the path. Figure 10-6

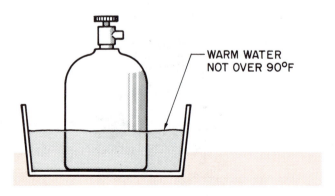

Figure 10-4. Refrigerant drum in warm water (not greater than 90°F).

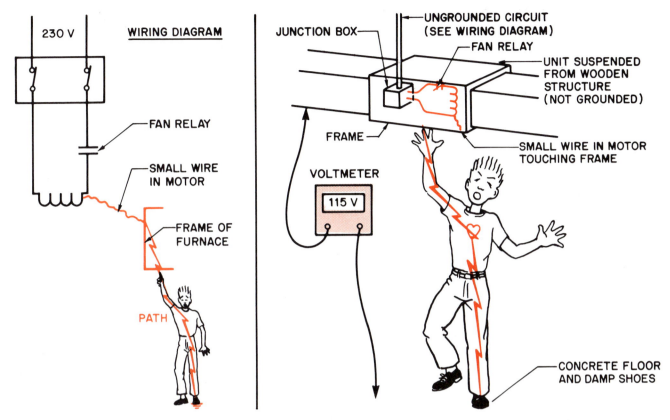

Figure 10-6. Ways for the technician to become part of the electrical circuit and receive an electrical shock.

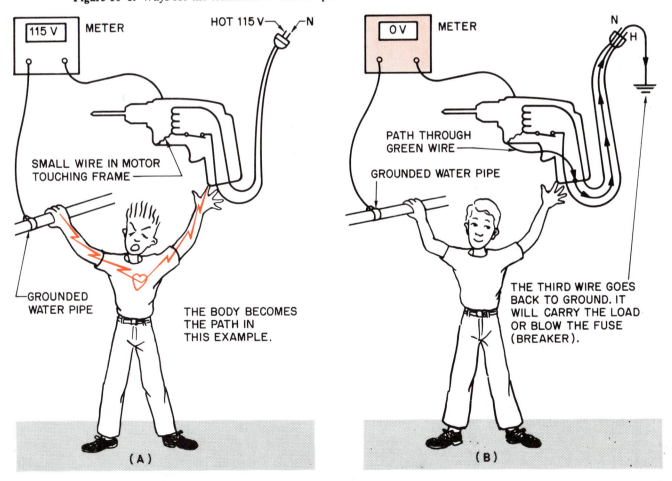

Figure 10-7. Electrical circuit to ground from metal frame of drill.

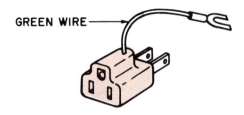

Figure 10–8. A three- to two-prong adapter.

Figure 10–10. A double-insulated drill motor. *Photo by Bill Johnson*

is a wiring diagram showing situations where the technician is part of a circuit.

The technician should use properly grounded power tools connected to properly grounded circuits.
Caution should be taken when using electric drill motors. These are hand-held devices with potential electrical energy inside just waiting for a path to flow through. Some drills are constructed with metal frames. These should all have a grounding wire in the power cord. The grounding wire protects the operator. The drill motor will work without it, but it is not safe. If the motor inside the drill develops a loose connection and the frame of the drill motor becomes electrically hot, the third wire, rather than your body, will carry the current, and a fuse or breaker will interrupt the circuit, Figure 10–7.

Many technicians use the three- to two-prong adapters at job sights because the wall receptacle only has two connections and their drill motor has a three-wire plug, Figure 10–8. This adapter has a third wire that

must be connected to a ground for the circuit to give protection. If this wire is fastened under the wall plate screw and this screw terminates in an ungrounded box nailed to a wooden wall, you are not protected, Figure 10–9.

An alternative to the metal drill motor is a plastic cased drill motor that has the motor and electrical connections insulated within the tool. These are called double-insulated and are very safe, Figure 10–10.

Electrical Burns

Do not wear jewelry (rings and watches) while working on live electrical circuits because they can cause shock and possible burns.

Other electrical burns can come from an electrical

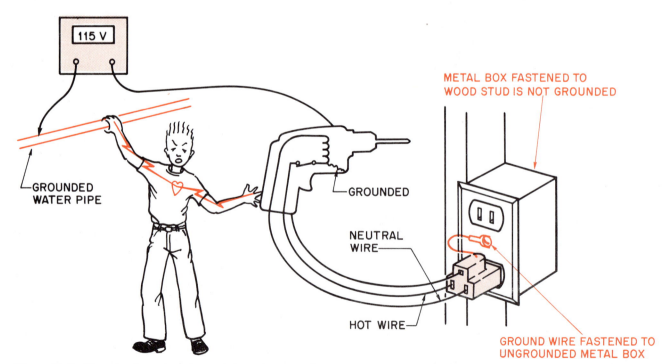

Figure 10–9. The wire from the adapter is intended to be fastened under the screw in the duplex wall plate. However, this will provide no protection if the outlet box is not grounded.

arc, such as in a short circuit to ground, when uncontrolled electrical energy flows. For example, you may let a screwdriver slip while working in a panel and the blade may complete a circuit to ground. The potential flow of electrical energy is tremendous from this type of accident. An example using Ohm's law can be used to show what the potential is. When a circuit has a resistance of 10 Ω and is operated on 120 V, it would have a current flow of

$$I = E/R = 120 \text{ V}/10 \text{ } \Omega = 12 \text{ A}$$

If this example is calculated again with less resistance, the current will be greater because the voltage is divided by a smaller number. Suppose the resistance is lowered to 1 Ω; the current flow is then

$$I = E/R = 120/1 = 120 \text{ A}$$

If the resistance is reduced to 0.1 Ω, the current flow will be 1200 A. By this time the circuit breaker will trip, but you may have already incurred burns, Figure 10–11. You will learn more about this in Unit 11, Basic Electricity.

10.3   HEAT

The use of heat requires special care. A high concentration of heat comes from torches. Torches are used for many things, including soldering or brazing. Many combustible materials may be in the area where soldering is required. For example, the refrigeration system in a completed restaurant may need repair. The upholstered restaurant furniture, grease, and other flammable materials must be treated as carefully as possible. *When soldering or using concentrated heat, a fire extinguisher should always be close by, and you should know exactly where it is and how to use it.* Learn to use a fire extinguisher *before* the fire occurs.

Figure 10–12. A typical fire extinguisher. *Photo by Bill Johnson*

A fire extinguisher should be a part of the service tool setup on a well-stocked truck, Figure 10–12.
When a solder connection must be made next to combustible materials or a finished surface, use a shield of noncombustible materials for insulation. This is often necessary when soldering in packaged equipment, for example, when an ice-maker compressor is changed or when a drier must be soldered in the line, Figure 10–13.

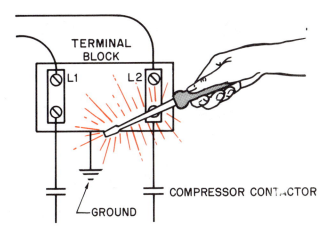

Figure 10–11. Wiring illustration showing short circuit caused by a slip of a screwdriver.

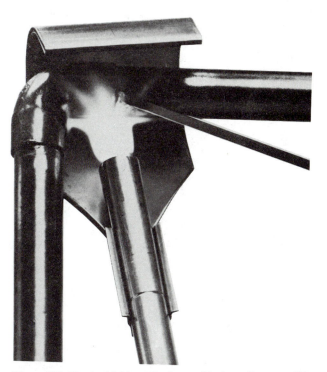

Figure 10–13. A shield used when soldering. *Courtesy Wengaersheek*

SAFETY PRECAUTION: *Hot refrigerant lines, hot heat exchangers, and hot motors can burn your skin and leave a permanent scar. Care should be used while handling them.*

10.4 COLD

Cold can be as harmful as heat. Liquid refrigerant can burn your skin instantaneously. But long exposure to cold is also harmful. Working in cold weather can cause frostbite. Wear proper clothing and waterproof boots, which also help protect against electrical shock. A cold wet technician will not make decisions based on logic. Make it a point to stay warm. *Waterproof boots not only protect your feet from water and cold, but help to protect you from electrical shock hazard when working in wet weather.*

Low-temperature freezers are just as cold in the middle of the summer as in the winter. Cold-weather gear must be used when working inside these freezers. For example, an expansion valve may need changing, and you may be in the freezer for more than an hour. It is a shock to the system to step into a room that is 0°F from the outside where it may be 95° or 100°F. If you are on call for any low-temperature applications, carry a coat and gloves and wear them in cold environments.

10.5 MECHANICAL EQUIPMENT

Rotating equipment can damage body and property. Motors that drive fans, compressors, and pumps are among the most dangerous because they have so much power. *If a shirt sleeve or coat were caught in a motor drive pulley or coupling, severe injury could occur. Loose clothing should never be worn around rotating machinery. Even a small electric hand drill can wind a necktie up before the drill can be shut off, Figure 10–14.*

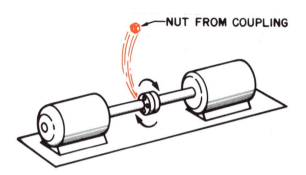

NUT FROM COUPLING

Figure 10-15. *Insure that all nuts are tight on couplings and other components.*

When starting an open motor, stand well to the side of the motor drive mechanism. If the coupling or belt were to fly off the drive, it would fly outward in the direction of rotation of the motor. All set screws or holding mechanisms must be tight before starting a motor, even if the motor is not connected to a load. All wrenches must be away from a coupling or pulley. A wrench or nut thrown from a coupling can be a lethal projectile, Figure 10-15.

When a large motor, such as a fan motor, is coasting to a stop, don't try to stop it. If you try to stop the motor and fan by gripping the belts, the momentum of the fan and motor may pull your hand into the pulley and under the belt, Figure 10–16.

10.6 MOVING HEAVY OBJECTS

Heavy objects must be moved from time to time. Think out the best and safest method to move these objects. Don't just use muscle power. Special tools can help you move equipment. When equipment must be installed on top of a building, a crane or even a helicopter can be used, Figure 10–17. A technician with a weak back is limited. Do not take a chance, get help.

When you must lift heavy equipment, use your legs not your back, Figure 10–18. Some available tools are a pry bar, a lever truck, a refrigerator hand truck, a lift gate on the pickup truck, and a portable dolly, Figure 10–19.

Figure 10–14. *Never wear a necktie or loose clothing when using or working around rotating equipment.*

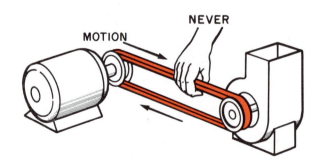

NEVER

MOTION

Figure 10-16. *Never attempt to stop a motor or other mechanism by gripping the belt.*

Figure 10-17. Helicopter lifting air conditioning equipment to roof.

Figure 10-18. *Use legs, not back, to lift objects. Keep back straight.*

When moving large equipment, such as a reach-in cooler, across a carpeted floor, or a package unit across a gravel-coated roof, first lay some plywood down. Keep the plywood in front of the equipment as it is moved along. When equipment has a flat bottom, such as a package air conditioner, short lengths of pipe may be used to move the equipment across a solid floor. Figure 10-20.

10.7 USING CHEMICALS

Chemicals are often used to clean equipment such as air-cooled condensers and evaporators. They are also used for water treatment. The chemicals are normally simple and mild, except for some harsh cleaning products used for water treatment. *These chemicals should be handled according to the manufacturer's directions. Do not get careless. If you spill chemicals*

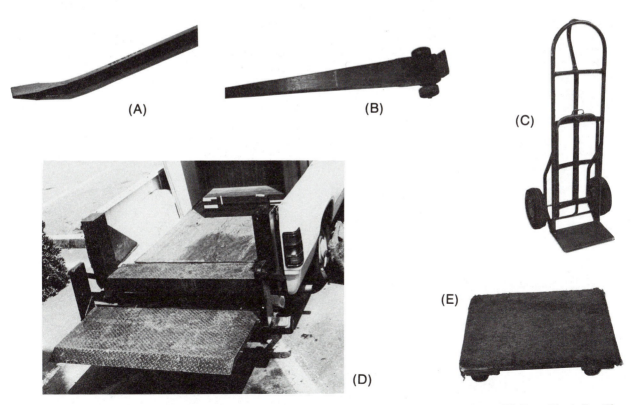

(A)   (B)   (C)

(E)

(D)

Figure 10-19. (A) Pry bar. (B) Lever truck. (C) Hand truck. (D) Lift gate on a pickup truck. (E) Portable dolly. *Photos by Bill Johnson*

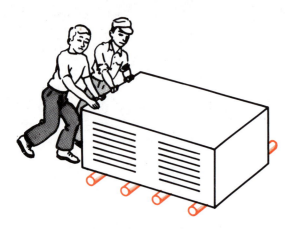

Figure 10–20. Moving equipment using short lengths of pipe as rollers.

on your skin or splash them in your eyes, follow the manufacturer's directions and go to a doctor. It is a good idea to read the entire label before starting a job. It's hard to read the first aid treatment for eyes after they've been damaged.✱

SUMMARY

- ✱The technician must use every precaution when working with pressures, electrical energy, heat, cold, rotating machinery, chemicals, and moving heavy objects.✱
- ✱Safety situations involving pressure are encountered while working with pressurized systems and vessels.✱
- ✱Electrical energy is encountered while troubleshooting energized electrical circuits.✱
- ✱Heat is encountered while soldering and working on heating systems.✱
- ✱Liquid R-22 refrigerant boils at −41°F at atmospheric pressure and will cause frostbite.✱
- ✱Rotating equipment such as fans and pumps can be dangerous and should be treated with caution.✱
- ✱When moving heavy equipment, use correct techniques.✱
- ✱Chemicals are used for cleaning and water treatment and must be handled with care.✱

REVIEW QUESTIONS

1. Where would a technician encounter freezing liquid refrigerant?
2. Why should the technician not use his or her hands to stop liquid refrigerant from escaping?
3. What part of the body does electrical shock harm?
4. What two types of injury are caused by electrical energy?
5. How can the technician prevent electrical shock?
6. How can the technician prevent paint burning on equipment while soldering?
7. What safety precaution should be taken before starting a large motor with the coupling disconnected?
8. What is the third wire used for on an electric drill motor?
9. How can heavy equipment be moved across a rooftop?
10. What special precautions should be taken before using chemicals to clean an air-cooled condenser?

BASIC ELECTRICITY AND MAGNETISM

OBJECTIVES

After studying this unit, you should be able to

- describe the structure of an atom.
- identify atoms with a positive charge and atoms with a negative charge.
- explain the characteristics that make certain materials good conductors.
- describe how magnetism is used to produce electricity.
- state the differences between alternating current and direct current.
- list the units of measurement for electricity.
- explain the differences between series and parallel circuits.
- state Ohm's law.
- state the formula for determining electrical power.
- describe a solenoid.
- explain inductance.
- describe the construction of a transformer and the way that a current is induced in a secondary circuit.
- describe how a capacitor works.
- state the reasons for using proper wire sizes.
- describe procedures for making electrical measurements.

11.1 STRUCTURE OF MATTER

To understand the theory of how an electric current flows, you must understand something about the struc-ture of matter. Matter is made up of atoms. Atoms are made up of protons, neutrons, and electrons. Protons and neutrons are located at the center (or nucleus) of the atom. Protons have a positive charge. Neutrons have no charge and have little or no effect as far as electrical characteristics are concerned. Electrons have a negative charge and travel around the nucleus in orbits. The number of electrons in an atom is the same as the number of protons. Electrons in the same orbit are the same distance from the nucleus but do not follow the same orbital paths, Figure 11–1.

The hydrogen atom is a simple atom to illustrate because it has only one proton and one electron, Figure 11–2. Not all atoms are as simple as the hydrogen atom. Most wiring used to conduct an electrical current is made of copper. Figure 11–3 illustrates a copper atom.

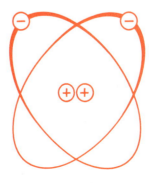

Figure 11–1. Orbital paths of electrons.

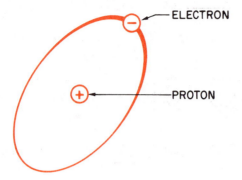

Figure 11-2. Hydrogen atom with one electron and one proton.

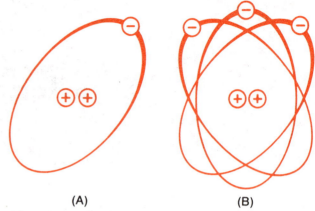

(A)   (B)

Figure 11-4 (A) This atom has two protons and one electron. It has a shortage of electrons and thus a positive charge. (B) This atom has two protons and three electrons. It has an excess of electrons and thus a negative charge.

There are 29 protons and 29 electrons. Some electron orbits are farther away from the nucleus than others. As can be seen, 2 travel in an inner orbit, 8 in the next, 18 in the next, and 1 in the outer orbit. It is this single electron in the outer orbit that makes copper a good conductor.

11.2 MOVEMENT OF ELECTRONS

When sufficient energy or force is applied to an atom, the outer electron (or electrons) becomes free and moves. If it leaves the atom, the atom will contain more protons than electrons. Protons have a positive charge. This means that this atom will have a positive charge, Figure 11-4A. The atom the electron joins will contain more electrons than protons, so it will have a negative charge, Figure 11-4B.

Like charges repel each other, and unlike charges attract each other. An electron in an atom with a surplus of electrons (negative charge) will be attracted to an atom with a shortage of electrons (positive charge). An electron entering an orbit with a surplus of electrons will tend to repel an electron already there and cause it to become a free electron.

11.3 CONDUCTORS

Good conductors are those with few electrons in the outer orbit. Three common metals—copper, silver, and gold—are good conductors, and each has one electron in the outer orbit. These are considered to be free electrons because they move easily from one atom to another.

11.4 INSULATORS

Atoms with several electrons in the outer orbit are poor conductors. These electrons are difficult to free, and materials made with these atoms are considered to be insulators. Glass, rubber, and plastic are examples of good insulators.

11.5 ELECTRICITY PRODUCED FROM MAGNETISM

Electricity can be produced in many ways, for example, from chemicals, pressure, light, heat, and magnetism. The electricity that air conditioning and heating technicians are most involved with is produced by a generator using magnetism.

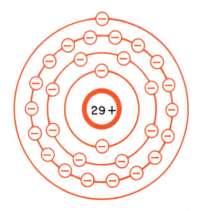

Figure 11-3. Copper atom with 29 protons and 29 electrons.

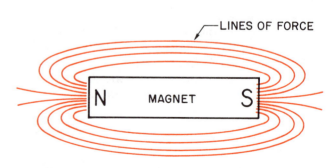

Figure 11-5. Permanent magnet with lines of force.

Magnets are common objects with many uses. Magnets have poles usually designated as the north (N) pole and the south (S) pole. They also have fields of force. Figure 11–5 shows the lines of the field of force around a permanent bar magnet. This field causes the like poles of two magnets to repel each other and the unlike poles to attract each other.

If a conductor, such as a copper wire, is passed through this field and cuts these lines of force, the outer electrons in the atoms in the wire are freed and begin to move from atom to atom. They will move in one direction. It does not matter if the wire moves or if the magnetic field moves. It is only necessary that the conductor cut through the lines of force, Figure 11–6.

This movement of electrons in one direction produces the electric current. The current is an impulse transferred from one electron to the next. If a golf ball were pushed into a tube already filled with golf balls, one would be ejected instantly from the other end, Figure 11–7. Electric current travels in a similar manner at a speed of 186,000 miles per second. The electrons do not travel through the wire at this speed, but the repelling and attracting effect causes the current to do so.

An electrical generator has a large magnetic field and many turns of wire cutting the lines of force. A large magnetic field or one with many turns of wire produces more electricity than a smaller field or a field with few turns of wire. The magnetic force field for generators is usually produced by electromagnets. Electromagnets have similar characteristics to permanent magnets and will be discussed later in this unit. Figure 11–8 shows a simple generator.

11.6 DIRECT CURRENT (DC)

Direct current travels in one direction. Because electrons have a negative charge and travel to atoms with a positive charge, direct current is considered to flow from negative to positive.

Figure 11-7. Tube filled with golf balls.

11.7 ALTERNATING CURRENT (AC)

Alternating current is continually and rapidly reversing. The charge at the power source (generator) is continually changing direction; thus the current continually reverses itself. Most electrical energy generated for public use is alternating current. There are several reasons for this. It is much more economical to transmit electrical energy long distances in the form of alternating current. The voltage of this type of electrical current flow can be readily changed so that it has many more uses. There are still many applications for direct current, but it is usually obtained by changing AC to DC or by producing the DC locally where it is to be used.

11.8 ELECTRICAL UNITS OF MEASUREMENT

Electromotive force (emf) or voltage (V) is used to indicate the difference of potential in two charges. When there is an electron surplus built up on one side of a circuit and a shortage of electrons on the other side, there is a difference of potential or electromotive force. The unit used to measure this force is the *volt*.

The *ampere* is the unit used to measure the quantity of electrons moving past a given point in a specific period of time (electron flow).

All materials oppose or resist the flow of an electrical current to some extent. In good conductors this opposition or resistance is low. In poor conductors the

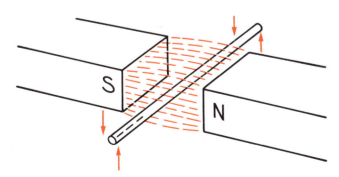

Figure 11-6. Movement of wire up and down, and cutting the lines of force causes an electric current to flow in the wire.

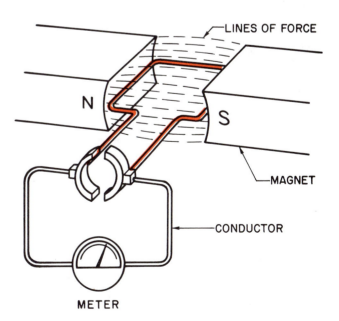

Figure 11-8. Simple generator.

resistance is high. The unit used to measure resistance is the *ohm*. A conductor has a resistance of one ohm when a force of one volt causes a current of one ampere to flow.

Volt = Electrical force or pressure (V)
Ampere = Quantity of electron flow (A)
Ohm = Resistance to electron flow (Ω)

11.9 THE ELECTRICAL CIRCUIT

An electrical circuit must have a power source, a conductor to carry the current, and a load or device to use the current. There is also generally a means for turning the electrical current flow on and off. Figure 11-9 shows an electrical generator for the source, a wire for the conductor, a small heater for the load, and a switch for opening and closing the circuit.

The generator produces the current by passing many turns of wire through a magnetic field. If it is a DC generator, the current will flow in one direction. If it is an AC generator, the current will continually reverse itself. However, the effect on this circuit will generally be the same whether it is AC or DC.

The wire or conductor provides the means for the electricity to flow to the heater and complete the circuit. The electrical energy is converted to heat and light energy at the heater element.

The switch is used to open and close the circuit. When the switch is open, no current will flow. When it is closed, the heater element will produce heat.

11.10 MAKING ELECTRICAL MEASUREMENTS

In the circuit illustrated in Figure 11-9 electrical measurements can be made to determine the voltage (emf) and amperes (current). In making the measurements, Figure 11-10, the voltmeter is connected across the terminals of the heater without interrupting the circuit. The ammeter is connected directly into the

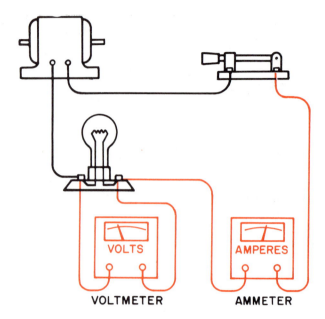

Figure 11-10. Voltage is measured across the resistance. Amperage is measured in series.

circuit so that all the current flows through it. Figure 11-11 illustrates the same circuit using symbols.

Often a circuit will contain more than one resistance or load. These resistances may be wired in series or in parallel, depending on the application or use of the circuit. Figure 11-12 shows three loads in series. This is shown pictorially and by symbols. Figures 11-13A and 11-13B illustrate these loads wired in parallel.

In circuits where devices are wired in series, all of the current passes through the device. When two or more devices are wired in parallel, the current is divided among the devices. This will be explained in more detail later. Power-passing devices such as switches are wired in series. Most resistances or loads (power-consuming devices) air conditioning and heating technicians work with are wired in parallel.

Figure 11-14 illustrates how a voltmeter is con-

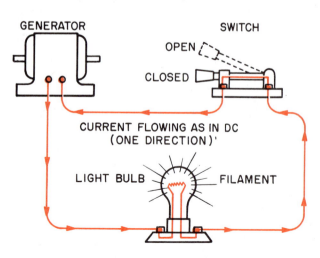

Figure 11-9. Electric circuit.

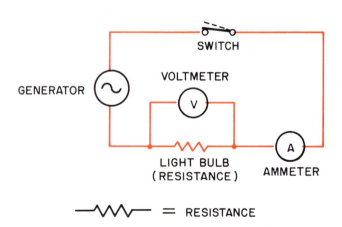

Figure 11-11. Same circuit as Figure 11-10, illustrated with symbols.

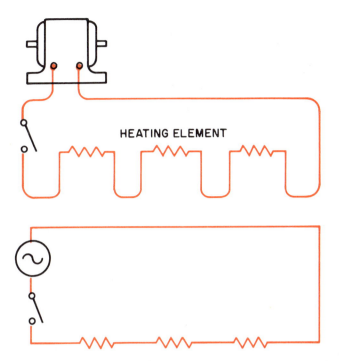

HEATING ELEMENT

Figure 11-12. Multiple resistances (small heating elements) in series.

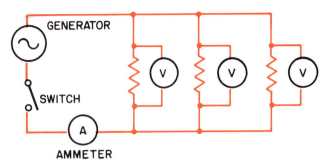

Figure 11-14. Voltage readings are taken across the resistances in the circuit.

nected for each of the resistances. The voltmeter is in parallel with each resistance. The ammeter is also shown and is in series in the circuit. An ammeter has been developed that can be clamped around a single conductor to measure amperes, Figure 11-15. This is convenient because it is often difficult to disconnect the circuit to connect the ammeter in series. This type of ammeter, usually called a "clamp-on" type, will be discussed later in this unit.

11.11 OHM'S LAW

During the early 1800s the German scientist Georg S. Ohm did considerable experimentation with electrical circuits and particularly with regard to resistances in these circuits. He determined that there is a relationship between each of the factors in an electrical circuit. This relationship is called Ohm's law. The following describes this relationship. Letters are used to represent the different electrical factors.

E or V = Voltage (electromotive force)
I = Amperage (current)
R = Resistance (load)

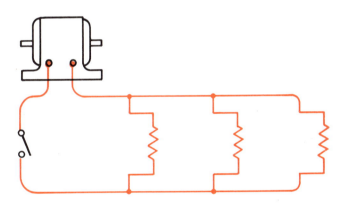

Figure 11-13A. Multiple resistances (small heating elements) in parallel.

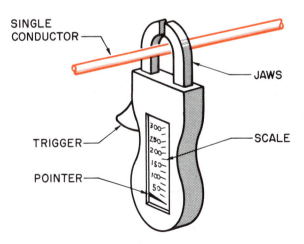

Figure 11-15. Clamp-on ammeter. *Courtesy Amprobe Instrument Division, Core Industries*

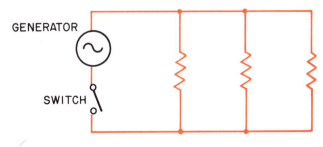

Figure 11-13B. Three resistances in parallel using symbols.

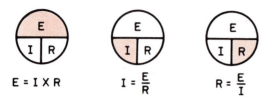

$$E = I \times R \qquad I = \frac{E}{R} \qquad R = \frac{E}{I}$$

Figure 11-16. To determine the formula for the unknown quantity, cover the letter representing the unknown.

The voltage equals the amperage times the resistance:

$$E = I \times R$$

The amperage equals the voltage divided by the resistance:

$$I = \frac{E}{R}$$

The resistance equals the voltage divided by the amperage:

$$R = \frac{E}{I}$$

Figure 11-16 shows a convenient way to remember these formulae. The symbol used for ohms is Ω.

In Figure 11-17 the resistance of the heating element can be determined as follows:

$$R = \frac{E}{I} = \frac{120}{1} = 120 \ \Omega$$

In Figure 11-18 the voltage across the resistance can be calculated as follows:

$$E = I \times R = 2 \times 60 = 120 \ V$$

Figure 11-19 indicates the voltage to be 120 V and the resistance to be 20 Ω. The formula for determining the current flow is

$$I = \frac{E}{R} = \frac{120}{20} = 6 \ A$$

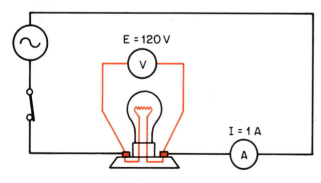

Figure 11-17

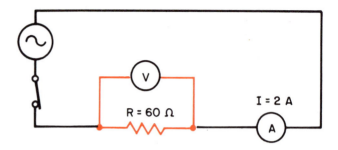

Figure 11-18

In series circuits with more than one resistance, simply add the resistances together as if they were one. In Figure 11-20 there is only one path for the current to follow (a series circuit), so the resistance is 40 Ω (20 Ω + 10 Ω + 10 Ω). The amperage in this circuit will be

$$I = \frac{E}{R} = \frac{120}{40} = 3 \ A$$

The ohms in the individual resistances can be determined by disconnecting the resistance to be measured from the circuit and reading the ohms from an ohmmeter, as illustrated in Figure 11-21.

SAFETY PRECAUTION: *Do not use any electrical measuring instruments without specific instructions from a qualified person. The use of electrical measuring instruments will be discussed later in this unit.*

11.12   CHARACTERISTICS OF SERIES CIRCUITS

In series circuits:

- The voltage is divided across the different resistances.
- The total current flows through each resistance or load.
- The resistances are added together to obtain the total resistance, Figure 11-20.

11.13   CHARACTERISTICS OF PARALLEL CIRCUITS

In parallel circuits:

- The total voltage is dropped across each resistance.

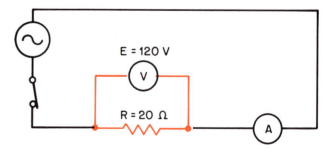

Figure 11-19

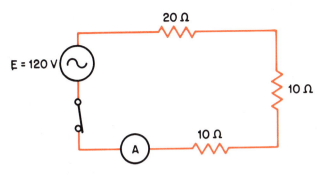

Figure 11-20

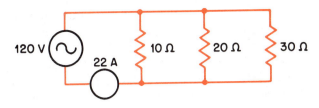

Figure 11-22

- The current is divided between the different loads, or the total current is equal to the sum of the currents in each branch.
- The total resistance is less than the value of the smallest resistance.

Calculating the resistances in a parallel circuit requires a different procedure than simply adding them together as in a series circuit. A parallel circuit allows current flow along two or more paths at the same time. This type of circuit applies equal voltage to all loads. The general formula used to determine *total resistance* in a parallel circuit is as follows:

$$R_{total} = \frac{1}{\dfrac{1}{R_1} + \dfrac{1}{R_2} + \dfrac{1}{R_3} + \cdots}$$

The total resistance of the circuit in Figure 11-22 is determined as follows:

$$R_{total} = \frac{1}{\dfrac{1}{10} + \dfrac{1}{20} + \dfrac{1}{30}}$$
$$= \frac{1}{0.1 + 0.05 + 0.033}$$
$$= \frac{1}{0.183}$$
$$= 5.46 \ \Omega$$

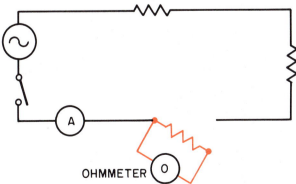

Figure 11-21. To determine resistance, disconnect the resistance from the circuit and check with an ohmmeter.

To determine the total current draw use Ohm's law:

$$I = \frac{E}{R} = \frac{120}{5.46} = 22 \ \text{A}$$

11.14 ELECTRICAL POWER

Electrical power (P) is measured in watts. A *watt* (W) is the power used when one ampere flows with a potential difference of one volt. Therefore, power can be determined by multiplying the voltage times the amperes flowing in a circuit.

$$\text{Watts} = \text{Volts} \times \text{Amperes}$$

or

$$P = E \times I$$

The consumer of electrical power pays the electrical utility company according to the number of kilowatts used. A kilowatt is equal to 1000 W. To determine the power being consumed, divide the number of watts by 1000:

$$P \ (\text{in kilowatts}) = \frac{E \times I}{1000}$$

11.15 MAGNETISM

Magnetism was briefly discussed previously in this unit to point out how electrical generators are able to produce electricity. Magnets are classified as either permanent or temporary. Permanent magnets are used in only a few applications that air conditioning and refrigeration technicians would work with. Electromagnets, a form of temporary magnet, are used in many components of air conditioning and refrigeration equipment.

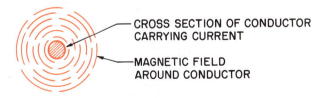

CROSS SECTION OF CONDUCTOR CARRYING CURRENT

MAGNETIC FIELD AROUND CONDUCTOR

Figure 11-23. Cross section of a wire showing magnetic field around conductor.

A magnetic field exists around a wire carrying an electrical current, Figure 11–23. If the wire or conductor is formed in a loop, the magnetic field will be increased, Figure 11–24. If the wire is wound into a coil, a stronger magnetic field will be created, Figure 11–25. This coil of wire carrying an electrical current is called a *solenoid*. This solenoid or electromagnet will attract or pull an iron bar into the coil, Figure 11–26.

If an iron bar is inserted permanently in the coil, the strength of the magnetic field will be increased even more.

This magnetic field can be used to generate electricity and to cause electric motors to operate. The magnetic attraction can also cause motion, which is used in many controls and switching devices, such as solenoids, relays, and contactors, Figure 11–27A. Figure 11–27B is a cutaway view of a solenoid.

Figure 11–24. Magnetic field around loop of wire. This is a stronger field than that around a straight wire.

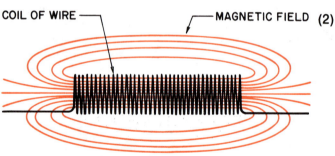

Figure 11–25. There is a stronger magnetic field surrounding wire formed into a coil.

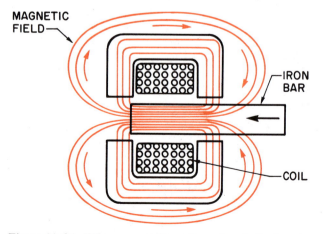

Figure 11–26. When current flows through coil, the iron bar will be attracted into it.

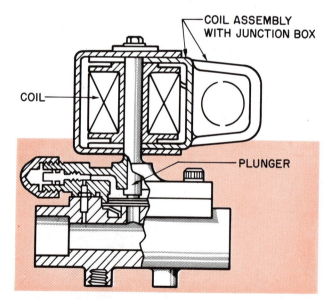

Figure 11–27A. (1) Solenoid. (2) Relay. (3) Contactor. *Courtesy (1) Sporlan Valve Company. (3) Honeywell, Inc., Residential Division. Photo (2) by Bill Johnson*

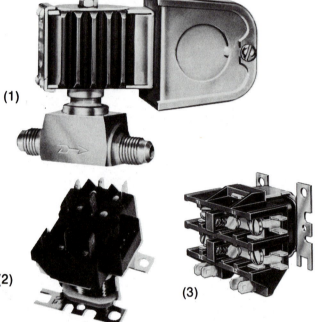

Figure 11–27B. Cutaway view of solenoid. *Courtesy Parker Hannifin Corporation*

11.16 INDUCTANCE

As mentioned previously, when voltage is applied to a conductor and current flows, a magnetic field is produced around the conductor. In an AC circuit the current is continually changing direction. This causes the magnetic field to continually build up and immediately collapse. When these lines of force build up and collapse, they cut through the wire or conductor and produce an emf or voltage. This voltage opposes the existing voltage in the conductor.

In a straight conductor this induced voltage is very small and is usually not considered, Figure 11–28. However, if a conductor is wound into a coil, these lines of force overlap and reinforce each other, Figure 11–29. This does develop an emf or voltage that is strong enough to provide opposition to the existing voltage. This opposition is called *inductive reactance* and is a type of resistance in an AC circuit. Coils, chokes, and transformers are examples of components that produce inductive reactance. This will be covered in more detail later in this section. See Figure 11–30 for the electrical symbol.

11.17 TRANSFORMERS

Transformers are electrical devices that produce an electrical current in a second circuit through electromagnetic induction.

In Figure 11–31A a voltage applied across terminals A-A will produce a magnetic field around the steel or iron core. This is alternating current causing the magnetic field to continually build up and collapse as the current reverses. This will cause the magnetic field around the core in the second winding to cut across the conductor wound around it. An electrical current is induced in the second winding.

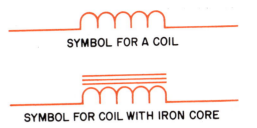

Figure 11–30. Symbols for a coil.

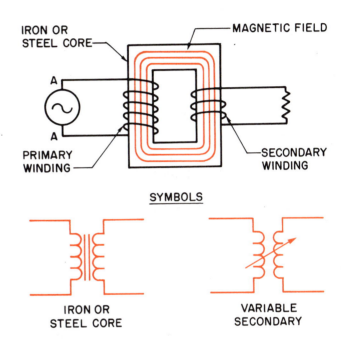

Figure 11–31A. Voltage applied across terminals produces a magnetic field around an iron or steel core.

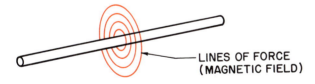

Figure 11–28. Straight conductor with magnetic field.

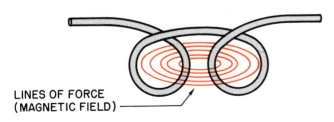

Figure 11–29. Conductor formed into coil with lines of force.

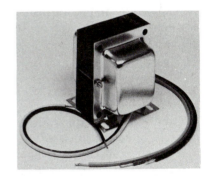

Figure 11–31B. Transformers. *Reprinted with permission of Motor and Armatures, Inc.*

Transformers, Figure 11–31B, have a primary winding, a core usually made of thin plates of steel laminated together, and a secondary winding. There are step-up and step-down transformers. A step-down transformer contains more turns of wire in the primary winding than in the secondary winding. The voltage at the secondary is directly proportional to the number of turns of wire in the secondary as compared to the primary windings. For example, Figure 11–32 is a transformer with 1000 turns in the primary and 500 turns in the secondary. A voltage of 120 V is applied, and the voltage induced in the secondary is 60 V. Actually the voltage is slightly less due to some loss into the air of the magnetic field and because of resistance in the wire.

A step-up transformer has more windings in the secondary than in the primary. This causes a larger voltage to be induced into the secondary. In Figure 11–33, with 1000 turns in the primary, 2000 in the secondary, and an applied voltage of 120 V, the voltage induced in the secondary is doubled, or approximately 240 V.

The same power (watts) is available at the secondary as at the primary (except for a slight loss). If the voltage is reduced to one half that at the primary, the current nearly doubles.

Step-up transformers are used at generating stations to increase the voltage to produce more efficiency in delivering the electrical energy over long distances to substations or other distribution centers. At the substation the voltage is reduced for further distribution. To reduce the voltage, a step-down transformer is used. At a residence the voltage may be reduced to 240 V or 120 V. Further step-down transformers may be used with air conditioning and heating equipment to produce the 24 V commonly used in thermostats and other control devices.

11.18   CAPACITANCE

A device in an electrical circuit that allows electrical energy to be stored for later use is called a *capacitor*. A very simple capacitor is composed of two plates with insulating material between them, Figure 11–34. The capacitor can store a charge of electrons on one plate. When the plate is fully charged in a DC circuit, no current will flow until there is a path back to the positive plate, Figure 11–35. When this path is available, the electrons will flow to the positive plate until the negative plate no longer has a charge, Figures 11–36 and 11–37. At this point both plates are neutral.

In an AC circuit the voltage and current are continuously changing direction. As the electrons flow in one direction, the capacitor plate on one side becomes charged. As this current and voltage reverses, the charge on the capacitor becomes greater than the

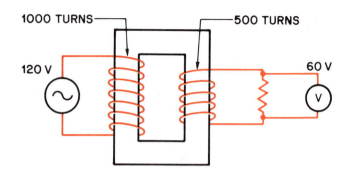

Figure 11–32. Step-down transformer.

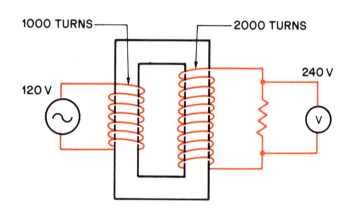

Figure 11–33. Step-up transformer.

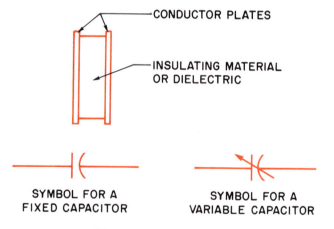

Figure 11–34. Capacitor.

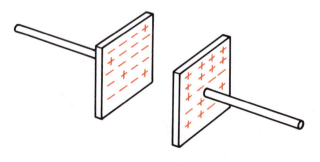

Figure 11–35. Charged capacitor.

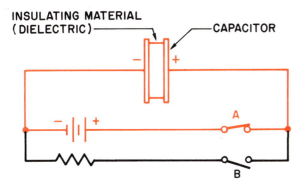

Figure 11-36. Electrons will flow to negative plate from battery. Negative plate will charge until capacitor has the same potential difference as the battery.

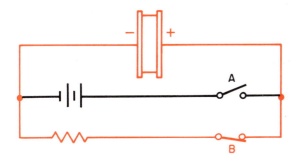

Figure 11-37. When capacitor is charged, switch A opened, and switch B closed, the capacitor discharges through the resistor to the positive plate. The capacitor then has no charge.

source voltage, and the capacitor begins to discharge. It is discharged through the circuit, and the opposite plate becomes charged. This continues through each AC cycle.

A capacitor has *capacitance*. This is the amount of charge that can be stored. The capacitance is determined by the following physical characteristics of the capacitor:

1. Distance between the plates
2. Surface area of the plates
3. Dielectric material between the plates

A capacitor opposes current flow in an AC circuit similar to a resistor or to inductive reactance. This opposition or resistance is called *capacitive reactance*. The capacitive reactance depends on the frequency of the voltage and the capacitance of the capacitor. Capacitors will be studied in more detail later in this section.

11.19 ELECTRICAL MEASURING INSTRUMENTS

A multimeter is an instrument that measures voltage, current, resistance, and, on some models, temperature. It is a combination of several meters and can be used for making AC or DC measurements in several ranges. It is the instrument used most often by heating, refrigeration, and air conditioning technicians.

A multimeter often used is the volt-ohm-milliammeter (VOM), Figure 11-38. This meter is used to measure AC and DC voltages, direct current, resistance, and alternating current when used with an AC ammeter adaptor, Figure 11-39. The meter has two main switches: the *function* switch and the *range* switch. The function switch, located at the left side of

Figure 11-38. Volt-ohm-milliammeter (VOM). *Courtesy of Simpson Electric Co., Elgin, Illinois*

Figure 11-39. Ammeter adaptor to VOM. *Courtesy of Simpson Electric Co., Elgin, Illinois*

the lower front panel, Figure 11–40, has −DC, +DC, and AC positions.

The range switch is in the center of the lower part of the front panel, Figure 11–40. It may be turned in either direction to obtain the desired range. It also selects the proper position for making alternating current measurements when using the AC "clamp-on" adapter.

The zero ohms control on the right of the lower panel is used to adjust the meter to compensate for the aging of the meter's batteries, Figure 11–41.

When the meter is ready to operate, the pointer must read zero. If the pointer is off zero, use a screwdriver to turn the screw clockwise or counterclockwise until the pointer is set exactly at zero, Figure 11–41.

The test leads may be plugged into any of eight jacks. In this unit the common (−) and the positive (+) jacks will be the only ones used, Figure 11–42. Only a few of the basic measurements will be discussed. Other measurements will be described in detail in other units.

The following instructions are for familiarization with the meter and procedures only. *Do not make any measurements without instructions and approval from an instructor or supervisor.* Insert the black test lead into the common jack. Insert the red test lead into the + jack.

Figure 11–43 is a DC circuit with 15 V from the battery power source. To check this voltage with the VOM, set the function switch to +DC. Set the range switch to 50 V, Figure 11–43. If you are not sure about the magnitude of the voltage, always set the range switch to the highest setting. After measuring, you can set the switch to a lower range if necessary to obtain a more accurate reading. Be sure the switch in the circuit is open. Now connect the black test lead to the negative side of the circuit, and connect the red test lead to the positive side, as indicated in Figure 11–43. Note

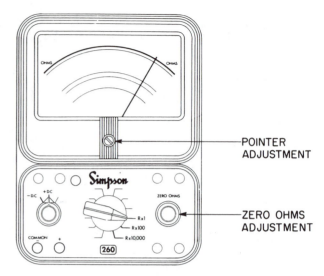

Figure 11–41. Zero ohms adjustment and pointer adjustment. *Courtesy of Simpson Electric Co., Elgin, Illinois*

that the meter is connected across the load (in parallel). Close the switch and read the voltage from the DC scale.

To check the voltage in the circuit in Figure 11–44, follow the steps listed:

1. Turn off the power.
2. Set the function switch to AC.
3. Set the range switch to 500 V.
4. Plug the black test lead into the − common jack and the red test lead into the + jack.
5. Connect the test leads across the load as shown in Figure 11–43.
6. Turn on the power. Read the red scale marked AC. Use the 0–50 figures and multiply the reading by 10.

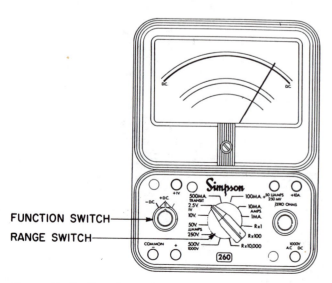

Figure 11–40. Function switch and range switch on VOM. *Courtesy of Simpson Electric Co., Elgin, Illinois*

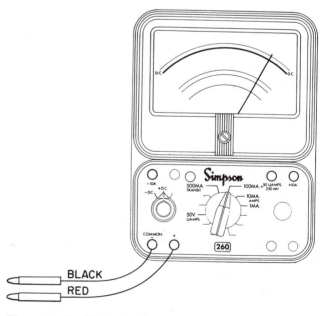

Figure 11–42. VOM showing the − and + jacks. *Courtesy of Simpson Electric Co., Elgin, Illinois*

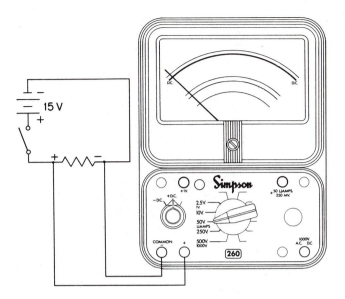

Figure 11-43. VOM with function switch set at +DC. Range switch set at 50 V. *Courtesy of Simpson Electric Co., Elgin, Illinois*

7. Turn the range switch to 250 V. Read the red scale marked AC and use the black figures immediately above the scale.

To determine the resistance of a load, disconnect the load from the circuit. Make sure all power is off while doing this.

1. Make the zero ohms adjustment in the following manner:
   a. Turn the range switch to the desired ohms range.
      Use R × 1 for 0–200 Ω
      Use R × 100 for 200–20 000 Ω
      Use R × 10 000 for above 20 000 Ω
   b. Plug the black test lead into the − common jack and the red test lead into the + jack.
   c. Connect the test leads to each other.

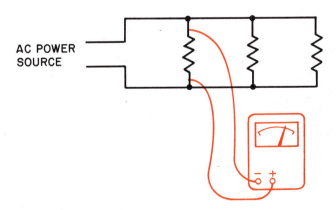

Figure 11-44

d. Rotate the zero ohms control until the pointer indicates zero ohms. (If the pointer cannot be adjusted to zero, replace one or both batteries.)
2. Disconnect the ends of the test leads and connect them to the load being tested.
3. Set the function switch at either −DC or +DC.
4. Observe the reading on the ohms scale at the top of the dial. (Note that the ohms scale reads from right to left.)
5. To determine the actual resistance, multiply the reading by the factor at the range switch position.

Many meters have an AC current attachment to allow current measurements without breaking the circuit. To use this attachment, connect the leads to the − common and + jacks.

1. Set the function switch to AC and the range switch to 2.5 V.
2. Set the current range selector on the clamp-on attachment to a range that covers the probable current being measured.
3. Place the clamp end over a single conductor as illustrated in Figure 11-45.
4. Read the amperage.

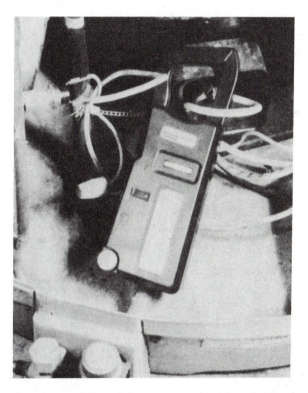

Figure 11-45. Measuring amperage by clamping meter around conductor. From Herman and Sparkman, *Electricity and Controls for Heating, Ventilating and Air Conditioning,* © *1986 by Delmar Publishers Inc.*

VOLTAGE

1 volt = 1000 millivolts (mV)
1 volt = 1 000 000 microvolts (μV)

AMPERAGE

1 ampere = 1000 milliamperes (mA)
1 ampere = 1 000 000 microamperes (μA)
Note that the symbol for micro or millions is μ

Figure 11–46. Units of voltage and amperage.

SAFETY PRECAUTION: *Do not perform any of these readings without approval from an instructor or supervisor. These instructions are simply a general orientation to meters. Be sure to read the operator's manual for the particular meter available to you.*

It is often necessary to determine voltage or amperage readings to a fraction of a volt or ampere. See Figure 11–46.

There are many styles and types of meters for making electrical measurements. The VOM is one type, and several companies manufacture them. Figure 11–47 shows some of these meters. Many modern meters come with digital readouts, Figure 11–48.

11.20 WIRE SIZES

All conductors have some resistance. The resistance depends on the conductor material, the cross-sectional area of the conductor, and the length of the conductor. A conductor with low resistance carries a current more easily than a conductor with high resistance.

Always use the proper wire (conductor) size. The size of a wire is determined by its diameter or cross section, Figure 11–49. A large diameter means there

Figure 11–48. Typical VOM with digital readout. *Courtesy Simpson Electric Co., Elgin, Illinois*

are more free electrons at any point, so the wire will carry more current. *If a wire is too small for the current passing through, it will overheat and possibly burn the insulation and could cause a fire.* Standard copper wire sizes are identified by American Standard Wire Gauge numbers and measured in circular mils. A circular mil is the area of a circle $\frac{1}{1000}$ in. in diameter. Temperature is also considered because resistance increases as temperature increases. Increasing wire-size numbers indicate *smaller* wire diameters and greater resistance. Check the tables in the National Electrical Code to determine proper wire size.

11.21 CIRCUIT PROTECTION DEVICES

SAFETY PRECAUTION: *Electrical circuits *must* be protected from current overloads. If too much cur-

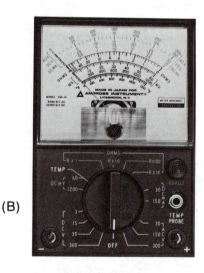

Figure 11–47. Meters used for electrical measurements. (A) DC millivoltmeter. (B) Multimeter (VOM). (C) Digital clamp-on ammeter. *Courtesy (A) Simpson Electric Co., Elgin, Illinois. (B and C) Amprobe Instrument Div., Core Industries*

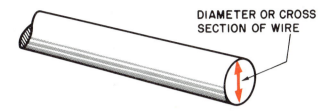

Figure 11-49. Cross section of a wire.

Figure 11-51. Dual-element plug fuse. *Reprinted with permission by Bussmann Division, McGraw-Edison Company*

rent flows through the circuit, the wires and components will overheat, resulting in damage and possible fire. Circuits are normally protected with fuses or circuit breakers.*

Fuses

A *fuse* is a simple device. Most fuses contain a strip of metal that has a higher resistance than the conductors in the circuit. This strip also has a relatively low melting point. Because of its higher resistance, it will heat up faster than the conductor. When the current exceeds the rating on the fuse, the strip melts and opens the circuit.

Plug Fuses. Plug fuses have either an Edison base or a Type S base, Figure 11-50A. Edison-base fuses are used in older installations and can be used for replacement only. Type S fuses can be used only in a Type S fuse holder specifically designed for the fuse; otherwise an adapter must be used. Each adapter is designed for a specific ampere rating, and these fuses cannot be interchanged. The amperage rating determines the size of the adapter, Figure 11-50B. Plug fuses are rated up to 125 volts and 30 A.

Dual-element Plug Fuses. Many circuits have electric motors as the load or part of the load. Motors draw more current when starting and can cause a plain fuse to burn out or open the circuit. Dual-element fuses are frequently used in this situation, Figure 11-51. One element in the fuse will melt when there is a large overload, such as a short circuit. The other element will melt and open the circuit when there is a smaller current overload lasting more than a few seconds. This allows for the larger starting current of an electric motor.

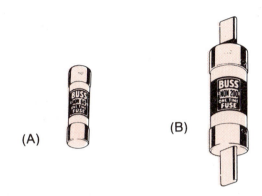

Figure 11-52. (A) Ferrule-type cartridge fuses. (B) Knife-blade cartridge fuse. *Reprinted with permission by Bussmann Division, McGraw-Edison Company*

Cartridge Fuses. For 230 V-600 V service up to 60A the ferrule cartridge fuse is used, Figure 11-52A. From 60 A to 600 A knife-blade cartridge fuses can be used, Figure 11-52B. A cartridge fuse is sized according to its ampere rating to prevent a fuse with an inadequate rating from being used. Many cartridge fuses have an arc-quenching material around the element to prevent damage from arcing in severe short-circuit situations, Figure 11-53.

Circuit Breakers

A circuit breaker can function as a switch as well as a means for opening a circuit when there is a current overload. Most modern installations in houses and many commercial and industrial installations use circuit breakers rather than fuses for circuit protection.

Circuit breakers use two methods to protect the circuit. One is a bimetal strip that heats up with a current overload and trips the breaker, opening the circuit. The other is a magnetic coil that causes the

Figure 11-50. (A) 1. Edison base fuse. 2. Type S base plug fuse. (B) Type S fuse adapter. *Reprinted with permission by Bussmann Division, McGraw-Edison Company*

Figure 11-53. Knife-blade cartridge fuse with arc-quenching material.

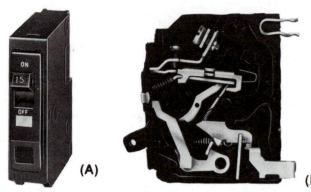

Figure 11–54. (A) Circuit breaker. (B) Cutaway. *Courtesy Square D Company*

breaker to trip and open the circuit when there is a short circuit or other excessive current overload in a very short period of time, Figure 11–54.

Ground Fault Circuit Interrupters

SAFETY PRECAUTION: *Ground fault circuit interrupters (GFCI) help protect individuals against shock, in addition to providing current overload protection. The GFCI, Figure 11–55, detects even a very small current leak to a ground. Under certain conditions this leak may cause an electrical shock. This small leak, which may not be detected by a conventional circuit breaker, will cause the GFCI to open the circuit.*

Figure 11–55. Ground fault circuit interrupter. *Courtesy Square D Company*

SUMMARY

- Matter is made up of atoms.
- Atoms are made up of protons, neutrons, and electrons.

- Protons have a positive charge, and electrons have a negative charge.
- Electrons travel in orbits around the protons and neutrons.
- Electrons in outer orbits travel from one atom to another.
- When there is a surplus of electrons in an atom, it has a negative charge. When there is a deficiency of electrons, the atom has a positive charge.
- Conductors have fewer electrons in the atom's outer orbit. They conduct electricity easily by allowing electrons to move easily from atom to atom.
- Insulators have more electrons in the outer orbits, which makes it difficult for the electrons to move from atom to atom. Insulators are poor conductors of electricity.
- Electricity can be produced by using magnetism. A conductor cutting magnetic lines of force produces electricity.
- Direct current is an electrical current moving in one direction.
- Alternating current is an electrical current that is continually reversing.
- Volt = electrical force or pressure.
- Ampere = quantity of electron flow.
- Ohm = resistance to electron flow.
- An electrical circuit must have an electrical source, a conductor to carry the current, and a resistance or load to use the current.
- Resistances or loads may be wired in series or in parallel.
- Voltage (E) = Amperage (I) × Resistance (R). This is Ohm's law.
- In series circuits the voltage is divided across the resistances, the total current flows through each resistance, and the resistances are added together to obtain the total resistance.
- In parallel circuits the total voltage is dropped across each resistance, the current is divided between the resistances, and the total resistance is less than that of the smallest resistance.
- Electrical power is measured in watts. $P = E \times I$.
- Inductive reactance is the resistance caused by the magnetic field surrounding a coil in an AC circuit.
- A coil with an electric current flowing through the loops of wire will cause an iron bar to be attracted into it. Electrical switching devices can be designed to use this action. These switches are called solenoids, relays, and contactors.
- Transformers use the magnetic field to step up or step down the voltage. A step-up transformer increases the voltage and decreases the current. A step-down transformer decreases the voltage and increases the current.
- A capacitor in a DC circuit collects electrons on one plate. These collect until they are equal to the source voltage. When a path is provided for these electrons

to discharge, they will do so until the capacitor becomes neutral.

- A capacitor in an AC circuit will continually charge and discharge as the current in the circuit reverses.
- A capacitor has capacitance. This is the amount of charge that can be stored.
- A multimeter is a common electrical measuring instrument used by air conditioning, heating, and refrigeration technicians. A multimeter often used is the VOM (volt-ohm-milliammeter).
- *Properly sized conductors must be used. Larger wire sizes will carry more current than smaller wire sizes without overheating.*
- *Fuses and circuit breakers are used to interrupt the current flow in a circuit when the current is excessive.*

REVIEW QUESTIONS

1. Describe the structure of an atom.
2. What is the charge on an electron? a proton? a neutron?
3. Describe the part of an atom that moves from one atom to another.
4. What effect does this movement have on the losing atom? on the gaining atom?
5. Describe the electron structure in a good conductor.
6. Describe the electron structure in an insulator.
7. Describe how electricity is generated through the use of magnetism.
8. State the differences between direct current and alternating current.
9. State the electrical units of measurement and describe each.
10. What components make up an electrical circuit?
11. Describe how a meter would be connected in a circuit to measure the voltage at a light bulb.
12. Describe how a meter would be connected in a circuit to measure the amperage.
13. Describe how an amperage reading would be made using a clamp-on or clamp-around ammeter.
14. Describe how the resistance in a DC circuit is determined.
15. Write Ohm's law for determining voltage, amperage, and resistance.
16. Sketch three loads wired in parallel in a circuit.
17. Illustrate how three loads would be wired in series.
18. Describe the characteristics of the voltage, amperage, and resistances when there is more than one load in a series circuit.
19. Describe the characteristics of the voltage, amperage, and resistances when there is more than one load in a parallel circuit.
20. What is the formula for the total resistance of three loads in a parallel circuit?
21. What is the unit of measurement for electrical power?
22. What is the formula for determining electrical power?
23. Explain inductance.
24. Explain how a solenoid operates.
25. Describe how a transformer operates.
26. Sketch a step-up transformer.
27. How does a step-down transformer differ from a step-up transformer?
28. Describe how a transformer is constructed.
29. Describe a capacitor.
30. How does a capacitor work in a DC circuit?
31. How does a capacitor work in an AC circuit?
32. What electrical measurements will a multimeter make?
33. What do the letters VOM stand for when referring to an electrical measuring instrument?
34. What are the two main switches on a VOM?
35. Why is there a zero ohms adjustment on a VOM?
36. Why is it important to use a properly sized wire in a particular circuit?
37. What is a circular mil?
38. Describe two kinds of plug fuses.
39. Describe two reasons for using a circuit breaker.
40. What force opens a circuit breaker?

INTRODUCTION TO AUTOMATIC CONTROLS

OBJECTIVES

After studying this unit, you should be able to

- **define bimetal.**
- **make general comparisons between different bimetal applications.**
- **describe the rod and tube.**
- **describe fluid-filled control.**
- **describe partial liquid, partial vapor filled control.**
- **distinguish between the bellows, diaphragm, and Bourdon tube.**
- **discuss the thermocouple.**
- **explain the thermistor.**

12.1   TYPES OF AUTOMATIC CONTROLS

The heating, air conditioning, and refrigeration field requires many types and designs of automatic controls to stop or start equipment. Modulating controls that vary the speed of a motor or modulate a valve are found less frequently in the residential and light commercial range of equipment. Controls also provide protection to people and equipment.

Controls can be classified in the following categories: electrical, mechanical, electromechanical, and electronic. Pneumatic and hydraulic controls will not be covered because they normally do not apply to residential and light commercial equipment.

Electrical controls are electrically operated and normally control electrical devices. Mechanical controls are operated by pressure and temperature to control fluid flow. Electromechanical controls are driven by pressure or temperature to provide electrical functions, or they are driven by electricity to control fluid flow. Electronic controls use electronic circuits and devices to perform the same functions that electrical and electromechanical controls perform.

The automatic control of a system is intended to maintain stable or constant conditions with a controllable device. This can involve protection of people and equipment. The system must regulate itself within the design boundaries of the equipment. If the system's equipment is allowed to operate outside of its design boundaries, the equipment components may be damaged.

In this industry the job is to control space or product condition by controlling temperature, humidity, and cleanliness.

12.2   DEVICES THAT RESPOND TO THERMAL CHANGE

Automatic controls in this industry usually provide some method of controlling temperature. Temperature control is used to maintain space or product temperature and to protect equipment from damaging itself. When used to control space or product temperature, the control is called a *thermostat;* when used to protect equipment, it is known as a *safety device.* A good example of both of these applications can be found in a household refrigerator. The refrigerator maintains the space temperature in the fresh food section at about 35°F. When food is placed in the box and stored for a long time, it becomes the same temperature as the space, Figure 12–1. If the space temperature is allowed

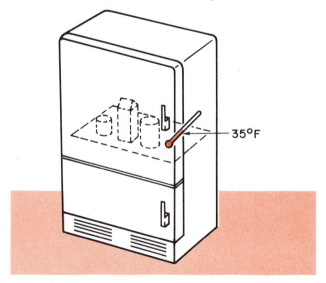

Figure 12–1. The household refrigerator maintains a specific temperature inside the box.

to go much below the 35°F mark, the food begins to freeze. Foods such as eggs, tomatoes, and lettuce do not serve well after freezing.

The refrigerator is often able to maintain this fresh food compartment condition for 15 or 20 years without failure. The frozen food compartment is another situation and will be covered later. For now think of what it would be like without automatic controls that keep food cold to preserve it but not cold enough to freeze and ruin it. The owner of the refrigerator would have to anticipate the temperature in the food compartment and get up in the middle of the night and turn it on or off to maintain the temperature. It is hard to imagine how many times the thermostat stops and starts the refrigeration cycle in 20 years to maintain the proper conditions in the refrigerator.

The refrigerator's compressor has a protective device to keep it from overloading and damaging itself, Figure 12–2. This overload is an automatic control designed to function on the rare occasion that an overload or power problem may cause damage to the compressor. One such occasion is when the power goes off and comes right back on while the refrigerator is running. The overload will stop the compressor for a cool-down period until it is ready to go back to work again without overloading and hurting itself.

You are exposed to many automatic controls that you are probably unaware of because usually there is no need to know about them. You need an understanding of these controls because they are just as valuable as the equipment they control. After all, if the control fails, the equipment will probably fail. You must have a very thorough understanding of automatic controls to choose, install, or repair equipment or systems.

Some common automatically controlled devices are

- Household refrigerator's fresh and frozen food compartments
- Residential and office cooling and heating air conditioning
- Water-heater temperature control
- Electric oven temperature control
- Garbage disposal overload control

Figure 12–2. Compressor overload device. *Photo by Bill Johnson*

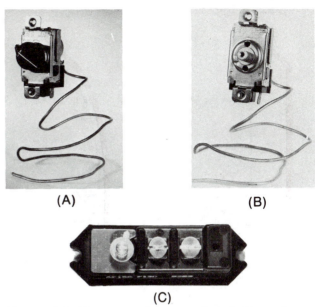

Figure 12–3. These controls operate household appliances. (A) Refrigerator thermostat. (B) Window air conditioner thermostat. (C) Water heater thermostat. *Photos by Bill Johnson*

- Fuses and circuit breakers that control current flow in electrical circuits in a home.

See Figure 12–3 for examples of automatic controls.

Automatic controls used in the air conditioning and refrigeration industry are devices that monitor temperature and its changes. Some controls respond to temperature changes and are used to monitor electrical overloads by temperature changes in the wiring circuits. This response is usually a change of dimension or electrical characteristic in the control-sensing element.

12.3 THE BIMETAL DEVICE

The *bimetal* device is probably the most common device used to detect thermal change. In its simplest form the device consists of two unlike metal strips, attached back to back, that have different rates of expansion, Figure 12–4. Brass and steel are commonly used. When the device is heated, the brass expands faster than the steel, and the device is warped out of shape. This warping action is a known dimensional change that can be attached to an electrical component or valve to stop, start, or modulate electrical current or fluid flow, Figure 12–5. This control is limited in its application by the amount of warp it can accomplish with a temperature change. For instance, when the bimetal is fixed on one end and heated, the other end moves a certain amount per degree of temperature change, Figure 12–6.

To obtain enough travel to make the bimetal practical over a wider temperature range, add length to the

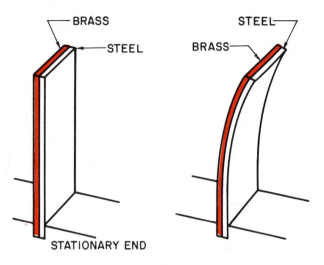

Figure 12-4. Basic bimetal strip made of two unlike metals such as brass and steel fastened back to back.

bimetal strip. When adding length, the bimetal strip is normally coiled in a circle, shaped like a hair pin, wound in a helix fashion, or formed in a worm shape, Figure 12-7. The movable end of the coil or helix can be attached to a pointer to indicate temperature, a switch to stop or start current flow, or a valve to modulate fluid flow. One of the basic control applications is shown in Figure 12-8.

Rod and Tube.   The *rod and tube* is another type of control that uses two unlike metals and the difference in thermal expansion. The rod and tube can be described more accurately as the rod in tube. It has an outer tube of metal with a high expansion rate and an internal rod of metal with a low expansion rate, Figure 12-9. This control has been used for years in the residential gas water heater. The tube is inserted into the tank and gives very accurate sensing of the water temperature. As the tank water temperature changes, the tube pushes the rod and opens or closes the gas

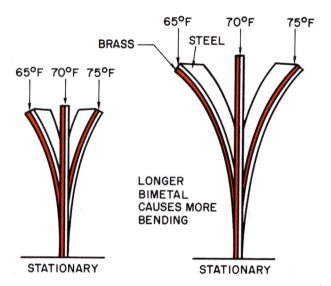

Figure 12-6. This bimetal is straight at 70°F. The brass side contracts more than the steel side when cooled causing a bend to the left. The brass side expands faster than the steel side on a temperature rise causing a bend to the right. This bend is a predictable amount per degree of temperature change. The longer the strip the more the bend.

valve to start or stop the heat to the water in the tank, Figure 12-10.

Snap-disc.   The *snap-disc* is another application of the bimetal used in some applications to sense temperature changes. This control is treated apart from the bimetal because of its snap characteristic that gives it a quick open-and-close feature. Some sort of snap-action feature has to be incorporated into all controls that stop and start electrical loads, Figure 12-11.

12.4   CONTROL BY FLUID EXPANSION

Fluid expansion is another method of sensing temperature change. Earlier, a mercury thermometer was

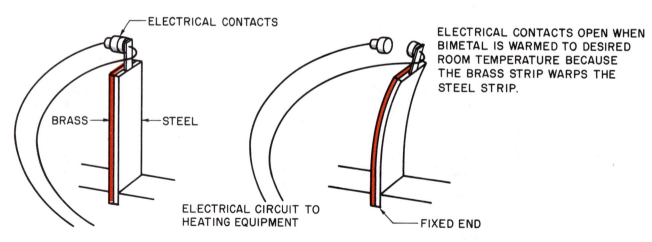

Figure 12-5. Basic bimetal strip used for a heating thermostat.

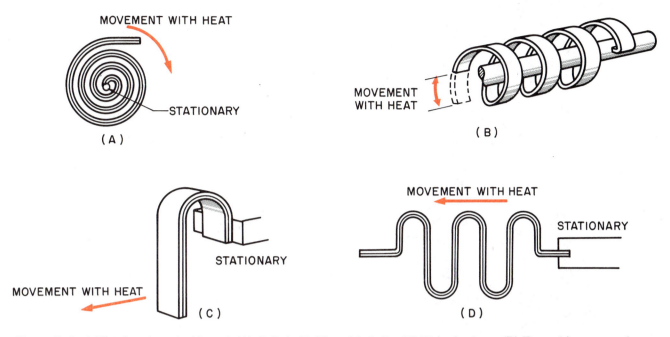

Figure 12-7. Adding length to the bimetal. (A) Coiled. (B) Wound in helix. (C) Hair pin shape. (D) Formed in a worm shape.

described as a bulb with a thin tube of mercury rising up a stem. As the mercury in the bulb is heated or cooled, it expands or contracts and either rises or falls in the stem of the thermometer. The level of the mercury in the stem is based on the temperature of the mercury in the bulb, Figure 1-1. The reason for the rising and falling is because the mercury in the tube has no place else to go. When the mercury in the bulb is heated and expands, it has to rise up in the tube. When it is cooled, it naturally falls down the tube. This

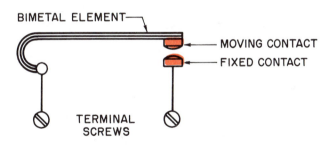

Figure 12-8. Movement of the bimetal due to changes in temperature opens or closes electrical contacts.

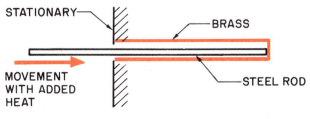

Figure 12-9. Rod and tube.

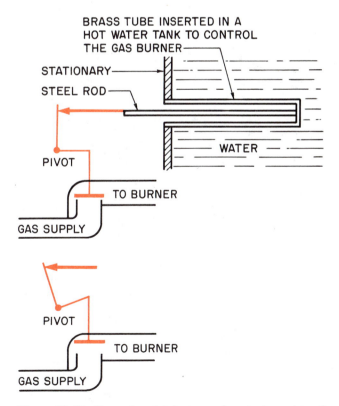

Figure 12-10. The rod and tube type of control consists of two unlike metals with the fastest expanding metal in a tube normally inserted in a fluid such as a hot-water tank.

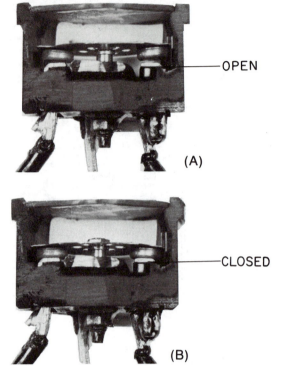

—OPEN

(A)

—CLOSED

(B)

Figure 12–11. The snap-disc is another variation of the bi-metal concept. The snap-disc is usually round and fastened on the outside. When heated, the disc snaps to a different position. (A) Open circuit. (B) Closed circuit. *Photos by Bill Johnson*

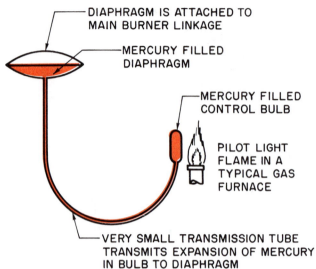

DIAPHRAGM IS ATTACHED TO MAIN BURNER LINKAGE

MERCURY FILLED DIAPHRAGM

MERCURY FILLED CONTROL BULB

PILOT LIGHT FLAME IN A TYPICAL GAS FURNACE

VERY SMALL TRANSMISSION TUBE TRANSMITS EXPANSION OF MERCURY IN BULB TO DIAPHRAGM

Figure 12–13. The bulb in or near the flame is mercury filled. The heated bulb causes the mercury to expand and move up the transmission tube and flex out the diaphragm. This proves that the pilot flame is present to ignite the main burner.

same idea can be used to transmit a signal to a control that a temperature change is taking place.

The liquid rising up the transmitting tube has to act on some device to convert the rising liquid to usable motion. One device used is the diaphragm. A *diaphragm* is a thin, flexible metal disc with a large area. It can move in and out with pressure changes underneath it, Figure 12–12.

When a bulb is filled with a liquid and connected to a diaphragm with piping, the bulb temperature can be transmitted to the diaphragm by the expanding liquid, Figure 12–13. In Figure 12–13 the bulb is filled with mercury and placed in the pilot-light flame on a gas furnace to insure that a pilot light is present to ignite the gas burner before the main gas valve is opened. The

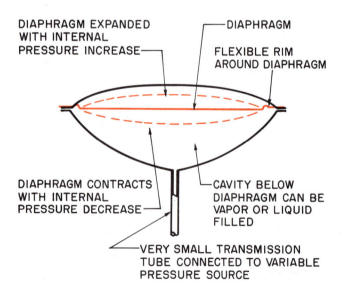

DIAPHRAGM EXPANDED WITH INTERNAL PRESSURE INCREASE

DIAPHRAGM

FLEXIBLE RIM AROUND DIAPHRAGM

DIAPHRAGM CONTRACTS WITH INTERNAL PRESSURE DECREASE

CAVITY BELOW DIAPHRAGM CAN BE VAPOR OR LIQUID FILLED

VERY SMALL TRANSMISSION TUBE CONNECTED TO VARIABLE PRESSURE SOURCE

Figure 12–12. The diaphragm is a thin flexible movable membrane (brass-steel or other metal) used to convert pressure changes to movement. This movement can stop and start controls or modulate controls.

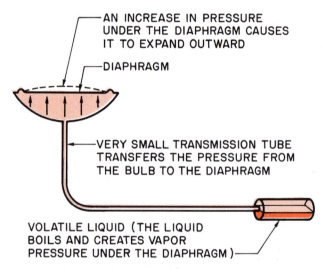

AN INCREASE IN PRESSURE UNDER THE DIAPHRAGM CAUSES IT TO EXPAND OUTWARD

DIAPHRAGM

VERY SMALL TRANSMISSION TUBE TRANSFERS THE PRESSURE FROM THE BULB TO THE DIAPHRAGM

VOLATILE LIQUID (THE LIQUID BOILS AND CREATES VAPOR PRESSURE UNDER THE DIAPHRAGM)

Figure 12–14. A large bulb partially filled with a volatile liquid, one that boils and creates vapor pressure when heated. This causes an increase in vapor pressure, which forces the diaphragm to move outward. When cooled, the vapor condenses, and the diaphragm moves inward.

entire mercury-filled tube and mercury-filled diaphragm are sensitive to temperature changes, and because the pilot light is much hotter than either the inner connecting tube or the diaphragm, the sensing bulb is located in the pilot-light flame.

To maintain more accurate control at the actual bulb location, you can use a bulb partially filled with a liquid that will boil and make a vapor, which is then transmitted to the diaphragm at the control point, Figure 12–14. You should realize that the liquid will respond to temperature change much more than will the vapor, which is used to transmit the pressure.

The following explanation describes how this can work using R-12. See Figure 3–15 for the pressure and temperature chart for R-12. Our example is a walk-in refrigerated box. The inside temperature is maintained by cutting the refrigeration machine off when the box temperature reaches 35°F and starting the refrigeration machine when the box temperature reaches 45°F. A control with a remote bulb is used to regulate the space temperature. The bulb is located inside the box, and the control is located outside the box so it can be adjusted.

For illustration purposes a pressure gage is installed

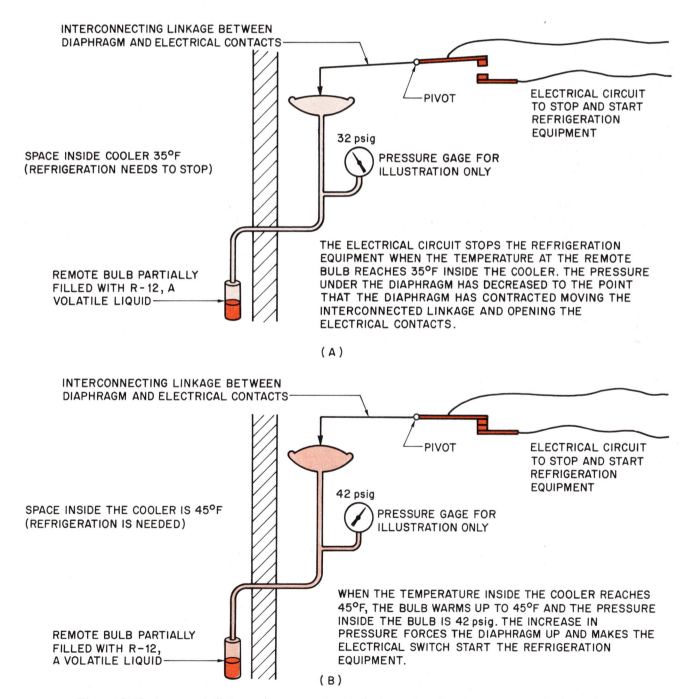

Figure 12–15. A remote bulb transmits pressure to the diaphragm based on the temperature in the cooler.

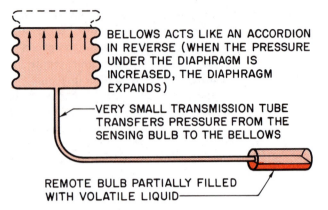

BELLOWS ACTS LIKE AN ACCORDION IN REVERSE (WHEN THE PRESSURE UNDER THE DIAPHRAGM IS INCREASED, THE DIAPHRAGM EXPANDS)

VERY SMALL TRANSMISSION TUBE TRANSFERS PRESSURE FROM THE SENSING BULB TO THE BELLOWS

REMOTE BULB PARTIALLY FILLED WITH VOLATILE LIQUID

Figure 12–16. The bellows is applied where more movement per degree is desirable. This control would normally have a partially filled bulb with vapor pushing up in the bellows section.

in the bulb to monitor the pressures inside the bulb as the temperatures change. See Figure 12–15 for a progressive explanation of this example. At the point that the refrigeration machine needs to be cycled off, the bulb temperature is 35°F. This corresponds to a pressure of 32 psig for R-12. A control mechanism can be designed to open an electrical circuit and stop the refrigeration machine at this point. When the cooler temperature rises to 45°F, it is time for the refrigeration machine to restart. At 45°F for R-12, the pressure inside the control is 42 psig, and the same mechanism can be designed to close the electrical circuit and start the refrigeration machine.

The diaphragm also has a travel problem like the bimetal strip. The diaphragm has a limited amount of travel but a great deal of power during this short travel. The travel of the liquid-filled control is limited to the expansion of the liquid in the bulb for the temperature

range it is working within. When more travel is needed, another device, called *bellows,* can be used. The bellows is very much like an accordion. It has a lot of internal volume with a lot of travel, Figure 12–16. The bellows is normally used with a vapor instead of a liquid inside it.

The remote bulb partially filled with liquid may also be used to indicate temperature using a Bourdon tube by driving a needle on a calibrated dial, Figure 12–17.

The partially filled bulb control is widely used in this industry because it is reliable, simple, and economical. This control has been in the industry since it began, and it takes many configurations. See Figure 12–18 for some remote bulb thermostats. The Bourdon tube is often used in the same fashion as the diaphragm and the bellows to monitor fluid expansion.

12.5   THE THERMOCOUPLE

The *thermocouple* differs from other methods of controlling with thermal change because it does not use expansion; instead, it uses electrical principles. The thermocouple consists of two unlike metals formed together on one end (usually wire made of unlike metals such as iron and constantan). When heated on the fastened end, an electrical current flow is started due to the difference in temperature in the two ends of the device, Figure 12–19. Thermocouples can be made of many different unlike metal combinations, and each one has a different characteristic. Each different thermocouple has a *hot* junction and a *cold* junction. The hot junction, as the name implies, is at a higher temperature level than the cold junction. This difference in temperature is what starts the current flowing. Heat will cause an electrical current to flow in one direction in one metal and in the opposite direction in the other. When these metals are connected they make an electri-

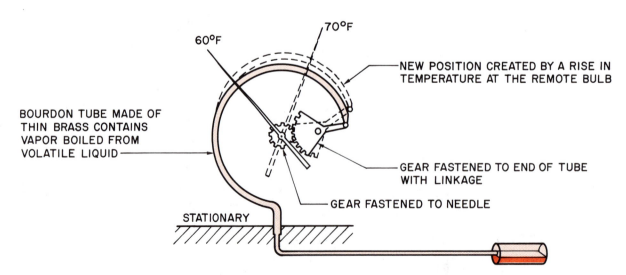

60°F

70°F

NEW POSITION CREATED BY A RISE IN TEMPERATURE AT THE REMOTE BULB

BOURDON TUBE MADE OF THIN BRASS CONTAINS VAPOR BOILED FROM VOLATILE LIQUID

GEAR FASTENED TO END OF TUBE WITH LINKAGE

GEAR FASTENED TO NEEDLE

STATIONARY

Figure 12–17. This remote bulb is partially filled with liquid. When heated, the expanded vapor is transmitted to a Bourdon tube that straightens out with an increase in vapor pressure. A decrease in pressure causes the Bourdon tube to curl inward.

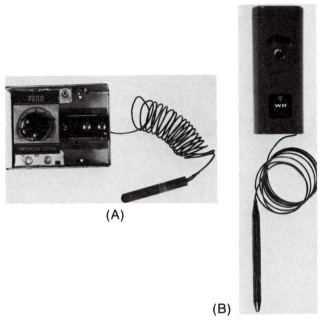

(A)

(B)

Figure 12-18. Remote-bulb refrigeration temperature controls. *Photos by Bill Johnson*

Figure 12-19. Thermocouple used to detect a pilot light in a gas-burning appliance. *Courtesy Robertshaw Controls Company*

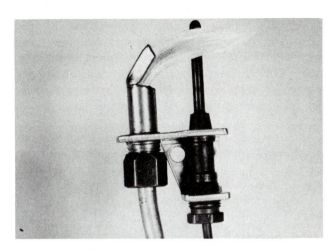

Figure 12-21. Thermocouple senses whether the gas furnace pilot light is on. *Photo by Bill Johnson*

cal circuit, and current will flow when heat is applied to one end of the device.

Tables and graphs for various types of thermocouples show how much current flow can be expected from a thermocouple under different conditions for hot and cold junctions. Figure 12–20 illustrates a thermocouple. The current flow in a thermocouple can be monitored by an electronic circuit and used for many temperature-related applications, such as a thermometer, a thermostat to stop or start a process, or a safety control, Figure 12–21.

The thermocouple has been used extensively for years in gas furnaces to detect the pilot-light flame for safety purposes. This application is beginning to be phased out because it works best with standing pilot-light systems. A gradual change in design using inter-

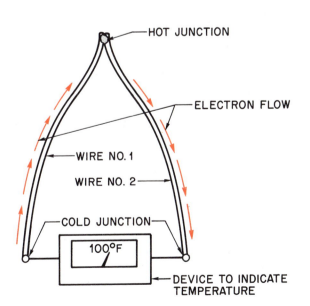

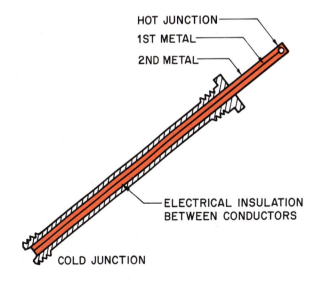

Figure 12-20. The thermocouple on the left is made of two wires of different metals welded on one end. It is used to indicate temperature. The thermocouple on the right is a rigid device and used to detect a pilot light.

mittent pilots is causing this phase out (these are pilot lights that are extinguished each time the burner goes out and relit each time the room thermostat calls for heat). This thermocouple application has an output voltage of about 20 millivolts, all that is needed to control a safety circuit to prove there is flame, Figure 12–22.

Thermocouples ganged together to give more output are called *thermopiles*. The thermopile is used on some gas-burning equipment as the only power source. This type of equipment has no need for power other than the control circuit, so the power supply is very small (about 500 millivolts). The thermopile has also been used to operate radios from the sun or from a small fire in remote areas.

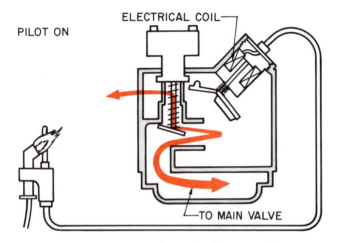

Figure 12–22. Thermocouple and control circuit used to detect a gas flame. When the flame is lit, the thermocouple generates an electrical current. This energizes an electromagnet that holds the gas valve open. When the flame is out, the thermocouple stops generating electricity, and the valve closes. Gas is not allowed to flow.

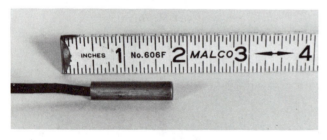

Figure 12–23. A thermometer probe using a thermistor to measure temperature. *Photo by Bill Johnson*

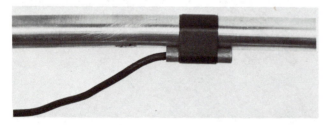

Figure 12–24. Thermistor application. *Photo by Bill Johnson*

12.6 ELECTRONIC TEMPERATURE SENSING DEVICES

The *thermistor* is an electronic solid-state device known as a semiconductor and requires an electronic circuit to utilize its capabilities. It is a *semiconductor;* that is, it is a better conductor than an insulator but not as good as a conductor. Based on its temperature, the thermistor allows varied amounts of current to flow. A temperature rise provides less resistance to current flow.

The thermistor can be very small and will respond to small temperature changes. The changes in current flow in the device are monitored by special electronic circuits that can stop, start, and modulate machines or provide a temperature readout, Figure 12–23 and Figure 12–24.

SUMMARY

- A bimetal element is two unlike metal strips such as brass and steel fastened back to back.
- Bimetal strips warp with temperature changes and can be used to stop, start, or modulate electrical current flow and fluid flow when used with different mechanical, electrical, and electronic helpers.
- The travel of the bimetal can be extended by coiling it. The helix and the coil are the names given to the extended bimetal.
- The rod and tube is another version of the bimetal.
- Fluid expansion is used in the thermometer to indicate temperature and to operate controls that are totally liquid filled.
- The pressure–temperature relationship studied earlier is applied to some controls that are partially filled with liquid.
- The diaphragm is used to move the control mechanism when either liquid or vapor pressure is applied to it.
- The diaphragm has very little travel but much power.
- The bellows is used for more travel and is normally vapor filled.
- The Bourdon tube is sometimes used like the diaphragm or bellows.
- The thermocouple generates electrical current flow when heated at the hot junction.
- This current flow can be used to monitor temperature changes to stop, start, or modulate electrical circuits.
- The thermistor is an electronic device that varies its resistance to electrical current flow based on temperature changes.

REVIEW QUESTIONS

1. Describe the bimetal strip.
2. What are some applications of the bimetal strip?
3. How can the bimetal strip be extended to have more stroke per degree of temperature change?
4. What is a rod and tube?
5. What two metals can be used in the bimetal or the rod and tube?
6. Define a diaphragm.
7. Name two characteristics of a diaphragm.
8. What fluid can be used in a totally liquid-filled bulb-type control?
9. What is one application for the totally liquid-filled control?
10. Define a bellows.
11. What is the difference between a bellows and a diaphragm?
12. Which type of fluid is normally found in a bellows?
13. How is the pressure–temperature relationship used to understand the partially filled remote-bulb-type of control?
14. What is a thermocouple?
15. How can a thermocouple be used to verify that a gas flame is present?
16. What is used with a thermocouple to control mechanical devices?
17. What are thermocouples normally made of?
18. Define a thermistor.
19. What must be used with a thermistor to control a mechanical device or machine?
20. What does a thermistor do that makes it different from a thermocouple?

SYSTEM CONTROL COMPONENTS

OBJECTIVES

After studying this unit, you should be able to

- **define space-temperature control.**
- **choose the most efficient space-temperature control device.**
- **describe the mercury control bulb.**
- **define the difference between high- and low-voltage controls.**
- **name the components of low- and high-voltage controls.**

13.1 RECOGNITION OF CONTROL COMPONENTS

This unit describes in words and pictures how various controls look and work. Recognizing a control and understanding its function and what component it influences is vitally important and will eliminate much confusion when reading a diagram or troubleshooting a system. This familiarization could take years of trial and error without some explanation. The ability to see a control on a circuit diagram and then recognize it on the equipment will come easier after a description and picture is studied.

Fortunately there are some similarities and categories that can make control recognition easier. The previous unit showed how thermal change is detected and then transferred to action and how pressure changes are converted to action. Temperature will now be covered in more detail using actual controls as examples.

13.2 TEMPERATURE CONTROLS

Temperature is controlled in many ways for many reasons. For instance, space temperature is controlled for comfort. The motor-winding temperature in a compressor motor is controlled to prevent overheating and damage to the motor. The motor could overheat and do damage to itself from the very power that operates it. These are two examples that are far apart in application but perform the same type of function. They both require some device that will sense temperature rise with a sensing element and make a known response. The space temperature example uses the control as an operating control, whereas the motor temperature example serves as a safety device.

The space temperature application takes two different directions, depending on whether winter heating or summer cooling is needed. In winter the control must break a circuit to stop the heat when conditions are satisfied. In summer the conditions are reversed: The control must make a circuit and start the air conditioner, based on a temperature rise. The terminology goes like this: **The heating thermostat opens on a rise in temperature, and the cooling thermostat closes on a rise in temperature.** *Note:* in both cases the control was described as *functioning on a rise*. This terminology is important because it is used in industry.

The motor-temperature cutout takes the same circuit action as the heating thermostat. It opens the circuit on a rise in motor temperature and stops the motor. The heating thermostat and a motor-winding thermostat may make the same move under the same conditions, but they do not physically resemble each other, Figure 13–1.

Another difference between the two thermostats is the medium to be detected. The motor-winding thermostat must be in close contact with the motor winding. It is fastened to the winding itself. The space-temperature thermostat is merely suspended in air under the decorative cover and relies on random air currents passing over it. These different concepts must be considered for each control designed.

Another important design concept is the current-carrying characteristics of the various controls. In the space-temperature application the stopping or starting of a popular heating system (the gas or oil furnace) involves stopping and starting low-voltage (24 V) components and high-voltage (115 V or 230 V) components. The gas or oil furnace normally has a low-voltage gas valve or relay and a high-voltage fan motor.

There seems to be no firm rule saying that one voltage or the other is all that is used in any specific application. However, it can easily be seen that the stopping and starting of a 3-hp compressor requires a larger switch mechanism than a simple gas valve does. A 3-hp compressor could require a running current of

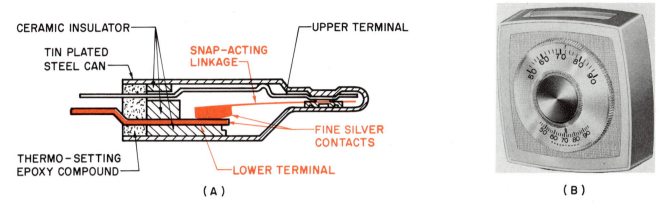

Figure 13–1. Both thermostats open on a rise in temperature, but they serve two different purposes. (A) The motor-winding thermostat measures the temperature of iron and copper while in close contact with motor windings. (B) The space thermostat measures air temperature by random air currents. *Courtesy (B) Robertshaw Controls Company*

18 A and a starting current of 90 A, whereas a simple gas valve might use only $\frac{1}{2}$ A. If the bimetal were large enough to carry the current for a 3-hp compressor, the control would be so large that it would be slow to respond to air temperature changes. This is one reason for using low-voltage controls to stop and start high-voltage components.

Residential systems usually have low-voltage control circuits. There are four reasons for this:

1. Economy.
2. Safety.
3. More precise control of relatively still air temperature.
4. In most states a technician does not need an electrician's license to install and service low-voltage wiring.

The low-voltage thermostat is energized from the residence power supply and transformed down to 24 V with a small transformer usually furnished with the equipment, Figure 13–2.

13.3 SPACE-TEMPERATURE CONTROLS, LOW VOLTAGE

The low-voltage space-temperature control (thermostat) normally regulates other controls and does not carry much current—seldom more than 2 A. The thermostat consists of the following components.

Electrical Contact Type. The mercury bulb is probably the most popular component used to make and break the electrical circuit in low-voltage thermostats. The mercury bulb is inside the thermostat, Figure 13–3. It consists of a glass bulb filled with an inert gas (a gas that will not support oxidation) with a small puddle of mercury free to move from one end to the other. The principal is to be able to make and break a small electrical current in a controlled atmosphere. When an

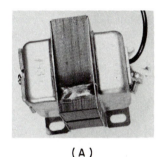

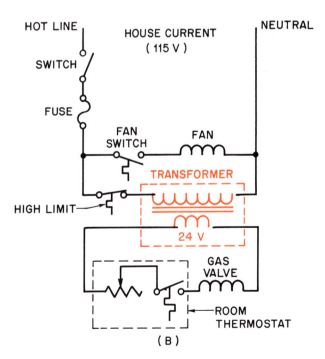

Figure 13–2. The typical transformer used in residences and light commercial buildings to change 115 V to 24 V (control voltage). *Photo by Bill Johnson*

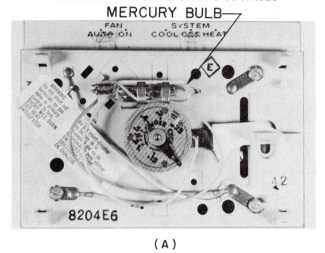

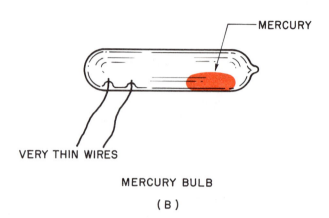

(A)   **(B)**

Figure 13-3. (A) A wall thermostat with the cover off and the mercury bulb exposed. (B) A detail of the mercury bulb. Note the very fine wire that connects the bulb to the circuit. The bulb is attached to the movable end of a bimetal coil. When the bimetal tips the bulb, the mercury flows to the other end and closes the circuit by providing the contact between the two wires. The wires are fastened to contacts inside the glass bulb, which is filled with an inert gas. This inert gas helps keep the contacts from pitting and burning up. *Photo by Bill Johnson*

electrical current is either made or broken, a small arc is present. The arc is hot enough to cause oxidation in the vicinity of the arc. When this arc takes place inside the bulb with an inert gas, where there is no oxygen, there is no oxidation.

The mercury bulb is fastened to the movable end of the bimetal, so it is free to rotate with the movement of the bimetal. The wire that connects the mercury bulb to the electrical circuit is very fine to prevent drag on the movement of the bulb. The mercury cannot be in both ends of the bulb at the same time, so when the bimetal rolls the mercury bulb to a new position, the mercury rapidly makes or breaks the electrical current flow. This is called *snap* or *detent action.*

Two other types of contacts in the low-voltage thermostat use conventional contact surfaces of silver-coated steel contacts. One is simply an open set of contacts, usually with a protective cover, Figure 13-4. The other is a set of silver-coated steel contacts enclosed in a glass bulb. Both of these contacts use a magnet mounted close to the contact to achieve detent or snap action.

Heat Anticipator. The *heat anticipator* is a small resistor, usually adjustable, used to cut off the heating equipment prematurely, Figure 13-5. For a moment, imagine that an oil or a gas furnace has heated a home to just the right cutoff point. The furnace itself could weigh several hundred pounds and be hot from running a long time. When the fire is stopped in the furnace, the furnace still contains a great amount of heat. The fan is allowed to run to dissipate this heat. The heat left in the furnace is enough to drive the house temperature past the comfort point.

The heat anticipator is located inside the thermostat to fool the furnace into cutting off early to dissipate

this heat, Figure 13-6. The resistor gives off a small amount of heat and causes the bimetal to think the room temperature is warmer than it actually is. *Note:* **The current flow through the heat anticipator must be accurately matched with the heat anticipator setting.** Directions come with the thermostat to show how to do this. **The heat anticipator is wired in series with the mercury-bulb heating contacts.**

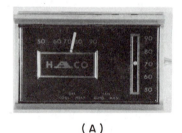

(A)

Figure 13-4. (A) Low-voltage thermostat. (B) Low-voltage thermostat illustrated with open contacts. *Photo by Bill Johnson*

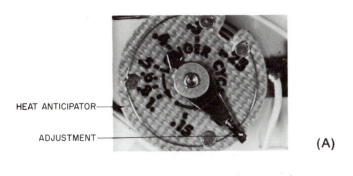

HEAT ANTICIPATOR

ADJUSTMENT

(A)

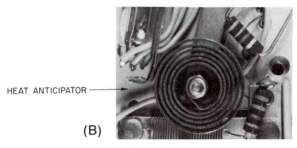

HEAT ANTICIPATOR

(B)

Figure 13–5. Heat anticipators as they appear in actual thermostats. *Photos by Bill Johnson*

The Cold Anticipator. The cooling system needs to be started up just a few minutes early to allow the air conditioning system to get up to capacity when needed. If the air conditioning system were not started up until it was needed, it would be 5 to 15 minutes before it would be producing to capacity. This is

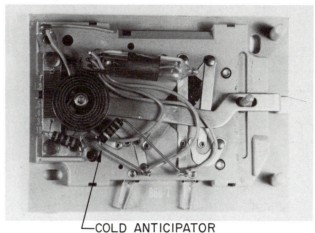

COLD ANTICIPATOR

Figure 13–7. The cold anticipator is normally a fixed resistor similar to resistors found in electronic circuitry. They are small round devices with colored bands to denote the resistance and wattage. *Photo by Bill Johnson*

enough time to cause a temperature drop in the conditioned space. The cold anticipator fools the system into starting up a few minutes early, so capacity is reached on time. This is normally a fixed resistor that is not adjustable in the field, Figure 13–7. **The cold anticipator is wired in parallel with the mercury-bulb cooling contacts,** Figure 13–8.

Thermostat Cover. The thermostat cover is intended to be decorative and protective. A thermometer is

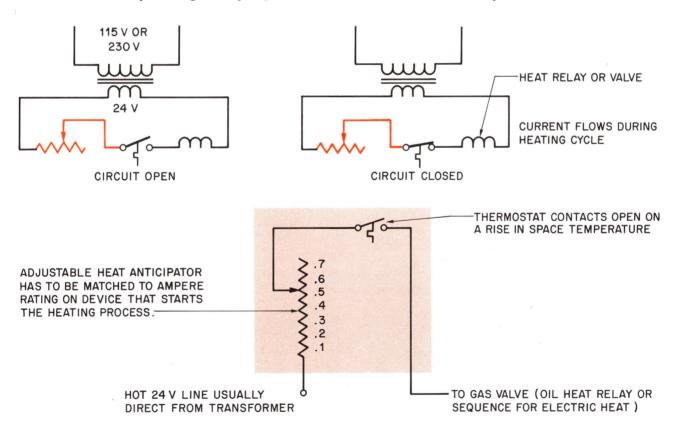

115 V OR
230 V

24 V

CIRCUIT OPEN

HEAT RELAY OR VALVE

CURRENT FLOWS DURING
HEATING CYCLE

CIRCUIT CLOSED

THERMOSTAT CONTACTS OPEN ON
A RISE IN SPACE TEMPERATURE

ADJUSTABLE HEAT ANTICIPATOR
HAS TO BE MATCHED TO AMPERE
RATING ON DEVICE THAT STARTS
THE HEATING PROCESS.

.7
.6
.5
.4
.3
.2
.1

HOT 24 V LINE USUALLY
DIRECT FROM TRANSFORMER

TO GAS VALVE (OIL HEAT RELAY OR
SEQUENCE FOR ELECTRIC HEAT)

Figure 13–6. The heat anticipator is usually a wirewound, slide-bar type of variable resistor.

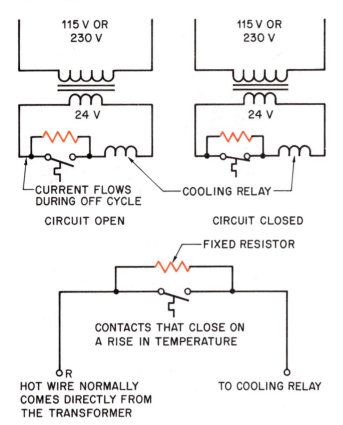

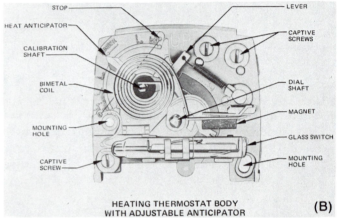

Figure 13-8. The cold anticipator is wired in parallel with the cooling contacts on the thermostat. This allows the current to flow through it during the off cycle. This means that the cold anticipator and the relay to start the cooling cycle are in the circuit when the thermostat is open.

Figure 13-10. Thermostat assemblies. *Courtesy (B) Robertshaw Controls Company. Photo (A) by Bill Johnson*

usually mounted in it to indicate the surrounding (ambient) temperature. The thermometer is functionally separate from any of the controls and would serve the same purpose if it were hung on the wall next to the thermostat. Thermostats come in many shapes. Each manufacturer tries to have a distinctive design, Figure 13-9.

Thermostat Assembly. The *thermostat assembly* houses the thermostat components already mentioned and is normally mounted on a subbase fastened to the wall. This assembly could be called the brain of the system. In addition to mercury bulbs and anticipators

the thermostat assembly includes the movable levers that adjust the temperature. These levers or indicators normally point to the set point or desired temperature, Figure 13-10. When the thermostat is functioning correctly, the thermometer on the front will read the same as the set point.

The Subbase. The *subbase,* which is usually separate from the thermostat, contains the selector switching levers, such as the fan-on switch or the heat-off-cool

Figure 13-9. Thermostat decorative covers. *Photos by Bill Johnson*

switch, Figure 13–11. The subbase is important because the thermostat mounts on it. The subbase is first mounted on the wall, then the interconnecting wiring is attached to the subbase. **When the thermostat is attached to the subbase, the electrical connections are made between the two components.**

13.4   SPACE-TEMPERATURE CONTROLS, HIGH (LINE) VOLTAGE

Sometimes it is desirable to use line-voltage thermostats to stop and start equipment. Some equipment that is self-contained does not need a remote thermostat. To add a remote thermostat would be an extra expense and would only result in more potential trouble.

The window air conditioner is a good example of this. It is self-contained and needs no remote thermostat. If it had a remote thermostat, it would not be a plug-in device but would require an installation of the thermostat. Note that the remote-bulb-type of thermostat is normally used with the bulb located in the return

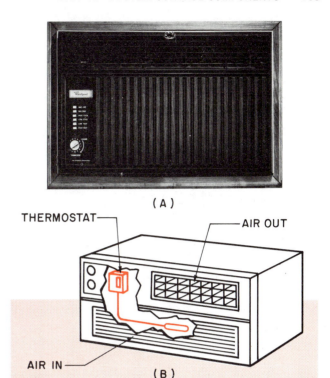

Figure 13–12. The window air conditioner has a line-voltage thermostat that stops and starts the compressor. When the selector switch is turned to cool, the fan comes on and stays on. The thermostat cycles the compressor only. *Courtesy (A) Whirlpool Corporation*

air stream. The fan usually runs all of the time and keeps a steady stream of room return air passing over the bulb. This gives more sensitivity to this type of application, Figure 13–12.

The concept of the line-voltage thermostat is used in many types of installations. The household refrigerator, reach-in coolers, and free-standing package air conditioning equipment are just a few examples. All of these have something in common, the thermostat is a heavy-duty type and may not be as sensitive as the low-voltage type. The line-voltage thermostat must be chosen with care. Most of the time it is chosen by the equipment supplier.

The line-voltage thermostat must be matched to the voltage and current that the circuit is expected to use. For example, a reach-in cooler for a convenience store has a compressor and fan motor to be controlled (i.e., stopped and started). The combined running-current draw for both is 16.2 A at 115 V (15.1 A for the compressor and 1.1 A for the fan motor). The locked-rotor amperage for the compressor is 72 A. (Locked-rotor amperage is the inrush current that the circuit must carry until the motor starts turning. The inrush was not considered for the fan because it is so small.) This is a $\frac{3}{4}$-hp compressor motor. A reliable, long-lasting thermostat must be chosen to operate this equip-

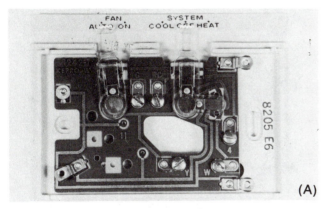

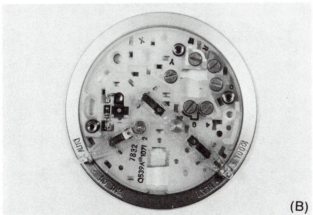

Figure 13–11. The subbase normally mounts on the wall and the wiring is fastened inside the subbase on terminals. These terminals are designed in such a way as to allow easy wire makeup. When the thermostat is screwed down onto the subbase, electrical connections are made between the two. The subbase normally contains the selector switches, such as the fan on–auto and heat–off–cool. *Photos by Bill Johnson*

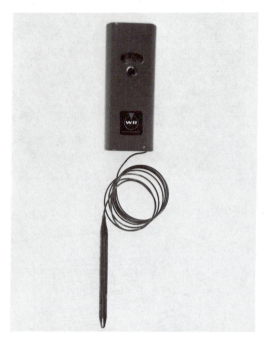

Figure 13–13. Line-voltage thermostat used to stop and start a compressor that draws up to 20 A full-load current and has an inrush current up to 80 A. *Photo by Bill Johnson*

ment. A control rated at 20 A running current and 80 A locked-rotor current is selected, Figure 13–13.

Usually, thermostats are not rated at more than 25 A because the size becomes prohibitive. This limits the size of the compressor that can be started directly with a line-voltage thermostat to about 1 hp on 115 V or 2 hp on 230 V. Remember, the same motor would draw exactly half of the current when the voltage is doubled.

When larger current-carrying capacities are needed, a motor starter is normally used with a line-voltage thermostat or a low-voltage thermostat.

Line-voltage thermostats usually consist of the following components.

The Switching Mechanism. Some line-voltage thermostats use mercury as the contact surface, but most switches are a silver-coated base metal. The silver helps conduct current at the contact point. This silver contact point takes the real load in the circuit and is the component in the control that wears first, Figure 13–14.

The Sensing Element. The sensing element is normally a bimetal, a bellows, or a remote bulb, Figure 13–15, located where it can sense the space temperature. If sensitivity is important, a slight air velocity should cross the element. The levers or knobs used to change the adjustment of the thermostat are attached to the main thermostat.

The Cover. Because of the line-voltage inside, the cover is usually attached with some sort of fastener to

discourage easy entrance, Figure 13–16. If the control is applied to room space temperature, such as a building, a thermometer may be mounted in the cover to read the room temperature. If the control is used to control a box space temperature, such as a reach-in cooler, the cover might be just a plain protective cover.

Subbase. When the thermostat is used for room space-temperature control, there must be some way to mount it to the wall. A *subbase* that fits on the electrical outlet box is usually used. The wire leading into a line-voltage control is normally a high-voltage wire routed between points in conduit. The conduit is connected to the box that the thermostat is mounted on. ＊If an electrical arc due to overload or short circuit occurs, it is enclosed in the conduit or box. This reduces fire hazard, Figure 13–17.＊

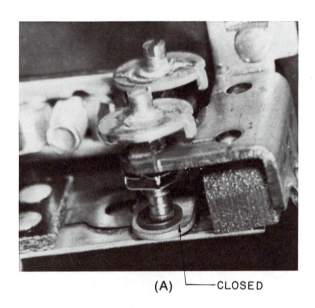

(A) ⌐CLOSED

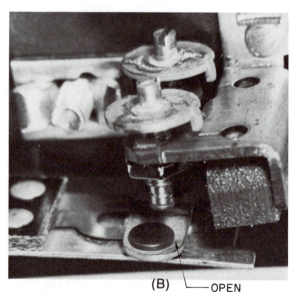

(B) ⌐OPEN

Figure 13–14. Silver-contact type line-voltage thermostat. (A) Closed. (B) Open. *Photos by Bill Johnson*

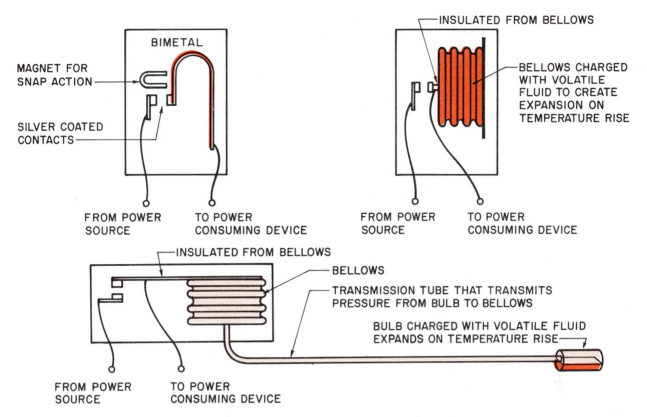

Figure 13–15. Bimetal, bellows, and remote-bulb sensing elements used in line-voltage thermostats. The levers or knobs used to adjust these controls are arranged to apply more or less pressure to the sensing element to vary the temperature range.

Figure 13–16. The cover for a line-voltage thermostat is usually less decorated than the cover on a low-voltage thermostat. *Photo by Bill Johnson*

SUMMARY

- Controls for regulating temperature are used throughout this industry.
- Temperature controls are either operating controls or safety controls.
- Space-temperature controls can either be low-voltage or line-voltage types.
- Low voltage is normally applied to all residential heating and cooling controls.
- Temperature is controlled for several reasons, comfort and product temperatures being among the most important.

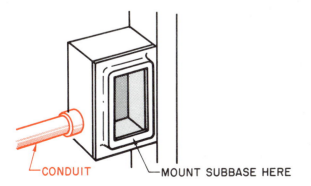

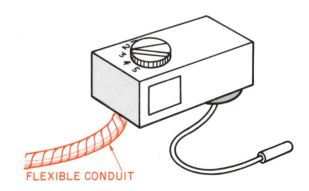

Figure 13–17. (A) Box for line-voltage, wall-mounted thermostat. (B) Remote-bulb thermostat with flexible conduit. In both cases the interconnected wiring is covered and protected.

- Temperature controls may carry the full load of the appliance or be pilot-operated or low-voltage operated.
- Some temperature-control devices may have the electric contacts enclosed in a bulb and use mercury as the contact surface, or they may have open contacts.
- Heating thermostats normally have a heat anticipator circuit in series with the thermostat contacts.
- Cooling thermostats may have a cooling anticipator in parallel with the thermostat contacts.
- The typical space-temperature thermostat has three parts: the subbase, the sensing elements, and the decorative cover.
- Line-voltage thermostats are normally rated up to 2 A of current because they are pilot operated.
- Low-voltage thermostats normally will carry only 2 A of current because they are pilot operated.

REVIEW QUESTIONS

1. Why is it important to be able to recognize the various controls?
2. Why is a low-voltage thermostat normally more accurate than a high-voltage thermostat?
3. Name three kinds of switching mechanisms in a low-voltage thermostat.
4. What is an inert gas?
5. What does a heat anticipator do?
6. What does a cold anticipator do?
7. What does a bimetal do?
8. In a residential system what voltage is considered low voltage?
9. What component steps down the voltage to the low-voltage value?
10. Name four reasons that low voltage is desirable for residential control voltage.
11. What is the maximum amperage usually encountered by a low-voltage thermostat?
12. A heating thermostat is known as a _____ on a rise device.
13. A cooling thermostat is known as a _____ on a rise device.
14. What instrument is often found in the cover of a room thermostat? What is it used for?
15. What two types of switches are normally found in the subbase of a low-voltage thermostat?
16. When is a line-voltage thermostat used?
17. Name three sensing elements that can be used with a line-voltage thermostat?
18. What material usually coats the contacts on a line-voltage thermostat? Why?
19. What is the maximum amperage generally encountered in a line-voltage thermostat?
20. Why is a line-voltage thermostat mounted on an electrical box that has conduit or flexible conduit connected to it?

CONTROLLING THE TEMPERATURE OF MASS

OBJECTIVES

After studying this unit, you should be able to

- define motor temperature.
- name two ways motors are protected from high temperature.
- make general comparisons between devices used to protect against overheating in motors.
- describe temperature sensing.

As mentioned earlier, a room thermostat for the purpose of controlling a space-heating system is the same as a motor-temperature thermostat as far as action is concerned. They both open or interrupt the electrical circuit upon a rise in temperature. However, the difference in the two controls from an appearance standpoint is considerable. The reason for this is that one control is designed to sense the temperature of slow-moving air, and the other is designed to sense the temperature of the motor winding. The winding has much more mass, and the control must be in much closer contact to get the response needed. Thus, substances other than air also need to be temperature-sensed, and controls to do this must be designed. A breakdown of types of substances will help you understand why controls have particular design features.

14.1   SENSING THE TEMPERATURE OF SOLIDS

A key point to remember is that any sensing device indicates or reacts to the temperature of the sensing element. **Remember that a mercury-bulb thermometer indicates the temperature of the bulb at the end of the thermometer, not the temperature of the substance it is submerged in. If it stays in the substance long enough to attain the temperature of the substance, an accurate reading will be achieved.** This statement will help you understand control sensing-element design. See Figure 14–1 for an illustration of the mercury thermometer.

To accurately determine the temperature of solids, **the sensing element must assume the temperature of the**

substance to be sensed as soon as practical. It can be quite difficult to get a round mercury bulb close enough to a flat piece of iron so that it senses only the temperature of the iron. Only a fractional part of the bulb can be made to touch the flat iron at any time, Figure 14–2. This leaves most of the area of the bulb exposed to the surrounding (ambient) air. How can the thermometer tell the true temperature of the iron plate?

Placing an insulator over the thermometer to hold it tightly on the plate and to shield the bulb from the ambient air helps give a more accurate reading. Sometimes a gum-type substance is used to hold the thermometer's bulb flat against the surface and insulate it from the ambient air, Figure 14–3. A well to insert the thermometer in may be necessary for a more permanent installation, Figure 14–4.

Any sensing element is going to have the same difficulties as the mercury bulb in the previous example in reaching the same temperature as the substance to be sensed. Some sensing elements are designed to fit the surface to be sensed. The external motor-temperature

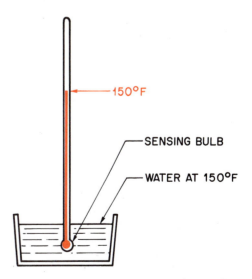

Figure 14–1. The mercury thermometer rises or falls based on expansion or contraction of the mercury in the bulb. The bulb is the sensing device.

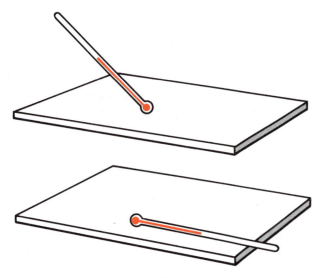

Figure 14-2. Only a fractional part of the mercury bulb thermometer can touch the surface of the flat iron plate at one time. This leaves most of the bulb exposed to the surrounding (ambient) temperature.

sensing element is a good example. It is manufactured flat to fit close to the motor housing, Figure 14-5. One reason it can be made flat is because it is a bimetal. This control is normally mounted inside the terminal box of the motor or compressor to shield it from the ambient temperature.

The protection of electrical motors is one of the most important subjects in this industry because motors are the prime movers of refrigerants, air, and water. Motors, especially the compressor motor, are the most expensive component in the system. Motors are normally made of steel and copper and need to be protected from heat and from overload that will lead to heat.

All motors build up heat as a normal function of work. The electrical energy that passes to the motor is intended to be converted to magnetism and work, but some of it is converted to heat. All motors have some means of detecting and dissipating this heat under

SOFT INSULATING GUM ABOUT THE CONSISTENCY OF BUBBLE GUM

Figure 14-3. Example of sensing temperature quite often used for field readings. The gum holds the bulb against the metal and insulates it from the ambient air.

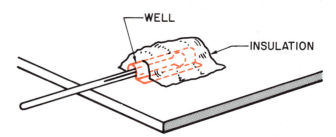

WELL — INSULATION

Figure 14-4. Example of how a well for a thermometer is designed. The well is fastened to the metal plate so that heat will conduct both into and out of the bulb.

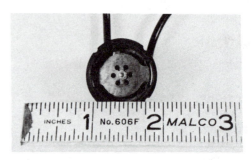

Figure 14-5. Motor-temperature sensing thermostat. *Photo by Bill Johnson*

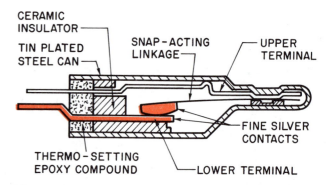

CERAMIC INSULATOR — SNAP-ACTING LINKAGE — UPPER TERMINAL

TIN PLATED STEEL CAN — FINE SILVER CONTACTS

THERMO-SETTING EPOXY COMPOUND — LOWER TERMINAL

Figure 14-6. Bimetal motor-temperature protection device.

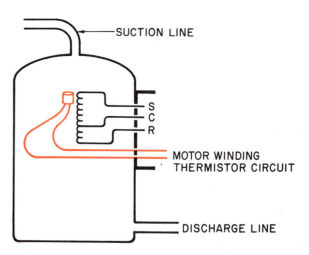

SUCTION LINE

S C R

MOTOR WINDING THERMISTOR CIRCUIT

DISCHARGE LINE

Figure 14-7. Terminal box on the side of the compressor. It has more terminals than the normal common, run, and start of the typical compressor. The extra terminals are for the internal motor protection.

normal or design conditions. Sometimes the heat will begin to build up. This heat must be detected and dealt with if it gets excessive. Otherwise, the motor will overheat and be damaged.

Motor high-temperature protection is usually accomplished with some variation of the bimetal or the thermistor. The bimetal device can either be mounted on the outside of the motor, usually in the terminal box, or embedded in the windings themselves, Figure 14–6. The section on motors will cover in detail how these devices actually protect the motors.

The thermistor type is normally embedded into the windings. This close contact with the windings gives fast, accurate response, but it also means that the wires must be brought to the outside of the compressor. This involves extra terminals in the compressor terminal box, Figure 14–7. Figures 14–8 and 14–9 show wiring diagrams of temperature-protection devices.

We need to mention some things about troubleshooting motor-temperature problems. Because of its size and weight, a motor can take a long time to cool after overheating. If the motor is an open type, a fan or moving airstream can be devised to cool it more quickly. If the motor is inside a compressor shell, it may be suspended from springs inside the shell. This means that the actual motor and the compressor may

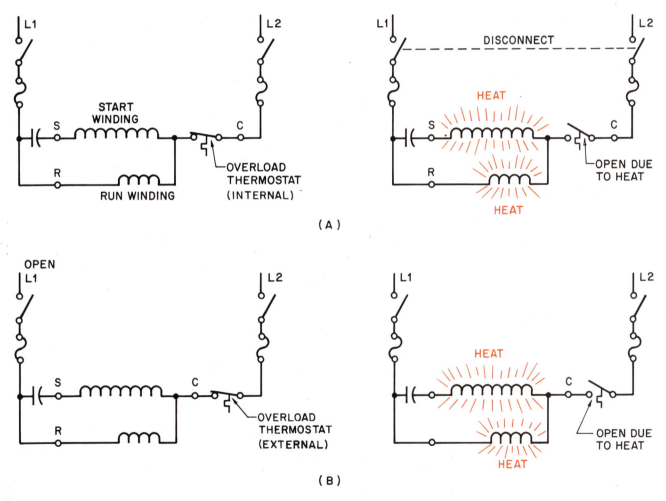

Figure 14–8. Line-voltage bimetal sensing device under hot and normal operating conditions. (A) Permanent split capacitor motor with internal protection. Note that the motor would indicate an open circuit if the ohm reading were taken at the C terminal to either start or run if the overload thermostat were to open. There is still a measurable resistance between start and run. (B) The same motor, except the motor protection is on the outside. Note that it is easier to troubleshoot the overload, but it is not as close to the windings for fast response.

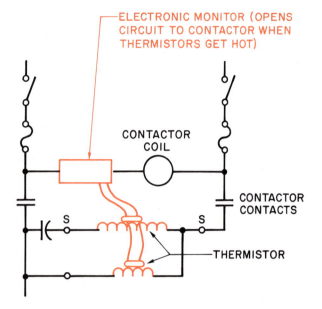

Figure 14-9. The thermistor type of temperature-monitoring device uses an electronic monitoring circuit to check the temperature at the thermistor. When the temperature reaches a predetermined high, the monitor interrupts the circuit to the contactor coil and stops the compressor.

be hot and hard to cool even though the shell does not feel hot. Figure 14-10 shows a method often used to cool a hot compressor. There is a vapor space between the outside of the shell and the actual heat source, Figure 14-11. *The unit must have time to cool. If you are in a hurry, set up a fan, or even cool the compressor with water, but don't allow water to get into the electrical circuits. Turn the power off and cover electrical circuits with plastic before using water to cool a hot compressor. Be careful when restarting. Use an ammeter to look for overcurrent, and gages to determine the charge level of the equipment.* Remember, that

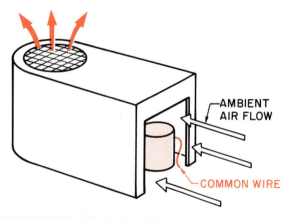

THE COMPRESSOR COMPARTMENT DOOR IS REMOVED AND THE FAN IS STARTED. THIS MAY BE ACCOMPLISHED BY REMOVING THE COMPRESSORS COMMON WIRE AND SETTING THE THERMOSTAT TO CALL FOR COOLING.

Figure 14-10. Ambient air is used to cool compressor motor.

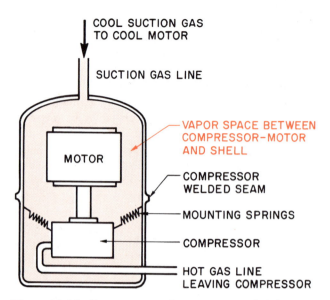

Figure 14-11. Compressor and motor suspended in vapor space in compressor shell. The vapor conducts heat slowly.

most hermetic compressors are cooled by the suction gas. If there is an undercharge, there will be undercooling (or overheating) of the motor.

The best procedure when a hot compressor is encountered is to shut off the compressor with the space-temperature thermostat "off" switch and return the next day. This gives it ample time to cool and keeps the crank-case heat on. Many service technicians have diagnosed an open winding in a compressor that was only hot, and they later discovered that the winding was open because of internal thermal protection.

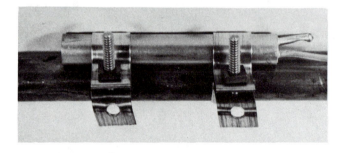

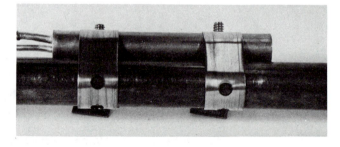

Figure 14-12. Correct way to get good contact with a sensing bulb and the suction line. Bulb is mounted on a straight portion of the line with a strap that holds it secure against the line. *Photo by Bill Johnson*

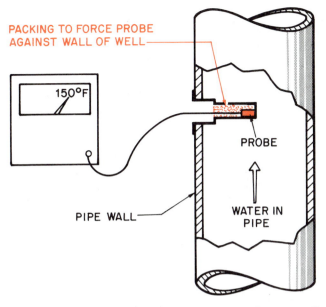

Figure 14-13. Well packed to get the probe against the wall of the well.

14.2 MEASURING THE TEMPERATURE OF FLUIDS

Measuring the temperature of fluids is quite similar to measuring the temperature of solids. The term *fluid* applies to both the liquid and vapor states of matter. Liquids are heavy and change temperature very slowly. Whatever the sensing element might be, it must be able to reach the temperature of the medium to be measured as soon as practical. Since liquids are contained in vessels (or pipes), the measurement can be made either by contact with the vessel or by some kind of immersion. When a temperature is detected from the outside of the vessel by contact, care must be taken that the ambient temperature does not affect the reading.

A good example of this is the sensing bulb used in refrigeration work for measuring the performance of the thermostatic expansion valve. The technician is trying to sense refrigerant gas temperature very accurately to keep liquid from entering the suction line. Quite often the technician will strap the sensing bulb to the suction line in the correct location but will fasten it to the line incorrectly. The technician may forget to insulate the bulb from the ambient when it needs to be insulated. Therefore the bulb is sensing the ambient temperature and averaging it in with the line temperature. Make sure when mounting a sensing bulb to a line for a contact reading that the bulb is in the very best possible contact, Figure 14-12.

When a temperature reading is needed from a larger pipe in a permanent installation, different arrangements are made. A well can be welded into the pipe during installation so that a thermometer or controller sensing bulb can be inserted into it. The well must be matched to the sensing bulb to get a good contact fit, or you won't get an accurate reading, Figure 14-13.

Sometimes it is desirable to remove the thermometer from the well and insert an electronic thermometer with a small probe for troubleshooting purposes. The well inside diameter is much larger than the probe. The well can be packed so that the probe is held firm against the well for an accurate reading.

Another method for obtaining an accurate temperature reading in a water circuit for test purposes is to use one of the valves in a water line to give a constant bleed. For instance, if the leaving-water temperature of a home boiler is needed and the thermometer in the well is questionable, try the following procedure. Place a small container under the drain valve in the leaving-water line. (A coffee cup will work.) Allow a small amount of water from the system to run continuously into the cup. An accurate reading will be obtained from this water. This is not a long-term method for detecting temperature because it requires a constant bleeding of water, but it is an effective field method, Figure 14-14.

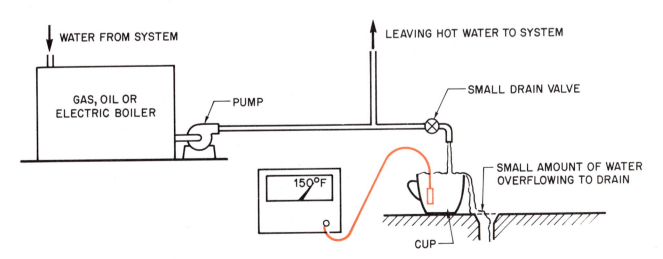

Figure 14-14. Method to obtain a leaving-water temperature reading from a small boiler.

Figure 14–15. Devices used to sense air temperatures in duct work and furnaces. *Photo by Bill Johnson*

14.3 SENSING TEMPERATURE IN AN AIRSTREAM

Sensing temperature in fast-moving airstreams such as duct work and furnace heat exchangers is usually done by inserting the sensing element into the actual airstream. The bimetal of one type or another is usually used for this job. Sometimes the flat-type snap-disc or the helix coil may be used, Figure 14–15.

14.4 THINGS TO REMEMBER ABOUT SENSING DEVICES

Sensing devices are not mysterious; something reacts to the temperature change, such as a bimetal device or a thermistor. Look over any temperature-sensing control and study it, and the working concept will normally come to you. If you are still confused, consult a catalog or the supplier of the control.

SUMMARY

- A heating thermostat has the same action as a motor-temperature cutout protector; it opens on a rise in temperature.
- A mercury-bulb thermometer indicates the temperature of the bulb, not necessarily of the medium to be checked.
- If the mercury bulb is left in the medium to be checked long enough that the mercury bulb reaches the temperature of the medium, an accurate reading has been achieved.
- To get a correct temperature reading of a flat or round surface, the sensing element (either mercury bulb, remote bulb, or bimetal) must be in good contact and insulated from the ambient.
- Some installations have wells in the substance to be sensed.
- All motors build up heat from electrical energy converted to heat energy.

- Motors have both internal and external types of motor-temperature sensing devices.
- The internal type of motor-temperature sensing device can be either a bimetal or a thermistor inserted inside the motor windings.
- When a motor is used in a compressor, it can take a long time to cool because the motor could be suspended in the vapor space of the compressor shell.
- When a well is provided for temperature measurement and an electronic thermometer probe is to be used, the probe can be inserted into the well and packed to hold it tight against the side of the well.
- Some sensing elements are inserted into the fluid stream. Examples are the fan or limit switch on a gas, oil, or electric furnace.
- All temperature-sensing elements change in some manner with a change in temperature: the bimetal warps, the thermistor changes resistance, and the thermocouple (covered in an earlier section) changes voltage.

REVIEW QUESTIONS

1. Name two differences when sensing the temperature of solids as compared to air.
2. What method is used to get a mercury thermometer to indicate accurately on a flat or round surface?
3. What does a mercury-bulb thermometer indicate the temperature of?
4. Why must a mercury-bulb thermometer be left in the substance to be measured for a long time?
5. Do other temperature-sensing devices have to be left in the substance to be measured?
6. Name a method not mentioned in Question 2 to get the sensing element in contact with the substance to be sensed.
7. Why do motors build up heat?
8. How is motor heat dissipated?
9. What are the two types of motor-temperature sensing devices?
10. What is the main precaution for an externally mounted motor-temperature protector?
11. What is the principal of operation of most externally mounted overload protection devices?
12. What are the two principal types of operation of the internal-type motor-temperature protector?
13. Name a method to speed up the cooling of an open motor.
14. Name two methods to cool an overheated compressor.
15. How can an electronic thermometer be used in a well to obtain an accurate response?
16. How can the temperature be checked in a water circuit if no temperature well is provided?

17. Name a temperature-sensing device commonly used in measuring airflow.
18. How does a bimetal change with temperature change?
19. How does the thermistor change with a temperature change?
20. How does a thermocouple change with a temperature change?

PRESSURE-SENSING DEVICES

OBJECTIVES

After studying this unit, you should be able to

- **define pressure-sensitive devices.**
- **describe the difference between a diaphragm and a bellows control.**
- **state the uses of pressure-sensitive controls.**
- **explain the difference between low-pressure controls and high-pressure controls.**
- **describe a high-pressure control.**
- **describe a low-pressure control.**
- **describe a pressure relief valve.**

15.1 INTRODUCTION TO PRESSURE-SENSING DEVICES

Pressure-sensing devices are normally applied to refrigerant, air, gas, and water. Notice that the terms "pressure control" and "pressure switch" are frequently used. Quite often the pressure control as a component has a switch action but is called a "pressure control" instead of a "pressure switch." This is how the terminology is used in the field. It can be confusing, because in conversation one cannot always tell if a switch or a valve is being discussed. The actual application of the component should indicate if an electrical circuit is to be broken or if fluid is to be controlled.

1. Pressure switches are used to stop and start electrical loads, such as motors, Figure 15–1.
2. Pressure controls contain a bellows, a diaphragm or a Bourdon tube to create movement when the pressure inside it is changed. Pressure controls may be attached to switches or valves, Figure 15–2.
3. When used as a switch, the bellows, Bourdon tube or diaphragm is attached to the linkage that operates the electrical contacts. When used as a valve, they are normally attached directly to the valve.
4. The electrical contacts are the component that actually open and close the electrical circuit.
5. The electrical contacts either open or close with snap action on a rise in pressure, Figure 15–3.
6. The pressure control can either open or close on a rise in pressure. This opening and closing action can control water or other fluids, depending on the type, Figure 15–4.
7. The pressure control can sense a pressure differential and be designed to open or close a set of electrical contacts.
8. The pressure control can be an operating-type control or a safety-type control, Figure 15–5.
9. It can either operate at low pressures (even below atmospheric pressure) or high pressures, depending on the design of the control mechanism.
10. Pressure controls can sometimes be recognized by the small pipe running to them for measuring fluid pressures.
11. Pressure switches are manufactured to handle control voltages or line currents to start a compressor up to about 3 hp maximum. The refrigeration industry is the only industry that uses the high-current draw controls.
12. The high-pressure and the low-pressure controls in refrigeration and air conditioning equipment are the two most widely used pressure controls in this industry.
13. Some pressure switches are adjustable, and some are not, Figure 15–6.
14. Some controls are automatic reset, and some are manual reset, Figure 15–7.
15. Some pressure controls have the high-pressure and the low-pressure controls built into one housing. They are called *dual-pressure controls,* Figure 15–8.
16. Pressure controls are usually located near the compressor on air conditioning and refrigeration equipment.
17. *When used as safety controls, pressure controls should be installed in such a manner that they are not subject to being valved off (by the service valves).*
18. The point or pressure setting at which the control interrupts the electrical circuit is known as the *cutout.* The point or pressure setting at which the electrical circuit is made is known as the *cut-in.* The difference in the two settings is known as the *differential.*

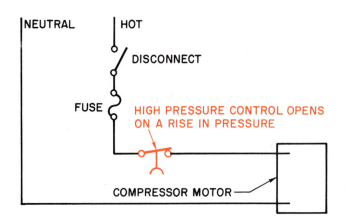

Figure 15–1. Electrical circuit of a refrigeration compressor with a high-pressure control. This control has a normally closed contact that opens on a rise in pressure.

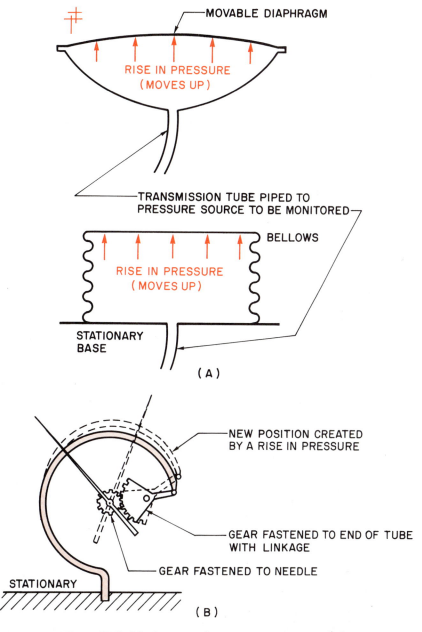

Figure 15–2. Moving part of most pressure-type controls.

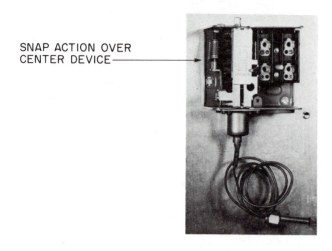

SNAP ACTION OVER
CENTER DEVICE

Figure 15-3. Snap action over center device. *Photo by Bill Johnson*

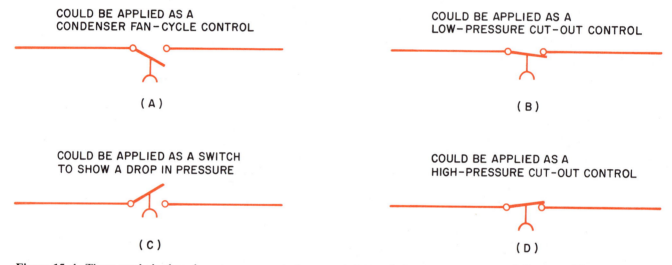

COULD BE APPLIED AS A
CONDENSER FAN-CYCLE CONTROL

(A)

COULD BE APPLIED AS A
LOW-PRESSURE CUT-OUT CONTROL

(B)

COULD BE APPLIED AS A SWITCH
TO SHOW A DROP IN PRESSURE

(C)

COULD BE APPLIED AS A
HIGH-PRESSURE CUT-OUT CONTROL

(D)

Figure 15-4. These symbols show how pressure controls connected to switches appear on control diagrams. The two symbols, A and B, indicate that the circuit will make on a rise in pressure. A indicates the switch is normally open when the machine does not have power to the electrical circuit. B shows that the switch is normally closed without power. The two symbols, C and D, indicate that the circuit will open on a rise in pressure. C is normally open and D is normally closed.

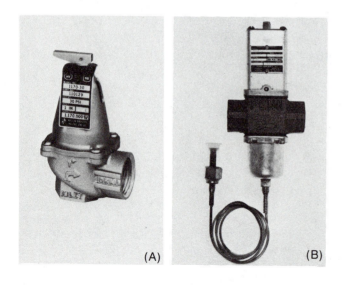

(A) (B)

Figure 15-5. (A) A safety control (boiler relief valve) and (B) an operating control (water regulating valve). *Photos by Bill Johnson*

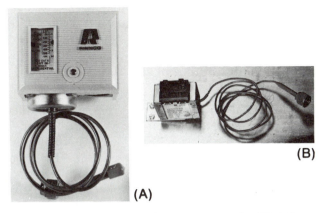

Figure 15-6. Commonly used pressure controls. (A) Adjustable. (B) Nonadjustable. *Photos by Bill Johnson*

Figure 15-7. The control on the left is a manual reset control, the one on the right is automatic reset. Note the push lever on the left hand control. *Photo by Bill Johnson*

Figure 15-8. The control has two bellows acting on one set of contacts. Either control can stop the compressor. *Photo by Bill Johnson*

15.2 HIGH-PRESSURE CONTROLS

The high-pressure control (switch) on an air conditioner stops the compressor if the pressure on the high-pressure side becomes excessive. This control appears in the wiring diagram as a normally closed control that opens on a rise in pressure. The manufacturer may determine the upper limit of operation for a particular piece of equipment and furnish a high-pressure cutout control to insure that the equipment does not operate above these limits, Figure 15-9.

A compressor is known as a positive displacement device. When it has a cylinder full of vapor, it is going to pump out the vapor or stall. If a condenser fan motor on an air-cooled piece of equipment burns out and the compressor continues to operate, very high pressures will occur. ✱The high-pressure control is one method of insuring safety for the equipment and the surroundings. Some compressors are strong enough to burst a pipe or a container. The overload in the compressor offers some protection in this event, but it is really a secondary device because it is not directly responding to the pressures. The overload may also be a little slow to respond.✱

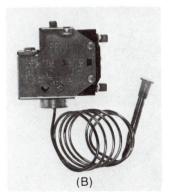

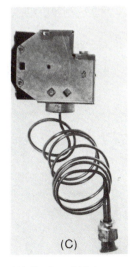

Figure 15-9. High-pressure controls. (A) Enclosed high-pressure control. (B) Manual reset. (C) Automatic reset. *Photos by Bill Johnson*

15.3   LOW-PRESSURE CONTROLS

The recent popularity of the capillary tube as a metering device has caused the low-pressure control to be reconsidered as a standard control on all equipment. The capillary tube metering device equalizes pressure during the off cycle and causes the low-pressure control to short cycle the compressor if it is not carefully applied. Remember, the capillary tube is a fixed-bore metering device and has no shutoff valve action. To prevent this short cycle, some equipment comes with a time-delay circuit that will not allow the compressor to restart for a predetermined time period.

There are two good reasons why it is undesirable to operate a system without an adequate charge.

1. The motors used most commonly in this industry, particularly in the air conditioning field, are cooled by the refrigerant. Without this cooling action, the motor will build up heat when the charge is low. The motor-temperature cutout discussed in the previous unit is used to detect this condition. It quite often takes the place of the low-pressure cutout by sensing motor temperature.
2. If the refrigerant escaped from the system through a leak in the low side of the system, the system can operate until it goes into a vacuum. Remember that when a vesssel is in a vacuum, the atmosphere's pressure is greater than the vessel's pressure. This causes the atmosphere to be pushed into the system. Technicians often say "pull air into the system." The reference point the technician uses is atmospheric

pressure. This air in the system is sometimes hard to detect; if it is not removed, it causes acid to form in the system.

15.4   OIL-PRESSURE SAFETY CONTROLS

The oil-pressure safety control (switch) is used to insure that the compressor has oil pressure when operating, Figure 15–11. This control is used on larger compressors and has a different sensing arrangement than the high- and low-pressure controls. The high- and low-pressure controls are single diaphragm or single bellows controls because they are comparing atmospheric pressure to the pressures inside the system. Atmospheric pressure can be considered a constant for any particular locality because it does not vary more than a small amount.

The oil-pressure safety control is a pressure differential control. The reason is that this control actually does measure a difference in pressure to establish that positive oil pressure is present. A study of the compressor will show that the compressor crankcase (this

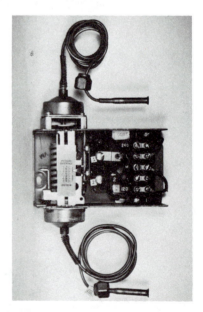

Figure 15–11. Two views of an oil-pressure safety control. This control satisfies two requirements: how to measure net oil pressure effectively; and how to get the compressor started to build oil pressure. *Photos by Bill Johnson*

Figure 15–10. Low-pressure control. *Photo by Bill Johnson*

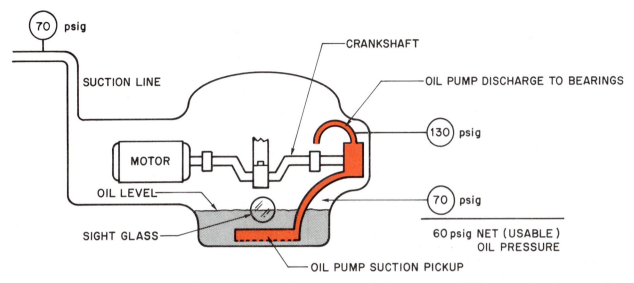

Figure 15-12. The oil pump suction is actually the suction pressure of the compressor. This means that the true oil pump pressure is the oil pump discharge pressure less the compressor suction pressure. For example, if the oil pump discharge pressure is 130 psig and the compressor suction pressure is 70 psig, the net oil pressure is 60 psig. This is usable oil pressure.

is the oil pump suction inlet) is the same as the compressor suction pressure, Figure 15-12. The suction pressure will vary from the off or standing reading to the actual running reading, not to mention the reading that would occur when a low charge is experienced. For example, when a system is using R-22 as the refrigerant, the pressures may be similar to the following: 125 psig while standing; 70 psig while operating, and 20 psig during a low-charge situation.

A plain low-pressure cutout control would not function at all of these levels, so a control had to be devised that would sensibly monitor pressures at all of these conditions.

Most compressors need at least 30 psig of actual oil pressure for proper lubrication. This means that whatever the suction pressure is operating at, the oil pressure has to be at least 30 lb above the oil pump

inlet pressure, because the oil pump inlet pressure is the same as the suction pressure. For example, if the suction pressure is 70 psig, the oil pump outlet pressure must be 100 psig for the bearings to have a net oil pressure of 30 psig. This difference in the suction pressure and the oil pump outlet pressure is called the *net oil pressure*.

The basic low-pressure control has the pressure under the diaphragm or bellows and the atmosphere's pressure on the other side of the diaphragm or bellows. The atmosphere's pressure is considered a constant because it doesn't vary more than a small amount. The oil-pressure control uses a double bellows—one bellows opposing the other to detect the net or actual oil pressure. The pump inlet pressure is under one bellows, and the pump outlet pressure is under the other bellows. These bellows are opposite each other either

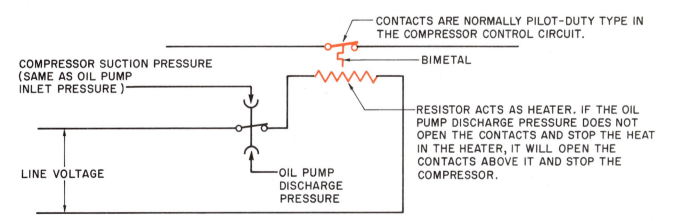

Figure 15-13. The oil pressure control is used for lubrication protection for the compressor. The oil pump that lubricates the compressor is driven by the compressor crankshaft. Therefore, there has to be a time delay to allow the compressor to start up and build oil pressure. This time delay is normally 90 seconds. The time delay is accomplished either with a heater circuit heating a bimetal or an electronic circuit.

physically or by linkage. The bellows with the most pressure is the oil pump outlet, and it overrides the bellows with the least amount of pressure. This override reads out in net pressure and is attached to a linkage that can stop the compressor when the net pressure drops for a predetermined time.

Because the control needs a differential in pressure to allow power to pass to the compressor, it must have some means to allow the compressor to get started. Remember, there is no pressure differential until the compressor starts to turn, because the oil pump is attached to the compressor crankshaft. There is a time delay built into the control to allow the compressor to get started and to prevent unneeded cutouts when oil pressure may vary for only a moment. This time delay is normally about 90 seconds. It is accomplished with a heater circuit and a bimetal device or electronically. *The manufacturer's instructions should be consulted when working with any compressor that has an oil safety control, Figure 15–13.*

15.5 AIR-PRESSURE CONTROLS

Air-pressure controls (switches) are used in the following applications.

1. Heat pumps sense air-pressure drop across the outdoor coil due to ice buildup. When the coil has a predetermined pressure drop across it, there is an ice buildup that justifies a defrost. When used in this manner, the control is an operating control. All manufacturers do not use air pressure as an indicator for defrost, Figure 15–14.
2. When electric heat is used in a remote duct as terminal heat, an air switch (sail switch) is sometimes used to insure that the fan is passing air through the duct before the heat is allowed to be energized. This is a safety switch and should be treated as such, Figure 15–15.

Air-pressure switches are usually large in diameter because they need the large diaphragm to sense the

Figure 15–15. A sail switch detects air movement. *Photo by Bill Johnson*

very low pressures in air systems. Mounted on the outside of the diaphragm is a small switch (microswitch) capable only of stopping and starting control circuits.

15.6 GAS-PRESSURE SWITCHES

Gas-pressure switches are similar to air-pressure switches but usually smaller. They detect the presence of gas pressure in gas-burning equipment before burners are allowed to ignite. *This is a safety control and should never be bypassed except for troubleshooting and then only by experienced service technicians.*

15.7 DEVICES THAT CONTROL FLUID FLOW AND DO NOT CONTAIN SWITCHES

The pressure relief valve can be considered a pressure-sensitive device. It is used to detect excess pressure in any system that contains fluids (water or refrigerant). Boilers, water heaters, hot-water systems, refrigerant systems, and gas systems all use pressure relief valves, Figure 15–16. This device can either be pressure and temperature sensitive (called a P & T

Figure 15–14. An air-pressure differential switch used to detect ice on a heat pump outdoor coil. *Photo by Bill Johnson*

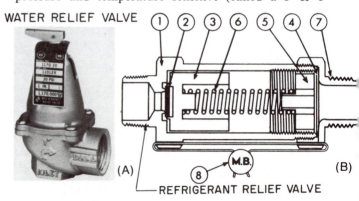

Figure 15–16. Pressure relief valves. *Courtesy (B) Superior Valve Co. (A) Photo by Bill Johnson*

valve) or just pressure sensitive. The P & T valves are normally associated with water heaters. The pressure-sensitive type will be the only one considered here.

The pressure relief valve can usually be recognized by its location. It can be found on all boilers at the high point on the boiler. *It is a safety device that must be treated with the utmost respect.* Some of these valves have levers on top that can be raised to check the valve for flow. Most of the valves have a visible spring that pushes the seat downward toward the system's pressure. A visual inspection will usually reveal that the principle of the valve is for the spring to hold the seat against the valve body. The valve body is connected to the system, so, in effect, the spring is holding the valve seat down against the system pressure. When the system pressure becomes greater than the spring tension, the valve opens and relieves the pressure and then reseats. *The pressure setting on a relief valve is factory set and should never be tampered with. Normally the valve setting is sealed or marked in some manner so that it can be seen if it is changed.*

Most relief valves are automatically reset. After they relieve, they automatically seat back to the original position. A seeping or leaking relief valve should be replaced. *Never adjust the valve or plug the valve. Extreme danger could be encountered.*

15.8   WATER PRESSURE REGULATORS

The water pressure regulator is a common device used in this industry to control water. Two types of valves are commonly used: the pressure regulating valve for system pressure, and the pressure regulating valve for head-pressure control on water cooled refrigeration systems. Air conditioning and refrigeration systems operate at lower pressures and temperatures when water cooled, but the water circuit needs a special kind of maintenance. At one time water-cooled equipment was widely used. Air-cooled equipment has now become the dominant type because it is easy to main-

tain. Water valves are therefore not as common as they were.

Water pressure regulating valves are used in two basic applications:

1. They reduce city water pressure to the operating pressure of the system. This type of valve has an adjustment screw on top that increases the tension on the spring regulating the pressure. A great many boilers in hydronic systems in homes or businesses use circulating hot water at about 15 psig, Figure 15–17. The city water may have a working pressure of 75 psig or more and have to be reduced to the system working pressure. If the city pressure were to be allowed into the system, the boiler pressure relief valve would relieve its pressure. This valve is installed in the city water makeup line that adds water to the system when some leaks out. Most water regulating valves have a manual valve arrangement that allows the service technician to remove the valve from the system and service it without stopping the system. A manual feedline is also furnished with most systems to allow the system to be filled by bypassing the water regulating valve. *Care must be taken not to overfill the system.*

2. The water regulating valve controls water flow to the condenser on water-cooled equipment for head-pressure control. This valve takes a pressure reading from the high-pressure side of the system and uses this to control the water flow to establish a predetermined head pressure, Figure 15–18. For example, if an ice maker were to be installed in a restaurant where the noise of the fan would be objectionable, consider a water-cooled ice maker. This might very well be a "wastewater type of system" where the water is allowed to go down the drain. In the winter the water may be very cool, and in the summer the water may get warm. In other words, the ice maker

Figure 15–17. Adjustable water regulating valve. *Photo by Bill Johnson*

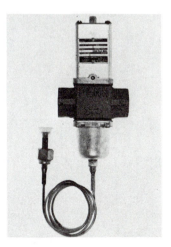

Figure 15–18. This valve maintains a constant head pressure for a water-cooled system during changing water temperatures and pressures. *Photo by Bill Johnson*

may have a winter need and a summer need. The water regulating valve would modulate the water in both cases to maintain the required head pressure. There is an added benefit because when the system is off for any reason, the head pressure reduces and the water flow is stopped until needed again. All of this is accomplished without an electric solenoid.

15.9 GAS-PRESSURE REGULATORS

The gas-pressure regulator is used in all gas systems to reduce the gas transmission pressure to usable pressure at the burner. The gas pressure at the street in a natural-gas system could be 5 psig, and the burners are manufactured to burn gas at 3.5 in. of water column (this is a pressure measuring system that indicates how high the gas pressure can push a column of water as the indicator of pressure). *This pressure must be reduced to the burner's design, and this is done with a pressure-reducing valve that acts like the water pressure regulating valve in the boiler system, Figure 15–19.* There is an adjustment on top of the valve for qualified personnel to adjust the valve.

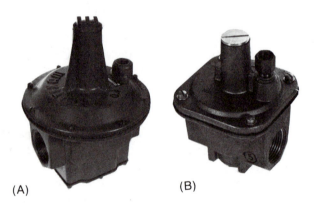

Figure 15–19. Gas-pressure regulators. (A) Gas pressure regulator at the meter. (B) Gas pressure regulator at the appliance. *Photos by Bill Johnson*

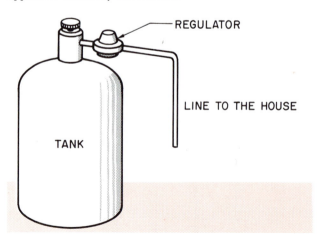

Figure 15–20. Gas-pressure regulator on a bottled-gas system. This regulator is at the tank.

When bottled gas is used, the tank can have as much as 150 psig pressure to be reduced to the burner design pressure of 11 in. of water column. The regulator is normally located at the tank for this pressure reduction, Figure 15–20.

SUMMARY

- Pressure-sensing devices can be applied to all fluids.
- This industry usually uses pressure-sensing devices on water, air, and refrigerant.
- Pressure controls (switches) electrical ratings are either pilot duty or line duty.
- The refrigeration industry uses the line-duty type more than the air conditioning industry does.
- Pressure controls normally are either diaphragm, Bourdon tube or bellows operated.
- Linkage connects the diaphragm, Bourdon tube or bellows to the electrical contacts.
- Pressure controls can operate at high pressures or low pressures, even below atmospheric pressure, in a vacuum.
- Pressure controls can also detect differential pressures.
- Pressure controls usually have a small pipe that goes straight to the diaphragm, Bourdon tube or bellows.
- Some pressure controls are field adjustable. Some have a seal that would show evidence of adjustment.
- Controls applied to electrical circuits are either manual reset or automatic reset.
- Pressure-sensitive controls can control fluid flow. These are sometimes modulating-type controls.
- *Gas pressure must be reduced before it enters a house or place of business to be burned in an appliance.*

REVIEW QUESTIONS

1. What is a pressure-sensitive device?
2. Name two methods used to convert pressure changes into action.
3. Name two actions that can be obtained with a pressure change.
4. Can pressures below atmospheric pressure be detected?
5. Name one function of the low-pressure control.
6. Name one function of the high-pressure control.
7. What is the function of the small pipe leading to most pressure controls?
8. Do all pressure controls have switches?
9. Name two types of water pressure control.
10. Why should a safety control not be adjusted if it has a seal to prevent tampering?

TROUBLESHOOTING
BASIC CONTROLS

OBJECTIVES

After studying this unit, you should be able to

- define and identify power- and non-power-consuming devices.
- describe how a voltmeter is used to troubleshoot electrical circuits.
- identify some typical problems in an electrical circuit.
- describe how an ammeter is used to troubleshoot an electrical circuit.
- recognize the components in a heat–cool electrical circuit.
- follow the sequence of electrical events in a heat–cool electrical circuit.
- differentiate between a pictorial and a line-type electrical wiring diagram.

16.1   INTRODUCTION TO TROUBLESHOOTING

Each control must be evaluated as to its function (major or minor) in the system. Recognizing the control and its purpose requires understanding. The ability to stop and study a control before taking action will save a lot of trouble in the future. When observing a control for the first time, see what it actually does. Look for pressure lines going to pressure controls and temperature-activated elements on temperature controls. See if the control stops and starts a motor, opens and closes a valve, or provides some other function. As mentioned previously, controls are either electrical, mechanical, or a combination of electrical and mechanical. (Electronic controls will be considered electrical for now. An explanation of electronic controls will be included in another unit.)

Electrical devices can be considered as power-consuming or non-power-consuming as far as their function in the circuit is concerned. Power-consuming devices use power and have either a magnetic coil or a resistance circuit. They are wired in parallel with the power source.

Non-power-consuming devices pass power. These terms can be described with a simple light-bulb circuit with a switch. The light bulb in the circuit actually consumes the power, and the switch passes the power to the light bulb. The object is to wire the light bulb to both sides (hot and neutral) of the power supply. This will then insure that the light bulb is in parallel with the power supply, and a complete circuit will be established. The switch is a power-passing device and is wired in series with the light bulb. To understand this troubleshooting method, you must realize that for any power-consuming device to consume power it has to have a potential voltage. A potential voltage is the voltage indicated on a voltmeter between two power legs (such as line 1 to line 2) of a power supply. This can be thought of as electrical pressure. In a home, for example, the light bulb has to be wired from the hot leg (the wire that has a fuse or breaker in it) to the neutral leg (the grounded wire that actually is wired to the earth ground), Figure 16–1.

Service technicians have tried to express which way the electrons flow in electrical circuits in many different ways. Confusion soon consumes the technicians' thinking if they are not careful. Some liberty will be used in explaining the circuits in this unit. These liberties should help straighten out some confusion without creating the problems of understanding the pure theory.

The following simplified explanation of a circuit should clarify what we have said.

1. It doesn't make any difference which way the electrons are flowing in an AC circuit when you keep in mind that the object is to get the electrons (current) to a power-consuming device and complete the circuit.
2. The electrons may pass through many non-power-consuming devices before getting to the power-consuming device.
3. Devices that don't consume power pass power.
4. Devices that pass power can either be safety devices or operating devices.

Let's use the light-bulb example again. Suppose the light bulb were to be used to add heat to a well pump to keep it from freezing in the winter. To prevent the bulb from burning all the time, a thermostat is installed. A fuse must also be installed in the line to

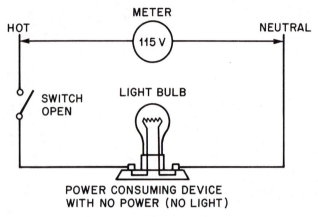

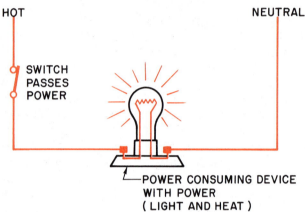

Figure 16-1. Power consuming device (light bulb) and non-power-consuming device (switch).

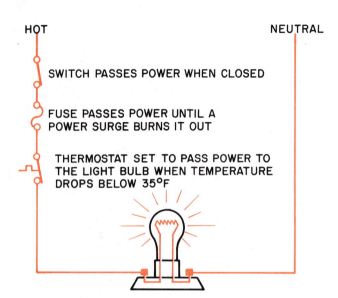

Figure 16-2. When closed, the switch passes power to the fuse.

protect the circuit, and a switch must be wired to allow the circuit to be serviced, Figure 16-2. Note that the power now must pass through three non-power-consuming devices to reach the light bulb. The fuse is a safety control, and the thermostat turns the light bulb on and off. The switch is not a control but a service convenience device. Notice that all switches are on one side of the circuit. Let's suppose that the light bulb does not light up when it is supposed to. Which component is interrupting the power to the bulb: the switch, the fuse, or the thermostat? Or is the light bulb itself at fault? In this example the thermostat (switch) is open.

16.2 TROUBLESHOOTING A SIMPLE CIRCUIT

To troubleshoot the circuit in the previous paragraph, a voltmeter may be used in the following manner. Turn the voltmeter selector switch to a voltage setting higher than the voltage supply. In this case the supply voltage should be 115 V, so the 250-V scale is a good choice. Follow the diagram in Figure 16-3 as you read the following.

1. Place the red lead of the voltmeter on the hot line and the black lead on the neutral. The meter will read 115 V.

2. Place the red lead, the lead being used to find and detect power in the hot line, on the load side of the switch. (The "load" side of the switch is the side of the switch that the load is connected to. The other side of the switch, where the line is connected, is the "line" side of the switch.) The black lead should remain in contact with the neutral line. The meter will read 115 V.

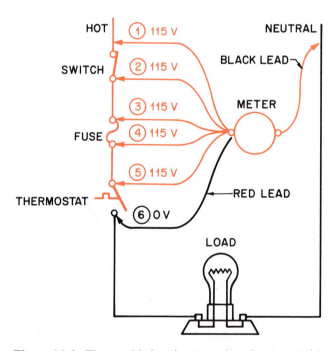

Figure 16-3. The troubleshooting procedure is to establish the main power supply of 115 V from the hot line to the neutral line. When this power supply is verified, the lead on the hot side is moved down the circuit toward the light bulb. The voltage is established at all points. When point 6 is reached, no voltage is present. The thermostat contacts are open.

3. Place the red lead on the line side of the fuse. The meter will read 115 V.
4. Place the red lead on the load side of the fuse. The meter will read 115 V.
5. Place the red lead on the line side of the thermostat. The meter will read 115 V.
6. Place the red lead on the load side of the thermostat. The meter will read 0 V. There is no power available to energize the bulb. The thermostat contacts are open. Now ask the question, is the room cold enough to cause the thermostat contacts to make? If the room temperature is below 35° F, the circuit should be closed; the contacts should be made.

Note: The red lead was the only one moved. It is important to note that if the meter had read 115 V at the light-bulb connection when the red lead was moved to this point, then further tests should be made.

Another step is necessary to reach the final conclusion. Let's suppose that the thermostat is good and 115 V is indicated at point 6.

7. The meter lead can now be moved to the terminal on the light bulb, Figure 16–4. Suppose it reads 115 V. Now, move the black lead to the light-bulb terminal on the right. If there is no voltage, the neutral wire is open between the source and the bulb.

   If there is voltage at the light bulb and it will not burn, the bulb is defective, Figure 16–5. Remember Ohm's law: $I = E/R$. This is another way of saying that *when there is a power supply and a path to flow through, then current will flow.* The light-bulb fila-

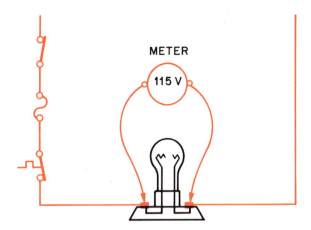

Figure 16–5. Power is available at the light bulb. The hot line on the left side and the neutral on right side complete the circuit. The light bulb filament is burned out, so there is no path through the bulb.

ment should be the path for the current to flow through in the form of a measurable resistance.

16.3   TROUBLESHOOTING A COMPLEX CIRCUIT

The following example is a progressive circuit that would be typical of a combination heating–cooling unit. This circuit is not standard because each manufacturer may design its circuits differently. They may vary, but will be similar in nature.

The unit chosen for this example is a package air conditioner with $1\frac{1}{2}$ tons cooling and 5 kW of electric heat. This unit resembles a window air conditioner because all of the components are in the same cabinet. The unit can be installed through a wall or on a roof, and the supply and return air can be ducted to the conditioned space. The reason for using this unit is that it comes in many sizes—from $1\frac{1}{2}$ tons of cooling (18,000 Btu/hr) to very large systems. The unit is popular with shopping centers because a store could have several roof units to give good zone control. If one unit were off, other units would help to hold the heating–cooling conditions. The unit also has all of the control components, except the room thermostat, within the unit's cabinet. The thermostat is mounted in the conditioned space. The unit can be serviced without disturbing the conditioned space, Figure 16–6.

The first thing we will consider is the thermostat. The thermostat is not standard but is a version used for illustration purposes. This is a usable circuit designed to illustrate troubleshooting. The diagram for the thermostat is in Figure 16–10.

1. This thermostat is equipped with a selector switch for either HEAT or COOL. When the selector switch is in the HEAT position and heat is called for, the *electric heat relay* is energized. The fan must

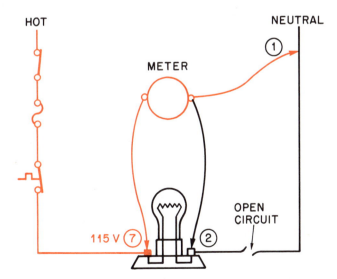

Figure 16–4. Thermostat contacts are closed, and there is a measurement of 115 V when one lead is on the neutral and one lead on the light-bulb terminal. (See position 7.) When the black lead is moved to the light bulb in position 2, there is no voltage reading. The neutral line is open.

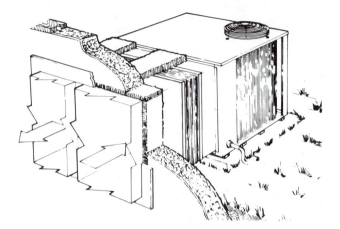

Figure 16-6. Small package air conditioner. *Courtesy Climate Control*

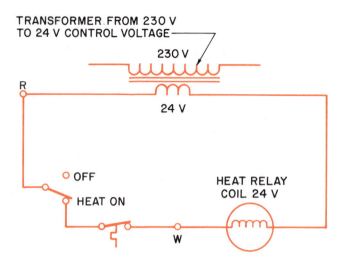

Figure 16-7. Simple thermostat with a HEAT OFF/HEAT ON position. There is no heat anticipator. The selector switch is closed, and the thermostat circuit is closed, calling for heating. When the heat relay is energized, the heating system starts.

run in the heating cycle. It will be started and run in the high-voltage circuit. This will be covered later.

2. When the fan is switched to ON, the indoor fan will run all the time. This thermostat is equipped with a FAN ON or FAN AUTO position, so the fan can be switched on manually for air circulation when desired. When the fan switch is in the AUTO position, the fan will come on upon a call for cooling.

3. When the selector switch is in the cooling mode, the cooling system will start upon a call for cooling. The indoor fan will start through the AUTO mode on the fan selector switch. The outdoor fan must run also, so it is wired in parallel with the compressor.

Figure 16-7 is the beginning of the control explanation. This is the simplest of thermostats, having only a set of heating contacts and a selector switch. Follow the power from the R terminal through the selector switch to the thermostat contacts and on to the W terminal. The W terminal designation is not universal but is quite common. When these contacts are closed, we can say that the thermostat is calling for heat. These contacts pass power to the heat relay coil. Power for the other side of the coil comes directly from the control transformer. When this coil is energized (24 V), the heat should come on. This coil is going to close a set of contacts in the high-voltage circuit to pass power to the electric heat element.

In Figure 16-8 the heat anticipator is added to the circuit. The heat anticipator is in series with the heat relay coil and will have current passing through it at any time the heat relay coil is energized. This current passing through the anticipator creates a small amount of heat in the vicinity of the thermostat bimetal sensing element. This causes the bimetal to break the thermostat contacts early to dissipate the heat in the heater. If the contacts to the heater did not open until the space temperature was actually up to temperature, the extra heat would overheat the space. The fan to move the

heat is stopped and started in the high-voltage circuit. This will be discussed later.

The heat anticipator is often a variable resistor and must be adjusted to the actual system. The current must be matched to the current used by the heat relay coil. All thermostat manufacturers explain how this is done in the installation instructions for the specific thermostat.

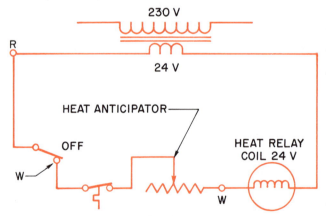

NOTE: THE HEAT ANTICIPATOR IS A VARIABLE RESISTANCE THAT IS DESIGNED TO CREATE A SMALL AMOUNT OF HEAT NEAR THE BIMETAL. THIS FOOLS THE THERMOSTAT INTO CUTTING OFF EARLY. THE FAN WILL CONTINUE TO RUN AND DISSIPATE THE HEAT REMAINING IN THE HEATERS.

Figure 16-8. Same diagram as Figure 16-7 with a heat anticipator in series with the thermostat contacts. The selector switch is set for heat, and the thermostat contacts are closed, calling for heat. The fan in this application will be operated with the high-voltage circuit when in the heating mode.

Figure 16–9 illustrates how a fan starting circuit can be added to a thermostat. Notice the addition of the G terminal. It is used to start the indoor fan. The G terminal designation is not universal but is quite common. Follow the power from the R terminal through the fan selector switch to the G terminal and on to the indoor fan relay coil. Power is supplied to the other side of the coil directly from the control transformer. When this coil has power on it (24 V), it closes a set of contacts in the high-voltage circuit and starts the indoor fan.

Figure 16–10 completes the circuitry in this thermostat by adding a Y terminal for cooling. Again, this is not the only letter used for cooling, but it is also quite common. Follow the power from the R terminal down to the cool side of the selector switch through to the contacts and on to the Y terminal. When these contacts are closed, power will pass through them and go two ways. One path will be through the fan AUTO switch to the G terminal and on to start the indoor fan. It has to run in the cooling cycle. The other way is straight to the cool relay. When the cool relay is energized (24 V), it closes a set of contacts in the high-voltage circuit to start the cooling cycle. *Note:* The cooling anticipator is in parallel with the cooling contacts. This means that current will flow through this anticipator when the thermostat is satisfied (or when the thermostat's circuit is open). The cool anticipator is normally a fixed, nonadjustable type.

16.4 TROUBLESHOOTING THE THERMOSTAT

The thermostat is an often misunderstood, frequently suspected component during equipment mal-

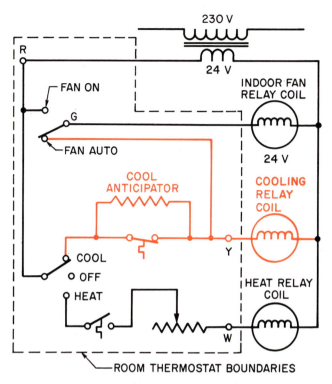

NOTE: THE CIRCUIT THROUGH FAN-AUTO INSURES THE FAN WILL COME ON DURING COOLING.

Figure 16–10. Same diagram as Figure 16–9 with cooling added. The cooling circuit has a cooling anticipator in parallel with the cooling contacts, so current flows through it during the off cycle.

function. But the thermostat is really not mysterious. It does a straightforward job of monitoring temperature and distributing the power leg of the 24-V circuit to the correct component to control the temperature. Service technicians should remind themselves as they approach the job that "one power leg enters the thermostat and it distributes power where called." Every technician needs a technique for checking the thermostat for circuit problems.

One way to troubleshoot a thermostat is to first turn the fan selector switch to the FAN ON position and see if the fan starts. If the fan does not start, there may be a problem with the control voltage. Control voltage has to be present for the thermostat to operate. Assume for a moment that the thermostat will cause the fan to come on when turned to FAN ON but will not operate the heat or cooling cycles. The next step may be to take the thermostat off the subbase and jump the circuit out manually with an *insulated jumper*. Jump from R to G, and the fan should start. Jump from R to W, and the heat should come on. Jump from R to Y, and the cooling should come on. If the circuit can be made to operate without the thermostat but not with it, the thermostat is defective, Figure 16–11. One does not have to be afraid to jump these circuits out one at a time because only one leg of power comes to

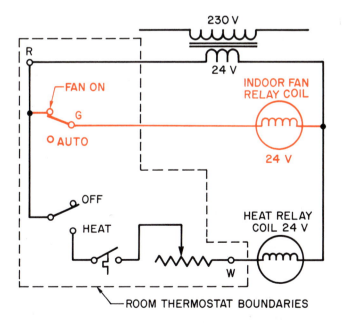

Figure 16–9. The addition of the fan relay to the thermostat allows the owner to switch the fan on for continuous operation.

NOTE: THE FAN SELECTOR SWITCH IS NORMALLY
A PART OF THE SUBBASE.

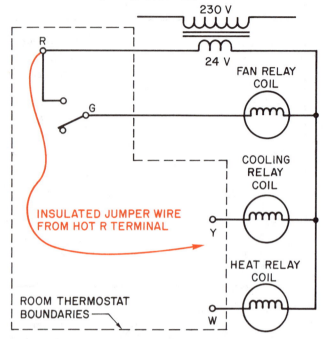

Figure 16-11. Terminal designation in a heat-cool thermostat's subbase. The letters R, G, Y, and W are common designations. A simple jumper wire from the R terminal to G should start the fan; from R to Y should start the cooling; from R to W should start the heat.

the thermostat. If all circuits were jumped out at one time in this thermostat, the only thing that would happen is that the heating and cooling would run at the same time. Of course, this should not be allowed to continue.

SAFETY PRECAUTION: *This should be done only under supervision of the instructor. Do not restart the air conditioner until five minutes have passed. This allows the system to equalize.*

We have just covered what happens in the basic thermostat. The next step is to move into the high-voltage circuit and see how the thermostat's actions actually control the fan, cooling, and heating. The progressive circuit will again be used.

The first thing to remember is that the high-voltage (230 V) is the input to the transformer. Without high voltage there is no low voltage. When the service disconnect is closed, the potential voltage between line 1 and line 2 is the power supply to the primary of the transformer. The primary input then induces power to the secondary.

See Figure 16-12 for the high-voltage operation of the fan circuit. The power-consuming device is the fan motor in the high-voltage circuit. The fan relay contact is in the line 1 circuit. It should be evident that the fan relay contacts have to be closed to pass power to the

fan motor. These contacts close when the fan relay coil is energized in the low-voltage control circuit.

See Figure 16-13 for the addition of the electric heat element and the electric heat relay. The electric heat element and the fan motor are the power-consuming devices in the high-voltage circuit. In the low-voltage circuit when the HEAT-COOL selector switch is moved to the HEAT position, power is passed to the thermostat contacts. When there is a call for heat, these contacts close and pass power through the heat anticipator to the heat relay coil. This closes the two sets of contacts in the high-voltage circuit. One set starts the fan, and the other set passes power to the limit switch, and then through the limit switch to the line 1 side of the heater. The circuit is completed through the fuse link in line 2.

See Figure 16-14 for the addition of cooling or air conditioning to the circuit. For this illustration the heating circuit was removed to reduce clutter and confusion. Three components have to operate in the cooling mode: the indoor fan, the compressor, and the outdoor fan. The compressor and the outdoor fan are wired in parallel and, for practical purposes, can be thought of as one component.

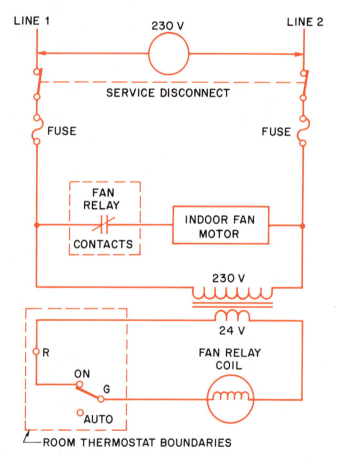

Figure 16-12. When the fan switch is turned to the ON position, a circuit is completed to the fan relay coil. When this coil is energized, it closes the fan relay contacts and passes power to the indoor fan motor.

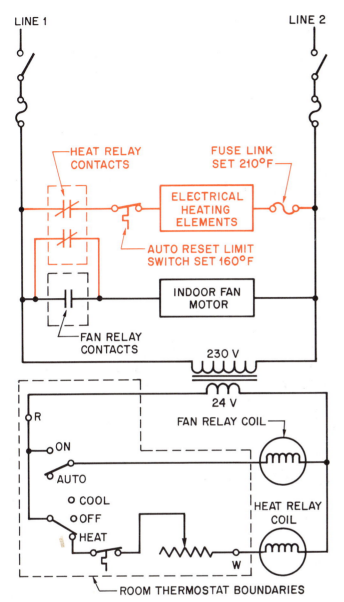

Figure 16-13. When the selector switch is in the HEAT position and the thermostat contacts close, the heat relay coil is energized. This action closes two sets of contacts in the heat relay. One set starts the fan because they are wired in parallel with the fan relay contacts. The other set passes power to the auto reset limit switch (set at 160°F). Power reaches the other side of the heat element through the fuse link. When power reaches both sides of the element, the heating system is functioning.

The line 2 side of the circuit goes directly to the three power-consuming components. Power for the line 1 side of the circuit comes through two different relays. The cooling relay contacts start the compressor and outdoor fan motor; the indoor fan relay starts the indoor fan motor. In both cases the relays pass power to the components upon a call from the thermostat.

When the thermostat selector switch is set to the COOL position, power is passed to the contacts in the thermostat. When these contacts are closed, power

passes to the cooling relay coil. When the cooling relay coil is energized, the cooling relay contacts in the high-voltage circuit are closed. This passes power to the compressor and outdoor fan motor.

The indoor fan must run whenever there is a call for cooling. When power passes through the contacts in the thermostat, a circuit is made through the AUTO side of the selector switch to energize the indoor fan relay coil and start the indoor fan. Note that if the fan selector switch is in the FAN ON position, the fan would run all the time and would still be on for

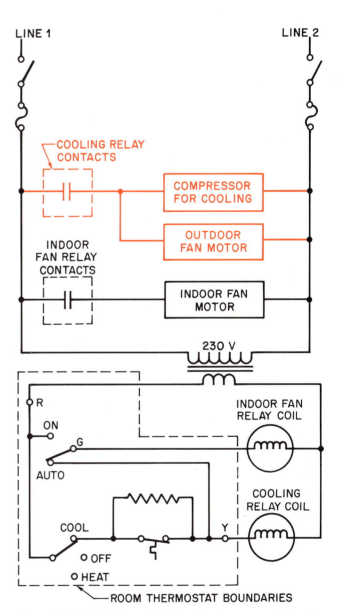

Figure 16-14. Addition of cooling to the system. When the thermostat selector switch is set to COOL and the thermal contacts close, the cooling relay coil is energized, closing the high-voltage contacts to the compressor and outdoor fan motor. Notice the circuit in the thermostat that starts the indoor fan through the AUTO switch. If this switch were ON, the fan would have been running.

cooling. The following situations will be used as examples of actual service situations. For the sake of knowing where the discussion is going, the actual problem is stated at the beginning of each service call.

16.5 SERVICE TECHNICIAN CALLS

Service Call 1

The system is not cooling. The control transformer has an open circuit in the primary, Figure 16–15.

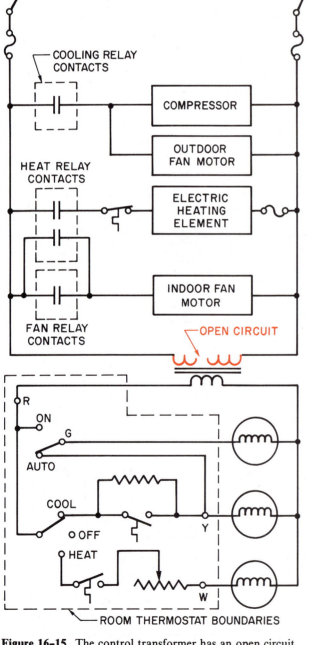

Figure 16–15. The control transformer has an open circuit.

1. Go to the indoor thermostat and turn the indoor fan switch to the FAN ON position. The fan will not start. This is an indication that you possibly have no control voltage. You must have high voltage before you can have low voltage. Go to the outdoor unit and check for high voltage.
2. With your meter set on 250 V, check for high voltage. You find that you have power.
3. Take the cover off the control compartment (usually this compartment can be found because the low-voltage wires enter close by).
4. Check for the high-voltage power supply at the primary of the transformer. You find that you have power there.
5. Check for power at the secondary of the low-voltage transformer. You will find that there is no power. The transformer must be defective. Now to prove it, perform the next step.
6. *Turn off the main power supply.* Take one lead off the low-voltage transformer and check the transformer for continuity with the ohmmeter. *Here is where you will find that the primary circuit is open.* A new transformer must be installed.

Service Call 2

The system is not cooling. The cooling relay coil is open, Figure 16–16.

1. Again, go to the thermostat and try the fan circuit. When the fan switch is moved to FAN ON, the indoor fan will start. Now switch the thermostat to the COOL position, and you will notice that the fan will run. This tells you that the power is passing through the thermostat, so the problem is probably not there. Leave the thermostat calling for cooling. Go to the outdoor unit.
2. Since the primary for the low-voltage transformer comes from the high voltage, we know that we have high voltage.
3. A look at the diagram shows that the only requirement for the cooling relay to close its contacts is that its coil be energized.
4. Check for 24 V at the cooling relay coil. You will find that you have it.
5. *Turn off the main power and remove one lead from the cooling relay coil.* Use the ohmmeter to check for continuity. There is none. A new relay is needed.

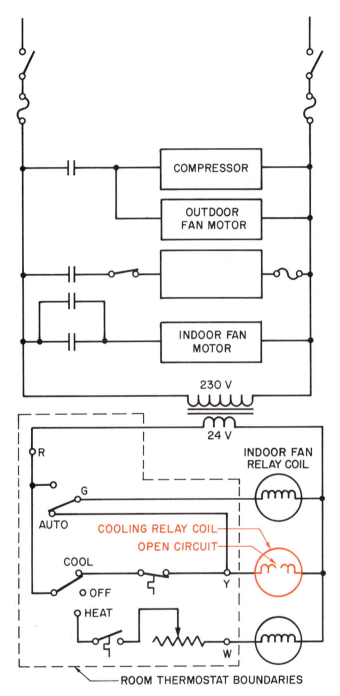

Figure 16–16. The cooling relay has an open circuit and will not close the cooling relay contacts. Note that the indoor fan is operable.

Service Call 3

There is no heat. Heat relay has shorted coil that overloaded the transformer and burned it out, Figure 16–17.

1. Again, go to the thermostat and try the indoor fan. You will find that it won't run.
2. Go to the outdoor unit and check for high voltage. You will find that you have power.

3. Check for low-voltage power. You will find that you do not have 24 V. Check the transformer as before. The transformer will have to be changed.
4. When the transformer is changed and the system is switched to HEAT, notice that the system did not come on. You may notice that the transformer is getting hot.
5. Go back to the thermostat and switch the heat off and see if the indoor fan will run. When it starts, switch back over to HEAT and the fan will probably stop because the current flow through the shorted electric heat relay may pull the voltage down. If this happens, turn the heat off at once, or the heat anticipator in the thermostat may be damaged.
6. Go to the outdoor unit and check the ohm reading across the coil. Compare it to the reading across the cooling contactor coil. If there is considerably less resistance across the heat relay coil, it may very well be shorted. Change it. *You must remember that all power-consuming devices must have a measurable resistance. If this resistance is reduced, the current flow increases.*
7. Another check that will verify the problem is to shut off the power to the outdoor unit and remove one wire from the coil on the heat relay. Go into the house and make the thermostat call for heat. Check voltage at the two wires, one still on the other side of the coil and the other hanging loose. If you have 24 V, touch the wire on the other side of the coil and see if the voltage reduces (maybe to 12 V). If this happens, it proves that the coil is pulling the voltage down.

16.6 TROUBLESHOOTING AMPERAGE IN LOW-VOLTAGE CIRCUITS

Transformers are rated in *volt-amperes,* commonly called VA. This rating can be used to determine if the transformer is underrated or drawing too much current.

For example, it is quite common to use a 40-VA transformer for the low-voltage power source on a combination cooling and heating piece of equipment. This tells the technician that at 24 V, the maximum amperage that the transformer can be expected to carry is 1.66 A. This was determined as follows:

$$\frac{40 \text{ VA}}{24 \text{ V}} = 1.66 \text{ A}$$

Clamp-on ammeters will not readily measure such a low current with any degree of accuracy, so arrange-

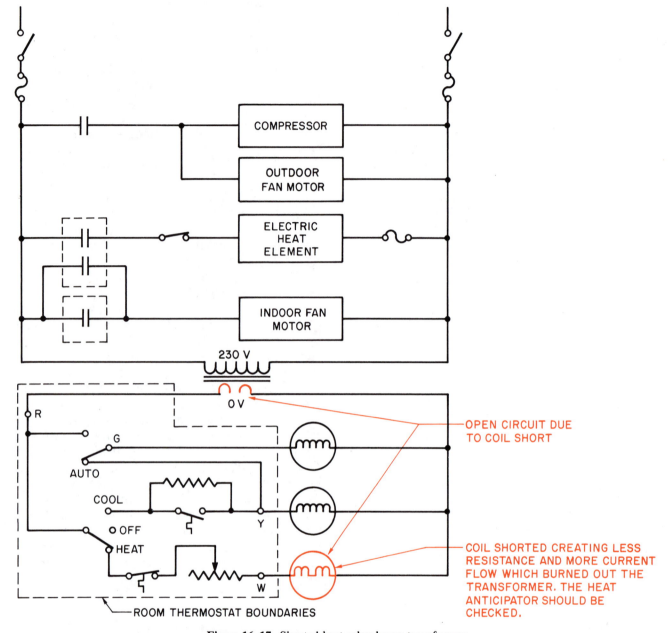

Figure 16–17. Shorted heat relay burns transformer.

ments have to be made to determine an accurate current reading. Use a jumper wire and coil it ten times (called a 10-wrap amperage multiplier). Place the ammeter's jaws in the 10-wrap loop and place the jumper in series with the circuit. **The reading on the ammeter will have to be divided by 10.** For instance, if the heat relay coil amperage reads 7 A, it is really only carrying 0.7 A, Figure 16–18.

Some ammeters have attachments to read in ohms and volts. The volt attachments can be helpful for voltage readings, but the ohm attachment should be checked to make sure that it will read in the range of ohms that is needed. Some of the ohmmeter attachments will not read very high resistance. For instance,

it may show an open circuit on a high-voltage coil that has a considerable amount of resistance.

16.7   TROUBLESHOOTING VOLTAGE IN THE LOW-VOLTAGE CIRCUIT

The voltmeter or VOM has the capability of checking continuity, milliamps, and volts. The most common applications are checking voltage and continuity. The volt scale can be used to check for the presence of voltage, and the ohmmeter scale can be used to check for continuity. A high-quality meter will give accurate readings for each measurement.

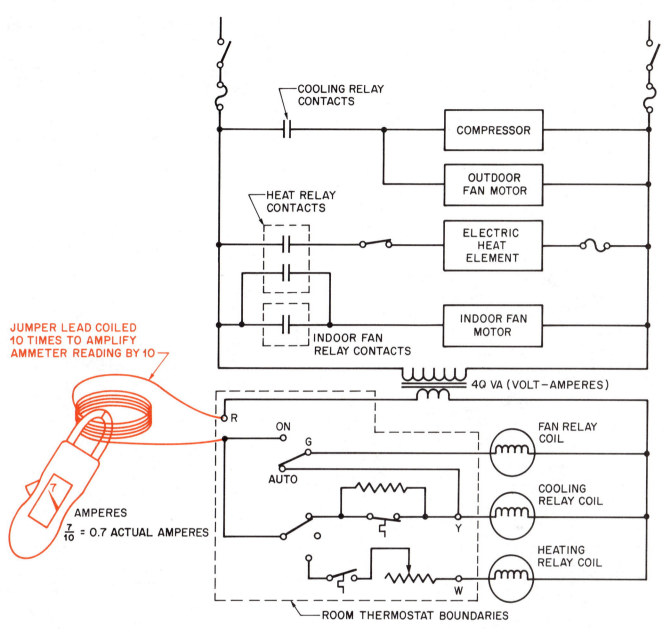

Figure 16–18. Clamp-on ammeter to measure current draw in the 24-V control circuit.

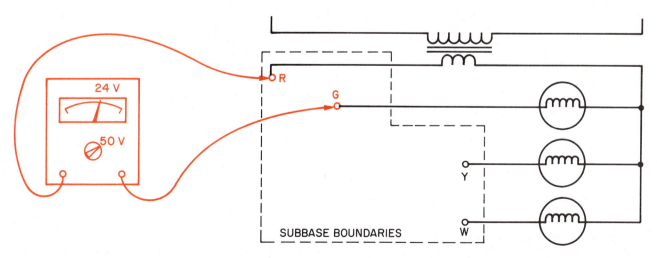

Figure 16–19. The volt-ohmmeter can be used at the thermostat location with the thermostat removed from the subbase.

Remember that any power-consuming device in the circuit must have the correct voltage to operate. The voltmeter can be applied to any power-consuming device for a voltage check by placing one probe on one side of the coil and the other probe on the other side of the coil. When power is present at both points, the coil should function. If not, a check of continuity through the coil is the next step, using the ohmmeter. Again keep in mind that when voltage is present and a path is provided, current will flow. If the flowing current does not cause the coil to do its job, the coil is defective and will have to be changed.

For example, the coil in the last service call was shorted. Current flowed, but the coil was shorted, and so much current flowed that the circuit voltage was pulled down.

The voltmeter cannot be used to much advantage at the actual thermostat because the thermostat is a closed device. When the thermostat is removed from the sub-base, the method discussed earlier of jumping the ther-

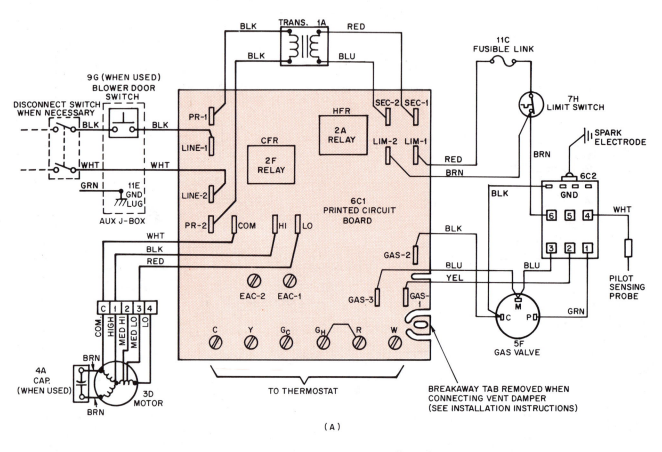

(A)

(B)

Figure 16–20. (A) A pictorial diagram showing the relative positions of the components as they actually appear. Note the terminal designation along the bottom. (B) Photo shows the actual circuit board. *Courtesy (A) BDP Company, a part of United Technologies Corporation. Photo (B) by Bill Johnson*

mostat terminals is usually more effective. With the thermostat removed from the subbase and the subbase terminals exposed, the voltmeter can be used in the following manner.

Turn the voltmeter selector switch to the scale just higher than 24 V. Attach the voltmeter lead to the hot leg feeding the thermostat, Figure 16–19. This terminal is sometimes labeled R or V. There is no standard letter or number used. When the other lead is placed on the other terminals one at a time, the circuits will be verified.

For example, with one lead on the R terminal and one on the G terminal, the terminal normally used for the fan circuit, a voltage reading of 24 V will be read. The meter is reading one side of the line straight from the transformer and the other side of the line through the coil on the fan relay. When the meter probe is moved to the circuit assigned to cooling (this terminal could be lettered Y), 24 V is read. When the meter probe is moved to the circuit assigned to heating (sometimes the W terminal), 24 V again appears on the meter. The fact that the voltage reads through the coil in the respective circuits is evidence that a complete circuit is present.

16.8 PICTORIAL AND LINE DIAGRAMS

The previous examples were all on troubleshooting the low-voltage circuit. This is recommended to start with. The control circuits that use high voltage (115 V or 230 V) work the same way. The same rules apply with more emphasis on safety. The service technician has to have a good mental picture of the circuit or a good diagram to work from. There are two distinct types of diagrams furnished with equipment: the *pictorial* diagram and the *line* diagram. Some equipment has only one diagram, some has both.

The pictorial diagram is used to locate the different components in the circuit, Figure 16–20. The example in the figure is a gas furnace with a printed circuit board, which includes an electronic air cleaner, a time delay, an electronic pilot-light ignition, and a vent

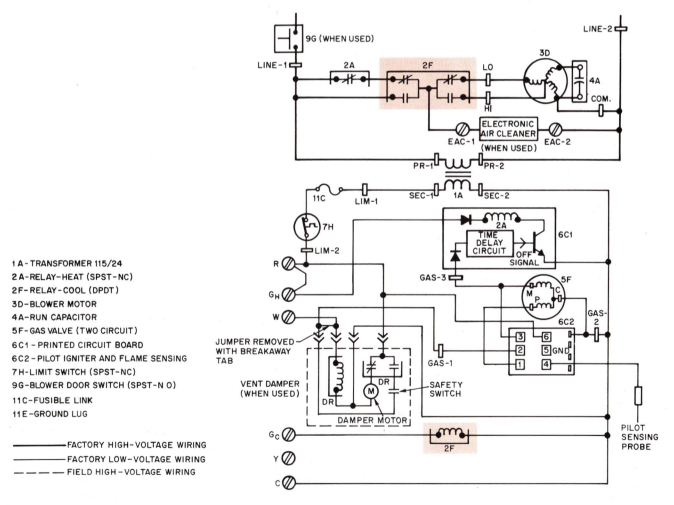

1 A – TRANSFORMER 115/24
2 A – RELAY-HEAT (SPST-NC)
2 F – RELAY-COOL (DPDT)
3 D – BLOWER MOTOR
4 A – RUN CAPACITOR
5 F – GAS VALVE (TWO CIRCUIT)
6 C1 – PRINTED CIRCUIT BOARD
6 C2 – PILOT IGNITER AND FLAME SENSING
7 H – LIMIT SWITCH (SPST-NC)
9 G – BLOWER DOOR SWITCH (SPST-N O)
11 C – FUSIBLE LINK
11 E – GROUND LUG

—————— FACTORY HIGH-VOLTAGE WIRING
——————— FACTORY LOW-VOLTAGE WIRING
– – – – – FIELD HIGH-VOLTAGE WIRING

Figure 16–21. Line diagram of Figure 16–20. Notice the arrangement of the components. The right side has no switches. The power goes straight to the power-consuming devices. Notice also that the components have to be separated to illustrate them in this manner. *Courtesy BDP Company, a part of United Technologies Corporation*

damper motor. This diagram is organized just as you would see it with the panel door open. For instance, if the diagram illustrates the control transformer in the upper portion of the picture, it will be in the upper portion of the control panel when the door is opened. This is very useful when you don't know what a particular component looks like. Study the diagram until you find the control and then locate it in the corresponding place in the control panel. The diagram also gives the wire color to further verify the component. The general outline of components is the same. In this diagram the transformer is shown at the top with four prominent wires leading to it. In the actual picture the transformer is off to the side with the four prominent wires leading to it.

The line diagram, sometimes called the *ladder* diagram, is the easier diagram to use to follow the circuit, Figure 16–21. This diagram can normally be studied briefly, and the circuit function should become obvious. All power-consuming devices are between the lines. Most manufacturers try to make the right side of the diagram a common line to all power-consuming devices. The right side of the diagram will normally have no switches in it. This can make troubleshooting the circuit more practical.

Notice that the components have to be separated to illustrate them in the line diagram style. For example, the fan relay contacts 2F at the top of the diagram in Figure 16–21 are actually operated by the 2F coil illustrated at the bottom of the diagram; follow the Gc wire to the right.

Pictorial and line diagrams are an example of the way most manufacturers illustrate the wiring in their equipment. Each manufacturer has its own way to illustrate their points of interest. The only standard that industry seems to have established is the symbols used to illustrate the various components.

Anyone studying electrical circuits could benefit by first using a colored pencil for each circuit. Every diagram can be divided into circuits by a skilled person. The colored pencil will allow an unskilled person to make a start in dividing the circuits into segments.

SUMMARY

- Each control has to be evaluated as to its purpose in a circuit.
- Electrical controls are divided into two categories: power-consuming and non-power-consuming.
- Non-power-consuming controls are known to pass power.
- One method of understanding a circuit basically is that the potential power supply is between two different power legs of different potential.

- Devices or controls that pass power are known as safety devices or operating devices.
- The light bulb controlled by a thermostat with a fuse in the circuit is an example of an operating device and a safety device in the same circuit.
- The voltmeter may be used to follow the circuit from the beginning to the power-consuming device.
- There are three separate power-consuming circuits in the low-voltage control of a typical heating and cooling fan unit: heat circuit, cool circuit, and fan circuit. The selector switch for heating and cooling decides which function will operate.
- The fan relay, cooling relay, and heat relay are all power-consuming devices.
- The low-voltage relays start the high-voltage power-consuming devices.
- The voltmeter is used to trace the actual voltage at various points in the circuit.
- The ohmmeter is used to check for continuity in a circuit.
- The ammeter is used to detect current flow.
- The pictorial diagram has wire colors and destinations printed on it. It shows the actual locations of all components.
- The line or ladder diagram is used to trace the circuit to understand its purpose.

REVIEW QUESTIONS

1. Name three types of automatic controls.
2. Name two categories of electric controls.
3. Name the circuit that is energized when a thermostat terminal is designated Y.
4. Name the circuit that is energized when a thermostat terminal is designated G.
5. Name the circuit that is energized when a thermostat terminal is designated W.
6. When a thermostat terminal is designated R, it is known by what name?
7. Is the heat anticipator in parallel or in series with the heat contacts in a thermostat?
8. Is the cool anticipator in parallel or in series with the cooling contacts in a thermostat?
9. Which anticipator is adjustable?
10. True or False: The indoor fan must always run in the cooling cycle.
11. True or False: When the control transformer is not working, the indoor fan will not run in the FAN ON position.
12. What switch should be turned off when working on the compressor circuit?
13. True or False: The pictorial wiring diagram is used to locate components from the diagram to the actual unit.

14. True or False: The line diagram is used to follow and understand the intent of the circuit.

15. How can amperage be measured in a low-voltage circuit with a clamp-on ammeter?

TROUBLESHOOTING MECHANICAL AND ELECTROMECHANICAL CONTROLS

OBJECTIVES

After studying this unit, you should be able to

- **define mechanical controls.**
- **describe electromechanical controls.**
- **distinguish between mechanical and electromechanical controls.**
- **locate information to understand a particular control.**

17.1 MECHANICAL AND ELECTROMECHANICAL CONTROLS

Mechanical and electromechanical controls are controls that have some driving method other than electricity. A high-pressure switch is an example of an electromechanical control. The switch contacts are the electrical part, and the bellows or diaphragm is the mechanical part. The mechanical action of the bellows is transferred to the switch to stop a motor when high pressures occur, Figure 17–1.

An example of a mechanical control is a water pressure regulating valve. This valve is used to maintain a preset water pressure in a boiler circuit and has no electrical contacts or connections, Figure 17–2. It acts independently from any of the other controls, yet the system depends on it as part of the team of controls that make the system function trouble free.

17.2 MECHANICAL CONTROLS

The starting point for troubleshooting mechanical controls is the same as for electrical controls: recognition and understanding—know the description and function of each control in the system. Doing this is more difficult than for electrical controls because the control diagram does not always describe mechanical controls as well as it does electrical controls.

17.3 SERVICE TECHNICIAN CALLS

Service Call 1

Water regulator valve (boiler water feed) out of adjustment. Water is seeping from the

Figure 17–1. High-pressure control. It is considered an electromechanical control. *Photo by Bill Johnson*

Figure 17–2. A water pressure regulating valve for boiler. *Photo by Bill Johnson*

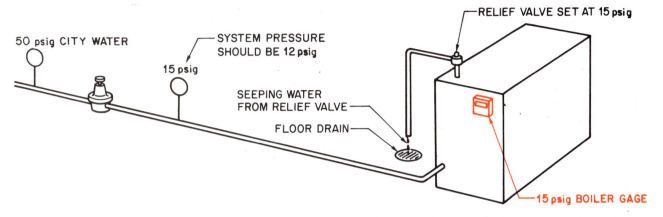

Figure 17-3. This illustration indicates how one control can cause another control to look defective. The water pressure regulating valve adjustment has been changed by mistake allowing more than operating pressure into the system. The water pressure regulating valve is designed to regulate the water pressure down to 12 psi. This valve is necessary to allow water to automatically keep the system full should a water loss develop due to a leak.

boiler's pressure relief valve, Figure 17-3. Examination of the control in the system will often reveal the purpose of the control. Notice that the water regulating valve is in the city water line entering the boiler, so it must have something to do with the feedwater system. Actually, this valve keeps the city pressure from reaching the boiler and feeds water into the system when there is a loss of water due to a possible leak. City water pressure could easily be 75 psig or more. The boiler could have a working pressure of 30 psig with a relief valve setting of 15 psig. If city water were allowed into the boiler, it would push the pressure past the design pressure of the system. The design pressure of the system may be 12 psig. This would cause the pressure relief valve to relieve the pressure and dump water from the system until the pressure was correct. In this example one mechanical automatic control is causing another control to show that the system has problems. If you did not examine the problem, you might think that the relief valve is bad because water is seeping out of it. This is not the case; the relief valve is doing its job. The water regulating valve is at fault.

The first thing to do is find out what the boiler pressure is actually supposed to be. Now check the actual pressure. There should be a reliable gage on the boiler, but you may need to add one. When you see that the pressure is above the normal working pressure, it will be evident that the relief valve is acting normally. The relief pressure stamped on the valve indicates the system's maximum working pressure. The pressure must be coming from somewhere else. The freshwater makeup could

be the suspect because the water is seeping not surging from time to time. Constant seeping indicates a constant pressure.

If you check the water regulating valve, you will notice that the adjusting screw has been tampered with. After talking with maintenance personnel, you discover that the adjustment was altered accidently. When the adjustment screw is returned to its normal operating position, the relief valve stops relieving, and the system is back to normal.

Service Call 2

Gas valve will not close. The relief valve is relieving water periodically, Figure 17-4. The boiler is a gas boiler serving an apartment

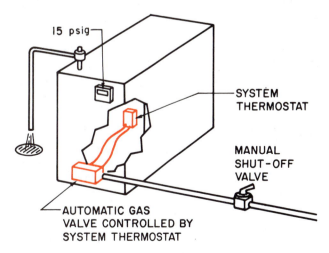

Figure 17-4. This problem is similar to the one in Figure 17-3, except that the relieving of the relief valve is not continuous.

house. The season is mild, so the boiler has a small load on it.

Because the relief valve is relieving intermittently, you may believe that there is a periodic pressure buildup in the boiler. If the problem were the same as Service Call 1, where a constant, bleeding relief occurred, the water regulating valve or the relief valve might be the suspect. This is not the case.

Notice that when the relief valve relieves, the pressure in the boiler is actually going above the set point of the relief valve, 15 psig. Examination of the boiler during the off cycle shows that the gas valve is not completely shutting off. The gas burner is still burning when the boiler's temperature control satisfies. When the gas valve is replaced, the burner shuts off completely, and the relief valve starts functioning normally.

These service calls are examples of how a trained technician might approach these service problems. Every technician develops their own methods of approach to problems. The point is that no diagram was consulted. A technician either must have prior knowledge of the system or must be able to look at the system and make accurate deductions based upon a knowledge of basic principles.

Each control has distinguishing features to give it a purpose. There are hundreds of controls and dozens of manufacturers (some of whom are out of business). If you find a control with no information as to what it does, consult the manufacturer. If this is not possible, try to find someone who has experience with the equipment. Otherwise, you'll just have to use your imagination. In general, when the control is examined in the proper perspective, the design parameters will help. For example: What is the maximum temperature or pressure that is practical? What is the minimum temperature or pressure that is practical? Ask yourself these questions when you run across a strange control; the answer may be obvious.

17.4   ELECTROMECHANICAL CONTROLS

Electromechanical controls convert a mechanical movement into some type of electrical activity. As we said earlier, these controls normally appear on the electrical print with a symbol adjacent to the electrical contact describing what the control does, Figure 17–5. These symbols are supposed to be standard, however, old equipment, installed long before standardization was considered, is still in service. You need imagination and experience to understand the intentions of the

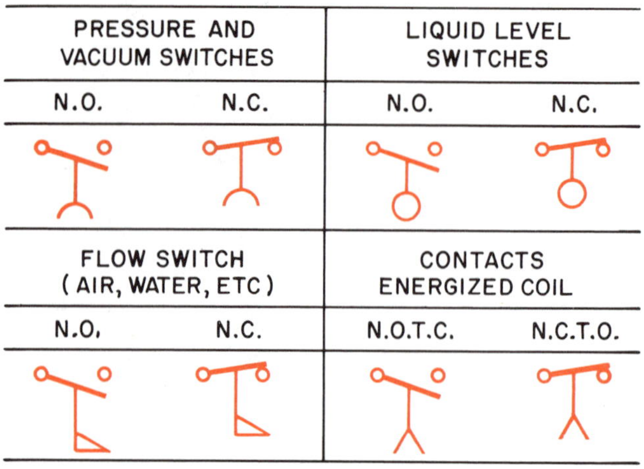

Figure 17–5. Table of electromechanical control symbols.

manufacturer. Even with standardization, it is easy to get confused.

17.5   SERVICE TECHNICIAN CALLS

Service Call 1

A residential gas furnace will not heat. The fan motor has an open winding, Figure 17–6. The technician first checks to see if there is

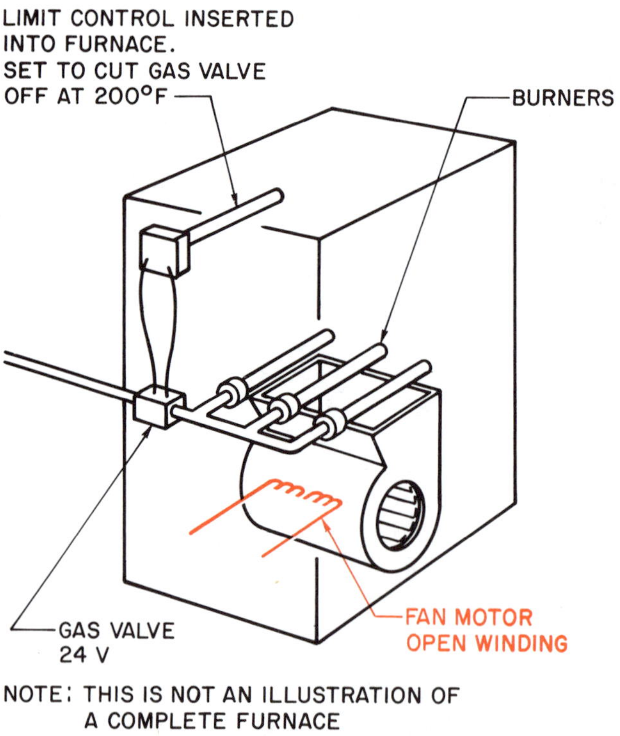

LIMIT CONTROL INSERTED
INTO FURNACE.
SET TO CUT GAS VALVE
OFF AT 200°F

NOTE: THIS IS NOT AN ILLUSTRATION OF
A COMPLETE FURNACE

Figure 17–6. When the thermostat calls for heat, the gas valve opens, and gas is ignited by a pilot light. When the furnace gets warm (approximately 140°F) the fan comes on through a temperature-operated fan switch. In this case, the fan motor has an open circuit and will not start. The temperature in the furnace continues to rise to 200°F. At this temperature the limit control cuts off the main burner.

power at the furnace and notices that although the furnace is hot to the touch, no heat is coming out of the duct. Further investigation shows that the fan is not running. A quick check shows that the fan has an open circuit and will not start up. The furnace has a *limit* control that allows the furnace to get up to about 200°F before cutting off the gas supply.

The limit control is an electromechanical control because it uses a bimetal to detect temperature. Through linkage the bimetal movement opens a set of electrical contacts and stops the heat source before dangerous temperatures occur.

Service Call 2

A central air conditioning system in a small office building will not cool because it is low on refrigerant causing the low pressure control to stop the unit, Figure 17–7. The service technician first turns the thermostat to COOL and turns the thermostat indicator down to call for cooling. The indoor fan motor starts. This indicates that the control voltage is operating and will send power on a request for cooling. Only

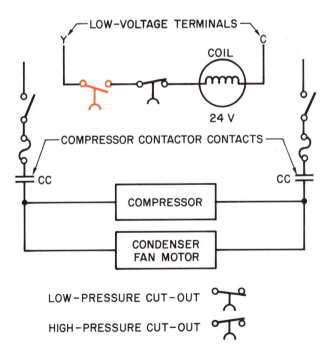

Figure 17–7. A troubleshooting problem. There is no cooling. The indoor fan will run when there is a call for cooling. This indicates that the problem is outside at the cooling unit. When the panel is removed, the wiring diagram reveals that two controls could be the problem: either the low-pressure cutout or the high-pressure cutout. Notice that both of these controls are normally closed in the symbols, but the low-pressure cutout is open in the system. This indicates that the system is low on refrigerant.

recirculated air is coming out of the ducts. The technician checks the outside unit and notices that it is not running. Obviously no cooling can be obtained without the outside unit running. Power is checked to make sure there is power to the unit.

The control panel is removed to see what controls will actually prevent the unit from starting. This gives some indication as to what the possible problem may be. There is a high-pressure cutout switch and a low-pressure cutout switch. Since the unit did not even start, it is unlikely that it is off because of high pressure. Pressure gages are applied and show that there is not enough pressure in the system to make the low-pressure cutout contacts. (Most air conditioning systems use R-22. If the system were standing at 80°F, the system should have 144 psig of pressure, according to the P & T chart for R-22 at 80°F.) If the system has a pressure of less than the standing pressure for the temperature conditions, it is low on refrigerant. Note, care should be used in the standing pressure comparison. Part of the unit is inside, at 75°F air, and part of the unit is outside. Some of the liquid refrigerant will go to the coolest place if the system valves will allow it to. *The gage readings may correspond to a temperature between the indoor and outdoor temperature but not less than the cooler temperature.* Many modern systems use fixed-opening metering devices that allow the system to equalize and the refrigerant to migrate to the coolest place in the system.

In this example let's say that the pressure in the system is 60 psig, the refrigerant is R-22, and the temperature is about 80°F. A typical pressure control setup may call for the control to interrupt the compressor when the system pressure gets down to 20 psig and to make the control and start the compressor when the pressure rises back to 100 psig. It can easily be determined that the low-pressure cutout is keeping the compressor off; no cooling is possible until the condition is corrected. The low-pressure cutout is an electromechanical control that takes a pressure reading and uses linkage to stop an electrical operation.

The foregoing examples are used as typical service problems to show how a service technician might arrive at the correct diagnosis, using the information at hand. For more detailed explanations of specific equipment, see the section on troubleshooting.

Always save and study the manufacturer's literature. There is no substitute for being aware of the manufacturer's intentions.

SUMMARY

- Mechanical controls perform mechanical functions without electricity.
- Electromechanical controls have both electrical and mechanical functions.
- Mechanical controls quite often do not appear on the unit diagrams.
- Electromechanical controls appear in the unit wiring diagram.
- Quite often there is no description of mechanical controls with the unit diagram.
- Some purely mechanical controls are water regulating valve, pressure relief valve, and expansion valve.
- Some electromechanical controls are low-pressure cutout, high-pressure cutout, high limit, thermostat.
- *When information about a control cannot be found from the manufacturer, consult an experienced person.*
- Keep and study manufacturer's literature.

REVIEW QUESTIONS

1. True or False: The term *mechanical controls* applies to controls that have no electrical components.
2. True or False: The term *electromechanical controls* applies to controls that have electrical components as well as mechanical components.
3. True or False: Mechanical controls always appear on the wiring diagram.
4. How can you find information about a control when there is no description with the unit?

Use M or E to indicate whether the following controls are mechanical or electromechanical.

5. Low-pressure cutout.
6. High-pressure cutout.
7. Water regulating valve.
8. Pressure relief valve.
9. Limit control.
10. Expansion valve.

ELECTRONIC AND PROGRAMMABLE CONTROLS

OBJECTIVES

After studying this unit, you should be able to

- describe several applications for electronic controls.
- describe why electronic controls are more applicable to some applications than are electromechanical controls.
- describe the electronic control boards used for air conditioning circuits, oil burners, and gas furnaces.
- recognize and troubleshoot a basic electronic control circuit board.
- describe programmable thermostats.

18.1 ELECTRONIC CONTROLS

Electronic controls are being used more frequently for automatic control of equipment. For many years electronic controls have been used to control larger equipment. The development of these controls has now reached the point that they are feasible for use in residential and light commercial equipment. These controls are economical, reliable, and allow efficient energy management.

Electronic controls serve the same purposes as many electrical and electromechanical controls. Operating and safety functions are the main uses, but energy management applications are becoming more popular each year. These controls normally come in circuit boards with terminal strips for external circuit connections. Some manufacturer's design these circuit boards as though they were individual controls. For troubleshooting purposes this control board can normally be treated as an individual component. Another method used by some manufacturers is to furnish a special module that can be plugged into the control circuit to locate problems. When a module is required, the technician is basically tied to this specific manufacturer for an orderly repair.

Following are examples of applications of electronic controls in residential and light commercial systems.

18.2 GAS-FURNACE PILOT LIGHTS

Pilot lights are gradually changing from the standing pilot to intermittent ignition to shut the pilot-light flame off when the furnace is off. (The summer off period can be quite lengthy in mild climate areas, and the air conditioning system must overcome the pilot light's heat.) This intermittent ignition is normally accomplished with an electronic circuit. The manufacturers furnish the best troubleshooting guide for these controls because each manufacturer has its own method of applying these controls. See Figure 18-1 for an example of an electronic circuit board used in a gas furnace.

The circuit board has some extra features. Other components may easily be added to the board, such as controls for an electronic air cleaner, a humidifier, and a vent damper shutoff. The component's wires can be connected to the board without a lot of control circuit wiring. For example, wiring an electronic air cleaner or a humidifier requires that neither component be energized unless the indoor furnace fan is running. This is called *interlocking* the components. The circuit board has connections that interlock the electronic air cleaner or the humidifier with the furnace fan in both summer (high-speed) and winter (low-speed) applications. No wiring decisions have to be made by the installing technician. The application and troubleshooting of intermittent ignition will be covered in more detail in the unit on gas heat.

18.3 OIL FURNACES

The oil-furnace industry is currently using an electronic circuit to prove the flame in oil-fired fur-

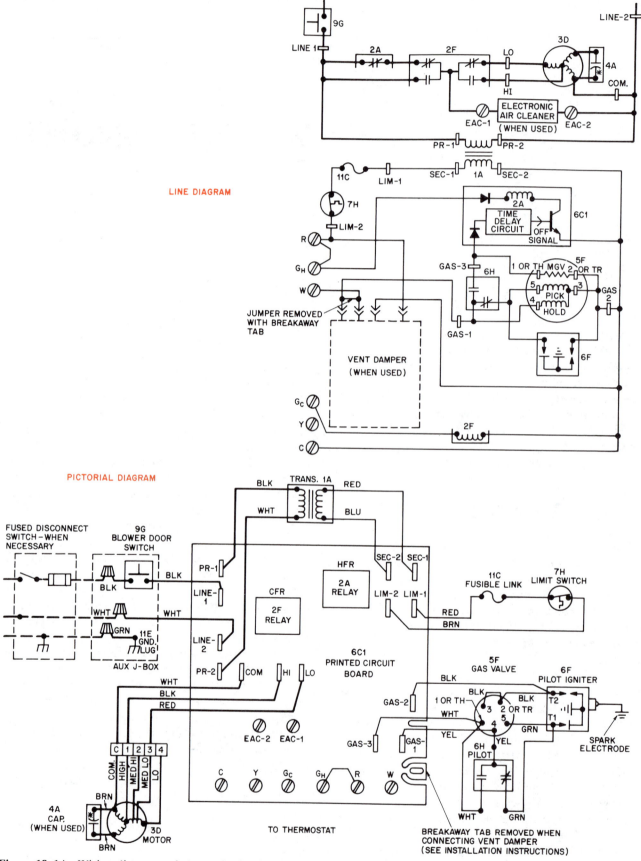

Figure 18-1A. Wiring diagrams of electronic circuit board. The terminals on the pictorial diagram are placed exactly like the ones in the actual unit. Different components can be added. This board automatically starts these components at the correct time when the fan is running. *Courtesy BDP Company, a part of United Technologies Corporation.*

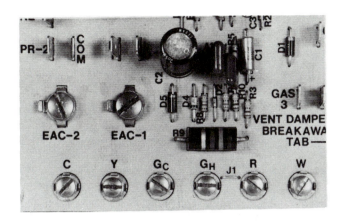

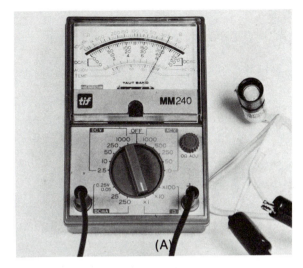

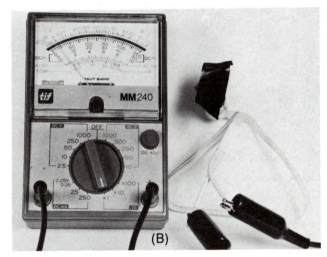

Figure 18–1B. Photos of the electronic circuit board. *Photos by Bill Johnson*

Figure 18–2. The cad cell changes resistance with light changes. (A) The cad cell with light shining in and a resistance of 20 Ω. (Note: The meter is on the R × 1 scale.) (B) The cell eye covered with electrical tape and the reading changes to 620 Ω. This cad cell is used to "see" the flame in an oil furnace. *Photos by Bill Johnson*

naces. This control is known as a *cad cell,* which stands for cadmium sulfide cell. Basically the same as the light meter in a camera, it varies in resistance with more or less light, Figure 18–2. This varying resistance is monitored by an electronic circuit to stop the oil from being injected into the furnace when there is no flame to ignite it. See Figure 18–3 for an example of the circuit board for an oil furnace. This control is called a primary control when applied to an oil furnace. The cad cell is not used on gas furnaces because of the quality of light. The oil furnace burns with a yellow flame, and the gas furnace burns with a blue flame. The cad cell does not see blue well. There is more on the cad cell in the unit on oil heat.

18.4 AIR CONDITIONING APPLICATIONS

The electronic control board is used by some manufacturers of air conditioning equipment to provide electrical protection for the equipment. For example, the electronic control can monitor the voltage being

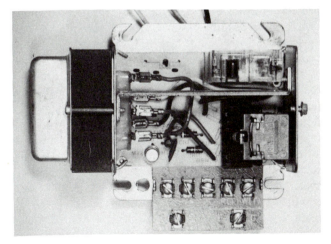

Figure 18–3. An actual oil-burner circuit board. *Photo by Bill Johnson*

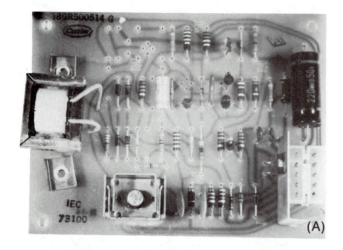

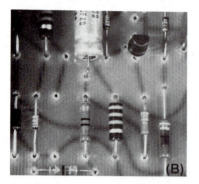

Figure 18-4. (A) Circuit board used in residential air conditioning. It has the following features: time delay, low-voltage monitor, high-voltage monitor, and amperage protection. (B) The electronics in this board resemble those in a radio or TV. The service technician cannot repair the components but will have to replace the board if the components fail. (C) Notice the pin connectors for field wiring connections. *Photos by Bill Johnson*

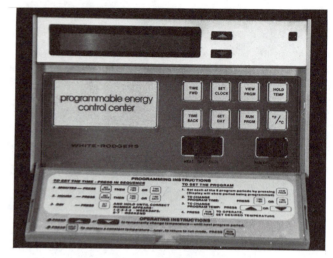

Figure 18-5. Thermostat programmable for energy management. It can be used in any 24-V circuit for one-stage cooling and one-stage heating. This is typical of one that could be installed in a residence or small business. *Photo by Bill Johnson*

supplied to a unit. While measuring the voltage, it can measure the current draw of the compressor motor to protect it from overload, Figure 18-4. Overload protection can be readily obtained with electromechanical devices, but low- and high-voltage monitoring and protection cannot.

The electromechanical controls can be connected to the circuit board in such a manner that the time-delay feature is used to keep the unit from short cycling. The time delay before a compressor restart could be something like five minutes. For example, most residential air conditioning units use fixed-bore metering devices, such as the capillary tube, that equalize during the off cycle. (The thermostatic expansion valve does not equalize completely during the off cycle, so the short cycle was not as much of a problem when that valve was used.) The unit could come right back on after a low-pressure cutout if not for the time-delay circuit. The time delay can be easily obtained with electronics.

18.5 ELECTRONIC THERMOSTATS

Electronic thermostats can be manufactured with programs for various needs at a reasonable price. This makes them attractive to the small user who wants economy and total control. They can be easily programmed to stop and start the heating and cooling equipment at many predetermined times, Figure 18-5. For example, homeowners that work may want to lower the heating or cooling while they are at work, then turn on the heat or air conditioner before arriving home. The thermostat can also cut the system back at night and bring the home back to temperature just before the residents get up in the morning. The program schedule can be altered when necessary. It will automatically return to the original program at the next scheduled change.

There are many features available that concern programming time. The temperature control itself has a faster response time than the bimetal type because of the mass of the sensing element. It could easily be a thermistor element, which is lighter than the bimetal. From a troubleshooting standpoint this thermostat is the same as a regular thermostat. The terminal designations in the subbase are basically the same. We can say that the thermostat either works or it doesn't. If it doesn't work, install a new one. The directions furnished by the manufacturer show how to check the program to make sure that the problem is not an operator's error. If the program is correct and the thermostat will not function as it was designed to, the technician can't do much to repair the electronic parts.

18.6   DIAGNOSTIC THERMOSTATS

The future is going to bring thermostats that are diagnostic centers in themselves, which will be the message centers for the computer. Sensors will be mounted in strategic locations on the unit to monitor pressures and temperatures. When the unit is beginning to experience difficulty, the computer will monitor the messages from the sensors and send a signal to the thermostat, which will display the problems that the unit is experiencing.

To anticipate trouble, the technician could call the system computer and get a readout of potential problems. Of course, cost is the big factor as to when this will happen. More elaborate commercial and industrial installations are already using this feature. As more and more homes get personal computers, the idea will become more popular.

18.7   TROUBLESHOOTING ELECTRONIC CONTROLS

Troubleshooting electronic circuit boards currently manufactured is much like troubleshooting a single component in the circuit. See Figure 18–6 for an electronic thermostat that is programmable along with the subbase. The low-voltage power supply is doing the same thing as in a conventional circuit in that it is still working to supply power to the power-consuming control circuit. The hot leg of the transformer goes directly to the thermostat, and the thermostat distributes this hot leg of power to the individual components just as in a conventional thermostat. There is still a compressor contactor, a fan relay, and the same heating components to energize.

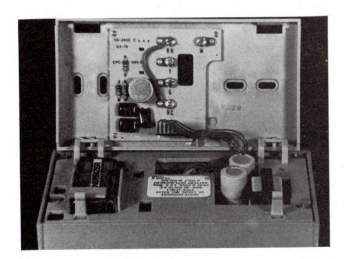

Figure 18–6. Photo showing the back of the thermostat. *Photo by Bill Johnson*

18.8   TROUBLESHOOTING THE ELECTRONIC THERMOSTAT

The following is an example of how a typical programmable thermostat for combination heating and cooling can be diagnosed for some typical problems. This could be considered a replacement thermostat for a standard heat–cool thermostat. A review of the unit on thermostats may be helpful in understanding this example.

Remember that the hot terminal feeding the thermostat provides the power to the other terminals. See Figures 18–7, 18–8 and 18–9 for circuit diagrams. As you can see from the manufacturer's literature, the installation is quite simple.

When the existing unit has two different power supplies, the transformers must be *in phase,* or they will oppose each other, Figure 18–10. Isolation relays are used to prevent stray unwanted electrical feedback that can cause erratic operation. This manufacturer recommends *isolation relays* in certain applications that have electronic spark ignition and a vent damper shutoff, Figure 18–11.

This thermostat has many features that can save money in the cooling and heating seasons by turning on the equipment only when really needed. The conventional thermostat maintains a constant temperature in the conditioned space unless it is manually changed. The electronic thermostat can be set up to maintain minimum or maximum temperatures when no one is present and then automatically change to the correct conditions just before the people return. The typical homeowner or business manager can also override the program in special cases such as a Saturday night party or Thursday night business meeting. When the program has been altered in this manner, it will automatically go back to the program at the next scheduled change. See Figure 18–12 for a list of the manufacturer's features and specifications.

18.9   PROGRAMMING THE ELECTRONIC THERMOSTAT

The programming of this thermostat is typical of most thermostats on the market, and is intended for the average person to be able to set the basic functions. Each manufacturer has its own method of programming their thermostat but most of them have similar features and similar methods to achieve them. Most of the thermostats have explicit instructions for the operator to follow. These instructions should be studied and the operator should become very aware of the various operating modes. See Figure 18–13 for some of the programming information on this thermostat. It is very flexible for the owner who does not have a predictable schedule.

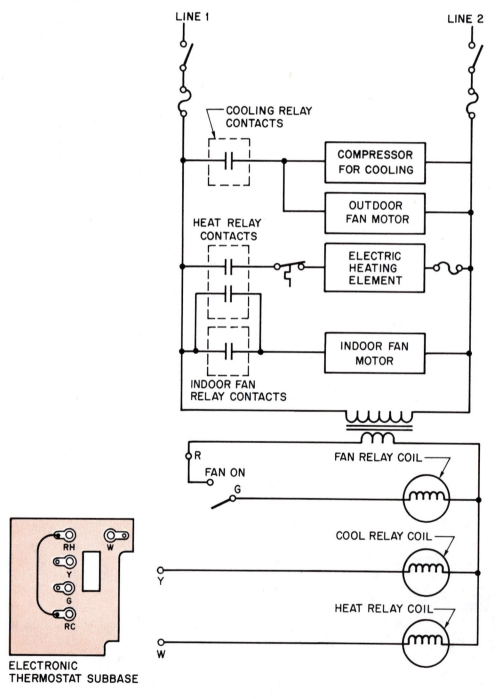

Figure 18–7. Same application as was shown in Figure 16–11, only on this diagram there are two R terminals with a jumper between them. This is a split-base thermostat that can use two 24-V power supplies. When only one is used, a jumper is applied.

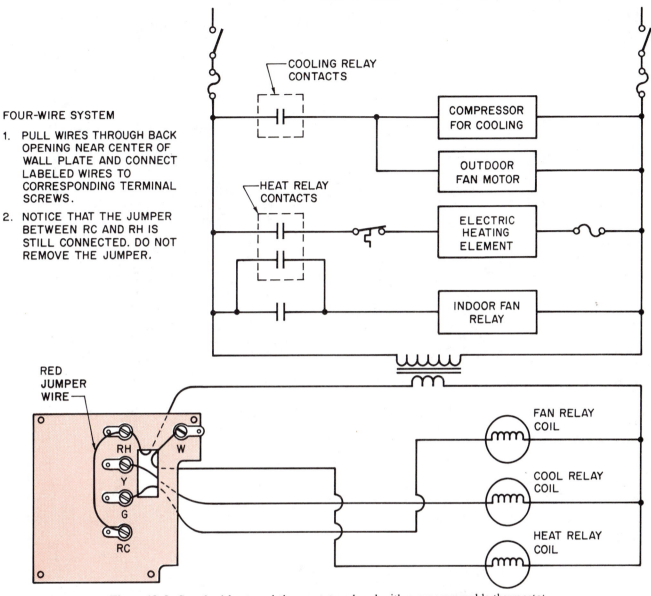

FOUR-WIRE SYSTEM

1. PULL WIRES THROUGH BACK OPENING NEAR CENTER OF WALL PLATE AND CONNECT LABELED WIRES TO CORRESPONDING TERMINAL SCREWS.

2. NOTICE THAT THE JUMPER BETWEEN RC AND RH IS STILL CONNECTED. DO NOT REMOVE THE JUMPER.

COOLING RELAY CONTACTS

HEAT RELAY CONTACTS

COMPRESSOR FOR COOLING

OUTDOOR FAN MOTOR

ELECTRIC HEATING ELEMENT

INDOOR FAN RELAY

RED JUMPER WIRE

RH
W
Y
G
RC

FAN RELAY COIL

COOL RELAY COIL

HEAT RELAY COIL

Figure 18-8. Standard heat-cool thermostat replaced with a programmable thermostat.

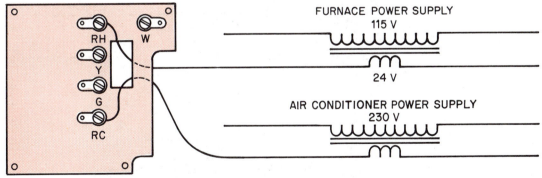

RH
W
Y
G
RC

FURNACE POWER SUPPLY
115 V

24 V

AIR CONDITIONER POWER SUPPLY
230 V

FIVE-WIRE SYSTEM WHERE SEPARATE TRANSFORMERS ARE USED ON HEATING SYSTEM AND COOLING SYSTEM.
1. REMOVE AND DISCARD THE RED JUMPER WIRE THAT CONNECTS TERMINALS RH AND RC ON WALL PLATE.
2. PULL WIRES THROUGH BACK OPENING NEAR CENTER OF WALL PLATE AND CONNECT LABELED WIRES TO CORRESPONDING TERMINAL SCREWS.

Figure 18-9. The above illustrates an application with two 24-V power supplies. One power supply comes to R (heat), and the other to R (cool). The thermostat and subbase keep the two supplies from coming together. (If the two power supplies are phase checked, it is permissible to wire them together. Phase checking involves making sure that the primaries of the transformers are parallel and that the secondaries are parallel, in phase.)

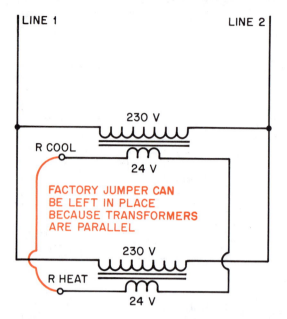

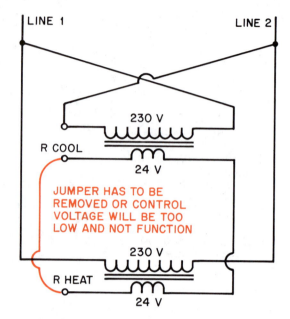

EXAMPLE OF TWO LOW-VOLTAGE
TRANSFORMERS THAT ARE PHASED TOGETHER
BECAUSE THEY ARE EXACTLY PARALLEL ON
PRIMARY AND SECONDARY.

(A)

EXAMPLE OF TWO LOW-VOLTAGE
TRANSFORMERS THAT ARE NOT PHASED
TOGETHER IN PARALLEL.

(B)

Figure 18-10. The second example can be corrected by changing the primary or secondary wiring. Sparks will be emitted if the secondary wiring is wrong.

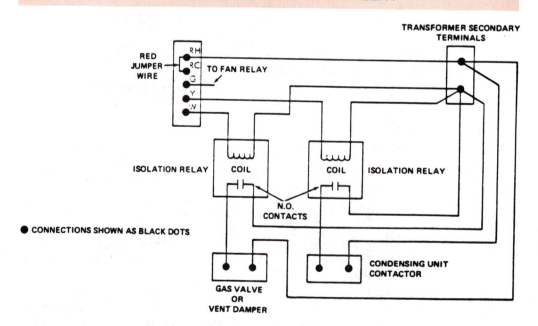

Figure 18-11. Wiring diagram showing use of an isolation relay.

SPECIAL FEATURES

- Separate set-back programming for five day week and two day weekend.
- Four separate time/temperature settings per 24 hour period.
- LCD displays continuous set point, time and room temperature alternately.
- Manual temperature control for overriding program.
- Fahrenheit or Celsius temperature display.
- Sensor for 100° F protection.

- Independently adjustable anticipation for heating and cooling.
- Compressor short cycle protection.
- Pre-programmed thermostat.
- Heating set point resets to 68° F (power loss).
- Cooling set point resets to 78° F (power loss).
- Indicators for "Hold Temp" — "System Cycle".
- Battery backup option.

SPECIFICATIONS

Electrical Rating: 17 to 30 volts 60 Hz
0.15 to 1.5 Amperes
WARNING: Do not use on circuits exceeding 30 volts.
Higher voltage will damage control—could
cause shock or fire hazard.
Temperature Range: 40° F to 90° F
4° C to 32° C
Rated Differential: 0.3° to 1.8° F—Heating
0.8° to 1.8° F—Cooling
Mountings: Wiring wall plate mounts on wall.
Dimensions: 6-3/8″ W × 1-3/4″ D × 3-1/2″ H

Figure 18–12. An example of features and specifications as shown by one manufacturer. *Courtesy White-Rodgers Division, Emerson Electric Co.*

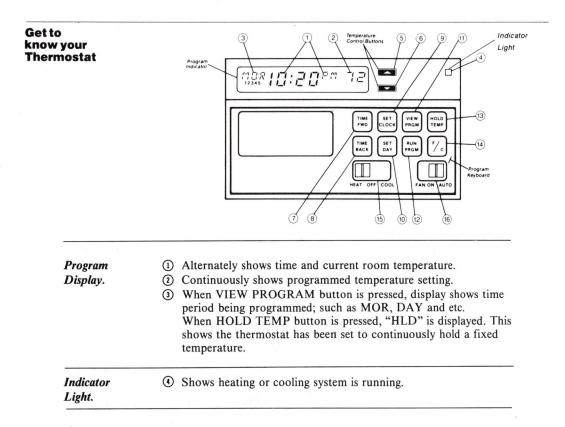

Program Display.	① Alternately shows time and current room temperature.
	② Continuously shows programmed temperature setting.
	③ When VIEW PROGRAM button is pressed, display shows time period being programmed; such as MOR, DAY and etc.
	When HOLD TEMP button is pressed, "HLD" is displayed. This shows the thermostat has been set to continuously hold a fixed temperature.
Indicator Light.	④ Shows heating or cooling system is running.

Figure 18–13. Manufacturer's instructions for a programmable thermostat. *Courtesy White-Rodgers Division, Emerson Electric Co.*

Temperature Control Buttons.	⑤ ▲	Red arrow raises temperature setting. (90°F or 32°C maximum)
	⑥ ▼	Blue arrow lowers temperature setting. (40°F or 4°C minimum)

Program Control Buttons.	⑦ TIME FWD	For advancing time to change program setting.
	⑧ TIME BACK	For moving back time to change program setting.
	⑨ SET CLOCK	For setting current time.
	⑩ SET DAY	For setting current day of the week.
	⑪ VIEW PRGM	Press and hold to review weekday/weekend programming for all 8 program time/temperature schedules.
	⑫ RUN PRGM	Push to start your thermostat when programming is complete.
	⑬ HOLD TEMP	To temporarily override all program temperatures without changing the weekly program.
	⑭ F/C	For setting the display to show either Fahrenheit or Celsius temperature.
	⑮ HEAT OFF COOL	Sets the thermostat to control either your furnace or your air conditioner.
	⑯ FAN ON AUTO	For selecting continuous or automatic fan operation.

Planning Your Personal Schedule	Before you begin to program the thermostat you must plan your program. To do this, fill out the **Personal-Use Chart** on the next page as you answer the following questions:

Now, let's plan your heating season schedule for weekdays:	1. Morning a. What time does the first member of your household usually get up in the morning? (We suggest that you program the thermostat 1/2 hour before this time, so the house reaches the temperature you want at the desired time.) b. What temperature would you like the house, at this time? 2. Day a. What time does the last person usually leave the house for the day (for work, school, etc.)? b. What temperature would you like the house, at this time? (**NOTE:** If someone stays home all day, you can program this temperature to be the same as in the morning.) 3. Evening a. What time does your first family member return home? (We suggest that you program the thermostat 1/2 hour before this time, so the house reaches the temperature you want at the desired time.) b. What temperature would you like the house, at this time? (**NOTE:** If someone stays home all day, you can program this temperature to be the same as in the morning). 4. Night a. What time does the last member of your family usually go to bed? b. What temperature would you like the house, at this time?

Planning your weekend and cooling season schedules:	WHEN PLANNING YOUR INDIVIDUAL WEEKEND & COOLING SEASON SCHEDULES, SIMPLY REPEAT STEPS 1 THROUGH 4. Remember, for the greatest energy savings:

Figure 18-13. *(continued)*

—when *heating* your home, program the temperature to be cooler when you are gone.

—when *cooling* your home, program the temperature to be warmer when you are gone.

Pre-Programmed Time and Temperature

Your thermostat is pre-programmed with Time and Temperatures for Heating and Cooling Programs that are typical of the average residential user's lifestyle as follows:

HEATING Program for All Days of the Week:			COOLING Program for All Days of the Week:		
STEP	**TIME**	**TEMP**	**STEP**	**TIME**	**TEMP**
MOR	5:00 A.M.	70	MOR	5:00 A.M.	78
DAY	9:00 A.M.	70	DAY	9:00 A.M.	82
EVE	4:00 P.M.	70	EVE	4:00 P.M.	78
NTE	10:00 P.M.	62	NTE	10:00 P.M.	78

NOTE: If "System Switch" is in "HEAT" or "OFF" position, when thermostat is installed, the pre-program shown will be for **Heating Program Only.** (Program will be shown when view program button is pressed.)

NOTE: If "System Switch" is in "COOL" position when thermostat is installed, the pre-program shown will be for **Cooling Program Only.** (Program will be shown when view program button is pressed.)

PERSONAL-USE CHART

Weekdays (1–5)

	HEATING		COOLING	
	TIME	**TEMP**	**TIME**	**TEMP**
1. Morning	a._____	b._____	a._____	b._____
2. Day	a._____	b._____	a._____	b._____
3. Evening	a._____	b._____	a._____	b._____
4. Night	a._____	b._____	a._____	b._____

Weekends (6 & 7)

	TIME	TEMP	TIME	TEMP
1. Morning	a._____	b._____	a._____	b._____
2. Day	a._____	b._____	a._____	b._____
3. Evening	a._____	b._____	a._____	b._____
4. Night	a._____	b._____	a._____	b._____

Figure 18–13. *(continued)*

18.10 POWER OUTAGE AND THE ELECTRONIC THERMOSTAT

If the power goes off, is the program ruined? No. Manufacturers use various methods to prevent this. For example, in Figure 18–14 the thermostat is backed up with a battery for up to three days. When a long power failure occurs, the program *is* likely to be lost. The owner's manual shows how to set up the program again. Some thermostats maintain an acceptable set temperature until the program is reestablished. This protects the premises from overheating or overcooling (maybe below freezing).

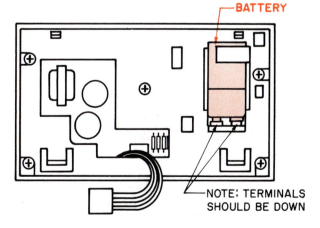

Figure 18-14. An optional 9-V battery can be installed as a backup for power failure. *Photo by Bill Johnson*

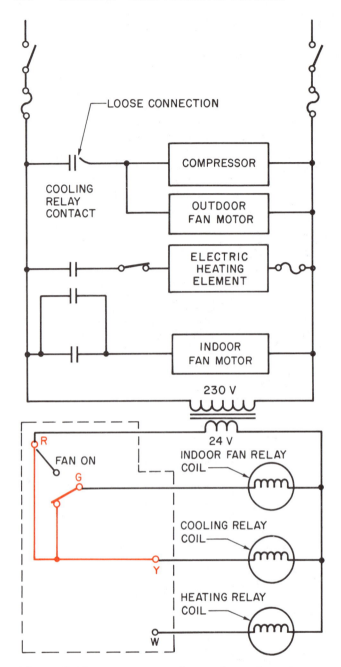

Figure 18–15. A loose connection at the cooling relay terminal does not allow power to get to the compressor and outdoor fan motor. The coil is energized because it is humming. This represents what happens inside the electronic circuit on a call for cooling. R passes power to G and Y.

18.11   SERVICE TECHNICIAN CALLS

Service Call 1

The unit is not cooling. A wire is burned off the cooling relay contacts (compressor) due to a bad connection. Again, the package unit will be used as in the example in Unit 17, Figure 18–15.

1. Go to the indoor thermostat, which is a programmable type. Switch the indoor fan to ON. It runs.
2. Turn the thermostat to call for cooling. Note that the indoor fan comes on again. This proves that the control voltage and thermostat are trying to operate the equipment.
3. Go to the outdoor unit. The cooling relay is humming. This means that the low-voltage call for cooling is reaching the outdoor unit.
4. *Turn off the power and remove the cover.* Turn on the power. Notice that the cooling relay is energized. Line power has to be reaching the unit because it is the power source for the control transformer. So check the line power into the relay and out.
5. Power is going in but not coming out. The relay has a bad connection. *Turn off the power and repair the connection.*

Service Call 2

The system will not cool. The thermostat will not pass power, Figure 18–16.

1. Go to the indoor thermostat and switch on the indoor fan relay. The indoor fan starts up. This proves that the line voltage is reaching the outdoor unit because the 24-V control transformer is located in the outdoor unit.
2. Turn the thermostat to call for cooling. Neither the indoor fan nor the outdoor unit starts.
3. Go to the outdoor unit, remove the panel, and check for voltage at the cooling relay coil. There is no voltage.
4. Return to the indoor thermostat and check the program according to the manufacturer's instructions. The program is correct. Remove the thermostat from the subbase. Place a jumper between the R terminal and the G terminal. The indoor fan starts. Leave the jumper on the fan circuit and place another jumper between the R terminal and the Y terminal. The outdoor fan and compressor start and run with the indoor fan. Obviously the thermostat is not passing power and needs to be replaced.

As with all equipment and controls, the manufacturer has certain intentions. These intentions are usually printed in the literature. Get acquainted with the written material before starting any installation or service job. The manufacturer or his distributor will also be helpful if you have questions.

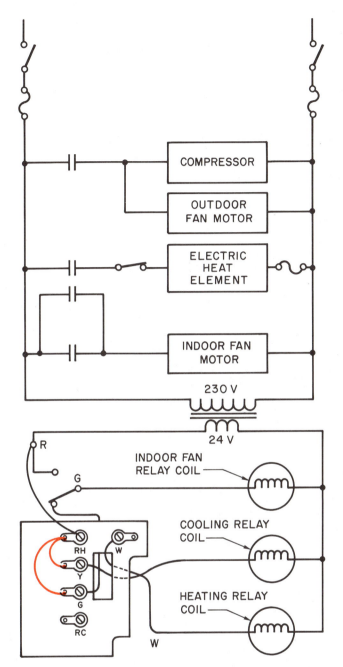

Figure 18–16. The 24-V control voltage is going into the thermostat but will not come out upon a call for cooling. The thermostat is removed from the subbase and a jumper attached from R to G, and the indoor fan starts. Another jumper is attached from R to Y, and the outdoor unit starts.

SUMMARY

● Electronic controls are being used more frequently because they are reliable and economical.
● Electronic controls are basically used to replace electrical and electromechanical controls where they are applicable.

● Electronic controls are basically used in a safety and operating capacity, with more emphasis on energy management.
● Electronic controls can monitor high and low voltages and easily add time delays and sequence of operation to the control system.
● The residential and light commercial air conditioning control circuit uses electronics to monitor high and low voltages, time delay, and current draw of the compressor. It has the connection board for electromechanical controls where applicable.
● The oil furnace is using the electronic control circuit to achieve flame-proving methods.
● Oil-burner equipment uses the cad cell to see the flame and report to the electronic circuit.
● The gas furnace can use electronic controls to monitor gas flame for flame-proving methods.
● The gas furnace uses several different methods to prove the flame and report to the electronic circuit board.
● The electronic programmable thermostat is becoming one of the most attractive electronic devices.
● Troubleshooting electronic controls is similar to troubleshooting electromechanical controls because each control is normally a component; a circuit goes in and comes out.

REVIEW QUESTIONS

Answer questions 1–9 either true or false.

1. Electronic controls are normally more economical than electromechanical controls.
2. Electronic thermostats react faster than bimetal thermostats.
3. The cad cell sees darkness and shuts off the oil burner when there is no flame.
4. The gas furnace can use electronic flame detection.
5. Electronic programmable thermostats can be programmed for only one program.
6. Electronic programmable thermostats can normally be installed in place of a typical thermostat.
7. Can electronic circuits generally be repaired in the field?
8. Electronic programmable thermostats are designed so the owner can reprogram them.
9. A qualified technician should be called to troubleshoot electronic circuit problems.
10. Which of the following control objectives are best obtained with electronic controls: time delay, low-pressure monitoring, high-voltage protection, low-voltage protection, vibration protection, temperature sensing?

Section Four
Commercial Refrigeration

THE COMPRESSION CYCLE, EVAPORATORS

OBJECTIVES

After studying this unit, you should be able to

- define refrigeration.
- define high-, medium-, and low-temperature refrigeration.
- identify the boiling temperature in an evaporator.
- work with different types of evaporators.
- describe multiple- and single-circuit evaporators.

19.1 REFRIGERATION

The word *refrigeration* describes the intentional moving of heat. When heat is allowed to move naturally, it moves from a warm place to a colder place. Heat does not naturally move from a cold place to a warm place. It has to be physically moved. To remove heat from a refrigerator at a temperature of 35°F and release this heat into a kitchen at 75°F, the heat has to be forcibly moved. This movement requires energy. Heat is moved *up the temperature hill* from one place to another, Figure 19–1. **Refrigeration is transferring heat from a place where it is not wanted to a place where it makes no difference.**

The household refrigerator absorbs heat from the food products and releases this heat into the kitchen where it makes no difference. The amount of heat released is so small it is not even noticed. Heat added to the inside of the refrigerator (e.g., the conduction heat gained through the wall of a refrigerator) is moving naturally from a warm room to the cooler refrigerator. In the refrigerator the heat is absorbed into the system, pumped to the outside of the box, and released into the room. The heat is actually absorbed into the refrigerant while the refrigerant is changing from a liquid to a gas. This gas is then piped to the outside of the box where the heat is released, Figure 19–2. Pumping this heat-laden gas up the temperature hill is what makes refrigeration different from a typical exchange in heat between two substances. It takes work to move the heat up the temperature scale. An understanding of Unit 3 is vital at this point.

When observing any refrigeration system, keep in mind that **when heat is being absorbed into a refrigeration system, it also has to be rejected from the system.** Three noticeable components make a working refrigerating system: the *evaporator* for absorbing the heat, the *compressor* for pumping the heat-laden vapor, and the *condenser* for rejecting the heat. See Figure 19–3 for an illustration of the absorbing-pumping-rejection of heat using a sponge as an example. While any study of a refrigeration system is taking place, these components have to be kept in mind. The fourth component, the *expansion device,* is important, but it does not have the active noticeable functions of the other three components.

207

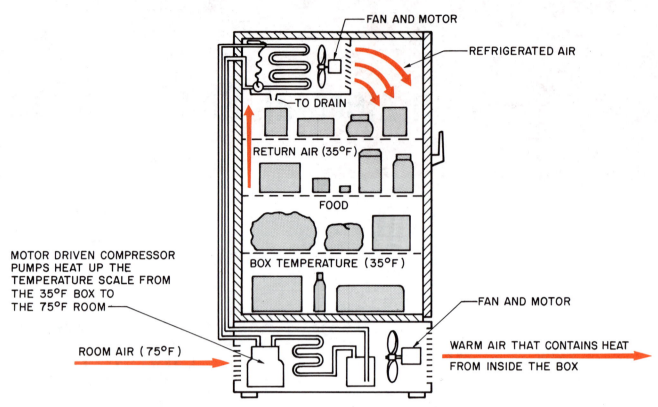

Figure 19–1. Heat normally flows from a warm place to a colder place. When it is desirable for heat to be moved from a colder to a warmer place, the heat has to be moved by force. The compressor in the refrigeration system is the pump that pumps the heat up the temperature hill by force. These compressors are normally electric motor driven.

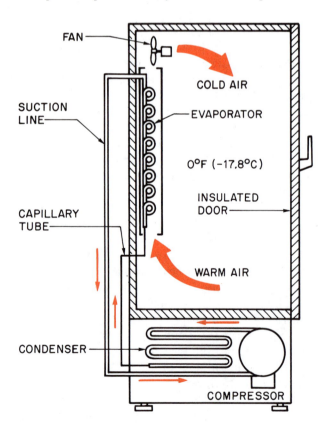

Figure 19–2. Illustration of how the refrigeration compression cycle is used to pump heat in the refrigerant given up by the foodstuff inside the refrigerator.

Figure 19–3. How a sponge absorbs water. The water can then be carried in the sponge to another place. When the sponge is squeezed, the water is rejected to another place. The squeezing of the sponge is considered the energy that it takes to pump the water.

19.2 REFRIGERATION AND FOOD PRESERVATION

The word "refrigeration" is used in conjunction with the food industry. As we said earlier, when heat is moved up the temperature scale it is called refrigeration. In the strictest sense this is true. However, in the industry the people involved in comfort cooling are called air conditioning people. Refrigeration practices and theory are applied to comfort cooling, but it is called *air conditioning*. One outstanding difference in refrigeration and air conditioning is the fact that refrigeration has to operate year around. The food or beverage compartment is normally in a conditioned space such as a store or restaurant. This space is kept the same temperature all year, so the refrigeration has about the same job all year.

19.3 REFRIGERATION AS AIR CONDITIONING

Refrigeration normally is associated with the responsibility of cooling substances to temperatures lower than the air conditioning industry systems. The air conditioning design temperature is normally in the range of a 40°F evaporator coil. The 40°F evaporator coil cools some of the room air down to about 55°F and mixes it with the remaining room air to maintain room conditions of about 75°F, Figure 19–4. Refrigeration temperature ranges normally start at 35°F evaporator temperatures to cool the air or substances down to about 50°F; this is called *high-temperature* refrigeration. Many components are interchangeable between air conditioning systems and high-temperature refrigeration systems. Some high-temperature system applications would be the storage of flowers or candy. It is desirable to keep these products cool but not cold.

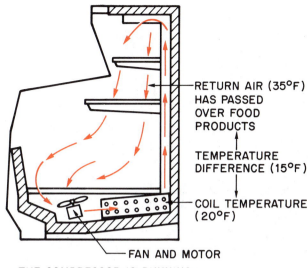

RETURN AIR (35°F) HAS PASSED OVER FOOD PRODUCTS

TEMPERATURE DIFFERENCE (15°F)

COIL TEMPERATURE (20°F)

FAN AND MOTOR

THE COMPRESSOR IS RUNNING — THIS LOWERS COIL TEMPERATURE

Figure 19–5. Relationship of the coil's boiling temperature to the air passing over the coil while operating in the design range.

19.4 TEMPERATURE RANGES OF REFRIGERATION

When the evaporator temperature is operated below 35°F, it is approaching the freezing temperature of water. The coil has to operate at a colder temperature than the medium to be cooled for the heat to transfer into the evaporator coil, Figure 19–5. If the water freezes on the coil, some method of defrost has to be devised to remove the ice. When the evaporator is operated near or below the freezing point, it is called *medium-temperature* refrigeration. Most medium-temperature refrigeration applications have some means to remove the ice from the coil, the most common is

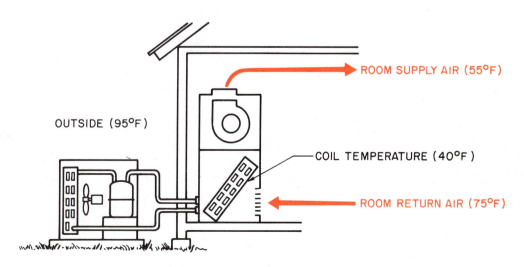

ROOM SUPPLY AIR (55°F)

OUTSIDE (95°F)

COIL TEMPERATURE (40°F)

ROOM RETURN AIR (75°F)

Figure 19–4. Air conditioning example of refrigeration.

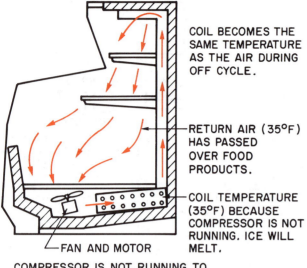

COIL BECOMES THE SAME TEMPERATURE AS THE AIR DURING OFF CYCLE.

RETURN AIR (35°F) HAS PASSED OVER FOOD PRODUCTS.

COIL TEMPERATURE (35°F) BECAUSE COMPRESSOR IS NOT RUNNING. ICE WILL MELT.

FAN AND MOTOR

COMPRESSOR IS NOT RUNNING TO LOWER THE COIL TEMPERATURE.

Figure 19–6. Relationship of the refrigerant and the air in the cooler during the off cycle.

called *off-cycle defrost*. Off-cycle defrost uses the space-temperature air to do the defrosting. The temperature of the conditioned space is above freezing. The temperature of a medium-temperature cooler only goes down to about 35°F air or box temperature. See Figure 19–6 for the temperature relationships between the coil and the air during the off cycle. The coil temperature would be about 15° to 20°F for a 35°F box (or air) temperature.

Refrigeration that operates at a temperature of around 0°F is called *low-temperature* refrigeration. Ice removal (defrost) has to come from heat added to the system from external sources, such as electric heat, Figure 19–7. Low-temperature refrigeration has different components than high- and medium-temperature refrigeration. One difference is the fin spacing in the evaporator coil, which allows some ice buildup. Another is the defrost heaters and controls, Figure 19–8.

The temperature ranges used by the industry are as

follows: High-temperature-range evaporators–35° to 50°F; Medium-temperature-range evaporators–10° to 35°F; Low-temperature-range evaporators–below 10°F. These are actual coil temperatures. Box air temperatures are normally 15° to 20°F higher than the coil temperature.

19.5 BOILING AND CONDENSING

The two main points used in understanding refrigeration are (1) boiling temperature and (2) condensing

(A)

(B)

Figure 19–8. Fin spacing of (A) a low-temperature evaporator and (B) a medium-temperature evaporator. *Photos by Bill Johnson*

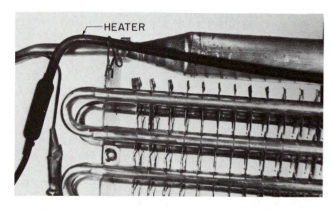

HEATER

Figure 19–7. Heaters used for the defrost of evaporators that operate below freezing. *Photo by Bill Johnson*

temperature. The boiling temperature and its relationship to the system takes place in the evaporator and will be discussed first. The condensing temperature takes place in the condenser. These temperatures can be followed by using the pressure and temperature chart in conjunction with a set of refrigeration pressure gages, Figures 19–9, 19–10.

19.6   THE EVAPORATOR AND BOILING TEMPERATURE

The *boiling temperature* of the liquid refrigerant determines the coil operating temperature. In the air conditioning example in Unit 3 a 40°F evaporator coil with 75°F air passing over it gave conditions used for air conditioning or high-temperature refrigeration. Boiling is normally associated with high temperatures and water. Unit 3 discussed the fact that water boils at 212°F at atmospheric pressure. It also discussed the fact that water boils at other temperatures, depending

on the pressure. When the pressure is reduced, water will boil at 40°F. This is still boiling—changing a liquid to a vapor. In a refrigeration system the refrigerant may boil at 20°F by absorbing heat from the 35°F food.

The service technician has to be able to determine what operating pressures and temperatures are correct for the various systems being serviced under different load conditions. This comes from experience. When the thermometers and gages are observed, the readings have to be evaluated. There can be as many different readings as there are changing conditions.

There are some guidelines that can help the technician know the pressure and temperature ranges at which the equipment should operate. Relationships exist between the entering air temperature and the evaporator for each system. These relationships are similar from installation to installation. For example, in the air conditioning example in Unit 3, the boiling

TEMPERATURE °F	REFRIGERANT 12	22	502	TEMPERATURE °F	REFRIGERANT 12	22	502	TEMPERATURE °F	REFRIGERANT 12	22	502
−60	19.0	12.0	7.2	12	15.8	34.7	43.2	42	38.8	71.4	83.8
−55	17.3	9.2	3.8	13	16.4	35.7	44.3	43	39.8	73.0	85.4
−50	15.4	6.2	0.2	14	17.1	36.7	45.4	44	40.7	74.5	87.0
−45	13.3	2.7	1.9	15	17.7	37.7	46.5	45	41.7	76.0	88.7
−40	11.0	0.5	4.1	16	18.4	38.7	47.7	46	42.6	77.6	90.4
−35	8.4	2.6	6.5	17	19.0	39.8	48.8	47	43.6	79.2	92.1
−30	5.5	4.9	9.2	18	19.7	40.8	50.0	48	44.6	80.8	93.9
−25	2.3	7.4	12.1	19	20.4	41.9	51.2	49	45.7	82.4	95.6
−20	0.6	10.1	15.3	20	21.0	43.0	52.4	50	46.7	84.0	97.4
−18	1.3	11.3	16.7	21	21.7	44.1	53.7	55	52.0	92.6	106.6
−16	2.0	12.5	18.1	22	22.4	45.3	54.9	60	57.7	101.6	116.4
−14	2.8	13.8	19.5	23	23.2	46.4	56.2	65	63.8	111.2	126.7
−12	3.6	15.1	21.0	24	23.9	47.6	57.5	70	70.2	121.4	137.6
−10	4.5	16.5	22.6	25	24.6	48.8	58.8	75	77.0	132.2	149.1
−8	5.4	17.9	24.2	26	25.4	49.9	60.1	80	84.2	143.6	161.2
−6	6.3	19.3	25.8	27	26.1	51.2	61.5	85	91.8	155.7	174.0
−4	7.2	20.8	27.5	28	26.9	52.4	62.8	90	99.8	168.4	187.4
−2	8.2	22.4	29.3	29	27.7	53.6	64.2	95	108.2	181.8	201.4
0	9.2	24.0	31.1	30	28.4	54.9	65.6	100	117.2	195.9	216.2
1	9.7	24.8	32.0	31	29.2	56.2	67.0	105	126.6	210.8	231.7
2	10.2	25.6	32.9	32	30.1	57.5	68.4	110	136.4	226.4	247.9
3	10.7	26.4	33.9	33	30.9	58.8	69.9	115	146.8	242.7	264.9
4	11.2	27.3	34.9	34	31.7	60.1	71.3	120	157.6	259.9	282.7
5	11.8	28.2	35.8	35	32.6	61.5	72.8	125	169.1	277.9	301.4
6	12.3	29.1	36.8	36	33.4	62.8	74.3	130	181.0	296.8	320.8
7	12.9	30.0	37.9	37	34.3	64.2	75.8	135	193.5	316.6	341.2
8	13.5	30.9	38.9	38	35.2	65.6	77.4	140	206.6	337.2	362.6
9	14.0	31.8	39.9	39	36.1	67.1	79.0	145	220.3	358.9	385.0
10	14.6	32.8	41.0	40	37.0	68.5	80.5	150	234.6	381.5	408.4
11	15.2	33.7	42.1	41	37.9	70.0	82.1	155	249.5	405.1	432.9

Vacuum — Red Figures
Gage Pressure — Bold Figures

Figure 19–9. Pressure and temperature chart in in. Hg vacuum or psig.

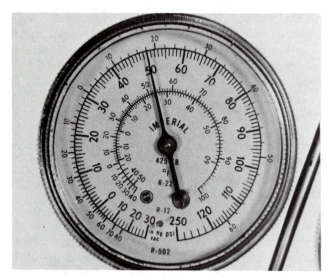

Figure 19-10. Pressure gages have pressure and temperature relationship printed on the gage. *Photo by Bill Johnson*

refrigerant of 40°F and a return air of 75°F gave a leaving air of about 55°F. In this example the return air is 75°F, and the coil temperature is 40°F. The difference of 35°F between the entering air and the boiling refrigerant is characteristic of the air conditioning industry and is good for removing moisture from the air. The coil to boiling temperature relationships in refrigeration coils is normally 10° to 20°F difference between the boiling refrigerant and the entering air temperature.

19.7 REMOVING MOISTURE

Dehumidifying the air means to remove the moisture, and this is frequently desirable. Moisture removal is similar from one air conditioning system to another.

Knowing this relationship can help the technician know what the suction pressure gage should be reading at various conditions, Figure 19-11. In the figure the load on the coil would rise or fall accordingly as the return air temperature rises or falls. For example, if the house got very warm before the air conditioning system started, the coil would have more heat to remove. It has more load on it. It would be very much like boiling water in an open pan on the stove. The water boils at one rate with the burner on medium and at an increased rate with the burner on high. The boiling pressure stays the same in the boiling water in a pan because the pan is open to the atmosphere. When this same boiling process takes place in an enclosed coil, the pressures will rise when the boiling takes place at a faster rate. This causes the operating pressure of the whole system to rise, Figure 19-12.

The refrigeration evaporator is a component that absorbs heat for the system. This heat must be rejected from the system by the condenser. The evaporator can be thought of as the *sponge* of the system. It is responsible for a heat exchange between the conditioned space or product and the refrigerant inside the system. Some evaporators absorb heat more efficiently than others. See Figure 19-13 for an illustration of this heat exchange between air and refrigerant.

19.8 HEAT EXCHANGE CHARACTERISTICS OF THE EVAPORATOR

Conditions that govern the rate of heat exchange are as follows:

1. The evaporator *material* through which the heat has to be exchanged. Evaporators may be manufactured from copper, steel, brass, stainless steel, or aluminum. Corrosion is one factor that determines what material is used. For instance, when acidic materials

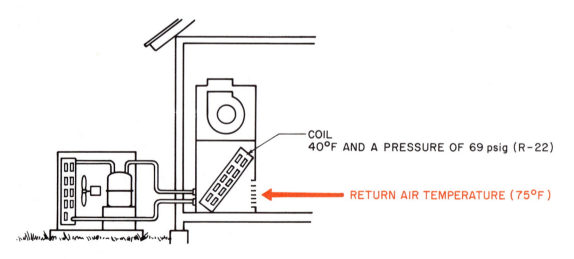

Figure 19-11. Coil-to-air temperature relationship under normal working conditions.

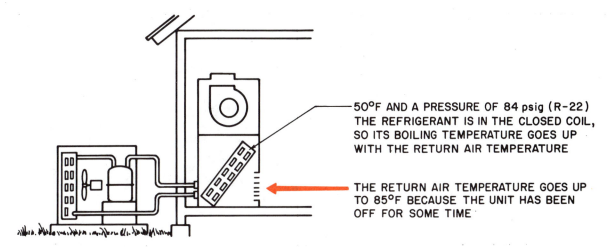

50°F AND A PRESSURE OF 84 psig (R-22)
THE REFRIGERANT IS IN THE CLOSED COIL,
SO ITS BOILING TEMPERATURE GOES UP
WITH THE RETURN AIR TEMPERATURE

THE RETURN AIR TEMPERATURE GOES UP
TO 85°F BECAUSE THE UNIT HAS BEEN
OFF FOR SOME TIME

Figure 19-12. Coil-to-air temperatures under changing conditions. As the load on a coil goes up, the pressure in the coil goes up.

need to be cooled, copper, or aluminum coils would be eaten away. Stainless steel may be used instead, but stainless steel does not conduct heat as well as copper.

2. The *medium* to which the heat is exchanged. Giving heat up from air to refrigerant is an example. The best heat exchange takes place between two liquids, such as water to liquid refrigerant. However, this is not always practical because heat frequently has to be exchanged between air and vapor refrigerant. The vapor to vapor exchange is slower than the liquid to liquid exchange, Figure 19-14.

3. The *film factor*. This is a relationship between the medium giving up heat or absorbing heat and the heat-exchange surface. The film factor relates to the velocity of the medium passing over the exchange surface. When the velocity is too slow, the film between the medium and the surface becomes an insulator and slows the heat exchange. The velocity keeps the film to a minimum, Figure 19-15. The correct velocity is chosen by the manufacturer.

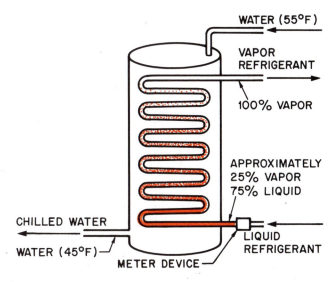

WATER (55°F)

VAPOR
REFRIGERANT

100% VAPOR

APPROXIMATELY
25% VAPOR
75% LIQUID

CHILLED WATER

WATER (45°F)

METER DEVICE

LIQUID
REFRIGERANT

Figure 19-14. Heat exchange between a liquid in a heat exchanger and the refrigerant inside the coil.

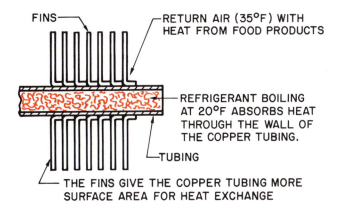

FINS

RETURN AIR (35°F) WITH
HEAT FROM FOOD PRODUCTS

REFRIGERANT BOILING
AT 20°F ABSORBS HEAT
THROUGH THE WALL OF
THE COPPER TUBING.

TUBING

THE FINS GIVE THE COPPER TUBING MORE
SURFACE AREA FOR HEAT EXCHANGE

Figure 19-13. Relationship of the air to refrigerant heat exchange.

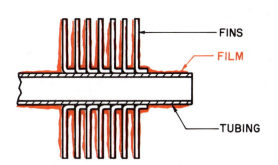

FINS

FILM

TUBING

Figure 19-15. One of the deterring factors in a normal heat exchange. The film factor is the film of air or liquid next to the tube in the heat exchange.

19.9 TYPES OF EVAPORATORS

Numerous types of evaporators are available, and each has its purpose. The first evaporators for cooling air were natural convection type. This evaporator was used in early walk-in coolers and was mounted high in the ceiling. It relied on the air being cooled, falling to the floor, and setting up a natural air current. The evaporator had to be quite large for the particular application because the velocity of the air passing over the coil was so slow. It is still occasionally used today.

The use of a blower to force or induce air over the coil improved the efficiency of the heat exchange. This meant that smaller evaporators could be used to do the same job. Design trends in industry have always been to smaller, more efficient equipment, Figure 19–16.

The expansion of the evaporator surface to a surface larger than the pipe itself gives a more efficient heat exchange. The *stamped evaporator* is a result of the first designs to create a larger pipe surface. The stamped evaporator is two pieces of metal stamped with the impression of a pipe passage through it, Figure 19–17.

Pipe with fins attached, called a *finned tube evaporator,* are used today more than any other heat exchange between air and refrigerant. This heat exchanger is quite efficient because the fins are in good contact with the pipe carrying the refrigerant. See Figure 19–18 for an example of a finned tube evaporator.

Multiple circuits improve evaporator performance and efficiency by reducing pressure drop inside the evaporator. The pipes inside the evaporator can be polished smooth, but they still offer resistance to the flow of both liquid and vapor refrigerants. The shorter the evaporator is, the less resistance there is to this flow. When an evaporator becomes quite long, it is common to cut it off and run another circuit in parallel next to it, Figure 19–19.

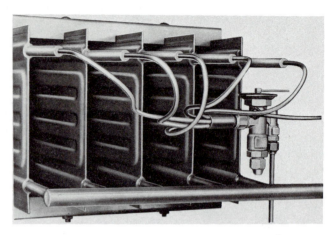

Figure 19–17. Stamped evaporator. *Courtesy Sporlan Valve Company*

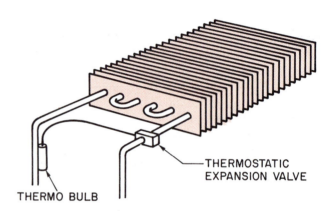

Figure 19–18. Finned evaporator.

The evaporator for cooling liquids or making ice operates under the same rules as one for cooling air but is designed differently. It may be strapped on the side of a drum with liquid inside, submerged inside the liquid container, or be a double-pipe system with the refrigerant inside one pipe and the liquid to be cooled circulated inside an outer pipe, Figure 19–20.

19.10 EVAPORATOR EVALUATION

Knowing the design considerations helps in *evaporator evaluation.* When the service technician arrives at the job, it may be necessary to evaluate whether or not a

Figure 19–16. Forced draft evaporator. *Courtesy Bally Case and Cooler, Inc.*

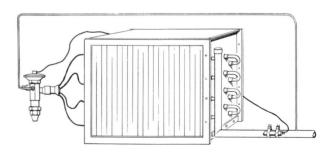

Figure 19–19. Multicircuit evaporator. *Courtesy Sporlan Valve Company*

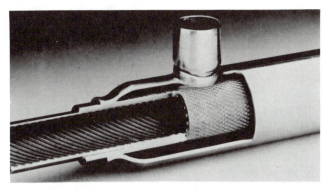

Figure 19-20. Liquid heat exchanger. *Courtesy Noranda Metal Industries Inc.*

particular evaporator is performing properly. This can be considered one of the starting points in organized troubleshooting. The evaporator has to absorb heat for the condenser to reject it or the compressor to pump it. The example used here will be a medium-temperature walk-in box. However, the evaporator evaluation would be about the same for any typical application.

Evaporator Specifications

1. Copper pipe coil
2. Aluminum fins attached to the copper pipe coil
3. Forced draft with a prop-type fan
4. One continual refrigerant circuit
5. R-12
6. Evaporator to maintain space temperature at 35°F
7. Evaporator is clean and in good working condition

The first consideration will be to describe how the evaporator functions when it is working correctly.

Entering the evaporator is a partial liquid–partial vapor mixture at 20°F and 21 psig, it is approximately 75% liquid and 25% vapor. Approximately 25% of the liquid entering the expansion device at the evaporator is changed to a vapor and cools the remaining 75% of the liquid to the evaporator's boiling condition (20°F). This is accomplished with the pressure drop across the expansion device. When the warm liquid passes through the small opening in the expansion device into the low pressure (21 psig) of the evaporator side of the device, some of the liquid flashes to a gas, Figure 19-21.

As the partial liquid–partial vapor mixture moves through the evaporator, more of the liquid changes to a vapor. This is called *boiling* and is a result of heat absorbed into the coil from whatever medium the evaporator is cooling. Finally, near the end of the evaporator the liquid is all boiled away to a vapor. At this point the refrigerant is known as *saturated vapor*. This means that the refrigerant vapor is saturated with heat. If any more heat is added to it, it will rise in temperature. If any heat is taken away from it, it will start changing back to a liquid. This vapor is saturated with heat, but it is still at the evaporating temperature corresponding to the boiling point, 20°F. **This is a most important point in the workings of an evaporator because the evaporator has to boil all of the liquid away as close to the end of the coil as possible to (1) keep the coil efficiency up and (2) insure that liquid refrigerant does not leave the evaporator and move into the compressor.** For the evaporator to run

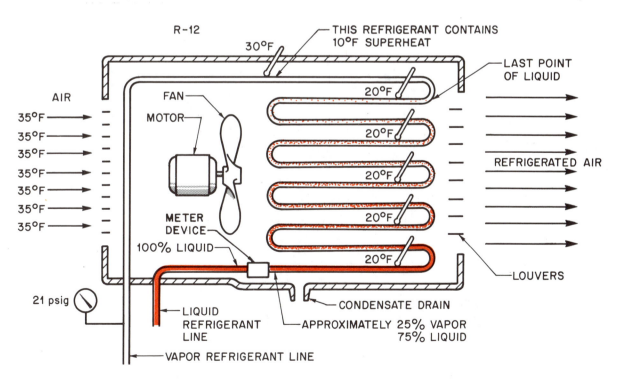

Figure 19-21. Evaporator operating under normal working conditions.

efficiently, it has to operate as full of liquid as possible without boiling over because the best heat exchange is between the liquid refrigerant and the air passing over the coil.

19.11 LATENT HEAT IN THE EVAPORATOR

The **latent heat absorbed during the change of state is much more concentrated than the sensible heat that would be added to the vapor leaving the coil.** Refer to the example of heat in Unit 1 that showed how it only takes 1 Btu to change the temperature of 1 lb of 211°F water to 212°F water. It also showed that it takes 970 Btu to change 1 lb of 212°F water to 212°F steam. This is true for water or refrigerant. The change of state is where the great amount of heat is absorbed into the system.

19.12 THE FLOODED EVAPORATOR

To get the maximum efficiency from the evaporator heat exchange, some evaporators are operated full of liquid or flooded, with a device to keep the liquid refrigerant from passing to the compressor. These *flooded evaporators* are specially made and normally use a float metering device to keep the liquid level as high as possible in the evaporator. This text will not go into detail about this system because it is not a device encountered often. Manufacturer's literature should be consulted for any special application.

If the evaporator is not flooded—that is, the refrigerant starts out as a partial liquid and boils away to a vapor in the heat exchange pipes—it is known as a *dry-type* or *direct expansion evaporator*.

19.13 DRY-TYPE EVAPORATOR PERFORMANCE

To check the performance of a dry-type evaporator, the service technician would first make sure that the refrigerant coil is operating with enough liquid inside the coil. This is generally done by comparing the boiling temperature inside the coil to the line temperature leaving the coil. The difference in temperatures is usually 8° to 12°F. For example, in the coil pictured in Figure 19–21, the boiling temperature in the coil was arrived at by converting the coil pressure (suction pressure) to temperature. In this example, the pressure was 21 psig, which corresponds to 20°F.

19.14 SUPERHEAT

The difference in temperature between the boiling refrigerant and the suction line temperature is known as *superheat*. Superheat is the sensible heat added to the vapor refrigerant after the change of state has taken place. Superheat is the best method of checking to see when a refrigerant coil has a proper level of refrigerant. When a metering device is not feeding enough refrigerant to the coil, the coil is said to be a *starved coil*, and the superheat is greater, Figure 19–22.

19.15 HOT PULL DOWN (EXCESSIVELY LOADED EVAPORATOR)

When the refrigerated space has been allowed to warm up considerably, the system must go through a hot pull down. On a hot pull down the evaporator and metering device are not expected to act exactly as they would in a typical design condition. For instance, if a

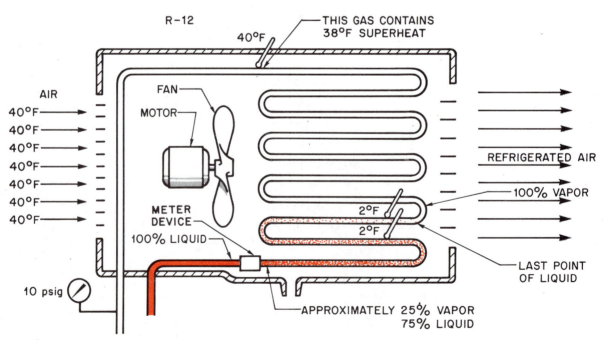

Figure 19–22. Starved evaporator.

walk-in cooler were supposed to maintain 35°F and it were allowed to warm up to 60°F and have some food or beverages inside, it would take an extended time to pull the air and product temperature down. The coil may be boiling the refrigerant so fast that the superheat may not come down to 8° to 12°F until the cooler has cooled down to closer to the design temperature. **A superheat reading on a hot pull down should be interpreted with caution,** Figure 19–23.

When a dry-type coil is fed too much refrigerant, the refrigerant does not all change to a vapor. This coil is thought of as a *flooded* coil. **Do not confuse this with a coil flooded by design,** Figure 19–24.

19.16 PRESSURE DROP IN EVAPORATORS

Multiple-circuit evaporators are used when the coil would become too long for a single circuit, Figure 19–25. The same evaluating procedures hold true for a multiple-circuit evaporator as for a single-circuit evaporator. The dry-type evaporator will again be used as the example.

The evaporator has to be as full as possible to be efficient. Each circuit should be feeding the same amount of refrigerant. If this needs to be checked, the service technician can check the common pressure tap for the boiling pressure. This pressure can be converted to temperature. Then the temperature will have to be checked at the outlet of each circuit to see if any circuit is overfeeding or starving, Figures 19–26 and 19–27.

Some reasons for uneven feeding for a multicircuit evaporator are

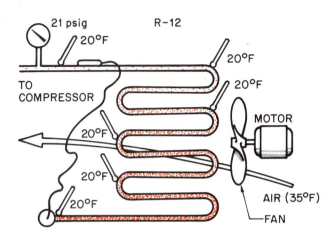

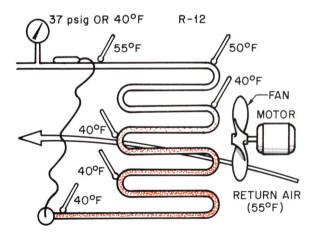

Figure 19–23. Hot pull down with a coil. This is a medium-temperature evaporator that should be operating at 21 psig (R-12 20°F). The return air is 55°F instead of 35°F. This causes the pressure in the coil to be higher. The warm box boils the refrigerant at a faster rate. The expansion valve is not able to feed the evaporator fast enough to keep the superheat at 10°F. The evaporator has 15°F of superheat.

Figure 19–24. Evaporator flooding because expansion device is not controlling refrigerant flow.

1. Stopped-up distribution system
2. Dirty coil
3. Uneven air distribution
4. Coil circuits of different lengths

19.17 LIQUID COOLING EVAPORATORS

A different type of evaporator is required for liquid cooling. It acts much the same as that for cooling air. They are normally of the dry type of expansion evaporator in the smaller systems, Figure 19–28. These evaporators have more than one refrigerant circuit to prevent pressure drop in the evaporator. The use of refrigeration gages and some accurate method of checking temperature of the suction line are very important. These evaporators sometimes have to be checked out for performance to see if they are absorbing heat like they should. They have a normal superheat range similar to air-type evaporators (8° to 12°F). When the superheat is within this range and all circuits in a multicircuit evaporator are performing alike, the evaporator is doing its job on the refrigerant side. However, this does not mean that it will cool properly. The liquid side of the evaporator has to be clean so that the liquid will come in proper contact with the evaporator.

Typical problems on the liquid side of the evaporator are

1. Mineral deposits that may build up on the liquid side and cause a poor heat exchange. They would act like an insulator.
2. Poor circulation of the liquid to be cooled where a circulating pump is concerned.

When the superheat is found to be correct and the coil is feeding correctly on a multicircuit system, the

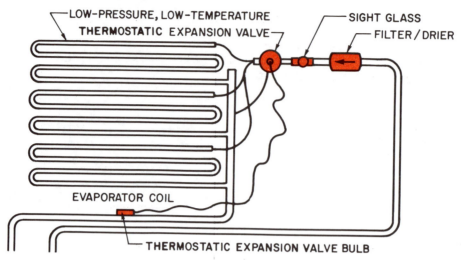

Figure 19–25. Multicircuit evaporator.

technician should consider the temperature of the liquid. The superheat may not be within the prescribed limits if the liquid to be cooled is not cooled down close to the design temperature. On a hot pull down of a liquid product, the heat exchange can be such that the coil appears to be starved for refrigerant because the coil is so loaded up that it is boiling the refrigerant faster than normal. The answer to this is time. You cannot rush a pull down, Figure 19–29. Air-to-refrigerant evaporators do not have quite the pro-

nounced difference in pull down that liquid heat exchange evaporators do.

19.18 EVAPORATORS FOR LOW-TEMPERATURE APPLICATION

Low-temperature evaporators that are used for cooling space or product to below freezing are designed differently. They require the coil to operate below freezing.

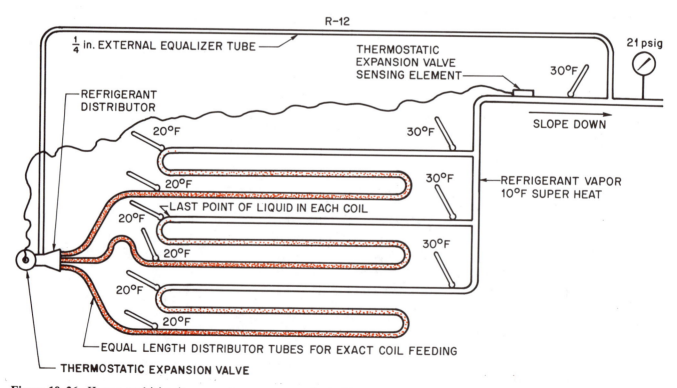

Figure 19–26. How a multicircuit evaporator appears on the inside when it is feeding correctly. It is like several evaporators piped in parallel.

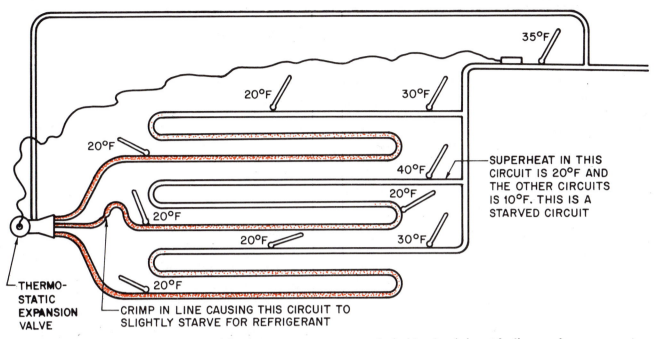

Figure 19-27. How a multicircuit evaporator appears on the inside when it is not feeding evenly.

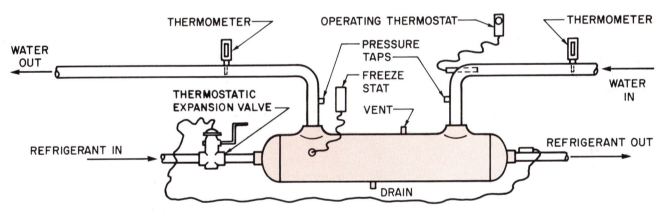

Figure 19-28. Evaporators used for exchanging heat between liquids and refrigerant. Most of these evaporators are of the direct-expansion type.

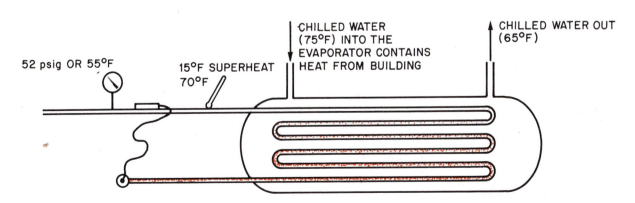

Figure 19-29. A hot pull down on a liquid evaporator giving up its heat to refrigerant. This evaporator normally has 55°F water-in and 45°F water out. The hot pull down with 75°F water instead of 55°F water boils the refrigerant at a faster rate. The expansion valve may not be able to feed the evaporator fast enough to maintain 10°F superheat. No conclusions should be made until the system approaches design conditions.

In an airflow application the water that accumulates on the coil will freeze and will have to be removed. The design of the fin spacing must be carefully chosen. A very small amount of ice accumulated on the fins will restrict the airflow. Other than the airflow blockage due to ice buildup, these low temperature evaporators perform much the same as medium-temperature evaporators. They are normally dry-type evaporators and have one or more fans to circulate the air across the coil. The defrosting of the coil has to be done by raising the coil temperature above freezing to melt the ice. Then the condensate water has to be drained off and kept from freezing. Defrost is sometimes accomplished with heat from outside the system. Electric heat can be added to the evaporator to melt the ice, but this heat adds to the load of the system and needs to be pumped out after defrost, Figure 19–30.

19.19 DEFROST OF ACCUMULATED MOISTURE

Defrost can be accomplished with heat from inside the system using the hot gas that is discharged from the refrigerant. This system is economical because power does not have to be purchased for electric heaters that will heat the evaporator. It is accomplished with a magnetic solenoid valve that routes the hot gas to the evaporator entrance on the evaporator side of the expansion valve. For either system the evaporator fan has to be turned off during defrost, or two things will happen: (1) The heat from defrost will be transferred directly to the conditioned space. (2) The cold conditioned air would slow down the defrost process.

Evaporators that are applied to some ice making processes have similar defrost methods. They have to have some method of applying heat to the evaporator to melt the ice. Sometimes the heat is electric or hot gas. When the evaporator is being used to make ice, the makeup water for the ice maker is sometimes used for defrost.

In summation, when checking an evaporator remember that its job is to absorb heat into the refrigeration system.

SUMMARY

- Refrigeration is defined as the process of "removing heat from a place where it is not wanted and depositing it in a place where it makes no difference."
- Heat travels normally from a warm substance to a cool substance.
- For heat to travel from a cool substance to a warm substance, work must be performed. The motor that drives the compressor in the refrigeration cycle does this work.
- The evaporator is the component that absorbs the heat into the refrigeration system.
- The evaporator has to be cooler than the medium to be cooled to have a heat exchange.
- The refrigerant boils to a vapor in the evaporator and absorbs heat because it is boiling at a low pressure and low temperature.
- The boiling temperature of the refrigerant in the evaporator determines the evaporator (low-side) pressure.
- The pressure–temperature relationship is printed directly on some pressure gages.
- Medium temperature systems can use off-cycle defrost (the product is above freezing, and the heat from it can be used to defrost).
- Low-temperature refrigeration must have heat added to the evaporator to melt the ice.
- Evaporators have the same characteristics for the same types of installations regardless of location.
- The first evaporator coils were bare pipe.
- Stamped plate steel evaporators were among the first attempts to extend the surface area of a bare pipe.
- Fins were later added to further extend the bare tube and give more efficiency.
- Most refrigeration coils are copper with aluminum fins.
- The starting point in organized troubleshooting is to determine if the evaporator is operating efficiently.
- Checking the superheat is the best method the service technician has for evaluating evaporator performance.
- Some evaporators are called dry-type because they use a minimum of refrigerant.
- Dry-type evaporators are also called direct expansion evaporators and maintain a constant superheat when operating correctly within their design range.

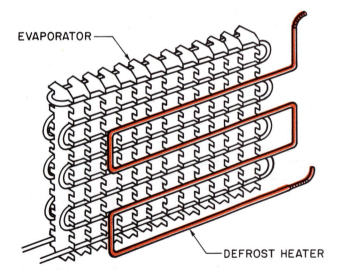

EVAPORATOR

DEFROST HEATER

Figure 19–30. Heaters used for electric defrost of low-temperature evaporators.

- Some evaporators are flooded and use a float to meter the refrigerant. *Superheat checks on these evaporators should be interpreted with caution.*
- Some evaporators have a single circuit, and some have multiple circuits.
- Multicircuit evaporators keep excessive pressure drop from occurring in the evaporator.
- There is a relationship between the boiling temperature of the refrigerant in the evaporator and the temperature of the medium (air) being cooled.
- The coil normally operates from 10° to 20°F colder than the air passing over it.

REVIEW QUESTIONS

1. What is the responsibility of the evaporator in the refrigeration system?
2. What happens to the refrigerant in the evaporator?
3. True or False: Some evaporators are made of bare pipe.
4. What is the heat called that is added to the vapor after the liquid is boiled away?
5. When there is superheat, there is no _____.
6. What determines the pressure on the low-pressure side of the system?
7. What is considered a typical superheat for a refrigeration system evaporator?
8. What does a high superheat indicate?
9. What does a low superheat indicate?
10. Why is a multiple-circuit evaporator used?
11. What expansion device do flooded evaporators use?
12. An evaporator that is not flooded is thought of as what type of evaporator?
13. When the evaporator experiences a load increase, what happens to the suction pressure?
14. What is used to defrost the ice from a medium-temperature evaporator?
15. What is commonly used to defrost the ice from a low-temperature evaporator?
16. A medium-temperature refrigeration box operates within what temperature range?
17. A low-temperature refrigeration device operates within what temperature range?

20

CONDENSERS

OBJECTIVES

After studying this unit, you should be able to

- explain the condenser in a refrigeration system.
- distinguish the operating characteristics of water-cooled and air-cooled systems.
- describe the basics of exchanging heat in a condenser.
- explain the difference between a tube in a tube-coil type of condenser and a tube in a tube-serviceable condenser.
- describe the difference between a shell and coil condenser and a shell and tube condenser.
- describe a wastewater system.
- describe a recirculated water system.
- describe a cooling tower.
- explain the relationship between the condensing refrigerant and the condensing medium.

20.1 THE CONDENSER

The *condenser* is a heat exchange device similar to the evaporator that rejects the heat from the system absorbed by the evaporator. This heat is in the form of hot gas that has to be cooled down to the point where it will condense. When heat was being absorbed into the system, we pointed out that it is at the point of change of state (liquid to a vapor) of the refrigerant that the greatest amount of heat is absorbed. The same thing, in reverse, is true in the condenser. The point where the change of state (vapor to a liquid) takes place is where the greatest amount of heat is rejected.

The condenser is operated at higher pressures and temperatures than the evaporator is and is often located outside. The same laws apply to heat exchange

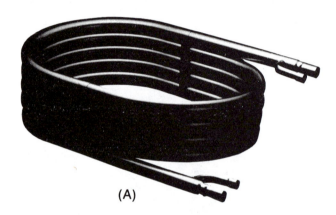

(A)

(B)

Figure 20-2. Two types of tube within a tube condensers. (A) A pipe within a pipe. (B) A flanged type of condenser. The flanged condenser can be cleaned by removing the flanges. Removal of the flanges opens only the water circuit not the refrigerant circuit. *Courtesy (A) Noranda Metal Industries Inc. (B) Photo by Bill Johnson*

Figure 20-1. An early water-cooled condensing unit. *Courtesy Tecumseh Products Company*

in the condenser as in the evaporator. The materials a condenser is made of and the medium used to transfer heat into make a difference in the efficiency of the heat exchanger.

20.2 WATER-COOLED CONDENSERS

Unlike the first evaporators, the first commercial refrigeration condensers were water cooled. These condensers were crude compared to modern water-cooled devices, Figure 20–1. Water-cooled condensers are quite efficient compared to air-cooled condensers and operate at much lower condensing temperatures. Water-cooled equipment comes in several styles: The tube within a tube, the shell and coil, and the shell and tube are the most common.

20.3 TUBE WITHIN A TUBE CONDENSERS

The *tube within a tube* condenser comes in two styles: the coil type, and the cleanable type with flanged ends, Figure 20–2.

The tube within a tube that is fabricated into a coil is manufactured by slipping one pipe inside another and sealing the ends in such a manner that the outer tube becomes one container and the inner tube becomes another container, Figure 20–3. The two pipes are then formed into a coil to save space. The heat exchange takes place between the fluid inside the outer pipe and the fluid inside the inner pipe, Figure 20–4.

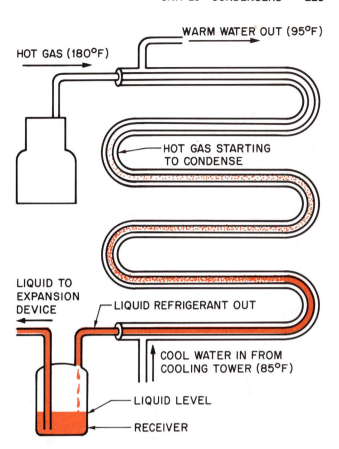

Figure 20–4. Fluid flow through the condenser. The refrigerant is flowing in one direction, water in the other.

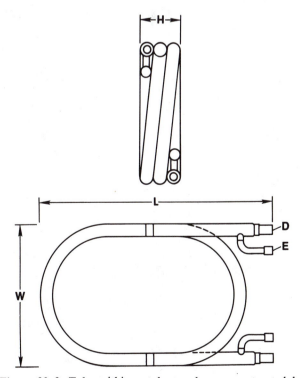

Figure 20–3. Tube within a tube condenser constructed by sliding one tube through the other tube. The tubes are sealed in such a manner that the inside tube is separate from the outside tube. *Courtesy Noranda Metal Industries Inc.*

Figure 20–5. Method of adding water treatment to a cooling tower. Treatment is being metered at a rate that will last about a month so that the operator will not have to be in attendance at the tower all of the time. There is a constant bleed of the tower water to the drain to keep the water from being overconcentrated with minerals. *Courtesy Calgon Corporation*

20.4 MINERAL DEPOSITS

Since water flows through one of the tubes, mineral deposits and scale will form even in the best water. The heat in the vicinity of the discharge gas has a tendency to cause any minerals in the water to deposit onto the tube surface. This is a slow process, but it will happen in time to any water-type condenser. These mineral deposits act as an insulator between the tube and the water and have to be kept to a minimum. Water treatment can be furnished to help prevent this buildup of mineral scale. This treatment is normally added at the tower or injected into the water by chemical feed pumps. See Figure 20-5 for an example of treatment being added to a tower. See Figure 20-6 for an example of water treatment being pumped into the water piping.

In some mild cases of scale buildup more water circulation to improve the heat exchange will occur. Later in this unit, variable water-flow controls are introduced. These controls will step up the water flow upon an in-crease in head pressure. This type of control causes more water to flow through the condenser automatically when the mineral deposits cause an increase in head pressure because of a poor heat exchange. In a water-cooled condenser a dirty condenser is like a dirty air-cooled condenser. If the water is wasted instead of cooled and used again, the water bill would go up before the operator would notice there was a condenser problem.

The tube within a tube condenser that is made into a coil cannot be cleaned mechanically with brushes. It has to be cleaned with chemicals designed not to harm the metal in the condenser. Professional help from a chemical company that specializes in water treatment is recommended when a condenser has to be cleaned with chemicals. Condensers of this type are normally made from copper or steel; some special condensers are made of stainless steel or copper and nickel.

20.5 CLEANABLE CONDENSERS

The tube within a tube condenser that is fabricated with flanges on the end can be mechanically cleaned.

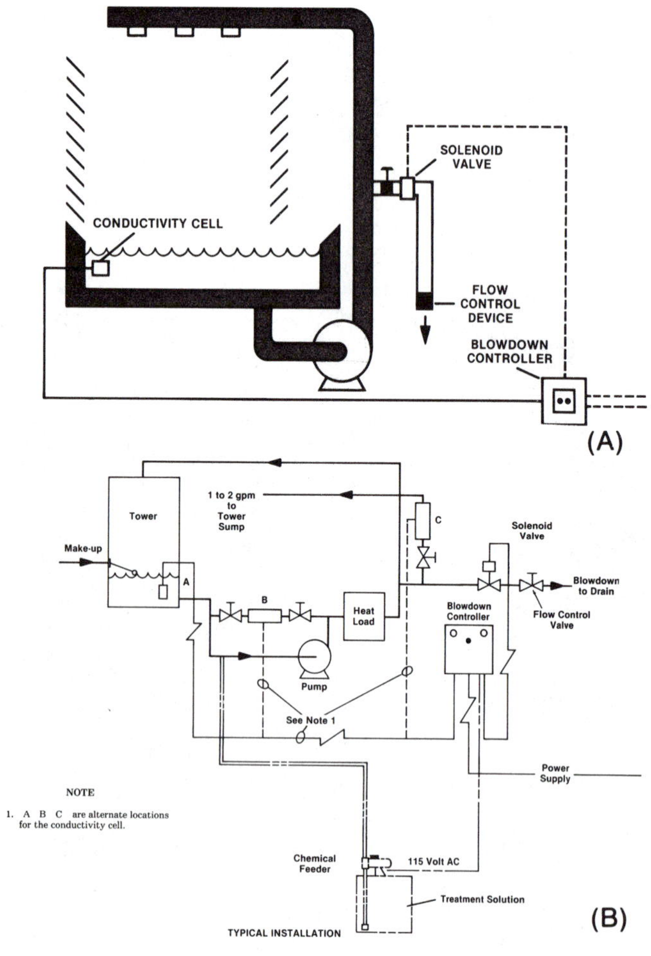

Figure 20-6. Automatic system of feeding t the water system. It includes an automatic monitoring system that determines when the system actually needs chemicals added. This type of system is normally used on larger systems because of the economics of the total system cost. *Courtesy Calgon Corporation*

Figure 20-7. Condenser is flanged for service. When the flanges are removed, the refrigerant circuit is not disturbed. *Photo by Bill Johnson*

The flanges can be removed, and the tubes can be examined and brushed with an approved brush, Figure 20–7. This condenser has the gaskets on the water circuit with the refrigerant flowing around the tubes, so the refrigerant circuit is not opened to clean the tubes, Figure 20–8. Consult the manufacturer for the correct brush. Fiber is usually preferable. This is a more expensive type of condenser, but it is serviceable.

20.6 SHELL AND COIL

The shell and coil condenser is similar to the tube within a tube coil. It is a coil of tubing packed into a shell that is then closed and welded. Normally the refrigerant gas is discharged into the shell, and the water is circulated in the tube packed into the shell. The shell of the condenser serves as a receiver storage tank for the extra refrigerant in the system. This condenser is not mechanically serviceable because the coil is not straight, Figure 20–9. It must be cleaned chemically.

20.7 SHELL AND TUBE CONDENSERS

Shell and tube condensers are more expensive than shell and coil condensers, but they can be cleaned mechanically with brushes. They are constructed with the tubes fastened into an end sheet in the shell. The refrigerant is discharged into the shell, and the water is circulated through the tubes. The ends of the shell are like end caps (known as water boxes) with the water circulating in them, Figure 20–10. These end caps can be removed, so the tubes can be inspected and brushed out if needed. The shell acts as a receiver storage tank for extra refrigerant. This is the most expensive condenser and is normally used in larger applications.

The water-cooled condenser is used to remove the heat from the refrigerant. When it is removed, the heat is in the water. There are two things that can be done at this point:

(1) waste the water, or
(2) take the water to a remote place, remove the heat, and reuse the water.

(A)

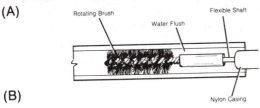

(B)

Figure 20–8. Brushes actually being pushed through the water side of the condenser. Use only approved brushes. *Courtesy Goodway Tools Corporation*

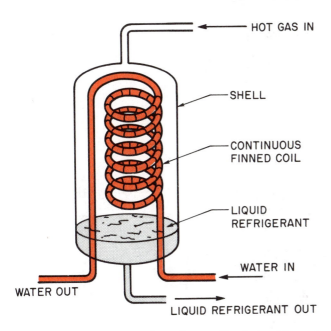

Figure 20–9. Shell and coil condenser. The hot refrigerant gas is piped into the shell, and the water is contained inside the tubes.

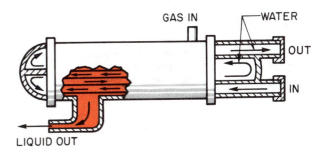

Figure 20–10. How water can be circulated back and forth through the condenser by using the end caps to give the water the proper direction.

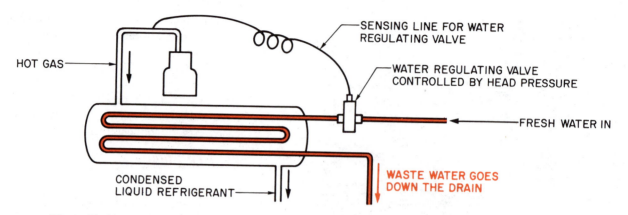

Figure 20–11. Wastewater system used when water is plentiful at a low cost, such as from a well or lake.

20.8   WASTEWATER SYSTEMS

Wastewater systems are just what the name implies. The water is used once, then wasted down the drain, Figure 20–11. This is worthwhile if water is free, or if only a small amount is used. Where large amounts of water are used, it is probably more economical to save the water, cool it in an outside water tower, and reuse it.

The water supplied to systems that use the water only once and waste it has a broad temperature range.

For instance, the water in summer may be 75°F out of the city mains and as low as 40°F in the winter, Figure 20–12. Water that runs through a building with long pipe runs may have warm water in the beginning from standing in the pipes. When the standing water is run through, the main water temperature may be quite low. This change in water temperature has an effect on the head pressure (condensing temperature). The head pressure has to be about 100 psig higher than the suction pressure for the expansion device to function properly.

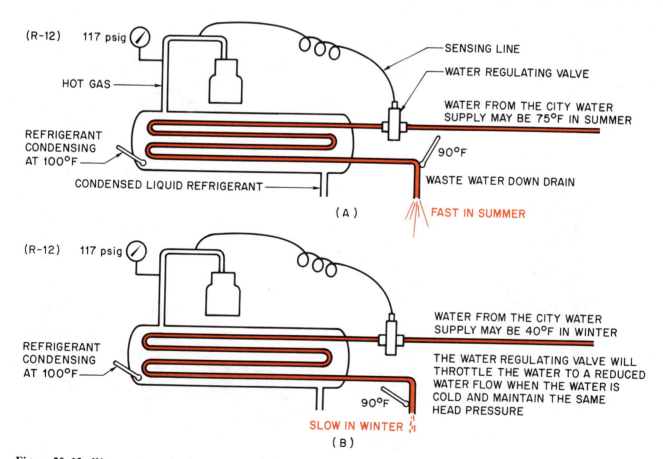

Figure 20–12. Wastewater condenser system at two sets of conditions: (A) summer with warm water entering the system, (B) winter when the water in the city mains is colder.

20.9   REFRIGERANT-TO-WATER TEMPERATURE RELATIONSHIP

There is a relationship between the condensing refrigerant and the water used to condense the refrigerant. Normally the condensing temperature is about 10°F higher than the leaving water. For example, in summer the water from the mains may be 80°F. This would step up the water flow to the point that the leaving water may be 90°F. This 90°F can be used as a reference point to find out what the head pressure should be (90°F plus 10°F = 100°F); this would be the approximate condensing temperature. The head pressure corresponding to this 100°F condensing temperature is 117 psig. In winter when the water temperature in the main is down to 40°F, the leaving water will still be about 90°F due to the automatic water valve that throttles the water to maintain a constant head pressure, Figure 20-12.

In a wastewater system the water flow can be varied to suit the need by means of a regulating valve for the water. This valve has a pressure tap that fastens the bellows in the control to the high-pressure side of the system. When the head pressure goes up, the valve opens and allows more water to flow through the condenser to keep the head pressure in line, Figure 20-13.

20.10   RECIRCULATED WATER SYSTEMS

When the system gets large enough that saving water is a concern, then a system that will recirculate the water is considered. This system uses the condenser to absorb the heat out of the system into the water just as in the wastewater system. The water is then pumped to an area away from the condenser where the heat is removed from the water, Figure 20-14. There is a relationship between the water and the refrigerant temperatures. The refrigerant will normally condense at a temperature of about 10°F higher than the leaving-water temperature, Figure 20-15.

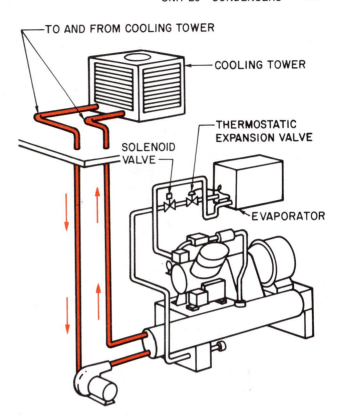

Figure 20-14. Water-cooled condenser that absorbs heat from the refrigerant and pumps the water to a cooling tower at a remote location. The condenser is located close to the compressor, and the tower is on the roof outside the structure.

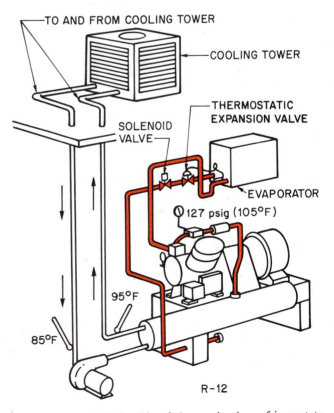

Figure 20-15. Relationship of the condensing refrigerant to the leaving-water temperature.

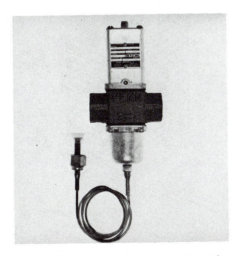

Figure 20-13. Water regulating valve to control water flow during different demands. *Photo by Bill Johnson*

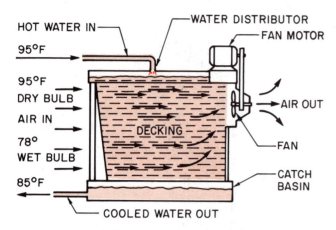

Figure 20-16. Relationship of a forced-draft cooling tower to the ambient air. Cooling tower performance depends on the wet-bulb temperature of the air. This relates to the humidity and the ability of the air to absorb moisture.

20.11 COOLING TOWERS

The *cooling tower* is a device that passes the outside air over the water to remove the system heat from the water. Any water tower is limited in capacity to the amount of evaporation that occurs. The evaporation rate is linked to the wet-bulb temperature of the outside air (humidity). Usually a cooling tower can cool the water that returns back to the condenser to within 7°F of the wet-bulb temperature of the outside air, Figure 20-16. This tower arrangement comes in sizes of about 2 tons of refrigeration and up. Towers can either be

(1) natural draft,
(2) forced draft, or
(3) evaporative.

20.12 NATURAL-DRAFT TOWERS

The *natural-draft tower* does not have a blower to move air through the tower. It is customarily made of some material that the weather will not deteriorate, such as redwood, fiberglass, or galvanized sheet metal.

Since natural-draft towers rely on the natural prevailing breezes to blow through them, they need to be located in the prevailing wind. The water is sprayed into the top of the tower through spray heads, and some of the water evaporates as it falls to the bottom of the tower. This evaporation takes heat from the remaining water and adds to the capacity of the tower. The evaporated water has a makeup system using a float assembly connected to the city water supply that makes up evaporated water automatically, Figure 20-17.

The tower location must be chosen carefully. If it is located in a corner between two buildings where the breeze cannot blow through it, higher than normal

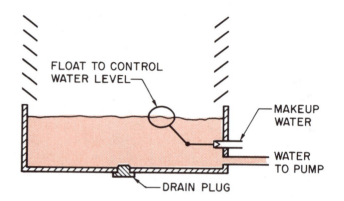

Figure 20-17. Makeup water system in a cooling tower. Since the cooling tower performance depends partly on evaporation of water from the tower, this makeup is necessary.

water temperatures will occur, which will cause higher than normal head pressures, Figure 20-18.

These towers have two weather-related conditions that have to be dealt with: (1) The tower must be operated in the winter on refrigeration systems, and the water will freeze in some climates if freeze protection is not provided. Heat can be added to the water in the basin of the tower; antifreeze will also prevent this from happening, Figure 20-19. (2) The water can get cold enough to cause a head pressure drop. A water regulating valve can be installed to prevent this from happening. Natural-draft towers can be seen on top of buildings as structures that look like they are made of slats. These slats keep the water from blowing out of the tower, Figure 20-20.

20.13 FORCED-DRAFT TOWERS

Forced-draft cooling towers differ from natural-draft towers because they have a fan to force air

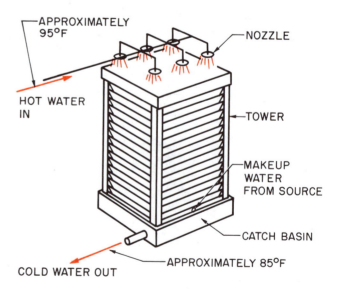

Figure 20-18. Natural-draft cooling tower. (Must be located in prevailing winds.)

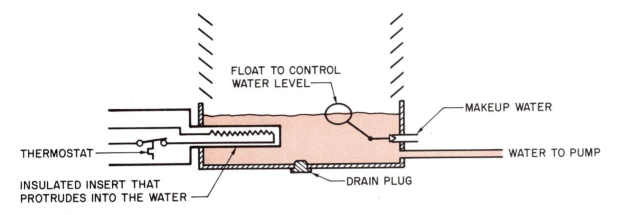

FLOAT TO CONTROL
WATER LEVEL

MAKEUP WATER

WATER TO PUMP

THERMOSTAT

DRAIN PLUG

INSULATED INSERT THAT
PROTRUDES INTO THE WATER

Figure 20–19. Type of heat that may be applied to keep the water in the basin from freezing in winter. This heat can be thermostatically controlled to prevent it from being left on when not needed.

over a wetted surface, Figure 20–21. They are customarily designed with the warm water from the condenser being pumped into a flat basin in the top of the tower. This basin has calibrated holes drilled in it to allow a measured amount of water to pass downward through the fill material, Figure 20–22. The fill material is usually redwood or manmade fiber and gives the water surface area for the fan to blow air over to evaporate and cool the water, Figure 20–23. As the water is evaporated, it is replaced with a water makeup system using a float, similar to the natural-draft tower.

Forced-draft towers can be located almost anywhere because the fan can move the air. They can even be located inside buildings, where the air is brought in and out through ducts, Figure 20–24. The tower is fairly enclosed as far as the prevailing winds are concerned, so no water regulating valve is normally necessary. The

fan can be cycled off and on to control the water temperature and thus control the head pressure. The mass of the water in the tower gives a long cycle between the time the fan comes on and off for this type of tower. Forced-draft towers are small compared to natural-draft towers. They are quite versatile because of the forced movement of air. *Note:* An induced draft tower is similar to a forced draft tower only the air is pulled, not pushed across the wetted surface.

20.14 EVAPORATIVE CONDENSERS

Evaporative condensers are a different type altogether because the refrigerant condenser is actually located inside the tower. They are often confused with cooling towers, Figure 20–25. In the previous water towers the condenser containing the refrigerant was remote from the tower, and the water was piped

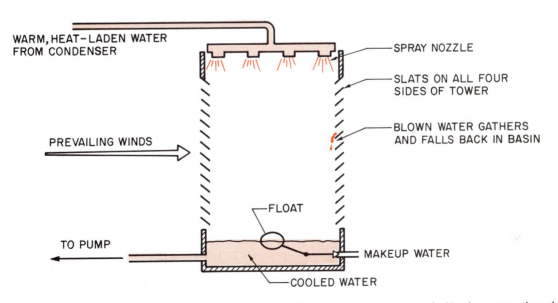

WARM, HEAT–LADEN WATER
FROM CONDENSER

SPRAY NOZZLE

SLATS ON ALL FOUR
SIDES OF TOWER

BLOWN WATER GATHERS
AND FALLS BACK IN BASIN

PREVAILING WINDS

FLOAT

TO PUMP

MAKEUP WATER

COOLED WATER

Figure 20–20. How the slats on the sides of the natural-draft cooling tower keep the water inside the tower when the wind is blowing.

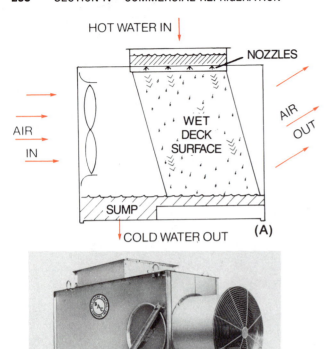

HOT WATER IN

NOZZLES

AIR OUT

WET DECK SURFACE

AIR IN

SUMP

COLD WATER OUT

(A)

Figure 20-21. Forced-draft tower. *Courtesy of Baltimore Aircoil Company Inc.*

(B)

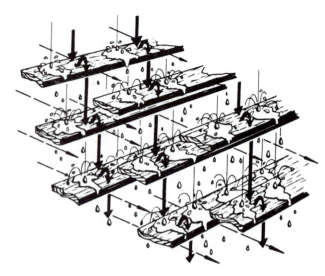

Figure 20-23. How the water trickles down through the fill material. *Courtesy Marley Cooling Tower Company*

20.15   AIR-COOLED CONDENSERS

Air-cooled condensers use air as the medium instead of water to reject heat into. This can be very advantageous where it is difficult to handle water. The first air-cooled condensers were bare pipe with air from the compressor's flywheel blowing over the condenser. The compressors were open drive at this time. To improve the efficiency of the condenser and to make it smaller, the surface area was then extended with fins. The condensers at this time were normally steel with steel fins, Figure 20-26. These condensers resembled radiators and were sometimes referred to as radiators.

Steel air-cooled condensers are still used in many small installations on refrigeration units. This is another trend that is characteristic of the refrigeration industry because steel-fin condensers are not used at all

through the condenser to the tower. The evaporative condenser uses the same water over and over with a pump located at the tower. As the water is evaporated, it is replaced with a makeup system using a float, as the other towers. When the evaporative condenser is used in cold climates, freeze protection must be provided in winter.

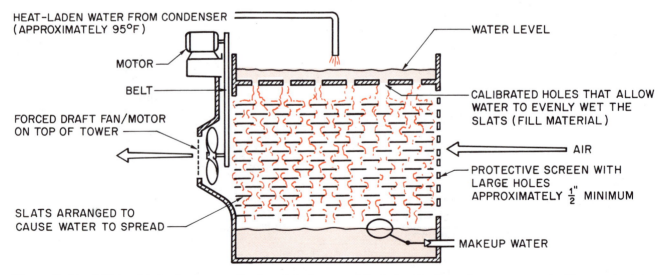

HEAT-LADEN WATER FROM CONDENSER (APPROXIMATELY 95°F)

MOTOR

BELT

FORCED DRAFT FAN/MOTOR ON TOP OF TOWER

SLATS ARRANGED TO CAUSE WATER TO SPREAD

WATER LEVEL

CALIBRATED HOLES THAT ALLOW WATER TO EVENLY WET THE SLATS (FILL MATERIAL)

AIR

PROTECTIVE SCREEN WITH LARGE HOLES APPROXIMATELY $\frac{1}{2}$" MINIMUM

MAKEUP WATER

Figure 20-22. Calibrated holes in the top of a forced-draft tower. The holes distribute the water over the fill material below and are covered by the returning warm water from the system.

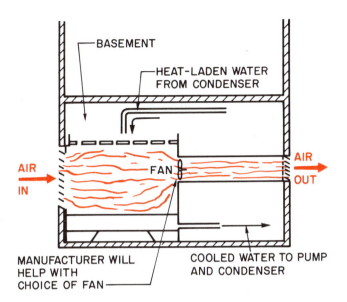

Figure 20-24. Forced-draft tower located inside a building with air ducted to the outside.

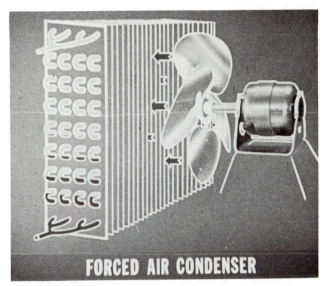

FORCED AIR CONDENSER

Figure 20-26. Fins designed to give the coil more surface area. *Reproduced courtesy of Carrier Corporation*

in the air conditioning industry. Possibly this is due to weight. The larger refrigeration systems use the same condenser styles that the air conditioning industry uses, Figure 20-27.

Air-cooled condensers come in a variety of styles. Some are horizontal with the air blowing through them and are subject to prevailing winds, Figure 20-28. Some air-cooled condensers are vertical in the airflow pattern. They take air into the bottom and discharge air out of the top. The prevailing winds do not effect

these condensers to any extent, Figure 20-29. Another style of air-cooled condenser takes the air in the sides and discharges it out of the top. This condenser can be affected by prevailing winds, Figure 20-30.

The smaller refrigeration systems are often located in the conditioned space, such as a restaurant or store. These air-cooled condensers normally have widely spaced steel fins on a steel coil, which allow more time before the coil will stop up with dust and other airborne material.

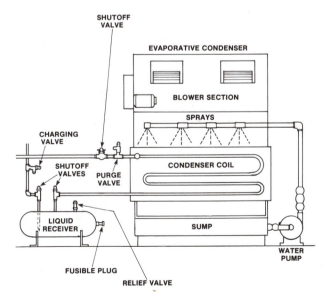

Figure 20-25. Evaporative condenser. The refrigerant coil is located inside the actual tower, and the refrigerant is piped to the tower instead of piping the water to the condenser. *Courtesy of Trane Company*

Figure 20-27. Larger refrigeration air-cooled condenser. It resembles an air conditioning condenser. *Courtesy Bohn Heat Transfer*

Figure 20-28. Horizontal air-cooled condenser subject to prevailing winds blowing through it. *Courtesy Copeland Corporation*

Figure 20-29. Condenser with vertical airflow pattern. Air enters the bottom and blows out the top. It is unaffected by prevailing winds. *Courtesy Bohn Heat Transfer*

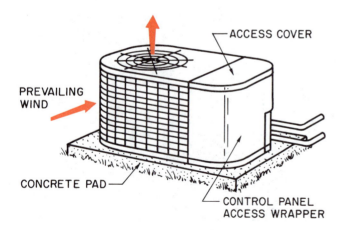

Figure 20-30. Condenser takes the air in from the sides and discharges it out the top. The prevailing winds could affect it. *Reproduced Courtesy of Carrier Corporation*

20.16 INSIDE THE CONDENSER

The hot gas normally enters the air-cooled condenser at the top. The beginning tubes of the condenser will be receiving the hot gas straight from the compressor. This gas will be highly superheated. (Remember that superheat is heat that is added to the refrigerant after the change of state in the evaporator.) When the superheated refrigerant from the evaporator reaches the compressor and is compressed, more heat is added to the gas. Part of the energy applied to the compressor transfers into the refrigerant in the form of heat energy instead of work. This additional heat added by the compressor causes the refrigerant leaving the compressor to be heavily heat laden. For a more thorough understanding of this, read Unit 21 on compressors. For now, just imagine that the refrigerant entering the condenser is heat laden. On a hot day (95°F) the hot gas leaving the compressor could easily reach 200°F. The condenser has to remove this heat down to the condensing temperature before any condensation can take place.

See Figure 20-31 for an illustration of the following description of an air-cooled, R-12 condenser located outside a store. It is responsible for rejecting the heat absorbed inside a medium-temperature walk-in cooler. This cooler has reach-in doors typical of those in a convenience store. This box has the beverages on shelves, and store personnel can stock the shelves from the walk-in portion behind the shelves. The cooler is maintained at 35°F. The outside air temperature is 95°F. The refrigeration system must absorb heat at 35°F and reject that same heat to the outside where the condensing medium is 95°F.

20.17 CONDENSING REFRIGERANT AND THE AMBIENT AIR RELATIONSHIP

Air-cooled condensers have a relationship to the air passing over them just like the evaporator did in the previous text. The relationship can be stated like this: The refrigerant inside the coil will normally condense at 35°F higher temperature than the air passing over the condenser (also known as the ambient air). This statement is true for most standard-efficiency condensers that have been in service long enough to have a typical dirt deposit on the fins and tubing. The relationship can be improved by adding condenser surface. With an outside air temperature of 95°F, the condensing temperature will be about 130°F. With the refrigerant R-12 condensing at 130°F, the head-pressure or high-pressure gage should read 180 psig. (See the pressure and temperature chart for R-12.) This is important because it helps the service technician establish what the head pressure should be. Read the following example carefully.

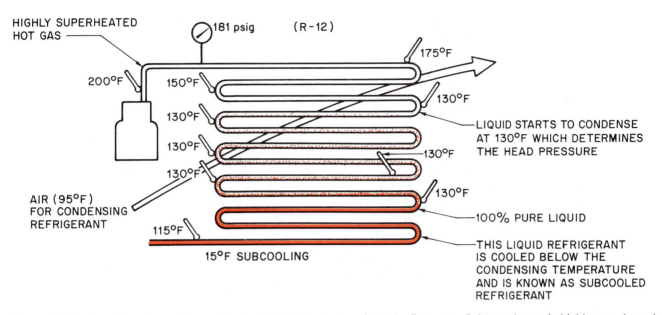

Figure 20-31. The following points are illustrated: (1) The hot gas into the first part of the condenser is highly superheated. The condensing temperature is 130°F, and the discharge gas has to be cooled from 200° to 130°F before any condensing will take place. (2) The best heat exchange is between the liquid and the air on the outside of the coil. More heat is removed during the change of state than while desuperheating the vapor. (3) When the liquid is all condensed and there is pure liquid in the coil, the liquid can be subcooled to below the condensing temperature.

1. The hot gas is entering the condenser at 200°F. The condensing temperature is going to be 35°F warmer than the outside (ambient) air.

2. The outside air temperature is 95°F. The condensing temperature is 95°F plus 35°F = 130°F. The refrigerant has to be cooled to 130°F before any actual condensing takes place. Thus the condenser has to lower the hot-gas temperature 70°F in the first part of the coil. This is called *desuperheating*. It is the first job of the condenser.

3. Part way down the coil the superheat is removed down to the actual condensing temperature of 130°F, and liquid begins to form in the coil. From the discussion of the evaporator, recall that the best heat exchange takes place not between the vapor and the air outside the coil but between liquid in the coil and the outside air. The forming liquid speeds up the heat exchange process.

4. When the refrigerant is down near the last few turns of the coil, the tubes will be nearly full of liquid at the 130°F condensing temperature.

5. When the refrigerant gets to the end of the coil, the condenser tubes will be full of liquid and then drain into the receiver. If the condenser is long enough, the liquid may even cool below the condensing temperature of 130°F. This is called *subcooling*.

6. The liquid in the bottom of the condenser draining into the receiver may cool to 115°F.

20.18 THE CONDENSER AND LOW AMBIENT CONDITIONS

The foregoing example is how an air-cooled condenser operates on a hot day. An example of a condenser operating under different conditions is a supermarket with a small package-display case located inside the store. The condenser is located inside the store, Figure 20-32. The inlet air to the case is in the store itself and may be quite cool if the air conditioner is operating. The store temperature may be 70°F at times. This reduces the operating pressure on the high-pressure side of the system. When the condenser relationship temperature rule is applied, we see that the new condensing temperature would be 70°F plus 35°F = 105°F. The head pressure would go down to 127 psig. This may be enough to effect the performance of the expansion device. This will be covered in more detail later, but for now it is necessary to know that the head pressure has to be about 100 psig higher than the suction or low-side pressure for the expansion device to operate correctly. The low-side pressure for a medium-temperature fixture will be in the neighbor-

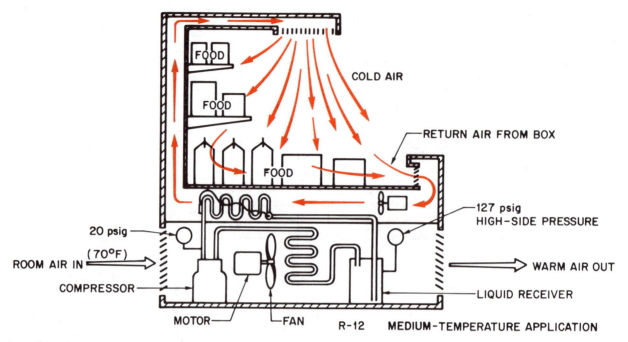

Figure 20–32. Package-display case located inside the store with the compressor and the condenser located inside the cabinet. This is a plug-in device; no piping is required.

hood of 20 psig at the lowest point. When the relationship of the low-side pressure of 20 psig is subtracted from the high-side pressure of 127 psig for a condenser with 70°F air passing over it, the difference is 107 psig. This is close to the 100 psig minimum difference. Should this fixture be exposed to temperatures of lower than 70°F, difficulties with a starved evaporator coil may result.

If the medium-temperature box were moved outside the store, the fixture may quit working to capacity during winter. For example, when the air temperature over the coil drops to 50°F outside, the head pressure is going to drop to 50°F plus 35°F = 85°F condensing temperature, which corresponds to 92 psig. Thus pressure difference of 92 psig − 20 psig = 72 psig isn't sufficient to feed enough liquid refrigerant through the expansion device to properly feed the coil, Figure 20–33. Notice in the figure the pressure and temperature difficulties encountered. The head pressure must be regulated.

Another example of a condenser operating outside the design parameter would be an ice-holding box like those found at service stations and convenience stores. These fixtures hold ice made at another location. They have to keep the ice hard in all types of weather and often operate at about 0°F inside the box. They do not make ice, they only hold or store it.

It may be 30°F outside the box where the small air-cooled condenser is rejecting the heat. The condenser would be operating at about 30°F plus 35°F = 65°F, or at a head pressure of 64 psig. The evaporator should be operating at about −15°F, or at a suction

pressure of 2.5 psig. This gives a pressure difference of 64 psig − 2.5 psig = 61.5 psig, which will starve the evaporator, Figure 20–34. This unit has to run to furnish ice, so something has to be done to get the head pressure up to a value at least 100 psig higher than the suction pressure of 2.5 psig.

One thing that prevents problems for some equipment in low-ambient conditions is the load is reduced. The ice-holding box, for example, does not have to run as much in 30°F weather to keep the ice frozen at 0°F.

20.19 HEAD-PRESSURE CONTROL

Practical methods to maintain the correct workable head pressure automatically and not cause equipment wear are fan cycle control, dampers, and condenser flooding.

Fan Cycling Devices

The air-cooled condenser normally has a small fan that passes the air over the condenser. When this fan is cycled off, the head pressure will go up in any kind of weather. This is provided the prevailing winds do not take over and do the fan's job. To cycle off the fan, we will use a pressure control piped into the high-pressure side of the system that will close a set of electrical contacts upon a preset rise in pressure, Figure 20–35. The electrical contacts will stop and start the fan motor upon pressure changes.

A common setup for R-12 may call for the fan to cut off when the head pressure falls to 125 psig and to re-

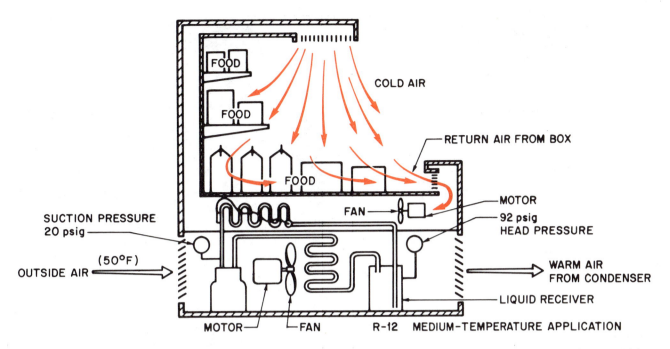

Figure 20–33. The fixture in Figure 20–32 is relocated outside the store. Now the condenser is subject to the winter conditions. The performance will fall off if some type of head pressure control is not furnished. The head pressure is so low that it cannot push enough liquid refrigerant through the expansion valve. The evaporator is starved. The unit has a reduced capacity that would be evident if a load of warm food were placed in the box. It may not pull the food temperature down. The coil would ice-up because there would not be any off cycle defrost.

start the fan when the head pressure reaches 175 psig. This setting will not interfere with the summer operation of the system and will give good performance in the winter. The settings are far enough apart to keep the fan from short cycling any more than it has to. When the thermostat calls for the compressor to come on and the ambient air is cold, the head pressure would be so low that the system would never get up to good running capacity with the condenser fan blowing more cold air over the condenser. The condenser fan cycling

control would keep the condenser fan off until the head pressure is within the correct operating range.

The fan cycling device is one method of maintaining a correct operating head pressure. From an economical standpoint it is very attractive. It can be added to the system without much expense for the control, and the piping normally does not have to be altered. One problem with this device is that it has a tendency to cause the head pressure to swing up and down as the fan is stopped and started. This can also affect the expansion

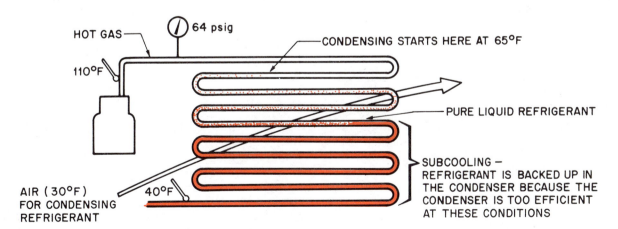

Figure 20–34. How the condenser acts in low ambient condition. Notice that this condenser has much more liquid refrigerant in it and the head pressure is very low. It may be so low that it may starve the expansion device because it relies on this pressure to force liquid refrigerant through its metering device.

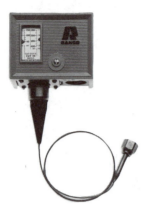

Figure 20–35. Condenser fan cycling device. This control has the same action (make on a rise in pressure) as the low-pressure control except that it operates at a much higher range. *Courtesy Ranco*

Figure 20–37. Condenser with air shutter. With one fan it is the only control. With multiple fans the other fans can be cycled by temperature with the shutter controlling the final fan. *Courtesy of Trane Company*

device operation, because the pressure may be 175 psig for part of the cycle and then, when the fan comes on, drop to 125 psig. This pressure swing causes the expansion device to operate erratically. There are other means to control the head pressure at a steady state on air-cooled condensers.

When a condenser has more than one fan, one fan can be put in the lead, and the other fans can be cycled off by temperature, Figure 20–36. The lead fan can be cycled off by pressure just like a single fan. This can help prevent the pressure swings from being so close together like the single-fan application. For example, when three fans are used, the first will cycle off at approximately 70°F, and the second will cycle off at approximately 60°F. The remaining fan will be controlled by the pressure.

Air Shutters or Dampers

These shutters may be located either at the inlet to the condenser or at the outlet. The air shutter has a pressure-operated motor that pushes a shaft to open the shutters when the head pressure rises to a

predetermined pressure, Figure 20–37. This pressure-operated motor is actually a piston in a cylinder. When the pressure rises, the piston arm extends to open the shutters, Figure 20–38.

When a single fan is used, the shutter is installed over the inlet or over the outlet to the fan. When there are multiple fans, the shutter-covered fan is operated all the time, and the other fans can be cycled off by pressure or temperature. This arrangement can give good head-pressure control down to low temperatures.

Condenser Flooding

Flooding the condenser with liquid refrigerant causes the head pressure to rise just as though the condenser were covered with a plastic blanket. It is accomplished by having enough refrigerant in the system to flood the condenser with liquid refrigerant in both mild and cold weather. This calls for a large refrigerant charge and a place to keep it. In addition to the charge there must be a valve arrangement to allow the refrigerant liquid to fill the condenser during both mild and cold weather. This condenser flooding method is designed to maintain the correct head pressure in the coldest weather during start up and while operating, Figure 20–39.

Figure 20–36. Multiple-fan condenser. *Courtesy Bohn Heat Transfer*

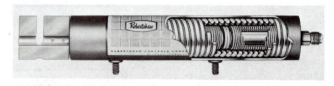

Figure 20–38. Piston-type shutter operator. This piston has the high-pressure discharge gas on one side of the bellows and the atmosphere on the other side. When the head pressure rises to a predetermined point, the shutters begin to open. In summer the shutters will remain wide open during the running cycle. *Courtesy Robertshaw Controls Company*

Figure 20–39. Head-pressure control for condenser flooding. This valve allows the refrigerant to flood the condenser during both mild and cold weather. This method requires enough refrigerant to flood the condenser and has a large receiver to hold the refrigerant during the warm season when it is not needed to keep the head pressure up. *Courtesy Sporlan Valve Company*

20.20 UTILIZING THE CONDENSER'S SUPERHEAT

Air-cooled condensers have the characteristic of high discharge line temperatures even in winter. This can be used to advantage, because the heat can be captured in winter and redistributed as heat for the structure or to heat hot water. The refrigeration system is rejecting heat out of the refrigerated box, heat that leaked into the box from the store itself. This heat has to be rejected to a place that it is unobjectionable. In summer the heat should be rejected outside the store or possibly into a domestic hot-water system. The domestic hot-water system can be used for cleaning around the store, Figure 20–40.

20.21 HEAT RECLAIM

In winter the store needs heat. If heat could be rejected inside the store, it could reduce the heating

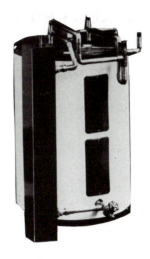

Figure 20–40. Heat exchanger to capture the heat from the highly superheated discharge line and use it to heat domestic hot water. This line can easily be 200°F and can furnish 140°F water in a supply limited by the size of the refrigeration system. The larger the system, the more heat available. *Courtesy Noranda Metal Industries Inc.*

cost. Any heat that is recovered from the system is heat that does not have to be purchased, Figure 20–41.

With air-cooled equipment heat recovery can be accomplished easily. The discharge gas can be offered an option of passing to the rooftop condenser or to a coil mounted in the ductwork that supplies heat to the store. The condensing temperature of air-cooled equipment is high enough to be used as heat, and the quantity of heat is sizable enough to be important. Some stores in moderate climates will be able to extract enough heat from the refrigeration system to supply the full amount of heat to the store.

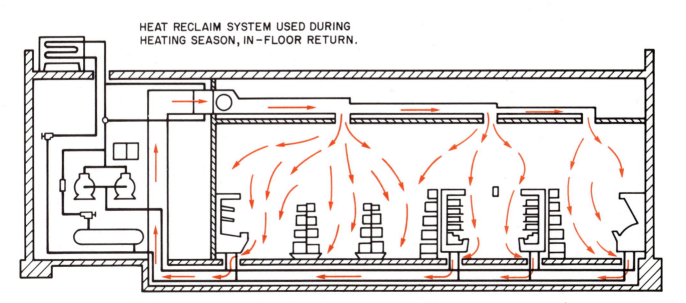

HEAT RECLAIM SYSTEM USED DURING HEATING SEASON, IN-FLOOR RETURN.

Figure 20–41. System that can supply heat to the store.

20.22 CONDENSER EVALUATION

A final note: Don't get lost in the details of the equipment. Every compression system has a condenser to reject the heat from the system. Examination of the equipment will disclose the condenser, whatever type it may be. The condenser will be hot on air-cooled equipment and warm on water-cooled equipment.

SUMMARY

- The condenser is the component that rejects the heat from the refrigeration system.
- The refrigerant condenses to a liquid in the condenser and gives up heat.
- Water was the first medium that heat was rejected into through the water-cooled condenser.
- There are three types of water-cooled condensers: the pipe within a pipe, the shell and coil, and the shell and tube.
- The pipe within a pipe with flanges and the shell and tube condensers can be cleaned with brushes.
- The pipe within a pipe and the shell and coil condensers have to be cleaned with chemicals.
- *When condenser cleaning is desirable, consult the manufacturer.*
- The greatest amount of heat is given up from the refrigerant while the condensing process is taking place.
- The refrigerant normally condenses about 10°F higher than the condensing medium in a water-cooled condenser.
- The first job of the condenser is to desuperheat the hot gas flowing from the compressor.
- After the refrigerant is condensed to a liquid, the liquid can be further cooled below the condensing temperature. This is called subcooling.
- When the condensing medium is cold enough to reduce the head pressure to the point that the expansion device will starve the evaporator, head-pressure control must be used, whether air or water is used as the condensing medium.
- Water-cooled equipment is more efficient than air-cooled equipment.
- Water-cooled equipment has two places to deposit heat absorbed by the condenser: (1) down the drain, in the wastewater system; (2) into the atmosphere, in a cooling tower.
- There are three types of cooling towers: (1) natural draft, (2) forced draft, (3) evaporative condenser.
- Recirculated water uses evaporation to help the cooling process.
- *When water is evaporated, it has a tendency to overconcentrate the minerals, and water must be added to the system to keep this from happening.*
- Several types of common head-pressure controls are fan cycling, shutters, and condenser flooding.
- There is a relationship of the condensing temperature to the air passing over a condenser that can help the service technician determine what the high-pressure reading should be on an air-cooled condenser. Normally the refrigerant condenses at about 35°F higher than the air entering the condenser.
- The heat from a refrigeration system can be captured and used to heat water or to add heat to the conditioned space.

REVIEW QUESTIONS

1. What is the responsibility of the condenser in the refrigeration system?
2. Name three types of water-cooled condensers.
3. Why do some condensers have to be cleaned with brushes and others with chemicals?
4. Name three materials that condensers are normally made of.
5. Who should be consulted when condenser cleaning is needed?
6. When a water-cooled condenser is operating, the refrigerant normally condenses at _____°F higher than the leaving water.
7. The first job of the condenser is to _____ the gas before condensing can take place.
8. When is the most heat removed from the refrigerant in the condensing process?
9. After the refrigerant is condensed to a liquid, the remaining liquid can be further cooled down. This further cooling is called _____.
10. After heat is absorbed into a condenser medium in a water-cooled condenser, the heat can be deposited in one of two places. What are they?
11. Why does a water-cooling tower overconcentrate in mineral content?
12. A water-cooling tower capacity is governed by what aspect of the ambient air?
13. When an air-cooled condenser is used, the condensing refrigerant will normally be _____°F higher in temperature than the entering air temperature.
14. Name three methods of controlling head pressure in an air-cooled condenser.
15. The high pressure normally has to be _____psig higher than the suction pressure for the expansion device to feed the evaporator correctly.
16. The prevailing winds can effect which of the air-cooled condensers?
17. Which condenser is more efficient: air-cooled or water-cooled?
18. Which type of condenser has the lower operating head pressures: air-cooled or water-cooled?
19. True or False: Water can be heated with the hot discharge gas from a compressor.
20. Name two methods to use the heat from an air-cooled condenser.

COMPRESSORS

OBJECTIVES

After studying this unit, you should be able to

- **explain the function of the compressor in a refrigeration system.**
- **identify reciprocating and rotary compressors.**
- **state specific conditions under which a compressor is expected to operate.**
- **explain the difference between a hermetic compressor and a semihermetic compressor.**
- **describe the various working parts of reciprocating and rotary compressors.**

21.1 THE FUNCTION OF THE COMPRESSOR

The *compressor* is considered the heart of the refrigeration system. It is a pump, like the heart in the circulatory system of the human body. However, the compressor only pumps vapor. The simplest definition of a compressor is *vapor pump*. The compressor ac-

tually lifts (increases) the pressure in the system from the suction pressure level to the discharge pressure level. For example, in a low-temperature system the suction pressure for a system that has R-12 as the refrigerant may have a suction pressure of 3 psig and a discharge pressure of 181 psig. The compressor lifts the pressure 178 psig (181 − 3 = 178), Figure 21–1. A system next to the low-temperature system may have a different lift. It may be a medium-temperature system and have a suction pressure of 21 psig with a discharge pressure of 181 psig. This system has a lift of 160 psig (181 − 21 = 160), Figure 21–2. An understanding of Unit 1 is vital at this point.

The compressor has cool refrigerant entering the suction valve to fill the cylinders. This cool vapor contains the heat absorbed in the evaporator. When the refrigerant was boiled (changed to a vapor), it absorbed heat. The compressor is now pumping this heat-laden vapor to the condenser so that it can be rejected from the system.

R-12   LOW-TEMPERATURE APPLICATION

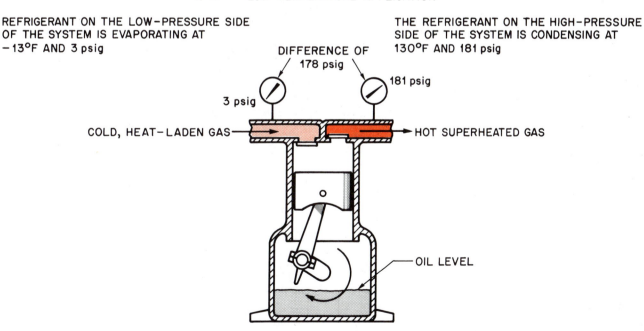

Figure 21-1. Pressure difference between the suction and discharge side of the compressor.

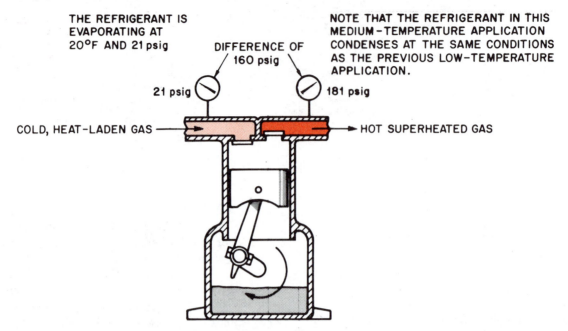

Figure 21-2. Combination of suction and discharge pressures. This is R-12 applied to a medium-temperature system. The compressor only has to lift the suction gas 160 psi.

The vapor leaving the compressor can be very warm. With a discharge pressure of 181 psig, the discharge line at the compressor could easily be 200°F. The vapor is compressed with the heat from the suction gas concentrated in the gas leaving the compressor, Figure 21-3.

21.2   TYPES OF COMPRESSORS

Two types of compressors are commonly used in small and medium refrigeration systems: (1) reciprocating and (2) rotary. Reciprocating compressors are more numerous than rotary compressors. They are used almost exclusively in the smaller range of compressors. The reciprocating compressor and the rotary compressor will be covered in this text as they apply to refrigeration.

The Reciprocating Compressor

Reciprocating compressors are categorized by the compressor housing and by the drive mechanisms. The two housing categories are *hermetic* and *open* compressors, Figure 21-4. Hermetic refers to the type of housing the compressor is contained in and is divided into two types: fully welded and serviceable, Figure 21-5. The drive mechanisms may either be enclosed inside the shell or outside the shell. When the compressor is hermetic, the drive mechanism is direct. The compressor and motor shaft are the same shaft.

Fully Welded Hermetic Compressors. The motor and compressor are contained inside a single shell that is welded closed when a welded hermetic compressor is manufactured. This unit is sometimes called the *tin can* compressor because it cannot be serviced without cut-

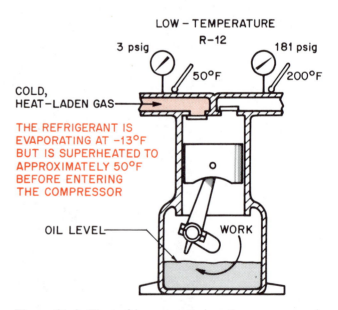

Figure 21-3. The refrigerant entering the compressor is called heat laden because it contains the heat that was picked up in the evaporator from the boiling process. The gas is cool but full of heat that was absorbed at a low pressure and temperature level. When this gas is compressed in the compressor, the heat concentrates. In addition to the heat that was absorbed in the evaporator, the act of compression converts some energy to heat.

Figure 21-4. (A) Open-drive compressor. (B) Hermetic compressor. *Courtesy (A) Trane Company (B) Copeland Corporation*

Figure 21-5. (A) Serviceable hermetic compressor is used in some smaller refrigeration systems and exclusively in the larger hermetic applications. (B) The welded hermetic compressor is used in the smaller compressor sizes, up to about 7.5 hp. *Courtesy (A) Copeland Corporation (B) Bristol Compressors Inc.*

Figure 21-6. Welded hermetic compressor. Most welded hermetic compressors have a few things in common. The suction line usually pipes directly into the shell and is open to the crankcase. The discharge line normally is piped from the compressor inside the shell to the outside of the shell. The compressor shell is typically thought of as a low-side component. *Courtesy Bristol Compressors Inc.*

ting open the shell, Figure 21-6. Characteristics of the fully welded hermetic compressor are listed here:

1. There is no access to the inside of the shell except by cutting the shell open.
2. They are only opened by a very few companies that specialize in this type of work. Otherwise, unless the manufacturer wants the compressor back for examination, it is a throw-away compressor.
3. The motor shaft and the compressor crankshaft are one shaft.
4. It is usually considered a low-side device because the suction gas is vented to the crankcase, the whole inside of the shell. The discharge (high-pressure) line is normally piped to the outside of the shell so that the shell only has to be rated at the low-side working-pressure value.
5. Generally, they are cooled with suction gas.

6. They usually have a pressure lubrication system.

7. The combination motor and crankshaft are customarily in a vertical position with a bearing at the bottom of the shaft next to the oil pump. The second bearing is located about halfway on the shaft between the compressor and the motor.

8. The pistons and rods work outward from the crankshaft, so they are working at a 90° angle in relation to the crankshaft, Figure 21–7.

Serviceable Hermetic Compressors. When a serviceable hermetic compressor is manufactured, the motor and compressor are contained inside a single shell that is bolted together. This unit can be serviced by removing the bolts and opening the shell at the appropriate place, Figure 21–8. Characteristics of the serviceable hermetic compressor are listed here:

1. The unit is bolted together at locations that will be conducive to service and repair.

2. The housing is normally cast iron and may have a steel housing fastened to the cast iron compressor. They are normally heavier than the fully welded type.

3. The motor and crankshaft combination are similar to that in the fully welded type except that the crankshaft is usually horizontal.

4. They generally use a splash-type lubrication system in the smaller compressors and a pressure lubrication system in the larger compressors.

5. They are often air cooled and can be recognized by the fins in the casting or extra sheet metal on the outside of the housing to give the shell more surface area.

Figure 21–8. Serviceable hermetic compressor designed in such a manner that it can be serviced in the field. *Courtesy Copeland Corporation*

6. The piston heads are normally at the top or near the top of the compressor and work in and out from the center of the crankshaft, Figure 21–9.

Open-Drive Compressors. Open-drive compressors are manufactured in two styles: belt drive and direct drive, Figure 21–10. Any compressor with the drive on the outside of the casing has to have a shaft seal to keep the refrigerant from escaping to the atmosphere. This seal arrangement has not changed much in many years.

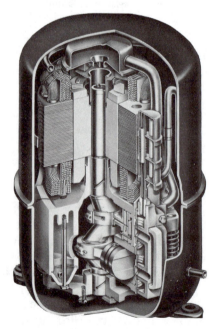

Figure 21–7. Internal workings of a welded hermetic compressor. *Courtesy Tecumseh Products Company*

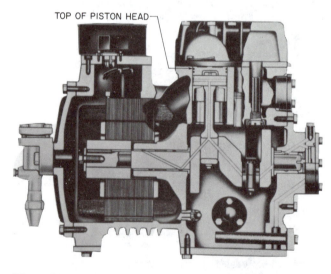

Figure 21–9. Working parts of the serviceable hermetic compressor. The crankshaft is in the horizontal position, and the rods and pistons move in and out from the center of the shaft. The oil pump is on the end of the shaft and draws the oil from the crankcase at the bottom of the compressor. *Courtesy Copeland Corporation*

Figure 21–10. (A) Belt-drive compressor. (B) Direct-drive compressor. The belt-drive may have different speeds, depending on the pulley sizes. The direct-drive turns at the speed of the motor because it is attached directly to the motor. 1750 rpm and 3450 rpm are common speeds. The coupling between the motor and the compressor is slightly flexible. *Reproduced courtesy of Carrier Corporation*

Belt-Drive Compressors. The belt-drive compressor was the first type of compressor and is still used to some extent. With the belt-drive unit the motor and its shaft are parallel with the compressor's shaft. The motor is beside the compressor. Notice that since the compressor and motor shaft are parallel, there is a sideways pull on both shafts to tighten the belts. This strains both shafts and requires the manufacturer to compensate for this in the shaft bearings, Figure 21–11.

Direct-Drive Compressors. The direct-drive compressor differs from the belt-drive in that the compressor shaft is end to end with the motor shaft. These shafts have a coupling between them with a small amount of flexibility. The two shafts have to be in very good alignment to run correctly, Figure 21–12.

21.3 COMPRESSOR COMPONENTS

Regardless of the type of compressor, they all have similarities in the method of pumping gas.

Crankshafts

The *crankshaft* of a reciprocating compressor transmits the circular motion to the rods, and the motion is changed to back and forth (reciprocating) for the pistons, Figure 21–13. Crankshafts are normally manufactured of cast iron or soft steel. The crankshaft can be cast (molten metal poured into a mold) into a general shape and machined into the exact size and shape. In this case the shaft has to be cast iron, Figure 21–14. This machining process is very critical because the throw (the off-center part where the rod fastens) does not turn in a circle (in relationship to the center of the shaft) when placed in a lathe. The machinist has to know how to work with this type of setup.

These off-center shafts normally have two main-bearing surfaces in addition to the off-center rod-bearing surfaces: One is on the motor end of the shaft, and one is on the other end. The bearing on the motor end is normally the largest because it carries the greatest load.

Some shafts are straight and have a cam-type arrangement called an *eccentric*. This allows the shaft to be manufactured straight and from steel. The shaft may not be any more durable, but it is easier to machine. The eccentrics can also be machined straight with an off-center hole to accomplish the reciprocating action, Figure 21–15. Notice that the rod has to be different for the eccentric shaft because the end of the rod has to fit over the large eccentric on the crankshaft.

All of these shafts have to be lubricated. The smaller compressors using the splash system may have a catch basin to catch oil and cause it to flow down the center of the shaft, Figure 21–16. It is then slung to the outside of the eccentric when the compressor runs, which causes the oil to move to the other parts, such as the rods on both ends.

Some of the shafts are drilled and lubricated with a pressure lubrication system. These compressors have an oil pump mounted on the end of the crankshaft that turns with the crankshaft, Figure 21–17. *Note:* When the compressor starts, there is no lubrication until the compressor is running.

Since some compressors have vertical shafts and some have horizontal, manufacturers have been challenged to provide proper lubrication where needed. Consult the compressor manufacturer for any question you may have about a specific application.

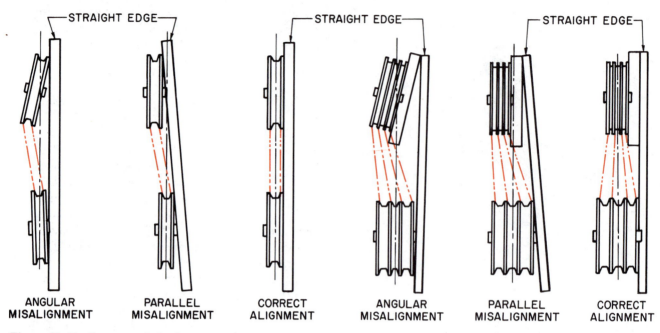

Figure 21-11. Correct and the incorrect alignments of belt-drive compressor with its motor. The belts used on a multibelt installation are matched at the factory for the correct length.

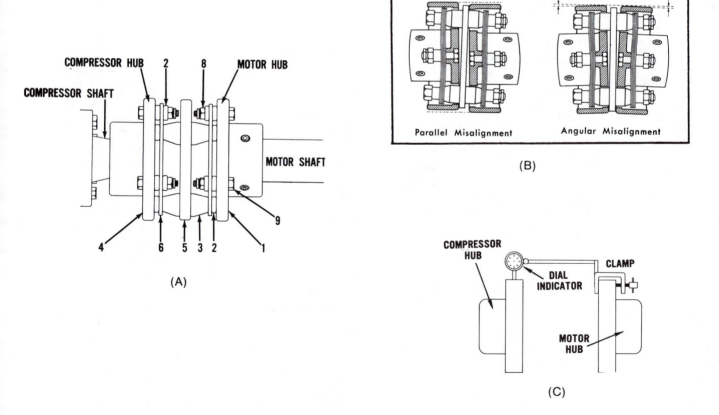

Figure 21-12. Alignment of a direct-drive compressor and motor. This alignment has to be within the compressor manufacturer's specifications. These compressors turn so fast that if the alignment is not correct, there will be a seal or bearing failure very soon. When the correct alignment is attained, the compressor and motor are normally fastened permanently to the base they are mounted on. The motor and the compressor can be rebuilt in place on larger installations. *Courtesy Trane Company*

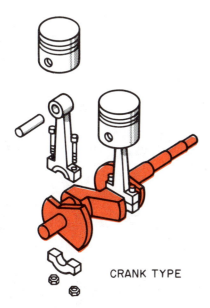

Figure 21-13. Relationship of the pistons, rods, and crankshaft. *Reproduced courtesy of Carrier Corporation*

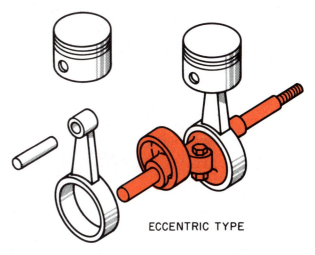

Figure 21-15. This crankshaft obtains the off-center action with a straight shaft and an eccentric. The eccentric is much like a cam lobe. The rods on this shaft have large bottom throws and slide off the end of the shaft. This means that to remove the rods, the shaft must be taken out of the compressor. *Reproduced courtesy of Carrier Corporation*

Connecting Rods

Connecting rods connect the crankshaft to the piston. These rods, as they are called in industry, are normally made in two styles: the type that fit the crankshaft with off-center throws, and the type to fit the eccentric crankshaft, Figure 21-18. Rods can be made of several different metals. Iron, brass, and aluminum are common. The rod design is very important because it takes a lot of the load in the compressor. If the crankshaft is connected directly to the motor and the motor is running at 3450 revolutions per minute (rpm) the piston at the top of the rod is changing directions 6900 times per minute. The rod is the connection between this piston and the crankshaft and is the link between all of this changing of direction.

The rods with the large holes in the shaft end are for eccentric shafts. They cannot be taken off with the shaft in place. The shaft has to be removed to take the piston out of the cylinder. The rods with the small holes are for the off-center shafts; these are split and have rod bolts, Figure 21-19. These rods can be separated at the crankshaft, and the rod and piston can be removed with the crankshaft in place.

The rod is small on the piston end and fastens to the piston by a different method. The rod normally has

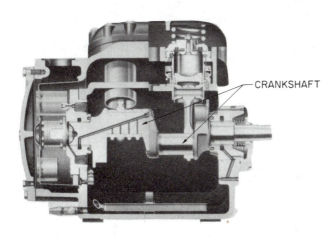

Figure 21-14. Crankshafts are cast into the general shape and machined to the correct shape in a machine shop. *Courtesy Trane Company*

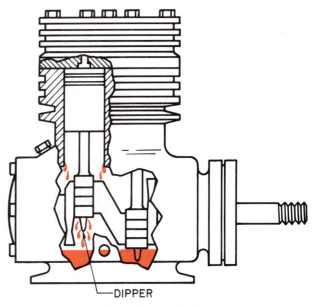

Figure 21-16. Splash lubrication system that splashes oil up onto the parts. *Reproduced courtesy of Carrier Corporation*

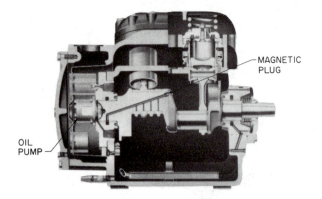

Figure 21-17. Crankshaft drilled for the oil pump to force the oil up the shaft to the rods, then up the rods to the wrist pins. Magnetic elements are sometimes placed along the passage to capture iron filings. *Courtesy Trane Company*

Figure 21-19. Service technician removing the rods from a compressor. The rods have the split bottoms. *Courtesy Trane Company*

a connector called a *wrist pin* that slips through the piston and the upper end of the rod. This almost always has a *snap ring* to keep the wrist pin from sliding against the cylinder wall, Figure 21-20.

The Piston

The *piston* is the part of the cylinder assembly exposed to the high-pessure gas during compression. Pistons have high-pressure gas on top and suction or

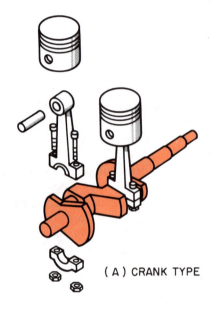

(A) CRANK TYPE

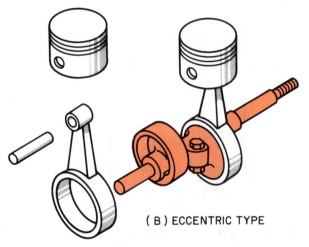

(B) ECCENTRIC TYPE

Figure 21-18. Different types of rods. The rod that has bolts fits the off-center crankshaft, and the rods with the large holes in the bottom are for the eccentric type of crankshaft. *Reproduced courtesy of Carrier Corporation*

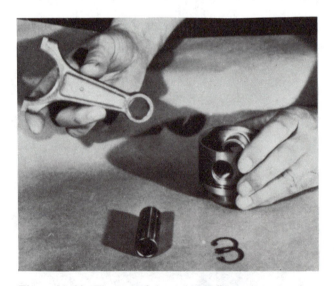

Figure 21-20. The end of the rod that fits up into the piston. The wrist pin holds the rod to the piston while allowing the pivot action that takes place at the top of the stroke. The wrist pin is held secure in the piston with snap rings. *Photo by Bill Johnson*

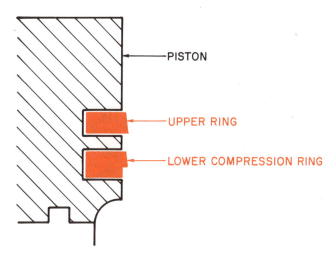

Figure 21-21. Piston rings for a refrigeration compressor resemble the rings used on automobile pistons. There are oil and compression rings. *Courtesy Trane Company*

low-pressure gas on the bottom during the upstroke. They have to slide up and down in the cylinder in order to pump. They must have some method of preventing the high-pressure gas from slipping by to the crankcase. Piston rings are used on the larger piston just as in automobile engines. As in an automobile, these rings are of two types: compression and oil. The smaller compressors use the oil on the cylinder walls as the seal. An example of these rings can be seen in Figure 21-21.

Refrigerant Cylinder Valves

The valves in the top of the compressor determine the direction in which the gas entering the compressor will flow. A cutaway of a compressor cylinder is shown in Figure 21-22. These valves are made of very hard

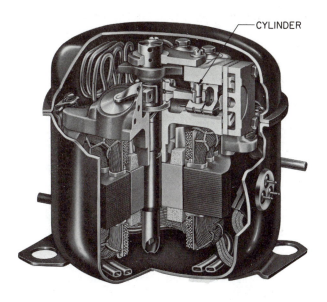

Figure 21-22. Cutaway of compressor showing typical cylinder. *Courtesy Tecumseh Products Company*

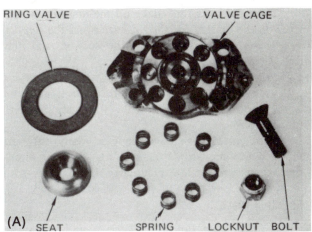

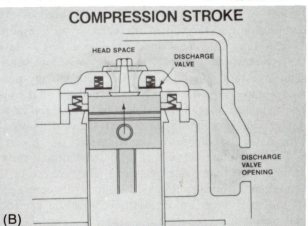

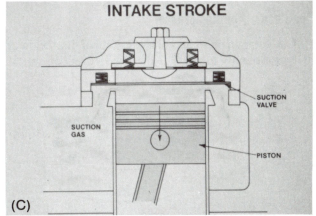

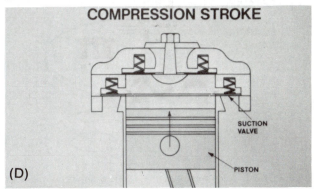

Figure 21-23. Ring valves. They normally have a set of small springs to close them. *Courtesy Trane Company*

steel. The two styles that make up the majority of the valves on the market are the *ring valve* and the *flapper (reed) valve*. They serve both the suction and the discharge ports of the compressor.

The ring valve is made in a circle with springs under it. If ring valves are used for the suction and the discharge, the larger one will be the suction valve, Figure 21–23.

Flapper valves have been made in many different shapes to accomplish the same thing. Each manufacturer has its own version, Figure 21–24.

The Valve Plate

The *valve plate* holds the suction and discharge flapper valves. It is located between the head of the compressor and the top of the cylinder wall, Figure 21–25. Many different methods have been used to hold the valves in place without taking up any more space than necessary. The bottom of the plate actually protrudes into the cylinder. Any volume of gas that cannot be pumped out of the cylinder because of the valve design will reexpand on the downstroke of the piston. This makes the compressor less efficient.

There are other versions of the crankshaft and valve arrangements than those listed here. They are not used enough in the refrigeration industry to justify coverage. If you need more information, contact the specific manufacturer.

The Head of the Compressor

The component that holds the top of the cylinder and its assembly together is the *head*. It sets on top of

Figure 21–25. Valve plate typical of those used to hold the valves. They can be replaced or rebuilt if not badly damaged. There is a gasket on both sides. *Reproduced courtesy of Carrier Corporation*

the cylinder and contains the high-pressure gas from the cylinder until it moves into the discharge line. They often contain the suction chamber, separated from the discharge chamber by a partition and gaskets. These heads have many different design configurations and need to accomplish two things. They hold the pressure in and hold the valve plate on the cylinder. They are made of steel in some welded hermetic compressors and of cast iron in a serviceable hermetic type. The cast iron heads may be in the moving airstream and have fins on them to help dissipate the heat from the top of the cylinder, Figure 21–26.

Mufflers

Mufflers are used in many fully hermetic compressors to muffle compressor pulsation noise. There are audible suction and discharge pulsations that can be transmitted into the piping if they are not muffled. Mufflers must be designed to have a low pressure drop and still muffle the discharge pulsations, just as in automobiles, Figure 21–27.

The Compressor Housing

The *housing* holds the compressor and sometimes the motor. It is made of stamped steel for the welded hermetic and of cast iron for the serviceable hermetic.

The welded hermetic compressor is designed so the compressor shell is under low-side pressure and will normally have a working pressure of from 150 to 250 psig, depending on the application. The compressor is mounted inside the shell, and the discharge line is normally piped to the outside of the shell. This frees the shell from needing to have a test pressure as high as the high-side pressure. A cutaway of a hermetic compressor inside a welded shell and the method used to weld the shell together are shown in Figure 21–28.

Two methods are used to mount the compressor inside the shell: rigid and springs.

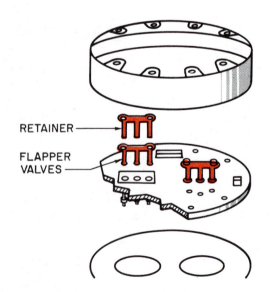

RETAINER

FLAPPER VALVES

Figure 21–24. Reed or flapper valves held down on one end. This provides enough spring action to close the valve when reverse flow occurs. *Reproduced courtesy of Carrier Corporation*

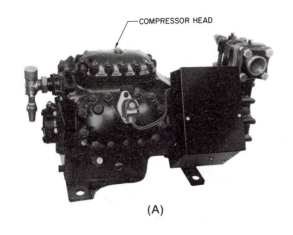

Figure 21-26. Typical compressor heads (A) suction-cooled compressor and (B) air-cooled compressor. The compressors that have air-cooled motors are located in a moving airstream, or overheating will occur. *Courtesy Copeland Corporation*

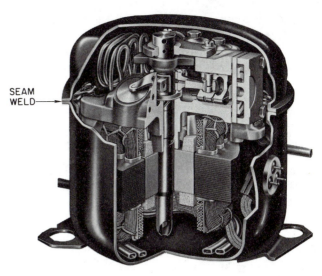

Figure 21-28. Compressor shell fastened with a seam and then welded. *Courtesy Tecumseh Products Company*

Rigid-mounted compressors were used for many years. The compressor shell was mounted on external springs that had to be bolted tightly for shipment. The springs were supposed to be loosened when installed, they often were not, and the compressor vibrated because, without the springs, it was mounted rigidly to the condenser casing. External springs can also rust, especially where there is a lot of salt in the air, Figure 21-29.

The internal spring-mounted compressors actually suspend the compressor from springs inside the shell. These compressors have methods of keeping the compressor from moving too much during shipment.

Figure 21-27. Compressor muffler. *Courtesy Trane Company*

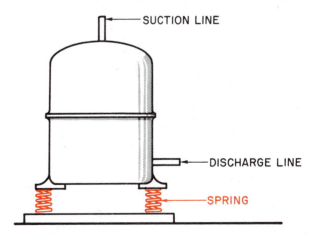

Figure 21-29. Compressor motor pressed into its cast iron shell. It requires experience to remove the motor. The compressor has springs under the mounting feet to help eliminate vibration. This compressor is shipped with a bolt tightened down through the springs. This bolt must be loosened by the installing contractor.

Sometimes a compressor will come loose from one or two of the internal springs. When this happens, the compressor will normally run and pump just like it is supposed to but will make a noise on startup or shutdown or both. If the compressor comes off the springs and they are internal, there is nothing that can be done to repair it in the field, Figure 21-30.

Compressor Motors in a Refrigerant Atmosphere

The compressor motor operating inside the refrigerant atmosphere must have special consideration. Motors for hermetic compressors differ from standard electric motors. They actually run in a refrigerant vapor. The materials used in a hermetic motor are not the same materials that would be used in a fan or pump motor that would run in air. Hermetic motors have to be manufactured of materials compatible with the system refrigerants. For instance, rubber cannot be used because the refrigerant would dissolve it. The motors are assembled in a very clean atmosphere and kept dry. When a hermetic motor malfunctions, take it to a qualified motor repair shop if it must be repaired.

Motor Electrical Terminals. There must be some conductor to carry the power from the external power supply to the internal motor. The power to operate the compressor has to be carried through the compressor housing without the refrigerant leaking. The connection also has to insulate the electric current from the compressor shell. These terminals are sometimes fused

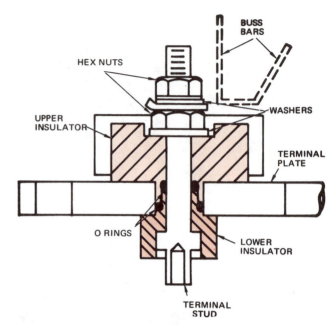

Figure 21-31. Motor terminals. **The power to operate the** compressor is carried through the compressor shell but must be insulated from the shell. This is a fiber block used as the insulator. O rings keep the refrigerant in. *Courtesy Trane Company*

glass with a terminal stud through the middle on the smaller compressors. When larger terminals are required, the terminals are sometimes placed in a fiber block with an O ring-type seal, Figure 21-31.

Care has to be taken with these terminals (due to loose electrical connections) to prevent overheating. Should the terminal overheat, a leak could occur. If the terminal block is a fused-glass type, it would be very hard to repair. The fused-glass type will stand more heat, but there is a limit to how much heat it can take. Less heat can be tolerated with the O ring and fiber-type of terminal board. When the O ring and fiber board are damaged, they can be replaced with new parts. Refigerant loss can result before the problem is discovered.

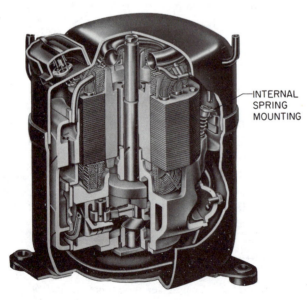

Figure 21-30. Compressor mounted inside the welded hermetic shell on springs. The springs have guides that only allow them to move a certain amount during shipment. *Courtesy Tecumseh Products Company*

Internal Overloads. Internal overloads in hermetic motors protect the motor from overheating. These overloads are embedded in the windings and are wired in two different ways. One style of overload breaks the line circuit inside the compressor. Since it is internal and carries the line current, it is limited to smaller compressors. It has to be enclosed to prevent the electrical arc from affecting the refrigerant. If contacts of this line type remained open, the compressor cannot be restarted. The compressor would have to be replaced.

Another type of overload breaks the control circuit. This is wired to the outside of the compressor to the control circuit. If the pilot-duty type wired to the outside of the compressor were to remain open, an external overload could be substituted, Figure 21-32.

The Serviceable Hermetic Compressor. The serviceable hermetic normally has a cast iron shell and is considered a low-side device. Again, because of the piping arrangement in the head, the discharge gas is contained either under the head or out of the discharge line. The motor is rigidly mounted to the shell, and the compressor must be externally mounted on springs or other flexible mounts to prevent vibration. The serviceable hermetic is used exclusively in larger compressor sizes because it can be rebuilt. The compressor components are much the same as the components in the welded hermetic and will not be covered again, Figure 21-33.

Open-Drive Compressor. Open-drive compressors are manufactured with the motor external to the compressor shell. The shaft protrudes through the casing to the outside where either a pulley or a coupling is attached. This compressor is normally very heavy duty in nature. It has to be able to be mounted down tightly to a foundation. The motor is either mounted end to end with the compressor shaft or beside the compressor and belts are used to turn the compressor.

The Shaft Seal

The pressure inside the compressor crankcase can be either in a vacuum (below atmosphere) or a positive

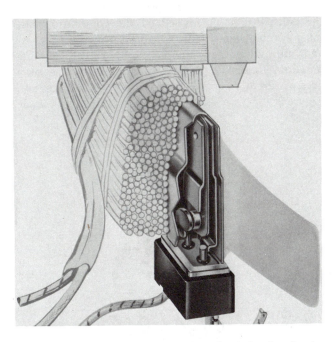

Figure 21-32. Internal compressor overload that breaks the line circuit. Since this set of electrical contacts is inside the refrigerant atmosphere, they are contained inside a hermetic container of their own. If the electrical arc were allowed inside the refrigerant atmosphere, the refrigerant would deteriorate in the vicinity of the arc. *Courtesy Tecumseh Products Company*

Figure 21-33. Serviceable hermetic compressors. *Courtesy (A) Reproduced courtesy of Carrier Corporation (B) Trane Company (C) Copeland Corporation*

pressure. If the unit were an extra-low-temperature unit using R-12 as the refrigerant, the crankcase pressure could easily be in a vacuum. If the shaft seal were to leak, the atmosphere would enter the crankcase. When the compressor is sitting still, it could have a high positive pressure on it. For example, when R-502 is used and the system is off for extended periods (the whole system may get up to 100°F in a hot climate), the crankcase pressure may go over 200 psig. The crankcase shaft seal has to be able to hold refrigerant inside the compressor under all of these conditions. The shaft is turning at high speed during any running conditions, Figure 21-34.

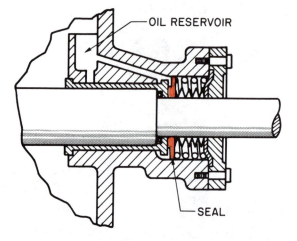

Figure 21–34. Shaft seal is responsible for keeping the refrigerant inside the crankcase and allowing the shaft to turn at high speed. This seal must be installed correctly. If the seal is installed on a belt-drive compressor, the belt tension is important. If it is installed on a direct-drive compressor, the shaft alignment is important. *Reproduced courtesy of Carrier Corporation*

The shaft seal has a rubbing surface to keep the refrigerant and the atmosphere separated. This surface is normally a carbon material rubbing against a steel surface. If assembled correctly, these two surfaces can rub together for years and not wear out. This correct assembly normally consists of the shafts being aligned correctly. The belts have to have the correct tension if the unit is a belt drive. If the unit is direct drive, the shafts have to be aligned according to manufacturer's instructions.

21.4 BELT-DRIVE MECHANISM CHARACTERISTICS

The belt-drive compressor has the motor mounted to the side of the compressor and has a pulley on the motor as well as on the compressor. The pulley on the motor is called the *drive pulley,* and the pulley on the compressor is called the *driven pulley.* The drive pulley is sometimes adjustable in a fashion that will allow the compressor speed to be adjustable. The drive pulley can also be changed to a different size to vary the compressor speed. This can be advantageous when a compressor is too large for a job (too much capacity) and needs to be slowed down to compensate. Different pulleys are shown in Figure 21–35.

The compressor can also be speeded up for more capacity if the motor has enough horsepower in reserve (if the motor is not already running at maximum horsepower). If a pulley size change is needed, consult the compressor manufacturer to be sure that the design limits of the compressor are not exceeded. Most compressors are not designed to turn over a certain rpm, and the compressor manufacturer has this information. When a particular pulley size is needed, the pulley supplier will help you choose the correct size. Belt sizes also have to be calculated. Choosing the correct belts and pulleys is very important and should be done by an experienced person.

The compressor and motor shafts have to be in correct alignment with the proper tension applied to the belts. The motor base and the compressor base must be tightened rigidly so that no dimensions will vary during the operation. There are several belt combinations that may have to be considered. The compressor drive mechanism may have multiple belts, Figure 21–36, or a single belt if the compressor is small.

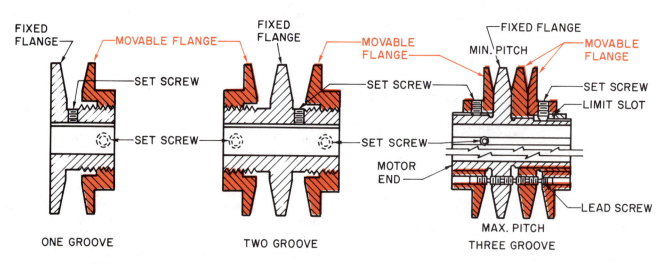

Figure 21–35. Different pulleys used on belt drive compressors. They are single groove and multiple groove pulleys with different belt widths. Some pulleys are adjustable for changing the compressor speed.

Figure 21–36. Multiple-belt compressor. *Courtesy Tecumseh Products Company*

Figure 21–37. Flexible coupling. There is an extensive procedure to obtain the correct shaft alignment. This must be done or bearings and seal will fail prematurely. *Courtesy Lovejoy, Inc.*

Belts also come in different types. The width of the belt, the grip type, and the material have to be considered.

When multiple belts are used, they have to be bought as matched sets. For example, if a particular compressor and motor drive has four V belts of B width and 88 in. long, you'll be in trouble if you buy four 88-in. belts because the four belts will vary slightly in length and will not pull correctly. The proper method is to order four 88-in. belts of B width that are factory matched for the exact length. These are called *matched belts* and have to be used on multiple-belt installations.

21.5   DIRECT-DRIVE COMPRESSOR CHARACTERISTICS

The direct-drive compressor is limited to the motor speed that the drive motor is turning. This installation has the motor and compressor shafts end to end. These shafts have a slightly flexible coupling between them but have to have very close alignment, or the bearings and seal will fail prematurely, Figure 21–37. This compressor and motor combination are mounted on a common rigid base.

Both the motor and compressor are customarily manufactured so that they can be rebuilt in place. Thus once the shafts are aligned, the shell of the motor and the shell of the compressor can be fastened down and always remain in place. If the motor or compressor has to be rebuilt, then the shafts will automatically line back up when reassembled.

The reciprocating compressor has not changed appreciably for many years. Manufacturers continuously try to improve the motor and the pumping efficiencies. The future will bring about different concepts to turn the circular motion of the motor into a more efficient and longer-lasting device. The valve arrangements can

make a difference in the pumping efficiency and are being studied for improvements. New valve concepts will be developed in the future. It seems simple to say design it and produce it, but food spoilage is at stake, and manufacturers want their products to be reliable.

21.6   THE ROTARY COMPRESSOR

Rotary compressors have been used to some extent in small refrigeration for household units. They are not used to any great extent for medium-sized applications. They are actually a more efficient compressor because they do not have the clearance volume at the top of the stroke. Manufacturers are continuously looking at methods to improve the compressor and will certainly be exploring further use of the rotary compressor.

The rotary compressor is like the reciprocating compressor in that it can be used in a shell as a hermetic or out of the shell as an open-drive unit. When the motor is in the shell, it has to be cooled because it builds up heat just as the reciprocating compressor does. Suction gas is often the choice for this cooling; however, it is not the only choice.

The rotary compressor has two basic design types: the stationary vane and the rotary vane.

Stationary Vane Rotary Compressor

The *stationary vane* compressor has one vane located in the housing of the compressor. The rotor (the movable part) is an off-center drum-shaped device that turns like the cam lobe. The tolerances between the rotor and the casing have to be very close for good efficiency. As the rotor turns, the vane slides in and out to trap gas between the rotor and the case. The gas is then pushed out the suction line. A valve is needed to keep the discharge gas from leaking back into the suction side during the off cycle, Figure 21–38.

Rotary Vane Rotary Compressor

The *rotary vane* compressor has several vanes that slide back and forth in the rotor. The rotor is fitted to a shaft exactly in the center of the rotor. The vanes

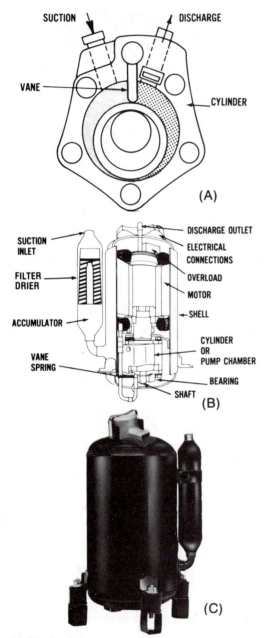

Figure 21–38. Stationary vane compressor has the compressor cylinder mounted off center on the shaft with the vane in the housing. As the cylinder of the compressor turns, it traps the refrigerant in the chamber and forces it out into the discharge line. This compressor has valves to prevent the high-pressure gas from coming back down the discharge line through the compressor and into the suction side of the system during the off cycle. *Reprinted with permission of Motors and Armatures, Inc.*

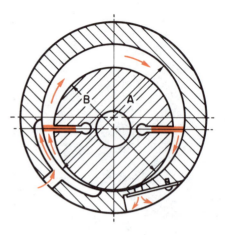

Figure 21–39. This compressor works much the same as the compressor in Figure 21–38 except there is more than one vane and they turn in the compressor's rotary cylinder. The cylinder on this compressor has the shaft drilled straight through the center. The trapping action of the gas comes from the sliding vanes.

trap the gas and force it out the discharge port. A suction valve is also needed to keep the discharge gas from leaking back to the suction side of the system during the off cycle, Figure 21–39.

21.7 RECIPROCATING COMPRESSOR EFFICIENCY

A compressor's efficiency is determined by the design of the compressor. The efficiency of a compressor starts with the filling of the cylinder. The following sequence of events takes place inside a reciprocating compressor during the pumping action.

A medium-temperature application will be used as the example for the pumping sequence. The suction pressure is 20 psig, and the discharge pressure is 180 psig.

1. **Piston at the top of the stroke and starting down.** When the piston has moved down far enough to create less pressure in the cylinder than is in the suction line, the flapper valve will open and the cylinder will start to fill with gas, Figure 21–40.
2. **Piston continues to the bottom of the stroke.** At this point the cylinder is nearly as full as it is going to get. There is a very slight time lag at the bottom of the stroke as the crankshaft carries the rod around the bottom of the stroke, Figure 21–41.
3. **Piston is starting up.** The rod throw is past dead center, and the piston starts up. When the cylinder is as full as it is going to get, the suction flapper valve closes.
4. **The piston proceeds to the top of the stroke.** When the piston reaches a point that is nearly at the top, the pressure in the cylinder becomes greater than the pressure in the discharge line. If the discharge pres-

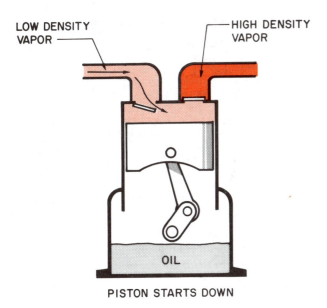

LOW DENSITY VAPOR

HIGH DENSITY VAPOR

OIL

PISTON STARTS DOWN

Figure 21–40. What happens inside the reciprocating compressor while it is pumping. When the piston starts down, a low pressure is formed under the suction reed valve. When this pressure becomes less than the suction pressure and the valve spring tension, the cylinder will begin to fill. Gas will rush into the cylinder through the suction reed valve.

sure is 180 psig, the pressure inside the cylinder may have to reach 190 psig to overcome the discharge valves and spring tension, Figure 21–42.

5. **The piston is at exactly top dead center.** (As close to the top of the head as it can go. There has to be a certain amount of clearance in the valve assemblies and between the piston and the head or they would touch. This clearance is known as

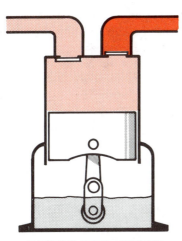

BOTTOM DEAD-CENTER

Figure 21–41. When the piston gets near the bottom of the stroke, the cylinder is nearly as full as it is going to get. There is a short time lag as the crankshaft circles through bottom dead center, during which a small amount of gas can still flow into the cylinder.

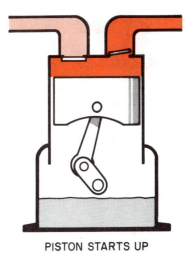

PISTON STARTS UP

Figure 21–42. When the piston starts back up and gets just off the bottom of the cylinder, the suction valve will have closed and pressure will begin to build in the cylinder. When the piston gets close to the top of the cylinder, the pressure will start to approach the pressure in the discharge line. When the pressure inside the cylinder is greater than the pressure on the top side of the discharge reed valve, the valve will open, and the discharge gas will empty out into the high side of the system.

clearance volume.) The piston is going to push as much gas out of the cylinder as time and clearance volume will allow. There will be a small amount of gas left in the clearance volume. This gas will be at the discharge pressure mentioned earlier. When the piston starts back down, this gas will reexpand, and the cylinder will not start to fill until the cylinder pressure is lower than the suction pressure of 20 psig. This reexpanded refrigerant is part of the reason that the compressor is not 100% efficient. Valve design and the short period of time the cylinder has to fill at the bottom of the stroke are other reasons the compressor is not 100% efficient, Figure 21–43.

21.8 LIQUID IN THE COMPRESSOR CYLINDER

The piston-type reciprocating compressor is known as a positive displacement device. This means that when the cylinder starts on the upstroke, it is going to empty itself or stall. *If the cylinder is filled with liquid refrigerant that does not compress (liquids are not compressible), something is going to break. Piston breakage, valve breakage, and rod breakage can all occur if a large amount of liquid reaches the cylinder, Figure 21–44.*

Liquid in compressors can be a problem from more than one standpoint. Large amounts of liquid, called a *slug* of liquid or *liquid slugging,* usually cause immediate damage. Small amounts of liquid floodback

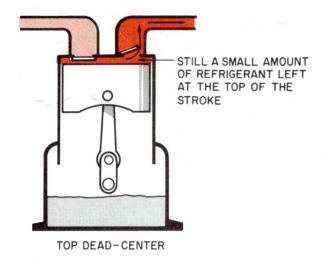

Figure 21-43. A reciprocating compressor cylinder cannot completely empty because of the clearance volume at the top of the cylinder. The manufacturers try to keep this clearance volume to a minimum but cannot completely do away with it.

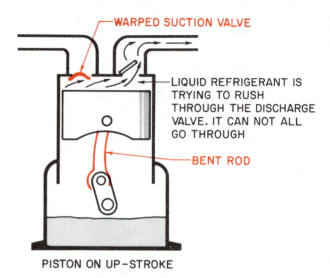

Figure 21-44. Cylinder trying to compress liquid. Something has to give.

can be just as detrimental but with a slower action. When small amounts of liquid enter the compressor, they can cause oil dilution. If the compressor has no oil lubrication protection, this may not be noticed until the compressor fails. One of the tricky parts of this failure is that marginal oil pressure may cause the compressor to throw a rod. Throwing a rod would be called mechanical failure. This may cause motor damage and burn out the motor. The technician may diagnose this as an electrical problem. Not so! If the compressor is a welded hermetic, the technician may not be aware of the problem until another failure occurs. *Never take a compressor failure for granted. Give the system a thorough checkout on startup and actively look for a problem.*

21.9   SYSTEM MAINTENANCE AND COMPRESSOR EFFICIENCY

The compressor's overall efficiency can be improved by maintaining the correct working conditions. This involves keeping the suction pressure as high as practical and the head pressure as low as practical within the design parameters.

A dirty evaporator will cause the suction pressure to drop. When the suction pressure goes below normal, the vapor that the compressor is pumping becomes less dense and gets thin, sometimes called rarified vapor. The compressor performance decreases.

A dirty condenser makes the head pressure rise. This causes the amount of refrigerant in the clearance volume (at the top of the cylinder) to be greater than the design conditions allow. This makes the compressor efficiency drop. If there is a dirty condenser (high head pressure) and or a dirty evaporator (low suction pressure), the compressor will run longer to keep the re-

frigerated space at the design temperature. The overall efficiency drops. A customer with a lot of equipment may not be aware of this if only part of the equipment is not efficient.

When the efficiency of a compressor drops, the owner is paying more money for less refrigeration. A good maintenance program is very economical in the long run.

SUMMARY

- The compressor is a vapor pump.
- The compressor cannot compress liquid refrigerant.
- The compressor lifts the low-pressure gas from the suction side of the system to the discharge side of the system.
- The discharge gas can be quite hot because the heat contained in the cool suction gas is concentrated when compressed in the compressor.
- Additional heat is added to the gas as it passes through the compressor because some of the work energy does not convert directly to compression but converts to heat.
- A discharge line on a compressor on a hot day can be as hot as 200°F and still be normal.
- Two methods are usually used to achieve compression in refrigeration compressors: the reciprocating compressor and the rotary compressor.
- Hermetic and open-drive are two types of reciprocating compressors.
- Hermetic compressors are manufactured as welded hermetic and serviceable hermetic.
- The welded hermetic compressor has to have the shell cut open to be serviced, and this work is done by special rebuilding shops only.
- The serviceable hermetic compressor has the shell bolted together and can be disassembled in the field.

- Reciprocating compressors have similar components: crankshafts, oil pumps, rods, pistons, valve plates, heads, and shells.
- Most reciprocating compressor motors are cooled by suction gas. Some are air cooled.
- *In all hermetic compressors the motor is operating in the refrigerant atmosphere and special precautions have to be taken in the manufacture and service of these motors.*
- Hermetic compressors have a shaft with the motor on one end and the compressor on the other end.
- Special overloads called internal overloads are used on hermetic motors that operate inside the refrigerant atmosphere.
- One type internal overload breaks the actual line current. If this overload does not close back when it should, the compressor is defective.
- The other internal overload is a pilot-duty type that breaks the control voltage. If something were to happen to this overload, an external type could be installed.
- The rotary compressor has vanes inside it that trap the gas for pumping purposes.
- The rotary compressor is more efficient than the reciprocating compressor but not as widely used.
- Both reciprocating and rotary compressors are positive displacement pumps, meaning that when they have a cylinder full of gas or liquid, the cylinder is going to be emptied or damage will occur.
- Open compressors have the motor on the outside of the shell.
- The motor can either be found mounted beside the compressor with the compressor and motor shafts side by side, or the motor may be mounted at the end of the compressor shaft with a flexible coupling between them.
- In either case shaft-to-motor alignment is very important.
- Belts for belt-drive applications come in many different types. The manufacturer's supplier should be consulted for advice.
- Compressor efficiencies in the reciprocating compressor depend mostly on the clearance volume and the motor efficiency.
- Compressor efficiencies for both reciprocating and rotary compressors depend on a good maintenance program.

REVIEW QUESTIONS

1. Describe the operation of a compressor.
2. Name two types of compressors with respect to the method of compression.
3. Will a compressor compress a liquid?
4. Why is the discharge gas leaving the compressor so hot?
5. What would be considered a high normal temperature for a discharge gas line on a compressor?
6. How is the compressor motor normally cooled in a welded hermetic compressor?
7. Which compressor type uses pistons to compress the gas?
8. Which compressor type uses vanes to trap the gas for compression?
9. What type of compressor uses belts to turn the compressor?
10. Define piston, rod, crankshaft, valves, valve plate, head, shaft seal, internal overload, pilot-duty overload, coupling.
11. Name two things in the design of the reciprocating compressor that control the efficiency of the compressor.
12. Which compressor is more efficient?
13. At what speed does a hermetic compressor normally turn?
14. What effect does a slight amount of liquid refrigerant have on a compressor if it runs for a long time?
15. What lubricates the refrigeration compressor?
16. What can the service technician do to keep the refrigeration machine operating at peak efficiency?

EXPANSION DEVICES

OBJECTIVES

After studying this unit, you should be able to

- describe the three most popular types of expansion valves.
- describe the operating characteristics of the three most popular valves.
- describe how the three valves respond to load changes.

22.1 EXPANSION DEVICES

The *expansion device,* often called the *metering device,* is the fourth component necessary for the compression refrigeration cycle to function. The expansion device is not as visible as the evaporator, the condenser, or the compressor. Generally, the device is concealed inside the cabinet and not obvious to the casual observer. It can either be a valve or a fixed-bore device. Figure 22–1 shows a thermostatic expansion valve installed inside the evaporator cabinet.

The expansion device is one of the division lines between the high side of the system and the low side of the system (the compressor is the other), Figure 22–2. An understanding of Unit 3 is important at this time. The expansion device is responsible for metering the correct amount of refrigerant to the evaporator. In Unit 3 it was explained that the evaporator performs best when it is as full of liquid refrigerant as possible without any running over into the suction line. Any refrigerant that enters the suction line may reach the compressor because there should not be any appreciable heat added to the refrigerant in the suction line to boil the liquid to a vapor. Later in this unit a suction line heat exchanger for special applications will be discussed, which is used to boil away liquid that may be in the suction line. Usually, liquid in the suction line is a problem.

The expansion device is normally installed in the liquid line between the condenser and the evaporator. The liquid line may be warm to the touch on a hot day and can be followed quite easily to the expansion device where there is a pressure drop and an accom-

panying temperature drop. For example, on a hot day the liquid line entering the expansion device may be 110°F. If this is a low-temperature cooler using R-12, the low-side pressure on the evaporator side may be 3 psig at a temperature of −15°F. This is a dramatic temperature drop and can be easily detected when found. The device may be warm on one side and frosted on the other, Figure 22–3. Since some expansion devices are valves and some are fixed-bore devices, this change can occur in a very short space—less than an inch on a valve, or a more gradual change on some fixed-bore devices.

Expansion devices come in five different types, but only three are currently being furnished with equipment, Figure 22–4, and they only will be covered in this book: (1) high-side float, (2) low side float, (3) thermostatic expansion valve, (4) automatic expansion valve, and (5) fixed bore, such as the capillary tube. The high-side float and the low-side float are not cur-

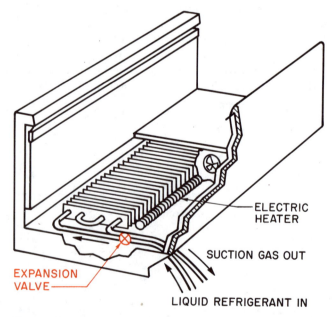

Figure 22–1. Metering device installed on a refrigerated case. The valve is not out in the open and is not as visible as the compressor, condenser, or evaporator.

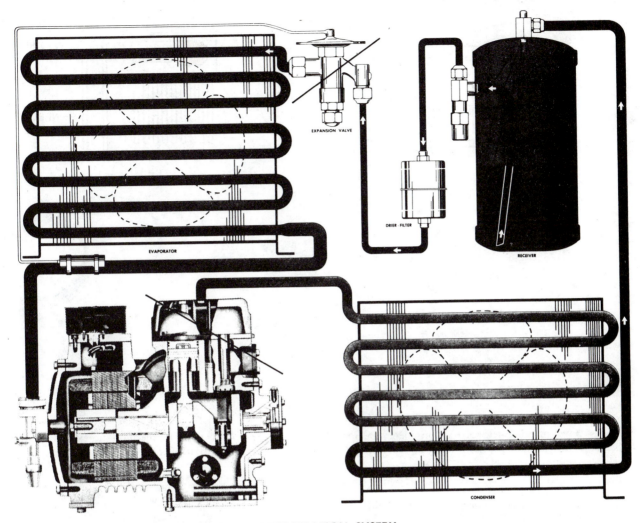

TYPICAL COMPRESSION REFRIGERATION SYSTEM

Figure 22–2. Two division lines in the refrigeration system. Each has the low-side pressure on one side and the high-side pressure on the other side. The compressor is one divider, and the expansion device the other. *Courtesy Copeland Corporation*

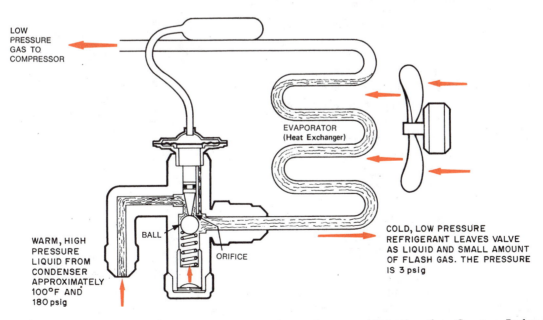

Figure 22–3. The expansion device has a dramatic temperature change from one side to the other. *Courtesy Parker Hannafin Corporation*

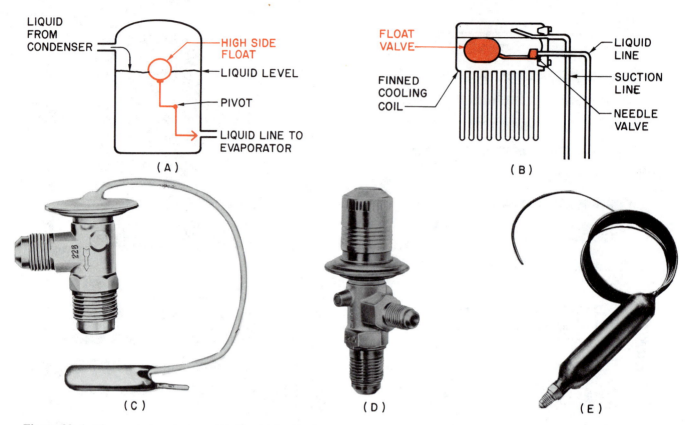

Figure 22–4. Five metering devices. (A) The high side float. (B) The low side float. (C) The thermostatic expansion valve. (D) The automatic expansion valve. (E) The capillary tube. *Courtesy (C and D) Singer Controls Division, Schiller Park, Illinois, (E) Parker Hannafin Corporation*

rently being used on typical equipment and should not be encountered in the field.

22.2   THERMOSTATIC EXPANSION VALVE

The *thermostatic expansion valve* meters the refrigerant to the evaporator using a thermal sensing element to monitor the superheat. The word "valve" means there is some characteristic about this device that changes in dimension. This valve changes in dimension in the seat area in response to a thermal element. The thermostatic expansion valve maintains a constant superheat in the evaporator. Remember, when there is superheat, there is no liquid refrigerant. Excess superheat is *not* desirable, but a small amount is necessary with this valve to assure that no liquid refrigerant leaves the evaporator.

22.3   THERMOSTATIC EXPANSION VALVE COMPONENTS

The thermostatic expansion valve consists of the (1) valve body, (2) diaphragm, (3) needle and seat, (4) spring, (5) adjustment and packing gland, and (6) the sensing bulb and transmission tube, Figure 22–5.

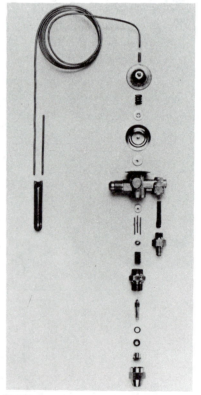

Figure 22–5. Exploded view of thermostatic expansion valve. All parts are visible in the order in which they go together in the valve. *Courtesy Singer Controls Division, Schiller Park, Illinois*

22.4 THE VALVE BODY

On common refrigerant systems the *valve body* is an accurately machined piece of solid brass or stainless steel that holds the rest of the components and fastens the valve to the refrigerant piping circuit, Figure 22–6. Notice that the valves have several different configurations. Some of them are one piece and cannot be disassembled, and some are made so that they can be taken apart.

These valves may be fastened to the system by three methods: (a) flare, (b) solder, or (c) flange. When a valve is installed in a refrigeration system, future service should be considered, so a flare connection or a flange-type valve should be used, Figure 22–7. If a solder connection is used, a valve that can be disassembled and rebuilt in place is desirable, Figure 22–8. The valve often has an inlet screen with a very fine mesh to strain out any small particles that may stop up the needle and seat, Figure 22–9.

Some valves have a third connection called an *external equalizer*. This connection is normally a $\frac{1}{4}$-in.

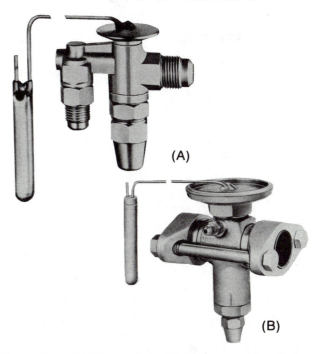

(A)

(B)

Figure 22–7. (A) Flare and the (B) flange-type valve. It can be removed from the system and replaced easily when it is installed where it can be reached with wrenches. *Courtesy Singer Controls Division, Schiller Park, Illinois*

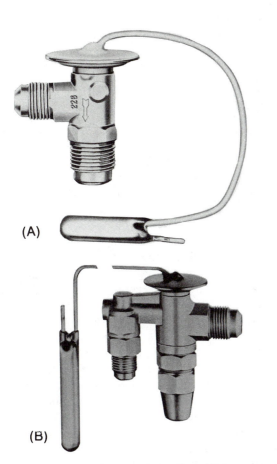

(A)

(B)

Figure 22–6. Thermostatic expansion valves. All have remote sensing element. (A) Some of these valves are one-piece valves that are thrown away when defective, and (B) some can be rebuilt in place. This feature is particularly good when the valve is soldered into the system. *Courtesy Singer Controls Division, Schiller Park, Illinois*

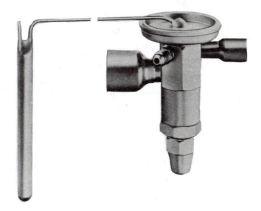

Figure 22–8. Solder-type of valve that can be disassembled and rebuilt without taking it out of the system. This valve can be quite serviceable in some situations. *Courtesy Singer Controls Division, Schiller Park, Illinois*

Figure 22–9. Most valves have some sort of inlet screen to strain any small particles out of the liquid refrigerant before they reach the very small opening in the expansion valve. *Courtesy Singer Controls Division, Schiller Park, Illinois*

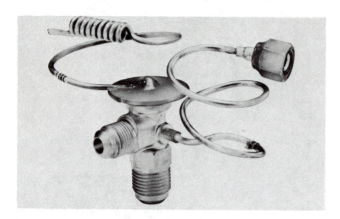

Figure 22–10. The third connection on this expansion valve is called the external equalizer. *Courtesy Singer Controls Division, Schiller Park, Illinois*

flare or $\frac{1}{4}$-in. solder and is on the side of the valve close to the diaphragm, Figure 22–10. As will be discussed later, the evaporator pressure has to be represented under the diaphragm of the expansion valve. When an evaporator has a very long circuit and pressure drop in the evaporator may occur, then an external equalizer is used. A pressure connection is made at the end of the evaporator that supplies the evaporator pressure under the diaphragm. Some evaporators have several circuits and a method of distributing the refrigerant that will cause pressure drop between the expansion valve outlet and the evaporator inlet. This installation has to have an external equalizer for the expansion valve to have correct control of the refrigerant, Figure 22–11. This distributor will be discussed in more detail under refrigerant components.

22.5   THE DIAPHRAGM

The *diaphragm* is located inside the valve body and moves the needle and seat in response to system load

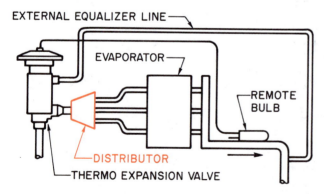

Figure 22–11. Evaporator with multiple circuits and an external equalizer line to keep pressure drops to a minimum. When an evaporator becomes so large that the length would create pressure drop, the evaporator is divided into circuits, and each circuit must have the correct amount of refrigerant.

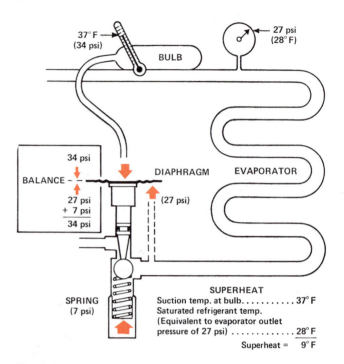

Figure 22–12. The diaphragm in the expansion valve is a thin membrane that has a certain amount of flexibility. It is normally made of a hard metal such as stainless steel. *Courtesy Parker Hannafin Corporation*

changes. The diaphragm is made of thin metal and is under the round dome-like top of the valve, Figure 22–12.

22.6   NEEDLE AND SEAT

The *needle and seat* control the flow of refrigerant through the valve. They are normally made of some type of very hard metal, such as stainless steel, to prevent the refrigerant passing through from eroding the seat. The needle and seat are used in a metering device so that close control of the refrigerant can be obtained, Figure 22–13. Some valve manufacturers have needle and seat mechanisms that can be changed for different capacities or to correct a problem.

The size of the needle and seat determines how much liquid refrigerant will pass through the valve with a specific pressure drop. For example, when the pressure is 180 psig on one side of the valve and 3 psig on the other side of the valve, a measured and predictable amount of liquid refrigerant will pass through the valve. If this same valve were used when the pressure is 100 psig and 3 psig, the valve would not be able to pass as much refrigerant. The conditions that the valve is going to have to operate under must be considered when selecting the valve. The manufacturer's manual for a specific valve is the best place to get the proper information to make these decisions. The pressure difference from one side of the valve to the other is not

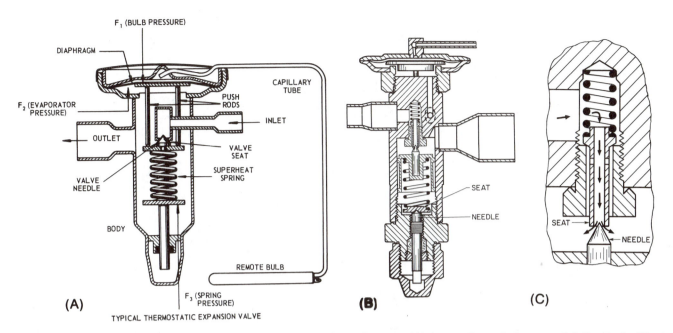

Figure 22-13. Needle and seat devices used in **expansion valves.** *Courtesy (A) Singer Controls Division, Schiller Park, Illinois (B, C) Sporlan Valve Company*

necessarily the discharge versus the suction pressure. Pressure drop in the condenser and interconnecting piping may be enough to cause a problem if not considered, Figure 22-14. Notice in the figure that the actual discharge and suction pressure are not the same as the pressure drop across the expansion valve.

Thermostatic expansion valves are rated in tons of refrigeration at a particular pressure drop condition. The capacity of the system and the working conditions of the system must be known. For example, using the manufacturer's catalog in Figure 22-15, we see that a 1-ton valve is needed in a medium-temperature cooler

with a 20°F evaporator and a 1-ton capacity, which is going to operate inside the store. The inside of the store is expected to stay at 70°F year around. The head pressure is expected to remain at a constant 125 psig. The 1-ton valve has a capacity of 1.3 ton at these conditions. If the same cooler is moved outside where the temperature is warmer and a head-pressure control is used, the same valve will have a capacity of 1.6 ton. This is not enough to create a problem. The outside ambient temperature will vary from 0° to 95°F. The head-pressure control will vary the head pressure from 125 psig at the low end to 180 psig at the high end.

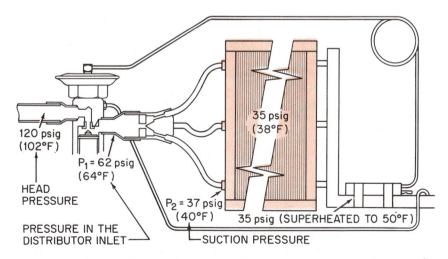

Figure 22-14. Real pressures as they would appear in a system. The pressure drop across the expansion valve is not necessarily the head pressure versus the suction pressure. The pressure drop through the refrigerant distributor is one of the big pressure drops that has to be considered. The refrigerant distributor is between the expansion valve and the evaporator coil. *Courtesy Sporlan Valve Company*

		EVAPORATOR TEMPERATURE (°F)														
		+50								+40						
	Nominal Capacity	PRESSURE DROP ACROSS VALVE (psi)														
Model	(tons)	40	60	80	100	125	150	175	200	40	60	80	100	125	150	175	200
128	1/4	0.32	0.38	0.43	0.48	0.55	0.60	0.65	0.72	0.30	0.34	0.38	0.43	0.49	0.53	0.55	0.57
223	1/2	0.50	0.67	0.79	0.86	0.95	1.2	1.3	1.4	0.45	0.65	0.75	0.84	0.90	1.0	1.1	1.3
226	1	0.95	1.1	1.2	1.5	1.7	1.8	2.1	2.3	0.90	1.0	1.2	1.4	1.6*	1.7	2.0	2.2
228	1-1/2	1.5	1.6	1.9	2.1	2.4	2.5	2.6	2.7	1.4	1.5	1.7	1.9	2.1	2.2	2.3	2.4
326	2	2.0	2.3	2.6	2.9	3.3	3.4	3.5	3.6	1.9	2.2	2.4	2.6	3.0	3.1	3.2	3.3
328	3	3.1	3.5	4.0	4.5	4.9	5.0	5.1	5.2	3.0	3.4	3.7	3.9	4.0	4.2	4.3	4.4
	4	4.0	5.0	5.9	6.8	8.0	8.5	9.4	10.0	3.5	4.5	5.2	6.0	6.6	7.4	8.0	8.5
426/428	5	4.6	5.7	6.5	7.3	8.5	9.3	10.1	10.8	4.4	5.0	5.3	6.5	7.1	7.9	8.6	9.5
407	7-1/2	7.4	9.1	10.5	11.7	13.0	13.5	14.0	14.4	7.0	8.0	9.3	10.3	11.5	11.7	12.0	12.4
	10	9.4	11.5	13.3	14.8	16.6	17.0	17.6	17.9	9.0	10.2	11.8	13.2	14.6	14.9	15.3	15.6
419	12-1/2	10.0	12.3	14.1	15.8	17.6	18.4	19.0	19.5	10.0	10.8	12.5	14.0	15.6	17.1	18.2	19.3
420	16	13.8	15.7	18.1	20.2	22.6	23.1	24.2	25.0	13.0	13.9	16.0	17.9	20.0	21.9	22.9	24.3
	19	16.2	18.6	21.5	23.9	26.6	27.3	28.4	29.3	15.9	16.5	19.0	21.2	23.6	24.5	25.6	27.5
	25	22.0	24.5	28.3	31.5	34.8	35.5	36.7	37.4	21.0	21.7	25.0	28.0	30.9	31.9	33.7	34.9

Condition 1

Condition 2

		EVAPORATOR TEMPERATURE (°F)														
		+20								0							
	Nominal Capacity	PRESSURE DROP ACROSS VALVE (psi)														
Model	(tons)	60	80	100	125	150	175	200	225	60	80	100	125	150	175	200	225
128	1/4	0.24	0.28	0.31	0.34	0.39	0.44	0.50	0.53	0.22	0.25	0.28	0.31	0.34	0.36	0.39	0.40
223	1/2	0.50	0.63	0.70	0.79	0.92	1.0	1.2	1.3	0.48	0.55	0.63	0.70	0.88	0.97	1.1	1.2
226	1	1.0	1.1	1.2	1.3	1.5	1.6	1.7	1.9	0.89	1.0	1.1	1.2	1.4	1.5	1.6	1.8
228	1-1/2	1.5	1.6	1.7	1.8	2.0	2.1	2.2	2.4	1.3	1.4	1.5	1.7	1.9	2.0	2.1	2.2
326	2	1.9	2.1	2.3	2.5	2.7	2.9	3.0	3.2	1.7	2.0	2.1	2.3	2.5	2.6	2.7	3.1
328	3	3.0	3.3	3.5	3.7	3.9	4.0	4.2	4.4	2.8	2.9	3.0	3.2	3.3	3.4	3.5	3.6
	4	3.2	4.0	4.5	5.6	6.3	7.0	7.6	8.2	3.0	3.5	4.0	4.8	5.9	6.6	7.0	7.8
426/428	5	3.7	4.2	4.7	6.0	7.1	7.6	8.2	8.6	3.5	4.0	4.3	5.2	6.3	7.0	7.5	8.0
407	7-1/2	5.9	6.8	7.6	8.4	9.3	9.6	9.9	10.1	4.6	5.0	5.4	6.0	6.6	7.2	7.8	8.4
	10	7.5	8.6	9.6	10.7	11.6	12.1	12.4	12.7	5.8	6.3	6.9	7.7	8.4	9.2	9.9	10.3
419	12-1/2	7.9	9.2	10.2	11.6	12.4	13.1	14.0	14.9	6.0	6.6	7.3	8.2	9.0	9.7	10.5	11.2
420	16	10.1	11.7	13.1	14.7	16.0	17.2	18.1	19.3	7.9	8.4	9.4	10.5	11.5	12.4	13.1	14.2
	19	12.0	13.9	15.6	17.6	19.0	20.5	21.8	23.0	9.4	10.0	11.1	12.4	13.6	14.7	15.3	16.4
	25	15.9	18.3	20.5	23.0	25.0	26.3	27.7	29.1	12.2	13.0	14.6	16.4	17.9	19.4	20.6	22.0

Figure 22–15. Manufacturer's table that shows the capacity of valves at different pressure drops. Notice that the same valve has different capacities at different pressure drops. The more the pressure drop, the more capacity a valve has. This is a partial table and not intended for use in design of a system. *Courtesy Singer Controls Division, Schiller Park, Illinois*

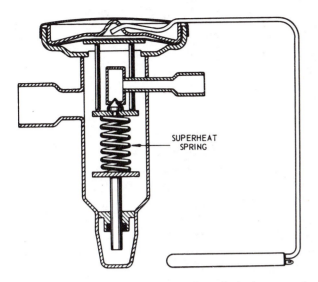

Figure 22-16. Spring used in the thermostatic expansion valve. *Courtesy Singer Controls Division, Schiller Park, Illinois*

22.7 THE SPRING

The *spring* is one of the three forces that act on the diaphragm. It raises the diaphragm and closes the valve by pushing the needle into the seat. When a valve has an adjustment, the adjustment applies more or less pressure to the spring to change the tension for different superheat settings. The spring tension is factory set for a predetermined superheat of 8° to 12°F, Figure 22-16.

The adjustment part of this valve can either be a screw slot or a square-headed shaft. Either type is normally covered with a cap to prevent water, ice, or other foreign matter from collecting on the stem. The cap also serves as a backup leak prevention. Most adjustment stems on expansion valves have a packing gland that can be tightened to prevent refrigerant from leaking. The cap would cover the stem and the gland, Figure 22-17. Normally one complete turn of the stem can change the superheat reading from 0.5 to 1°F of superheat, depending on the manufacturer.

22.8 THE SENSING BULB AND TRANSMISSION TUBE

The *sensing bulb* and *transmission tube* are extensions of the valve diaphragm. The bulb detects the temperature at the end of the evaporator on the suction line and transmits this temperature, converted to pressure, to the top of the diaphragm. The bulb contains a fluid, such as refrigerant, that responds to a pressure and temperature relationship chart just like R-12 or R-22 would. When the suction line temperature goes up, this temperature change takes place inside the bulb. When there is a pressure change, the transmission line (which is nothing more than a small-diameter hollow tube) allows the pressure between the bulb and diaphragm to equalize back and forth, Figure 22-18.

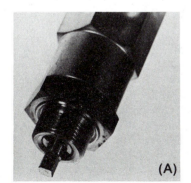

Figure 22-17. Adjustment stems on expansion valves. Some of them are adjusted with a screwdriver, and some with a valve wrench. One full turn of the stem will normally change the superheat 0.5°F to 1°F. *Photos by Bill Johnson*

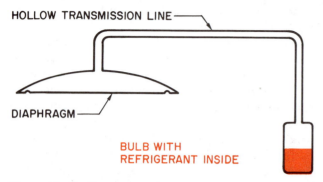

HOLLOW TRANSMISSION LINE

DIAPHRAGM

BULB WITH REFRIGERANT INSIDE

Figure 22–18. Illustration of the diaphragm, the bulb, and the transmission tube.

The seat in the valve is stationary in the valve body, and the needle is moved by the diaphragm. One side of the diaphragm gets its pressure from the bulb, and the other side gets its pressure from the evaporator. The diaphragm is movable and moves up and down in response to three different pressures. These three different pressures act at the proper time to open, close, or modulate the valve needle between open and closed. These three pressures are the bulb pressure, the evaporator pressure, and the spring pressure. They all work as a team to position the valve needle at the correct position for the load conditions at any particular time. Figure 22–19 is a series of illustrations that show how the thermostatic expansion valve functions under different load conditions.

22.9   TYPES OF BULB CHARGE

The fluid that comes inside the expansion valve bulb is known as the *charge* for the valve. There are four types of charge that can be obtained with the thermostatic expansion valve: liquid charge, cross liquid charge, vapor charge, and cross vapor charge.

22.10   THE LIQUID CHARGE BULB

The *liquid charge bulb* is a valve and bulb charged with a fluid characteristic of the refrigerant in the system. The diaphragm and bulb are not actually liquid full. They have enough liquid, however, that they always have some liquid inside them. The liquid will not all boil away. The pressure and temperature relationship is almost a straight line on a graph. When the temperature goes up a degree, the pressure goes up a specific amount and can be followed on a pressure and temperature chart. When this pressure–temperature concept is carried far enough, it can easily be seen that when high temperatures are encountered, high pressures will exist. During defrost the expansion valve bulb may reach high temperatures. This can cause two things to happen: The pressures inside the bulb can

cause excessive pressures over the diaphragm, and the valve will open wide when the bulb gets warm. This will overfeed the evaporator and can cause liquid to flood the compressor when defrost is terminated, Figure 22–20. This can cause the service technician or manufacturer to change to the cross liquid charge bulb.

22.11   THE CROSS LIQUID CHARGE BULB

The *cross liquid charge bulb* has a fluid that is different from the system fluid. It does not follow the pressure and temperature relationship. It has a flatter curve and will close the valve faster upon a rise in evaporator pressure. This valve closes during the off cycle when the compressor shuts off and the evaporator pressure rises. This will help prevent liquid refrigerant from flooding over into the compressor at start up, Figure 22–21.

22.12   THE VAPOR CHARGE BULB

The *vapor charge bulb* is actually a valve that has only a small amount of liquid refrigerant in the bulb, Figure 22–22. It is sometimes called a *critical charge bulb*. When the bulb temperature rises, more and more of the liquid will boil to a vapor until there is no more liquid. When this point is reached, an increase in temperature will no longer bring an increase in pressure. The pressure curve will be flat, Figure 22–23.

When this valve is applied, care must be taken that the valve body does not get colder than the bulb, or the liquid that is in the bulb will condense above the diaphragm. When this happens, the control at the end of the suction line by the bulb will be lost to the valve diaphragm area, Figure 22–24. The valve will be controlled by the temperature of the liquid that is at the diaphragm and the valve will lose control. Small heaters have been installed at the valve body to keep this from happening.

22.13   THE CROSS VAPOR CHARGE BULB

The *cross vapor charge bulb* has a similar characteristic to the vapor charged valve, but it has a fluid different from the refrigerant in the system. This gives a different pressure and temperature relationship under different conditions. These special valves are applied to special systems. Manufacturers or suppliers should be consulted when questions are encountered.

22.14   FUNCTIONING EXAMPLE OF A TXV (THERMOSTATIC EXPANSION VALVE)

When all of these components are assembled, the expansion valve will function as follows. Note: This is a liquid filled bulb.

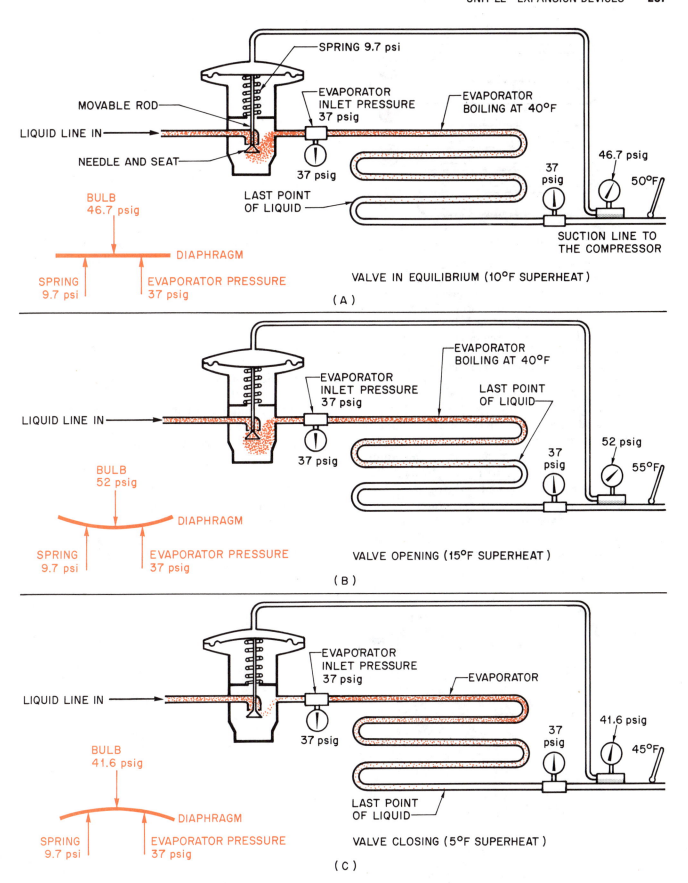

Figure 22-19. All components work together to hold the needle and seat in the correct position to maintain stable operation with the correct superheat. (A) Valve in equilibrium. (B) Valve opening. (C) Valve closing.

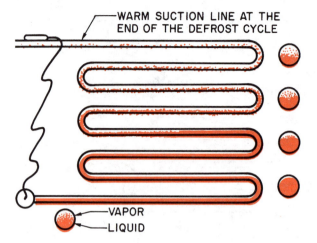

Figure 22–20. This can happen at the end of the defrost cycle. The valve is open because the evaporator suction line is warm after the defrost cycle. This allows the liquid in the evaporator to spill over into the suction line before the expansion valve can gain control. The bulb can become very hot because it is fastened to the end of the suction line. The coil can become quite hot in an extended defrost cycle, and the remaining liquid can move on to the suction line.

1. **Normal load conditions.** The valve is operating in equilibrium (no change, stable), Figure 22–25. The evaporator is operating at a medium-temperature application and just prior to the cutoff point. The suction pressure is 21 psig, and the refrigerant, R-12, is boiling (evaporating) in the evaporator at 20°F. The expansion valve is maintaining 10°F of superheat, so the suction line temperature is 30°F at the bulb location. The bulb has been on the line long enough that it is the same temperature as the line,

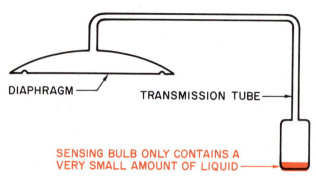

SENSING BULB ONLY CONTAINS A
VERY SMALL AMOUNT OF LIQUID

Figure 22–22. This gas charged bulb is actually a liquid charged bulb with a critical charge. When the bulb reaches a point, the liquid is boiled away and the pressure won't rise any more. When this valve is used, care must be taken that the valve body does not get colder than the bulb, or the small amount of liquid in the valve will condense in the diaphragm area. The control is where the liquid is.

30°F. For now suppose that the liquid in the bulb will be 28.5 psig, corresponding to 30°F. The spring is exerting a pressure equal to the difference in pressure to hold the needle at the correct position in the seat to maintain this condition. The spring pressure in this example is 7.5 psig.

2. **Load changes with food added to the cooler.** See Figure 22–26. When food that is warmer than the inside of the cooler is added to the cooler, the load on the evaporator changes. The warmer food warms the air inside the cooler, and the load is added to the refrigeration coil by the air. This warmer air passing over the coil causes the liquid refrigerant inside the coil to boil faster. The suction pressure will also rise. The net effect of this condition will be

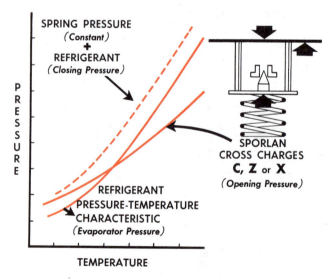

Figure 22–21. This cross charged valve has a flatter pressure and temperature curve. The pressure rise after defrost or at the end of a cycle will have a tendency to close the valve and prevent liquid problems. *Courtesy Sporlan Valve Company*

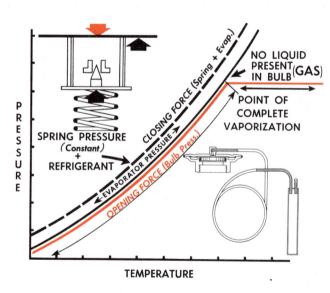

Figure 22–23. Graph showing pressures inside the valve when the sensing bulb is in a hot area. *Courtesy Sporlan Valve Company*

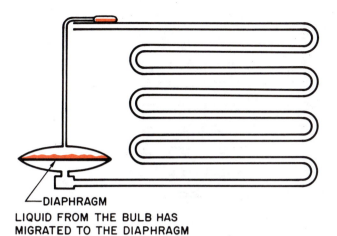

LIQUID FROM THE BULB HAS
MIGRATED TO THE DIAPHRAGM

Figure 22-24. How liquid can migrate to the valve body when the body becomes cooler than the bulb. When this happens, the valve body can be heated with a warm cloth, and the liquid will move back to the bulb.

that the last point of liquid in the coil will be farther from the end of the coil than when the coil was under normal working conditions. The coil will start to starve for liquid refrigerant. The thermostatic expansion valve will start to feed more refrigerant to compensate for this shortage. When this condition of increased load has gone on for an extended period of time, the thermostatic expansion valve will stabilize in the feeding of refrigerant and reach a new point of equilibrium, where no adjustment takes place.

3. **Load changes with food removed from the cooler.** See Figure 22-26. When a large portion of the food is removed from the cooler, the load will decrease on the evaporator coil. There will no longer be enough load to boil the amount of refrigerant that the expansion valve is feeding into the coil, so the expansion valve will start overfeeding the coil. The coil is beginning to flood with liquid refrigerant. The thermostatic expansion valve needs to throttle the refrigerant flow. When the thermostatic expansion valve has operated at this condition for a period of time, it will stabilize and reach another point of equilibrium. When the condition exists for a long enough time, the thermostat will stop the compressor.

22.15 TXV RESPONSE TO LOAD CHANGES

The TXV responds to a change in load in the following manner. When the load is increased, for example when a load of food warmer than the cooler (refrigerated box) is placed inside the cooler, the thermostatic expansion valve opens, allowing more refrigerant into the coil. The evaporator needs more refrigerant at this time because the increased load is evaporating the refrigerant in the evaporator faster. The suction pressure increases, Figure 22-26. When there is a load decrease (e.g., when some of the food is removed from the cooler), the liquid in the evaporator evaporates slower and the suction pressure decreases. At this time the thermostatic expansion valve will throttle back by

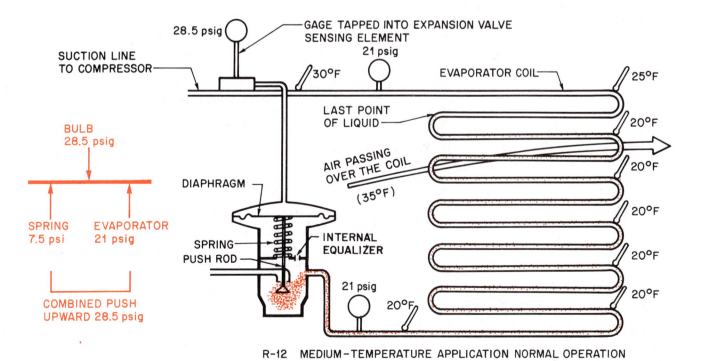

R-12 MEDIUM-TEMPERATURE APPLICATION NORMAL OPERATION

Figure 22-25. Thermostatic expansion valve under a normal load condition. The valve is said to be in equilibrium. The needle is stationary.

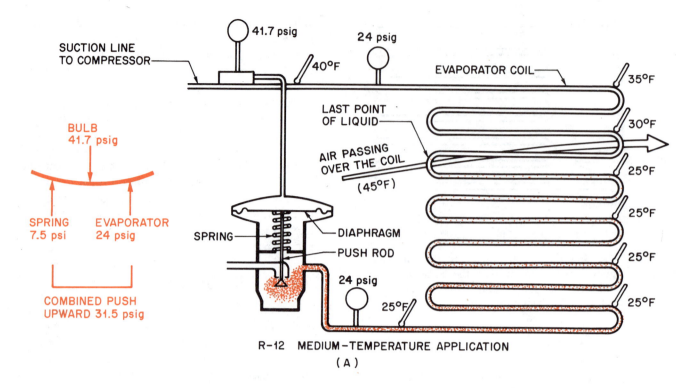

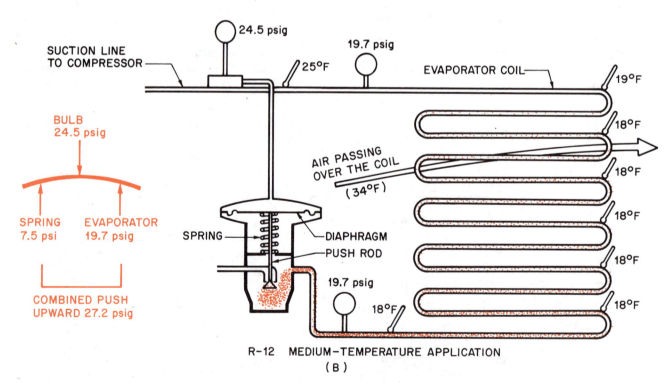

Figure 22–26. (A) When food is added to the cooler, the load on the coil goes up. The extra Btu added to the air in the cooler are transferred into the coil, which causes a temperature increase of the suction line. This causes the valve to open and feed more refrigerant. If this condition is prolonged, the valve will reach a new equilibrium point with the needle being steady and not moving. (B) When there is a load decrease, such as when some of the food is removed, the refrigeration machine now has increased capacity. This causes a decrease in suction pressure, and the valve will begin to close. When this goes on for a prolonged time, the valve will reach a new point of equilibrium at a low load condition. The thermostat will cut off the unit if this condition goes on long enough. The equilibrium point in (A) and (B) is the valve superheat set point of 10°F superheat. The valve should always return to this 10°F set point with prolonged steady state operation.

slightly closing the needle and seat to maintain the correct superheat. **The thermostatic expansion valve responds to a load increase by metering more refrigerant into the coil. This causes an increase in suction pressure.**

22.16   TXV VALVE SELECTION

The TXV must be carefully chosen for a particular application. **Each thermostatic expansion valve has its own refrigerant and has to be selected for that refrigerant.** One valve on the market uses a spring change to change refrigerant, but the refrigerant still has to be considered in the selection of the valve.

The capacity of the system is very important. If the system calls for a $\frac{1}{2}$-ton expansion valve and a 1-ton valve is used, the valve will not control correctly because the needle and seat are too large. If a $\frac{1}{4}$-ton valve were used, it would not pass enough liquid refrigerant to keep the coil full, and a starving coil condition would occur.

22.17   PRESSURE LIMITING TXV

The pressure limiting TXV has another bellows that will only allow the evaporator to build to a predetermined pressure, and then the valve will shut off the flow of liquid. This valve is desirable on low-temperature applications because it keeps the evaporator pressure down during a hot pull down that could overload the compressor. For example, when a low-temperature cooler is started with the inside of the box hot, the compressor will operate under an overloaded condition until the box cools down. The pressure-limiting thermostatic expansion valve will prevent this from happening, Figure 22–27.

Figure 22–27. Pressure-limiting type expansion valve. When the pressure is high in the evaporator, such as during a hot pull down, this valve will override the thermostatic element with a pressure element and throttle the refrigerant. *Courtesy Sporlan Valve Company*

22.18   SERVICING THE TXV

When any thermostatic expansion valve is chosen, care should be taken that the valve is serviceable and will perform correctly. There are several things that should be considered: (a) type of fastener (flare, solder, or flange), (b) location of the valve for service and performance, and (c) expansion valve bulb location. This valve has moving parts that are subject to wear. When a valve has to be replaced, an exact replacement is usually best. When this is not possible, a supplier can furnish you with the information needed to change to another valve.

22.19   SENSING-ELEMENT INSTALLATION

Particular care should be taken when installing the expansion valve sensing element. Each manufacturer

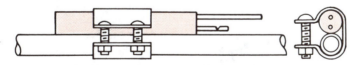

EXTERNAL BULB ON SMALL SUCTION LINE

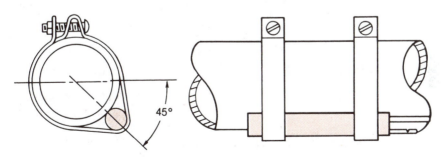

EXTERNAL BULB ON LARGE SUCTION LINE

Figure 22–28. Best positions for mounting the expansion valve sensing bulb. *Courtesy ALCO Controls Division, Emerson Electric Company*

has a recommended method for this installation, but they are all similar. The valve sensing bulb has to be mounted at the end of the evaporator on the suction line. The best location is near the bottom of the line on a horizontal run where the bulb can be mounted flat and not be raised by a fitting, Figure 22–28. The bulb should not be located on the bottom of the line because there will be oil returning that will act as an insulator to the sensing element. The object of the sensing element is to sense the temperature of the suction line. To do this the line should be very clean, and the bulb fastened to this line very securely. Normally the manufacturer suggests that the bulb be insulated from the ambient temperature if the ambient is much warmer than the suction line temperature because the bulb will be influenced by the ambient temperature instead of by the line temperature.

22.20 THE SOLID-STATE EXPANSION VALVE

The solid-state expansion valve uses a thermistor as a sensing element to vary the voltage to a heat motor operated valve (a valve with a bimetal element). This valve normally uses 24 V as the control voltage to operate the valve, Figure 22–29.

When the voltage is cut off at the end of the cycle, the valve will shut off and the system can be pumped down. If the voltage is allowed to remain on the element, the valve will remain open during the off cycle, and the pressures will equalize.

The thermistor is inserted into the vapor stream at the end of the evaporator. It is very small in mass and will respond very quickly to temperature changes.

The electronic expansion valve responds to the change in temperature of the sensing element just like the typical TXV except that it does not have a spring. When the thermistor is suspended in dry vapor, it is heated by

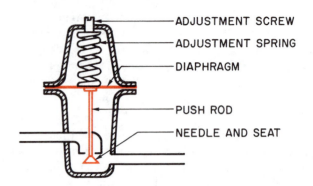

Figure 22–30. The automatic expansion valve uses the diaphragm as the sensing element and maintains a constant pressure in the evaporator but does not control superheat.

the current passing through it. This creates a faster response than does merely measuring the vapor temperature. When the valve opens and saturated vapor reaches the element, the valve begins to close slightly. This valve controls to a very low superheat, which allows the evaporator to utilize maximum surface area.

22.21 THE AUTOMATIC EXPANSION VALVE

The *automatic expansion valve* is an expansion device that meters the refrigerant to the evaporator by using a pressure-sensing device. This device is also a valve that changes in inside dimension in response to its sensing element. **The automatic expansion valve maintains a constant pressure in the evaporator. Notice that superheat was not mentioned.** This device has a needle and seat just like the TXV that is fastened to a diaphragm, Figure 22–30. The diaphragm has one side vented to the evaporator, and the other side to the atmosphere. When the evaporator pressure drops for any reason, the valve begins to open and feed more refrigerant into the evaporator.

The automatic expansion valve is built much the same as the TXV except that it does not have a sensing bulb. The body is normally made of machined brass. The adjustment of this valve is normally at the top of the valve. There may be a cap to remove, or there may just be a cap to turn. This adjustment changes the spring tension that supports the atmosphere in pushing down on the diaphragm. When the tension is increased, the valve will feed more refrigerant and increase the suction pressure, Figure 22–31.

22.22 AUTOMATIC EXPANSION VALVE RESPONSE TO LOAD CHANGES

The automatic expansion valve responds differently than the TXV to load changes. It actually acts in reverse. When a load is added to the coil, the suction pressure starts to rise. The automatic expansion valve will start to throttle the refrigerant by closing enough

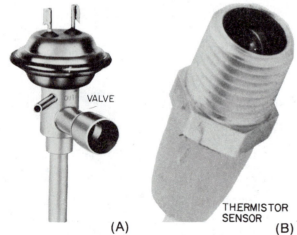

Figure 22–29. Expansion valve controlled with a thermistor and a heat motor. *Courtesy Singer Controls Division, Schiller Park, Illinois*

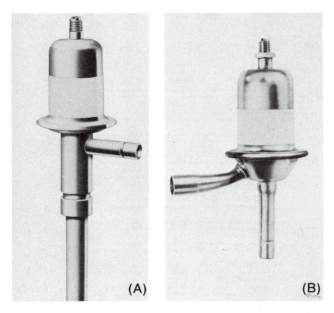

Figure 22–31. Automatic expansion valves. They resemble the thermostatic expansion valve, but they do not have the bulb for sensing temperature at the end of the suction line. *Courtesy Singer Controls Division, Schiller Park, Illinois*

to maintain the suction pressure at the set point. This has the effect of starving the coil slightly. A large increase in load will cause more starving. When the load is decreased and the suction pressure starts to fall, the automatic expansion valve will start to open and feed more refrigerant into the coil. If the load reduces too much, liquid could actually leave the evaporator and proceed down the suction line, Figure 22–32.

We can see from these examples that this valve responds in reverse to the load. The best application for this valve is where there is a fairly constant load. One of its best features is that it can hold a constant pressure. When this valve is applied to a water-type evaporator, freezing will not occur.

22.23 SPECIAL CONSIDERATIONS FOR THE TXV AND AUTOMATIC EXPANSION VALVE

The TXV and the automatic expansion valve both are expansion devices that use more or less refrigerant, depending on the load. Both need a storage device (receiver) for refrigerant when it is not needed. The receiver is a small tank located between the condenser and the expansion device. Normally the condenser is close to the receiver. It has a *king* valve that functions

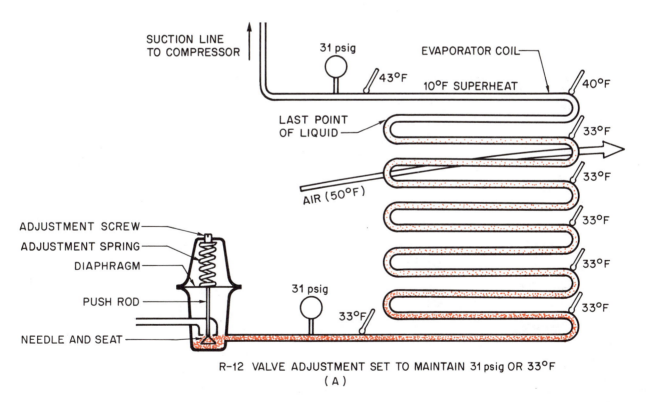

Figure 22–32. Automatic expansion valve under varying load conditions. This valve responds in reverse to a load change. (A) Normal operation. (B) When the load goes up the valve closes down and starts to starve the coil slightly to keep the evaporator pressure from rising. (C) When the load goes down, the valve opens up to keep the evaporator pressure up. This valve is best applied where the load is relatively constant. When applied to a water-type evaporator, freeze protection can be a big advantage with this valve.

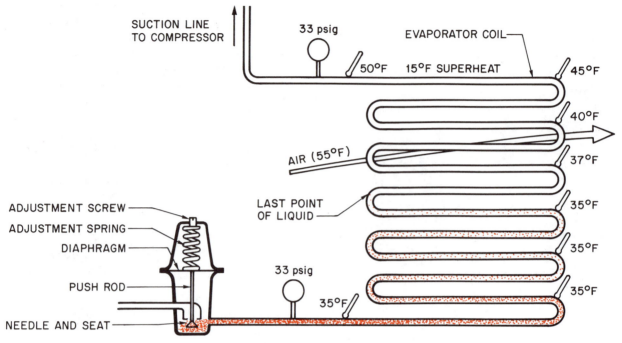

R-12 VALVE ADJUSTMENT SET TO MAINTAIN 31 psig OR 33°F
(B)

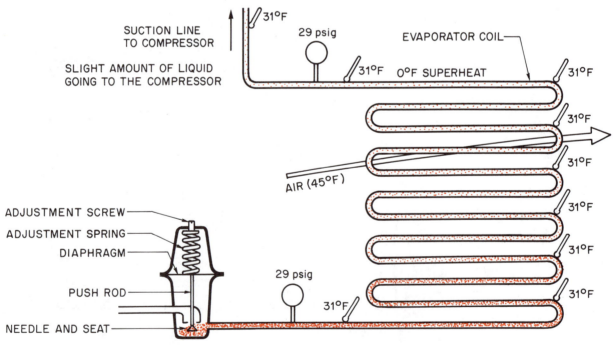

R-12 VALVE ADJUSTMENT SET TO MAINTAIN 31 psig OR 33°F
(C)

Figure 22–32. (*Continued*)

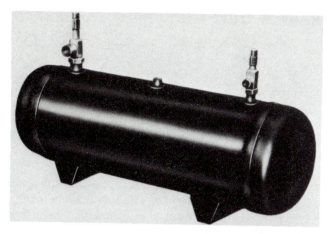

Figure 22–33. Refrigerant receiver. When the load increases, more refrigerant is needed and moves from the receiver into the system. *Courtesy Refrigeration Research*

Figure 22–34. Capillary tube metering device. *Courtesy Parker Hannafin Corporation*

as a service valve; this valve stops the refrigerant from leaving the receiver when the low side of the system is serviced. This can serve both as a storage tank for different load conditions and as a tank into which the refrigerant can be pumped for service, Figure 22–33. We will discuss the receiver later in this unit.

22.24 THE CAPILLARY TUBE EXPANSION DEVICE

The *capillary tube* metering device controls refrigerant flow by pressure drop. It is a copper tube with a very small calibrated inside diameter. The diameter and the length of the tube determine how much liquid will pass through the tube at any given pressure drop. The capillary tube can be quite long on some installations and may be wound in a coil to store the extra tubing length. The capillary tube does not control superheat or pressure. It is a fixed-bore device with no moving parts. Since this device cannot adjust to load change, it is usually used where the load is relatively constant with no large fluctuations, Figure 22–34.

The capillary tube is a very inexpensive device for the control of refrigerant and is used often in small equipment. This device does not have a valve and does not stop the liquid from moving to the low side of the system during the off cycle. This reduces the motor starting torque requirements for the compressor because the pressures equalize during the off cycle, Figure 22–34.

22.25 OPERATING CHARGE FOR THE CAPILLARY TUBE SYSTEM

The capillary tube requires only a small amount of refrigerant because it does not modulate or feed more and less refrigerant according to the load. When the refrigerant charge is analyzed, we notice that when the unit is operating at the design conditions there is a

specific amount of refrigerant in the evaporator and a specific amount of refrigerant in the condenser. This is the amount of refrigerant required for proper refrigeration. Any other refrigerant that is in the system is in the pipes for circulating purposes only.

The amount of refrigerant in the system is very critical in capillary tube systems. It is easy for technicians to overcharge the system if they are not very careful and familiar with the system. Most capillary tube systems have the charge for the system printed on the nameplate of the equipment. The manufacturer always recommends measuring the refrigerant into these systems either by scales or by liquid charging cylinders.

SUMMARY

- The expansion device is one of the dividing points between the high and low sides of the system.
- The thermostatic expansion valve maintains a constant *superheat* in the evaporator.
- The TXV is composed of a body, a diaphragm, a needle and seat, a spring, an adjustment, and a bulb and transmission tube.
- The TXV has three forces acting on its needle and seat: the bulb, the evaporator, and the spring pressure.
- The bulb pressure is the only force that acts to open the valve.
- The forces inside the expansion valve all work together to hold the needle and seat in the correct position so that the evaporator will have the correct amount of refrigerant under all load conditions.

- The bulb and diaphragm are charged with one of four different charges: the liquid charge, the cross liquid charge, the vapor charge, or the cross vapor charge.
- The TXV bulb has to be mounted securely to the suction line to sense accurately the suction line temperature.
- The TXV responds to a load increase by feeding more refrigerant into the evaporator.
- The automatic expansion valve maintains a constant *pressure* in the evaporator.
- The automatic expansion valve responds in reverse to a load change; when the load increases, the automatic expansion valve throttles the refrigerant instead of feeding more refrigerant as the thermostatic expansion valve does.
- The capillary tube expansion device is a fixed-bore metering device usually made of copper, with a very small inside diameter, and no moving parts.

REVIEW QUESTIONS

1. What are the three forces acting on the TXV diaphragm?
2. What material is the needle and seat normally made of?
3. The TXV maintains what condition in the evaporator?
4. Name the four types of charge that the TXV uses to control refrigerant flow.
5. Where is the bulb of the TXV mounted?
6. When do some TXVs require an external equalizer?
7. How does a TXV respond to an increase in load?
8. What condition does the automatic expansion valve maintain in the evaporator?
9. How does the automatic expansion valve respond to a load increase?
10. What determines the amount of refrigerant that flows through a capillary tube metering device?

SPECIAL REFRIGERATION SYSTEM COMPONENTS

OBJECTIVES

After studying this unit, you should be able to

- **define the categories of system components.**
- **distinguish between mechanical and electrical controls.**
- **explain that a control stops, starts, or modulates.**
- **explain how and why mechanical controls function.**
- **define low ambient operation.**
- **describe electrical controls that apply to refrigeration.**
- **describe off-cycle defrost.**
- **describe random and planned defrost.**
- **explain temperature-terminated defrost.**
- **describe the various refrigeration accessories.**
- **describe the low-side components.**
- **describe the high-side components.**

23.1 THE FOUR BASIC COMPONENTS

The compression refrigeration cycle must have the four basic components to function: the compressor, the condenser, the evaporator, and the expansion device. However, there are many more devices and components that can enhance the performance and the reliability of the refrigeration system. Some of these protect the components, and some improve the reliability under various conditions. The components in this unit will be described in different combinations as needed.

System components divide themselves into two broad categories: controls and accessories. Control components are divided into mechanical, electrical, and electromechanical. Electromechanical controls are those that may be both electrical and mechanical in nature. This control may be discussed in the category of the feature that is most prominent.

23.2 MECHANICAL CONTROLS

Mechanical controls generally stop, start, or modulate fluid flow and can be operated by pressure, temperature, or electricity. These controls can usually be

identified because they are almost always found in the piping.

23.3 TWO-TEMPERATURE CONTROLS

Two-temperature operation is desirable when more than one evaporator is used with one compressor. This occurs if two or more evaporators are designed to operate in different temperature ranges, such as when an evaporator operating at 30°F is used on the same compressor with an evaporator operating at 20°F, Figure 23–1. Two-temperature application is normally accomplished with a purely mechanical valve or a temperature-controlled valve.

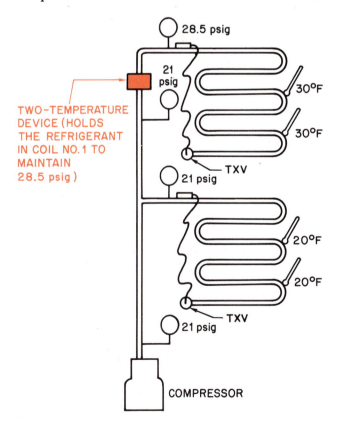

Figure 23–1. Two-temperature operation. Two evaporators are operating on one compressor. One evaporator operates at 20°F (21 psig), the other at 30°F (28.5 psig).

277

23.4 EVAPORATOR PRESSURE CONTROL

The *evaporator pressure regulator valve* is a mechanical control that keeps the refrigerant in the evaporator from going below a predetermined point. The evaporator pressure regulator (*EPR valve* as it will be called from now on) is installed in the suction line at the evaporator outlet. The bellows in the EPR valve senses evaporator pressure and throttles (modulates) the suction gas to the compressor. This will then allow the evaporator pressure to go as low as the pressure setting on the valve. When the EPR valve is used with the thermostatic expansion valve (TXV), the system now has the characteristics of maintaining a constant superheat because of the TXV and the advantage of the automatic expansion valve by keeping the pressure from going too low, Figure 23–2.

The EPR valve can be applied to a system that cools water. The evaporator will not go below the predetermined point, which could be freezing, 32°F. For example, when the system is started, with a load on it, the EPR valve would be wide open. The TXV would be throttling the liquid refrigerant into the evaporator. When the refrigerant is cooled to the point at which the EPR valve is set, it will begin to throttle a slow flow of refrigerant. If this setting is just above freezing, the valve will throttle off enough to keep the evaporator from freezing until the thermostat responds and shuts the system off, Figure 23–3.

23.5 MULTIPLE EVAPORATORS

When more than two evaporators are used with one compressor, one of them will sometimes be at a different temperature and pressure range. For example,

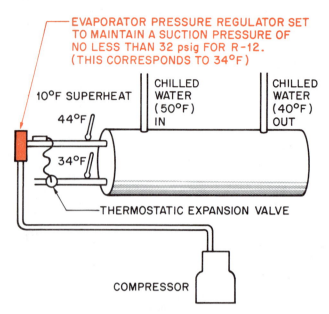

Figure 23–3. Thermostatic expansion valve keeps the evaporator at the proper level of refrigerant while the EPR valve keeps the evaporator from getting too cold. This combination of controls has all of the advantages of the thermostatic expansion valve and the automatic expansion valve in one system. The EPR valve may be used on a water-type evaporator to keep it from operating below freezing. The thermostat should cut off the compressor soon after the EPR valve starts to throttle.

when an evaporator that needs to operate at 20 psig (15°F) is piped with another evaporator of 25 psig (25°F) to the same compressor, an EPR valve is needed in the highest-temperature evaporator's suction line. The true suction pressure for the system will be the low value of 20 psig, but the other evaporator will operate at the correct pressure of 25 psig. Several evaporators of different pressure requirements can be piped together using this method, Figure 23–4.

It is desirable to know the actual pressure in the evaporator because the evaporator pressure is not the same as the true suction pressure. Normally there is a gage port, known as a Schrader valve port, permanently installed in the EPR valve body that allows the service technician to take gage readings on the evaporator side of the valve. The true suction pressure can be obtained at the compressor service valve, Figure 23–5.

23.6 TWO-TEMPERATURE VALVE

The two-temperature valve operates in much the same way as the EPR valve except it is more definitive in its action. It generally has some method to open or close the valve instead of modulating it. This can be accomplished either with an overcenter movement in-

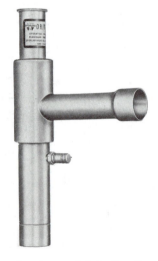

Figure 23–2. EPR valve to modulate the flow of vapor refrigerant leaving the evaporator. It limits the pressure in the evaporator and keeps it from dropping below the set point. *Courtesy Sporlan Valve Company*

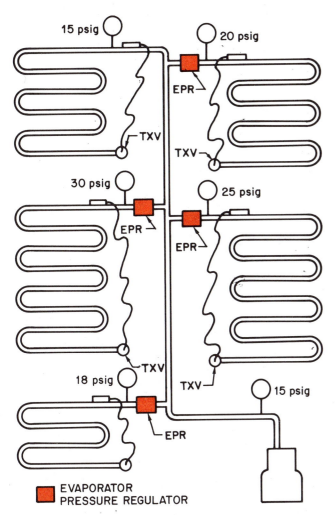

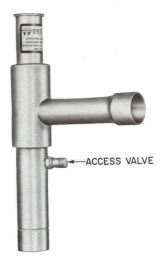

Figure 23–5. Pressure tap provided on the evaporator side of the EPR valve so that the technician can observe the actual evaporator pressure. The true suction pressure can be obtained at the compressor suction service valve. *Courtesy Sporlan Valve Company*

Figure 23–4. Evaporators piped together and one compressor used to maintain a suction pressure for the lowest pressure evaporator. Evaporators that have pressures higher than the true suction pressure of the lowest evaporator all have EPR valves set to their individual needs.

side the valve or with a thermostat and an electric valve.

The mechanical type of two-temperature valve uses the overcenter device and gives a distinct temperature spread in the evaporator. It has low- and high-temperature settings that can be seen. The high-temperature setting can be set above freezing for the coil to assure defrost before the valve will open up and allow refrigerant to flow out of the coil, Figure 23–6.

The electric type of device could be a solenoid valve placed either at the evaporator outlet or the evaporater inlet before the expansion valve. If it is placed at the outlet, it would act much as the mechanical type except it is electrically controlled with a thermostat in the box. If the valve is placed in the liquid line before the expansion valve, when the valve closes the liquid would be pumped out of the coil into the receiver. Each type has its advantage. An example of a solenoid valve can be seen in Figure 23–17.

The thermostatic type of the two-temperature valve acts much like the thermostatic expansion valve except it controls the vapor leaving the evaporator. The sensing bulb can be located either in the air inlet or in the air outlet to the evaporator. This would be a modulating type of valve, Figure 23–7.

23.7 CRANKCASE PRESSURE REGULATOR

The crankcase pressure regulator (*CPR valve*) looks much the same as the EPR valve, but it has a different function. The CPR valve is in the suction line also, but

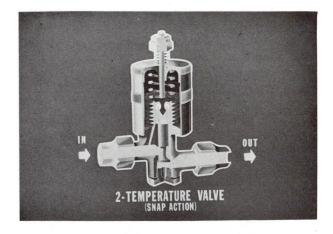

Figure 23–6. This two-temperature valve uses an overcenter device that actually gives two different temperatures inside the evaporator. This control has a low-temperature event and a high-temperature event. The high-temperature event can be set above freezing and used as defrost. This is not a modulating type of control. *Reproduced courtesy Carrier Corporation*

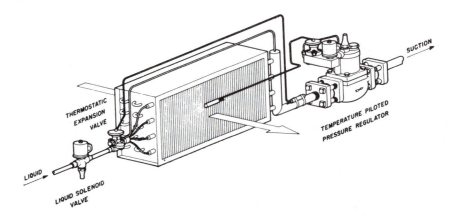

Figure 23-7. Thermostatic type two-temperature device. This valve may have its sensing bulb in the entering airstream or the leaving airstream of the evaporator. *Courtesy ALCO Controls Division, Emerson Electric Company*

it is usually located close to the compressor rather than at the evaporator outlet. The CPR valve sensing bellows is on the true compressor suction side of the valve and would normally have a gage port on the evaporator side of the valve, Figure 23-8.

This valve is used to keep the low-temperature compressor from overloading on a hot pull down. A hot pull down would occur (a) when the compressor has been off for a long-enough time and the foodstuff has a rise in temperature, or (b) on startup with a hot box. In either case the temperature in the refrigerated box influences the suction pressure. When the temperature is high, the suction pressure is high. When the suction pressure goes up, the density of the suction gas goes up. The compressor is a constant-volume pump and does not know when the gas it is pumping is dense enough to create a motor overload.

When a compressor is started up and the refrigerated cooler is warmer than the design range of the compressor, an overload occurs. The refrigeration compressor motor can be operated at only 10% over its rated capacity without harm during this pull down. When a motor has a full-load amp rating of 20 A, it could run at 22 A for an extended pull down, and no harm would come to the motor. If the cooler temperature were to be 75° or 80° F, the motor would be overloaded to the point that the motor overcurrent protection would shut it off. This protection is automatically reset, so the compressor would try to restart immediately, causing it to short cycle. This would go on until the manual reset motor control, the fuse or breaker, stopped the compressor for good, or the system would slowly pull down during the brief running times.

The CPR valve would throttle the suction gas into the compressor to keep the compressor current at no more than the rated value. Before CPR valves were used, the service technician had to start the system manually by throttling the suction service valve on a hot startup, Figure 23-9.

23.8 ADJUSTING THE CPR VALVE

The setting of the CPR valve should be accomplished on a hot pull down or at least when the compressor has enough load that it is trying to run overloaded. This can be done by shutting off the unit until the box or cooler is warm enough to create a load on the evaporator. This loads up the compressor enough to cause it to run at a high current. Use the ammeter on the compressor while adjusting the CPR valve to the full-load amperage of the compressor. For example, if the compressor is suppose to draw 20 A, throttle the CPR valve back until the compressor amperage is 20 A and the valve is set.

Figure 23-8. CPR valve to keep the compressor from running in an overloaded condition during a hot pull down. *Courtesy Sporlan Valve Company*

THIS IS A LOW-TEMPERATURE FREEZER WITH THE COMPRESSOR DESIGNED TO OPERATE AT 2 psig AND THE EVAPORATOR AT -16°F

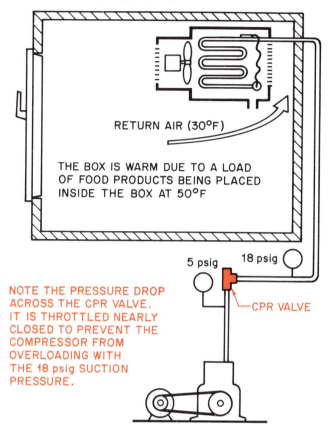

RETURN AIR (30°F)

THE BOX IS WARM DUE TO A LOAD OF FOOD PRODUCTS BEING PLACED INSIDE THE BOX AT 50°F

5 psig   18 psig

NOTE THE PRESSURE DROP ACROSS THE CPR VALVE. IT IS THROTTLED NEARLY CLOSED TO PREVENT THE COMPRESSOR FROM OVERLOADING WITH THE 18 psig SUCTION PRESSURE.

CPR VALVE

COMPRESSOR MOTOR FULL-LOAD AMPERES IS 20 A WHEN THE SUCTION PRESSURE IS 5 psig

Figure 23-9. The CPR valve throttles the vapor refrigerant entering the compressor. There will be a pressure drop across the valve when the evaporator is too warm. This valve is adjusted and set up using the compressor full load amperage on a hot pull down. The CPR valve can be set to throttle the suction gas to the compressor and maintain a suction pressure on the compressor side of the valve that will not allow the compressor to overload. When the refrigerated box temperature is pulled down to the design range, the suction gas in the evaporator will be the same as the pressure at the compressor because the valve is wide open.

23.9 RELIEF VALVES

Relief valves were discussed in general in Unit 15. Relief valves used in refrigeration are much like those explained in the previous unit except that they hold back refrigerant. Refrigerant costs much more than water that may be lost from a water system. Refrigerant relief valves come in two different types: the spring-loaded type that will reset, and the one-time type that does not close back.

The *spring-loaded* type of relief valve is normally brass with a neoprene seat. This valve is piped so that

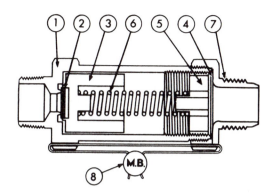

Figure 23-10. Spring-loaded type of relief valve used to protect the system from very high pressures that might occur if the condenser fan failed or if a water supply failed on a water-cooled system. This relief valve is designed to reseat after the pressure is reduced. There are threads on the valve outlet, so the valve can be piped to the outside if needed. *Courtesy Superior Valve Company*

it is in the vapor space of the condenser or receiver. The relief valve must be in the vapor space, not in the liquid space, for it to relieve pressure. The object is to let vapor off to vaporize some of the remaining liquid and lower its temperature, Figure 23-10.

The top of the valve normally has threads, so the refrigerant can be piped to the outside of the building. *When relief valves are used to protect vessels in a fire, it is desirable for the refrigerant to be removed from the area. Refrigerant gives off a noxious gas when it is burned.*

The *one-time* relief valve is relieved by temperature. These valves, often called fusible plugs, are designed with one of the following methods: a fitting filled with a low-melting-temperature solder; a patch of copper soldered on a drilled hole in the copper line; or a fitting that has been drilled out at the end with a spot of solder over the hole. Sometimes the melting temperature is printed on the fitting. The melting temperature will be very low, about 220°F. When soldering around the compressor make sure that the solder in the fusible plug is not melted away. This type of device will normally never relieve unless there is a fire. It can often be found on the suction side of the system close to the compressor to protect the system and the public. This will keep the compressor shell from experiencing high pressures and rupturing during a fire, Figure 23-11.

23.10 LOW AMBIENT CONTROLS

Low ambient controls are very important in refrigeration systems because refrigeration is needed year around. When the condenser is located outside, the head pressure will go down in the winter to the point that the expansion valve will not have enough

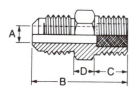

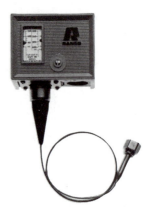

Figure 23–11. Fusible relief. It may be a low-temperature solder patch that covers a hole drilled in a fitting or pipe. Care should be taken when soldering in the vicinity of this relief device, or it will be melted loose. *Courtesy Mueller Brass Company*

Figure 23–12. Pressure control with the same action as a low-pressure control; it makes on a rise in head pressure to start the condenser fan. This pressure control operates at a higher pressure range than the low-pressure control normally used on the low side of the system. Fan cycle devices will cause the head pressure to fluctuate up and down. This can affect the operation of the expansion device. *Courtesy Ranco*

pressure drop across it to feed refrigerant correctly. When this happens, some method has to be used to keep the head pressure up to an acceptable level. The most common methods are

1. Fan cycling using a pressure control.
2. Fan speed control.
3. Air volume control using shutters and fan cycling.
4. Condenser flooding devices.

23.11 FAN CYCLING HEAD-PRESSURE CONTROL

Fan cycling has been used for years because it is simple. When a unit has one small fan, this is a simple and reliable method because only one fan is cycled. When there is more than one fan, the extra fans may be cycled by temperature, and the last fan can be cycled by head pressure. The control used is a pressure control that closes on a rise in pressure to start the fan and opens on a fall in pressure to stop the fan when the head pressure falls. When fan cycling is used, the fan will cycle in the winter but will run constantly in hot weather, Figure 23–12.

This type of control can be hard on a fan motor because of all the cycling. The fan motor needs to be a type that does not have a high starting current. The control has to have enough contact surface area to be able to start the motor many times to be reliable. The motor current should be compared to the control capabilities very carefully.

Fan cycling can vary the pressure to the expansion device a great deal. The technician must choose whether to have the control set points close together for the best expansion device performance or far apart to keep the fan from short cycling. The best application for fan cycling devices is in the use of multiple fans,

where the last fan has a whole condenser to absorb the fluctuations. Vertical condensers may be affected by the prevailing winds. If these winds are directed into the coil, they can act just like the fan. Sometimes a shield has to be installed to prevent this, Figure 23–13.

23.12 FAN SPEED CONTROL FOR CONTROLLING HEAD PRESSURE

Fan speed control devices have been used very successfully in some installations. This device can be used with multiple fans, where the first fans are cycled off by temperature and the last fan is controlled by head

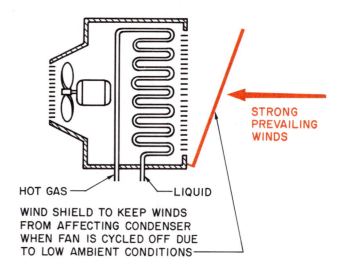

HOT GAS — LIQUID

WIND SHIELD TO KEEP WINDS FROM AFFECTING CONDENSER WHEN FAN IS CYCLED OFF DUE TO LOW AMBIENT CONDITIONS

STRONG PREVAILING WINDS

Figure 23–13. If fan cycling controls are used, care must be taken if the condenser is vertical. The prevailing winds can cause the head pressures to be too low even with the fan off. A shield can be used to prevent the wind from affecting the condenser.

motor speed increases. At some predetermined point the additional fans are started. On a hot day all fans will be running, with the variable fan running at maximum speed. Some fan speed controls use a temperature sensor to monitor the condenser's temperature, Figure 23–14.

23.13 AIR VOLUME CONTROL FOR CONTROLLING HEAD PRESSURE

Air volume control using shutters is accomplished with a damper motor driven with the high-pressure refrigerant. When there are multiple fans, the first fans can be cycled off using temperature, and the shutter can be used on the last fan, as in the previous two systems. This system results in a steady head pressure, as with fan speed control. The expansion valve inlet pressure does not go up and down like it does in pure fan cycling. The shutter can be located either on the fan inlet or on the fan outlet, Figure 23–15. Be careful that the fan motor is not overloaded with a damper.

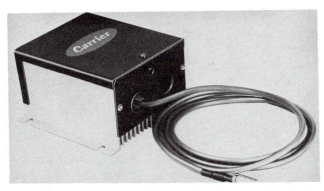

Figure 23–14. Control used to vary the fan speed on a special motor based on condenser temperature. *Reproduced courtesy of Carrier Corporation*

pressure or condensing temperature. The device that controls this fan is normally a transducer that converts pressure to motor speed control or a temperature sensor. As the temperature drops on a cool day, the motor speed is reduced. As the temperature rises, the

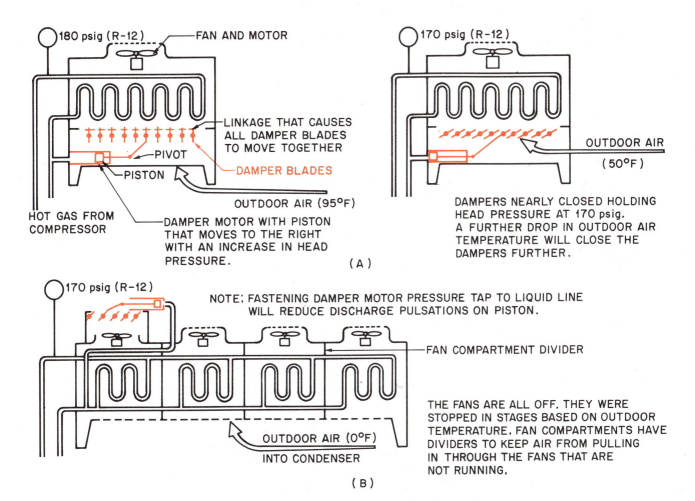

Figure 23–15. Installation using dampers to vary the air volume instead of fan speed. This is accomplished with a refrigerant-operated damper motor that operates the shutters. This application modulates the airflow and gives steady head-pressure control. If more than one fan is used, the first fans can be cycled off based on ambient temperature. Care should be taken that the condenser fan will not be overloaded when the shutters are placed at the fan outlet.

23.14 CONDENSER FLOODING FOR CONTROLLING HEAD PRESSURE

Condenser flooding devices are used in both mild and cold weather to cause the refrigerant to move from an oversized receiver tank into the condenser. This excess refrigerant in the condenser acts like an overcharge of refrigerant and causes the head pressure to be much higher than it normally would be on a mild or cold day. The head pressure will remain the same as it would on a warm day. This method gives very steady control with no fluctuations while running. This system requires a large amount of refrigerant because in addition to the normal operating charge, there must be enough refrigerant to flood the condenser in the winter. A large receiver is used to store the refrigerant in the summer when the valves will divert the excess liquid into the receiver.

Condenser flooding has one added benefit in the winter. Since the condenser is nearly full of refrigerant,

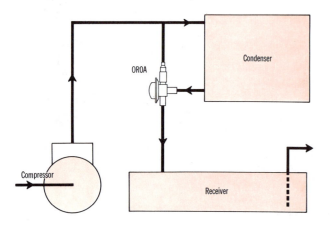

Figure 23-16. Low ambient head-pressure control accomplished with refrigerant. A large receiver stores the refrigerant when it is not needed (the summer cycle). The valve in the system diverts the hot gas to the condenser during summer operation, and the system acts like a typical system with a large receiver. In the winter cycle part of the gas goes to the condenser, depending on the outdoor temperature, and part of the gas to the top of the receiver. The gas going to the receiver keeps the pressure up for correct expansion valve operation. The gas going to the condenser is changed to a liquid and subcooled in the condenser. It then returns to the receiver to subcool the remaining liquid. *Courtesy Sporlan Valve Company*

the liquid refrigerant that is furnished to the expansion devices is well below the condensing temperature. Remember, this is called subcooling and will help improve the efficiency of the system. The liquid may be subcooled to well below freezing for systems operating in a cold climate, Figure 23–16.

23.15 ELECTRICAL CONTROLS

Electrical controls are used to stop, start, or modulate electron flow to control motors and fluid flow. Modulation is not used as much in electrical applications as it is in mechanical controls. The reason for this is that electrical components are not easy to modulate. It is difficult to get a motor to stop at a certain point. Special motors used for this are often too expensive for simple tasks. These motors use a resistance-type of thermostat. The thermostat is one circuit of a Wheatstone bridge electronic circuit. An electronic manual should be consulted for more information. The technician normally checks for voltage and if the motor doesn't turn, it should be changed.

The very nature of mechanical controls makes them appropriate to modulate devices, because most of them have sensing elements that change in dimension with temperature and pressure changes. This dimension change can easily be linked to a valve to modulate fluid flow.

23.16 THE SOLENOID VALVE

The *solenoid valve* is the most frequently used component to control fluid flow. This valve has a magnetic coil that when energized will lift a plunger into the coil, Figure 23–17. This valve can either be normally open (NO) or normally closed (NC). The NC valve is closed until energized; then it opens. The NO valve is normally open until energized; then it closes.

Solenoid valves are a snap-acting type of valve that open and close very fast with the electrical energy to the coil. This valve can be used to control either liquid or vapor flows. The snap action can cause liquid hammer when installed in a liquid line, so be careful when locating the valve. Liquid hammer occurs when the fast-moving liquid is shut off abruptly by the solenoid valve, causing the liquid to stop abruptly.

The solenoid valve is responsible for stopping and starting fluid flow. Two things can prevent the solenoid valve from doing this: the direction in which the valve is mounted, and the position in which the valve is installed. The fluid flow has to be in the correct direction with a solenoid valve, or the valve may not close tightly, Figure 23–18. The valve is mounted in the correct direction when the fluid helps to close the valve. If the high pressure is under the valve seat, the valve may have a tendency to lift off its seat. The valve will have an arrow to indicate the direction of flow. While

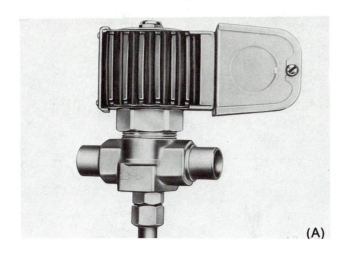

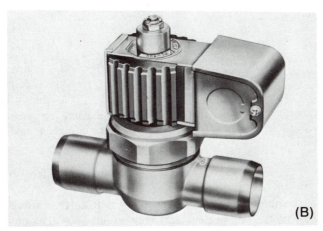

Figure 23-17. Electrically operated solenoid valves. The valve seat is moved by the plunger attached to the seat. The plunger moves into the magnetic coil when the coil is energized to either open or close the valve. *Courtesy Sporlan Valve Company*

placing the solenoid valve in the correct direction, the position of the valve has to be considered. Most solenoid valves have a heavy plunger that is open and lifted. When the plunger is not magnetized, the weight of this plunger holds the valve on its sealing seat. If this type of valve is installed on its side or upside down, the valve will remain in the magnetized position when it is not magnetized, Figure 23-19.

The solenoid valve must be fastened to the refrigerant line so that it will not leak refrigerant. It can be fastened by flare, flange, or solder connections. Most valves have to be serviced at some point in time. The valves that are soldered in the line can be serviced easily if they can be disassembled.

A special solenoid valve known as *pilot acting* is available. This valve uses a very small valve seat to divert high-pressure gas that causes a larger valve to change position. This type of solenoid valve uses this difference in pressure to cause the large movement while the solenoid's magnetic coil only has to lift a small seat. These valves are used when large vapor or

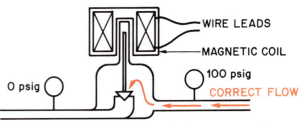

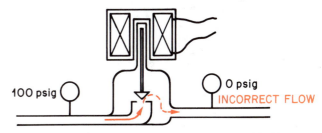

Figure 23-18. The fluid flow in a solenoid valve must be in the correct direction, or the valve will not close tight. If the high-pressure fluid is under the seat, it will have a tendency to raise the valve off the seat because the fluid helps to hold the valve on the seat.

liquid lines have to be switched; they can have more than one inlet and outlet. Some are known as *four-way* valves, others as *three-way* valves. These have special functions. If a coil had to be designed to make the switching action of a valve that is large, the coil would become very large and draw a large current, Figure 23-20.

23.17 PRESSURE SWITCHES

Pressure switches are used to stop and start electrical current flow to refrigeration components. They play

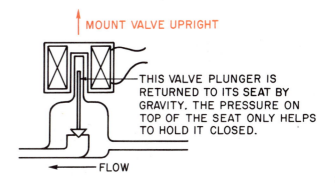

Figure 23-19. Solenoid valve installed in the correct position. Most valves use gravity to seat the valve, and the magnetic force raises the valve off its seat. If the valve is installed on its side or upside down, the valve will not seat when deenergized.

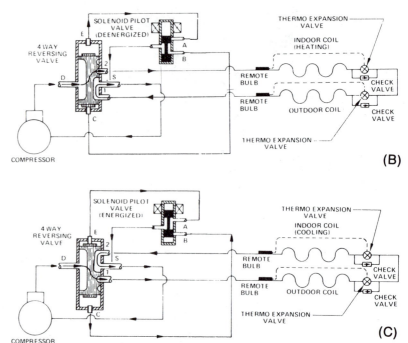

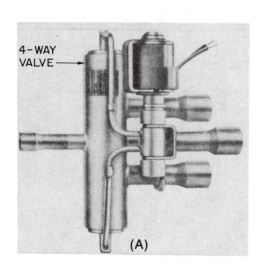

Figure 23–20. Pilot-operated solenoid valve. The magnetic coil only controls a small line that directs the high-pressure refrigerant to one end of a sliding piston. The magnetic coil only has to do a small amount of work, and the difference in pressure does the rest. *Courtesy ALCO Controls Division, Emerson Electric Company*

a very important part in the function of refrigeration equipment and will be discussed here as they apply to refrigeration. The typical pressure switch can be a

1. Low-pressure switch—closes on a rise in pressure.
2. High-pressure switch—opens on a rise in pressure.
3. Low ambient control—closes on a rise in pressure.
4. Oil safety switch—has a time delay; closes on a rise in pressure.

23.18   THE LOW-PRESSURE SWITCH

The low-pressure switch is used in two applications in refrigeration: low charge protection, and control of space temperature.

The low-pressure control can be used as a low-charge protection by setting the control to cut out at a value that is below the typical evaporator operating pressure. For example, in a medium-temperature cooler the air temperature may be expected to be no lower than 34°F. When this cooler uses R-12 for the refrigerant, the lowest pressure at which the evaporator would be expected to operate would be around 20 psig because the coil would normally operate at about 15°F colder than the air temperature in the coil (34°F − 15°F = 19°F). The refrigerant in the coil should not operate below this temperature of 19°F, which converts to 20 psig.

The low-pressure cutout should be set below the expected operating condition of 20 psig and above the

atmospheric pressure of 0 psig. A setting that would cut off the compressor at 5 psig would keep the compressor from going into a vacuum in the event of loss of refrigerant. This setting would be well below the typical operating condition. This control would normally be automatically reset. When a low-charge condition existed, the compressor would be cut off and on with this control and maintain some refrigeration. The store owner may call with a complaint that the unit is cutting off and on but not cooling the cooler properly. The cut-in setting for this application should be a pressure that is below the highest temperature the cooler is expected to experience in a typical cycle and just lower than the thermostat cut-in point. For example, the thermostat may be set at 45°F, which would mean that the pressure in the evaporator may rise as high as 42 psig. The low-pressure control should be set to cut in at about 40 psig. The previous example of a low-pressure control set to cut out at 5 psig and to cut in at 40 psig means that the control differential is 35 psig. The wide differential helps to keep the compressor from coming back on too soon or from short cycling, Figure 23–21.

23.19   THE LOW-PRESSURE CONTROL APPLIED AS A THERMOSTAT

The low-pressure control setup described in the previous subsection is for low-charge protection only,

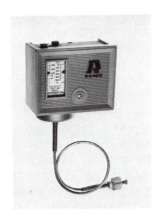

Figure 23–21. Low-pressure control. This control's contacts make on a rise in pressure (open on a fall). If the low-side pressure goes down below the control set point for any reason, the control stops the compressor. *Courtesy Ranco*

to keep the system from going into a vacuum. The same control can be set up to operate the compressor to maintain the space temperature in the cooler and to serve as a low-charge protection. Using the same temperatures as in the preceding paragraph, 34°F and 45°F, as the operating conditions, the low-pressure control can be set up to cut out when the low pressure reaches 21 psig. This corresponds to a coil temperature of 20°F. When the air in the cooler reaches 34°F, the coil temperature should be 19°F, with a corresponding

pressure of 20 psig. This system has a room-to-coil temperature difference of 15°F (34°F room temperature −19°F coil temperature = 15°F temperature difference). When the compressor cuts off, the air in the cooler is going to raise the temperature of the coil to 34°F and a corresponding pressure of 32 psig. As the cooler temperature goes up, it will raise the temperature of the refrigerant. When the temperature of the air increases to 45°F, the coil temperature should be 45°F and have a corresponding pressure of 42 psig. This could be the cut-in point of the low-pressure control. The settings would be cut out at 20 psig and cut in at 42 psig. This is a differential of 22 psig.

One of the advantages of this type of control arrangement is that there are no interconnecting wires between the inside of the cooler and the condensing unit. If a thermostat is used to control the air temperature in the cooler, a pair of wires has to be run between the condensing unit and the inside of the cooler. Some installations have the condensing unit a great distance from the cooler, which makes this impractical. With the temperature being controlled at the condensing unit, the owner is less likely to turn the control and cause problems.

There are as many low-pressure control settings as there are applications. Different situations call for different settings. Figure 23–22 is a chart of recommended settings by one manufacturer.

APPROXIMATE PRESSURE/CONTROL SETTINGS

	Refrigerant					
	12		22		502	
Application	Out	In	Out	In	Out	In
Ice Cube Maker—Dry Type Coil	4	17	16	37	22	45
Sweet Water Bath—Soda Fountain	21	29	43	56	52	66
Beer, Water, Milk Cooler, Wet Type	19	29	40	56	48	66
Ice Cream Trucks, Hardening Rooms	2	15	13	34	18	41
Eutectic Plates, Ice Cream Truck	1	4	11	16	16	22
Walk In, Defrost Cycle	14	34	32	64	40	75
Reach In, Defrost Cycle	19	36	40	68	48	78
Vegetable Display, Defrost Cycle	13	35	30	66	38	77
Vegetable Display Case—Open Type	16	42	35	77	44	89
Beverage Cooler, Blower Dry Type	15	34	34	64	42	75
Retail Florist—Blower Coil	28	42	55	77	65	89
Meat Display Case, Defrost Cycle	17	35	37	66	45	77
Meat Display Case—Open Type	11	27	27	53	35	63
Dairy Case—Open Type	10	35	26	66	33	77
Frozen Food—Open Type	7	5	4	17	8	24
Frozen Food—Open Type—Thermostat	2°F	10°F	—	—	—	—
Frozen Food—Closed Type	1	8	11	22	16	29

Figure 23–22. Table to be used as a guide for setting low-pressure controls for the different applications. *Courtesy C.C. Dickson*

Low-pressure controls are rated by their pressure range and the current draw of the contacts. A low-pressure control that is suitable for R-12 may not be suitable for R-502 because of pressure range. For the same application in the previous paragraph, the cutout would be 53 psig and the cut-in would be 88 psig. A control of the correct pressure range has to be chosen. Some of these controls are single-pole–double-throw. They can make or break on a rise. This control can serve as one component for two different jobs.

The contact rating for a pressure control has to do with the size of an electrical load that the control can carry. If the pressure control is expected to start a small compressor, the inrush current should also be considered. Normally a pressure control used for refrigeration is rated so it can directly start up to a 3-hp single phase compressor. If the compressor is any larger, or three phase, a contactor is normally used. The pressure control can then control the contactor's coil, Figure 23–23.

23.20 THE HIGH-PRESSURE CONTROL

The high-pressure control is normally not as complicated as the low-pressure switch. It is used to keep the compressor from operating with a high head pressure. This control is very necessary on water-cooled equipment because an interruption of water is more likely than an interruption of air. This control opens on a rise in pressure and should be set above the typical high pressure that the machine would normally en-

counter. The high-pressure control may be either automatic reset or manual reset.

When an air-cooled condenser is placed outside, the condenser can be expected to operate at no more than 35°F warmer than the ambient air. This condition is true after the condenser has run long enough to have a coat of dirt built up on the coil. It is true that a clean condenser is important, but they are more often slightly dirty than clean. If the ambient air is 95°F, the condenser would be operating at about 180 psig if the system used R-12 (95°F + 35°F = 130°F condensing temperature, which corresponds to 180 psig). The same system using R-502 would have a pressure of 318 psig. The high-pressure control should be set well above 180 psig for the R-12 system. If the control were set to cut out at 250 psig, there should be no interference with normal operation, and it still would give good protection, Figure 23–24.

The control cut-in point has to be above the ambient temperature of 95°F. If the compressor cuts off and the outdoor fan continues to run, the temperature inside the condenser will quickly reach the ambient temperature. For example if the ambient is 95°F, the pressure will quickly fall to 108 psig. If the high-pressure control were set to cut in at 125 psig, the compressor could come back on with a safe differential of 125 psig (cut-out 250 − cut-in 125 = 125 psig differential).

Some manufacturers specify a manual reset high-pressure control. When this control cuts out, someone must press the reset button to start the compressor. This calls attention to the fact that there is a problem. The

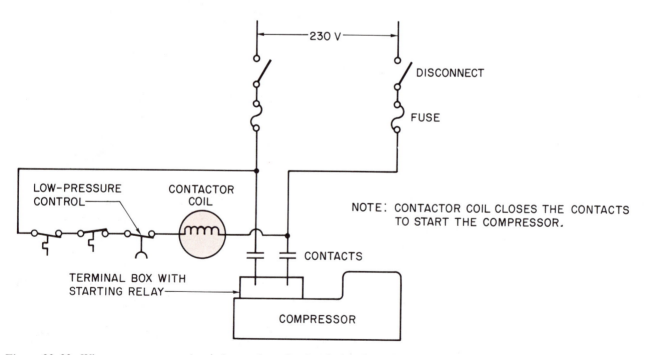

Figure 23–23. When a compressor that is larger than the electrical rating of the pressure control's contacts is encountered, the compressor has to be started with a contactor. The wiring diagram shows how this is accomplished.

Figure 23–24. High-pressure switch to keep the compressor from running when high system pressures occur. *Courtesy Ranco*

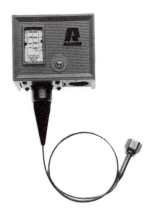

Figure 23–25. Low ambient controls used to open the contacts on a drop in head pressure to stop the condenser fan. *Courtesy Ranco*

manual reset control is better equipment protection, but the automatic reset control may save the food by allowing the compressor to run at short intervals. An observant owner or operator should notice the short cycle of the automatic control if the compressor is close by in the workspace.

23.21 THE LOW AMBIENT FAN CONTROL

The low ambient fan control has the same switch action as the low-pressure control but operates at a higher pressure range. This control stops and starts the condenser fan in response to head pressure. This control has to be coordinated with the high-pressure control to keep them from working against each other. The high-pressure control stops the compressor when the head pressure gets too high, and the low ambient control starts the fan when the pressure gets to a predetermined point before the high-pressure control cuts the compressor off.

When a low ambient control is used, the high-pressure cut-out has to be checked to make sure that it is higher than the cut-in point of the low ambient control. For example, if the low ambient control is set up to maintain the head pressure between 125 psig and 175 psig, a high-pressure control setting of 250 psig should not interfere with the low ambient control setting. This can easily be verified by installing a gage and stopping the condenser fan to make sure that the high-pressure control is cutting out where it is supposed to. With the gage on the high side, the fan action can be observed also to see that the fan is cutting off and on as it should. The fan will operate all of the time on high ambient days, and the fan control will not stop the fan. The low ambient pressure control can be identified by the terminology on the control's action. It is described as a "close on a rise" in pressure, Figure 23–25.

23.22 THE OIL SAFETY SWITCH

The oil safety cut-out switch is used on larger installations with larger compressors, normally above 5 hp. Most of the larger compressors discussed in this book have pressure lubrication systems. The oil pump is attached to the crankshaft and turns when the compressor is turning. There is no oil pressure until the compressor crankshaft turns. The compressor has to be allowed some running time at the beginning of the cycle to build oil pressure, so a plain pressure switch will not meet the requirements. Usually there is a 90-sec time delay built into the oil safety control to allow the compressor to build up a minimum of 30 psig of net oil pressure. Net oil pressure is the pressure registered above the crankcase pressure (the same as suction pressure). See Figure 23–26 for an oil safety control.

23.23 THE DEFROST CYCLE

The defrost cycle in refrigeration is divided into medium-temperature and low-temperature ranges, and the components that serve the defrost cycle are different.

23.24 MEDIUM TEMPERATURE REFRIGERATION

The medium-temperature refrigeration coil normally operates below freezing and rises above freezing during the off cycle. A typical temperature range would be from 32° to 45°F space temperature inside the cooler. As mentioned before, the coil temperature is normally 10° to 15°F cooler than the space temperature in the refrigerated box. This means that the coil temperatures would normally operate as low as 17°F (32° − 15° = 17°). The air temperature inside the box will always rise above the freezing point during the off cycle and can be used for the defrost. This is called *off-cycle* defrost and can either be random or planned.

Figure 23–26. Oil safety controls have a bellows on each side of the control. They are opposed in their forces and serve to measure the net oil pressure; the difference in the suction pressure (oil pump inlet) and the oil pump outlet pressure. This control has a 90 second time delay to allow the compressor to get up to speed and establish oil pressure before it shuts down. *Courtesy Ranco*

23.25 RANDOM OR OFF-CYCLE DEFROST

Random defrost will occur when the refrigeration machine has enough reserve capacity to cool more than the load requirement. When the refrigeration machine has reserve capacity, it will be shut down from time to time by the thermostat and the air in the cooler can defrost the ice from the coil. When the compressor is off, the evaporator fans will continue to run, and the air in the cooler will defrost the ice from the coil. When the refrigeration machine does not have enough capacity or the refrigerated box has a constant load, there may not be enough off time to accomplish defrost. This is when it has to be planned.

23.26 PLANNED DEFROST

Planned defrost is accomplished by forcing the compressor to shut down for short periods of time so that the air in the cooler can defrost the ice from the coil. This is accomplished with a timer that can be programmed. Normally the timer stops the compressor during times that the refrigerated box is under the least amount of load. For example, a restaurant unit may defrost at 2 A.M. and 2 P.M. to avoid the rush hours, Figure 23–27.

23.27 LOW-TEMPERATURE EVAPORATOR DESIGN

Low-temperature evaporators all operate below freezing and must have planned defrost. Since the air inside the refrigerated box is well below freezing, heat

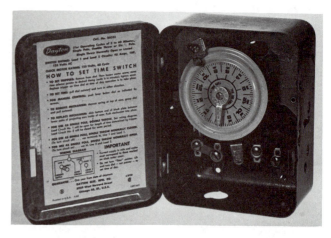

Figure 23–27. Timer to program off-cycle defrost. This is a 24-hr timer that can have several defrost times programmed for the convenience of the installation. *Photo by Bill Johnson*

has to be added to the evaporator for defrost. This defrost is normally accomplished with *internal heat* or *external heat.*

23.28 DEFROST USING INTERNAL HEAT (HOT GAS DEFROST)

The internal heat method of defrost normally uses the hot gas from the compressor. This hot gas can be introduced into the evaporator from the compressor discharge line at the inlet and allowed to flow until the evaporator is defrosted. A portion of the energy used for hot gas defrost is available in the system. This makes it attractive from an energy-saving standpoint.

Injecting hot gas into the evaporator is rather simple if the evaporator is a single circuit type because a T in the expansion valve outlet is all that is necessary. When a multiple-circuit evaporator is used, the hot gas has to be injected between the expansion valve and the refrigerant distributor. This gives an equal distribution of hot gas to defrost all of the coils equally.

The defrost cycle is normally always started with a timer in space-temperature applications, where forced air is used to cool the product. When the defrost cycle is started, some method has to be used to terminate it. Defrost can be terminated by time or temperature. The amount of time it takes to defrost the coil has to be known before time alone can be used efficiently to terminate defrost. Since this time can vary from one situation to another, the timer could be set for too long a time, and the unit would run in defrost when it is not desirable.

Defrost can be started with time and terminated with temperature. When this is done, a temperature-sensing device is used to determine that the coil is above freezing. The hot gas entering the evaporator is stopped, and the system goes back to normal operation, Figure 23–28.

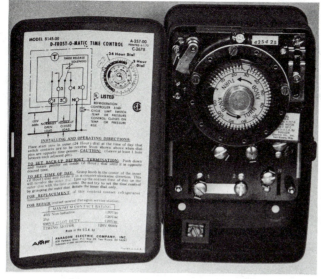

(A)

Figure 23-28A. Timer with a mechanism that can stop defrost with an electrical signal from a temperature-sensing element. When defrost is over, the coil temperature will rise above freezing. There is no reason for the defrost to continue after the ice has melted. *Photo by Bill Johnson*

Figure 23-28B. Wiring diagram of a circuit to control the defrost cycle. The events happen like this: When there is a defrost call (the timer's contacts close), the solenoid valve opens and the compressor continues to run and pump the hot gas into the evaporator. The coil gets warm enough to cause the thermostat to change from the cold contacts to the hot contacts; the defrost will be terminated by the X terminal on the timer. When the coil cools off enough for the thermostat to change back to the cold contacts, the fan will restart. This is another method of keeping the compressor from running overloaded on a hot pull down.

During the hot-gas defrost cycle, several things have to happen at one time. The timer is used to coordinate the following functions:

1. The hot gas solenoid must open.
2. The evaporator fans must stop, or cold air will keep defrost from occurring.
3. The compressor must continue to run.
4. A maximum defrost time must be determined and programmed into the timer.
5. Drain pan heaters may be energized.

23.29   EXTERNAL HEAT TYPE OF DEFROST

The external heat method of defrost is often accomplished with electric heating elements that are factory mounted next to the evaporator coil. This type of defrost is also a planned defrost that is controlled by a timer. The external heat method is not as efficient as the internal heat method because energy has to be purchased for defrost. When electric defrost is used, it

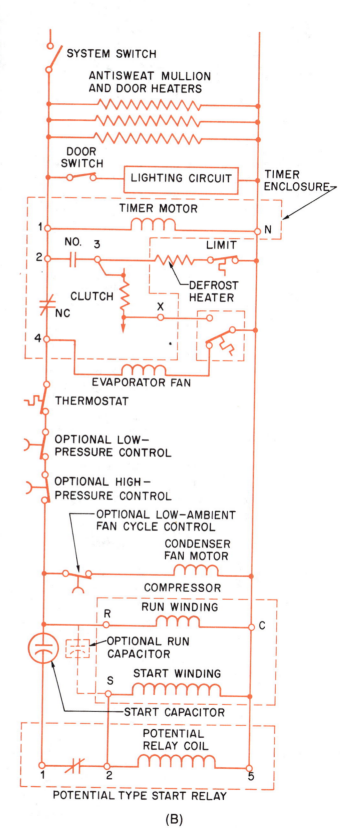

(B)

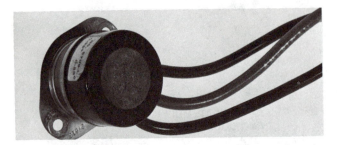

Figure 23-29. Sensor used with the timer in Figure 23–28. Sensor has three wires; it is a single-pole–double-throw device. It has a hot contact (made from common to the terminal that is energized on a rise in temperature) and a cold contact (made from common to the terminal that is energized when the coil is cold). This control is either in the hot or the cold mode. *Photo by Bill Johnson*

is more critical that the defrost be terminated at the earliest possible time. The timer controls the following events for electric defrost:

1. The evaporator fan stops.
2. The compressor stops. (There may be a pump-down cycle to pump the refrigerant out of the evaporator to the condenser and receiver.)
3. The electric heaters are energized.
4. Drain pan heaters may be energized.

Note: A temperature sensor may be used to terminate defrost when the coil is above freezing. A maximum defrost time should be programmed into the timer, Figure 23–29.

23.30   REFRIGERATION ACCESSORIES

Accessories that are in the refrigeration cycle are devices that improve the system performance and service functions. This text will start at the condenser, where the liquid refrigerant leaves the coil, and add various accessories as they are encountered in systems. Each system does not have all the accessories.

23.31   THE RECEIVER

The *receiver* is located in the liquid line and is used to store the liquid refrigerant after it leaves the condenser. The receiver should be lower than the condenser so the refrigerant has an incentive to flow into it naturally. This is not always possible. The receiver is a tank like device that can either be upright or horizontal, depending on the installation. See Figure 23–30 for an example. Receivers can be quite large on systems that need to store large amounts of refrigerant. Figure 23–31 is a photo of a large receiver that will hold several hundred pounds of refrigerant.

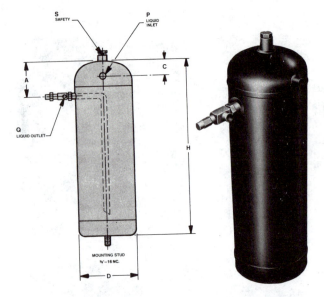

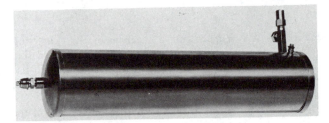

Figure 23-30. Vertical and horizontal receivers. *Courtesy Refrigeration Research*

Figure 23-31. Large receivers store the charge on a condenser flooding system. It may hold more than 100 lb of refrigerant. *Photo by Bill Johnson*

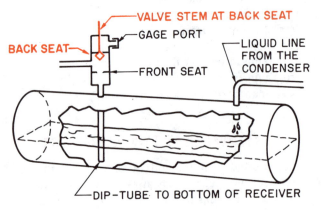

Figure 23-32. King valve piped in the circuit. The back seat is open to the gage port when the valve is front seated. This valve has to be back seated for the gage to be removed if there is refrigerant in the receiver.

The receiver's inlet and outlet connection can be at almost any location on the outside of the tank body. On the inside of the receiver, however, the refrigerant must enter the receiver at the top in some manner. The refrigerant that is leaving the receiver must be taken from the bottom. This is accomplished with a dip tube if the line is at the top. Figure 23–32 is an illustration that shows how this is done and why.

23.32 THE KING VALVE ON THE RECEIVER

The *king* valve is located in the liquid line between the receiver and the expansion valve. It is quite often fastened to the receiver tank as the outlet. The king valve is important in service work because when it is valved off, no refrigerant can leave the receiver. If the compressor is operated with this valve closed, the refrigerant will all be pumped into the condenser and receiver. Most of it will flow into the receiver. The other valves in the system can then be closed and the low-pressure side of the system can be opened for service. When service is complete and the low side of the system is ready for operation, the king valve can be opened, and the system can be put back into operation.

The king valve may have a pressure service port that a gage manifold can be fastened to. When the valve stem is turned away from the back seat, this port will give a pressure reading in the liquid line between the expansion valve and the compressor high-side gage port. When the valve is front seated and the system is being pumped down, this gage port is on the high-pressure side of the system common to the receiver. Gage line removal may be difficult until the king valve is back seated after the repair is completed, Figure 23–34.

The next component that can be found in many systems is a solenoid valve. This was covered earlier. It is a valve that stops and starts the liquid flow.

23.33 FILTER-DRIERS

The *refrigerant filter-drier* can be found at any point on the liquid line after the king valve. The drier is a device that removes foreign matter from the refrigerant. This foreign matter can be dirt, flux from soldering, solder beads, filings, moisture, parts, and acid, caused by moisture. Filter-driers can remove construction dirt (filter only), moisture, and acid.

These filter-drying operations are accomplished with a variety of materials that are packed inside the drier. Some of the manufacturers furnish beads of chemicals and some use a porous block made from the drying agent. The most common agents found in the filter-driers are activated alumina, molecular sieve, or silica gel. The component has a fine screen at the outlet to catch any fine particles, Figure 23–33.

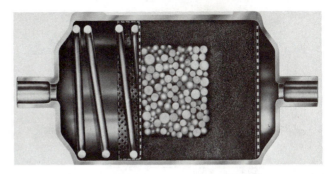

Figure 23–33. Filter-drier. This device is for a liquid line application and removes particles and moisture. The moisture removal is accomplished with a desiccant inside the drier shell. This desiccant material will be in bead or block form. *Courtesy Parker Hannafin Corporation*

The filter-drier comes in two styles: permanent and replaceable core. Both types of driers can be used in the suction line when chosen for that application. This will be discussed in more detail when we get to the suction line components.

The filter-drier can be fastened to the liquid line by a flare connection in the smaller sizes up to $\frac{5}{8}$ in. Solder connections can be used for a full range of sizes from $\frac{1}{4}$ in. to $1\frac{5}{8}$ in. The larger solder filter-driers are all replaceable core types. The replaceable core filter-drier can be very useful for future service.

The filter-drier can be installed anywhere in the liquid line and is found in many locations. The closer it is to the metering device, the better it cleans the refrigerant before it enters the tiny orifice in the metering device. The closer it is to the king valve, the easier it is to service. Different service technicians and engineers choose their own placement, based on experience and judgment.

23.34 REFRIGERANT SIGHT GLASSES

The *refrigerant sight glass* is normally located anywhere that it can serve a purpose. When it is installed just prior to the expansion device, the technician can be assured that a solid column of liquid is reaching the expansion device. When it is installed at the condensing unit, it can help with troubleshooting. More than one sight glass, one at each place, is sometimes a good investment.

Sight glasses come in two basic styles: plain glass, and glass with a moisture indicator. The plain glass type is used to observe the refrigerant as it moves along the line. The sight glass with a moisture indicator in it can tell the technician what the moisture content is in the system. It has a small element in it that changes color when moisture is present. An example of both of these can be seen in Figure 23–34.

Figure 23–34. (A) Sight glass with an element that indicates the presence of moisture in the system. (B) Sight glass used only to view the liquid refrigerant to be sure there are no vapor bubbles in the liquid line. *Courtesy (A) Superior Valve Company (B) Henry Valve Company*

23.35 LIQUID REFRIGERANT DISTRIBUTOR

The *refrigerant distributor* is fastened to the outlet of the expansion valve. It distributes the refrigerant to each individual evaporator circuit. This is a precision-machined device that assures that the refrigerant is divided equally to each circuit. This is not a simple task because the refrigerant is not all liquid or all vapor. It is a mixture of liquid and vapor. A mixture has a tendency to stratify and feed more liquid to the bottom and the vapor to the top, Figure 23–35. Some distributors have a side inlet that hot gas can be injected into for hot-gas defrost, Figure 23–36.

23.36 THE HEAT EXCHANGER

A *heat exchanger* is often placed in the suction line leaving the evaporator. This heat exchange is between the suction and the liquid line. It

1. Improves the capacity of the evaporator by subcooling the liquid refrigerant entering it, which can allow a smaller evaporator in the refrigerated box.
2. Prevents liquid refrigerant from moving out of the evaporator to the suction line and into the compressor. Most of these heat exchangers are simple and straightforward.

The heat exchanger has no electrical circuit or wires. It can be recognized by the suction line and the liquid line piped to the same device. Some small refrigeration devices have the capillary tube soldered to the suction line. This accomplishes the same thing that a larger heat exchanger does, Figure 23–37.

The next two components that could be installed in the suction line are the EPR valve (evaporator pressure

Figure 23–35. Multicircuit refrigerant distributor. Distributor to feed equal amounts of liquid refrigerant to the different refrigerant circuits. The combination of liquid and vapor refrigerant has a tendency to stratify with the liquid moving to the bottom. This is a precision machined device to separate the mixture evenly. *Courtesy Sporlan Valve Company*

Figure 23–36. Same type of distributor as in Figure 23–35 except that it has a side inlet for allowing hot gas to enter the evaporator evenly during defrost. *Courtesy Sporlan Valve Company*

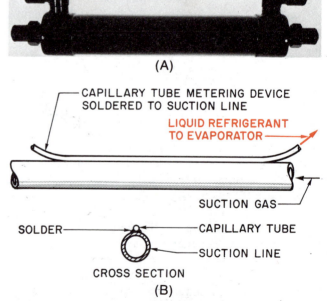

Figure 23–37. Heat exchangers. (A) Plain liquid-to-suction type. (B) Capillary tube fastened to the suction line that accomplishes the same thing. *Photo by Bill Johnson*

regulator) and the CPR valve (crankcase pressure regulator). These were covered earlier in this unit as control components.

23.37 SUCTION LINE ACCUMULATOR

The *suction line accumulator* can be located in the suction line to prevent liquid refrigerant from passing into the compressor. Under certain conditions liquid refrigerant may leave the evaporator. The suction line is insulated and will not evaporate much refrigerant. The suction line accumulator collects this liquid and gives it a place to boil off (evaporate) to a gas before continuing on to the compressor.

The suction line accumulator will also collect oil from the suction line. After it is collected, the oil is moved on to the compressor by velocity or by a small hole in the piping close to the bottom of the accumulator, Figure 23-38. The accumulator is usually located close to the compressor. It should not be insulated so that any liquid that is in it may have a chance to boil off.

23.38 THE SUCTION LINE FILTER-DRIER

The *suction line filter-drier* is similar to the drier in the liquid line. It is rated for suction line use. A drier is placed in the suction line to protect the compressor and it is good insurance in any installation. It is essential to install a filter-drier after any failure that

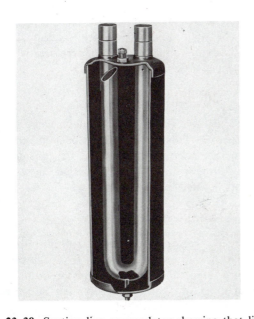

Figure 23-38. Suction line accumulator showing that liquid refrigerant returning down the suction line can be trapped and allowed to boil away before entering the compressor. The oil bleed hole allows a small amount of any liquid in the accumulator to be returned to the compressor. If the liquid is refrigerant, there will not be enough of it getting back through the bleed hole to cause damage. *Courtesy AC and R Components, Inc.*

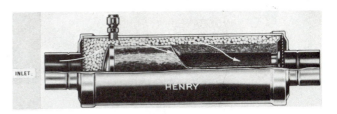

Figure 23-39. Suction line filter-drier. This device cleans any vapor that is moving toward the compressor. *Courtesy Henry Valve Company*

contaminates a system. A motor burnout in a compressor shell usually moves acid and contamination into the whole system. When a new compressor is installed, a suction line filter-drier can be installed to clean the refrigerant and oil before it reaches the compressor, Figure 23-39.

23.39 THE SUCTION SERVICE VALVE

The *suction service valve* is normally attached to the compressor. Equipment used for refrigeration installations usually has service valves. This is not always the case on air conditioning equipment. The suction service valve can never be totally closed because of the valve's seat design. When a service valve is mentioned, the terms back seat, front seat, and midseat are used.

The suction service valve is used

1. As a gage port.
2. To throttle the gas flow to the compressor.
3. To valve off the compressor from the evaporator for service.

An example of these functions is shown in Figure 23-40.

The service valve consists of the valve cap, valve stem, packing gland, inlet, outlet, and valve body, Figure 23-41.

The *valve cap* is used as a backup to the packing gland to prevent refrigerant from leaking around the stem. It is normally made of steel and should be kept dry on the inside. An oil coating on the inside will help prevent rust.

The *valve stem* has a square head for the valve wrench to fasten to. This stem is normally steel and should be kept rust free. If rust does build up on the stem, a light sanding and a coat of oil will help. The valve turns in and out through the valve packing gland. Rust on the stem will destroy the packing in the gland.

The *packing gland* can either be a permanent type or an adjustable type. If the gland is not adjustable and a leak is started, it can only be stopped with the valve cap. If the gland is adjustable, the packing can be replaced. Normally it is graphite rope, Figure 23-42. Any time the valve stem is turned, the gland should be first loosened if it is adjustable.

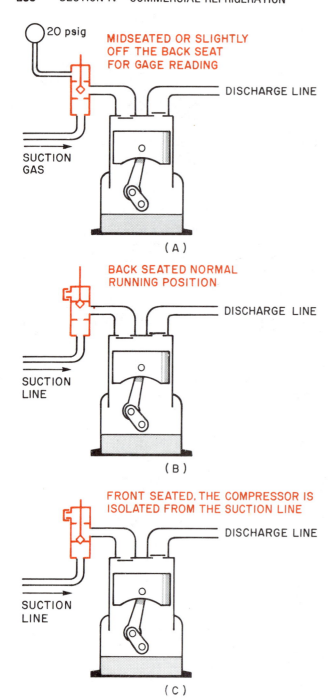

Figure 23–40. Three positions for the suction service valve.

The suction service valve attaches to the compressor on one side. The refrigerant piping from the evaporator fastens to the other side of the valve. The piping can either be flared or soldered.

23.40   THE DISCHARGE SERVICE VALVE

The *discharge service valve* is the same as the suction service valve except that it is located in the discharge line. This valve can be used as a gage port, and to valve off the compressor for service. *The compressor can-

not be operated with this valve front seated except for closed loop capacity checks under experienced supervision. Extremely high pressures will result if the test equipment is not properly applied.*

The *compressor service valves* are used for many service functions. One of the most important is to change the compressor. When both service valves are front seated, the compressor is isolated and can be removed. When a new one is installed, the only part of the system that must be evacuated is the new compressor. A compressor can be changed in this fashion with little loss of refrigerant. All that is lost would be the vapor between the service valves.

23.41   REFRIGERATION SERVICE VALVES

Refrigeration service valves are normally hand-operated specialty valves used for service purposes. These valves can be used in any line that may have to be valved off for any reason. They come in two types: the diaphragm valve and the ball valve.

23.42   THE DIAPHRAGM VALVE

The *diaphragm valve* has the same internal flow pattern that a "globe" valve does. The fluid has to rise up and over a seat, Figure 23–43. There is a measurable pressure drop through this type of valve. The valve can be tightened by hand enough to hold back high pressures. This valve can be fastened either by a flare connection or soldered into the system. When it is soldered, care should be taken that it is not over-heated. Most of these valves have seats made of materials that would melt at low temperatures when being soldered into a line.

23.43   THE BALL VALVE

The *ball valve* is a straight-through valve with little pressure drop. This valve can also be soldered into the line, but the temperature has to be considered. All manufacturers furnish directions that show how to install their valve, Figure 23–44.

23.44   THE OIL SEPARATOR

The *oil separator* is a component installed in the discharge line to separate the oil from the refrigerant and return the oil to the compressor crankcase. All reciprocating and rotary compressors allow a small amount of oil to pass through into the discharge line. Once the oil leaves the compressor, it would have to go through the complete system to get back to the compressor crankcase. The oil separator has a float to allow this oil to shortcut and return to the crankcase. The oil separator should be kept warm to keep liquid refrigerant from condensing in it during the off cycle. The float does not distinguish between oil and liquid

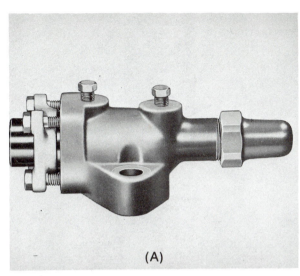

(A)

PCE. NO.	DESCRIPTION	QUAN.
1	BODY	1
2	STEM	1
3	SEAT DISC	1
4	RETAINER RING	1
5	DISC SPRING	1
6	GASKET	1
7	CAPSCREW	4
8	PACKING WASHER	1
9	PACKING	1
10	PACKING GLAND	1
11	CAP	1
12	CAP GASKET	1
13	FLANGE	1
14	ADAPTER	1

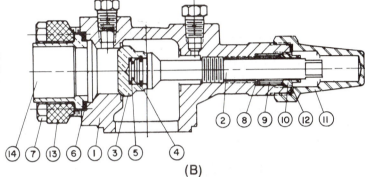

(B)

Figure 23–41. Components of a service valve. The suction and discharge valves are made the same except the suction valve is often larger. *Courtesy Henry Valve Company*

PCE. NO.	DESCRIPTION	QUAN.
1	BODY	1
2	STEM	1
3	SEAT DISC ASS'Y	1
4	DISC SPRING	1
5	DISC PIN	4
6	RETAINER RING	1
7	PACKING WASHER	1
8	PACKING	2 *
9	PACKING GLAND	1
10	CAP	1
11	CAP GASKET	1
12	FLANGE	1
13	ADAPTER	1
14	GASKET	1
15	CAPSCREW	4
16	PIPE PLUG	2

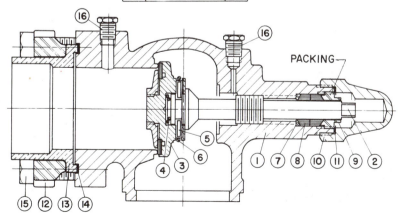

PACKING

Figure 23–42. Service valve showing packing material. This can be replaced if it leaks. *Courtesy Henry Valve Company*

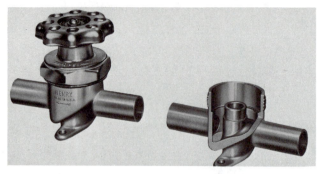

Figure 23–43. Diaphragm-type hand valve used when servicing a system. This valve is either open or closed, unlike the suction and discharge valve with a gage port. The valve has some resistance to fluid flow, called pressure drop. It can be used anywhere a valve is needed. The larger sizes of this valve are soldered in the line. Care should be used that the valve is not overheated when soldering to the valve. *Courtesy Henry Valve Company*

refrigerant. If liquid refrigerant were in the separator, it would return the refrigerant to the compressor crankcase. This would dilute the oil and cause marginal lubrication, Figure 23–45.

23.45 PRESSURE ACCESS PORTS

Pressure access ports are a method of taking pressure readings at places that do not have service ports. Several types can be used effectively. Some can be attached to a line while the unit is operating. This can be very helpful when a reading is needed in a hurry and the machine needs to keep on running. There are two types that can be fastened while running. One type is bolted on the line and has a gasket. When this valve is bolted in place, a pointed plunger is forced through the pipe. A very small hole is pierced in the line, just enough to take pressure readings and transfer small amounts of refrigerant if needed, Figure 23–46.

The other type of valve that can be installed while running requires the valve to be soldered on the line with a low-temperature solder. This can only be done on a vapor line, because a liquid line will not heat up as long as there is liquid in it. The manufacturers claim that there is no damage to the refrigerant in the line for this short soldering job. This valve may be more leak free because it is soldered. After the valve is

(A)

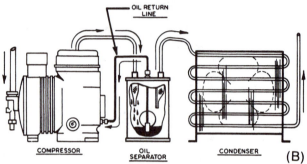

COMPRESSOR   OIL SEPARATOR   CONDENSER   (B)

Figure 23–45. Oil separator used in the discharge line of a compressor to return some of the oil to the compressor before it gets out into the system. This is a float action valve and will return liquid refrigerant as well as oil. It must be kept warm to keep refrigerant from condensing in it during the off cycle. *Courtesy AC and R Components, Inc.*

soldered on the line, a puncture is made similar to the valve just described, Figure 23–47.

Other valves that are used as ports must have a hole drilled in the refrigerant line. The valve stem is inserted into the line and soldered. This must be done when there is no pressure in the line, Figure 23–48.

Attaching a gage hose to the valves can be done in two ways. Some of the valves have handles that shut off the valve to the atmosphere. The others normally use a *Schrader* connection, which is like a tire valve on a car or bicycle except that it has threads that accept the $\frac{1}{4}$-in. gage hose connector from a gage manifold, Figure 23–49.

Figure 23–44. Ball valve. This valve is open straight through and creates very little pressure drop or resistance to the refrigerant flow. *Courtesy Henry Valve Company*

Figure 23–46. Pressure tap devices used to obtain pressure readings when there are no gage ports. *Courtesy J/B Industries*

Figure 23–47. This valve can be soldered on a vapor line while the system has pressure in it. Low-temperature solder is used. It cannot be soldered to a line with liquid in it because the liquid will not allow the line to get hot enough. *Courtesy J/B Industries*

Figure 23–48. This valve port must have a hole drilled in the line and can only be installed with the system at atmospheric pressure. *Courtesy J/B Industries*

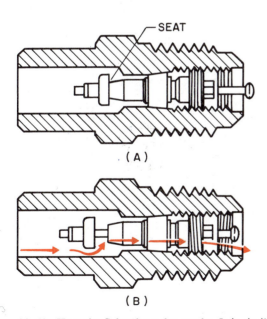

Figure 23–49. How the Schrader valve works. It is similar to the valve on a car tire except that it has threads to accept the service technician's gage line threads. *Courtesy J/B Industries*

SUMMARY

- Four basic components of the compression cycle are the compressor, condenser, evaporator, and expansion device.
- Two other types of components that enhance the refrigeration cycle are control components and accessories.
- Control components can be classified as mechanical, electrical, and electromechanical.
- Mechanical components are used to stop, start, or modulate either fluid flow or electrical energy.
- Two-temperature operation (evaporator pressure regulator, EPR valve, two-temperature-valve, solenoid valve) is used for multiple evaporators at different temperatures.
- Crankcase pressure regulator limits the amount of gas that a compressor can pump under an overloaded condition, such as a hot pull down.
- Relief valves prevent high pressures from occurring in the system.
- Low ambient controls maintain a proper working head pressure at low ambient conditions on air-cooled equipment (fan cycling, fan speed control, air volume control, and condenser flooding).
- Electrical controls stop, start, or modulate electron flow for the control of motors and fluid flow.
- When energized, the solenoid valve, a valve with a magnetic coil, will open or close a valve to control fluid flow.
- Pressure switches stop and start system components.
- The low-pressure switch can be used for low-charge protection and as a thermostat.
- The high-pressure switch protects the system against high operating pressures. This control can be either manual or automatic reset.
- The low ambient control (switch) maintains the correct operating head pressures on air-cooled equipment in both mild and cold weather by cycling the condenser fan.
- The oil safety switch insures that the correct oil pressure is available 90 sec after startup on larger compressors (normally above 5 hp). This control is manual reset.
- Defrost cycle controls are either for medium-temperature or low-temperature applications.
- Medium-temperature applications either use random defrost, where the air during the off cycle is used for defrost when the machine cuts off with the thermostat, or planned off-cycle defrost, which is accomplished with a timer.
- Low-temperature defrost must be accomplished with heat, from either within or without the system. The evaporator fan must be stopped in both cases because the air in the cooler is so cold that defrost cannot occur with the fans on.

- Defrost with internal heat can be accomplished with hot gas from the compressor.
- Defrost with external heat is normally done with electric heaters in the vicinity of the evaporator. The compressor must be stopped during this defrost.
- Refrigeration accessories normally *do not* automatically change the flow of the refrigerant but enhance the operation of the system.
- These accessories can be service valves, filter-driers, sight glasses, refrigerant distributors, heat exchangers, storage tanks, oil separators, or pressure taps.
- No system has to have all of these components, but all systems will have some of them.

REVIEW QUESTIONS

1. Name the four basic components necessary for the compression cycle.
2. Define modulating fluid flow.
3. What is the purpose of the two-temperature valve?
4. What is the purpose of a crankcase pressure regulator?
5. What does a relief valve protect the system from?
6. Name two types of relief valves.
7. Why is low ambient control necessary?
8. Name four methods used for low ambient control on air-cooled equipment.

Describe the devices in Questions 9–12.

9. Low-pressure switch
10. High-pressure switch
11. Low ambient control (switch)
12. Oil safety switch
13. What is off-cycle defrost?
14. Name two types of off-cycle defrost.
15. Name two methods for accomplishing defrost in low-temperature refrigeration systems.

Describe how the following components enhance the refrigeration cycle:

16. Filter-drier
17. Heat exchanger
18. Suction accumulator
19. Receiver
20. Pressure taps

APPLICATION OF REFRIGERATION SYSTEMS

OBJECTIVES

After studying this unit, you should be able to

- **distinguish between different types of equipment.**
- **discuss different applications of refrigeration systems.**
- **describe the different types of display equipment.**
- **discuss heat reclaim.**
- **describe package versus remote condensing applications.**
- **describe merchandising techniques in refrigeration.**
- **describe mullion heat.**
- **describe the various defrost methods.**
- **discuss walk-in refrigeration applications.**
- **describe ice-making equipment.**
- **describe basic beverage cooling.**

24.1 APPLICATION DECISIONS

Someone must make the decisions as to what the needs are in any potential installation. This same person may also be the one to decide what the equipment choices are to be. The factors that enter into this decision process are the first cost of the installation, the conditions to be maintained, the operating cost of the equipment and the long-term intentions of the installation. When first cost alone is considered, the equipment may not perform acceptably. For example, if the wrong equipment choice is made for storing meat, the humidity may be too low in the cooler, and dehydration may occur. When too much moisture is taken out of the meat, there will be a weight loss. This weight loss actually will go down the drain as condensate during defrost.

This unit discusses some of the different options to consider when equipment is installed. The service technician has to know these options to make service decisions.

24.2 REACH-IN REFRIGERATION MERCHANDISING

Retail stores use reach-in refrigeration units for merchandising their products. Customers can go from one section of the store to another and choose items to purchase. These reach-in display cases are available in high-, medium-, and low-temperature ranges. Each display case has open display and closed display types of cases to choose from. These open or closed styles may be chest type, upright type with display shelves or upright type with doors. The display boxes can be placed end to end for a continuous length of display. When this is done, the frozen food is kept together, and fresh foods (medium-temperature applications) are grouped, Figure 24–1.

All of these combinations of reach-in display add to the customer's convenience and enhance the sales appeal of the products. The two broad categories of open and closed displays are purely for merchandising the product. A closed case is more efficient than an open case. The open case is more appealing to the customer because the food is more visible and easier to reach. Open display cases are able to maintain conditions because the refrigerated air is heavy and settles to the bottom. These display cases can even be upright and store low-temperature products. This is accomplished by designing and maintaining air patterns to keep the

Figure 24–1. Display cases used to safely store food displayed for sale. Some display cases are open to allow the customer to reach in without opening a door; some have doors. Closed cases are the most energy efficient. *Courtesy Hill Refrigeration*

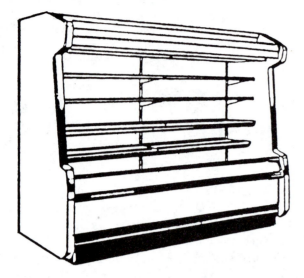

Figure 24-2. Open display cases accomplish even, low-temperature refrigeration because cold air can be controlled. It falls naturally. When these cases have shelves that are up high, air patterns are formed to create air curtains. These air curtains must not be disturbed by the air conditioning system's air discharge. All open cases have explicit directions as to how to load them with load lines conspicuously marked. *Courtesy Tyler Refrigeration Corporation*

Figure 24-3. Self-contained refrigerated box has its condensing unit built in. Refrigeration tubing does not have to be run. This box rejects the heat it absorbs back into the store. It is movable because it only needs an electrical outlet and, sometimes, a drain, depending on the box. *Courtesy Hill Refrigeration*

refrigerated air from leaving the case, Figure 24-2. When reach-in refrigerated cases are used as storage, such as in restaurants, they do not display products, although they have the same components.

24.3 SELF-CONTAINED REACH-IN FIXTURES

Reach-in refrigeration fixtures can be either self-contained with the condensing unit built inside the box or they can have a remote condensing unit. Making the best decision as to which type of fixture to purchase depends on several factors.

Self-contained equipment rejects the heat at the actual case. The compressor and condenser are located at the fixture, and the condenser rejects its heat back into the store, Figure 24-3. This is good in winter. In summer it is desirable in warmer climates to reject this heat outside. Self-contained equipment can be moved around to new locations without much difficulty. It normally plugs into either a 115-V or 230-V electrical outlet. Only the outlet may have to be moved. Figure 24-4 shows some examples of wiring diagrams of medium- and low-temperature fixtures.

When service problems arise, only one fixture is affected, and the food can be moved to another cooler. Self-contained equipment is located in the conditioned space, and the condenser is subject to any airborne particles in the conditioned space. Keeping the several condensers clean at a large job may be difficult. For example, in a restaurant kitchen large amounts of

grease will deposit on the condenser fins and coil, which will then collect dust. The combination of grease and dust is not easy to clean in the kitchen area without contaminating other areas. The kitchen may like the versatility of self-contained equipment, but remote condensing units may be easier to maintain. The warm condensers are also a natural place for pests to locate.

All refrigeration equipment has to have some means to dispose of the condensate that is gathered on the evaporator. This condensate must either be piped to a drain or be evaporated. Self-contained equipment can use the heat that the compressor gives off to evaporate this condensate, provided the condensate can be drained to the area of the condenser. This means that the condenser has to be lower than the evaporator. This is not easy to do when the condenser is on top of the unit, Figure 24-5.

24.4 REMOTE CONDENSING UNIT EQUIPMENT

Remote equipment can be designed in two ways: an individual condensing unit for each fixture, or one compressor or multiple compressors manifolded together to serve several cases.

24.5 INDIVIDUAL CONDENSING UNITS

When individual condensing units are used and trouble is encountered with the compressor or a

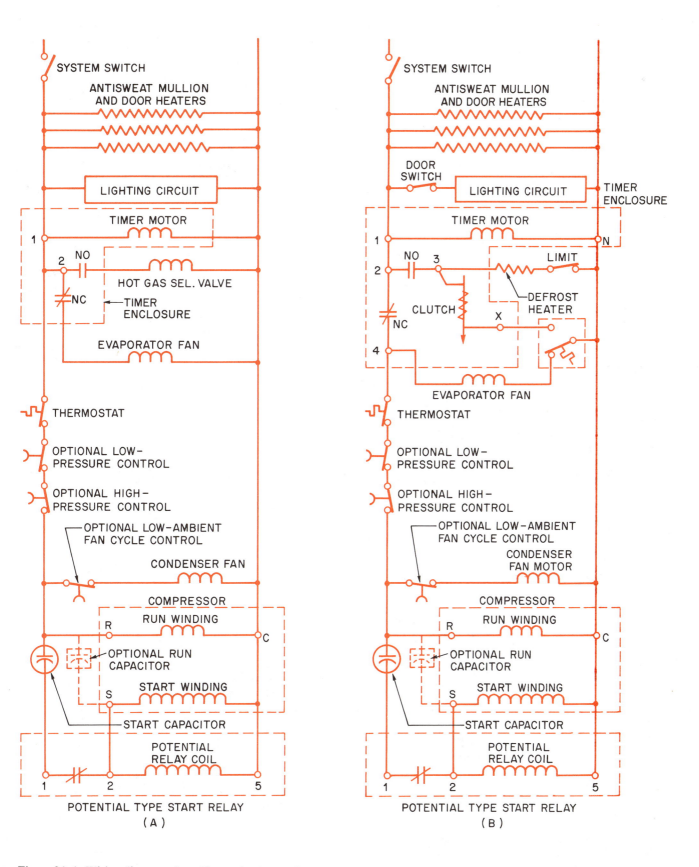

Figure 24-4. Wiring diagrams for self-contained reach-in boxes. (A) Low-temperature with hot gas defrost. (B) Low-temperature with electric defrost. (C) Medium-temperature with planned off-cycle defrost. (D) Medium- or high-temperature with random defrost.

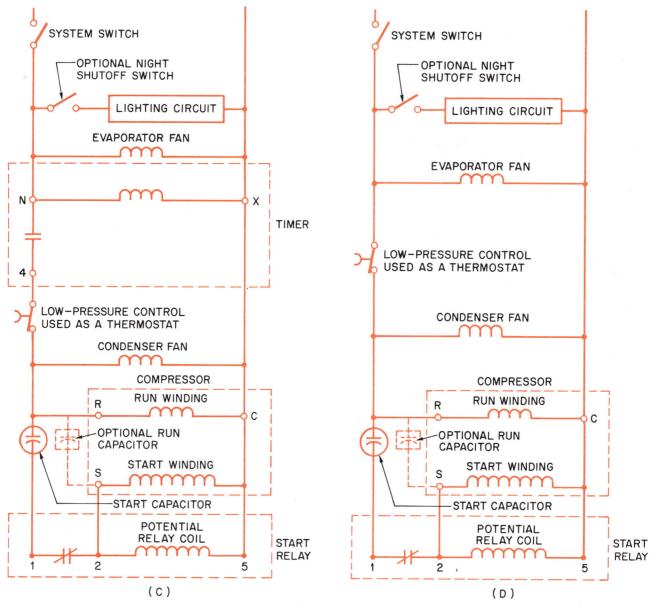

Figure 24-4. (*Continued*)

refrigerant leak occurs, only one system is affected. The condensing unit can be located outside, with proper weather protection, or in a common equipment room, where all equipment can be observed at a glance for routine service. The air in the room can be controlled with dampers for proper head-pressure control. The equipment room can be arranged so that it recirculates store air in the winter and reuses the heat that it took out of the store through the refrigeration system. This saves buying heat from a local utility, Figure 24-6.

24.6 MULTIPLE EVAPORATORS AND SINGLE-COMPRESSOR APPLICATIONS

The method of using one compressor on several fixtures has its own advantages:

1. The compressor motors are more efficient because they are larger.
2. The heat from the equipment can be more easily captured for use in heating the store or hot water for store use, Figure 24-7.

These installations are designed in two basic ways. One method uses one compressor for several cases. A 30-ton compressor may serve 10 or more cases. This particular application may have cycling problems at times when the load varies unless the compressor has a cylinder unloading design, a method to vary the capacity of a multicylinder compressor by stopping various cylinders from pumping at predetermined pressures, Figure 24-8.

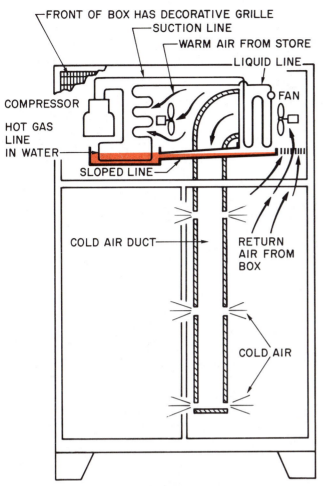

Figure 24–5. The evaporator can be high enough on a self-contained cooler so that the condensate can drain into a pan and be evaporated by the hot-gas line. No drain is needed in this application.

24.7   MULTIPLE MEDIUM-SIZED COMPRESSORS

Another method is to use more than one compressor manifolded on a rack. They may have a common suction and discharge line and a common receiver. The compressors can be cycled on and off as needed for capacity control. For example, there may be four compressors that each have a capacity of $7\frac{1}{2}$ ton of refrigeration. Some compressor racks have different capacity compressors, Figure 24–9. In this arrangement the compressors are responsible for maintaining a suction pressure in the suction manifold that will serve the coldest evaporator. There may be 10 or more cases piped to this common suction line. Each case would have its own expansion valve and liquid line from a common receiver. Either method can be used and has advantages and disadvantages. For example, when a single compressor is used, the total system depends on this one component.

If multiple compressors are used and one fails, the other compressors can continue to run. However, a bad motor burn on one compressor will contaminate the other three compressors. The system may be cleaned using filter driers and may prevent damage to the remaining compressors if accomplished quickly. In either case a refrigerant leak will affect the whole system. Design decisions will have to be left to the designer. Input from service personnel would be helpful to determine the best arrangement for each installation.

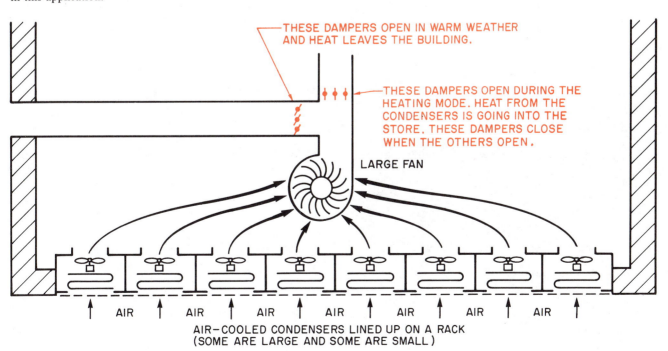

Figure 24–6. An equipment room with multiple compressors and condensers can provide the heat to be circulated throughout the store in the winter.

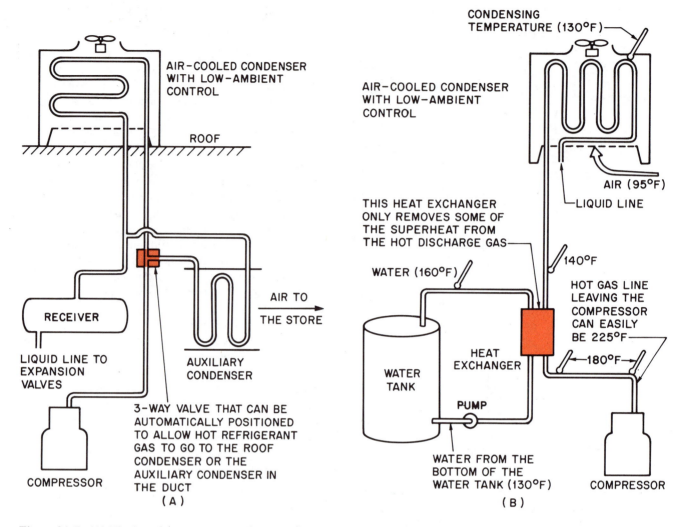

Figure 24-7. (A) The heat from a common discharge line can be captured and reused. The heat can go out on the roof or into a coil mounted in the duct work. (B) The other device is a hot-water heat-reclaim device. It uses the heat from the hot-gas line to heat water.

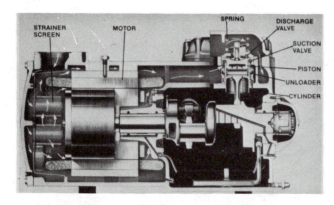

Figure 24-8. Compressor with cylinder unloading. This system is furnished with certain compressors to vary the capacity at reduced loads. This is accomplished by preventing various cylinders from pumping upon demand. For example, a 4-cylinder compressor with 40 ton capacity has 10 ton capacity per cylinder. With the proper controls this compressor may be a 10-, 20-, 30-, or 40-ton. *Courtesy Trane Company*

24.8 EVAPORATOR TEMPERATURE CONTROL

When multiple evaporators are used, they may not all be the same temperature rating. The coldest evaporator may require a suction pressure of 20 psig, and the warmest evaporator may require a suction pressure of 28 psig. Evaporator pressure regulator (EPR) valves can be located in each of the higher-temperature evaporators. When the load varies, the compressors can be cycled on and off to maintain a suction pressure of 20 psig. The compressors can be cycled using low-pressure control devices.

This type of installation is used in some larger installations such as supermarkets. An advantage of this system is that the heat being removed from several fixtures is concentrated into one discharge line. This heat can be piped in such a manner that it can be rejected to the atmosphere or back into the store. This

Figure 24–9. These compressors have varying capacities but are manifolded together with a common suction and common discharge line. These compressors can be cycled on and off individually on demand. *Courtesy Tyler Refrigeration Corporation*

can be a big supplement to the store's heating system in the winter, Figure 24–10.

24.9 INTERCONNECTING PIPING IN MULTIPLE-EVAPORATOR INSTALLATIONS

When the fixtures are located in the store and the compressors are in a common equipment room, the liquid line must be piped to the fixture, and the suction line must be piped back to the compressor area. The suction line should be insulated to prevent it from picking up heat on the way back to the compressor. This can be accomplished by preplanning and by providing a pipe

chase in the floor. Preferably the pipe chase should be accessible in case of a leak, Figure 24–11.

24.10 TEMPERATURE CONTROL OF THE FIXTURE

Control of this type of remote medium-temperature application can be accomplished without interconnecting wiring being installed between the fixture and the equipment room. A power supply for the fixture has to be located at the fixture. Where the application is medium temperature, planned off-cycle defrost can be accomplished at the equipment room with a time clock and a liquid line solenoid valve. The clock and solenoid valve can be located at the case, but the equipment room gives a more central location for all controls. The clock can cut off the solenoid valve for a predetermined time. The refrigerant will be pumped out of the individual fixture. The evaporator fan will continue to run, and the air in the fixture will defrost the coil. When the proper amount of time has passed, the solenoid valve will open and the coil will go back to work. The compressor will continue to run during the defrost of the individual cases. Their defrost times can be staggered by offsetting their defrost times. This is sometimes accomplished with a master time clock with many circuits.

Low-temperature installations must have a more extensive method of defrost because heat must be furnished to the coil in the fixture, and the fan has to be stopped. This can be accomplished at the fixture with a time clock and heating elements. The power supply for the fixture is in the vicinity of the fixture. It can also be accomplished with hot gas, but a hot-gas line must then be run from the compressor to the evaporator. This is a third line to be run to each case. It must also be insulated to keep the gas hot until it reaches the evaporator. The defrost of each case can

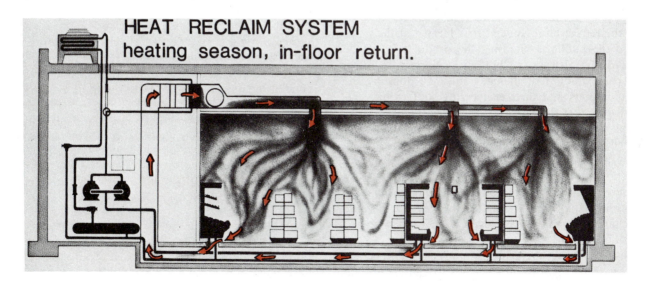

Figure 24–10. Piping arrangement for heat reclaim for space-heating purposes. *Courtesy Hill Refrigeration*

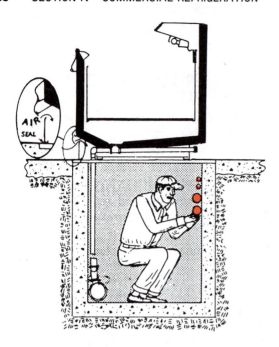

Figure 24–11. Pipe chase is one method of running piping from the fixture to the equipment room and is quite effective. When a chase like this is used, the piping can be serviced if needed. *Courtesy Tyler Refrigeration Corporation*

be staggered with different time-clock settings, Figure 24–12. This type of defrost is the most efficient because the heat from defrost is coming from the other cases.

These methods of defrost are typical, but by no means the only methods. Different manufacturers devise their own methods of defrost to suit their equipment.

24.11   THE EVAPORATOR AND MERCHANDISING

The evaporator is the device that absorbs the heat into the refrigeration system. It is located at the point where the public is choosing the product. It can be built in several ways. At best an evaporator is bulky. A certain amount of planning must be done by the manufacturers and their engineering staff to have attractive fixtures that are also functional, Figure 24–13.

The buying appeal of the customer must be considered in the choice of equipment. There is a saying in the retail business: "All equipment can be stopped, but the cash registers must continue to run to stay in business." The service technician may not understand why some equipment is installed the way it is because merchandising may have played a major part in the decision.

Display fixtures are available as (1) chest type (open, open with refrigerated shelves, closed) and (2) upright (open with shelves, with doors).

24.12   CHEST-TYPE DISPLAY FIXTURES

Chest-type reach-in equipment can be designed with an open top or lids. Vegetables, for example, can easily be displayed in the open type. The vegetables can be stocked, rotated, and kept damp with a sprinkler hose, Figure 24–14. The product can be covered at night with plastic lids or film. The customer can see the product because the fixture may have its own lights. Meat that is stored in the open is normally packaged in clear plastic.

This type of fixture may have the evaporator in the bottom of the box. Fans blow the cold air to the correct grilles to give good air circulation. Service for the coil components and fan is usually through removable panels under the vegetable storage or on the front side of the fixture. The appliance is normally placed with a wall on the back side, Figure 24–15.

The chest-type fixture can be furnished with the condensing unit built in or designed to be located at a remote location, such as an equipment room. If the condensing unit is furnished with the cabinet, it is usually located underneath the front and can be serviced by removing a front panel. When this fixture has the condensing unit furnished with it, it is called self-contained.

24.13   REFRIGERATED SHELVES

Some chest-type fixtures have refrigerated shelves located at the top. These shelves must have either air flow or plate type evaporators to maintain the food conditions. When there are evaporators in the top, they are normally piped in series with the evaporators in the bottom, Figure 24–16.

24.14   CLOSED-TYPE CHEST FIXTURES

The closed-type chest is normally low temperature and can store ice cream or frozen foods. The lids may be lifted off or slide from side to side.

The upright closed display normally has doors that the customer can look through to see the product. This type of cooler may have a self-contained or remote condensing unit. It can also be piped into a system with one compressor. This display is sometimes used as one side of the wall for a walk-in cooler. The display shelves can then be loaded from inside the walk in cooler, Figure 24–17.

24.15   CONTROLLING SWEATING ON THE CABINET OF FIXTURES

Display cases that have any cabinet surfaces that may operate below the dew-point temperature of the room (the temperature at which moisture will form) must have some means (usually small heaters) to keep this moisture from forming. This cabinet sweating nor-

LATENT HEAT DEFROST AS USED ON A COMBINED SYSTEM

(Do not confuse this with Reverse Cycle Defrost with all cases being defrosted at once.)
Only one-third to one-fourth of the total Btuh load can be defrosted at any one time.

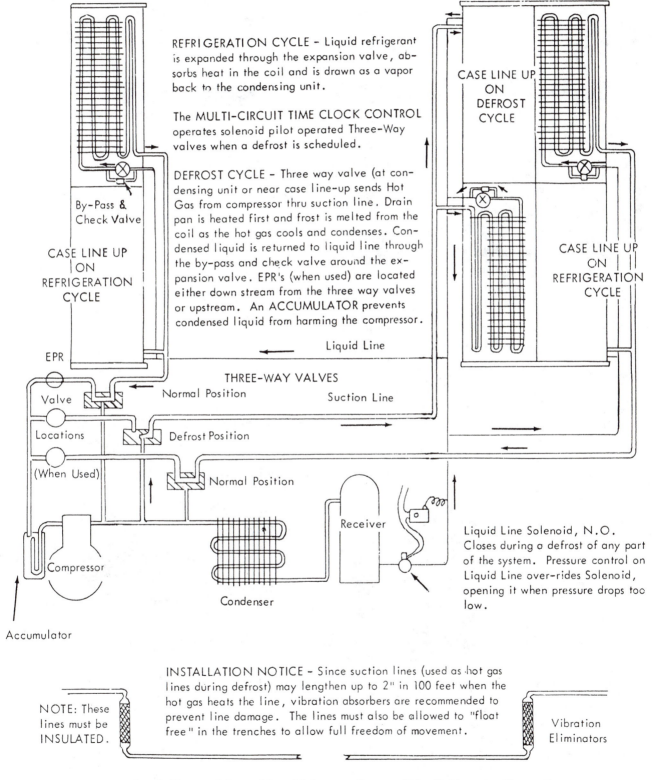

REFRIGERATION CYCLE - Liquid refrigerant is expanded through the expansion valve, absorbs heat in the coil and is drawn as a vapor back to the condensing unit.

The MULTI-CIRCUIT TIME CLOCK CONTROL operates solenoid pilot operated Three-Way valves when a defrost is scheduled.

DEFROST CYCLE - Three way valve (at condensing unit or near case line-up sends Hot Gas from compressor thru suction line. Drain pan is heated first and frost is melted from the coil as the hot gas cools and condenses. Condensed liquid is returned to liquid line through the by-pass and check valve around the expansion valve. EPR's (when used) are located either down stream from the three way valves or upstream. An ACCUMULATOR prevents condensed liquid from harming the compressor.

CASE LINE UP ON DEFROST CYCLE

CASE LINE UP ON REFRIGERATION CYCLE

By-Pass & Check Valve

CASE LINE UP ON REFRIGERATION CYCLE

Liquid Line

THREE-WAY VALVES

EPR

Valve

Locations

(When Used)

Normal Position   Suction Line

Defrost Position

Normal Position

Receiver

Compressor

Condenser

Accumulator

Liquid Line Solenoid, N.O. Closes during a defrost of any part of the system. Pressure control on Liquid Line over-rides Solenoid, opening it when pressure drops too low.

INSTALLATION NOTICE - Since suction lines (used as hot gas lines during defrost) may lengthen up to 2" in 100 feet when the hot gas heats the line, vibration absorbers are recommended to prevent line damage. The lines must also be allowed to "float free" in the trenches to allow full freedom of movement.

NOTE: These lines must be INSULATED.

Vibration Eliminators

Figure 24-12. Hot-gas defrost with multiple cases. *Courtesy Tyler Refrigeration Corporation*

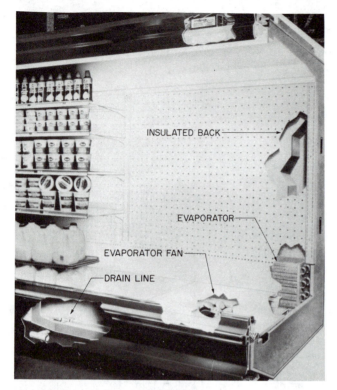

Figure 24–13. Dairy case showing the location of the evaporator. *Courtesy Hill Refrigeration*

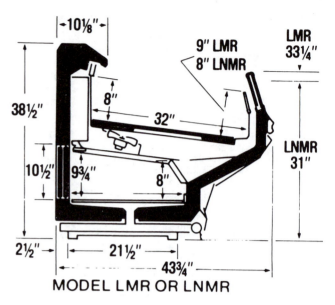

MODEL LMR OR LNMR

Figure 24–15. Cutaway side view of a chest-type display case. The fans can be serviced from inside the case. The case is normally located with its back to a wall or another fixture. *Courtesy Hill Refrigeration*

mally occurs around doors on closed type equipment. The colder the refrigerated fixture, the more need there is for the protection. The heaters are made with a resistance-type wire that is run just under the surface of the cabinet and are called *mullion* heaters. Mullion means division between panels; for refrigeration it is between the inside panels that are cold and the out-

Figure 24–14. Open display case used to store fresh vegetables. The case has a drain, so the vegetables can be wet down with a hand-operated sprinkler. *Courtesy Tyler Refrigeration Corporation*

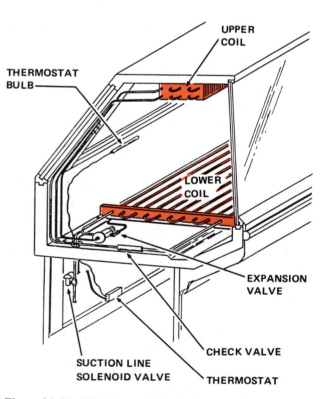

Figure 24–16. Multiple evaporators in a single case because refrigeration is needed in more than one place. *Courtesy Tyler Refrigeration Corporation*

Figure 24–17. Display case with walk-in cooler as its back wall. The display case can be supplied from inside the cooler. *Courtesy Hill Refrigeration*

side panels that may be located in a humid place. There can be a large network of these heaters in some equipment, Figure 24–18. *Note:* Some of them can be thermostatically controlled or controlled by a humidistat based on the humidity in the store. The air conditioning systems are used for controlling the humidity in the stores more than ever. Owners realize that with a low humidity in the store, they do not have to use as much mullion heat and do not need as much defrosting of coils operating below freezing.

24.16   MAINTAINING STORE AMBIENT CONDITIONS

The humidity is taken out of the store in the summer with the air conditioning system and the display cases. The more humidity removed by the air conditioning system, the less removed by the refrigeration fixtures. This means less defrost time. Some stores keep a positive air pressure in the store with makeup air that is conditioned through the air conditioning system. This method is different from the random method of taking in outside air just when the doors are opened by

customers. It is a more carefully planned approach to the infiltration of the outside air that is going to get in somewhere, Figure 24–19. You'll notice this system when the front doors are opened and a slight volume of air blows in your face.

The doors on display cases are usually double-pane glass sealed around the edges to keep moisture from entering between the panes. These doors must be rugged because any fixture that the public uses is subject to abuse, Figure 24–20.

The cabinets of these fixtures have to be made of tough material that is easy to clean. Stainless steel, aluminum, porcelain, and vinyl, as well as painted surfaces, are commonly used. The most expensive boxes are stainless steel, but they are the longest lasting.

24.17   WALK-IN REFRIGERATION

Walk-in refrigeration equipment is either permanently erected or of the knock-down type. The permanently erected equipment cannot be moved. Very large installations are permanent.

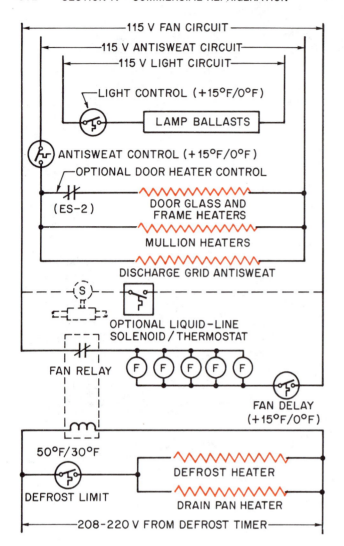

Figure 24-18. Wiring diagram of mullion heaters on a closed display case shows that there can be several circuits that apply just to the heaters. Note, this unit has 2 power supplies, 115 V and 208 or 230 V.

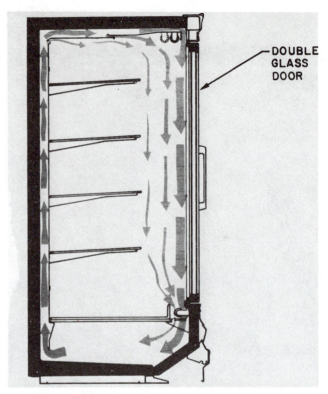

Figure 24-20. Doors are rugged and reliable. They have double glass sealed on the edges to keep them from sweating between the glass. *Courtesy Hill Refrigeration*

24.18 KNOCK-DOWN WALK-IN COOLERS

Knock-down walk-in coolers are constructed of panels from 1 to 4 in. thick, depending on the temperature inside the cooler. They are a sandwich type of construction with metal on each side and foam insulation between. The metal in the panels may be galvanized sheet metal or aluminum. The panels are strong

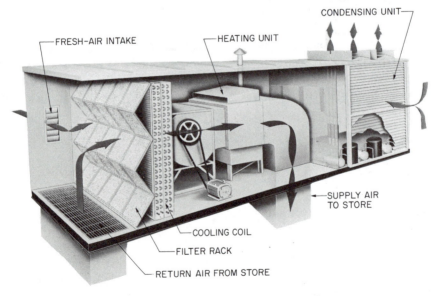

Figure 24-19. Planned infiltration known as ventilation. *Courtesy Tyler Refrigeration Corporation*

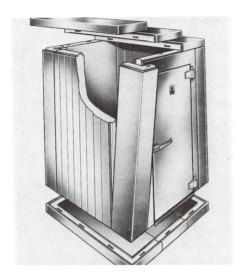

Figure 24-21. Knock-down walk-in cooler that can be assembled on the job. It can be moved at a later time if needed. The panels are foam with metal on each side, creating its own structure that needs no internal braces. This prefabricated box can be located inside or outside. It is weatherproof. *Courtesy Bally Case and Cooler, Inc.*

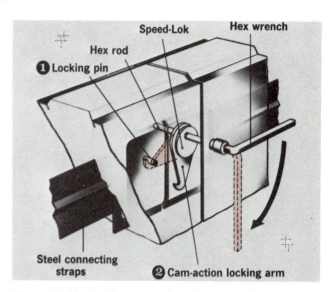

Figure 24-22. Locking mechanism for prefabricated walk-in cooler panels. They can be unlocked for moving the cooler when needed. *Courtesy Bally Case and Cooler, Inc.*

enough that no internal support steel is needed for small size coolers. The panels can be interlocked together. They are shipped disassembled on flats and can be assembled at the job site. This cooler can be moved from one location to another and can be reassembled, Figure 24-21.

Walk-in coolers come in a variety of sizes and applications. One wall of the cooler can be a display, and the shelves can be filled from inside the cooler. The coolers are normally waterproof and can be installed outdoors. The outside finish may be aluminum or galvanized sheet metal. When the panels are pulled together with their locking mechanism, they become a prefabricated structure, Figure 24-22.

24.19   WALK-IN COOLER DOORS

Walk-in cooler doors are very durable and must have a safety latch on the inside to allow anyone trapped on the inside to get out, Figure 24-23.

24.20   EVAPORATOR LOCATION IN A WALK-IN COOLER

Refrigerating a walk-in cooler is much like cooling any large space. The methods that are used today take advantage of evaporators that have fans to improve the air circulation and make them compact. Systems used to refrigerate these coolers are listed here:

1. Evaporators piped to condensers using field assembled pipe.
2. Evaporators with precharged piping.
3. Package units, wall-hung, or top-mounted with condensing units built in.

The evaporators should be mounted in such a manner that the air currents blowing out of them do not blow all of the air out the door when the door is opened. They can be located on a side wall or in a corner. These evaporators are normally in aluminum cabinets that have the expansion device and electrical connection accessed at the end panel or through the bottom of the cabinet, Figure 24-24.

Figure 24-23. Walk-in cooler doors are very rugged. They have a safety latch and can be opened from the inside. *Courtesy Bally Case and Cooler, Inc.*

(A)

(B)

Figure 24-24. Fan coil evaporators are used in walk-in coolers. *Courtesy Bally Case and Cooler, Inc.*

24.21 CONDENSATE REMOVAL

The bottom of the cabinet on the evaporator contains the condensate drain piped to the outside of the cooler. When the inside of the cooler is below freezing, heat has to be provided to keep the condensate in the line from freezing, Figure 24-25. This heat is normally provided by an electrical resistance heater that can be field installed. The line is piped to a drain and must have a trap to prevent the atmosphere from being pulled into the cooler. The line and trap must be heated if the line is run through below-freezing surroundings. These drain line heaters sometimes have their own thermostats to keep them from using energy during warm weather.

24.22 REFRIGERATION PIPING

One of two methods may be used to install refrigeration piping. One method is to pipe the cooler in the conventional manner. The other is to use precharged tubing, called a *line set*. When the conventional method is used, the installing contractor usually furnishes the copper pipe and fittings as part of the agreement. This piping can have straight runs and factory elbows for the corners. When the piping is completed, the installing contractor has to leak check, evacuate, and charge the system. This requires the service of an experienced technician.

When precharged tubing is used, the tubing is furnished by the equipment manufacturer. It is sealed on both ends and has quick-connect fittings. This piping has no fittings for corners and must be handled with care when bends are made. The condenser has its operating charge in it, and the evaporator has its operating charge in it. The tubing has the correct operating charge for the particular length chosen for the job. This has some advantage because the installation crew does not have to balance the operating charge in the system. The system is a factory-sealed system with field connections, Figure 24-26. No soldering or flare connections have to be made. The system can be installed by someone with limited experience.

If, because of miscalculation, the piping is too long for the installation, you can obtain a new line set from the supplier, or you can cut the existing line set to fit. If the existing line set is altered, it can be leak checked, evacuated, and charged to specifications. If the line set is too long, the extra coils should be coiled up in a horizontal method to assure good oil return. Figure 24-27 shows a step-by-step illustration of how the line set can be altered.

24.23 PACKAGE REFRIGERATION FOR WALK-IN COOLERS

Wall-hung or ceiling-mount units are package-type units, Figure 24-28. They can be installed by personnel that do not understand refrigeration. These units are

DRAIN-PAN HEATER

Figure 24-25. Drain pan heaters to keep drain pans from freezing. *Courtesy Larkin Coils, Inc.*

Figure 24-26. Quick-connect fittings, which allow the customer to have a system that is factory sealed and charged. *Courtesy Aeroquip Corporation*

factory assembled and require no field evacuation or charging. They come in both high-, medium-, and low-temperature ranges. Only the electrical connections need to be made in the field.

24.24 ICE-MAKING EQUIPMENT

Ice-making equipment is a distinctly different application of refrigeration. Package-type ice-making equipment will be discussed here. This equipment is manufactured to be either plug-in or wired to the electrical circuit. A water makeup line and a drain will also have to be furnished. This equipment is factory assembled and needs no field refrigeration installation.

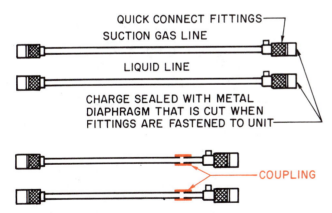

Figure 24-27. Altering quick-connect lines. Note: The suction line is charged with vapor. The liquid line has a liquid charge. It contains about as much liquid as may be pulled into it under a deep vacuum. Vent the charge in each tube and cut to the desired length. Fasten together with couplings. *Leak check, evacuate and charge: (1) the suction line with vapor (2) the liquid line with liquid. You now have a short line set with the correct charge.* See Units 7 and 8 on System Evacuation and System Charging.

Figure 24-28. This wall-hung unit, "saddle unit", actually hangs on the wall of a cooler. The weight is distributed on both sides. This is a package unit that only has to be connected to the power supply to be operable. *Courtesy Bally Case and Cooler, Inc.*

The temperature range of ice-making equipment is above 0°F and below freezing. The ice that is made will be about 32°F. Ice from package machines can be cube ice, flake ice, or cylindrical ice with a hole in the middle. (Note, ice-making equipment should not be confused with ice-holding equipment. Ice-holding equipment is equipment that may be located at various convenient locations for the public to purchase the ice. Ice-holding equipment may well operate at 0°F. The ice is made somewhere else, normally at a bulk ice plant, where it is bagged and moved to the ice-holding machine for purchase.)

24.25 MAKING CUBE ICE

Cube ice is made in at least two ways. One way is to make an ice sheet on an inclined evaporator. When the sheet is the correct thickness, defrost occurs, and the sheet slides on to a grid of wires. This grid of wires has a low-voltage circuit that prevents electrical shock. It produces enough heat to cut the ice sheet into cubes. The cubes are then dropped into a storage bin. The sheet thickness is determined with sensors, Figure 24-29.

Another method of making cubes is to spray water against an inverted or a vertical refrigerated evaporator with cups. The water freezes in the cups. When the correct cube size is formed (usually determined by time or compressor suction pressure), defrost occurs. The cubes drop down on a chute that delivers them to a storage bin. These cups can be square or round to give square, round, or round flat ice, Figure 24-30.

24.26 MAKING FLAKE ICE

Flake ice is normally made on an evaporator that is a refrigerated cylinder. As the ice freezes to the walls, an auger scrapes it off. The flakes then drop down into the storage bin. The auger is a heavy-duty gear-driven device, Figure 24-31.

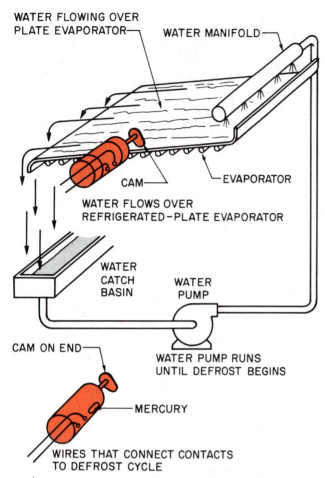

Figure 24-29. Sensors that control ice thickness in this machine. The ice thickness switch rotates when ice begins to form on the evaporator plate. The cam touches the ice during rotation (approximately 1 rpm). The switch has mercury contacts in the rear. As the cam touches the ice it causes the mercury to roll to the back. The contacts are made and defrost begins.

24.27 MAKING CYLINDER ICE

Cylinder ice that has a hole in it is normally made with a tube within a tube evaporator. The refrigerant is circulated in the outer tube, and the water is circulated in the inner tube. The refrigeration in the outer tube causes the water to start to freeze inside the inner tube. As more ice forms, a pressure drop occurs because the hole inside the ice keeps getting smaller. This pressure drop can be used to signal defrost. The ice is then pushed out of the tube, broken to length, and dropped into the storage bin.

24.28 PACKAGE-ICE MACHINES

Most package-ice-making equipment lets the ice in the bin do the refrigerating for holding purposes. The bin is well insulated, and only a small amount of ice will melt to keep the rest of the ice cold. This ice will be at 32°F and easy to use because it is not stuck together. The melted ice must be piped to a drain, Figure 24-32.

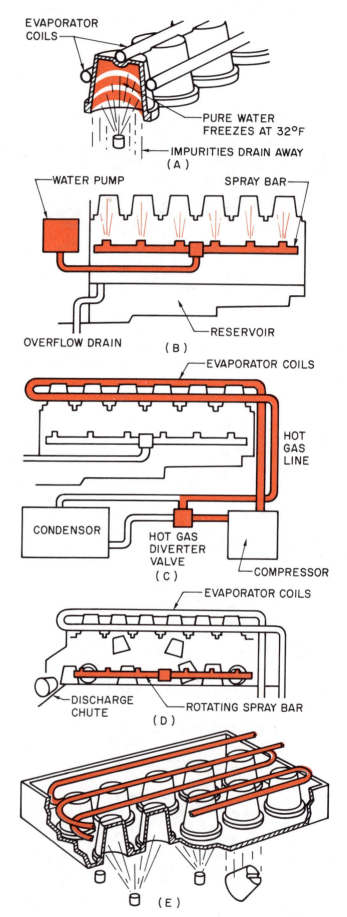

Figure 24-30. Ice cubes and the evaporator they are made on.

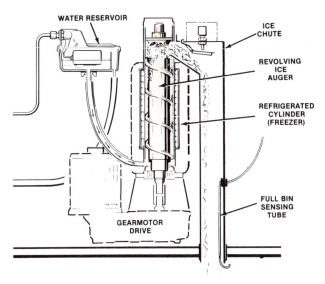

Figure 24-31. A flake ice maker, its evaporator, and auger. The auger turns and shaves the ice off the evaporator. The tolerances are very close. *Courtesy Scotsman*

WATER RESERVOIR

ICE CHUTE

REVOLVING ICE AUGER

REFRIGERATED CYLINDER (FREEZER)

GEARMOTOR DRIVE

FULL BIN SENSING TUBE

Figure 24-32. Ice-holding bin. It melts some of the ice to hold the rest at 32° F. This ice is wet and easy to handle because it does not stick together. *Courtesy Scotsman*

The condensing units furnished with ice-making equipment are generally air cooled and typical of the other condensing units. Some ice makers are water cooled, in which case they usually waste the water and use water regulating valves to help with water consumption.

24.29 BEVERAGE COOLERS

Beverage coolers are used to cool either canned or bulk beverages. Canned beverage coolers can be self-dispensing or self-help. Medium-temperature refrigeration is applied to them. The beverages must be maintained above freezing. Most small beverage coolers are package units with air-cooled condensers. When the beverages are in cans, the evaporators can be sized small and operate at low temperatures because evaporation of the product does not have to be considered. The countertop units that are seen at soda fountains are very small air-cooled units that are normally taken into the shop for service.

In conclusion, it should be mentioned that each manufacturer has its own design ideas. They are all usually correct in their applications for their own reasons. The service technician has to be familiar with these different applications to know what the original intention of the manufacturer was. In general, when exposed food or floral products are stored, the atmosphere inside the cooler is considered. If the humidity is too low, these products will dry out or dehydrate. Food products that are correctly packaged do not have this problem. Every opportunity should be taken by the service technician to understand the manufacturer's intentions. There are times when there is no substitute for knowing what the original intentions were, because the technician must often improvise when replacing components that are not available. As equipment

becomes outdated, the components may not be available for exact replacement.

SUMMARY

- Items to consider when choosing refrigeration equipment for a job are (1) first cost, (2) conditions to be maintained, (3) operating cost of the equipment, and (4) long-range intentions of the equipment.
- Product dehydration can be a factor in the choice of equipment.
- Reach-in fixtures allow customers to help themselves.
- The three temperature ranges are high, medium, and low.
- Display equipment may be package, or it may have remote condensers.
- With package or self-contained equipment, the condenser rejects the heat back into the store when located inside.
- The condensing unit is normally on top or underneath the fixture for packaged equipment.
- All fixtures have condensate that must be drained away or evaporated.
- Individual small compressors are not as efficient as larger compressors because of motor efficiency.
- Fixtures may also be piped to a common equipment room with either individual compressors or single-compressor units (the single-compressor unit may be two or more large compressors or one large compressor).
- When a single large compressor is used, capacity control is desirable.
- When two or more large compressors are used, they are manifolded together with a common suction and discharge line.
- Capacity control is accomplished by cycling compressors on the multiple-compressor racks.
- Several evaporators are piped together to a common suction line.
- The liquid lines are manifolded together after the liquid leaves the receiver.

- Each fixture has its own expansion valve.
- There must be a defrost cycle on both medium- and low-temperature fixtures.
- Heat must be added to the coil for defrost in low-temperature refrigeration systems. The fan has to be stopped during defrost. Hot-gas defrost must have a hot-gas line run from the equipment room to the fixture. Hot gas defrost is more efficient because it uses the heat from the other cases to accomplish defrost.
- Common discharge lines concentrate the heat that was absorbed into the system.
- This heat can be used to heat the store or heat hot water with heat-recovery devices.
- Fixtures can either be open or closed and either chest-type or upright.
- Cold air falls to the bottom of a refrigerated box.
- Air patterns are used to cool open display boxes when they have high shelves.
- When doors are used, they are rugged, usually two panes of glass with an air space between.
- Heaters are used around the doors to keep the cabinet around the doors above the dew-point temperature of the room.
- These heaters are called mullion heaters and are normally electric resistance heat that is sometimes thermostatically controlled.
- Some stores are controlling humidity by introducing fresh air from the outside through the air conditioning system.
- Humidity (usually in the summer) leaves the store by two means, the refrigeration equipment and the air conditioning equipment.
- It is less expensive to remove humidity with the air conditioning equipment than with the refrigeration equipment's defrost cycle.
- Some walk-in coolers are permanent structures, and some are knock-down type that can be assembled in the field and even moved.
- They can be assembled into a structure that can be located inside or outside.
- All coolers must have drains for the condensate; if the piping is below freezing, heat has to be applied to the drain lines.
- Drain heaters are normally electric resistance heat and may have thermostats to keep them from operating above freezing.
- Some coolers have remote condensing units, and some have wall-hung or roof-mount package equipment.
- Some of the remote units are field piped, and some are piped with quick-connect tubing.
- The quick-connect and the package systems may not require a skilled technician to start them.
- Ice-making equipment makes flake, cube, and cylindrical ice.
- Most small applications are package equipment that need only a water supply, electrical supply, and drain.

- An ice-holding machine and an ice-making machine are different.
- Ice-holding machines may operate at 0° F.
- Ice-making machines normally make ice at above 0° F and hold the ice in a bin at about 32° F.
- Cube ice is normally made in inverted or vertical evaporators with water spraying or running up into the evaporator or on flat plate evaporators.
- Flake ice is normally made in a circular evaporator, and the ice is scraped off with a turning auger.
- Most package ice makers are air cooled; when water cooled, they are normally wastewater systems.
- Package beverage coolers can be dispensing or self-serve types.
- Most small bulk beverage coolers are taken into the shop for service.
- *There is no substitute for knowing the manufacturer's intentions. Don't improvise without this knowledge.*

REVIEW QUESTIONS

1. Name three things that should be considered when choosing equipment for a refrigeration installation.
2. Why should a service technician be familiar with application?
3. What are the two broad categories of display cases?
4. How are conditions maintained in open display cases when there are high shelves?
5. What type of doors do closed display cases have?
6. What are mullion heaters?
7. What are the three temperature ranges of refrigeration systems?
8. Name two methods of rejecting heat from refrigerated cases.
9. How are multiple cases applied with the same compressor?
10. Name two advantages of having multiple cases applied to one compressor.
11. Where is the piping routed when the compressors are in the equipment room?
12. When one large compressor is used, what desirable feature should this compressor have?
13. How is defrost accomplished when the equipment room is in the back and the fixtures are in the front on medium-temperature fixtures?
14. Name two types of defrost used with low-temperature fixtures.
15. What materials are normally used in manufacturing display fixtures?
16. What materials are used in manufacturing walk-in coolers?
17. What two special precautions should be taken with drain lines in walk-in coolers?
18. What holds up the structure of a walk-in cooler?
19. Name three types of ice made by package ice makers.
20. Why is product dehydration not a concern with canned beverages?

TROUBLESHOOTING AND TYPICAL OPERATING CONDITIONS FOR COMMERCIAL REFRIGERATION

OBJECTIVES

After studying this unit, you should be able to

- **list the typical operating temperatures and pressures for the low-pressure side of a refrigeration system for high, medium, and low temperatures.**
- **list the typical operating pressures and temperatures for the high-pressure side of the systems.**
- **state how R-12 and R-502 compare on the high-pressure and low-pressure sides of the system.**
- **diagnose an inefficient evaporator.**
- **diagnose an inefficient condenser.**
- **diagnose an inefficient compressor.**
- **bench test a compressor for capacity.**

25.1 ORGANIZED TROUBLESHOOTING

To begin troubleshooting in any area, you need some idea of what the conditions are supposed to be. In commercial refrigeration we may be dealing with air temperatures inside a box or outside at the condenser. We may need to know the current draw of the compressor or fan motors. The pressures inside the system may add to the puzzle. There are many conditions that reflect on the system from outside and from within.

It is helpful to keep in mind that when a piece of equipment has been running well for some period of time without noticeable problems that normally only one problem will trigger any sequence of events that we will encounter. For instance, it is common for the technician to approach a piece of equipment and think that two or three components have failed at one time. The chance of two parts failing at one time is remote, unless one part causes the other to fail. These failures can almost always be traced to one original cause.

Knowing how the equipment is supposed to be functioning helps. You should know how it is supposed to sound, where it should be cool or hot, and when a particular fan is supposed to be operating. Knowing the correct operating pressures for a typical system can help you get started. **Before you do anything, look over the whole system for obvious problems.**

These troubleshooting procedures will be divided into the temperature application ranges of high, medium, and low with some typical pressures for each. Each system has its own temperature range, depending on what it is supposed to be refrigerating. The table in Figure 25–1 shows the temperature ranges for some refrigerated storage applications. Notice that there is a different temperature requirement for almost every food under more than one situation. There may be a different temperature requirement for long-term storage than for short-term storage. The first subject to be covered is the *space* or *product temperature*. The temperatures and pressures on the low-pressure side of the system are a result of the product load. Remember that **an evaporator acts the same whether there is a single compressor close to the evaporator or a large compressor in the equipment room.** The examples given relate to evaporators. The compressor's job is to lower the pressure in the evaporator to the correct boiling point to achieve a desired condition in the evaporator.

25.2 TROUBLESHOOTING HIGH-TEMPERATURE APPLICATIONS

High-temperature applications start at about 45°F and go up to about 60°F refrigerated box temperature. Normally a product temperature will be at one end of the range or the other depending on the product (flowers may be at 60°F and candy at 45°F). The intention is to give you the low and high temperature and pressure conditions that may be encountered for high-temp refrigeration systems. The coil temperature is normally 10° to 20°F cooler than the box temperature. This difference is referred to as temperature difference or TD. This means that the coil will normally be operating at 25 psig (45 − 20 = 25°F coil temperature) at the lowest temperature for an R-12 machine. The pressure would be 58 psig (60°F coil temperature) at the end of the cycle when the compressor is off. Figure 25–2 is a chart of some typical operating conditions for high, medium, and low temperature applications. These figures are for R-12 and R-502 for their typical applica-

COMMODITY	STORAGE TEMP. °F	RELATIVE HUMIDITY %	APPROXIMATE STORAGE LIFE
Carrots			
Prepackaged	32	80–90	3–4 weeks
Topped	32	90–95	4–5 months
Celery	31–32	90–95	2–4 months
Cucumbers	45–50	90–95	10–14 days
Dairy products			
Cheese	30–45	65–70	
Butter	32–40	80–85	2 months
Butter	0 to −10	80–85	1 year
Cream (sweetened)	−15	—	several months
Ice cream	−15	—	several months
Milk, fluid whole			
Pasteurized Grade A	33	—	7 days
Condensed, sweetened	40	—	several months
Evaporated	Room Temp	—	1 year, plus
Milk, dried			
Whole milk	45–55	low	few months
Non-fat	45–55	low	several months
Eggs			
Shell	29–31	80–85	6–9 months
Shell, farm cooler	50–55	70–75	
Frozen, whole	0 or below	—	1 year, plus
Frozen, yolk	0 or below	—	1 year, plus
Frozen, white	0 or below	—	1 year, plus
Oranges	32–34	85–90	8–12 weeks
Potatoes			
Early crop	50–55	85–90	—
Late crop	38–50	85–90	—

(A)

COIL-TO-AIR TEMPERATURE DIFFERENCES TO MAINTAIN PROPER BOX HUMIDITY

TEMPERATURE RANGE	DESIRED RELATIVE HUMIDITY	TD (REFRIGERANT TO AIR)
25° F to 45° F	90%	8° F to 12° F
25° F to 45° F	85%	10° F to 14° F
25° F to 45° F	80%	12° F to 16° F
25° F to 45° F	75%	16° F to 22° F
10° F and Below	—	15° F or less

(B)

Figure 25–1. (A) Chart showing some storage requirements for typical products. (B) Coil-to-air temperature difference chart to select desired relative humidity. *Courtesy (A) Used with permission from ASHRAE, Inc. 1791 Tuller Circle NE, Atlanta, GA 30329 (B) Copeland Corporation*

This table is intended to show the typical operating pressures for commercial refrigeration systems. Column 1 is with the compressor off and the box temperature just at the cut-in point. Column 2 is just after the compressor comes on. Column 3 is just before the compressor cuts off. This is not intended to be the only operating pressures, but the upper and lower limits as applied to high-, medium- and low-temperature typical applications. A coil-to-inlet air temperature of 15° F (TD) was used as an average. Many systems will use 10° F or 20° F TD. See Figure 25–1 (A) for different TD applications.

	R-12		
	Column 1 Compressor Off	Column 2 Compressor On	Column 3 Compressor On
HIGH-TEMPERATURE			
Box Temperature	60° F	60° F	45° F
Coil Temperature	60° F	40° F	25° F
Temperature Difference	0° F	20° F	20° F
Suction Pressure	58 psig	37 psig	25 psig
MED-TEMPERATURE			
Box Temperature	45° F	45° F	30° F
Coil Temperature	45° F	25° F	10° F
Temperature Difference	0° F	20° F	20° F
Suction Pressure	42 psig	25 psig	15 psig
LOW-TEMPERATURE			
Box Temperature	5° F	5° F	−20° F
Coil Temperature	5° F	−15° F	−40° F
Temperature Difference	0° F	20° F	20° F
Suction Pressure	12 psig	2.5 psig	11 in. Hg
	R-502		
LOW-TEMPERATURE			
Box Temperature	5° F	5° F	−20° F
Coil Temperature	5° F	−15° F	−40° F
Temperature Difference	0° F	20° F	20° F
Suction Pressure	36 psig	19 psig	4.1 psig
	R-12		

ICE MAKING
Ice begins to form at 20° F coil temperature
Suction pressure 21#
End of Cycle, Ice about to Harvest 5° F coil temperature
Suction Pressure 11#

When a service technician installs the gage manifold, the readings for a typical installation should not exceed or go below these readings. For instance, if the gages were applied to a medium-temperature vegetable box, the highest suction pressure that should be encountered on a typical box should be 25 psig with the compressor running and a low of 15 psig at the end of the cycle. Any readings above or below these will be out of range for a box that has a 20° F temperature difference between the air temperature entering the coil and the coil's refrigerant boiling temperature.

Figure 25–2. This chart can be used for typical operating temperatures and pressures for refrigerating systems using forced-air evaporators.

tions. If any other refrigerant or application is used, the temperature readings and pressure readings may be converted to the different refrigerant.

When the technician suspects problems and installs the gages for pressure readings, the following readings should be found on a 20°F TD system:

1. With the compressor running just before the cutout point, a reading below 25 psig would be considered too low.
2. With the compressor running just after it started up with a normal box temperature, 60°F (60 − 20 = 40°F coil temperature) a reading above 37 psig would be considered too high.
3. When the compressor is off and the box temperature is up to the highest point, the coil pressure would correspond to the air temperature in the box.

If a 60°F box is the application, the suction pressure could be 58 psig. This would be just before the compressor starts in a normal operation.

These pressures are fairly far apart but serve as reference points for this particular application. The preceding illustration is for a TD of 20°F. There will be times when a 10°F TD is considered normal for a particular application. The best procedure for the technician is to consult the chart in Figure 25–1 for the correct application and determine what the pressures should be by converting the box temperatures to pressure. This chart shows what the storage temperatures are.

The correct box humidity conditions may also be found by studying Figure 25–1. The box humidity conditions will determine the coil-to-air temperature difference for a particular application. For example, you may want to store cucumbers in a cooler. The chart shows that cucumbers should be stored at a temperature of 45° to 50°F and a humidity of 90% to 95% to prevent dehydration. Examining the second part of the chart indicates that to maintain these conditions the coil-to-air temperature difference should be 8° to 12°F; probably 10°F should be used.

The pressures from one piece of equipment to another may vary slightly. Do not forget that any time the compressor is running the coil will be 10° to 20°F cooler than the fixture air temperature.

25.3   TROUBLESHOOTING MEDIUM-TEMPERATURE APPLICATIONS

Medium-temperature refrigerated box temperatures range from about 30° to 45°F in the refrigerated space. Many products will not freeze as low as 30°F box temperature. Using the same methods that were used in the high-temperature examples to find the low- and high-pressure readings that would be considered normal, we find the following. The lowest pressures that would

normally be encountered would be at the lowest box temperatures of 30°F. If the coil were to be boiling the refrigerant at 20°F below the inlet air temperature, the refrigerant would be boiling at 10°F (30 − 20 = 10). This corresponds to a pressure for R-12 of 15 psig. If the service technician encounters pressures below 15 psig for an R-12 medium-temperature system, there is most likely a problem. The highest pressures that would normally be encountered while the system is running would be at the boiling temperature just after the compressor starts at 45°F. This is 45 − 20 = 25°F, or a corresponding pressure of 25 psig, Figure 25–2. The pressure in the low side while the compressor is off would correspond to the air temperature inside the cooler. When 45°F is considered the highest temperature before the compressor comes back on, the pressure would be 42 psig. Consult the chart in Figure 25–1 for any specific application because each is different. The chart is used as a guideline for the product type. The equipment operating conditions may vary slightly from one manufacturer to another.

25.4   TROUBLESHOOTING LOW-TEMPERATURE APPLICATIONS

Low-temperature applications start at freezing and go down in temperature. There is not much application at just below freezing. The first usable application is the making of ice, which normally occurs from 20°F evaporator temperature to about 5°F evaporator temperature. Ice making has many different variables. These variables are associated with the type of ice being made. Flake ice is a continuous process so the pressures are about the same regardless of the manufacturer. The manufacturer of the specific piece of equipment should be consulted for exact temperatures and pressures for flake ice machines.

Normally for cube ice makers you will find that a suction pressure of 21 psig (20°F) at the beginning of the cycle will begin to make ice using R-12. The ice will normally be harvested before the suction pressure reaches the pressure corresponding to 5°F or 11 psig, Figure 25–2. If you discover that the suction pressure would not go below 21 psig for an R-12 machine, suspect a problem. If the suction pressure goes below 11 psig, suspect a problem. These pressures are for the normal running cycle.

The next low-temperature application to be discussed will be frozen foods. It is generally considered that frozen food starts at 5°F space temperature and goes down from there. Foods require different temperatures to be frozen hard. Some are hard at 5°F, others at −10°F. The designer or operator will use the most economical design or operation. For instance, ice cream may be frozen hard at −10°F, but it will not hurt it to go to −30°F. The economics of cooling it to a cooler temperature may not be wise. As the temperature goes down, the compressor has less capacity because the suction gas becomes thinner. The compressor may not even cut off at the very low temperatures if the box thermostat is set this low.

Using the same guidelines as in high- and medium-temperature applications, we find the highest normal refrigerant boiling temperature to be $5°F - 20°F = -15°F$ for R-12. This has the refrigerant boiling at 20°F below the space temperature because the compressor is running and creating a suction pressure of 2.5 psig. The lowest suction pressure may be anything down to the pressure corresponding to −20°F space temperature. This would be $-20°F - 20°F = -40°F$ or 11 in. Hg vacuum for an R-12 system, Figure 25–2. With the compressor off the evaporator would be 5°F, the same as the space temperature or 12 psig.

Earlier in the unit on basic refrigeration, we discussed the idea of using a different refrigerant for low-temperature applications to keep the pressures positive, above atmospheric. If we used R-502 for this application, we would find a new set of figures. 5°F corresponds to 36 psig when the space temperature is 5°F with the compressor off. When the space temperature is 5°F and the compressor is running, the refrigerant will be boiling at about 20°F below the space temperature or $5°F - 20°F = -15°F$. This −15°F corresponds to a pressure of 19 psig. The lowest temperature normally encountered in low-temperature refrigeration is a space temperature of about −20°F. This would make the suction pressure about $-20°F - 20°F = -40°F$ or 4.1 psig, Figure 25–2. We can easily see that by using R-502 in this application the system is going to operate well above a vacuum. R-502 boils at −50°F at 0 psig, so the cooler would get down to −30°F air temperature before the suction would get down to atmospheric pressure or 0 psig with a 20°F TD.

Temperatures below −20°F space or product temperature are obtainable and used. This text will not consider these extra-low-temperature ranges because the equipment is not typical for the scope of this text.

The chart in Figure 25–2 can be used as a guideline for typical operating conditions for the foregoing systems on the low side of the system.

25.5 TYPICAL HIGH-PRESSURE CONDITIONS

Typical high-pressure side operating conditions can either be applied to air-cooled equipment or water-cooled equipment. Troubleshooting is very different for each.

25.6 TYPICAL AIR-COOLED CONDENSER OPERATING CONDITIONS

Air-cooled condensers have an operating temperature range from cold to hot as they are exposed to outside temperatures. The equipment may even be located inside a conditioned building. The condenser has to maintain a pressure that will create enough pressure drop across the expansion device. This pressure drop across the expansion device will push enough refrigerant through the device for proper evaporator refrigerant levels. Expansion valves for R-12 are normally sized at a 60 to 80 psig pressure drop. For example, when a valve manufacturer says a valve has a capacity of 1 ton at an evaporator temperature of 20°F, this is for a pressure drop of 80 psig. The same valve will have less capacity if the pressure drop goes to 60 psig. Controlling head pressure was discussed in detail in Unit 20.

Since any piece of equipment located outside must have head-pressure control, the minimum values at which the head-pressure control is set would be the minimum expected head pressure. For R-12 this is

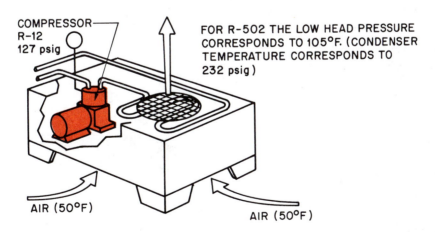

COMPRESSOR R-12 127 psig

FOR R-502 THE LOW HEAD PRESSURE CORRESPONDS TO 105°F. (CONDENSER TEMPERATURE CORRESPONDS TO 232 psig)

AIR (50°F)

AIR (50°F)

Figure 25–3. Refrigeration system showing the relationship of the air-cooled condenser to the entering air temperature on a cool day. This unit has head-pressure control for mild weather. The outside air entering the condenser is 50°F. This would normally create a head pressure of about 85 psig. This unit has a fan cycle device to hold the head pressure up to 125 psig minimum.

normally about 127 psig for air-cooled equipment. This corresponds to 105°F condensing temperature. R-502 has a head pressure of 232 psig at 105°F, Figure 25–3. When the discharge pressure is less than 127 psig for an R-12 machine, the condenser is operating at the minimum value and may not be feeding the expansion valve correctly.

25.7 CALCULATING THE CORRECT HEAD PRESSURE FOR AIR-COOLED EQUIPMENT

The maximum normal high-side pressure would correspond to the maximum ambient temperature in which the condenser is operated. Most air-cooled con-

densers will condense the refrigerant at a temperature of about 35°F above the ambient temperature when the ambient is above 70°F. See Figure 25–4 for an example of a refrigeration system operating under low ambient conditions without head pressure control. This would be considered the high end of the condenser operation. When the ambient temperature drops below 70°F, the relationship changes to a lower value.

If the condenser is a higher-efficiency condenser or oversized, the condenser may condense at 25°F above the ambient. Using the 35°F figure, we see that if a unit were located outside and the temperature were 95°F, the condensing temperature would be 95 + 35 = 130°F. This corresponds to 181 psig for R-12 and 321 psig for R-502, Figure 25–5. These are the two most common refrigerants for commercial refrigeration.

CONDENSER: CONDENSING AT 70°F WITH A HEAD PRESSURE OF 70 psig (R-12)

COMPRESSOR — 70 psig — AIR OUT

5 psig

AIR IN (40°F)   AIR IN (40°F)

EVAPORATOR IS OPERATING AT 5 psig. IT IS STARVED FOR REFRIGERANT. THE SUPERHEAT AT THE END OF THE EVAPORATOR IS HIGH. ICE WILL FORM ON THE BEGINNING OF THE COIL.

EVAPORATOR

70 psig

FANS

EXPANSION VALVE
SIGHT GLASS

FULL OF LIQUID AT THE LOW PRESSURE OF 70 psig. THE PRESSURE WILL NOT FEED ENOUGH REFRIGERANT TO THE EXPANSION VALVE.

MEDIUM–TEMPERATURE EVAPORATOR SHOULD BE OPERATING AT 21 psig (20°F BOILING TEMPERATURE)

Figure 25–4. Refrigeration system operating at an ambient temperature that caused the head pressure to be too low for normal operation. The expansion valve is not feeding correctly because there is not enough head pressure to push the liquid refrigerant through the valve.

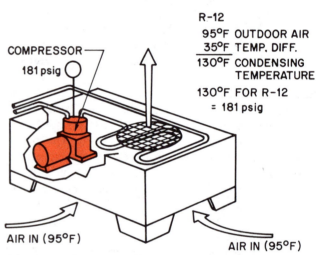

R-12

95°F OUTDOOR AIR
35°F TEMP. DIFF.
130°F CONDENSING TEMPERATURE

130°F FOR R-12 = 181 psig

COMPRESSOR — 181 psig

AIR IN (95°F)   AIR IN (95°F)

STANDARD EFFICIENCY AIR-COOLED CONDENSER CONDITIONS WITH A TYPICAL LIGHT FILM OF AIRBORNE DIRT CONDENSING AT 35°F ABOVE THE AMBIENT ENTERING AIR TEMPERATURE.

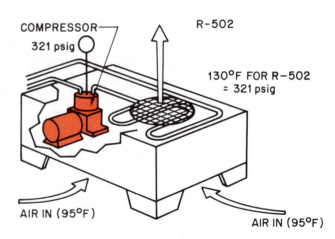

R-502

COMPRESSOR — 321 psig

130°F FOR R-502 = 321 psig

AIR IN (95°F)   AIR IN (95°F)

Figure 25–5. Air-cooled refrigeration system with condenser outside. Typical pressures are shown.

25.8 TYPICAL OPERATING CONDITIONS FOR WATER-COOLED EQUIPMENT

Water-cooled condensers are used in many systems in two ways. Some are wastewater, and some reuse the same water by extracting the heat with a cooling tower. Water has to be added in a system that reuses the water to make up for water that evaporates. Makeup water is also necessary for the water that is wasted to keep the remaining water from becoming too concentrated with minerals. See Figure 25–6 for a wastewater system.

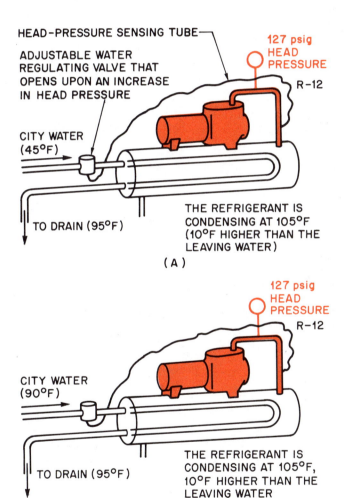

Figure 25–6. Water-cooled wastewater refrigeration system. The water flow is adjusted by the water regulating valve, which keeps too much water from flowing, maintains a constant head pressure and shuts the water off at the end of the cycle. There is a relationship between the condensing refrigerant and the leaving-water temperature. This is normally a 10°F temperature difference.

25.9 TYPICAL OPERATING CONDITIONS FOR WASTEWATER CONDENSER SYSTEMS

Wastewater systems use the same condensers as the cooling tower applications, but the water is wasted down the drain. Normally a water regulating valve is used to regulate the water flow for economy and to regulate the head pressure. The condensers are either cleanable condensers or coil type that cannot be cleaned. With either type condenser it is advantageous to know from the outside how the condenser is performing on the inside.

When a water regulating valve is used to control the water flow, the water flow will be greater if the condenser is not performing correctly. When the head pressure goes up, the water will start to flow faster to compensate. This will take place until the capacity of the valve opening is reached. Then the head pressure will increase with maximum water flow. The head pressure has a more consistent relationship with the outlet water temperature because the outlet water temperature is more of a constant than the inlet water. Sometimes the inlet water is colder, such as in winter. It would not be unusual for the inlet water to be 45°F in the winter. 90°F may be the high value if the water travels through a hot ceiling in the summer, Figure 25–6.

The condensing temperature is normally 10°F higher than the water temperature leaving the condenser. If the system water valve were set to maintain 127 psig head pressure for an R-12 system, the condensing temperature of the refrigerant would be 105°F, and the leaving water should be about 95°F, Figure 25–6. If the condensing temperature were much above this, you would suspect a dirty condenser. If the gages indicated 220 psig (145°F condensing temperature) and the leaving water temperature were to be 95°F, the difference in temperature of the condensing refrigerant is 145°F − 95°F = 50°F. This indicates that the condenser is not removing the heat; the coil is dirty, Figure 25–7. It can

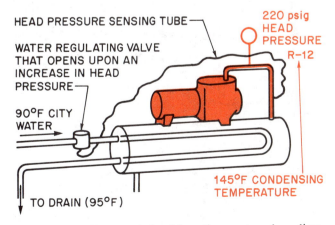

Figure 25–7. Water-cooled refrigeration system has dirty condenser tubes, and the water is not taking the heat out of the system.

be cleaned chemically or with brushes if it is a cleanable condenser. Whenever a noticeable increase in water flow takes place, a dirty condenser is suspected.

25.10 TYPICAL OPERATING CONDITIONS FOR RECIRCULATED WATER SYSTEMS

Water-cooled condensers that use a cooling tower to remove the heat from the water normally do not use water regulating valves to control the water flow. It is normally a constant volume of water that is pumped by a pump. The volume is customarily designed into the system in the beginning and can be verified by checking the pressure drop across the water circuit at the condenser inlet to outlet. There has to be some pressure drop for there to be water flow. The original specifications of the job should include the engineer's intentions with regard to the water flow, but these may not be obtainable on an old job.

Most systems that reuse the same water with a cooling tower have a standard 10°F temperature rise across the condenser. For example, if the water from the tower were to be 85°F entering the condenser, the water leaving the condenser should be 95°F, Figure 25–8. If the difference were to be 15 or 20°F, you might think that the condenser is doing its job of removing the heat from the refrigerant, but the water flow is insufficient, Figure 25–9. If the water temperature entering the condenser were to be 85°F and the leaving water were to be 90°F, it may be that there is too much water flow. If the head pressure is not high, the condenser is removing the heat, and there is too much water flow. If the head pressure is high and there is the

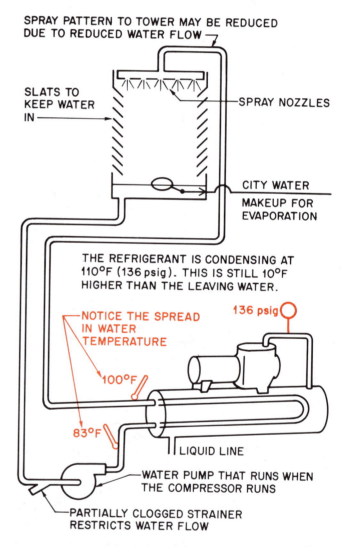

Figure 25–8. A water-cooled refrigeration system reusing the water after the heat is rejected to the atmosphere. There is a constant bleed of water to keep the system from overconcentrating with the minerals left behind when the water is evaporated. The difference in the incoming water and the leaving water is 10°F. This is the typical temperature rise across a water tower system.

Figure 25–9. This system has too much temperature rise, indicating there is not enough water flow. The water strainers may be stopped up. The condensing temperature is higher than normal, which causes the head pressure to rise. A decrease in water flow may be detected by water pressure drop if it is known what it is supposed to be.

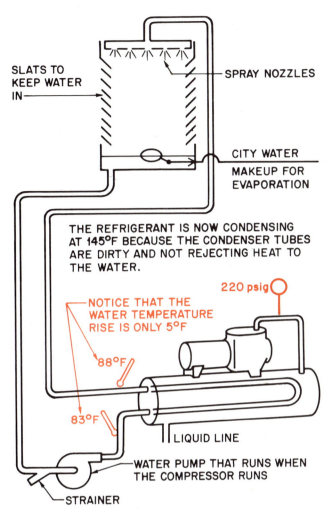

Figure 25–10. System has dirty condenser tubes.

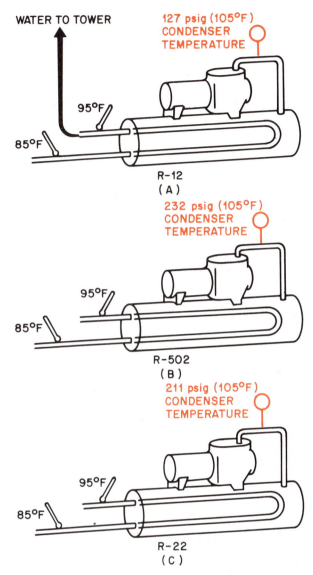

Figure 25–11. The water tower system has a relationship with the leaving water and the condensing refrigerant just like the wastewater system. Normally the refrigerant condenses at 10°F warmer than the water leaving the condenser.

right amount of water, the condenser is dirty. See Figure 25–10 for an example of a dirty condenser.

The condenser has a relationship of heat exchange with the condenser water like the wastewater system. The refrigerant normally condenses about 10°F warmer than the leaving water. If the leaving water is 95°F, a properly operating condenser should be condensing at 105°F or a discharge pressure of 127 psig for R-12 and 232 psig for R-502, Figure 25–11. If the condensing temperature is much more than 10°F above the leaving water temperature, the condenser is not removing the heat from the refrigerant. It could be dirty.

In this type of system the condenser is getting its inlet water from the cooling tower. A cooling tower has a heat exchange relationship with the ambient air. Usually the cooling tower is located outside and may be natural draft or forced-draft (i.e., a fan forces the air through the tower). Either tower will be able to supply water temperature according to the humidity or moisture content in the outside air. The cooling tower can normally cool the water to within 7°F of the wet-bulb temperature of the outside air, Figure 25–12. If the outside wet-bulb temperature (taken with a psy-

chrometer) is 78°F, the leaving cooling tower water will be about 85°F if the tower is performing correctly. Wet-bulb temperature concerns the moisture content in air and is discussed in more detail in Unit 37.

25.11   SIX TYPICAL PROBLEMS

Six typical problems that can be encountered by any refrigeration system are

1. Low refrigerant charge
2. Excess refrigerant charge
3. Inefficient evaporator
4. Inefficient condenser
5. Restriction in the refrigerant circuit
6. Inefficient compressor

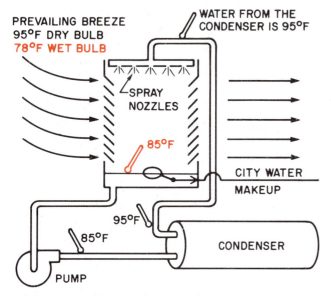

COOLED WATER IN BASIN OF TOWER IS 85°F. IT
CAN NORMALLY BE COOLED TO WITHIN 7°F OF
THE OUTDOOR WET-BULB TEMPERATURE DUE
TO EVAPORATION.
NOTICE THAT THE FINAL WATER TEMPERATURE
IS MUCH COOLER THAN THE OUTDOOR DRY-BULB
TEMPERATURE.

Figure 25–12. Cooling tower has a temperature relationship
with the air that is cooling the water. Most cooling towers
can cool the water that goes back to the condenser to within
7°F of the wet-bulb temperature of the ambient air. For
example, if the wet-bulb temperature is 78°F, the tower
should be able to cool the water to 85°F. If it does not, a
tower problem should be suspected.

25.12 LOW REFRIGERANT CHARGE

A low refrigerant charge affects most systems in
about the same way, depending on the amount of the
refrigerant needed to be correct. The normal symptoms
are low capacity. The system has a starved evaporator
and cannot absorb the rated amount of Btu or heat.
The suction gage will read low, and the discharge gage
will read low. The exception to this is a system with
an automatic expansion valve, Figure 25–13. It will be
discussed later in this unit.

If the system has a sight glass, it will have bubbles
in it that look like air but which are actually vapor
refrigerant. Figure 25–14 is a typical liquid-line sight
glass. Remember, a sight glass that is full of vapor or
liquid may look the same. If there is only vapor in the
glass, a slight film of oil may be present. This is a
good indicator of vapor only.

When a system has a sight glass, it will generally
have a thermostatic or automatic expansion valve and
a receiver. These valves will hiss when a partial
vapor–partial liquid mixture is going through the valve.
If the system has a capillary tube, it will probably not
have a sight glass. The technician needs to know how

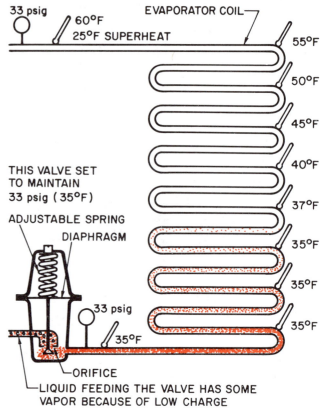

THE VALVE IS WIDE OPEN AND IS STILL
MAINTAINING THE SET POINT FOR PRESSURE.
COIL HAS A HIGH SUPERHEAT.

Figure 25–13. Low refrigerant charge system characteristics
when the metering device is an automatic expansion valve.

the system feels at different points to determine the gas
charge level without using gages.

The low charge affects the compressor by not supplying the cool suction vapor to cool the motor. Most
compressors are suction cooled, so the result is a hot
compressor motor. It may even be off because of the
motor-winding thermostat, Figure 25–15. If the compressor is air cooled, the suction line coming back to

Figure 25–14. A system with a sight glass to indicate when
pure liquid is in the liquid line. *Courtesy Henry Valve Company*

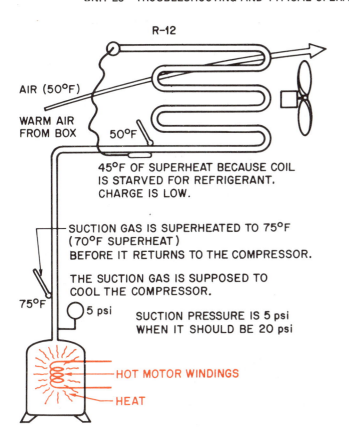

Figure 25-16. An air-cooled compressor. The compressor motor will not normally get hot as a result of a low refrigerant charge, but the discharge gas leaving the compressor may get too warm because the refrigerant entering the compressor is too warm. Most compressor manufacturers require the suction gas not to be over 65°F for continuous operation. *Courtesy Tyler Refrigeration Company*

Figure 25-15. System with a suction-cooled hermetic compressor. The compressor is hot enough to cause it to cut off because of motor temperature. It is hard to cool off a hot hermetic compressor from the outside because the motor is suspended inside a vapor atmosphere.

the compressor will not be as cool as a suction-cooled compressor. It may be warm by comparison, Figure 25–16.

25.13 REFRIGERANT OVERCHARGE

A refrigerant overcharge also acts much the same way from system to system. The discharge pressure is high, and the suction pressure may be high. The automatic expansion valve system will not have a high suction pressure because it maintains a constant suction pressure, Figure 25–17. The thermostatic expansion valve may have a slightly higher suction pressure if the head pressure is excessively high because the system capacity may be down, Figure 25–18.

The capillary tube will have a high suction pressure because the amount of refrigerant flowing through it depends on the difference in pressure across it. The more head pressure, the more liquid it will pass. The capillary tube will allow enough refrigerant to pass so that it will allow liquid into the compressor. When the compressor is sweating down the side or all over, it is a sign of liquid refrigerant in the compressor, Figures 25–19 and 25–20.

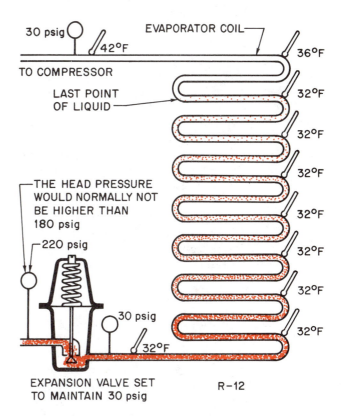

Figure 25-17. System has an automatic expansion valve for a metering device. This device maintains a constant suction pressure. The head pressure is higher than normal, but the suction pressure remains the same.

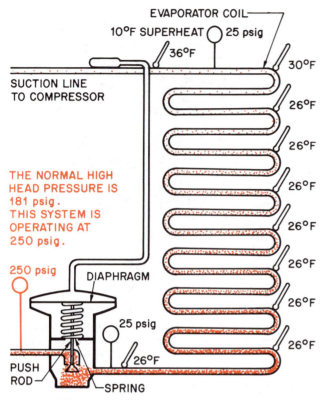

THE NORMAL HIGH HEAD PRESSURE IS 181 psig. THIS SYSTEM IS OPERATING AT 250 psig.

THE NORMAL SUCTION PRESSURE IS 20 psig

Figure 25–18. A thermostatic expansion valve system that has a higher than normal head pressure.

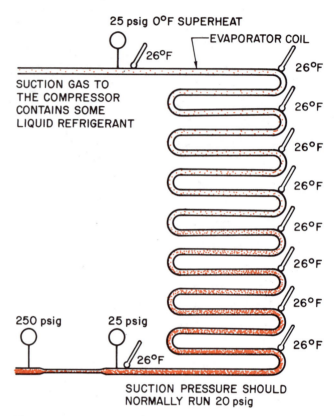

SUCTION PRESSURE SHOULD NORMALLY RUN 20 psig

Figure 25–19. The capillary tube system has an overcharge of refrigerant, and the head pressure is higher than normal. This has a tendency to push more refrigerant than normal through the metering device.

The compressor should only have vapor entering it. Vapor will rise in temperature as soon as it touches the compressor shell. When liquid is present, it will not rise in temperature and will cool the compressor shell. **Liquid refrigerant still has its latent heat absorption capability and will absorb a great amount of heat without changing temperature. A vapor absorbs only sensible heat and will change in temperature quickly,** Figure 25–20. Another reason that liquid may get back to the compressor is poor heat exchange in the evaporator in a capillary tube system. If liquid is getting back to the compressor, the evaporator heat exchange should be checked before removing refrigerant. See the next subsection.

25.14 INEFFICIENT EVAPORATOR

An inefficient evaporator does not absorb the heat into the system and will have a low suction pressure. The suction line may be sweating or frosting back to the compressor. This can be caused by a dirty coil, a fan running too slow, an expansion valve starving the coil, recirculated air, ice buildup, or product interference causing blocked airflow, Figure 25–21.

All of these can be checked with an evaporator performance check. This check can be performed by making sure that the evaporator has the correct

amount of refrigerant with a superheat check, Figure 25–22. The heat exchange surface should be clean. The fans should be blowing enough air and not recirculating it from the discharge to the inlet of the coil.

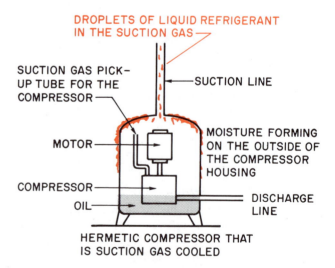

HERMETIC COMPRESSOR THAT IS SUCTION GAS COOLED

Figure 25–20. When liquid refrigerant gets back to the compressor, the latent heat that is left in the refrigerant will cause the compressor to sweat more than normal. When vapor only gets back to the compressor, the vapor changes in temperature quickly and does not sweat a lot.

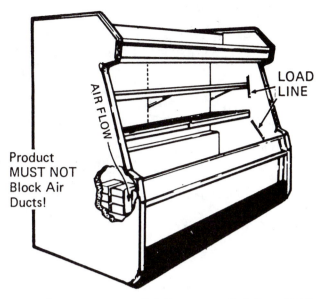

Canned goods and other non-refrigerated items are great for imposing piles and eye-catching displays. Don't try the same gimmicks with perishables. The case will fail to refrigerate any of the merchandise when the air ducts are blocked.

Low temperature multi-shelf cases are particularly sensitive to air pattern changes as well as extremes of humidity and temperature in the store.

Jumble displays may have some sales benefits, but they really foul up the protective layer of cold air in the case. Observe the LOAD LINE stickers!

Figure 25–21. Evaporator that cannot absorb the required Btu because of product interference. There is a load line on the inside where the product is stored. *Courtesy Tyler Refrigeration Company*

The refrigerant boiling temperature should not be more than 20°F colder than the entering air on an air evaporator coil, Figure 25–2. A water coil should have no more than a 10°F difference in the boiling refrigerant and the leaving water, Figure 25–23. When the boiling refrigerant relationship to the medium being cooled starts increasing, the heat exchange is decreasing.

25.15 INEFFICIENT CONDENSER

An inefficient condenser acts the same whether it is water cooled or air cooled. If the condenser cannot

remove the heat from the refrigerant, the head pressure will go up. The condenser does three things and has to be able to do them correctly, or excessive pressures will occur.

1. Desuperheat the hot gas from the compressor. This gas may be 200°F or hotter on a hot day on an air-cooled system. This is accomplished in the beginning of the coil.

2. Condense the refrigerant. This is done in the middle of the coil. The middle of the coil where the condensing is taking place is the only place that the coil

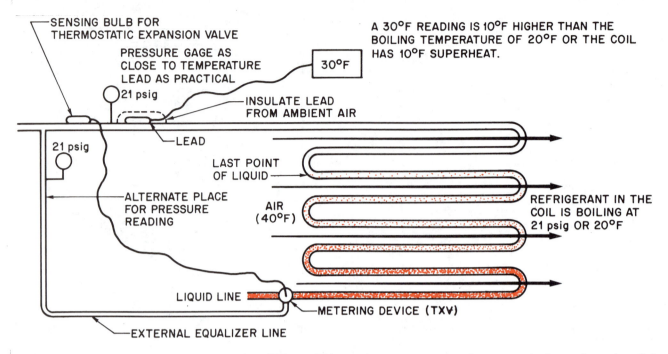

Figure 25–22. How a coil can be analyzed for efficiency. This requires temperature and pressure checks at the outlet of the evaporator. When a coil has the correct refrigerant level, checked by superheat and the correct air-to-refrigerant heat exchange, the coil will absorb the correct amount of heat. The correct heat exchange is taking place if the refrigerant is boiling at 10°F to 20°F cooler than the entering air. Note, this occurs when the cooler temperature is within the design range, 30°F to 45°F for medium-temperature for example. Coils with multiple circuits must have every circuit checked for even distribution.

temperature will correspond to the head pressure. You could check the temperature against the head pressure if a correct temperature reading can be taken, but the fins are usually in the way.

3. Subcool the refrigerant before it leaves the coil. This subcooling is cooling the refrigerant to a point below the actual condensing temperature. A subcooling of 5° to 20°F is typical. The subcooling can be checked just like the superheat, only the temperature is checked at the liquid line and compared to the high-side pressure converted to condensing temperature, Figure 25–24.

The condenser must have the correct amount of cooling medium (air or water). This medium must **not** be recirculated without being cooled (mixed with the incoming medium). It is easy to locate air-cooled equipment in such a manner that the air leaving the condenser circulates right back into the inlet. This air is hot and will cause the head pressure to go up in proportion to the amount of recirculation, Figure 25–25. An air-cooled condenser should not be located down low, close to the roof, even though the air comes in the side. The temperature is higher at the roof level than it is a few inches higher. A clearance of about 18 in. will give better condenser performance, Figure 25–26. An air-cooled condenser that has a vertical coil may be influenced by prevailing winds. If the fan is trying to discharge its air into a 20 mph wind, it may

not move the correct amount of air, and a high head pressure may occur, Figure 25–27.

25.16 REFRIGERANT FLOW RESTRICTIONS

Restrictions in the refrigeration circuit are either partial or full. A partial restriction will be in a gas line

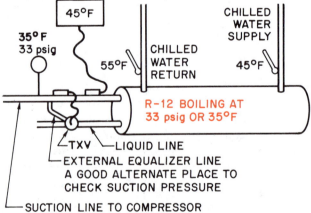

Figure 25–23. Water coils can be analyzed in much the same way as air coils. The refrigerant temperature and pressure at the end of the evaporator will indicate the refrigerant level. The refrigerant boiling temperature should not be more than 10°F cooler than the leaving water.

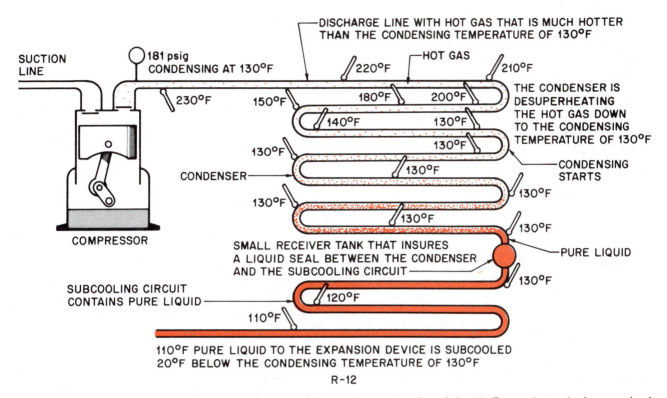

Figure 25-24. Checking the subcooling on a condenser. The condenser does three jobs: (1) Desuperheats the hot gas, in the first part of the condenser. (2) Condenses the vapor refrigerant to a liquid, in the middle of the condenser. (3) Subcools the refrigerant at the end of the condenser. The condensing temperature corresponds to the head pressure. Subcooling is the temperature of the liquid line subtracted from the condensing temperature. A typical condenser can subcool the liquid refrigerant 5°F to 20°F cooler than the condensing temperature.

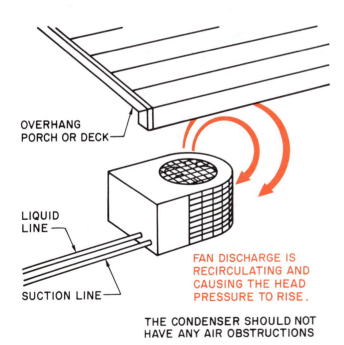

Figure 25-25. The air-cooled condenser is located too close to an obstacle, and the hot air leaving the condenser is recirculated back into the inlet of the coil.

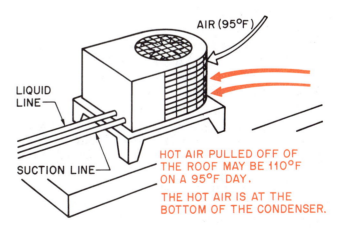

Figure 25-26. An air-cooled condenser should not be located close to a roof. The temperature of the air coming directly off the roof is much warmer than the ambient air because the roof acts like a solar collector.

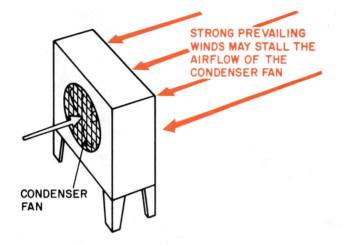

STRONG PREVAILING WINDS MAY STALL THE AIRFLOW OF THE CONDENSER FAN

CONDENSER FAN

Figure 25–27. A condenser that is located in such a manner that it is discharging its air into a strong prevailing wind may not get enough air across the coil.

(high or low side) or the liquid line. A restriction always causes pressure drop at the point of the restriction. Different conditions will occur, depending on where the restriction is. Pressure drop can always be detected with gages, but they cannot always be applied to the correct place. Gage ports sometimes have to be installed for pressure testing.

If a restriction happens due to something outside the system, it is usually physical damage, such as flattened or bent tubing. These can be hard to find if they are in hidden places such as under the insulation or behind a fixture.

If a partial restriction occurs in a liquid line, it will be evident because the refrigerant will have a pressure drop and will start to act like an expansion device at the point of the restriction. **When there is a pressure drop in a liquid line, there is always a temperature change.** A temperature check on each side of a restriction will locate the place. Sometimes when the drop is across a drier, the temperature difference from one side to the other may not be enough to feel with bare hands, but a temperature tester will detect it, Figure 25–28.

If a system has been running for a long time and a restriction occurs, physical damage may have occurred. If the restriction occurs soon after start, a filter or drier may be stopping up. When this occurs in the liquid line drier, bubbles will appear in the sight glass when the drier is located before the sight glass, Figure 25–29.

Another occurrence that may create a partial restriction could be valves that do not open all the way. Normally the thermostatic expansion valve either works or it doesn't. It will function correctly, but if it loses its charge in the thermal bulb, it will close and cause a full restriction, Figure 25–30.

There is a strainer at the inlet to most expansion devices that can trap particles and stop up slowly. If the device is a valve that can be removed, it can be

inspected and cleaned if necessary. If it is a capillary tube, it will be soldered into the line and not easy to inspect, Figure 25–31.

Water circulating in any system that operates below freezing will freeze at the first cold place it passes through. This would be in the expansion device. One drop of water can stop a refrigeration machine. Sometimes a piece of equipment that has just been serviced will show signs of moisture on the first hot day, Figure 25–32. This is because the drier in the liquid line will have more capacity to hold moisture when it is cool. The first hot day, the drier may turn a drop or two of water loose, and it will freeze in the expansion device, Figure 25–33. When you suspect this, apply heat to thaw the water to a liquid. *Care must be used when applying heat. A hot wet cloth is a good source.* If applying a hot cloth to the metering device causes the system to go back to refrigerating, the problem is free water in the system. Change the drier and evacuate the system.

Other components that may cause restrictions are any automatic valves in the lines, such as the liquid line solenoid, crankcase pressure regulator, or the evaporator pressure regulator. These valves may easily be checked with gages applied to both sides of them where pressure taps are provided, Figure 25–34.

25.17 INEFFICIENT COMPRESSOR

Inefficient compressor operation can be one of the most difficult troubles to find. When a compressor will

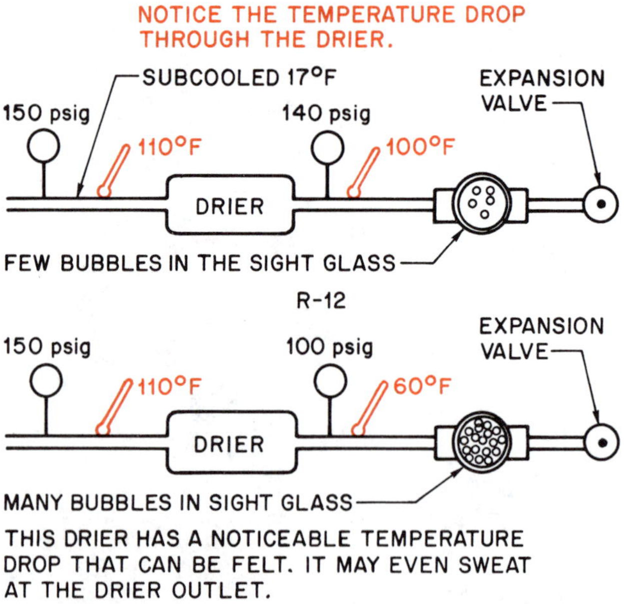

NOTICE THE TEMPERATURE DROP THROUGH THE DRIER.

SUBCOOLED 17°F

150 psig   110°F   DRIER   140 psig   100°F   EXPANSION VALVE

FEW BUBBLES IN THE SIGHT GLASS

R-12

150 psig   110°F   DRIER   100 psig   60°F   EXPANSION VALVE

MANY BUBBLES IN SIGHT GLASS

THIS DRIER HAS A NOTICEABLE TEMPERATURE DROP THAT CAN BE FELT. IT MAY EVEN SWEAT AT THE DRIER OUTLET.

Figure 25–28. The driers each have a restriction. One of them is very slight. Where there is pressure drop in a liquid line, there is temperature drop. If the temperature drop is very slight, it can be detected with a temperature tester. Sometimes gages are not easy to install on each side of a drier to check for pressure drop.

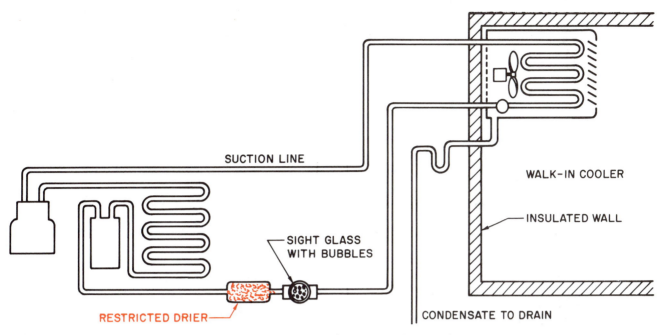

Figure 25–29. The restriction in a system may occur shortly after startup. This indicates that solid contaminants from installing the system must be in the drier. If the restriction occurs after the system has been running for a long time, the restriction may be physical damage, such as a bent pipe. Normally, loose contamination will make its way through the system to the drier in a matter of hours.

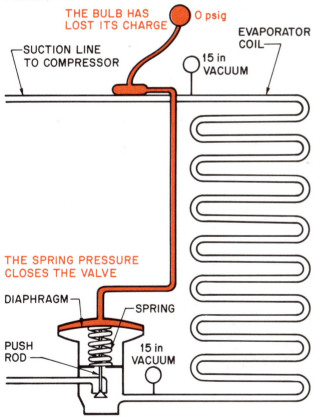

Figure 25–30. What happens when a thermostatic expansion valve loses its charge. The charge in the bulb is the only force that opens the valve, so when it loses its pressure, the valve closes. The system can go into a vacuum if there is no low-pressure control to stop the compressor. A partial restriction can occur if the bulb loses part of the charge or if the inlet strainer stops up.

not run, it is evident where the problem is. Motor troubleshooting procedures are covered in a separate unit. When a compressor is pumping at slightly less than capacity, it is hard to determine the problem. It helps at this point to remember that a compressor is a vapor pump. It should be able to create a pressure from the low side of the system to the high side of the system under design conditions and operate under the design energy requirements of the compressor.

Figure 25–31. Capillary tube metering device with the strainer at the inlet. This strainer is soldered into the line and is not easy to service. *Courtesy Parker Hannafin Corporation*

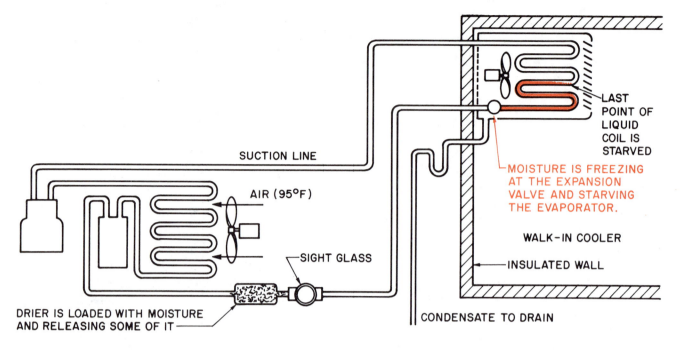

SUCTION LINE

AIR (95°F)

LAST POINT OF LIQUID COIL IS STARVED

MOISTURE IS FREEZING AT THE EXPANSION VALVE AND STARVING THE EVAPORATOR.

WALK-IN COOLER

INSULATED WALL

SIGHT GLASS

DRIER IS LOADED WITH MOISTURE AND RELEASING SOME OF IT

CONDENSATE TO DRAIN

Figure 25–32. The system has had some moisture in it. The drier has all the moisture it can hold at mild temperatures. When the weather gets warm, the drier cannot hold all of the moisture and turns some of it loose. The moisture will freeze at the expansion device, where the first refrigeration is experienced if the system operates below freezing.

The following methods are all used by service technicians to discover the compressor problems.

25.18 COMPRESSOR VACUUM TEST

The compressor vacuum test is usually performed on a test bench with the compressor out of the system. This test may be performed in the system when the system has service valves. *Care should be taken not to pull air into the system while in a vacuum.* All reciprocating compressors should immediately go into a vacuum if the suction line is valved off when the compressor is running. This test proves that the suction valves are seating correctly on at least one cylinder. This test is *not satisfactory* on a multicylinder compressor. If one cylinder will pump correctly, a vacuum will be pulled. A reciprocating compressor should pull 26 to 28 in. Hg vacuum with the atmosphere as the discharge pressure, Figure 25–35. The compressor should pull about 24 in. Hg vacuum against 100 psig discharge pressure, Figure 25–36. When the compressor has pumped a differential pressure and is stopped off, the pressures should not equalize. For example, if a compressor has been operated until the suction pressure is 24 in. Hg vacuum and the head pressure is 100 psig then it is stopped. These pressures should stay the same when off. When refrigerant is used for this pumping test, the 100 psig will drop some because of the condensing refrigerant. Nitrogen is a better choice to pump in the test because the pressure will not drop.

*Care should be taken when operating a hermetic compressor in a vacuum because the motor is subject to damage. Also, most compressor motors are cooled with suction gas and will get hot if operated for any length of time. This vacuum test should not take more than three to five minutes. The motor will not overheat in

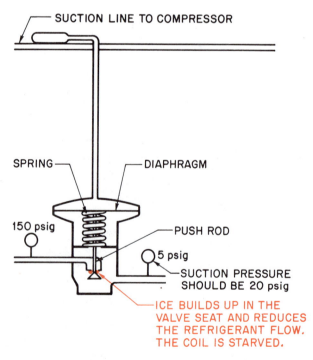

SUCTION LINE TO COMPRESSOR

SPRING

DIAPHRAGM

150 psig

PUSH ROD

5 psig

SUCTION PRESSURE SHOULD BE 20 psig

ICE BUILDS UP IN THE VALVE SEAT AND REDUCES THE REFRIGERANT FLOW. THE COIL IS STARVED.

Figure 25–33. If it is suspected that moisture is frozen in the metering device, mild heat can be added. Heat from a hot wet cloth will normally thaw the moisture out, and the system will start refrigerating again.

this period of time. The test should only be performed by experienced technicians.*

25.19 CLOSED LOOP COMPRESSOR RUNNING BENCH TEST

Running bench testing the compressor can be accomplished by connecting a line from the discharge to the suction of the compressor and operating the compressor in a closed loop. A difference in pressure can be obtained with a valve arrangement or gage manifold. This will prove the compressor will pump. **When the compressor is hermetic, it should operate at close to full-load current in the closed loop when design pressures are duplicated.** Nitrogen or refrigerant can be used as the gas to compress. Typical operating pressures will have to be duplicated for the compressor to operate at near the full-load current rating. For example, for a medium-temperature compressor, a suction

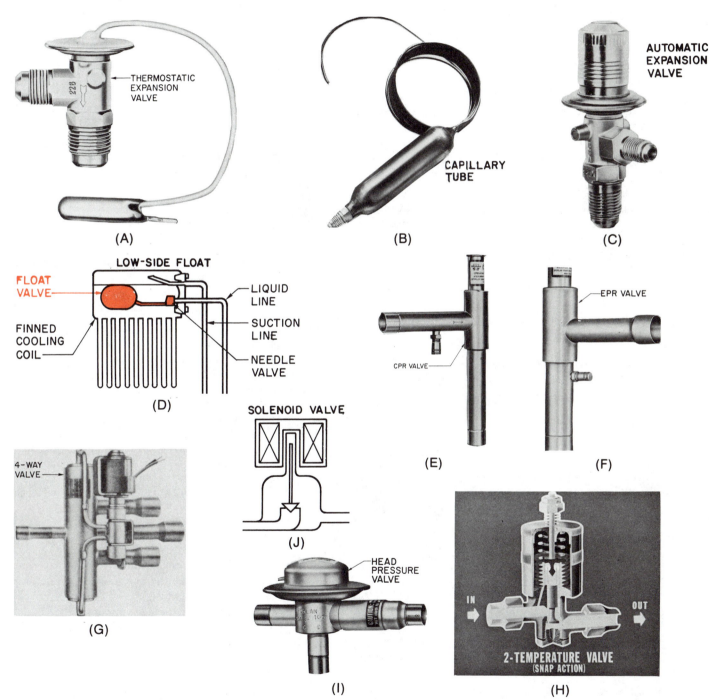

Figure 25–34. The components are all subject to closing and causing a restriction. Any valve that can close is subject to do so. *Courtesy (A) Parker Hannafin Corporation (B, C, E) Singer Controls Division, Schiller Park, Illinois (F, I) Sporlan Valve Company (G) ALCO Controls Division, Emerson Electric Company (H) Carrier Corporation*

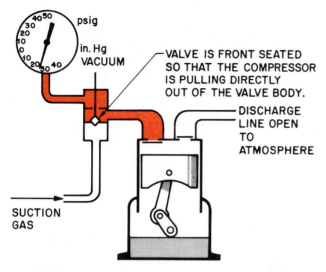

Figure 25–35. Compressor pulling a vacuum with the atmosphere as the head pressure. Most reciprocating compressors can pull 26 to 28 in. Hg vacuum with the atmosphere as the discharge pressure.

pressure of 20 psig will duplicate a typical condition for R-12 on a hot day. The compressor should operate at near to nameplate full-load current when the design voltage is supplied to the motor.

The following is a step-by-step procedure for performing this test. Use this procedure with the information in Figure 25–37. *This test should be performed only under the close supervision of an instructor or by a qualified person and on equipment under 3 hp. Safety goggles must be worn.*

1. Use the gage manifold and fasten the suction line to the gage line in such a manner that the compressor is pumping only from the gage line.

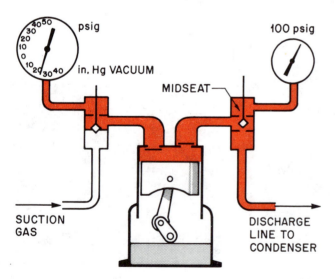

Figure 25–36. Compressor check by pulling a vacuum while pumping against a head. A reciprocating compressor can normally pull 24 in. Hg vacuum against 100 psig head pressure.

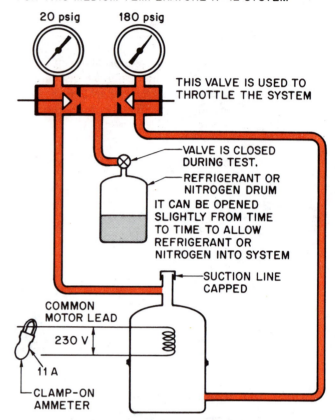

Figure 25–37. Checking a compressor's pumping capacity by using a closed loop. The test is accomplished by routing the discharge gas back into the suction with a piping loop. The discharge gage manifold valve is gradually throttled towards closed *(do not close entirely)*. *Note:* It can never be closed, or tremendous pressure will occur. When the design suction and discharge pressures are reached, the compressor should be pulling close to nameplate full-load amps. *Note:* A suction-cooled compressor cannot be run for long in this manner, or the motor will get hot. The refrigerant characteristic to the compressor or nitrogen can be used to circulate for pumping. Nitrogen will not produce the correct amperage, but it will be close enough. *This test should only be performed by experienced technicians.*

2. Fasten the discharge line to the discharge gage line in such a manner that the compressor is pumping only into the gage manifold.
3. Plug the center line of the gage manifold.
4. *Open both gage manifold valves wide open, counterclockwise.*
5. *Start the compressor, keep your hand on the off switch.*
6. The compressor should now be pumping out the discharge line and back into the suction line. The discharge gage manifold valve may be slowly throttled *(do not close entirely)* towards closed until the

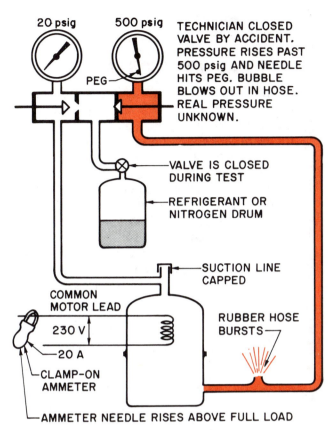

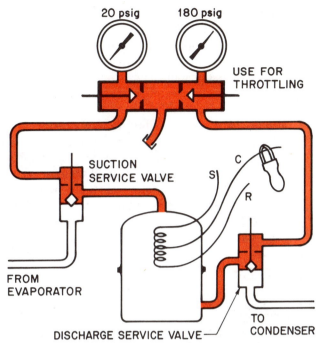

Figure 25–38. What happens if a compressor is started up in a closed loop with no place for the discharge gas to go (discharge gage manifold closed by accident). One cylinder full of gas could be enough to build tremendous pressures.

discharge pressure rises to the design level and the suction pressure drops to the design level.

7. When the desired pressures are reached, the amperage reading on the compressor motor should compare closely to the full load amperage of the compressor. If the correct pressures cannot be obtained with the amount of gas in the compressor, a small amount of gas may be added to the loop system by attaching the center line to a refrigerant drum and slightly opening the drum valve.

8. The amperage may vary slightly from full load because of the input voltage. For example, a voltage above the nameplate will cause an amperage below full load and vice versa.

Figure 25–38 illustrates a situation where a technician accidently closed the wrong valve.

25.20 COMPRESSOR CLOSED LOOP RUNNING FIELD TEST

When a compressor has service valves, this test can be performed in place in the system using a gage manifold as the loop, Figure 25–39. The compressor is started with the compressor service valves turned all the way to the front seat and the gage manifold valves open. The center line on the gage manifold must be plugged. *This can be a dangerous startup and should

Figure 25–39. Compressor test being performed in a system using a gage manifold as a closed loop. Notice that this test can be performed in the system because the compressor has service valves.

only be performed under supervision of an experienced person. The compressor has to have a place to pump the discharge gas because reciprocating compressors are positive displacement pumps. When the compressor cylinder is full of gas, it is going to pump it somewhere or stall. In this test the gas goes around through the gage manifold and back into the suction. Should the compressor be started before the gage manifold is open for the escape route through the gage manifold, tremendous pressures will result before you can stop the compressor. It only takes one cylinder full of vapor to fill the gage manifold to more than capacity. This test should only be performed on compressors under 3 hp. Safety goggles must be worn.*

1. Turn the unit OFF and fasten the suction line of the gage manifold to the suction service valve. Plug the center line of the gage manifold.

2. Turn the suction service valve stem to the front seat.

3. Fasten the discharge line to the discharge valve and turn the discharge service valve stem to the front seat.

4. *Open both gage manifold valves all the way, counterclockwise.*

5. *Start the compressor and keep your hand on the switch.*

6. The compressor should now be pumping out through the discharge line, through the gage manifold and back into the suction line.

7. The discharge gage manifold valve may be throttled toward the seat to restrict the flow of refrigerant and create a differential in pressure. Throttle *(do not close entirely)* the valve until the design head and suction pressure for the system is attained and the compressor should then be pulling near to full load amperage. As in the bench test the amperage may vary slightly because of the line voltage.

25.21 COMPRESSOR RUNNING TEST IN THE SYSTEM

Running test in the system can be performed by creating typical design conditions in the system. Typically a compressor will operate at a *high suction* and a *low head pressure* when it is not pumping to capacity. This will cause the compressor to operate at a low current. When the technician gets to the job, the conditions are not usually at the design level. The fixture is usually not refrigerating correctly—this is what instigated the call to begin with. The technician may not be able to create design conditions, but the following approach should be tried if the compressor capacity is suspected.

1. Install the high- and low-side gages.
2. Make sure that the charge is correct (not over nor under), using manufacturer's recommendations.
3. Check the compressor current and compare to full load.
4. Block the condenser airflow and build up the head pressure.

If the compressor will not pump the head pressure up to the equivalent of a $95°$ F $(95 + 35 = 130°$ F condensing temperature or 181 psig for R-12 and 318 psig for R-502) **day and draw close to nameplate full-load current, the compressor is not pumping. When the compressor is a can type of hermetic, it may whistle when it is shut down. This whistle is evidence of an internal leak from the high side to the low side.**

If the compressor has service valves a closed loop test can be performed on the compressor using the methods we have explained, while it is in the system. *Make sure no air is drawn into the system while it is in a vacuum.*

The compressor's temperature should be monitored at any time these tests are being conducted. If the compressor gets too warm, it should be stopped and allowed to cool. If the compressor has internal motor-temperature safety controls, don't operate it when the control is trying to stop the compressor.

25.22 SERVICE TECHNICIAN CALLS

In addition to the six typical problems encountered in the refrigeration system, there are many more not so typical problems. The following service situations will help you understand troubleshooting. Most of these service situations have already been described, although not as an actual troubleshooting procedure. Sometimes the symptoms do not describe the problem and a wrong diagnosis is made. Remember, do not draw any conclusion until the whole system has been examined. Become a system doctor. Examine the system and say, "This needs further examination," then do it.

Remember, refrigeration systems often cool large amounts of food so they are slow to respond. The temperature may drop very slowly on a hot or warm pull down.

Service Call 1

Customer calls and complains that a medium-temperature walk-in cooler with a remote condensing unit has a compressor that is short cycling and not cooling correctly. *The evaporator has two fans, and one is burned up. The unit is short cycling on the low-pressure control because there is not enough load on the coil.*

On the way to the job the technician goes over the possibilities of the problem. This is where it helps to have some familiarity with the job. The technician remembers that the unit has a low-pressure control, a high-pressure control, a thermostat, an overload, and an oil safety control. Defrost is accomplished by off cycle using the refrigerated space air with the fans running. Process of elimination helps make the decision that the thermostat and the oil safety control are not at fault. The thermostat has a $10°$ F differential, and the cooler temperature should not vary $10°$ F in a short cycle. The oil safety control is manual reset and will not short cycle. This narrows the possibilities down to the overload and the high- or low-pressure controls.

Upon arrival, the technician looks over the whole job before doing anything and notices that one evaporator fan is not running and the coil is iced up so the suction is too low. This causes the low-pressure control to shut off the compressor. When the examination is complete, the fan motor is changed, and the sys-

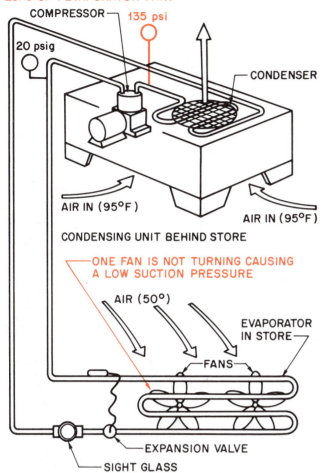

NOTICE THAT THE HEAD PRESSURE WOULD
NORMALLY BE 181 psig (95°F + 35°F = 130°F)
BUT IT IS 135 psig BECAUSE OF THE REDUCED
LOAD OF 1 EVAPORATOR FAN.

COMPRESSOR — 135 psi

20 psig

CONDENSER

AIR IN (95°F)

AIR IN (95°F)

CONDENSING UNIT BEHIND STORE

ONE FAN IS NOT TURNING CAUSING
A LOW SUCTION PRESSURE

AIR (50°)

EVAPORATOR
IN STORE

FANS

EXPANSION VALVE

SIGHT GLASS

MEDIUM-TEMPERATURE EVAPORATOR SHOULD BE
OPERATING AT 21 psig (20°F BOILING
TEMPERATURE) BUT IT IS OPERATING ALL OF THE
TIME AND THE COOLER TEMPERATURE IS 50°F.

Figure 25–40. Symptoms of a medium-temperature system with one of the two evaporator fans burned out. This system has a thermostatic expansion valve.

tem is put back in operation. The compressor is shut off, and the fans allowed to run long enough to defrost the coil. The temperature inside the cooler is 50°F, and because of the amount of food it will take a long while for the cooler to pull down to the cutout point of 35°F. The technician cautions the owner to watch the thermometer in the cooler to make sure that it is pulling down. A call later in the day will confirm that the unit is working properly, Figure 25–40.

Service Call 2

Customer calls. The unit is not cooling, and the compressor is cutting on and off. The

system is a medium-temperature reach-in cooler with the condensing unit on top. *The unit has a thermostatic expansion valve and a low charge. Two tubes rubbed together and created a leak.*

The technician arrives at the job and finds the unit is short cycling on the low-pressure control. The sight glass has bubbles in it, indicating a low charge. The technician discovers that the small tube leading to the low-pressure control has rubbed a hole in the suction line. There is oil around the point of the leak. The system is pumped down by closing the king valve, and the leak is repaired. The system has not operated in a vacuum because of the low-pressure control, so an evacuation is not necessary, Figure 25–41.

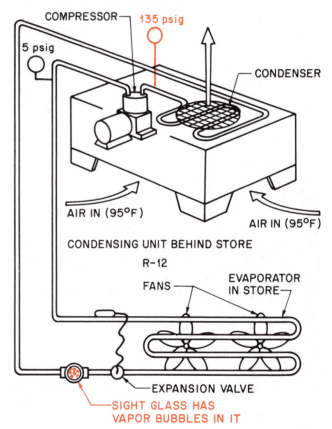

NOTICE THAT THE HEAD PRESSURE IS
DOWN DUE TO REDUCED LOAD.

COMPRESSOR — 135 psig

5 psig

CONDENSER

AIR IN (95°F)

AIR IN (95°F)

CONDENSING UNIT BEHIND STORE

R-12

FANS

EVAPORATOR
IN STORE

EXPANSION VALVE

SIGHT GLASS HAS
VAPOR BUBBLES IN IT

MEDIUM-TEMPERATURE EVAPORATOR SHOULD BE
OPERATING AT 21 psig (20°F BOILING
TEMPERATURE) BUT IT IS OPERATING AT 5 psig
BECAUSE OF A LOW CHARGE CAUSED BY A LEAK.

Figure 25–41. Symptoms of a medium-temperature system operating in a low-charge situation; thermostatic expansion valve.

Since the system had a low-pressure control that cut off the compressor well before the low side went into a vacuum, no atmosphere was pulled into the system. The king valve is closed, the system is pumped down into the condenser, and the leak is repaired. After the repair a leak check is performed. The system is then started back up and charged to the correct charge. A call later in the day will verify the system is functioning correctly.

Service Call 3

Customer calls and reports that the unit is off, and the food is thawing out in the low-temperature walk-in cooler. This cooler lost its charge in the winter, the leak was repaired and the unit recharged. *The unit has a thermostatic expansion valve and an overcharge of refrigerant. This is the first hot day, and the unit is off because of a manual reset high-pressure control.*

The service technician remembers that this unit was serviced by a new technician when the loss-of-charge incident happened. The technician suspects that the unit may be off because of the manual reset high-pressure control. This is the only control on the unit that will keep it off, unless there was a power problem. The technician arrives at the job and finds that the unit is off on the high-pressure control. Before resetting the control, the technician installs a gage manifold so that the pressures can be observed. When the high-pressure control is reset, the compressor starts up, but the head pressure is going up to 375 psig and cutting the compressor off. The system uses R-502. The ambient temperature is 90°F, so the head pressure should not be more than the pressure corresponding to 125°F. This was arrived at by adding 35°F to the ambient temperature: 35 + 90 = 125. The head pressure corresponding to 125°F is 298 psig. It is obvious the head pressure is too high. The condenser seems clean enough, and air is not recirculating back to the condenser inlet, so too much refrigerant is suspected. Refrigerant is let out of the system until the head pressure is down to 298 psig. The sight glass remains full, so the technician knows that the unit still has enough refrigerant. The technician asks the customer to watch the cooler temperature during the next few hours to be sure that it is going down. A call later in the day proves the repair worked because the temperature is going down in the cooler, Figure 25–42.

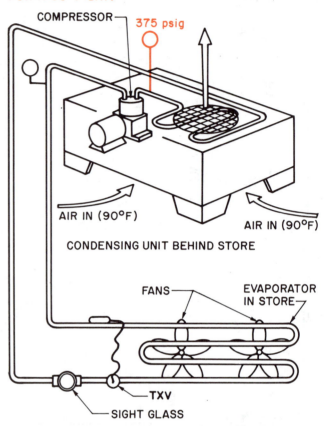

COMPRESSOR — 375 psig

AIR IN (90°F) AIR IN (90°F)

CONDENSING UNIT BEHIND STORE

FANS EVAPORATOR IN STORE

TXV

SIGHT GLASS

LOW-TEMPERATURE SYSTEM USING R-502 WOULD HAVE A TYPICAL LOW-SIDE PRESSURE OF 18 psig (−16°F) AND A HIGH-SIDE PRESSURE OF 301 psig (90°F AMBIENT + 35°F = 125°F)

Figure 25–42. Symptoms of a low-temperature system that has an overcharge of refrigerant. The unit is off on the manual reset high-pressure control.

Service Call 4

Customer calls and says the medium-temperature unit that was just installed is running all the time and not cooling the box. *The thermostatic expansion valve sensing bulb is not secure to the suction line, and the valve is overfeeding the evaporator. This is keeping the suction pressure up and the box will not pull down.*

The technician goes to the job and looks it over. This is a new installation of a walk-in medium-temperature freezer and was just started up yesterday. The hermetic compressor is sweating down the side of the housing. The evaporator fan is operating, and the box temperature is not down to the cutoff point, so there is enough load to boil the refrigerant to a vapor. The expansion valve is overfeeding the evaporator. Before the superheat is adjusted, the sensing bulb location should be checked. Examination shows the bulb is not down tight on the suction line. When the bulb is secured to the suction line and the system is operated for a short period of time, the sweat starts to dry on the side of the compressor. A temperature tester is fastened to the suction line, and gages are installed. The superheat is found to be 10°F at the suction line leaving the evaporator. The unit is left running. A call late in the day verifies the cooler temperature is down to normal, Figure 25–43.

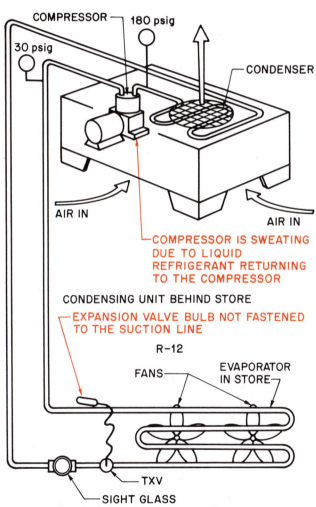

COMPRESSOR — 180 psig

30 psig

— CONDENSER

AIR IN

AIR IN

— COMPRESSOR IS SWEATING DUE TO LIQUID REFRIGERANT RETURNING TO THE COMPRESSOR

CONDENSING UNIT BEHIND STORE

— EXPANSION VALVE BULB NOT FASTENED TO THE SUCTION LINE

R–12

FANS —

EVAPORATOR IN STORE —

— TXV

— SIGHT GLASS

MEDIUM–TEMPERATURE EVAPORATOR SHOULD BE OPERATING AT 21 psig (20°F BOILING TEMPERATURE) BUT IT IS OPERATING AT 30 psig. THIS YIELDS A COIL TEMPERATURE OF 32°F. NOT LOW ENOUGH TO PULL THE REFRIGERATOR BOX TEMPERATURE DOWN.

Figure 25–43. Symptoms of a medium-temperature system with the thermostatic expansion valve overfeeding due to a loose sensing bulb.

Service Call 5

Customer calls and says that the low-temperature reach-in cooler that had a burned compressor last week is not cooling properly. The food is beginning to thaw. *The system has a thermostatic expansion valve and the liquid line drier is stopping up with sludge from the motor burn. A suction line filter-drier was not installed at the time of the compressor change-out.*

The technician arrives and examines the whole system. The evaporator seems to be starving for refrigerant. When the technician approaches the condensing unit in the back of the store, it is noticed that the liquid line is sweating where it leaves the drier. This indicates a restriction at this point. The drier is acting like an expansion valve and flashing some of the liquid refrigerant to a vapor. This starves the expansion device, and it starves the evaporator. The unit is pumped down, and a new liquid line filter-drier is installed. At the same time a suction line drier is installed close to the compressor. The low side of the system is evacuated after a leak test. The system is started back up. It is now realized that the system had much more contamination than was planned for at the compressor changeout. The suction drier could have been installed at the compressor changeout and saved a call back out. Remember, call backs cost the company. The customer is called later in the day and verifies that the cooler temperature is going down, Figure 25–44.

Service Call 6

Customer calls and says the reach-in cooler in the lunch room is not cooling and is running all the time. This is a medium-temperature cooler with the condensing unit under the bottom of the fixture. It has not had a service call in a long time. *The evaporator is dirty and icing over. The unit has a thermostatic expansion valve and does not get any off time for defrost because it never can get the box cool enough.*

The technician arrives and sees that the system is iced up. The compressor is sweating down the side; liquid is slowly coming back to the compressor—not enough to cause a noise from slugging. The thermostatic expansion valve is supposed to maintain a constant superheat, but it may lose control if the pressure drops too low. The first thing that has to be done is to defrost the evaporator. This is done

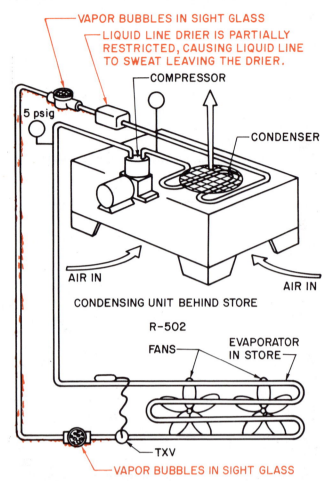

COMPRESSOR

5 psig

CONDENSER

AIR IN

AIR IN

CONDENSING UNIT BEHIND STORE

R-502

FANS

EVAPORATOR IN STORE

TXV

VAPOR BUBBLES IN SIGHT GLASS

LOW-TEMPERATURE COOLER USING R-502 SHOULD HAVE A SUCTION PRESSURE OF ABOUT 18 psig. IT IS MUCH LOWER BECAUSE THE EXPANSION VALVE IS GETTING PART VAPOR AND PART LIQUID INSTEAD OF PURE LIQUID.

Figure 25-44. Symptoms of a low-temperature system with a partially stopped-up liquid line drier; thermostatic expansion valve.

by stopping the compressor and using a heat gun (like a high-powered hair drier). The evaporator has a lot of dirt on it. *The evaporator is cleaned with a coil cleaner that is approved for food-handling areas.* When the evaporator is cleaned, the system is started back up. The unit is now operating with a full sight glass, and the suction line leaving the evaporator is cold. From this point it will take time for the unit to pull the cooler temperature down. The service technician leaves and calls later in the day to confirm the repair, Figure 25-45.

SAFETY PRECAUTIONS: *During this series of service situations, gages will only be installed when necessary. A capillary tube system has a critical charge. A

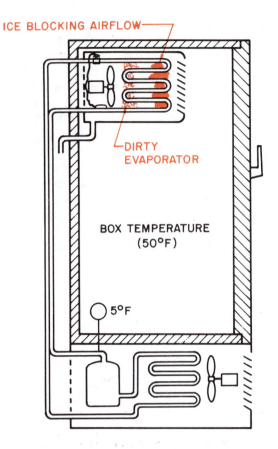

ICE BLOCKING AIRFLOW

DIRTY EVAPORATOR

BOX TEMPERATURE (50°F)

5°F

Figure 25-45. Symptoms of a medium-temperature cooler with a dirty evaporator coil; thermostatic expansion valve.

service technician can install gages on a system with a correct charge, and the amount of refrigerant that enters the gage lines can be enough to affect the charge adversely. The high-side gage in particular will cause high pressure refrigerant to condense in the line. A 3-ft gage line full of liquid may be enough to cause the unit to be undercharged. If a suction gage only is used, it will not alter the charge significantly because it only has low-pressure vapor in it.*

Service Call 7

The customer calls and says the reach-in cooler that stores the dairy products is not cooling properly. The temperature is 55°F. It has been cooling well until early this morning. *There is a leak due to a stress crack (caused by age and vibration) in the suction line near the compressor. The compressor is vibrating because the customer has moved the unit, and the condensing unit has fallen down in the frame. This system has a capillary tube metering device.*

The technician's examination discloses that the compressor is vibrating because the condensing unit is not setting straight in the frame. While securing the condensing unit, the technician notices an oil spot on the bottom of the suction line. Also, the compressor shell is hot. Gages are installed, and it is discovered that the suction pressure is operating in a vacuum. *Care must be used when installing gages if a vacuum is suspected or air may be drawn into the system.* The compressor is stopped, and the low-side pressure is allowed to rise. There is not enough pressure in the system to accomplish a good leak test, so refrigerant is added from a refrigerant tank. A leak check in the vicinity of the suction line reveals a leak. This appears to be a stress crack due to the vibration. The refrigerant is exhausted from the system to repair the leak. **Air must have been pulled in while operating in a vacuum, so the refrigerant charge is wasted. This system does not have a low-pressure control.** A short length of pipe is installed where the stress crack was found. A new liquid line drier is installed because the old one may not have any capacity left. The system is leak checked, triple evacuated, and charged, using a measured charge that the manufacturer recommended. The technician calls the owner later in the day and hears that the unit is functioning correctly, Figure 25-46.

Service Call 8

Customer calls and says the reach-in freezer that is used for ice storage is running all the time. This system was worked on by a competitor in the early spring. A leak was found, repaired, and the unit was recharged. Hot weather is here, and the unit is running all the time. *This system has a capillary tube metering device and there is an overcharge of refrigerant. The other service technician did not measure the charge into the system.*

The service technician examines the fixture and notices that the compressor is sweating down the side of the can. The condenser feels hot for the first few rows and then warm. This appears to be an overcharge of refrigerant. The condenser should be warm near the bottom where the condensing is taking place. The evaporator fan is running, and the evaporator looks clean, so the evaporator must be doing its job. Gages are installed, and the head pressure is 400 psig with an outside temperature of 95°F; the refrigerant is R-502. The head pressure should be no more than 318 psig on a 95°F day (95 + 35 = 130°F condensing temperature or 318 psig). This system calls for

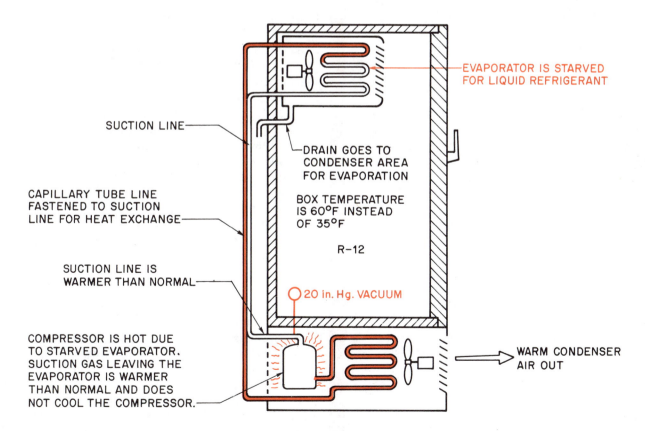

Figure 25-46. Symptoms of a medium-temperature system with a low charge; capillary tube.

a measured charge of 2 lb 8 oz. There are two approaches that can be taken: (1) alter the existing charge; or (2) exhaust the charge and measure a new charge into the unit while the unit is in a deep vacuum. It is a lot of trouble to exhaust the charge and connect the vacuum pump.

The technician chooses to alter the existing charge. This is a plain capillary tube system with a heat exchanger (the capillary tube is soldered to the suction line after it leaves the evaporator). A temperature tester lead is fastened to the suction line **after the evaporator** but **before the heat exchanger.** The suction pressure is checked for the boiling point of the refrigerant and balanced against the temperature tester temperature. The superheat is 0° F with the existing charge of refrigerant. Refrigerant is exhausted until the superheat is 5° F at this point. *Note:* The heat exchanger will allow a lower superheat than normal. When the system charge is balanced, the technician leaves the job and will call the operator later in the day for a report on how the system is functioning, Figure 25–47.

Service Call 9

The customer calls and says the reach-in medium-temperature beverage cooler is running all the time. The suction line is frosting back to the compressor. *A small drier is stopped up with sludge from a compressor changeout after a motor burnout. A suction line drier should have been installed.*

The technician looks the system over closely and sees that the suction line is frosting and that the compressor has a frost patch on it. A first glance might indicate that the fan at the evaporator is off or that the coil is dirty. Further examination shows that the liquid line is frosting starting at the outlet of the drier. This means the drier is partially stopped up. The pressure drop across the drier makes the drier act like an expansion device. This effectively means there are two expansion devices in series because the drier is feeding the capillary tube. The drier is sweated into the liquid line, and there are no service valves. The refrigerant charge has to be exhausted, and the drier replaced. While the system is open, the technician solders a suction line drier in the suction line close to the compressor.

The reason the unit was frosting instead of sweating is that the evaporator was starved for liquid refrigerant. The suction pressure went down to a point below freezing. The unit's

capacity was reduced to the point where it was running constantly and had no defrost. The frost will become denser and block the air through the coil and act as an insulator. This will cause the coil to get even colder with more frost. This condition will continue with the frost line moving on to the compressor. Some people have the mistaken opinion that an iced-up coil will refrigerate. The ice or frost acts as an insulator and an air blockage. Air has to circulate across the coil for the unit to have the rated capacity, Figure 25–48.

Service Call 10

The customer calls and says the reach-in frozen food cooler for ice cream is running. but the temperature is rising. *The system has a capillary tube metering device and the evaporator fan is burned. The customer hears the compressor running and thinks the whole unit is running.*

The technician has never been to this job and has to learn it from the beginning. Upon examining the system, the technician discovers that the frost line on the suction line goes all the way back to the compressor and down the side of the can. The first thought is that the system is not going through the proper defrost cycle. The coil is iced and has to be defrosted before anything can be done. The defrost timer is advanced until the unit goes through defrost. The defrost cycle clears the coil of ice. When defrost is over, the fan does not restart as it should. The technician takes off the panel to the fan compartment and checks the voltage. There is voltage, but the fan will not run. When the motor is checked for continuity through the windings, it is discovered that the fan motor winding is open. The fan motor is changed, and the system is started again. It will take several hours for the fixture to pull back down to the normal running temperature of −10° F. A call later in the day verifies that the cooler's temperature is going down, Figure 25–49.

Service Call 11

Customer calls and says that his reach-in dairy case used for milk storage is running constantly, and the temperature is 48° F. This unit has never had a service call in 10 years of service. *The evaporator is dirty and this unit uses a capillary tube metering device. These coils never have filters, and years of dust will accumulate on the coils.*

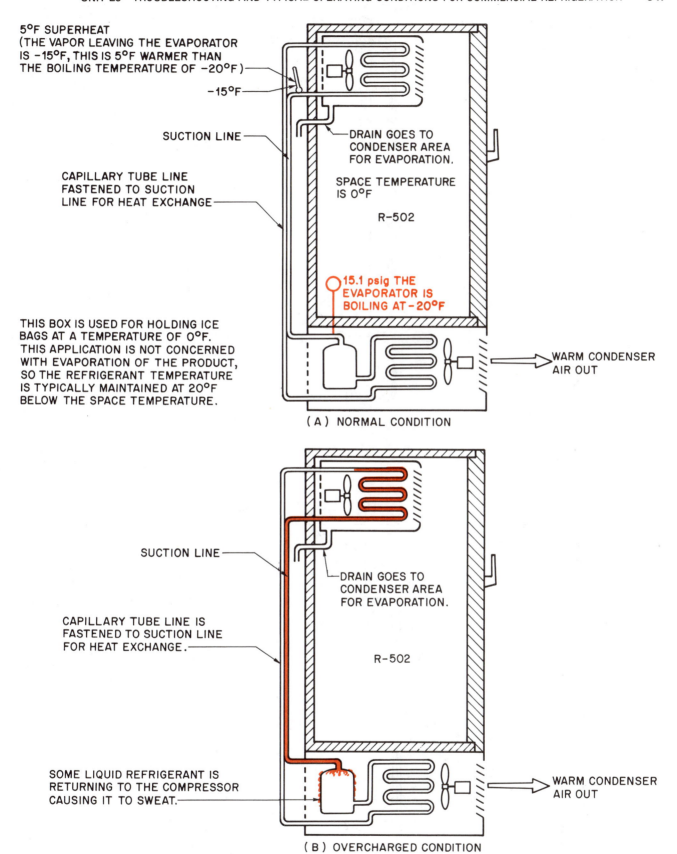

5°F SUPERHEAT
(THE VAPOR LEAVING THE EVAPORATOR
IS -15°F, THIS IS 5°F WARMER THAN
THE BOILING TEMPERATURE OF -20°F)

-15°F

SUCTION LINE

CAPILLARY TUBE LINE
FASTENED TO SUCTION
LINE FOR HEAT EXCHANGE

DRAIN GOES TO
CONDENSER AREA
FOR EVAPORATION.

SPACE TEMPERATURE
IS 0°F

R-502

15.1 psig THE
EVAPORATOR IS
BOILING AT -20°F

THIS BOX IS USED FOR HOLDING ICE
BAGS AT A TEMPERATURE OF 0°F.
THIS APPLICATION IS NOT CONCERNED
WITH EVAPORATION OF THE PRODUCT,
SO THE REFRIGERANT TEMPERATURE
IS TYPICALLY MAINTAINED AT 20°F
BELOW THE SPACE TEMPERATURE.

WARM CONDENSER
AIR OUT

(A) NORMAL CONDITION

SUCTION LINE

DRAIN GOES TO
CONDENSER AREA
FOR EVAPORATION.

CAPILLARY TUBE LINE IS
FASTENED TO SUCTION LINE
FOR HEAT EXCHANGE.

R-502

SOME LIQUID REFRIGERANT IS
RETURNING TO THE COMPRESSOR
CAUSING IT TO SWEAT.

WARM CONDENSER
AIR OUT

(B) OVERCHARGED CONDITION

Figure 25–47. Symptoms of a reach-in freezer with an overcharge of refrigerant; capillary tube.

WHEN THE SYSTEM STAYS ON FOR LONG
PERIODS OF TIME WITH A STARVED
COIL, THE ICE KEEPS MOVING DOWN
THE COIL UNTIL THE WHOLE COIL IS
ICED. THEN THE SUCTION LINE BACK
TO THE COMPRESSOR WILL BECOME
COLD AND MAY EVEN BUILD ICE.

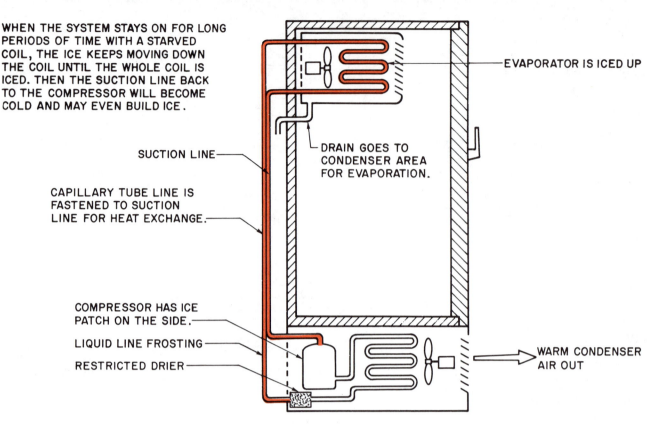

EVAPORATOR IS ICED UP

DRAIN GOES TO
CONDENSER AREA
FOR EVAPORATION.

SUCTION LINE

CAPILLARY TUBE LINE IS
FASTENED TO SUCTION
LINE FOR HEAT EXCHANGE.

COMPRESSOR HAS ICE
PATCH ON THE SIDE.

LIQUID LINE FROSTING

RESTRICTED DRIER

WARM CONDENSER
AIR OUT

Figure 25-48. Symptoms of a medium-temperature reach-in cooler with a partially stopped-up liquid line drier.

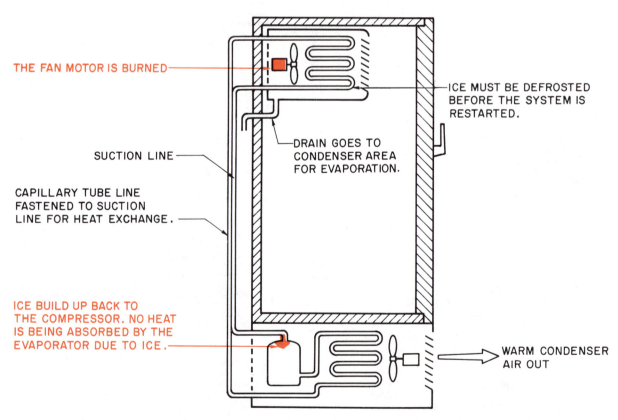

THE FAN MOTOR IS BURNED

ICE MUST BE DEFROSTED
BEFORE THE SYSTEM IS
RESTARTED.

DRAIN GOES TO
CONDENSER AREA
FOR EVAPORATION.

SUCTION LINE

CAPILLARY TUBE LINE
FASTENED TO SUCTION
LINE FOR HEAT EXCHANGE.

ICE BUILD UP BACK TO
THE COMPRESSOR. NO HEAT
IS BEING ABSORBED BY THE
EVAPORATOR DUE TO ICE.

WARM CONDENSER
AIR OUT

Figure 25-49. Symptoms of a reach-in freezer with a burned evaporator fan motor; capillary tube.

The technician finds the suction line very cold. The compressor is sweating down the side of the can. This is evidence of an overcharge, or else the refrigerant is not boiling to a vapor in the evaporator. A close examination of the evaporator indicates the evaporator is not exchanging heat with the air in the cooler. *The evaporator is cleaned with a special detergent approved for evaporator cleaning and for use in food-handling areas.* Areas that have dairy products are particularly difficult because dairy products absorb odors easily. The system is started back up, and the sweat gradually moves off the compressor, and the suction line feels normally cool. The technician leaves the job and will call back later to ask the operator if the unit is performing, Figure 25-50.

Remember, an automatic expansion valve maintains a constant pressure. It responds in reverse to a load change. If a load of additional product is added to the cooler, the rise in suction pressure will cause the automatic expansion valve to throttle back and slightly starve the evaporator. If the load is reduced, such as with a dirty coil, the valve will overfeed to keep the refrigerant pressure up.

Service Call 12

Customer calls and says that a small beverage cooler is not cooling the drinks down to the correct temperature. *A small leak at the flare nut on the outlet to the automatic expansion valve has caused a partial loss of refrigerant.*

The technician arrives and hears the expansion valve hissing. This means the expansion valve is passing vapor along with the liquid it is supposed to pass. The sight glass shows some bubbles. Gages are installed; the suction pressure is normal, and the head pressure is low. The suction pressure is 26 psig; this corresponds to 25°F boiling temperature. The boiling refrigerant normally is about 10°F cooler than the liquid to be cooled. The liquid beverage in the cooler is 50°F and should be 35°F. The head pressure should be about 136 psig, the pressure corresponding to 110°F. The ambient is 75°F, and the condensing temperature should be 110°F (75 + 35 = 110°F). The head pressure is 100 psig. All of these signs point to an undercharge. The technician turns the cooler off and allows the low-side

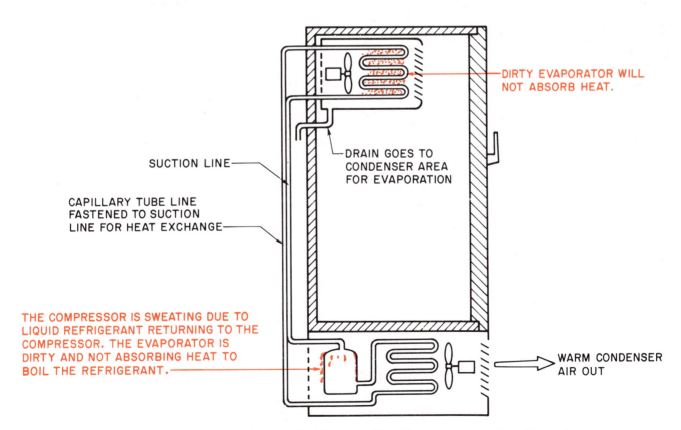

DIRTY EVAPORATOR WILL NOT ABSORB HEAT.

DRAIN GOES TO CONDENSER AREA FOR EVAPORATION

SUCTION LINE

CAPILLARY TUBE LINE FASTENED TO SUCTION LINE FOR HEAT EXCHANGE

THE COMPRESSOR IS SWEATING DUE TO LIQUID REFRIGERANT RETURNING TO THE COMPRESSOR. THE EVAPORATOR IS DIRTY AND NOT ABSORBING HEAT TO BOIL THE REFRIGERANT.

WARM CONDENSER AIR OUT

Figure 25-50. Symptoms of a reach-in dairy case with a dirty evaporator coil; capillary tube.

pressure to rise so that there will be a better chance for a leak to show up. A leak is found at the flare nut leaving the expansion valve. The nut is tightened, but it doesn't stop the refrigerant from leaking. The flare connection must be defective.

The charge is exhausted, and the flare nut is removed. There is a crack in the tubing at the base of the flare nut. The flare is repaired, and the system is leak checked and triple evacuated. A new charge of refrigerant is measured into the system from a vacuum. This gives the most accurate operating charge. The system is started back up. A call back later is recommended, Figure 25–51.

Service Call 13

Customer calls and says the compressor in the pie case is cutting off and on and sounds like it is in a strain while it is running. This unit was charged after a leak, then shut down and put in storage. It has not been operated in several months. *The unit has an automatic expansion valve and has an overcharge of refrigerant. When the unit was charged it was in the back room in the winter, and it was cold in the room. The unit was charged to a full sight glass with a cold condenser.*

The service technician remembers the unit was started up and charged in a cold ambient and may have an overcharge of refrigerant. When the technician arrives, the unit is started up in the location where it is going to stay; the ambient is warm. Gages are installed, and the head pressure is 250 psig. The pressure should be 147 psig at the highest because the ambient

temperature is 80°F. This should create a condensing temperature of no more than 115°F (80 + 35 = 115°F) condensing temperature or 147 psig. The compressor is cutting off because of high pressure at 260 psig. Refrigerant is exhausted from the machine until the head pressure is down to 147 psig, the correct head pressure for the ambient temperature. A call back to the owner later in the day verifies the system is working correctly, Figure 25–52.

Service Call 14

Customer calls and says the reach-in cooler is rising in temperature, and the unit is running all the time. It is medium temperature and should be operating between 35° and 45°F and uses an automatic expansion valve. This unit has not had a service call in three years. *The evaporator fan motor is not running. This system has off-cycle defrost.*

The service technician arrives at the job and examines the system. The compressor is on top of the fixture and is sweating down the side of the can. Since this uses an automatic expansion valve, the sweating is a sign that the evaporator is not boiling the refrigerant to a vapor. The evaporator could be dirty or the fan may not be moving enough air. After removing the fan panel, the technician sees that the evaporator fan motor is not running. The blades are hard to turn, indicating the bearings are tight. The bearings are lubricated, and the fan is started up and seems to run like it should. This is a nonstandard motor and cannot be purchased locally. The system is left in

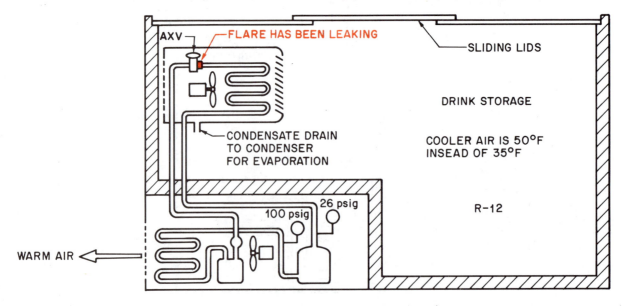

Figure 25–51. Symptoms of a reach-in beverage cooler with a low charge; automatic expansion valve.

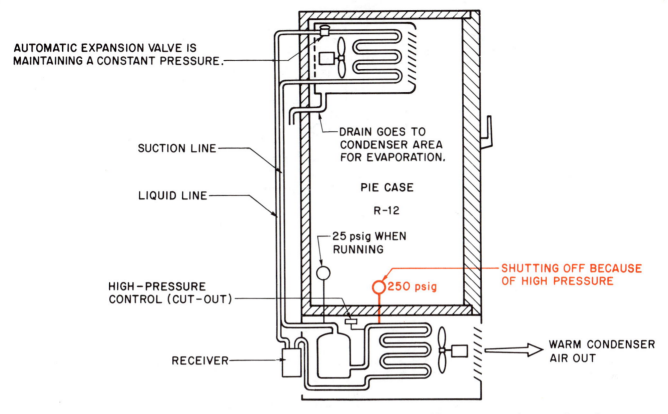

Figure 25–52. Symptoms of a reach-in pie case with an overcharge of refrigerant; automatic expansion valve.

running condition until a new fan motor can be obtained.

The technician returns in a week with the correct motor and exchanges it for the old one. The system has been working correctly all week. The old motor would fail again because the bearings are scored. This is the reason for exchanging a working motor for a new one, Figure 25–53.

Service Call 15

Customer calls and says that a reach-in medium-temperature cooler is rising in temperature. This unit has been performing satisfactorily for several years. *A new stockclerk has loaded the product too high, and the product is interfering with the airflow of the evaporator fan. The system has an automatic expansion valve.*

The technician arrives at the job and notices the product is too high in the product area. The extra product is moved to another cooler. The stock clerk is shown the load level lines in the cooler. A call back later in the day proves the cooler is now working correctly, Figure 25–54.

Service Call 16

Customer calls on the first hot day in the spring and says the walk-in freezer is rising in temperature. The food will soon start to thaw if something is not done soon. *The condenser is dirty. This is an air-cooled condenser in the back of the store.*

The service technician arrives, examines the whole system, and notices that the liquid line is hot, not warm. This is a thermostatic expansion valve system and has a sight glass. It is full, indicating a full charge of refrigerant. The compressor is cutting off because of high pressure and then restarting periodically. This system uses R-502, and the head of the compressor is painted purple to signify the refrigerant type. Gages are installed, and the head pressure is starting out at 345 psig and rising to the cutout point of 400 psig. The head pressure should be 318 psig or a condensing temperature of 130°F because it is 95°F (95 + 35 = 130°F or 318 psig for R-502). The condenser coil appears to have a lot of lint and dust built up between the fins. The unit is stopped long enough to clean the condenser and then it is restarted. When the unit is restarted and the condenser has time to dry from

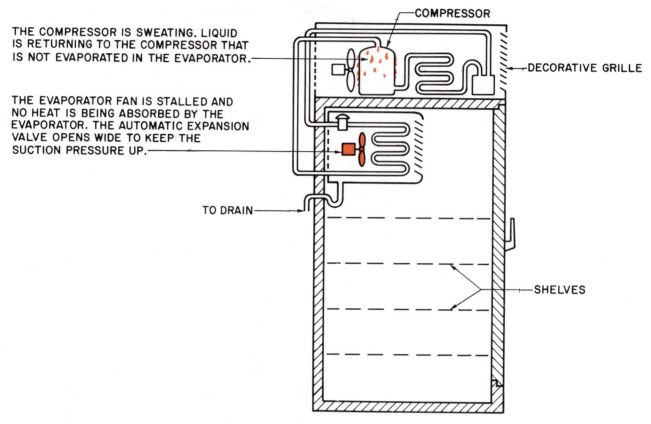

Figure 25-53. Symptoms of a reach-in cooler with a defective fan motor; automatic expansion valve.

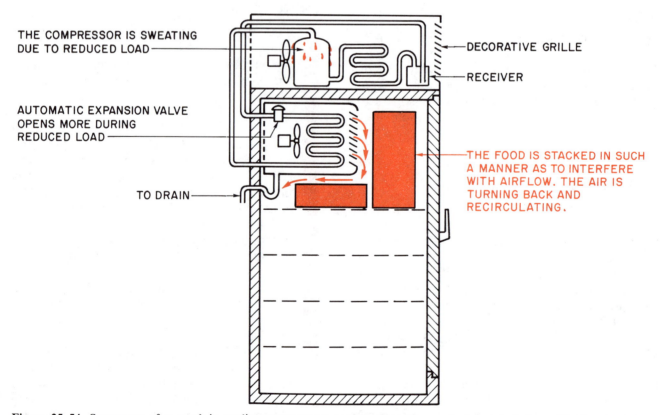

Figure 25-54. Symptoms of a reach-in medium-temperature cooler where the product is interfering with the air pattern; automatic expansion valve.

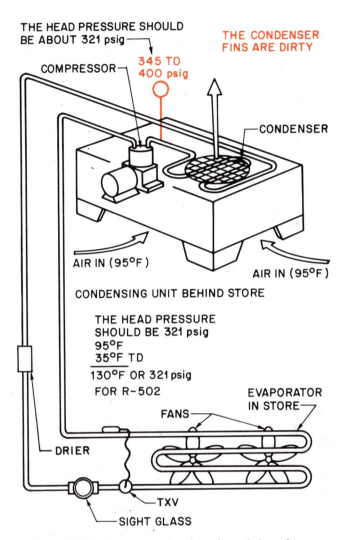

THE HEAD PRESSURE SHOULD BE ABOUT 321 psig

THE CONDENSER FINS ARE DIRTY

345 TO 400 psig

COMPRESSOR

CONDENSER

AIR IN (95°F)

AIR IN (95°F)

CONDENSING UNIT BEHIND STORE

THE HEAD PRESSURE SHOULD BE 321 psig
95°F
35°F TD
130°F OR 321 psig
FOR R-502

FANS

EVAPORATOR IN STORE

DRIER

TXV

SIGHT GLASS

Figure 25–55. Symptoms of a dirty air-cooled condenser.

the cleaning, the head pressure drops back to 318 psig, the correct head pressure for the conditions. The moisture on the coil after cleaning will give a false indication of head pressure because of the evaporation on the hot coil. A call back to the operator later confirms that the system is dropping in temperature, Figure 25–55.

Service Call 17

Customer says the condensing unit on the medium-temperature walk-in cooler is running all the time, and the cooler is rising in temperature. *This is the first hot day, and some boxes have been stacked too close to the condenser outlet. The hot air is leaving the condenser and recirculating back into the fan inlet.*

The service technician arrives and looks over the system. The temperature is 52°F in-

side the cooler. The liquid line temperature is hot instead of warm. This is a thermostatic expansion valve system, and the sight glass is clear. The unit appears to have a full charge and running a high head pressure. Gages are installed, and the head pressure is 275 psig. The system is using R-12 as the refrigerant. The head pressure should be no more than 170 psig, which is a condensing temperature of 125°F; it is 90°F (90 + 35 = 125°F or 170 psig). Further examination of the job indicates that air is leaving the condenser and recirculating back to the condenser inlet. There are several boxes that have been stored in front of the condenser. These are moved, and the head pressure drops to 170 psig. The reason there has not been a complaint up to now is that the air has been cold enough in the previous mild weather to keep the head pressure down even with the recirculation problem. A call back later in the day shows that the cooler temperature is back to normal, Figure 25–56.

Service Call 18

Customer says a water-cooled compressor is cutting off from time to time, and the walk-in cooler is losing temperature. This is a medium-temperature cooler. *A dirty water-cooled condenser. This condenser has been in service for several years and has not been cleaned.*

The service technician arrives to find a waste-water condenser (the water is regulated by a water regulating valve to control the head pressure and then goes down the drain). The water is coming out of the condenser and going down the drain at a rapid rate. The liquid line is hot, not warm. The water is not taking the heat out of the refrigerant like it should. Gages are installed on the compressor, and the head pressure is 200 psig. The compressor is cutting off and on because of the high-pressure control. The head pressure should be about 125 psig with a condensing temperature of 105°F. This condenser has removable heads, so the water circuit can be cleaned. The heads are removed, and the condenser is found to be dirty. The tubes are brushed with a nylon brush to remove the film. The heads are fastened back in place, and the water is turned on. The system is restarted. The customer is called later in the day, and the cooler is working properly, Figure 25–57.

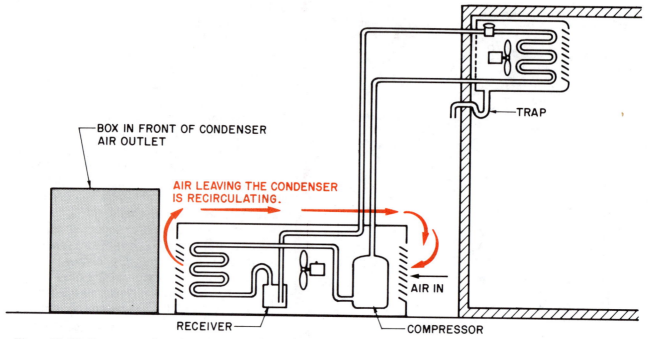

Figure 25–56. Symptoms of an air-cooled condenser with the hot discharge air recirculating back into the condenser inlet.

Service Call 19

Customer calls and says the temperature is going up in the low-temperature freezer. This is a walk-in cooler. *The defrost time-clock motor is burned, and the system will not go into defrost.*

The service technician arrives, examines the system, and sees that the evaporator coil is iced up with thick ice. It is evident that defrost has not been working. The first thing to do is to force a defrost. The technician goes to the timer and when the clock dial is examined, it is found that the time indicator says 4:00 A.M.,

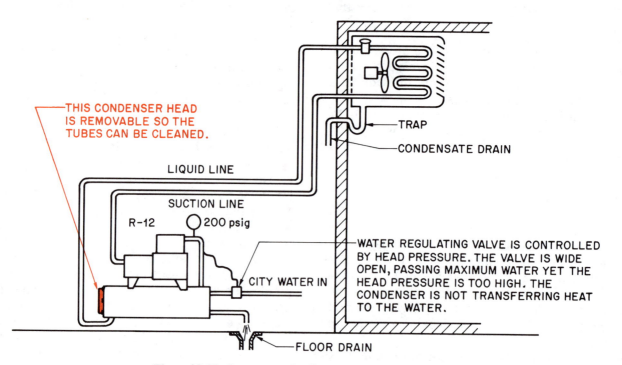

Figure 25–57. Symptoms of a dirty water-cooled condenser.

THE SUCTION PRESSURE SHOULD BE ABOUT 18 psig (−16°F) FOR A 0°F FREEZER TEMPERATURE. THIS FREEZER IS WARMER THAN 0°F AND THE SUCTION PRESSURE IS LOW.

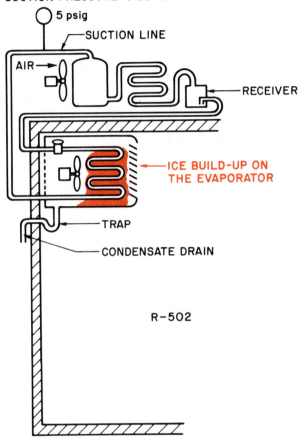

Figure 25–58. Symptoms of a defrost problem with a defective time clock.

but it is 2:00 P.M. Either the power has been off, or the timer motor is not advancing the timer. The technician advances the timer by hand until the defrost cycle starts and then marks the time. After the defrost is terminated by the temperature-sensing device, the system goes into normal operation. The technician gives the clock about one-half hour and sees no advancement in the time. A new time clock is installed. The customer is cautioned to look out for heavy ice buildup. A call the next day verifies that the system is working correctly, Figure 25–58.

Service Call 20

Customer calls and says the defrost seems to be lasting too long in the low-temperature walk-in cooler. It used to defrost, and the fans would start back up in about 10 minutes. It is now taking 30 minutes. *The defrost termination switch is not terminating defrost with the*

temperature setting. This is an electric defrost system. The system is staying in defrost until the timed override in the timer takes it out of defrost. This also causes the compressor to operate at too high a current because the fans are not supposed to come on until the coil cools to below 30°F. With the defrost termination switch stuck in the cold position, the fans will come on when the timer terminates defrost, and the compressor will be overloaded by the heat left in the coil from the defrost heaters.

The technician arrives at the job and examines the system. The coil is free of ice so defrost has been working. The cooler is cold, −5°F. The technician advances the time clock to the point that defrost starts. The termination setting on the defrost timer is 30 minutes. This may be a little long unless the cooler is only defrosted once a day. The timer settings are to defrost twice a day, at 2:00 P.M. and 2:00 A.M. The coils are defrosted in a matter of 5 minutes, but the defrost continues. The technician allows defrost to continue for 15 minutes to allow the coil to get to maximum temperature, so the defrost termination switch will have a chance to make. After 15 minutes the defrost is still going on, so the timer is advanced to the end of the timed cycle. When the system goes out of defrost, the fans start back up. They should not start until the coil gets down to below 30°F. The defrost termination switch is removed. It has three terminals, and the wires are marked for easy replacement of a new control. A new control is acquired and installed. The system is started up. The owner calls the next day and says that he observed a normal defrost, and the system is acting normally. The old defrost termination switch is checked out with an ohmmeter after having been room temperature for a long time, and it is found that the cold contacts are still closed. The control is definitely defective, Figure 25–59.

Service Call 21

Customer calls and says the walls between the doors are sweating on the reach-in freezer. This has never happened. The freezer is about 15 years old. *The mullion heater in the wall of the cooler that keeps the panel above the dew-point temperature of the room is not functioning.*

The technician knows what the problem is before going to the job. After arriving at the

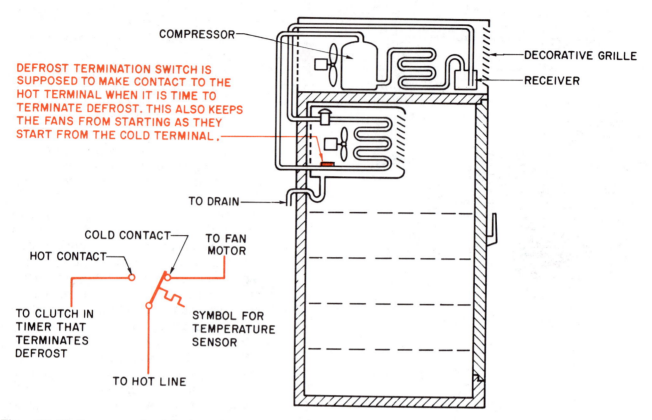

COMPRESSOR

DEFROST TERMINATION SWITCH IS SUPPOSED TO MAKE CONTACT TO THE HOT TERMINAL WHEN IT IS TIME TO TERMINATE DEFROST. THIS ALSO KEEPS THE FANS FROM STARTING AS THEY START FROM THE COLD TERMINAL.

DECORATIVE GRILLE

RECEIVER

TO DRAIN

COLD CONTACT

HOT CONTACT

TO FAN MOTOR

TO CLUTCH IN TIMER THAT TERMINATES DEFROST

SYMBOL FOR TEMPERATURE SENSOR

TO HOT LINE

Figure 25–59. Symptoms of a defective defrost termination thermostat (stuck with cold contacts closed) in a low-temperature freezer.

job, the technician examines the fixture for the reason the mullion heater is not getting hot. If the heater is burned out, the panel needs to be removed. The circuit is traced to the back of the cooler, and the wires to the heater are found. An ohm check proves the heater is still in the circuit; it has continuity. A voltage check shows there is no voltage going to the heater. After further tracing by the technician, a loose connection is located in a junction box. The wires are fastened together, and a current check is performed to prove the heater is working, Figure 25–60.

Service Call 22

Customer calls and says the medium-temperature walk-in cooler is off. The breaker was off, and they tried resetting it, but it tripped again. *The compressor motor is burned.*

The technician arrives at the job and examines the system. The food is warming up, and the cooler is up to 55°F. Before resetting the breaker, the technician uses an ohmmeter and finds the compressor has an open circuit through the run winding. There is nothing that can be done except change the compressor. A refrigerant line is opened slightly, and the

refrigerant is smelled to see if the burn is bad or mild. The refrigerant has a high acid odor, indicating a bad motor burn. The owner is told that it will be several hours before the system

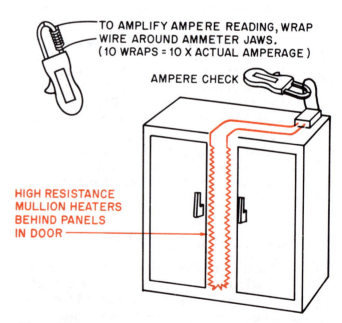

TO AMPLIFY AMPERE READING, WRAP WIRE AROUND AMMETER JAWS. (10 WRAPS = 10 X ACTUAL AMPERAGE)

AMPERE CHECK

HIGH RESISTANCE MULLION HEATERS BEHIND PANELS IN DOOR

Figure 25–60. Symptoms of a defective mullion heater on a low-temperature freezer.

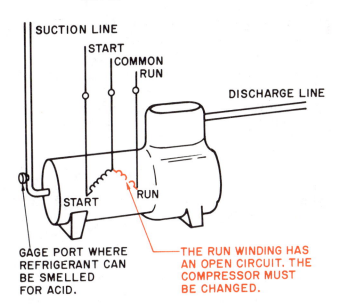

SUCTION LINE

START

COMMON

RUN

DISCHARGE LINE

START   RUN

GAGE PORT WHERE
REFRIGERANT CAN
BE SMELLED
FOR ACID.

THE RUN WINDING HAS
AN OPEN CIRCUIT. THE
COMPRESSOR MUST
BE CHANGED.

Figure 25–61. Symptoms of a system with a burned compressor motor.

can be put back into service, so he moves the food to another cooler.

The technician goes to the supply house and picks up the required materials, including a suction line filter-drier with high acid-removing qualities. This drier has removable cores for easy changeout. The compressor is changed, and the suction line drier is installed. The system is purged with dry nitrogen to push as much of the free contaminants as possible out of the system. The system is leak checked, and a triple vacuum is pulled. The system is then charged and started up. After the compressor runs for an hour, an acid test is taken on the crankcase oil. It shows a slight acid count. The unit is run for another four hours, and an acid check shows even less acid. The system is pumped down, and the cores are changed in the suction line drier. The liquid line drier is changed, and the system is allowed to operate overnight.

The technician returns the next day and takes another acid check. It shows no sign of acid. The system is left to run in this condition wih the drier cores still in place in case some acid turns loose at a later date. *The motor burn is attributed to a random motor failure,* Figure 25–61.

Service Call 23

The customer calls and says the walk-in freezer for the frozen food is going up in temperature. Although it tries, the compressor will not start. *The start relay coil is burned. The*

compressor is trying to start, and the overload is cutting it off.

The service technician hears the compressor try to start, and the overload cuts it off, making a clicking sound. The compressor is hot, too hot to start many times before the internal winding thermostat will cut it off and keep it off for a long period of time. This is because every start the motor goes through makes it hotter. The technician needs to make sure that it starts and stays on the line the next time it is started. This is a single-phase compressor and has a starting relay and starting capacitor. The problem can be in several places: (1) the starting capacitor, (2) the starting relay, or (3) the compressor. The technician uses an ohmmeter to check for continuity in the compressor, from the common terminal to the run terminal, and from the common terminal to the start terminal. The compressor has continuity and should start, provided it is not stuck.

An ohm check of the starting capacitor shows that it will charge and discharge. (This test is accomplished by first shorting the capacitor terminals out and then using the R × 100 scale. The capacitor should cause the meter to rise and then fall. Reversing the leads will cause it to rise and fall again.) *Note:* This check does not indicate the actual capacity of a capacitor. It indicates that the capacitor will charge and discharge.

The starting relay is checked for continuity from terminal 1 to terminal 2. This is the circuit the starting capacitor is energized through. The circuit shows no resistance; the contacts are good. The coil of the relay is checked and found to be an open circuit (the coil circuit is from terminal 2 to terminal 5). This coil will not pick up the contacts and take the starting capacitor and starting winding out of the circuit when the motor gets up to speed. The compressor is drawing high current and cutting off because of an overload. A new starting relay is installed, and the compressor is turned on again. An ammeter is clamped on the wire leading to the common terminal to make sure the compressor is operating at the correct current. It does. A call later in the day proves the compressor is stopping and starting correctly, Figure 25–62.

If a starting relay had not been available to get the compressor back on the line, the technician could have started the compressor by using an approved start accessory cable. This can be done to keep the food from thawing out. *The starting winding has to be discon-*

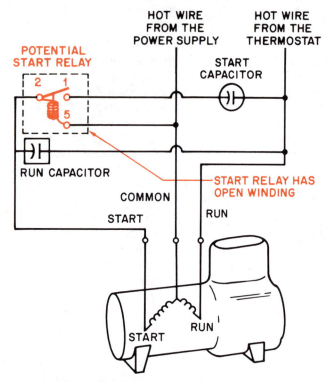

Figure 25-62. Symptoms of a system with a defective starting relay coil.

nected from the circuit in less than 5 sec. This will work in an emergency. If the compressor stops for any reason, it will not restart. There are times when the compressor is going to run for a long period after startup because the refrigerated box is warm. If you know that the compressor is going to have a long running period, leave it running, and the operator can watch the system until a relay can be obtained. *This type of service must only be performed by a skilled technician or under the supervision of a skilled technician. It should only be performed in emergency situations.*

Service Call 24

The customer calls and says the medium-temperature reach-in cooler is not cooling properly. The temperature is 55°F inside, and it should be no higher than 45°F. The compressor sounds like it is trying to start but then cuts off. *The starting capacitor is defective. It has an open circuit. The compressor will not start and is going off because of overload.*

The technician arrives at the job in time to hear the compressor try to start. As in Problem

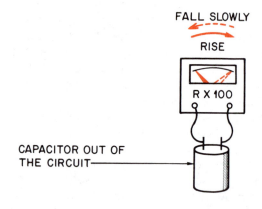

GOOD CAPACITOR

THE CAPACITOR HAS BEEN CHECKED TO SEE THAT IT IS NOT CHARGED BY PLACING A SCREWDRIVER BLADE FROM TERMINAL TO TERMINAL. (A RESISTOR IS RECOMMENDED FOR THIS BUT MOST TECHNICIANS DO NOT CARRY RESISTORS).

1. TOUCH ONE LEAD, THEN THE OTHER TO THE CAPACITOR TERMINALS. THE NEEDLE SHOULD RISE TO THE 0 RESISTANCE SIDE THEN FALL BACK.

2. IF THERE IS A RESISTOR BETWEEN THE TERMINALS THE NEEDLE WILL FALL BACK TO THE VALUE OF THE RESISTOR (KNOWN AS A BLEED RESISTOR).

3. TO PERFORM THE TEST AGAIN, THE LEADS MUST BE REVERSED. THE METER'S BATTERY IS DIRECT CURRENT. YOU WILL NOTICE AN EVEN FASTER RISE IN THE NEEDLE THE SECOND TIME BECAUSE THE CAPACITOR HAS THE METER'S BATTERY CHARGE.

DEFECTIVE CAPACITOR

BLEED THE CAPACITOR WITH A SCREWDRIVER

1. TOUCH THE METER LEADS TO THE CAPACITOR TERMINALS. IF IT IS DEFECTIVE, IT WILL NOT RISE ANY HIGHER THAN THE VALUE OF THE BLEED RESISTOR.

Figure 25-63. Symptoms of a system with a defective starting capacitor.

24, there are several things that can keep the compressor from starting. It is best to give the compressor the benefit of the doubt and assume it is good. Check the starting components first and then check the compressor. The starting capacitor is removed from the circuit for checking. The capacitor has discoloration around the vent at the top, and the vent is pushed upward. These are signs the capacitor is bad. An ohm test shows the capacitor open. The capacitor is replaced with a similar capacitor, and the compressor is started. The compressor is allowed to run for several minutes to allow the suction gas to cool the motor; then it is stopped and restarted to make sure that it is operating correctly. A call back the next day proves that the compressor is still stopping and starting correctly, Figure 25–63.

Service Call 25

Customer calls and says the compressor in the low-temperature walk-in freezer is not starting. The box temperature is 0°F, and it normally operates at −10°F. *The compressor is stuck. This is a multiple-evaporator installation with four evaporators piped into one suction line. An expansion valve on one of the evaporators has been allowing a small amount of liquid refrigerant to get back to the compressor. This has caused marginal lubrication, and the compressor has scored bearings that are bad enough to lock the compressor.*

The service technician arrives and examines the job. The compressor is a three-phase compressor and does not have a starting relay or starting capacitor. It is important that the compressor be started within 5 hours, or the food must be moved. The technician turns off the power to the compressor contactor. The leads are then removed from the load side of the contactor. The motor is checked for continuity with an ohmmeter. The motor windings appear to be normal. *Note:* Some compressor manufacturers furnish data that tells what the motor-winding resistances should be. The meter is then turned to the voltage selection for a 230-V circuit. The contactor is then energized to test the load side of the starter to make sure that each of the three phases has the correct power. *Note:* This is a no-load test. The leads have to be connected to the motor terminals and the voltage applied to the leads while the motor is trying to start before the technician knows for sure there is a full 230-V under the starting load.

The technician connects the motor leads back up and gets two more meters for checking voltage. This allows the voltage to be checked from phase 1 to phase 2, phase 1 to phase 3 and phase 2 to phase 3 at the same time while the motor is being started. When the power is applied to the motor, there is 230 V from each phase to the other, and the motor will not start. **This is conclusive. The motor is stuck.** The motor can be reversed (by changing any two leads, such as L1 and L2) and it may start, but it is not likely that it will run for long.

The technician now has four hours to either change the compressor or move the food products. A call to the local compressor supplier shows that a compressor is in stock in town. The technician calls the shop for help. Another technician is dispatched to pick up the new compressor and a crane is ordered to set the new compressor in and take the old one out. The compressor is 30 hp and weighs about 800 lb.

When the crane gets to the job, the old compressor is disconnected and ready to set out. This was accomplished by front seating the suction and discharge service valves to isolate the compressor. When the new compressor is installed, all that will have to be done after evacuating the compressor is open the service valves and start it. The original charge is still in the system. By the time the old compressor is removed, the new compressor arrives. The new compressor is set in place and the service valves are connected. While a vacuum is being pulled, the motor leads are connected. In an hour the new compressor has been triple evacuated and is ready to start.

When the new compressor is started, it is noticed that the suction line is frosting on the side of the compressor. Liquid is getting back to the compressor. A temperature tester lead is fastened to each of the evaporators at the evaporator outlet. It is discovered that one of the evaporator suction lines is much colder than the others. The expansion valve bulb is examined and is found to be loose and not sensing the suction line temperature. Someone has taken the screws out of the mounting strap and used them somewhere else.

The expansion valve bulb is then secured to the suction line, and the system is allowed to operate. The frost line moves back from the side of the compressor, and the system begins to function normally. A trip by the job the next day shows the frost line to the compressor is correct, not frosting the compressor, Figure 25–64.

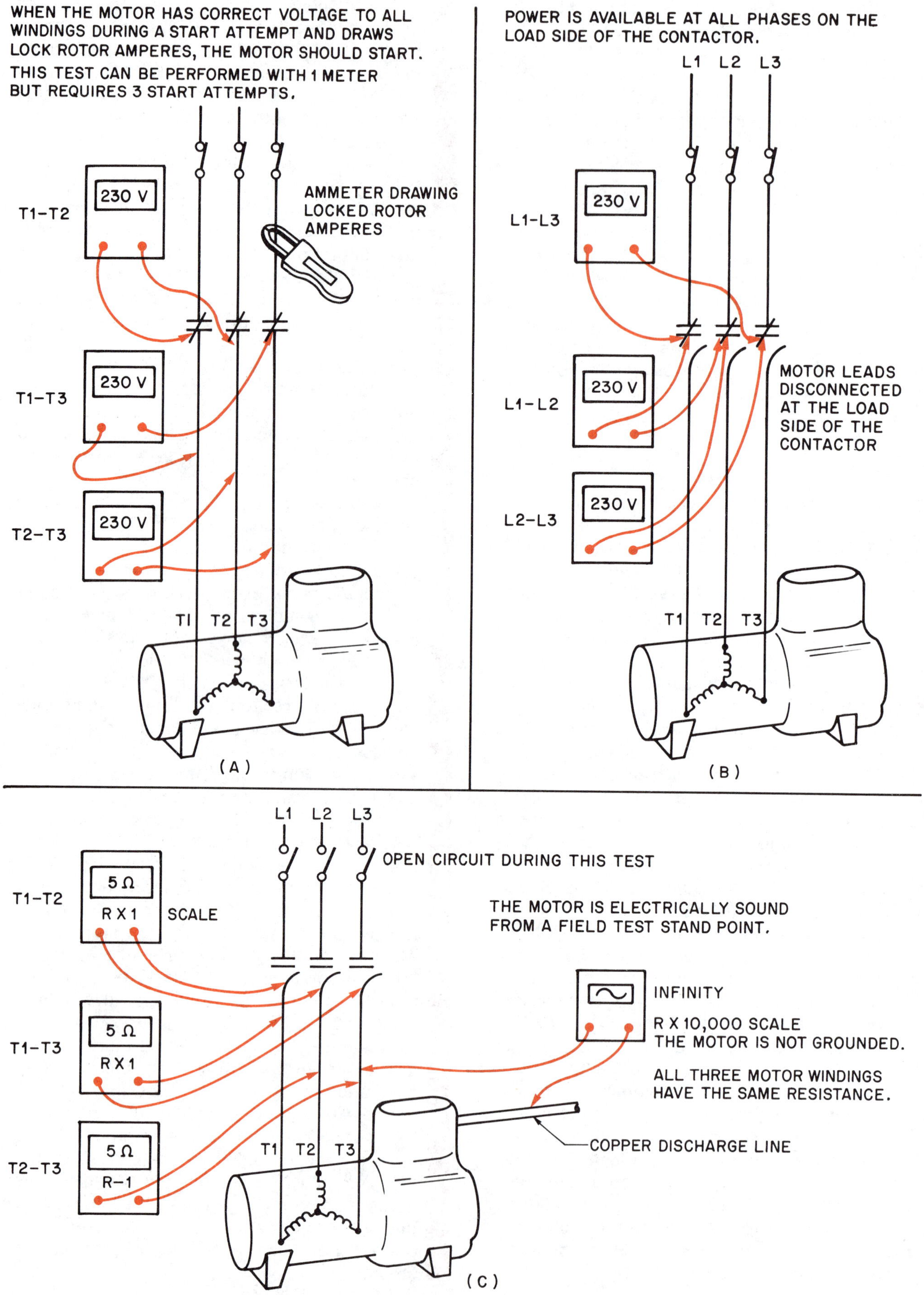

WHEN THE MOTOR HAS CORRECT VOLTAGE TO ALL
WINDINGS DURING A START ATTEMPT AND DRAWS
LOCK ROTOR AMPERES, THE MOTOR SHOULD START.
THIS TEST CAN BE PERFORMED WITH 1 METER
BUT REQUIRES 3 START ATTEMPTS.

POWER IS AVAILABLE AT ALL PHASES ON THE
LOAD SIDE OF THE CONTACTOR.

T1–T2   **230 V**

AMMETER DRAWING
LOCKED ROTOR
AMPERES

T1–T3   **230 V**

T2–T3   **230 V**

T1   T2   T3

(A)

L1 L2 L3

L1–L3   **230 V**

MOTOR LEADS
DISCONNECTED
AT THE LOAD
SIDE OF THE
CONTACTOR

L1–L2   **230 V**

L2–L3   **230 V**

T1   T2   T3

(B)

L1 L2 L3

T1–T2   **5 Ω**
R X 1   SCALE

OPEN CIRCUIT DURING THIS TEST

THE MOTOR IS ELECTRICALLY SOUND
FROM A FIELD TEST STAND POINT.

T1–T3   **5 Ω**
R X 1

∿   INFINITY

R X 10,000 SCALE
THE MOTOR IS NOT GROUNDED.

ALL THREE MOTOR WINDINGS
HAVE THE SAME RESISTANCE.

T2–T3   **5 Ω**
R–1

T1   T2   T3

COPPER DISCHARGE LINE

(C)

Figure 25-64. Symptoms of a system with a three-phase compressor that is stuck.

SUMMARY

- To begin to troubleshoot you need to know what the equipment performance is supposed to be.
- Usually only one component fails at a time. If there is more than one component defective, it is because one component caused the others to fail. There is normally a sequence of events.
- The technician should know how the equipment should sound, where it should be cool, and where it should be warm. This comes with experience.
- *The technician should look the whole system over before doing anything.*
- All evaporators that cool air have a relationship with the air they are cooling. The coil will generally boil the refrigerant between 10°F and 20°F colder than the air entering the evaporator.
- When higher humidities are desired inside the cooler to prevent product dehydration, a 10°F coil-to-air relationship is customarily used.
- High-temperature evaporators normally operate between 25°F and 40°F boiling temperature of the liquid refrigerant.
- Medium-temperature evaporators normally operate between 10°F and 25°F boiling refrigerant temperature of the liquid refrigerant.
- Low-temperature applications are in two areas: Ice making and food storage.
- Ice making normally starts at a coil temperature of about 20°F.
- Low-temperature food-storage systems normally operate from −15°F down to about −40°F refrigerant boiling temperature.
- There are tables that show the various storage temperatures for different products for both short- and long-term storage.
- There are three temperatures that should be considered when any evaporator is evaluated: (1) When the compressor is off, just before the thermostat calls for the compressor to start back up. The coil temperature is the same as the box temperature at this point. (2) Just after the compressor starts back up before the box begins to cool down. The compressor is running, so the coil will be cooler than the box temperature. (3) While the compressor is running, just before the thermostat calls for the compressor to stop. The coil temperature will be cooler than the box temperature.
- Typical conditions for the high-pressure side of the system are divided into two categories: air cooled and water cooled.
- There is a relationship between the entering air and the condensing refrigerant. Usually the refrigerant should not condense at more than 35°F higher than the entering air.
- High-efficiency condensers may have lower relationships, down to about 25°F higher than the entering air.
- When the refrigerant is condensing at a temperature greater than 35°F higher than the entering air, something is wrong.
- Water-cooled condensers have a relationship with the leaving water. The condensing temperature is usually 10°F warmer than the leaving water.
- A low refrigerant charge is indicated by a low suction pressure and a low head pressure when the metering device is a thermostatic expansion valve or a capillary tube.
- When a low charge is encountered and the metering device is an automatic expansion valve, low head pressure and bubbles in the sight glass will occur.
- Hissing can be heard during low charge operation at the expansion valve when a thermostatic or automatic expansion valve is the metering device.
- The automatic expansion valve maintains a constant low-side pressure. This is the reason that the low-side pressure does not go down during low charge operation as with a capillary tube and a thermostatic expansion valve.
- An overcharge of refrigerant always causes an increase in head pressure.
- An increase in head pressure causes an increase in suction pressure when the thermostatic expansion valve or the capillary tube is the metering device.
- *An increase in head pressure causes more refrigerant to flow through a capillary tube. Liquid refrigerant may flood into the compressor with an overcharge.*
- The automatic expansion valve responds in reverse to load changes: When the load is increased, the valve throttles back and will slightly starve the evaporator. When the load is decreased excessively, the valve is subject to overfeed the coil and may allow small amounts of liquid refrigerant to enter the compressor.
- When an inefficient evaporator is encountered, the refrigerant boiling temperature will be too low for the entering air temperature.
- When an inefficient condenser is encountered, the condensing refrigerant temperature will be too high for the heat rejection medium. For air-cooled equipment the condensing refrigerant should be no more than 35°F higher than the entering air. For water-cooled equipment the condensing refrigerant temperature should be no more than 10°F higher than the leaving water.
- The best check for an inefficient compressor is to see if it will pump to capacity.
- Design load conditions can almost always be duplicated by building the head pressure up and duplicating the design suction pressure.

REVIEW QUESTIONS

1. What is the first thing a technician should know before beginning troubleshooting?
2. Name the refrigerated space temperatures that apply to the three temperature ranges.
3. What is the coil-to-air temperature relationship for a refrigeration system designed for minimum food dehydration?
4. When food dehydration is not a factor, what is the coil-to-air temperature relationship?
5. How is minimum food dehydration accomplished?
6. Why is minimum dehydration not used on every job?
7. What is the coil-to-air temperature relationship for an average coil where dehydration is not a factor?
8. At what temperature does an ice maker usually begin to make ice?
9. What is the evaporator coil-to-air temperature relationship when the compressor is not running?
10. What is the coil-to-air temperature relationship between the entering air and the condensing temperature for an air-cooled condenser?
11. What is the leaving-water temperature to the condensing temperature relationship for a water-cooled condenser?
12. How can an evaporator be tested for efficiency?
13. How can a condenser be tested for efficiency?
14. How can a hermetic compressor be tested for efficiency?

How do the following systems respond to an overcharge?

15. Thermostatic expansion valve
16. Automatic expansion valve
17. Capillary tube

How do the following systems respond to an undercharge?

18. Thermostatic expansion valve
19. Automatic expansion valve
20. Capillary tube

TYPES OF ELECTRIC MOTORS

OBJECTIVES

After studying this unit, you should be able to

- describe the different types of open single-phase motors used to drive fans, compressors, and pumps.
- describe the applications of the various types of motors.
- state which motors have high starting torque.
- list the components that cause a motor to have a higher starting torque.
- describe a multispeed PSC motor and describe how the different speeds are obtained.
- explain the operation of a three-phase motor.
- describe a motor used for a hermetic compressor.
- explain the motor terminal connections in various compressors.
- describe the different types of compressors that use hermetic motors.

26.1   USES OF ELECTRIC MOTORS

Electric motors are used to turn the prime movers of air, water, and refrigerant. The prime movers are the fans, pumps, and compressors, Figure 26–1. Several types of motors, each with its particular use, are available. For example, some applications need motors that will start under heavy loads and still develop their rated work horsepower at a continuous running condition, whereas others are used in installations that don't need much starting torque but must develop their rated horsepower under a continuous running condition. Some motors run for years in dirty operating condi-

tions, and others operate in a refrigerant atmosphere. These are a few of the typical applications of motors in this industry. The technician must understand which motor is suitable for each job so that effective troubleshooting can be accomplished and, if necessary, the motor replaced by the proper type. But the basic operating principles of an electric motor must be first understood. Although there are many types of electric motors, most motors operate on similar principles.

26.2   PARTS OF AN ELECTRIC MOTOR

Electric motors have a *stator* with windings, a *rotor, bearings, end bells,* and some means to hold these parts in the proper position, Figures 26–2 and 26–3.

26.3   ELECTRIC MOTORS AND MAGNETISM

It would be beneficial to review the paragraphs on magnetism in Unit 11 before proceeding with this section. Unlike poles of a magnet attract each other, and like poles repel each other. If a stationary horseshoe magnet were placed with its two poles (north and south) at either end of a free-turning magnet as in Figure 26–4, one pole of the free rotating magnet would line up with the opposite pole of the horseshoe magnet. If the horseshoe magnet were an electromagnet and the wires on the battery were reversed, the poles of this magnet would reverse and the poles on the free magnet would be repelled, causing it to rotate until the unlike poles again were lined up. This is the basic principle of how an electric motor operates. The horseshoe magnet is the stator, and the free rotating magnet the rotor.

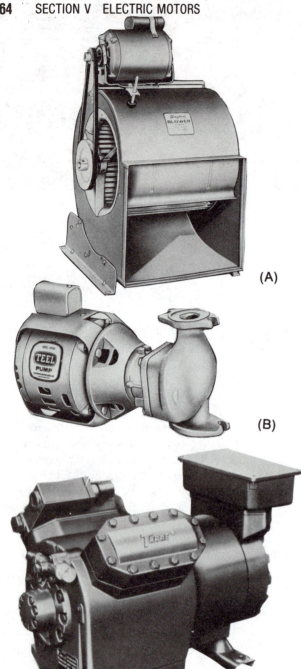

Figure 26–2. Cutaway of an electric motor. *Courtesy Century Electric, Inc.*

netic field is produced in the windings and a magnetic field is also induced in the rotor. The bars in the rotor actually form a coil. This is similar to the field induced in a transformer secondary by the magnetic field in the transformer primary. The field induced in the rotor has a polarity opposite that in the running windings.

The attracting and repelling action between the poles of the running windings and the rotor sets up a rotating magnetic field and causes the rotor to turn. Since this is alternating current reversing 60 times per second, the rotor turns, in effect "chasing" the changing polarity in the running windings.

26.4 STARTING WINDINGS

Starting windings placed between the running windings insure that the rotor starts properly and that it turns in the desired direction, Figure 26–6. The starting windings have more turns than the running windings

Figure 26–1. (A) Fans move air. (B) Pumps move water. (C) Compressors move vapor refrigerant. *Courtesy (A and B) W. W. Grainger, Inc. (C) Trane Company*

In a two-pole split-phase motor the stator has two poles with insulated wire windings called the *running windings*. When an electrical current is applied, these poles become an electromagnet with the polarity changing constantly. In normal 60-cycle operation the polarity changes 60 times per second.

The rotor may be constructed of bars, Figure 26–5. This type is called a *squirrel cage rotor*. When the rotor shaft is placed in the bearings in the bell type ends, it is positioned between the running windings. When an alternating current is applied to these windings, a mag-

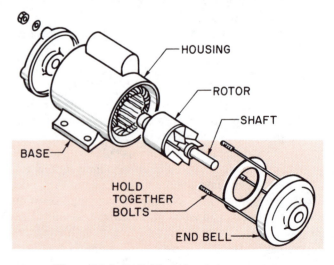

Figure 26–3. Individual electric motor parts.

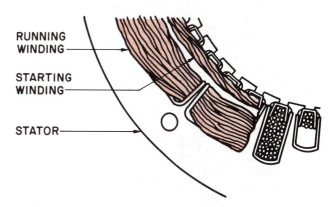

Figure 26-6. Placement of starting and running windings inside stator.

ings. This is due to a larger resistance in the smaller diameter wire and more turns in the start windings.

We have just described a two-pole split-phase *induction* motor, which is rated to run at 3600 revolutions per minute (rpm). It actually turns at a slightly slower speed when running under full load. When the motor reaches approximately 75% of its normal speed, a centrifugal switch opens the circuit to the starting windings and the motor continues to operate on only the running windings. Many split-phase motors have four poles and run at 1800 rpm.

26.5   DETERMINING A MOTOR'S SPEED

The following formula can be used to determine the synchronous speed (without load) of motors.

$$S \text{ (rpm)} = \frac{\text{frequency} \times 120}{\text{number of poles}}$$

Frequency is the number of cycles per second (also called *hertz*). *Note:* The magnetic field builds and collapses twice each second (each time it changes direction), 120 is used in the formula instead of 60.

$$\text{Speed of two-pole split-phase motors} = \frac{60 \times 120}{2}$$
$$= 3600$$

$$\text{Speed of four-pole split-phase motors} = \frac{60 \times 120}{4}$$
$$= 1800$$

The speed under load of each motor will be approximately 3450 rpm and 1750 rpm.

26.6   STARTING AND RUNNING CHARACTERISTICS

Two major considerations of electric motor applications are the starting and running characteristics. A motor applied to a refrigeration compressor must have a high starting torque—it must be able to start under heavy starting loads. A refrigeration compressor may

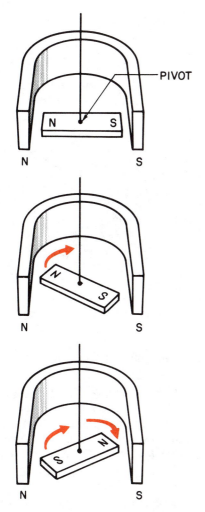

Figure 26-4. Poles (north and south) on rotating magnet will line up with opposite poles on stationary magnet.

and are wound with a smaller-diameter wire. This produces a larger magnetic field and greater resistance, which helps the rotor to start turning and determines the direction in which it will turn. This happens as a result of these windings being located between the running windings. It acts to change the phase angle between the voltage and the current in these wind-

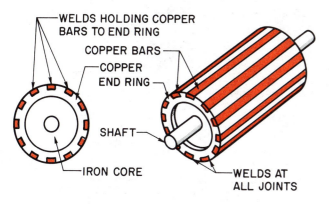

Figure 26-5. Simple sketch of squirrel cage rotor.

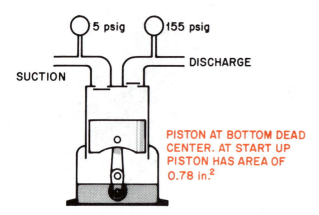

Figure 26–7. Compressor with high-side pressure 155 psig and low-side pressure 5 psig.

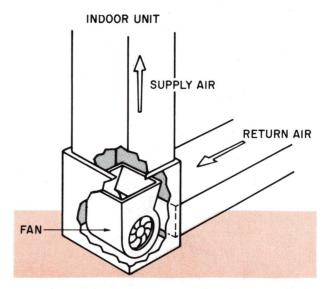

Figure 26–8. This fan has no pressure difference to overcome while starting. When it stops, the air pressure equalizes.

have a head pressure of 155 psig and a suction pressure of 5 psig and still be required to start. The pressure difference of 150 psig is the same as saying that the compressor has a starting resistance of 150 psi of piston displacement. If this compressor has a 1-in. diameter piston, the area of the piston is 0.78 in.^{2} ($A = \pi r^2 = 3.14 \times 0.5 \times 0.5$). This area multiplied by the pressure difference of 150 psi is the starting resistance for the motor (117 lb). This is similar to a 117-lb weight resting on top of the piston when it tries to start, Figure 26–7.

To start a small fan, a motor does not need as much starting torque. The motor must simply overcome the friction needed to start the fan moving. There is no pressure difference because a fan equalizes the pressures when not running, Figure 26–8.

26.7 POWER SUPPLIES FOR ELECTRIC MOTORS

One other basic difference in electric motors is the power supply that is used to operate them. For example, only single-phase power is available to homes. Motors designed to operate on single-phase power must therefore be used. For large loads in factories, for example, single-phase power is inadequate, so three-phase power is used to run the motors. The difference in the two power sources changes the starting and running characteristics of the motors.

We now describe some motors currently used in the heating, air conditioning, and refrigeration industry. We emphasize the electrical characteristics, not the working conditions. The working conditions will be described later. These motors are the motors that are being manufactured today. There are still some older motors in operation, but will not be discussed in this text.

26.8 SINGLE-PHASE OPEN MOTORS

The power supply for most *single-phase* motors is

either 125 V or 208–230 V. A home furnace has a power supply of 125 V, whereas the air conditioner outside uses a power supply of 230 V, Figure 26–9. A commercial building may have either 230 V or 208 V, depending on the power company. Some single-phase motors are dual voltage. The motor has two run-windings and one start-winding. The two run-windings have the same resistance and the start-winding has a high resistance. The motor will operate with the two run-windings in parallel in the low-voltage mode. When it is required to run in the high-voltage mode, the technician changes the numbered motor leads to the manufacturer's suggested pattern. This wires the run-windings in series with each other and delivers an effective voltage of 125 V to each winding. It can be said that

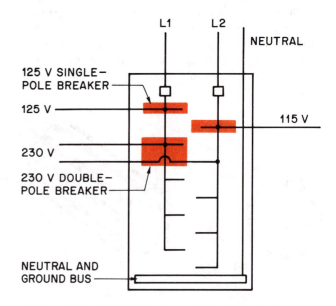

Figure 26–9. Main breaker panel for a typical residence.

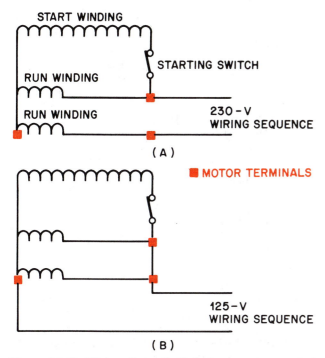

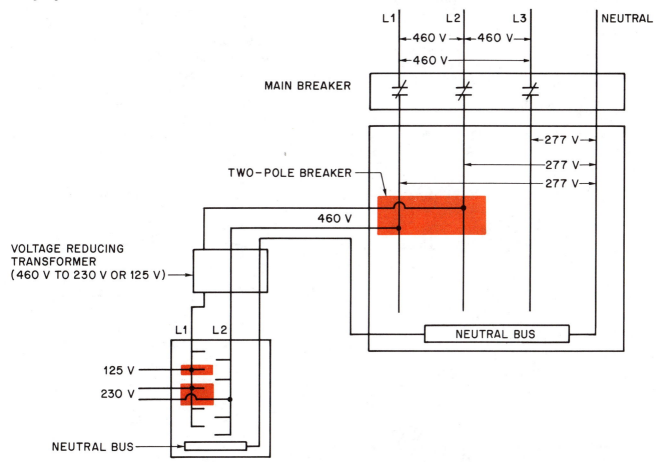

Figure 26-10. Wiring diagram of dual-voltage motor. It is made to operate using 125 V or 230 V, depending on how the motor is wired. (A) 230-V wiring sequence. (B) 125-V wiring sequence.

the motor windings are actually only 125 V because they only operate on 125 V, no matter which mode they are in. The technician can change the voltage at the motor terminal box, Figure 26-10.

Some commercial and industrial installations may use a 460-V power supply for large motors. The 460 V may be reduced to a lower voltage to operate the small motors. The smaller motors may be single phase and must operate from the same power supply, Figure 26-11.

A motor can rotate clockwise or counterclockwise. Some motors are reversible from the motor terminal box, Figure 26-12.

26.9   SPLIT-PHASE MOTORS

Split phase motors have two distinctly different windings, Figure 26-13. They have a medium amount of starting torque and a good operating efficiency. The split-phase motor is normally used for operating fans in the fractional horsepower range. Its normal operating ranges are 1800 rpm and 3600 rpm. An 1800-rpm motor will normally operate at 1725-1750 rpm under a load. The difference in the rated rpm and the actual rpm is called *slip*. If the motor is loaded to the point

Figure 26-11. 460-V commercial building power supply. Normally when a building has a 460-V power supply, it will have a step-down transformer to 125 V for office machines and small appliances.

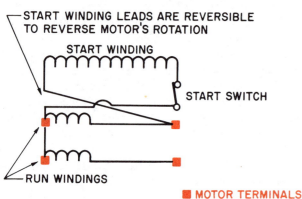

Figure 26–12. Single-phase motor can be reversed by changing the connections in the motor terminal box. The direction the motor turns is determined by the start winding. This can be shown by disconnecting the start-winding leads and applying power to the motor. It will hum and will not start. The shaft can be turned in either direction and the motor will run in that direction.

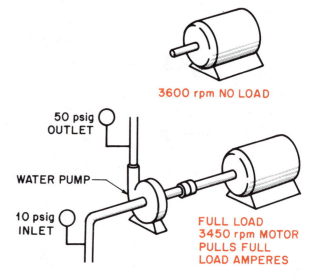

Figure 26–14. How motor revolutions per minute change under motor load.

where the speed falls below 1725 rpm, the current draw will climb above the rated amperage. Motors rated at 3600 rpm will normally slip in speed to about 3450–3500 rpm, Figure 26–14. Some of these motors are designed to operate at either speed, 1750 or 3450. The speed of the motor is determined by the number of motor poles and by the method of wiring the motor poles. The technician can change the speed of a two-speed motor at the motor terminal box.

26.10 THE CENTRIFUGAL SWITCH

The *centrifugal switch* is used to disconnect the starting winding from the circuit when the motor reaches three-fourths of the rated speed. We are discussing now motors that run in the atmosphere. (Hermetic motors that run in the refrigerant environment are discussed later.) When a motor is started in the air, the arc from the centrifugal switch will not harm the atmosphere. (It will harm the refrigerant, so there must be no arc in a refrigerant atmosphere.)

The centrifugal switch is a mechanical device attached to the end of the shaft with weights that will sling outward when the motor reaches three-quarter speed. For example, if the motor has a rated speed of 1725 rpm, at 1294 rpm (1725 × 0.75) the centrifugal weights will change position and open a switch to remove the starting winding from the circuit. This switch is under a fairly large current load, so a spark

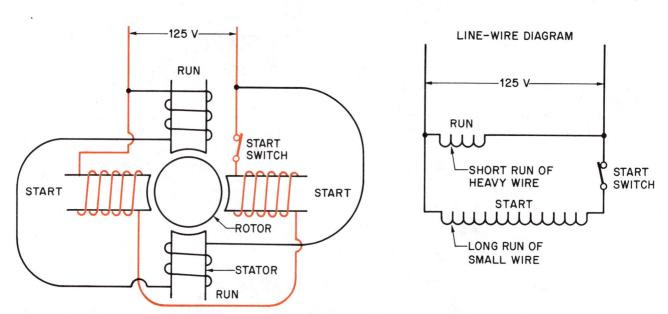

Figure 26–13. Diagram showing difference in the resistance in start and run windings.

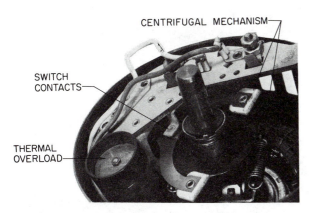

Figure 26-15. Centrifugal switch located in the end of the motor. *Photo by Bill Johnson*

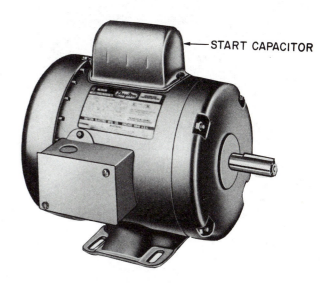

Figure 26-16. Capacitor-start motor. *Courtesy W. W. Grainger, Inc.*

will occur. If the switch fails to open its contacts and remove the starting winding, the motor will draw too much current and will stop because of the overload.

The more the switch is used, the more its contacts will burn from the arc. If this type of motor is started many times, the first thing that will likely fail will be the centrifugal switch. This switch makes an audible sound when the motor starts and stops, Figure 26-15.

26.11 THE ELECTRONIC RELAY

The *electronic relay* is used with some motors to open the starting windings after the motor has started. This is a solid-state device designed to open the starting winding circuit when the design speed has been obtained. Other devices are also used to perform this function; they are described with hermetic motors.

26.12 CAPACITOR-START MOTOR

The capacitor-start motor is the same basic motor as the split-phase motor, Figure 26-16. It has the two distinctly different windings for starting and running. The previously mentioned methods may be used to interrupt the power to the starting windings while the motor is running. A start capacitor is wired in series

with the starting windings to give the motor more starting torque. Figure 26-17 shows voltage and current cycles in an induction motor. In an inductive circuit the current *lags* the voltage. In a capacitive circuit the current *leads* the voltage. The amount by which the current leads or lags the voltage is the *phase angle*. A capacitor is chosen to make the phase angle such that it is most efficient for starting the motor, Figure 26-18. This capacitor is not designed to be used while the motor is running, and it must be switched out of the circuit soon after the motor starts. This is done at the same time the starting windings are taken out of the circuit.

26.13 CAPACITOR-START, CAPACITOR-RUN MOTOR

Capacitor-start, capacitor-run motors are much the same as the split-phase motor. The run capacitor is wired into the circuit to provide the most efficient phase angle between the current and voltage when the motor is running. The run capacitor is in the circuit at

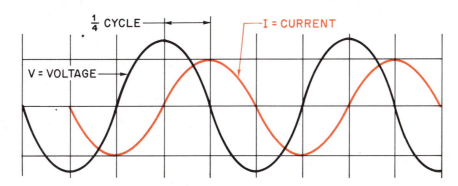

Figure 26-17. AC cycle, voltage and current in an induction circuit. The current lags the voltage.

Figure 26-18. A start capacitor. *Photo by Bill Johnson*

any time the motor is running. If a run capacitor fails because of an open circuit within the capacitor, the motor may start, but the running amperage will be about 10% too high and the motor will get hot if operated at full load, Figure 26–19. The capacitor-start, capacitor-run motor is one of the most efficient motors used in refrigeration and air conditioning equipment. It is normally used with belt-drive fans and compressors.

26.14 PERMANENT SPLIT-CAPACITOR MOTOR

The permanent split-capacitor (PSC) motor has windings very similar to the split-phase motor, Figure 26–20, but it does not have a start capacitor. Instead it uses one run capacitor wired into the circuit in a similar way to the run capacitor in the capacitor-start, capacitor-run motor. This is the simplest split-phase motor. It is very efficient and has no moving parts for the starting of the motor, however, the starting torque

Figure 26-20. Permanent split-capacitor motor. *Courtesy Universal Electric Company*

is very low so the motor can only be used in low-starting-torque applications, Figure 26–21.

The permanent split-capacitor (PSC) motor may have several speeds. A multispeed motor can be identified by the many wires at the motor electrical connections, Figures 26–22 and 26–23. PSC motors are explained later in the unit. As the resistance of the motor winding decreases, the speed of the motor increases. When more resistance is wired into the circuit, the motor speed decreases. Most manufacturers use this motor in the fan section in air conditioning and heating systems. Earlier systems used a capacitor-start, capacitor-run motor and a belt drive, and air volumes were adjusted by varying the drive pulley diameter. Motor speed can be changed by switching the wires.

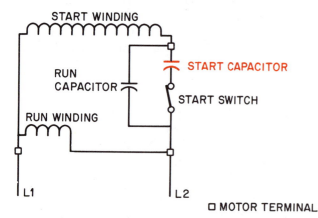

Figure 26-19. Wiring diagram of capacitor-start, capacitor-run motor. The start capacitor is only in the circuit during starting, and the run capacitor is in the circuit during starting and running of the motor.

Figure 26-21. Open PSC motor that may be used to turn a fan. *Courtesy Universal Electric Company*

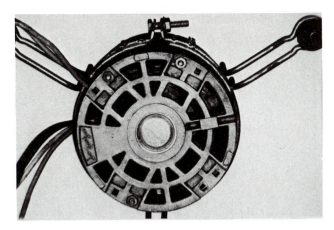

Figure 26–22. Multispeed PSC motor. *Photo by Bill Johnson*

The permanent split-capacitor motor may be used to obtain slow fan speeds during the winter heating season for higher leaving air temperatures with gas, oil, and electric furnaces. The fan speed can be increased by switching to a different resistance in the winding using a relay. This will provide more airflow in summer to satisfy cooling requirements, Figure 26–24.

26.15   THE SHADED-POLE MOTOR

The shaded-pole motor has very little starting torque and is not as efficient as the PSC motor, so it is only used for light-duty applications. These motors have small starting windings at the corner of each pole that help the motor start by providing an induced current and a rotating field, Figure 26–25. It is a very economical motor from a first-cost standpoint. The shaded-pole motor is normally manufactured in the fractional horsepower range. For years it has been used in air-cooled condensers to turn the fans. Figure 26–26.

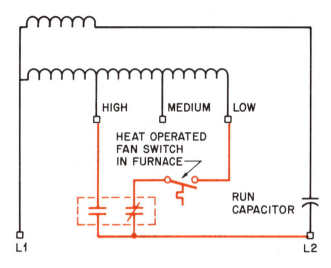

FAN RELAY: WHEN ENERGIZED, SUCH AS IN COOLING, THE FAN CANNOT RUN IN THE LOW-SPEED MODE. WHEN DEENERGIZED THE FAN CAN START IN THE LOW-SPEED MODE THROUGH THE CONTACTS IN THE HEAT OPERATED FAN SWITCH.

IF THE FAN SWITCH AT THE THERMOSTAT IS ENERGIZED WHILE THE FURNACE IS HEATING, THE FAN WILL MERELY SWITCH FROM LOW TO HIGH. THIS RELAY PROTECTS THE MOTOR FROM TRYING TO OPERATE AT 2 SPEEDS AT ONCE.

Figure 26–24. Diagram of a PSC motor showing how the motor can be applied for high air volume in the summer and low air volume in the winter.

26.16   THREE-PHASE MOTOR

Three-phase motors are used on commercial equipment. The building power supply must have three-phase power available. (Three-phase power is seldom found in a home.) Three-phase motors have no starting windings or capacitors. They can be thought of as having three single-phase power supplies, Figure 26–27.

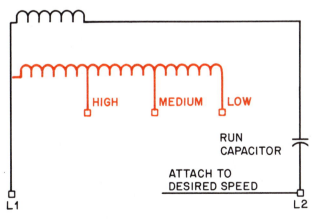

Figure 26–23. Diagram showing how a three-speed motor may be wired to run at a slow speed in the winter and high speed in the summer.

Figure 26–25. Shaded-pole motor. *Courtesy Universal Electric Company*

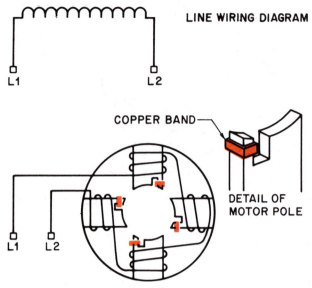

Figure 26-26. Wiring diagram of a shaded-pole motor.

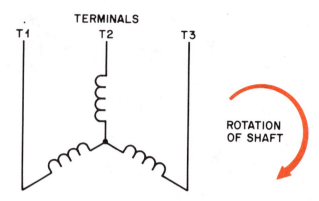

Figure 26-28. Diagram of a three-phase motor.

Each of the phases can have either two or four poles. A 3600-rpm motor will have three sets, each with two poles (total of six), and an 1800-rpm motor will have three sets, each with four poles (total of 12). Each phase changes direction of current flow at different times but always in the same order. A three-phase motor has a high starting torque because of the three phases of current that operate the motor. At any given part of the rotation of the motor, one of the windings is in position for high torque. This makes starting large fans and compressors very easy, Figure 26-28.

The three-phase motor rpm also slips to about 1750 and 3450 rpm when under full load. The motor is not normally available with dual speed; it is either an 1800-rpm or a 3600-rpm motor.

The rotation of a three-phase motor may be changed by switching any two motor leads, Figure 26-29. This rotation must be carefully observed when three-phase fans are used. If a fan rotates in the wrong direction it will move only about half as much air. If this occurs, reverse the motor leads and the fan will turn in the correct direction.

All of the motors we have described are considered to be *open* motors and are used for fans and pumps.

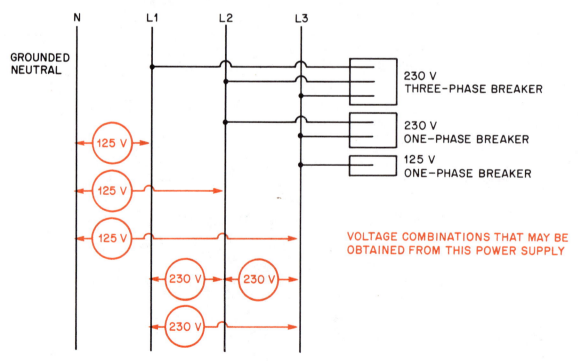

Figure 26-27. Diagram of three-phase power supply.

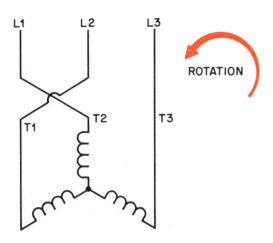

Figure 26–29. Wiring diagram of a three-phase motor. The rotation of this motor can be reversed by changing any two motor leads.

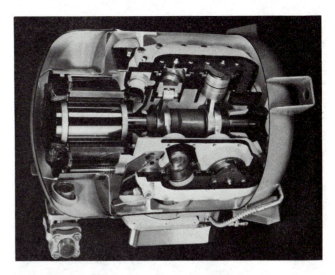

Figure 26–30. Typical motor for a hermetic compressor. *Courtesy of Trane Company*

These motors have other characteristics that must be considered when selecting a motor for a particular job: for example, the motor mounting. Is the motor solidly mounted to a base, or is there a flexible mount to minimize noise?

The bearing choice is another factor. Is the motor going to be used where ball bearings would make too much noise? If so, sleeve bearings should be used.

Still another factor is the operating temperature of the motor surroundings. A condenser fan motor that pulls the air over the condenser coil and then over the motor requires a motor that will operate in a warmer atmosphere. The best advice is to replace any motor with an exact replacement.

26.17   SINGLE-PHASE HERMETIC MOTOR

The single-phase hermetic motor is similar to the split-phase motor from a wiring standpoint. It has start and run windings, each with a different resistance. The motor runs with the run winding and uses a potential relay to open the circuit to the start windings. A run capacitor is often used to improve running efficiency. A hermetic motor is designed to operate in a refrigerant, usually vapor, atmosphere. It is undesirable for liquid refrigerant to enter the shell, say by an overcharge. Single-phase hermetic compressors usually are manufactured up to 5 hp. If more capacity is needed, multiple systems or larger three-phase units are used.

Hermetic compressor motor materials must be compatible with the refrigerant and oil circulating in the system. The coatings on the windings, the materials used to tie the motor windings, and the papers used as wedges must be of the correct material. The motor is assembled in a dry clean atmosphere because it will be used in compressors, Figure 26–30.

Hermetic motors are started in much the same way as the other motors described. The start windings must be removed from the circuit when the motor gets to about 75% of its normal operating speed. The start windings are not removed in the same way as for an open motor because the windings are in a refrigerant atmosphere. Open single-phase motors are operated in air, and a spark is allowed when the start winding is disconnected. This cannot be allowed in a hermetic motor because the spark makes the refrigerant deteriorate. Special devices determine when the compressor motor is running at the correct speed to disconnect the start winding.

Because the hermetic motor is enclosed in refrigerant, the motor leads must pass through the compressor shell to the outside. A terminal box on the outside houses the three motor terminals, Figure 26–31: one for the run winding, one for the start winding, and one for the line common to the run and start windings. See Figure 26–32 for a wiring diagram of a three-terminal compressor. The start winding has more resistance than the run winding.

The motor leads are insulated from the steel compressor shell. Neoprene was the most popular insulating material for years. However, if the motor terminal becomes too hot, due to a loose connection, the neoprene may eventually become brittle and possibly leak, Figure 26–33. Many compressors now use a ceramic material to insulate the motor leads.

26.18   THE POTENTIAL RELAY

The *potential relay* is often used to break the circuit to the start winding when the motor reaches approximately 75% of its normal speed. This relay has a normally closed set of contacts. The coil is designed to operate at a slightly higher voltage than the applied

Figure 26-31. Motor terminal box on the outside of the compressor. *Photo by Bill Johnson*

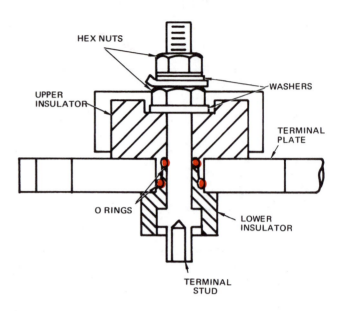

Figure 26-33. These motor terminals use neoprene O rings as the insulator between the terminal and the compressor housing. *Courtesy of Trane Company*

(line) voltage. When the rotor begins to turn, a transformer action takes place at the start winding; as the rotor approaches 75% of its design speed, the voltage exceeds the applied voltage and is sufficient to energize the coil. This opens the contacts, which open the start winding circuit, Figure 26-34.

26.19   THE CURRENT RELAY

The *current relay* also breaks the circuit to the start winding. It uses the inrush current of the motor to determine when the motor is running up to speed. A motor draws locked-rotor current during the time the power is applied to the windings and the motor has not started turning. As the motor starts turning, the current peaks; it begins to reduce as the motor starts turning. The current relay has a set of normally open contacts that close when the inrush current flows through its coil, energizing the start windings. When the motor

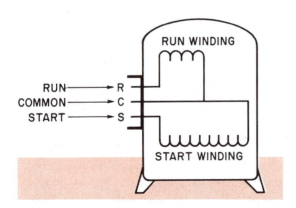

Figure 26-32. Wiring diagram of what is behind the three terminals on a single-phase compressor.

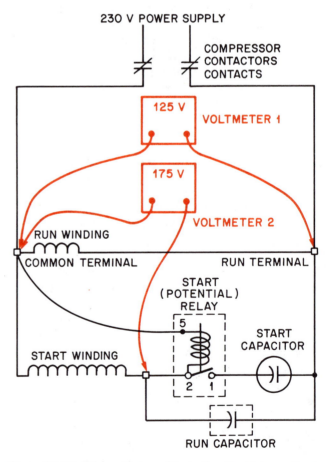

Figure 26-34. Wiring diagram illustrating the higher voltage of the start winding in a typical motor.

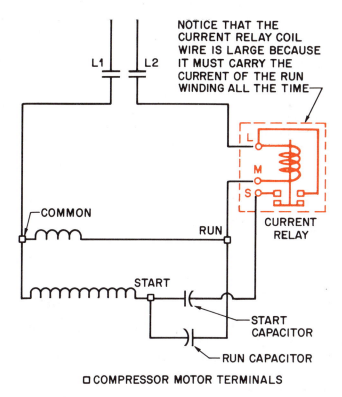

NOTICE THAT THE CURRENT RELAY COIL WIRE IS LARGE BECAUSE IT MUST CARRY THE CURRENT OF THE RUN WINDING ALL THE TIME

Figure 26-35. Wiring diagram of a current relay.

speed reaches about three fourths of the rated rpm, the current relay opens its contacts, either by gravity or by a spring, Figure 26-35. The coil is wired in series with the run winding of the motor. The full current of the motor must flow through the coil of the current relay. **The current relay may always be identified by the size of the wire in the relay coil. This wire is unusually large because it must carry the full-load current of the motor, Figure 26-36.**

The starting methods we have talked about are used on many compressors with split-phase motors that need high starting torque. If a system has a capillary tube metering device or a fixed-bore orifice metering device, the pressures will equalize during the off cycle and a high-starting-torque compressor may not be necessary. A PSC (permanent split-capacitor) motor

may be used for this application. PSC motors are often used in residential air conditioning and heat pumps with the equalizing type refrigeration cycle. In the PSC motor the starting and running windings are both energized whenever there is power to the motor. The motor does utilize a running capacitor wired between the run and start terminals, so line voltage is not applied directly to the start windings.

26.20   POSITIVE-TEMPERATURE-COEFFICIENT START DEVICE

The permanent split capacitor (PSC) motor may not need any start assistance when conditions are well within the design parameters. If it does need start assistance, a potential relay and start capacitor may be added to provide additional torque or a positive-temperature-coefficient (PTC) device may be added. The PTC is a thermistor (a thermistor changes resistance with a change in temperature) that has no resistance to current flow when the unit is off. When the unit is started, the current flow through the PTC causes it to heat very fast and create a high resistance in its circuit. This changes the phase angle of the start windings. It will not give a motor the starting torque that a start capacitor will, but it is advantageous because it has no moving parts. The PTC is wired in parallel with the run capacitor and acts just like a short across the run capacitor during starting. This provides full line voltage to the start windings during starting, Figure 26-37.

26.21   TWO-SPEED COMPRESSOR MOTORS

Two-speed compressor motors are used by some manufacturers to control the capacity required from small compressors. For example, a residence or small office building may have a 5-ton air conditioning load at the peak of the season and a $2\frac{1}{2}$-ton load as a

Figure 26-36. Current relay is identified by the size of the wire in the holding coil. *Photo by Bill Johnson*

Figure 26-37. A positive-temperature-coefficient device (PTC). *Photo by Bill Johnson*

minimum. Capacity control is desirable for this application. A two-speed compressor may be used to accomplish capacity control. Two-speed operation is obtained by wiring the compressor motor to operate as a two-pole motor or a four-pole motor. The automatic changeover is accomplished with the space-temperature thermostat and the proper compressor contactor for the proper speed. For all practical purposes, this can be considered two motors in one compressor housing. One motor turns at 1800 rpm, the other at 3600 rpm. The compressor uses either motor, based on the capacity needs. This compressor has more than three motor terminals to operate the two motors in the compressor.

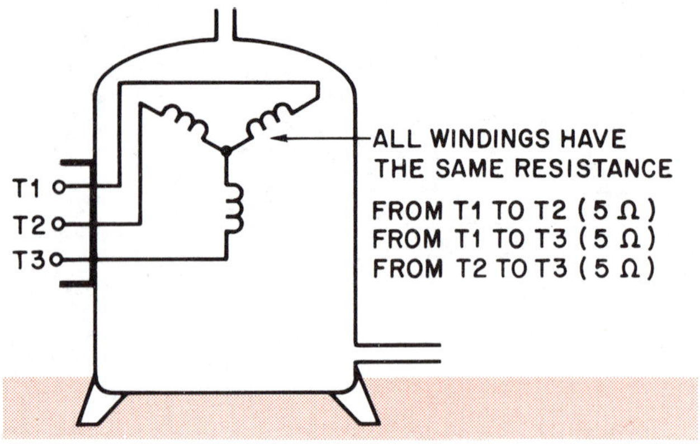

Figure 26-39. Three-phase compressor with three leads for the three windings. The windings all have the same resistance.

26.22 SPECIAL-APPLICATION MOTORS

There are some special-application single-phase motors, which may have more than three motor terminals, that are not two-speed motors. Some manufacturers design an auxiliary winding in the compressor to give the motor more efficiency. These motors are normally in the 5 hp and smaller range. Other special motors may have the winding thermostat wired through the shell as two extra motor terminals. The winding thermostat can be wired out of the circuit if it should fail with its circuit open, Figure 26-38.

26.23 THREE-PHASE MOTOR COMPRESSORS

Large commercial and industrial installations will have three-phase power for the air conditioning and refrigeration equipment. Three-phase compressor motors normally have three motor terminals, but the resistance across each winding is the same, Figure 26-39. As explained earlier, three-phase motors have a high starting torque and, consequently, should experience no starting problems.

Welded hermetic compressors were limited to $7\frac{1}{2}$ tons for many years but are now being manufactured in

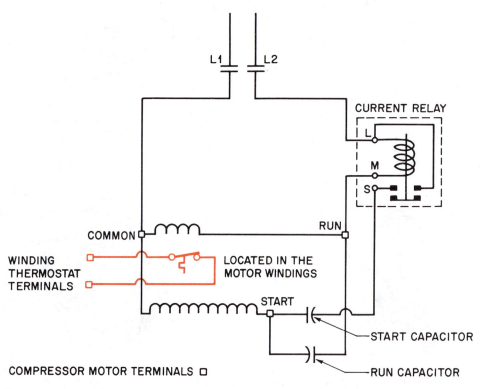

Figure 26-38. Compressor with five terminals in the terminal box, two of which are wired to the winding thermostat.

Figure 26–40. Large welded hermetic compressor. *Courtesy of Trane Company*

Figure 26–41. Serviceable hermetic compressors. *Courtesy of Trane Company*

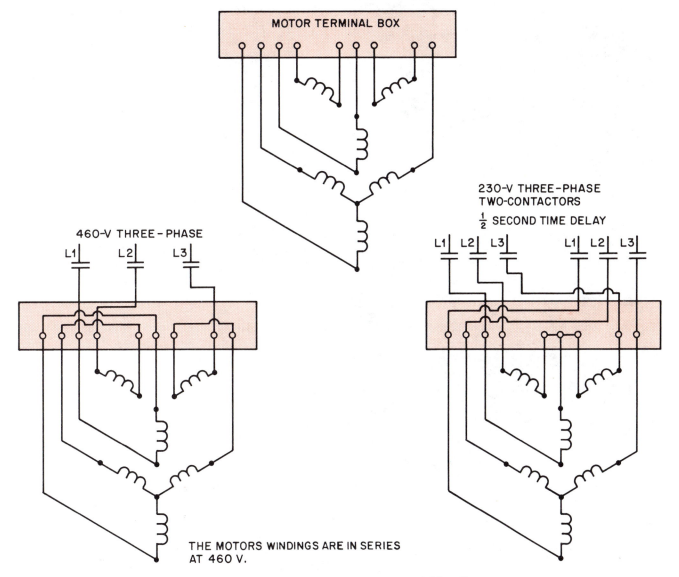

Figure 26–42. Dual-voltage compressor motor wiring diagram.

sizes up to about 50 tons. The larger welded hermetic compressors are traded to the manufacturer when they fail and are remanufactured. These must be cut open for service, Figure 26–40.

Serviceable hermetic compressors of the reciprocating type are manufactured in sizes up to about 125 tons, Figure 26–41. These compressors may have dual-voltage motors for 208–230 V or 460-V operation, Figure 26–42. These compressors are normally rebuilt or remanufactured when the motor fails, so an overhaul is considered when a motor fails. The compressors may be rebuilt in the field or traded for remanufactured compressors. Most large metropolitan areas will have companies that can rebuild the compressor to the proper specifications. The difference in rebuilding and remanufacturing is that one is done by the original manufacturer or an authorized rebuilder, and the other by an independent rebuilder.

(A)

(B)

Figure 26–43. All motors must be cooled or they will overheat. (A) This compressor is cooled by the refrigerant gas passing over the motor winding. (B) This compressor is cooled by air from a fan. The air-cooled motor has fins on the compressor to help dissipate the heat. *Courtesy Copeland Corporation*

26.24   COOLING ELECTRIC MOTORS

All motors must be cooled. Most open motors are air cooled. Hermetic motors may be air-, water-, or refrigerant-gas cooled, Figure 26–43. The small-to-medium-sized motors are normally refrigerant-gas cooled. Only very large motors are water cooled. An air-cooled motor has fins on the surface to help give the surface more area for dissipating heat. These motors must be located in a moving air stream. To cool properly, refrigerant-gas-cooled motors must have an adequate refrigerant charge.

SUMMARY

- Motors turn fans, compressors, and pumps.
- Some of these applications need high starting torque and good running efficiencies; some need low starting torque with average or good running efficiencies.
- Compressors applied to refrigeration normally require high-starting-torque motors.
- Small fans normally need motors that have low starting torque.
- The voltage supplied to a particular installation will determine the motor's voltage. The common voltage for furnace fans is 125 V; 230 V is the common voltage for home air conditioning systems.
- Common single-phase motors are split phase, PSC, and shaded pole.
- When more starting torque is needed, a start capacitor is added to the motor.
- A run capacitor improves the running efficiency of the split-phase motor.
- A centrifugal switch breaks the circuit to the start winding when the motor is up to running speed. The switch changes position with the speed of the motor.
- An electronic switch may be used to interrupt power to the start winding.
- The common rated speed of a single-phase motor is determined by the number of poles or windings in the motor. The common speeds are 1800 rpm, which will slip in speed to about 1725 rpm, and 3600 rpm, which will slip to about 3450 rpm.
- Three-phase motors are used for all large applications. They have a high starting torque and a high running efficiency. Three-phase power is not available at most residences, so these motors are limited to commercial and industrial installations.
- The power to operate hermetic motors must be conducted through the shell of the compressor by way of motor terminals.
- Since the winding of a hermetic compressor is in the refrigerant atmosphere, a centrifugal switch may not be used to interrupt the power to the start winding.
- A potential relay takes the start winding out of the circuit using back emf.

- A current relay breaks the circuit to the start winding using the motor's run current.
- The PSC (permanent split capacitor) motor is used when high starting torque is not required. It needs no starting device other than the run capacitor.
- The PTC (positive temperature coefficient) device is used with some PSC motors to give small amounts of starting torque. It has no moving parts.
- When compressors are larger than 5 ton, they are normally three phase.
- Dual-voltage three-phase compressors are built with two motors wired into the housing.
- Three-phase compressors come in sizes up to about 125 ton.

REVIEW QUESTIONS

1. Name the two popular operating voltages in residences.
2. What device takes the start winding out of the circuit when an open motor gets up to running speed?
3. Describe the difference in resistances of the starting and running windings of a split-phase motor.
4. What device may be wired into the starting circuit to improve the starting torque of a compressor?
5. What device may be wired into the circuit to improve the running efficiency of a compressor?
6. What is back emf?
7. Why does a hermetic compressor need special materials?
8. Name the two types of motors used for single-phase hermetic compressors.
9. How does the power pass through the compressor shell to the motor windings?
10. Name the two types of relays used to start the single-phase hermetic compressor.
11. What is a PTC device?
12. How are some small compressors operated at two different speeds?
13. Why are two speeds desirable?
14. How are some compressor motors operated at different voltages?
15. Name two methods to cool most hermetic compressors.

APPLICATION OF MOTORS

OBJECTIVES

After studying this unit, you should be able to

- identify the proper power supply for a motor.
- describe the application of three-phase versus single-phase motors.
- describe other motor applications.
- explain how the noise level in a motor can be isolated from the conditioned space.
- describe the different types of motor mounts.
- identify the various types of motor drive mechanisms.

27.1 MOTOR APPLICATIONS

Because electric motors perform so many different jobs, choosing the proper motor is crucial to safe and effective performance. It is usually the manufacturer or design engineer for a particular job that chooses the motor for each piece of equipment. However, as a technician you often will need to substitute a motor when an exact replacement is not available, so you should understand the reasons for choosing a particular motor for a job. For example, when a fan motor burns out in an air conditioning condensing unit the correct motor must be obtained or another failure may occur. *In an air-cooled condenser the air is normally pulled through the hot condenser coil and passed over the fan motor. This hot air is used to cool the motor. You must be aware of this, or you will install the wrong motor.* The motor must be able to withstand the operating temperatures of the condenser air, which may be as high as 130°F, Figure 27–1.

Open motors are discussed in this unit because they are the only ones a technician can select from. The following are some design differences that influence the application.

- The power supply
- The work requirements
- The motor insulation type or class
- The bearing types
- The mounting characteristics

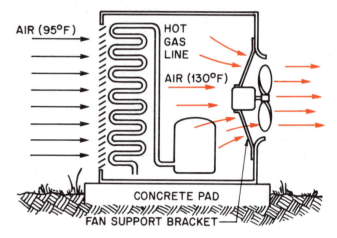

Figure 27–1. Motor operating in the hot airstream after the air has been pulled through the condenser.

27.2 THE POWER SUPPLY

The power supply was discussed to some extent in Unit 26. The power supply must provide the correct voltage and sufficient current. For example, the power supply in a small shop building may be capable of operating a 5-hp air compressor. But suppose air conditioning is desired. If the air conditioning contractor prices the job expecting the electrician to use the existing power supply, the customer is going to be in for a surprise. The electrical service for the whole building will have to be changed. The motor equipment nameplate information and the manufacturer's catalogs provide the needed information for the additional service, but someone must put the whole project together. The installing air conditioning contractor may have that responsibility. See Figure 27–2 for a typical motor nameplate and Figure 27–3 for a part of a page from manufacturer's catalog with the electrical data included. Figure 27–4 is an example of a typical electrical panel rating.

The power supply data contains

A. The voltage (120 V, 208 V, 230 V, 460 V)
B. The current capacity in amperes
C. The frequency in hertz or cycles per second (60 cps in the United States and 50 cps in many foreign countries)
D. The phase (single or three phase)

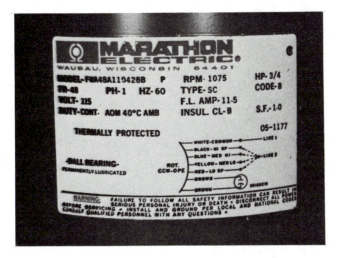

Figure 27-2. Motor nameplate. *Photo by Bill Johnson*

Motors and electrical equipment must fit within the system's total electrical capacity, or failures may occur. Voltage and current are most often the inadequate characteristic encountered. The technician can check them, but it is best to have a licensed electrician make the final calculations.

Voltage

The voltage of an installation is important because every motor operates within a specified voltage range—usually within ±10%. Figure 27-5 gives the upper and lower limits of common voltages. If the voltage is too low, the motor will draw a high current. For example, if a motor is supposed to operate on 230 V but the supply voltage is really 200 V, the motor's current will go up. The motor is trying to do its job, but it lacks the power and it will overheat, Figure 27-6.

If the applied voltage is too high, the motor may develop local hot spots within its windings, but it will not always experience high amperage. The high voltage will actually give the motor more power than it can use. A 1-hp motor with a voltage rating of 230 V that is

TYPE CS • NEMA SERVICE FACTOR • 60 HERTZ • CUSHION BASE

HP	RPM	Bearings	Overload Protector	Full Load (12) Amps	Frame
$\frac{1}{6}$	1800	Sleeve	Auto	4.4	K48
$\frac{1}{4}$	1800	Sleeve	Auto	2.7	K48
		Ball	Auto	2.7	K48
$\frac{1}{3}$	3600	Sleeve	Auto	3.1	L48
	1800	Sleeve	Auto	2.9	L48
		Sleeve	Auto	2.6	J56
		Ball	No	2.9	L48
		Ball	No	2.6	J56
		Ball	Auto	2.6	J56
	1200	Sleeve	No	3.0	K56
$\frac{1}{2}$	3600	Sleeve	Auto	4.0	L48
	1800	Sleeve	No	3.6	J56
		Sleeve	Auto	3.6	J56
		Ball	No	3.6	J56
		Ball	Auto	3.6	J56
$\frac{3}{4}$	3600	Sleeve	Auto	4.6	J56
	1800	Sleeve	No	5.2	K56
		Sleeve	Auto	5.2	K56
		Ball	No	5.2	K56
		Ball	Auto	5.2	K56
1	3600	Sleeve	Auto	6.0	K56
		Ball	No	6.0	K56
	1800	Sleeve	Auto	6.5	L56
		Ball	No	6.5	L56
		Ball	Auto	6.5	L56
$1\frac{1}{2}$	3600	Ball	Auto	8.0	L56
	1800	Ball	Auto	7.5	M56
2	3600	Ball	No	9.5	L56

Figure 27-3. Part of a page from a manufacturer's catalog. *Courtesy Century Electric Inc.*

Figure 27-4. Nameplate from an electrical panel used for a power supply. *Photo by Bill Johnson*

208 VOLT-RATED MOTOR	+10%	228.8 VOLTS
	−10%	187.2 VOLTS
230 VOLT-RATED MOTOR	+10%	253 VOLTS
	−10%	207 VOLTS
208–230 VOLT-RATED MOTOR	+10%	253 VOLTS
	−10%	187.2 VOLTS

Figure 27–5. Table showing the maximum and minimum operating voltages of typical motors.

operating at 260 V is running above its 10% maximum. This motor may be able to develop $1\frac{1}{4}$ hp at this higher-than-rated voltage, but the windings are not designed to operate at that level. The motor can overheat and eventually burn out if it continually runs overloaded. This can happen *without* drawing excessive current, Figure 27–7.

Current Capacity

There are two current ratings of a motor. The full-load amperage (FLA) is the current the motor draws while operating at a full-load condition at the rated voltage. This is also called the run-load amperage (RLA). For example, a 1-hp motor will draw approximately 16 A at 115 V or 8 A at 230 V in a single-phase circuit. See Figure 27–8 for a chart that shows approximate amperages for some typical motors.

The other amperage rating that may be given for a motor is the locked-rotor amperage (LRA). These two current ratings are available for every motor and are stamped on the motor nameplate for an open motor.

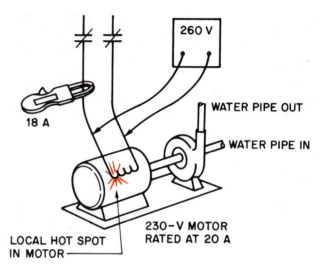

Figure 27–7. Motor rated at 230 V and operating at 260 V.

Some compressors do not have both ratings printed. The LRA or the FLA (RLA) tells the technician if the motor is operating outside its design parameters. Normally the LRA is about five times the FLA. For example, a motor that has an FLA of 5 A will normally have an LRA of about 25 A. If the LRA is given on the nameplate and the FLA is not given, divide the LRA by 5 to get an approximate FLA or RLA. For example, if a compressor nameplate shows an LRA of 80 A the approximate FLA is 80/5 = 16 A.

Every motor has a service factor that may be listed in the manufacturer's literature. This service factor is

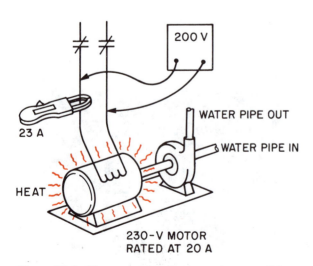

Figure 27–6. Motor operating at low-voltage condition.

APPROXIMATE FULL LOAD AMPERAGE VALUES FOR ALTERNATING CURRENT MOTORS

Motor	Single Phase		3-Phase-Squirrel Cage Induction		
HP	115V	230V	230V	460V	575V
$\frac{1}{6}$	4.4	2.2			
$\frac{1}{4}$	5.8	2.9			
$\frac{1}{3}$	7.2	3.6			
$\frac{1}{2}$	9.8	4.9	2	1.0	0.8
$\frac{3}{4}$	13.8	6.9	2.8	1.4	1.1
1	16	8	3.6	1.8	1.4
$1\frac{1}{2}$	20	10	5.2	2.6	2.1
2	24	12	·6.8	3.4	2.7
3	34	17	9.6	4.8	3.9
5	56	28	15.2	7.6	6.1
$7\frac{1}{2}$			22	11.0	9.0
10			28	14.0	11.0

Does not include shaded pole.

Figure 27–8. Chart showing approximate full load amperage values. *Courtesy BDP Company*

actually reserve horsepower. A service factor of 1.15 applied to a motor means that the motor can operate at 15% over the nameplate horsepower before it is out of its design parameters. A motor operating with a variable load and above normal conditions for short periods of time has a larger service factor. If the voltage varies at a particular installation, a motor with a high service factor may be chosen. The service factor is standardized by the National Electrical Manufacturer's Association (NEMA). Figure 27-9 is a typical manufacturer's chart showing service factors.

Frequency

The frequency in cycles per second (cps) is the frequency of the electrical current the power company supplies. The technician has no control over this. Most motors are 60 cps in the United States but could be 50

THREE PHASE ● DRIPPROOF

Type SC Squirrel Cage ● Fractional HP
- 60 Hertz
- Ball Bearing
- 40°C Ambient
  Class B Insulation

- NEMA Service Factor
  1/20 thru 1/8 HP—1.40
  1/6 thru 1/3 HP—1.32
  1/2 thru 3/4 HP—1.25
  1 thru 200 HP—1.15

- Versatile 208-430/460 volt motors available in many ratings.

HP	RPM	Volts	Full Load (5) Amps	Frame
Rigid Base				
$\frac{1}{4}$	1800	200-230/460	0.8	K48
	1200	230/460	0.6	H56
$\frac{1}{3}$	3600	200-230/460	0.7	B56
	1800	200-230/460	0.8	K48
		208-230/460	0.8	B56
	1200	200-208	1.7	J56
		230/460	0.8	J56
$\frac{1}{2}$	3600	208-230/460	0.9	B56
	1800	200-208	2.4	B56
		230/460	1.1	B56
		208-230/460	1.1	B56
	1200	200-208	2.0	J56
		230/460	1.0	J56
$\frac{3}{4}$	3600	208-230/460	1.2	J56
	1800	200-208	3.2	H56
		230/460	1.3	H56
		200-230/460	1.3	H56
	1200	200-208	3.3	J56
		200-208	3.3	M143T
		230/460	1.6	J56
		230/460	1.6	M143T
1	3600	200-208	3.2	J56
		230/460	1.5	J56
	1800	200-208	3.8	J56
		200-230/460	1.7	L143T
		200-230/460	1.7	J56
		575	1.4	L143T
	1200	200-208	3.8	N145T
		230/460	1.9	K56
		230/460	1.9	N145T

Figure 27-9. Chart showing service factors for motors. *Courtesy Century Electric, Inc.*

cps in a foreign country. Most 60-cps motors will run on 50 cps, but they will develop only five sixths of their rated speed (50/60). If you believe that the supply voltage is not 60 cps, contact the local power company. When motors are operated with local generators as the power supply, the generator's speed will determine the frequency. A cps meter is normally mounted on the generator and can be checked to determine the frequency.

Phase

The number of phases of power supplied to a particular installation is determined by the power company. They make this determination based on the amount of electrical load and the types of equipment that must operate from their power supply. Normally, single-phase power is supplied to residences and three-phase power is supplied to commercial and industrial installations. Single-phase motors will operate on two phases of three-phase power, Figure 27–10. Three-phase motors will not operate on single-phase power, Figure 27–11. The technician must match the motors to the number of phases of the power supply, Figure 27–12.

27.3 ELECTRICAL MOTOR WORKING CONDITIONS

The motor's working conditions determine which motor is the most economical for the particular job. For example, an open motor with a centrifugal starting switch (single phase) for the air conditioning fan may not be used in a room with explosive gases. When the motor's centrifugal switch opens to interrupt the power

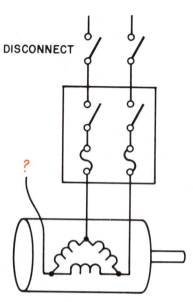

Figure 27–11. Three-phase motor with three leads that cannot be wired into single-phase circuit.

to the start winding, the gas may ignite. An explosion-proof motor enclosed in a housing must be used. Local codes should be checked and adhered to, Figure 27–13. The explosion-proof motor is too expensive to be installed in a standard office building, so a proper choice of motors should be made. A motor operated in a very dirty area may need to be enclosed, giving no ventilation for the motor windings. This motor must have some method to dissipate the heat from the windings, Figure 27–14.

A drip-proof motor should be used where water can fall on it. It is designed to shed water, Figure 27–15.

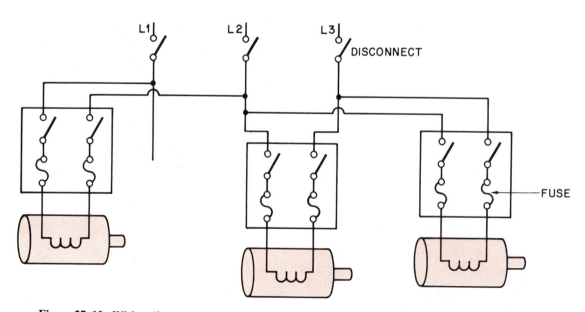

Figure 27–10. Wiring diagram showing how single-phase motors are wired into a three-phase circuit.

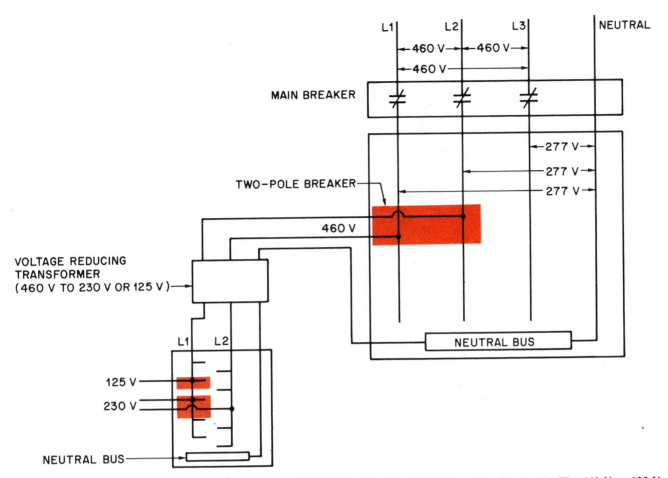

Figure 27–12. 460-V panel power supply for a commercial building showing distribution of various loads. The 460-V to 125-V step-down transformer is used to operate appliances and office machines.

27.4 THE INSULATION TYPE OR CLASS

The insulation type or class describes how hot the motor can safely operate in a particular ambient temperature condition. The example of the motor used earlier in this unit in an air conditioning condensing unit is typical. This motor must be designed to operate in a high ambient temperature. Motors are classified by the maximum allowable operating temperatures of the motor windings, Figure 27–16.

For many years motors were rated by the allowable temperature rise of the motor above the ambient temperature. Many motors still in service are rated this way. A typical motor has a temperature rise of 40°C (Celsius). If the maximum ambient temperature is 40°C (104°F), the motor could have a winding temperature of 40°C + 40°C = 80°C (176°F). The technician that cannot convert Celsius to Fahrenheit will have dif-

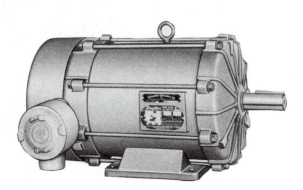

Figure 27–13. Explosion-proof motor. *Courtesy W. W. Grainger, Inc.*

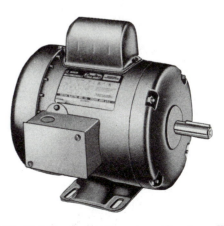

Figure 27–14. Totally enclosed motor. *Courtesy W. W. Grainger. Inc.*

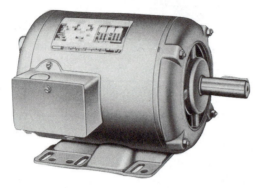

Figure 27–15. Drip-proof motor. *Courtesy W. W. Grainger, Inc.*

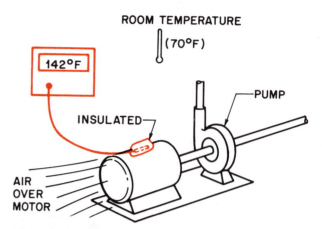

NOTE: CARE MUST BE TAKEN THAT TEMPERATURE TESTER LEAD IS TIGHT ON MOTOR AND NEXT TO MOTOR WINDINGS. THE LEAD MUST BE INSULATED FROM SURROUNDINGS. THE TEMPERATURE CHECK POINT MUST BE AS CLOSE TO THE WINDING AS POSSIBLE.

Figure 27–17. This motor is being operated in an ambient that is less than the maximum allowable for its insulation.

ficulty when troubleshooting these motors. If the temperature of the motor winding can be determined, the technician can tell if a motor is running too hot for the conditions. For example, a motor in a 70°F room is allowed a 40°C rise on the motor. The maximum winding temperature is 142°F (70°F = 21°C; 21°C + 40°C rise = 61°C). This is 142°F, Figure 27–17.

27.5   TYPES OF BEARINGS

Load characteristics and noise level determine the type of bearings that should be selected for the motor. Two common types are the *sleeve* bearing and the *ball* bearing, Figure 27–18.

The sleeve bearing is used where the load is light and noise must be low (e.g., a fan motor on a residential furnace). A ball-bearing motor would probably make excessive noise in the conditioned space. Metal duct work is an excellent sound carrier. Any noise in the system is carried throughout the entire system. Sleeve bearings are normally used in smaller applications, such as in residential and light commercial air conditioning systems for this reason. They are quiet and dependable, but they cannot stand great pressures (e.g., if fan belts are too tight). These motors have either vertical or horizontal shaft applications. The typical air-cooled condenser has a vertical motor shaft and is pushing the air out the top of the unit, Figure 27–19. A furnace fan has a horizontal motor shaft, Figure 27–20. These two types may not look very different, but they are. The vertical condenser fan is trying to fly downward into the unit to push air out the top. This

puts a real load on the end of the bearing (called the *thrust surface*), Figure 27–21.

Sleeve bearings are made from material that is softer than the motor shaft. The bearing must have lubrication—an oil film between the shaft and the bearing

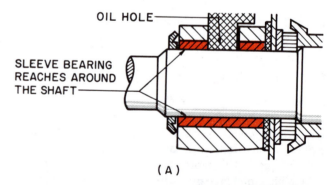

OIL HOLE

SLEEVE BEARING REACHES AROUND THE SHAFT

(A)

(B)

Figure 27–18. (A) Sleeve bearings and (B) ball bearings. *Courtesy (B) Century Electric, Inc.*

Class A	221°F	(105°C)
Class B	266°F	(130°C)
Class F	311°F	(155°C)
Class H	356°F	(180°C)

Figure 27–16. Temperature classification of typical motors.

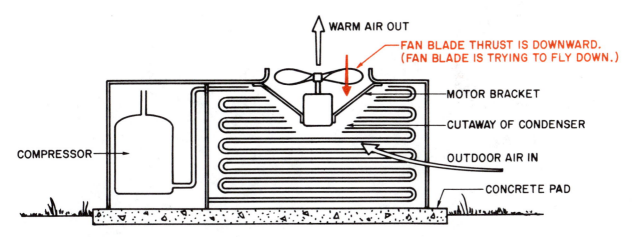

Figure 27-19. Motor working against two conditions: the normal motor load and the fact that the fan is trying to fly downward while pushing air upward.

surface. The shaft actually floats on this oil film and never should touch the bearing surface. The oil film is supplied by the lubrication system. Two types of lubrication systems for sleeve bearings are the *oil port* and the *permanently lubricated* bearing.

The oil port bearing has an oil reservoir that is filled from the outside by means of an access port. This bearing must be lubricated at regular intervals with the correct type of oil, which is usually 20-weight nondetergent motor oil or an oil specially formulated for electric motors. If the oil is too thin, it will allow the shaft to run against the bearing surface. If the oil is too thick, it will not run into the clearance between the shaft and the bearing surface. The correct interval for lubricating a sleeve bearing depends on the design and use of the motor. Manufacturer's instructions will indicate the recommended interval. Some motors have large reservoirs and do not need lubricating for years. This is good if there is limited access to the motor.

The permanently lubricated sleeve bearing is constructed with a large reservoir and a wick to gradually feed the oil to the bearing. This bearing truly does not need lubrication until the oil deteriorates. If the motor has been running hot for many hours, the oil will deteriorate and fail. Shaded-pole motors operating in the heat and weather have these bearing systems and many have operated without failure for years.

Ball-bearing motors are not as quiet as sleeve-bearing motors and are used in locations where their noise levels will not be noticed. Large fan motors and pump motors are normally located far enough from the conditioned space that the bearing noises will not be noticed. These bearings are made of very hard material and usually lubricated with grease rather than oil. Motors with ball bearings generally have permanently lubricated bearings or grease fittings.

Permanently lubricated bearings are similar to sleeve bearings, but they have reservoirs of grease sealed in the bearing. They are designed to last for years with the lubrication furnished by the manufacturer, unless working conditions are really bad.

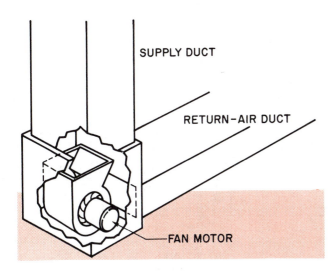

Figure 27-20. Furnace fan motor mounted in a horizontal position.

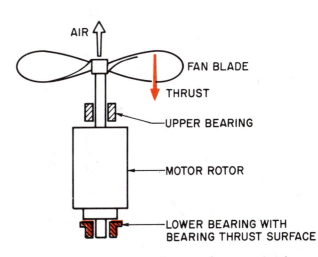

Figure 27-21. Thrust surface on a fan motor bearing.

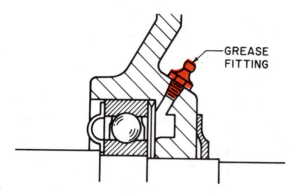

Figure 27–22. Fitting through which the bearing is greased.

Bearings needing lubrication have grease fittings, so a grease gun can force grease into the bearing. This is often done by hand, Figure 27–22. Only an approved grease must be used. In Figure 27–23 the slotted screw at the bottom of the bearing housing is a *relief* screw. When grease is pumped into the bearing, this screw must be removed, or the pressure of the grease may push the grease seal out and grease will leak down the motor shaft.

Large motors use a type of ball bearing called a *roller* bearing, which has cylindrically shaped rollers instead of balls.

27.6 MOTOR MOUNTING CHARACTERISTICS

Mounting characteristics of a motor determine how it will be secured during its operation. Noise level must be considered when mounting a motor. Two primary means are *rigid* mount and *resilient* or rubber mount: Rigid-mount motors are bolted, metal to metal, to the frame of the fan or pump and will transmit any motor noise into the piping or duct work. (The motor hum is an electrical noise, which is different from bearing noise, and must also be isolated in some installations.)

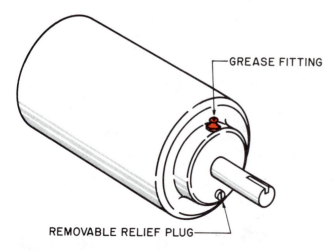

Figure 27–23. Motor bearing using grease for lubrication. Notice the relief plug.

Figure 27–24. Grounding strap to carry current from the frame if the motor has a grounded winding. *Courtesy Universal Electric Company*

Resilient-mount motors use different methods of isolating the motor noise and bearing noise from the metal frame work of the system. Notice the ground strap on the resilient mount motor. This motor is electrically and mechanically isolated from the metal frame, Figure 27–24. If the motor were to have a ground (circuit from the hot line to the frame of the motor), the motor frame would be electrically hot without the ground strap. *When replacing a motor always connect the ground strap or the motor could become dangerous, Figure 27–25.*

The four basic mounting styles are *cradle mount, rigid-base mount, end mount* (with tabs or studs or flange mount) and *belly-band-strap mount,* all of which fit standard dimensions established by NEMA and

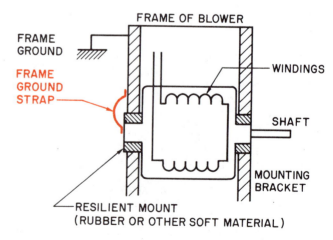

Figure 27–25. How the grounding strap works.

MOTOR DIMENSIONS FOR NEMA FRAMES

Standardized motor dimensions as established by the National Electrical Manufacturers Association (NEMA) are tabulated below and apply to all base-mounted motors listed herein which carry a NEMA frame designation.

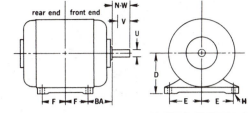

NEMA FRAME	All Dimensions in Inches							V(§) Min.	Key			NEMA FRAME
	D(*)	2E	2F	BA	H	N-W	U		Wide	Thick	Long	
42	2⅝	3½	1¹¹⁄₁₆	2¹⁄₁₆	⁹⁄₃₂ slot	1⅛	⅜	—	—	²¹⁄₆₄ flat	—	42
48	3	4¼	2¾	2½	¹¹⁄₃₂ slot	1½	½	—	—	²⁹⁄₆₄ flat	—	48
56	3½	4⅞	3	2¾	¹¹⁄₃₂ slot	1⅞(†)	⅝(†)	—	³⁄₁₆(†)	³⁄₁₆(†)	1⅜(†)	56
56H			3&5(‡)									56H
56HZ	3½	**	**	**	**	2¼	⅞	2	³⁄₁₆	³⁄₁₆	1⅜	56HZ

Figure 27–26. Dimensions of typical motor frames. *Courtesy W. W. Grainger, Inc.*

which are distinguished from each other by *frame numbers.* Figure 27–26 shows some typical examples.

Cradle-Mount Motors

Cradle-mount motors are used for either direct-drive or belt-drive applications. They have a cradle that fits the motor end housing on each end. The end housing is held down with a bracket, Figure 27–27. The cradle is fastened to the equipment or pump base with machine screws, Figure 27–28. Cradle-mount motors are available only in the small horsepower range. A handy service feature is that the motor can easily be removed.

Rigid-Mount Motors

Rigid-base-mount motors are similar to cradle mount except that the base is fastened to the motor body, Figure 27–29. The sound isolation for this motor is in the belt, if one is used, that drives the prime mover. The belt is flexible and dampens motor noise. This motor is often used as a direct drive to turn a compressor or pump. A flexible coupling is used between the motor and prime mover, Figure 27–30. We will discuss the different types of drives later in the text.

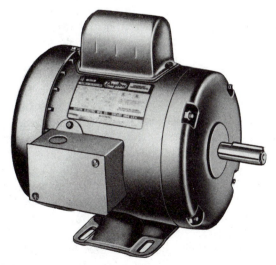

Figure 27–28. Cradle fastened to base of pump. *Courtesy W. W. Grainger, Inc.*

Figure 27–27. Cradle-mount motor. *Courtesy W. W. Grainger, Inc.*

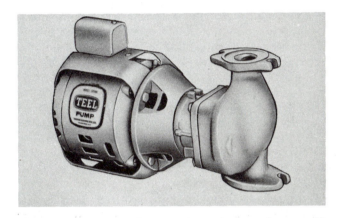

Figure 27–29. Rigid-mount motor. *Courtesy W. W. Grainger, Inc.*

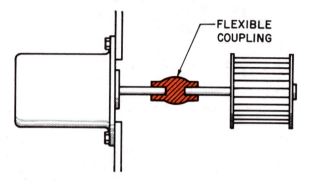

Figure 27-30. Direct-drive motor with flexible coupling.

Figure 27-31. Motors end mounted with tabs and studs. *Courtesy W. W. Grainger, Inc.*

End-Mount Motors

End-mount motors are very small motors mounted to the prime mover with tabs or studs fastened to the motor housing, Figure 27-31. Flange-mounted motors have a flange as a part of the motor housing, for example, an oil-burner motor, Figure 27-32.

Belly-Band-Mount Motors

Belly-band-mount motors have a strap that wraps around the motor to secure it with brackets mounted to the strap. These motors are often used in air conditioning air handlers. Several universal types of motor kits are belly-band mounted and will fit many different applications. These motors are all direct drive, Figure 27-33.

27.7 MOTOR DRIVES

Motor drives are devices or systems that connect a motor to the driven load. For instance, the motor is a driving device, and a fan is a driven component. All motors drive their loads by belts, direct drives through couplings, or the driven component may be mounted on the motor shaft. Gear drives are a form of direct drive and will not be covered in this text because they are used mainly in large industrial applications.

The drive mechanism is intended to transfer the motor's rotating power or energy to the driven device. For example, a compressor motor is designed to transfer the motor's power to the compressor to pump refrigerant from the low side to the high side of the system. Efficiency, speed to the driven device, and noise level are some factors involved in this transfer. It takes energy to turn the belts and pulleys on a belt-drive system in addition to the compressor load. Therefore a direct drive motor may be better suited for this application. Figure 27-34 is an example of both direct and belt drives.

Belt-drive applications have been used for years to drive both fans and compressors. Pulley sizes can be changed, and the speed of the driven device may be

Figure 27-32. Motor for an oil burner with a flange on the end to hold the motor to the equipment that is being turned. *Courtesy W. W. Grainger, Inc.*

Figure 27-33. Belly-band-mount motors. *Courtesy W. W. Grainger, Inc.*

Figure 27-34. Motor drive mechanisms. *Reproduced courtesy of Carrier Corporation*

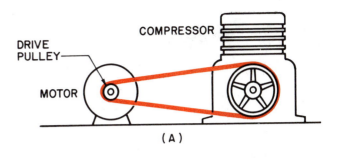

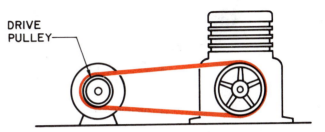

SAME MOTOR–LARGER DRIVE PULLEY WILL
CAUSE THE COMPRESSOR TO TURN FASTER

(B)

Figure 27-35. Belt drive.

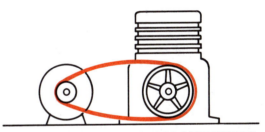

MOTOR IS ADJUSTED TOWARD COMPRESSOR
FOR BELTS TO BE INSTALLED.

Figure 27-36. Correct method for installing belts over a pulley. The adjustment is loosened to the point that the belts may be passed over the pulley side.

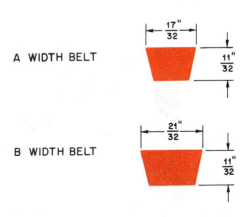

Figure 27-37. A and B width belts.

changed. This is a versatility of the belt-drive type of system, Figure 27-35. This can be a great advantage if the capacity of a compressor or a fan speed needs to be changed. However, the changes must be made within the capacity of the drive motor.

Belts are manufactured in different types and sizes. Some have different fibers inside to prevent stretching. *Handle belts carefully during installation. A belt designed for minimum stretch must not be installed by forcing it over the side of the pulley because it may not stretch enough. Fibers will break and weaken the belt, Figure 27-36.*

Belt widths are denoted by "A" and "B." An A width belt must not be used with a B width pulley nor vice versa, Figure 27-37.

Belts can have different grips, Figure 27-38.

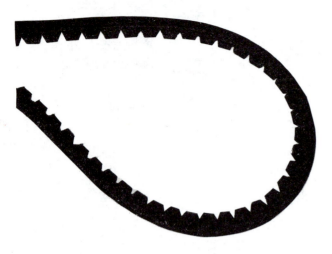

Figure 27–38. Belt with grooves has a tractor type grip. *Photo by Bill Johnson*

When a drive has more than one belt, the belts must be matched. Two belts with the same length marked on the belt are not necessarily matched. They may not be exactly the same length. A *matched* set of belts means the belts are *exactly* the same length. A set of 42-in. belts *marked* as a matched set means each belt is exactly 42 in. If the belts are not marked as a matched set, one may be $42\frac{1}{4}$ in., and the other may be $41\frac{3}{4}$ in. Thus the belts will not pull evenly—one belt will take most of the load and will wear out first.

Belts and pulleys wear like any moving or sliding surface. When a pulley begins to wear, the surface roughens and wears out the belts. Normal pulley wear is caused by use or running time. Belt slippage will cause premature wear. Pulleys must be inspected occasionally, Figure 27–39.

Belts must have the correct tension, or they will cause the motor to operate in an overloaded condition. A belt-tension meter should be used to correctly adjust belts to the proper tension. This meter is used in conjunction with a chart that gives the correct tension for different types of belts of various lengths.

Direct-drive motor applications are normally used with drive motors for fans, pumps, and compressors. Small fans and hermetic compressors are direct drive, but the motor shaft is actually an extended shaft that has the fan or compressor on the end, Figure 27–40. The technician can do nothing to alter these. When this type is used in an open-drive application, some sort of coupling must be installed between the motor and the driven device, Figure 27–41. Some couplings have springs that connect the two coupling halves together to absorb small amounts of vibration from the motor or pump.

A more complicated coupling is used between the motor and a larger pump or a compressor, Figure 27–42. This coupling and shaft must be in very close alignment, or vibration will occur. This was discussed to some extent in Unit 21. The alignment must be checked to see that the motor shaft is parallel with the compressor or pump shaft. Alignment is a very precise operation and is done by experienced technicians. If two shafts are aligned to within tolerance while the motor and driven mechanism is at room temperature, the alignment must be checked again after the system is run long enough to get the system up to operating temperature. The motor may not expand and move the

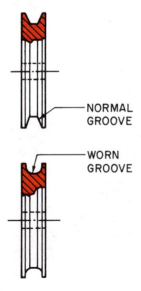

Figure 27–39. Normal and worn pulley comparison.

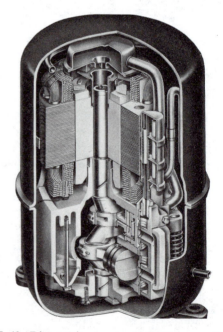

Figure 27–40. Direct-drive compressor. *Courtesy Tecumseh Products Company*

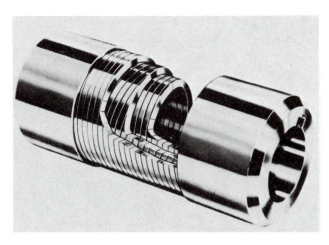

Figure 27–41. Small flexible coupling. *Courtesy Lovejoy, Inc.*

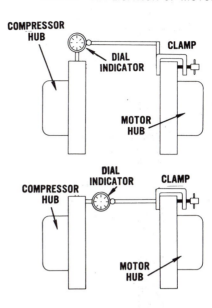

Figure 27–43. The alignment of the two shafts must be very close for the system to operate correctly. *Courtesy Trane Company*

same distance as the driven mechanism and the alignment may need to be adjusted to the warm value, Figure 27–43.

When a new motor must be installed to replace an old one, try to use an exact replacement. Sometimes the motor must be obtained from the original equipment manufacturer; sometimes an exact replacement can be found at a motor supply house. When the motor is not a normally stocked motor, you can save much time by taking the old motor to the distributor and asking for one "just like this."

SUMMARY

- In many installations only one type of motor can be used.
- The power supply determines the applied voltage, the current capacity, the frequency, and the number of phases.
- The working conditions (duty) for a motor deal with the atmosphere in which the motor must operate (i.e., wet, explosive, or dirty).
- Motors are also classified according to the insulation of the motor windings and the motor temperature under which they operate.
- Each motor has sleeve, ball, or roller bearings. Sleeve bearings are the quietest but will not stand heavy loads.
- *Motors must be mounted in the fashion designed for the installation.*
- *An exact motor replacement should be obtained whenever possible.*
- The drive mechanism transfers the motor's energy to the driven device (the fan, pump, or compressor).

Figure 27–42. More complicated coupling used to connect larger motors to large compressors and pumps. *Courtesy Lovejoy, Inc.*

REVIEW QUESTIONS

1. Name four items to consider in a power supply for a system and its motors.
2. What is the allowable voltage variation for typical motors?
3. What are the two main power-supply characteristics that the technician has some control over?

4. What is meant by the service factor of a motor?
5. What two categories of electric motors concern the power supply?
6. Name some of the typical conditions a motor must operate in.
7. How does the insulation class of a motor affect its use?
8. Name two types of bearings commonly used on small motors.

9. Name four types of motor mounts.
10. What is the best replacement motor to use for a special application?
11. What does the drive mechanism do?
12. What is a matched set of belts?
13. Name the different types of belts.
14. Why must direct-drive couplings be aligned?
15. Why are springs used in a small coupling?

MOTOR STARTING

OBJECTIVES

After studying this unit, you should be able to

- describe the differences between a relay, a contactor, and a starter.
- state how the locked-rotor current of a motor affects the choice of a motor starter.
- list the basic components of a contactor and starter.
- compare two types of external motor overload protection.
- describe conditions that must be considered when resetting devices to restart electric motors.

28.1 INTRODUCTION TO MOTOR CONTROL DEVICES

Unit 27 dealt with motor starting and running after the power was distributed to the motor terminals. This unit concerns those components used to close or to open the power-supply circuit to the motor—relays, contactors, and starters.

For example, a compressor in a residential air conditioner is controlled in the following manner. The thermostat contacts close on a temperature rise in the space to be conditioned. These contacts pass low voltage and energize the coil on the compressor relay. This closes the contacts on this device, allowing the applied or line voltage to pass to the compressor's motor windings. Relays, contactors, and starters do the same job. Figure 28–1 is a diagram of a motor starting relay. They all have coils that, when energized, close contacts and pass the line current to the motor.

The size of the motor and the application usually determines the type of switching device (relay, contactor or starter). For example, a small manual switch can start and stop a hand-held drier, and a person operates the switch. A large 100-hp motor that drives an air conditioning compressor must start, run, and stop unattended. It will also consume much more current than the hair drier. The components that start and stop large motors must be more elaborate than those that start and stop small motors.

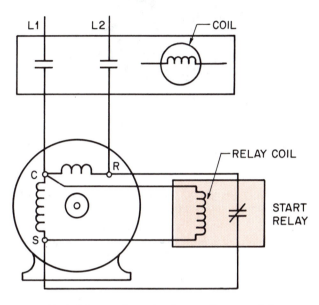

Figure 28–1. The starting relay is actually part of the motor circuit.

28.2 RUN LOAD AND LOCKED-ROTOR CURRENT

Electric motors have two current (amperage) ratings: the *run load amperage* (RLA), sometimes referred to as the *full load amperage* (FLA), and the *locked-rotor amperage* (LRA). The RLA or FLA is the current drawn while the motor is running. The LRA is the current drawn by the motor just as it begins to start. Both currents must be considered when choosing the component (relay, contactor, starter) that passes the line voltage to the motor.

28.3 THE RELAY

The *relay* has a magnetic coil that closes one or more sets of contacts, Figure 28–2. It is a throw-away device because parts are not available for rebuilding it.

Relays are designed for light duty. *Pilot* relays can switch (on and off) larger contactors or starters. Pilot relays for switching circuits are very light duty and are not designed to directly start motors. Relays designed for starting motors are not really suitable as switching relays because they have more resistance in the contacts.

Figure 28-2. Small relay for starting a motor. It has a magnetic coil that closes the contacts when the coil is energized. *Photo by Bill Johnson*

The pilot-duty relay contacts are often made of a fine silver alloy and designed for low-level current switching. Use on a higher load would melt the contacts. Heavier-duty motor switching relays are often made of silver cadmium oxide with a higher surface resistance and are physically larger than pilot-duty relays.

If a relay starts the indoor fan in the cooling mode in a central air conditioning system, it must be able to withstand the inrush current of the fan motor on startup, Figure 28-3. (Remember, a motor normally has a starting current of five times the running current.) Relays are usually rated in horsepower: If a relay is rated for a 3-hp motor, it will be able to stand the inrush or locked-rotor current of a 3-hp motor.

Figure 28-3. Fan relay that may be used to start an evaporator fan on a central air conditioning system. This is a throw-away relay. *Photo by Bill Johnson*

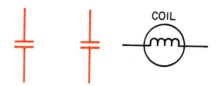

Figure 28-4. Double-pole, single-throw relay with two sets of contacts that close when the coil is energized.

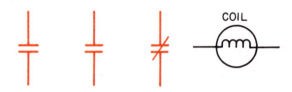

Figure 28-5. Relay with two sets of contacts that close and one set that opens when the coil is energized. It has two normally open contacts (NO) and one normally closed (NC).

A relay may have more than one type of contact configuration. It could have two sets of contacts that close when the magnetic coil is energized, Figure 28-4, or it may have two sets of contacts that close and one set that opens when the coil is energized, Figure 28-5. A relay with a single set of contacts that close when the coil is energized is called a *single-pole–single-throw, normally open relay* (spst,NO). A relay with two contacts that close and one that opens is called a *triple-pole–single-throw,* with two *normally open* contacts, and one *normally closed contact, TPST.* Figure 28-6 shows some different relay contact arrangements.

CONTACT ARRANGEMENT DIAGRAMS

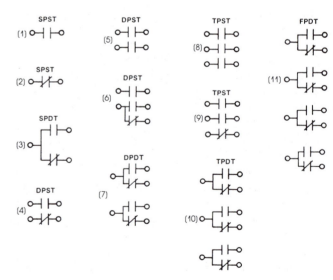

Figure 28-6. Some of the more common combinations of contacts supplied with relays. *Reproduced courtesy of Carrier Corporation*

Figure 28–7. Large contactor. *Courtesy Honeywell, Inc.— Residential Division*

28.4 CONTACTOR

The *contactor* is a larger version of a relay, Figure 28–7. It can be rebuilt if it fails, Figure 28–8. A contactor has movable and stationary contacts. The holding coil can be designed for various operating voltages; 24-V, 208–230-V, or 460-V operation.

Contactors can be very small or very large, but all have similar parts. A contactor may have many configurations of contacts—from the most basic, which has just one contact, to more elaborate types with several. The single-contact type contactor is used on many residential air conditioning units. It interrupts power to one leg of the compressor, which is all that is necessary to stop a single phase compressor. Sometimes a single-contact contactor is needed to provide crankcase heat to the compressor. A trickle charge of electricity passes through this contact to the motor windings. If you use a contactor with two contacts for a replacement, the compressor would not have crankcase heat. Once again, an exact replacement is the best choice. Figure 28–9

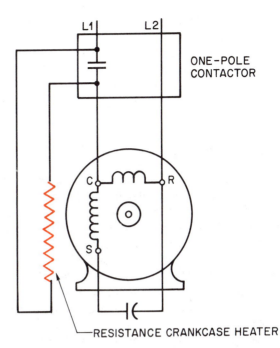

ONE-POLE CONTACTOR

RESISTANCE CRANKCASE HEATER

Figure 28–9. This contactor only has one set of contacts. Only one line of the power needs to be broken in order to stop a motor like the one in the diagram. This is a method of supplying crankcase heat to the compressor only during the off cycle. The current during the running cycle (with the contacts closed) bypasses the heater and goes to the run winding.

Figure 28–8. Parts that may be replaced in a contactor: the contacts, both movable and stationary; the springs that hold the contacts; and the holding coil. *Photo by Bill Johnson*

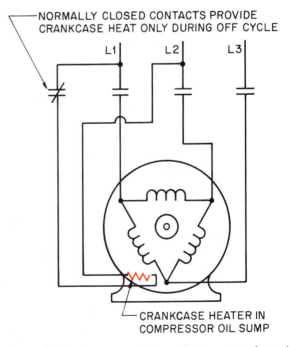

NORMALLY CLOSED CONTACTS PROVIDE CRANKCASE HEAT ONLY DURING OFF CYCLE

CRANKCASE HEATER IN COMPRESSOR OIL SUMP

Figure 28–10. Contactor with auxiliary contacts for switching other circuits.

shows another method of supplying crankcase (oil sump) heat only during the off cycle.

Some contactors have as many as five or six sets of contacts. Generally for larger motors there will be three heavy-duty sets to start and stop the motor; the rest (*auxiliary contacts*) can be used to switch auxiliary circuits, Figure 28–10.

28.5 MOTOR STARTERS

The *starter,* or motor starter as it is sometimes called, is similar to the contactor. In some cases a contactor may be converted to a motor starter. The motor starter differs from a contactor because it has motor overload protection built into its framework, Figures 28–11 and 28–12. A motor starter may be rebuilt and ranges in size from small to very large.

＊Motor protection will be discussed more fully later in this unit, but we should know that overload protection protects that particular motor. See Figure 28–13 for a melting alloy-type overload heater. The fuse or circuit breaker cannot be wholly relied upon for protection because it protects the entire circuit, which may have many components.＊ In some cases the motor starter's protection is a better indication of a motor problem than the motor winding thermostat protection will provide. Choosing the correct size of overload device will be discussed later also.

The contact's surfaces become dirtier and more pitted with each motor starting sequence, Figure 28–14. Some technicians believe that these contacts may be buffed or cleaned with a file or sandpaper. ＊Contacts may be cleaned, but this should be done as a temporary measure only. Filing or sanding exposes base metal under the silver plating and speeds its deterioration. Replacing the contacts is the only recommended repair.＊ If this device is a relay, the complete relay must be replaced. If the device is a contactor or a starter, the contacts may be replaced with new ones, Figure 28–14. There are both movable and stationary contacts and springs that hold tension on the contacts.

You can use a voltmeter to check resistance across a set of contacts under full load. When the meter leads are placed on each side of a set of contacts and there is a measurable voltage, there is resistance in the contacts. When the contacts are new and have no resistance, no voltage should be read on the meter. An old set of contacts will have some slight voltage drop and will produce heat, due to resistance in the contact's surface, where the voltage drop occurs. Figure 28–15 shows an example of how contacts may be checked with an ohmmeter.

Each time a motor starts, the contacts will be exposed to the inrush current (the same as LRA) for the moment the motor starts turning. These contacts are under a tremendous load for this moment. When the contacts open, there is an arc caused by the breaking

Figure 28–11. Motor starter. *Courtesy Square D Company*

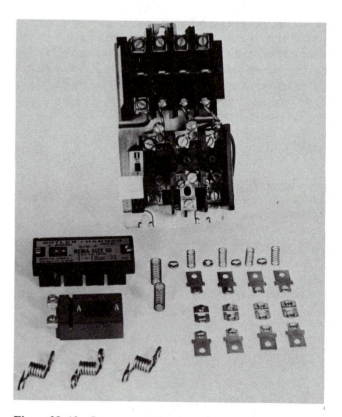

Figure 28–12. Components that may be changed on a starter: contacts, both movable and stationary; the springs; the coil; and the overloads. *Photo by Bill Johnson*

Figure 28–13. Melting alloy type overload heater. *Courtesy Square D Company*

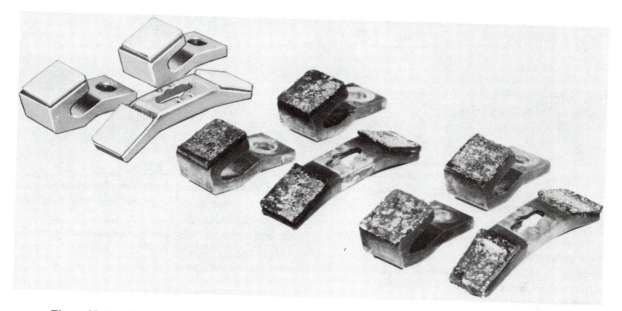

Figure 28–14. Clean contacts contrasted with a set of dirty pitted contacts. *Courtesy Square D Company*

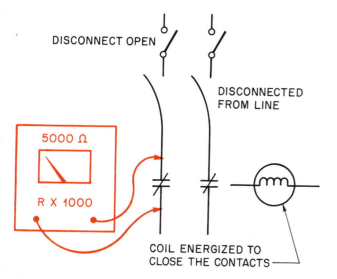

Figure 28–15. Resistance in contacts due to dirty surface.

of the electrical circuit. These contacts must make and break as fast as possible. The contacts have a magnet that pulls them together with a mechanism to take up any slack. For example, there may be three large movable contacts in a row that must all be held equally tight against the stationary contacts. The springs behind the movable contacts keep this pressure even, Figure 28–16. If there is a resistance in the contact surface, the contacts may get hot enough to take the tension out of the springs. This will make the contact-to-contact pressure even less and make the heat greater. The tension in these springs must be maintained.

28.6 MOTOR PROTECTION

The electric motors used in air conditioning, heating, and refrigeration equipment are the most expensive single components in the system. These motors con-

LINE "L" AND LOAD "T" TERMINALS

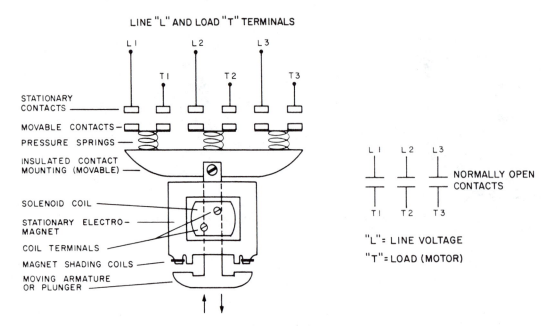

Figure 28–16. Springs holding the three movable contacts tightly against the stationary contacts. From Alerich, *Electric Motor Control,* © 1983 by Delmar Publishers Inc.

sume large amounts of electrical energy, and there is considerable stress on the motor windings. Therefore they deserve the best protection possible within the economic boundaries of a well-designed system. The more expensive the motor, the more expensive and elaborate the protection should be.

Fuses (see Unit 11) are usually used as circuit (not motor) protectors. *The conductor wiring in the circuit must not be allowed to pass too much current or it will overheat and cause conductor failures or fires.* A motor may be operating at an overloaded condition that would not cause the conductor to be overloaded; hence the fuse will not open the circuit. Let's use a central air conditioning system as an example. There are two motors in the condensing unit: the compressor (the largest) and the condenser fan motor, (the smallest). In a typical unit the compressor may have a current draw of 23 A, and the fan motor may use 3 A. The fuse protects the total circuit. If one motor is overloaded, the fuse may not open the circuit, Figure 28–17. Each motor should be protected within its own operating range.

Motors can operate without harm for short periods at a slight overcurrent condition. The overload protection is designed to disconnect the motor at some current draw value that is slightly more than the FLA value, so the motor can be operated at its full-load design condition. Time is involved in this value in such a manner that the higher the current value above FLA, the more quickly the overload should react. The amount of the overload and the time are both figured into the design of the particular overload device.

Overload current protection is applied to motors in different ways. Overload protection, for example, is not needed for some small motors that will not cause circuit overheating or will not damage themselves. Some small motors do not have overload protection because they will not consume enough power to damage the motor unless shorted from winding to wind-

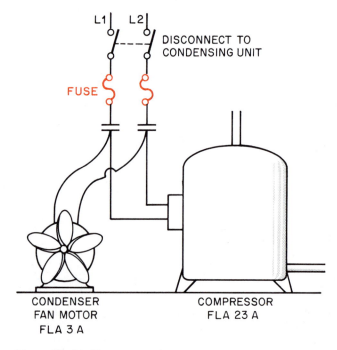

Figure 28–17. Two motors in the same circuit served by the same conductor.

Figure 28-18. Small condenser fan motor does not pull enough amperage at locked rotor to create enough heat to be a problem. This is known as impedance motor protection. *Courtesy W. W. Grainger, Inc.*

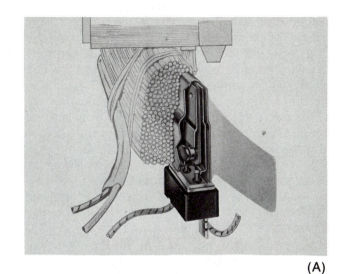

(A)

(B)

Figure 28-19. (A) Inherent overload protector. (B) External overload protector. *Courtesy (A) Tecumseh Products Company (B) Photo by Bill Johnson*

ing or from the winding to the frame (to ground). See Figure 28-18 for an example of a small condenser fan motor that does not draw enough amperage at the LRA condition to overheat. This motor does not have overload protection. It is not a very expensive motor. If the motor fails because of a burn out, the current draw will be interrupted by the circuit protector.

Overload protection is divided into inherent protection and external protection.

28.7 INHERENT MOTOR PROTECTION

Inherent protection is that provided by internal thermal overloads in the motor windings or the thermally activated snap-disc (bimetal). These were discussed in Unit 14, pertaining to compressors, Figure 28-19. The same types of devices are used with open motors.

28.8 EXTERNAL MOTOR PROTECTION

External protection is applied to the device passing power to the motor contactor or starter. These are normally devices actuated by current overload and break the circuit to the contactor coil. The contactor stops the motor. When a motor is started with a relay, the motor is normally small and only has internal protection, Figure 28-20. Contactors are used to start larger motors, and either inherent protection or external protection is used. Large motors (above 5 hp in air conditioning, heating, and refrigeration systems) use starters and overload protection either built into the starter or in the contactor's circuit.

The value (trip point) and type of the overload protection is normally chosen by the system design engineer or by the manufacturer. The technician checks the overload devices when there is a problem, such as random shutdowns because of an overload tripping. The technician must be able to understand the designer's intentions as well as the motor's operation and the overload

device operation because they are closely related in a working system.

28.9 SERVICE FACTOR

All motors have an FLA rating at which the motor is designed to operate for continuous periods. Some

Figure 28-20. Fan motor that is normally started with a relay. Motor protection is internal. *Courtesy W. W. Grainger, Inc.*

motors have a service factor, which is the reserve capacity of the motor. The motor can operate above the FLA and within the service factor range for short periods without harm. Typical service factors are 1.15 to 1.25. For example, a motor with an FLA of 10 A can operate from 11.5 to 12.5 A, depending on the service factor, for short periods without damaging the motor. The overload protection for a particular motor takes the service factor into account.

Some motors are protected at a value above the service factor. For example, protection for a hermetic compressor motor may be at a value of 140% over the FLA rating.

28.10 NATIONAL ELECTRICAL CODE STANDARDS

The National Electrical Code (NEC) sets the standard for all electrical installations, including motor overload protection. The code book published by NEMA should be consulted for any overload problems or misunderstandings that may occur regarding correct selection of the overload device.

The purpose of the overload protection device is to disconnect the motor from the circuit when an overloaded condition occurs. Detecting the overload condition and opening the circuit to the motor can be done in several ways: by an overload device mounted on the motor starter, or by a separate overload relay applied to a system with a contactor. See Figure 28–21 for an example of a thermal overload relay.

28.11 TEMPERATURE-SENSING DEVICES

Various sensing devices are used for overcurrent situations. A review of Unit 14 would help at this time because most of these devices have been discussed in

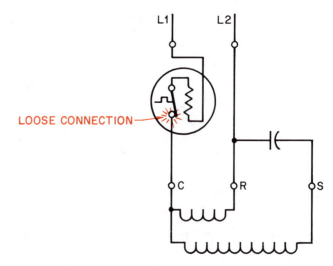

Figure 28–22. Overload tripping because of a loose connection.

detail. The most popular devices are those that are sensitive to temperature changes.

The bimetal element is an example. The line current of the motor passes through a heater (that can be changed to suit a particular motor amperage) that heats a bimetal strip. When the current is excessive, the heater warps the bimetal, opening a set of contacts that interrupt power to the contactor's coil circuit. Many bimetal devices are designed as snap-discs to avoid excessive arcing. These thermal-type overloads are sensitive to any temperature conditions around them. High ambient temperature and loose connections are frequent problems encountered. See Figure 28–22 for an example of a thermal overload with a loose connection.

A low-melting solder may be used in place of the bimetal. This is called a *solder pot*. The solder will melt from the heat caused by an overcurrent condition. The overload heater is sized for the particular amperage draw of the motor it is protecting. The overload's control circuit will interrupt the power to the motor's contactor coil and stop the motor in case of overload. The solder melts and the overload's mechanism turns because it is spring loaded. It can be reset when it cools, Figure 28–21.

Both of these overload protection devices are sensitive to temperature. The temperature of the heater causes them to function. Heat from any source, even if it has nothing to do with motor overload, makes the protection devices more sensitive. For example, if the overload device is located in a metal control panel in the sun, the heat from the sun may affect the performance of the overload protection device. A loose connection on one of the overload device leads will cause local heat and may cause it to open the circuit to the motor even though there is actually no overload, Figure 28–23.

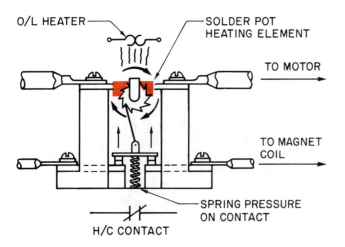

Figure 28–21. Overload using a resistor-type of heater that heats a low-temperature solder. From Alerich, *Electric Motor Control* © 1983 by Delmar Publishers Inc.

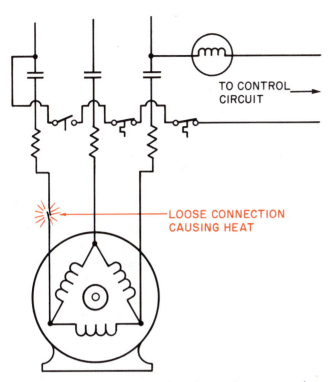

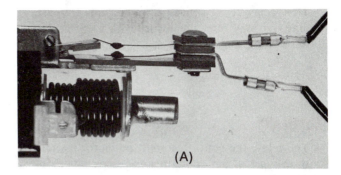

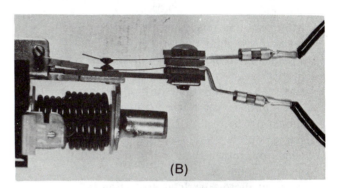

Figure 28–23. Local heat, such as from a loose connection, will influence the thermal type of overload.

Figure 28–24. Magnetic overload. (A) Contacts open. (B) Contacts closed. (C) Overload enclosure. *Photos by Bill Johnson*

28.12 MAGNETIC OVERLOAD DEVICES

Magnetic overload devices are separate components and are not attached to the motor starter. This component is very accurate and not affected by ambient temperature, Figure 28–24.

28.13 RESTARTING THE MOTOR

When a motor is stopped for safety reasons, don't restart it immediately. When a motor stops because it is overloaded, the overload condition at the instant the motor is stopped, is reduced to 0 A. This does *not* mean that the motor should be immediately restarted. The cause of the overload may still exist. Also, the motor could be too hot and may need time to cool.

There are various ways of restarting a motor after an overload condition has occurred. Some manufacturers design their control circuits with a manual reset to keep the motor from restarting, and some use a time delay to keep the motor off for a predetermined time. Others use a relay that will keep the motor off until the thermostat is reset. The units that have a manual reset at the overload device may require someone to go to the roof to reset the overload if the unit is located on the roof. See Figure 28–25 for an example of a manual reset. When the reset is in the thermostat circuit, the protection devices may be reset from the room thermostat. This is very convenient, but several controls may be reset at the same time and the technician may not know which control is being reset.

Figure 28–25. Manually reset overload. *Courtesy Square D Company*

Time-delay reset devices keep the unit from short cycling but may reset themselves with a problem condition still existing.

SUMMARY

- The relay, the contactor, and the motor starter are three types of motor starting and stopping devices.
- The relay is used for switching circuits and motor starting.
- Motor starting relays are used for heavier-duty jobs than are switching relays.
- Contactors are large relays that may be rebuilt.
- Starters are contactors with motor overload protection built into the framework of the contactor.
- The contacts on relays, contactors, and starters should not be filed or sanded.
- Large motors should be protected from overload conditions other than by normal circuit overload protection devices.
- Inherent motor overload protection is provided by sensing devices within the motor.
- External motor overload protection is applied to the current-passing device: the relay, contactor, or starter.
- The technician must have an understanding of the motor operation and the overload device operation.
- The service factor is the reserve capacity of the motor.
- Bimetal and solder-pot devices are thermally operated.
- Magnetic overload devices are very accurate and not affected by ambient temperature.

● *Most motors should not be restarted immediately after shutdown from an overloaded condition because they may need time to cool. When possible, determine the reason for the overloaded condition before restarting the motor.*

REVIEW QUESTIONS

1. What is the recommended repair policy for a relay?
2. What components can be changed on a contactor and a starter for rebuilding purposes?
3. What are the two types of relays?
4. What two amperages influence the choice for replacing a motor starting component?
5. What is the difference between a contactor and a starter?
6. Can a contactor ever be converted to a starter?
7. What are the contact surfaces of relays, contactors, and starters made of?
8. What causes an overload protection device to function?
9. What are typical coil voltages used for relays, contactors, and starters?
10. Why is it not a good idea to file or sand the contactor's contacts?
11. Why is it not a good idea to use circuit protection devices to protect large motors from overload conditions?
12. Under what conditions are motors allowed to operate with slightly higher than design loads?
13. Describe the difference between inherent and external overload protection.
14. What is the purpose of overload protection at the motor?
15. State reasons why a motor should not be restarted immediately.

TROUBLESHOOTING ELECTRIC MOTORS

OBJECTIVES

After studying this unit, you should be able to

- **describe different types of electric motor problems.**
- **list common electrical problems in electric motors.**
- **identify various mechanical problems in electric motors.**
- **describe a capacitor checkout procedure.**
- **explain the difference in troubleshooting a hermetic motor problem and an open motor problem.**

29.1 ELECTRIC MOTOR TROUBLESHOOTING

Electric motor problems are divided into mechanical and electrical problems. Mechanical problems may appear to be electrical. For example, a bearing dragging in a small PSC fan motor may not make any noise. The motor may not start, and it appears to be an electrical problem. The technician must know how to correctly diagnose the problem. This is particularly true with open motors, because if the driven component is stuck a motor may be changed unnecessarily. If the stuck component is a hermetic compressor, the whole compressor must be changed; if it is a serviceable hermetic compressor, you can change the motor or rebuild the compressor running gear.

29.2 MECHANICAL MOTOR PROBLEMS

Mechanical motor problems normally occur in the bearings or the shaft where the drive is attached. The bearings can be tight or worn due to lack of lubrication. Grit can easily get into the bearings of some open motors and cause them to wear.

Motor problems are not usually repaired by heating, air conditioning, and refrigeration technicians. They are handled by technicians trained in motor and rotating equipment rebuilding. A motor vibration may require you to seek help from a qualified balancing technician. Explore every possibility to insure that the vibration is not caused by a field problem, such as a fan loaded with dirt or liquid flooding into a compressor.

When motor bearings fail, they can be replaced. If the motor is small, the motor is normally replaced because it would cost more to change the bearings than to purchase and install a new motor. The labor involved in obtaining bearings and disassembling the motor can take too much time to make a profit. This is particularly true for fractional horsepower fan motors. These small motors almost always have sleeve bearings pressed into the end bells of the motor, and special tools may be needed to remove and to install new bearings, Figure 29–1.

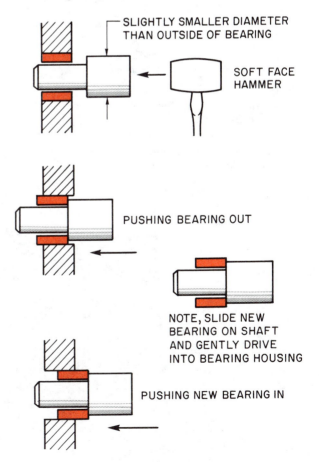

Figure 29–1. Special tool to remove bearings.

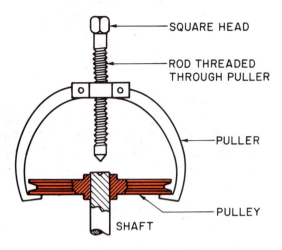

Figure 29-2. Pulley puller.

29.3 REMOVING DRIVE ASSEMBLIES

To remove the motor, you need to remove the pulley, coupling, or fan wheel from the motor shaft. The fit between the shaft and whatever assembly it is fastened to may be very tight. *Removing the assembly from the motor shaft must be done with care.* The assembly may have been running on this shaft for years, and it may have rust between the shaft and the assembly. You must remove the assembly without damaging it. Special pulley pullers will help, but other tools or procedures may be required, Figure 29-2.

Most assemblies are held to the motor shaft with set screws threaded through the assembly and tightened against the shaft. A flat spot is usually provided on the shaft for the set screw to be seated and to keep it from damaging the shaft surface. The set screw is very hard steel, much harder than the motor shaft, Figure 29-3. When larger motors with more torque are used, a matching keyway is normally machined in the shaft and assembly. This keyway provides a better bond between the motor and assembly, Figure 29-4. A set

Figure 29-3. Flat spot on motor shaft where pulley set screw is tightened. *Photo by Bill Johnson*

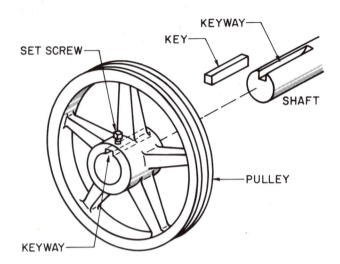

Figure 29-4. Pulley with a groove cut in it that matches a groove in the shaft. A key is placed in these grooves and a set screw is tightened down on top of the key.

screw is then often tightened down on the top of the key to secure the assembly to the motor shaft.

Many technicians make the mistake of trying to drive a motor shaft out of the assembly fastened to the shaft. In doing so, they blunt or distort the end of the motor shaft. When it is distorted, the motor shaft will never go through the assembly without damaging it, Figure 29-5. The shaft is made from mild steel and can be easily damaged. If the shaft must be driven, use a similar shaft with a slightly smaller diameter as the driving tool, Figure 29-6.

29.4 BELT TENSION

Many motors fail because of overtightened belts and incorrect alignment. The technician should be aware of the specifications for the motor belt-tension on belt-drive systems. A belt-tension meter will insure properly adjusted belts. Belts that are too tight strain the bearings so that they wear out prematurely, Figure 29-7.

29.5 PULLEY ALIGNMENT

Pulley alignment is very important. If the motor shaft and drive shaft are not parallel, a strain is imposed on the shafts' drive mechanisms. The pulleys may be made parallel with the help of a straight edge, Figure 29-8. There is a certain amount of adjustment tolerance built into the motor base on small motors, and this may be enough to allow the motor to be out of alignment if the base becomes loose. Aligning shafts on a furnace may not be easy to do under a house by flashlight, but it must be done or the motor or drive mechanism will not last.

When mechanical problems occur with a motor, the motor is normally either replaced or taken to a motor

Figure 29-5. Motor shaft damaged from trying to drive it through the pulley. *Photo by Bill Johnson*

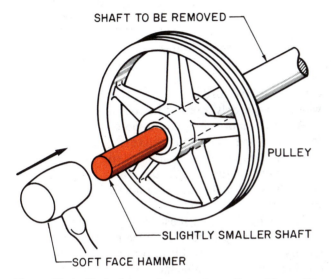

Figure 29-6. Shaft driven through the pulley with another shaft as the contact surface. The shaft that is used as the contact surface is smaller in diameter than the original shaft.

repair shop. Bearings can be replaced in the field by a competent technician, but it is generally better to leave this type of repair to motor experts. When the problem is the pulley or drive mechanism, the air conditioning, heating, and refrigeration technician may be responsible for the repair. *Proper tools must be used for motor repair, or shaft and motor damage may occur.*

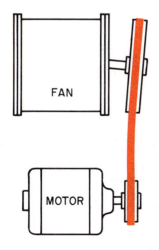

Figure 29-7. Belt that is too tight.

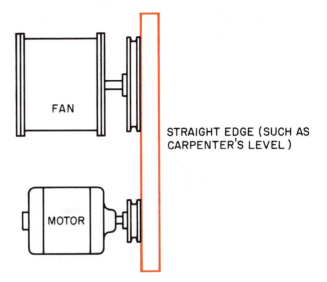

Figure 29-8. Pulleys must be in proper alignment or belt and bearing wear will occur. The pulleys may be aligned using a straight edge.

29.6 ELECTRICAL PROBLEMS

Electrical motor problems are the same for hermetic and open motors. Open motor problems are a little easier to understand or diagnose because they can often be seen. When an open motor burns up, this can often be easily diagnosed because the winding can be seen through the end bells on many motors. With a hermetic motor, instruments must be used because they are the only means of diagnosing problems. There are three common electrical motor problems: (1) an open winding, (2) a short circuit from the winding to ground, and (3) a short circuit from winding to winding.

29.7 OPEN WINDINGS

Open windings in a motor can be found with an ohmmeter. There should be a known measurable resistance from terminal to terminal on every motor for it to run when power is applied to the windings. Single-phase motors must have the applied system voltage at the run winding to run and at the start winding during starting, Figures 29-9. Figure 29-10 is an illustration of a motor with an open start winding.

29.8 SHORTED MOTOR WINDINGS

Short circuits in windings occur when the conductors in the winding touch each other. This creates a short path for the electrical energy to flow through. This path has a lower resistance and increases the current flow in the winding. Although motor windings appear to be made from bare copper wire, they are coated with an insulator to keep the copper wires from touching each other. The measurable resistance men-

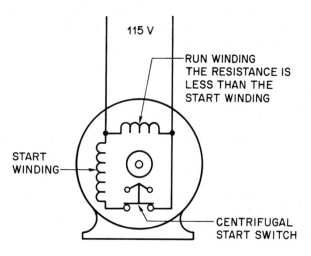

Figure 29–9. Wiring diagram with run and start windings.

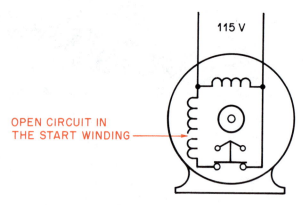

Figure 29–10. Motor with open winding.

tioned in the previous paragraph is known for all motors. Some motors have a published resistance for their windings. The best way to check a motor for electrical soundness is to *know what the measurable resistance should be for a particular winding and verify it with a good ohmmeter,* Figure 29–11. This measurable resistance will normally be less than the rated value when a motor has short-circuit problems. The decrease in resistance causes the current to rise, which causes motor overload devices to trip and possibly even the circuit overload protection may trip. If the resistance does not fall within these tolerances, there is a problem with the winding. This table is quite helpful when you troubleshoot a hermetic compressor. Tables may not be easy to obtain for open motors, and the windings are not as easy to check because the individual windings do not all come out to terminals as they do on a hermetic compressor.

If the decrease in resistance in the windings is in the start winding, the motor may not start. If it is in the run winding, the motor may start and draw too much current while running. If the motor winding resistance cannot be determined, then it is hard to know whether

a motor is overloaded or whether it has a defective winding when only a few of its coils are shorted.

When the motor is an open motor, the load can be removed. For example, the belts can be removed or the coupling can be taken apart, and the motor can be started without the load. If the motor starts and runs correctly without the load, the load may be too great, Figure 29–12.

Three-phase motors must have the same resistance for each winding. (There are three identical windings.) Otherwise there is a problem. An ohmmeter check will quickly reveal an unbalance in winding resistance, Figure 29–13.

29.9   SHORT CIRCUIT TO GROUND (FRAME)

A short circuit from winding to ground or the frame of the motor may be detected with a good ohmmeter. No circuit should be detectable from the winding to ground. The copper suction line is a good source for checking to the ground. *"Ground" or "frame" are interchangeable terms because the frame is supposed to be grounded to the earth ground through the building's electrical system, Figure 29–14.*

To check a motor for a ground, use a good ohmmeter with an $R \times 10\ 000$ ohm scale. Special

Compressor Model	Voltage	MOTOR AMPS				FUSE SIZE		Winding Resistance in Ohms
		Full Winding		1/2 Winding		Recommended Max		
		Rated Load	Locked Rotor	Rated Load	Locked Rotor	Fusetron	Std.	
9RA - 0500 - CFB	230/1/60	27.5	125.0			FRN-40	50	Start 1.5 Run 0.40
9RB TFC	208-230/3/60	22.0	115.0			FRN-25	40	0.51-0.61
9RJ TFD	460/3/60	12.1	53.0			FRS-15	15	2.22-2.78
9TK TFE	575/3/60	7.8	42.0			FRS-10	15	3.40-3.96
MRA FSR	200-240/3/50	17.0	90.0	8.5	58.0	FRN-25	35	0.58-0.69
MRB FSM	380-420/3/50	9.5	50.0	4.8	32.5	FRS-15	20	1.80-2.15
MRF								

Figure 29–11. Resistances for some typical hermetic compressors. *Courtesy Copeland Corporation*

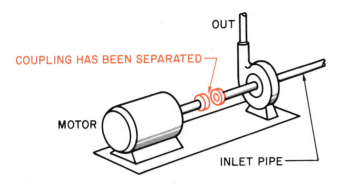

OUT

COUPLING HAS BEEN SEPARATED

MOTOR

INLET PIPE

Figure 29-12. The coupling was disconnected between this motor and pump because it was suspected that the motor or pump was locked up and would not turn.

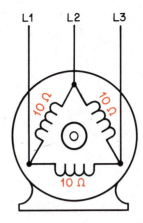

L1 L2 L3

10 Ω 10 Ω

10 Ω

Figure 29-13. Wiring diagram of three-phase motor. The resistance is the same across all three windings.

instruments for finding very high resistances to ground are used for larger, more sophisticated motors. Most technicians use ohmmeters. Top-quality instruments can detect a ground in the 10 000 000 Ω and higher range. A *megger* even has an internal high-voltage DC supply to help create conditions to detect the ground, Figure 29-15. The term "megger" means megohms or

1 000 000 ohms. It has to do with the capacity of the meter to detect very high resistances.

A typical ohmmeter will detect a ground of about 1 000 000 Ω or less. The rule of thumb is that if an ohmmeter set to the $R \times 10\ 000$ scale will even move the needle, with one lead touching the motor terminal

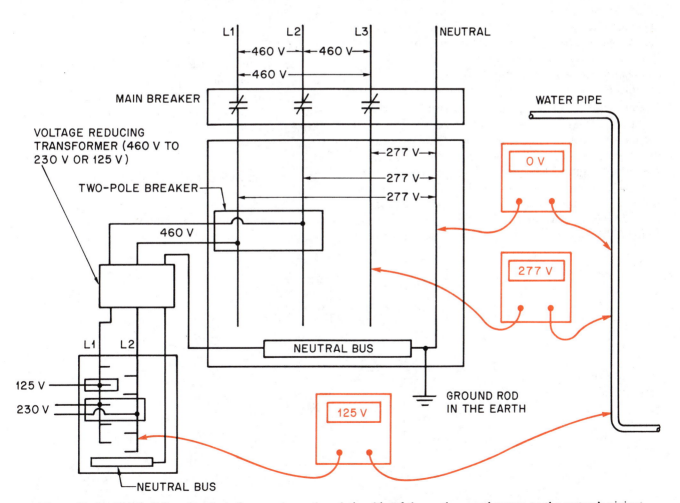

Figure 29-14. This building electrical diagram shows the relationship of the earth ground system to the system's piping.

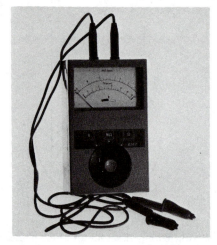

Figure 29-15. A megger. *Photo by Bill Johnson*

and the other touching a ground (such as a copper suction line), the motor should be started with care. *If the meter's needle moves to the midscale area, do not start the motor, Figure 29-16.* When a meter reads a very slight resistance to ground, the windings may be dirty and damp if it is an open motor. Clean the motor and the ground will probably be eliminated. Some motors operating in a dirty, damp atmosphere may indicate a slight circuit to ground in damp weather. Air-cooled condenser fan motors are an example. When the motor is started and allowed to run long enough to get warm and dry, the ground circuit may disappear.

Hermetic compressors may occasionally have a slight ground due to the oil and liquid refrigerant in the motor splashing on the windings. The oil may have dirt suspended in it and show a slight ground. Liquid refrigerant causes this condition to be worse. If the ohmmeter shows a slight ground but the motor starts, run the motor for a little while and check again. If the ground persists, the motor is probably going to fail soon if the system is not cleaned up. A suction line

filter-drier may help remove particles that are circulating in the system and causing the slight ground.

For troubleshooting electric motors, the ammeter and the voltmeter are the main instruments used. Remember the rules of current flow. When there is a power supply, a path for the power to flow on (conductor) and a resistance, current will flow. If the resistance is correct, the motor is electrically sound. When all three of these conditions are met, the motor should start, provided there is not a mechanical problem.

29.10   MOTOR STARTING PROBLEMS

Symptoms of electric motor starting problems are

1. The motor hums and then shuts off.
2. The motor runs for short periods and shuts off.
3. The motor will not try to start at all.

The technician must decide whether there is a motor mechanical problem, a motor electrical problem, a circuit problem, or a load problem. If the motor is an open motor, turn the power off and try to rotate the motor by hand. If it is a fan or a pump, it should be easy to turn, Figure 29-17. If it is a compressor, the shaft may be hard to turn. Use a wrench to grip the coupling when trying to turn a compressor. *Be sure the power is off.*

If the motor and the load turn freely, examine the motor windings and components. If the motor is humming and not starting, the starting switch may need replacing or the windings may be burned. If the motor is open, you may be able to visually check them; if you can't, remove the motor end bell, Figure 29-18.

29.11   CHECKING CAPACITORS

Motor capacitors may be checked to some extent with an ohmmeter in the following manner. Turn off the power to the motor and remove one lead of the

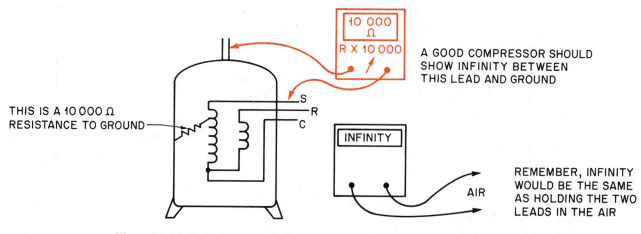

Figure 29-16. Volt-ohmmeter detecting a circuit to ground in a compressor winding.

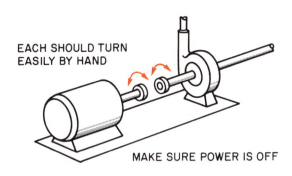

EACH SHOULD TURN
EASILY BY HAND

MAKE SURE POWER IS OFF

Figure 29-17. Example of how a pump or compressor coupling can be disconnected and the component's shaft turned by hand to check for a hard to turn shaft, or a stuck component.

capacitor. *Short from one terminal to the other with an insulated screwdriver to discharge the capacitor in case the capacitor has a charge stored in it. Some start capacitors have a resistor across these terminals to bleed off the charge during the off cycle, Figure 29-19. Short across the capacitor anyway. The resistor may be open and won't bleed the charge from the capacitor. If you place ohmmeter leads across a charged capacitor, the meter movement may be damaged.*

Set the ohmmeter scale to the $R \times 10$ scale and touch the leads to the capacitor terminals, Figure 29-20. If the capacitor is good, the meter needle will go to 0 and begin to fall back toward infinite resistance. If the leads are left on the capacitor for a long time, the needle will fall back to infinite resistance. If the needle falls part of the way back and will drop no more, the capacitor has an internal short. The electrolyte is touching the other pole of electrolyte. If the needle will not rise at all, try the $R \times 100$ scale. Try reversing the leads. The ohmmeter is charging the capacitor with its internal battery. This is DC voltage.

Figure 29-18. The end bell has been removed from this motor to examine the start switch and the windings. *Photo by Bill Johnson*

Figure 29-19. Start capacitor with bleeder resistor across the terminals to bleed off the charge during the off cycle. *Photo by Bill Johnson*

When the capacitor is charged in one direction, the meter leads must be reversed for the next check. If the capacitor has a bleeder resistor, the capacitor will charge to 0 Ω, then drop back to the value of the resistor.

This simple check with an ohmmeter will not give the capacitance. You need a capacitor tester to find the actual capacitance of a capacitor, Figure 29-21. It is not often that a capacitor will change in value, so this meter is used primarily by technicians that need to perform many capacitor checkouts.

29.12 IDENTIFICATION OF CAPACITORS

A run capacitor is contained in a metal can and is oil filled. If this capacitor gets hot due to overcurrent, it will often swell, Figure 29-22. The capacitor should then be changed. *Run capacitors have an identified terminal to which the lead that feeds power to the capacitor should be connected. When the capacitor is wired in this manner, a fuse will blow if the capacitor is shorted to the container. If the capacitor is not wired in this manner and a short occurs, current can flow through the motor winding to ground during the off cycle and overheat the motor, Figure 29-23.*

The start capacitor is a dry type and may be contained in a shell of paper or plastic. Paper containers are no longer used, but you may find one in an older motor. If this capacitor has been exposed to overcurrent, it may have a bulge at a vent at the top of the container, Figure 29-24. The capacitor should then be changed.

29.13 WIRING AND CONNECTORS

The wiring and connectors that supply the power to a motor must be in good condition. When a connection

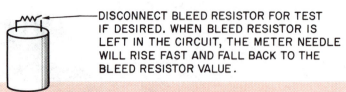

DISCONNECT BLEED RESISTOR FOR TEST IF DESIRED. WHEN BLEED RESISTOR IS LEFT IN THE CIRCUIT, THE METER NEEDLE WILL RISE FAST AND FALL BACK TO THE BLEED RESISTOR VALUE.

FOR BOTH RUN AND START CAPACITORS

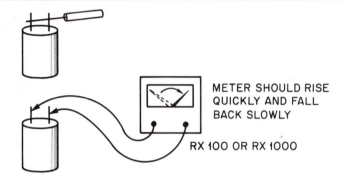

1. FIRST SHORT THE CAPACITOR FROM POLE TO POLE USING SCREWDRIVER OR INSULATED WIRE.

2. USING THE RX 100 OR RX 1000 SCALE TOUCH THE METER'S LEADS TO THE CAPACITOR'S TERMINALS. METER NEEDLE SHOULD RISE FAST AND FALL BACK SLOWLY. IT WILL EVENTUALLY FALL BACK TO INFINITY IF THE CAPACITOR IS GOOD. (PROVIDED THERE IS NO BLEED RESISTOR).

METER SHOULD RISE QUICKLY AND FALL BACK SLOWLY

RX 100 OR RX 1000

3. YOU CAN REVERSE THE LEADS FOR A REPEAT TEST, OR SHORT THE CAPACITOR TERMINALS AGAIN. IF YOU REVERSE THE LEADS, THE METER NEEDLE MAY RISE EXCESSIVELY HIGH AS THERE IS STILL A SMALL CHARGE LEFT IN THE CAPACITOR.

4. FOR RUN CAPACITORS THAT ARE IN A METAL CAN: WHEN ONE LEAD IS PLACED ON THE CAN AND THE OTHER LEAD ON A TERMINAL INFINITY SHOULD BE INDICATED ON THE METER USING THE RX 10 000 OR RX 1000 SCALE

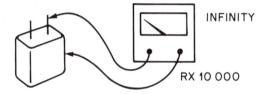

INFINITY

RX 10 000

Figure 29–20. Procedure to field check a capacitor.

Figure 29–21. Capacitor analyzer. *Photo by Bill Johnson*

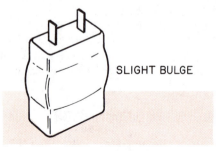

SLIGHT BULGE

Figure 29–22. Capacitor swelled because of internal pressure.

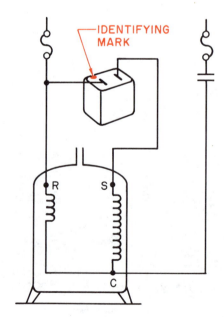

IDENTIFYING MARK

R   S

C

Figure 29–23. How a run capacitor should be wired for proper fuse protection of the circuit.

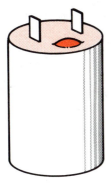

Figure 29-24. When a start capacitor has overheated, the small rubber diaphragm in the top will bulge.

becomes loose, oxidation of the copper wire occurs. The oxidation acts as an electrical resistance and causes the connection to heat more, which in turn causes more oxidation. This condition will only get worse. Loose connections will result in low voltages at the motor and in overcurrent conditions. Loose connections will appear the same as a set of dirty contacts and may be located with a voltmeter, Figure 29-25. If a connection is loose enough to create an over-current condition, it can often be located by a temperature rise at the loose connection.

29.14 TROUBLESHOOTING HERMETIC MOTORS

Diagnosing hermetic compressor motor problems differs from diagnosing open motor problems because the motor is enclosed in a shell and cannot be seen; the motor sound level may be dampened by the compressor shell. Motor noises that are obvious in an open motor may become hard to hear in a hermetic compressor.

The motor inside a hermetic compressor can only be checked electrically from the outside. It will have the

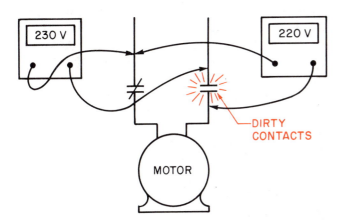

Figure 29-25. When voltmeter leads are applied to both sides of a dirty contact, the meter will read voltage drop across the contact.

same problems as an open motor—open circuit, short circuit, or grounded circuit. The technician must give the motor a complete electrical checkout from the outside of the shell; if it cannot be made to operate, the compressor must be changed. A motor checkout includes the starting and running components for a single-phase compressor, such as the run and start capacitors and the start relay.

As mentioned earlier, compressors operate in remote locations and are not normally attended. A compressor can throw a rod, tearing it up, and the motor may keep running until damaged. The compressor parts damage the motor, which may then burn out. The compressor damage is not detected because the damage cannot be seen. When a compressor is changed because of a bad motor, you should suspect mechanical damage.

29.15 SERVICE TECHNICIAN CALLS

Service Call 1

The manager of a retail store calls. *They have no air conditioning and the piping going into the unit is frozen solid. The air handler is in the stock room. The evaporator motor winding is open, and the outdoor unit has continued to run without air passing over the evaporator, freezing it solid.*

Upon arriving at the store and noticing that the fan is not running the technician turns the cooling thermostat to OFF to stop the condensing unit, then turns the fan switch to ON. *The technician then turns off the power and removes the fan compartment cover.* The fan motor body is cool to the touch, indicating the motor has been off for some time. The technician then turns on the power and checks power at the motor terminal block—the motor has power. The power is turned off and the technician checks the motor windings with an ohmmeter—the motor windings are open.

The fan motor is changed, but this is not the end of the problem. The unit cannot be started up until the coil is allowed to thaw. The technician instructs the store manager to allow the fan to run until air is felt at the outlet in the mens room. This insures the airflow has started. Wait 15 minutes, and set the thermostat to COOL and the fan switch to AUTO. The technician leaves. A call back later in the day proves the system is now cooling correctly.

Service Call 2

A residential customer calls. *There is no heat. The fan motor on the gas furnace has a* defective run capacitor. The furnace burner is cycling on and off with the high-limit control. The fan will run for a few minutes each cycle before shutting off from over temperature. The defective capacitor causes the fan motor current to run about 15% too high.

The technician arrives and goes under the house where the horizontal furnace is located. The furnace is hot. The technician hears the burner ignite and watches it for a few minutes. The burner shuts off, and about that time the fan comes on. The technician removes the fan compartment door, and sees the fan turning very slowly. *The power is shut off.* The motor is very hot. It seems to turn freely, so the bearings must be normal. The technician removes the run capacitor and uses an ohmmeter for a capacitor check, Figure 29–20. The capacitor is open—no circuit or continuity. The capacitor is replaced. The fan compartment door is shut and power is restored. It is evident from the sound of the motor that it is turning faster than before the capacitor was changed.

Before leaving the job, the technician turns off the power, oils the motor, and replaces the air filters.

Service Call 3

A retail store customer calls. *There is no air conditioning. The compressor, located outside, starts up, then shuts off. The fan motor is not running. The dispatcher tells the customer to shut the unit off until the technician arrives. The PSC (permanent split capacitor) motor has bad bearings. The motor feels free to turn the shaft but if power is applied to the motor when it is turning, the motor will stop.*

The technician arrives and talks to the store manager. The technician goes to the outdoor unit and disconnects the power so that the system can be controlled from outside. The room thermostat is set to call for cooling.

The technician goes outside and turns the disconnect on and observes. The compressor starts up, but the fan motor does not. *The power is turned off and the compressor leads are disconnected at the load side of the contactor so that the compressor will not be in a strain.* The technician makes the disconnect switch again and, using the clamp-on ammeter, checks the current going to the fan motor. It is drawing current and trying to turn. It is not known at this point if the motor or the capacitor is bad. The power is shut off. The technician spins the motor and turns the power on while the motor is turning. The motor acts

like it has breaks and stops. This indicates bad bearings or an internal electrical motor problem.

The motor is changed and the new motor performs normally even with the old capacitor. The compressor is reconnected, and power is resumed. The system now cools normally. The technician changes the air filter and oils the indoor fan motor before leaving.

Service Call 4

An insurance office customer calls. *There is no air conditioning. This system has an electric furnace with the air handler (fan section) mounted above the suspended ceiling in the stock room. The condensing unit is on the roof. The low-voltage power supply is at the air handler. The condenser fan motor is grounded and has tripped the breaker at the outdoor unit.*

The technician arrives and goes to the room thermostat. It is set on cooling, and the indoor fan is running. The technician goes to the roof and discovers the breaker is tripped. Before resetting the breaker, the technician decides to find out why the breaker tripped. The cover to the electrical panel to the unit is removed. A voltage check shows that the breaker has all the voltage off. The contactor's 24-V coil is the only thing energized because its power comes from downstairs at the air handler.

The technician checks the motors for a ground circuit by placing one lead on the load side of the contactor and the other on the ground with the meter set on $R \times 10\,000$. The meter shows a short circuit to ground (no resistance to ground). See Figure 29–16 for an example of a motor ground check. From this test, the technician does not know if the problem is the compressor or the fan motor. The wires are disconnected from the line side of the contactor to isolate the two motors. The compressor shows normal, no reading. The fan motor shows 0 Ω resistance to ground; it is defective.

The motor is changed, and the disconnect is closed. The unit starts up normally and runs. The technician changes the air filters and oils the indoor fan motor before leaving.

Service Call 5

The manager of a local restaurant calls. *The air conditioning is not cooling the dining room. The air filters where just changed and the panel was left off the air handler while changing the filters. The indoor fan is pulling in too much air and causing the fan motor to pull too much current.*

The technician arrives and discovers the evaporator coil has an ice build up; it is not frozen solid. The fan is running. The technician suspects a low refrigerant charge at first; then looks at the air handler and notices the door is not in place. About that time, the fan motor shuts off due to overload. The panel is replaced and the motor current is checked. It is normal with the panel in place. To verify what was suspected, the fan panel is opened while looking at the ammeter. The amperage rises to above the nameplate rating. Before leaving, the technician checks the belts, greases the motor, and cautions the owner about leaving the panel off.

Service Call 6

A commercial customer calls. *There is no air conditioning in the upstairs office. This is a three-story building with an air handler on each floor and a chiller in the basement. The fan motor on the top floor is three phase and has a fuse blown on one phase. The fuse has a loose connection at the load-side terminal. The motor is said to be "single phasing" with the fuse blown. This motor has a motor starter, and the overload is tripped.*

The technician arrives and goes to the fan room on the second floor where the complaint is. There would be complaints from the other floors if the main chiller were not running. The technician notices the fan motor is not running and feels the chilled-water coil piping. It is cool; the chiller is definitely running. *Since the motor may have electrical problems, the technician proceeds with caution by opening the electrical disconnect.* The technician pushes the reset button on the fan motor starter and hears the ratchet mechanism reset. The unit must have been pulling too much current.

Using an ohmmeter, the technician checks for a ground by touching one lead to a ground terminal and the other to one of the motor leads on the load side of the starter. The meter is set on $R \times 10\,000$ and will detect a fairly high resistance to ground. See Figure 29-16 for an example of a motor ground check. *Any movement of the meter's needle when touched to the motor lead would indicate a ground so caution is necessary. If the meter's needle moves as much as one fourth of its scale, the*

circuit should not be energized until the ground is cleared up, or physical damage may occur.*

The motor is not grounded. The resistance between each winding is the same, so the motor appears to be normal. The technician then turns the motor over by hand to see if the bearings are too tight; the motor turns normally.

The technician shuts the fan compartment door and then fastens a clamp-on ammeter to one of the motor leads at the load side of the starter. The technician then pushes the disconnect switch closed and tries to start the fan motor. It will not start and pulls a high amperage. The motor seemed to be normal from an electrical standpoint and turned freely, so the power supply is now suspected.

The technician quickly pulls the disconnect to the off position and gets an ohmmeter. Each fuse is checked with the ohmmeter. The fuse in L2 is open. The fuse is replaced and the motor is started again. The motor starts and runs normally with normal amperage on all three phases. The question is, why did the fuse blow?

The technician decides to stay with the unit for a few minutes to see if it is going to continue to run. All of the air filters are changed in all three fan rooms, all motor and fan shaft bearings are oiled, and all belts are checked for correct tension. When the technician returns to the disconnect where the fuse was blown, everything seems normal except that the wire fastened to the load side of the fuse that burned feels warm and the insulation is changing color. The technician shuts the disconnect off and holds a hand close to the wire and can feel the heat. The screw is tightened two full turns. This loose connection has caused the heat that opened the fuse.

Service Call 7

A customer that owns a truck stop calls. *There is no cooling in the main dining room. The customer does some of the maintenance. The motor on the indoor air handler is tripping off because of an overloaded condition. The customer overtightened the belts when servicing the system.*

The technician arrives and can hear the condensing unit on the roof running; it shuts off about the time the technician gets out of the truck. The customer had just serviced the system and the fan motor then began shutting off. It could be reset at the fan contactor, but it would not run long.

The technician and the customer go to the stock room where the air handler is located. The fan is not running, and the condensing unit is shut down from low pressure. The technician fastens the clamp-on ammeter on one of the motor leads and restarts the motor. The motor is pulling too much current on all three phases. *The technician shuts off the power and removes the fan motor compartment door.* The technician turns over the fan and motor by hand and can tell by the sound that the belts seem extra tight. The belts are touched to see how easy they depress, but they hardly press down. Obviously, the belts are too tight. The adjustments are loosened until a gentle push will depress the belt.

The technician then places the fan door back temporarily and starts the fan. The current draw drops back to normal. The belt tension is very important: if too loose, they will slip; if too tight, they will wear out the bearings and draw high current while doing it.

SUMMARY

- Motor problems are divided into mechanical problems and electrical problems.
- Mechanical problems are bearing or shaft problems.
- Bearing problems are often caused by belt tension.
- Shaft problems are often caused by the technician while removing pulleys or couplings.
- Motor balancing problems are normally not handled by the heating, air conditioning, and refrigeration technician.

- *Make sure that vibration is not caused by the system.*
- Most electrical problems are open windings, short-circuited windings, or grounded windings.
- The laws of current flow must be used while troubleshooting motors. For current to flow, there must be a power supply, a path (conductor), and a measurable resistance.
- If the motor is receiving the correct voltage and is electrically sound, check the motor components.
- Troubleshooting hermetic compressor motors is different from troubleshooting open motors because hermetic motors are enclosed.

REVIEW QUESTIONS

1. What are the two main categories of motor problems?
2. In what category is a motor shaft problem?
3. Who normally works with motor balancing problems?
4. How is a stuck hermetic compressor repaired?
5. What can be done to check an open motor to see if it is locked up or if the load component is locked up?
6. What three electrical conditions must be met to have current flow?
7. What are the two electrical test instruments most used for motor problems?
8. Where is a convenient place to check a motor for an electrical ground?
9. If the motor is electrically sound, what should then be checked?
10. How does checking a compressor motor differ from checking an open motor?

Section Six
Air Conditioning (Heating and Humidification)

30

ELECTRIC HEAT

OBJECTIVES

After studying this unit, you should be able to

- discuss the efficiency, relative purchase and installation costs, and operating cost of electric heat.
- list types of electric heaters and state their uses.
- describe how sequencers operate in electric forced-air furnaces.
- trace the circuitry in a diagram of an electric forced-air furnace.
- perform basic tests in troubleshooting electrical problems in an electric forced-air furnace.

30.1 INTRODUCTION

Electric heat is produced by converting electrical energy to heat. This is done by placing a known resistance of a particular material in an electrical circuit. The resistance has relatively few free electrons and does not conduct electricity easily. The resistance to electron flow produces heat at the point of resistance. This resistance is obtained with a special wire called nichrome, for nickle chromium.

Electric heat is very efficient but is more expensive to operate compared to other sources of heat. It is efficient because there is very little loss of electrical energy from the meter to the heating element. It is expensive because it takes large amounts of electrical energy to produce the heat, and the cost of electrical energy in

most areas of the country can be expensive compared to fossil fuels (coal, oil, and gas).

The purchase price of electrical heating systems is usually less than other systems. The installation and maintenance is also usually less expensive. This makes electric heating systems attractive to many purchasers.

When electric heat is used as the primary heat source, a high value of insulation is normally used throughout the structure to lower the amount of heat required.

This unit briefly describes several types of electric heating devices, particularly central forced-air electric heat, because it is frequently serviced by technicians in this trade.

30.2 PORTABLE ELECTRIC HEATING DEVICES

Portable or small space heaters are sold in many retail stores, Figure 30–1. Some have glowing coils (due to the resistance of the wire to electron flow). These transfer heat by radiation (infrared rays) to the solid objects in front of the heater. The radiant heat travels in a straight line and is absorbed by solid objects, which warm the space around them. Radiant heat also provides heating comfort to individuals. The heat concentration decreases by the square of the distance and is soon dissipated into the space. Quartz and glass panel heaters are also used to heat small spaces by radiation, Figure 30–2.

Other space heaters use fans to move air over the heating elements and into the space. This is called

Figure 30–1. Portable electric space heater. *Courtesy W. W. Grainger, Inc.*

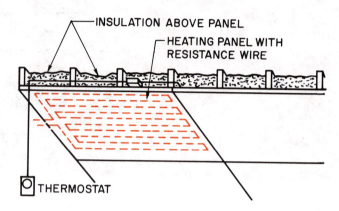

Figure 30–3. Radiant ceiling heating panel.

forced-convection heat because the heat-laden air is moved mechanically. The units may be designed to move enough air across the heating elements so that they do not glow.

Radiant spot heating can be effectively used at doorways, warehouses, work areas, and even outdoors. The effect is much like a sunlamp pointing at the heated area. As mentioned previously, distance has a great bearing on the effectiveness of radiant heating.

30.3   RADIANT HEATING PANELS

Radiant electric heating panels are used in residential and light commercial buildings. The panel is made of gypsum board with wire heating circuits running throughout. The panels are usually installed in ceilings and controlled with individual room thermostats, Figure 30–3. This provides room or zone control. *Be careful that nails or other objects driven into or mounted on the panels do not damage the electrical circuits.*

Heating panels must have good insulation behind (or above) to keep heat from escaping. The electrical

Figure 30–2. Quartz heater. *Courtesy Fostoria Industries, Inc.*

connections are also on the back, so the junctions are easily accessible from an attic.

The heat produced is even, easy to control, and tends to keep the mass of the room warm by its radiation. This makes items in the room pleasingly warm to the touch.

30.4   ELECTRIC BASEBOARD HEATING

Baseboard heaters are popular convection heaters used for whole-house, spot, or individual room heating, Figure 30–4. They are economical to install and can be controlled by individual room thermostats. Unused rooms can be closed off and the heat turned down, which makes them economical in certain applications.

Baseboard units are thin and mounted on the wall just above the floor or carpet. Outside walls are usually used. The heater is a natural-draft unit. Air enters near the bottom, passes over the electric element, is heated, and rises in the room. As the air cools, it settles, setting up a natural-convection air current.

Baseboard heat is easy to control, safe, quiet, and evenly distributed throughout the house.

30.5   UNIT HEATERS

Unit heaters are suspended from the ceiling and use a fan to force the air across the elements into the space to be heated, Figure 30–5. *Any heater designed to have air forced across the element should not be operated without the fan.*

30.6   ELECTRIC HYDRONIC BOILERS

The small electric hydronic (hot water) boiler system is used for some residential and light commercial applications. The boiler is somewhat like an electric domestic water heater in principle. It is slightly different because of the control arrangement and safety devices. It is also connected to a closed loop of piping and requires a pump to move water through the loop

(A)

(B)

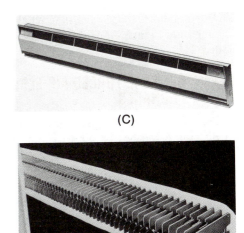

(C)

(D)

Figure 30-4. (A) Wall thermostat. (B) Built-in thermostat. (C) Electric baseboard heater. (D) Finned element. *Courtesy W. W. Grainger, Inc.*

consisting of the boiler and the terminal heating units, Figure 30–6.

The electric boiler is small and compact for easy location and installation. The first cost of a small electric boiler is comparable to that of an oil or gas boiler. It is relatively easy to troubleshoot and repair.

Any boiler handling water is subject to all the problems of a water circulating system. For instance, if the boiler shuts down and the room temperature drops below freezing, pipes will burst and the boiler

may be damaged; water treatment against scale and corrosion is also needed in a boiler installation.

The boiler itself is very efficient because it converts virtually all of its electrical energy input into heat energy and transfers it to the water. When the boiler is located in the conditioned space, any heat loss through the boiler's walls or fittings is not lost from the heated structure.

Figure 30-5. Electric unit heater. *Courtesy International Telephone and Telegraph Corporation—Reznor Division*

Figure 30-6. Electric hydronic boiler. *Courtesy Burnham Corporation*

This system is very quiet and reliable, but summer cooling and winter humidification cannot easily be added.

30.7 CENTRAL FORCED-AIR ELECTRIC FURNACES

Central forced-air furnaces are used with duct work to distribute the heated air to rooms or spaces away from the furnace. The heaters are factory installed in the furnace unit with the air-handling equipment (fan), or they can be purchased as duct heaters and installed within the ductwork, Figures 30–7, 30–8.

Central heaters are usually controlled by a single thermostat, resulting in one control point—there is no individual room or zone control. An advantage with duct heaters is that temperature in individual rooms or zones can be controlled with multiple heaters placed in the ductwork system. However, an interlock system must be incorporated to insure that the fan is running to pass air over the heater to prevent it from burning up.

The heating elements are made of nickel chromium (nichrome) resistance wire mounted on ceramic or mica insulation. They are enclosed in the ductwork or the furnace housing in an electric furnace.

Forced-air heating can usually have added summer air conditioning and humidification because of the air-handling feature. This system is more versatile then the others.

✳Be careful when servicing these systems because many exposed electrical connections are behind the inspection panels. Electric shock can be fatal.✳

30.8 AUTOMATIC CONTROL FOR FORCED-AIR ELECTRIC FURNACES

Automatic controls are used to maintain temperature at desired levels in given spaces and to protect the equipment and occupant. Three common controls used in electric heat applications are *thermostats, sequencers* (discussed later in this unit), and *contactors (or relays).* See Figure 30–9 for a typical thermostat.

30.9 THE LOW-VOLTAGE THERMOSTAT

The low-voltage thermostat is used for thermostats, sequencers, and contactors because it is compact, very

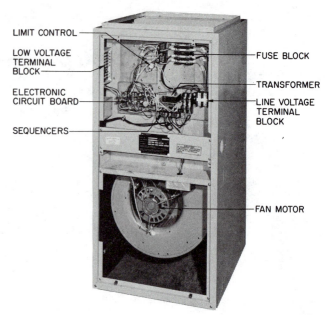

Figure 30–8. Central forced-air electric furnace with multiple sequencers. *Courtesy The Williamson Company*

responsive, safe, and easy to install and troubleshoot. The low-voltage wiring may be run without an electrical license in many localities. Figure 30–10 shows a diagram of a low voltage thermostat typical of those used with an electric forced air heating furnace.

The thermostat has an isolated subbase that allows two power supplies to be run to the thermostat. This isolated subbase may be needed when air conditioning is added after the furnace is installed. The furnace will have its low-voltage power supply, and the air conditioning unit may have its own low-voltage power supply. If one power supply is used for both, a jumper is required from the R terminal to the 4 terminal. This is shown in Figure 30–10 by a dotted line. The R terminal is the cooling terminal, and the 4 terminal is assigned to heating if the subbase has two power supplies.

The *heat anticipator* used in a thermostat with electric heat must be set at the time of the installation.

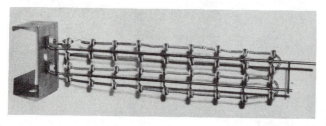

Figure 30–7. Electric duct heater. *Photo by Bill Johnson*

Figure 30–9. Low-voltage thermostat. *Courtesy Robertshaw Controls Company*

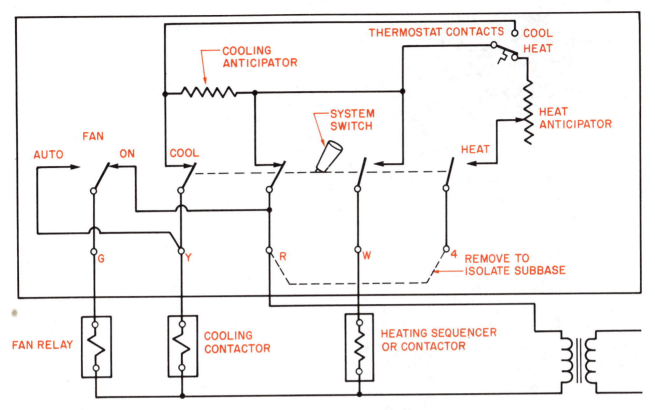

Figure 30–10. Diagram of low-voltage thermostat.

The setting is determined by adding all the current draw in the 24-V circuit that passes through the thermostat control bulb. When this current is determined, it is then set with the indicator on the heat anticipator. For example, suppose that the current passing through terminal 4 is 0.75 A. This number is used to set the heat anticipator in the low-voltage thermostat subbase. Most sequencers will have the ampere load of the low-voltage circuit printed on the sequencer. If there is more than 1, they can be added. For example, for three sequencers, each with a heater load of 0.3 A, the heat anticipator is set on $0.3 + 0.3 + 0.3 = 0.9$ A.

30.10   CONTROLLING MULTIPLE STAGES

Most electric heating furnaces have several heating elements activated in stages to avoid putting a high-power load in service all at once. Some furnaces may have as many as six heaters to be connected to the electrical load at the proper time. The *sequencer* is used to do this. It uses low-voltage control power to start and stop the electric heaters. A sequencer can be described as a heat-motor type of device. It uses a bimetal strip with a low-voltage wire wrapped around it. When the thermostat calls for heat, the low-voltage wire heats the strip and warps it out of shape in a known direction for a known distance. As it warps or bends, it closes electrical contacts to the electric heat circuit. This bending takes time, and each set of contacts closes

quietly and in a certain order with a time delay between the closings.

See Figure 30–11 for a diagram of a package sequencer. This sequencer can start or stop three stages of strip heat. Some sequencers have five circuits, Figure 30–12: three heat circuits, a fan circuit, and a circuit to pass low-voltage power to another sequencer for

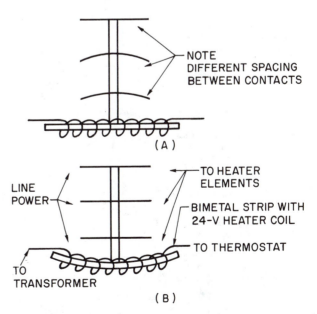

Figure 30–11. Sequencer with three contacts. (A) OFF position. (B) ON position.

Figure 30–12. Multiple-type sequencer. *Photo by Bill Johnson*

three more stages of heat. There are five sets of contacts, and none of them are in the same circuit.

Another sequencer design, called *individual sequencers,* has only a single circuit that can be used for starting and stopping an electric heat element, but it could have two other circuits: one to energize another sequencer through a set of low-voltage contacts, and one for starting the fan motor. Several stages of electric heat may be controlled with several of these sequencers, one for each heat strip.

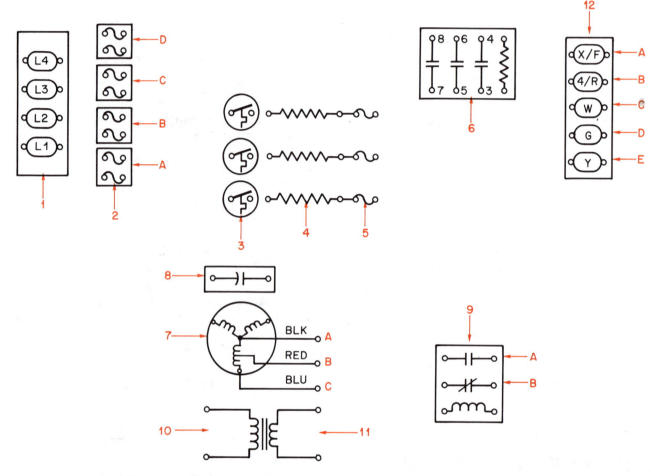

1. Line voltage terminal block from disconnect
2. Fuse block
    A. L1 Fuse block
    B. L2 Fuse block
    C. L3 Fuse block
    D. L4 Fuse block
3. Automatic reset limit switch
4. Electric heat element
5. Fusible link
6. Sequencer
7. Fan motor
    A. High speed motor lead
    B. Medium speed motor lead
    C. Low speed motor lead

8. Fan motor capacitor
9. Fan relay (double pole double throw)
    A. Normally open relay
    B. Normally closed relay
10. Primary side of transformer (line voltage)
11. Secondary side of transformer (low voltage—24 V)
12. Low voltage terminal block
    A. Common terminal from transformer
    B. Terminal for hot lead from transformer
    C. Heating
    D. Fan relay for cooling
    E. Cooling

Figure 30–13. Components on pictorial diagram.

30.11   WIRING DIAGRAMS

Individual manufacturers vary in how they illustrate electrical circuits and components. Some use pictorial and schematic illustrations. The *pictorial* type shows the location of each component as it actually appears to the person installing or servicing the equipment. When the panel door is opened, the components inside the control box are in the same location as on the pictorial diagram.

The *schematic* (sometimes called *ladder type*) shows the current path to the components. Schematic diagrams help the technician to understand and follow the intention of the design engineer. Both pictorial and schematic diagrams are discussed in Unit 16.

Figure 30–13 contains a legend of electrical symbols in an electric heating circuit. Such a legend is vital in following a wiring diagram. Many symbols are standardized throughout the industry, even throughout the world.

30.12   CONTROL CIRCUITS FOR FORCED-AIR ELECTRIC FURNACES

The low-voltage control circuit safely and effectively controls the heating elements that do the work and pull the most current. Low-voltage circuits control these large current flows safely and effectively. The circuit contains devices for safety and for control. For example, the limit switch, a safety device, shuts off the unit if high temperature occurs; the room thermostat,

control or operating device, stops and starts the heat based on room temperature.

Safety and operating devices can consume power and do work or pass power to a power-consuming device. For example, the sequencer's heater coil that operates the bimetal or a magnetic solenoid that moves the armature in a contactor are power-consuming devices in the low-voltage circuit. The contacts of a contactor or a limit control pass power to the power-consuming devices and are in the high-voltage circuit. **Power-passing devices are wired in series to the power-consuming devices. Power-consuming devices are wired in parallel to each other.**

Figure 30–14 is a diagram of a low-voltage control circuit. By tracing this circuit, you can see that when the thermostat contacts are closed, a circuit is completed. This in turn activates the contacts in the sequencer, which activate the heating elements in the high- or line-voltage circuit. The line to the G terminal is used for the cooling circuit and has no other purpose. The heating elements and the fan circuit have been omitted to simplify the circuit.

Single-heating-element control is illustrated in Figure 30–15. The electrical current from the L1 fuse block goes directly to the limit switch. This is a temperature-actuated switch (A) that opens under excessive heat and provides protection to the furnace. It is usually an automatic reset switch that closes when the temperature cools. It is wired to the heating element (B) and to a fusible link (C) that provides additional protection from overheating of the elements. Under higher temper-

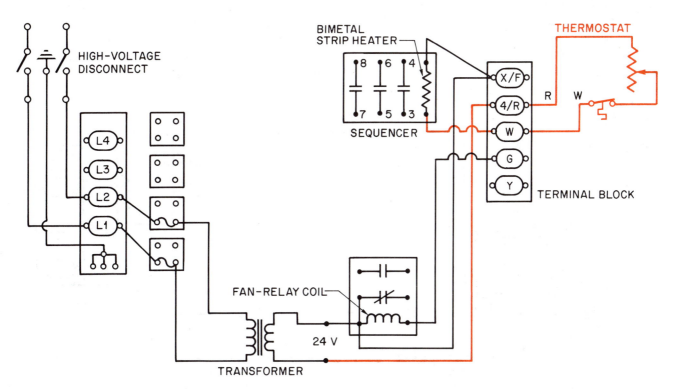

Figure 30–14. Diagram of low-voltage control circuit.

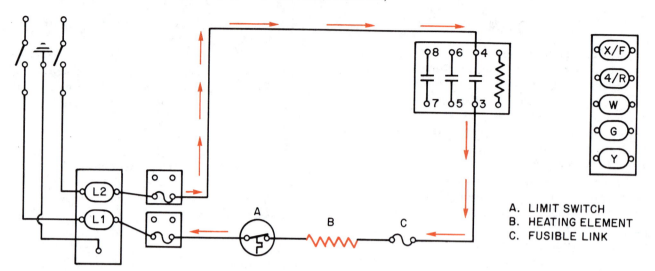

Figure 30–15. Single-heating-element circuit with sequencer (high voltage).

atures this link melts and must be replaced. The link is wired to terminal 3 of the sequencer with the circuit completed through L2 when the sequencer contact is closed.

See Figure 30–16 for a diagram of a furnace with two heating elements. See Figure 30–17 for an example of one with three elements. Each of these examples is using the same package sequencer as shown in Figure 30–15.

30.13   FAN MOTOR CIRCUITS

The fan motor, a power-consuming device, that forces the air over the electric heat elements must be started and stopped at the correct time. See Figure 30–18 for an example of the fan wiring circuit. *It must run before the furnace gets too hot and continue to run until the furnace cools down.*

Note that the L1 terminal is wired directly to the fan motor and that the L2 terminal is wired directly to terminal 4 on the sequencer. From terminal 4 a circuit is made to the NO contact on the fan relay. This circuit could have been made directly to L2. This is the high-speed fan circuit used to start the fan in the cooling mode.

Power is passed through terminal 4 to terminal 3 when the sequencer is energized long enough for the bimetal to bend and close the contacts. The power then passes to the NC terminal on the fan relay and on to the slow-speed winding required for heating.

Remember that the sequencer contacts and the relay contacts *pass* power and do not consume power. They are wired in series.

Figure 30–19 is an example legend of terminal designations used on electric forced-air furnaces.

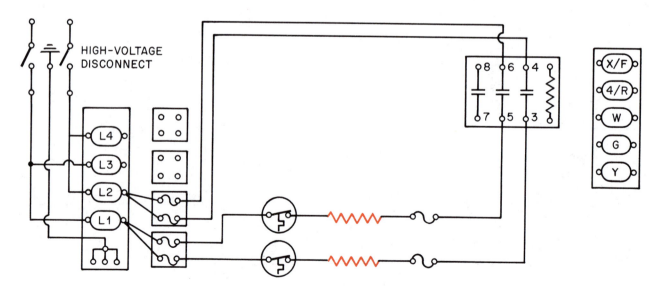

Figure 30–16. Two-heating element circuits.

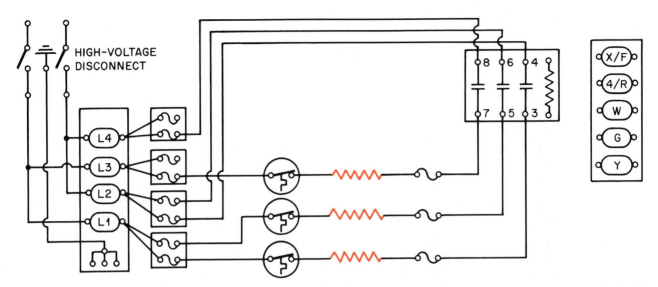

Figure 30–17. Three-heating element circuits.

Figure 30–20 is a wiring diagram for a forced-air furnace with individual sequencers. Note the low-voltage circuit wiring sequence;

1. Power is wired directly from the transformer to terminal C and on to all power-consuming devices.

There are no power-passing devices in this circuit. This can be called the *common* circuit.

2. Power is wired directly from the other side of the transformer to terminal R and then on to the R terminal on the room thermostat. The room thermostat passes (or distributes) power to the respec-

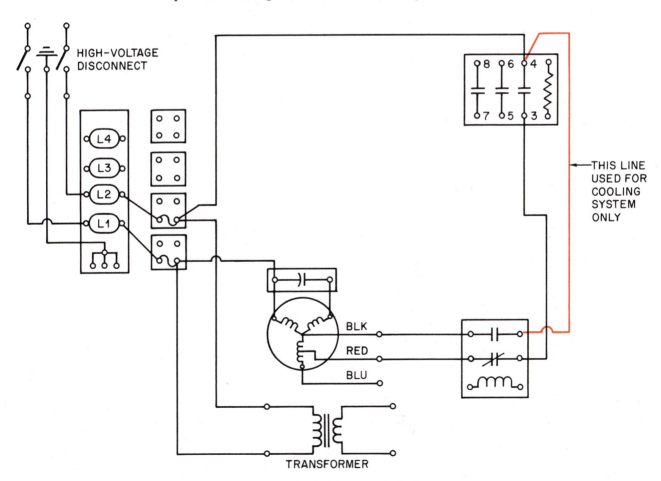

Figure 30–18. Fan control circuit.

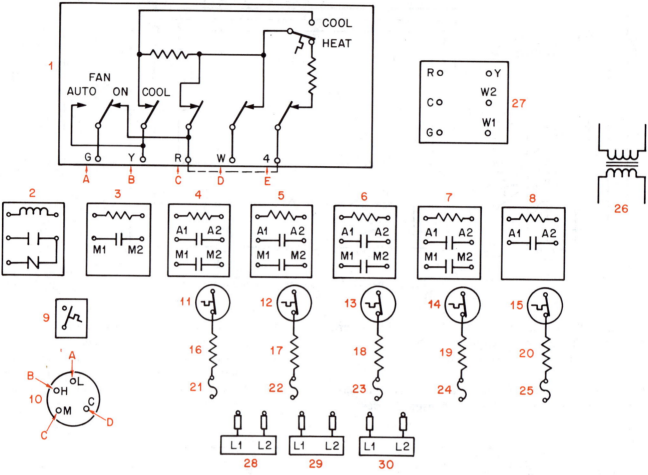

Figure 30–19. Legend for Figure 30–20.

1. Thermostat
   A. G terminal (to fan relay)
   B. Y terminal (to cooling contactor)
   C. R terminal (from transformer secondary)
   D. W terminal (to heating sequencer)
   E. 4 terminal (red or jumped)
2. Fan relay (for cooling)
3. Timed fan control
4. Heat Sequencer I
5. Heat Sequencer II
6. Heat Sequencer III
7. Heat Sequencer IV
8. Heat Sequencer V
   (A1 and A2 terminals for 24 volts only. These are for auxiliary sequencer heater strips which energize next

sequencer. M1 and M2 are terminals for heating elements.)
9. Temperature actuated fan control
10. Fan motor
    A. Low speed
    B. High speed
    C. Medium speed
    D. Common
11–15. Temperature actuated limit switches (automatic reset)
16–20. Electric heating elements
21–25. Fusible links
26. Primary side of transformer
27. Low voltage terminal board
28–30. Line terminal blocks

tive power-consuming-device circuits upon a call for HEAT, COOL, or FAN ON.

3. Upon a call for cooling, the Y and G terminals are energized and start the air conditioning and the indoor fan in high speed.

4. Upon a call for heat, the thermostat energizes the W terminal and passes power to the timed fan control and the first sequencer. The first sequencer passes power to the next sequencer heater coil, then the next.

See Figure 30–20 for the high-voltage sequence.

1. Although there are six fuses, they are labeled either L1 or L2, meaning that there are only two lines supplying power to the unit. The six fuses break the power-consuming heaters and fan motor down into smaller-amperage increments.

2. Upon a call for heat, the contacts in sequencer 1 close and pass power from L2 to the limit switch and on to heater 1. At the same time the timed fan

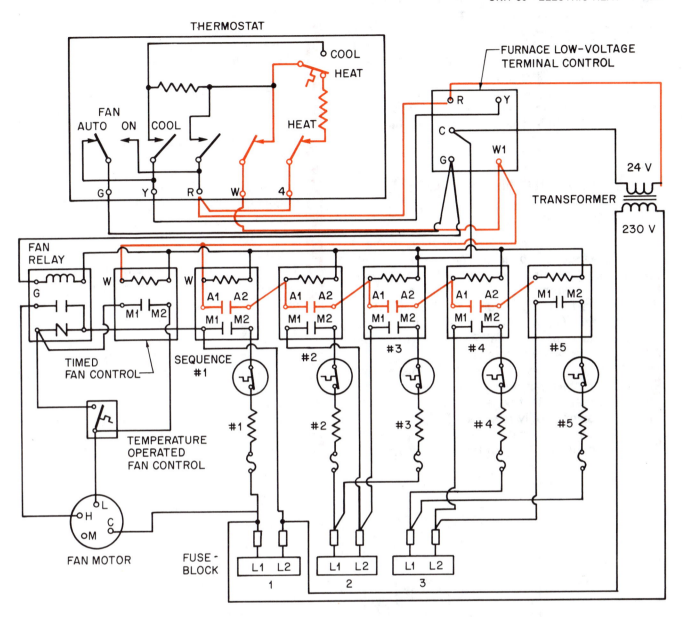

Figure 30–20. Wiring diagram for electric furnace.

control is energized and its contacts close, starting the fan in low speed.

3. When the contacts close for the high-voltage load, the low-voltage contacts also close and pass power in the low-voltage circuit to the next sequencer's operating coil, which sets up the same sequence for the second heater. This continues down the line until all sequencer contacts are closed and all the heat is on. *Note:* This sequence may take 20 sec or more per sequencer, and there are five heat sequencers for the heat to be completely on. When the room thermostat is satisfied, it only takes about 20 sec to turn off the heat because the power to all sequencers is interrupted by the room thermostat. The temperature operated fan switch takes command of the fan at this time and keeps the fan on until the heaters cool.

The fan circuit is a circuit by itself. The fan will start in low speed if the fan relay is not energized. If the fan is running in slow speed and someone turns the fan switch to ON, the fan will switch to high speed. It cannot run in both speeds at the same time due to the relay with NO and NC contacts.

30.14 CONTACTORS TO CONTROL ELECTRIC FURNACES

Using contactors (or relays) to control electric heat means that all of the heat will be started at one time unless separate time delay relays are used. Contactors are magnetic and make noises that a magnetic device makes. They snap in and may hum. Contactors are normally used on commercial systems where they may not be directly attached to the duct work. This prevents the noise from traveling down the system, Figure 30–21. The contactor's magnetic coils are 24 V and 230 V. The 24-V coils are controlled by the room thermostat and they energize the 230-V coil. There are four contactors and their coils for the electric heat and one for the fan. The 24-V coils are energized at one time and pull in together.

30.15 SERVICE TECHNICIAN CALLS

Service Call 1

A customer with a 20-kW electric furnace calls. *There is no heat. The customer says that the fan comes on, but the heat does not. This is a system that has individual sequencers with a fan-starting sequencer. The first-stage sequencer has a burned-out coil and will not close its contacts. The first stage starts the rest of the heat, so there is none.*

The technician is familiar with this system and had started it after installation. The technician realizes that because the fan starts the 24-V power supply is working. Upon arriving at the job, the thermostat is set to call for heat. The technician goes under the house with a spare sequencer, a volt-ohmmeter, and an ammeter.

When approaching the electric furnace, the fan can be heard running. The panel covering the electric heat elements and sequencers is removed. The ammeter is used to check to see if any of the heaters are using current; none are. *The technician observes electrical safety precautions.* The voltage is checked at the coils of all sequencers. See Figure 30–20 for a diagram. It is found that 24 V is present at the timed fan control and the first-stage sequencer, but the contacts are not closed. The electrical disconnect is opened, and the continuity is checked across the first-stage sequencer coil. The circuit is open. When the sequencer is changed and power is turned on, the electric heat comes on.

Service Call 2

A residential customer calls. *There is no heat, but the customer smells smoke. The company dispatcher advises that the customer should turn the system off until the technician arrives. The fan motor has an open circuit and will not run. The smoke smell is coming from the heating unit cycling on the limit control.*

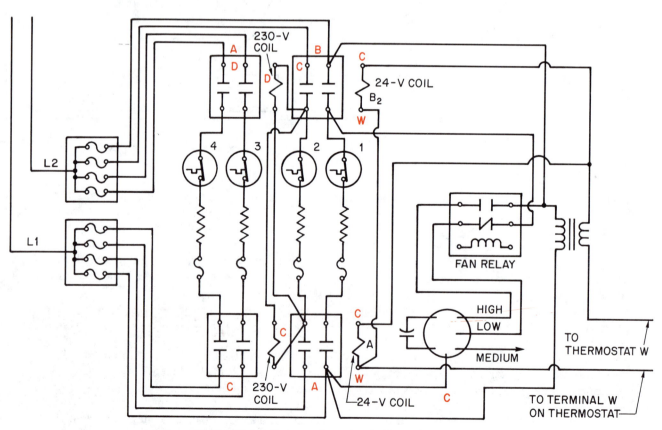

Figure 30–21. Wiring diagram of an electric furnace using contactors.

The technician arrives and goes to the room thermostat and turns the system to heat. This is a system that does not have cooling or the thermostat would be turned to FAN ON to see if the fan will start. If it does not start, this helps to solve the problem. When the technician goes to the electric furnace in the hall closet, it is noticed that a smoke smell is present. When an ammeter is used to check the current at the electric heater, it is found that it is pulling 40 A, but the fan is not running. *The technician observes electrical safety precautions.*

The technician looks at the wiring diagram and discovers that the fan should start with the first-stage heat from a set of contacts on the sequencer. The voltage is checked to the fan motor; it is getting voltage but is not running. The unit power is turned off and a continuity check of the motor proves that the motor winding is open. The technician changes the motor, and the system operates normally.

Service Call 3

A small business calls. *Their heating system is not putting out enough heat. This is the first day of very cold weather, and the space temperature is only getting up to 65°F with the thermostat set at 72°F. The system is a package air conditioning unit with 30 kW (six stages of 5 kW each) of strip heat located on the roof. The fuse links in two of the heaters are open due to previously dirty air filters.*

The technician arrives, goes to the room thermostat, checks the setting of the thermostat, and finds that the space temperature is 5°F lower than the setting. The technician goes to the roof with a volt-ohmmeter and an ammeter. *After removing the panels on the side of the unit where the strip heat is located, the amperage is carefully checked.* Two of the six stages are not pulling any current. It looks like a sequencer problem at first. A voltage check of the individual sequencer coils shows that all of the sequencers should have their contacts closed; there are 24 V at each coil.

A voltage check at each heater terminal shows that stages 4 and 6 have voltage but are not pulling any current. The technician shuts off the power, pulls the stage 4 heater element from the unit for examination, and finds that the fuse link is open. The same is true of stage 6. The fuse links are replaced, and the heater restarted. Stages 4 and 6 now draw current.

The technician knows that the heaters have been too hot for some reason. A check of the air filters shows that they are clean, too clean. The technician looks in the trash can and finds the old filters and they are nearly completely stopped up. The building management is informed about changing the filters on a regular basis whether they look like they need it or not.

Service Call 4

The service technician is on a routine service contract call and inspection. *The terminals on the electric heat units have been hot. The insulation on the wire is burned at the terminals.*

When removing the panels on the electric heat panel, the technician sees the burned insulation on the wires and shows this to the store manager. It is suggested that these wires and connectors be replaced. The store manager asks what the consequences of waiting for them to fail would be. The technician explains that they may fail on a cold weekend and allow the building water to freeze. If the overhead sprinkler freezes and thaws, all of the merchandise will get wet; it is not worth the chance. *The technician turns off the power and completes the job, using the correct wire size, high-temperature insulation, and connectors.*

SUMMARY

- Electric heat is a convenient way to heat individual rooms and small spaces.
- Central electric heating systems are usually less expensive to install and maintain than other types.
- Operational costs may be more than with other types of fuels.
- The low-voltage thermostat, sequencer, and relays are control mechanisms used in central electric forced-air systems.
- The sequencer is used to activate the heating elements in stages. This avoids putting heavy kilowatt loads in service at one time. If this were done, it could cause fluctuations in power, resulting in voltage drop, flickering lights, and other disturbances.
- Sequencers have bimetal strips with 24-V heaters. These heaters cause the bimetal strip to bend, closing contacts, and activating the heating elements.
- *The fan in a central system must operate while the heating elements are on. The systems must be wired to insure that this occurs.*
- Systems are protected with limit switches and fusible links (temperature controlled).

REVIEW QUESTIONS

1. What are two economic advantages of electric heat?
2. What is an economic disadvantage of electric heat?
3. List four types of electric heating devices or systems.
4. What is a disadvantage in the control of central electric forced-air heating systems?
5. List three types of controls used in central electric forced-air heating systems.
6. Of what material are the heating elements made in these systems?
7. Briefly describe how a sequencer operates.
8. Why are sequencers used in electrical forced-air furnaces?
9. Why is a limit switch used in the heating element circuit?
10. Power-consuming devices are wired in _____. Power-passing devices are wired in _____.

GAS HEAT

OBJECTIVES

After studying this unit, you should be able to

- describe each of the major components of a gas furnace.
- list three fuels burned in gas furnaces and describe characteristics of each.
- discuss gas-pressure measurement in inches of water column and describe how a manometer is used to make this measurement.
- discuss gas combustion.
- describe a solenoid, diaphragm, and heat-motor gas valve.
- list the functions of an automatic combination gas valve.
- describe the standing-pilot, electric spark-to-pilot, and direct-spark ignition systems.
- state conditions under which glow-coil reignition systems might be used.
- list three flame-proving devices and describe the operation of each.
- discuss reasons and the systems used for the delay in starting and stopping the furnace fan.
- state the purpose of a limit switch.
- describe flue-gas venting systems.
- discuss the types of vent dampers.
- sketch a gas piping system as it should be installed immediately upstream from the gas valve.
- describe procedures used to check leaks in gas piping systems.
- sketch a basic wiring diagram for a gas furnace.
- sketch a basic wiring diagram of a glow-coil reignition circuit.
- describe flame rectification when used with electric spark-to-pilot ignition systems.
- compare the designs of a high-efficiency gas furnace and a conventional furnace.
- describe procedures for taking flue-gas CO_2 and temperature readings.

31.1 INTRODUCTION TO GAS-FIRED FORCED-HOT-AIR FURNACES

Gas-fired forced-hot-air furnaces have a heat-producing system and a heated air distribution sys-

tem. The heat-producing system includes the manifold and controls, burners, heat exchanger, and the venting system. See Figure 31-1 for a modern high efficiency condensing gas furnace.

The heated-air distribution system consists of the blower that moves the air in the duct work. The manifold and controls meter the gas to the burners where the gas is burned, which creates flue gases in the heat exchanger and heats the air in and surrounding the heat exchanger. The venting system allows the flue gases to be exhausted into the atmosphere. The blower distributes the heated air through duct work to the areas where the heat is wanted.

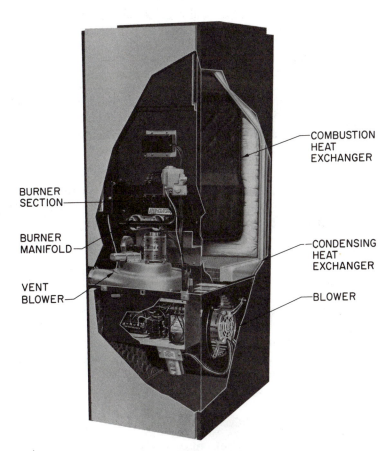

Figure 31-1. A modern condensing gas furnace. *Courtesy Heil-Quaker Corporation*

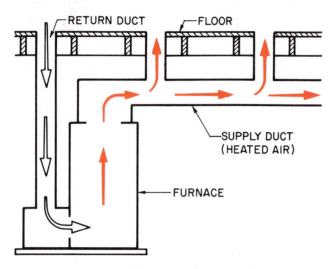

Figure 31-2. Upflow gas furnace airflow.

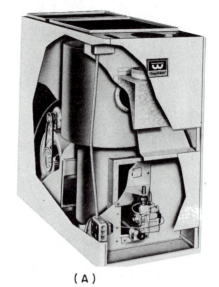

(A)

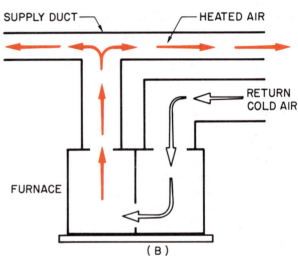

(B)

Figure 31-3. (A) Low-boy gas furnace. (B) Airflow for low-boy furnace. *Courtesy (A) The Williamson Company*

31.2   TYPES OF FURNACES

Upflow

The *upflow* furnace stands vertically and needs headroom. It is designed for first floor installation with the duct work in the attic or for basement installation with the duct work between or under the first floor joists. The furnace takes in cool air from the rear, bottom, or sides near the bottom. It discharges hot air out the top, Figure 31-2.

"Low-boy"

The *low-boy* furnace is approximately four feet high. It is used primarily in basement installations with low head room and the duct work under the first floor. Air intake and discharge are both at the top, Figure 31-3.

Downflow

The *downflow* furnace, sometimes referred to as a *counterflow* furnace, looks like the upflow furnace. The duct work may be in a concrete slab floor or in a crawl space under the house. The air intake is at the top, and the discharge is at the bottom, Figure 31-4.

Horizontal

The *horizontal* furnace is positioned on its side. It is installed in crawl spaces, in attics or suspended from floor joists in basements. In these installations it takes no floor space. The air intake is at one end, the discharge at the other. They are designed for air to flow from right to left or left to right, Figure 31-5.

31.3   GAS FUELS

Natural gas, manufactured gas, and liquified petroleum (LP) are commonly used in gas furnaces.

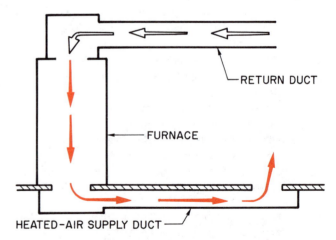

Figure 31-4. Airflow for a downflow or counterflow gas furnace.

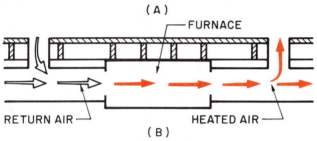

Figure 31–5. (A) Horizontal gas furnace with air conditioning. (B) Airflow for horizontal gas furnace. *Courtesy (A) BPD Company*

Natural Gas

Natural gas was formed along with oil millions of years ago from dead plants and animals. This organic material accumulated for years and years and gradually was washed or deposited into hollow spots in the earth, Figure 31–6. This material accumulated to great depths, causing tremendous pressure and high temperature from its own weight. This caused a chemical reaction, changing this organic material to oil and gas, which accumulated in pockets and porous rocks deep in the earth.

Natural gas is composed of 90–95% methane and other hydrocarbons almost all of which are combustible, which makes this gas efficient and clean burning. Natural gas has a specific gravity of 0.60, and dry air

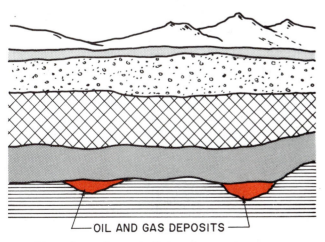

Figure 31–6. Gas and oil deposits deep in the earth.

has a specific gravity of 1.0, so natural gas weighs 60% as much as dry air. Therefore, natural gas rises when discharged into the air. The characteristics of natural gas will vary somewhat from one location to another. When burned, 1 ft³ of natural gas provides approximately 1050 Btu of heat. It takes 10 ft³ of air for this combustion. Contact the utility company providing the gas for its specific characteristics in your area.

Gas from these pockets deep in the earth (more generally called wells) may contain moisture and other gases that must be removed before distributing the gas.

Natural gas by itself has neither odor nor color and is not poisonous. *But it is dangerous, because it can displace oxygen in the air, causing suffocation, and when it accumulates it can explode.*

Sulfur compounds, called *odorants,* which have a garlic smell are added to the gas to make leak detection easier. There is no odor left, however, when the gas has been burned.

Manufactured Gas

Manufactured gas is produced in many ways, often as a by-product of burning coal to make coke. The average cubic foot of manufactured gas produces 530+ Btu/ft³. Manufactured gas is used primarily for special applications in particular localities. Its specific gravity is also 0.60, so it is also lighter than air.

Liquified Petroleum

Liquified petroleum (LP) is liquified propane, butane, or a combination of propane and butane. When in the vapor state these gases may be burned as a mixture of one or both of these gases with air. The gas is liquified by keeping it under pressure until ready to be used. A regulator at the tank reduces the pressure as the gas leaves the tank. As the vapor is drawn from the tank, it is replaced by a slight boiling or vaporizing of the remaining liquid. LP gas is obtained from natural gas or as a by-product of the oil-refining process. The boiling point of propane is −44° F, which makes it feasible to store it in tanks for use in low temperatures during northern winters.

When burned, 1 ft³ of *propane* produces 2500 Btu of heat. However, it requires 24 ft³ of air to support this combustion. The specific gravity of propane gas is 1.52, which means it is 1½ times heavier than air, so it sinks when released in air. *This is dangerous because it will replace oxygen and cause suffocation; it will also collect in low places to make pockets of highly explosive gas.*

Butane produces approximately 3200 Btu/ft³ of gas when burned. LP gas contains more atoms of hydrogen and carbon than natural gas does, which causes the Btu output to be greater. Occasionally, butane and propane are mixed with air to alter its characteristics. It is then called propane or butane air gas. This reduces the Btu/

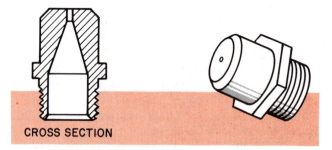

CROSS SECTION

Figure 31–7. Spud showing the orifice through which the gas enters the burner.

ft³ and the specific gravity. The characteristics may be changed with air to closely duplicate natural gas. This allows it to be substituted and burned in natural gas burners without adjustment for standby purposes. LP gas is used primarily where natural gas is not available.

The specific gravity of a gas is important because it affects the gas flow through the piping and through the orifice of the furnace. The orifice is a small hole in a fitting (called a *spud*) through which the gas must flow to the burners, Figure 31–7. The gas flow rate also depends on the pressure and on the size of the orifice. At the same pressure more light gas than heavy gas will flow through a given orifice.

SAFETY PRECAUTION: ✳LP gas alone must not be used in a furnace that is set up for natural gas because the orifice will be too large. This results in overfiring.✳ When the proper mixture of LP gas and air is used, it will operate satisfactorily in natural gas furnaces because the proper mixture will match the orifice size and the burning rate will be the same.

Manifold Pressures

The manifold pressure at the furnace should be set according to the manufacturer's specifications because they relate to the characteristics of the gas to be burned. The manifold gas pressure is much lower than 1 psi and is expressed in inches of water column (in. W. C.). *One inch of water column* pressure is the pressure required to push a column of water up one inch.

SOME COMMON MANIFOLD PRESSURES	
Natural gas and propane/air mixture	3–3½ in. W. C.
Manufactured gas (below 800 Btu quality)	2½ in. W. C.
LP gas	11 in. W. C.

A *water manometer* is used to measure gas pressure. It is a tube of glass or plastic formed into a U. The tube is about half full of plain water, Figure 31–8. The standard manometer used in domestic and light commercial installations is graduated in in. W. C. The pressure is determined by the difference in levels in the two

Figure 31–8. A manometer used for measuring pressure in inches of water column. *Courtesy Robinair Division—Sealed Power Corporation*

columns. Gas pressure is piped to one side of the tube, and the other side is open to the atmosphere to allow the water to rise, Figure 31–9. The instrument often has a sliding scale, so it can be adjusted for easier reading. Some common conversions follow.

WEIGHTS PER SQUARE INCH
1 lb = 27.71 in. W.C.
1 oz = 1.732 in. W.C.
2.02 oz = 3.5 in. W.C. (standard pressure for natural gas)
6.35 oz = 11 in. W.C. (standard pressure for LP gas)

SAFETY PRECAUTION: ✳Turn the gas off while connecting the water manometer.✳

31.4 GAS COMBUSTION

To properly install and service gas heating systems, the heating technician must know the fundamentals of combustion. These systems must be installed to operate safely and efficiently.

Combustion needs fuel, oxygen, and heat. It is a reaction between the fuel, oxygen, and heat known as *rapid oxidation* or the process of burning.

The fuel in this case is the natural gas (methane), propane, or butane; oxygen comes from the air, and the heat comes from a pilot flame. The fuels contain hydrocarbons that unite with the oxygen when heated to the ignition temperature. The air contains approximately 21% oxygen. Enough air must be supplied

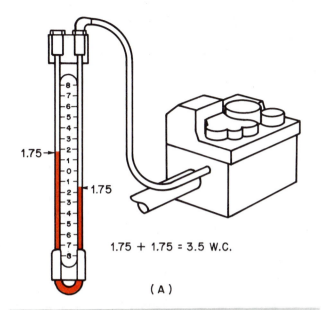

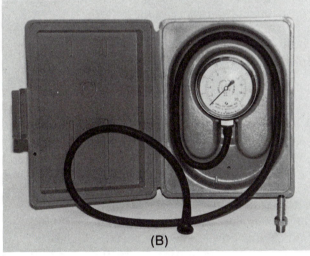

Figure 31-9. (A) Manometer connected to a gas valve to check the manifold pressure. (B) Gauge calibrated in inches of water column. *Photo (B) by Bill Johnson*

to furnish the proper amount of oxygen for the combustion process. The ignition temperature for natural gas is $1100°$–$1200°$ F. The pilot flame must provide this heat, which is the minimum temperature for burning. When the burning process takes place, the chemical formula is

$$CH_4 \text{ (methane)} + 2O_2 \text{ (oxygen)} \longrightarrow$$
$$CO_2 \text{(carbon dioxide)} + 2H_2O \text{ (water vapor)} + heat$$

This formula represents perfect combustion. The results of this process produce carbon dioxide (a harmless gas), water vapor, and heat. Although perfect combustion seldom takes place, slight variations create no danger. *The technician must see to it that the combustion is as close to perfect as possible. By-products of poor combustion are carbon monoxide

(CO)—a poisonous gas—soot, and minute amounts of other products. It should be obvious that carbon monoxide production must be avoided.* Soot lowers furnace efficiency because it collects on the heat exchanger and acts as an insulator, so it also must be kept at a minimum.

31.5 THE GAS BURNER

Most gas furnaces and appliances use an atmospheric burner because the gas and air mixtures are at atmospheric pressure during burning. The gas is metered to the burner through the orifice. The velocity of the gas pulls in the primary air around the orifice, Figure 31-10A.

The burner tube diameter is reduced where the gas is passing through it to induce the air. This burner diameter reduction is called the *venturi*, Figure 31-10A. The gas-air mixture moves on through the venturi into the mixing tube where the gas and air are mixed for better combustion. The mixture is forced by its own velocity through the burner ports or slots where it is ignited as it leaves the port, Figure 31-10B.

When the gas is ignited at the port, secondary air is drawn in around the burner ports to support combustion. The flame should be a well-defined blue with slightly orange, *not yellow*, tips, Figure 31-10C. *Yellow tips indicate an air-starved flame emitting poisonous carbon monoxide.* Orange streaks in the flame are not to be confused with yellow. Orange streaks are dust particles burning.

Other than pressure to the gas orifice, the primary air is the only adjustment that can be made to the flame. Modern furnaces have very little adjustability; this is intentional design to maintain minimum standards of flame quality. The only thing that will change the furnace combustion characteristics enough to matter, except gas pressure, is dirt or lint drawn in with the primary air. If the flame begins to burn *yellow*, primary air restriction should be suspected.

For a burner to operate efficiently, the gas flow rate must be correct and the proper air must be supplied. Modern furnaces use a primary and a secondary air supply, and the gas supply or flow rate is determined by the orifice size and the gas pressure. A normal pressure at the manifold for natural gas is 3–$3\frac{1}{2}$ in. W.C. It is important, however, to use the manufacturer's specifications when adjusting the gas pressure.

A gas-air mixture that is too lean (not enough gas) or too rich (too much gas) will not burn. If the mixture contains 0% to 4% natural gas, it will not burn. If the mixture contains from 4% to 14% gas, it will burn. *However, if the mixture is allowed to accumulate, it will explode when ignited.* If the gas in the mixture is 15–100%, the mixture will not burn or explode. The burning mixtures are known as the *limits of flammability* and are different for different gases.

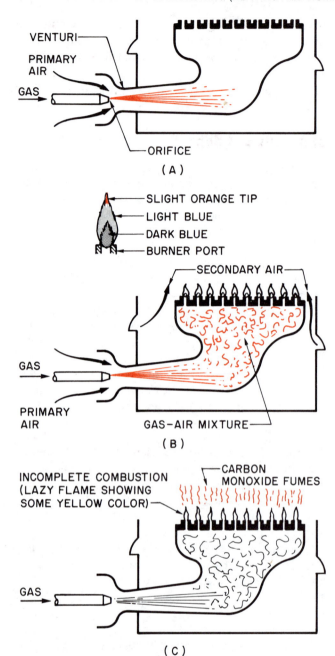

Figure 31-10. (A) Primary air is induced into the air shutter by the velocity of the gas stream from the orifice. (B) Ignition of the gas is on top of the burner. (C) Incomplete combustion yields yellow "lazy" flame. Any orange color indicates dust particles drawn in with primary air.

The rate at which the flame travels through the gas and air mixture is known as the *ignition or flame velocity* and is determined by the type of gas and the air to gas ratio. Natural gas (CH_4) has a lower burning rate than butane or propane because it has fewer hydrogen atoms per molecule. Butane (C_4H_{10}) has 10 hydrogen atoms, whereas natural gas has 4. The speed is also increased in higher gas-air mixtures within the limits of flammability; therefore, the burning speed can be changed by adjusting the air flow.

Perfect combustion requires two parts of oxygen to one part methane. The atmosphere consists approximately of one-fifth oxygen and four-fifths nitrogen with very small quantities of other gases. Approximately 10 ft^3 of air is necessary to obtain 2 ft^3 of oxygen to mix with 1 ft^3 of methane. This produces 1050 Btu and approximately 11 ft^3 of flue gases, Figure 31-11.

The air and gas never mix completely in the combustion chamber. Extra air is always supplied so that the methane will find enough oxygen to burn completely. This excess air is normally supplied at 50% over what would be needed if it were mixed thoroughly. This means that to burn 1 ft^3 of methane, 15 ft^3 of air is supplied. There will be 16 ft^3 of flue gases to be vented from the combustion area, Figure 31-10A. There will also be additional air added at the draft hood. This will be covered later.

Flue gases contain approximately 1 ft^3 of oxygen (O_2), 12 ft^3 of nitrogen (N_2), 1 ft^3 of carbon dioxide (CO_2), and 2 ft^3 of water vapor (H_2O).

It is essential that the furnace be vented properly so that these gases will all be dissipated into the atmosphere.

Cooling the flame will cause inefficient combustion. This happens when the flame strikes the sides of the combustion chamber (due to burner misalignment) and is called *flame impingement*. When the flame strikes the cooler metal of the chamber, the temperature of that part of the flame is lowered below the ignition temperature. This results in poor combustion and produces carbon monoxide and soot.

The percentages and quantities indicated above are for the burning of methane only. These figures vary considerably when propane, butane, propane-air, or butane-air are used. Always use the manufacturer's specifications for each furnace when making adjustments.

SAFETY PRECAUTION: *Care should be taken while taking flue gas samples not to touch the hot vent pipe.*

31.6 GAS REGULATOR

Natural gas pressure in the supply line does not remain constant and is always at a much higher pressure than required at the manifold. The *gas regulator* drops the pressure to the proper level (in. W.C.) and maintains this constant pressure at the outlet where the gas is fed to the gas valve. Many regulators can be adjusted over a pressure range of several inches of water column, Figure 31-12. The pressure is increased when the adjusting screw is turned clockwise; it is decreased when the screw is turned counterclockwise. Some regulators have limited adjustment capabilities; others have no adjustment. Such regulators are either fixed permanently or sealed so that an adjustment

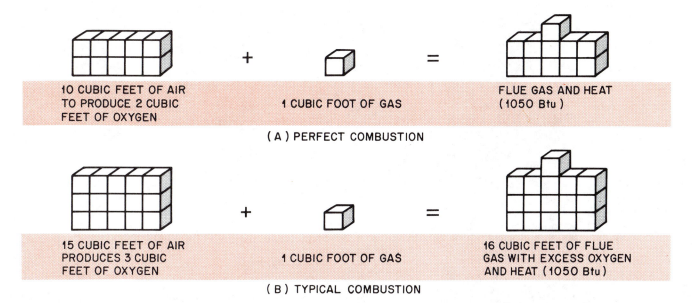

10 CUBIC FEET OF AIR TO PRODUCE 2 CUBIC FEET OF OXYGEN + 1 CUBIC FOOT OF GAS = FLUE GAS AND HEAT (1050 Btu)

(A) PERFECT COMBUSTION

15 CUBIC FEET OF AIR PRODUCES 3 CUBIC FEET OF OXYGEN + 1 CUBIC FOOT OF GAS = 16 CUBIC FEET OF FLUE GAS WITH EXCESS OXYGEN AND HEAT (1050 Btu)

(B) TYPICAL COMBUSTION

Figure 31-11. Quantity of air for combustion.

cannot be made in the field. The natural gas utility company should be contacted to determine the proper setting of the regulator. In most modern furnaces the regulator is built into the gas valve, and the manifold pressure is factory set at the most common pressure: $3\frac{1}{2}$ in. W.C. for natural gas.

LP gas regulators are located at the supply tank. These regulators are furnished by the gas supplier. Check with the supplier to determine the proper pressure at the outlet from the regulator. The outlet pressure range is normally 10–11 in. W.C. The gas distributor would normally adjust this regulator. In some localities a higher pressure is supplied by the distributor. The installer then provides a regulator at the appliance and sets the W.C. pressure to the manufacturer's specifications, usually 11 in. W.C. LP gas installations normally *do not* use gas valves with built-in regulators.

SAFETY PRECAUTION: * Only experienced technicians should adjust gas pressures.*****

31.7 GAS VALVE

From the regulator the gas is piped to the *gas valve* at the manifold, Figure 31-13. There are several types

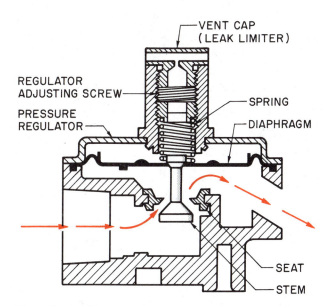

VENT CAP (LEAK LIMITER)

REGULATOR ADJUSTING SCREW

PRESSURE REGULATOR

SPRING

DIAPHRAGM

SEAT

STEM

Figure 31-12. Diagram of a standard gas-pressure regulator.

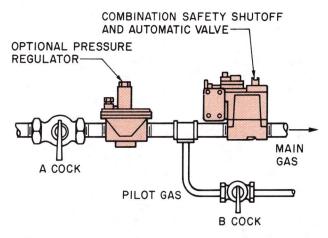

COMBINATION SAFETY SHUTOFF AND AUTOMATIC VALVE

OPTIONAL PRESSURE REGULATOR

A COCK

PILOT GAS

B COCK

MAIN GAS

Figure 31-13. Natural gas installation where the gas passes through a separate regulator to the gas valve and then to the manifold.

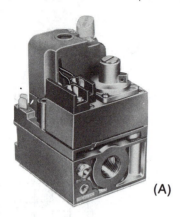

(A)

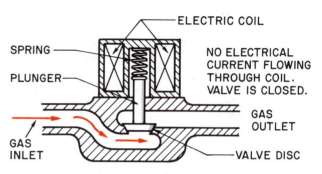

Figure 31–15. Solenoid gas valve in normally closed position.

(B)

Figure 31–14. Two gas valves with pressure regulator combined. *Courtesy (A) Honeywell, Inc., Residential Division. (B) Robertshaw Controls Company*

of gas valves. Many are combined with pilot valves and then called *combination* gas valves, Figure 31–14. We will first consider the gas valve separately and then in combination.

Valves are generally classified as solenoid, diaphragm, and heat motor.

31.8 THE SOLENOID VALVE

The gas type *solenoid* valve is a normally closed (NC) valve, Figure 31–15. The plunger in the solenoid is attached to the valve or is in the valve. When an electric current is applied to the coil, the plunger is pulled into the coil. This opens the valve. The plunger is spring loaded so that when the current is turned off the spring forces the plunger to its NC position, shutting off the gas, Figure 31–16.

31.9 THE DIAPHRAGM VALVE

The *diaphragm* valve uses gas pressure on one side of the diaphragm to open the valve. When there is gas pressure above the diaphragm and atmospheric pressure below it, the diaphragm will be pushed down and the valve port will be closed, Figure 31–17. When the gas is removed from above the diaphragm the pressure from below will push the diaphragm up and open the valve, Figure 31–17. This is done by a very small valve, called a *pilot-operated* valve because of its small size. It has two ports—one open while the other is closed. When the port to the upper chamber is closed and not allowing gas into the chamber above the diaphragm, the port to the atmosphere is opened. The gas already in this chamber is vented or bled to the pilot where it is burned. The valve controlling the gas into this upper chamber is operated electrically by a small magnetic coil, Figure 31–18.

When the thermostat calls for heat in the thermally operated valve, a bimetal strip is heated, which causes it to warp. A small heater is attached to the strip, or a resistance wire is wound around it, Figure 31–19. When the strip warps, it closes the valve to the upper chamber and opens the bleed valve. The gas in the upper chamber is bled to the pilot, reducing the pres-

ELECTRIC CURRENT APPLIED TO COIL PULLS PLUNGER INTO COIL, OPENING VALVE AND ALLOWING GAS TO FLOW. THE SPRING MAKES THE VALVE MORE QUIET AND HELPS START THE PLUNGER DOWN WHEN DEENERGIZED AND HOLDS IT DOWN.

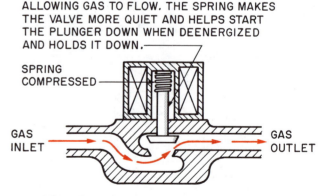

Figure 31–16. Solenoid valve in open position.

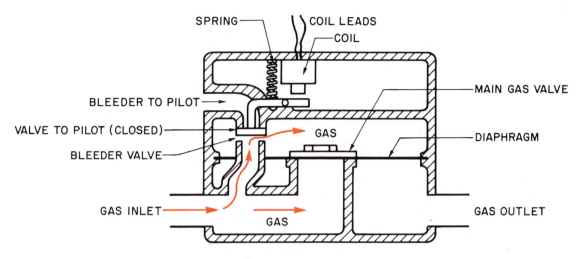

Figure 31-17. Electrically operated magnetic diaphragm valve.

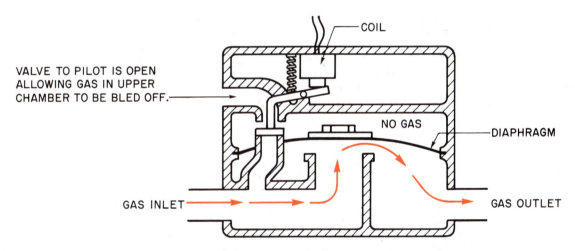

Figure 31-18. When an electric current is applied to the coil, the valve to the upper chamber is closed as the lever is attracted to the coil. The gas in the upper chamber bleeds off to the pilot reducing the pressure in this chamber. The gas pressure from below the diaphragm pushes the valve open.

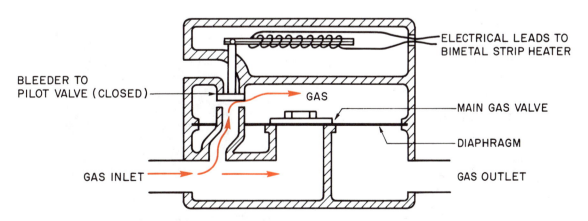

Figure 31-19. Thermally operated diaphragm gas valve.

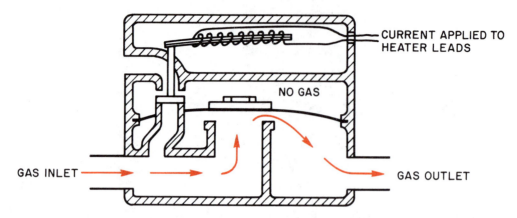

Figure 31-20. When an electric current is applied to the leads of the bimetal strip heater, the bimetal warps, closing the valve to the upper chamber, opening the valve to bleed the gas from the upper chamber. The gas pressure is then greater below the diaphragm, pushing the valve open.

sure above the diaphragm. The gas pressure below the diaphragm pushes the valve open, Figure 31–20.

31.10 HEAT-MOTOR-CONTROLLED VALVE

In a heat-motor-controlled valve an electric heating or resistance wire is wound around a rod attached to the valve, Figure 31–21. When the thermostat calls for heat, this heating coil or wire is energized and produces heat, which expands the rod. When expanded, the rod opens the valve, allowing the gas to flow. As long as heat is applied to the rod, the valve remains open.

When the heating coil is deenergized by the thermostat, the rod contracts. A spring will close the valve.

It takes time for the rod to expand and then contract. This varies with the particular model but the average time is 20 sec to open the valve and 40 sec to close it.

SAFETY PRECAUTION: *Be careful while working with heat-motor gas valves because of the time delay. Because there is no audible click, you cannot determine the valve's position. Gas may be escaping.*

31.11 AUTOMATIC COMBINATION GAS VALVE

Most modern furnaces designed for residential and light commercial installations use an automatic combination gas valve (ACGV), Figure 31–22. These valves incorporate a manual control, the gas supply for the pilot, the adjustment and safety shutoff features for the pilot, the pressure regulator, and the controls to operate the main gas valve. They also combine the

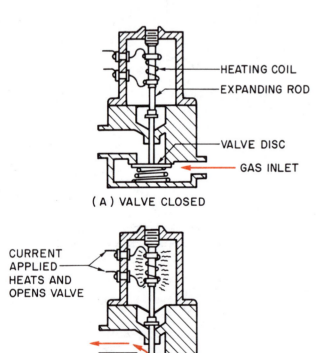

Figure 31-21. Heat-motor-operated valve.

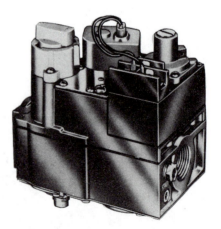

Figure 31-22. Automatic combination gas valve. *Courtesy Honeywell, Inc., Residential Division*

Figure 31-23. Aerated pilot. *Courtesy Robertshaw Controls Company*

Figure 31-24. Nonaerated pilot. *Courtesy Robertshaw Controls Company*

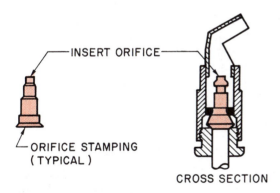

Figure 31-25. Nonaerated pilot burner.

features described earlier relating to the control and safety shutoff of the gas.

31.12 PILOTS

Pilot flames ignite the gas at the burner on most conventional gas furnaces. Pilot burners can be *aerated* or *nonaerated*. In the aerated pilot the air is mixed with the gas before it enters the pilot burner, Figure 31-23. The air openings, however, often clog and require periodic cleaning if there is dust or lint in the air. Nonaerated pilots use only secondary air at the point where combustion takes place. Little maintenance is needed with these, so most furnaces are equipped with nonaerated pilots, Figure 31-24.

The pilot is actually a small burner, Figure 31-25. It has an orifice, similar to the main burner, through which the gas passes. If pilots go out or do not perform properly, safety devices stop the gas flow.

Standing pilots burn continuously; other pilots are ignited by an electric spark when the thermostat calls for heat. In furnaces without pilots the electric spark ignites the gas at the burner. In furnaces with pilots, the pilot must be ignited and burning, and this must be proved before the gas valve to the main burner will open.

The pilot burner must direct the flame so that there will be ignition at the main burners, Figure 31-26. The pilot flame also provides heat for the safety device that shuts off the gas flow if the pilot flame goes out.

31.13 SAFETY DEVICES AT THE STANDING PILOT

Three main types of safety devices, called *flame-proving* devices, keep the gas from flowing through the main valve if the pilot flame goes out: the thermocouple or thermopile, the bimetallic strip, and the liquid-filled remote bulb.

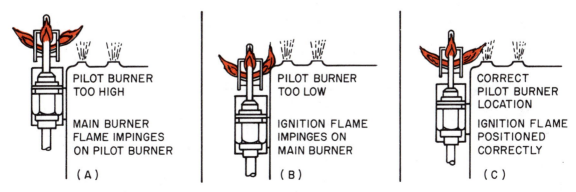

Figure 31-26. The pilot flame must be directed at the burners and adjusted to the proper height.

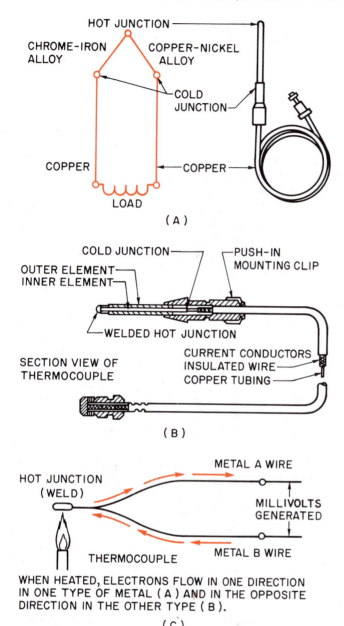

Figure 31-27. Thermocouple. *Courtesy Robertshaw Controls Company*

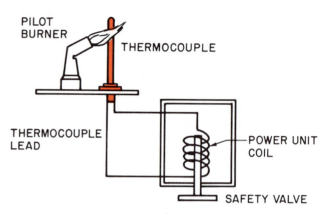

Figure 31-28. The thermocouple generates electrical current when heated by the pilot flame. This induces a magnetic field in the coil of the safety valve holding it open. If flame is not present, the coil will deactivate, closing the valve, and the gas will not flow.

31.14   THERMOCOUPLES AND THERMOPILES

The *thermocouple* consists of two dissimilar metals welded together at one end, Figure 31–27, called the "hot junction." When this junction is heated, it generates a small voltage (several millivolts) across the two wires or metals at the other end. The other end is called the "cold junction." The thermocouple is connected to a shutoff valve, Figure 31–28. As long as the electrical current in the thermocouple energizes a coil, the gas can flow. If the flame goes out, the thermocouple will cool off in about 30 sec, and no current will flow and the gas valve will close. A *thermopile* consists of several thermocouples wired in series to increase the voltage. If a thermopile is used, it performs the same function as the thermocouple, Figure 31–29.

31.15   THE BIMETALLIC SAFETY DEVICE

In the *bimetallic* safety device the pilot heats the bimetal strip, which closes electrical contacts wired to the gas safety valve, Figure 31–30(A). As long as the

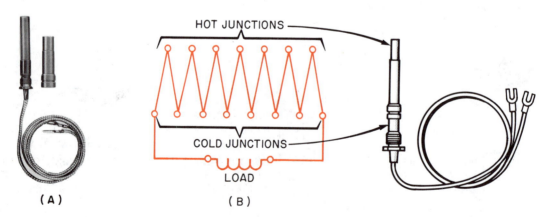

Figure 31-29. A thermopile consists of a series of thermocouples in one housing. *Photo courtesy Honeywell, Inc., Residential Division*

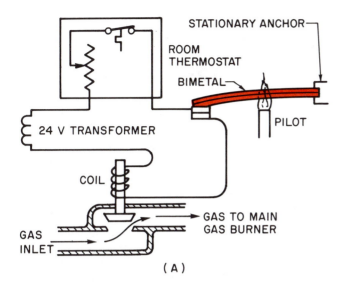

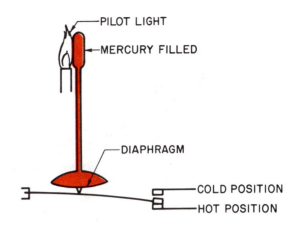

Figure 31–31. Liquid-filled remote bulb.

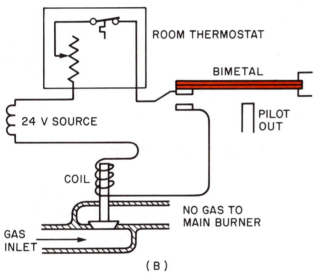

Figure 31–30. (A) When the pilot is lit, the bimetal warps, causing the contacts to close. The coil is energized, pulling the plunger into the coil opening the valve. This allows the gas to flow to the main gas valve. (B) When the pilot is out, the bimetal straightens, opening the contacts. The safety valve closes. No gas flows to the furnace burners.

pilot flame heats the bimetal, the gas safety valve remains energized and gas will flow when called for. When the pilot goes out, the bimetal strip cools within about 30 sec and straightens, opening contacts and causing the valve to close, Figure 31–30(B).

31.16 LIQUID-FILLED REMOTE BULB

The *liquid-filled remote bulb* includes a diaphragm, a tube, and a bulb, all filled with a liquid, usually mercury. The remote bulb is positioned to be heated by the pilot flame, Figure 31–31. The pilot flame heats the liquid at the remote bulb. The liquid expands, causing the diaphragm to expand, which closes contacts wired to the gas safety valve. As long as the pilot

flame is on, the liquid is heated and the valve is open, allowing gas to flow. If the pilot flame goes out, the liquid cools in about 30 sec and contracts, opening the electrical contacts and closing the gas safety valve.

31.17 THE MANIFOLD

The *manifold* in the gas furnace is a pipe through which the gas flows to the burners and on which the burners are mounted. The gas orifices are threaded into the manifold and direct the gas into the venturi in the burner. The manifold is attached to the outlet of the gas valve, Figure 31–32.

31.18 THE ORIFICE

The *orifice* is a precisely sized hole through which the gas flows from the manifold to the burners. The orifice is located in the spud, Figure 31–7. The spud is

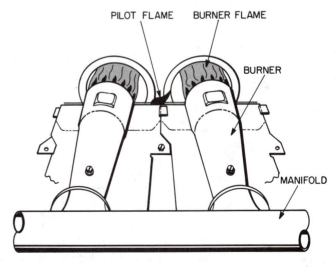

Figure 31–32. Manifold. *Courtesy BDP Company*

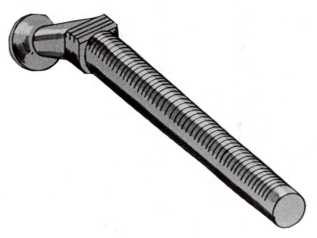

Figure 31–33. A cast iron burner with slotted ports. *Reproduced courtesy Carrier Corporation*

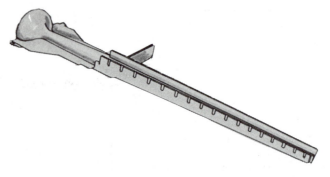

Figure 31–35. A ribbon burner. *Reproduced courtesy Carrier Corporation*

screwed into the manifold. The orifice allows the correct amount of gas into the burner.

31.19   THE BURNERS

Gas combustion takes place at the burners. Combustion uses primary and secondary air. Primary air enters the burner from near the orifice, Figure 31–10. The gas leaves the orifice with enough velocity to create a low-pressure area around it. The primary air is forced into this low-pressure area and enters the burner with the gas. The procedure for adjusting the amount of primary air entering the burner will be explained later in this unit.

The primary air is not sufficient for proper combustion. Additional air, called secondary air, is available in the combustion area. Secondary air is vented into this area through ventilated panels in the furnace. Both primary and secondary air must be available in the correct quantities for proper combustion. The gas is ignited at the burner by the pilot flame.

The drilled port burner is generally made of cast iron with the ports drilled. The slotted port burner is similar to the drilled port except the ports are slots, Figures 31–33 and 31–34. The ribbon burner produces a solid flame down the top of the burner, Figure 31–35. The single port burner is the simplest and has, as the name implies, one port. This is often called the *inshot* or *upshot* burner, Figure 31–36. All of these burners are known as *atmospheric* burners because the air for the burning process is at atmospheric pressure. In some larger gas burners, air is forced in with blowers.

31.20   THE HEAT EXCHANGER

The burners are actually located at the bottom of the heat exchanger, Figure 31–37. The *heat exchanger* is divided into sections with a burner in each section.

Heat exchangers are made of sheet steel and designed to provide rapid transfer of heat from the hot combustion products through the steel to the air that will be distributed to the space to be heated.

Poor combustion can corrode the heat exchanger so good combustion is preferred. The steel in the exchangers may be coated or bonded with aluminum, glass, or ceramic material. These materials are more corro-

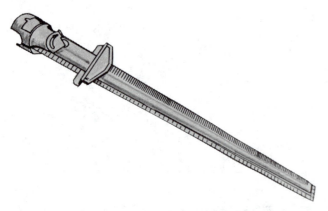

Figure 31–34. A stamped steel slotted burner. *Reproduced courtesy Carrier Corporation*

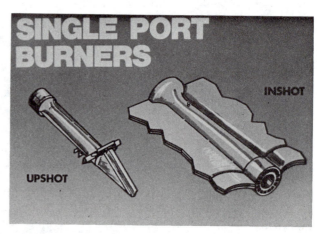

Figure 31–36. An inshot burner. *Reproduced courtesy Carrier Corporation*

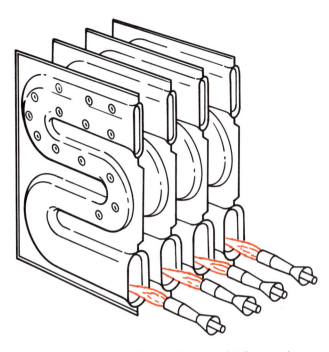

Figure 31–37. Modern heat exchanger with four sections.

sion resistant, and the unit will last longer. Some exchangers are made of stainless steel, which is more expensive but resists corrosion extremely well. *The exchanger must not be corroded or pitted enough to leak since one of its functions is to keep combustion gases separated from the air to be heated and circulated throughout the building.*

31.21 THE FAN SWITCH

The fan switch automatically turns the blower on and off. The blower circulates the heated air to the conditioned space. The switch can be temperature controlled or time delay. In either instance the heat exchanger is given a chance to heat up before the fan is turned on. This delay keeps cold air from being circulated through the duct system before the heat exchanger gets hot at the beginning of the cycle. The heat exchanger must be hot in a conventional furnace because the heat provides a good draft for properly venting the combustion gases.

There is also a delay in shutting off the fan. This allows the heat exchanger time to cool off and dissipates the furnace heat at the end of the cycle.

The temperature-sensing element of the switch, usually a bimetal helix, is located in the airstream near the heat exchanger, Figure 31–38. When the furnace comes on, the air is heated, which expands the bimetal and closes the contacts, thus activating the blower motor. This is called a *temperature on–temperature off* fan switch.

The fan switch could activate the blower with a time delay and shut it off with a temperature-sensing device.

When the thermostat calls for heat and the furnace starts, a small resistance-heating device is activated. This heats a bimetal strip that will close electrical contacts to the blower when heated. This provides a time delay and allows the furnace to heat the air before the blower comes on. The bimetal helix in the airstream keeps the contacts closed even after the room thermostat is satisfied and the burner flame goes out. When the furnace shuts down, the bimetal cools at the same time as the heat exchanger and turns off the blower. This fan switch provides a positive starting of the fan with temperature stopping it and is called *time on–temperature off*.

A third type, called a *time on–time off*, switch uses a small heating device, such as the one used in the time on–temperature off switch. The difference is that this switch is not mounted in such a way that the heat exchanger heat will influence it. The time delay is designed into the switch and is not adjustable. Most models of the other two switches are adjustable. Procedures to make these adjustments will be discussed later in this unit.

31.22 THE LIMIT SWITCH

The *limit switch* is a safety device. If the fan does not come on or if there is another problem causing the heat exchanger to overheat, the limit switch will open its contacts, which closes the gas valve. *Almost any circumstance causing a restriction in the airflow to the conditioned space can make the furnace overheat: for

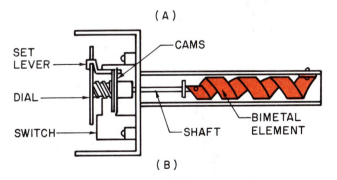

(A)

SET LEVER

DIAL

SWITCH

CAMS

SHAFT

BIMETAL ELEMENT

(B)

Figure 31–38. Temperature on–temperature off fan switch. *Photo by Bill Johnson*

example, dirty filters, a blocked duct, dampers closed, fan malfunctioning, or a loose or broken fan belt. The furnace may also be overfired due to an improper setting or malfunctioning of the gas valve. It is extremely important that the limit switch operate as it is designed to do.✱

The limit switch has a heat sensing element. When the furnace overheats this element opens contacts closing the main gas valve. This switch can be combined with a fan switch, Figure 31–39.

31.23 VENTING

Conventional gas furnaces use a hot flue gas and natural convection to get the products of combustion to vent. The hot gas is vented quickly primarily to prevent cooling of the flue gas, which produces condensation and other corrosive actions. However, these furnaces lose some efficiency because considerable heat is lost up the flue.

High-efficiency furnaces recirculate the flue gases through a special extra heat exchanger to keep more of the heat available for space heating in the building. The gases are then pushed out the flue by a small fan. This causes condensation and thus more corrosion in the flue. Plastic pipe is often used because it is not damaged by the corrosive materials. These furnaces can have efficiencies of up to 97%.

✱Regardless of the type, a gas furnace must be properly vented.✱ The venting must provide a safe and effective means of moving the flue gases to the outside air. In a conventional furnace it is important for the flue gases to be vented as quickly as possible. Conven-

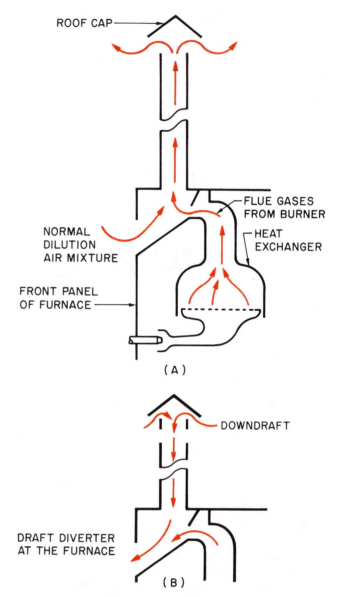

Figure 31–40. Draft hood.

tional furnaces are equipped with a draft hood that blends some room air with the rising flue gases, Figure 31–40. The products of combustion enter the draft hood and are mixed with air from the area around the furnace. Approximately 100% additional air (called *dilution* air) enters the draft hood at a lower temperature than the flue gases. The heated gases rise rapidly and create a draft, bringing the dilution air in to mix and move up the vent.

If wind conditions produce a downdraft, the opening in the draft hood provides a place for the gases and air to go, diverting it away from the pilot and main burner flame. This *draft diverter* helps reduce the chance of the pilot flame being blown out and the main burner flame being altered, Figure 31–40.

✱Make sure replacement air is available in the furnace area.✱ Remember it takes 10 ft³ of air for each 1 ft³ of natural gas to support combustion. To this is

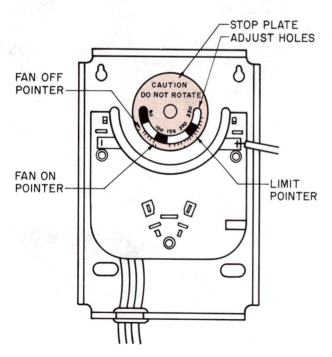

Figure 31–39. Fan on-off and limit control.

Figure 31–41. Section of type B gas vent.

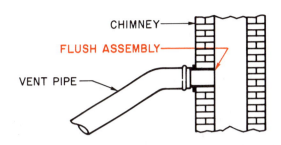

Figure 31–43. The vent pipe should not extend beyond the inside of the flue lining.

added 5 ft³ more (excess air) to insure enough oxygen. Another 15 ft³ is added at the draft hood. All of this air must be replaced in the furnace area and also must be vented as flue gases and dilution air rise up the vent. If the air is not replaced in the furnace area, it will become a negative pressure area, and air will be pulled down the flue. Products of combustion will then fill the area.

Type B vent or approved masonry materials are required for conventional gas-furnace installations. *Type B* venting systems consist of metal vent pipe of the proper thickness, which is approved by a recognized testing laboratory, Figure 31–41. The venting system must be continuous from the furnace to the proper height above the roof. Type B vent pipe is usually of a double-wall construction with air space between. The inner wall may be made of aluminum and the outer wall constructed of steel.

The vent pipe should be at least the same diameter as the vent opening at the furnace. The horizontal run should be as short as possible. Horizontal runs should always be sloped upward as the vent pipe leaves the furnace. A slope of $\frac{1}{4}$ in. per foot of run is the minimum recommended, Figure 31–42. Use as few elbows as possible. Long runs lower the temperature of the flue gases before they reach the vertical vent or chimney,

and they reduce the draft. Do not insert the vent connector beyond the inside of the wall of the chimney when the gases are vented into a masonry type chimney, Figure 31–43.

If two or more gas appliances are vented into a common flue, the common flue size should be calculated based on recommendations from the National Fire Protection Association (NFPA) or local codes.

The top of each gas furnace vent should have an approved vent cap to prevent the entrance of rain and debris and to help prevent wind from blowing the gases back into the building through the draft hood, Figure 31–44.

Metal vents reach operating temperatures quicker than masonry vents do. Masonry chimneys tend to cool the flue gases. Masonry chimneys must be lined with a glazed type of tile or vent pipe installed in the chimney. If an unlined chimney is used, the corrosive materials from the flue gas will destroy the mortar joints. Condensation occurs at startup and during short running times in mild weather. It takes a long running cycle to heat a heavy chimney. The warmer the gases, the faster they rise and the less damage there is from condensation and other corrosive actions. Vertical vents can be masonry with lining and glazing or prefabricated metal chimneys approved by a recognized testing laboratory.

When the furnace is off, heated air can leave the structure through the draft hood. Automatic vent dampers that close when the furnace is off and open when the furnace is started can be placed in the vent to prevent air loss.

Several types of dampers are available. One type uses a helical bimetal with a linkage to the damper. When

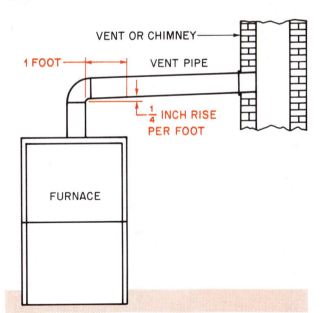

Figure 31–42. The minimum rise should be $\frac{1}{4}$ in. per foot of horizontal run.

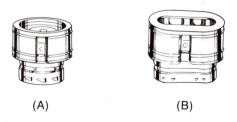

(A)   (B)

Figure 31–44. Gas vent caps.

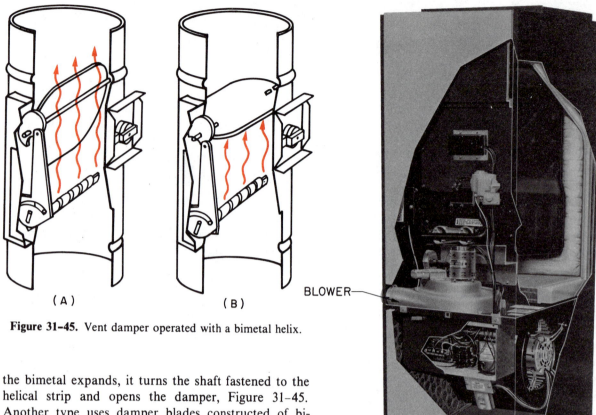

(A)   (B)

Figure 31–45. Vent damper operated with a bimetal helix.

the bimetal expands, it turns the shaft fastened to the helical strip and opens the damper, Figure 31–45. Another type uses damper blades constructed of bimetal. When the system is off, the damper is closed. When the vent is heated by the flue gases, the bimetal action on the damper blades opens them, Figure 31–46.

In high-efficiency systems a small blower is installed in the vent system near the heat exchanger. This blower operates only when the furnace is on so that heated room air is not being vented 24 hr a day. This blower will vent gases at a lower temperature than is possible with natural draft, allowing the heat exchanger to absorb more of the heat from the burner for distribution to the conditioned space, Figure 31–47. When blowers are used for furnace venting, the room thermostat normally energizes the blower. Then when blower air is established, the burner ignites. Blower air

BLOWER

Figure 31–47. Forced venting with a vent blower. *Courtesy Heil Quaker Corporation*

can be established by air-flow switches or a centrifugal switch on the motor.

31.24   GAS PIPING

Installing or replacing gas piping can be a very important part of a technician's job. The first thing a

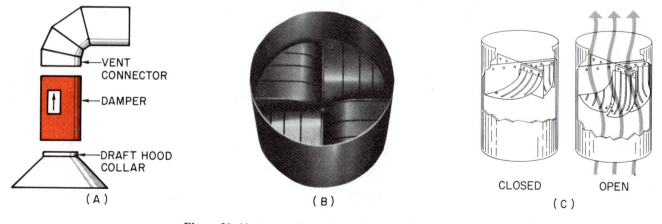

VENT CONNECTOR

DAMPER

DRAFT HOOD COLLAR

(A)   (B)   CLOSED   OPEN   (C)

Figure 31–46. Automatic vent damper with bimetal blades.

technician should do is to become familiar with the national and local codes governing gas piping. Local codes may vary from national codes. A technician should also be familiar with the characteristics of natural gas and LP gas in the particular area in which he or she works. Pipe sizing and furnace Btu ratings will vary from area to area due to varying gas characteristics.

The piping should be kept as simple as possible with the pipe run as direct as possible. Distributing or piping gas is similar to distributing electricity. The piping must be large enough, and there should be as few fittings as possible in the system because each of these fittings creates a resistance. Pipe that is too small with other resistances in the system will cause a pressure drop, and the proper amount of gas will not reach the furnace. The specific gravity of the gas must also be taken into consideration when designing the piping. Systems should be designed for a maximum pressure drop of 0.35 in. W.C. In designing the piping system the amount of gas to be consumed by the furnace must be determined. The gas company should be contacted to determine the heating value of the gas in that area. They can also furnish you with pipe sizing tables and helpful suggestions. Most natural gas will supply 1050 Btu/ft³ of gas. To determine the gas to be consumed in 1 hr for a typical natural gas, use the following formula:

$$\frac{\text{Furnace Btu input per hour rating}}{\text{Gas heating rating}} = \text{Cubic feet of gas needed per hour}$$

Suppose the Btu rating of the furnace were 100,000 and the gas heating rating were 1050. Then

$$\frac{100,000}{1050} = 95.2 \text{ ft}^3 \text{ of natural gas needed per hour}$$

Tables give the size (diameter) of pipe needed for the length of the pipe to provide the proper amount of gas needed for this furnace. The designer of the piping system would have to know whether natural gas or LP

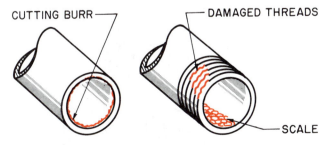

Figure 31-48. Insure that threads are not damaged. Deburr pipe. Clean all scale, dirt, and other loose material from pipe threads and inside of pipe.

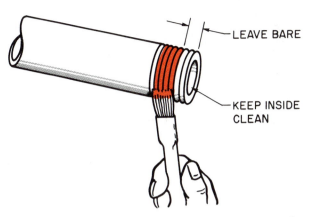

Figure 31-49. Pipe dope should be used, but do not apply to last two threads at end of pipe.

gas were going to be used, since the pipe sizing would be different due to the differences in specific gravities of the gases.

Steel or wrought iron pipe should be used. Aluminum or copper tubing may be used, but they will have to be treated to prevent corrosion. They are used only in special circumstances.

SAFETY PRECAUTION: *Insure that all piping is free of burrs and that threads are not damaged, Figure 31-48. All scale, dirt, or other loose material should be cleaned from pipe threads and from the inside of the pipe. Any loose particles can move through the pipe to the gas valve and keep it from closing properly. It may also stop up the small pilot-light orifice.*

When assembling threaded pipe and fittings, use a joint compound, commonly called *pipe dope*. Do not apply this compound on the last two threads at the end of the pipe because it could get into the pipe and plug the orifice or prevent the gas valve from closing, Figure 31-49. Also be careful when using teflon tape to seal the pipe threads since it can also get into the pipe and cause similar problems.

At the furnace the piping should provide for a drip trap, a shutoff valve, and a union, Figure 31-50. A manual shutoff valve within 2 ft of the furnace is required by most local codes. The drip trap is installed to catch dirt, scale, or condensate (moisture) from the supply line. A union between the tee and the gas valve allows the gas valve or the entire furnace to be removed without disassembling other piping. Piping should be installed with a pitch of ¼ in. for every 15 ft of run in the direction of flow. This will prevent trapping of the moisture that could block the gas flow, Figure 31-51. Small amounts of moisture will move to the drip leg and slowly evaporate. Use pipe hooks and/or straps to support the piping adequately, Figure 31-52.

*When completed, test the piping assembly for leaks. There are several methods, one is to use the

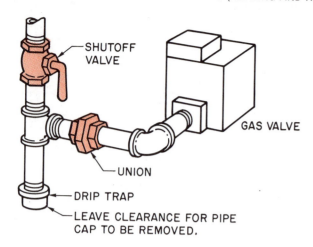

Figure 31-50. A shutoff valve, drip trap, and union should be installed ahead of the gas valve.

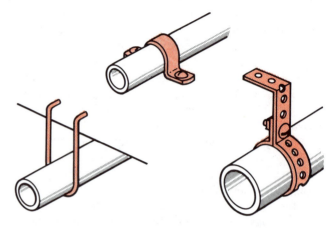

Figure 31-52. Piping should be supported adequately with pipe hooks or straps.

gas in the system. Do not use other gases, and especially *don't* use oxygen.* Turn off the manual shutoff to the furnace. Make sure all joints are secure. Turn the gas on in the system. Watch the gas meter dial to see if it moves, indicating that gas is passing through the system. A check for 5–10 min will indicate a large leak if the meter dial continues to show gas flow. An overnight standing check is better. If the dial moves, indicating a leak, check each joint with soap and water, Figure 31-53. A leak will make the soap bubble. When you find the leak and repair it, repeat the same procedure to insure that the leak has been repaired and that there are no other leaks.

Leaks can also be detected with a manometer. Install the manometer in the system to measure the gas pressure. Turn the gas on and read the pressure in in. W.C. on the instrument. Now turn the gas off. There should be no loss of pressure. Some technicians allow an over-

night standing period with no pressure drop as their standard. Use soap and water to locate the leak as before. *Note:* If a standing-pilot system that is not 100% shut off is being checked, the pilot-light valve *must be closed* because it will indicate a leak.

A high pressure test is more efficient. For a high-pressure leak check, use air pressure from a bicycle tire pump. Ten pounds of pressure for 10 min with no leakdown will prove there are no leaks. *Don't let this pressure reach the automatic gas valve. The manual valve must be closed during the high-pressure test.*

If there are no gas leaks, the system must be *purged;* that is the gas must be bled off to rid the system of air and/or other gases. *Purging should take place in a well-ventilated area. It can often be done by disconnecting the pilot tubing or loosening the union in the piping between the manual shutoff and the gas valve. Do not purge where the gas will collect in the combustion area.* After the system has been purged, allow it to set for at least 15 min to allow any accumulated gas to dissipate. If you are concerned about gas collecting in the area, wait for a longer period. Moving the

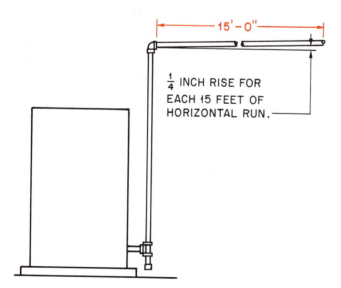

Figure 31-51. Piping from furnace should have a rise of $\frac{1}{4}$ in. for each 15 ft of horizontal run.

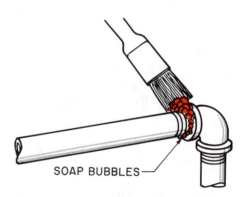

Figure 31-53. Apply liquid soap to joints. Bubbling indicates a leak.

air with a hand-operated fan can speed this up. When it is evident that no gas has accumulated, the pilot can be lit.

31.25   GAS-FURNACE WIRING DIAGRAMS

Figure 31–54 is a wiring diagram of a gas furnace. It does not show the safety features necessary in a furnace for proving that the pilot flame is lit or for a glow coil or intermittent spark ignition. These will be discussed later.

The path for the 115-V electrical current (white wire) goes from N to the primary transformer, to the limit switch (which is normally closed), and back to the L1 terminal on the red wire.

The transformer reduces the voltage to 24 V. The thermostat and the gas valve are in this circuit (blue wire). It goes from the transformer secondary to the thermostat, to the gas valve, and back to the other terminal on the transformer secondary.

The fan circuit is shown with the black wire with part of it in red. It goes from L1 through the fan switch to the blower motor and back to the disconnect to N.

The transformer is usually energized all the time. The current path can be traced through the red wire and the NC limit switch. When there is a call for heat, the contacts in the thermostat will close. This will cause a current to flow in the secondary (blue wire). A coil in the gas valve will be energized, opening the gas valve. The pilot safety and flame-proving device are in another circuit. When enough heat is produced, a bimetal control will warp, closing the contacts to the fan. This will cause the fan to operate, distributing the heat to the space where it is needed.

The limit switch is a safety device. If the furnace overheats, a bimetal will warp, opening this switch in the 115-V circuit. This will stop the current flow to the transformer, closing the gas valve. This will not interrupt the power to the fan, which will continue to operate and distribute the heat and cool the heat exchanger.

31.26   TROUBLESHOOTING TECHNIQUES

To troubleshoot these circuits, first set the selector switch on a VOM (volt-ohm-milliammeter) to 250 V to be sure that the VOM will not be damaged by over-

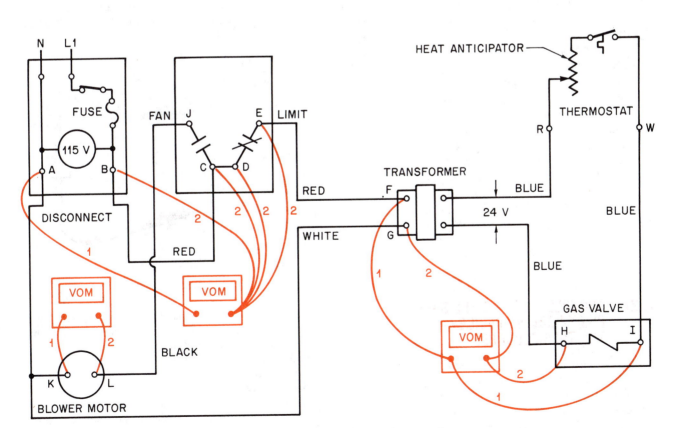

Figure 31–54. Basic wiring diagram for a gas furnace.

voltage. To check the 115-V circuit, place probe 1 on terminal A or the neutral wire connection and probe 2 at B. A reading of 115 V indicates that power is being supplied to the unit. No reading would indicate that the fuse, circuit breaker, or switch was defective.

Leave probe 1 at a neutral terminal and move probe 2 to *C:* no power indicates that the conductor is defective; *D:* no reading indicates that the jumper in the limit switch is open; *E:* no reading indicates that the limit switch is open. (Note that the limit switch is heat operated. It should not open until the temperature in the heat exchanger is greater than 200° F.)

To check power to the transformer, place probe 1 at F and probe 2 at G. 115 V should be read. This shows that the correct voltage is supplied to the transformer. If there is no reading here, there must be a break in the conductor.

On a service call check the voltage at the transformer first. If there is no voltage, then follow the other steps to determine where the interruption is.

To check the fan circuit, do the following: Place probe 1 at A and probe 2 at B. If you read 115 V, place probe 2 at C. If there is no power, a conductor must be open. If there is power, move probe 2 to J. (*Note:* This is a temperature-actuated switch and will not close until the temperature in the heat exchanger reaches approximately 140° F.) If there is no power at J after proper heat has been achieved, the fan switch is defective and should be replaced. If there is power at J, move probe 1 to K and probe 2 to L. If there is a 115-V reading here but the motor does not run, the motor is defective.

To troubleshoot the output voltage from the transformer, place probe 1 on I and probe 2 on H. If the meter reads 24 V, the gas valve should open. If the valve does not open, turn the power off. Take one lead off the gas valve, adjust the meter to measure ohms, place one probe on H and the other at I. If there is no resistance, the coil in the valve is defective and the valve should be replaced. If there is a resistance the valve itself is stuck and needs to be replaced. If there is no voltage at H and I, place a jumper across R and W at the thermostat. If there is voltage at H and I, the thermostat should be replaced. If there is still no voltage, a wire must be broken.

31.27 TROUBLESHOOTING THE SAFETY PILOT-PROVING DEVICE— THE THERMOCOUPLE

The thermocouple generates a small electrical current when one end is heated. When the pilot flame is lit and heating the thermocouple, it generates a current that energizes a coil holding a safety valve open. This valve is manually opened when lighting the pilot, the

coil only holds it open. If the pilot flame goes out, the current would no longer be generated and the safety valve would close, shutting off the supply of gas.

To light the pilot, turn the gas valve control to PILOT position. Depress the control knob and light the pilot. Hold the knob down for approximately 45 sec, release, and turn to ON. The thermocouple may be defective if the pilot goes out when the knob is released. Figure 31-55 describes the no-load test for a thermocouple. It is important to remember that the thermocouple only generates enough current to *hold* the coil armature in. The armature is pushed in when the valve handle is depressed.

To check under load conditions, unscrew the thermocouple from the gas valve. Insert the thermocouple testing adapter into the gas valve. The testing adapter allows you to take voltage readings while the thermocouple is operating. Screw the thermocouple into the top of this adapter, Figure 31-56. Relight the pilot and check the voltage produced with a millivoltmeter, using the terminals on the adapter. The pilot flame must cover the entire top of the thermocouple rod. If the thermocouple produces at least 9 mV under load (while connected to its coil), it is good. Otherwise replace it. Manufacturer's specifications must be checked to determine the acceptable voltage for holding in different valve coils. Thermopiles produce a greater voltage. If the thermocouple functions properly but the flame will not continue to burn when the knob is released, the coil in the safety valve must be replaced. This is normally done by replacing the entire gas valve.

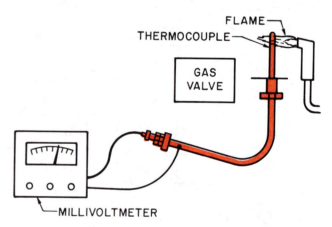

Figure 31-55. Thermocouple no load test. Operate the pilot flame for at least 5 min with the main burner off. (Hold the gas cock knob in the pilot position and depressed.) Disconnect the thermocouple from the gas valve while still holding the knob down to maintain the flame on the thermocouple. Attach the leads from the millivoltmeter to the thermocouple. Use DC voltage. Any reading below 20 mV indicates a defective thermocouple or poor pilot flame. If pilot flame is adjusted correctly, test thermocouple under load conditions.

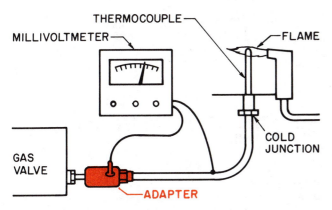

Figure 31-56. Thermocouple test under load conditions. Disconnect thermocouple from gas valve. Screw thermocouple test adapter into gas valve. Screw thermocouple into test adapter. Light the pilot and main burner and allow to operate for 5 min. Attach one lead from the millivoltmeter to either connecting post of the adapter and the other lead to the thermocouple tubing. Any reading under 9 mV would indicate a defective thermocouple or insufficient pilot flame. Adjust pilot flame. Replace thermocouple if necessary. Insure that pilot, pilot shield, and thermocouple are positioned correctly. Too much heat at the cold junction would cause a satisfactory voltage under no load conditions but unsatisfactory under a load condition.

31.28 GLOW-COIL IGNITION CIRCUIT

The glow coil is used to automatically reignite a pilot light if it goes out. This happens often on equipment located in a drafty area. The glow coil is normally applied to equipment with a standing pilot that is *not* 100% shut off. A thermal type of main gas safety valve is used to allow the glow coil enough time to heat up and light the pilot. The pilot has gas going to it at all times; if the pilot goes out, the circuit to the glow coil will be energized at the same time that the circuit to the gas valve is deenergized, Figure 31-57. If the glow coil fails to ignite the gas, the small amount of gas from the pilot will go up the flue. Notice the door switch in the circuit. This switch will not allow power to pass when the front panel is removed. It will have to be held shut to check voltage inside the furnace. It will appear like a light switch in a refrigerator door, Figure 31-58.

The power supply in this glow-coil circuit is from the low-voltage transformer, however, it is not 24 V but 12 V. It may be taken from a third tap or wire from the middle of the control transformer's coil. This is not the only method used to obtain the low voltage for a glow coil. Sometimes, the full transformer 24 V may go

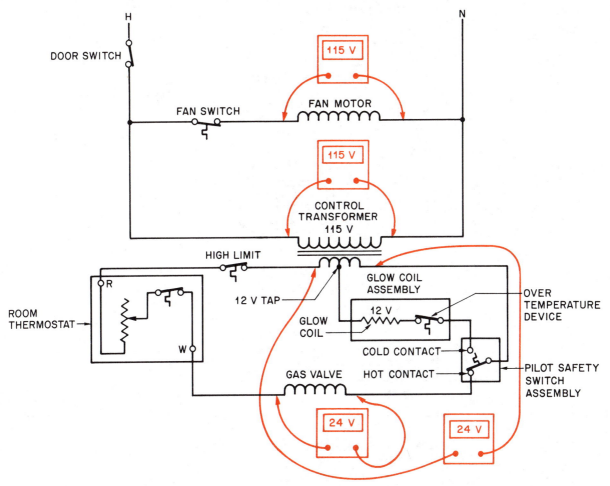

Figure 31-57. Normal operating voltages.

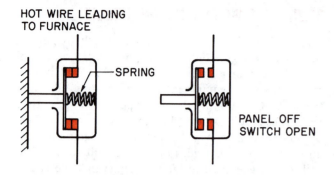

Figure 31-58. Panel switch used to interrupt power to the furnace circuit for safety purposes.

to the glow-coil component. When this is the case, the glow coil will more than likely still be a 12-V glow coil and have another method for reducing the voltage (such as using a resistor), Figure 31–59.

The glow coil as a component may also have a high-temperature device to keep the coil from glowing in the event that the pilot does not light. The glow coil itself is normally made of platinum and would burn up if left on indefinitely. Stopping the glow coil is often accomplished with a high-temperature cutout furnished with the glow-coil assembly. The cutout shuts off the glow coil after about 15 or 20 sec if the pilot does not light.

31.29 TROUBLESHOOTING THE GLOW-COIL CIRCUIT

Troubleshooting this circuit is very much like troubleshooting any other circuit. There has to be a power supply, the transformer, interconnecting wiring, and a load—the glow coil. To get these in proper perspective, let's use a customer complaint. The customer calls and says there is no heat. The problem is that the pilot has gone out and the pilot safety valve did not make to the cold contact to provide power to the glow coil. See Figure 31–60 for the voltmeter readings that should appear at the various points.

Procedures to use:

1. Make sure that the thermostat is calling for heat.

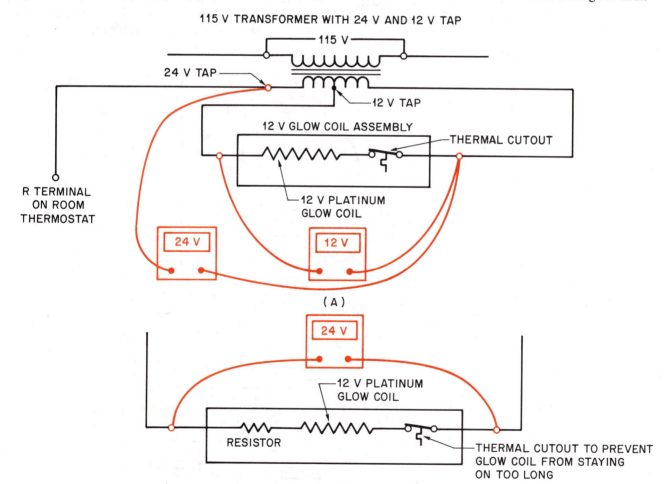

Figure 31-59. (A) Voltage reduced using third wire on transformer. (B) Resistor used in glow-coil circuit.

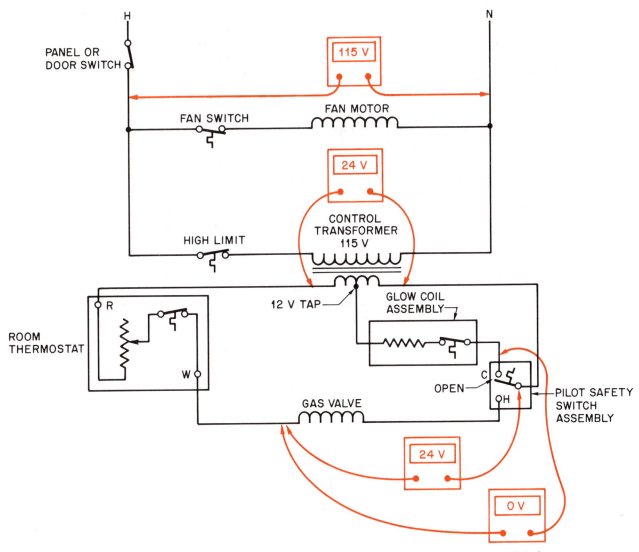

Figure 31–60. The pilot safety switch is not made to the cold contact when the pilot light is out.

2. Check to see that there is power to the unit (approximately 115 V).
3. Check for 24 V at the control transformer on the secondary side.
4. Check for 24 V at the pilot safety terminal on the line (inlet side) with the other probe at the glow coil terminal.
5. Check for voltage at the glow-coil's other terminal. Suppose you get 0 V. This indicates that the pilot safety switch is not making contact on the cold contact. It needs to be replaced. *Note:* The glow-coil component should have power to it whenever there is no pilot. This power is supplied through the pilot safety switch.

   If you had 12 V to the glow coil and it would not glow, wait for a few minutes to allow the thermal element to cool. If it did not cool and come on, then the glow-coil component needs to be replaced. When you get the pilot component out of the unit, check it with an ohmmeter. Sometimes the

thermal contacts will close back while removing the glow coil assembly. Since this is a standing-pilot-type of device, the pilot can be lit with a match until a replacement can be obtained.

The glow coil has to be positioned in such a fashion that the gas from the pilot is directed toward the glow coil. If the pilot light is dirty or the gas is not hitting the glow coil, it may not light. If the glow coil glows and will not light the pilot, then check

1. The pilot gas stream hitting the glow coil
2. If the glow coil is hot enough (check for full voltage)

The glow coil can be used for applications other than just the standing pilot. For instance, it can be used for intermittent pilot. This is a system where the pilot is extinguished at the end of each cycle and relit upon a call for heat. This will be covered later in the unit. The glow coil can be used to ignite gas in applications that are designed for it. The same troubleshooting procedures would be used. Voltage must be at the coil for it to glow.

31.30 TROUBLESHOOTING THE GAS-VALVE CIRCUIT

This is much the same as troubleshooting the glow-coil circuit. We again use the previous example, but this time the pilot light is burning. The problem is that the pilot safety switch is not making to the hot contact. The following procedure will determine the problem. See Figure 31–61 for the voltmeter readings.

1. Check to see that the thermostat is calling for heat.
2. Check for 115 V at the unit.
3. Check for 24 V at the secondary of the transformer. *Note:* If there is voltage to the unit and not to the transformer, the limit control may be open. If a limit control suddenly becomes open, be sure to check the airflow. The limit control normally *never*

moves from the made position. See Figure 31–62 for volt readings with an open limit.

4. Check for voltage at the gas valve. Here we find that we have no voltage. There are three possibilities: (1) The thermostat contacts are open; (2) the pilot safety switch is defective; (3) the conductors leading to the gas valve from the 24-V power supply are open.

Figure 31–62 shows still another problem that can occur. The fan switch is not making. The fan will not come on to dissipate the heat in the furnace, so the limit control shuts the gas off. Notice that voltage is absent everywhere except at the line coming in. This will be true when the limit switch is open. When the limit switch closes, the furnace will light for short periods of time. *Note:* When approaching a furnace in this condition, see if it is hot to the touch. If the furnace is hot and the fan is not running, there is a fan or fan circuit problem.

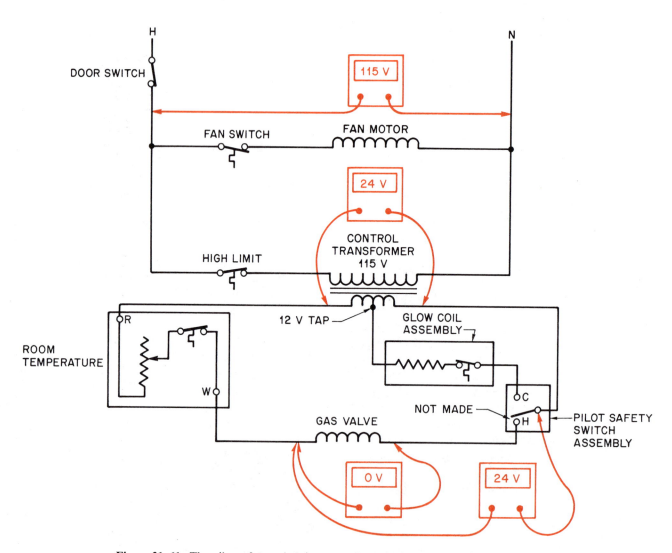

Figure 31–61. The pilot safety switch is not made to the hot contact after a pilot outage.

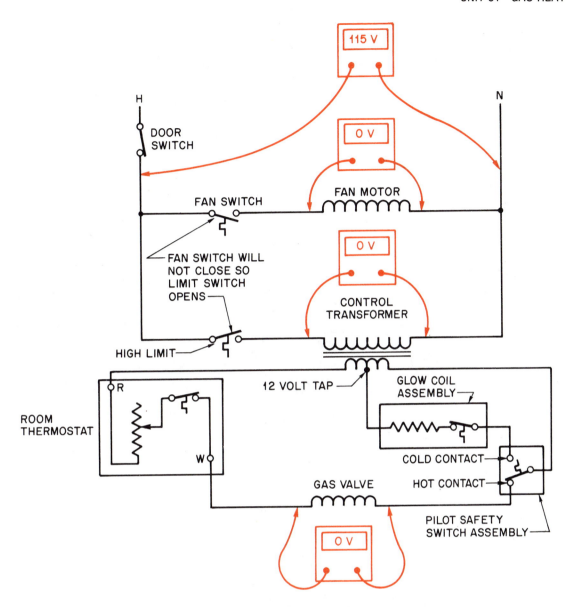

Figure 31–62. The pilot light is lit—the limit switch is open.

31.31 SPARK-TO-PILOT IGNITION

In the spark-to-pilot-type of gas ignition system, a spark ignites the pilot, which ignites the main gas burners, Figure 31–63. The pilot only burns when the thermostat calls for heat. This system is popular because fuel is not wasted with the pilot burning when not needed. Two types of gas valves are used with this system. One is used with natural gas and is not considered a 100% shutoff system. If the pilot does not ignite, the pilot valve will remain open, the spark will continue, and the main gas valve will not open. The other type is used with LP gas and some natural gas applications and is a 100% shutoff system. If there is no pilot ignition, the pilot gas valve will close after approximately 45 sec and the spark will stop. This

Figure 31–63. Spark-to-pilot ignition system. *Courtesy Robertshaw Controls Company*

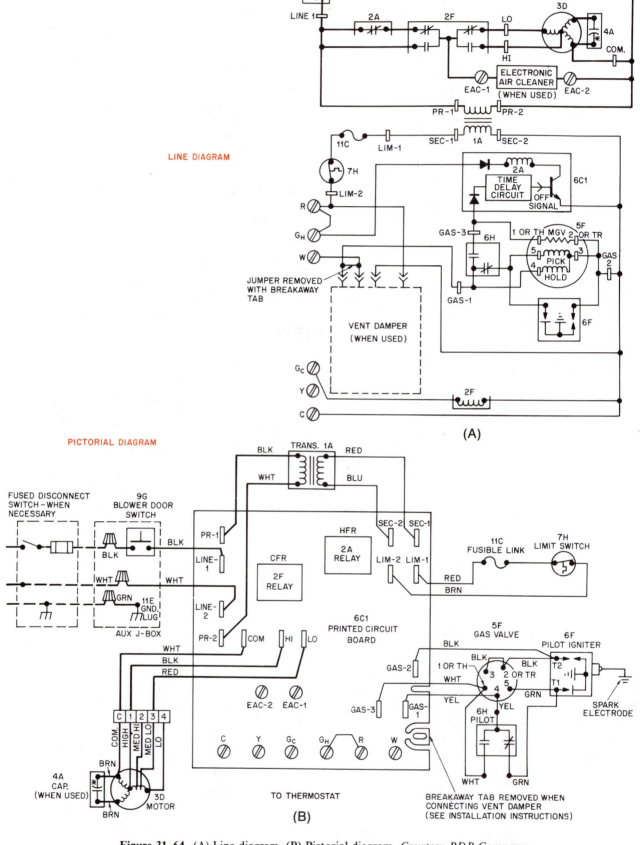

Figure 31–64. (A) Line diagram. (B) Pictorial diagram. *Courtesy BDP Company*

system must be manually reset, usually at the thermostat.

When there is a call for heat in the natural gas system, contacts will close in the thermostat, providing 24-V to the pilot igniter and to the pilot valve coil. The coil opens the pilot valve, and the spark ignites the pilot. Two types of systems are used to open the main gas valve: a mercury vapor tube and a flame rectification system. In the *mercury vapor tube* system the remote bulb is located at the pilot. When this bulb is heated by the pilot flame, it expands a bellows or diaphragm that makes a switch and opens the main gas valve. The pilot flame then ignites the main burners.

In the *flame rectification* system the heat from the pilot flame changes the normal alternating current to direct current. The electronic components in the system will only energize and open the main gas valve with a direct current. It is important that the pilot flame quality be correct to insure proper operation. If there is not enough heat, insufficient DC voltage will be produced. If there is too much heat, the AC will not change to DC. Consequently, the main gas valve will open only with the proper pilot flame quality.

The spark is intermittent and arcs approximately 30 times per minute. It must be a high-quality arc or the pilot will not ignite. The main gas valve should open 50–70 sec after the thermostat contacts close. This delay may vary depending on the manufacturer of the furnace controls.

31.32 TROUBLESHOOTING SPARK-IGNITION PILOT LIGHTS

Most spark-ignition pilot-light assemblies have internal circuits on printed circuit boards. The technician can only troubleshoot the circuit to the board, not circuits within the board. If the trouble is in the circuit board, the board must be changed. Figure 31–64 shows how one manufacturer accomplishes spark ignition using the flame rectification method. Flame rectification uses the flame to change AC to DC. This current is picked up by an electronic circuit to prove that a pilot light is present before the main gas valve is opened. There are two diagrams, one is a pictorial and the other is a line type of diagram. Notice that the manufacturer has the terminal board on the bottom for the pictorial and on the side for the line diagram. Be aware that all components do not remain in the same position from one diagram to another.

The spark-ignition board in the previous example has terminals provided for adding other components, such as an electronic air cleaner and a vent damper shutoff motor. This is quite handy because the installing technician only has to follow the directions to wire these components when they are used in conjunction with this furnace. The electronic air cleaner is properly interlocked to operate only when the fan is running.

This is not always easy to wire into the circuit to accomplish the proper sequence, particularly with two-speed fan applications.

The following reference points can be used for this circuit board:

1. *When the front panel is removed, everything stops until the switch is blocked closed. This should only be done by a qualified service technician.*
2. The fan motor is started through the 2A contacts and is single speed.
3. The 2A contacts are held open during the off cycle by energizing the 2A coil. This coil is deenergized during the on cycle.
4. If the control transformer is not functioning, the fan will run all the time. You must remember this. If the fan runs constantly but nothing else works, suspect a bad control transformer.
5. The fan starts and stops through the time delay in the heating cycle.

The following description of the electrical circuit can be used for an orderly troubleshooting procedure. See Figures 31–65 and 31–66 for the voltages in the pictorial and line diagrams.

1. Line power should be established at the primary of the control transformer.
2. 24 V should be detected from the C terminal (common to all power-consuming devices) to the LIM-1 terminal. See the meter lead on the 24-V meter labeled 1.
3. Leaving the probe on the C terminal, move the other probe to positions 2, 3, and 4. 24 V should be detected at each terminal, or there is no need to proceed.
4. The key to this circuit is to have 24 V between the C terminal and the GAS-1 terminal. At this time the pilot should be trying to light. The spark at the pilot should be noticeable. It makes a ticking sound with about one-half second between sounds.
5. If the pilot light is lit and the ticking has stopped, the 6H relay is supposed to have changed over. The NC contacts in the 6H relay should be open and the NO contacts should be closed. This means that you should have 24 V between the C terminal, which is the same as the GAS-2 terminal, and the GAS-3 terminal. The 6H relay is inside the circuit board. If it will not pass power, the board must be replaced.
6. When the circuit board will pass power to the GAS-3 terminal (probe position 5) and the burner will not light, the gas valve is the problem. Before changing it, be sure that its valve handle is turned to the correct direction and the interconnecting wire is good. The procedure of leaving one probe on the C terminal and moving the other probe can be done

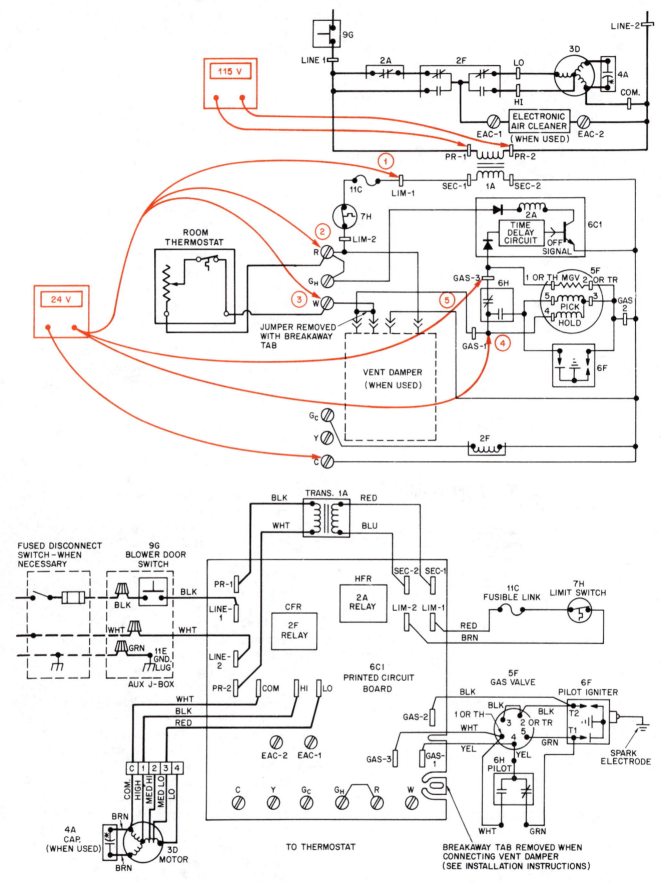

Figure 31-65. Note voltages indicated. *Courtesy BDP Company*

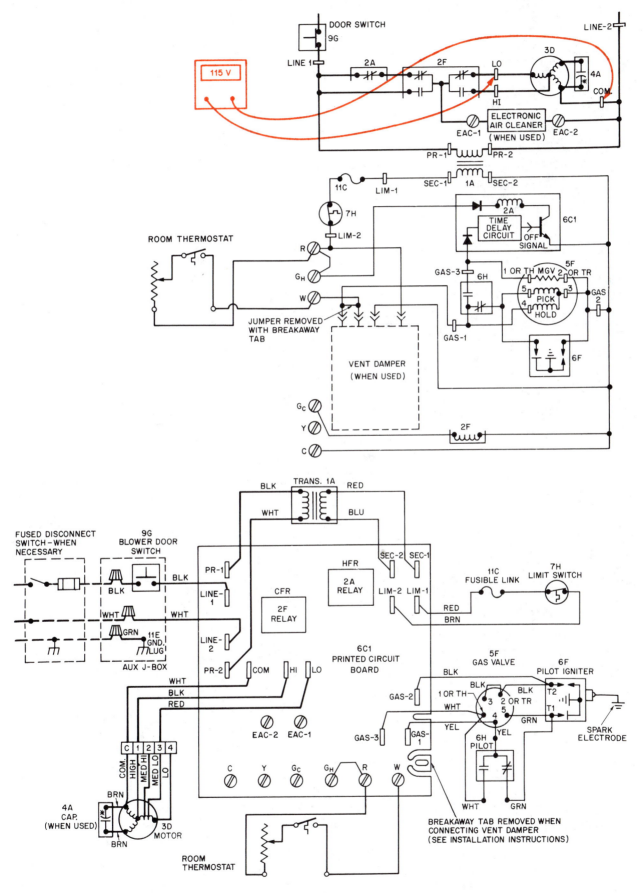

Figure 31–66. Note voltages indicated. *Courtesy BDP Company*

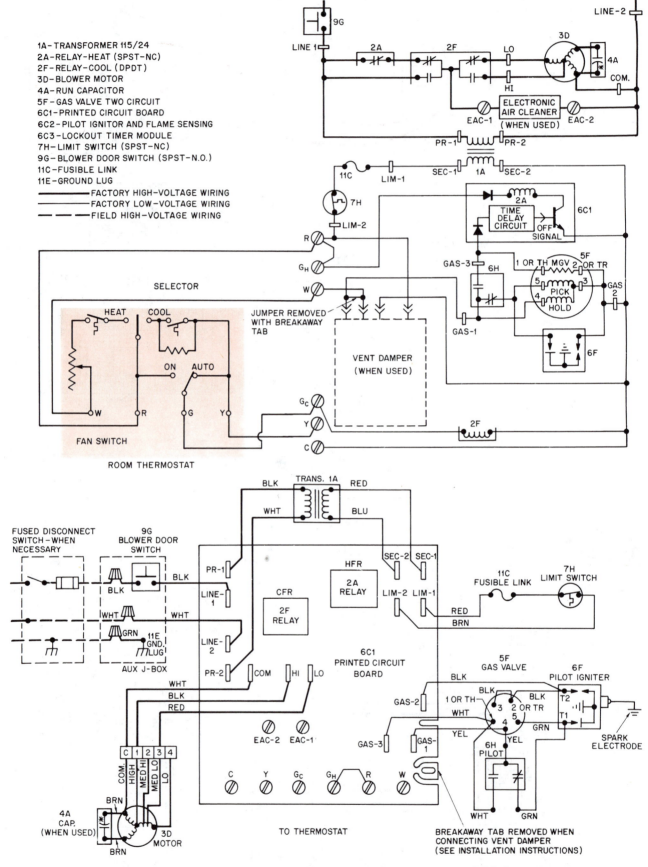

Figure 31-67. Furnace circuit board with thermostat diagram. *Courtesy BDP Company*

very quickly when a furnace has a circuit board with terminals. This makes troubleshooting easier.

7. If you notice that the R terminal has 24 V and the W terminal does not, the thermostat and the interconnecting wiring are the problem. You can place a jumper from R to W for a moment; if the spark starts to arc, the problem is the wiring or the thermostat.

8. The fan is started through the circuit board and is part of the sequence of operation also. The fan is started when the 2A relay is deenergized through the time-delay circuit. This is accomplished when the 6H relay in the circuit board changes position. When 24 V is detected from terminal C to terminal GAS-3, the time-delay circuit starts timing. When the time is up, the contacts open and deenergize the fan relay coil; the contacts close, and the fan starts.

9. The voltmeter can be relocated to the COM terminal and the LO terminal. When relay 2A is passing power, 115 V should be detected here. If there is power to the fan motor and it will not start, open up the fan section and check the motor for continuity. Check the capacitor. See if the motor is hot. If the motor cannot be started with a good capacitor, replace the motor.

See Figure 31–67 for an example of the same furnace circuit board that has been expanded to include more features. This board has added cooling and a two-speed fan. The high-speed fan is desirable in some applications where the quantity of air for cooling is more than that for heating. Notice how the electronic air cleaner is started in either cooling or heating through a different relay application. This board has two relays: the same 2A relay as before, and the double-pole–double-throw (dpdt) relay 2F. The 2F relay is energized by the thermostat in the cooling mode through the Gc terminal.

The key to troubleshooting these circuits is to keep in mind which components are controlling other components. A study of the previous diagrams will help you decide which component is actually defective—the board or the component that the board is controlling.

31.33 DIRECT-SPARK IGNITION (DSI)

Many modern furnaces are designed with a spark ignition direct to the main burner. No pilot is used in this system. Components in the system are the igniter/sensor assembly and the DSI module, Figure 31–68. The sensor rod verifies that the furnace has fired and sends a microamp signal to the DSI module confirming this. The furnace will then continue to operate. This system goes into a "safety lockout" if the flame is not established within the "trial for ignition" period (approximately 6 sec). The system can then only be reset by turning the power off to the system control

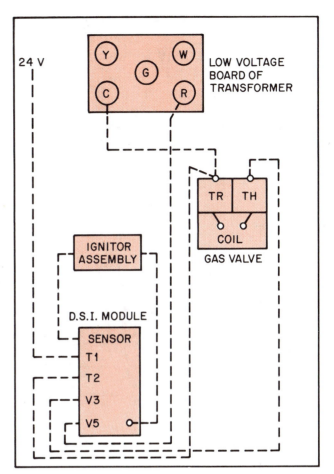

Figure 31–68. Diagram of components in a direct-spark ignition system. *Courtesy Heil-Quaker Corporation*

and waiting one minute before reapplying the power. This is a "typical" system. The technician should follow the wiring diagram and manufacturer's instructions for the specific furnace being installed or serviced.

Most ignition problems are caused from improperly adjusted spark gap, igniter positioning, and bad grounding, Figure 31–69. The igniter is centered over the left port. Most manufacturers also provide specific troubleshooting instructions. Intermittent sparking after the furnace is lit is normal in most systems. Continued sparking may be due to excessive resistance in the igniter wire or to a defective DSI module. The continual sparking may not be harmful to the system, but it is noisy and often can be heard in the living or working area. Replace the igniter wire. If sparking continues, replace the DSI module.

31.34 HIGH-EFFICIENCY GAS FURNACES

Gas furnaces with high-efficiency ratings have been developed and are being installed in many homes and businesses. The U. S. Federal Trade Commission requires manufacturers to provide an annual fuel utilization efficiency rating (AFUE). This rating allows the

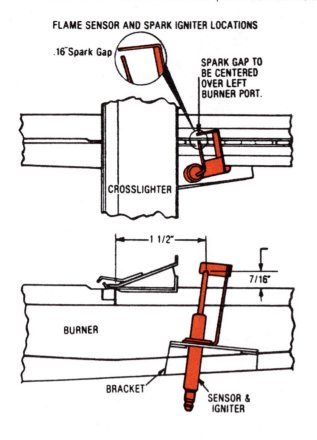

Figure 31–69. Spark gap and igniter position for a direct-spark ignition system. *Courtesy Heil-Quaker Corporation*

The system shown in Figure 31–70 reduces the flue temperature by as much as 150°. Efficiencies of 97+% have been attained with some modern furnaces, such as the *pulse furnace* and the *condensing furnace*.

The pulse furnace ignites minute quantities of gas 60 to 70 times per second in a closed combustion chamber. The process begins when small amounts of natural gas and air enter the combustion chamber. This mixture is ignited with a spark igniter. This ignition forces the combustion materials down a tailpipe to an exhaust decoupler. The pulse is reflected back to the combus-

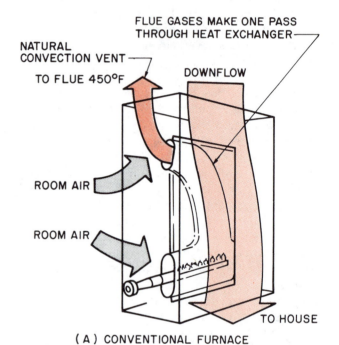

(A) CONVENTIONAL FURNACE

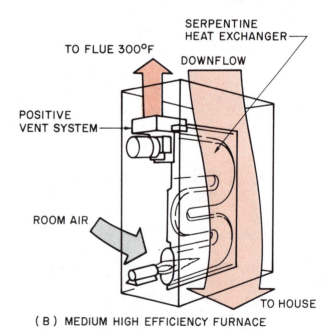

(B) MEDIUM HIGH EFFICIENCY FURNACE

Figure 31–70. Comparison of flue-gas temperatures between a standard furnace and a medium-high-efficiency furnace.

consumer to compare furnace performances before buying. Efficiency ratings have been increased in some instances from 65% to 97+% by keeping excessive heat from being vented to the atmosphere. Stack temperatures in conventional furnaces are kept high to provide a good draft for proper venting. They also prevent condensation and therefore corrosion in the vent.

Modern furnaces are now designed with a forced draft so that flue temperatures can be reduced. Figure 31–70 shows a furnace with an efficiency of about 85%.

Heat exchangers are designed with baffles to absorb more of the heat, providing more usable heat for the conditioned space. Some installations have bimetal vent dampers to control the flue gases and keep them in the furnace until heated to a predetermined temperature. Other vent dampers are opened during running time and closed when the furnace stops holding in the heat until the fan dissipates it. This damper is always closed when the furnace is off. In older systems some heated room air is vented up the flue 24 hr a day whether or not the furnace is operating.

tion chamber, igniting another gas and air mixture, and the process is repeated. Once this process is started, the spark igniter can be turned off because the pulsing will continue on its own 60–70 times per second. This furnace also uses heat exchangers to absorb much of the heat produced. Air is circulated over the exchanger and then to the space to be heated. This furnace also utilizes a condensing component.

The condensing furnace pipes the flue gas through a heat exchanger in the return air and cools the flue gas to the point that the moisture actually condenses out of it. This moisture is very corrosive and at a low temperature. The flue gas can be piped away from the furnace with PVC pipe, and the moisture can be piped down the drain, Figure 31–71. These condensing furnaces may reach efficiencies of 97%.

(A)

(B)

Figure 31–71. (A) **High-efficiency pulse furnace.** Courtesy Lennox Industries Inc. (B) **High-efficiency condensing furnace.** *Courtesy Heil-Quaker Corporation*

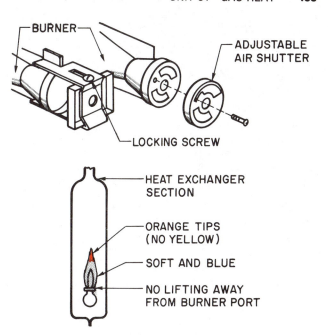

Figure 31–72. Main burner flame adjustment. The air shutter is turned to produce the proper flame. To check the flame, turn the furnace on at the thermostat. Wait a few minutes to insure that any dust or other particles have been burned and no longer have an effect on the flame. The flames should be stable, quiet, soft and blue with slightly orange tips. They should not be yellow. They should extend directly upward from the burner ports without curling downward, floating or lifting off the ports and they should not touch the side of the heat exchanger.

The manufacturer must be consulted for troubleshooting procedures because they are highly specialized.

31.35 COMBUSTION EFFICIENCY

With the high and ever-increasing cost of fuel it is essential that adjustments be made to produce the most efficient but safe combustion possible. It is also necessary to insure that an overcorrection is not made that would produce carbon monoxide.

Atmospheric gas burners use primary air, which is sucked in with the gas, and secondary air, which is pulled into the combustion area by the draft produced when the ignition takes place. ✱Incomplete combustion produces carbon monoxide (CO). Enough secondary air must be supplied so that carbon monoxide will not be produced.✱

Primary air intake can be adjusted with shutters near the orifice. This adjustment is made so that the flame has the characteristics specified by the manufacturer. Generally, however, the flame is blue with a small orange tip when it is burning efficiently, Figure 31–72.

Gas-designed furnaces, that is furnaces designed and built at the factory to burn gas, generally have no secondary air adjustment. These furnaces are designed at the factory for maximum efficiency.

Conversion units do normally have secondary air adjustments, and these must be adjusted in the field for maximum efficiency. The CO_2 test is used to get the necessary information from the flue gases to make the secondary air adjustment. CO_2 in the flue gases is actually a measurement of the excess secondary air supplied. As the percentage of CO_2 in the flue gases increases, the secondary air supplied decreases and the chance of producing carbon monoxide becomes greater.

If the exact amount of air is supplied and perfect combustion takes place, a specific amount of CO_2 by volume will be present in the flue gases. This is called the ultimate CO_2 content. The following table indicates the ultimate CO_2 content for various gases: (*These figures are for perfect combustion and are not used as the CO_2 reading for a gas-burning appliance.*)

Natural gas	11.7–12.2
Butane gas	14.0
Propane gas	13.7

A gas furnace is never adjusted to produce the ultimate CO_2 because there is too much danger of producing carbon monoxide. Therefore, the accepted practice is to adjust the secondary air to produce excess air resulting in less CO_2 by volume in the flue gases. The adjustment should be made to produce from 65% to 80% of the ultimate CO_2. This would produce 8% to $9\frac{1}{2}\%$ CO_2 in the flue. It is very important that manufacturer's specifications be followed when making any adjustments.

The flue-gas temperature should also be taken if the percent of combustion efficiency is to be determined. The percent combustion efficiency is an index of the useful heat obtained. It is not possible to obtain 100%

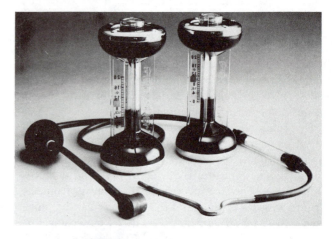

(A)

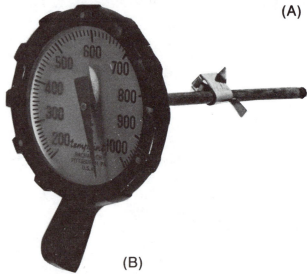

(B)

Figure 31–74. (A) O_2 and CO_2 analyzer. (B) Stack thermometer.

efficiency since heated air must be allowed to escape through the flue to provide a means for the products of the combustion to leave the building.

The CO_2 and flue-gas temperature are taken in the flue between the draft diverter and the furnace. A sampling tube can be inserted through the draft diverter projecting into the inlet side in a gas-burning furnace, Figure 31–73. An O_2 and CO_2 analyzer and a stack thermometer are used for making these tests, Figure 31–74. You can refer to a chart provided by the manufacturer of the test equipment to determine the percent efficiency. Average gas-burning furnaces manufactured before 1982 should produce 75% to 80% efficiency.

Test equipment is also available to check CO content in the flue gases. CO free combustion is defined as less than 0.04% CO in an air-free sample of the flue gas. CO can be produced by flame impingement on a cool surface, insufficient primary air (yellow flame) as well as insufficient secondary air.

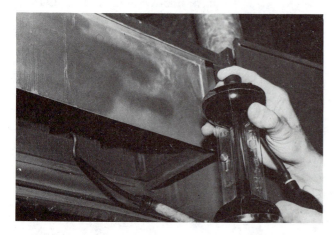

Figure 31–73. Sampling tube in draft diverter. *Photo by Bill Johnson*

Other adjustments that can be checked and compared with manufacturer's specifications are suction at the draft hood, draft at the chimney, gas pressure, and gas input.

31.36 SERVICE TECHNICIAN CALLS

Service Call 1

A residential customer calls. *The furnace stopped in the middle of the night. The furnace is old, and the thermocouple is defective.*

The technician arrives and goes directly to the thermostat. The thermostat is set correctly for heating. The thermometer in the house showed a full 10°F below the thermostat's setting. This unit has air conditioning, so the technician turned the fan switch to ON to see if the indoor fan would start; it did. This proved that the system's low-voltage power supply (transformer) was working.

The furnace was in an upstairs closet. The technician saw that the pilot light was not burning. This system has 100% shutoff, so there is no gas to the pilot unless the thermocouple holds the pilot-valve solenoid open. The technician positioned the main gas valve to the PILOT position and pressed the red button to allow gas to the pilot light. The pilot light burned when a lit match (an igniter may be used) was placed next to it. The technician then held the red button down for 30 sec and slowly released it. The pilot light went out. This was a sure sign that either the thermocouple or the pilot solenoid coil was defective.

The technician used an adapter to check the thermocouple voltage output and determined that it will only generate 2 mV when connected to the coil (this is its load). Again this can mean either that the coil was pulling too much current due to a short or that the thermocouple was bad. When the thermocouple was disconnected from the coil, it still only generated 7 mV. It should have easily generated 20 mV, so it needed replacing.

When the thermocouple was replaced, the pilot light was lit and stayed lit when the red button was released. The gas valve was turned ON, and the furnace started. While at the job, the technician changed the furnace filter and oiled the indoor fan motor.

Service Call 2

A residential customer calls. *The furnace is not heating the house. The furnace is located in the basement and is very hot, but no heat is moving into the house. The dispatcher tells the customer to turn the furnace off until a service technician can get to the job. The furnace fan motor is defective and will not run, so the burner is cycling on the high limit control.*

The technician has some idea that the fan motor is bad from the symptoms. The technician knows what kind of furnace it is from previous service calls, so a fan motor is brought along. When the technician arrives, the room thermostat is set to call for heat. From the service request it is obvious that the low-voltage circuit is working because the customer says the burner will come on. The technician then goes to the basement and hears the burner operating. When the furnace has had enough time to get warm and the fan has not started, the technician takes the front control panel off the furnace and notices that the temperature-operated fan switch dial (circular dial type, Figure 31–39) has rotated as if it were sensing heat. *The technician carefully checks the voltage entering and leaving the fan switch.* (This is done by placing one meter lead on the neutral wire and the other first to the wire going into the fan switch and then to the wire leaving the fan switch and going to the fan motor.) There are 118 V going to the fan motor, but it is not turning. The technician then shuts off the power and removes the panel to the fan motor. The motor housing is cool to the touch; it seems it has not even been trying to run. If the motor was locked up or had a bad start winding, it would be cycling on overload due to high current and therefore would be warm or hot to the touch.

The technician then checks the motor for continuity with the ohmmeter and finds that the motor has an open circuit in the windings. The motor is changed and power is resumed. The burner lights, the fan switch begins to move as it senses heat, and the fan motor starts. The system is back to normal.

Service Call 3

A customer in a retail store calls. *There is no heat. This is an upflow gas furnace with a standing pilot and air conditioning. The low-voltage transformer is burned out because the gas valve coil is shorted. This causes excess current for the transformer. The furnace is located in the stockroom.*

The technician arrives at the job, goes to the room thermostat and places the fan switch to ON to see if the indoor fan will start. It will not, so the technician suspects that the low-voltage power supply is not working. The thermostat is set to call for heat. The technician goes to the stockroom where the furnace is located. The voltage is checked at the transformer secondary, and it is 0 V. The power is turned off and the ohmmeter is ued to check continuity of the transformer. The secondary (low-voltage) coil has an open circuit, so the transformer is changed. Before connecting the secondary wires, the continuity of the low-voltage circuit is checked. There is only 2 Ω of resistance in the gas valve coil. This is so low that it will cause a high current flow in the secondary circuit. (The resistance should be at least 20 Ω.) The technician goes to the truck and checks the continuity of another gas valve and finds it to be 50 Ω, so the gas valve is also changed.

The system is started with the new gas valve and transformer. A current check of the gas valve circuit shows that it is only pulling 0.5 A, which is not overloaded.

The technician changes the air filters and calls the store manager back to the stockroom for a conference before leaving. The store manager is informed that the boxes of inventory must be kept away from the furnace because they may present a fire hazard.

Service Call 4

A customer in a residence calls. *Fumes can be smelled and are probably coming from the furnace in the hall closet. The furnace has not been operated for the past two weeks because the weather has been mild. Since the fumes might be harmful, the dispatcher tells the customer to shut the furnace off until a technician arrives. The flue is stopped up with a shingle laid on top of the brick chimney by a roofer who had been making repairs.*

The technician arrives and starts the furnace. A match is held at the draft diverter to see if the flue has a negative pressure. The flue-gas fumes are not rising up the flue for some reason; the match flame blows away from the flue pipe and draft diverter. The technician turns off the burner and examines it and the heat exchanger area with a flashlight. There is no soot that would indicate the burner had been burning incorrectly. The technician goes to the roof to check the flue and notices a shingle on top of the chimney. The shingle is lifted off and heat then rises out of the chimney.

The technician goes back to the basement and starts the furnace again. A match is held at the draft diverter, and the flame is drawn towards the flue. The furnace is now operating correctly. The furnace filter is changed, and the fan motor is oiled before the technician leaves the job.

Service Call 5

A residential customer calls. *There is no heat. The furnace is in the basement, and there is a sound that sounds like a clock ticking. The electrode in the pilot has shifted position, and the unit is arcing and trying to light. It cannot light, but the arc can be heard.*

The service technician arrives and goes to the basement. The furnace door is removed. The technician hears the arcing sound. The shield in front of the burner is removed. The technician sees the arc. A match is lit near the pilot to see if there is gas at the pilot and to see if the pilot will light: It does. The arcing stops, as it should. The burner lights after the proper time delay. Because there is a gas stream, the problem must be the electrode alignment.

The technician turns off the power and removes the pilot-light assembly and finds the screw holding the electrode to be loose, and the electrode slightly out of alignment. The technician looks at the installation and startup manual left with the furnace to find the correct alignment and tolerance. See Figure 31–69 for a typical alignment procedure.

The electrode assembly is aligned and replaced in the furnace, and power is restored. The electronic ignition arcs five times and lights. The technician starts and stops the ignition sequence several times with good results. The filters are changed, and the motor is oiled before leaving the job.

Service Call 6

A residential customer calls. *There is no heat. The main gas valve coil is shorted, and the low-voltage fuse is blown. This is an electronic intermittent ignition system, Figure 31–66. This symptom can lead the technician to believe that the printed circuit board is bad.*

The technician arrives and notices the indoor fan is running. The technician cannot stop the fan from running by turning the room thermostat to OFF. (Remember, the low voltage circuit holds the fan motor off in this system.) A low-voltage problem is suspected.

The technician goes to the furnace and checks the output of the low-voltage transformer; the meter reads 24 V at the transformer but 0 V after the circuit goes through the fuse. The fuse is blown. *The technician turns off the power and replaces the fuse.* When the power is restored, the electronic ignition circuit lights the pilot light. Everything seems normal, but the fuse blows again.

The technician turns off the power and changes the electronic circuit board and 1-A fuse. When power is restored, the pilot lights and everything is normal for a moment, but the fuse again blows. In such a short time it is not normal to have two bad components that act alike.

The technician turns off the power and fastens a clamp-on ammeter in the low-voltage system, using ten wraps of wire to amplify the ampere reading, Figure 16–18. The control fuse is changed. When power is restored, the pilot lights again and the amperage is normal (about 0.5 A). When the pilot proves and it is time for the main gas valve (MGV) to be energized, the current goes up to 3 A and the fuse blows again.

The technician turns off the power and uses the ohmmeter to check the resistance of all power-consuming components in the circuit. When the MGV coil in the gas valve is checked, it shows 2 Ω of resistance.

The gas valve is replaced and the fuse changed. Power is restored. The pilot light is lit and proved. The main gas valve opens and the burner ignites. The system operates normally.

The technician turns off the power and changes the circuit board back to the original board and then restores the power. The furnace goes through a normal startup.

Service Call 7

A customer calls and wants an efficiency check on a furnace. *This customer thinks that the gas bill is too high for the conditions and wants the system checked out from one end to the other.*

The technician arrives and meets the customer, who wants to watch the complete procedure of the service call. The technician will run an efficiency test on the furnace burners at the very beginning and then at the end, so the customer can see the difference.

The furnace is in the basement and easily accessible. The technician turns the thermostat to 10°F above the room temperature setting to insure that the furnace will not shut off in the middle of the test. The technician then goes to the truck and gets the flue-gas analyzer kit. *Being careful not to touch the stack, the stack temperature is taken at the heat exchanger side of the draft diverter.* A sample of the flue gas is drawn into the sample chamber after the flue-gas temperature has stopped rising and the indoor fan has been running for about 5 min. This insures that the furnace is up to maximum temperature.

The flue-gas reading shows that the furnace is operating at 80% efficiency. This is normal for a standard efficiency-type furnace. After the test is completed and recorded, the technician shuts off the gas to the burner and allows the fan to run to cool down the furnace.

When the furnace has cooled down enough to allow the burners to be handled, the technician takes them out. The burners are easily removed in this furnace by removing the burner shield and pushing the burners forward one at a time while raising up the back. They will then clear the gas manifold and can be removed.

There is a small amount of rust, which is normal for a furnace in a basement. The draft diverter and flue pipe are removed so that the technician and customer can see the top of the heat exchanger. All is normal—no rust or scale.

With a vacuum cleaner the technician removes the small amount of dirt and loose rust from the heat exchanger and burner area. The burners are taken outside where they are blown out with a compressed air tank at the truck. *The technician wears eye protection.* After the cleaning is complete, the technician assembles the furnace and tells the customer that his furnace was in good shape and that no difference in efficiency will be seen. Modern gas burners do not stop up as badly as the older ones, and the air adjustments will not allow the burner to get out of adjustment more than 2% or 3% at the most.

The fan motor is oiled, and the filter is changed. The system is started up and allowed to get up to normal operating temperature. While the furnace is getting up to temperature, the technician checks the fan current and finds it to be running at full load. The fan is doing all the work that it can. The efficiency check is run again, and the furnace is still running about 80%. It is now clean, and the customer has piece of mind. The thermostat is set before leaving.

Service Call 8

A customer calls. *The pilot light will not stay lit. The pilot light goes out after it has been lit a few minutes. The heat exchanger has a hole in it very close to the pilot light. The pilot light will light, but when the fan starts it blows out the pilot light.*

The technician arrives and checks the room thermostat to see that it is set above the room temperature. The technician then goes to the basement. The standing pilot light is not lit, so the technician lights it and holds the button down until the thermocouple will keep the pilot light lit. When the gas valve is turned to ON, the main burner lights. Everything is normal until the fan starts, then the flame starts to wave around and the pilot light goes out. The thermocouple cools, and the gas to the main burner goes out. The technician shows the customer the hole in the heat exchanger. The gas valve is turned off and it is explained to the customer that the furnace cannot be operated in this condition because of the potential danger of gas fumes.

The technician explains to the customer that this furnace is 18 years old, and the customer really should consider getting a new one. A heat exchanger can be changed in this furnace, but it requires much labor plus the price of the heat exchanger. The decision is for a new furnace.

SUMMARY

- Forced-hot-air furnaces are normally classified as upflow, downflow (counterflow), horizontal, or low-boys. Low-boys take the return air in and discharge the heated air at the top.
- Furnace components consist of the cabinet, gas valve, manifold, pilot, burners, heat exchangers, blower, electrical components, and venting system.
- Gas fuels are natural gas, manufactured gas (used in few applications), and LP gas (propane, butane, or a mixture of the two).
- Inches of water column is the term used when determining or setting gas fuel pressures.
- A water manometer measures gas pressure in inches of water column.
- For combustion to take place there must be fuel, oxygen, and heat. The fuel is the gas, the oxygen comes from the air, and the heat comes from the pilot flame or other igniter.
- Gas burners use primary and secondary air. Excess air is always supplied to insure as complete combustion as possible.

- Gas valves control the gas flowing to the burners. The valves are controlled automatically and allow gas to flow only when the pilot is lit or when the ignition device is operable.
- Some common gas valves are classified as solenoid, diaphragm, or heat-motor valves.
- Ignition at the main burners is caused by heat from the pilot or from an electric spark. There are standing pilots that burn continuously, and intermittent pilots that are ignited by a spark when the thermostat calls for heat. There is also a direct-spark ignition, in which the spark ignites the gas at the burners.
- The thermocouple, the bimetal, and the liquid-filled remote bulb are three types of safety devices (flame-proving devices) to insure that gas does not flow unless the pilot is lit.
- The manifold is a pipe through which the gas flows to the burners and on which the burners are mounted.
- The orifice is a precisely sized hole in a spud through which the gas flows from the manifold to the burners.
- The burners have holes or various designs of slots through which the gas flows. The gas burns immediately on the outside of the burners at the top or end depending on the type.
- The gas burns in an opening in the heat exchanger. Air passing over the heat exchanger is heated and circulated to the conditioned space.
- A blower circulates this heated air. The blower is turned on and off by a fan switch, which is controlled either by time or by temperature.
- The limit switch is a safety device. If the fan does not operate or if the furnace overheats for another reason, the limit switch causes the gas valve to close.
- *Venting must provide a safe and effective means of moving the flue gases to the outside atmosphere. Flue gases are mixed with other air through the draft hood. Venting may be by natural draft or by forced draft. Flue gases are corrosive.*
- Gas piping should be kept simple with as few turns and fittings as possible. It is important to use the correct size pipe.
- *All piping systems should be tested carefully for leaks.*
- A glow coil is a reignition device that ignites the pilot if it goes out.
- High-efficiency gas furnaces have been developed. More heat is retained in the furnace for distribution to the conditioned space.
- The combustion efficiency of furnaces should be checked, and adjustments made when needed.

REVIEW QUESTIONS

1. List the four types of gas furnaces and describe each.

2. What two gas fuels are most commonly used in residential furnaces?
3. What is the specific gravity of each type of gas fuel? How does the specific gravity affect each type?
4. Describe a manometer.
5. What is an inch of water column?
6. What percentage of air is oxygen?
7. Why is the percentage of oxygen in air a factor when determining proper air supply to support gas combustion?
8. What is a typical pressure in inches of water column for natural gas at the manifold of the furnace?
9. Why is excess air supplied to support combustion in a gas furnace?
10. Why is a gas regulator necessary before the gas reaches the valve?
11. List three types of gas valves.
12. Describe an automatic combination gas valve.
13. Describe two types of pilots.
14. When is a glow coil reignition system used?
15. Describe how a thermocouple flame-proving safety device works.
16. Describe the bimetallic flame-proving safety device.
17. Describe the liquid-filled remote-bulb flame-proving safety device.
18. Why are the devices in Questions 15–17 considered safety devices?
19. What does the manifold do?
20. What is an orifice?
21. List four types of gas burners.
22. Describe the function of a heat exchanger.
23. Why is it important that a heat exchanger not be corroded?

24. How does a temperature on–temperature off fan switch operate?
25. How does a time-delay fan switch differ from a temperature on–temperature off switch?
26. What is a limit switch? How does it work?
27. Sketch a vent hood and describe how it functions. What is dilutant air?
28. How does a vent hood or draft diverter operate in downdraft conditions?
29. What two types of venting systems are used for gas-fuel installation?
30. Why is a metal vent often preferred over a masonry chimney?
31. Describe how at least one type of automatic vent damper operates.
32. What kind of metal is most gas piping made from?
33. Why does the size (diameter) of gas piping have to be considered when designing a piping system?
34. Describe the proper procedures for making a threaded joint in gas piping.
35. What three types of fittings are needed in a piping system just before the gas valve?
36. Describe the procedure to test for leaks in a piping system.
37. Sketch a wiring diagram for a basic gas furnace.
38. Describe the load and no-load thermocouple test.
39. Why is a spark-to-pilot ignition system often preferred over a standing-pilot system?
40. How does a flame rectification system operate?
41. Describe a direct-spark ignition system.
42. List differences between modern high-efficiency gas furnaces and older models.
43. Why is a CO_2 test made on a gas furnace?

OIL HEAT

OBJECTIVES

After studying this unit, you should be able to

- describe how the fuel oil and air are prepared and mixed in the oil-burner unit for combustion.
- list products produced as a result of combustion of the fuel oil.
- list the components of gun-type oil burners.
- describe basic service procedures for oil-burner components.
- sketch wiring diagrams of the oil-burner primary control system and the fan circuit.
- state tests used to determine oil-burner efficiency.
- explain corrective actions that may be taken to improve burner efficiency, as indicated from the results of each test.

32.1 INTRODUCTION TO OIL-FIRED FORCED-WARM-AIR FURNACES

The oil-fired forced-warm-air furnace has two main systems: a heat-producing system and a heat-distribution system. The *heat-producing system* consists of the oil burner, the fuel supply components, combustion chamber, and heat exchanger. The *heat distribution system* is composed of the blower fan, which moves the air through the duct work, and other related components, Figure 32–1.

Following is a brief description of how such a furnace operates. When the thermostat calls for heat, the oil burner is started. The oil–air mixture is ignited. After the mixture has heated the combustion chamber and the heat exchanger reaches a certain temperature, the fan comes on and distributes the heated air through the duct work to the space to be heated. When this space has reached a predetermined temperature, the thermostat will cause the burner to shut down. The fan will continue to run until the heat exchanger cools to a set temperature.

32.2 PHYSICAL CHARACTERISTICS

The physical appearance and characteristics of forced-air furnaces vary to some extent. The *low-boy* is

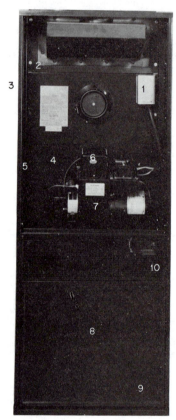

1 – FAN SWITCH
2 – HEAT EXCHANGER
3 – CABINET PANEL
4 – COMBUSTION CHAMBER
5 – INSULATION
6 – TRANSFORMER
7 – OIL BURNER
8 – BLOWER
9 – BASE PAN
10 – PRIMARY CONTROL

Figure 32–1. Forced-warm-air oil furnace. *Courtesy Ducane Corporation*

often used when there is not much headroom, Figure 32–2. Low-boys may have a cooling coil on top to provide air conditioning.

An *upflow* furnace is a vertical furnace in which the air is taken in at the bottom and is forced across the heat exchanger and out the top, Figure 32–2. This furnace is installed with the duct work above it (e.g., in the attic). It could also be installed in a basement with the duct work above it under the floor.

A *downflow* furnace looks very similar to an upflow except that the air is drawn in from the top and forced out the bottom, Figure 32–3. The duct work in this case would be just below or in the floor below the furnace.

A *horizontal* furnace is usually installed in a crawl

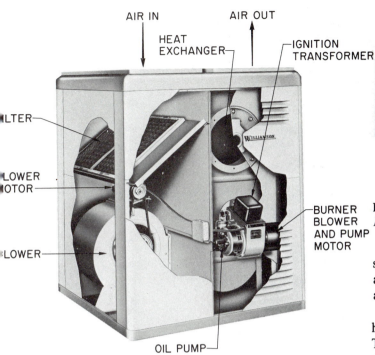

AIR IN — AIR OUT

HEAT EXCHANGER

IGNITION TRANSFORMER

FILTER

BLOWER MOTOR

BLOWER

OIL PUMP

BURNER BLOWER AND PUMP MOTOR

Figure 32–2. Low-boy forced-air furnace. *Courtesy The Williamson Company*

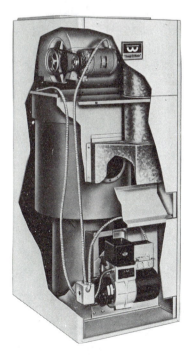

Figure 32–3. A downflow or counterflow furnace. *Courtesy The Williamson Company*

Figure 32–4. Horizontal forced-air furnace. *Courtesy Dornback Furnace and Foundry Company*

space under a house or in an attic, Figure 32–4. They are available with right-to-left airflow or left-to-right airflow.

Oil-fired forced-air heating is a popular method of heating residences and light commercial buildings. There are several million furnaces of this type used in the country at the present time. It is essential that the air conditioning and heating technician understand proper installation and servicing techniques involved with oil burners. Safety is always a concern of technicians but with the high cost of fuel oils, efficiency in the combustion process is also extremely important.

32.3 FUEL OILS

Six grades of fuel oil are used as heating oils, with the No. 2 grade the one most commonly used to heat residences and light commercial buildings. Fuel oil is obtained from petroleum pumped from the ground. The lighter fuel oils are products of a distillation process at the oil refinery during which the petroleum is vaporized, condensed, and the different grades separated.

Fuel oil is a combination of liquid hydrocarbons containing hydrogen and carbon in chemical combination. Some of these hydrocarbons in the fuel oil are light, and others are heavy. Combustion takes place when the hydrocarbons unite rapidly with oxygen. This is when the fuel oil burns. Heat, carbon dioxide, and water vapor are produced when the combustion takes place.

32.4 PREPARATION OF FUEL OIL FOR COMBUSTION

Fuel oil must be prepared for combustion. It must first be converted to a gaseous state by forcing the fuel oil, under pressure, through a nozzle. The nozzle breaks the fuel oil up to form tiny droplets. This process is called *atomization,* Figure 32–5. The oil droplets are then mixed with air which contains oxygen. The lighter hydrocarbons form gas covers around the droplets. Heat is introduced at this point

Figure 32–5. Atomized fuel oil droplets and air leaving burner nozzle. *Courtesy Delavan Corporation*

with a spark. The vapor ignites (combustion takes place), and the temperature rises, causing the droplets to vaporize and burn.

Perfect combustion exists in theory only. In reality, products other than carbon dioxide and water vapor are produced in this burning process. These other products are carbon monoxide, soot, and quantities of unburned fuel.

High-pressure gun-type oil burners are generally used to achieve this combustion. Oil is fed under a pressure of 100 psi to a nozzle. Air is forced through a tube that surrounds this nozzle. Usually the air is swirled in one direction and the oil in the opposite direction. The oil forms into tiny droplets combining with the air. The ignition transformer furnishes a high-voltage spark between two electrodes located near the front of the nozzle, and combustion takes place.

32.5 GUN-TYPE OIL BURNERS

The main parts of a gun-type oil burner are the burner motor, blower or fan wheel, pump, nozzle, choke, air tube, electrodes, transformer, and primary controls, Figures 32–6 and 32–7.

Burner Motor

The oil-burner *motor* is usually a split-phase fractional horsepower motor that provides power for both the fan and the fuel pump. A flexible coupling is used to connect the shaft of the motor to the shaft of the pump, Figure 32–7. The motor speeds may be either 1750 or 3450 rpm. The pump should always match the motor rpm.

PRIMARY SAFETY CONTROL

IGNITION TRANSFORMER

NOZZLE LINE ELECTRODE ASSEMBLY

CADMIUM CELL

BLOWER WHEEL

FLAME-RETENTION HEAD

FUEL UNIT

DRIVE MOTOR

BURNER HOUSING

Figure 32–6. An oil burner. *Courtesy R. W. Beckett Corporation*

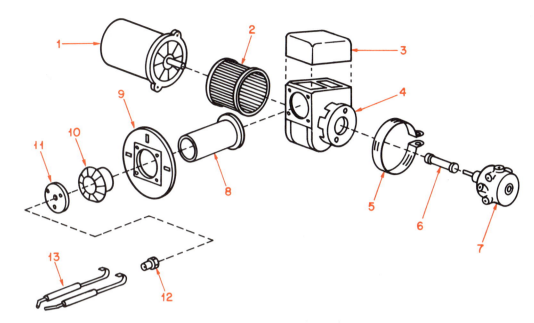

Figure 32-7. Exploded view of an oil burner assembly.

1 - MOTOR
2 - BLOWER WHEEL
3 - TRANSFORMER
4 - BLOWER HOUSING
5 - ADJUSTABLE AIR INLET COLLAR
6 - FLEXIBLE COUPLER
7 - FUEL PUMP
8 - AIR TUBE
9 - MOUNTING FLANGE
10 - END CONE
11 - STATIC DISC
12 - NOZZLE
13 - ELECTRODES

Burner Fan or Blower

The burner *fan* or blower is a squirrel cage type with adjustable air inlet openings in a collar attached to the blower housing. This provides a means for regulating the volume of air being drawn into the blower. The fan forces air through the air tube to the combustion chamber where it is mixed with the atomized fuel oil to provide the necessary oxygen to support the combustion, Figure 32-8.

Fuel Oil Pumps

Several types of oil pumps are used in gun-type oil burners, Figure 32-9. A *single-stage pump* is used when

(A)

(B)

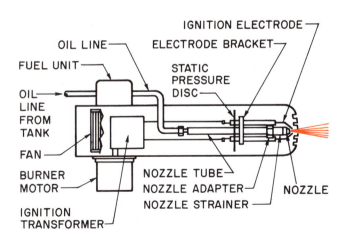

Figure 32-8. Schematic drawing (top view) of a typical gun-type oil burner. *Courtesy Honeywell, Inc., Residential Division*

Figure 32-9. (A) Single-stage fuel oil pump. (B) Two-stage fuel oil pump. *Courtesy (A) Webster Electric Company (B) Suntec Industries Incorporated, Rockford, Illinois*

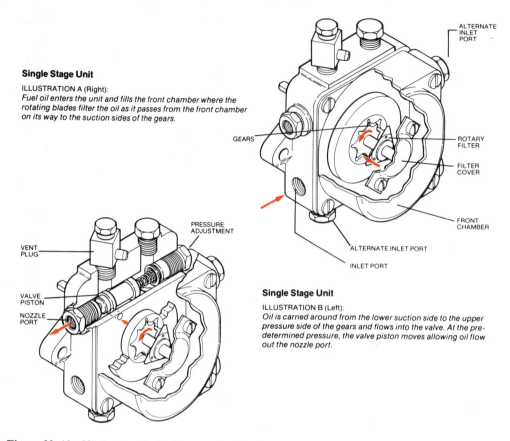

Single Stage Unit

ILLUSTRATION A (Right):
Fuel oil enters the unit and fills the front chamber where the rotating blades filter the oil as it passes from the front chamber on its way to the suction sides of the gears.

Single Stage Unit

ILLUSTRATION B (Left):
Oil is carried around from the lower suction side to the upper pressure side of the gears and flows into the valve. At the predetermined pressure, the valve piston moves allowing oil flow out the nozzle port.

Figure 32–10. Single-stage fuel oil pump showing flow of oil. *Courtesy Webster Electric Company*

the fuel oil storage tank is above the burner. The fuel oil flows to the burner by gravity, and the pump provides oil pressure to the nozzle, Figure 32–10.

A *one-pipe* supply system may be used with a single-stage pump. This means that there is one pipe from the tank to the burner. In a normal operation a surplus of oil is pumped to the nozzle. The nozzle cannot handle this surplus, and the excess fuel is returned to the low-pressure or inlet side of the pump, Figure 32–11.

When installing a one-pipe system, make sure that the bypass plug is not in place so that the surplus oil can return to the inlet side of the pump, Figure 32–12. If the fuel oil gets too low in the storage tank or if the piping is opened for any reason in a one-pipe system, the air must be bled from the system at the pump. This will be discussed in more detail later.

A *dual- or two-stage pump* is used when the oil is stored below the burner. One stage of the pump lifts

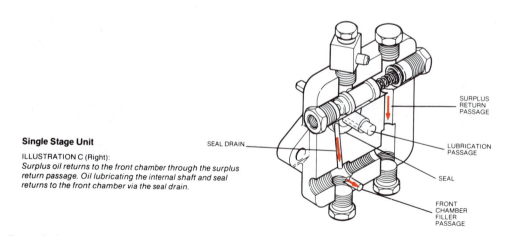

Single Stage Unit

ILLUSTRATION C (Right):
Surplus oil returns to the front chamber through the surplus return passage. Oil lubricating the internal shaft and seal returns to the front chamber via the seal drain.

Figure 32–11. Part of single-stage fuel oil pump illustrating return of surplus or excess oil pumped to the nozzle. *Courtesy Webster Electric Company*

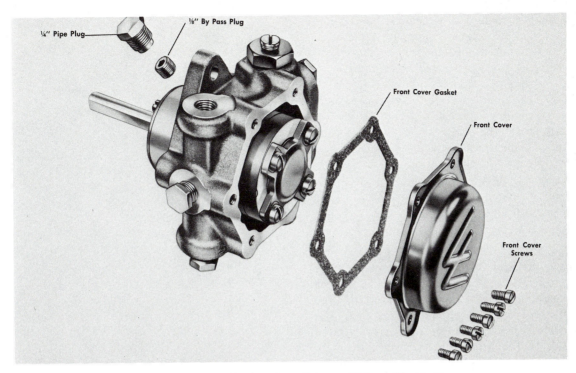

Figure 32–12. Bypass plug location. *Courtesy Webster Electric Company*

the oil to the pump inlet, and the other stage provides oil pressure to the nozzle. This dual-stage unit does the same job as the single-stage unit. However, it has an extra pump gear, called a suction pump gear. It produces a greater vacuum for lifting the fuel oil than can be obtained from a single-stage unit. The suction pump supplies the fuel oil needed by the pressure pump, Figure 32–13.

If the burner has a dual- or two-stage pump, the supply system should have two pipes. Many single-stage pumps are also installed with a two-pipe system. Two pipes are required with a dual-stage pump when the

supply tank is below the pump because there must be a vacuum on the tank side of the pump to lift the oil. The surplus oil pumped to the nozzle in this case is returned to the storage tank through the second pipe, Figure 32–14. The bypass plug must be inserted in the pump. Two-pipe systems are self-venting (they do not have to be bled) and are preferred in all installations.

The pumps used in high-pressure gun-type burners are rotary pumps using either a cam system or gears or a combination of the two to provide the pressure,

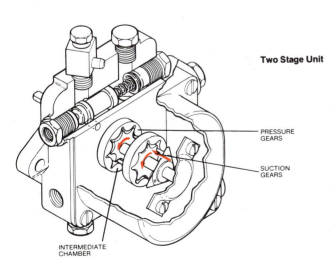

Figure 32–13. Two-stage fuel oil pump. *Courtesy Webster Electric Company*

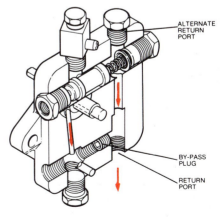

Figure 32–14. Excess oil pumped to the nozzle is returned to the storage tank through the return port. Bypass plug is in place forcing the oil through the return port. *Courtesy Webster Electric Company*

Figure 32–13. The pump itself should not be repaired by the technician. Pumps are normally replaced if defective, so it is not really important to the technician whether cams or gears are used in the pump.

Built in to each pump is a pressure regulating valve, Figure 32–10. The pump provides excessive pressure. The pressure regulating valve can be set so that the fuel oil being delivered to the nozzle is under a set pressure of 100 psi. As mentioned previously, oil is also delivered in greater quantities than the nozzle can handle. This excess oil is diverted back to the inlet side of the pump in one pipe systems and back to the storage tank in two pipe systems.

Nozzle

The *nozzle* prepares the fuel oil for combustion by atomizing, metering, and patterning the fuel oil. It does this by separating the fuel into tiny droplets. The smallest of these droplets will be ignited first. The larger droplets (there are more of these) provide more heat transfer to the heat exchanger when they are ignited. The atomization of the fuel oil is a complex process. The straight lateral movement of the fuel oil must be changed to a circular motion. This circular movement opposes a similar airflow. The fuel enters the orifice of the nozzle through the swirl chamber. Figure 32–15 is a typical nozzle.

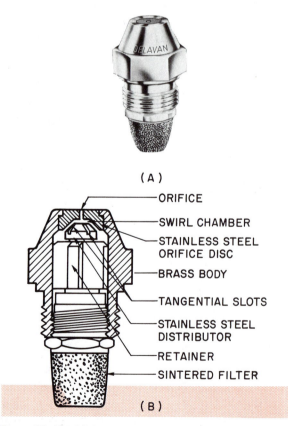

(A)

— ORIFICE
— SWIRL CHAMBER
— STAINLESS STEEL ORIFICE DISC
— BRASS BODY
— TANGENTIAL SLOTS
— STAINLESS STEEL DISTRIBUTOR
— RETAINER
— SINTERED FILTER

(B)

Figure 32–15. Oil-burner nozzle. *Courtesy Delavan Corporation*

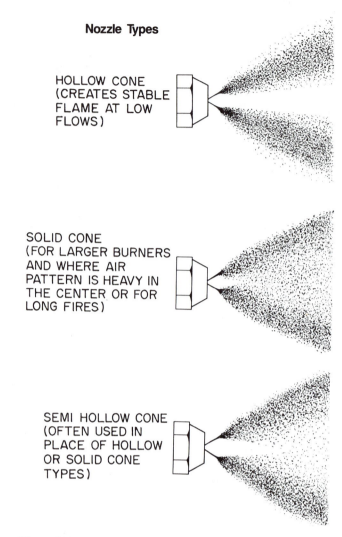

Nozzle Types

HOLLOW CONE (CREATES STABLE FLAME AT LOW FLOWS)

SOLID CONE (FOR LARGER BURNERS AND WHERE AIR PATTERN IS HEAVY IN THE CENTER OR FOR LONG FIRES)

SEMI HOLLOW CONE (OFTEN USED IN PLACE OF HOLLOW OR SOLID CONE TYPES)

Figure 32–16. Spray patterns. *Courtesy Delavan Corporation*

The bore size of the nozzle is designed to allow a certain amount of fuel through at a given pressure to produce the Btu desired. Each nozzle is marked as to the amount of fuel it will deliver. A nozzle marked 1.00 will deliver one gallon of fuel oil per hour (gph) if the input pressure is 100 psi. With No. 2 fuel oil at a temperature of 60°F approximately 140,000 Btu would be produced each hour. A 0.8 nozzle would deliver 0.8 gph, and so on.

The nozzle must be designed so that the spray will ignite smoothly and provide a steady quiet fire that will burn clean and efficiently. It must provide a uniform spray pattern and angle that is best suited to the requirements of the specific burner. There are three basic spray patterns: hollow, semihollow, and solid with angles from 30° to 90°, Figure 32–16.

Hollow cone nozzles generally have a more stable spray angle and pattern than do solid cone nozzles of the same flow rate. These nozzles are often used where the flow rate is under 1 gph.

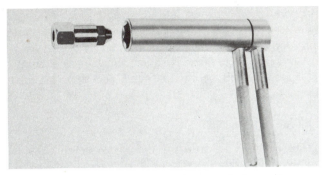

Figure 32–17. A nozzle changer. *Courtesy Delavan Corporation*

Solid cone nozzles distribute droplets fairly evenly throughout the pattern. These nozzles are often used in larger burners.

Semihollow cone nozzles are often used in place of the hollow or solid cone nozzles. The higher flow rates tend to produce a more solid spray pattern, and the lower flow rates tend to produce a hollower spray pattern.

The high-pressure oil-burner nozzle is a precision device that must be handled carefully. *Do not attempt to clean these nozzles. Metal brushes or other cleaning devices will distort or otherwise damage the precision machined surfaces. When the nozzle is not performing properly, replace it.* Use a nozzle changer designed specifically for this purpose, to remove and install a nozzle, Figure 32–17.

To avoid an after-fire drip at the nozzle, install a solenoid-type cutoff valve. This valve reduces smoking after the burner is shut off, Figure 32–18.

Air Tube

Air is blown into the combustion chamber through the *air tube*. Within this tube the straight or lateral movement of the air is changed to a circular motion. This circular motion is opposite that of the circular motion of the fuel oil. The air moving in a circular motion in one direction mixes with the fuel oil moving in a circular motion in the opposite direction in the combustion chamber. The air tube of some burners contains a stationary disc, Figure 32–19. This disc increases the static air pressure and reduces the air volume. The circular air motion is achieved by vanes located in the air tube after the nozzle. The head of the air tube chokes down the air and fuel oil mixture, causing a higher velocity.

The amount of air required for the fuel oil preignition treatment is greater than the amount required for combustion. Prior to ignition approximately 2000 ft³ of air is needed per gallon of fuel oil. Combustion requires 1540 ft³ of air per gallon of fuel oil. The air blown into the combustion chamber is greater than what would be needed for the theoretical perfect com-

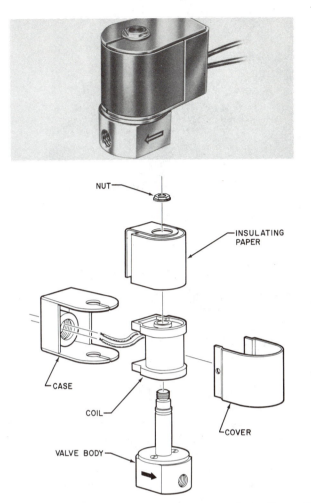

Figure 32–18. Solenoid positive automatic fuel oil cutoff control to avoid after-fire drip. *Courtesy Honeywell, Inc., Residential Division*

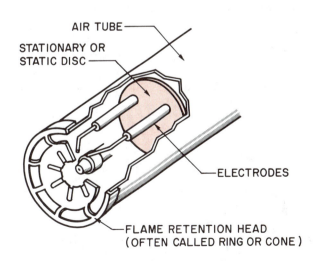

Figure 32–19. Oil-burner air tube.

Figure 32–20. Flame retention device. *Courtesy Ducane Corporation*

bustion. Should the amount of air exceed the 2000 ft³, the burner flame will be long and narrow and possibly impinge on or strike the rear of the combustion chamber. Air adjustments are often made by eye. But poor judgment can severely lower burner efficiency and affect the life of the combustion chamber. Airflow adjustments should be made only after analyzing the product of combustion, the flue gases. This will be discussed in detail later in this unit.

Some burners are equipped with a *flame retention ring* or *cone* located at the front of the air tube. These rings or cones are designed to provide greater burner efficiency by creating more turbulence within the air–oil mixture, providing a stable flame front. The flame is locked to the burner cone, Figures 32–19 and 32–20.

Electrodes

Two *electrodes* are located within the air tube of each oil burner unit, Figure 32–21. The electrodes are metal rods insulated with a ceramic material to prevent an electrical ground. The rear portions of the electrodes are made of a flat brass alloy that must make firm contact with the transformer terminals. The position of the electrodes is adjustable. This will be discussed later in this unit.

There are two types of ignition. Modern furnaces are designed so that the electrodes are providing the high-voltage spark during the entire burning cycle. This is

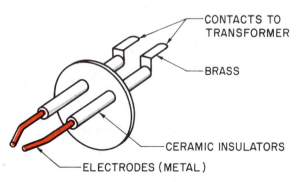

Figure 32–21. Electrode assembly.

continuous or constant ignition. Some older furnaces were designed so that the electrodes provide the spark only for a short period of time during the beginning of the burning cycle. This is called *intermittent* ignition. This system will be rarely found and only in older furnaces.

Ignition Transformer

Oil burners use a *step-up transformer* to provide high voltage to the electrodes, which produce the spark for ignition. The voltage is increased from 110 V to 10,000 V. The transformer is located on the top portion of the burner assembly and is normally hinged for service purposes, Figure 32–22. The transformer terminals are often held in firm contact with the electrodes with springs.

Transformers cannot be serviced in the field. When a weak or defective transformer is detected, replace it with one that meets the manufacturer's specifications, Figure 32–23.

Primary Control Unit

The oil-burner *primary controls* provide a means for operating the burner and a safety function whereby the burner is shut down in the event that combustion does not take place.

A primary control must turn the burner on and off in response to the low-voltage operating controls (thermostat). When this thermostat closes its contacts, low voltage is supplied to a coil that energizes and closes a switch by magnetic force. This switch transfers line voltage to the burner, and it begins to operate.

If the burner does not ignite or if it flames out during the combustion cycle, the safety function of the primary control must shut down the burner to prevent large quantities of unburned oil from accumulating in the combustion chamber.

Older installations may use a bimetal-actuated switch, called a *stack switch* or *stack relay,* for this safety feature. This switch is installed in the flue between the heat exchanger and the draft damper, Figure 32–24. This device is a heat-sensing component and is wired to shut down the burner if heat is not detected in the stack.

Modern furnaces are designed with a light-sensing device made of cadmium sulfide, called a *cad cell.* The cad cell works with a special primary control, Figure 32–25, and is located in the oil-burner unit beneath the ignition transformer, Figures 32–26 and 32–27. It must be located so that it can "sight" the flame through the air tube. When not sensing light from the flame, it offers a very high resistance to an electrical current. When the flame is on, the cad cell resistance drops, permitting a greater current flow. Electrical circuits and switching devices are designed so that this change in resistance (flame on or flame off) allows the burner to remain on or shut down.

Figure 32–22. Transformer mounted on oil-burner assembly. *Courtesy Ducane Corporation*

Figure 32–23. Assorted transformers. *Courtesy Webster Electric Company*

Figure 32–24. Stack switch. *Honeywell, Inc., Residential Division*

Figure 32–25. This primary control works in conjunction with the cad cell for flame detection. *Photo by Bill Johnson*

Figure 32–26. Cad cell. *Courtesy Honeywell, Inc., Residential Division*

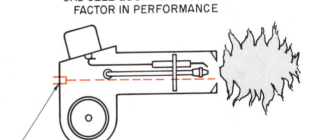

CAD CELL LOCATION CRITICAL FACTOR IN PERFORMANCE

CAD CELL MUST BE POSITIONED TO SIGHT FLAME

1. CELL REQUIRES A DIRECT VIEW OF FLAME.
2. ADEQUATE LIGHT FROM THE FLAME MUST REACH THE CELL TO LOWER ITS RESISTANCE SUFFICIENTLY.
3. CELL MUST BE PROTECTED FROM EXTERNAL LIGHT.
4. AMBIENT TEMPERATURE MUST BE UNDER 140°F.
5. LOCATION MUST PROVIDE ADEQUATE CLEARANCE. METAL SURFACES MUST NOT AFFECT CELL BY MOVEMENT, SHIELDING OR RADIATION.

Figure 32–27. Cad cell location.

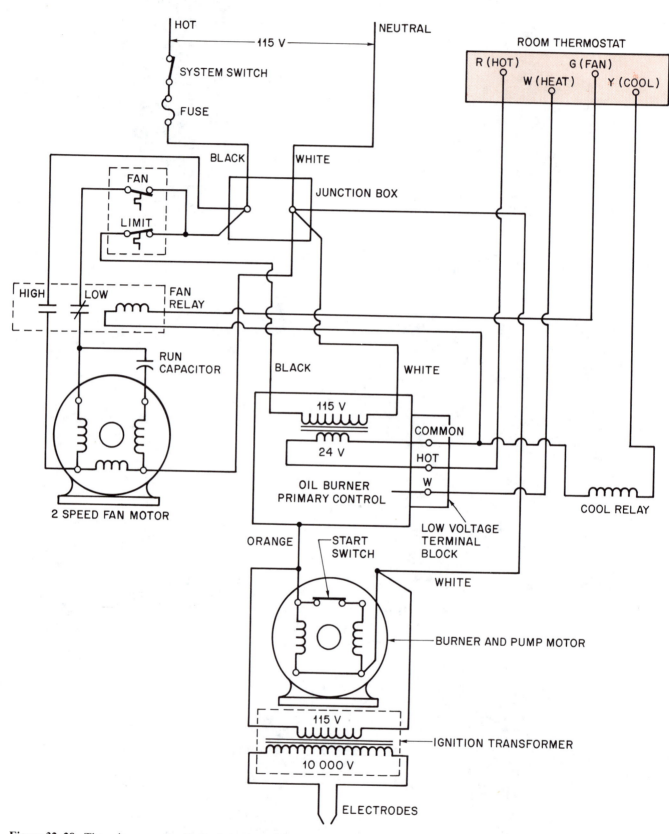

Figure 32–28. The primary control has a hot wire (black) and a neutral wire (white) feeding in. When a call from the thermostat for heat occurs, the orange wire is energized and the burner motor starts at the same time as the arc at the electrodes.

32.6 OIL FURNACE WIRING DIAGRAMS

Figure 32–28 is a wiring diagram for a typical forced-warm-air oil furnace. This diagram is designed to be used with an air conditioning system as well. It includes wiring for the fan circuit for the air distribution, the oil-burner primary control circuit, and the 24-V control circuit. Note that this diagram illustrates the circuitry for a multispeed blower motor.

Figure 32–29 illustrates the wiring for the blower fan motor operation only. The fan relay Ⓐ protects the blower motor from current flowing in the high- and low-speed circuits at the same time. If the normally open (NO) contacts on the high-speed circuit close, the NC contacts on the low-speed circuit will open. The NO fan limit switch Ⓑ is operated by a temperature-sensing bimetal device. When the temperature at the heat exchanger reaches approximately 110°, these contacts close, providing current to the low-speed terminal on the fan motor. The high-speed circuit can be activated only from the thermostat. The high speed on the fan motor is *not* used for heating. It will operate only when the thermostat is set to FAN ON manually or for cooling (air conditioning).

The fan circuit for heating would then be from the power source through the NC fan relay, to the temperature-actuated fan limit switch, to the low-speed fan motor terminal, and back through the neutral wire.

Figure 32–30 shows the oil-burner wiring diagram. The burner is wired through a NC limit switch Ⓐ. This limit switch is temperature actuated to protect the furnace and the building. If there is an excessive temperature buildup from the furnace overheating, this switch will open, causing the burner to shut down. This overheating could result from the burner being over-fired, from a fan problem, from airflow restriction, or from similar problems. The wiring can then be followed from the limit switch to the primary control Ⓑ.

The black wire and orange wire at the primary pass the same current, but the orange is wired through the 24-V primary high-voltage contacts Ⓒ. When the thermostat calls for heat, these contacts close, providing current to the ignition transformer, burner motor, and fuel valve, if one is used. The return is through the neutral (white wire) to the power source. The transformer in the primary control reduces the 120-V line voltage to 24 V for the thermostat control circuit.

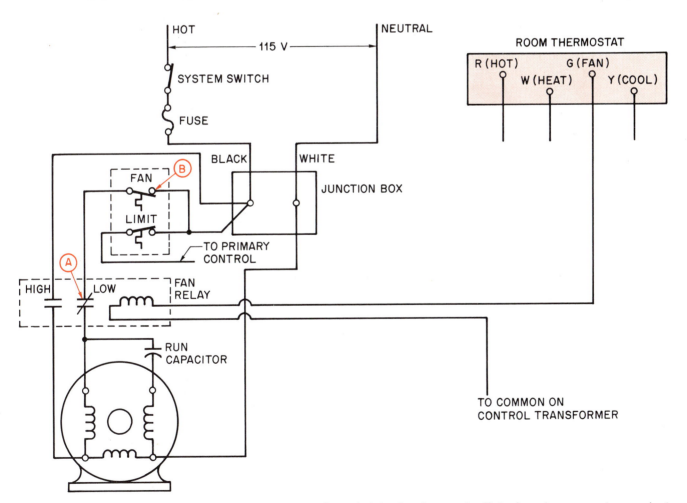

Figure 32–29. The fan runs using the thermally activated fan switch in the winter cycle. If the fan relay were to be energized while in the HEAT mode, the relay would switch from low to high speed.

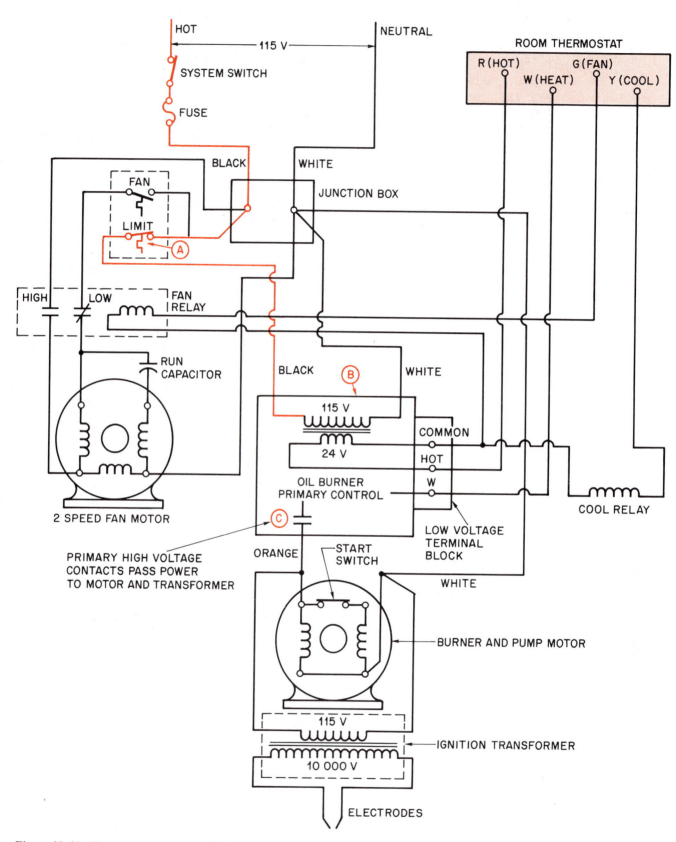

Figure 32–30. The power to operate the burner motor and transformer passes through the NC limit switch to the primary. On a call for heat, the thermostat energizes an internal 24-V relay in the primary. If the primary safety circuit is satisfied, power will pass through the orange wire to the burner motor and ignition transformer.

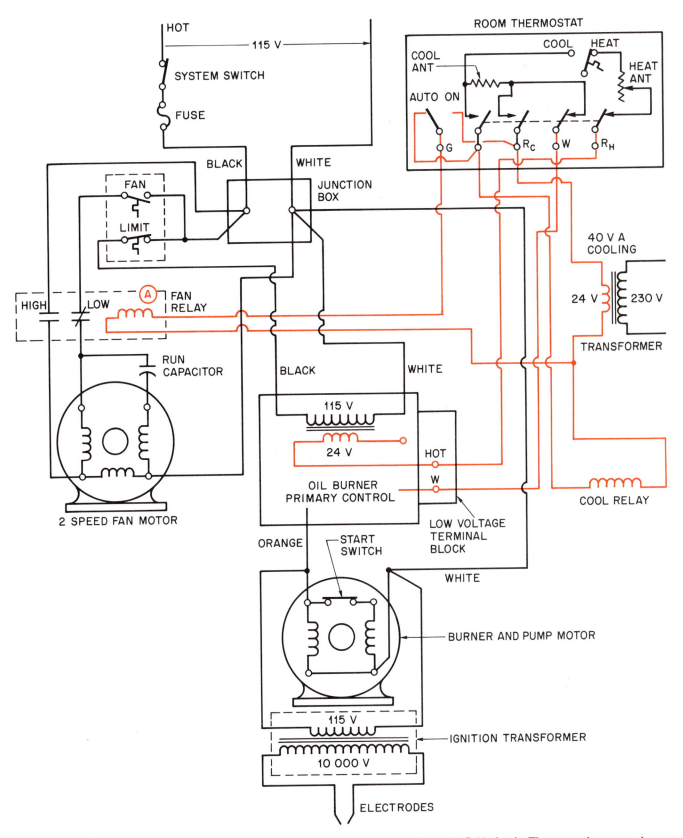

Figure 32-31. The common wire in the primary transformer does not extend into the field circuit. The room thermostat has a split subbase with the hot wire from two different transformers feeding it. These two hot circuits do not come in contact because of the split subbase.

If there is a problem in the startup of the burner, the safety device (stack switch or cad cell) will shut down the burner. This is explained in the following paragraphs.

Figure 32-31 illustrates the wiring for the 24-V thermostat control circuit. The letter designations in this diagram indicate what the wire colors would normally be: R (red), "hot" leg from the transformer; W (white), heat; Y (yellow), cooling; G (green), manual fan relay.

It will occasionally be necessary to convert a system from one designed for heating only, to one that will also accommodate air conditioning. A heating–cooling system requires a larger 24-V control transformer than one normally supplied with a heating only system. To make this conversion, a 40-VA transformer is required in addition to the one supplied in the normal primary control circuit. The thermostat in this conversion should also have an isolating subbase so that one transformer is not connected electrically with the other. A fan relay Ⓐ is also needed in the fan circuit to activate the high-speed fan.

Power-consuming devices must be connected to both legs (i.e., connected in parallel). Power-passing devices will be wired in series through the hot leg.

32.7 STACK SWITCH SAFETY CONTROL

The bimetal element of the stack switch is positioned in the flue pipe. When the bimetal element is heated, proving that there is ignition, it expands and pushes the drive shaft in the direction of the arrow, Figure 32-32. This closes the hot contacts and opens the cold contacts.

Figure 32-33 is a wiring diagram showing the current flow during the initial startup of the oil burner. The 24-V room thermostat calls for heat. This will energize the 1K coil, closing the 1K1 and 1K2 contacts. Current will flow through the safety switch heater and cold contacts. The hot contacts remain open. The closing of the 1K1 contacts provides current to the oil-burner motor, oil valve, and ignition transformer. Under normal conditions there will be ignition and heat produced from the combustion of the oil–air mixture. This will provide heat to the stack switch in the flue pipe, causing the hot contacts to close and the cold contacts to open. The current flow then is shown in Figure 32-34. The circuit is completed through the 1K2 contact, the hot contacts to the 1K coil. The 1K1 coil remains closed, and the furnace continues to run as in a normal safe startup. The safety switch heater is no longer in the circuit.

If there is no ignition and consequently no heat in the stack, the safety switch heater remains in the circuit. In approximately 90 sec the safety switch heater opens the safety switch, which causes the burner to shutdown. To start the cycle again, depress the manual

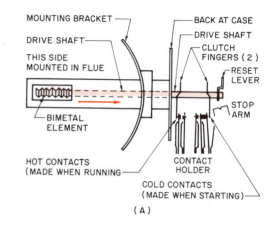

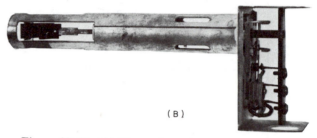

Figure 32–32. (A) Illustration of a stack switch. (B) Photo of a stack switch. *Courtesy (A) Honeywell, Inc., Residential Division (B) Photo by Bill Johnson*

reset button. Allow approximately 2 min for the safety switch heater to cool, before attempting to restart.

32.8 CAD CELL SAFETY CONTROL

The cad cell is designed to offer a low resistance when it senses light. When there is ignition (flame), the cad cell has a low resistance. It has a very high resistance when there is no light (no flame), Figure 32-35. The cad cell must be positioned properly to sense sufficient light for the burner to operate efficiently.

In a standard modern primary circuit the cad cell may be coupled with a triac, which is a form of a solid-state device designed to conduct current when the cad cell circuit resistance is high (no flame).

In Figure 32-36 the circuit shows the current flow in a normal startup. There is low resistance in the cad cell. The triac does not conduct readily; therefore the current bypasses the safety switch heater, and the burner continues to run.

In Figure 32-37 there is no flame and a high resistance in the cad cell. By design the triac will then conduct current that will pass through the safety switch heater. If this current continues through the heater, it will open the safety switch and cause the burner to shut down. To recycle, depress the reset button.

In most modern installations the cad cell is preferred over the thermal stack switch. It acts faster, has no mechanical moving parts, and consequently is considered more dependable.

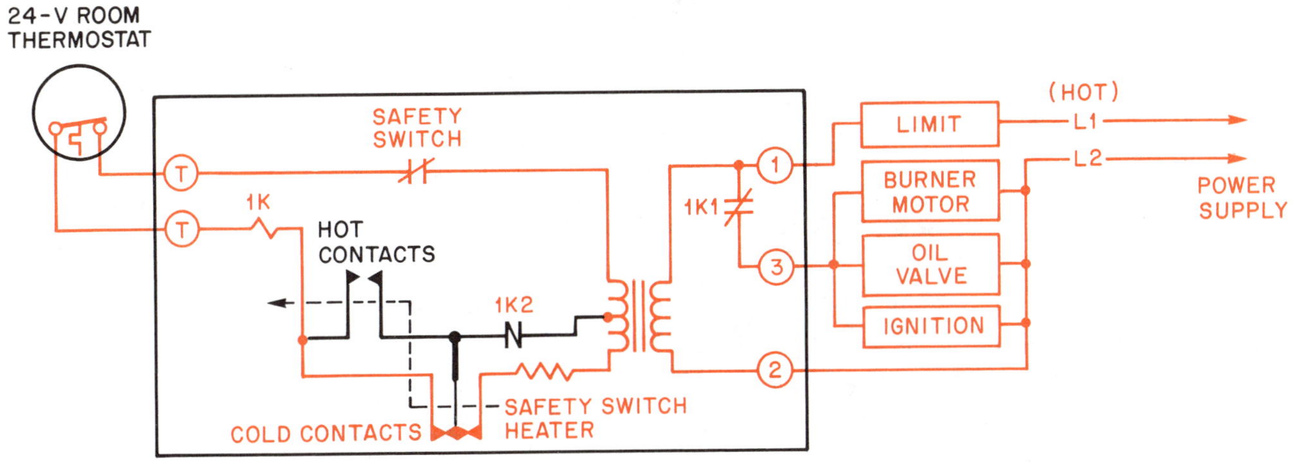

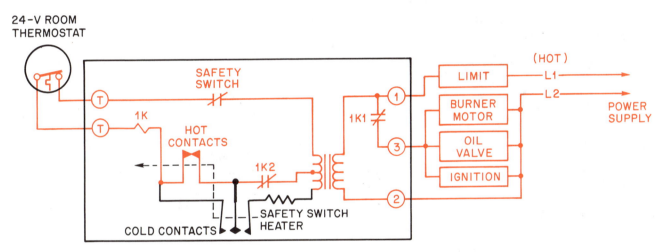

Figure 32–33. Stack switch circuit when burner first starts. Note current flow through safety switch heater.

Figure 32–34. Typical stack switch circuitry with flame on, hot contacts closed, cold contacts open. Safety switch heater no longer is in the circuit.

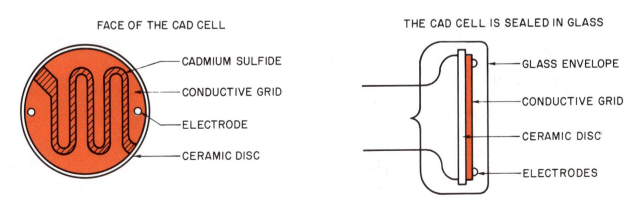

Figure 32–35. Diagram of the face and side of a cad cell.

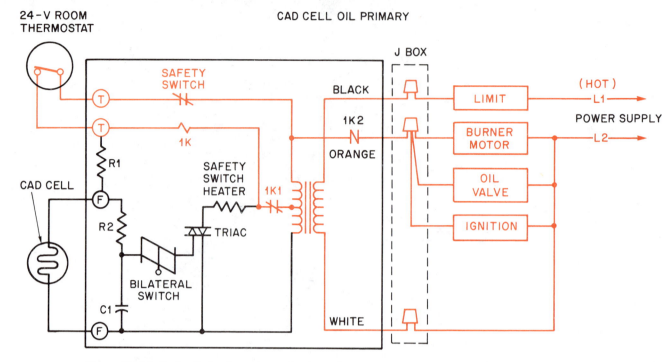

Figure 32–36. Cad cell circuitry when cell "sights" flame. Safety switch heater is not in the circuit.

SAFETY PRECAUTION: *Care should be taken not to reset any primary control too many times because unburned oil may accumulate after each reset. If a puddle of oil is ignited, it will burn intensely. If this should happen, the technician should stop the burner motor but should allow the furnace fan to run. The air shutter should be shut off to reduce the air to the burner. Notify the fire department. Do *not* try to open the inspection door to put the fire out. Let the fire burn itself out with reduced air.*

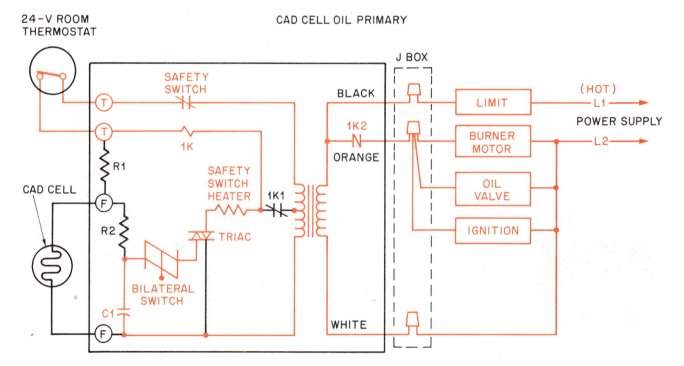

Figure 32–37. Cad cell primary circuitry when cell does not "sight" flame. Safety switch heater will heat and open safety switch.

If the tank is above the burner level, a one-pipe system may be used. The flow of the fuel oil is then by gravity through the fuel filter, Figure 32–38, then to the burner, Figure 32–39. When the tank is below the burner level, a two-pipe system is used. Fuel oil is drawn or pumped from the tank, Figure 32–40.

Whenever the fuel oil line filter or pump is serviced in a one-pipe system, all air in the fuel line and the oil-burner unit must be bled off to insure proper fuel oil flow. Bleeding the system may also be necessary if the fuel oil tank is allowed to become empty. Air bleeding is automatic in the two-pipe system. Because of this a two-pipe system is preferred.

The *viscosity* or thickness of the fuel oil is determined to a great extent by its temperature. The viscosity controls the rate at which the fuel oil flows. The lower the temperature of the oil, the slower it flows. In cold climates the fuel oil tank should be installed in the basement or buried underground to insure that the oil does not thicken to the extent that its flow rate would be reduced below that needed for a normal operation.

Auxiliary Fuel Supply Systems

In some installations oil-burner units will be located above the fuel oil supply tank at a height that will be beyond the lift capabilities of the burner pump. This

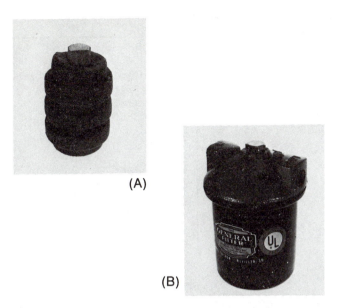

Figure 32–38. Filter cartridge and housing. *Photos by Bill Johnson*

32.9   FUEL OIL SUPPLY SYSTEMS

The fuel oil supply system for a residential or light commercial system uses one or two fuel lines. The location of the fuel storage tank in relation to the burner determines whether one or two lines are used.

ITEM	DESCRIPTION
1	OIL BURNER
2	ANTIPULSATION LOOP
3	OIL FILTER
4	SHUTOFF VALVE (FUSIBLE)
5	INSTRUCTION CARD
6	OIL GAGE
7	VENT ALARM
8	VENT
9	FILL LINE COVER

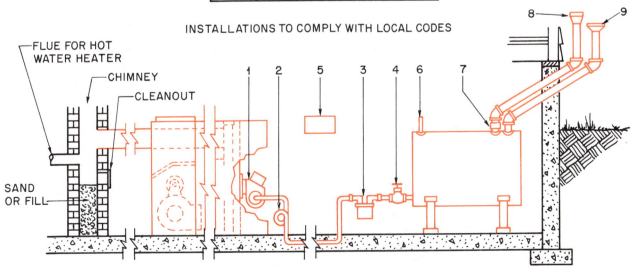

Figure 32–39. One-pipe system from tank to burner.

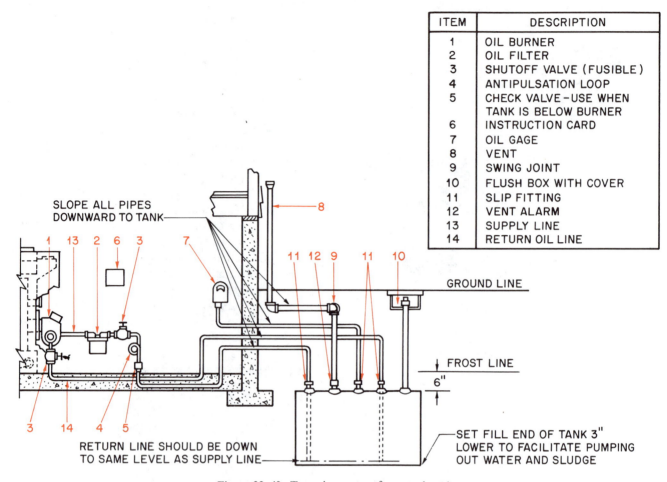

ITEM	DESCRIPTION
1	OIL BURNER
2	OIL FILTER
3	SHUTOFF VALVE (FUSIBLE)
4	ANTIPULSATION LOOP
5	CHECK VALVE – USE WHEN
	TANK IS BELOW BURNER
6	INSTRUCTION CARD
7	OIL GAGE
8	VENT
9	SWING JOINT
10	FLUSH BOX WITH COVER
11	SLIP FITTING
12	VENT ALARM
13	SUPPLY LINE
14	RETURN OIL LINE

Figure 32–40. Two-pipe system from tank to burner.

occurs primarily in commercial applications. Heights exceeding 15 ft from the bottom of the tank are beyond the capabilities of even the two-stage systems. In such installations an auxiliary fuel system must be used to get the fuel supply to the burner.

In this instance a fuel oil booster pump is used to pump the fuel to an accumulator or reservoir tank, Figure 32–41. This booster pump is wired separately from the burner unit and pumps the fuel oil through a separate piping system. The accumulator tank is kept full. Check valves are used to maintain the prime of the booster pump when it is not operating. An adjustable pressure regulator valve is a part of the fuel oil booster pump, and this regulator must be set to maintain a fuel oil pressure of 5 psi measured at the accumulator tank. Failure to properly adjust this regulator could result in seal damage to the fuel unit.

Fuel Line Filters

A filter should be located between the tank and the pump to remove many fine solid impurities from the fuel oil before it reaches the pump, Figures 32–38 and 32–40. Filters may also be located in the fuel line at the pump outlet to further reduce the impurities that may reach the nozzle, Figure 32–42.

32.10   COMBUSTION CHAMBER

The atomized oil and air mixture is blown from the air tube into the combustion chamber where it is ignited by a spark from the electrodes. The atomized oil must be burned in suspension; that is, it must be burned while in the air in the combustion chamber. If the flame hits the wall of the combustion chamber before it is totally ignited, the cooler wall will cause the oil vapor to condense, and efficient combustion will not take place. Combustion chambers should be designed to avoid this condition. These chambers may be built of steel or a refractory material. A silicon refractory is used in many modern furnaces, Figure 32–43. Figure 32–44 shows a flame in a combustion chamber.

32.11   HEAT EXCHANGER

A heat exchanger in a forced-air system takes heat caused by the combustion in the furnace and transfers it to the air that is circulated to heat the building, Figure 32–45. The heat exchanger is made of material that will cause a rapid transfer of heat. Modern exchangers, particularly in residential furnaces, are made of sheet steel, which is frequently coated with special substances to resist corrosion. Acids produced by combustion cause corrosion.

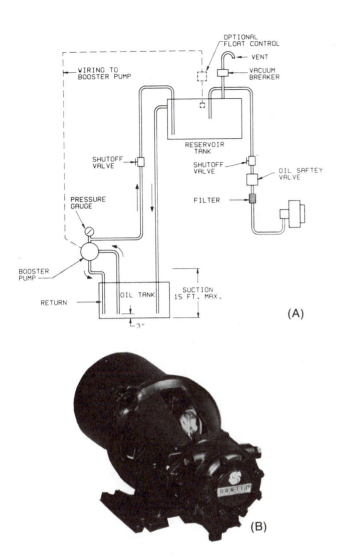

(A)

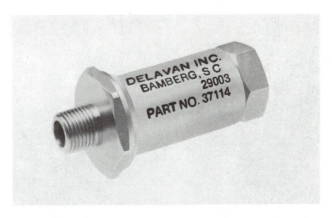

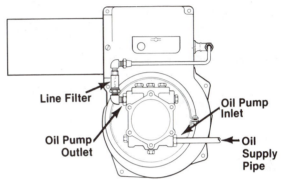

Figure 32–42. Fuel-line filter at pump outlet. *Courtesy Delavan Corporation*

Figure 32–41. (A) Diagram of system using auxiliary booster pump and reservoir tank. (B) Booster pump. *Courtesy (B) Suntec Industries Incorporated, Rockford, Illinois*

The heat exchanger is also designed to separate flue gases and other combustion materials from the air circulated throughout the building. *The flue gases from combustion should never mix with the circulating air.*

Heat exchangers and combustion chambers are often inspected by the technician. A mirror made of chrome-plated steel is an ideal tool for this job, Figure 32–46.

32.12 SERVICE PROCEDURES

Pumps

The performance of the fuel oil system from the tank to the nozzle in residential and commercial units without booster pumps can be determined with a vacuum gage and a pressure gage. Before the vacuum and pressure checks are made, however, the following should be determined:

1. The tank has sufficient fuel oil.
2. The tank shutoff valve is open.
3. The tank location is noted (above or below burner).
4. The type of system is noted (one or two pipe).

Connect the vacuum and pressure gages to the fuel oil unit as shown in Figure 32–47. If the tank is above the burner level, the supply system may be either a one- or two-pipe system. If the tank is above the burner and

Figure 32–43. An alumna silicon refractory combustion chamber. *Courtesy Ducane Corporation*

Figure 32–44. Inside combustion chamber. Note retention ring with flame locked to it. *Courtesy R. W. Beckett Corporation*

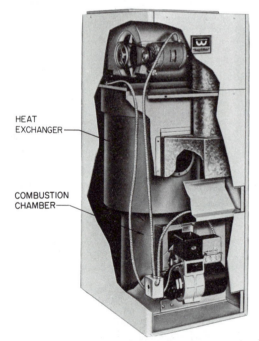

Figure 32–45. Heat exchanger in oil-fired forced-air furnace. *Courtesy The Williamson Company*

Figure 32–46. Mirror used by the heating technician to view the heat exchanger, the inside of the combustion chamber, and the flame. The handles on some models telescope and the mirror is adjustable. *Courtesy Delavan Corporation*

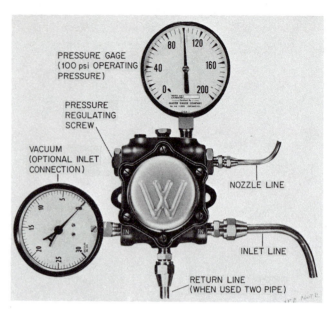

Figure 32–47. Pump with pressure and vacuum gauge in place for tests and servicing. *Courtesy Webster Electric Company*

with the unit in operation, the vacuum gage should read 0 in. Hg. If the vacuum gage indicates a vacuum, one or more of the following may be the problem:

- The fuel line may be kinked.
- The line filter may be clogged or blocked.
- The tank shutoff valve may be partially closed.

If the supply tank is below the burner level, the vacuum gage should indicate a reading. Generally, the gage should read 1 in. Hg for every foot of vertical lift and 1 in. Hg for every 10 ft of horizontal run. The lift and horizontal run combination should not total more than 17 in. Hg. (Remember, the lift capability of a dual-stage pump is approximately 15 ft.) Vacuum readings in excess of this formula may indicate a tubing size too small for the run. This may cause the oil to separate into a partial vapor and become milky.

The pressure gage indicates the performance of the pump and its ability to supply a steady even pressure to the nozzle. The pressure regulator should be adjusted to 100 psi. See Figure 32–48 for an illustration of a pressure adjusting valve. To adjust the regulator, with the pressure gage in place turn the adjusting screw until the gage indicates 100 psi. With the pump shown in Figure 32–48, the valve-screw cover screw should be removed and the valve screw turned with a $\frac{1}{8}$ in. Allen wrench.

With the regulator set and the unit in operation, the pressure gage should indicate a steady reading. If the gage pulsates, one or more of the following may be the problem:

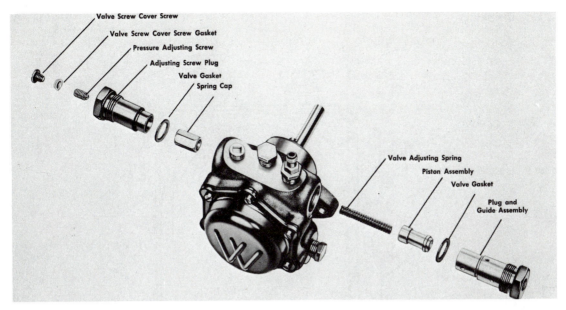

Figure 32–48. Fuel pump illustrating pressure adjusting valve. *Courtesy Webster Electric Company*

- Partially clogged supply filter element
- Partially clogged unit filter or screen
- Air leak in fuel oil supply line
- Air leak in fuel oil pump cover

A pressure check of valve differential is necessary when oil enters the combustion chamber after the unit has been shut down. To make this check, insert the pressure gage in the outlet for the nozzle line, Figure 32–49. With the gage in place and the unit running, record the pressure. With the unit shutdown, the pressure gage should drop 15 psi and hold. If the gage drops to 0 in. psi, the fuel oil cutoff inside the pump is defective and the pump may need to be replaced. If the pressure gage indicates a hold of 15 psi differential, but there is still an indication of improper fuel oil shut off the problem may be

- Possible air leak in fuel unit (pump)
- Possible air leak in fuel oil supply system
- Nozzle strainer may be clogged

Burner Motor

If the burner motor does not operate

1. Press the reset button (*Note:* Reset may not function if motor is too hot. Wait for motor to cool down.)
2. Check the electrical power to the motor. This can be done with a voltmeter across the orange and white or black and white leads to the motor. If there is power to the motor and there is nothing binding it physically and it still does not operate, it should be replaced.

Bleeding a One-Pipe System

When starting an oil burner with a one-pipe system for the first time or whenever the fuel-line filter or pump is serviced, the system piping will have air in it. The system must be bled to remove this air. The system should also be bled if the supply tank is allowed to become empty.

To bleed the system, a ¼-in. flexible transparent tubing should be placed on the vent or bleed port, and the free end of the tube placed in a container, Figure 32–50. Turn the bleed port counterclockwise. Follow the manufacturer's specifications as to how much to turn this port. Usually from ⅛ to ¼ turn is sufficient. The burner should then be started, allowing

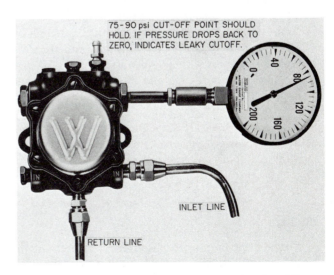

Figure 32–49. Pressure gage in place to check valve differential. *Courtesy Webster Electric Company*

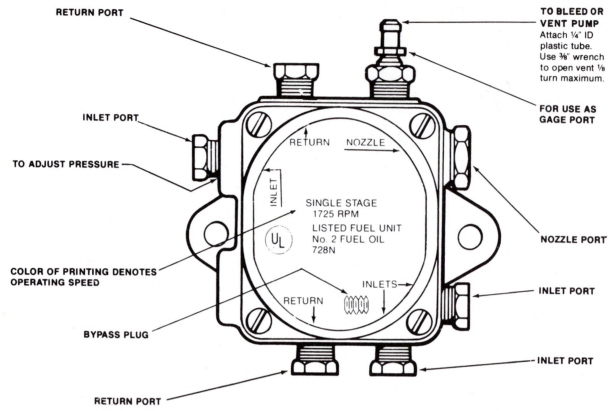

Figure 32–50. Diagram of fuel oil pump. *Courtesy Webster Electric Company*

the fuel oil to flow into the container. Continue until the fuel flow is steady, and there is no further evidence of air in the system.

Converting a One-Pipe System to a Two-Pipe System

Most residential burners are shipped to be used with a one-pipe system. To convert these units to a two-pipe system, do the following:

1. Shut down all electrical power to the unit.
2. Close the fuel oil supply valve at the tank.
3. Remove inlet port plug from the unit, Figure 32–51.
4. Insert bypass plug shipped with unit into the deep seat of inlet port with an Allen wrench.
5. Replace inlet port plug and install flare fitting and copper line returning to the tank.

SAFETY PRECAUTION: * Do not start the pump with the bypass plug in place unless the return line to the tank is in place. Without this line, there is no place for the excess oil to go except to possibly rupture the shaft seal.*****

Nozzles

Nozzle problems may be discovered by observing the condition of the flame in the combustion chamber, taking readings on a pressure gage, and analyzing the flue gases. (Flue-gas analysis is discussed later.)

Common conditions that may relate to the nozzle are

- Pulsating pressure gage
- Flame changing in size and shape
- Flame impinging (striking) on sides of combustion chamber
- Sparks in flame
- Low CO_2 (carbon dioxide) reading in flue gases (less than 8%)
- Delayed ignition
- Odors present

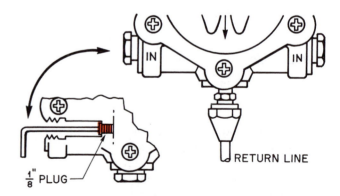

Figure 32–51. Installation of bypass plug for conversion to two-pipe system. *Courtesy Webster Electric Company*

When nozzle problems are apparent, the nozzle should be replaced. The nozzle is a delicate finely machined component, and wire brushes and other cleaning tools should not be used on it. Generally, a nozzle should be replaced annually as a normal servicing procedure. *Do *not* attempt to clean the nozzle strainer.*

If the nozzle strainer is clogged, the cause should be determined. Clean the fuel oil nozzle line and the fuel unit (pump) strainer. Change the supply filter element. There may also be water in the supply tank, and the fuel oil may otherwise be contaminated.

Carbon formation at the nozzle or fuel oil burning at the nozzle may be caused by bent or distorted nozzle features. *Nozzles should be removed and replaced with a special nozzle wrench. Never use adjustable pliers or a pipe wrench.*

When the nozzle has been removed, keep the oil from running from the tube by plugging the end of the nozzle assembly with a small cork. If oil leaves this tube, air will enter and cause an erratic flame when the burner is put back into operation. It may also cause an after-drip or after-fire when the burner is shut down.

It is important that nozzles do not overheat. Overheating causes the oil within the swirl chamber to break down and form a varnish-like substance to build up. Major causes of overheating and varnish buildup are

- Fire burning too close to the nozzle
- Firebox too small
- Nozzle too far forward
- Inadequate air-handling components on burner
- Burner overfired

If the fire burns too close to the nozzle, the blast tube opening may be narrowed. This will increase the air velocity as it leaves the tube and move the flame away from the nozzle. Increasing the pressure of the pump will increase the velocity of the oil droplets leaving the nozzle and will help alleviate this problem.

A nozzle that is too far forward will clog from reflected heat from the firebox from the gum formation caused by the cracking (overheated) oil. The nozzle may be moved back into the air tube by using a short adapter or by shortening the oil line.

Do not allow an after-drip or after-fire to go unchecked. Any leakage or after-fire will result in carbon formation and clogging of the nozzle. Check for an oil leak, air in the line, or a defective cutoff valve if there is one. An air bubble behind the nozzle will not always be pushed out during burning. It may expand when the pump is shut off and cause an after-drip.

Ignition System

Of all phases of oil-burner servicing, ignition problems are among the easiest to recognize and solve. Much ignition service consists of making the proper spark gap adjustment, cleaning the electrodes, and making all connections secure.

To check the ignition transformer,

1. Turn off power to the oil-burner unit.
2. Swing back ignition transformer.
3. Shut off fuel supply or disconnect burner motor lead.
4. Turn power back on to the oil burner. With a voltmeter measure the voltage of the transformer output. This reading should be 10 000 V. If a lesser voltage is read, the transformer is defective and should be replaced. *Follow the manufacturer's instructions and keep your distance from the 10 000 V leads.*

To check the electrodes,

1. Insure that the three spark gap settings of the electrodes are set properly. Check Figure 32–52 for the settings of the gap, height above the center of the nozzle, and distance of the electrode tips forward from the nozzle center.

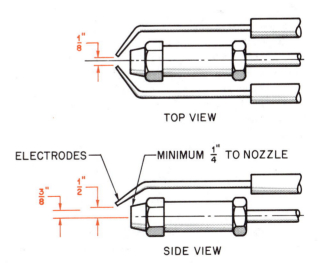

TOP VIEW

ELECTRODES — — MINIMUM $\frac{1}{4}$" TO NOZZLE

SIDE VIEW

HEIGHT ADJUSTMENT

$\frac{1}{2}$" RESIDENTIAL INSTALLATION

$\frac{3}{8}$" COMMERCIAL INSTALLATION

POSITION OF ELECTRODES IN FRONT OF NOZZLE IS DETERMINED BY SPRAY ANGLE OF NOZZLE

Figure 32–52. Electrode adjustments. Electrodes cannot be closer than $\frac{1}{4}$ in. to any metal part.

2. Make sure the tips of the electrodes are in back of the oil spray. This can be checked by using a flame mirror, Figure 32–46. If this is not possible, remove the electrode assembly and spray the fuel oil into an open container making sure the electrodes are not in the fuel oil path.

Electrode Insulators. Wipe the insulators clean with a cloth moistened with a solvent. If they remain discolored after a good cleaning, they are filled with carbon throughout their porous surfaces and should be replaced.

To check the insulator,

1. Disconnect burner motor lead.
2. Remove one side of the high-tension lead and hold it close to the discolored portion of the other insulator, which is still connected to its lead.
3. Turn on the burner switch.
4. If any spark jumps to the insulator from the lead, the electrode should be replaced. Check both sides, using this procedure.

SAFETY PRECAUTION: *This is 10 000 V and can cause electrical shock.*

32.13 COMBUSTION EFFICIENCY

Until a few years ago the heating technician was primarily concerned with insuring that the heating equipment operated cleanly and safely. Technicians made adjustments by using their eyes and ears. High costs of fuel, however, now make it necessary for the technician to make adjustments using test equipment to ensure efficient combustion.

Fuels consist mainly of hydrocarbons in various amounts. In the combustion process new compounds are formed and heat is released. The following are simplified formulas showing what happens during a perfect combustion process:

$$C + O_2 \longrightarrow CO_2 + heat$$
$$H_2 + \tfrac{1}{2}O_2 \longrightarrow H_2O + heat$$

During a normal combustion process, carbon monoxide, soot, smoke, and other impurities are produced along with heat. Excess air is supplied to insure that the oil has enough oxygen for complete combustion, but even then the air and fuel may not be mixed perfectly. The technician must make adjustments so that near perfect combustion takes place, producing the most heat and reducing the quantity of unwanted impurities. On the other hand, too much excess air will absorb heat, which will be lost in the stack (flue gas) and reduce efficiency. Air contains only about 21% oxygen, so the remaining air does not contribute to the heating process.

The following tests can be made for proper combustion:

1. Draft
2. Smoke
3. Net temperature (flue stack)
4. Carbon dioxide

Technicians develop their own procedures in combustion testing. Making an adjustment to help correct one problem will often correct or help to correct others. Compromises have to be made also. When correcting one problem, another may be created. In some instances problems may be caused by the furnace design.

Some technicians use individual testing devices for each test, Figure 32–53. Others use electronic combustion analyzers with digital readouts, Figure 32–54.

To make these tests with the individual testing devices, a hole must be drilled or punched in the flue pipe 12 in. from the furnace breeching, on the furnace side of the draft regulator, and at least 6 in. away from it. The hole should be of the correct size so that the stem or sampling tube of the instrument can be inserted into it. ***The manufacturer's instructions furnished with the instruments should be followed carefully. Procedures indicated here are very general and should not take the place of the manufacturer's instructions.***

Draft Test

Correct draft is essential for efficient burner operation. The draft determines the rate at which combus-

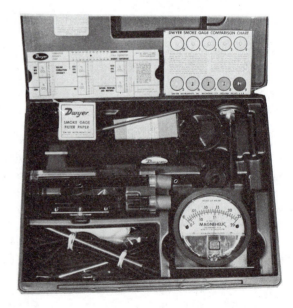

Figure 32–53. Combustion efficiency testing equipment. *Photo by Bill Johnson*

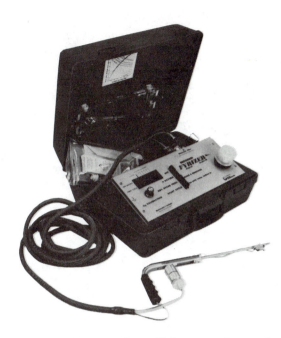

Figure 32–54. Combustion efficiency analyzer. *Courtesy United Technologies Bacharach*

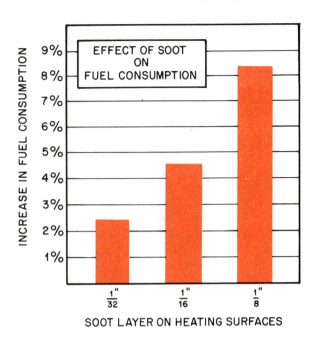

Figure 32–56. Effect of soot on fuel consumption. *Courtesy United Technologies Bacharach*

tion gases pass through the furnace, and it governs the amount of air supplied for combustion. The draft is created by the difference in temperatures of the hot flue gases and is negative pressure in relation to the atmosphere. Excessive draft can increase the stack temperature and reduce the amount of CO_2 in the flue gases. Insufficient draft may cause pressure in the combustion chamber, resulting in smoke and odor around the furnace. Adjust the draft before you make other adjustments to obtain maximum efficiency.

To make the test,

1. Drill a hole into the combustion area for the draft tube. (A bolt may be removed on some furnaces

and the bolt hole used for this access.) This is necessary to determine the overfire draft, Figure 32–55.
2. Place the draft gage on a level surface near the furnace and adjust to 0 in.
3. Turn the burner on and let run for at least 5 min.
4. Insert the draft tube into the combustion area to check the overfire draft, Figure 32–55.
5. Insert the draft tube into the flue pipe to check the flue draft.

The overfire draft should be at least −0.02 in. water column.

The flue draft should be adjusted to maintain the proper overfire draft. Most residential oil burners require a flue draft of −0.04 to −0.06 in. water column to maintain the proper overfire draft. These drafts are updrafts and are negative in relationship to atmospheric pressure. Longer flue passages require a higher flue draft than shorter flue passages.

Smoke Test

Excessive smoke is evidence of incomplete combustion. This incomplete combustion can result in a fuel waste of up to 15%. A 5% fuel waste is not unusual. Excessive smoke also results in a soot buildup on the heat exchanger and other heat-absorbing areas of the furnace. Soot is an insulator. This results in less heat being absorbed by the heat exchanger and increased heat loss to the flue. A $\frac{1}{16}$-in. layer of soot can cause a $4\frac{1}{2}\%$ increase in fuel consumption, Figure 32–56.

The smoke test is accomplished by drawing a prescribed number of cubic inches of smoke-laden flue

Figure 32–55. Checking overfire draft. *Photo by Bill Johnson*

products through a specific area of filter paper. The residue on this filter paper is then compared with a scale furnished with the testing device. The degree of sooting can be read off the scale. A smoke tester such as the one illustrated in Figure 32–57 may be used.

To make the smoke test, Figure 32–58,

1. Turn on the burner and let run for at least 5 min or until the stack thermometer stops rising.
2. Insure that filter paper has been inserted in tester.
3. Insert sampling tube of test instrument into hole in flue.
4. Pull the tester handle the number of times indicated by the manufacturer's instructions.
5. Remove the filter paper and compare with the scale furnished with the instrument.

Excessive smoke can be caused by

- Improper fan collar setting (burner air adjustment)
- Improper draft adjustment (draft regulator may be required or need adjustment)
- Poor fuel supply (pressure)
- Oil pump not functioning properly
- Nozzle defective or of incorrect type
- Excessive air leaks in furnace (air diluting flame)
- Improper fuel-air ratio
- Firebox defective
- Improper burner air-handling parts

Net Stack Temperature

The net stack temperature is important because an abnormally high temperature is an indication that the furnace may not be operating as efficiently as possible. The net stack temperature is determined by subtracting the air temperature around the furnace from the measured stack or flue temperature. For instance, if the flue temperature reading was 650°F and the basement air temperature where the furnace was located was 60°F, the net stack temperature is 650° − 60° = 590°F. Manufacturer's specifications should be con-

Figure 32–58. Making smoke test. *Photo by Bill Johnson*

sulted to determine normal net stack temperatures for particular furnaces.

To determine the stack temperature, Figure 32–59.

1. Turn on the burner and allow to run for at least 5 min or until the stack thermometer stops rising in temperature.
2. Insert the thermometer stem into the hole in the flue.

Figure 32–59. Making stack gas temperature test. *Photo by Bill Johnson*

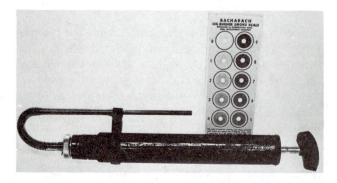

Figure 32–57. Smoke tester. *Photo by Bill Johnson*

3. Subtract the basement or ambient temperature from the stack temperature reading.
4. Compare with manufacturer's specifications.

A high stack temperature may be caused by one or more of the following:

- Excessive draft through the combustion chamber
- Dirty or soot-covered heat exchanger
- Lack of baffling
- Undersized furnace
- Incorrect or defective combustion chamber
- Overfiring

CO_2 Test

The CO_2 test is an important combustion efficiency test. A high CO_2 reading is good. If the CO_2 test reading is low, it indicates that the fuel oil has not burned efficiently or completely. The CO_2 reading should be considered with all of the test readings. Under most normal conditions a $10+\%$ CO_2 reading should be obtained. If problems exist that would be very difficult to correct but the furnace was considered safe to operate and had a net stack temperature of $400°F$ or less, a CO_2 reading of 8% could be acceptable. However, if the net stack temperature is over $500°F$, a reading of at least 9% CO_2 should be obtained.

To take the CO_2 test, Figure 32–60,

1. Turn on the burner and operate for at least 5 min.
2. Insert thermometer and wait for the temperature to stop rising.
3. Insert the CO_2 sampling tube into the hole previously made in the flue pipe.
4. Remove a test sampling, using the procedure provided by the manufacturer of the test instrument.

Figure 32–60. Making CO_2 test. *Photo by Bill Johnson*

5. Mix the fluid in the test instrument with the sample gases from the flue according to instructions.
6. Read the percent CO_2 from the scale on the instrument.

A low percent CO_2 reading may be caused by one or more of the following:

- High draft or draft regulator not working properly
- Excess combustion air
- Air leakage into combustion chamber
- Poor oil atomization
- Worn, clogged, or incorrect nozzle
- Oil-pressure regulator set incorrectly

Electronic Combustion Analyzers

The technician may have an electronic combustion analyzer. This equipment provides information similar to that provided by the test instruments already described. These instruments may indicate carbon monoxide and oxygen concentrations rather than carbon dioxide percentages, but the technician can determine acceptable levels and necessary corrections to be made on the furnace from the manufacturer's instructions.

Generally, the tests are made from samplings from the same position in the flue, as with the other instruments. These electronic instruments are easier to use and save time. The readings are provided by a digital readout system that is convenient to use, Figure 32–54.

32.14   SERVICE TECHNICIAN CALLS

Service Call 1

A new customer calls. *The customer wants a complete checkup of the oil furnace, including an efficiency test. This customer will stay with the service technician and watch the complete procedure.*

The technician arrives and explains to the customer that the first thing to do is run the furnace and perform an efficiency test. By running a test before and after adjustment, the technician can report results. The thermostat is set to about $10°F$ above the room temperature to allow time for it to warm up while the technician sets up. This will also ensure that the furnace will not shut off during the test. The technician gets the proper tools and goes to the basement where the furnace is located. The customer is already there and asks if the technician would mind an observer. The technician explains that a good technician should not mind being watched.

The technician inserts the stack thermometer in the flue and observes the temperature; it is no longer rising. The indoor fan is operating. A sample of the combustion gas is taken and checked for efficiency. A smoke test is also performed. The unit is operating with a slight amount of smoke, and the test shows the efficiency to be 65%. This is about 5 to 10% lower than normal. See Figure 32–53 for an example of a flue gas kit.

The technician then removes a low-voltage wire from the primary, which shuts off the oil burner and allows the fan to continue to run and cool the furnace. The technician then removes the burner nozzle assembly and replaces the nozzle with an exact replacement. Before returning the nozzle assembly to its place, the technician also sets the electrode spacing and then changes the oil filter, the air filter and oils the furnace motor. The furnace is ready to start again.

The technician starts the furnace and allows it to heat up. When the stack temperature has stopped rising, the technician pulls another sample of flue gas. The furnace is now operating at about 67% efficiency and still a little smokey in the smoke test. The technician checks the furnace nameplate and notices that this furnace needs a 0.75 gallon per hour (gph) nozzle but has a 1 gph nozzle. Evidently the last service technician put in the wrong nozzle. It is common (but poor practice) for some technicians to use what they have, even if it is not exactly correct.

This technician removes the nozzle just installed as an exact replacement and then gets more serious about the furnace because it has been mishandled. The technician installs the correct nozzle, starts the unit, and checks the oil pressure. The pressure has been reduced to 75 psig to correct for the oversized nozzle. The technician changes the pressure to 100 psig (the correct pressure), standard for all residential gun-type burners.

The air to the burner is adjusted, meanwhile the stack temperature is reading 660°F (this is a net temperature of 600°F since the room temperature is 60°F). The technician runs another smoke test, which now shows minimum smoke. The efficiency is now 73%. This is much better.

The customer has been kept informed all through the process and is surprised to learn that oil burner service is so exact. The technician sets the room thermostat back to normal and leaves a satisfied customer.

Service Call 2

A customer calls. *The oil furnace in the basement under the family room is making a noise when it shuts off. The oil pump is not shutting off the oil fast enough.*

The technician arrives, goes to the room thermostat, and turns it up above the room temperature to keep the furnace running. This furnace has been serviced in the last 60 days, so a nozzle and oil filter change is not needed. The technician goes to the furnace and removes a low voltage wire from the primary control to stop the burner. The fire does not extinguish immediately but shuts down slowly with a rumble.

The technician installs a solenoid in the small oil line that runs from the pump to the nozzle and wires the solenoid coil in parallel with the burner motor so that it will be energized only when the burner is operating. The technician disconnects the line where it goes into the burner housing and turns it into a bottle to catch any oil that may escape. The burner is then turned on for a few seconds. This clears any air that may be trapped in the solenoid out of the line leading to the nozzle.

The technician reconnects the line to the housing and starts the burner. After it has been running for a few minutes, the technician shuts down the burner. It has a normal shutdown. The startup and shutdown is repeated several times. The furnace operates correctly. The technician then sets the room thermostat to the correct setting before leaving.

Service Call 3

A customer calls. *The customer smells smoke when the furnace starts up. This customer does not have the furnace serviced each year. The oil filter and nozzle are in such poor condition that the furnace has soot buildup in the heat exchanger and flue. This creates a restriction to the flue gas leaving the furnace. When the burner starts, it puffs smoke out around the burner until the draft is established.*

The technician goes to the furnace in the basement, and the customer follows. The technician turns the system switch off and goes back upstairs and sets the thermostat 5°F above the room temperature so the furnace can be started from the basement. When the technician gets back to the basement, the furnace is started and the puff of smoke is observed. The technician inserts a draft gage in the burner door, the draft is +0.01 in. W.C. positive pressure. The burner is not venting. *The furnace is shut off before it gets too hot to work with.* The technician explains to the customer that the furnace must have a complete service and cleaning and that it is hard on the furnace to allow it to get in such poor condition before having it serviced. The technician wants the customer to understand that it is a good investment to have an oil furnace serviced each year—the nozzle and filter changed and the electrodes adjusted. The technician proves this by letting the customer see the inside of the furnace.

Breathing protection should be worn for the following procedure. The technician removes the burner and opens the inspection covers to the heat exchanger and the flue pipe. The canister vacuum and a set of small brushes is brought in from the truck. The technician starts by vacuuming out the combustion chamber, then the flue area. Next small brushes are used and passed between the heat exchanger walls to loosen the soot. The soot is then vacuumed out. Coat hanger wire is pushed into every conceivable crevice of the flue passages of the furnace. The customer cannot believe what comes out of the furnace.

The technician then takes the flue pipe loose at the chimney and cleans the horizontal run of the flue pipe and the bottom of the chimney. It has soot buildup that closes the connection by nearly half. The burner nozzle is changed, the burner motor is oiled, and the electrodes are aligned. Then the oil filter and the air filter are changed. The furnace is then reassembled. This has all taken about $1\frac{1}{2}$ hr. It would even take longer if the furnace were a horizontal furnace in the crawl space under a house.

The technician installs a pressure gage to monitor the oil pressure, starts the furnace and points out that the draft over the fire is now −0.02 in. W. C. The furnace is now venting correctly. There is no smoke. When the furnace has warmed up, a flue-gas analysis shows the furnace is operating at 72% efficiency.

The customer now knows that it is not worth it to let a furnace get in poor condition. There is no way of knowing how long the furnace has been operating inefficiently. The technician sets the room thermostat to the original setting and leaves the job.

Service Call 4

A customer from a duplex apartment calls. *There is no heat. The customer is out of fuel. The customer had fuel delivered last week, but the driver filled the wrong tank. This system has an underground tank and it is sometimes hard to get the oil pump to prime (pull fuel to the pump). The technician will be fooled for some time, thinking that there is fuel and that the pump is bad.*

The technician arrives and goes to the room thermostat. It is set at 10°F above the room temperature setting. The technician goes to the furnace, which is in the garage at the end of the apartment. The furnace is off because the primary control has tripped. The technician takes a flashlight and examines the heat exchanger for any oil buildup that may have accumulated if the customer had been resetting the primary. (If a customer had repeatedly reset the primary trying to get the furnace to fire and the technician then starts the furnace, this excess oil would be dangerous.) There is no excess oil in the heat exchanger, so the reset button is pushed. The burner does not fire.

The technician suspects that the electrodes are not firing correctly or that the pump is not pumping. The first thing to do is to fasten a hose to the bleed port on the side of the oil pump. The hose is placed in a bottle to catch any oil that may escape, then the bleed port is opened and the primary control is reset. This is a two-pipe system and when the correct oil quality is found at the bleed port the oil supply is verified. The burner and pump motor start. No oil is coming out of the hose, so the pump must be defective. Before changing the pump, the technician decides to check to make sure there is oil in the line. The line entering the pump is removed. There is no oil in the line. The technician opens the filter housing and finds very little oil in the filter. The technician now decides to check the tank and borrows the stick the customer uses to check the oil level. The stick is pushed through the fill hole in the tank. There is no oil.

The technician tells the customer, who calls the oil company. The driver that delivered the oil is close by and dispatched to the job. The driver points to the tank that was filled, the wrong one, and then fills the correct tank.

The technician starts the furnace and bleeds the pump until a solid column of liquid oil flows, then closes the bleed port. The burner ignites and goes through a normal cycle. This furnace is not under contract for maintenance, so the technician suggests to the occupant that a complete service call with nozzle change, electrode adjustment, and filter change be done. The occupant says to do the service call. The technician completes the service, turns the thermostat to the normal setting, and leaves.

Service Call 5

A customer calls. *There is no heat in a small retail store. The cad cell is defective, and the furnace will not fire.*

The technician arrives, goes to the room thermostat, and discovers it set at 10°F higher than the room temperature. The thermostat is calling for heat. The technician goes to the furnace in the basement and discovers it needs resetting for it to run. The technician examines the heat exchanger with a flashlight and finds no oil accumulation. The primary control is then reset. The burner motor starts, and the fuel ignites. It runs for 90 sec and shuts down. There is something wrong in the flame-proving circuit, the primary, or the cad cell.

The technician turns off the power, removes the burner assembly, and examines the cad cell. It is not dirty, so the resistance through it should be checked. The cad cell is checked open with room light shining on it. The resistance through a cad cell should be about 200 Ω with room light shining on it. The technician replaces the cad cell. While the burner assembly is out, the technician replaces the nozzle, and sets the electrodes. Before firing the furnace again, the oil filter and the air filters are changed and the pump and furnace fan motor are oiled.

When the furnace is started, the burner comes on and stays on. The technician runs an efficiency check as a routine part of the call. The furnace is running at 72% efficiency, so the technician sets the thermostat at the original setting and leaves.

Service Call 6

A customer calls. *There is no heat. The primary control is bad. The furnace will not start when the reset button is pushed.*

The technician arrives and checks to see that the thermostat is calling for heat. The set point is much higher than the room temperature. The technician goes to the garage where the furnace is located and presses the reset; nothing happens. The primary is carefully checked to see if there is power to the primary control. It shows 115 V. Next the circuit leaving the primary, the orange wire, is checked. (The technician realizes that power comes into the primary on the white and black wires, the white being neutral, and leaves on the orange wire.) There is power on the white to black but none on white to orange, Figure 32–30. The primary is not allowing power to pass. The technician shuts off the power and replaces the primary control. Power is resumed, and the furnace starts. This furnace has had a complete checkup within the last 60 days, so the technician sees no reason to do anything else.

SAFETY PRECAUTION: *Be careful not to spill fuel oil while changing the nozzle and filter. The oil is not highly flammable, but it has a smell that will stay for a long time. Be extra cautious while troubleshooting the electrical circuit. The control transformer has a secondary voltage of 10 000 V that may cause electrical shock.

Always be very cautious when starting any oil furnace that the customer has restarted by pushing the reset button. Every time the reset button is pushed, oil is pumped into the combustion chamber. If, after many attempts, this oil is ignited the furnace will get excessively hot and may rumble and vibrate from the excess fire. You may not be able to extinguish the fire: it will have to burn itself out. If this happens, shut off the burner and shut off all the air to the burner that can be shut off. Make sure you shut off the burner in such a fashion that the in-door fan is allowed to run to dissipate the heat. Turn the air adjustment on the burner to allow minimum air. Call the fire department.

Avoid burns when taking flue-gas efficiency tests.*

SUMMARY

- Number 2 grade fuel oil is most commonly used in heating residences and light commercial buildings.
- Fuel oil is composed primarily of hydrogen and carbon in chemical combination.
- Fuel oil is prepared for combustion by converting it to a gaseous state. This is done by atomization—breaking it up into tiny droplets and mixing with air.
- Gun-type oil-burner parts are the burner motor, blower or fan wheel, pump, nozzle, choke, air tube, electrodes, transformer, and primary controls.

- The motor is a split-phase fractional horsepower motor.
- The fan is a squirrel cage type.
- The pump can be single or dual stage. Single stage or dual stage is used when the fuel supply is above the burner. Dual stage is used when the fuel supply is below the burner, and the fuel must be lifted to the pump. Pumps have pressure regulating valves that must be set at 100 psi.
- The nozzle atomizes, meters, and patterns the fuel oil.
- Air is blown through the air tube into the combustion chamber and mixed with the atomized fuel oil.
- The electrodes provide the spark for the ignition of the fuel oil and air mixture.
- The transformer provides the high voltage, producing the spark across the electrodes.
- The primary controls provide the means for operating the furnace and the means for shutting down the furnace if there is no ignition.
- Fuel storage and supply systems can be one-pipe or two-pipe design. A two-pipe system is preferred in all installations. A one-pipe system may be used in installations where the tank is above the burner. A two-pipe system is used in all installations where the tank is below the burner.
- An auxiliary or booster supply system must be used where the burner is more than 15 ft above the storage tank.
- The atomized oil and air mixture is ignited in the combustion chamber. Modern chambers are often built of or lined with a refractory material.
- The heat exchanger takes heat caused by the combustion and transfers it to the air that is circulated to heat the building.
- The performance of the pump can be checked with a vacuum gage and a pressure gage.
- One-pipe systems must be bled before the burner is started for the first time or whenever the fuel supply lines have been opened.
- To convert a pump from a one-pipe to a two-pipe system, insert a bypass plug in the return or inlet port. The manufacturer's specifications must be checked to determine this procedure.
- *Nozzles should not be cleaned or unplugged. They should be replaced.*
- Electrodes should be clean, connections should all be secure, and the spark gap should be adjusted accurately.
- Combustion efficiency tests should be made when servicing oil burners, and corrective action taken to obtain maximum efficiency.

REVIEW QUESTIONS

1. How many grades of fuel oil are normally considered as heating oils?
2. What grade is most commonly used in heating residences and light commercial buildings?
3. Describe how the fuel oil is prepared for combustion.
4. What products would be produced if pure combustion took place?
5. List other products that may be produced in the burning process in an oil burner.
6. What is the standard pressure under which oil is fed to the nozzle of an oil burner?
7. List the main components of a gun-type oil burner.
8. Describe the two types of pumps used in gun-type oil burners.
9. Describe the three functions of the nozzle.
10. Sketch and label the three basic spray patterns.
11. When should air adjustments be made?
12. What is the purpose of the electrodes?
13. What is the purpose of the ignition transformer?
14. What is the normal output voltage of the transformer?
15. What are the two functions of the primary control unit?
16. Where is the stack switch located? How does it operate?
17. Where is the cad cell located? How does it operate?
18. When can a one-pipe supply system be used?
19. What is the disadvantage in a one-pipe supply system?
20. When are two-pipe systems used?
21. Describe an auxiliary fuel supply system.
22. When are auxiliary fuel supply systems used?
23. What will happen in the combustion process if the combustion chamber is not designed properly and the flame hits the wall before it is fully ignited?
24. Describe the purpose of the heat exchanger.
25. What are the two gages used when servicing oil burner pumps?
26. What should the vacuum gage reading be with the pump in operation in a gravity fed system?
27. List the conditions considered in determining the vacuum gage reading if the supply tank is located below the burner.
28. List possible problems if the pressure gage installed on the pump pulsates.
29. List five possible problems that may relate to a defective nozzle.
30. Describe the purpose of the flame retention ring.
31. Sketch a wiring diagram of a primary control circuit using a cad cell.
32. Sketch a wiring diagram for an air-handling system.
33. List four tests to determine oil-burner efficiency.
34. List two corrective actions that might be taken as a result of readings on each of the four tests in Question 33.

HYDRONIC HEAT

OBJECTIVES

After studying this unit, you should be able to

- describe a basic hydronic heating system.
- describe reasons for a hydronic heating system to have more than one zone.
- list the three heat sources commonly used in hydronic systems.
- state the reason a boiler is constructed in sections or tubes.
- discuss the purposes for eliminating air from the system.
- describe the purpose of limit controls and low-water cutoff devices.
- state the purpose of a pressure relief valve.
- describe the two purposes of an air cushion or expansion tank.
- state the purpose of a zone control valve.
- explain centrifugal force as it applies to hydronic circulating pumps.
- sketch a finned-tube baseboard unit.
- sketch a one-pipe hydronic heating system.
- describe a two-pipe reverse return system.
- describe a tankless domestic hot-water heater used with a hydronic space-heating system.

33.1 INTRODUCTION TO HYDRONIC HEATING

Hydronic heating systems are systems where water or steam carries the heat through pipes to the areas to be heated. This text will cover only hot-water systems since they are generally used in residential and light commercial installations.

In hydronic systems water is heated in a boiler and circulated through pipes to a heat transfer component called a *terminal unit,* such as a radiator or finned-tube baseboard unit. Here heat is given off to the air in the room. The water is then returned to the boiler. There is no forced moving air in these systems in most residential installations, and if properly installed there will be no hot or cold spots in the conditioned space. Hot water stays in the tubing and heating units even when the boiler is not running, so there are no sensations of rapid cooling or heating, which might occur with forced-warm-air systems.

These systems are designed to include more than one zone when necessary. If the system is heating one small home or area, it may have one zone. If the house is a long ranch type or multilevel, there may be several zones. Separate zones are often installed in bedrooms so that these temperatures can be kept lower than those in the rest of the house, Figure 33–1.

The water is heated at the boiler, using an oil, gas, or electric heat source. These burners or heating elements are very similar to those discussed in other units in this section of the book. Sensing elements in the boiler start and stop the heat source according to the boiler temperature. The water is circulated with a centrifugal pump. A thermostatically controlled zone control valve allows the heated water into the zone

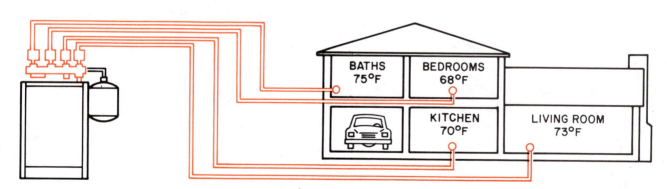

Figure 33–1. Four-zone hydronic heating system.

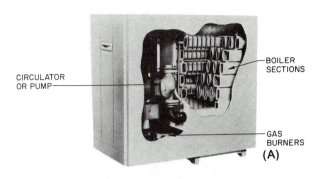

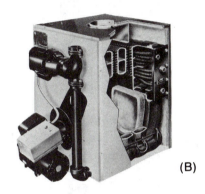

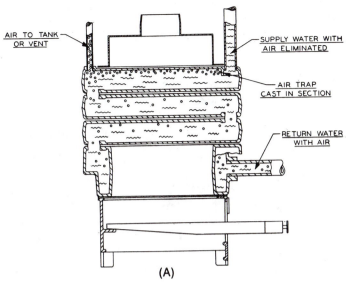

Figure 33-2. (A) A gas-heated hydronic boiler. (B) An oil-heated hydronic boiler. *Courtesy (A) The Peerless Heater Company (B) Burnham Corporation*

TOP OUTLET BOILER

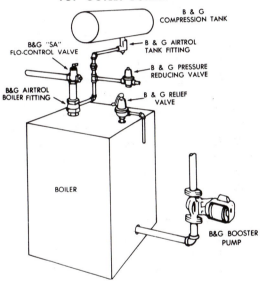

Hot water heating boilers having large internal passages are often excellent low velocity points for air separation. The dip tube which should be pushed down into the boiler (always install ABF with a short nipple) permits only bubble free water from a point well below the top of the boiler to circulate out to the system.

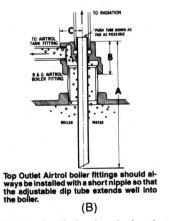

Top Outlet Airtrol boiler fittings should always be installed with a short nipple so that the adjustable dip tube extends well into the boiler.

(B)

Figure 33-3. An electrically heated boiler. *Courtesy Burnham Corporation*

Figure 33-4. (A) An air-eliminating device that traps the air at the top, venting it to the expansion tank. (B) Another type air eliminator. *Courtesy (A) The Peerless Heater Company (B) ITT Fluid Handling Division*

needing the heat. Most residential installations use finned-tube baseboard units to transfer heat from the water to the air.

The design or sizing of the system will not be covered in detail in this unit, but it is necessary that the boiler, piping, and terminal units be the proper size. All components including the pump and valves should also be sized properly for the correct water flow.

33.2 BOILER

A *boiler* is, in its simplest form, a furnace that heats water, using oil, gas, or electricity as the heat source. Some larger commercial boilers use a combination of two fuels. When one fuel is more available than another, it can be used, or the boiler can be easily changed over or converted to use the other fuel. The part of the boiler containing the water is usually constructed in sections or in tubes when oil or gas heat sources are used, Figure 33–2. This provides more surface area for the burner to heat and is more efficient than heating a tank of water. See Figure 33–3 for an example of an electric boiler.

It is important to eliminate air from the water in a boiler for several reasons. Air in water at normal atmospheric temperatures causes a significant amount of corrosion. As the temperature of the water is increased, the corrosive factor of this air in the water is increased many times. Air pockets can also form in the system, causing a blockage of the water circulation. Air in the system can also cause undesirable noise. Many boilers are designed to speed up elimination of air by trapping it and forcing it to be vented into the outside air or into an expansion tank, Figure 33–4. Once the system has been operating and most air has been eliminated, the only new air to be concerned with is that entering with the makeup water. This is usually a small amount of water and a small amount of air that with proper venting will not be a problem. Figure 33–5 is a photo of a manual type and an automatic air vent. These vents are placed on top of the boiler, at high points in the piping system, and at the finned-tube baseboard units.

Boilers are also equipped with several safety controls:

33.3 LIMIT CONTROL

A *limit control* in a hot-water boiler shuts down the heating source if the water temperature gets too high, Figure 33–6.

33.4 WATER REGULATING VALVE

Water heating systems should have an automatic method of adding water back to the system if water is lost due to leaks. City water pressure is too great for the system. A *water regulating valve* is installed in the water makeup line leading to the boiler and is set to maintain the pressure on its leaving side (entering the boiler) less than the relief valve on the boiler.

Figure 33–5. (A) Manual air vent. (B) Automatic air vent. *Courtesy ITT Fluid Handling Division*

Figure 33–6. High limit control. This is set so that under normal conditions the boiler shuts down before the relief valve opens. This avoids the problem of water being released from the furnace under pressure causing damage.

Figure 33–7. An automatic water regulating valve with a system pressure relief valve. *Courtesy ITT Fluid Handling Division*

A system pressure relief valve is sometimes coupled with this water regulating valve to assure that the system pressure does not exceed the boiler's pressure relief setting, Figure 33–7.

33.5 PRESSURE RELIEF VALVE

American Society of Mechanical Engineers (ASME) Boiler and Pressure Vessel Codes require that each hot-water heating boiler have at least one officially rated *pressure relief valve* set to relieve at or below the maximum allowable working pressure of the boiler. This valve discharges excessive water when pressure is created by expansion. It also releases excessive steam pressure if there is a runaway overfiring emergency, Figure 33–8.

33.6 AIR CUSHION TANK OR EXPANSION TANK

When water is heated, it expands. A hot-water heating system operates with all components and piping full of water. When water is heated from its source temperature to over 200°F, it will expand considerably, and some provision must be made for this expansion. Thus an *air cushion tank* or *expansion tank* is used, Figure 33–9. This is an airtight tank located above the boiler that provides space for air initially trapped in the system and for the expanded water when it is heated. This tank should provide the only air space within the system. Often when initially filling the system too much air is trapped in the tank. Figure 33–10 illustrates a system using a vent tube through which air can be released one time. An expansion tank may have a flexible diaphragm to keep the air and water separated.

33.7 ZONE CONTROL VALVES

Zone control valves are thermostatically controlled. Many are heat-motor operated, Figure 33–11. When the thermostat calls for heat, a resistance wire around

Figure 33–8. Safety relief valve. *Courtesy ITT Fluid Handling Division*

Figure 33–9. Air cushion or expansion tank. *Courtesy ITT Fluid Handling Division*

the valve heats and causes the valve's bimetal element to expand and open slowly. Other types are electric-motor operated, which allows slow opening and closing of the valve. Slow opening and closing reduces expansion noise and prevents water hammer.

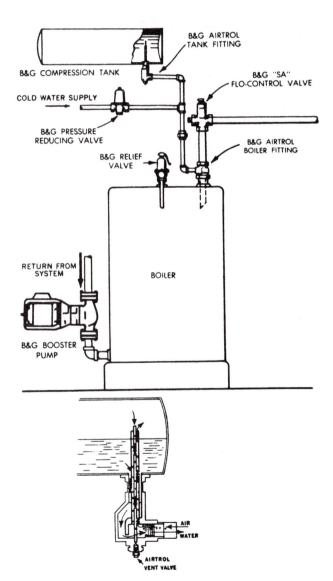

Figure 33–10. A system with a vent tube that will release air one time after initially starting up the boiler. This air is vented out the valve at the bottom. *Courtesy ITT Fluid Handling Division*

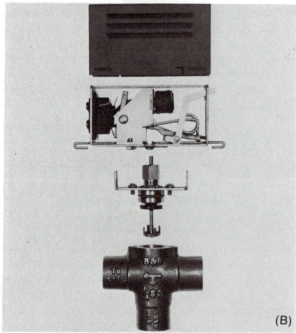

Figure 33-11. (A) A heat motor zone control valve. (B) Exploded view. *Courtesy ITT Fluid Handling Division*

33.8 CENTRIFUGAL PUMPS

Centrifugal pumps, also called *circulators,* force the hot water from the boiler through the piping to the heat transfer units and back to the boiler. These pumps use centrifugal force to circulate the water through the system. Centrifugal force is generated whenever an object is rotated around a central axis. The object or matter being rotated tends to fly away from the center due to its velocity. This force increases proportionately with the speed of the rotation, Figure 33-12.

The *impeller* is that part of the pump that spins and forces water through the system. The proper direction of rotation of the impeller is essential. The vanes or blades in the impeller must "slap" and then throw the water, Figure 33-13. Impellers used on circulating pumps in hot-water heating systems usually have sides

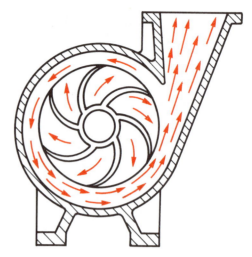

Figure 33-12. When the impeller of the pump is rotated it "throws" the water away from the center of the rotation and out through the opening.

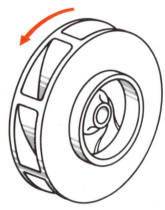

Figure 33-13 Pump impeller. Note the direction of the rotation indicated by the arrow. This is an example of a closed impeller.

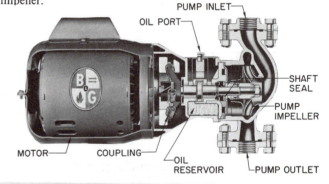

Figure 33-14. Centrifugal pump. *Courtesy of ITT Fluid Handling Division*

enclosing the vanes. Such types are called *closed* impellers, Figure 33–13. Many pumps, Figure 33–14, used in closed systems where some makeup water is used (which will cause corrosion) are called *bronze-fitted* pumps. They generally have a cast iron body with the impeller and other moving parts made of bronze or nonferrous metals. Others have stainless steel or all bronze parts.

33.9 FINNED-TUBE BASEBOARD UNITS

Most residences with modern hydronic heating systems use finned-tube baseboard heat transfer for terminal heating units, Figure 33–15. Air enters the bottom of these units and passes over the hot fins. Heat is given off to the cooler air, causing it to rise by convection. The heated air leaves the unit through the damper area. The damper can be adjusted to regulate the heat flow. Baseboard units are generally available in lengths from 2 ft to 8 ft and are relatively easy to install following the manufacturer's instructions. Figure 33–16 shows a radiator and fan coil terminal units.

When installing baseboard terminal units, provide for expansion. The hot water passing through the unit makes it expand. Expansion occurs toward both ends of the unit so the ends should not be restricted. It is a good practice to install expansion joints in longer units, Figure 33–17A. If one expansion joint is used,

place it in the center. If two are used, space them evenly at intervals one third of the length of the unit. When piping drops below the floor level, the vertical risers should each be at least 1 ft long. The table in Figure 33–17B shows the diameter of the holes needed through the floor for the various pipe sizes. The table in Figure 33–18 lists the maximum lengths that can be installed if both ends of the baseboard unit are free to expand when the water temperature is raised from 70°F to the temperature shown.

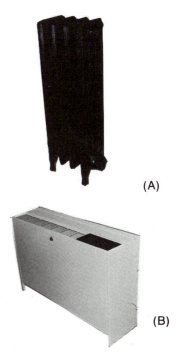

(A)

(B)

Figure 33–16. Terminal hydronic heating units. (A) Water in the radiator radiates heat out into the room. (B) The fan coil units are similar to the baseboard units but have a fan which blows the hot air into the room. *Photos by Bill Johnson*

Figure 33–17A. An expansion joint. *Courtesy Edwards Engineering Corporation*

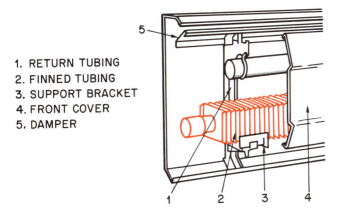

1. RETURN TUBING
2. FINNED TUBING
3. SUPPORT BRACKET
4. FRONT COVER
5. DAMPER

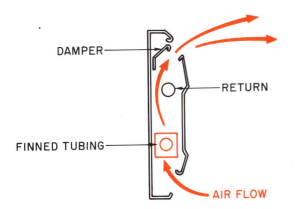

Figure 33–15. Two-pipe finned-tube baseboard unit.

TUBE SIZE		RECOMMENDED MINIMUM HOLE (INCHES)
NOMINAL (INCHES)	O.D. (INCHES)	
$\frac{1}{2}$	$\frac{5}{8}$	1
$\frac{3}{4}$	$\frac{7}{8}$	$1\frac{1}{4}$

Figure 33–17B. This table indicates the diameter of the holes required for each pipe size to allow for expansion.

AVERAGE WATER TEMPERATURE °F	MAXIMUM LENGTH OF STRAIGHT RUN (FEET)
220	26
210	28
200	30
190	33
180	35
170	39
160	42
150	47

Figure 33–18. Maximum lengths of baseboard unit that can be installed when the temperature is raised from 70°F to the temperature indicated in the table.

Figure 33–19. Balancing valve. *Courtesy ITT Fluid Handling Division*

33.10 BALANCING VALVES

When designing a hot water heating system, consideration must be given to the flow rate of the water and the friction in the system. Friction is caused by the resistance of the water flowing through the piping, valves, and fittings in the system. The flow rate is the number of gallons of water flowing each minute through the system. A system is considered to be in balance when the resistance to the water flow is the same in each flow path. A means for balancing the system should be provided in all installations. One method is to install a *balancing valve,* Figure 33–19, on the return side of each heating unit. This valve is adjustable. *The valve adjustment is generally determined by the system designer, and the installer sets them accordingly. Be sure to read the manufacturer's instructions for the correct procedure to set the valve.*

33.11 FLOW CONTROL VALVES

Flow control valves or *check valves* are necessary so that the water will not flow by gravity when the circulator is not operating, Figure 33–20.

33.12 HORIZONTAL AND VERTICAL FORCED-AIR-DISCHARGE UNIT HEATERS

Horizontal and vertical forced-air-discharge unit heaters (used generally in commercial and industrial applications) have fans to blow the air across the heat transfer elements. The heating elements are normally made of heavy copper tubing through which the hot

water is circulated. The tubing is surrounded with aluminum fins to more efficiently give off heat to the air blown across them, Figure 33–21.

33.13 HYDRONIC HEATING PIPING SYSTEMS

Most residential systems use a *series loop* or a *one pipe system* layout for each zone. Figure 33–22 illustrates a series loop system. In this system, all of the hot water flows through all the heating units. Neither the temperature nor the amount of flow can be varied from one unit to the next without affecting the entire system. The water temperature in the last unit will be lower than in the first since heat is given off as the water moves through the system. This system is simple and economical but without much flexibility.

(A)

(B)

Figure 33–20. Flow control valves. *Courtesy ITT Fluid Handling Division*

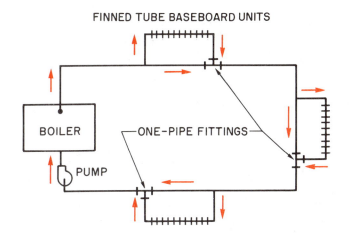

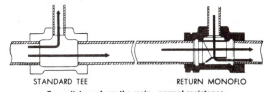

Figure 33–23. One-pipe hydronic heating system.

HOW TO USE B & G MONOFLO FITTINGS

- Be sure the RING is between the risers OR
- Be sure the SUPPLY arrow on the supply riser and the RETURN arrow on the return riser point in the direction of flow.

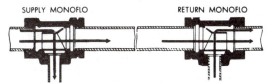

STANDARD TEE   RETURN MONOFLO
For radiators above the main—normal resistance
For most installations where radiators are above the main, only one Monoflo Fitting need be used for each radiator.

SUPPLY MONOFLO   RETURN MONOFLO
For radiators above the main—high resistance
Where characteristics of the installation are such that, resistance to circulation is high, two Fittings will supply the diversion capacity necessary.

SUPPLY MONOFLO   RETURN MONOFLO
For radiators below the main
Radiators below the main require the use of both a Supply and Return Monoflo Fitting, except on a ¾" main use a single return fitting.

Figure 33–24. A tee for one-pipe systems. Partial water flow is diverted through the tee to the terminal unit. *Courtesy ITT Fluid Handling Division*

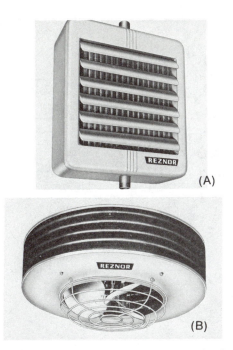

Figure 33–21. (A) Horizontal and (B) vertical air discharge unit heaters. *Courtesy of ITT Reznor Division*

A basic one-pipe system is illustrated in Figure 33–23. It has a one-pipe main supply with branches to each of the heating units. The piping in the branches is smaller in diameter than the main pipe. One of the tees to each of the units is a special tee called a *one-pipe fitting,* Figure 33–24. This fitting allows some of the water in the line to go through the heating unit and the rest to continue in the main supply line.

The *two-pipe reverse return system,* Figure 33–25, has two pipes running parallel with each other, and the water flowing in the same direction in each pipe. Notice that the water going into the first heating unit is the last to be returned to the boiler. The water going into the last heating unit is the first to be returned to the boiler. This system equalizes the distance of the water

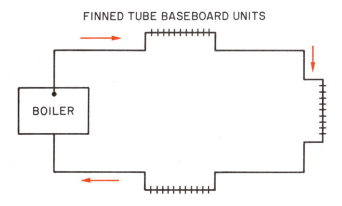

Figure 33–22. Series loop hydronic heating system.

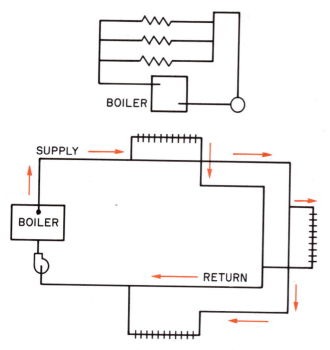

Figure 33-25. Two-pipe reverse return system.

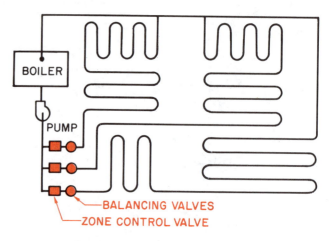

Figure 33-27. A radiant panel system.

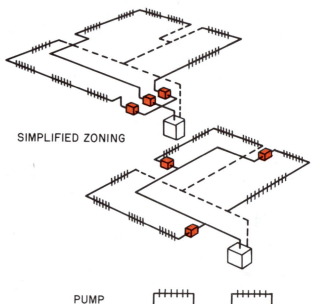

SIMPLIFIED ZONING

flow in the system. It is used in light commercial installations but seldom in residences.

In the *two-pipe direct return system*, Figure 33-26, the water flowing through the nearest heating unit to the boiler is the first back to the boiler. It therefore has the shortest run. The water flowing through the unit farthest away has the longest run. It is much more

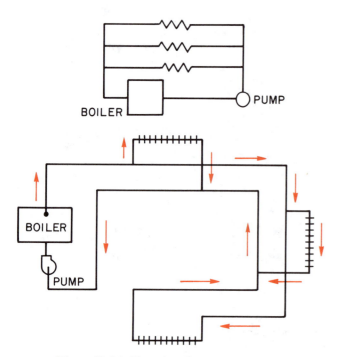

Figure 33-26. Two-pipe direct return system.

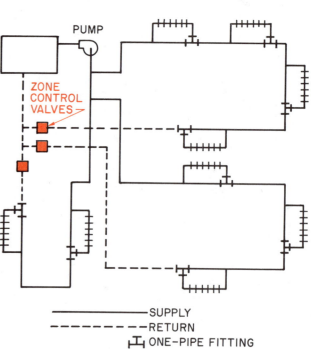

Figure 33-28. Multizone installation.

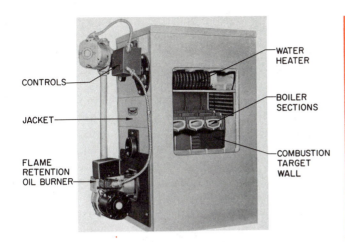

Figure 33–29. Tankless domestic water heater installed in a boiler. *Courtesy The Peerless Heater Company*

difficult to balance this type of system, and it is seldom used in residential applications.

Hot-water heating systems installed in floors and ceilings are known as *panel systems.* They are normally designed with each coil as an individual zone or circuit and are each balanced individually, usually at a place where they come together, Figure 33–27.

Most installations will be multizone applications, Figure 33–27. Each zone is treated as a separate system, Figure 33–28. The zone control valve controls water circulated through that zone. Each unit is balanced at the terminal unit.

33.14 TANKLESS DOMESTIC HOT-WATER HEATERS

Most hot-water heating furnaces (oil and gas fired) can be furnished with a domestic hot-water heater consisting of a coil inserted into the boiler. The domestic hot-water is contained within the coil and heated by the boiler quickly, which eliminates the need for a storage tank. This system can meet most domestic hot water needs without using a separate tank, Figure 33–29. This is generally a very efficient way to produce hot water. This heater is installed with a flow control valve to allow only the proper amount of cold water into the system to be heated.

33.15 SERVICE TECHNICIAN CALLS

Service Call 1

A customer calls. *There is no heat at a motel. The problem is on the top two floors of a four-story building. The system has just been started for the first time this season, and it has air in the circulating hot water. The automatic vent valves have rust and scale in them due to lack of water treatment and proper maintenance.*

The technician arrives and goes to the boiler room. The boiler is hot, so the heat source is working. The water pump is running, because the technician can hear water circulating. The technician decides to go to the top floor to one of the room units and removes the cover; the coil is at room temperature. The technician listens to the coil very closely and can hear nothing circulating. There may be air in the system and the bleed ports should be located. The technician knows they must be at the high point in the system and normally would be on the top of the water risers from the basement.

The technician goes to the basement where the water lines start up through the building and finds a reference point. The technician can then go to the top floor and find the top portions of these pipes. The pipes rise through the building next to the elevator shaft. The technician goes to the top floor to the approximate spot and finds a service panel in the hall in the ceiling. Using a ladder, the automatic bleed port for the water supply is found over the panel. This is the pump discharge line. See Figure 33–5 for an example of an automatic bleed port.

The technician removes the rubber line from the top of the automatic vent. The rubber line carries any water that bleeds to a drain. There is a valve stem in the automatic vent (as in an automobile tire). The technician presses the stem, but nothing escapes. There is a hand valve under the automatic vent, so the valve is closed and the automatic vent is removed. The technician then carefully opens the hand valve and air begins to flow out. Air is allowed to bleed out until water starts to run. The water is very dirty, much like dirty water in a drainage ditch. *Care should be used around hot water.*

The technician then bleeds the return pipe in the same manner until all the air is out, and water only is at the top of both pipes. The technician then goes to the heating coil where the cover was previously removed. The coil now has hot water circulating.

The technician then takes the motel manager to the basement and drains some water from the system and shows the manager the dirty water. The water should be clear or slightly colored with water treatment. The technician suggests that the manager call a water treatment company that specializes in boiler water treatment, or the system will have some major troubles in the future.

The technician then replaces the two automatic vents with new ones and opens the hand valves so that they can operate correctly. The technician tells the manager again that the system needs help from a water treatment expert and then leaves.

Service Call 2

A customer in a small building calls. *There is no heat in one portion of the building. The system is heating well in most of the building. The building has four hot-water circulating pumps, and one of them is locked up—the bearings are seized. These pumps are each 230-V, three-phase, and 2 hp.*

The technician arrives at the building and consults the customer. They go to the part of the building where there is no heat. The thermostat is calling for heat. They go to the basement where the boiler and pumps are located. They can tell from examination that the number 3 pump, which serves the part of the building that is cool, is not turning. The technician carefully touches the pump motor; it is cool and has not been trying to run. Either the motor or the pump has problems.

The technician chooses to check the voltage to the motor first, then goes to the disconnect switch on the wall and measures 230 V across phases 1 to 2, 2 to 3, and 1 to 3. This is a 230-V three-phase motor with a motor starter. The motor's overload protection is in the starter, and it is tripped.

The technician still does not know if the motor or the pump has problems. *The power is shut off with the disconnect and a resistance test is performed on the motor.* All three windings have equal resistance, and there is no ground circuit in the motor. The power is turned on, an ammeter is clamped on one of the motor leads, and the overload reset is pressed. The technician can hear the motor try to start, and it is pulling locked-rotor current. The disconnect is pulled to stop the power from continuing to the motor. It is evident that either the pump or the motor is stuck.

The technician returns to the pump. It is on the floor next to the others. The guard over the pump shaft coupling is removed. The technician tries to turn the pump over by hand—remember, the power is off. The pump is very hard to turn. The question now is whether it is the pump or the motor that is tight.

The technician disassembles the pump coupling and tries to turn the motor by hand. It is free. The technician then tries the pump; it is too tight. The pump bearings must be defective. The technician gets permission from the owner to disassemble the pump. The technician shuts off the valves at the pump inlet and outlet, removes the bolts around the pump housing, then removes the impeller and housing from the main pump body. The technician takes the pump impeller housing and bearings to the shop and replaces them, installs a new shaft seal, and returns to the job.

While the technician is assembling the pump, the customer asks why the whole pump was not taken instead of just the impeller and housing assembly. The technician shows the customer where the pump housing is fastened to the pump with bolts and dowl pins. The relationship of the pump shaft and motor shaft have been maintained by not removing the complete pump. They will not have to be aligned after they are reassembled. The pump is manufactured to be rebuilt in place to avoid this.

After completing the pump assembly, the technician turns it over by hand, assembles the pump coupling, and affirms that all fasteners are tight. Then the pump is started. It runs and the amperage is normal. The technician turns off the power and assembles the pump coupling guard and leaves.

Service Call 3

A customer calls. *One of the pumps beside the boiler is making a noise. This is a large home with three pumps, one for each zone. The pump coupling is defective. It is a spring type coupling and is wearing out fast.*

The technician arrives and goes to the basement, hears the coupling, and notices there are metal filings. See Figure 33–14 for an example of this kind of coupling. *The technician shuts off the pump, replaces the coupling, and restarts the pump.* It now sounds good. The technician writes on the report that this pump is several years old and that if another coupling fails, the pump cradle mount (rubber) or neoprene may need replacing. If the rubber or neoprene is soft, misalignment will occur and coupling failure will result. The technician lubricates the motors and pump bearings before leaving the job.

Service Call 4

A customer calls from an apartment house. There is no heat in one of the apartments. This apartment house has 25 fan coil units with a zone control valve on each and a central boiler. One of the zone control valves is defective.

The technician arrives and goes to the apartment with no heat. The room thermostat is set above the room setting. The technician goes to the fan coil unit located in the hallway ceiling and opens the fan coil compartment door. It drops down on hinges. The technician stands on a short ladder to reach the controls. There is no heat in the coil, and there is no water flowing.

This unit has a zone control valve with a small heat motor. The technician checks for power to the valve's coil; it has 115 V, as it should. See Figure 33–11 for an example of this type of valve. This valve has a manual open feature, so the valve is opened by hand. The technician can hear the water start to flow and can feel the coil get hot. The valve's heat motor or valve assembly must be changed. The technician chooses to change the assembly because the valve and its valve seat are old.

The technician obtains a valve assembly from the stock of parts at the apartment and proceeds to change it. *The technician shuts off the power and the water to the valve and water coil. A plastic drop cloth is then spread to catch any water that may fall. Then a bucket is placed under the valve before the top assembly is removed.* The old top is removed, and the new top installed. The electrical connections are made, and the valves are opened.

The technician turns on the power and can see the valve begin to move after a few seconds (remember this is a heat-motor valve, and it responds slowly). The technician can now feel heat in the coil. The compartment door is closed. The plastic drop cloth is removed, the room thermostat is set, and the technician leaves.

Service Call 5

A homeowner calls. There is water in the floor around the boiler in the basement. The relief valve is relieving because the gas valve is not shutting off tight and is overheating the boiler.

The technician goes to the basement and sees the boiler relief valve seeping. The boiler gage reads 15 psig, which is the rating of the relief valve. The system normally operates at 12 psig. The technician looks at the flame in the burner section and sees the burner burning at a low fire. A voltmeter check shows there is no voltage to the gas valve's operating coil. This means the valve is stuck partially open.

The technician shuts off the gas and the power and replaces the valve. When this is accomplished, power and gas are restored. The boiler control is set higher, the gas valve opens, and the burner lights. When the technician resets the boiler control to normal, the burner goes out. The relief valve has stopped leaking, so the technician leaves the job.

Care should be taken when working with hot water systems to ensure that all controls function normally. An overheated boiler has great explosion potential.

SUMMARY

- Hydronic heating systems use hot water or steam to carry heat to the areas to be heated.
- Water is heated by gas, oil, or electricity and circulated through pipes to terminal units where the heat is exchanged to the air of the conditioned space.
- A boiler is a furnace that heats water. In gas and oil systems the part of the boiler containing the water is constructed in sections or tubes.
- Air should be eliminated from the water to prevent corrosion, noise, and water-flow blockage.
- A limit control is used to shut the heat off if the water temperature gets too high.
- A low-water cutoff shuts down the heat source if the water level gets too low. Some of these devices have an automatic water feeder that allows water to flow into the system. If the proper water level still cannot be maintained, the low-water cutoff activates a switch that cuts off electrical power to the heat source.
- A pressure relief valve is required to discharge water when excessive pressure is created by expansion due to overheating.
- An air cushion tank or expansion tank is necessary to provide space for trapped air and to allow for expansion of the heated water.
- Zone control valves are thermostatically operated and control the flow of hot water into individual zones.

- A centrifugal pump circulates the water through the system.
- Finned-tube baseboard terminal units are commonly used in residential hydronic heating systems.
- A balancing valve is used to equalize the flow rate throughout the system.
- One-pipe systems and two-pipe reverse return piping systems are most commonly used in residential and light commercial installations.
- A one-pipe system requires a specially designed tee to divert some water to the baseboard unit while allowing the rest to flow through the main pipe.
- Tankless domestic hot-water heaters are commonly used with hot-water heating systems. The water is contained within a coil and heated by the boiler.

REVIEW QUESTIONS

1. Describe a basic hydronic heating system.
2. Why do hot-water heating systems often have more than one zone?
3. List the three heat sources commonly used with hydronic systems.
4. Why are gas- or oil-heated boilers constructed in sections or with tubes?
5. List three reasons for eliminating air from the system.
6. Where is this air vented?
7. What is the purpose of a limit control on a boiler?
8. Describe the purpose of a low-water cutoff used with an automatic water feeder.
9. What is the purpose of a pressure relief valve?
10. What is the purpose of an air cushion tank or expansion tank?
11. Describe two types of zone control valves.
12. What is the purpose of zone control valves?
13. Describe centrifugal force as it applies to hydronic circulating pumps.
14. What is the purpose of the impeller on a circulating pump?
15. Make a sketch of an impeller and show the direction of rotation.
16. What is the purpose of the fins on a finned-tube baseboard unit?
17. Make a sketch of a finned-tube baseboard unit and show the circulation of the air.
18. Why must expansion be considered when installing baseboard units?
19. Why are balancing valves necessary?
20. Sketch a one-pipe hydronic heating system.
21. What is the purpose of the one-pipe fitting (tee) used in these systems?
22. How does a two-pipe reverse return system operate?
23. Describe a radiant panel system.
24. How does a tankless domestic hot-water system operate?

SOLAR HEAT

OBJECTIVES

After studying this unit, you should be able to

- **describe the difference between passive and active solar systems.**
- **discuss direct and diffuse radiation.**
- **describe the declination angle and the effect it has on the sun's radiation during winter and summer.**
- **explain the construction of air and liquid solar collectors.**
- **list the typical components in an air-based solar system and describe the function of each.**
- **list the typical components in a liquid based system and describe the function of each.**
- **describe the difference between sensible-heat storage and latent-heat storage.**
- **describe an air-based rock storage bed.**
- **list the types of liquid-based storage tanks and describe advantages and disadvantages of each.**
- **describe the operation of various solar space-heating systems.**

34.1   STORED SOLAR ENERGY

Stored solar energy is the energy from fossil fuels such as coal, gas, and oil. Fossil fuels have formed over thousands of years from decayed plants and animals, which, when alive depended on the sun for life. This supply is being rapidly used and becoming increasingly expensive.

34.2   DIRECT SOLAR ENERGY

The sun furnishes the earth tremendous amounts of direct energy each day. It is estimated that two weeks of the sun's energy reaching the earth are equal to all of the known deposits of coal, gas, and oil. The challenge facing scientists, engineers, and technicians is to better utilize this energy. We know the sun heats the earth, which heats the air immediately above the earth. One of the challenges is to learn how to collect this heat, store it, and distribute it to provide heat and hot water for homes and business.

Many advances have been made, but there are still those who feel that the design and installation of solar systems has progressed very slowly. Many others feel that rapid progress has been made compared to the progress in the design of other technologies over the years. It is assumed, though, that as the resources are further depleted and as economics or political actions cause energy crises, it will be only a matter of time before direct energy of the sun is used extensively.

34.3   PASSIVE SOLAR DESIGN

Many structures being constructed presently are using *passive solar* designs. These designs utilize non-moving parts of a building or structure to provide heat or cooling, or they eliminate certain parts of a building that help cause inefficient heating or cooling. Some examples follow.

- Place more windows on the east, south, or west sides of homes and fewer on the north side. This allows warming from the morning sun in the east and from the sun throughout the rest of the day from the south and west. By eliminating windows on the north side, this coldest side of the house can be better insulated.
- Place greenhouses, usually on the south side, to collect heat from the sun to help heat the house, Figure 34–1.

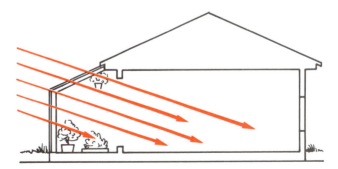

Figure 34–1. A greenhouse will allow the sun to shine into the house. It may also have a masonry floor or barrels of water may be located in the greenhouse to help store the heat until evening when the sun goes down. This heat then may be circulated throughout the house.

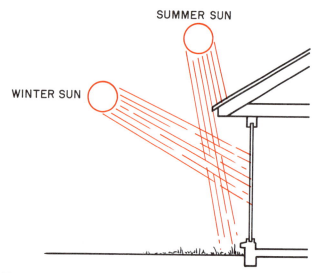

Figure 34–2. An overhang may be constructed to allow the sun to shine into the house in the winter when it is lower in the sky and yet shade the window in the summer.

- Design roof overhangs to shade windows from the sun in the summer but to allow sun to shine through the windows in the winter, Figure 34–2.
- Provide a large mass such as a concrete or brick wall to absorb heat from the sun and temper the inside environment naturally.
- Place latent-heat storage tubes containing phase-change materials in windows where they can collect heat to be released at a later time, Figure 34–3.

34.4 ACTIVE SOLAR SYSTEMS

This unit will be limited to *active solar systems*. These systems use electrical and/or mechanical devices to help collect, store, and distribute the sun's energy. The components to collect and store the heat from the sun are relatively new. The distribution of this heat to the conditioned space is by means of the same type of equipment used in fossil-fuel furnaces.

Figure 34–3. Latent-heat storage tubes may be placed in windows where heat is stored when the sun is shining. *Courtesy Calortherm Associates*

34.5 DIRECT AND DIFFUSE RADIATION

The sun is a star often called the "daystar." A very small amount of the sun's energy reaches the earth. Much of the energy that does reach the earth's atmosphere is reflected into space or absorbed by moisture

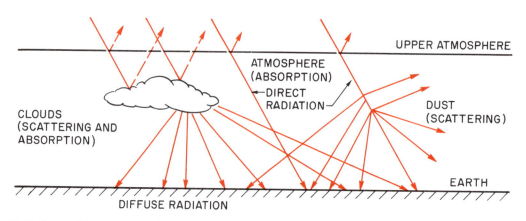

Figure 34–4. Radiation striking the earth directly from the sun is considered direct radiation. If it reaches the earth after it has been deflected by clouds or dust particles it is called diffuse radiation.

and pollutants before reaching the earth. The energy reaching the earth directly is *direct* radiation. The reflected or scattered energy is *diffuse* radiation, Figure 34–4.

34.6 SOLAR CONSTANT

The rate of solar energy reaching the outer limits of the earth's atmosphere is the same at all times. It has been determined that the radiation from the sun at these outer limits produces 429 $Btu/ft^2/hr$ on a surface perpendicular (90°) to the direction of the sun's rays. This is known as the *solar constant*. The energy from the sun is often called *insolation*.

34.7 DECLINATION ANGLE

The earth revolves once each day around an axis that passes through the north and south poles. This axis is tilted, so the intensity of the sun's energy reaching the northern and southern atmospheres varies as the earth orbits around the sun. This tilt or angle is called the *declination angle* and is responsible for differences during the year in the distribution or the intensity of the solar radiation, Figure 34–5.

The amount of radiation reaching the earth also varies according to the distance it travels through the atmosphere. The shortest distance is when the sun is perpendicular (90°) to a particular surface. This is when the greatest energy reaches that section of the earth, Figure 34–6. The angle of the sun's rays with regard to a particular place on the earth plays an important part in the collection of the sun's energy.

34.8 SOLAR COLLECTORS

Solar collectors are used in active solar systems to collect this energy from the sun. Most solar collectors are flat and are positioned and tilted to catch as many

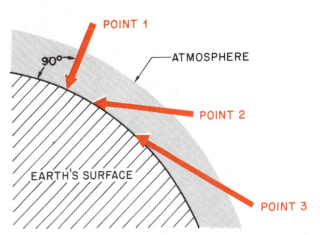

Figure 34–6. Radiation from the sun will be warmer at point 1 as it travels the shortest distance through the atmosphere.

of the sun's rays as close to 90° as possible in winter, Figure 34–7. For the greatest efficiency these collectors could be mounted on a device that follows the sun's path throughout the day and throughout the season. At the present time this type of device is too expensive, so the positioning of the collectors is done to collect the most energy possible. These collectors will collect direct radiation and, to a lesser extent, diffuse radiation. This means that some heat can be generated on hazy days and, to some extent, through clouds.

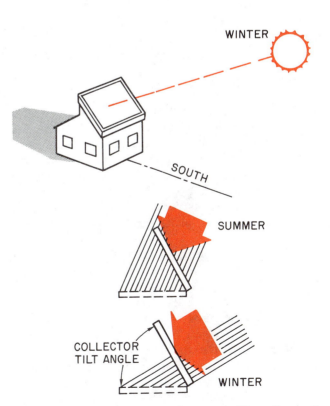

Figure 34–7. Collectors should face south. The collector tilt angle should be so that the sun's rays are 90° to the collector in the winter when the heat is needed the most.

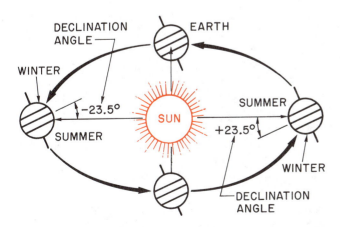

Figure 34–5. The angle of declination (23.5°) tilts the earth so that the angle from the sun north of the equator is greater in the winter. The sun's rays are not as direct, and the normal temperatures are colder than during the summer.

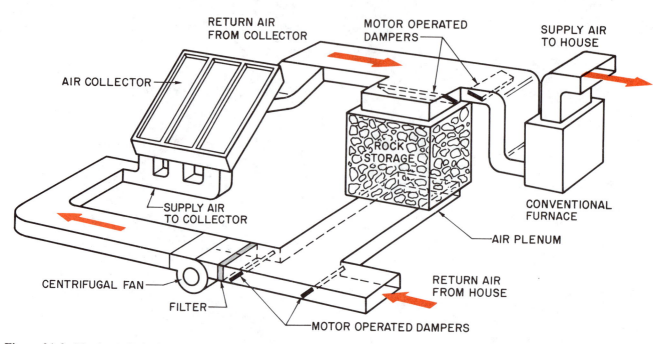

Figure 34–8. The heated air from the collector is blown through the duct to the conventional furnace where it is distributed throughout the house.

34.9 ACTIVE SOLAR AIR SPACE HEATING

In an *active solar air space heating system,* air is circulated through the collector where it is heated by the sun's radiation. The blower forces this heated air into a storage unit or to the space to be heated. Automatic dampers control whether this air is blown into storage or into the conditioned space, Figure 34–8.

34.10 AIR SOLAR COLLECTORS

Air solar collectors take in cold air from a manifold duct. The air is circulated through the collector, often through baffles to contain it in the collector longer so that it will be heated to the desired or maximum temperature before being blown to the storage unit or space to be heated, Figure 34–9. Air collectors are usually constructed with one or two tempered glass panels, an absorber plate, insulation, and a metal frame, Figure 34–10.

The sun radiates through the one or two panels of glass. There may be some reflection and loss of heat, but the glass is designed to allow the maximum radiation to pass through. These glass panels, usually constructed of a low-iron tempered glass, are also designed to keep the heat from being radiated from the absorber plate out of the collector. Low-iron glass is used because it allows more of the sun's radiation to pass through. The absorber plate is painted or treated with a special coating to absorb as much energy as possible. Often a selective coating designed to absorb certain wavelengths of the sun's radiation is used. It may not absorb as much of the heat as a normal black-

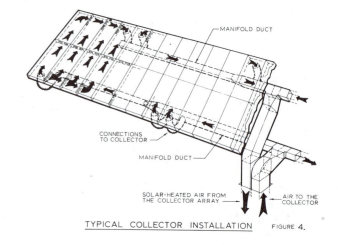

Figure 34–9. Solar air collector showing manifold ducts. *Courtesy Solaron Corporation*

painted coating would, but it absorbs those longer wavelengths that do not escape back through the glass. Thus more heat is trapped in the collector. The coated surface is often rough, allowing more heat to be absorbed.

The air to be heated is circulated underneath the absorber plate. This cool air absorbs heat from the underside of the absorber plate as it moves through the collector and is returned either to storage or to the heated space. It is important that the panels be insulated properly to keep unwanted heat loss from the collector to a minimum. Remember that the collector is located outside where the air is often very cold. This air could absorb significant amounts of the heat from the collector if it were not properly insulated.

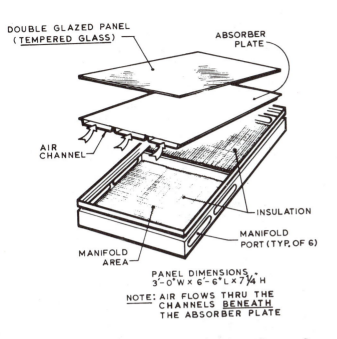

Figure 34-10. Construction of an air solar collector. *Courtesy Solaron Corporation*

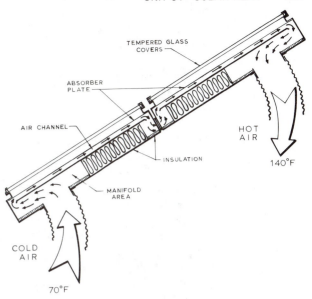

Figure 34-11. Two collectors are mounted side by side. Arrows show the air flow. *Courtesy Solaron Corporation*

As many collectors as needed can be mounted side by side to obtain sufficient surface area to give the desired heating results, Figure 34-11. Note, in the figure, that the air passes under the absorber plate.

In most areas these solar heating systems are either auxiliary systems to a conventional heating system, or the conventional system is a backup for the solar system.

34.11 LIQUID SOLAR COLLECTORS

Liquid solar collectors are flat-plate collectors similar to air collectors except that a liquid (either water or antifreeze solution) is passed through it in copper tubing. The liquid is heated as it passes through the collector. There are many designs for a system using liquid collectors. A basic one is illustrated in Figure 34-12. The water in the collector's tubing is heated and pumped to the storage tank. Water from this storage tank is pumped to a coil in a conventional

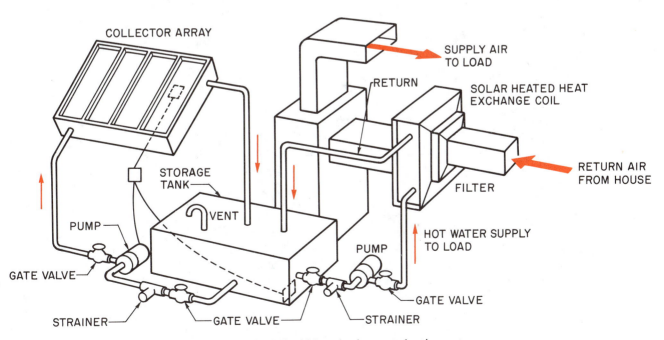

Figure 34-12. A liquid-based solar space-heating system.

furnace where air is blown across it. This air absorbs heat from the coil, and this heated air is distributed to the space to be heated.

Liquid collectors also utilize one or two panels of glass. Again, this glass may vary in quality and design, but a low-iron tempered glass is recommended.

The absorber plate is usually made of copper and formed tightly around the copper tubing, Figure 34–13. For this type of design the tubing should be soldered the entire length to the absorber plate. Some modern designs have the tubes and plates mechanically pressed or fastened together to achieve maximum contact. The absorber plate is painted or coated, generally with a selective coating similar to that described for the air collectors. Copper is generally used for the absorber plate and tubing because it has very good heat transfer characteristics. The absorber plate absorbs the radiation from the sun, and this heat is transferred to the tubing and then to the liquid by conduction. The continuous soldered joints, if used, help with this conduction transfer. The fluid is pumped through the collector, removing heat as it passes through. These collectors must also be insulated very well to keep heat loss from the collector to a minimum.

34.12   COLLECTOR LOCATION AND POSITIONING

Flat-plate collectors are usually positioned permanently to receive the maximum energy from the sun. This location and positioning is determined according to the position of the sun at midday during the winter when most heat is needed. The collectors should face south. The tilt will vary from one location to another because of the location's latitude. The *latitude* is the distance from the equator and is expressed in degrees. The collector should be tilted so that it will be at a 90° angle to the sun's rays at midday during the coldest season. Figure 34–14 and Figure 34–15 indicate the latitude of various cities. The rule of thumb for the angle of

DEGREES LATITUDE	LOCATION
48.2	Glasgow, Montana
43.6	Boise, Idaho
40.0	Columbus, Ohio
35.4	Oklahoma City, Oklahoma
40	Salt Lake City, Utah
29.5	San Antonio, Texas
32.8	Fort Worth, Texas
40.3	Grand Lake, Colorado
42.4	Boston, Massachusetts
27.9	Tampa, Florida
33.4	Phoenix, Arizona
33.7	Atlanta, Georgia
35.1	Albuquerque, New Mexico
40.8	State College, Pennsylvania
42.8	Schenectady, New York
43.1	Madison, Wisconsin
33.9	Los Angeles, California
45.6	St. Cloud, Minnesota
36.1	Greensboro, North Carolina
36.1	Nashville, Tennessee
39.0	Columbia, Missouri
30.0	New Orleans, Louisiana
32.5	Shreveport, Louisiana
42.0	Ames, Iowa
42.4	Medford, Oregon
44.2	Rapid City, South Dakota
38.6	Davis, California
38.0	Lexington, Kentucky
42.7	East Lansing, Michigan
40.5	New York, New York
41.7	Lemont, Illinois
46.8	Bismark, North Dakota
39.3	Ely, Nevada
31.9	Midland, Texas
34.7	Little Rock, Arkansas
39.7	Indianapolis, Indiana

Figure 34–14. U.S. cities with their degrees latitude north.

the collector is to add 15° to the degrees of latitude for that particular location. In other words, for the latitude at Albuquerque, New Mexico of 35.1°, the collector tilt for a system in this city would be 35° + 15° = 50°. This angle is not critical so the 0.1° can be disregarded, Figure 34–16. An *inclinometer* is often used to position the collector at the correct angle, Figure 34–17.

Collectors can be mounted on roofs or on the ground. They should be mounted close to the rest of the system, Figure 34–18. This is particularly essential for air systems. Many times the roof may not face the south, or the roof may be shaded. In these cases the collectors may have to be located on the ground. The collector should not be in the shade from 9 A.M. to 3 P.M. during the heating season.

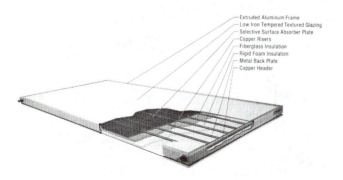

Extruded Aluminum Frame
Low Iron Tempered Textured Glazing
Selective Surface Absorber Plate
Copper Risers
Fiberglass Insulation
Rigid Foam Insulation
Metal Back Plate
Copper Header

Figure 34–13. A liquid solar collector. *Courtesy Solaron Corporation*

Figure 34–15. Map to determine the approximate latitude north.

34.13   STORAGE

The collectors will absorb heat only during the day when the sun is shining. They will collect some heat on cloudy days but probably not enough to heat the space to the temperature desired in cold weather. On bright sunny days they will collect more than enough to heat the conditioned space. It is therefore necessary to store the heat generated on sunny days to help heat the structure at night and on cloudy days. *Sensible-heat storage* and *latent-heat storage* are the two general types of storage systems.

34.14   SENSIBLE HEAT STORAGE

In sensible-heat storage the heat flowing into the storage raises the temperature of the storage material. When heating the space in a home or business, heat is removed from the storage material and its temperature drops. Most sensible-heat storage systems use either an air system with a rock bed or a liquid-based system. Regardless of the design or type of the system, a storage material, a container, and provisions for adding and removing heat will always be needed.

34.15   LATENT HEAT STORAGE

In latent-heat storage systems the material used for the storage changes state. It absorbs heat without

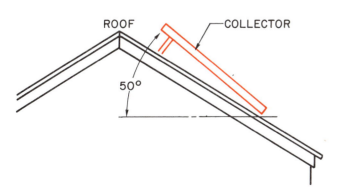

Figure 34–16. Collector tilt angle for Albuquerque, NM.

Figure 34–17. An inclinometer. *Photo by Bill Johnson*

Figure 34–18. Typical collector installations.

changing temperature. Heat flowing into the storage melts the material. When heat is taken out of the material, it solidifies or refreezes. (You may wish to review Unit 1 on latent heat.) This type of material is called phase-change material (PCM). Many types of PCM have been investigated. There are types that change by melting–freezing, boiling–condensing, changing from one solid crystalline structure to another, and changing from one liquid structure to another. The only PCM that is practical to use at this time is the melting–freezing type.

34.16 AIR-BASED SENSIBLE-HEAT STORAGE

Air-based systems generally use rocks as the storage medium or material. The rocks are placed in a container or bin. Air from the collectors that is not used directly for heating the building is forced into the bin and heats the rocks. Heat is stored in the rocks until needed and then air is blown through the rock bin and distributed to the space to be heated.

Rocks should be of medium size (1- to 3-in. diameter). They should be rounded if possible and should be all approximately the same size. A rule of thumb is that they should not be smaller than three-fourths the average size and should not be larger than $1\frac{1}{2}$ times the average size. If rocks are too large, only the outside will be heated, and the space taken up by the rest of the rocks will be wasted. If they are too small, they will fill up all space in the bin and reduce air circulation.

Some rocks are very soft; they break and cause dust. Others, particularly of volcanic origin, may have an odor. Avoid all of these. The rocks should be washed thoroughly to insure that there is no dust. Mix them (small and large) and shovel them into the container or bin evenly. *The bins must be strong enough to withstand the tremendous weight of the rocks.*

The design of the rock bed depends on the space available. Bins may be vertical or horizontal. Vertical beds provide the best characteristics because they take advantage of the natural tendency of hot air to rise. Air from the collectors is forced in from the top to the bottom. When heat is removed, air is forced in at the bottom and out the top. Most of the heat is at the top and therefore closest to the space to be heated, Figures 34–19 and 34–20.

Horizontal rock beds are used when a high space is not available. They may be located in a crawl space or similar area. Horizontal bins do have some disadvantages: (1) The airflow is from side to side rather than from bottom to top or top to bottom (when being charged with heat). The rocks will settle, particularly with the movement from expansion and contraction from heat and cooling, and leave an air space at the top. The air will flow horizontally through this space, bypassing the rocks and not picking up sufficient heat. A floating top will eliminate this air space. (2) As air from the collectors flows into the bin, it will rise and heat the rocks at the top more than the rocks at the

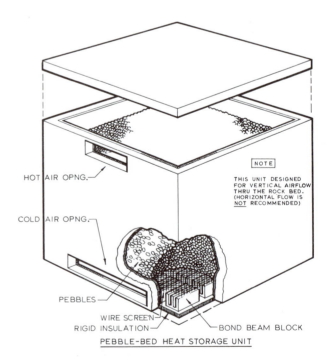

Figure 34–19. A rock-bed air-based storage unit. Note the air space at the bottom between the beam blocks and at the top. These air spaces are called plenums and are necessary to allow the proper air flow through the unit. *Courtesy Solaron Corporation*

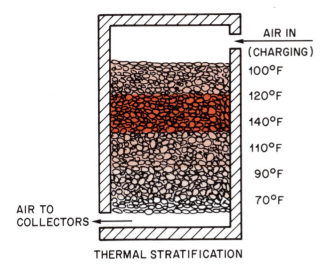

AIR IN
(CHARGING)
100°F
120°F
140°F
110°F
90°F
70°F

AIR TO
COLLECTORS

THERMAL STRATIFICATION

Figure 34–20. Vertical storage rock bed showing heat stratification.

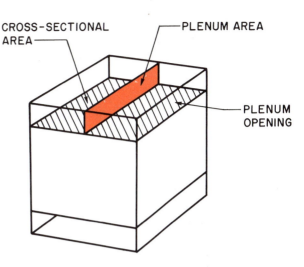

CROSS-SECTIONAL AREA

PLENUM AREA

PLENUM OPENING

Figure 34–22. The plenum area should be approximately 8% of the bin cross-section area.

bottom. When air is forced through for space heating, it will blow through the cool and warm rocks. When the air is mixed, its overall temperature will be lower than air taken out the top, Figure 34–21.

The rock bin should have *plenums* (air spaces) at the top and at the bottom of a vertical bed or at the ends of a horizontal bed. The plenum allows the air to circulate across the entire top and bottom, permitting air to circulate by all the rocks. The efficiency would drop significantly if there were obstructions in the plenum. Each plenum cross section should be approximately 8% of the bin cross section, Figure 34–22.

The airflow from the collector to the bed and back to the collector, and from the bed to the house and back to the bed is controlled by dampers. Dampers will be discussed in more detail later in this unit.

34.17   LIQUID-BASED STORAGE SYSTEMS

Liquid based storage systems usually use water as the storage medium. Many things must be considered when designing this type of system. The designer must determine the most appropriate size of the storage container or tank, its shape, material, and location. In addition, leak protection methods, temperature limits,

insulation and pressurization methods (if any) must be determined.

Storage tanks may be steel, fiberglass, concrete, or wood with a plastic lining. Normally a single tank is used for space-heating applications, but if the proper size is not available or if there are space limitations more than one tank may be used.

Direct solar space heating usually requires storage temperatures of 160°F or less. Space heating with solar-assisted heat pumps usually requires a storage temperature of 100°F or less—the lower the required storage temperature, the less insulation needed for the storage container.

34.18   STEEL TANKS

Steel tanks have been used for years in similar applications and considerable information concerning them is available. They are easy to manufacture, and piping and fittings can be attached easily. However, they are expensive, subject to rusting and corrosion, and difficult to install indoors.

Steel tanks should be purchased with a baked-on phenolic epoxy on the inside. If at all possible limit the temperature to 160°F because corrosion doubles with every 20° rise in temperature. The outside of buried tanks should be protected with a coal-tar epoxy, aboveground tanks with primer and enamel.

If a steel tank is used, steps may have to be taken to prevent excessive electrochemical corrosion, oxidation (rusting), galvanic corrosion, and pitting. Manufacturers of these tanks should be consulted for information as to how to prevent these conditions.

34.19   FIBERGLASS TANKS

Fiberglass tanks should be designed specifically for solar storage. They should be factory insulated.

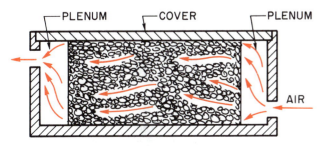

PLENUM   COVER   PLENUM

AIR

Figure 34–21. Horizontal rock storage bin.

Fiberglass tanks usually have two shells, an inner and outer shell, with insulation between. These tanks do not rust or corrode and may be installed above ground or in the ground. *The maximum temperature is limited with these tanks, and they normally cannot be pressurized. They are easily damaged, so inspect them before installing to insure that there will be no leaks.*

34.20 CONCRETE TANKS

Concrete tanks can be purchased precast, or they may be cast in place and constructed to fit many different space limitations. The concrete itself becomes part of the storage, providing some additional capacity. They do not corrode or rust. The weight of this type of tank must be considered when locating it, and enough support must be provided. The weight can be an advantage for buried tanks because they will not be easily floated up or out of the ground by groundwater. They should be insulated. Because heating and cooling will cause expansion and possibly result in leaks, some provision, such as a spray-on butyl rubber coating or a plastic liner, should be used to help prevent leaks.

34.21 INSULATION REQUIREMENTS

Indoor tanks located in heated areas require the least insulation. Tanks in unheated areas and outdoor installations require more. Simple covers are sufficient to protect the insulation in indoor installations, but underground tanks require waterproof insulation.

34.22 LATENT-HEAT STORAGE SYSTEMS

Latent-heat storage material melts when it absorbs heat, and it refreezes or solidifies when it gives up heat. The material does this without changing temperature. One feature of this system is that it can store more heat in less space than either of the other systems. Therefore it requires less space, a smaller container, and less insulation. If the storage material is chosen so that the melting point is slightly above the temperature required by the system, the collector output would need to be only a few degrees warmer than this melting point. The material will increase in temperature only until it reaches its melting point. It then changes state until it is completely melted. It has more even temperatures than rock or liquid storage systems. The collector output for a rock- or liquid-storage system would need to operate at a much higher temperature when the storage system is partially or nearly heated to its maximum temperature.

The melting point of the phase change material (PCM) is one of the most important factors in choosing the material. There is a temperature drop in heat exchangers, and forced-air heating systems should distribute heated air at a considerably warmer temperature than the desired room temperature. Therefore, in an air-based system the minimum melting point of the PCM should be approximately 90°F. Because liquid-based systems go through an additional heat exchange process, the minimum melting point in these systems should be about 100°F. Solar-assisted heat pumps can use materials with a melting point between 50° and 90°F.

Some latent-heat systems can store a small amount of sensible heat by heating the material above the melting point. The amount of heat stored is generally very small compared to the latent heat stored.

Figure 34-23 illustrates a system using tubes filled with the PCM spaced apart in a container. Air is blown around the tubes to either charge the system from the collectors or to remove heat for circulation to the area to be heated. Figure 34-24 shows a system using flat trays containing the PCM. Again, air is circulated through the container to charge the system with heat from the collectors or to remove the heat for space heating.

Salts and waxes can be used in these systems. They have melting points ranging from approximately 80° to 120°F. Air-based and liquid-based storage systems are currently used more than the latent-heat storage systems. The initial purchase costs are usually higher for the latent-heat system, but as the technology develops it is assumed that these systems will be used more widely.

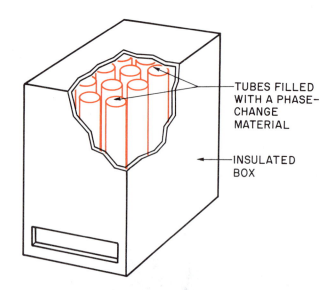

TUBES FILLED WITH A PHASE-CHANGE MATERIAL

INSULATED BOX

Figure 34-23. Tubes filled with a phase change material can be placed in an insulated container. Air can be forced into the top of the container from the collectors, charging the storage or it can be circulated from the bottom and out the top, extracting heat from the PCM and heating the conditioned space.

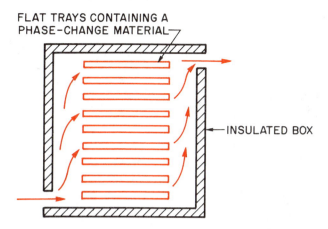

Figure 34-24. Trays with a phase change material can be placed in an insulated storage container. This functions similar to the PCM tube storage in Figure 34-23.

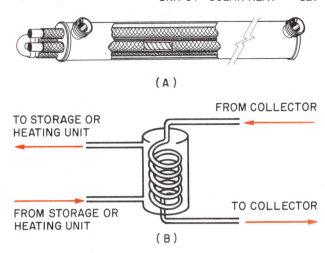

Figure 34-26. Liquid-to-liquid heat exchangers. *Courtesy Noranda Metal Industries, Inc.*

34.23 OTHER SYSTEM COMPONENTS

Heat Exchangers

Heat exchangers transfer heat from one medium to another, usually without mixing either of the materials. They may transfer air to air, air to liquid, liquid to liquid, or liquid to air.

In the air-based space-heating system a separate heat exchanger is not needed because the rock bed is the heat exchanger. The collectors heat the air, which charges the rock bed with heat. When the system is discharged or air is forced through the rock bed to the space to be heated, the rock bed is the place where the heat exchange takes place. The air being blown to the conditioned space absorbs heat from the rocks.

Air-to-liquid. *Air-to-liquid* heat exchangers are used in domestic hot-water heating. An exchanger, Figure 34-25, is placed in the heated air duct near the domestic hot-water heater. It absorbs heat from the air into the heat exchange fluid, which is pumped to the hot-water tank where it heats the water.

Liquid-to-liquid. In *liquid-to-liquid* exchangers the liquid-based solar collectors heat the fluid in the collectors. It is then piped into a coil in the storage heat exchanger, Figure 34-26. Heated fluid from the outlet of the heat exchanger is then piped either to the storage tank or to the water heating element, such as a hydronic baseboard heating unit.

Liquid-to-air. *Liquid-to-air* exchangers may employ a heat exchanger like that used in the air-to-liquid system. These exchangers usually are constructed with finned water tubing. The component is placed in the air distribution duct work, where air forced through it absorbs heat from the fins and tubing. The tubing may contain hot water directly from the solar collectors or from the storage tank.

Any system loses efficiency each time a heat exchanger is used. This is because of the difference in temperature between the heat exchange mediums. It is therefore necessary to limit the number of heat exchangers.

Filters

Filters should be installed at the rock-bed inlet and outlet. It is essential to filter as much dust as possible from the rock bed and from the conditioned space.

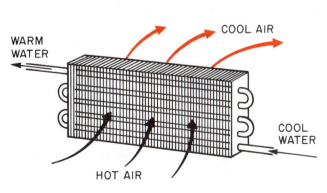

Figure 34-25. An air-to-water heat exchanger.

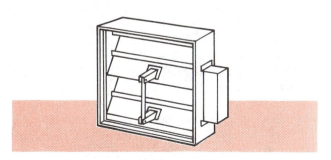

Figure 34-27. A motorized damper.

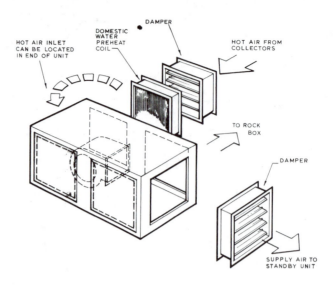

Figure 34–28. Diagram of air handler, showing possible locations of heating coil and dampers. The fan is shown with the dotted line inside the unit. *Courtesy Solaron Corporation*

Automatically Controlled Dampers

Dampers must be installed in solar air systems to control the air from the collectors to the storage or to the conditioned space and return to the storage bin and collector. Motor-driven, thermostatically controlled dampers are used, Figure 34–27. Back-draft dampers should also be installed between the rock bed and the collectors to prevent heat loss from the storage bed to the collectors when they are not absorbing heat from the sun.

Package Air Handlers

Air handlers that include a motor-driven fan and dampers are often used from the air collectors to the furnace or to the rock storage bin, Figure 34–28.

Controllers

Controllers turn pumps on and off, control dampers, and open and close valves, Figure 34–29. The differential thermostat is the main component in a controller. In a house heated by a conventional furnace, only one temperature setting is needed on the thermostat to turn the furnace on and off. In a solar-heated house the temperature at the collector, in the storage, and in the house must all be sensed so that the collector will operate the pumps, valves, and dampers for the system to function efficiently.

When the liquid temperature at the collectors is hotter than the temperature at the storage, the collector pump will be activated to circulate liquid from the collector to the storage tank. When the room thermostat calls for heat and the storage tank is hot enough,

Figure 34–29. A typical modern controller for a space-heating solar system. *Courtesy Heliotrope General*

the pump to the space heating coil is activated; when the room thermostat calls for heat and the storage is not warm enough, the furnace is activated. The controller must "know" the temperature at all these points to "determine" which circuits to activate, Figure 34–30.

Temperature Sensors

Temperature sensors are used in the rock bed to send the temperature reading at the top and bottom of the bed to the controller. Usually one sensor is installed 6 in. from the top and another 6 in. from the bottom

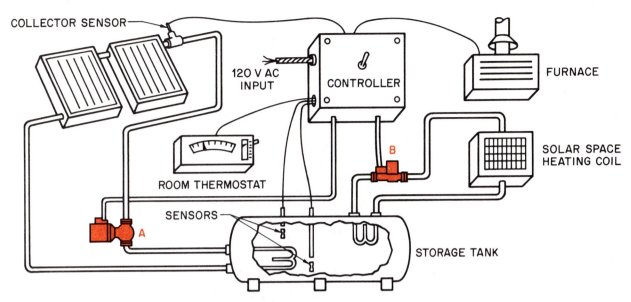

Figure 34–30. When the collectors are hotter by a predetermined amount than the storage temperature, pump A is activated, circulating hot liquid from the collectors to the storage. When the room thermostat calls for heat and the water in the storage tank is hot enough, pump B is activated circulating heated water to the solar space-heating coil. When the room thermostat calls for heat and the storage liquid is not hot enough, the furnace is activated.

of the rock bed. The sensors are connected to controllers that activate the auxiliary heat and/or dampers. Temperature sensors, located in the storage tank, are also used in liquid-based systems. The sensors tell the controller when the water at the top is hot enough for space heating or when the temperature at the bottom is such that the collector pump should be turned on or when the temperature is so hot that it is unsafe and should be cooled down. Sensors may be located in the tank, in wells in the tank, or outside the tank if it is metallic. The sensors should be protected from the water and, if installed on the tank, should be in good contact with the metal and insulated from the ambient. Sensors may be thermocouples, bimetal controls, liquid or vapor expansion tubes and bulbs, thermistors, or silicon transistors. Figure 34–31 and 34–32

are examples of typical thermistor sensors. Refer to Unit 12 for a review of many of these sensors. They should always be matched with the control. Figure 34–33 illustrates how a bulb type thermistor sensor may be installed in tanks.

Thermometers should also be installed at the top and bottom of the water storage tank. They are used to adjust the system's operation and to troubleshoot the system if it doesn't function properly.

Pumps

Liquid-based solar heating systems require *pumps* to move the liquid through the collectors to the storage and back to the collectors. The fluid may at times bypass the storage under certain conditions and be fed directly to the heat exchanger serving the heat distribution system. In this instance another pump is often used. This may be the same pump that pumps the water from storage to the heat exchanger.

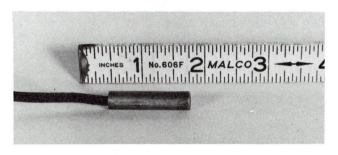

Figure 34–31. Thermistor sensor. *Photo by Bill Johnson*

Figure 34–32. Thermistor sensor in brass plug.

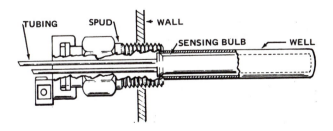

Figure 34–33. Thermistor sensor installed through wall of tank. *Courtesy Heliotrope General*

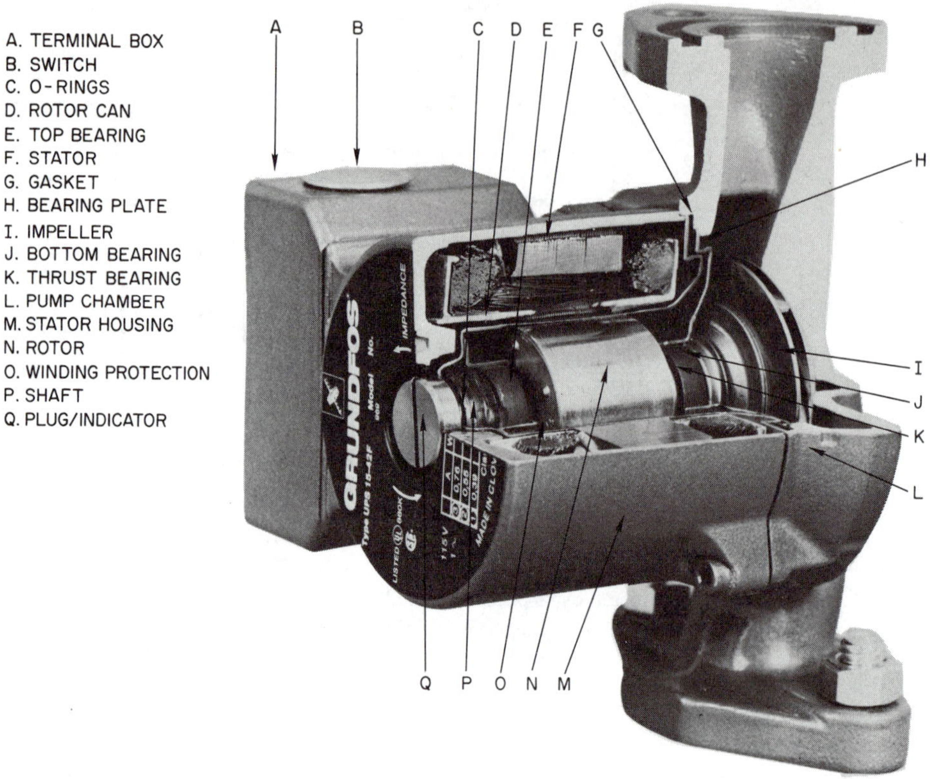

A. TERMINAL BOX
B. SWITCH
C. O-RINGS
D. ROTOR CAN
E. TOP BEARING
F. STATOR
G. GASKET
H. BEARING PLATE
I. IMPELLER
J. BOTTOM BEARING
K. THRUST BEARING
L. PUMP CHAMBER
M. STATOR HOUSING
N. ROTOR
O. WINDING PROTECTION
P. SHAFT
Q. PLUG/INDICATOR

Figure 34–34. A cutaway of a typical circulating pump used in a solar space-heating system. *Courtesy Grundfos Pumps Corporation*

Centrifugal pumps are generally used in solar systems. An impeller turns in a pump housing and forces the fluid through the system. These pumps produce a low pressure but can move a large volume of liquid. The term "head" is often used instead of pressure. You may read in some manufacturer's specifications or other materials that pumps have a low or high head. This simply means a low- or high-pressure difference that the pump can develop from the inlet to the outlet, Figure 34–34. Centrifugal pumps are also described in Unit 33.

34.24 VALVES

The following is a brief description of some valves in a liquid-based solar system.

Hand-operated valves. *Hand-operated valves* are used to isolate a particular section of the system. This may be necessary when the system is shut down for a period of time or when it is being serviced. The three common types of hand-operated valves are the *globe valve,* the *gate valve,* and the *ball valve,* Figure 34–35. The globe valve is used to shut off the system or to control the amount of flow. It offers resistance to the flow even when open, and it does not fully drain when installed horizontally. Gate valves are used as shutoff valves;

Figure 34–35. Hand-operated valves. (A) Globe valve. (B) Gate valve. (C) Ball valve. *Photos by Bill Johnson*

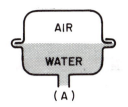

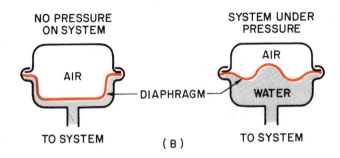

Figure 34–37. (A) No diaphragm in this type. (B) Expansion tank with a diaphragm to separate the water and air.

they will drain and offer little resistance. Gate valves are generally used in solar systems. Ball valves offer virtually no resistance. They may be used as shut-off valves, but may be more expensive.

Check valves. *Check valves* are installed between the pump outlet and the collector inlet to prevent siphoning back to the collector on cold nights when the collectors have been drained. Collectors using only water must be drained in cold weather at night and at other times when the water might freeze. Check valves must be installed properly. *Gravity operated check valves must be installed in the correct direction. If they are installed upside down or at any angle other than vertical, they will not operate properly; therefore siphoning, and possible freezing, will occur.*

Spring-operated check valves. *Spring-operated check valves* may be installed at any angle, but they must be installed in the line with the correct direction of flow. Although they are more dependable than gravity operated check valves, they offer more resistance, which causes a greater pressure drop.

Pressure relief valves. *Pressure relief valves* are necessary to relieve a closed system of excessive pressure to prevent damage to a system and to eliminate a hazardous condition. These valves are normally selected for the pressure and temperature at which they will open, Figure 34–36. *Always use exact replacement relief valves.*

34.25 EXPANSION TANKS

All closed systems must have space into which liquids can expand when heated to keep the system pressure under control. Therefore an *expansion tank* must be included in the system. The expansion tank

must be sized properly to avoid damage to the system. The kind of liquid in the system must be considered when determining the size of the tank. For instance, antifreeze solutions will expand more than water, so more space is needed for the antifreeze solutions.

Expansion tanks may be designed for the water or fluid to mix with the air in the tank, or they may be designed so that the liquid is kept separate from the air, Figure 34–37. This latter type of tank has a flexible diaphragm separating the liquid from the air, which prevents the liquid from absorbing the air.

34.26 INSULATION

Insulation is a key factor in efficient solar systems. Insulation should be properly installed wherever there is a chance for the heat to escape. This includes the sides and bottom of collectors, the storage container, and all heat transfer components such as piping, tubing, and duct work. Insulation designed specifically for duct and piping applications should be used rather than wrapping insulation around the duct and piping. Adapting insulation for applications for which it is not intended is inefficient because insulation should not be crushed, packed, or squeezed. This reduces the air space and lowers the efficiency. Insulation on exposed piping, such as on the roof, should be treated, painted, or of a material that resists deterioration from the sun's radiation.

Figure 34–36. A pressure–temperature relief valve. *Photo by Bill Johnson*

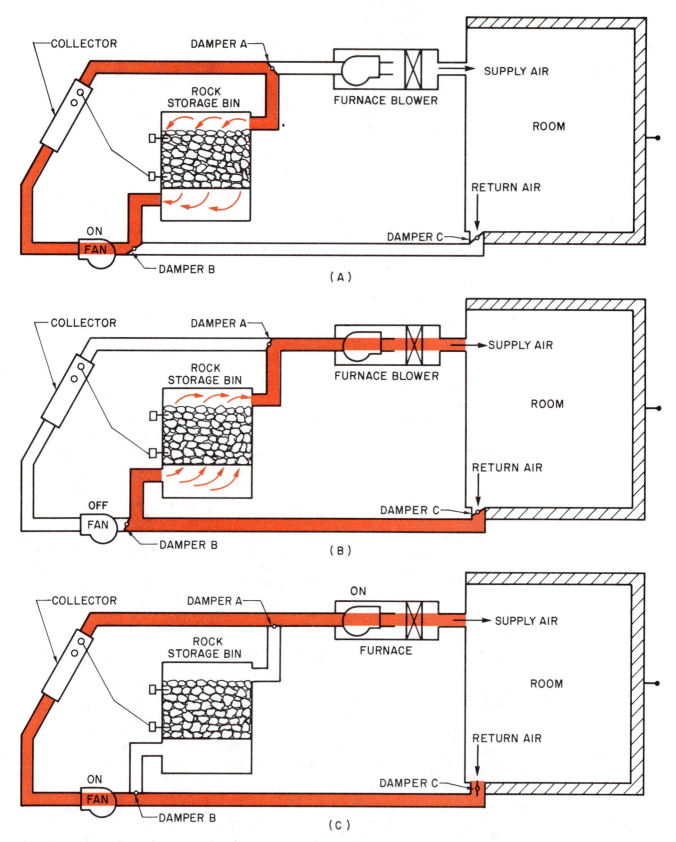

Figure 34-38. (A) When collectors are charging the storage bin, damper A closes airflow to the furnace and damper B allows the air to flow through the fan back to the collector. (B) When the room thermostat calls for heat and the collector temperature is not sufficient to supply the heat, the furnace blower will come on; damper A closes airflow from the collector allowing air to flow from the rock storage to the room and return to the storage. (C) When the house thermostat is calling for heat and the collectors are producing sufficient heat, dampers A and B will close to the storage, providing a path directly from the collectors to the house. Return air damper C will be open, providing a path for the air back to the collectors.

34.27 SOLAR SPACE-HEATING SYSTEMS

The following includes a few of the combinations possible in using solar radiation for space heating applications.

Air Collection–Rock-Bed Storage–Air Distribution to House–Auxiliary Heat by Conventional Forced-Warm-Air Furnace

This system utilizes the conventional warm air furnace to distribute the heat from the collectors and/or the storage bin to the house. The furnace is also used as the auxiliary heat supply when heat from the solar system is not available, Figure 34–38. When the collectors are absorbing heat and have reached a predetermined temperature and the house thermostat is not calling for heat, dampers A and B will open to the rock storage bin. The collector fan will start, and the heat will be blown into the top of the rock bed. This is called *charging* the storage bin. The rocks near the top will be heated first and will absorb the most heat.

When heat is not being collected and there is a call for heat from the house, dampers A and B will close to the collector, C will open to the house duct, and the furnace fan will activate. The airflow will be reversed; that is, it will now be from the bottom of the rock bed to the top, forcing out the hottest air, which will be distributed throughout the house by the furnace fan.

When heat is being collected and there is a simultaneous demand for heat in the house, dampers A and B will close to the storage, and the collector fan and furnace fan will start blowing heated air from the collectors directly to the furnace distribution system, returning through damper C and the solar duct system to the collector fan and back to the collectors. If the solar collectors and storage cannot produce enough heat, the conventional furnace will activate to satisfy the demand.

Liquid Collection–Water Storage–Air Distribution to House–Auxiliary Heat by Conventional Forced-Warm-Air Furnace

This system has collectors, a water storage tank, a liquid-to-air heat exchanger, centrifugal pumps, and uses a conventional forced-warm-air furnace to distribute the heat and for auxiliary heat. It also has a domestic hot-water tank with a heat exchanger in the storage tank, Figure 34–39.

When sensors indicate that heat is ready for collection at the collector and if the house thermostat is not calling for heat, the controller will activate pump A and circulate the water directly into the storage tank from the collectors. All liquids in this tank would be water as the collector liquid mixes with the storage liquid. Water is pumped through the collectors as long

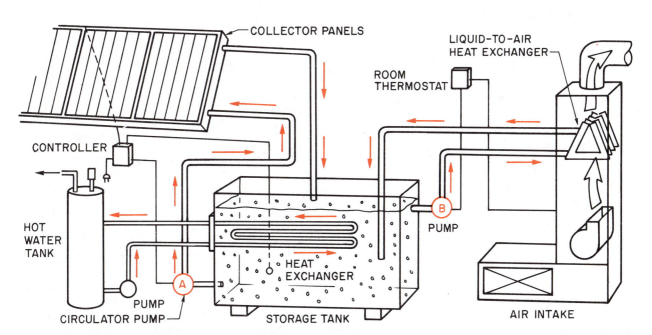

Figure 34–39. Collectors use water as the liquid. This is a drain-down system because the water in the collectors would freeze on cold nights. When collectors are producing heat at a higher temperature than the storage water, circulator pump A pumps water from the storage to be heated and from the collectors back to storage. When the house thermostat calls for heat, circulator pump B pumps water from storage to the heat exchanger in the furnace duct. The furnace fan blows air across the heat exchanger, warming the air, and distributes it to the house. Also shown is a domestic hot-water tank, heated through a heat exchanger in the water storage.

as the collectors are absorbing heat at a higher temperature than the temperature of the storage water. Should there be a call for heat from the house thermostat, the controller will activate circulator pump B and pump the heated storage water through the heat exchanger.

The liquid-to-air heat exchangers for these systems may be located in the return air duct, just as it enters the furnace, or above the heat source as shown in Figure 34–39. A two-stage thermostat is often used in these systems. If the solar heat cannot supply the heat needed, the second stage of the thermostat will activate the conventional furnace. It may be designed so that both systems are operating simultaneously because some heat from the solar radiation will help. When this solar heat drops below a predetermined temperature, this system will cut off and the auxiliary heat will take over.

This solar system is called a *drain-down* system. When circulator pump A shuts down, the collector water will all drain to a heated area to keep the water from freezing in cold weather. Any freezing can cause considerable damage to the system. Note that the pipe from the collector into the storage tank does not extend into the water. This provides venting and allows the collectors and piping to drain. If this space is not provided, other venting in the collector system is necessary.

An alternative design often preferred, particularly in colder climates, requires a closed collector piping system using antifreeze instead of water, Figure 34–40. Note that a heat exchanger is used to heat the storage water. The collector antifreeze fluid must be kept separated from the storage water. This system is less efficient than the drain-down system, which does not require the extra heat exchanger.

Many closed collector systems such as this are designed with a *purge coil*. On a hot day when the storage is fully charged with heat and the room thermostat is not calling for heat, the solution in the collector system may overheat. This could cause the safety valve to open, resulting in loss of solution and overheating, which will damage the equipment. When a predetermined temperature is reached, the controller diverts the solution through the three-way valve 2 to the purge coil. This is normally a finned-tube coil and dissipates the heat from the system to the air.

Liquid Collection–Water Storage–Water Distribution–Auxiliary Heat by Conventional Hot-Water Boiler

A liquid-based solar collector system may be paired with a hot-water finned-tube convector heating furnace. The collector system can be of the same design as an air distribution system. It can be a drain-down water design or one that uses antifreeze with a heat exchanger in the storage unit. Solar collector and storage temperatures are lower than those produced by a hot-water boiler. This can be compensated for to

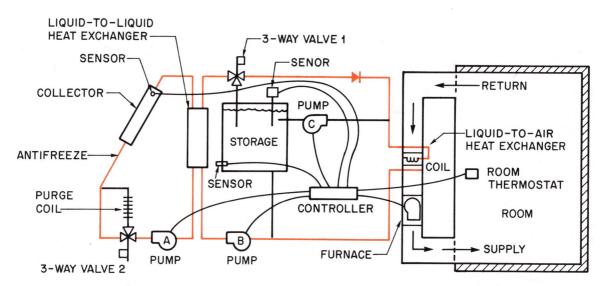

Figure 34–40. Closed liquid-collection, water storage system with air distribution to the house. The liquid in the collector system is an antifreeze and water solution. It is heated at the collectors and pumped through the coil at the heat exchanger and back to the collectors. This continues as long as the collectors are absorbing heat at a predetermined temperature. When the room thermostat is calling for heat, 3-way valve 1 allows the water to be circulated by pump B through the liquid to air heat exchanger and back to the liquid-to-liquid heat exchanger where it absorbs more heat. If the room thermostat is not calling for heat, 3-way valve 1 diverts the water to storage. If the storage temperature has reached a predetermined temperature, the collector solution is circulated through the purge coil. The solution is diverted to this purge coil through 3-way valve 2. (The heat is dissipated into the air outside through this coil.) This is a necessary safety precaution since it is possible for the collector systems to overheat and damage the equipment.

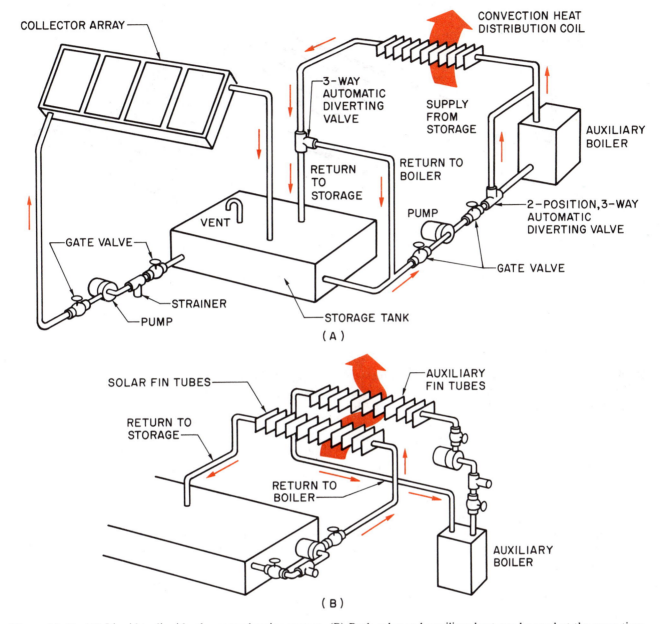

Figure 34–41. (A) Liquid-to-liquid solar space-heating system. (B) Both solar and auxiliary heat can be used at the same time.

some extent by using more fin tubing than would be used with a conventional hot-water system.

Figure 34–41 illustrates a design using a liquid-based solar collector with a hot-water boiler system. This design uses the auxiliary boiler when the storage water is not hot enough. If the solar storage is hot enough, the boiler does not operate. Water is pumped from the storage tank and bypasses the boiler. It goes directly to the baseboard fin tubing. The three-way *diverting valve* automatically causes this boiler bypass. When auxiliary heat is called for, the two three-way diverting valves cause the water to bypass the storage. It would be very inefficient to pump heated water through the storage tank.

This system can be designed as two separate systems in parallel as in Figure 34–41A. These separate systems can both be operated at the same time. With this design the solar system can help the auxiliary system when the temperature of the storage water is below that which would be satisfactory for it to operate alone. Sensors are used in either system to tell the controller when to start and stop the pumps and when to control the diverting valves.

Water Collection–Water Storage–Radiant Heat–Conventional Furnace Auxiliary Heat (any type)

This system is very appropriate for use in a solar heating application, Figure 34–42. The heating coils

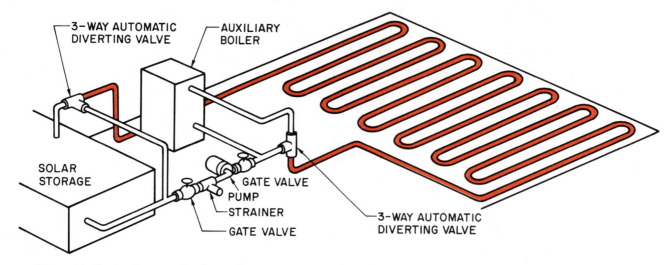

Figure 34–42. A solar space-heating radiant system. The radiant heating coils are normally imbedded in a concrete floor.

may be imbedded in concrete in the floor or in plaster in ceilings or walls. The normal surface temperature for floor heating is 85° F, for wall or ceiling panels 120° F. The same type controls, pumps, and valves are used with this system as with the baseboard hot-water system.

Floor installations are most common. The coils are imbedded in concrete approximately 1 in. below the surface. If this is a concrete slab on grade, it should be insulated underneath. In ceiling and wall applications there should be coils for each room to be heated. Centering the coils in walls, ceiling joists, or floor joists between rooms to heat more than one area is usually not satisfactory. When installations are on outside walls they should be insulated very well between the coils and outside surfaces to prevent extreme heat loss.

Radiant heating installed in the floor is very comfortable and there is little temperature variation from room to room. Figure 34–43 illustrates a polybutylene pipe being installed in a floor. The pipe is tied to a wire mesh and propped on blocks to keep it near the top of the concrete slab and the concrete is poured around it, covering it by approximately 1 in. There should be no joints in this pipe in the floor. Polybutylene pipe (plastic) is popular for this installation. Any joints necessary should be manifolded outside the slab. Almost any type of flooring material may be used over the slab. Common materials are tile, wood, and carpeting. A controller is installed with sensors at the collectors, the water storage, and outdoors. It is also connected to the room thermostat. The controller modulates the temperature of the water being circulated in the floor according to the outside temperature and the room temperature.

Water Collection–Water Storage–Water-to-Air Heat Pump Space Heating

The solar heating systems described previously are only effective when the storage is sufficiently charged with heat to provide the space-heating temperature desired. The heat required of the storage medium to be effective varies somewhat with the type system and equipment used. However, most systems are not effective when the storage temperature is below 90° F. In cold climates with cloudy weather it will not take long for the storage temperature to drop below 90° F.

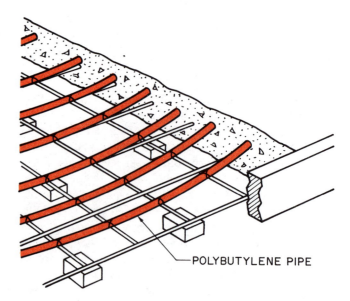

Figure 34–43. Polybutylene pipe being installed in a concrete slab to be used in a solar radiant heating system.

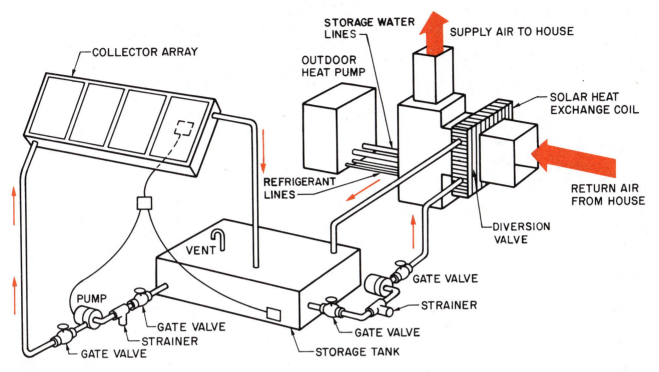

Figure 34-44. A liquid-based solar space-heating system.

A heat pump, however, can extract heat from temperatures far below 90° F. This makes it practical to connect a heat pump to a coil supplying heated storage water. If the storage water temperature is high enough, this will operate in a manner similar to the water-to-air system. If the temperature of the storage material drops below that which can effectively be used in this manner, the storage water will be diverted to the heat pump where it uses the heat available to help it provide the necessary heat for the conditioned space, Figure 34-44.

Solar Heated Domestic Hot Water

Many solar systems are designed and installed specifically to heat or to assist in heating domestic hot water. These use similar components to those used in space-heating systems, Figure 34-45. The heated collec-

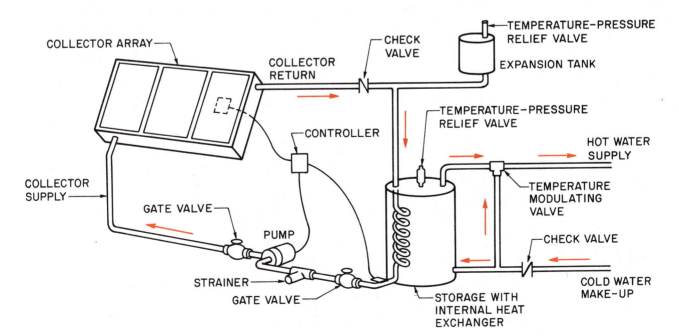

Figure 34-45. Solar domestic hot-water system.

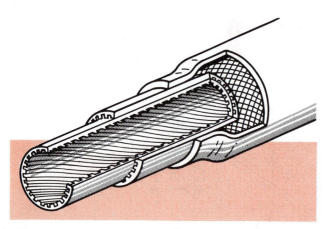

Figure 34–46. Heat exchanger piping has a double wall when an antifreeze solution is used in the collectors.

tor water or antifreeze is pumped through the heat exchanger where it provides heat to the water in the tank. This tank must have conventional heating components also since the solar heating will probably not meet the demand at all times. *If an antifreeze solution is used, the heat exchanger must have a double wall to prevent the toxic solution from mixing with the water used in the home or business, Figure 34–46.*

SAFETY PRECAUTION: *Ethylene glycol is normally used as an antifreeze. It is toxic and must be used with a double-wall heat exchanger. Propylene glycol is a "food grade" antifreeze and is recommended for use in a domestic hot-water system.*

Many systems are designed to use a solar preheat or storage tank in conjunction with a conventional hot-water tank. This gives additional solar-heated water to the system. Conventional heating (electric, gas, oil) will not have to be used as much with this extra storage, Figure 34–47.

When a space-heating system is installed, it will usually include domestic hot-water heating as well. There are several ways in which this can be designed, Figure 34–39. The cold makeup water is passed through the heat exchanger in the storage tank. It absorbs whatever heat is available before entering the hot-water tank, thus reducing the amount of heat that must be supplied by the conventional means.

SUMMARY

- Passive solar designs use nonmoving parts of a building to provide or enhance heating or cooling.
- Active solar systems use electrical and/or mechanical devices to help collect, store, and distribute the sun's energy.
- An active system includes collectors, a means for storing heat, and a system for distributing it.
- The tilt or angle of the earth on its axis is called the angle of declination. This angle is responsible for the differences in solar radiation during the seasons.
- The amount of radiation reaching the earth also varies according to the distance it travels through the atmosphere. The shortest distance is when the sun is perpendicular (90°) to a particular surface.
- Most solar collectors used in active solar heating systems are flat and positioned and tilted to catch as many of the sun's rays as close to 90° as possible.
- Air solar collectors heat the air, which is then circulated either to a storage component or directly to the conditioned space. The cooled air is then returned to the collector where the heating and circulation process is continued.
- Liquid solar collectors circulate water or an antifreeze solution through coils where the liquid is heated by the sun's radiation. It is then pumped to a storage tank or to a heat exchanger where the heat is distributed to the space to be heated.

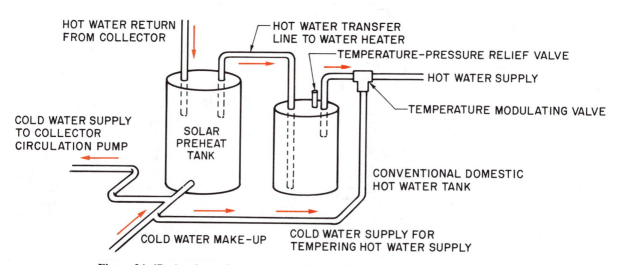

Figure 34–47. A solar preheat tank provides additional storage for a domestic hot water system.

- Collectors are positioned to face south. They are tilted to collect the sun's rays at a 90° angle in winter. The rule of thumb for the angle of this tilt is to add 15° to the degrees of latitude for the particular location of the collector.
- Heat is stored as sensible heat or as latent heat.
- Sensible heat is normally stored in rocks or in water.
- Latent heat is stored in a salt or wax material.
- Air-based systems normally store heat in rock beds.
- Liquid-based systems normally store heat in water storage tanks. These tanks may be made of steel, fiberglass, concrete, or wood with plastic liners.
- Latent-heat storage uses phase-change materials. These materials do not require as high a temperature as water or air and use less space, but they form a more expensive system.
- Other solar system components include heat exchangers, dampers, package air handlers, sensors, valves, pumps, and expansion tanks.
- Many combinations of components may be used for a complete solar space-heating system.

REVIEW QUESTIONS

1. List three types of passive solar designs.
2. How do active solar systems differ from passive designs?
3. What is the solar constant?
4. What is the declination angle?
5. How does the declination angle affect the sun's radiation during summer and winter?
6. Describe the difference between an active air-based solar heating system and an active liquid-based solar heating system.
7. Sketch a cutaway of an air-based collector.
8. Describe how air-based collectors function.
9. Sketch a diagram of a liquid-based collector.
10. Describe how liquid-based collectors function.
11. What type of glass is preferred for use in solar collectors?
12. What effect does the coating have on the absorber plate?
13. Why is a special selective coating often used on absorber plates?
14. How is the angle of the collector tilt determined?
15. Describe rock-bed air-based storage.
16. Describe the difference between sensible-heat and latent-heat storage systems.
17. Describe the types of tanks that may be used in liquid storage systems.
18. What are the advantages of latent-heat storage?
19. Describe one use of a heat exchanger in a solar space-heating system.
20. In what kind of system are air dampers used?
21. What is the purpose of using temperature sensors?
22. Why are hand-operated valves used?
23. Why are check valves used?
24. Why is an expansion tank necessary in a closed liquid-collection space-heating system?
25. Sketch a diagram of a typical air-based solar space-heating system.
26. Sketch a diagram of a typical liquid-based space-heating system.

WOOD HEAT

OBJECTIVES

After studying this unit, you should be able to

- describe what constitutes a cord of wood.
- list some reasons for the formation of creosote.
- describe methods to help prevent the formation of creosote.
- explain how stovepipe should be assembled and installed.
- describe the operation of a catalytic combustor.
- describe ways in which combustion air can be made up or replaced.
- list safety hazards that may be encountered with wood stoves.
- explain how a fireplace insert can improve on the heating efficiency of a fireplace.
- describe procedures for installing a stove to be vented through a fireplace chimney.
- describe a dual-fuel furnace.
- describe how a chimney should be cleaned.

Many people are burning wood as an alternative fuel for heating their homes. Some homes are heated entirely by wood; others use wood as supplemental heating. Wood must be readily available to make this type of heating practical. The heating technician should have an understanding of this type of heat if he/she lives and works in an area where wood is available.

35.1 ORGANIC MAKEUP AND CHARACTERISTICS OF WOOD

Wood plants manufacture glucose. Some of this glucose or sugar is turned into cellulose. Approximately 88% of wood is composed of cellulose and lignin in about equal parts. Cellulose is an inert substance and forms the solid part of wood plants. It forms the main or supporting structure of each cell in a tree. Lignin is a fibrous material that forms an essential part of woody tissue. The remaining 12% of the wood is composed of resins, gums, and a small quantity of other organic material.

Water in green or freshly cut wood constitutes from one third to two thirds of its weight. Thoroughly air-dried wood may have as little as 15% moisture content by weight.

Wood is classified as hardwood or softwood. Hickory, oak, maple, and ash are examples of hardwood. Pine and cedar are examples of softwood. Figure 35–1 lists some common types of wood and the heat value in each type in millions of Btu per cord. A cord of wood can be split or unsplit or mixed. It is important to know whether the wood is split because there is more wood in a stack that is split. Wood is sold by the cord which is a stack 4 ft × 4 ft × 8 ft or 128 ft^3.

Wood should be dry before burning. There is approximately 20% more heat available in dry wood than in green wood. Wood should be stacked off the ground on runners, well ventilated, and exposed to the sun when it is drying, Figure 35–2. When possible it should be covered, but air should be allowed to circulate through it. Wood splits easier when it is green or freshly cut and dries better when it has been split. When green wood is burned, combustion will be incomplete, resulting in unburned carbon, oils, and resins, which leave the fire as smoke.

During oxidation or burning, oxygen is added to the chemical process. This actually turns wood back into the products that helped it grow as a plant: primarily carbon dioxide (CO_2), water (H_2O), other miscellaneous

TYPE	WEIGHT CORD	BTU PER CORD AIR DRIED WOOD	EQUIVALENT VALUE #2 FUEL OIL, GALS.
White Pine	1800#	17,000,000	120
Aspen	1900	17,500,000	125
Spruce	2100	18,000,000	130
Ash	2900	22,500,000	160
Tamarack	2500	24,000,000	170
Soft Maple	2500	24,000,000	170
Yellow Birch	3000	26,000,000	185
Red Oak	3250	27,000,000	195
Hard Maple	3000	29,000,000	200
Hickory	3600	30,500,000	215

Figure 35–1. The table indicates the weight per cord, the Btu per cord of air dried wood, and the equivalent value of No. 2 fuel oil in gallons. *Courtesy Yukon Energy Corporation*

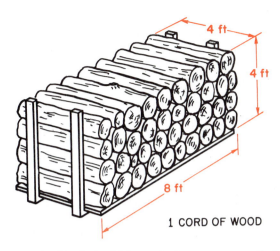

Figure 35–2. Wood should be stacked on runners so that air can circulate through it.

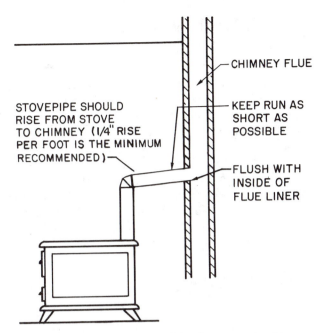

Figure 35–3. Stovepipe should rise from stove to chimney.

materials, and, of course, heat. Some woods produce more heat than others, Figure 35–1. Generally, dry hardwoods are the most efficient.

35.2 CREOSOTE

The combustion in a wood-burning appliance is never complete. The smoke contains the gases and other products of the incomplete combustion including a substance called *creosote*. Chemically, creosote is a mixture of unburned organic material. When it is hot, it is a thick dark-brown liquid. When it cools, it forms into a residue like tar. It then often turns into a black flakey substance inside the chimney or stovepipe. Some of the primary causes of excessive creosote are

- Smoldering lowheat fires
- Smoke in contact with cool surfaces in the stovepipe or chimney
- Burning green wood
- Burning softwood

✳Creosote buildup in the stovepipe and chimney is dangerous. It can ignite and burn with enough force to cause a fire in the building. Stovepipes have been known to be blown apart, and chimney fires are common with this excessive buildup.✳

To help prevent the formation of excessive creosote, burn dry hardwood. Fires should burn with some intensity. When the stovepipe or chimney flue temperature drops below 250°F, creosote will condense on the surfaces. Use well-insulated stovepipe with as little run to the chimney as possible. Use as few elbows or bends as possible and insure that the stovepipe has a rise at all points from the stove to the chimney. The minimum rise should be $\frac{1}{4}$ in. per foot of horizontal run. A rise of 30° is recommended, Figure 35–3. Anything that slows the movement of the gases allows them to cool, which will cause more creosote condensation. Stovepipes and chimneys should be cleaned regularly.

Assemble the stovepipe with the crimped end down, Figure 35–4. This will keep the creosote inside the pipe, allow it to drain to the stove, and be burned. Regardless of the preventive measures, proper design, and installation, creosote will form. Never let the creosote build up. Even a buildup of $\frac{1}{4}$ in. can burn with dangerous intensity.

35.3 TYPES OF WOOD-BURNING APPLIANCES

Wood can be burned in wood-burning stoves, fireplace inserts, wood-burning furnaces, and dual-fuel furnaces.

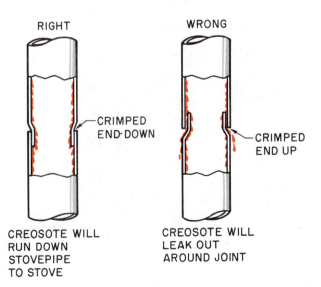

Figure 35–4. Stovepipe should be assembled with crimped end down so that creosote will not leak out of the joint.

35.4 WOOD-BURNING STOVES

Wood-burning stoves heat primarily by radiation. The heat radiates into the room until it is absorbed by some object(s), the walls, and floor of the room. After these have been heated they heat the air in the room. The combustion also heats the stove, which in turn heats the air around it. Some stoves are classified as radiant stoves. Others, however, have an outer jacket, usually with openings or grill work. This is to allow for air circulation around the stove. Some of these are designed for natural convection, others for forced-air movement with a blower, Figure 35-5. Good-quality stoves are built of a heavy-duty steel plate, $\frac{1}{4}$ in. or thicker, or of cast iron. Doors are often of cast iron to

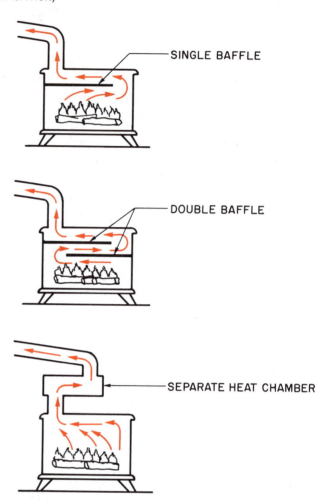

Figure 35-6. To help retain the heat within the stove for better efficiency, baffles and/or heat chambers are used.

prevent warping. Warped doors cause air leaks and uncontrolled combustion.

Better-quality stoves are airtight. No air can get to the fire except through the draft dampers. This allows good control of the combustion process. The more air that reaches the fire through cracks and other openings, the less control there is of the combustion process and the lower the efficiency of the stove. The primary source of air leaks is around the doors. The doors should be constructed of a heavy steel or cast iron that will resist warping. They should also have ceramic rope or other insulation where the door makes contact with the stove.

When wood stoves are used as the primary heat source in a home, they should be of the best quality, which will normally result in the greatest efficiency. It is necessary to keep the heat in the stove as long as possible to achieve the greatest efficiency. To do this, many stoves are designed with baffles or heat chambers, Figure 35-6. ✳Enough heat must leave the stove to provide a good draft condition to vent the combustion gases. This is also necessary to heat the stovepipe and chimney to prevent excessive creosote condensa-

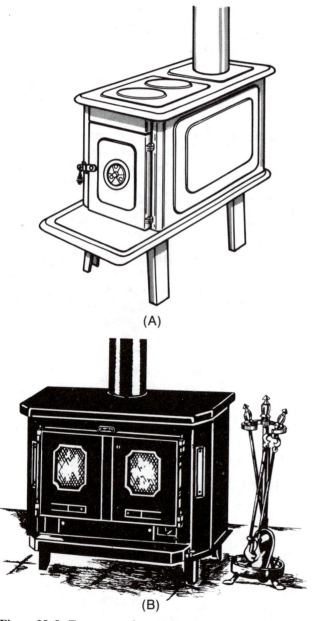

Figure 35-5. Two types of wood-burning stoves: (A) radiant and (B) forced convection. *Courtesy* (B) *Blue Ridge Mountain Stove Works*

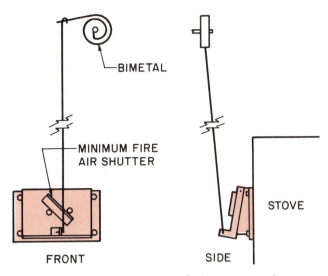

Figure 35–7. A thermostatic damper control.

tion.✱ Good-quality highly efficient stoves have a fire-brick lining to help retain heat and increase the life of the stove by providing added protection to the metal from the heat. Some stoves are designed with thermostatic controls. These controls open and close the damper with a bimetal device, Figure 35–7. Closing the dampers when the space is heated will result in better use of the fuel, thus providing more efficiency.

35.5   CATALYTIC COMBUSTORS

Catalytic combustors have been designed to improve the efficiency of the wood stove and to burn products present in the flue gases. ✱Some of these products are considered harmful air pollutants and, along with others, form creosote. As a result with catalytic combustors, combustion efficiency is improved, harmful air pollutants are reduced, and creosote buildup is reduced.✱

The combustor actually causes most of the products that make up the flue gases or smoke to burn. Smoke from a wood-burning stove contains products that will burn and consequently are a potential source for more

heat. For this smoke to burn, it must be heated to a temperature much higher than is possible or desirable in most stoves, even those with secondary combustion. These temperatures must reach 1000° to 1500°F; additional oxygen is needed, and the smoke must be exposed to these temperatures and the oxygen for a longer time.

The catalytic combustor is a form of an afterburner. Afterburners increase the burning of all by-products of wood, but an additional fuel is needed to do this. Afterburners are not practical devices. Catalytic combustors use a catalyst to cause combustion without the additional fuel. The catalyst causes a chemical reaction, which causes the flue gases to burn at about 500°F rather than the 1000° to 1500°F otherwise required.

Catalytic combustors come in various styles. Most are similar and use the same technology. They are cell-like structures, Figure 35–8, and consist of a substrate, washcoat, and catalyst. The *substrate* is a ceramic material formed into a honeycomb shape. Ceramic material is used because of its stability in extreme cold and hot conditions. The *washcoat,* usually made of an aluminum-based substance called *alumina* covers the ceramic material and helps disperse the catalyst across the combustor surface. The *catalyst* is made of a noble metal, usually platinum, palladium, or rhodium, which are chemically stable in extreme temperatures.

35.6   COMBUSTION IN THE CATALYTIC COMBUSTOR

For combustion to occur in the catalytic combustor, the fuel is the unburned flue gases or smoke, the oxygen is either excess from the main combustion chamber or supplied directly to the combustor, and the heat needed is 500° to 600°F furnished from the combustion of the wood. Once the combustor starts to function, however, the stove firebox temperature can be as low as 250° to 400°F because the combustor will generate enough of its own heat to maintain the combustion.

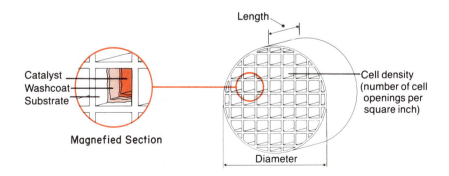

Figure 35–8. Catalytic combustor element. *Courtesy Corning Glassworks*

SAFETY PRECAUTION: *Very high temperatures can be generated by the burning of the flue gases in the combustor. Iron and steel stove parts can be subject to metal oxidation from this heat. These materials should be spaced at least two inches from the combustor; otherwise stainless steel should be used.*

35.7 A TYPICAL STOVE DESIGN WITH A CATALYTIC ELEMENT

Figure 35–9 shows a typical stove design with a catalytic element (F). Following are brief descriptions of each component of the stove:

Wood Combustion Chamber (Firebox).

Heat Exchanger Area. In the heat exchanger area, much of the heat from the firebox and combustor is transferred to the surface area and surrounding space before the gases are exhausted through the flue.

Baffle Plate. The baffle plate separates the firebox from the heat exchanger area. It contains the bypass damper (D), safety bypass (E), and combustor (F).

Bypass Damper. The bypass damper should be opened before opening the firebox door. The combustor restricts smoke exhaust. If this damper were not opened, smoke would come out the firebox door when opened. Most manufacturers recommend that this or a similar damper arrangement be opened when a hot fire exists. This allows the gases to escape and keeps the combustor from being damaged by the high fire and possible flame impingement.

Safety Bypass. The safety bypass is a relatively small hole that allows some smoke to escape by the combustor. This is necessary in case the combustor becomes plugged. Much of this smoke will still be burned as it passes over the combustor on its way to the flue.

Combustor. The combustor will normally be wrapped or enclosed in a stainless steel "can."

Secondary Air Inlet. Some stoves provide a preheated secondary air supply to the combustor through the secondary air distributor, which distributes the air uniformly.

Flame Guard. The flame guard protects the combustor from damage from flame impingement and from physically being damaged while the wood is loaded into the firebox.

Radiation Shield. The radiation shield is provided in many stove designs to protect the metal directly above the combustor from extreme heat. It also reflects radiant heat back to the combustor, causing it to operate hotter and thus more efficiently.

Barometric Damper. The barometric damper is helpful in controlling airflow and in keeping the combustor from overfiring or operating at too high a temperature.

View Port. Many manufacturers provide a viewport to actually see how the combustor is operating. It will glow when above 1000°F. Much of the time it will not be operating at a temperature this high, however. It does not have to glow to be operating properly.

35.8 OPERATING A STOVE WITH A CATALYTIC ELEMENT

When starting the stove, the gas temperatures must be raised to 500° to 600°F (260°–320°C). This is a medium to high firing rate, and this temperature should be maintained for approximately 20 min. This is called achieving *catalytic light-off*. When refueling a stove that has been burning below 500°F, you should let it be fired at a higher temperature for about 10 min to insure proper catalytic operation. Stoves with older combustors may require slightly higher temperatures to achieve light-off. The catalytic activity decreases as the combustor ages. The catalyst itself does not burn but promotes reactions in the hydrocarbons in the smoke. There are two separate reactions: a hydrogen reaction and a carbon monoxide reaction. Both cause burning or oxidation to take place more easily and at a lower temperature.

35.9 CAUSES OF CATALYTIC FAILURE

Failure of the ceramic structure or substrate can be caused by severe temperature cycling (extreme high temperatures at frequent intervals) and uneven air dis-

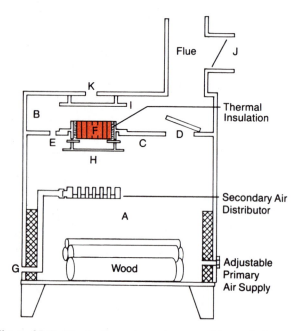

Figure 35–9. Illustration of wood stove with catalytic combustor. *Courtesy Corning Glassworks*

tribution, producing hot and cool zones, Figure 35–10. Flame impingement is another cause of failure.

SAFETY PRECAUTION: *Catalyst failure can be caused by extremely high temperatures (over 1800°F) and by what is known as *catalytic poisoning.* Only natural wood should be burned in a catalytic stove. Materials such as lead, zinc, and sulfur will form alloys with the noble metal used as the catalyst, or compounds will be formed to cover the catalyst.* Either of the above will result in the combustor being ineffective.

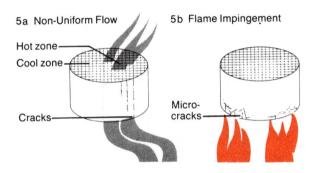

Figure 35–10. Uneven air distribution or flame impingement can cause catalytic failure. *Courtesy Corning Glassworks*

35.10 COMBUSTOR LIFE

Most combustors will last from two to five years, depending on the amount of use. Most stoves are designed for easy replacement of the combustor. An ineffective catalytic combustor will cause a stove to act sluggishly. It will be difficult to start and hard to maintain a good fire. When this happens, the catalytic element should be replaced. If not replaced, it should at least be removed. The stove should then operate properly (but without the benefit of the combustor). Figure 35–11 shows the catalytic element. It is placed in a stainless steel container.

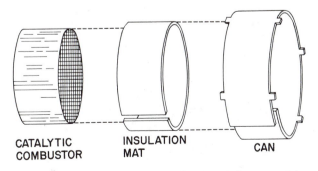

Figure 35–11. Catalytic element is placed in a container ("can") for protection. Elements are very fragile particularly after they have been used. *Courtesy Corning Glassworks*

35.11 AIR POLLUTION CONTROL

Woodsmoke contains both POM (polycyclic organic matter) and non-POM, which are considered to be health hazards and are of concern to many environmentalists. A properly operating catalytic combustor will enable many of these particles to be burned and eliminate most of this hazard.

Some states have enacted legislation and others are expected to do so which will require manufacturers to design and only build wood burning stoves and fireplace inserts which will control these particles from entering the atmosphere.

35.12 GENERAL OPERATING AND SAFETY PROCEDURES

*1. Check the combustor regularly to insure that it is not plugged and that it is otherwise operating properly.
2. Check the chimney and flue regularly for creosote buildup. When operating properly, the combustor will not eliminate all creosote, and normal safe practices should be followed.
3. Insure that safety bypasses are open. A blockage can result in smoke and flue-gas leakage.
4. Insure that all clearances from combustible materials are maintained when installing the stove.
5. Never burn materials other than natural wood. Never burn paper, plastic, coal, or painted wood. These materials will poison the catalyst.
6. Do not allow the flame to impinge on the combustor.*

35.13 LOCATION OF THE STOVE

Because heat rises, most people think the most efficient location for a stove is in the basement. This is true only if the basement area were the primary space to be heated. Remember, many stoves are radiant stoves. The heat will radiate through the air heating solid objects. Most of the heat will be absorbed by the basement walls and solid objects, which in turn will heat the basement space. The heated basement air will then move up through registers in the first floor. However, too much of the heat will be used in heating the basement.

The stove should be placed as near as practical to the space where the heat is desired. This is often near the center of the first floor area. This is not often practical, however, as there may not be a chimney located where the stove should be placed. The stove must be in the same room as the chimney inlet. An approved prefabricated chimney may be installed where it would be most appropriate for the stove and the room. Avoid long stovepipe runs connecting the stove to the chimney. Do not connect the stovepipe to a chimney flue used for another purpose. Therefore, it may be the location of the chimney that determines where the stove will be placed.

35.14 HEAT DISTRIBUTION

The radiant heat from the stove warms the objects and the walls in the room, which heat the air and make it rise. The heat reaches the ceiling area, cools, and descends to the floor area to be reheated and recirculated, Figure 35-12. To heat rooms above the stove, place registers in the ceiling to allow the heated air to pass through the registers and heat the room above. There should also be a register for the cooler air to descend to the lower floor, Figure 35-13.

A stairway near the stove is a good way for the heated air to rise to the upper floor. If the stairway is open to the room where the stove is located, the heated air can rise through the registers over the stove and the cooled air can return down the stairwell, Figure 35-14. If heat is desired in a room beside the one where the stove is located, registers should be cut through the wall next to the ceiling. The heated air rises to the ceiling and passes through these registers near the ceiling, Figure 35-15. It is difficult for the heat to pass from one room to another through doorways because the door openings do not go to the ceiling. The partial wall above the door blocks the circulation. The cooled air will return through the doorways, however. If there are no door openings, registers should be cut through the wall near the floor, also. Small two-level houses can be heated with wood stoves because heat will rise to the second floor. It is difficult to heat a ranch house with one stove. The heat can be ducted from the basement with forced air, but more than one stove is usually required for most satisfactory installations.

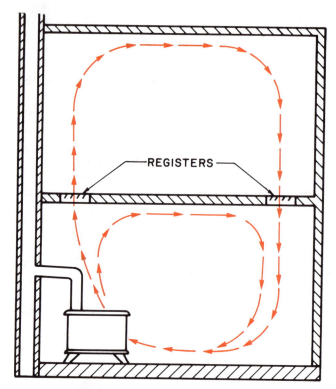

Figure 35-13. Convection circulation pattern for two rooms, one above the other.

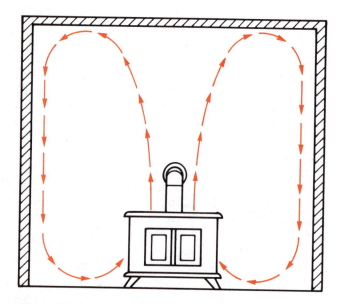

Figure 35-12. Convection circulation pattern for one room.

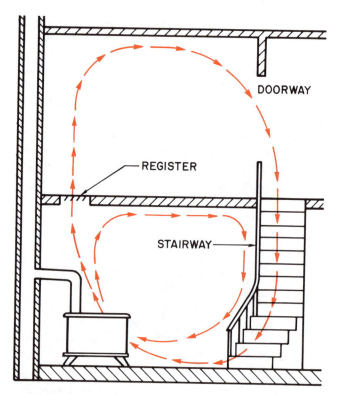

Figure 35-14. Convection circulation pattern using stairway to return cooled air.

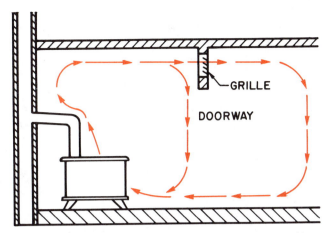

Figure 35–15. Convection circulation pattern with register near ceiling, cooled air returning to stove area through doorway.

35.15 MAKEUP AIR

Air used in combustion must be made up or resupplied from the outside, for example, with an inlet from outside directly to the stove, Figure 35–16. Often an older home has many air leaks. The hot air may leak out near the ceiling or through the attic. Cool air can leak in through cracks around windows, under doors, and elsewhere. Although this may disturb the normal heating cycle, it does help to make up air for that air leaving the chimney as a result of the combustion and venting. *In modern homes that are sealed and insulated well, some provision may have to be made to supply the makeup air. A door may have to be opened a crack to provide a proper draft. A stove could actually burn enough oxygen in a small home to make it difficult to get enough oxygen to breathe. Always make sure there is enough makeup air.*

35.16 SAFETY HAZARDS

*Live coals in the stove in the living area of a house along with creosote in stovepipes and chimneys make it absolutely necessary to install, maintain, and operate stoves safely. This cannot be emphasized enough.

Following are some of the safety hazards that may be encountered.

- A hot fire can ignite a buildup of creosote, resulting in a stovepipe and/or chimney fire.
- Radiation from the stove or stovepipe may overheat walls, ceilings, or other combustible materials in the house and start a fire.
- Sparks may get out of the stove, land on combustible materials, and ignite them. This could happen through a defect in the stove, while the door is left ajar, while the firebox is being filled or while ashes are removed.

- Flames could leak out through faulty chimneys, or heat could be conducted through cracks to a combustible material.
- Burning materials coming out of the top of the chimney can also start a fire at the outside of the house. These sparks or glowing materials can ignite roofing materials, leaves, brush, or other matter outside the house.*

35.17 INSTALLATION PROCEDURES

*A wood-burning stove or appliance should be approved by a national testing laboratory. Before installing a stove, stovepipe, or prefabricated chimney, be sure that all building and/or fire marshal's codes are followed as well as the instructions of the testing laboratory and manufacturer. If one code or set of instructions is more restrictive than another, it is absolutely necessary that the most restrictive instructions be followed.

These instructions should include the distance the stove should be located from any combustible material, such as a wall or the floor. They should indicate the minimum required protective material between the stove and the wall and between the stove and the floor. Excessive heat from the stove can heat the walls or floor to the point where a fire can be started.

The stove must be connected to the chimney with an approved stovepipe, often called a *connector* pipe. Single-wall or double-wall stovepipe may be used for this, Figure 35–17. The double-wall pipe has an air space between the two walls. Normally this type of stovepipe installation would require less spacing to the nearest combustible material than a single-wall pipe would. Codes and instructions must be followed.

A stove collar adapter should be used to install the stovepipe to the stove. The stovepipe should be placed on the outside of the stove collar and yet it must also be fitted inside the collar so that the creosote will run into, not on top of, the stove, Figure 35–18.

Figure 35–19 shows three different types of installations. The stovepipe must not run through any com-

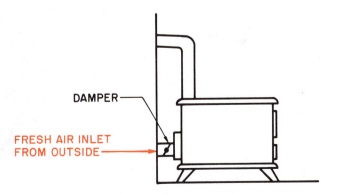

Figure 35–16. Stove with fresh-air tube from outside.

Figure 35–17. (A) Single-wall stovepipe section. (B) Double-wall stovepipe section.

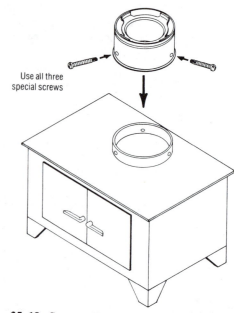

Use all three special screws

Figure 35–18. Stove collar adapter for double-wall stovepipe.

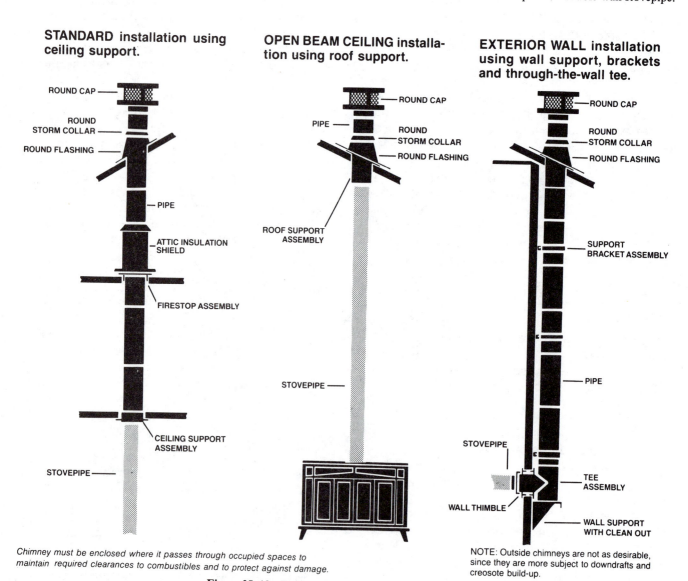

STANDARD installation using ceiling support.

ROUND CAP
ROUND STORM COLLAR
ROUND FLASHING
PIPE
ATTIC INSULATION SHIELD
FIRESTOP ASSEMBLY
CEILING SUPPORT ASSEMBLY
STOVEPIPE

Chimney must be enclosed where it passes through occupied spaces to maintain required clearances to combustibles and to protect against damage.

OPEN BEAM CEILING installation using roof support.

ROUND CAP
PIPE
ROUND STORM COLLAR
ROUND FLASHING
ROOF SUPPORT ASSEMBLY
STOVEPIPE

EXTERIOR WALL installation using wall support, brackets and through-the-wall tee.

ROUND CAP
ROUND STORM COLLAR
ROUND FLASHING
SUPPORT BRACKET ASSEMBLY
PIPE
STOVEPIPE
TEE ASSEMBLY
WALL THIMBLE
WALL SUPPORT WITH CLEAN OUT

NOTE: Outside chimneys are not as desirable, since they are more subject to downdrafts and creosote build-up.

Figure 35–19. Three different types of stove installations.

bustible material such as a ceiling or wall. Approved chimney sections with necessary fittings should be used. Figure 35–20 illustrates details for adapting the stovepipe to the "through the wall" chimney fittings. Remember to keep horizontal stovepipe runs to a minimum. A rise of $\frac{1}{4}$ in. per foot should be considered a minimum, Figure 35–3. Local codes or manufacturers may require a greater rise. Figure 35–21 shows various prefabricated chimney fittings and sections. Insure that approved thimbles, joist shields, insulation shields, and other necessary fittings are used where required. Insure that all materials are approved by a recognized national testing laboratory.

Chimneys may be double walled with insulation between the walls or triple walled with air space between the walls. This insulating material may be a ceramic fiber refractory blanket, Figure 35–22.

Only one stove can be connected to a chimney. For masonry chimneys with more than one flue, no more than one stove can be connected to each flue. A factory-built chimney used for wood stoves should be rated as residential and building heating appliance chimney. Check local codes.

Many stoves are vented through masonry chimneys. If a new chimney is being used, it should have been constructed according to applicable building codes and carefully inspected. If an old chimney is being used, it should be inspected by an experienced person to insure that it is safe to use. The mortar in the joints may be deteriorated and loose or the chimney may have been

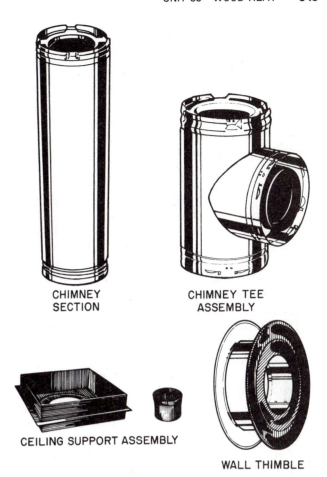

CHIMNEY SECTION   CHIMNEY TEE ASSEMBLY

CEILING SUPPORT ASSEMBLY

WALL THIMBLE

Figure 35–21. Prefabricated chimney fittings and sections

cracked by a serious chimney fire. All necessary repairs should be made to the chimney by a competent mason before connecting the stove to it.*

35.18   SMOKE DETECTORS

Smoke detectors should be used in all homes regardless of the type of heating appliance used. It is even more important to install smoke detectors when burning wood fuel. Many types of detectors are available. Either an AC-powered photoelectric type or an AC- or battery-powered ionization chamber detector are usually satisfactory.

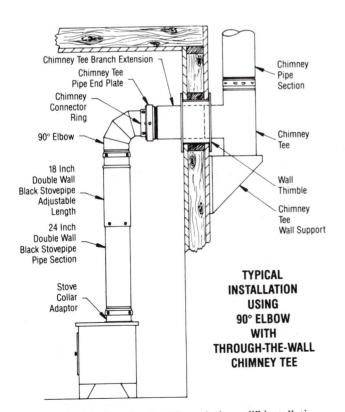

Chimney Tee Branch Extension
Chimney Tee Pipe End Plate
Chimney Connector Ring
90° Elbow
18 Inch Double Wall Black Stovepipe Adjustable Length
24 Inch Double Wall Black Stovepipe Pipe Section
Stove Collar Adaptor

Chimney Pipe Section
Chimney Tee
Wall Thimble
Chimney Tee Wall Support

TYPICAL INSTALLATION USING 90° ELBOW WITH THROUGH-THE-WALL CHIMNEY TEE

Figure 35–20. Details of a "through the wall" installation.

Figure 35–22. Double-walled chimney section with ceramic fiber refractory blanket.

35.19   FIREPLACE INSERTS

Fireplace inserts can convert a fireplace from a very inefficient heating source to one that is fairly efficient. Very little heat can be obtained from a fireplace without some device to get the heat from the fireplace into the room. The fireplace insert provides a way to retain the heat and blow it into the room. When fireplace fires are burned regularly without such a device, the heat is almost totally wasted, Figures 35–23, 35–24. *For an insert to be used, the fireplace and chimney must generally be of masonry construction. Local building codes and the manufacturer's specifications for the particular insert should be consulted to insure that the size and construction of the fireplace will accommodate the insert. Most factory-built or prefabricated fireplaces are not approved for inserts. Before an insert is installed, the chimney should be inspected. If there is creosote buildup, the chimney must be cleaned. Check for cracks or flaws in the chimney that could be a fire hazard. Inserts should never be used in a chimney flue used for another purpose.*

Instead of an insert many people wish to install a wood stove and exhaust the flue gases through the fireplace. There are two methods that may be used to make this conversion. An opening can be cut into the flue liner of the chimney above the fireplace opening, Figure 35–25. A flue adapter should be extended from the stove connector pipe into the chimney flue as in Figure 35–26. Connecting the pipe directly into the flue liner generally yields a better draft. *A mason should be employed to cut the opening and install the thimble. Be sure that there

Figure 35–24. Fireplace insert installed. *Courtesy Blue Ridge Mountain Stove Works*

is adequate clearance between the pipe and the ceiling, any wooden mantel, or other combustible material. The stovepipe should extend into the thimble as far as possible without protruding into the flue. The fireplace damper should be securely shut. An approved sealer should be used between the thimble and the masonry.*

To make the installation through the fireplace opening, first remove the damper. If it cannot be removed, fasten it in the open position. A prefabricated adapter should be purchased to provide the proper fitting and correct size to accept the connector pipe.

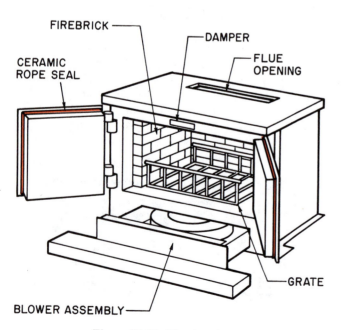

Figure 35–23. Fireplace insert.

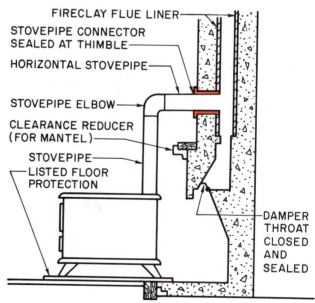

Figure 35–25. Flue opening above fireplace opening.

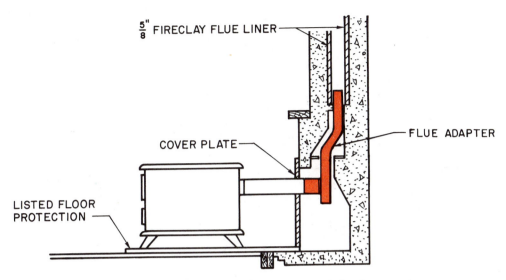

Figure 35-26. A flue adapter installed through fireplace opening.

35.20 ADD-ON WOOD-BURNING FURNACES

Wood-burning furnaces may be installed to operate in conjunction with an existing furnace, Figure 35-27. *Manufacturer's installation instructions and specifications must be followed carefully.* This add-on furnace may be used as the primary source of heat with the existing furnace supplementing it in extreme cold weather, or the wood add-on furnace may be used as the supplementary source of heat with the original furnace providing the primary heat.

35.21 DUAL-FUEL FURNACES

Dual-fuel furnaces provide the same service as the add-on and existing furnaces do except they are two furnaces in one unit. They are combination wood-burning furnaces that operate in conjunction with another type of fuel such as oil, gas, or electricity. Some of these furnaces are designed so that the wood is ignited by the combustion from the oil or gas side. Wood can be used as the primary fuel with the oil, gas, or electricity, or vice versa. Figure 35-28 is an illustration of a typical wood/oil dual furnace.

Thermostats for dual-fuel units are two-stage units. One stage controls the wood fire, and the second the other source of heat. If the stage controlling the wood furnace is set higher than the other, the thermostat will call for heat from the wood furnace first. When it calls for heat, the damper will open, allowing primary air to the wood fire. The wood will then increase its burning rate, and when the thermostat is satisfied the damper will close. The primary air will be shut way down, and

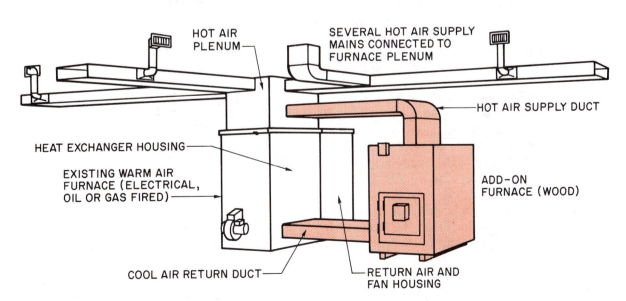

Figure 35-27. Add-on furnace. *Courtesy National Stove Works*

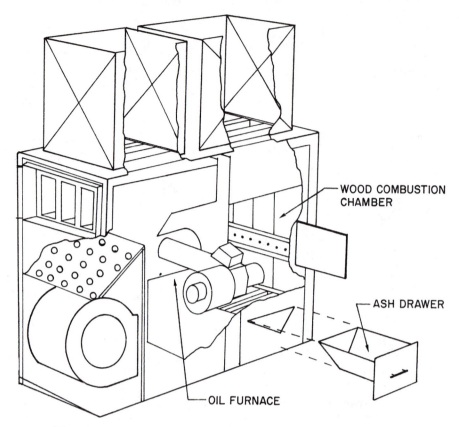

Figure 35–28. Dual-fuel furnace. *Courtesy Yukon Energy Corporation*

the fire will burn at a very slow rate. When the wood burns down to the point where it cannot produce the heat needed, the backup furnace will come on. The settings on the thermostat may be reversed if another fuel is the primary heat source.

Again, when installing the furnace insure that proper clearances to combustible materials are maintained. Proper clearances are also required for the duct work. The duct should also be large enough to handle natural convection airflow in the event there is an electrical power failure or a furnace fan failure. This guarantees heat to the house when power is off or the fan is not functioning properly and allows the heated air to pass through without overheating the duct.

It is very important to locate the draft regulator properly. It should be located in the connector pipe as close to the furnace as possible, Figure 35–29. It must be in the same room as the furnace.

Outside air to assure proper combustion must be available or provided. The fire in the furnace uses oxygen and must have a continuous supply. As with a wood-burning stove, the air in the house contains only enough oxygen to support the combustion for a short time. This makeup air should be provided through a fresh-air duct similar to that shown in Figure 35–30.

35.22   FACTORY-BUILT CHIMNEYS

*Factory-built chimneys (prefabricated) must be installed in accordance with the conditions stated by the listing laboratory. These conditions will be found in the manufacturer's instructions. Factory-built chimneys

BEST DRAFT REGULATOR LOCATIONS

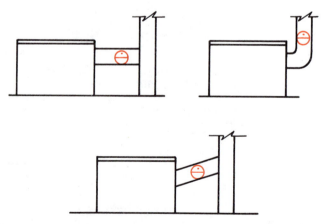

Figure 35–29. Suggested draft regulator locations. *Courtesy Yukon Energy Corporation*

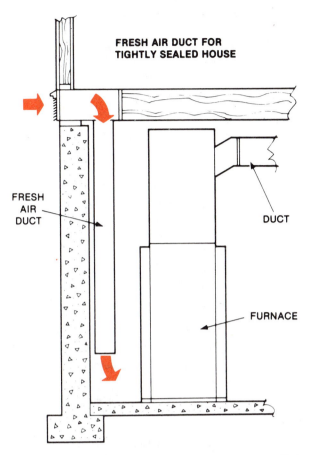

FRESH AIR DUCT FOR TIGHTLY SEALED HOUSE

FRESH AIR DUCT

DUCT

FURNACE

Figure 35-30. Typical makeup air duct. *Courtesy Yukon Energy Corporation*

used for wood furnaces must be listed as residential and building heating appliance chimneys. Masonry chimneys must be constructed according to local codes.*

35.23   CLEANING THE CHIMNEY, CONNECTOR PIPE, AND HEAT EXCHANGER

Chimney fires should be avoided at all costs. A check for creosote and soot buildup should be made twice a month. Creosote cannot only cause fire, but soot buildup on the heat exchanger in a furnace or stove can result in a great loss in efficiency.

Round steel brushes are the best for cleaning metal connector pipe and metal and clay flue chimneys. It is generally better to clean chimneys from the top. If the chimney is from a fireplace, it may be cleaned through the fireplace. An extendable handle should be used so that the brush can be inserted into the chimney, brushing and sweeping downward until the entire length of the chimney is cleaned. A weight may be placed under the brush, and the chimney can be cleaned by lowering the brush into the chimney. The brush will sweep the sides of the flue lining as it is raised.

The debris will be at the bottom of the chimney at the cleanout opening. Sweep the debris into a container. The connector pipe can be cleaned in a similar

manner with a brush of the proper size. A smaller flat steel brush can be used for cleaning the heat exchanger in the furnace. Be sure to provide a covering for flooring, carpeting or furniture where you may be working. You should have an industrial vacuum cleaner available when working where the soot and creosote can fall and settle in a living area.

Codes, regulations, and laws regarding wood heating appliances and components change frequently. It is essential that you be familiar with the appropriate current National Fire Protection Association information and local codes and laws.

SUMMARY

- Many people are burning wood as an alternative fuel for heating.
- Wood is classified as hardwood or softwood.
- Generally, hardwoods provide more heat than softwoods.
- Wood is sold by the cord which is a stack 4 ft $\times$ 4 ft $\times$ 8 ft or 128 ft^3.
- *Creosote is a product of incomplete combustion. It condenses in the stovepipe and chimney and can be hazardous. It can ignite and burn with intensity. The following conditions contribute to creosote buildup: smoldering low-heat fires, smoke in contact with cool surfaces in the stovepipe and chimney, burning green wood, burning softwood.*
- To help prevent creosote buildup, burn dry hardwood and insure that there is an adequate rise in the stovepipe.
- *Inspect and clean the stovepipe and chimney regularly.*
- Types of appliances in which wood can be burned are wood-burning stoves, fireplace inserts, wood-burning furnaces, and dual-fuel furnaces.
- Some wood-burning stoves heat primarily by radiation. The heat radiates into the room heating objects, walls, and floor. These in turn heat the air.
- Some stoves have a jacket that allows air to circulate around the stove. These are available in both natural-convection and forced-convection designs.
- Efficient stoves are airtight, which permits more control over the combustion.
- Catalytic combustors are used to improve heating efficiency and reduce creosote.
- Stoves should be located near the center of the area to be heated, but they must be located near a chimney and in the same room as the inlet to the chimney. Avoid long runs of stovepipe.
- Registers should be cut in the ceiling if room(s) above the stove are to be heated. Registers should be cut near the ceiling and floor in the wall if the room next to the room with a stove is to be heated.
- *Makeup air for air used in combustion should be provided through a fresh-air duct.

Possible safety hazards that may be encountered with stoves:

- Stovepipe and chimney fires may result from excessive creosote buildup.
- Radiation may overheat walls, ceilings, or other combustible materials.
- Sparks or glowing coals may land on combustible materials and ignite them.
- Flames could leak out through cracks in a faulty chimney.
- Burning materials coming out of the top of a chimney can ignite roofing materials or combustibles on the ground.

- Minimum clearances between stove and combustibles should be adhered to.
- Use only approved stovepipe, chimneys, and other components. Follow all manufacturer's instructions and specifications.
- Use smoke detectors.
- Install fireplace inserts only in approved fireplaces.
- Insure that proper clearances are maintained around fireplace inserts.
- If the stove is vented through the fireplace chimney, the damper opening must be completely sealed.*
- Typical thermostatic damper controls for stoves use a bimetal device to open and close the damper.
- Wood-burning furnaces may be installed to supplement an existing furnace. Each furnace can be controlled separately.
- Dual-fuel furnaces are two furnaces in one unit. They are normally wood furnaces with either an oil, gas, or electric furnace operating in conjunction with it.
- Two-stage thermostats are used for these furnaces. The stage for the primary heat (usually the wood) is set a few degrees above the other.
- *Adequate combustion makeup air should be provided.*
- *When installing the furnace and duct work, clearances to combustibles must be maintained.*
- Round steel brushes are best for cleaning connector pipe and chimney flues. An extendable handle should be used to clean the full length of the chimney from the top down. Heat exchangers in furnaces should be kept clean. A flat steel brush can be used for this.

REVIEW QUESTIONS

1. When is it practical to burn wood for home heating?
2. Describe the organic makeup of wood.
3. List three hardwoods.
4. List two softwoods.
5. Describe a cord of wood.
6. Describe how cut wood should be stacked for drying.
7. What are the primary causes of creosote buildup?
8. Why is creosote buildup dangerous?
9. What procedures can help prevent excessive creosote buildup?
10. Why should stovepipe be assembled with the crimped end down?
11. List other procedures to be followed when installing stovepipe.
12. Why are airtight stoves preferred over those with air leaks?
13. Why are some stoves designed with one or more internal baffles?
14. What factors should be considered when locating a stove in a house?
15. Why should registers be cut in a wall to a room beside the room where the stove is located?
16. Sketch the paths of air heated by (a) a stove in a room; (b) in a room with a room above it; (c) in a room with a room beside it.
17. Describe the operation of a catalytic combustor.
18. How can air used for combustion in a room be made up?
19. List safety hazards that may be encountered with stoves.
20. Where is a thimble used?
21. What is a fireplace insert?
22. How can a fireplace insert improve on the heating efficiency of a fireplace?
23. What are two methods for venting a stove through a fireplace or fireplace chimney?
24. What is an add-on furnace?
25. What is a dual-fuel furnace?
26. Explain how a dual-fuel furnace operates.
27. Sketch a design for getting combustion air makeup through a fresh-air duct.
28. Describe how a chimney should be cleaned.

HUMIDIFICATION

OBJECTIVES

After studying this unit, you should be able to

- explain relative humidity.
- list reasons for providing humidification in winter.
- discuss the differences between evaporative and atomizing humidifiers.
- describe bypass and under-duct-mount humidifiers.
- describe disc, plate, pad, and drum humidifier designs and the media used.
- explain the operation of the infrared humidifier.
- explain why a humidifier used with a heat pump or electric furnace may have its own independent heat source.
- describe the spray-nozzle and centrifugal atomizing humidifiers.
- state the reasons for installing self-contained humidifiers.
- list general factors used when sizing humidifiers.
- describe general procedures for installing humidifiers.

36.1   RELATIVE HUMIDITY

In fall and winter, homes very often are dry because cold air from outside infiltrates the conditioned space. The infiltration air in the home is artificially dried out when it is heated because it expands spreading out the moisture. The amount of moisture in the air is measured or stated by a term called *relative humidity*. It is the percentage of moisture in the air compared to the capacity of the air to hold moisture. In other words, if the relative humidity is 50%, each cubic foot of air is holding one-half the moisture it is capable of holding. The relative humidity of the air decreases as the temperature increases because air with higher temperatures can hold more moisture. When a cubic foot of 20° F outside air at 50% relative humidity is heated to room temperature (75° F), the relative humidity of that air drops.

For comfort, the dried-out air should have its moisture replenished. The recommended relative humidity for a home is between 35% and 50%. The dry warm air draws moisture from everything in the conditioned space, including carpets, furniture, woodwork, plants, and people. Furniture joints loosen, nasal and throat passages dry out, and skin becomes very dry. Dry air causes more energy consumption than necessary because the air gets moisture from the human body through evaporation from the skin. The person then feels cold and sets the thermostat a few degrees higher to become comfortable. With more humidity in the air a person is more comfortable at a lower temperature.

Static electricity is also much greater in dry air. It produces discomfort because of the small electrical shock a person receives when touching something after having walked across the room.

36.2   HUMIDIFICATION

Years ago people placed pans of water on radiators or on stoves. They even boiled water on the stove to make moisture available to the air. The water evaporated into the air and raised the relative humidity. Although this may still be done in some homes, efficient and effective equipment called *humidifiers* produce this moisture and make it available to the air by *evaporation*. The evaporation process is speeded up by using power, heat, or by passing air over large areas of water. The area of the water can be increased by spreading it over pads or by atomization.

36.3   HUMIDIFIERS

Evaporative humidifiers work on the principle of providing moisture on a surface called a *media* and exposing it to the dry air. This is normally done by forcing the air through or around the media and picking up the moisture from the media as a vapor or a gas. There are several types of evaporative humidifiers.

Bypass Humidifier.   The bypass humidifier relies on the difference in pressure between the supply (warm) side of the furnace and the return (cold) side. They may be mounted on either the supply plenum or duct or the cold-air return plenum or duct. Piping must be run from the plenum or duct where it is mounted to the other plenum or duct. If mounted in the supply duct, it must be piped to the cold-air return, Figures 36–1, 36–2. The difference in pressure between the two plenums draws some air through the humidifier to the supply duct and is distributed throughout the house.

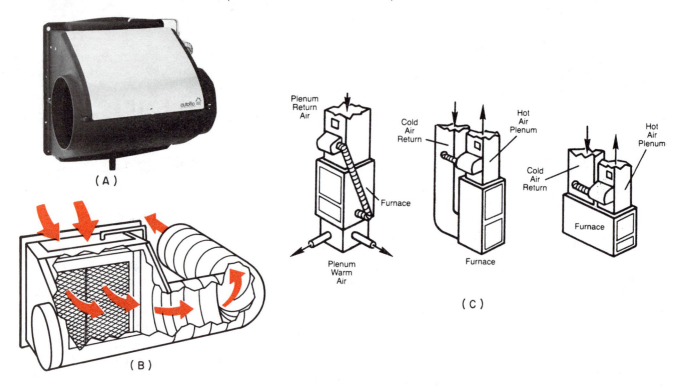

Figure 36-1. (A) Bypass humidifier. (B) Cutaway showing airflow from plenum through media to the return. (C) Typical installations. *Courtesy AutoFlo Company*

Plenum-Mount Humidifier. The plenum-mount humidifier is mounted in the supply plenum or the return-air plenum. The furnace fan forces air through the media where it picks up moisture. The air and moisture are then distributed throughout the conditioned space, Figure 36–3.

Under-Duct-Mount Humidifier. The under-duct-mount humidifier is mounted on the underside of the supply duct so that the media is extending into the airflow where moisture is picked up in the airstream. Figure 36–4 illustrates an under-duct humidifier.

36.4 HUMIDIFIER MEDIA

Humidifiers are available in several designs with various kinds of media. Figure 39–4 is a photo of a type using disc screens mounted at an angle. These discs are mounted on a rotating shaft that makes the slanted discs pick up moisture from the reservoir. The moisture is then evaporated into the moving airstream. The discs are separated to prevent electrolysis, which causes the minerals in the water to form on the media. The

Figure 36-2. Bypass humidifier used between plenums with a pipe to each. *Courtesy Aqua-Mist, Inc.*

Figure 36-3. A plenum-mounted humidifier with a plate type media that absorbs water from the reservoir and evaporates into the air in the plenum. *Courtesy AutoFlo Company*

Figure 36–4. Under-duct humidifier using disc screens as the media. *Courtesy Humid Aire Division, Adams Manufacturing Company*

Figure 36–6. Humidifier with electric heating elements. *Courtesy AutoFlo Company*

wobble from the discs mounted at an angle washes the minerals off and into the reservoir. The minerals can then be drained from the bottom of the reservoir.

Figure 36–5 illustrates a type of media in a drum design. A motor turns the drum, which picks up moisture from the reservoir. The moisture is then evaporated from the drum into the moving airstream. The drums can be screen or sponge types.

A plate- or pad-type media is shown in Figure 36–3. The plates form a wick that absorbs water from the reservoir. The airstream in the duct or plenum causes the water to evaporate from the wicks or plates.

Figure 36–6 shows an electrically heated water humidifier. In electric-furnace and heat pump installations, the temperature in the duct is not as high as in other types of hot-air furnaces. Media evaporation is not as easy with lower temperatures. The electrically heated humidifier heats the water with an electric ele-

ment, causing it to evaporate and be carried into the conditioned space by the airstream in the duct.

Another type is the infrared humidifier, Figure 36–7, which is mounted in the duct and has infrared lamps with reflectors to reflect the infrared energy onto the water. The water thus evaporates rapidly into the duct airstream and is carried throughout the conditioned space. This action is very similar to the sun's rays shining on a large lake and evaporating the water into the air.

Humidifiers are generally controlled by a *humidistat,* Figure 36–8. The humidistat controls the motor and/or the heating elements in the humidifier. The humidistat has a moisture-sensitive element, often made of hair or nylon ribbon. This material is wound around two or more bobbins and shrinks or expands,

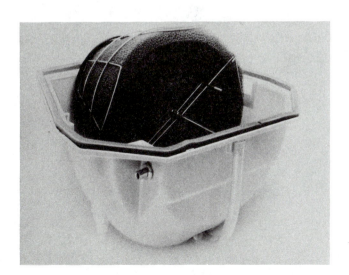

Figure 36–5. Under-duct humidifier using drum-style media. *Courtesy Herrmidifier Company, Inc.*

Figure 36–7. Infrared humidifier. *Courtesy Humid Aire Division, Adams Manufacturing Company*

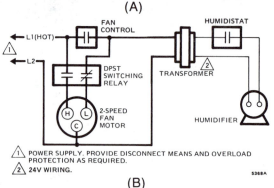

Figure 36-8. (A) Humidistat. (B) Wiring diagram. *Courtesy of Honeywell, Inc., Residential Division*

depending on the humidity. Dry air causes the element to shrink, which activates a snap-action switch and starts the humidifier. Many other devices are used, including electronic components which vary in resistance with the humidity.

36.5 ATOMIZING HUMIDIFIERS

Atomizing humidifiers discharge tiny water droplets (mist) into the air, which evaporate very rapidly into the duct airstream or directly into the conditioned space. These humidifiers can be *spray-nozzle* or *centrifugal* types, but they should not be used with hard water because it contains minerals (lime, iron, etc.) that leave the water vapor as dust and will be distributed throughout the house. Eight to 10 grains of water hardness is the maximum recommended for atomizing humidifiers.

The spray-nozzle type sprays water through a metered bore of a nozzle into the duct airstream where it is distributed to the occupied space. Another type sprays the water onto an evaporative media where it is absorbed by the airstream as a vapor. They can be mounted in the plenum, under the duct, or on the side

Figure 36-9. Combination spray-nozzle and evaporative pad humidifier. *Courtesy Aqua-Mist, Inc.*

of the duct. It is generally recommended that atomizing humidifiers be mounted on the hot-air or supply side of the furnace. Figures 36–9 and 36–10 illustrate two types.

The centrifugal atomizing humidifier uses an impeller or slinger to throw the water and break it into particles that are evaporated in the airstream, Figure 36–11.

SAFETY PRECAUTIONS: *Atomizing humidifiers should operate only when the furnace is operating, or moisture will accumulate and cause corrosion, mildew, and a major moisture problem where it is located.* Some models operate with a thermostat that controls

Figure 36-10. Atomizing humidifier. *Courtesy AutoFlo Company*

Figure 36-11. Centrifugal humidifier. *Courtesy Herrmidifier Company, Inc.*

a solenoid valve turning the unit on and off. The furnace must be on and heating before this type will operate. Others, wired in parallel with the blower motor, operate with it. Most are also controlled with a humidistat.

36.6 SELF-CONTAINED HUMIDIFIERS

Many residences and light commercial buildings do not have heating equipment with duct work through which the heated air is distributed. Hydronic heating systems, electric baseboard, or unit heaters, for example, do not utilize duct work.

To provide humidification where these systems are used, self-contained humidifiers may be installed. These generally employ the same processes as those used with forced-air furnaces. They may use the evaporative, atomizing, or infrared processes. These units may include an electric heating device to heat the water, or the water may be distributed over an evaporative media. A fan must be incorporated in the unit to

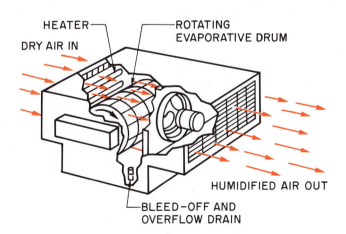

Figure 36-12. Drum-type self-contained humidifier.

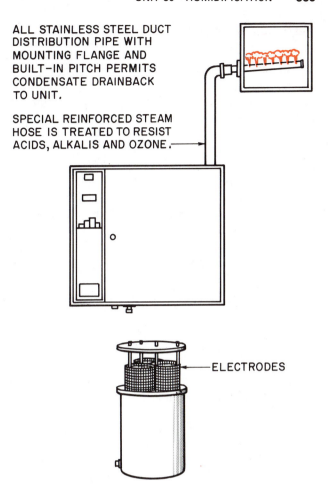

ALL STAINLESS STEEL DUCT DISTRIBUTION PIPE WITH MOUNTING FLANGE AND BUILT-IN PITCH PERMITS CONDENSATE DRAINBACK TO UNIT.

SPECIAL REINFORCED STEAM HOSE IS TREATED TO RESIST ACIDS, ALKALIS AND OZONE.

ELECTRODES

Figure 36-13. Self-contained steam humidifier.

distribute the moisture throughout the room or area. Figure 36-12 illustrates a drum type. A design using steam is shown in Figure 36-13. In this system the electrodes heat the water, converting it to steam. The steam passes through a hose to a stainless steel duct. Steam humidification is also used in large industrial applications where steam boilers are available. The steam is distributed through a duct system or directly into the air.

36.7 PNEUMATIC ATOMIZING SYSTEMS

Pneumatic atomizing systems use air pressure to break up the water into a mist of tiny droplets and disperse them.

SAFETY PRECAUTION: ＊These systems as well as other atomizing systems should only be applied where the atmosphere does not have to be kept clean or where the water has a very low mineral content because the minerals in the water are also dispersed in the mist throughout the air. The minerals fall out and accumulate on surfaces in the area. These are often used in manufacturing areas, such as textile mills. ＊

36.8 SIZING HUMIDIFIERS

The proper size humidifier should be installed. This text emphasizes installation and service, so the details for determining the size or capacity of humidifiers will not be covered. However, the technician should be aware of some very general factors involved in the sizing process:

1. The number of cubic feet of space to be humidified. This is determined by taking the number of heated square feet of the house and multiplying it by the ceiling height. A 1500-ft^2 house with an 8-ft ceiling height would have 1500 ft$^2 \times 8$ ft = 12,000 ft^3.
2. The construction of the building. This includes quality of insulation, storm windows, fireplaces, building "tightness," and so on.
3. The amount of air change per hour and the approximate lowest outdoor temperature.
4. The level of relative humidity desired.

36.9 INSTALLATION

The most important factor regarding installation of humidifiers is to follow the manufacturer's instructions. Evaporative humidifiers are often operated independent of the furnace. It is normally recommended that they be controlled by a humidistat, but it does no real harm for them to operate continuously, even when the furnace is not operating. Atomizing humidifiers, however, should not operate when the furnace (or at least the blower) is not operating. Moisture will accumulate in the duct if allowed to do so.

Particular attention should be given to clearances within the duct or plenum. The humidifier should not exhaust directly onto air conditioning coils, air filters, electronic air cleaners, blowers, or turns in the duct.

If mounting on a supply duct, choose one that serves the largest space in the house. The humid air will spread throughout the house, but the process will be more efficient when given the best distribution possible.

Plan the installation carefully, including locating the humidifer, as already discussed, and providing the wiring and plumbing (with drain). A licensed electrician and/or plumber must provide the service where required by code or law.

36.10 SERVICE, TROUBLESHOOTING, AND PREVENTIVE MAINTENANCE

Proper service, troubleshooting, and preventive maintenance play a big part in keeping humidifying equipment operating efficiently. Cleaning the components that are in contact with the water is the most important factor. The frequency of cleaning depends on the hardness of the water: the harder the water, the more minerals in the water. In evaporative systems these minerals collect on the media, on other moving parts, and in the reservoir. In addition, algae growth can cause problems, even to the extent of blocking the output of the humidifier. Algicides can be used to help neutralize algae growth. The reservoir should be drained regularly if possible, and components, particularly the media, should be cleaned periodically.

Humidifier Not Running. When the humidifier doesn't run, the problem is usually electrical, or a component is bound tight or locked due to a mineral buildup. A locked condition may cause a thermal overload protector to cut out. Using troubleshooting techniques described in other units in this text, check overload protection, circuit breakers, humidistat, and low-voltage controls if there are any. See if the motor is burned out.

Excessive Dust. If excessive dust is caused by the humidifier, the dust will be white due to mineral build-up on the media. Clean or replace the media. If excessive dust occurs in an atomizing humidifier, the wrong equipment has been installed.

Water Overflow. Water overflow indicates a defective float-valve assembly. It may need cleaning, adjusting, or replacing.

Moisture in or Around Ducts. Moisture in ducts is found only in atomizing humidifiers. *Remember, this equipment should operate only when the furnace operates.* Check the control to see if it operates at other times, such as COOL or FAN ON modes. A restricted airflow may also cause this problem.

Low or High Levels of Humidity. If the humidity level is too high or too low, check the calibration of the humidistat by using a sling psychrometer. If it is out of calibration, it may be possible to adjust it. Insure that the humidifier is clean and operating properly.

SUMMARY

- In fall and winter, homes are dry because colder air infiltrates into the home and is heated. When heated, the air is artificially dried out by expansion.
- The relative humidity is the percentage of moisture in the air compared to the capacity of the air to hold the moisture.
- Dry air draws moisture from carpets, furniture, woodwork, plants, and people and frequently has a detrimental effect.
- Humidifiers put moisture back into the air.
- Evaporative humidifiers provide moisture to a media and force air through it, evaporating the moisture.
- Evaporative humidifiers may be a bypass type mounted outside the duct work on a forced-air furnace with piping from the hot air through the

humidifier to the cold-air side of the furnace. They may be mounted in the plenum or under the duct.
- The media may be disc, drum, or plate types.
- For heat pump or electric heat installations the humidifier may have an independent electric heater to increase the temperature of the water for evaporation.
- Infrared humidifiers use infrared lamps and a reflector to cause the water to be evaporated into the airstream.
- Atomizing humidifiers may be spray nozzle or centrifugal types. They discharge a mist (tiny water droplets) into the air. They should not be used in hard-water conditions, and only where there is forced air movement.
- Self-contained humidifiers are used with hydronic, electric baseboard, or unit heaters, where forced air is not available. They have their own fan and often a heater.
- The number of cubic feet in the house, the type of construction, the amount of air change per hour, the lowest outdoor temperatures, and the relative humidity level desired determine the size of the humidifier.
- *Manufacturer's instructions should be followed carefully when installing humidifiers.*

REVIEW QUESTIONS

1. Why are homes drier in winter than in summer?
2. Explain relative humidity.
3. Why does relative humidity decrease as the temperature increases?
4. How does dry warm air dry out household furnishings?
5. Why does dry air cause a "cool feeling"?
6. How is moisture added to the air in a house?
7. Describe the differences between evaporative and atomizing humidifiers.
8. How does a bypass humidifier operate?
9. Is a plenum-mounted humidifier installed in the supply or return air plenum?
10. What is the purpose of the media in an evaporative humidifier?
11. Describe disc, plate, and drum humidifier designs. Describe the media that can be used on each.
12. Why do some humidifiers have their own water heating device?
13. Explain how an infrared humidifier operates.
14. What are the two types of atomizing humidifiers?
15. Describe how each type of atomizing humidifier operates.
16. Why is it essential for the furnace to be running when an atomizing humidifier is operating?
17. Why are self-contained humidifiers used?
18. Describe a self-contained humidifier.
19. What general factors are considered when determining the size or capacity of a humidifier?
20. Describe, in general terms, installation procedures for humidifiers used with forced-air furnaces.

37

COMFORT

OBJECTIVES

After studying this unit, you should be able to

- **recognize the four factors involved in comfort.**
- **explain the relationship of body temperature to room temperature.**
- **describe why one person is comfortable when another is not.**
- **define psychrometrics.**
- **define wet bulb and dry bulb.**
- **define dew-point temperature.**
- **explain vapor pressure of water in air.**
- **describe humidity.**
- **plot air conditions using a psychrometric chart.**

37.1   COMFORT

Comfort describes the delicate balance of feeling in the body in relationship to its surroundings. To be comfortable describes our surroundings when we are not aware of discomfort. Providing a comfortable atmosphere for people becomes the job of the heating and air conditioning profession. Comfort involves four things, (1) Temperature, (2) Humidity, (3) Air Movement and (4) Air Cleanliness.

The human body has a sophisticated control system for both protection and comfort. It can go from a warm house to 0°F outside, and it starts to compensate for the surroundings. It can go from a cool house to 95°F outside, and it will start to adjust to keep you comfortable and from overheating. Body adjustments are accomplished by the circulatory and respiratory systems. When the body is exposed to a climate that is too cold, it starts to shiver, an involuntary reaction, to warm the body. When the body gets too warm, the vessels next to the skin dilate to get the blood closer to the surrounding air in an effort to increase the heat exchange with the air. If this does not cool the body, it will break into a sweat. When this sweat is evaporated, it takes heat from the body and cools it.

37.2   FOOD ENERGY AND THE BODY

The human body is similar to a coal hot-water boiler. The coal is burned in the boiler to create heat. Heat is energy. Food to the human body is like coal to a hot-water boiler. The coal in a boiler is converted to heat for space heating. Some heat goes up the flue, some escapes to the surroundings, and some is carried away in the ashes. If fuel is added to the fire and the heat cannot be dissipated, the boiler will overheat, Figure 37–1.

The body uses food to create energy. Some energy is stored as fatty tissue, some goes off as waste, some as heat, and some as energy to keep the body functioning. If the body needs to dissipate some of its heat to the surroundings and cannot, it will overheat, Figure 37–2.

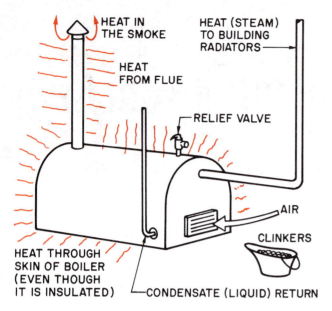

Figure 37–1. The boiler is much like the human body. It uses fuel for energy.

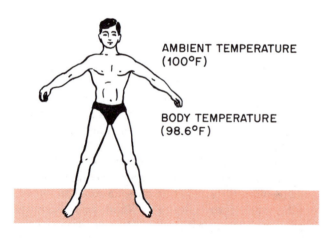

Figure 37–2. The human body must give off some of its generated heat to the surroundings or it will overheat.

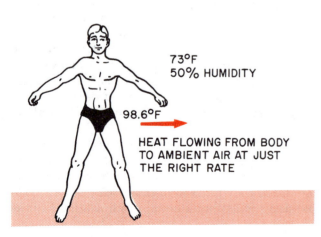

Figure 37–3. The body's normal temperature is 98.6°F.

37.3 BODY TEMPERATURE

Human body temperature is normally 98.6°F, Figure 37–3. We are comfortable when the heat level in our body due to food intake is transferring to the surroundings at the correct rate. But certain conditions must be met for this comfortable, or balanced, condition to exist.

The body gives off and absorbs heat by the three methods of heat transfer: conduction, convection, and radiation, Figure 37–4. Evaporation, in the form of perspiration, could be considered a fourth way, Figure 37–5. When the surroundings are at a particular comfort condition, the body is giving up heat at a steady rate that is comfortable. The surroundings must be cooler than the body for the body to be comfortable.

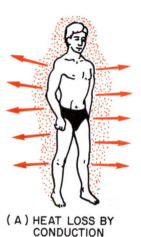

(A) HEAT LOSS BY CONDUCTION

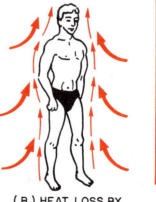

(B) HEAT LOSS BY CONVECTION

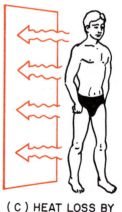

(C) HEAT LOSS BY RADIATION

Figure 37–4. The three direct ways the human body gives off heat. (A) Conduction. (B) Convection. (C) Radiation.

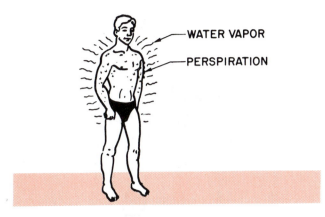

Figure 37-5. This could be called a fourth way the body gives off heat—perspiration or sweat.

Typically, when the body at rest (sitting) is in surroundings of 75°F and 50% humidity with a slight air movement, the body is very close to being comfortable during summer conditions. Notice that the room air at this condition is 23.6°F cooler than the human body, Figure 37-6. In winter a different set of conditions applies (e.g., we wear more clothing). The following statements can be used as guidelines for comfort.

1. In winter:
   A. Lower temperature can be offset with higher humidity.
   B. The lower the humidity is, the higher the temperature must be.
   C. Air movement is more noticeable.
2. In summer:
   A. When the humidity is high, air movement helps.
   B. Higher temperatures can be offset with lower humidity.
3. The comfort conditions in winter and in summer are different.

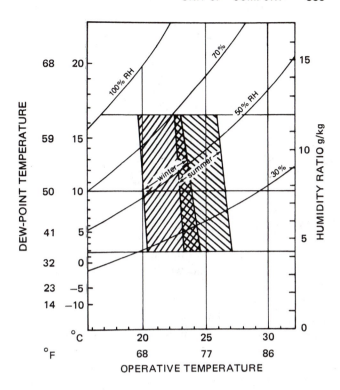

Figure 37-7. This chart is a generalized comfort chart, for different temperature and humidity conditions. This is for persons at rest or slightly active, normally clothed where there is low air movement. *Used with permission from ASHRAE, Inc., 1791 Tullie Circle, NE, Atlanta, GA 30329*

4. Styles of clothes in different parts of the country make a slight difference in the conditioned space-temperature requirements for comfort. For example, in Maine the styles would be warmer in the winter than in Georgia, so the inside temperature of a home or office will not have the same comfort level.
5. Body metabolism varies from person to person. Women, for example, are not as warm natured as men. The circulatory system generally does not work in older people as well as in younger people.

37.4   THE COMFORT CHART

The chart in Figure 37-7, often called a *generalized comfort chart,* can be used as a basis to compare one situation with another. It shows the different combinations of temperatures and humidity for both summer and winter. This table is for one air movement condition.

Cooling, heating, humidifying, and cleaning our air describes the heating and air conditioning profession. Air consists of approximately 78% nitrogen, 21% oxygen, and 1% other gases. Water, in the form of low-pressure vapor, is suspended in the air and called *humidity,* Figure 37-8. More on this later.

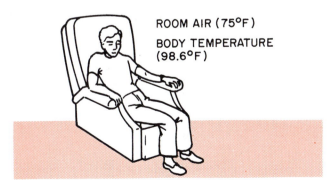

Figure 37-6. Relationship of the human body at rest with the atmosphere surrounding it.

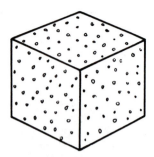

Figure 37–8. Moisture suspended in air.

37.5 PSYCHROMETRICS

The study of air and its properties is called *psychrometrics.* Dry air will be discussed first, then air with moisture added.

When we move through a room, we are not aware of the air inside the room, but the air has weight and takes up space just like water in a swimming pool. The water in a swimming pool is more dense than the air in the room, it weighs more per unit of volume. Air weighs 0.075 lb/ft^3 at 70°F at sea level. This weight or *density* changes slightly with temperature. Water weighs 62.4 lb/ft^3, Figure 37–9.

Air offers resistance to movement. To prove this, take a large piece of cardboard and try to swing it around with the flat side moving through the air, Figure 37–10. It is hard to do because of the resistance of the air—the larger the area of the cardboard, the more resistance there will be. For example, if you took a large piece of cardboard outside on a windy day, the wind will try to take the cardboard from you. The cardboard acts like a sail on a boat, Figure 37–11.

For another example, Figure 37–12, invert an empty glass and push it down in water. The air in the glass resists the water going up into the glass. This shows that air takes up space. Since a cubic foot of air at 70°F has a weight of 0.075 lb/ft^3, the weight of air in a room can be calculated by multiplying the room

Figure 37–10. This man is having a hard time swinging a piece of cardboard around because the air around him is taking up space also.

Figure 37–11. This man walks out of a building and into a breeze. He is pushed along by the breeze against the cardboard because the air has weight and takes up space.

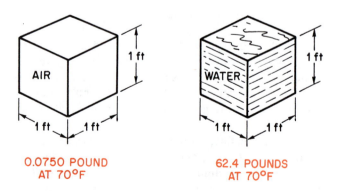

0.0750 POUND
AT 70°F

62.4 POUNDS
AT 70°F

Figure 37–9. Relative difference between the weight of air and the weight of water.

Figure 37–12. This illustration proves in another way that air takes up space.

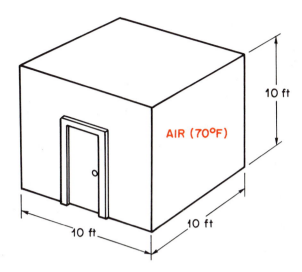

Figure 37–13. This room is 10 ft × 10 ft × 10 ft high. The room contains 1000 cubic feet of air that weighs 0.075 lb/ft³. 1000 ft³ × 0.075 lb/ft³ = 75.0 lb of air in the room.

volume by the weight of a cubic foot of air. In a 10-ft × 10-ft × 10-ft room, the volume is 1000 ft³, so the room air weighs 1000 ft³ × 0.075 lb/ft³ = 75 lb, Figure 37–13.

The number of cubic feet of air to make a pound of air can be obtained by taking the reciprocal of the density. The reciprocal of a number is 1 divided by that number. The reciprocal of the density of air at 70°F is 1 divided by 0.075 or 13.33 cubic feet per pound of air, Figure 37–14.

37.6 MOISTURE IN AIR

Air is not totally dry. Surface water and rain keep moisture in the atmosphere everywhere (even in a desert) at all times. (Remember, the earth's surface is approximately 65% water.) Moisture in the air is called *humidity*.

37.7 SUPERHEATED GASES IN AIR

Since air is made of several different gases, it is not a pure element or gas. Air is made up of nitrogen (78%), oxygen (21%) and approximately 1% other gases, Figure 37–15. These gases in the air are highly superheated. Nitrogen, for instance, boils at −319°F, and oxygen boils at −297°F at atmospheric pressure, Figure 37–16. Hence, nitrogen and oxygen in the atmosphere are *superheated gases*—they are superheated

---

1 ÷ 0.075 lb/ft³ = 13.33 ft³/lb
1 pound of air occupies 13.33 cubic feet.

Figure 37–14. This example shows how to find the number of cubic feet a pound of air occupies.

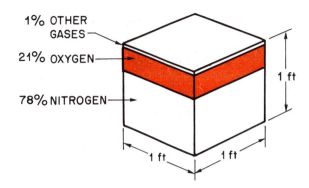

Figure 37–15. Relationship of nitrogen to oxygen in air.

several hundred degrees above absolute (0° Rankine). Each gas exerts pressure according to Dalton's Law of Partial Pressures. Simply stated, this law says that each gas in a mixture of gases acts independently of the other gases and the total pressure of a gas mixture is the sum of the pressures of each gas in the mixture. More than one gas can occupy a space at the same time.

Water vapor suspended in air is a gas that exerts its own pressure and occupies space with the other gases. Water at 70°F in a dish in the atmosphere exerts a pressure of 0.7392 in. Hg, Figure 37–17. If the water vapor pressure in the air is less than the water vapor pressure in the dish, the water in the dish will evaporate slowly to the lower pressure area of the water vapor in the air. For example, the room may be at a dry-bulb temperature of 70°F with a humidity of 30%. The vapor pressure for the moisture suspended in the air is 0.101 psia × 2.036 = 0.206 in. Hg, Figure 37–18. Vapor pressure for moisture in air can be found in some psychrometric charts and in saturated water tables. When reverse pressures occur, the action of the water vapor reverses. For example, if the water vapor pres-

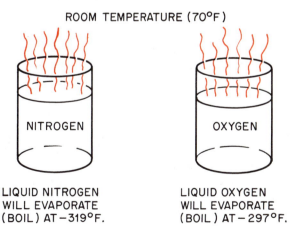

Figure 37–16. If you place a container of liquid nitrogen and a container of liquid oxygen in a room, they would start to evaporate or boil.

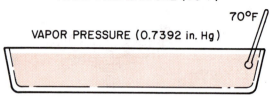

Figure 37-17. Vapor pressure at 70°F of water in an open dish in a room.

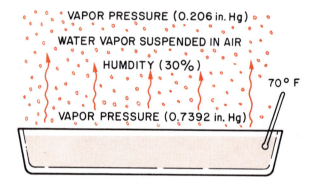

Figure 37-18. Moisture suspended in the air has a pressure controlled by the humidity in the room.

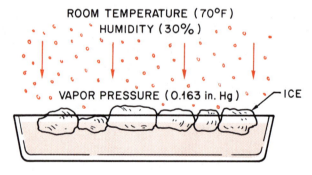

Figure 37-19. The moisture in the dish has ice in it, which lowers the vapor pressure to 0.163 in. Hg (0.08 psig × 2.036 = 0.163 in. Hg). The room temperature is still 70°F with a humidity of 30%, which has a vapor pressure of 0.206 in. Hg.

sure in the dish is less than the pressure of the vapor in the air, water from the air will condense into the water in the dish, Figure 37-19.

When water vapor is suspended in the air, the air is sometimes called "wet air." If the air has a large amount of moisture, the moisture can be seen (for example, fog or a cloud). Actually, the air is not wet because the moisture is suspended in the air. This could more accurately be called a nitrogen, oxygen, and water vapor mixture.

37.8 HUMIDITY

The moisture content in air (humidity) is measured by weight, expressed in pounds or grains (7000 grains per pound). Air can hold very little water vapor. 100% humid air at 29.92 in. Hg and 70°F can hold 110.5 grains (gr) of moisture (0.01578 lb) per cubic foot. Several methods are used to calculate the percentage of moisture content in the air. *Relative humidity* is the most practical and most used for field measurements. It is based on the weight of water vapor in a given volume of space compared to the weight of water vapor that the same volume could hold if it were 100% saturated, Figure 37-20.

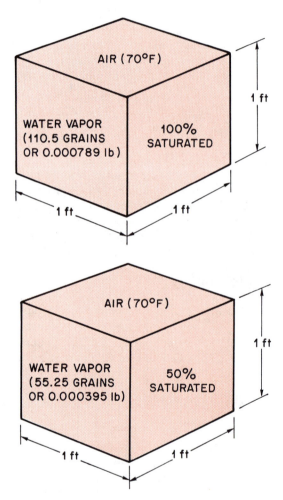

Figure 37-20. Relative humidity is based on the water vapor suspended in a given volume compared to the weight of water vapor this volume could hold if 100% saturated.

37.9 DRY-BULB AND WET-BULB TEMPERATURES

Temperature was discussed in Unit 1 as it applied to sensible and latent heats of substances. The moisture content of air can be checked by using a combination

DB TEMP.	WB DEPRESSION																													
	1	2	3	4	5	6	7	8	9	10	11	12	13	14	15	16	17	18	19	20	21	22	23	24	25	26	27	28	29	30
32	90	79	69	60	50	41	31	22	13	4																				
36	91	82	73	65	56	48	39	31	23	14	6																			
40	92	84	76	68	61	53	46	38	31	23	16	9	2																	
44	93	85	78	71	64	57	51	44	37	31	24	18	12	5																
48	93	87	80	73	67	60	54	48	42	36	34	25	19	14	8															
52	94	88	81	75	69	63	58	52	46	41	36	30	25	20	15	10	6	0												
56	94	88	82	77	71	66	61	55	50	45	40	35	34	26	24	17	12	8	4											
60	94	89	84	78	73	68	63	58	53	49	44	40	35	31	27	22	18	14	6	2										
64	95	90	85	79	75	70	66	61	56	52	48	43	39	35	34	27	23	20	16	12	9									
68	95	90	85	81	76	72	67	63	59	55	51	47	43	39	35	31	28	24	21	17	14									
72	95	91	86	82	78	73	69	65	61	57	53	49	46	42	39	35	32	28	25	22	19									
76	96	91	87	83	78	74	70	67	63	59	55	52	48	45	42	38	35	32	29	26	23									
80	96	91	87	83	79	76	72	68	64	61	57	54	54	47	44	41	38	35	32	29	27	24	21	18	16	13	11	8	6	1
84	96	92	88	84	80	77	73	70	66	63	59	56	53	50	47	44	41	38	35	32	30	27	25	22	20	17	15	12	10	8
88	96	92	88	85	81	78	74	71	57	64	61	58	55	52	49	46	43	41	38	35	33	30	28	25	23	21	18	16	14	12
92	96	92	89	85	82	78	75	72	69	65	62	59	57	54	51	48	45	43	40	38	35	33	30	28	26	24	22	19	17	15
96	96	93	89	86	82	79	76	73	70	67	74	61	58	55	53	50	47	45	42	40	37	35	33	31	29	26	24	22	20	18
100	96	93	90	86	83	80	77	74	71	68	65	62	59	57	54	52	49	47	44	42	40	37	35	33	31	29	27	25	23	21
104	97	93	90	87	84	80	77	74	72	69	66	63	61	58	56	53	51	48	46	44	41	39	37	35	33	31	29	27	25	24
108	97	93	90	87	84	81	78	75	72	70	67	64	62	59	57	54	52	50	47	45	43	41	39	37	35	33	31	29	28	26

Figure 37-21. Wet-bulb depression chart.

of dry-bulb and wet-bulb temperatures. *Dry-bulb* temperature is the sensible-heat level of air and is taken with an ordinary thermometer. Wet-bulb temperature is taken with a thermometer with a wick on the end that is soaked with distilled water. The reading from a wet-bulb thermometer takes into account the moisture content of the air. It reflects the total heat content of air. The thermometer will get cooler than the dry-bulb thermometer due to the evaporation of the distilled water. Distilled water is used because some water has undesirable mineral deposits. Some minerals will change the boiling temperature.

The difference between the dry-bulb reading and the wet-bulb reading is called the *wet-bulb depression*. See Figure 37-21 for a wet-bulb depression chart. As the amount of moisture suspended in the air decreases, the wet-bulb depression increases and vice versa. For example, a room with a dry-bulb temperature of 76°F and a wet-bulb temperature of 64°F has a wet bulb depression of 12°F and a relative humidity of 52%. If the 76°F dry-bulb temperature is maintained and moisture is added to the room so that the wet-bulb temperature rises to 74°F, the relative humidity increases to 91% and the new wet-bulb depression is 2°F. If the wet-bulb depression is allowed to go to 0°F. (for example, 76°F dry bulb and 76°F wet bulb), the relative humidity will be 100%. The air is holding all of the moisture it can—it is *saturated* with moisture.

37.10 DEW-POINT TEMPERATURE

The *dew-point temperature* is the temperature at which moisture begins to condense out of the air. For example, if you were to set a glass of warm water in a room with a temperature of 75°F and 50% relative humidity, the water in the glass would evaporate slowly to the room. If you gradually cool the glass with ice, then when the glass surface temperature becomes 55.5°F, water will begin to form on the surface of the glass, Figure 37-22. Moisture from the room will also

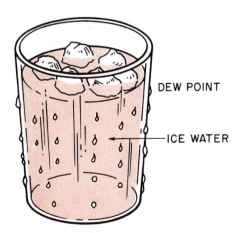

Figure 37-22. The glass was gradually cooled until beads of water began to form on the outside of the glass.

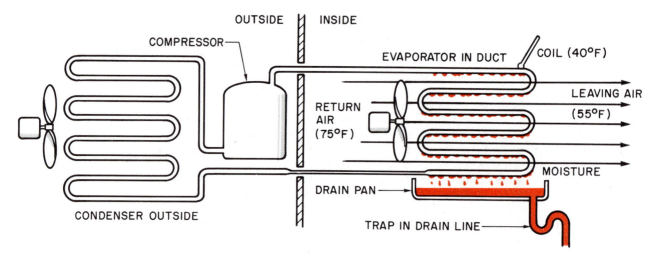

Figure 37-23. The cold surface on the air conditioning coil condenses moisture from the air passed over it.

collect in the water in the glass and the level will begin to rise. This temperature at which water forms is called the dew-point temperature of the glass. Air can be dehumidified by passing it over a surface that is below the dew-point temperature of the air; moisture will collect on the cold surface, for example, an air conditioning coil, Figure 37-23. The condensed moisture is drained. This is the moisture that you see running out the back of a window air conditioner, Figure 37-24.

37.11 THE PSYCHROMETRIC CHART

The foregoing description can all be plotted on a psychrometric chart, Figure 37-25A. The chart looks very complicated, but a clear plastic straightedge and a pencil will help you understand it. See Figure 37-25B through 37-25G for some examples of plottings of the different conditions on a psychrometric chart.

If you know any two conditions previously mentioned, you can plot any of the other conditions. The easiest conditions to determine from room air are the wet-bulb and the dry-bulb temperatures. For example, you can take an electronic thermometer and make a wet-bulb thermometer if you do not have one. Take two leads and tape them together with one lead about 2-in. below the other one, Figure 37-26. A simple wick can be made from a piece of white cotton, such as an undershirt. Make sure that it does not have any perspiration on it. Wet the lower bulb (with the wick on it) with distilled water that is warmer than the room air. Water from a clean condensate drain line may be used if real distilled water is not available. Water from the city system can be used but may give slightly wrong results. Hold the leads about 3 ft back from the element on the end and slowly spin them in the air. The wet lead will drop to a colder temperature than the dry lead will. Keep spinning them until the lower lead stops dropping in temperature but is still damp, Figure

Figure 37-24. The water dripping out of the back of this window air conditioner is moisture that was collected from the room onto the air conditioner's evaporator coil.

37-27. Quickly read wet-bulb and dry-bulb temperatures without touching the bulbs. Suppose the reading is 75°F DB (dry bulb) and 62.5°F WB (wet bulb). Put your pencil point at this place on the psychrometric chart, Figure 37-25, and make a dot. Draw a light circle around it so you can find the dot again. The following information can be concluded from this plot:

1. Dry-bulb temperature	75°F
2. Wet-bulb temperature	62.5°F
3. Dew-point temperature	55.5°F
4. Total heat content of 1 lb of air	28.2 Btu
5. Moisture content of 1 lb of air	65 gr
6. Relative humidity	50%
7. Density of the air	13.675 ft³/lb

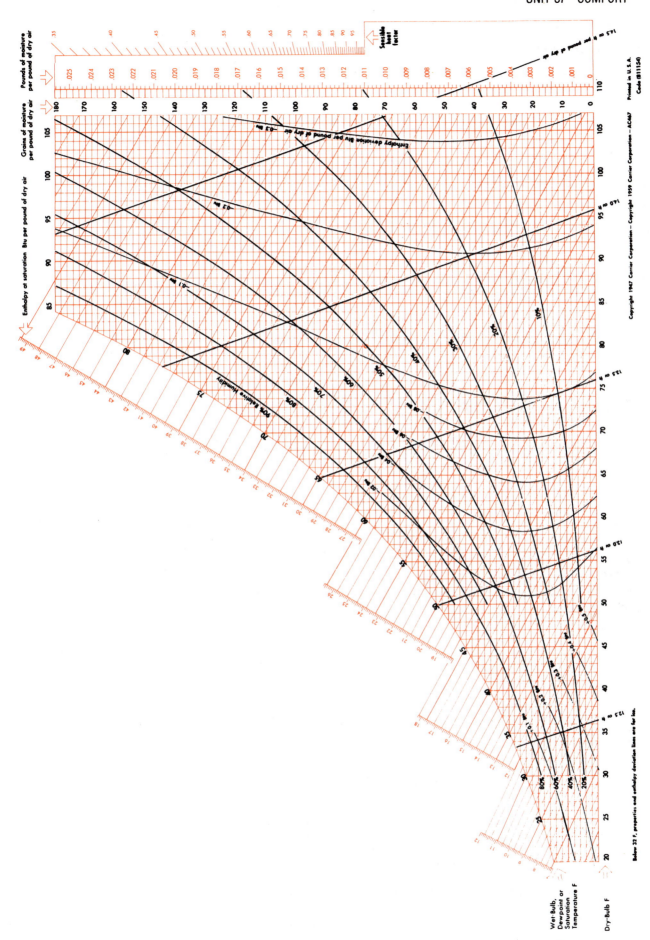

Figure 37-25A. Psychrometric chart. *Reproduced courtesy of Carrier Corporation*

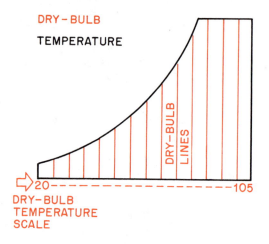

Figure 37–25B. Skeleton chart showing the dry-bulb temperature lines. From Lang, *Principles of Air Conditioning* © 1979 by Delmar Publishers Inc.

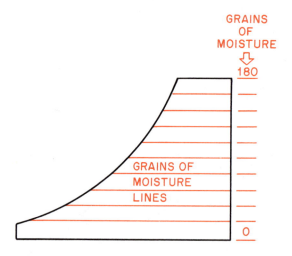

Figure 37–25E. Skeleton chart showing the moisture content of air expressed in grains per pound of air. From Lang, *Principles of Air Conditioning* © 1979 by Delmar Publishers Inc.

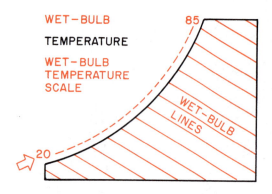

Figure 37–25C. Skeleton chart showing the wet-bulb lines. From Lang, *Principles of Air Conditioning* © 1979 by Delmar Publishers Inc.

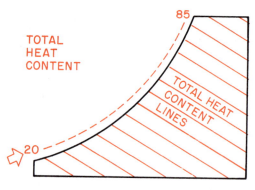

Figure 37–25F. Skeleton chart showing the total heat content of air in Btu/lb. These lines are almost parallel to the wet-bulb lines. From Lang *Principles of Air Conditioning* © 1979 by Delmar Publishers Inc.

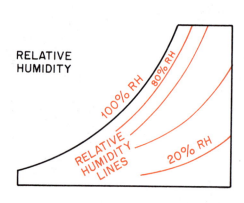

Figure 37–25D. Skeleton chart showing the relative humidity lines. From Lang, *Principles of Air Conditioning* © 1979 by Delmar Publishers Inc.

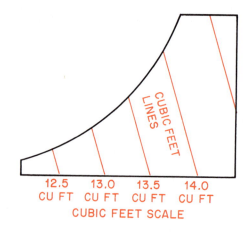

Figure 37–25G. Skeleton chart showing the specific volume of air at different conditions. From Lang, *Principles of Air Conditioning* © 1979 by Delmar Publishers Inc.

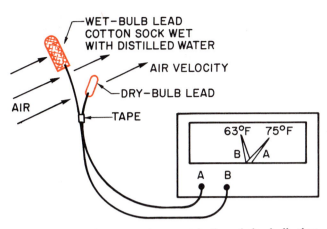

Figure 37–26.. How to make a wet-bulb and dry-bulb thermometer from an electronic thermometer.

Figure 37–27. Technician spinning a wet-bulb and dry-bulb thermometer, known as a sling psychrometer.

37.12 TOTAL HEAT

The capacity of a heating or cooling unit may be field checked with the total heat feature of the psychrometric chart. If the amount of air passing over a heat exchanger is known, the total heat can be checked where it enters and leaves the heat exchanger. This will give a fairly accurate account of the performance of the heat exchanger, Figure 37–28.

The air that surrounds us has to be maintained at the correct conditions for us to be comfortable. The air in our homes is treated by heating it, cooling it, dehumidifying it, humidifying it, and cleaning it so that our bodies will give off the correct amount of heat for comfort. Conditioned air is first picked up from the room air, conditioned, and blended with the room air. A small amount of air is induced from the outside into the conditioner to keep the air from becoming oxygen starved and stagnant. This is called fresh-air intake or *ventilation*. If a system has no ventilation, it is relying on air infiltrating the structure around doors and windows.

Modern energy-efficient homes can be built so tight that infiltration does not provide enough fresh air. If a home has odors, a fresh-air intake may have to be installed, Figure 37–29.

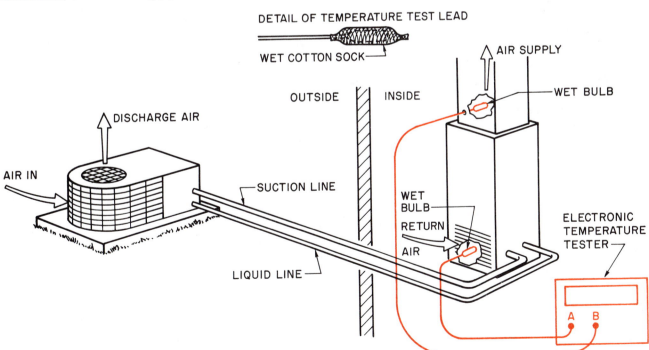

Figure 37–28. Wet-bulb reading can be taken on each side of an air heat exchanger.

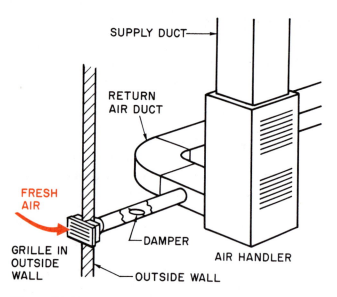

Figure 37-29. Fresh air drawn into return-air duct to improve the air quality inside the house.

SUMMARY

- *Comfort* describes the delicate balance of feeling in relationship to our surroundings.
- The body burns food and turns it to energy.
- The body stores energy, wastes it, consumes it in work, or gives the heat off to the surroundings.
- For the body to be comfortable, it has to be warmer than the surroundings, so it can give up excess heat to the surroundings.
- The body gives off heat in three conventional ways: conduction, convection, and radiation. Evaporation may be considered a fourth way.
- Air contains 78% nitrogen, 21% oxygen, 1% other gases, and suspended water vapor.
- The density of air at 70° F and 29.92 in. Hg is 0.0749 lb/ft³.
- The specific volume of air is the reciprocal of the density: 1/0.0749 = 13.35 ft³/lb.
- The moisture content of air can vary transfer of heat from the human body; therefore, different temperatures and moisture content can give the same relative comfort level.
- Slight air movement can help to offset higher temperatures in the summer.
- Dry-bulb temperature is registered with a regular thermometer.
- Wet-bulb temperature is registered with a thermometer that has a wet wick. The thermometer gets colder than the dry-bulb thermometer because the moisture evaporates on the wick.
- The difference between the wet-bulb reading and the dry-bulb reading is the wet-bulb depression. It can be used to determine the relative humidity of a conditioned space.
- The dew-point temperature of air is the point at which moisture begins to drop out of the air.
- Water vapor in the air creates its own vapor pressure.
- The psychrometric chart is used to plot various air conditions.
- The wet-bulb reading on a psychrometric chart shows the total heat content of a pound of air.
- When the cubic feet of air per minute is known, the wet-bulb reading in and out of an air heat exchanger can give the total heat being exchanged. This can be used for field calculating the capacity of a unit.

REVIEW QUESTIONS

1. Name the four comfort factors.
2. Name three ways the body gives off heat.
3. How does perspiration cool the body?
4. How can lower room temperatures be offset in winter?
5. What two conditions can offset higher room temperatures in summer?
6. How is the relative humidity of a conditioned space measured?
7. What is the name of the chart used to plot the various air conditions?
8. What two unknowns are the easiest to obtain in the field to plot air conditions?
9. Describe the dew point.
10. What is the density of air at 70° F?

REFRIGERATION APPLIED TO AIR CONDITIONING

OBJECTIVES

After studying this unit, you should be able to

- state two ways that air is conditioned for cooling.
- explain refrigeration as applied to air conditioning.
- describe an air conditioning evaporator.
- describe an air conditioning compressor.
- describe an air conditioning condenser.
- describe an air conditioning metering device.
- explain three ways in which heat gets into a structure.
- list different types of evaporator coils.
- identify different types of condensers.
- explain how "high-efficiency" is accomplished.
- describe package air conditioning equipment.
- describe split-system equipment.

38.1 REFRIGERATION

Refrigeration involves removing heat from a place where it is not wanted and depositing it in a place where it makes no difference. This description has already been applied in this book to both space and product cooling. This unit is concerned with comfort cooling. Some people never have air conditioning and seem to survive, but there are times when they are uncomfortable. When the nights are warm, above 75°F and the humidity is high, it is very hard to be comfortable enough to rest well.

38.2 STRUCTURAL HEAT GAIN

Summer heat leaks into a structure by conduction, infiltration and radiation (the sun's rays—solar load). The summer *solar load* on a structure is greater on the east and west sides because the sun shines for long periods of time on these parts of the structure, Figure 38–1. If a building has an attic, the air space can be ventilated to help relieve the solar load on the ceiling, Figure 38–2. If there is no attic, it is at the mercy of the sun unless it is well insulated, Figure 38–3.

Conduction heat enters through walls, windows, and doors. The rate depends on the temperature difference between the inside and the outside of the house, Figure 38–4.

Some of the warm air that gets into the structure is infiltrated through the cracks around the windows and doors. Air also leaks in when the doors are opened for entering and leaving. This infiltrated air is different in different parts of the country. Using the example in Figure 38–4, the typical design condition in Phoenix is 105°F dry bulb (DB) and 71°F wet bulb (WB). In Atlanta the air may be 90°F DB and 73°F WB. When the air leaks into the structure in Phoenix it is cooled to the space temperature and a certain amount of humidity will concentrate for each cubic foot of air that leaks in. In Atlanta, there will normally be more humidity involved with the infiltration than in Phoenix.

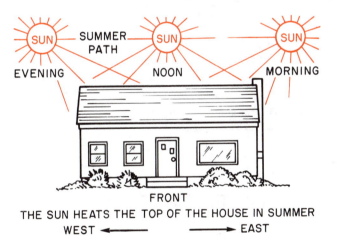

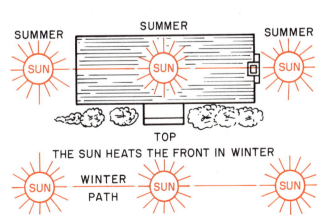

Figure 38–1. Solar load on a home.

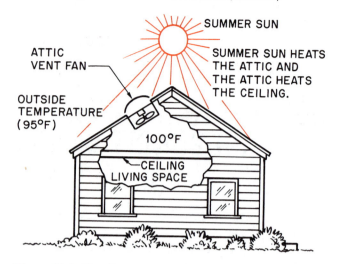

Figure 38–2. Ventilated attic to keep the solar heat from the ceiling of the house.

Figure 38–3. House has no attic. The sun shines directly on the ceiling of the living space.

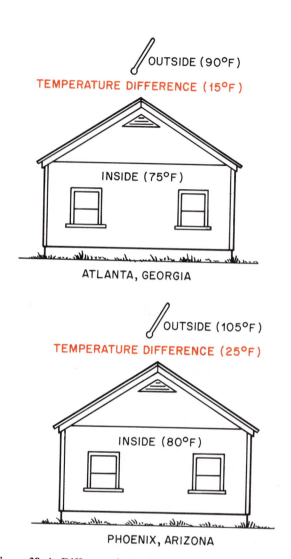

Figure 38–4. Difference in the inside and outside temperatures of a home in Atlanta, Georgia and a home in Phoenix, Arizona.

38.3 EVAPORATIVE COOLING

Air has been conditioned in more than one way to achieve comfort. In the climates where the humidity is low, a device called an *evaporative cooler,* Figure 38–5, has been used for years. This device uses a fiber mounted in a frame with water slowly running down the fiber as the cooling media. Fresh air is drawn through the water-soaked fiber and cooled by evaporation to a point close to the wet bulb (WB) temperature of the air. The air entering the structure is very humid but cooler than the dry bulb (DB) temperature. For example, in Phoenix, Arizona, the design dry bulb temperature in summer is 105°F. At the same time the dry bulb is 105°F, the wet bulb temperature may be 70°F. An evaporative cooler may lower the air temperature entering the room to 80°F DB. 80°F air is cool compared to 105°F, even if the humidity is high.

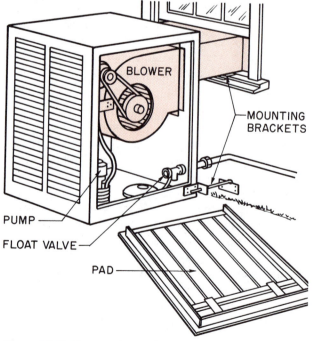

Figure 38–5. Evaporative cooler.

38.4 REFRIGERATED COOLING OR AIR CONDITIONING

Refrigerated air conditioning is similar to commercial refrigeration because the same components are used to cool the air: (1) the evaporator, (2) the compressor, (3) the condenser, and (4) the metering device. These components are assembled in several ways to accomplish the same goal—refrigerated air to cool space. Review Unit 3 if you are not familiar with the basics of refrigeration.

Package Air Conditioning

The four components are assembled into two basic types of equipment for air conditioning purposes: package equipment and split-system equipment. *Package equipment* has all of the components built into one cabinet; hence, it is also called *self-contained* equipment. Air is ducted to and from the equipment. Package equipment may be located beside the structure or on top of it. It sometimes has the heating equipment built into the same cabinet. See Figure 38–6 for an example of package air conditioning equipment.

Split-System Air Conditioning

In *split-system air conditioning* the condenser is remote from the evaporator and uses interconnecting refrigerant lines. The evaporator may be located in the attic, a crawl space, or a closet for upflow or downflow applications. The fan to blow the air across the

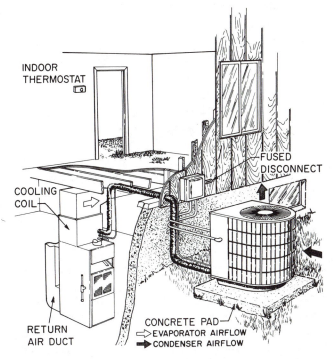

Figure 38–7. Split air conditioning system. *Courtesy Climate Control*

evaporator may be included in the heating equipment, or a separate fan may be used for the air conditioning system, Figure 38–7.

38.5 THE EVAPORATOR

The *evaporator* is the component that absorbs heat into the refrigeration system. It is a refrigeration coil made of aluminum or copper with aluminum fins on either type attached to the coil to give it more surface area for better heat exchange. The evaporator coil has several designs for airflow through the coil and draining the condensate water from the coil, depending on the installation. The different designs are known as the *A* coil, the *slant* coil and the *H* coil.

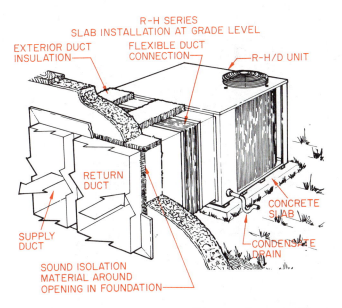

Figure 38–6. Package air conditioner. *Courtesy Climate Control*

Figure 38–8. An A coil. *Courtesy BDP Company*

Figure 38-9. A slant coil. *Courtesy BDP Company*

The A Coil

The *A coil* is used for upflow, downflow, and horizontal flow applications. It consists of two coils with their circuits side by side and spread apart at the bottom, Figure 38-8. When used for upflow or downflow, the condensate pan is at the bottom of the A pattern. When used for horizontal flow, a pan is placed at the bottom of the coil and the coil is turned on its side. The airflow through an A coil is through the core of the coil. It cannot be from side to side with the two coils in series. This is not the most popular method of applying the A coil. When horizontal

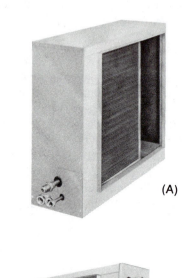

(A)

(B)

Figure 38-10. An H coil. *Courtesy BDP Company*

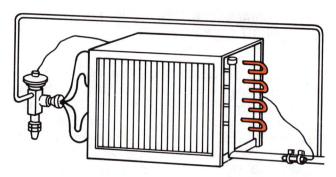

Figure 38-11. Multicircuit coil. *Courtesy Sporlon Valve Company*

airflow is needed, slant or H coils may be more desirable.

The Slant Coil

The *slant* coil is a one-piece coil mounted in the duct on an angle (usually 60°) or slant to give the coil more surface area and to drain the condensate water. The condensate pan is located at the bottom of the slant. The coil can be used for upflow, downflow or horizontal flow when designed for these applications, Figure 38-9.

The H Coil

The *H coil* is normally applied to horizontal applications, although it can be adapted to vertical applications by using special drain pan configurations. The drain is normally at the bottom of the H pattern, Figure 38-10.

Coil Circuits

All of the aforementioned coils may have more than one circuit for the refrigerant. As we said in Unit 3,

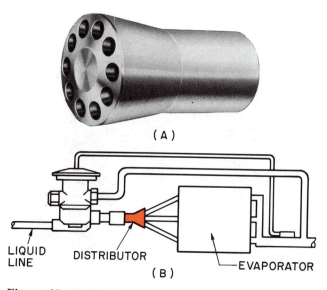

(A)

LIQUID LINE   DISTRIBUTOR   EVAPORATOR

(B)

Figure 38-12. Refrigerant distributor. *Courtesy Sporlon Valve Company*

State and Station	Col. 1	Col. 2 Latitude[b]		Col. 3 Longitude[b]		Col. 4 Elevation[c]	Winter[d] Col. 5 Design Dry-Bulb		Summer[e] Col. 6 Design Dry-Bulb and Mean Coincident Wet-Bulb			Col. 7 Mean Daily Range	Col. 8 Design Wet-Bulb		
		°	′	°	′	Ft	99%	97.5%	1%	2.5%	5%		1%	2.5%	5%
ARIZONA															
Douglas AP		31	3	109	3	4098	27	31	98/63	95/63	93/63	31	70	69	68
Flagstaff AP		35	1	111	4	6973	−2	4	84/55	82/55	80/54	31	61	60	59
Fort Huachuca AP (S)		31	3	110	2	4664	24	28	95/62	92/62	90/62	27	69	68	67
Kingman AP		35	2	114	0	3446	18	25	103/65	100/64	97/64	30	70	69	69
Nogales		31	2	111	0	3800	28	32	99/64	96/64	94/64	31	71	70	69
Phoenix AP (S)		33	3	112	0	1117	31	34	109/71	107/71	105/71	27	76	75	75
Prescott AP		34	4	112	3	5014	4	9	96/61	94/60	92/60	30	66	65	64
Tuscon AP (S)		32	1	111	0	2584	28	32	104/66	102/66	100/66	26	72	71	71
Winslow AP		35	0	110	4	4880	5	10	97/61	95/60	93/60	32	66	65	64
Yuma AP		32	4	114	4	199	36	39	111/72	109/72	107/71	27	79	78	77
CALIFORNIA															
Bakersfield AP		35	2	119	0	495	30	32	104/70	101/69	98/68	32	73	71	70
Barstow AP		34	5	116	5	2142	26	29	106/68	104/68	102/67	37	73	71	70
Blythe AP		33	4	114	3	390	30	33	112/71	110/71	108/70	28	75	75	74
Burbank AP		34	1	118	2	699	37	39	95/68	91/68	88/67	25	71	70	69
Chico		39	5	121	5	205	28	30	103/69	101/68	98/67	36	71	70	68
Los Angeles AP (S)		34	0	118	2	99	41	43	83/68	80/68	77/67	15	70	69	68
Los Angeles CO (S)		34	0	118	1	312	37	40	93/70	89/70	86/69	20	72	71	70
Merced-Castle AFB		37	2	120	3	178	29	31	102/70	99/69	96/68	36	72	71	70
Modesto		37	4	121	0	91	28	30	101/69	98/68	95/67	36	71	70	69
Monterey		36	4	121	5	38	35	38	75/63	71/61	68/61	20	64	62	61
Napa		38	2	122	2	16	30	32	100/69	96/68	92/67	30	71	69	68
Needles AP		34	5	114	4	913	30	33	112/71	110/71	108/70	27	75	75	74
Oakland AP		37	4	122	1	3	34	36	85/64	80/63	75/62	19	66	64	63
Oceanside		33	1	117	2	30	41	43	83/68	80/68	77/67	13	70	69	68
Ontario		34	0	117	36	995	31	33	102/70	99/69	96/67	36	74	72	71
GEORGIA															
Albany, Turner AFB		31	3	84	1	224	25	29	97/77	95/76	93/76	20	80	79	78
Americus		32	0	84	2	476	21	25	97/77	94/76	92/75	20	79	78	77
Athens		34	0	83	2	700	18	22	94/74	92/74	90/74	21	78	77	76
Atlanta AP (S)		33	4	84	3	1005	17	22	94/74	92/74	90/73	19	77	76	75
Augusta AP		33	2	82	0	143	20	23	97/77	95/76	93/76	19	80	79	78
Brunswick		31	1	81	3	14	29	32	92/78	89/78	87/78	18	80	79	79
Columbus, Lawson AFB		32	3	85	0	242	21	24	95/76	93/76	91/75	21	79	78	77
Dalton		34	5	85	0	720	17	22	94/76	93/76	91/76	22	79	78	77
Dublin		32	3	83	0	215	21	25	96/77	93/76	91/75	20	79	78	77
Gainesville		34	2	83	5	1254	16	21	93/74	91/74	89/73	21	77	76	75
NEW YORK															
Albany AP (S)		42	5	73	5	277	−6	−1	91/73	88/72	85/70	23	75	74	72
Albany CO		42	5	73	5	19	−4	1	91/73	88/72	85/70	20	75	74	72
Auburn		43	0	76	3	715	−3	2	90/73	87/71	84/70	22	75	73	72
Batavia		43	0	78	1	900	1	5	90/72	87/71	84/70	22	75	73	72
Binghamton AP		42	1	76	0	1590	−2	1	86/71	83/69	81/68	20	73	72	70
Buffalo AP		43	0	78	4	705r	2	6	88/71	85/70	83/69	21	74	73	72
Cortland		42	4	76	1	1129	−5	0	88/71	85/71	82/70	23	74	73	71
Dunkirk		42	3	79	2	590	4	9	88/73	85/72	83/71	18	75	74	72
Elmira AP		42	1	76	5	860	−4	1	89/71	86/71	83/70	24	74	73	71
Geneva (S)		42	5	77	0	590	−3	2	90/73	87/71	84/70	22	75	73	72

[a]Table 1 was prepared by ASHRAE Technical Committee 4.2, Weather Data, from data compiled from official weather stations where hourly weather observations are made by trained observers, See also Ref 1, 2, 3, 5 and 6.

[b]Latitude, for use in calculating solar loads, and longitude are given to the nearest 10 minutes. For example, the latitude and longitude for Anniston, Alabama are given as 33 34 and 85 55 respectively, or 33° 40, and 85° 50.

[c]Elevations are ground elevations for each station. Temperature readings are generally made at an elevation of 5 ft above ground, except for locations marked r, indicating roof exposure of thermometer.

[d]Percentage of winter design data shows the percent of the 3-month period, December through February.

[e]Percentage of summer design data shows the percent of 4-month period, June through September.

Figure 38–13. Excerpt from a table showing different design conditions for various parts of the United States. *Used with permission from ASHRAE, Inc. 1791 Tullie Circle, NE, Atlanta, GA 30329*

Figure 38-14. Evaporator with a coil case. *Courtesy BDP Company*

when a coil becomes too long and excessive pressure drop occurs it is advisable to have more than one coil in parallel, Figure 38-11. The coil may have as many circuits as necessary to do the job. However, when more than one circuit is used, a distributor must be used to distribute the correct amount of refrigerant to the individual circuits, Figure 38-12.

38.6   THE JOB OF THE EVAPORATOR

The evaporator is a heat exchanger that takes the heat from the room air and transfers it to the refrigerant. Two kinds of heat must be transferred: sensible heat and latent heat. *Sensible heat* lowers the air temperature, and *latent heat* changes the water vapor in the air to condensate. The condensate collects on the coil and runs through the drain pan to a trap (to stop air from pulling into the drain) and then it is normally piped to a drain.

Typically, room air may be 75°F DB and have a humidity of 50%, which is 62.5°F WB. The coil generally operates at a refrigerant temperature of 40°F to remove the required amount of sensible heat (to lower the air temeprature) and latent heat (to remove the correct amount of moisture). The air leaving the coil is approximately 55°F DB with a humidity about 95%, which is 54°F WB. These conditions are average for a climate with high humidity.

38.7   DESIGN CONDITIONS

A house built in Atlanta has less sensible-heat load and more latent-heat load than one constructed in Phoenix. The designer or engineer must be familiar with local design practices, Figure 38-13. The airflow across the same coil in the two different parts of the country may be varied to accomplish different air conditions.

38.8   EVAPORATOR APPLICATION

The evaporator may be installed in the airstream in several different ways. It may have a coil case that

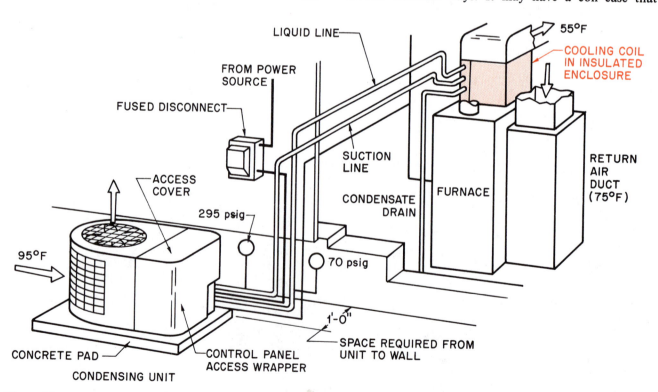

Figure 38-15. Operating conditions of an evaporator and a condenser. Notice the cooling coil is in an insulated enclosure to prevent sweating.

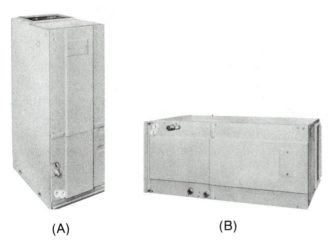

Figure 38–16. Evaporator mounted in an air handler. *Courtesy BDP Company*

encloses the coil, Figure 38–14, or it may be located in the duct work. The coil will normally operate below the dew-point temperature, and the coil enclosure should be insulated to keep it from absorbing heat from the surroundings, which makes it sweat, Figure 38–15. Some evaporators are built into the air handler by the manufacturer, Figure 38–16.

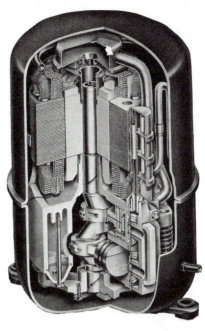

Figure 38–18. Suction gas-cooled compressor. *Courtesy Tecumseh Products Company*

38.9   THE COMPRESSOR

Compressors used for air conditioning are the same design as those used for refrigeration. The compressor is a *vapor pump* that pumps heat-laden vapor from the low-pressure side of the system in the evaporator to the high-pressure side of the system in the condenser. A refrigeration compressor rated for high-temperature application is sometimes found in an air conditioning

(A)

(B)

Figure 38–17. (A) Hermetic compressor. (B) Serviceable hermetic compressor. *Courtesy (A) Bristol Compressors, Inc. (B) Copeland Corporation*

Figure 38–19. Air-cooled compressor. Notice the suction line enters at the side of the cylinder. *Courtesy Copeland Corporation*

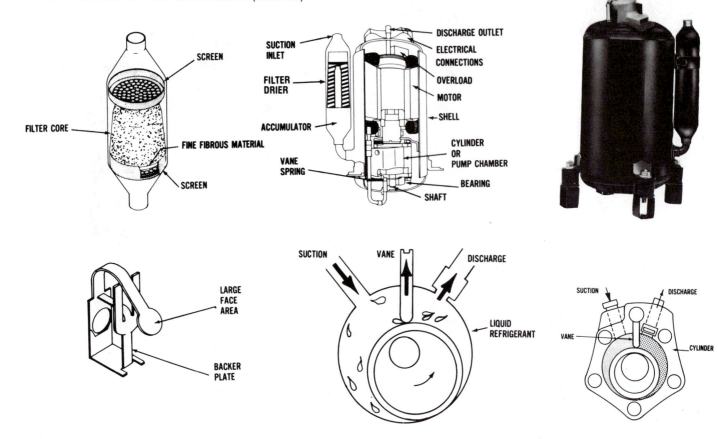

Figure 38–20. Rotary compressor. *Courtesy Rotorex*

system. Such a compressor is either the fully hermetic type or the serviceable hermetic type, Figure 38–17. They are either suction-gas cooled, Figure 38–18, or air cooled, Figure 38–19. Some of the compressors are rotary, Figure 30–20, or reciprocating, Figure 38–21. The rotary compressor may be cooled by discharge gas. A new compressor, called the *scroll compressor,* is being introduced for medium-sized air conditioning systems.

These compressors are all positive displacement and normally use R-22. Other refrigerants such as R-500 and R-12 have been used but are not common now. Unless the equipment was built before 1965, it will typically be an R-22 system, so R-22 will be discussed in this text.

38.10 COMPRESSOR SPEEDS (RPM)

Modern air conditioning compressors used in the small and medium size ranges are either fully hermetic or semihermetic and must turn standard motor speeds of 3450 rpm or 1750 rpm. Early compressors were 1750 rpm, used R-12, and were large and heavy. Present compressors use the faster motor and the more efficient refrigerant R-22, so equipment can be smaller and lighter.

The typical compressor is suction cooled. The suction-gas temperature is important because it cools the motor. Air-cooled compressors are also influenced by the suction pressure. The typical maximum high-suction gas temperature is 70° F.

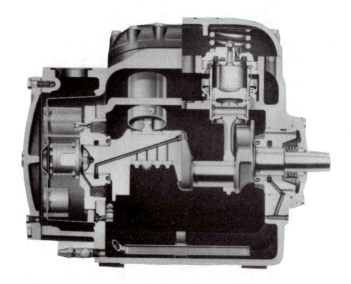

Figure 38–21. Reciprocating compressor. *Courtesy Trane Company*

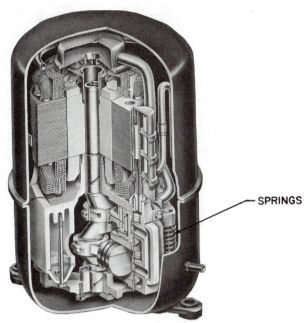

Figure 38–22. Hermetic compressor mounted internally on springs that suspend the compressor and motor from the shell. The shell is then mounted on rubber feet. *Courtesy Tecumseh Products Company*

38.11   COMPRESSOR MOUNTINGS

Hermetic reciprocating compressors all have rubber mounting feet on the outside, and the compressor is mounted on springs inside the shell, Figure 38–22.

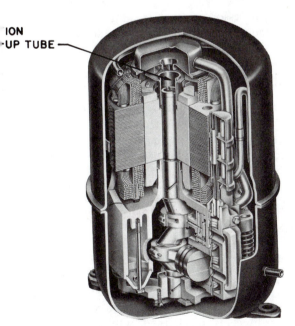

Figure 38–23. The suction pickup tube in this compressor is high in the shell so that liquid refrigerant or oil would have to be at the top of the tube before it could enter the cylinders of the compressor. *Courtesy Tecumseh Products Company*

Older compressors were mounted on springs outside the shell and the compressor was pressed into the shell. New compressors have a vapor space between the motor and the shell, so the motor-temperature sensor must be on the inside to sense the motor temperature quickly.

The suction gas dumps out into the shell, usually in the vicinity of the motor. Some compressors dump the suction gas directly into the rotor. The turning rotor tends to dissipate any liquid drops in the return suction gas. The suction pickup tube for the compressor, which is inside the shell, is normally located in a high position so that liquid refrigerant or foaming oil cannot enter the compressor cylinders, Figure 38–23.

The compressor for air conditioning equipment is normally located outside with the condenser. These compressors cannot be field serviced and often are not factory serviced. When one becomes defective, the manufacturer may authorize the technician to discard it or return it to the factory to determine what made it fail.

38.12   REBUILDING THE HERMETIC COMPRESSOR

Some manufacturers are remanufacturing hermetic compressors by opening the shell and repairing the compressor inside. The manufacturer will be the one to decide if this is economical or not.

The standard serviceable hermetic compressor is cast iron, and the manufacturer will want this compressor returned for remanufacturing. These compressors are not used widely in small air conditioning equipment because of the initial cost. They can either be suction-gas cooled, Figure 38–24, or air cooled, Figure 38–25. Suction-gas cooled serviceable compressors operate much like suction-gas cooled hermetic compressors be-

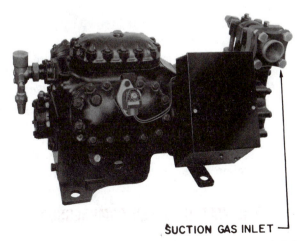

SUCTION GAS INLET

Figure 38–24. Serviceable hermetic compressor cooled by suction gas. *Courtesy Copeland Corporation*

(A)

Figure 38–26. Compressor similar to that in Figure 38–25, but it is applied to a water-cooled condenser. *Photo by Bill Johnson*

(B)

Figure 38–25. Serviceable hermetic compressor cooled by air passing over the ribs on the compressor body. *Courtesy (A) Tyler Refrigeration Company (B) Copeland Corporation*

cause the suction gas must be cool enough to cool the compressor motor. These compressors are designed so that the suction line is piped to the motor end of the compressor.

38.13 THE WATER-COOLED COMPRESSOR

The air-cooled compressor has ribs that must have air passing over them. It is essential that it be located

in a moving airstream. If this compressor is used on water-cooled equipment, the inlet water line is wrapped around the compressor motor body, Figure 38–26.

38.14 THE ROTARY COMPRESSOR

The rotary compressor is very small and light. It is sometimes cooled with compressor discharge gas, which makes the compressor appear to be running too hot. A warning may be posted in the compressor compartment that the housing will appear to be too hot. The rotary compressor is very efficient and will probably become more popular with manufacturers, Figure 38–20.

38.15 THE CONDENSER

The condenser for air conditioning equipment is the component that rejects the heat from the system. Most equipment is air cooled and rejects heat to the air. The coils are copper or aluminum and either type would have aluminum fins to add to the heat exchange surface area, Figure 38–27.

38.16 SIDE-AIR-DISCHARGE CONDENSING UNITS

Early condensers discharged the air out the side and were called side discharge, Figure 38–28. The advantage of this equipment is that the fan and motor are under the top panel, Figure 38–29. Any noise generated inside the cabinet is discharged into the leaving airstream and may be clearly heard in a neighbor's yard. *The heat from the condenser coil can be hot enough to kill plants that it blows on.* These condensers are still being used.

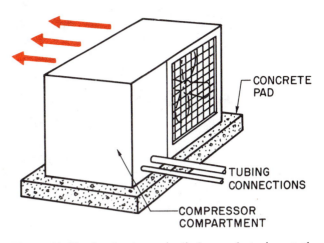

Figure 38-28. Condensing unit discharges hot air out the side.

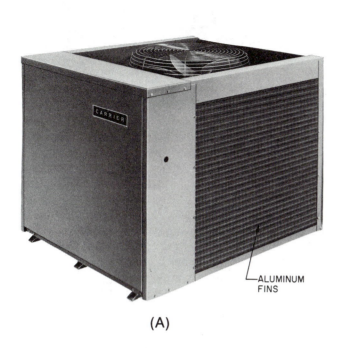

(A)

(B)

Figure 38-27. Condensers have either copper or aluminum tubes with aluminum fins. *Reproduced courtesy of Carrier Corporation*

Figure 38-29. Fan motor and all components are located under the top panel. *Courtesy Williamson Company*

38.17   TOP-AIR-DISCHARGE CONDENSERS

The modern trend in residential equipment is for the condenser to be a top-discharge type, Figure 38–30. In this unit the hot air and noise are discharged from the top of the unit into the air. This is advantageous as far as air and noise are concerned, but the fan and motor are on top of the equipment and rain, snow, and leaves can fall directly into the unit, so the fan motor should

Figure 38-30. Equipment with air discharged out the top of the cabinet. *Courtesy Coleman*

be protected with a rain shield, Figure 38–31. The fan motor bearings are in a vertical position, which means there is more thrust on the end of the bearing, so the bearing needs a thrust surface for this type of application, Figure 38–32.

38.18 CONDENSER COIL DESIGN

Some coil surfaces are vertical to the ground, and grass and dirt can easily get into the bottom of the coil. The coils must be clean for the condenser to operate efficiently. Some equipment uses the bottom few rows for a subcooling circuit to lower the condensed refrigerant temperature below the condensing temperature. It is common for the liquid line to be 10° to 15°F cooler than the condensing temperature. Each degree of superheat will add approximately 1% to the efficiency of the system. If the subcooling circuit is dirty, it could affect the capacity by 10 to 15%. This could mean the difference in a piece of equipment being able to cool a structure if the equipment were sized too close to the design cooling load.

Some manufacturers use horizontal or slant-type condensers to position the coil off the ground. These condensers are less likely to pick up leaves and grass at the ground level, Figure 38–33.

38.19 HIGH-EFFICIENCY CONDENSERS

More surface area in the condenser makes the compressor head pressure lower in the hottest weather. This means lower compressor current and less power consumed for the same amount of air conditioning. Some manufacturers with oversized condensers use two-speed

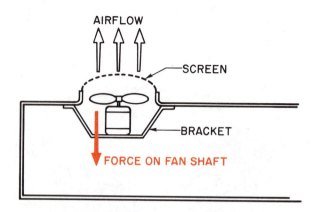

Figure 38–32. Top-air-discharge units have an additional load placed on the bearings. The fan is trying to fly down while it is running and pushing air out the top. During the starting and stopping of the fan, the bottom bearing has the fan blade and shaft resting on it. This bearing must have a thrust surface.

condenser fans—one speed for mild weather, one for hot. Without two-speed fans the condensers would be too efficient in mild weather and would make the head pressure too low, causing the expansion device to be starved and the capacity to be less. This would be similar to the low ambient problems discussed in Unit 20.

38.20 CABINET DESIGN

The condenser cabinet is usually located outside, so it needs weatherproofing. Most cabinets are galvanized and painted to give them years of life. Some cabinets are made of aluminum, which is lighter but may not last as long as in a salty environment.

Most small equipment is assembled with self-tapping sheet-metal screws. These screws are held by a drill screw holder during manufacturing and are threaded into the cabinet when turned with a drill motor. See Figure 38–34 for an example of a drill motor and holder. These screws should be made of weather-

Figure 38–31. The top-air-discharge units usually have the fan on top.

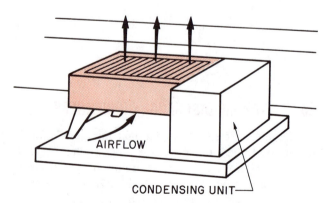

Figure 38–33. Some condensers are designed so that the coil is off the ground.

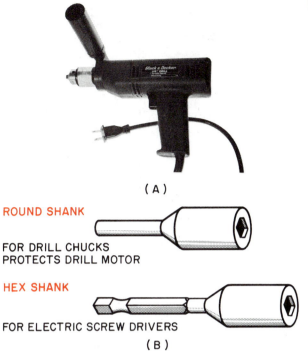

ROUND SHANK

FOR DRILL CHUCKS
PROTECTS DRILL MOTOR

HEX SHANK

FOR ELECTRIC SCREW DRIVERS
(B)

Figure 38–34. Drill motor and screw holder. *Photo by Bill Johnson*

Figure 38–35. Thermostatic expansion valve. *Courtesy ALCO Controls Division, Emerson Electric Company*

resistant material that will last for years out in the weather. In some locations, the weather may be salt air as in the coastal areas. In these locations, stainless steel is a good choice of metal for sheet-metal screws. When equipment that is assembled with drill screws is installed in the field, all of the screws should be fastened back into the cabinet tightly or the unit may rattle. After being threaded many times, the screw holes may become oversized; if so, just use the next size screw to tighten the cabinet panels.

38.21   EXPANSION DEVICES

The expansion device meters the refrigerant to the evaporator. The thermostatic expansion valve and the fixed-bore metering device (either a capillary tube or an orifice) are the types most often used.

Thermostatic expansion devices were described in Unit 22. This expansion valve is the same type except it has a different temperature range. Air conditioning expansion devices are in the high-temperature range, Figure 38–35. Thermostatic expansion valves are more efficient than fixed-bore devices because they allow the evaporator to reach peak performance faster. They allow more refrigerant into the evaporator coil during a hot pull down—when the conditioned space is allowed to get too warm before the unit is started. This results from equipment failure or because the operator waited too long before starting the equipment.

When a thermostatic expansion valve is used, the refrigerant pressures do not equalize during the off cycle unless the valve is made to equalize them. This means that the compressor must have a start capacitor

to start after the system has been off. Some valve manufacturers have a bleed port that always allows a small amount of refrigerant to bleed through. During the off cycle this valve will equalize pressures, and a compressor with a low-starting-torque motor can be used, Figure 38–35.

38.22   AIR-SIDE COMPONENTS

The air side of the air conditioning system consists of the supply air and the return-air systems. The airflow in an air conditioning system is normally 400 cfm/ton in the humid climates. This air is leaving the air handler at about 55°F in the average system. The duct work carrying this air must be insulated when it runs through an unconditioned space, or it will sweat and gain unwanted heat from the surroundings. The insulation should have a vapor barrier to keep the moisture from penetrating the insulation and collecting on the metal duct. All connections of the insulation must be fastened tight, and a vapor barrier must be used at all seams.

The return air will normally be about 75°F. If the return-air duct is run through the unconditioned space, it may not need insulation. If the unconditioned space is a crawl space or basement, the temperature may be 75°F, and no heat will exchange. Even if a small amount of heat did exchange, a cost evaluation should be made to see if it is more economical to insulate the return duct or allow the small heat exchange. If the duct is run through a hot attic, the duct must be insulated.

Unit 39 explains that cool air distributes better from high in the room because it falls. The final distribution point is where the cool air is mixed with the room air to arrive at a comfort condition. See Figure 38–36 for some examples of air conditioning diffusers.

38.23   INSTALLATION PROCEDURES

Package air conditioning systems were described earlier as systems where the whole air conditioner is built into one cabinet. A package air conditioner looks like a window air conditioner. Actually, a window unit

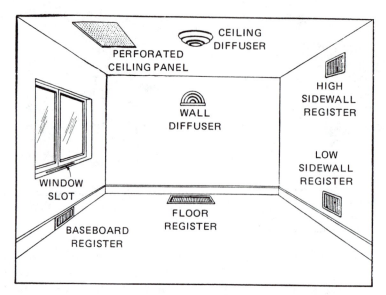

Figure 38–36. The final part of the air conditioning system is to get the refrigerated air properly distributed in the conditioned space. From Lang, *Principles of Air Conditioning,* © 1979 by Delmar Publishers Inc.

is a package air conditioner designed to blow freely into the conditioned space. Package air conditioners described in this text have two fan motors, one for the evaporator and one for the condenser, Figure 38–37. A window unit uses one fan motor with a double shaft that drives both fans.

The package air conditioner has the advantage that all equipment is located outside the structure, so all service can be performed on the outside. This equipment is probably more efficient than the split system because it is completely factory assembled and charged with refrigerant. The refrigerant lines are short, so there is less line loss.

The installation consists of mounting the unit on a firm foundation, fastening the package unit to the duct work, and connecting the electrical service to the unit. This can be quite attractive in some installations where the duct work can be readily attached to the unit. Some common installations are rooftop, beside the structure, and in the eaves of structures, Figure 38–38.

The split air conditioning system is used when the condensing unit must be located away from the evaporator. There must be interconnecting piping furnished by the installing contractor or the equipment supplier. When furnished by the equipment supplier, the tubing may be charged with its own operating charge or with dry nitrogen. These tubing sets are further discussed in Unit 40. See Figure 38–39 for an example of the two types of tubing.

The condensing unit should be located as close as possible to the evaporator to keep the interconnecting tubing short. The tubing consists of a cool gas line and a liquid line. The cool gas line is insulated and is the larger of the two lines. The insulation keeps unwanted

heat from conducting into the line and keeps the line from sweating and dripping. See Figure 38–40 for an example of a precharged line set.

A typical installation with pressures and temperatures is shown in Figure 38–41. This illustration gives some guidelines about the operating characteristics of a typical system.

SAFETY PRECAUTION: *Installation of equipment may require the technician to unload the equipment from a truck or trailer and move it to various locations. The condensing unit may be located on the other side

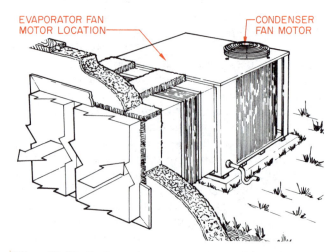

Figure 38–37. Package air conditioner with two fan motors— one for the evaporator coil and one for the condenser coil. *Courtesy Climate Control*

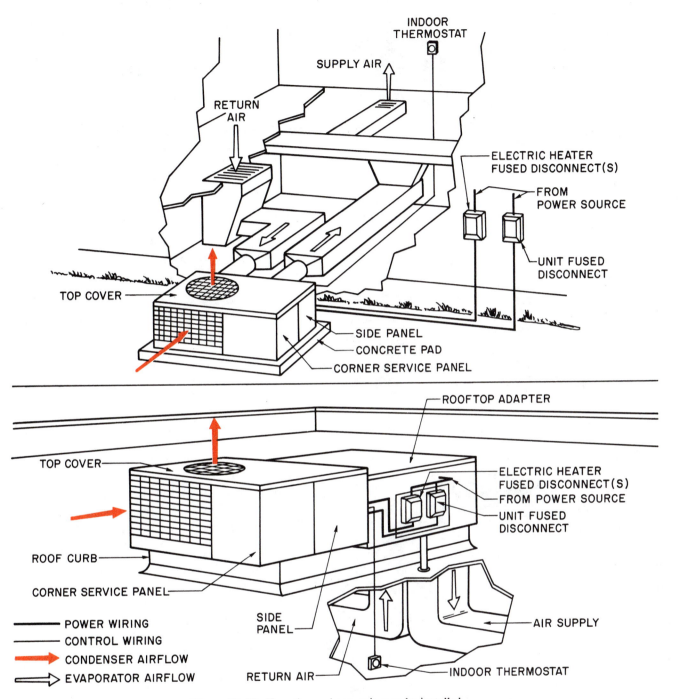

Figure 38-38. How the package unit may be installed.

of the structure, far away from the driveway, or it may even be located on a rooftop. Proper care should be taken in handling the equipment. Small cranes are often used for lifting. Lift gate trucks may be used to set the equipment down to the ground.

When installing the equipment, care should be taken while working under structures and in attics. Spiders and other types of stinging insects are often found in

these places and sharp objects such as nails are often left uncovered. The attic is the perfect place to step through a ceiling.

Care should be taken while handling the line sets during connection if they have refrigerant in them. Remember, liquid R-22 boils at −41°F and can inflict serious burns to the hands and eyes. Goggles should be worn when connecting line sets. ✱

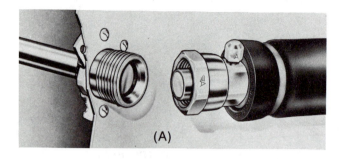

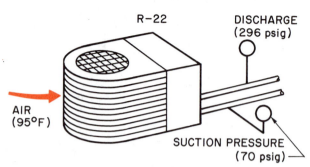

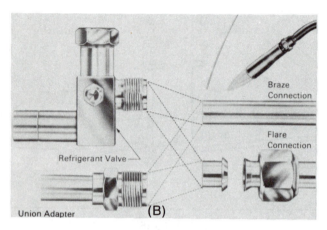

Figure 38–39. Tubing sets for precharged equipment. *Courtesy Aeroquip Corporation*

Figure 38–41. Installation showing the pressures and temperatures for a system in the humid southern part of the United States.

SUMMARY

- Refrigeration is the removal of heat from a place where it is not wanted and depositing it in a place where it makes no difference.
- Evaporative cooling and refrigerated air conditioning are two methods used for comfort air conditioning.
- Evaporative cooling may be used in areas where the temperature is high and the humidity is low.
- Refrigerated air conditioning cools the air and removes moisture.
- Refrigerated air conditioning is used in hot temperatures with high or low humidity.
- Refrigerated air conditioning systems have four major components: the evaporator, the compressor, the condenser, and the metering device.
- Evaporators are made in three types: the A coil, slant coil, and H coil. The type of coil depends on the position in the system and the particular manufacturer's design—an upflow, downflow, or horizontal flow.
- Evaporators operate at about 40°F when applied to air conditioning, and remove sensible heat and latent heat.
- Removal of sensible heat lowers the air temperature; removal of latent heat removes moisture.
- The compressor is the positive displacement pump that pumps the heat-laden vapor from the evaporator to the condenser.
- Compressors are cooled by suction gas, air, or discharge gas.

Figure 38–40. Suction (cool gas line) and liquid line set. *Photo by Bill Johnson*

- R-22 is the most commonly used refrigerant for air conditioning; R-500 and R-12 were commonly used in the past.
- Condensers are located outside to reject the heat to the outside.
- High efficiency in a condenser is achieved by increasing the condenser surface area. Two-speed fans may be used to keep the head pressure up during mild weather when the fan would run in the slow-speed mode.
- Package air conditioners are installed through the roof or through the wall at the end of the structure, wherever the duct can be fastened.
- The package unit has been charged at the factory and is factory assembled. It is probably more efficient than the split system.
- The split system has two interconnecting pipes. The large line is the insulated cool gas line; the small line is the liquid refrigerant line.

REVIEW QUESTIONS

1. Name two methods for cooling air for air conditioning.
2. What are the advantages of refrigerated air conditioning?
3. Name the four components of a refrigerated air conditioning system.
4. Name two types of refrigerated air conditioning equipment.
5. Name two types of compressors used in refrigerated air conditioning.
6. Name three methods used to cool the compressor motor.
7. List three types of metering devices used in air conditioning.
8. Which type of metering device requires a compressor motor with a high starting torque?
9. Why do the other metering devices not require a high-starting-torque motor?
10. What is the most popular refrigerant used in air conditioning?
11. Name three refrigerants that may be found in the air conditioning industry.
12. In what areas of the country is residential air conditioning popular?
13. What two types of heat must be removed with the air conditioning equipment?
14. Which component in the air conditioning system absorbs heat into the system?
15. Which component pumps heat?
16. In what substance is the heat absorbed, pumped, and rejected?
17. What component rejects heat from the system?
18. Define refrigeration.
19. What is the compressor mounted on?
20. Name the large and small interconnecting lines on an air conditioning system.

AIR DISTRIBUTION AND BALANCE

OBJECTIVES

After studying this unit, you should be able to

- **describe the prime mover of air for an air conditioning system.**
- **describe characteristics of the propeller and the centrifugal blowers.**
- **take basic air-pressure measurements.**
- **measure air quantities.**
- **list the different types of air-measuring devices.**
- **describe the common types of motors and drives.**
- **describe duct systems.**
- **explain what constitutes good airflow through a duct system.**
- **describe a return-air system.**
- **plot flow conditions on the friction chart.**

39.1 CONDITIONING EQUIPMENT

Unit 37 pointed out that air has to be conditioned for us to be comfortable. One way to condition air is to use a fan to move the air over the conditioning equipment. This conditioning equipment could be a cooling coil, a heating device, a device to add humidity, or a device to clean the air. This is known as a forced-air system because a fan is used to move the air. The forced-air system uses the same room air over and over again. Air from the room enters the system and is conditioned and returned to the room. Fresh air gets into the structure either by infiltration around the windows and doors or by ventilation from a fresh air inlet connected to the outside, Figure 39–1.

The forced-air system is different from a natural-draft system, where the air passes naturally over the conditioning equipment. Baseboard heat is an example of natural-draft heat. The warm water in the pipe heats the air in the vicinity of the pipe. The warmed air expands and rises. New air from the floor at a cooler temperature takes the place of the heated air, Figure 39–2. There is very little concern for the amount of air moving in a natural convection system.

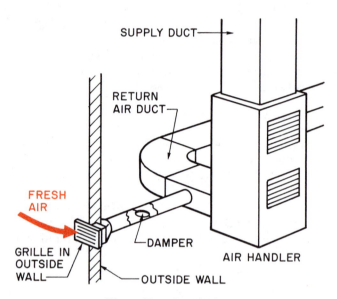

Figure 39–1. Ventilation.

39.2 CORRECT AIR QUANTITY

The object of the forced air system is to *deliver the correct quantity of conditioned air to the occupied space.* When this occurs, the air mixes with the room air and creates a comfortable atmosphere in that space.

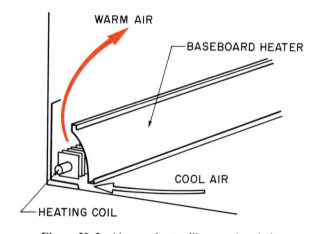

Figure 39–2. Air near heat will expand and rise.

592

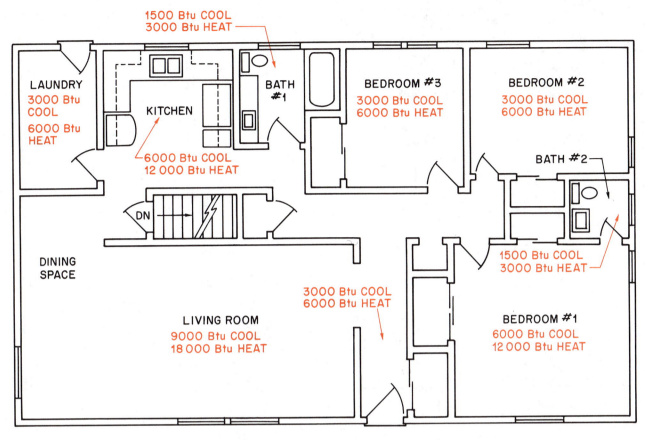

Figure 39–3. The floor plan has the heat and cooling requirements for each room indicated.

Different spaces have different air quantity requirements and it can easily be imagined that the same structure may have several different cooling requirements. For example, a house has rooms of different sizes with different requirements. A bedroom requires less heat and cooling than a large living room. Different amounts of air need to be delivered to these rooms to maintain comfort conditions, Figure 39–3 and Figure 39–4. Another example is a small office building with a high cooling requirement in the lobby and a low requirement in the individual offices. The correct amount of air must be delivered to each part of the building so that one area will not be overcooled while other areas are cooled correctly.

39.3   THE FORCED-AIR SYSTEM

The components that make up the *forced-air system* are the *blower* (fan), the *air supply system,* the *return-air system,* and the *grilles* and *registers* where the circulated air enters the room and returns to the conditioning equipment. See Figure 39–5 for an example of duct fittings. When these components are correctly chosen, they work together as a system with the following characteristics:

1. No air movement will be felt in the conditioned space that would normally be occupied.
2. No air noise will be noticed in the conditioned space.
3. No temperature swings will be felt by the occupants.
4. The occupants will not be aware that the system is on or off unless it stops for a long time and the temperature changes.

The lack of awareness of the conditioning system is important. It might be said that a well-designed comfort system is like a drummer in a symphony: you do not notice the drummer while he is playing, only when he stops.

39.4   THE BLOWER

The *blower* or fan provides the pressure to force the air into the duct system, through the grilles and registers, and into the room. Air has weight and has a resistance to movement. This means that it takes energy to move the air to the conditioned space. The fan may be required to push enough air through the evaporator and duct work for 3 tons of air conditioning. Typically 400 cubic feet of air must be moved per ton of air conditioning. This system would move

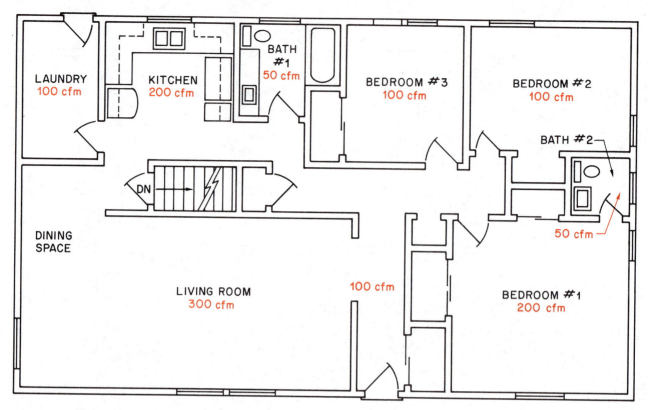

Figure 39–4. House plan of Figure 39–3 showing quantities of air delivered to the different rooms.

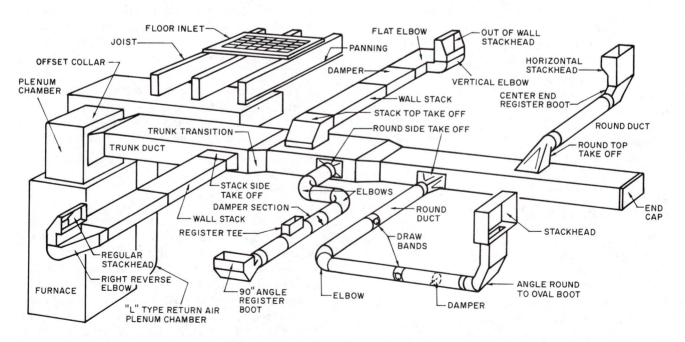

Figure 39–5. Duct fittings.

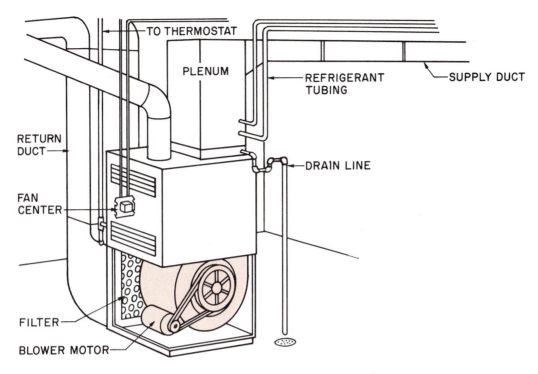

Figure 39-6. Fan and motor to move air.

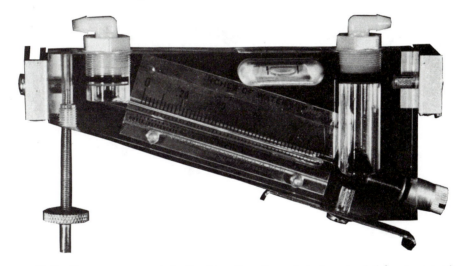

Figure 39-7. Water manometer is inclined to allow the scale to be extended for more graduations.

400 cfm × 3 tons or 1200 cubic feet of air per minute. Air has a weight of 1 pound per 13.35 cubic feet. This fan would be moving 90 pounds per minute (1200 cfm divided by 13.35 cubic feet per pound = 89.88). 90 pounds per minute × 60 minutes per hour means the fan moves 5400 pounds per hour. 5400 pounds per hour × 24 hours = 129,600 pounds per day. The motor is where the energy is consumed to move the air, Figure 39-6.

The pressure in a duct system for a residence or a small office building is too small to be measured in psi. It is measured in a unit of pressure that is still force per unit of area but in a smaller graduation. The pressure in duct work is measured in *inches of water column* (in. W.C.). A pressure of 1 in. W.C. is the pressure necessary to raise a column of water 1 in. Air pressure in a duct system is measured with a *water manometer*, which uses colored water that rises up a tube. Figure 39-7 shows a water manometer that is inclined for more accuracy at very low pressures. Figure 39-8 shows some other instruments that may be used to measure very low air pressures; they are graduated in inches of water even though they may not contain water.

The atmosphere exerts a pressure of 14.696 psi. 1 psi will raise a column of water 27.7 in. or 2.31 ft, Figure 39-9. The average duct system will not exceed a pressure of 1 in. W.C. A pressure of 0.05 psig will support

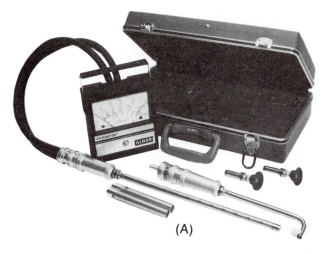

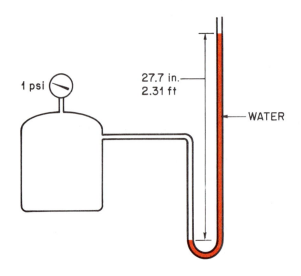

Figure 39–8. Other instruments used to measure air pressures. *Courtesy (A) Alnor Instrument Company (B) Photo by Bill Johnson*

a column of water 1.39 in. high (27.7 × 0.05 = 1.39 in.). Air flow pressure in duct work is measured in some very low figures.

39.5 SYSTEM PRESSURES

A duct system contains *static pressure, velocity pressure,* and *total pressure.* Static pressure is the same as the pressure on an enclosed vessel, such as a tank of refrigerant. This is like the pressure of the refrigerant in a drum that is pushing outward. Figure 39–10 shows a manometer for measuring static pressure. Notice the position of the sensing tube. The probe has a very small hole in the end, so the air rushing by the probe opening will not cause incorrect readings.

The air in a duct system is moving along the duct and therefore has velocity. The velocity of the air and the weight of the air create velocity pressure. Figure 39–11 shows a manometer for measuring velocity pressure in an air duct. Notice the position of the sensing tube. The air velocity goes straight into the tube inlet. The probe and manometer arrangement give the total pressure by canceling the static pressure with the second probe.

The total pressure of a duct can be measured with a manometer applied a little differently, Figure 39–12. Notice that the velocity component or probe of the manometer is positioned so that the air is directed into the end of the tube. This will register static and velocity pressures.

39.6 AIR-MEASURING INSTRUMENTS FOR DUCT SYSTEMS

The water manometer has been mentioned as an air pressure measuring instrument. An instrument used

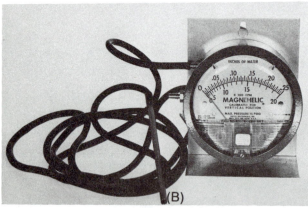

Figure 39–9. Vessel with a pressure of 1 psi inside. The water manometer has a column 27.7 in. high (2.31 ft).

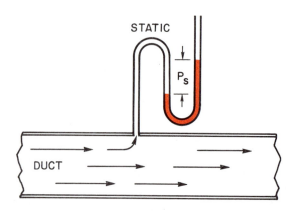

Figure 39–10. Manometer connected to measure the static pressure.

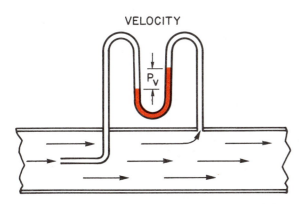

VELOCITY

P_V

THE STATIC PRESSURE PROBE CANCELS THE STATIC PRESSURE AT THE VELOCITY PRESSURE END AND INDICATES TRUE VELOCITY PRESSURE

Figure 39–11. Manometer connected to measure the velocity pressure of the air moving in the duct.

to measure the actual air velocity is the *velometer*. This instrument actually measures how fast the air is moving past a particular point in the system. See Figure 39–13 for an example of two different velometers. These instruments should be used as the particular manufacturer suggests.

A special device called a *pitot tube* was developed many years ago and is used with special manometers for checking duct air pressure at most pressure levels, Figure 39–14.

39.7 TYPES OF FANS

The blower or fan, as it is sometimes called, can be described as a device that produces airflow or movement. There are several different types of blowers that produce this movement, but all can be described as nonpositive displacement air movers. Remember, the compressor is a positive displacement pump. When the cylinder is full of refrigerant (or air for an air compressor), the compressor is going to empty that cylinder or

(A)

(B)

Figure 39–13. Two types of velometers. *Courtesy (B) Alnor Instrument Company (A) Photo by Bill Johnson*

break something. The fan is not positive displacement—it cannot build the kind of pressures that a compressor can. The fan has other characteristics that have to be dealt with.

The two fans that we will discuss in this text are the *propeller fan* and the *forward curve centrifugal fan* also called the *squirrel cage fan wheel*.

The propeller fan is used in exhaust fan and condenser fan applications. It will handle large volumes of air at low-pressure differentials. The propeller can be cast iron, aluminum, or stamped steel and is set into a housing called a *venturi* to encourage airflow in a

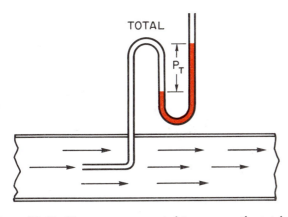

TOTAL

P_T

Figure 39–12. Manometer connected to measure the total air pressure in the duct.

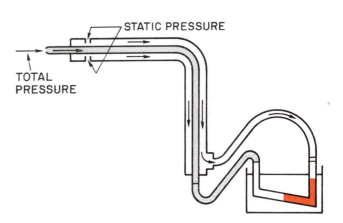

STATIC PRESSURE

TOTAL PRESSURE

Figure 39–14. Pitot tube.

straight line from one side of the fan to the other, Figure 39–15. The propeller fan makes more noise than the centrifugal fan so it is normally used where noise is not a factor.

The squirrel cage or centrifugal fan has characteristics that make it desirable for duct work. It builds more pressure from the inlet to the outlet and moves more air against more pressure. This fan has, what is called in the industry, a forward curved blade and a cutoff to shear the air spinning around the fan wheel. This air is thrown by centrifugal action to the outer perimeter of the fan wheel. Some of it would keep going around with the fan wheel if it were not for the shear that cuts off the air and sends it out the fan outlet, Figure 39–16. The centrifugal fan is very quiet when properly applied. It meets all requirements of duct systems up to very large systems that are considered high-pressure systems. High-pressure systems have pressures of 1 in. W.C. and more. Different types of fans, some of them similar to the forward curve centrifugal fan, are used in larger systems.

One characteristic that makes troubleshooting the centrifugal fan easier is the volume of air to horsepower requirement. This fan uses energy at the rate at which it moves air through the duct work. It can be said that the current draw of the fan motor is in proportion to the pounds of air it moves or pumps. The pressure the fan pumps against has little to do with the amount of energy used when operating at close to design conditions. For example, a fan motor that pulls full-load amperage at the rated fan capacity will pull less than full-load amperage at any value less than the fan capacity. If the fan is supposed to pull 10 A at maximum capacity of 3000 cfm, it will pull 10 A while moving only this amount of air. The weight of this volume of air can be calculated by dividing (3000 cfm ÷ 13.35 ft³/lb = 224.7 lb). If the fan inlet is stopped off, the suction side of the fan will be starved for air and the current will go down. If the discharge side of the fan is stopped off, the pressure will go up in the discharge of the fan and the current will go down because the fan is not handling as many pounds of air.

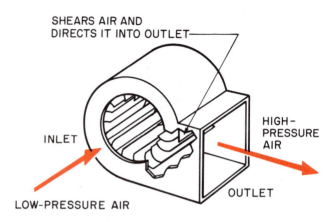

Figure 39–16. Centrifugal fan.

The air is merely spinning around in the fan and housing and not being forced into the duct work. This particular type of fan can be checked for airflow with an ammeter when making simple field measurements. If the current is down, the airflow is down. If the airflow is increased, the current will go up. For example, if the door on the blower compartment is opened, the fan current will go up because the fan will have access to the large opening of the fan compartment.

39.8 TYPES OF FAN DRIVES

The centrifugal blower must be turned by a motor. Two drive mechanisms are used: belt drive or direct drive. The belt-drive blower was used exclusively for many years. The motors were usually 1800 rpm. They actually run at 1750 rpm under load and operate very quietly. The motor normally had a capacitor and would go from a stopped position to 1750 rpm in about 1 sec. The motor may make more noise starting than running.

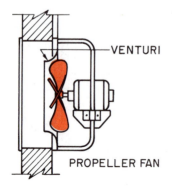

Figure 39–15. Propeller-type fan.

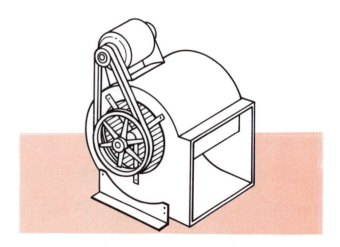

Figure 39–17. Blower driven with a motor using a belt and two pulleys to transfer the motor energy to the fan wheel.

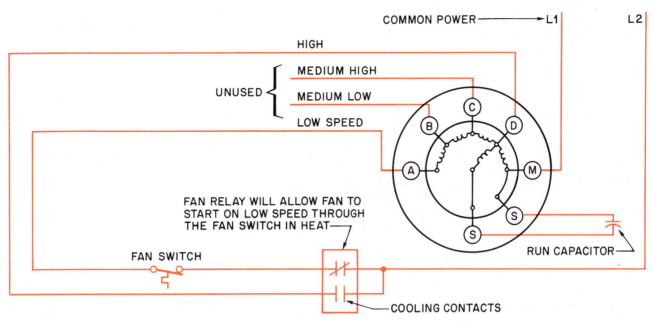

Figure 39–18. Wiring diagram of a multiple-speed motor.

Later, manufacturers began making equipment more compact and began using smaller blowers with 3600-rpm motors. These motors actually turn at 3450 rpm under load. They had to turn from 0 rpm to 3450 rpm in about 1 sec and could make quite a noise on startup. Sleeve bearings and resilient (rubber) mountings are used to keep bearing noise out of the blower section. Belt-drive blowers have two bearings on the fan shaft and two bearings on the motor. Sometimes these bearings are permanently lubricated from the manufacturer, so a technician cannot oil them.

The drive pulley on the motor, the driven pulley on the fan shaft, and the belt must be maintained. This motor and blower combination has many uses because the pulleys can be adjusted or changed to change fan speeds, Figure 39–17.

Recently, most manufacturers have been using a direct-drive blower. The motor is mounted onto the blower housing, usually with rubber mounts, with the fan shaft extending into the fan wheel. The motor is a PSC motor that starts up very slowly, taking several seconds to get up to speed. It is very quiet and does not have a belt and pulleys to wear out or to adjust.

Shaded-pole motors are used on some direct drive blowers.

With PSC motors the fan wheel bearing is located in the motor, which reduces the bearing surfaces from four to two. The bearings may be permanently lubricated at the factory. The front bearing in the fan wheel may be hard to lubricate if a special oil port is not furnished. There are no belts or pulleys to maintain or adjust. The fan turns at the same speed as the motor, so multispeed motors are common. The air volume may be adjusted with the different fan speeds instead of a pulley. The motor may have up to four different speeds that can be changed by switching wires at the motor terminal box. Common speeds are from about 1500 rpm down to about 800 rpm. The motor can be operated at a faster speed in the summer for more airflow for cooling, Figure 39–18.

39.9 THE SUPPLY DUCT SYSTEM

The supply duct system distributes air to the terminal units, registers, or diffusers into the conditioned space. Starting at the fan outlet, the duct can be fastened to the blower or blower housing directly or have a vibration eliminator between the blower and the duct work. The vibration eliminator is recommended on all jobs but is not always used. If the blower is quiet it may not be necessary, Figure 39–19.

The duct system must be designed to allow air moving towards the conditioned space to move as freely as possible, but the duct must not be oversized. Oversized duct is not economical. Duct systems can be *plenum, extended plenum, reducing plenum,* or *perim-*

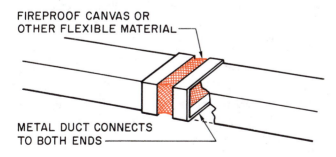

Figure 39–19. Vibration eliminator.

eter loop, Figure 39–20. Each system, of course, has its advantages and disadvantages.

39.10 THE PLENUM SYSTEM

The plenum system has an individual supply system that makes it well suited for a job where the room outlets are all very close to the unit. This system is very economical from a first-cost standpoint and can easily be installed by a beginning installer. It normally has the supply diffusers (where the air is diffused and blown into the room) located on the inside walls and is used for heating systems that have very warm or hot air as the heating source. Plenum systems work better on fossil-fuel (coal, oil or gas) systems than with heat pumps, because the leaving-air temperatures are much warmer in fossil-fuel systems.

When the supply diffusers are located on the inside walls, a warmer air is more desirable. The supply air temperature on a heat pump is rarely more than 100°F, whereas on a fossil-fuel system it could easily reach 130°F. The return-air system can be a single return, which makes materials very economical. We will further discuss the single-return system later. See Figure 39–21 for an example of a plenum system.

39.11 THE EXTENDED PLENUM SYSTEM

The extended plenum system can be applied to a long structure such as the ranch-style home. This system takes the plenum closer to the farthest point. The extended plenum is called the *trunk duct* and can be round, square, or rectangular, Figure 39–22. The system uses small ducts called *branches* to complete the connection to the terminal units. These small ducts can be round, square, or rectangular. In small sizes they are usually round because it is less expensive to manufacture and assemble round duct in the smaller sizes. An average home probably has 6-in. round duct for the branches.

39.12 THE REDUCING PLENUM SYSTEM

The reducing plenum system reduces the trunk duct size as branch ducts are added. This system has the advantage of saving materials and keeping the same pressure from one end of the duct system to the other, when properly sized. This ensures that each branch duct has approximately the same pressure pushing air into its takeoff from the trunk duct, Figure 39–23.

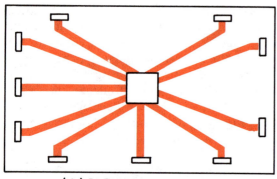

(A) PLENUM DUCT SYSTEM

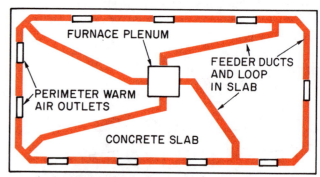

(C) REDUCING EXTENDED PLENUM SYSTEM

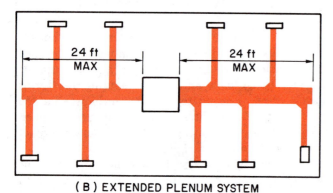

(B) EXTENDED PLENUM SYSTEM

(D) PERIMETER LOOP SYSTEM WITH FEEDER AND LOOP DUCTS IN CONCRETE SLAB

Figure 39–20. (A) Plenum system. (B) Extended plenum system. (C) Reducing plenum system. (D) Perimeter loop system.

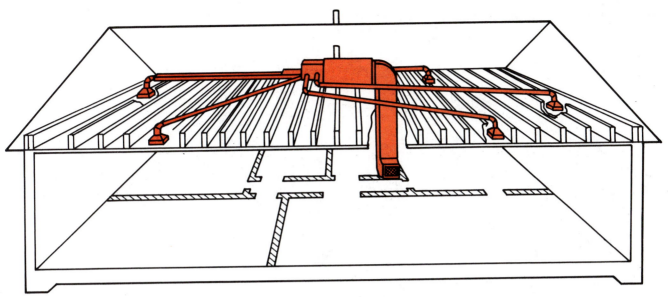

Figure 39-21. Using a plenum system.

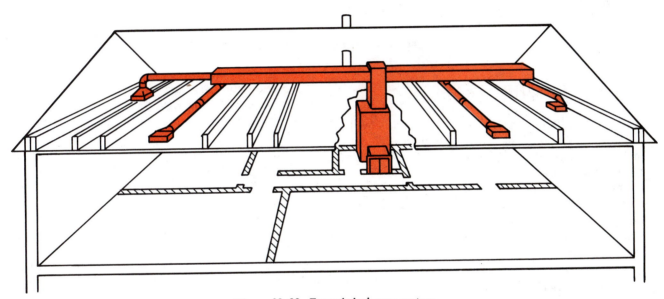

Figure 39-22. Extended plenum system.

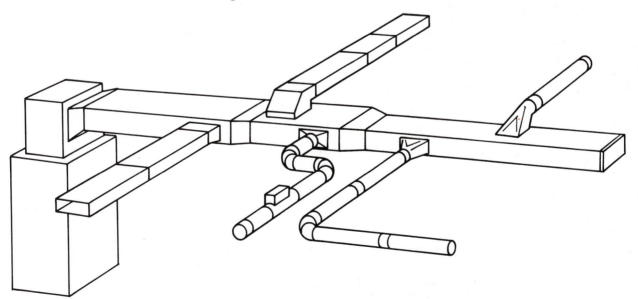

Figure 39-23. Reducing plenum system. *Courtesy Climate Control*

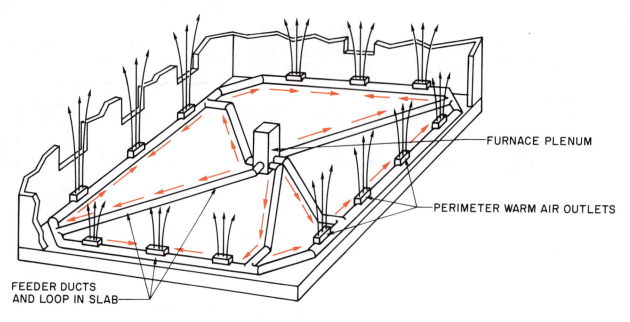

Figure 39-24. Perimeter loop system.

39.13   THE PERIMETER LOOP SYSTEM

The perimeter loop duct system is particularly well suited for installation in a concrete floor in a colder climate. The loop can be run under the slab close to the outer walls with the outlets next to the wall. There is warm air in the whole loop when the furnace fan is running, and this keeps the slab at a more even temperature. The loop has a constant pressure around the system and provides the same pressure to all outlets, Figure 39-24.

39.14   DUCT MATERIALS

The duct work for carrying the air from the fan to the conditioned space can be made of different materials. For many years galvanized sheet metal was exclusively used, but it is expensive to manufacture and assemble at the job. Galvanized metal is by far the most durable material. It can be used in walls where easy access for servicing is not available. Aluminum, fiberglass ductboard, spiral metal duct, and flexible duct have all been used successfully. *The duct material must meet the local codes for fire protection.*

39.15   GALVANIZED STEEL DUCT

Galvanized steel duct comes in several different thicknesses, called the *gauge* of the metal. When a metal is called 28 gauge, this means that it takes 28 thicknesses of the metal to make a piece 1-in. thick, or the metal is $\frac{1}{28}$ of an inch thick, Figure 39-25. The thickness of the duct can be less when the dimensions

of the duct work are small. When the duct work is larger, it must be more rigid or it will swell and make noises when the fan starts or stops. This action is similar to an oilcan end popping out when opened. Figure 39-26 shows a table to be used as a guideline for choosing duct thickness. Quite often the duct manufacturer will cross-break or make a seam from corner to corner on large fittings to make the duct more rigid.

Metal duct is normally furnished in lengths of 4 ft and can be round, square, or rectangular. Smaller round duct can be purchased in lengths up to 10 ft. Duct lengths can be fastened together with special fasteners, called S fasteners, and drive cleats if the duct is square or rectangular or self-tapping sheet-metal screws if the duct is round. These fasteners make a secure connection that is almost airtight at the low pressures at which the duct is normally operated. See Figure 39-27 for an example of these fasteners. If there is any question of the air leaking out, special tape can be applied to the connections to ensure that no air will

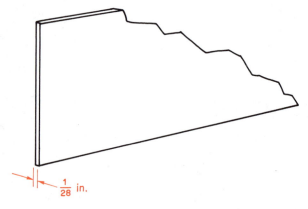

Figure 39-25. Duct thickness.

GAGES OF METAL DUCTS AND PLENUMS USED FOR COMFORT HEATING OR COOLING FOR A SINGLE DWELLING UNIT

	COMFORT HEATING OR COOLING			Comfort Heating Only
	Galvanized Steel			
	Nominal Thickness (In Inches)	Equivalent Galvanized Sheet Gage No.	Approximate Aluminum B & S Gage	Minimum Weight Tin-Plate Pounds Per Base Box
Round Ducts and Enclosed Rectangular Ducts				
14″ or less	0.016	30	26	135
Over 14″	0.019	28	24	—
Exposed Rectangular Ducts				
14″ or less	0.019	28	24	—
Over 14″	0.022	26	23	—

Figure 39–26. Table of recommended metal thicknesses for different sizes of duct.

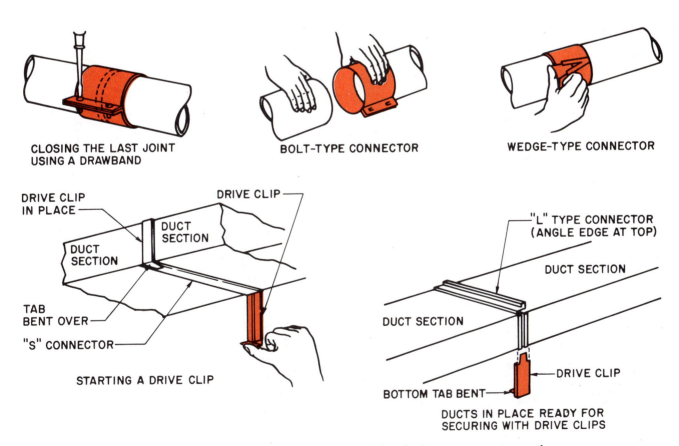

CLOSING THE LAST JOINT USING A DRAWBAND

BOLT-TYPE CONNECTOR

WEDGE-TYPE CONNECTOR

DRIVE CLIP IN PLACE

DRIVE CLIP

DUCT SECTION

DUCT SECTION

TAB BENT OVER

"S" CONNECTOR

STARTING A DRIVE CLIP

"L" TYPE CONNECTOR (ANGLE EDGE AT TOP)

DUCT SECTION

DUCT SECTION

DRIVE CLIP

BOTTOM TAB BENT

DUCTS IN PLACE READY FOR SECURING WITH DRIVE CLIPS

Figure 39–27. Fasteners for square and round duct for low-pressure systems only.

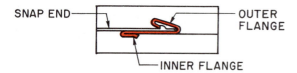

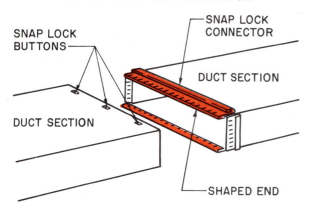

CROSS SECTION DETAIL OF
BUTTON SNAP LOCK CONNECTOR

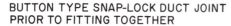

BUTTON TYPE SNAP-LOCK DUCT JOINT
PRIOR TO FITTING TOGETHER

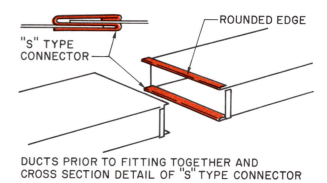

DUCTS PRIOR TO FITTING TOGETHER AND
CROSS SECTION DETAIL OF "S" TYPE CONNECTOR

Figure 39-27 *Continued*

leak out. See Figure 39–28 for a duct that is fastened together with self tapping sheet metal screws and a drill motor. See Figure 39–29 for a connection that has been taped after being fastened.

Aluminum duct follows the same guidelines as galvanized duct. The cost of aluminum prevents this duct from being used for many applications.

39.16 FIBERGLASS DUCT

Fiberglass duct is furnished in two styles: flat sheets for fabrication and round prefabricated duct. Fiberglass duct is normally 1 in. thick with an aluminum foil backing, Figure 39–30. The foil backing has a fiber reenforcement to make it strong. When the duct is fabricated, the fiberglass is cut to form the edges, and the reenforced foil backing is left intact to support the connections. These duct systems are easily transported and assembled in the field.

The ductboard can be made into duct work in several different ways. Special knives that cut the board in such a manner as to have overlapping connections

Figure 39–28. Fastening duct using a drill motor and sheet metal screws. *Photo by Bill Johnson*

Figure 39–29. Taped duct connection. *Photo by Bill Johnson*

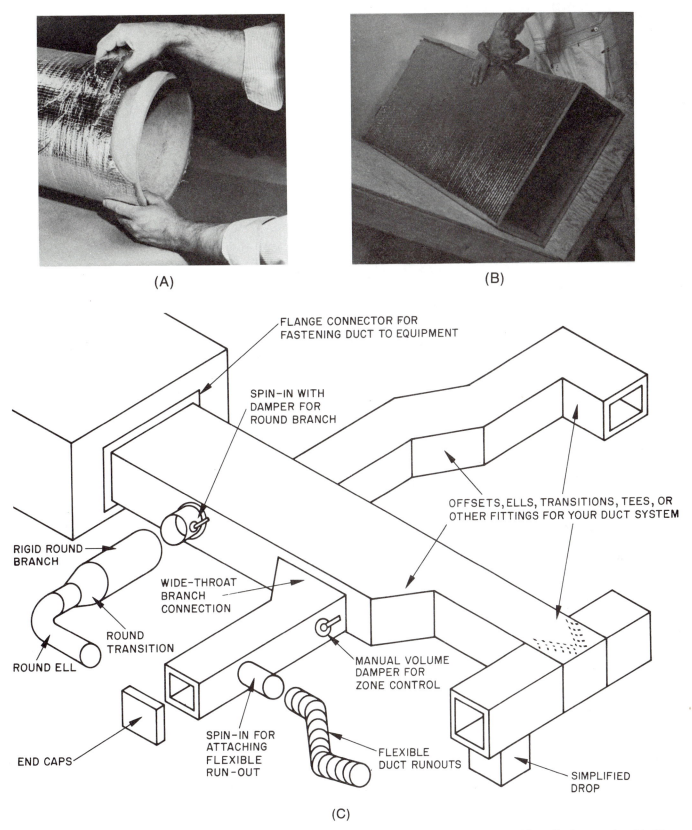

Figure 39-30. Ducts manufactured out of compressed fiberglass with a foil backing. *Courtesy (C) Manville Corporation (A, B) Photos by Bill Johnson*

Broad Blade Knife

Purple Tool

Orange Tool*

Grey Tool

*For V-Groove
cuts use Red Tool.

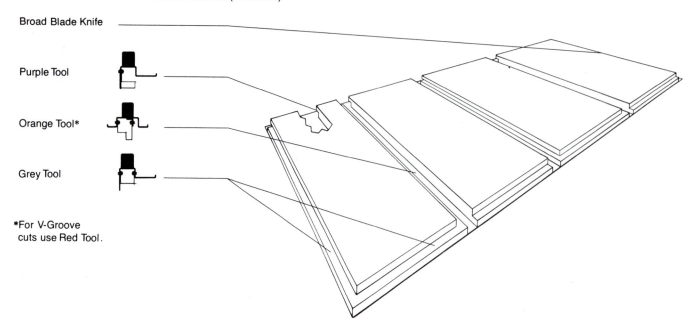

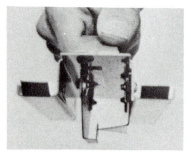

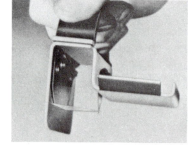

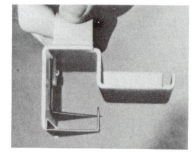

Orange Tool:
Cuts modified shiplap for forming corners.

Grey Tool:
The Grey Tool can also be used to make the closing corner joint. And, when making short sections, or when using square-edge board which does not have the premolded M/F edges, the Grey Tool cuts the female slip-joint used to join sections.

Purple Tool:
Cuts the male slip-joint used to join sections.

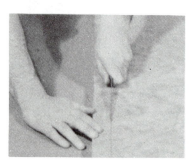

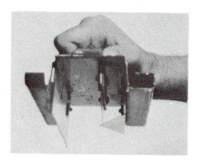

Broad Blade Knife:
Used in cutting the board to size and in cutting insulation from the staple flap.

Red Tool:
Cuts "V" or miter for forming corners.

Blue Tool:
Cuts shiplap for forming closing corner joints and also cuts the insulation to be stripped from the stapling flap.

Figure 39–31. Knives for fabricating fiberglass duct. *Courtesy Manville Corporation*

can be used in the field by placing the ductboard on any flat surface and using a straightedge to guide the knife, Figure 39–31. Special ductboard machines can be used to fabricate the duct in the field at the job sight or in the shop to be transported to the job. An operator has to be able to set up the machine for different sizes of ducts and fittings. When the duct is made in the shop, it can be cut and left flat and stored in the original boxes. This makes transportation easy. The pieces can be marked for easy assembly at the job site.

The machines or the knives cut away the fiberglass to allow the duct to be folded with the foil on the back side. When two pieces are fastened together, an overlap of foil is left so that one piece can be stapled to the other. A special staple is used that turns out and up on the ends. Then the connection is taped with special tape. The tape should be pressed on with a special iron, Figure 39–32. One of the advantages of fiberglass duct is that the insulation is already on the duct when it is assembled.

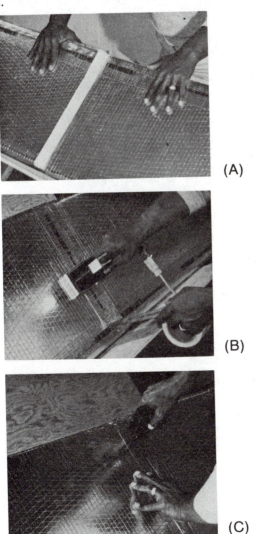

(A)

(B)

(C)

Figure 39–32. Connection assembly. *Courtesy Manville Corporation*

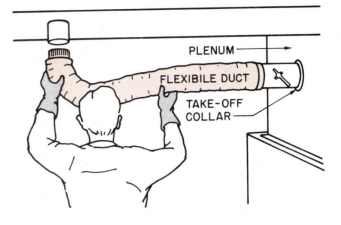

Figure 39–33. Flexible duct.

39.17 SPIRAL METAL DUCT

Spiral metal duct is used more on large systems. It is normally manufactured at the job sight with a special machine. The duct comes in rolls of flat narrow metal. The machine winds the metal off the spool and folds a seam in it. The length of a run of duct can be made very long with this method.

39.18 FLEXIBLE DUCT

Flexible round duct comes in sizes up to about 24 in. in diameter. Some of it has a reinforced aluminum foil backing and comes in a short box with the duct material compressed. It is hard to put the duct material back into the box. Without the insulation on it the duct looks like a coil spring with a flexible foil backing.

Some flexible duct comes with vinyl or foil insulation on it, in lengths of 25 ft to a box, Figure 39–33. It is compressed into the box and looks like a caterpillar when the box is opened and the duct is allowed to expand out of the box. Flexible duct has the advantage of going around corners easily. Keep the duct as short as practical and don't allow tight turns that may cause the duct to collapse. This duct has more friction inside it than metal duct does, but it also serves

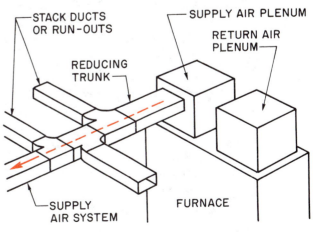

Figure 39–34. Square or rectangular duct.

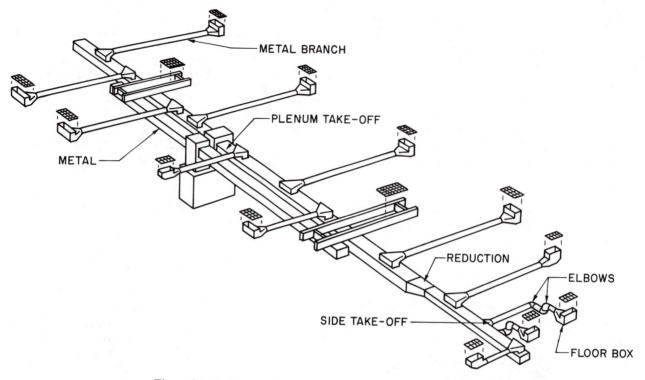

Figure 39-35. Rectangular metal trunk duct with round metal branch ducts.

as a sound attenuator to reduce blower noise down the duct. For best airflow, flexible duct should be stretched as tight as is practical.

39.19 COMBINATION DUCT SYSTEMS

Duct systems can be combined in various ways. For example:

1. All square or rectangular metal duct, the trunk line, and the branches are the same, Figure 39-34.
2. Metal trunk lines with round metal branch duct, Figure 39-35.
3. Metal trunk lines with round fiberglass branch duct, Figure 39-36.
4. Metal trunk lines with flexible branch duct, Figure 39-37.

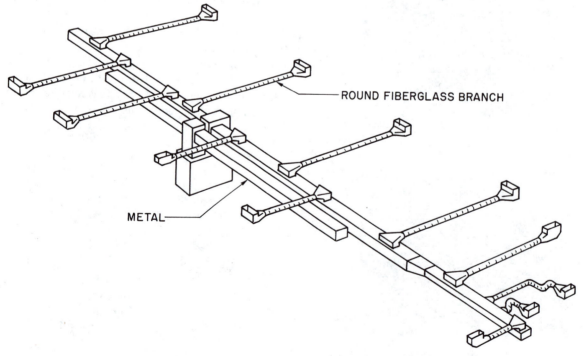

Figure 39-36. Metal trunk duct with round fiberglass branch ducts.

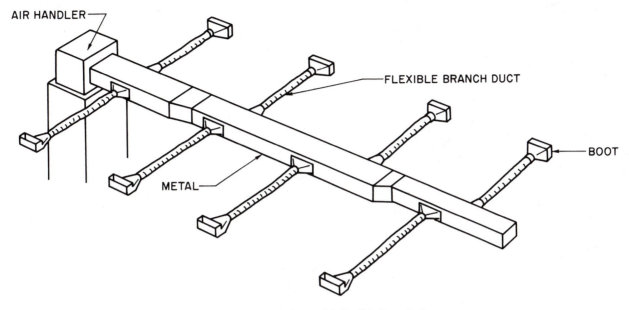

Figure 39–37. Metal trunk duct with flexible branch ducts.

5. Ductboard trunk lines with round fiberglass branches, Figure 39–38.
6. Ductboard trunk lines with round metal branches, Figure 39–39.
7. Ductboard trunk lines with flexible branches, Figure 39–40.
8. All round metal duct with round metal branch ducts, Figure 39–41.
9. All round metal trunk lines with flexible branch ducts, Figure 39–42.

39.20 DUCT AIR MOVEMENT

Special attention should be given to the point where the branch duct leaves the main trunk duct to get the correct amount of air into the branch duct. The branch duct must be fastened to the main trunk line with a *takeoff fitting*. Several fittings have been designed for this takeoff, Figure 39–43. The takeoff that has a larger throat area than the runout duct will allow the air to leave the trunk duct with a minimum of effort. This

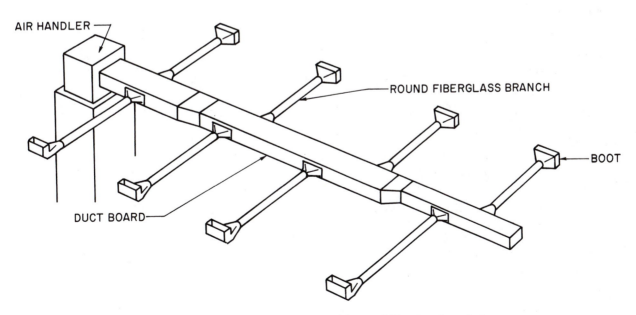

Figure 39–38. Fiberglass duct with round fiberglass branch ducts.

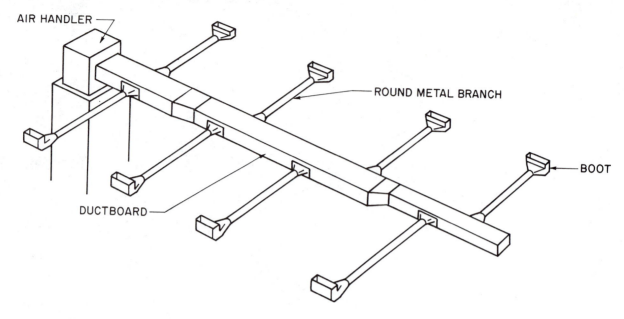

Figure 39-39. Ductboard trunk and round metal branches.

could be called a streamlined takeoff, Figure 39-43(B). The takeoff encourages the air moving down the duct to enter the takeoff to the branch duct.

Air moving in a duct has *inertia*—it wants to continue moving in a straight line. If air has to turn a corner, the turn should be carefully designed, Figure 39-44. For example, a square-throated elbow offers more resistance to airflow than a round-throated elbow does. If the duct is rectangular or square, turning vanes will improve the airflow around a corner, Figure 39-45.

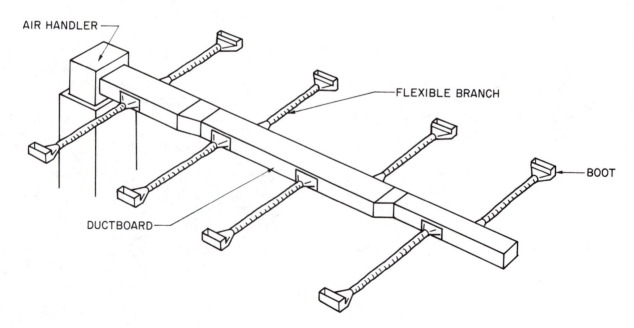

Figure 39-40. Fiberglass ductboard trunk and flexible branch ducts.

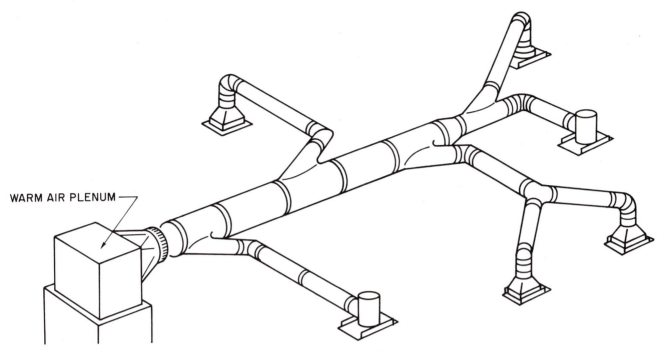

Figure 39–41. Metal and round duct.

39.21 BALANCING DAMPERS

A well-designed system will have *balancing dampers* in the branch ducts to balance the air in the various parts of the system. The dampers should be located as close as practical to the trunk line, with the damper handles uncovered if the duct is insulated. The place to balance the air is near the trunk, so if there is any air velocity noise it will be absorbed in the branch duct before it enters the room. A damper consists of a piece of metal shaped like the inside of the duct with a handle protruding through the side of the duct to the outside. The handle allows the damper to be turned at an angle to the airstream to slow the air down, Figure 39–46.

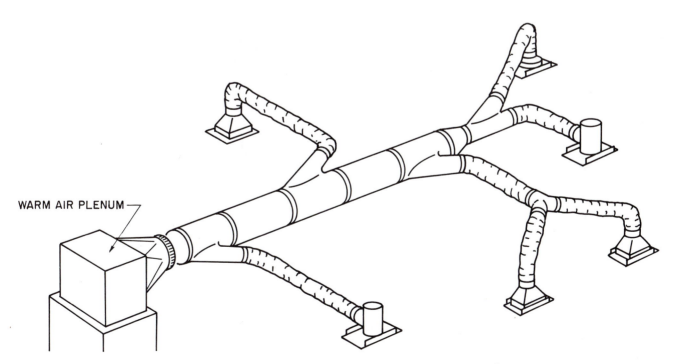

Figure 39–42. All round metal trunk line with flexible branch lines.

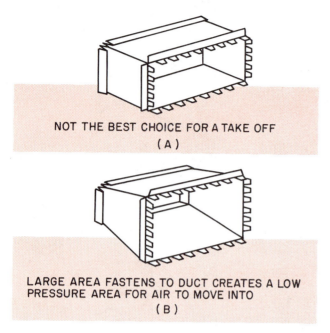

NOT THE BEST CHOICE FOR A TAKE OFF
(A)

LARGE AREA FASTENS TO DUCT CREATES A LOW
PRESSURE AREA FOR AIR TO MOVE INTO
(B)

Figure 39–43. (A) Standard takeoff fitting. (B) Streamlined takeoff fitting.

39.22 DUCT INSULATION

When duct work passes through an unconditioned space, heat transfer may take place between the air in the duct and the air in the unconditioned space. If the heat exchange adds or removes very much heat from the conditioned air, insulation should be applied to the duct work. A 15°F temperature difference from inside the duct to outside the duct is considered the maximum difference allowed before insulation is necessary.

The fiberglass duct has the insulation built into it from the manufacturer. Metal duct can be insulated in two ways: on the outside or on the inside. When applied to the outside, the insulation is usually a foil-

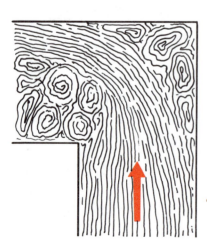

Figure 39–44. This illustration shows what happens when air tries to go around a corner.

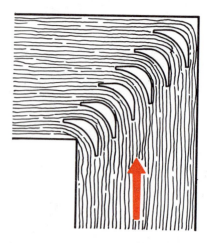

Figure 39–45. Square elbow with turning vanes.

or vinyl-backed fiberglass. It comes in several thicknesses, with 2-in. thickness the most common. The backing creates a moisture vapor barrier. This is important where the duct may operate below the dewpoint temperature of the surroundings and moisture would form on the duct. The insulation is joined by lapping it and stapling it. It is then taped to prevent moisture from entering the seams. External insulation can be added after the duct has been installed if the duct has enough clearance all around.

When applied to the inside of the duct, the insulation is either glued or fastened to tabs mounted on the duct by spot weld or glue. This insulation must be applied when the duct is being manufactured.

39.23 BLENDING THE CONDITIONED AIR WITH ROOM AIR

When the air reaches the conditioned space, it must be properly distributed into the room so that the room will be comfortable without anyone being aware that a conditioning system is operating. This means the final components in the system must place the air in the proper area of the conditioned space for proper air blending. Several guidelines are used for room air distribution. The following statements will help you understand some of the concepts:

1. When possible, air should be directed on the wall or load. For example, in winter air can be directed on the outside walls to cancel the load (cold wall) and keep the wall warmer. This will keep the wall from absorbing heat from the room air. In summer the same distribution will work; it will keep the wall cool and keep room air from absorbing heat from the wall. The *diffuser* spreads the air to the desired air pattern, Figure 39–47.

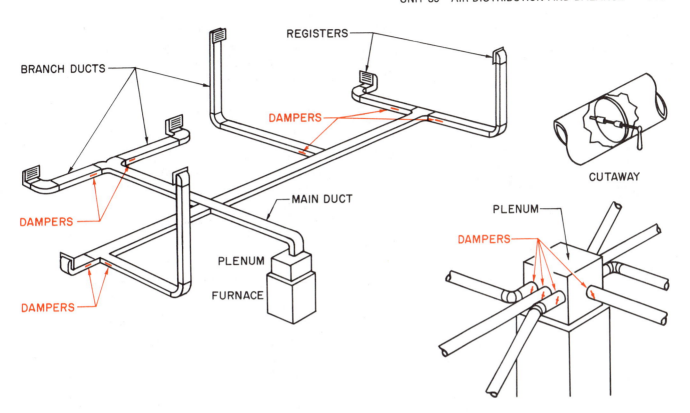

Figure 39-46. Balancing dampers.

2. Warm air for heating distributes better from the floor because it tends to rise, Figure 39–48.

3. Cool air for summer operation distributes better from the ceiling because it tends to fall, Figure 39–49.

4. The most modern concept for both heating and cooling is to place the diffusers next to the outside walls to accomplish this load-canceling effect, Figure 39–50.

5. The amount of throw (how far the air from the diffuser will blow away from the diffuser) depends on the air pressure behind the diffuser and the style

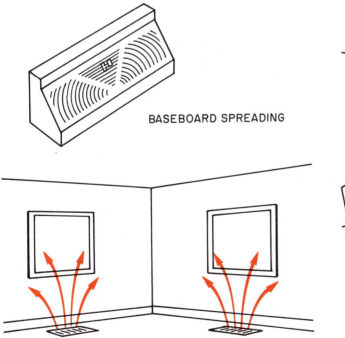

Figure 39-47. Diffuser.

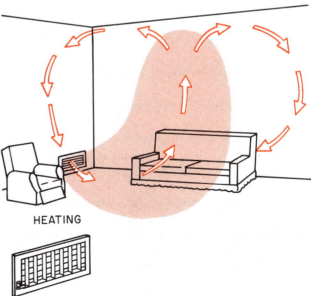

Figure 39-48. Warm air distribution.

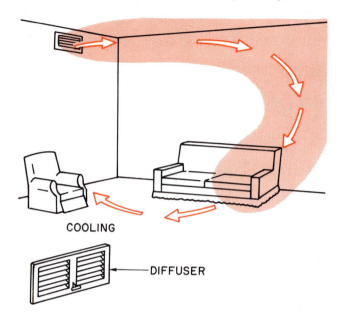

Figure 39–49. Cool air distribution.

of the diffuser blades. Air pressure in the duct behind the diffuser creates the velocity of the air leaving the diffuser.

Figure 39–51 shows some air registers and diffusers. The various types can be used for low side wall, high side wall, floor, ceiling, or baseboard.

39.24 THE RETURN-AIR DUCT SYSTEM

The return-air duct is constructed in much the same manner as the supply duct except that some jobs are built with central returns instead of individual room returns. Individual return-air systems have a return-air grille in each room that has a supply diffuser (with the exception of rest rooms and kitchens). The individual return system will give the most positive return-air system, but they are expensive. The return-air duct is normally sized at least slightly larger than the supply duct, so there is less resistance to the airflow in the return system than in the supply system. We will give more details when we discuss duct sizing. See Figure 39–52 for an example of a system with in-dividual room returns.

The central return system is usually satisfactory for a one level residence. Larger return-air grilles are located so that air from common rooms can easily get back to the common returns. For air to return to central returns, there must be a path, such as doors with grille work, open doorways, and undercut doors in common hallways. These open areas in the doors can prevent privacy that some people desire.

In a structure with more than one floor level, install a return at each level. Remember, cold air moves downward and warm air moves upward naturally

without encouragement. Figure 39–53 illustrates air stratification in a two-level house.

A properly constructed central return-air system helps to eliminate fan noise in the conditioned space. The return-air plenum should not be located on the furnace because fan running noise will be noticeable several feet away. The return-air grille should be around a corner from the furnace. If this cannot be done, the return-air plenum can be insulated on the inside to help deaden the fan noise.

Return-air grilles are normally large and meant to be decorative. They do not have another function unless there is a filter in them. They are usually made of stamped metal or have a metal frame with grille work, Figure 39–54.

39.25 SIZING DUCT FOR MOVING AIR

Moving air takes energy because (1) air has weight, (2) the air tumbles down the duct rubbing against itself and the duct work, and (3) fittings create resistance to the airflow. The friction part of the air movement will be discussed here.

Friction loss in duct work is due to the actual rubbing action of the air against the side of the duct and the turbulence of the air rubbing against itself while moving down the duct.

Friction due to rubbing the walls of the duct work cannot be eliminated but can be minimized with good design practices. Proper duct sizing for the amount of airflow helps the system performance. The smoother the duct surface is, the less friction there is. The slower the air is moving, the less friction there will be. It is beyond the scope of this text to go into details of

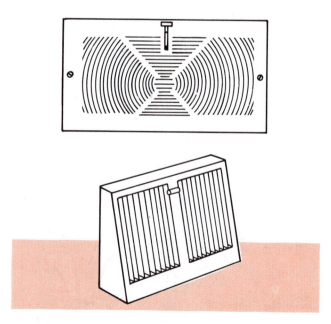

Figure 39–50. Diffusers used on the outside wall.

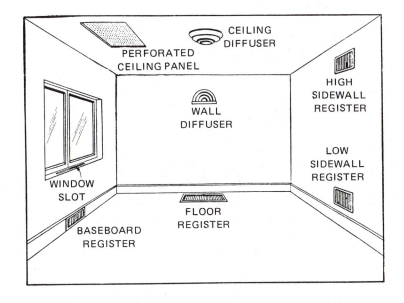

	FLOOR	BASEBOARD	LOW SIDEWALL	HIGH SIDEWALL	CEILING
COOLING PERFORMANCE	Excellent	Excellent if used with perimeter systems	Excellent if designed to discharge upward	Good	Good
HEATING PERFORMANCE	Excellent	Excellent if used with perimeter systems	Excellent if used with perimeter systems	Fair—should not be used to heat slab houses in Northern climates	Good—should not be used to heat slab houses in Northern climates
INTERFERENCE WITH DECOR	Easily concealed because it fits flush with the floor and can be painted to match	Not quite so easy to conceal because it projects from the baseboard	Hard to conceal because it is usually in a flat wall	Impossible to conceal because it is above furniture and in a flat wall	Impossible to conceal but special decorative types are available
INTERFERENCE WITH FURNITURE PLACEMENT	No interference— located at outside wall under a window	No interference— located at outside wall under a window	Can interfere because air discharge is not vertical	No interference	No interference
INTERFERENCE WITH FULL-LENGTH DRAPES	No interference— located 6 or 7 inches from the wall	When drapes are closed, they will cover the outlet	When located under a window, drapes will close over it	No interference	No interference
INTERFERENCE WITH WALL-TO-WALL CARPETING	Carpeting must be cut	Carpeting must be notched	No interference	No interference	No interference
OUTLET COST	Low	Medium	Low to medium, depending on the type selected	Low	Low to high— wide variety of types are available
INSTALLATION COST	Low because the sill need not be cut	Low when fed from below— sill need not be cut	Medium— requires wall stack and cutting of plates	Low on furred ceiling system; high when using under-floor system	High because attic ducts require insulation

Figure 39-51. Air registers and diffusers. From Lang, *Principles of Air Conditioning* © 1979 by Delmar Publishers Inc.

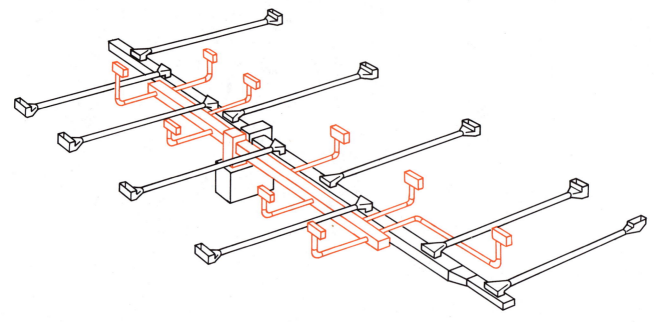

Figure 39-52. Duct plan of individual room return-air inlets. *Courtesy Carrier Corporation*

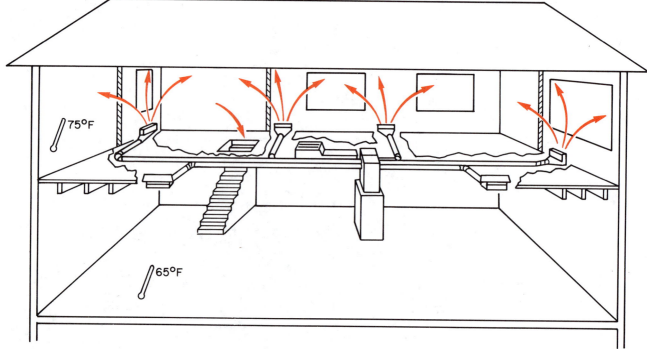

Figure 39-53. Air stratification.

(A) STAMPED LARGE-VOLUME
AIR INLET

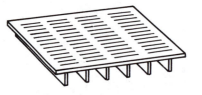

(B) FLOOR AIR INLET

(C) FILTER AIR INLET GRILLES

Figure 39-54. Return-air grilles.

duct design. The following text can be used as basic guidelines for a typical residential installation.

Each foot of duct offers a known resistance to airflow. This is called *friction loss*. It can be determined from tables and special slide calculators designed for this purpose. The following example will be used to explain friction loss in a duct system, Figure 39–55.

1. Ranch-style home requires 3 tons of cooling.
2. Cooling provided by a 3-ton cooling coil in the duct work.
3. The heat and fan are provided by a 100,000 Btu/hr furnace input, 80,000 Btu/hr output.
4. The fan has a capacity of 1360 cfm of air while operating against 0.40 in. W.C. static pressure with the system fan operating at medium to high speed. The system only needs 1200 cfm of air in the cooling mode. The system fan will easily be able to achieve this with a small amount of reserve capacity using a $\frac{1}{2}$-hp motor, Figure 39–56. Remember, the cooling mode usually requires more air than the heating mode. Cooling normally requires 400 cfm of air per ton; 3 ton × 400 = 1200 cfm.

5. The system has 11 outlets each requiring 100 cfm in the main part of the house and 2 outlets each requiring 50 cfm, located in the bathrooms. Most of these outlets are on the exterior walls of the house and distribute the conditioned air on the outside walls.
6. The return air is taken into the system from a common hallway, one return at each end of the hall.
7. While reviewing this system, think of the entire house as the system. The supply air must leave the supply registers and sweep the walls. It then makes its way across the rooms to the door adjacent to the hall. The air is at room temperature at this time and goes under the hall door to make its way to the return-air grille.
8. The return-air grille is where the duct system starts. There is a slight negative pressure (in relation to the room pressure) at the grille to give the air the incentive to enter the system. The filters are located in the return air grilles. The pressure on the fan side of the filter will be −0.03 in. W.C., which is less than the pressure in the room, so the room pressure pushes the air through the filter into the return duct.

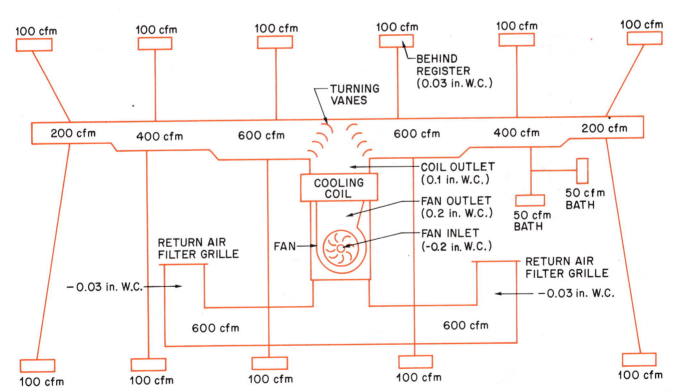

SYSTEM CAPACITY 3 TONS
cfm REQUIREMENT 400 cfm PER TON 400 X 3 = 1200 cfm
FAN STATIC PRESSURE (0.4 in. W.C.)

SUPPLY DUCT STATIC PRESSURE (0.2 in. W.C.)
RETURN DUCT STATIC PRESSURE (−0.2 in. W.C.)

Figure 39-55. Duct system.

SIZE	Blower Motor HP	Speed	External Static Pressure in. W.C.							
			0.1	0.2	0.3	0.4	0.5	0.6	0.7	0.8
048100	1/2 PSC	High	1750	1750	1720	1685	1610	1530	1430	—
		Med-High	1360	1370	1370	1360	1340	1315	—	—
		Med-Low	1090	1120	1140	1130	1100	—	—	—
		Low	930	960	980	980	965	945	—	—

Figure 39–56. Manufacturer's table for furnace airflow characteristics. *Courtesy BDP Company*

9. As the air proceeds down the duct toward the fan, the pressure continues to decrease. The lowest pressure in the system is in the fan inlet (−0.20 in. W.C.) below the room pressure.

10. The air is forced through the fan, and the pressure increases. The greatest pressure in the system is at the fan outlet, 0.20 in. W.C. above the room pressure. The pressure difference in the inlet and the outlet of the fan is 0.40 in. W.C.

11. The air is then pushed through the heat exchanger in the furnace where it drops. This new pressure is not useful to the service technician.

12. The air then moves through the cooling coil where it enters the supply duct system at a pressure of 0.10 in. W.C.

13. The air will take a slight pressure drop as it goes around the corner of the tee that splits the duct into two reducing plenums, one for each end of the house. This tee in the duct has turning vanes to help reduce the pressure drop as the air goes around the corner.

14. The first section of each reducing trunk has to handle an equal amount of air, 600 cfm each. Two branch ducts are supplied in the first trunk run, each with an air quantity of 100 cfm. This reduces the capacity of the trunk to 400 cfm on each side. A smaller trunk can be used at this point, and materials can be saved.

15. The duct is reduced to a smaller size to handle 400 cfm on each side. Because another 200 cfm of air is distributed to the conditioned space, another reduction can be made.

16. The last part of the reducing trunk on each side of the system need only handle 200 cfm for each side of the system.

This supply duct system will distribute the air for this house with minimal noise and maximal comfort. The pressure in the duct will be about the same all along the duct because as air was distributed off the trunk line, the duct size was reduced to keep the pressure inside the duct at the prescribed value.

At each branch duct dampers should be installed to balance the system's air supply to each room. This system will furnish 100 cfm to each outlet, but if a room did not need that much air, the dampers could be adjusted. The branch to each bathroom will need to be adjusted to 50 cfm each.

The return-air system is the same size on each side of the system. It returns 600 cfm per side and has the filters located in the return-air grilles in the halls. The furnace fan is located far enough from the grilles so that it won't be heard.

There are complete books written on duct sizing. Manufacturer's representatives may also help you with

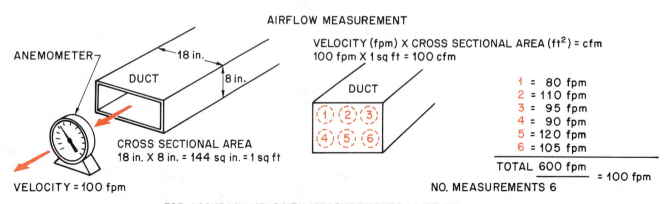

AIRFLOW MEASUREMENT

ANEMOMETER

DUCT 18 in. 8 in.

CROSS SECTIONAL AREA
18 in. X 8 in. = 144 sq in. = 1 sq ft

VELOCITY = 100 fpm

VELOCITY (fpm) X CROSS SECTIONAL AREA (ft²) = cfm
100 fpm X 1 sq ft = 100 cfm

DUCT

1   2   3
4   5   6

1 = 80 fpm
2 = 110 fpm
3 = 95 fpm
4 = 90 fpm
5 = 120 fpm
6 = 105 fpm

TOTAL 600 fpm
$\frac{}{}$ = 100 fpm
NO. MEASUREMENTS 6

FOR ACCURACY VELOCITY MEASUREMENTS MUST BE
AVERAGED OVER ENTIRE CROSS SECTION OF DUCT OUTLET

Figure 39–57. Cross section of duct with airflow shows how to measure duct area and air velocity.

specific applications. They also offer schools in duct sizing that use their techniques.

39.26   MEASURING AIR MOVEMENT FOR BALANCING

Air balancing is sometimes accomplished by measuring the air leaving each supply register. When one outlet has too much air, the damper in that run is throttled to slow down the air. This, of course, redistributes air to the other outlets and will increase their flow. See Figure 4–45 for an air instrument to determine airflow at an outlet.

The air quantity of an individual duct can be measured in the field to some degree of accuracy by using instruments to determine the velocity of the air in the duct. A velometer can be inserted into the duct to do this. The velocity must be measured in a cross section of the duct, and an average of the readings taken is used for the calculation. This is called *traversing* the duct. For example, if the air in a 1-ft^2 duct (12 in. × 12 in. = 144 in.^{2}), was traveling at a velocity of 1 ft/min, the volume of air passing a point in the duct would be 1 cfm. If the velocity were 100 ft/min, the volume of air passing the same point would be 100 cfm. The cross-sectional area of the duct is multiplied by the average velocity of the air to determine the volume of the moving air, Figure 39–57.

39.27   THE AIR FRICTION CHART

The previous system can be plotted on the friction chart in Figure 39–58. This chart is for volumes of air up to 500,000 cfm and is applied to all sizes of systems. Using the 400 cfm/ton mentioned earlier for air conditioning, we can use this chart for systems up to 1250 tons of cooling.

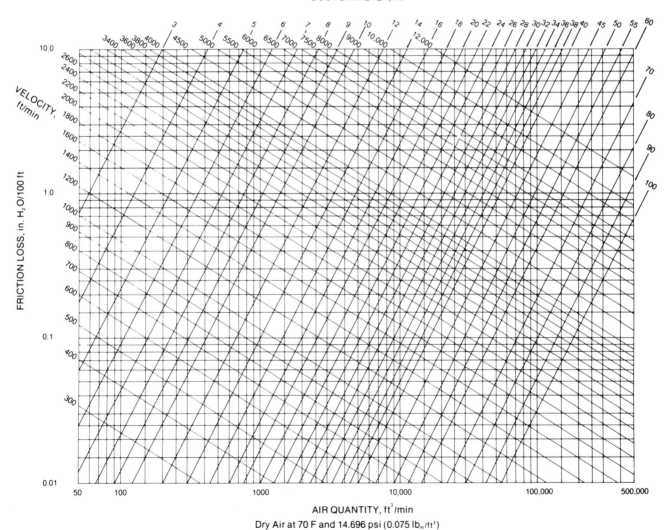

Figure 39–58. Air friction chart. *1985 chart used with permission from ASHRAE, Inc. 1791 Tullie Circle, NE, Atlanta, GA 30329*

The friction chart has cubic feet in the left column and round-pipe sizes angle from left to right toward the top on the page. These duct sizes are rated in round-pipe size on the chart and can be converted to square or rectangular duct by using the table in Figure 39–59. The round-pipe sizes are for air with a density of 0.075 lb/ft^3 using galvanized pipe with approximately 40 connections per 100 feet. The air velocity in the pipe is shown on the diagonal lines that run from right to left toward the top of the page. The friction loss in inches of water column per 100 feet of duct is shown along the bottom of the chart. For example, a run of pipe that carries 100 cfm (on the left side of the chart) can be plotted to the right of the intersection of the 6-in. pipe line. When this pipe is carrying 100 cfm of air, it has a velocity of just over 500 ft/min and a

Lgth Adj.[b]	6	7	8	9	10	11	12	13	14	15	16	17	18	19	20	22	24	26	28	30	Lgth Adj.[b]
																					Length of One Side of Rectangular Duct (a), in.
6	6.6																				6
7	7.1	7.7																			7
8	7.6	8.2	8.7																		8
9	8.0	8.7	9.3	9.8																	9
10	8.4	9.1	9.8	10.4	10.9																10
11	8.8	9.5	10.2	10.9	11.5	12.0															11
12	9.1	9.9	10.7	11.3	12.0	12.6	13.1														12
13	9.5	10.3	11.1	11.8	12.4	13.1	13.7	14.2													13
14	9.8	10.7	11.5	12.2	12.9	13.5	14.2	14.7	15.3												14
15	10.1	11.0	11.8	12.6	13.3	14.0	14.6	15.3	15.8	16.4											15
16	10.4	11.3	12.2	13.0	13.7	14.4	15.1	15.7	16.4	16.9	17.5										16
17	10.7	11.6	12.5	13.4	14.1	14.9	15.6	16.2	16.8	17.4	18.0	18.6									17
18	11.0	11.9	12.9	13.7	14.5	15.3	16.0	16.7	17.3	18.0	18.5	19.1	19.7								18
19	11.2	12.2	13.2	14.1	14.9	15.7	16.4	17.1	17.8	18.4	19.0	19.6	20.2	20.8							19
20	11.5	12.5	13.5	14.4	15.2	16.0	16.8	17.5	18.2	18.9	19.5	20.1	20.7	21.3	21.9						20
22	12.0	13.0	14.1	15.0	15.9	16.8	17.6	18.3	19.1	19.8	20.4	21.1	21.7	22.3	22.9	24.0					22
24	12.4	13.5	14.6	15.6	16.5	17.4	18.3	19.1	19.9	20.6	21.3	22.0	22.7	23.3	23.9	25.1	26.2				24
26	12.8	14.0	15.1	16.2	17.1	18.1	19.0	19.8	20.6	21.4	22.1	22.9	23.5	24.2	24.9	26.1	27.3	28.4			26
28	13.2	14.5	15.6	16.7	17.7	18.7	19.6	20.5	21.3	22.1	22.9	23.7	24.4	25.1	25.8	27.1	28.3	29.5	30.6		28
30	13.6	14.9	16.1	17.2	18.3	19.3	20.2	21.1	22.0	22.9	23.7	24.4	25.2	25.9	26.6	28.0	29.3	30.5	31.7	32.8	30
32	14.0	15.3	16.5	17.7	18.8	19.8	20.8	21.8	22.7	23.5	24.4	25.2	26.0	26.7	27.5	28.9	30.2	31.5	32.7	33.9	32
34	14.4	15.7	17.0	18.2	19.3	20.4	21.4	22.4	23.3	24.2	25.1	25.9	26.7	27.5	28.3	29.7	31.0	32.4	33.7	34.9	34
36	14.7	16.1	17.4	18.6	19.8	20.9	21.9	22.9	23.9	24.8	25.7	26.6	27.4	28.2	29.0	30.5	32.0	33.3	34.6	35.9	36
38	15.0	16.5	17.8	19.0	20.2	21.4	22.4	23.5	24.5	25.4	26.4	27.2	28.1	28.9	29.8	31.3	32.8	34.2	35.6	36.8	38
40	15.3	16.8	18.2	19.5	20.7	21.8	22.9	24.0	25.0	26.0	27.0	27.9	28.8	29.6	30.5	32.1	33.6	35.1	36.4	37.8	40
42	15.6	17.1	18.5	19.9	21.1	22.3	23.4	24.5	25.6	26.6	27.6	28.5	29.4	30.3	31.2	32.8	34.4	35.9	37.3	38.7	42
44	15.9	17.5	18.9	20.3	21.5	22.7	23.9	25.0	26.1	27.1	28.1	29.1	30.0	30.9	31.8	33.5	35.1	36.7	38.1	39.5	44
46	16.2	17.8	19.3	20.6	21.9	23.2	24.4	25.5	26.6	27.7	28.7	29.7	30.6	31.6	32.5	34.2	35.9	37.4	38.9	40.4	46
48	16.5	18.1	19.6	21.0	22.3	23.6	24.8	26.0	27.1	28.2	29.2	30.2	31.2	32.2	33.1	34.9	36.6	38.2	39.7	41.2	48
50	16.8	18.4	19.9	21.4	22.7	24.0	25.2	26.4	27.6	28.7	29.8	30.8	31.8	32.8	33.7	35.5	37.2	38.9	40.5	42.0	50
52	17.1	18.7	20.2	21.7	23.1	24.4	25.7	26.9	28.0	29.2	30.3	31.3	32.3	33.3	34.3	36.2	37.9	39.6	41.2	42.8	52
54	17.3	19.0	20.6	22.0	23.5	24.8	26.1	27.3	28.5	29.7	30.8	31.8	32.9	33.9	34.9	36.8	38.6	40.3	41.9	43.5	54
56	17.6	19.3	20.9	22.4	23.8	25.2	26.5	27.7	28.9	30.1	31.2	32.3	33.4	34.4	35.4	37.4	39.2	41.0	42.7	44.3	56
58	17.8	19.5	21.2	22.7	24.2	25.5	26.9	28.2	29.4	30.6	31.7	32.8	33.9	35.0	36.0	38.0	39.8	41.6	43.3	45.0	58
60	18.1	19.8	21.5	23.0	24.5	25.9	27.3	28.6	29.8	31.0	32.2	33.3	34.4	35.5	36.5	38.5	40.4	42.3	44.0	45.7	60
62		20.1	21.7	23.3	24.8	26.3	27.6	28.9	30.2	31.5	32.6	33.8	34.9	36.0	37.1	39.1	41.0	42.9	44.7	46.4	62
64		20.3	22.0	23.6	25.1	26.6	28.0	29.3	30.6	31.9	33.1	34.3	35.4	36.5	37.6	39.6	41.6	43.5	45.3	47.1	64
66		20.6	22.3	23.9	25.5	26.9	28.4	29.7	31.0	32.3	33.5	34.7	35.9	37.0	38.1	40.2	42.2	44.1	46.0	47.7	66
68		20.8	22.6	24.2	25.8	27.3	28.7	30.1	31.4	32.7	33.9	35.2	36.3	37.5	38.6	40.7	42.8	44.7	46.6	48.4	68
70		21.1	22.8	24.5	26.1	27.6	29.1	30.4	31.8	33.1	34.4	35.6	36.8	37.9	39.1	41.2	43.3	45.3	47.2	49.0	70
72			23.1	24.8	26.4	27.9	29.4	30.8	32.2	33.5	34.8	36.0	37.2	38.4	39.5	41.7	43.8	45.8	47.8	49.6	72
74			23.3	25.1	26.7	28.2	29.7	31.2	32.5	33.9	35.2	36.4	37.7	38.8	40.0	42.2	44.4	46.4	48.4	50.3	74
76			23.6	25.3	27.0	28.5	30.0	31.5	32.9	34.3	35.6	36.8	38.1	39.3	40.5	42.7	44.9	47.0	48.9	50.9	76
78			23.8	25.6	27.3	28.8	30.4	31.8	33.3	34.6	36.0	37.2	38.5	39.7	40.9	43.2	45.4	47.5	49.5	51.4	78
80			24.1	25.8	27.5	29.1	30.7	32.2	33.6	35.0	36.3	37.6	38.9	40.2	41.4	43.7	45.9	48.0	50.1	52.0	80
82				26.1	27.8	29.4	31.0	32.5	34.0	35.4	36.7	38.0	39.3	40.6	41.8	44.1	46.4	48.5	50.6	52.6	82
84				26.4	28.1	29.7	31.3	32.8	34.3	35.7	37.1	38.4	39.7	41.0	42.2	44.6	46.9	49.0	51.1	53.2	84
86				26.6	28.3	30.0	31.6	33.1	34.6	36.1	37.4	38.8	40.1	41.4	42.6	45.0	47.3	49.6	51.7	53.7	86
88				26.9	28.6	30.3	31.9	33.4	34.9	36.4	37.8	39.2	40.5	41.8	43.1	45.5	47.8	50.0	52.2	54.3	88
90				27.1	28.9	30.6	32.2	33.8	35.3	36.7	38.2	39.5	40.9	42.2	43.5	45.9	48.3	50.5	52.7	54.8	90
92					29.1	30.8	32.5	34.1	35.6	37.1	38.5	39.9	41.3	42.6	43.9	46.4	48.7	51.0	53.2	55.3	92
96					29.6	31.4	33.0	34.7	36.2	37.7	39.2	40.6	42.0	43.3	44.7	47.2	49.6	52.0	54.2	56.4	96

Figure 39–59. Chart to convert from round to square or rectangular duct. *1985 chart used with permission from ASHRAE, Inc. 1791 Tullie Circle, NE, Atlanta, GA 30329*

Structure	Supply Outlet	Return Openings	Main Supply	Branch Supply	Main Return	Branch Return
Residential	500–750	500	1,000	600	800	600
Apartments, Hotel Bedrooms, Hospital Bedrooms	500–750	500	1,200	800	1,000	800
Private Offices, Churches, Libraries, Schools	500–1,000	600	1,500	1,200	1,200	1,000
General Offices, Deluxe Restaurants, Deluxe Stores, Banks	1,200–1,500	700	1,700	1,600	1,500	1,200
Average Stores, Cafeterias	1,500	800	2,000	1,600	1,500	1,200

Figure 39–60. Chart of recommended velocities for different duct designs. From Lang, *Principles of Air Conditioning* © 1979 by Delmar Publishers Inc.

pressure drop of 0.085 in. W.C./100 feet. A 50-ft run would have half the pressure drop, 0.0425 in. W.C., Figure 39–61.

The friction chart can be used by the designer to size the duct system before the job price is quoted. This duct sizing gives the designer the sizes that will be used for figuring the duct materials. For all practical purposes, the duct should be sized using the chart in Figure 39–58, using the recommended velocities for various situations in the table in Figure 39–60. In the previous example a 4-in. pipe could have been used, but the velocity would have been nearly 1200 ft/min. This would be noisy, and the fan may not be big enough to push sufficient air through the duct.

High-velocity systems have been designed and used successfully in small applications. These systems will not be covered here since they may be considered the exception and not the rule. Such systems normally have a high air velocity in the trunk and branch ducts; the velocity is then reduced at the register to avoid drafts from the high-velocity air. If the air velocity is not reduced at the register, the register has a streamlined effect and is normally located in the corners of the room where someone is not likely to walk under it.

The friction chart can also be used by the field technician to troubleshoot airflow problems. Airflow problems come from system design, installation problems, and owner problems.

System design problems can be a result of a poor choice of fan or incorrect duct sizes. Any equipment supplier will help you choose the correct fan and equipment. They may even help with the duct design.

The installation of the duct system can make a difference.*The installers should protect the duct during the construction of the job. Stray material can make its way into the ducts and block the air, Figure 39–61. The installers must not collapse the duct and then insulate over the collapsed part. Duct work that

is run below a concrete slab can easily be damaged to the point that it will not pass the correct amount of air. When the insulation is applied to the inside of the duct, it must be fastened correctly or it may come loose and fall down into the airstream.*

Proper airflow techniques and design are a product of experience. The ability to design a system and get it installed and working within a prescribed budget takes time. Working with an experienced person is invaluable in this area.

SAFETY PRECAUTION: *The technician is responsible for obtaining air pressure and velocity readings in ductwork. This involves drilling and punching holes in metal duct. Use a grounded drill cord and be careful with drill bits and the rotating drill. Sheet metal can cause serious cuts and air blowing from the duct can blow chips in your eyes. Protective eye covering is necessary.*

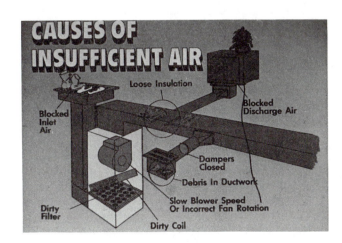

Figure 39–61. Obstructed airflow. *Courtesy Carrier Corporation*

SUMMARY

- Air is passed through conditioning equipment and recirculated into the room to condition it.
- The conditioning equipment may heat, cool, humidify, dehumidify, clean, or a combination of these to make air comfortable.
- Infiltration is air leaking into a structure.
- Ventilation is air being induced into the conditioning equipment and conditioned before it is allowed to enter the conditioned space.
- The duct system distributes air to the conditioned space. It consists of the blower (or fan), the supply duct, and the return duct.
- The blower or fan uses energy to move the air.
- Propeller and centrifugal fans are commonly used in residential and small commercial systems.
- The propeller type is used to move a lot of air against a small pressure. It can make noise.
- The centrifugal type is used to move large amounts of air in duct work, which offers resistance to the movement of air.
- Fans are turned either by belt-drive or direct-drive motors.
- A fan is not a positive displacement device.
- Small centrifugal fans use energy in proportion to the amount of air they move.
- Duct systems, both supply and return, are large pipes or tunnels that the air flows through.
- Duct can be made from aluminum, galvanized steel, flexible tubes, and fiberglass board.
- The pressure the fan creates in the duct is very small and is measured in inches of water column.
- 1 in. W.C. is the amount of pressure needed to raise a column of water, in a water manometer, 1 in.
- The atmosphere's pressure of 14.696 psia will support a column of water 34 ft high.
- 1 psi will support a column of water 27.1 in. or 2.31 ft high.
- Moving air in a duct system creates static pressure, velocity pressure, and total pressure.
- Static pressure is the pressure pushing outward on the duct.
- Velocity pressure is moving pressure created by the velocity of the air in the duct.
- Total pressure is the velocity pressure + the static pressure.
- The pitot tube is a probe device used to measure the air pressures.
- Air velocity (fpm) in a duct can be multiplied by the cross-sectional area (square feet) of the duct to obtain the amount of air passing that particular point in cubic feet per minute.
- Typical supply duct systems used are plenum, extended plenum, reducing plenum, and perimeter loop.
- The plenum system is economical and can easily be installed.
- The extended plenum system takes the trunk duct closer to the farthest outlets but is more expensive.
- The reducing plenum system reduces the trunk duct when some of the air volume has been reduced.
- The perimeter loop system supplies an equal pressure to all of the outlets. It is useful in a concrete slab floor.
- Branch ducts should always have balancing dampers to balance the air to the individual areas.
- When the air is distributed in the conditioned space, it is common practice to distribute the air on the outside wall to cancel the load.
- Warm air distributes better from the floor because it rises.
- Cold air distributes better from the ceiling because it falls.
- The amount of throw tells how far the air from a diffuser will reach into the conditioned space.
- Return-air systems commonly used are individual room return and common or central return.
- Each foot of duct, supply or return, has a friction loss that can be plotted on a friction chart for round duct.
- Round duct sizes can be converted to square or rectangular equivalents for sizing and friction readings.

REVIEW QUESTIONS

1. Name the five changes that are made in air to condition it.
2. Name the two ways that a structure obtains fresh air.
3. Name two types of blowers that move the air in a forced-air system.
4. Which type of blower is used to move large amounts of air against low pressures?
5. Which type of blower is used to move large amounts of air in duct work?
6. Name two reasons why duct work resists airflow.
7. Name four types of duct distribution systems.
8. Name two types of blower drives.
9. What is the pressure in duct work expressed in?
10. What is a common instrument used to measure pressure in duct work?
11. Why is pounds per square inch not used to measure the pressure in duct work?
12. Name the three types of pressures created in moving air in duct work.
13. Name two types of return air systems.
14. What component distributes the air in the conditioned space?
15. Where is the best place to distribute warm air? Why?
16. Where is the best place to distribute cold air? Why?
17. What chart is used to size duct work?
18. Name four materials used to manufacture duct.
19. What fitting leaves the trunk duct and directs the air into the branch duct?
20. The duct sizing chart is expressed in round duct. How can this be converted to square or rectangular duct?

OBJECTIVES

After completing this unit, you should be able to

- **list three crafts involved in air conditioning installation.**
- **identify types of duct system installations.**
- **describe the installation of metal duct.**
- **describe the installation of ductboard systems.**
- **describe the installation of flexible duct.**
- **recognize good installation practices for package air conditioning equipment.**
- **discuss different connections for package air conditioning equipment.**
- **describe the split air conditioning system installation.**
- **recognize correct refrigerant piping practices.**
- **state startup procedures for air conditioning equipment.**

40.1 THREE CRAFTS INVOLVED IN INSTALLATION

Installing air conditioning equipment requires three crafts: duct, electrical, and mechanical, which includes refrigeration. Some contractors use separate crews to carry out the different tasks. Others may do two of the crafts within their own company and subcontract the third to a more qualified contractor. Some small contractors do all three jobs with a few highly skilled people. The three job disciplines are often licensed at local and state levels; they may be licensed by different departments. The contractor could have all three licenses and therefore work in any of the areas.

Duct system components were discussed in Unit 39. Installing these components will be discussed here. The types of systems discussed are (1) all metal square or rectangular, (2) all metal round, (3) fiberglass ductboard, and (4) flexible duct.

40.2 SQUARE AND RECTANGULAR DUCTS

Square metal ducts are fabricated in a sheet-metal shop by qualified sheet-metal layout and fabrication personnel. The duct is then moved to the job sight and assembled to make a system. The duct must be as-sembled by qualified field craftsmen. Since the all-metal duct system is rigid, all dimensions have to be precise or the job will not go together. The duct must sometimes rise over objects or go beneath them and still measure out correctly so the takeoff will reach the correct branch location. The branch duct must be the correct dimension for it to reach the terminal point, the boot for the room register. See Figure 40–1 for an example of a duct system layout.

40.3 DUCT SECTION FASTENERS, SQUARE DUCT, AND RECTANGULAR DUCT

Square duct or rectangular duct is assembled with S fasteners and drive clips (see Unit 39), which make a duct connection nearly airtight. If further sealing is needed, the connection can be taped. While the duct is being assembled, it has to be fastened to the structure for support. This can be accomplished in several ways. It may lay flat in an attic installation or be hung by hanger straps. The duct should be supported so that it will be steady when the fan starts and will not transmit noise to the structure. The rush of air (which has weight) down the duct will move the duct if it is not fastened. Vibration eliminators under the fan section will prevent the transmission of fan vibrations down the duct. See Figure 40–2 for a flexible duct connector that can be installed between the fan and the metal duct. These are always recommended, but not always used. See Figure 40–3 for an example of a flexible connector that can be installed at the end of each run used to dampen sound.

Metal duct can be purchased in the popular sizes from some supply houses for small and medium systems. This makes metal duct systems available to the small contractor who may not have a sheet-metal shop. The assortment of standard duct sizes can be assembled with an assortment of standard fittings to build a system that appears to be custom-made for the job, Figure 40–4.

40.4 ROUND METAL DUCT SYSTEMS

Round metal duct systems are easy to install and are available from some supply houses in standard sizes for

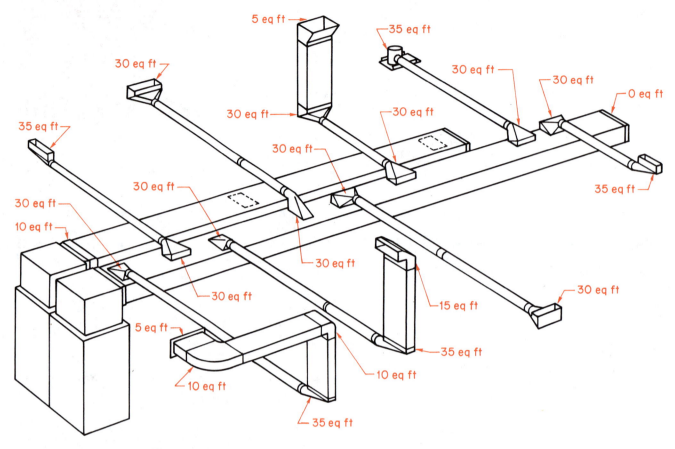

Figure 40–1. Duct system layout showing the equivalent feet of the fitting.

small and medium systems. These systems use reducing fittings from a main trunk line and may be assembled in the field. This type of system must be fastened together at all connections. Self-tapping sheet-metal screws are popular fasteners. The screws are held by a magnetized screwholder while an electric drill turns and starts the screw. Each connection should have a minimum of three screws spaced evenly to keep the duct steady. A good installer can fasten the connections as fast as the screws can be placed in the screwholder, Figure 40–5. A reversible variable-speed drill is the ideal tool to use, Figure 40–6.

Round metal duct takes more clearance space than square or rectangular duct. It must be supported and mounted at the correct intervals to keep it straight. When exposed, round metal duct does not look as good as square duct and rectangular duct, so it is often used in places that are out of sight.

SAFETY PRECAUTION: ✳Be careful when handling the sharp tools and fasteners. Use grounded or double insulated electrical tools.✳

Figure 40–2. Flexible duct connector.

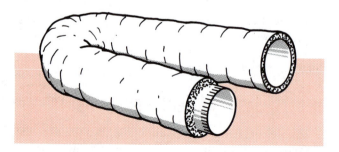

Figure 40–3. Round flexible duct connector.

Figure 40-4. Standard duct fittings.

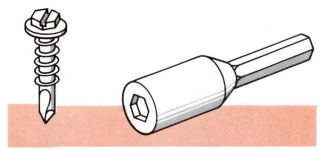

Figure 40-5. Self-tapping sheet metal screw and a magnetic screw holder.

40.5 INSULATION FOR METAL DUCT

Insulation for metal duct can be applied to the inside or the outside of the duct. When insulation is applied to the inside, the job is usually done in the fabrication shop. The insulation can be fastened with tabs, glue or both. The tabs are fastened to the inside of the duct and have a shaft that looks like a nail protruding from the duct wall. The liner and a washer are pushed over the tab shaft. The tabs have a base that the shaft is fastened to, Figure 40-7. This base is fastened to the inside of the duct by glue or spot welding. Spot welding is the most permanent method, but is difficult to do and is expensive. An electric spot welder must be used to weld the tab.

The liner can be glued to the duct, but it may come loose and block airflow, perhaps years from the installation date. This is difficult to find and repair, particularly if the duct is in a wall or framed in by the building structure. The glue *must* be applied correctly and even then may not hold forever. Using tabs with glue is a more permanent method.

The liner is normally fiberglass and is coated on the air side to keep the airflow from eroding the fibers. You may find it hard to believe that air can erode the glass fiber, but it can. Many pounds of air pass through the duct each season. An average air conditioning system handles 400 cfm/ton of air. A 3-ton system handles 1200 cfm or 72,000 cfh (400 cfm × 60 min/hr = 72,000 cfh). Then 72,000 cfh ÷ 13.35 ft³/lb = 5393

Figure 40-6. Electric hand drill. *Photo by Bill Johnson*

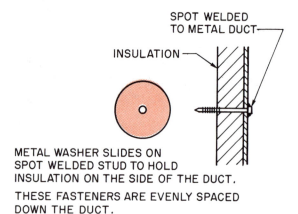

SPOT WELDED TO METAL DUCT

INSULATION

METAL WASHER SLIDES ON SPOT WELDED STUD TO HOLD INSULATION ON THE SIDE OF THE DUCT. THESE FASTENERS ARE EVENLY SPACED DOWN THE DUCT.

THIS TAB MAY BE GLUED ON THE DUCT WORK. CONTACT GLUE IS FURNISHED ON THE BACK OF TAB.

Figure 40-7. Tab to hold fiberglass duct liner to the inside of the duct.

lb/hr of air. This is more than $2\frac{1}{2}$ ton/hr of air. That's a lot of air in a year of operation.

SAFETY PRECAUTION: * Fiberglass insulation can irritate the skin. Gloves and goggles should be worn while handling it.*****

40.6 DUCTBOARD SYSTEMS

Fiberglass ductboard is very popular among many contractors because it requires little special training to construct a system. Special knives can be used to fabricate the duct in the field. When the knives are not available, simple connections can be made with a kitchen knife or a sheath knife. This duct has the insulation already attached to the outside skin. The skin is made of foil with a fiber running through it to give it strength. When assembling this duct, it is important to cut some of the insulation away so that the outer skin can lap over the surface that it is meeting. The two skins are strengthened by stapling them together and taped to make them airtight, Figure 40-8.

Fiberglass ductboard can be made into almost any configuration that metal duct can be made into. It is lightweight and easy to transport because the duct can be cut, laid out flat, and assembled at the job site. Metal duct fittings, on the other hand, are large and take up a lot of space in a truck. The original shipping boxes can be used to transport the ductboard and to keep it dry.

Round fiberglass duct is as easy to install as the ductboard because it can also be cut with a knife.

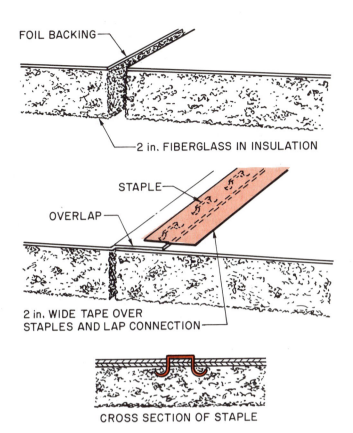

FOIL BACKING

2 in. FIBERGLASS IN INSULATION

STAPLE

OVERLAP

2 in. WIDE TAPE OVER
STAPLES AND LAP CONNECTION

CROSS SECTION OF STAPLE

Figure 40–8. Outer skin is lapped over, and the backing of one piece is stapled to the backing of the other piece.

Fiberglass duct must be supported just like metal duct to keep it straight. The weight of the ductboard itself will cause it to sag over long spans. A broad type of hanger that will not cut the ductboard's cover is necessary.

Fiberglass duct deadens sound because the inside of the duct has a coating (like the coating in the metal duct liner to keep the duct fibers from eroding) that helps to deaden any air or fan noise that may be transmitted into the duct. The duct itself is not rigid and does not carry sound.

40.7   FLEXIBLE DUCT

Flexible duct was discussed to some extent in Unit 39. This duct has a flexible liner and may have a fiberglass outer jacket for insulation if needed, for example, where a heat exchange takes place between the air in the duct and the space where the duct is routed. The outer jacket is held by a moisture-resistant cover made of fiber-reinforced foil or vinyl. The duct may be used for the supply or return and should be run in a direct path to keep bends from closing the duct. Sharp bends can greatly reduce the air flow and should be avoided.

When used in the supply system, flexible duct may be used to connect the main trunk to the boot at the room diffuser. The boot is the fitting that goes through

the floor. In this case, it has a round connection on one end for the flex duct and a rectangular connection on the other end where the floor register fastens to it. Its flexibility makes it valuable as a connector for metal duct systems. The metal duct can be installed close to the boot and the flexible duct can be used to make the final connection. Long runs of flexible duct are not recommended unless the friction loss is taken into account.

Flexible duct must be properly supported. A band, 1 in. or more wide, is the best method to keep the duct from collapsing and reducing the inside dimension. Some flexible ducts have built-in eyelet holes for hanging the duct. Tight turns should be avoided to keep the duct from collapsing on the inside and reducing the inside dimension. Flexible duct should be stretched to a comfortable length to keep the liner from closing and creating friction loss.

Flexible duct used at the end of a metal duct run will help reduce any noise that may be traveling through the duct. This can help reduce the noise in the conditioned space if the fan or heater is noisy.

40.8   ELECTRICAL INSTALLATION

The electrical installation of air conditioning equipment is a subject that should be discussed in an electrical textbook. However, there are some guidelines that the air conditioning technician should be familiar with to make sure that the unit has the correct power supply and that the power supply is safe for the equipment, the service technician, and the owner. The control voltage for the space thermostat is often installed by the air conditioning contractor even if the line-voltage power supply is installed by someone else.

40.9   THE POWER SUPPLY

For our purposes the power supply includes the correct voltage and wiring practices, including wire size. The law requires the manufacturer to provide with each electrical device a nameplate that gives the voltage requirements and the current the unit will draw, Figure 40–9. The applied voltage (the voltage that the unit will

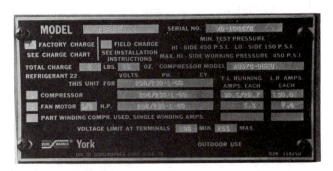

Figure 40–9. Air conditioning unit nameplate. *Photo by Bill Johnson*

actually be using) should be within ±10% of the rated voltage of the unit. That is, if the rated voltage of a unit is 230 V, the maximum operating voltage that the unit should be allowed to operate at would be 253 volts (230 × 1.10 = 253). The minimum voltage would be 207 volts (230 × 0.90 = 207). *If the unit operates for a long time beyond these limits, the motors and controls will be damaged.*

When package equipment is installed, there is one power supply for the unit. If the system is split, there are two power supplies, one for the inside unit and one for the outside unit. Both power supplies will go back to a main panel, but there will be a separate fuse or breaker at the main panel, Figure 40–10. *There should be a disconnect or cutoff switch within 25 ft of each unit and within sight of the unit.* This is a safety precaution for the technician working on the unit. If the disconnect is around a corner, someone may turn on the unit while the technician is working on the electrical system.

SAFETY PRECAUTION: *Care should be used while working with any electrical circuit. All safety rules *must* be adhered to. Particular care should be taken while working in the primary power supply to any building because the fuse that protects that circuit may be on the power pole outside. A screw driver touched across phases may throw off pieces of hot metal before the outside fuse responds.*

40.10 SIZING THE CONDUCTORS

Wire sizing tables specify the wire size that each component will need. Wire sizing is discussed in Unit 11. The National Electrical Code (NEC) provides installation standards, including workmanship, wire sizes, methods of routing wires, and types of enclosures for wiring and disconnects. Use it for all electrical wiring installation unless a local code prevails. The NEC can be very complicated for an inexperienced person.

40.11 LOW-VOLTAGE CIRCUITS

The control-voltage wiring in air conditioning and heating equipment is the line voltage reduced through a step-down transformer. It is installed with color-coded or numbered wires so that the circuit can be followed through the various components. For example, the 24-V power supply is often at the air handler, which may be in a closet, an attic, a basement, or a crawl space. The interconnecting wiring may have to leave the air handler and go to the room thermostat and the condensing unit at the back of the house. The air handler may be used as the junction for these connections, Figure 40–11.

The control wire is a light-duty wire because it carries a low voltage and current. The standard wire size is 18 gauge. A standard air conditioning cable has four wires each of 18 gauge in the same plastic-coated sheath and is called 18–4 (18 gauge, 4 wires). Wires

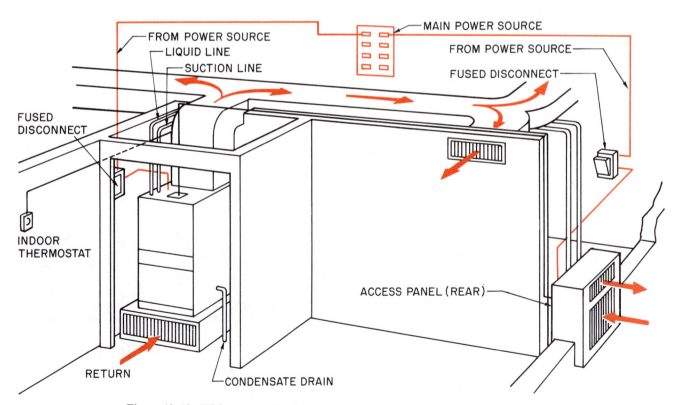

Figure 40–10. Wiring connecting indoor and outdoor units to the main power service.

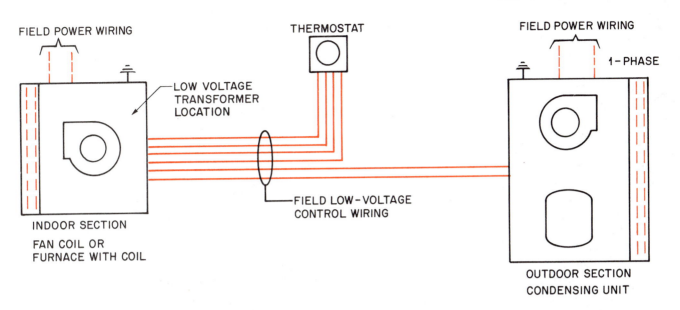

Figure 40-11. Pictorial diagram showing the relative position of the wiring as it is routed to the room thermostat, the air handler, and the condensing unit.

can have eight conductors and are called 18-8, Figure 40-12. There are red, white, yellow, and blue wires in the cable. The cable or sheath can be installed by the air conditioning contractor in most areas because it is low voltage. An electrical license may be required in some areas.

SAFETY PRECAUTION: *Although low-voltage circuits are generally harmless, you can be hurt if you are wet and touch live wires. Use all precautions for electric circuits.*

40.12   INSTALLATION OF THE REFRIGERATION SYSTEM

The mechanical or refrigeration part of the air conditioning installation is divided into package systems and split systems.

40.13   PACKAGE SYSTEMS

Package or self-contained equipment is equipment in which all components are in one cabinet or housing. The window air conditioner is a small package unit. Larger package systems can have 100 ton of air conditioning capacity. The smaller units will be discussed in this text, because larger units are usually nothing more than scaled-up smaller units with a different power supply.

The package unit is available in several different configurations to do different jobs. The following list describes some applications:

1. Air to air is similar to a window air conditioner except that it has two motors. This unit will be discussed because it is the most common. The term *air to air* is used because the refrigeration unit absorbs its heat from the air and rejects it into the air, Figure 40-13.

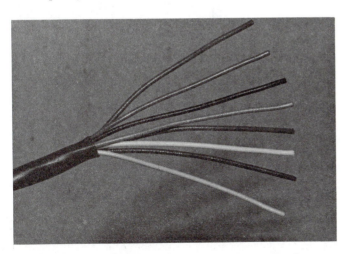

Figure 40-12. 18-8 thermostat wire. *Photo by Bill Johnson*

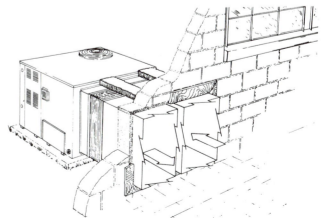

Figure 40-13. Air-to-air package unit. *Courtesy Climate Control*

2. Air to water describes the unit in Figure 40–14. This system absorbs heat out of the conditioned space air and rejects the heat into water. The water is wasted or passed through a cooling tower to reject the heat to the atmosphere. This system is sometimes called a water-cooled package unit.

3. Water to water describes the equipment in Figure 40–15. It has two water heat exchangers and is used in large commercial systems. The water is cooled and then circulated through the building to absorb the structure's heat, Figure 40–16. This system uses two pumps and two water circuits in addition to the fans to circulate the air in the conditioned space. To properly maintain this system, you need experience servicing pumps and water circuits.

4. Water to air describes the equipment in Figure 40–17. The unit absorbs heat from the water circuit and rejects the heat directly into the atmosphere. This equipment is used for larger commercial systems. *Note:* Air to air installations will be discussed in this book. The water portion of water installations is basically a plumbing job.

40.14 AIR-TO-AIR PACKAGE EQUIPMENT INSTALLATION

The air-to-air system installation requires that the unit be set on a firm foundation. The unit may be

Figure 40–15. Water-to-water package unit. *Reproduced courtesy of Carrier Corporation*

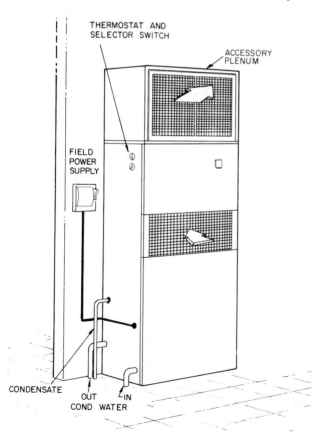

Figure 40–14. Air-to-water package unit. *Reproduced courtesy of Carrier Corporation*

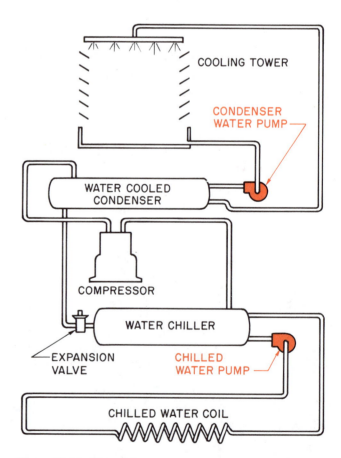

Figure 40–16. The total water-to-water system must have two pumps to move the water in addition to the fans to cool or heat the water.

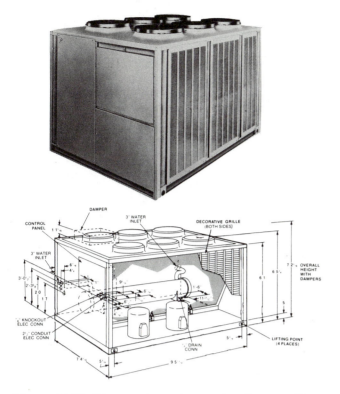

Figure 40-17. Water-to-air package unit. *Courtesy Trane Company*

furnished with a roof curb for rooftop installations. The roof curb will raise the unit off the roof and provide waterproof duct connections to the conditioned space below. When a unit is to be placed on a roof in new construction, the roof curb can be shipped separately, and a roofer can install it, Figure 40–18. The air conditioning contractor can then set the package unit on the roof curb for a watertight installation. The foundation for another type of installation

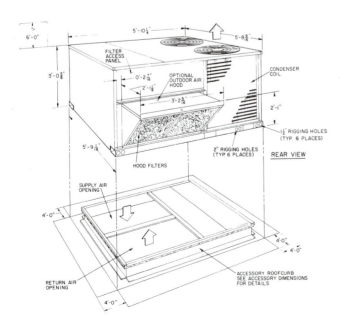

Figure 40-18. Unit will set on the roof curb. *Reproduced courtesy of Carrier Corporation*

may be beside the conditioned space and may be a high-impact plastic pad, a concrete pad, or a metal frame. See Figure 40–19 for an example of each. The unit should be placed where the water level cannot reach the unit.

40.15 VIBRATION ISOLATION

The foundation of the unit should be attached to the building in such a manner that the unit vibration is not transmitted into the building. The unit may need vibration isolation from the building structure. Two common methods for preventing vibration are: (1) rubber and cork pads and (2) spring isolators. Rubber and cork pads are the simplest and least expensive method for simple installations, Figure 40–20. The pads come in sheets that can be cut to the desired size. They are placed under the unit at the point where the unit rests on its foundation. If there are raised areas on the bottom of the unit for this contact, the pads may be placed there.

Spring isolators may also be used and need to be chosen correctly for the particular weight of the unit. The springs will compress and lose their effectiveness

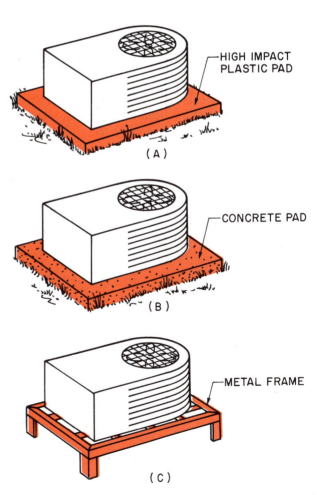

Figure 40-19. Various types of unit pads.

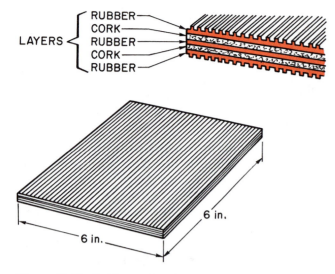

Figure 40-20. Rubber and cork pads may be placed under a unit to reduce the vibration.

under too heavy a load. They also may be too stiff for a light load if not chosen correctly.

40.16   DUCT CONNECTIONS FOR PACKAGE EQUIPMENT

The duct work entering the structure may be furnished by the manufacturer or made by the contractor. Manufacturers furnish a variety of connections for rooftop installations but do not normally have many factory-made fittings for going through a wall. See Figure 40–21 for a rooftop installation.

Package equipment comes in two different duct-connection configurations relative to the placement of the return and supply ducts, either side by side or over and under. The *side-by-side* pattern has the return duct connection beside the supply connection, Figure 40-22. The connections are almost the same size. There will

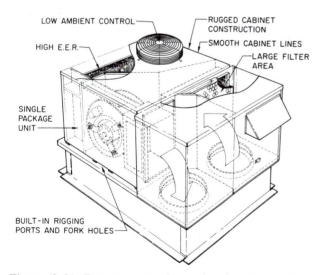

Figure 40-21. Duct connector for rooftop installation. *Courtesy Climate Control*

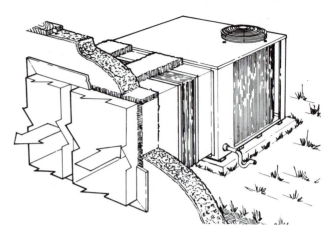

Figure 40-22. This air-to-air package unit has side-by-side duct connections. *Courtesy Climate Control*

be a great deal of difference in the connection sizes on an over and under unit. If the unit is installed in a crawl space, the return duct must run on the correct side of the supply duct if the crawl space is low. It is difficult to cross a return duct over or under a supply duct if there is not enough room. The side-by-side duct design makes it easier to connect the duct to the unit because of the sizes of the connections on the unit. They are closer to square in dimension than the connections on an over-and-under duct connection. If the duct has to rise or fall going into the crawl space, a standard duct size can be used.

The *over-and-under* duct connection is more difficult to fasten duct work to because the connections are more oblong, Figure 40-23. They are wide, because they extend from one side of the unit to the other, and are shallow, because they are on top of each other. A duct transition fitting is almost always necessary to connect this unit to the duct work. The transition is normally from wide shallow duct to almost square. When side-by-side systems are used, the duct system may be designed to be thin and wide to keep the duct transition from being so complicated.

The duct connection must be watertight and insulated, Figure 40-24. Some contractors cover the duct with a weatherproof hood.

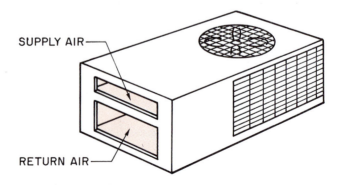

Figure 40-23. Over-and-under duct connections.

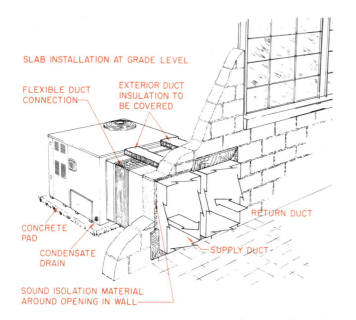

Figure 40–24. Air-to-air package unit installed through a wall. *Courtesy Climate Control*

At some time the package equipment will need replacing. The duct system usually outlasts two or three units. Manufacturers sometimes change their equipment style from over and under to side by side. This really complicates the new choice of equipment because an over-and-under unit may not be as easy to find as a side-by-side unit, and the duct already exists for an over-and-under connection, Figure 40–25.

Package air conditioners have no field run refrigerant piping. The refrigerant piping is done within the unit at the factory by the manufacturer. The refrigerant charge is included in the price of the equipment. The equip-

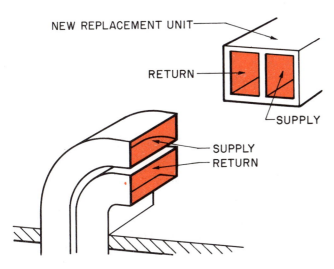

Figure 40–25. Air-cooled package unit with over-and-under duct connections. The manufacturer has changed design to a side-by-side duct connection and the installing contractor has a problem.

ment is ready to start up except for electrical and duct connections. One precaution must be observed, however. Most of these systems use R-22. The oil in the system has an affinity for R-22. The R-22 in the unit will all be in the compressor if it is not forced out by a crankcase heater. This was not mentioned in the unit on refrigeration because the refrigerants used in refrigeration equipment are not attracted to the oil like R-22. *All manufacturers that use crankcase heaters on their compressors supply a warning to leave the crankcase heater energized for some time, perhaps as long as 12 hours, before starting the compressor.* The installing contractor must coordinate the electrical connection and startup times closely, because the unit may not be started immediately when the power is connected.

40.17   INSTALLING THE SPLIT SYSTEM

In split-system air conditioning equipment installations, the condenser is separated from the evaporator. We will describe the installation beginning with the evaporator, moving to the condenser, and finishing with the interconnecting refrigerant lines. The startup of the equipment will be described separately at the end of the unit.

40.18   THE EVAPORATOR SECTION

The evaporator is normally located close to the fan section regardless of whether the fan is in a furnace or in a special air handler. The air handler (fan section) and coil must be located on a solid base or suspended from a strong support. Upflow and downflow equipment often has a base that may be fireproof and rigid for the air handler to rest on. Some vertical-mount air-handler installations have a wall-hanging support for the unit.

When the unit is installed horizontally, it may rest on the ceiling joists in an attic or on a foundation of

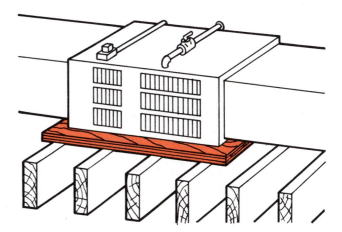

Figure 40–26. Furnace located in an attic crawl space. See manufacturer's recommendations for suggested base.

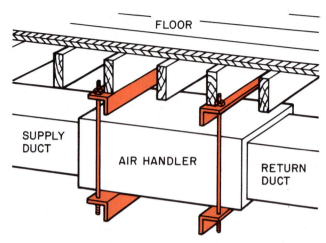

Figure 40–27. Air handler is hung from above, and the duct work is connected and hung at the same height.

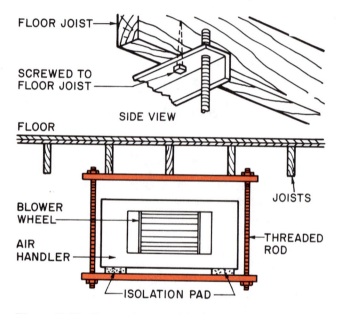

Figure 40–28. Trapeze hanger with vibration isolation pads.

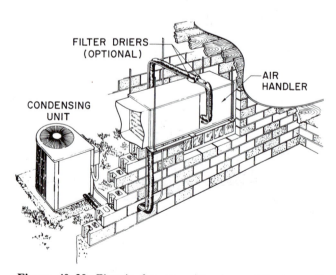

Figure 40–29. Electric furnace with air conditioning coil placed in the duct. *Courtesy Climate Control*

blocks or concrete in a crawl space, Figure 40–26. The air handler may be hung from the floor or ceiling joists in different ways, Figure 40–27. If the air handler is hung from above, vibration isolators are often located under it to keep fan noise or vibration from being transmitted into the structure. Figure 40–28 is a trapeze hanger using vibration isolation pads.

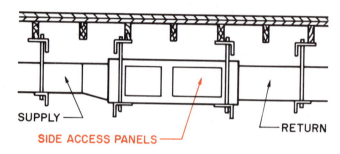

Figure 40–30. Side-access air handler in a crawl space.

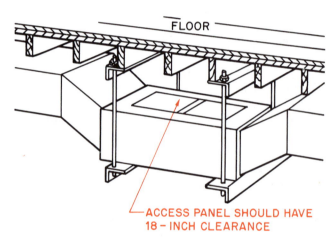

Figure 40–31. Top-access air handler.

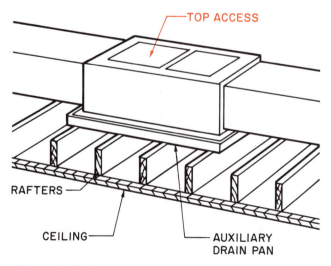

Figure 40–32. Evaporator installed in an attic crawl space.

40.19 SERVICE ACCESS

The air handler (fan section) should always be installed so that it will be easily accessible for future service. The air handler will always contain the blower and sometimes the controls and heat exchanger.

Most manufacturers have designed their air handlers or furnaces so that, when installed vertically, they are totally accessible from the front. This works well for closet installations where there is insufficient room between the closet walls and the side walls of the furnace or air handler. Figure 40–29 is an electric furnace used as an air conditioning air handler.

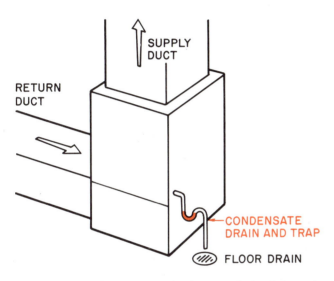

Figure 40–33. Condensate piped to a drain below the evaporator drain pan.

When a furnace or air handler is located in a crawl space, side access is important, so the air handler should be located off the ground next to the floor joists. The duct work will also be installed off the ground, Figure 40–30. If the air handler is top access, it will have to be located lower than the duct and the duct must make a transition downward to the air handler on the supply and return ends, Figure 40–31.

Top-access air handlers may be used in an attic crawl space where the air handler can be set on the joists. The technician can work on the unit from above, Figure 40–32.

40.20 CONDENSATE DRAIN PIPING

When the evaporator is installed, provisions must be made for the condensate that will be collected in the air conditioning cycle. An air conditioner in a climate with average humidity will collect about 3 pints of condensate per hour of operation for each ton of air conditioning. A 3-ton system would condense about 9 pt per hour of operation. This is more than a gallon of condensate per hour or more than 24-gal in a 24-hr operating period. This can add up to a great deal of water over a long time. If the unit is near a drain that is below the drain pan, simply pipe the condensate to the drain, Figure 40–33. A trap in the drain line will hold some water and keep air from pulling into the unit from the termination point of the drain. The drain may terminate in an area where foreign particles may be pulled into the drain pan. The trap will prevent this, Figure 40–34. If there is no drain close to the unit,

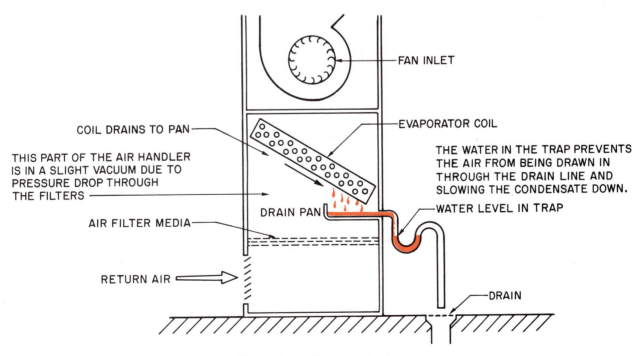

Figure 40–34. Cut-away of a trap.

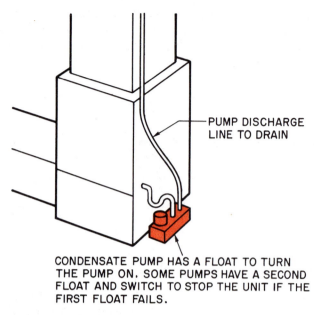

CONDENSATE PUMP HAS A FLOAT TO TURN THE PUMP ON. SOME PUMPS HAVE A SECOND FLOAT AND SWITCH TO STOP THE UNIT IF THE FIRST FLOAT FAILS.

Figure 40–35. The drain in this installation is above the drain connection on the evaporator, and the condensate must be pumped to a drain at a higher level.

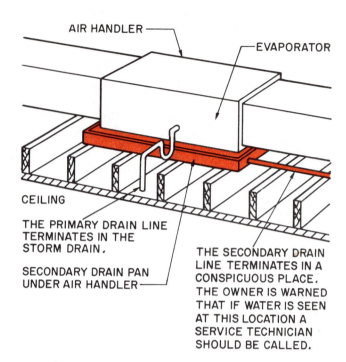

THE PRIMARY DRAIN LINE TERMINATES IN THE STORM DRAIN.

SECONDARY DRAIN PAN UNDER AIR HANDLER

THE SECONDARY DRAIN LINE TERMINATES IN A CONSPICUOUS PLACE. THE OWNER IS WARNED THAT IF WATER IS SEEN AT THIS LOCATION A SERVICE TECHNICIAN SHOULD BE CALLED.

Figure 40–37. Auxiliary drain pan installation.

the condensate must be drained or pumped to another location, Figure 40–35.

Some locations call for the condensate to be piped to a dry well. A dry well is a hole in the ground with gravel at the bottom. The condensate is drained into the well and absorbed into the ground, Figure 40–36. For this to be successful, the soil must be able to absorb the amount of water that the unit will collect.

When the evaporator and drain are located above the conditioned space, an auxiliary drain pan under the unit is recommended, Figure 40–37. Airborne particles, such as dust and pollen, can get into the drains. Algae

will also grow in the water in the lines, traps, and pans and will eventually plug the drain. If the drain system is plugged, the auxiliary drain pan will catch the overflow and keep the water from damaging whatever is below it. This auxiliary drain should be piped to a conspicuous place. The owner should be warned that if water ever comes from this drain line, a service call is necessary. Some contractors pipe this drain to the end of the house and out next to a driveway or patio so that if water ever came out it would be readily noticed, Figure 40–38.

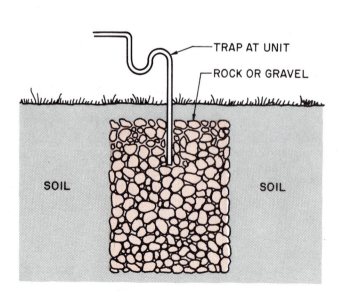

TRAP AT UNIT

ROCK OR GRAVEL

SOIL   SOIL

Figure 40–36. Dry well for condensate.

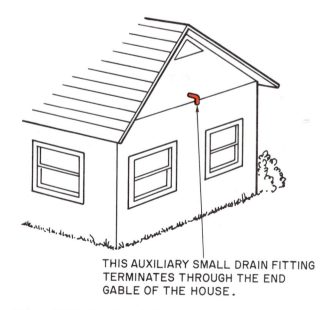

THIS AUXILIARY SMALL DRAIN FITTING TERMINATES THROUGH THE END GABLE OF THE HOUSE.

Figure 40–38. Auxiliary drain piped to the end of the house.

40.21 CONDENSING UNIT

The condensing unit location is remote from the evaporator. The following must be considered carefully when placing a condensing unit:

1. Proper air circulation
2. Convenience for piping and electrical service
3. Future service
4. Natural water and roof drainage
5. Solar influence
6. Appearance

40.22 AIR CIRCULATION AND INSTALLATION

The unit must have adequate air circulation. The air discharge from the condenser may be from the side or from the top. Discharged air must not hit an object and circulate back through the condenser. The air leaving the condenser is warm or hot and will create high head pressure and poor operating efficiency if it passes through the condenser again. Follow manufacturer's literature for minimum clearances for the unit, Figure 40–39.

40.23 ELECTRICAL AND PIPING CONSIDERATIONS

The refrigerant piping and the electrical service must be connected to the condensing unit. Piping is discussed later in this unit, but, for now, realize that the piping will be routed between the evaporator and the condenser. When the condenser is located next to a house, the piping must be routed between the house and the unit, usually behind or beside the house with the piping routed next to the ground. If the unit is placed too far from the house, the piping and electrical service become natural obstructions between the house and the unit. Children may jump on the piping and electrical conduit. If the unit is placed too close to the house, it may be difficult to remove the service panels. Before placing the unit, study the electrical and the refrigeration line connections and make them as short as practical, leave adequate room for service.

40.24 SERVICE ACCESSIBILITY

Unit placement may determine whether a service technician gives good service or barely adequate service. The technician must be able to see what is being worked on. A unit is often placed so that the technician can touch a particular component but be unable to see it, or, by shifting positions, can see the component but not touch it. The technician should be able to both see and touch the work at the same time.

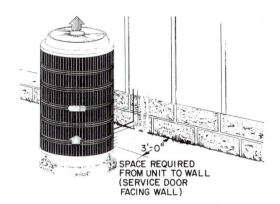

Figure 40–39. Condensing unit for a split system located so that it has adequate airflow and service room. *Reproduced Courtesy of Carrier Corporation*

40.25 WATER DRAINAGE FROM NATURAL SOURCES

The natural drainage of the groundwater and roof water should be considered in unit placement. *The unit should not be located in a low place where ground water will rise in the unit. If this happens, the controls may short to ground, and the wiring will be ruined.* All units should be placed on a base pad of some sort. Concrete and high-impact plastic are commonly used. Metal frames can be used to raise the unit when needed.

The roof drainage on a structure may run off in gutters. The condensing unit should not be located where the gutter drain or roof drainage will pour down on the unit. Condensing units are made to withstand rain water but not large volumes of drainage. If the unit is a top-discharge unit, the drainage from above will be directly into the fan motor, Figure 40–40.

40.26 SOLAR INFLUENCE

If possible, put the condensing unit on the shady side of the house, because the sun shining on the panels and coils will not help the efficiency. However, the difference is not crucial, and it would not pay to pipe the refrigerant tubing and electrical lines long distances just to keep the sun from shining on the unit.

Shade helps cool the unit, but it may also cause problems. Some trees have small leaves, sap, berries, or flowers that may harm the finish of the unit. For example, pine needles will fall into the unit, and the pitch that falls on the unit's cabinet will harm the finish .

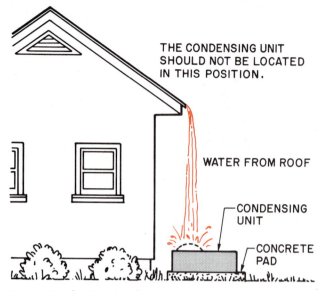

Figure 40-40. This unit is located improperly and allows the roof drain to pour down into the top of the unit.

THE CONDENSING UNIT SHOULD NOT BE LOCATED IN THIS POSITION.

WATER FROM ROOF

CONDENSING UNIT

CONCRETE PAD

The floor plan can be shown to the homeowners to help them understand the contractor's intentions.

40.28 INSTALLING REFRIGERANT PIPING

The refrigerant piping is always a big consideration when installing a split-system air conditioner. The choice of the piping system may make a difference in the startup time for the system. The piping should always be kept as short as practical. For an air conditioning installation there are three methods used to connect the evaporator and the condensing unit on almost all equipment under 5 tons (refrigeration systems for commercial refrigeration are not the same): (1) contractor-furnished piping, (2) flare or compression fittings, with the manufacturer furnishing the tubing (called a line set), and (3) precharged tubing with sealed quick-connect fittings, called a precharged line set. **All**

and outweigh the benefit of placing the unit in the shade of many types of trees.

40.27 PLACING A CONDENSING UNIT FOR BEST APPEARANCE

The condensing unit should be located where it will not be noticeable or will not make objectionable noise. When located on the side of a house, the unit may be hidden from the street with a low shrub. If the unit is a side-discharge type, the fan discharge must be away from the shrub or the shrub will not live. If the unit is top discharge, the shrub is unaffected but the unit's sound will rise. This noise may be objectionable in a bedroom located above the unit.

Locating the condensing unit at the back of the house usually places the unit closer to the evaporator and means shorter piping, but the back of the house is where patios and porches are located, and the homeowner, while sitting outside, may not want to hear the unit. In such a case, a side location at the end of the house where there are no bedrooms may be the best choice.

Each location has its considerations. The salesperson and the technician should consult with the owner about locating the various components. A floor plan of the structure is always a big help. Some companies use large graph paper to draw a rough floor plan to scale when estimating the job. The equipment can be located on the rough floor plan to help solve these problems.

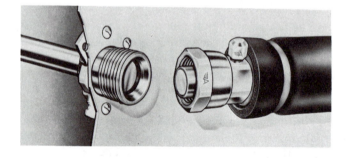

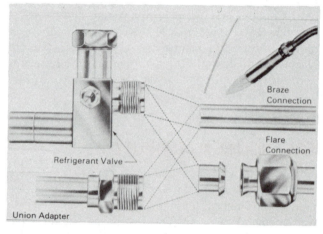

Braze Connection

Flare Connection

Refrigerant Valve

Union Adapter

Figure 40-41. Two basic piping connections for split-system air conditioning. *Courtesy Aeroquip Corporation*

of these systems have the operating charge for the system shipped in the equipment.

40.29 THE REFRIGERANT CHARGE

Regardless of who furnishes the connecting tubing, the complete operating charge for the system is normally furnished by the manufacturer. The charge is shipped in the condensing unit. The manufacturer furnishes enough charge to operate the unit with a predetermined line length, typically 30 ft. The manufacturer holds and stores the refrigerant charge in the condensing unit with service valves if the tubing uses flare or compression fittings. When quick-connect line sets are used, the correct operating charge for the line set is included in the actual lines. See Figure 40–41 for an example of both service valves and quick-connect fittings.

When service valves are used, the piping is fastened to the service valves by flare or compression fittings. Some manufacturers provide the option of soldering to the service valve connection. The piping is always hard-drawn or soft copper. In installations where the piping is exposed, straight pipe looks better. Hard-drawn tubing may be used for this with factory elbows used for the turns. When the piping is not exposed, soft copper is easily formed around corners where a long radius will be satisfactory.

40.30 THE LINE SET

When the manufacturer furnishes the tubing, it is called a *line set*. The suction line will be insulated, and the tubing will be charged with nitrogen, contained with rubber plugs in the tube ends. When the rubber plug is removed, the nitrogen rushes out with a loud hiss. This indicates that the tubing is not leaking.

When the tubing is used, uncoil it from the end of the coil. Place one end of the coil on the ground and roll it while keeping your foot on the tubing end on the ground, Figure 40–42.

Be careful not to kink the tubing when going around corners, or it may collapse. Because of the insulation on the pipe, you may not even see the kink.

40.31 TUBING LEAK TEST AND EVACUATION

When the piping is routed and in place, the following procedure is commonly used to make the final connections:

1. The tubing is fastened at the evaporator end.

Figure 40–42. Uncoiling tubing. *Photo by Bill Johnson*

When the tubing has flare nuts, a drop of oil applied to the back of the flare will help prevent the flare nut from turning the tubing when the flare connection is tightened. The suction line may be as large as $\frac{7}{8}$-in. OD tubing. Large tubing will have to be made very tight. Two adjustable wrenches are recommended, Figure 40–43. The liquid line will not be any smaller than $\frac{1}{4}$-in. OD tubing, and it may be as large as $\frac{1}{2}$-in. OD on larger systems.

2. The smaller (liquid) line is fastened to the condensing unit service valve.
3. The larger (suction) line is fastened hand tight.
4. The liquid-line service valve is slightly opened, and refrigerant is allowed to purge through the liquid

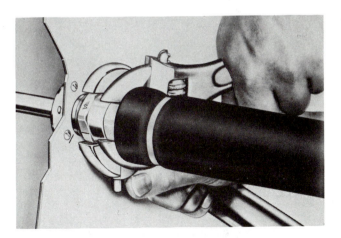

Figure 40–43. How to tighten fittings. *Courtesy Aeroquip Corporation*

line, the evaporator, and the suction line. This pushes the nitrogen out of the piping and the evaporator, and it escapes at the hand-tightened suction-line flare nut, Figure 40–44.

Line sets come in standard lengths of 10, 20, 30, 40 or 50 ft. If lengths other than these are needed, the manufacturer should be consulted. Some manufacturers have a maximum allowable line length, normally in the vicinity of 50 ft. If the line length has to be changed from the standard length, the manufacturer will also recommend how to adjust the unit charge for the new line length. Most units are shipped with a charge for 30 ft of line. If the line is shortened, refrigerant must be removed. If the line is longer than 30 ft, refrigerant must be added.

40.32   ALTERED LINE-SET LENGTHS

When line sets must be altered, they may be treated as a self-contained system of their own. The following procedures may be followed:

1. Alter the line sets as needed for the proper length. The nitrogen charge will escape during this alteration. **Do not remove the rubber plugs.**
2. When the alterations are complete, the lines may be pressured to about 25 psig to leak-check any connections you have made. The rubber plugs should stay in place. If you desire to pressure test with higher pressures, hook up the evaporator and condenser ends and you may safely pressurize to 150 psig with refrigerant or nitrogen. The line's service

port is common to the line side of the system. If the valves are not opened at this time, the line set and evaporator may be thought of as a sealed system of their own for pressure testing and evacuation, Figure 40–45.

3. After the leak test has been completed, evacuate the line set (and evaporator if connected). Break the vacuum to about 10 psig and go through the line-set purge just as though it is a factory installation. **After starting the unit, you will have to alter the charge to meet the manufacturer's guidelines.**

40.33   PRECHARGED LINE SETS (QUICK-CONNECT LINE SETS)

Precharged line sets with quick-connect fittings are shipped in most of the standard lengths. The difference is that the correct refrigerant operating charge is shipped in the line set. Refrigerant does not have to be added or taken out unless the line set is altered. The following procedures are recommended for connecting precharged line sets with quick-connect fittings.

1. Roll the tubing out straight.
2. Determine the routing of the tubing from the evaporator to the condenser and put it in place.
3. Remove the protective plastic caps in the tubing and the evaporator fittings. Place a drop of refrigerant oil (this is sometimes furnished with the tubing set) on the neoprene O rings on each line fitting. The O ring is used to prevent refrigerant from leaking out while the fitting is being connected, and it serves

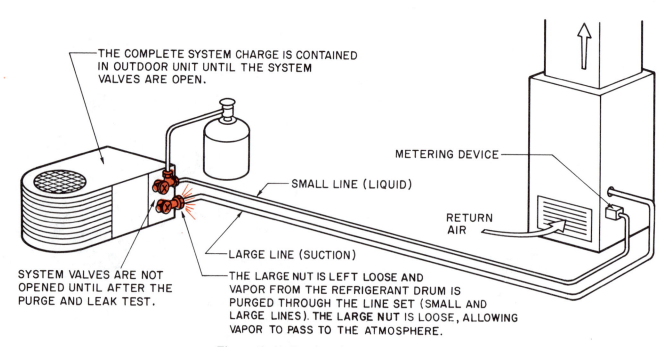

Figure 40–44. Purging nitrogen shipping charge.

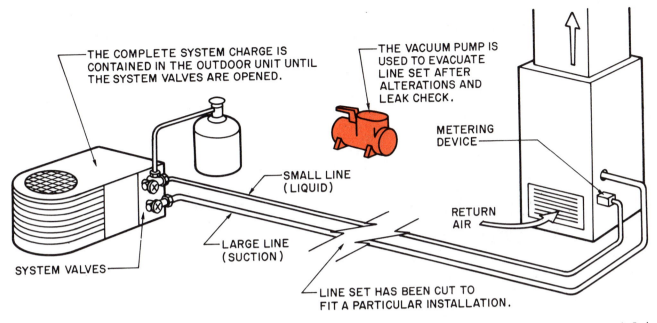

Figure 40-45. Altering the line set. After the line set has been altered, it is connected as in Figure 40–44 and purged. It is checked for leaks. The line set and evaporator are evacuated. After evacuation, a vapor holding charge is added, about 20 psig of the refrigerant, characteristic to the system.

no purpose after the connection is made tight. It may tear while making the connection if it is not lubricated.

4. Start the threaded fitting and tighten hand tight. Making sure that several threads can be tightened by hand will assure that the fitting is not cross-threaded.

5. When the fitting is tightened hand tight, finish tightening the connection with a wrench. You may hear a slight escaping of vapor while making the fitting tight; this purges the fitting of any air that may have been in the fitting. The O ring should be seating at this time. Once you have started tightening the fitting, do not stop until you are finished. If you stop in the middle, some of the system charge may be lost.

6. Tighten all fittings as indicated.

7. After all connections have been tightened, leak-check all connections that you have made.

40.34 ALTERED PRECHARGED LINE SETS

When the lengths of quick-connect line sets are altered, the charge will have to be altered. The line set may be treated as a self-contained system for alterations. The following is the recommended procedure.

1. *Before connecting the line set, bleed the refrigerant out of the line set. Do not just cut it because the liquid and suction lines contain some liquid refrigerant.*

2. Cut the line set and alter the length as needed.

3. Pressure test the line set.

4. Evacuate the line set to a low vacuum.

5. Valve off the vacuum pump and pressurize to about 10 psig on just the line set and connect it to the system using the procedures already given.

6. Read the manufacturer's recommendation as to the amount of charge for the new line lengths and add this much refrigerant to the system. If there are no recommendations, see Figure 40–46 for a table of liquid-line capacities. The liquid line should contain the most refrigerant, and the proper charge should be added to the liquid line to make up for what was lost when the line was cut. No extra refrigerant charge needs to be added for the suction line if it is filled with vapor before assembly.

40.35 PIPING ADVICE

The piping practices given for air conditioning equipment are typical of most manufacturers' recommendations. **The manufacturer's recommendation should always be followed.** Each manufacturer ships installation and startup literature inside the shipping crate. If it is not there, request it from the manufacturer and use it.

40.36 EQUIPMENT STARTUP

The final step in the installation is starting the equipment. The manufacturer will furnish startup in-

LIQUID LINE DIAMETER INCHES	OUNCES OF R-22 PER FOOT OF LENGTH OF LIQUID LINE
$\frac{3}{8}$	0.58
$\frac{5}{16}$	0.36
$\frac{1}{4}$	0.21

If 3 feet of liquid line must be added to a ($\frac{3}{8}$ in.) 30-foot line, an additional 1.74 ounces must be added to the system upon starting the unit. (3 × 0.58 = 1.74)

If the unit is shipped with a precharged line set, the complete charge for that particular line set is contained in the lines. If this set is altered, the complete liquid line length must be used. For example, if a 50-foot set ($\frac{1}{4}$ in.) is cut to 25 feet of length, the charge is exhausted from the lines, and 5.25 ounces must be added to the charge contained in the condenser when started. (25 × 0.21 = 5.25)

Figure 40–46. Table of liquid-line capacities.

structions. After the unit or units are in place, leak-checked, have the correct factory charge furnished by the manufacturer, and are wired for line and control voltage, follow these guidelines:

1. The line voltage must be connected to the unit and turned on. However, it should be arranged so that the compressor will not start. A disconnected low-voltage wire, such as the Y wire at the condensing unit, will prevent the compressor from starting. The line voltage allows the crankcase heater to heat the compressor crankcase. *This applies to any unit that has crankcase heaters. Heat must be applied to the compressor crankcase for the amount of time the manufacturer recommends (usually not more than 12 hr). Heating the crankcase boils any refrigerant out of the crankcase before the compressor is started. If the compressor is started with liquid in the crankcase, some of the liquid will reach the compressor cylinders and may cause damage. The oil will also foam and provide only marginal lubrication until the system has run for some time.*

2. It is normally a good idea to plan on energizing the crankcase heater in the afternoon and starting the unit the next day. Before you leave, make sure that the crankcase heater is hot.

3. If the system has service valves, open them. *Note:* Some units have valves that *do not* have back seats. *Do not try to back-seat these valves that do not have back seats or you may damage the valve.* Open the valve until resistance is felt, then stop, Figure 40–47.

4. Check the line voltage at the installation site and make sure it is within the recommended limits.

5. *Turn the power off when checking electrical connections, including the factory connections.*

6. Set the fan switch on the room thermostat to FAN ON and check the indoor fan for proper operation, rotation, and current draw. You should feel air at all registers, normally about 2 to 3 ft above a floor register. Make sure there are no air blockages at the supply and return openings.

7. Turn the fan switch to FAN AUTO and with the HEAT-OFF-COOL selector switch at the off position, replace the Y wire at the condensing unit. *The power at the unit should be turned off while making this connection.*

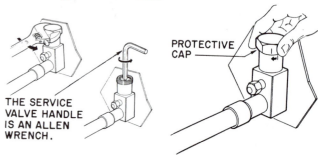

Figure 40–47. Service valves furnished with equipment may not have back seats. *Courtesy Aeroquip Corporation*

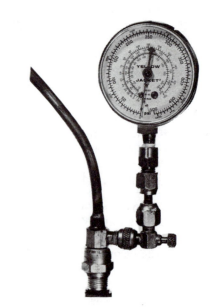

Figure 40–48. Short gauge line for high-pressure side of the system. *Photo by Bill Johnson*

8. Place your ammeter on the common wire to the compressor, have someone move the HEAT-OFF-COOL selector switch to COOL, and slide the temperature setting to call for cooling. The compressor should start. *Some manufacturer's literature will call for you to have a set of gages on the system at this time. Be careful of the line length on the gage you install on the high-pressure side of the system. If the system has a critical charge and a 6-ft gage line is installed. The line will fill up with liquid refrigerant and will alter the charge, possibly enough to affect performance. A short gage line is recommended, Figure 40-48.* Leak-check the gage port when the gages are removed. Replace the gage port cover.

If the manufacturer recommends that gages be installed for startup, install them before you start the system. If they do not recommend them, follow these recommendations:

9. When gages are not installed, there are certain signs that can indicate correct performance. The suction line coming back to the compressor should be cool, although the "coolness" can vary. Two things will cause the suction-line temperature to vary and still be correct: the metering device and the ambient temperature. Most modern systems use fixed-bore metering devices (capillary tube or orifice). When the outside temperature goes down to 75° or 80°F, for example, the suction line will not be as cool as on a hot day because the condenser becomes more efficient and liquid refrigerant is retained in the condenser, partially starving the evaporator. If the day is cool, 65° or 70°F, some of the air to the condenser may be blocked, which causes the head pressure to rise and the suction line to become cooler. The ambient temperature in the conditioned space causes the evaporator to have a large load, which makes the suction line not as cool until the inside temperature is reduced to near the design temperature, about 75°F.

The amperage of the compressor is a good indicator of system performance. If the outside temperature is hot and there is a good load on the evaporator, the compressor will be pumping at near capacity. The motor current will be near nameplate amperage. It is very rare for the compressor amperage to be more than the nameplate rating due to only the system load. It is also normal for the compressor amperage to be slightly below the nameplate rating.

When you are satisfied that the unit is running satisfactorily, inspect the installation. Check that

1. All air registers are open.
2. There are no air restrictions.
3. The duct is hung correctly, and all connections are taped.
4. All panels are in place with all screws.

5. The customer knows how to operate the system.
6. All warranty information is filled out.
7. The customer has the operation manual.
8. The customer knows how to contact you.

SAFETY PRECAUTION: *Before starting the system examine all electrical connections and moving parts. The equipment may have faults and defects that can be harmful (for example, a loose pulley may fly off when the motor starts to turn). Remember that vessels and hoses are under pressure. Always be mindful of potential electrical shock.*

SUMMARY

- The installation of air conditioning equipment normally involves three crafts: duct work, electrical, and mechanical or refrigeration.
- Duct systems are normally constructed of square, rectangular, or round metal, ductboard, or flexible material.
- Square- and rectangular-metal duct systems are assembled with S fasteners and drive clips.
- Round metal duct comes in many standard sizes with fittings.
- Round duct must be fastened with sheet-metal screws at each connection.
- The first fitting in a duct system may be a vibration eliminator to keep any fan noise or vibration from being transmitted into the duct.
- Insulation may be applied to the inside or outside of any metal duct system that may exchange heat with the ambient.
- When applied to the inside, insulation can be glued or glued and fastened with a tab.
- When the duct is insulated on the outside, a vapor barrier must be used to keep moisture from forming on the duct surface if the duct surface is below the dew-point temperature of the ambient.
- Fiberglass ductboard is compressed fiberglass with a reinforced foil backing that helps support it and create a vapor barrier.
- Flexible duct is a flexible liner that may have a cover of fiberglass that is held in place with vinyl or reinforced foil.
- *The electrical installation consists of choosing the correct enclosures, wire sizes, and fuses or breakers.*
- The electrical contractor will normally install the line-voltage wiring, and the air conditioning contractor will usually install the low-voltage control wiring. *Local codes should be consulted before any wiring is done.*
- The low-voltage control wiring is normally color coded.
- Air conditioning equipment is manufactured in package systems and split systems.
- Package equipment is completely assembled in one cabinet.

- Package equipment has two types of duct connections: over and under and side by side.
- Package units come in small to very large sizes; the larger sizes are normally scaled-up small sizes.
- Air-to-air package equipment installation consists of placing the unit on a foundation, connecting the duct work, and connecting the electrical service and control wiring.
- The duct connections may be made through a roof or through a wall at the end of a structure.
- Roof installations have waterproof roof curbs and factory-made duct systems.
- Isolation pads or springs placed under the equipment prevent equipment noise from travelling into the structure.
- Outside units may be installed on pads or frames; the pads are normally made of high-impact plastic or concrete.
- Split-system air conditioning equipment has the evaporator at an inside air handler (blower). The air handler may be an existing furnace or a separate blower package. The compressor is located outside with the condenser.
- Two refrigerant lines connect the evaporator and the condensing unit; the large line is the insulated suction line, and the small line is the liquid line.
- The air handler and condenser should be installed so that they are accessible for service.
- A condensate drain provision must be made for the evaporator.
- A secondary drain pan should be provided if the evaporator is located above the conditioned space.
- The condensing unit should be located as close as practical to the evaporator and the electrical service. It should not be located where natural water or roof water will drain directly into the top of the unit.
- The condensing unit should not be located where its noise will be bothersome.
- Line sets come in standard lengths of 10, 20, 30, 40, and 50 ft. The system charge is normally for 30 ft when the whole charge is stored in the condensing unit.
- Line sets may be altered in length. The refrigerant charge must be adjusted when the lines are altered.
- The startup procedure for the equipment is in the manufacturer's literature. Before startup, check the electrical connections, the fans, the airflow, and the refrigerant charge.

REVIEW QUESTIONS

1. Name the three crafts normally involved in an air conditioning installation.
2. Which craft normally installs the refrigerant lines and starts the system?
3. Which type of duct system is the most expensive and longest lasting?
4. What fasteners fasten square metal duct at the connections?
5. What is normally used to fasten round metal duct at the connections?
6. Name two methods for fastening insulation to the duct when it is insulated on the inside?
7. Why does insulation installed on the outside of duct have a vapor barrier when used for air conditioning?
8. What happens if insulation on the inside of duct comes loose?
9. What material is most duct insulation made from?
10. Why is a flexible connector installed in a metal duct system?
11. What happens if flexible duct is turned too sharply around a corner?
12. Why must flexible duct be pulled as tight as practical?
13. How should flexible duct be hung from above?
14. What are three advantages of package equipment?
15. What are some of the accessories that a manufacturer may supply to help the installing contractor have a simple waterproof installation in a rooftop installation with package equipment?
16. What are some advantages of a split air conditioning system?
17. What is the difference between a package unit and a split system?
18. Name the two duct connection configurations on a package unit.
19. Name the two refrigerant lines connecting the condensing unit to the evaporator.
20. Name three criteria of a good location for an air-cooled condensing unit.

CONTROLS

OBJECTIVES

After studying this unit, you should be able to

- **describe the control sequence for an air conditioning system.**
- **explain the function of the 24-V control voltage.**
- **describe the space thermostat.**
- **describe the compressor contactor.**
- **explain the operation of the high- and low-pressure controls.**
- **discuss the function of the overloads and motor winding thermostat.**
- **discuss the winding thermostat and the internal relief valve.**
- **identify operating and safety controls.**
- **compare the differences between modern control concepts and older concepts.**
- **describe how crankcase heat is applied in some modern equipment.**

41.1 CONTROLS FOR AIR CONDITIONING

Equipment control for maintaining correct air conditions involves the control of three components: the indoor fan, the compressor, and the outdoor fan. These components are used in air-cooled air conditioning equipment. Information on water-cooled equipment can be found in Unit 20. Water-cooled equipment for small air conditioning applications is seldom used.

The indoor fan, compressor, and outdoor fan must be operated in the correct sequence. They must be started and stopped automatically at the correct times. The normal sequence of operation is as follows:

1. The indoor fan must operate when the compressor operates.
2. The outdoor fan must operate when the compressor operates (except for units that may have a fan cycle device that allows the fan to be cycled off for short periods of time during cool weather for head pressure control).
3. The indoor fan may have a continuous-operation switch at the thermostat. In this position the indoor fan will run continuously, and the compressor and outdoor fan will cycle on and off upon demand.

The two basic functional categories of controls are operating and safety. A room thermostat, for example, is an operating control used to sense the space temperature to stop and start the compression system (compressor and outdoor fan) as needed, Figure 41–1. The high-pressure control is a safety control that keeps the unit from operating when the head pressure is too high, Figure 41–2. Both of these controls have something in common: neither consumes power. They pass power to other devices, such as the compressor's magnetic contactor.

41.2 PRIME MOVERS—COMPRESSORS AND FANS

The prime movers of the system are the fans (indoor and outdoor) and the compressor. These devices consume most of the power in the air conditioning process and are operated with high voltage, about 230 V in a residence. The controls are operated with 24 V for safety and convenience. The voltage reduction is accomplished by a transformer located in the condensing unit or air handler, Figure 41–3.

Figure 41–1. Room thermostat for controlling space temperature. *Photo by Bill Johnson*

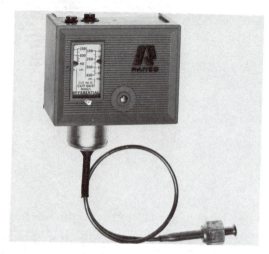

Figure 41-2. Commercial type of high-pressure control used in earlier residential equipment. *Courtesy Ranco*

Figure 41-3. Low-voltage transformer. *Reprinted with permission of Motors and Armatures, Inc.*

41.3 LOW-VOLTAGE CONTROLS

The low-voltage circuit operates the various power-consuming devices in the control system that start and stop the compressor and fan motors. These low-voltage devices do not use much current. The control transformer is usually a 40-VA transformer. This means that the highest amperage the control transformer will be able to produce on the secondary side is 1.666 A. This was arrived at by dividing the VA rating of 40 by the voltage, (40 VA ÷ 24V = 1.666 A). If the circuit current is greater than 1.666 A, the voltage will begin to drop and the transformer will get hot. In this situation a larger transformer is needed.

The contactor that actually starts and stops the compressor is considered one of the controls. It consumes power in its 24-V magnetic holding coil, Figure 41-4. The contacts of the contactor pass power when the magnetic coil is energized, Figure 41-5. The contact circuit to the compressor is 230 V.

41.4 SOME HISTORY OF RESIDENTIAL CENTRAL AIR CONDITIONING

Residential central air conditioning became popular in the late 1950s, and its popularity has continued to grow. This popularity has caused air conditioning equipment to become competitive in price. In the warm climates in the United States, central air conditioning is very desirable in all new construction, even if the owner or builder does not desire air conditioning, because of the resale value. Lending institutions may even require that a new home have air conditioning before they will finance the loan. The reason is that the lender may have to resell the house, and central air conditioning is a strong selling feature, which may not be easy to add in the future.

The first central air conditioning systems installed in residences were small commercial systems applied to homes. They were all water cooled and very efficient and reliable. Air-cooled equipment became popular

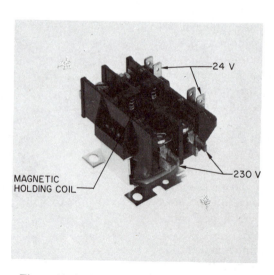

Figure 41-4. Contactor. *Photo by Bill Johnson*

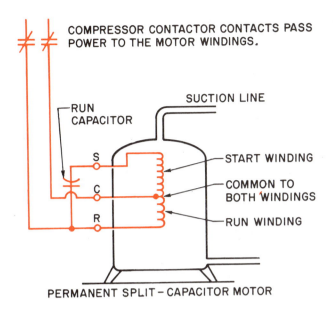

Figure 41-5. Diagram of circuit that passes power to the motor windings.

later. Air-cooled systems do not pump and handle water. Freezing water and mineral deposits are not problems.

The first air-cooled systems were very heavy duty, bulky, and hard-to-handle. Slow-speed hermetic compressors (1800 rpm) or belt-drive open compressors were used with R-12 as the refrigerant. The residential and light commercial equipment market began to grow to the point that price-conscious buyers began to look for less-expensive equipment. Speculating home builders began to build large developments and wanted to save on the air conditioning equipment. Manufacturers were thus forced to find more economical ways to build equipment. A more efficient refrigerant was also needed. When R-22 became available, the evaporator and the condenser tubing could be made smaller, making equipment more compact and lighter, which made storage and shipping much easier. In addition, the suction and liquid lines that connect the condensing unit and the evaporator are one size smaller and reduce the price of the installation.

41.5 ECONOMICS OF EQUIPMENT DESIGN

The compressors were then manufactured to turn at a higher speed, so a small compressor could pump more refrigerant and do the job of a larger compressor. Present compressors turn at 3600 rpm and are much smaller than earlier models. The faster compressors and the more efficient R-22 refrigerant means the equipment of today is smaller and more efficient than the earlier equipment. Every manufacturer constantly seeks to make equipment lighter, smaller, more efficient, and less expensive.

Air conditioning equipment manufactured to compete economically with other equipment will have only the essential controls yet may be very reliable. The following sections describe in detail typical controls for an air-cooled system built in the mid-1960s when central air conditioning systems were being installed in large numbers for the first time. The detailed description is important because many of these controls are no longer used. Another form of protection, an entirely new concept, has replaced them.

41.6 OPERATING CONTROLS FOR OLDER AIR-COOLED SYSTEMS

The room thermostat senses and controls the space temperature. The thermostat is not a power-consuming device, but it passes power to the power-consuming devices. It is sensitive to temperature changes because of the bimetal element under the cover. Early thermostats were larger than modern ones.

The fan relay starts and stops the indoor fan. The relay is a power-consuming device that receives power through the room thermostat to energize the magnetic coil and to close the contacts that start the fan.

The compressor contactor starts and stops the compressor and outdoor fan, which are normally wired in parallel, Figure 41-6. This control is a power-consuming device because of the magnetic coil. When the coil is energized, it closes the contacts and starts the compressor and the outdoor fan motor, Figure 41-7.

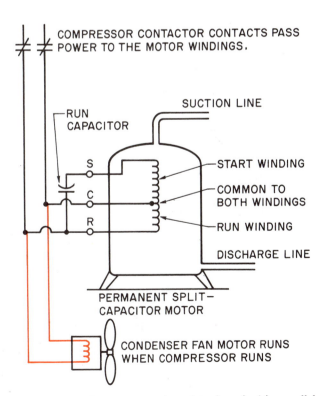

Figure 41-6. Compressor and outdoor fan wired in parallel.

Figure 41-7. Typical compressor contactor. *Photo by Bill Johnson*

The compressor starting and running circuits are actually not controls in the strictest sense, but most service technicians treat them as such. The compressors that were used for many years had capacitor-start, capacitor-run motors. These motors have a high starting torque and were used because thermostatic expansion valves were used as the metering devices. They do not equalize the high- and low-side pressures during the off cycle, and a high-torque compressor motor is required. The following components were part of the starting system: (1) a potential starting relay, (2) a start capacitor, and (3) a run capacitor. See Figure 41-8 for an example of each component. See Figure 41-9 for an example of the starting circuit diagram.

41.7 SAFETY CONTROLS FOR OLDER AIR-COOLED SYSTEMS

The high-pressure control stops the compressor when a high-pressure condition exists—for example, if the condenser fan stops when it should be running. The compressor has no way of knowing that the fan has stopped, and it keeps pumping refrigerant into the condenser. *The pressures can quickly rise to a dan-

RUN CAPACITOR   START CAPACITOR   POTENTIAL START RELAY

Figure 41-8. Components to start and run a capacitor-start, capacitor-run motor. *Photo by Bill Johnson*

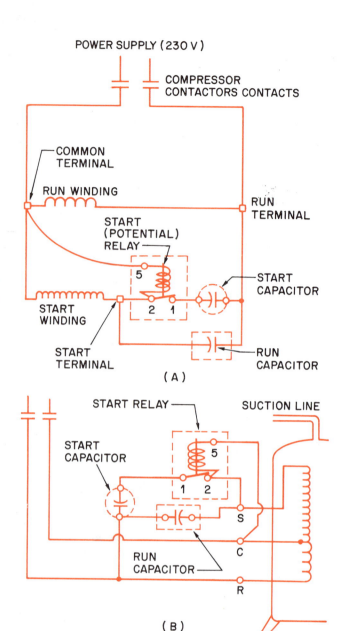

Figure 41-9. Starting circuit for the motor in Figure 41-8. (A) Schematic. (B) Pictorial.

gerous level, so the compressor must be stopped, Figure 41-10.*

For internal protection, the low-pressure control stops the compressor from pulling the suction pressure below a predetermined point. If the system loses all or part of its refrigerant charge, the low side of the system may pump into a vacuum without a low-pressure control. The low-pressure control can be set to shut off the compressor before a vacuum is reached, Figure 41-11. If the leak is on the low side of the system and the compressor pumps into a vacuum, air will be drawn into the system. Then there would be two problems (1) a leak, and (2) air in the system. The low-pressure control also provides freeze protection: When the air across the indoor coil is reduced, the suction pressure

sor contactor coil to stop the compressor before damage occurred, Figure 41–12.

Internal motor protection senses the actual motor temperature and stops the motor when it is too hot. Compressor internal protection may be the type that breaks the line power inside the compressor, or it may be the pilot type that interrupts the circuit to the contactor coil.

Short-cycle protection was used to prevent the compressor from short cycling when a safety or operating control would open the circuit and then make back in a short cycle. The overload is an example. Some types of overloads are current sensitive, Figure 28–24. These controls will open the circuit when the current is too high but will make back the instant it is reduced. If a compressor had a bad starting relay and stopped due to high current, it would try to start again immediately after stopping if there were no short-cycle control, Figure 41–13. This same control protects the compressor from short cycling if the homeowner turns the thermostat up and down several times while adjusting it.

41.8 OPERATING CONTROLS FOR MODERN EQUIPMENT

The room thermostat is now much smaller, See Figure 41–14. Some room thermostats are electronic and use a thermistor as the sensing element, Figure 41–15. These thermostats may have programs for night adjustments and daytime work schedule for both the heating and the cooling settings. They can be thought of as power-consuming because they must have a cir-

Figure 41–10. This commercial high-pressure control is much larger than currently used residential high-pressure controls. *Courtesy Ranco*

will be reduced to the point that a freeze condition may result. The condensate on the evaporator coil may turn to ice and block the airflow even more. The evaporator coil may turn to a solid block of ice. Reduced airflow may also be caused from dirty filters, closed air registers, blocked return-air grille or dirty evaporator coil.

The early systems had overload protection for the common and the run circuit of the compressor. This overload protection was usually a heat sensitive bimetal device with a current rating. If the motor current went above the rating on the overload device, the bimetal strip was heated and opened the circuit to the compres-

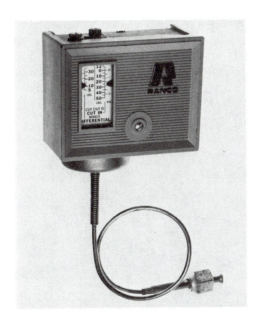

Figure 41–11. Type of low-pressure control used on commercial equipment and then used on early residential equipment. *Courtesy Ranco*

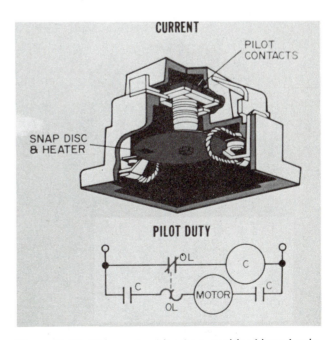

Figure 41–12. This overload is a heat sensitive bimetal strip. *Reproduced courtesy of Carrier Corporation*

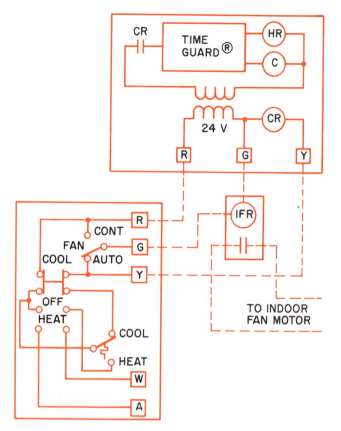

Figure 41-13. Timer to keep the compressor from short cycling.

Figure 41-14. This is a typical room thermostat. It is a combination heating and cooling thermostat with a bimetal element for sensing room temperature. *Photo by Bill Johnson*

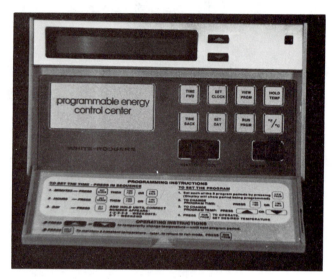

Figure 41-15. Electronic thermostat with a thermistor for the sensing element. *Photo by Bill Johnson*

cuit with a timer. They also pass power to various components (the fan relay and compressor contactor). The circuit that passes power is the contact circuit.

The fan relay for the modern equipment may be smaller than the older versions, but it does the same job, Figure 41-16.

The modern compressor contactor has one big difference on some manufacturers' equipment. Some contactors have only one set of contacts, Figure 41-17. Older contactors always had two sets of contacts that interrupted the power to the common and run circuits to the compressor. The modern contactor may only interrupt the power to one circuit. *Note:* The circuit with continuous power may feed power through a run capacitor to provide a small amount of current through the compressor windings to keep the compressor warm during the off cycle. This is a substitute for crank-

Figure 41-16. Fan relay. *Photo by Bill Johnson*

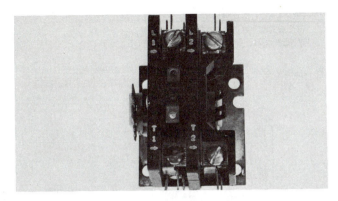

Figure 41-17. Compressor contactor with one set of contacts. *Photo by Bill Johnson*

case heat. If a contactor with two contacts is substituted, there will be no heat.

The starting circuit does not have as many components. If a retired service technician suddenly opened the panel of a modern piece of equipment, the conclusion would quickly be drawn that the equipment could not work correctly. It would seem unprotected because there are not as many components as before. The components used to start and run modern equipment are a run capacitor and possibly a start-assist unit such as a PTC (positive temperature coefficient) device. Remember, from Unit 26, the PTC device has no moving parts as a start relay does. It shunts the start terminal to the run terminal for a short time to get the compressor started. These devices are used on PSC (permanent split capacitor) motors as a start assist. The PSC motor has very little starting torque and is used with metering devices that equalize during the off cycle, Figure 41-18.

41.9 SAFETY CONTROLS FOR MODERN EQUIPMENT

Modern equipment may not have any visible safety controls. It may have only a motor-temperature control that senses the temperature of the motor windings. This control may be mounted in the windings and have no external terminals. A note on the compressor may read "this compressor is internally protected with a winding thermostat," Figure 41-19. The intent of the manufacturer is to let this control function cut off the compressor during the following situations:

A. When a system is operating with a low gas charge, the motor windings will overheat and the motor winding thermostat will stop the compressor. The suction gas is supposed to cool the compressor windings. It is functioning as low charge protection. When the motor gets warm or hot, the heat in the mass of the motor prevents the control from closing its contacts until the motor cools. This is a built-in short-cycle protection.

B. When the condenser is dirty, the compressor head pressure rises. This makes the current increase and the motor gets hot and cuts off on motor temperature cutout. The motor winding thermostat is functioning as a high-pressure control in this situation. It still has the same short-cycle protection.

C. *Internally in most compressors there is a pressure relief valve that will relieve hot gas from the compressor discharge onto the motor winding thermostat if a great pressure differential is experienced, Figure 41-20.* Suppose the condenser fan motor

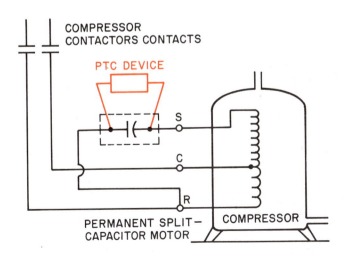

Figure 41-18. Wiring diagram showing how a PSC (permanent split capacitor) motor is wired when a PTC (positive temperature coefficient) device is used as a start assist for the compressor.

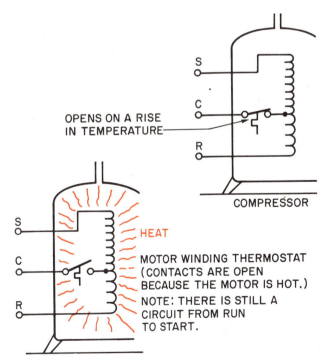

Figure 41-19. Motor winding thermostat.

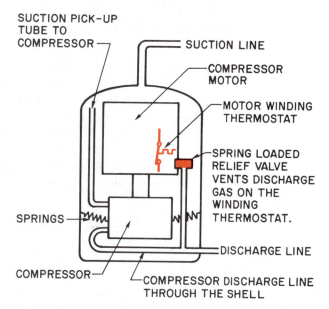

Figure 41-20. How the internal relief valve functions in a welded hermetic compressor.

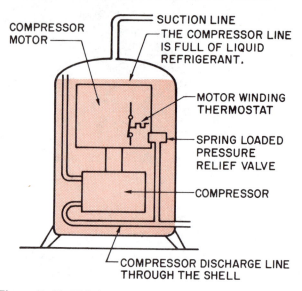

Figure 41-21. This is an example of one time when the relief valve and motor temperature combination may not give high pressure protection to stop the compressor.

burned and stopped. High head pressures would occur immediately. When these pressures exceed the relief valve setting, the valve relieves and blows hot gas on the winding thermostat to cut the compressor off. Internal relief valve settings may have a 450 psig differential. This means the head pressure has to be 450 psig higher than the suction pressure for the relief valve to open. If a condenser fan stopped, the head pressure could go to 540 psig, and the suction pressure could go to 90 psig before the internal relief valve opened (540 − 90 = 450 psig). *A reading of the high-pressure gage is not enough to determine that the relief valve is supposed to function.* The winding thermostat in this

application serves as the high-pressure cutout when fast action is needed.

D. The run capacitor may become defective, and the compressor will pull too much current trying to start. The motor will heat and cut out on winding thermostat. The winding thermostat is functioning as an overload in this situation.

E. The homeowner or children may fool with the space-temperature thermostat and turn it up and down several times. The compressor will try to start

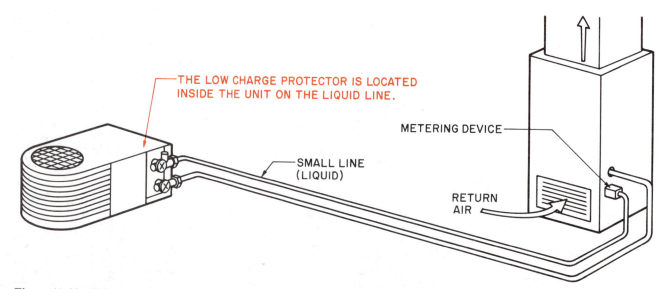

Figure 41-22. This loss-of-charge protector is located in the liquid line and will stop the compressor if the pressure goes as low as 5 psig.

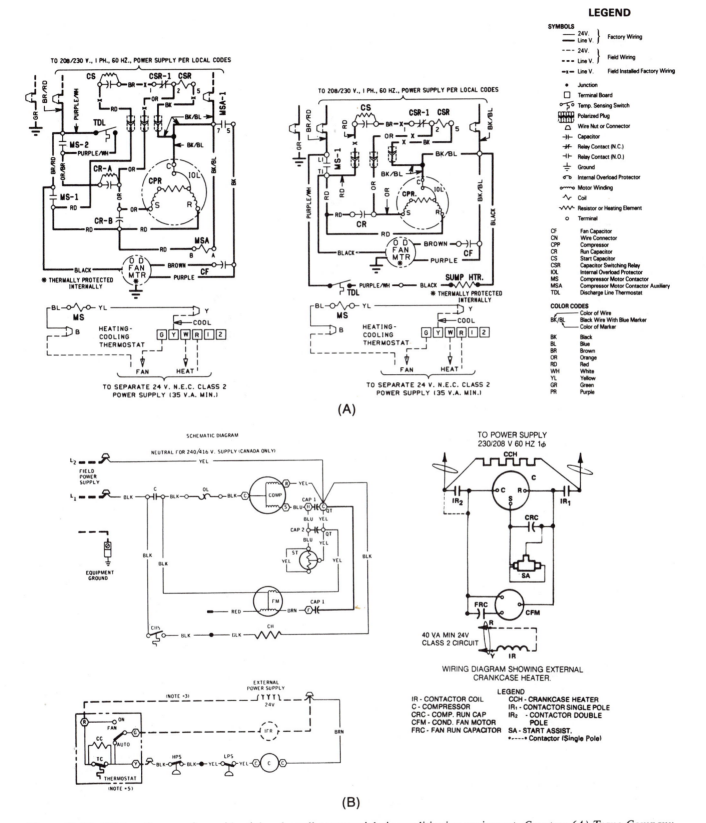

Figure 41-23. Wiring diagrams for residential and small commercial air conditioning equipment. *Courtesy (A) Trane Company (B) Climate Control*

several times, and the motor will get hot enough to cause the motor winding thermostat to cut it off. The motor winding thermostat is preventing the compressor from short cycling in this situation.

When an overcharge of refrigerant is great enough to cause the compressor windings to be cool, the winding thermostat may not give adequate protection, Figure 41–21. In this situation the only high-pressure protection the compressor has is the internal relief valve.

41.10 LOSS-OF-CHARGE PROTECTION

Some modern systems have a loss-of-charge protection in the form of a low-pressure cutout control located on the liquid line. It may have a setting as low as 5 psig and will stop the system only if there is complete loss of charge, Figure 41–22.

41.11 THE WORKING CONTROL PACKAGE

All controls must be assembled into a working assembly. Remember, manufacturers must be competitive, so they are always trying to provide a simple, effective control package that will protect the equipment and give it a long life span. See Figure 41–23 for some typical diagrams that apply to simple residential equipment.

Figure 41–24 summarizes the controls used on earlier equipment and newer equipment. The technician should be familiar with both because some manufacturers still use the older types of controls.

Some highly modern expensive equipment is very efficient, and part of the efficiency package is the method of controlling the equipment. Solid-state circuit boards are used by some manufacturers for some of the listed functions. For example, the short-cycle timer can easily be built into the electronic circuit, Figure 41–25.

41.12 ELECTRONIC CONTROLS AND AIR CONDITIONING EQUIPMENT

An electronic circuit board can monitor high voltage, whereas electromechanical controls cannot easily do this. Power companies often supply voltage that is higher than the equipment specification. For example, when a system is rated at 230 V, the equipment may be operated at ±10% of this voltage. If the voltage is too low, the motor will pull more than normal current and the overload (or winding thermostat) should cut off the motor. If the motor operates at a slightly high-voltage condition, the overloads will not help because the motor current will be below normal. The electronic circuit board may have a voltage monitor that will stop the compressor at a high voltage even if the current is low. The electronic circuit board can also react quickly to low voltage (known as "brownout"). It can monitor

OLDER EQUIPMENT	NEWER EQUIPMENT
Room thermostat	Room thermostat
Fan relay	Fan relay
Compressor contactor	Compressor contactor
Winding thermostat (maybe)	Winding thermostat
Run capacitor	Run capacitor
Internal relief valve (maybe)	Internal relief valve
Low-pressure control	No charge protection
High-pressure control	Winding thermostat
Short-cycle protection	Winding thermostat
Overloads (usually two)	Winding thermostat
Crankcase heater	Through the capacitor

Figure 41–24. Controls used on older and newer equipment.

and cut off the compressor before an overload has time to heat up and react.

Each manufacturer that uses a circuit board has a checkout procedure that they recommend. For practical troubleshooting, remember that a circuit board usually looks like a single control on the wiring diagram. The control circuit goes into the board and comes out of the board. However, sometimes the circuits on the board may be checked by using a jumper wire from one circuit to another to determine whether the board is defective.

SUMMARY

- The control of air conditioning equipment consists of controlling the various components to start and stop them in the correct sequence to maintain space temperature for comfort.
- The three main components controlled are the indoor fan, the compressor, and the outdoor fan.
- The control circuit uses low voltage (24 V) for safety.
- Low voltage is obtained from high voltage by a transformer located in the condensing unit or in the air handler.
- Some electronic thermostats use a thermistor as the sensing device and may incorporate other features, such as temperature set-back, for night and day.
- Air-cooled equipment is used more widely today than water-cooled equipment.
- Controls are either operating controls or safety controls.
- Some controls, such as contactors and electronic thermostats have more than one circuit. Some circuits may pass power and some may consume power.
- The compressor overloads are used to keep the compressor from drawing too much current and overworking.
- The winding thermostat shuts off the compressor when the compressor motor winding gets too hot.

COMPLETE CIRCUIT BOARD

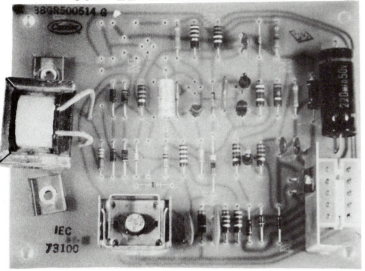

TYPICAL ELECTRONIC COMPONENTS

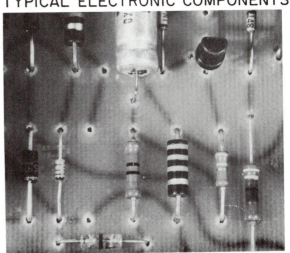

FIELD-CABLE CONNECTION ——

Figure 41-25. Electronic circuit board with voltage monitors built in. Also included is a timer circuit that is not evident. *Photos by Bill Johnson*

• The compressor internal relief valve vents high-pressure gas from the high-pressure side to the low-pressure side of the compressor internally. It can be used as a high-pressure indicator by directing this high-pressure, high-temperature gas on to the winding thermostat.

• Modern equipment may have only a few controls and components. It is common for the unit to have a single-pole contactor and a run capacitor as the only visible component.

• Electronic circuit boards are furnished with some equipment. The circuit boards may include short-cycle protection, as well as low- and high-voltage protection in their circuits.

REVIEW QUESTIONS

1. Name two types of controls.
2. Some controls _____ power, and some controls _____ power.

3. What type of control is the room thermostat?

What protection do the controls in Questions 4–8 offer?

4. High-pressure control

5. Low-pressure control

6. Winding thermostat

7. Internal relief valve

8. Electronic circuit board

9. Why can't you substitute a two-pole contactor for a single-pole contactor on some units?

10. What is the standard low-voltage control voltage for residential air conditioning?

11. What changes have manufacturers made in residential air conditioning equipment?

12. What component starts the compressor in a residential air conditioner?

13. What starts the indoor fan?

14. What starts the outdoor fan?

15. What stops the compressor in a low-charge condition if there is no low-pressure control?

TYPICAL OPERATING CONDITIONS

OBJECTIVES

After studying this unit, you should be able to

- **explain what conditions will vary the evaporator pressures and temperatures.**
- **define how the various conditions in the evaporator and ambient air affect condenser performance.**
- **state the relationship of the evaporator to the rest of the system.**
- **describe the relationship of the condenser to the total system performance.**
- **compare high-efficiency equipment and standard-efficiency equipment.**
- **establish reference points when working on unfamiliar equipment in order to know what the typical conditions should be.**
- **describe how humidity affects equipment suction and discharge pressure.**
- **explain three methods that manufacturers use to make air conditioning equipment more efficient.**

Air conditioning technicians must be able to evaluate both mechanical and electrical systems. The mechanical operating conditions are determined or evaluated with gages and thermometers; electrical conditions are determined with electrical instruments.

42.1 MECHANICAL OPERATING CONDITIONS

Air conditioning equipment is designed to operate at its rated capacity and efficiency at one set of design conditions. This design condition is generally considered to be at an outside temperature of 95°F and at an inside temperature of 80°F with a humidity of 50%. This rating is established by ARI (Air Conditioning and Refrigeration Institute). Equipment must have a rating in order for there to be a standard to work from. Equipment is also rated so that the buyer will have a common basis with which to compare one piece of equipment with another. All equipment in the ARI directory is rated under the same conditions. When an estimator or buyer finds that a piece of equipment is rated at 3 tons, it will perform at a 3-ton capacity or 36,000 Btu/hr under the stated conditions. When the

conditions are different, the equipment will perform differently. For example, most homeowners will not be comfortable at 80°F and a relative humidity of 50%. They will normally operate their system at about 75°F, and the humidity in the conditioned space will be close to 50%. The equipment will not have quite the capacity at 75°F that it had at 80°F. If the designer wants the system to have a capacity of 3 ton at 75°F and 50% humidity, the manufacturer's literature may be consulted to make the change in equipment choice. **This 75°F 50% humidity condition will be used as the design condition for this unit because it is a common operating condition of the equipment in the field.**

42.2 HUMIDITY AND THE LOAD

The inside humidity adds a significant load to the evaporator coil and has to be considered as part of the load. When conditions vary from the design conditions, the equipment will vary in capacity. The pressures and temperatures will also change.

42.3 SYSTEM COMPONENT RELATIONSHIPS UNDER LOAD CHANGES

It may be easy to understand that if the outside temperature increased, say to 100°F, that the equipment would be operating at a higher head pressure and would not have as much capacity. What is not so easy to understand is why the capacity varies when the space temperature goes up or down or when the humidity varies. Remember, there is a relationship between the various components in the system. The evaporator absorbs heat. When anything happens to increase the amount of heat that is absorbed into the system, the system pressures will go up. The condenser rejects heat. If anything happens to prevent the condenser from rejecting heat from the system, the system pressures will go up. The compressor pumps the heat-laden vapor. Vapor at different pressure levels and saturation points (in reference to the amount of superheat) will hold different amounts of heat. You must have a firm understanding of Units 3 and 19 to understand the following text.

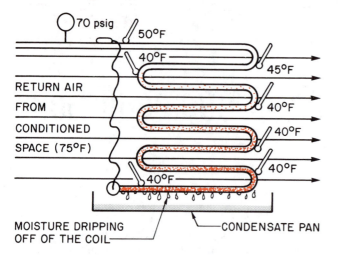

Figure 42-1. Evaporator operating at typical conditions. The refrigerant is boiling at 40°F in the coil. This corresponds to 68.5 psig typically rounded off to 70 psig for R-22.

42.4 EVAPORATOR OPERATING CONDITIONS

The evaporator will normally operate at a 40°F boiling temperature when operating at the 75°F 50% humidity condition. This will cause the suction pressure to be 70 psig for R-22. (The actual pressure corresponding to 40°F is 68.5 psig; for our purposes we will round this off to 70 psig.) This example is at design conditions and a steady-state load; in other words, the load is not changing. At this condition the evaporator is boiling the refrigerant exactly as fast as the expansion device is metering it into the evaporator. As an example, suppose the evaporator has a return-air temperature of 75°F, and the air has a relative humidity of 50%. The liquid refrigerant goes nearly to the end of the coil, and the coil has a superheat of 10°F. This

coil is operating as intended at this typical condition, Figure 42-1.

Late in the day, after the sun has been shining on the house, the heat load inside becomes greater. A new condition is going to be established. The example evaporator in Figure 42-2 has a fixed-bore metering device that will only feed a certain amount of liquid refrigerant. The space temperature in the house has climbed to 77°F and is causing the liquid refrigerant in the evaporator to boil faster, Figure 42-3. This causes the suction pressure and the superheat to go up slightly. The new suction pressure is 73 psig, and the new superheat is 13°F. This is well within the range of typical operating conditions for an evaporator. The system actually has a little more capacity at this point if the outside temperature is not too much above design. If the head pressure goes up because the outside temperature is 100°F, for example, the suction pressure will even go higher than 73 psig because head pressure will influence it, Figure 42-4.

From the previous discussion it can be seen that there can be many different conditions that will affect the operating pressures and air temperatures to the conditioned space. There can actually be as many different pressures as there are different inside and outside temperature and humidity variations. This can be very confusing, particularly to the new service technician. There are some common conditions, however, that the service technician can use in troubleshooting. This is necessary because there are very few times when the technician has a chance to work on a piece of equipment when the conditions are perfect. Most of the time when the technician gets to the job, the system has been off for some time or not operating correctly for long enough that the conditioned space temperature and humidity are higher than normal, Figure 42-5.

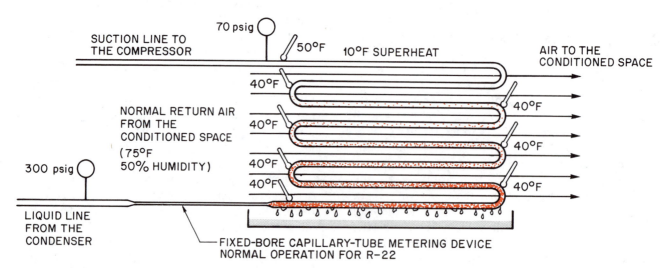

Figure 42-2. This fixed bore metering device will not vary the amount of refrigerant feeding the evaporator as much as the thermostatic expansion valve. It is operating at an efficient state of 10°F superheat.

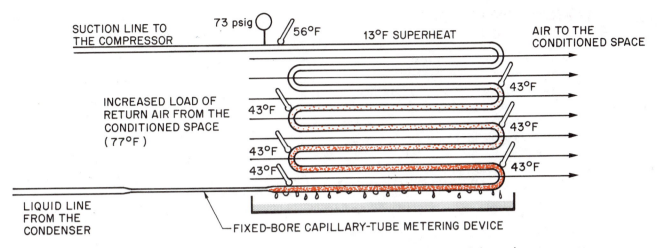

Figure 42-3. This is a fixed-bore metering device when an increase in load has caused the suction pressure to go up.

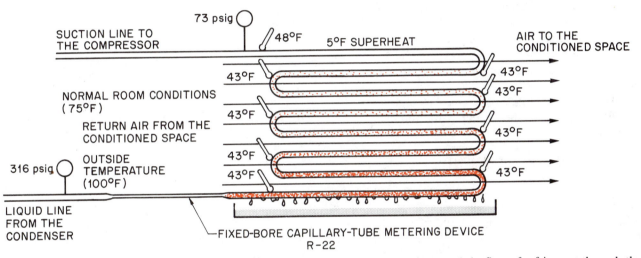

Figure 42-4. System with a high head pressure. The increase in head pressure has increased the flow of refrigerant through the fixed bore metering device and the superheat is decreased.

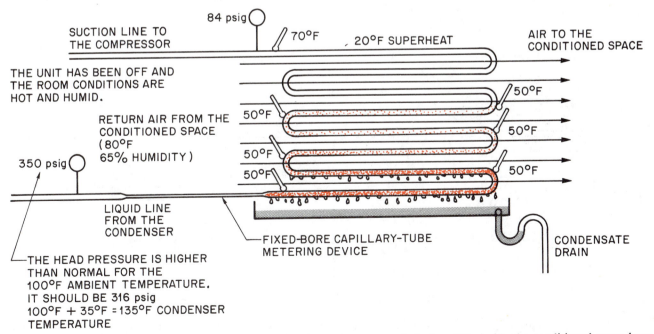

Figure 42-5. This system has been off long enough that the temperature and the humidity inside the conditioned space have gone up. Notice the excess moisture forming on the coil and going down the drain.

42.5 HIGH EVAPORATOR LOAD AND A COOL CONDENSER

High temperature in the conditioned space is not the only thing that will cause the system to have different pressures and capacity. The reverse can happen if the inside temperature is warm and the outside temperature is cooler than normal, Figure 42–6. For example, before going to work, a couple may turn the air conditioner off to save electricity. They may not get home until after dark to turn the air conditioner back on. By this time it may be 75°F outside but still be 85°F inside the structure. The air passing over the condenser is now cooler than the air passing over the evaporator. The evaporator may also have a large humidity load. The condenser becomes so efficient that it will hold some of the charge in the condenser. The reason is that the condenser starts to condense refrigerant in the first part of the condenser. Thus more of the refrigerant charge is in the condenser tubes, and this will slightly starve the evaporator. The system may not have enough capacity to cool the home for several hours. The condenser may hold back enough refrigerant to cause the evaporator to operate below freezing and freeze up before it can cool the house and satisfy the thermostat.

42.6 GRADES OF EQUIPMENT

Manufacturers have been constantly working on the design of air conditioning equipment to make it more efficient. There are normally three grades of equipment: economy grade, standard-efficiency grade, and high-efficiency grade. Some manufacturers manufacture all three grades and offer them to the supplier. Some manufacturers only offer one grade and may take offense at someone calling their equipment the lower grade. Economy grade and standard-efficiency grade are about equal in efficiency, but the materials they are made of and their appearances are different. High-efficiency equipment may be much more efficient and will not have the same operating characteristics, Figure 42–7. A condenser will normally condense the refrigerant at a temperature of about 30° to 35°F higher than the ambient temperature. For example, when the outside temperature is 95°F, the average condenser will condense the refrigerant at 125° to 130°F, and the head pressure for these condensing temperatures, would be 278 to 297 psig for R-22. High-efficiency air conditioning equipment may have a much lower operating head pressure. The high efficiency is gained by using a larger condenser surface and the same compressor or a more efficient compressor. The condensing temperature may be as low as 20°F greater than the ambient temperature. This would bring the head pressure down to a temperature corresponding to 115°F or 243 psig. The compressor will not use as much power at this condition, Figure 42–8.

42.7 DOCUMENTATION WITH THE UNIT

The technician needs to know what the typical operating pressures should be at different conditions. Some manufacturers furnish a chart with the unit to tell what the suction and discharge pressures should be at different conditions, Figures 42–9 and 42–10. Some publish a bulletin that lists all of their equipment along with the typical operating pressures and temperatures. Others furnish this information with the unit in the installation and startup manual, Figure 42–11. The homeowner may have this booklet, and it may be helpful to the technician.

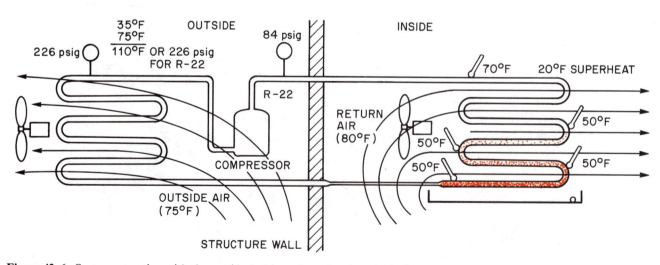

Figure 42–6. System operating with the outside ambient air cooler than the inside space temperature air. As the return air cools down, the suction pressure and boiling temperature will reduce.

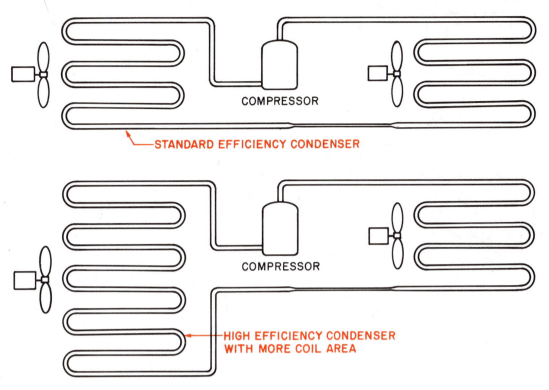

Figure 42-7. High-efficiency condenser and standard-efficiency condenser.

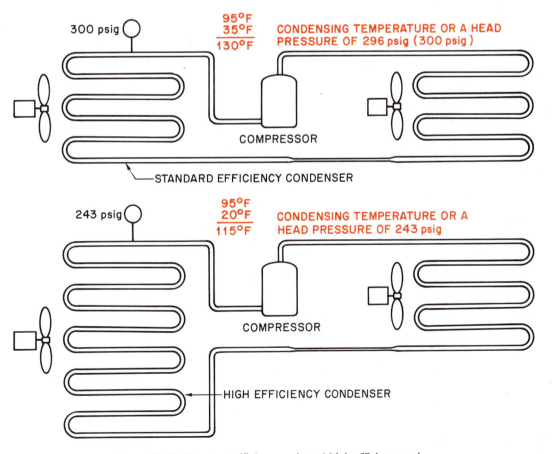

Figure 42-8. Standard-efficiency unit and high-efficiency unit.

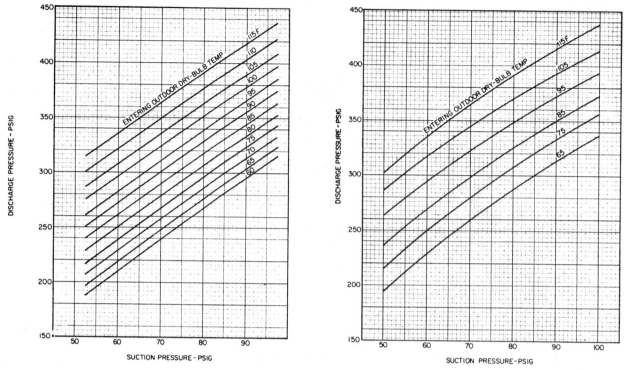

Figure 42–9. Charts furnished by some manufacturers to check unit performance. *Reproduced courtesy of Carrier Corporation*

42.8 ESTABLISHING A REFERENCE POINT ON UNKNOWN EQUIPMENT

When a technician arrives at the job and finds no literature and cannot obtain any, what does he or she do? The first thing is to try to establish some known condition as a reference point. For example, is the equipment standard efficiency or high efficiency? This will help establish a reference point for the suction and head pressure. The high-efficiency equipment is often larger than normal. The equipment is not always marked as high efficiency, and the technician may have to compare the size of the condenser to another one to determine the head pressure. It should be obvious that a larger or oversized condenser would have a lower head pressure. For example, a 3-ton compressor will have a full-load ampere (FLA) rating of about 17A. The ampere rating of the compressor may help to

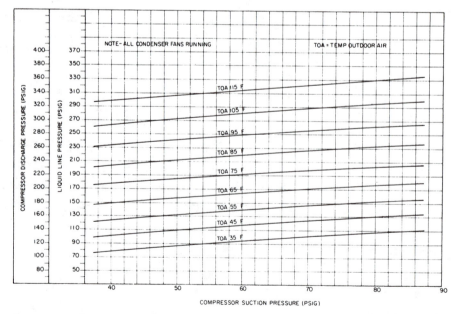

Figure 42–10. Page from a manufacturer's bulletin explaining how to charge one model of equipment. *Reproduced courtesy of Carrier Corporation*

REFRIGERANT CHARGING

SAFETY PRECAUTION: To prevent personal injury, wear safety glasses and gloves when handling refrigerant. Do not overcharge system. This can cause compressor flooding.

1. Operate unit a minimum of 15 minutes before checking charge.
2. Measure suction pressure by attaching a gage to suction valve service port.
3. Measure suction line temperature by attaching a service thermometer to unit suction line near suction valve. Insulate thermometer for accurate readings.
4. Measure outdoor coil inlet air dry-bulb temperature with a second thermometer.
5. Measure indoor coil inlet air wet-bulb temperature with a sling psychrometer.
6. Refer to table. Find air temperature entering outdoor coil and wet-bulb temperature entering indoor coil. At this intersection note the superheat.
7. If unit has higher suction line temperature than charted temperature, add refrigerant until charted temperature is reached.
8. If unit has lower suction line temperature than charted temperature, bleed refrigerant until charted temperature is reached.
9. If air temperature entering outdoor coil or pressure at suction valve changes, charge to new suction line temperature indicated on chart.
10. This procedure is valid, independent of indoor air quantity.

SUPERHEAT CHARGING TABLE
(SUPERHEAT ENTERING SUCTION SERVICE VALVE)

Outdoor Temp (°F)	INDOOR COIL ENTERING AIR °F WB													
	50	52	54	56	58	60	62	64	66	68	70	72	74	76
55	9	12	14	17	20	23	26	29	32	35	37	40	42	45
60	7	10	12	15	18	21	24	27	30	33	35	38	40	43
65	—	6	10	13	16	19	21	24	27	30	33	36	38	41
70	—	—	7	10	13	16	19	21	24	27	30	33	36	39
75	—	—	—	6	9	12	15	18	21	24	28	31	34	37
80	—	—	—	—	5	8	12	15	18	21	25	28	31	35
85	—	—	—	—	—	8	11	15	19	22	26	30	33
90	—	—	—	—	—	5	9	13	16	20	24	27	31
95	—	—	—	—	—	—	6	10	14	18	22	25	29
100	—	—	—	—	—	—	8	12	15	20	23	27
105	—	—	—	—	—	—	5	9	13	17	22	26
110	—	—	—	—	—	—	6	11	15	20	25
115	—	—	—	—	—	—	8	14	18	23

Figure 42–11. Chart furnished by manufacturer in the installation and startup literature. *Reproduced courtesy of Carrier Corporation*

determine the rating of the equipment, Figure 42–12. Although a 3-ton high-efficient piece of equipment will have a lesser ampere rating than a standard piece of equipment, the ratings will be close enough to compare and to determine the capacity of the equipment. If the condenser is very large for an ampere rating that should be 3 ton, the equipment is probably high efficiency, and **the head pressure will not be as high as in a standard piece of equipment.**

42.9 METERING DEVICES FOR HIGH-EFFICIENCY EQUIPMENT

High-efficiency air conditioning equipment often uses a thermostatic expansion valve rather than a fixed-bore metering device because a certain amount of efficiency may be gained. The evaporator may also be larger than normal. An oversized evaporator will add to the efficiency of the system. It is more difficult to make a determination regarding the evaporator size because, being enclosed in a casing or the duct work, it is not as easy to see.

The operating conditions of a system can vary so much that all possibilities cannot be covered; however,

a few general statements committed to memory may help. Standard-efficiency equipment and high-efficiency equipment will follow the general conditions listed in the following subsections.

Operating Conditions Near Design Space Conditions for Standard-Efficiency Equipment

1. The suction temperature is 40°F (70 psig for R-22 and 37 psig for R-12), Figure 42–13.
2. The head pressure should correspond to a temperature of no more than 35°F above the outside ambient temperature. This would be 296 psig for R-22 and 181 psig for R-12 when the outside temperature is 95°F (95° + 35° = 130°F condensing temperature or a pressure of 296 psig for R-22 and 181 psig for R-12), Figure 42–14.

Space Temperature Higher Than Normal for Standard-Efficiency Equipment

1. The suction pressure will be higher than normal. Normally, the refrigerant boiling temperature is

APPROXIMATE FULL LOAD AMPERAGE VALUES
FOR ALTERNATING CURRENT MOTORS

Motor	Single Phase		3-Phase-Squirrel Cage Induction		
HP	115 V	230 V	230 V	460 V	575 V
$\frac{1}{6}$	4.4	2.2			
$\frac{1}{4}$	5.8	2.9			
$\frac{1}{3}$	7.2	3.6			
$\frac{1}{2}$	9.8	4.9	2	1.0	0.8
$\frac{3}{4}$	13.8	6.9	2.8	1.4	1.1
1	16	8	3.6	1.8	1.4
$1\frac{1}{2}$	20	10	5.2	2.6	2.1
2	24	12	6.8	3.4	2.7
3	34	17	9.6	4.8	3.9
5	56	28	15.2	7.6	6.1
$7\frac{1}{2}$			22	11.0	9.0
10			28	14.0	11.0

Does not include shaded pole.

Figure 42-12. Table of current ratings for different sized motors at different voltages. *Courtesy BDP Company*

about 35°F cooler than the entering air temperature. (Recall the relationship of the evaporator to the entering air temperature described in the refrigeration section.) When the conditions are normal, the refrigerant boiling temperature would be 40°F when the return-air temperature is 75°F. This is when the humidity is normal. When the space temperature is higher than normal because the equipment has been off for a long time, the return-air temperature may go up to 85°F and the humidity may be high. The suction temperature may then go up to 50°F, with a corresponding pressure of 84 to 93 psig for R-22, Figure 42-15. This higher-than-normal suction pressure may cause the discharge pressure to rise also.

2. The discharge pressure is influenced by the outside temperature and the suction pressure. For example, the discharge pressure is supposed to correspond to a temperature of no more than 35°F higher than the ambient temperature under normal conditions. When the suction pressure is 80 psig, the discharge pressure

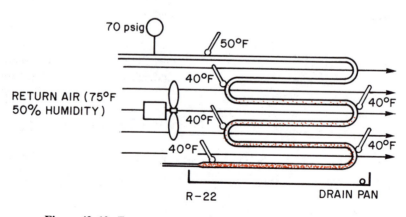

Figure 42-13. Evaporator operating at close to design conditions.

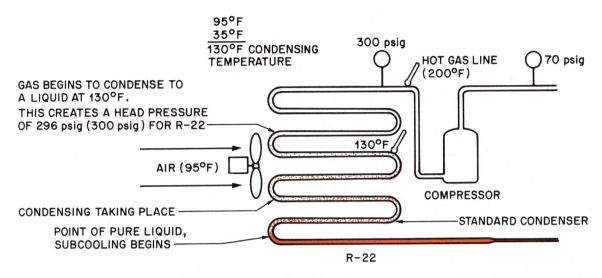

Figure 42–14. Normal conditions of a standard condenser operating on a 95°F day.

is going to rise accordingly. It may move up to a new condensing temperature 10°F higher than normal. This would mean that the discharge pressure for R-22 could be 337 psig (95°F + 35°F + 10°F = 140°F) while the suction pressure is high, Figure 42–16. When the unit begins to reduce the space temperature and humidity, the evaporator pressure will begin to reduce. The load on the evaporator is reduced. The head pressure will come down also.

Operating Conditions Near Design Conditions for High-Efficiency Equipment

1. The evaporator design temperature may in some cases operate at a slightly higher pressure and temperature on high-efficiency equipment because this

equipment is larger. The boiling refrigerant temperature may be 45°F with the larger evaporator at design conditions, and this would create a suction pressure of 76 psig for R-22 and be normal, Figure 42–17. R-12 is not mentioned in this coverage of high-efficiency equipment because it is not being used.

2. The refrigerant may condense at a temperature as low as 20°F more than the outside ambient temperature. For a 95°F day with R-22, the head pressure may be as low as 243 psig, Figure 42–17. If the condensing temperature were as high as 30°F above the ambient temperature, you should suspect a problem. For example, the head pressure should not be more than 277 psig on a 95°F day for R-22. This will be covered later.

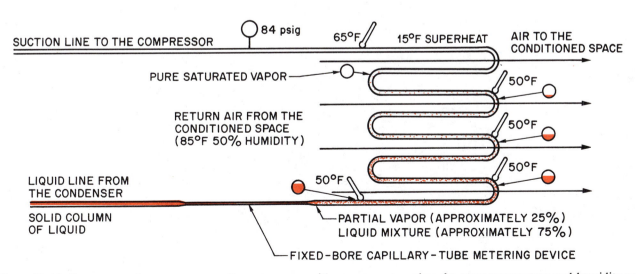

Figure 42–15. Pressures and temperatures as they may occur with an evaporator when the space temperature and humidity are above design conditions.

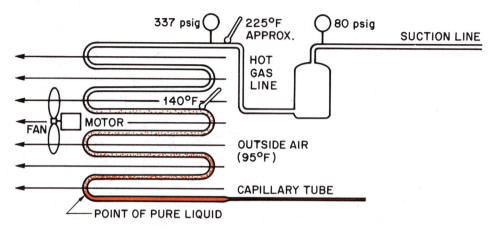

Figure 42–16. Condenser operating in design condition as far as the outdoor ambient air is concerned, but the pressure is high because the evaporator is under a load that is above design.

Other Than Design Conditions for High-Efficiency Equipment

1. When the unit has been off long enough for the load to build up, the space temperature and humidity are above design conditions, and the high-efficiency system pressures will be higher than normal, as would the standard-efficiency system. With standard equipment, when the return-air temperature is 75°F and the humidity is about 50%, the refrigerant boils at about 40°F. This is a temperature difference of 35°F. A high-efficiency evaporator is larger, and the refrigerant may boil at a temperature difference of 30°F, or around 55°F when the space temperature is 85°F, Figure 42–18. The exact boiling temperature relationship depends on the manufacturer and how much coil surface area was selected.

2. The high-efficiency condenser, like the evaporator, will operate at a higher pressure when the load is increased. The head pressure will not be as high as it would with a standard-efficiency condenser because the condenser has extra surface area.

The capacity of high-efficiency systems is not up to the rated capacity when the outdoor temperature is much below design. The earlier example of the family that shut off the air conditioning system before going to work and then turned it back on when they came home from work will be much worse with a high-efficiency system. The fact that the condenser became too efficient at night when the air was cooler with a standard condenser will be much more evident with the larger high-efficiency condenser. Most manufacturers have two-speed condenser fans so that a lower fan

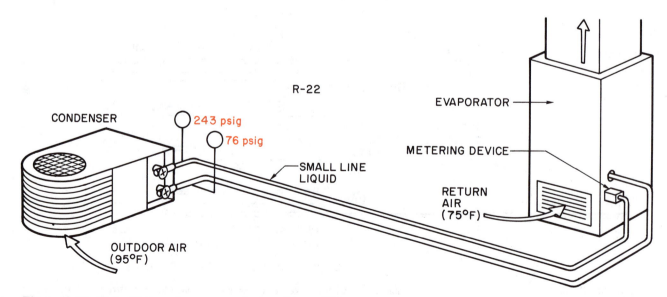

Figure 42–17. High efficiency system operating conditions, 45°F evaporator temperature and 115°F condensing temperature.

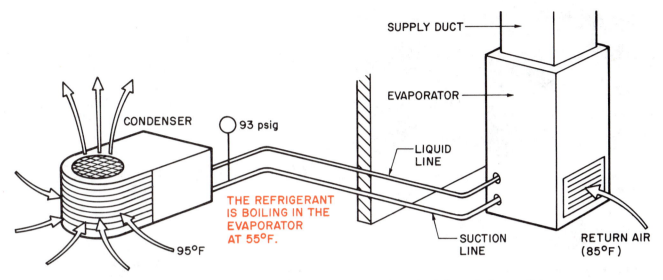

Figure 42-18. Evaporator in a high-efficiency system with a load above design conditions.

speed can be used in mild weather to help compensate for this temperature difference. The service technician that tries to analyze a component of high-efficiency equipment on a mild day will find the head pressure low. This will cause the suction pressure to also be low. Using the coil-to-air relationships for the condenser and the evaporator will help determine the correct pressures and temperatures.

Remember these two statements:

1. The evaporator absorbs heat, which is related to its operating pressures and temperatures.
2. The condenser rejects heat and has a predictable relationship to the load and ambient temperature.

42.10 TYPICAL ELECTRICAL OPERATING CONDITIONS

The electrical operating conditions are measured with a volt-ohmmeter and an ammeter. Three major power-consuming devices may have to be analyzed from time to time: the indoor fan motor, the outdoor fan motor, and the compressor. The control circuit is considered a separate function.

The starting point for considering electrical operating conditions is to know what the system supply voltage is supposed to be. For residential units, 230 V is the typical voltage. Light commercial equipment will nearly always use 208 V or 230 V single phase or 3 phase. Single and 3 phase may be obtained from a 3-phase power supply. The equipment rating may be 208/230 V. The reason for the two different ratings is that 208 V is the supply voltage that some power companies provide, and 230 V is the supply voltage provided by other power companies. Some light commercial equipment may have a supply voltage of 460 V three phase if the equipment is at a large commercial installation. For example, an office may have a 3- or 4-ton air conditioning unit that operates separately from the main central system. If the supply voltage is 460 V, the small unit may operate from the same power supply. When 208/230-V equipment is used at a commercial installation, the compressor may be three phase, and the fan motors single phase. The number of phases that the power company furnishes makes a difference in the method of starting the compressor. Single-phase compressors may have a start assist, such as the PTC device or a start relay and start capacitor. Three-phase compressors will have no start-assist accessories.

42.11 MATCHING THE UNIT TO THE CORRECT POWER SUPPLY

The typical operating voltages for any air conditioning system must be within the manufacturer's specifications. This is ±10% of the rated voltage. For the 208/230-V motor the minimum allowable operating voltage would be $208 \times 0.90 = 187.2$ V. The maximum allowable operating voltage would be $230 \times 1.10 = 253$ V. Notice that the calculations used the 208 V for the base for figuring the low voltage and the 230 V rating for calculating the highest voltage. This is because this application is 208/230 V. If the motor is rated at 208 V or 230 V alone, that value (208 V or 230 V) is used for evaluating the voltage. The equipment may be started under some conditions that are beyond the rated conditions. The technician must use some judgment. For example, if 180 V is measured, the motor should not be started, because the voltage will drop further. If the voltage reads 260 V, the motor may be started, because the voltage may drop slightly when the motor is started. If the voltage drops to within the limits, the motor is allowed to run.

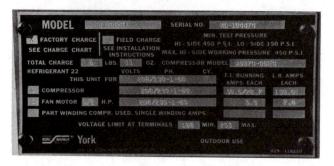

Figure 42-19. Condensing unit nameplate. *Photo by Bill Johnson*

42.12 STARTING THE EQUIPMENT WITH THE CORRECT DATA

When the correct rated voltage is known and the minimum and maximum voltages are determined, the equipment may be started if the voltages are within the limits. The three motors—indoor fan, outdoor fan, and the compressor—can be checked for the correct current draw.

The indoor fan is building the air pressure to move the air through the duct work, filters, and grilles to the conditioned space. By law the voltage characteristics must be printed on the motor in such a manner that they will not come off. This information may be printed on the motor, but the motor might be mounted so that they cannot be easily seen. In come cases the motor may be inside the squirrel cage blower. If so, removing the motor is the only way of determining the fan current. When the supplier can be easily contacted, you may obtain the information there. If the motor's electrical characteristics cannot be obtained from the nameplate, you might be able to get it from the unit nameplate. However, often the fan motor has been changed to a larger motor for more fan capacity.

42.13 FINDING A POINT OF REFERENCE FOR AN UNKNOWN MOTOR RATING

When a motor is mounted so that the electrical characteristics cannot be determined, you must improvise. We know that air conditioning systems normally move about 400 cfm of air per ton. This can help you determine the amperage of the indoor fan motor by comparing the fan amperage on an unknown system to the amperage of a known system. All you need is the approximate system capacity. As discussed earlier, you can find this by comparing compressor amperages of the unit in question and a known unit. For example, the compressor amperage of a 3-ton system is about 17 A. If you notice that the amperage of the compressor on the system you're checking is 17 A, you can assume that the system is close to 3 ton. The fan motor for a 3-ton system should be about $\frac{1}{3}$ hp for a typical duct system. The fan motor amperage for a $\frac{1}{3}$-hp PSC motor is 3.6 A. If the fan in question were pulling 5 A, suspect a problem.

Fan motors may be shipped in a warm-air furnace and may have been changed if air conditioning was added at a later date. In this case the furnace nameplate may not give the correct fan motor data. The condensing unit will have a nameplate for the condenser fan motor. This motor should be sized fairly close to its actual load and should pull close to nameplate amperage.

42.14 DETERMINING THE COMPRESSOR'S RUNNING AMPERAGE

The compressor current draw may not be as easy to determine as the fan motor current draw because compressor manufacturers do not all stamp the compressor full load amperage (FLA) rating on the compressor nameplate. There are so many different compressor sizes it is hard to state the correct full load amperage. For example, motors normally come in the following increments: 1, $1\frac{1}{2}$, 2, 3, and 5 hp. A unit rated at 34,000 Btu/hr is called a 3-ton unit, although it actually takes 36,000 Btu/hr to be a 3-ton unit. Ratings that are not completely accurate are known as *nominal ratings*. They are rounded off to the closest rating. A typical 3-ton unit would have a 3-hp motor. A unit with a rating of 34,000 Btu/hr does not need a full 3-hp motor, but it is supplied with one because there is no standard horsepower motor to meet its needs. If the motor amperage for a 3-hp motor were stamped on the compressor, it could cause confusion because the motor may never operate at that amperage. A unit nameplate shows complete electrical information, Figure 42-19. ✳The manufacturer may now stamp the compressor FLA on the unit nameplate. This amperage should not be exceeded.✳

42.15 COMPRESSORS OPERATING AT FULL-LOAD CURRENT

It is rare for a compressor motor to operate at its FLA rating. If design or above-design conditions were in effect, the compressor would operate at close to full load. When the unit is operating at a condition greater than design, such as when the unit has been off for some time in very hot weather, it might appear that the compressor is operating at more than FLA. However, there are usually other conditions that keep the compressor from drawing too much current. Remember, the compressor is pumping vapor and it is very light. It takes a substantial increase in pressure difference to create more work load.

42.16 HIGH VOLTAGE, THE COMPRESSOR, AND CURRENT DRAW

A motor operating at a voltage higher than the voltage rating of the motor is a condition that will prevent the motor from drawing too much current. A motor rated at 208/230 V has an ampere rating at some value between 208 and 230 V. Therefore, if the voltage is 230 V, the amperage may be lower than the nameplate amperage even during overload. The compressor motor may be larger than needed and may not reach its rated horsepower until it gets to the maximum rating for the system. It may be designed to operate at 105°F or 115°F outdoor ambient for very hot regions, but when the unit is rated at 3 tons, it would be rated down at the higher temperatures. The unit nameplate may contain the compressor's amperage at the highest operating condition that the unit is rated for, 115°F ambient temperature.

42.17 CURRENT DRAW AND THE TWO-SPEED COMPRESSOR

Some air conditioning manufacturers use two-speed compressors to achieve better seasonal efficiencies. These compressors may use a motor capable of operating as a two-pole or a four-pole motor. Remember that a four-pole motor runs at 1800 rpm, and a two-pole motor runs at 3600 rpm. A motor control circuit will slow the motor to half-speed for mild weather.

Variable speed may be obtained with electronic circuits. The system's efficiency is greatly improved because there is no need to stop the compressor and restart it at the beginning of the next cycle. However, this means more factors for a technician to consider. Manufacturer's literature must be consulted to establish typical running conditions under other than design conditions. The equipment should perform just as typical high-efficiency equipment would when operating at design conditions.

SAFETY PRECAUTION: *The technician must use all measure of caution while observing equipment for the operating conditions. Electrical, pressure and temperature readings are among the few things that must be explored while observing the equipment. Many times the readings must be made while the equipment is in operation.*

SUMMARY

- The technician must be familiar with mechanical and electrical operating conditions.
- Mechanical conditions deal with pressures and temperatures.
- The inside conditions that vary the load on the system are the space temperature and humidity.
- When the discharge pressure rises, the system capacity decreases.
- When suction pressure rises, head pressure also rises.
- When head pressure rises, suction pressure also rises.

- High-efficiency systems often use larger or oversized evaporators, and the refrigerant will boil at a temperature of about 45°F at typical operating conditions. This is a temperature difference of 30°F compared to the return air.
- High-efficiency equipment has a lower operating head pressure than standard-efficiency equipment partly because the condenser is larger. It has a different relationship to the outside ambient temperature, normally 20° to 25°F.
- Electrical instruments used to check working conditions are the ammeter and the voltmeter.
- The first thing the technician needs to know for electrical troubleshooting is what the operating voltage for the unit is supposed to be.
- Typical voltages are 230 V single phase for residential and 208/230 V single or three phase, or 460 V three phase for commercial installations.
- Equipment manufacturer's require operating voltages of ±10% of the rated voltage of the equipment.
- *The unit may have the compressor's FLA printed on it. It should **not** be exceeded.*
- Some manufacturers are now using two-speed motors and variable-speed motors for variable capacity.

REVIEW QUESTIONS

1. What is the standard condition at which air conditioning equipment is designed to operate?
2. What is the typical temperature relationship between a standard air-cooled condenser and the ambient temperature?
3. What is the temperature relationship between a high-efficiency condenser and the ambient temperature?
4. How is high-efficiency obtained with a condenser?
5. How does the temperature relationship between the condenser and the ambient temperature change as the ambient temperature changes?
6. What is the temperature relationship between the boiling refrigerant temperature and the entering air temperature with a standard-efficiency evaporator?
7. What does the head pressure do if the suction pressure rises?
8. What will cause the suction pressure to rise?
9. What happens to the suction pressure if the head pressure rises?
10. What can cause the head pressure to rise?
11. What two instruments are used to troubleshoot the system when an electrical problem is suspected?
12. What is the common voltage supplied to a residence?
13. How can an unknown current be found for a fan motor?
14. What are the typical motor horsepower sizes?
15. Name two methods that manufacturers use to vary the capacity of residential and small commercial compressors.

TROUBLESHOOTING

OBJECTIVES

After studying this unit, you should be able to

- determine the difference between mechanical and electrical problems.
- select the correct instruments for checking an air conditioning unit with a mechanical problem.
- determine the standard operating suction pressures for both standard- and high-efficiency equipment.
- calculate the correct operating suction pressures for both standard- and high-efficiency air conditioning equipment under conditions other than design.
- calculate the standard operating discharge pressures at various ambient conditions.
- select the correct instruments to troubleshoot electrical problems in an air conditioning system.
- check the line- and low-voltage power supplies.
- troubleshoot basic electrical problems in an air conditioning system.
- use an ohmmeter to check the various components of the electrical system.

43.1 TROUBLESHOOTING CATEGORIES

Troubleshooting air conditioning equipment involves troubleshooting mechanical and electrical systems since they have trouble symptoms that overlap. For example, if an evaporator fan motor capacitor fails, the motor will slow down and begin to get hot. It may even get hot enough to stop on internal overload. While it is running slowly, the suction pressure will go down and give symptoms of a restriction or low charge. If the technician diagnoses the problem based on suction pressure readings only, he or she will make the wrong decision.

43.2 MECHANICAL TROUBLESHOOTING

Mechanical troubleshooting is done with gages and temperature-testing equipment. The gage used is the same as that discussed in the units on refrigeration—namely, a *gage manifold,* Figure 43–1. This manifold has the suction or low-side gage on the left side of the manifold and the discharge gage or the

Figure 43–1. Low-side gage on the left and high-side gage on the right. *Courtesy Robinair Division-Sealed Power Corporation*

high-side gage on the right side. The most common refrigerant used in air conditioning is R-22. An R-22 temperature chart is printed on each gage for determining the saturation temperature for the low and high pressure sides of the system. Since these same gages are used for refrigeration, there is a temperature scale for R-12 and R-502 printed on the gage also, Figure 43–2. The R-12 scale is not used for air conditioning unless the equipment is so old that R-12 is the refrigerant. R-12 is the common refrigerant for automobile air conditioning but it is not discussed in this book.

43.3 GAGE MANIFOLD USAGE

The gage manifold determines the low- and high-side pressures while the unit is operating. These pressures can be converted to the saturation temperatures for the evaporating (boiling) refrigerant and the condensing refrigerant by using the pressure and temperature relationship. Remember, the low-side pressure can be converted to the boiling temperature. Let's say the boiling temperature for a system is supposed to be very close to 40°F, which converts to 69 psig for R-22. The

Figure 43-2. Gages have common refrigerant temperature relationships for R-12, R-22, and R-502 printed on them. *Photo by Bill Johnson*

superheat at the evaporator should be close to 10°F at this time. It is very hard to read the suction pressure at the evaporator because it normally has no gage port. Therefore, the technician takes the pressure and temperature readings at the condensing unit suction line to determine the system performance. Guidelines for checking the superheat at the condensing unit will be discussed later in this unit. If the suction pressure were 48 psig, the refrigerant would be boiling at about 24°F, which is cold enough to freeze the condensate on the evaporator coil and too low for continuous operation. The probable causes of low boiling temperatures are low charge or restricted airflow, Figure 43-3.

SAFETY PRECAUTION: *You must be extremely careful while installing manifold gages. High-pressure refrigerant will injure your skin and eyes. Wearing goggles and gloves are recommended. The danger from attaching the high-pressure gages can be reduced by shutting off the unit and allowing the pressures to equalize to a lower pressure. Liquid R-22 can cause serious frostbite since it boils at −44°F at atmospheric pressure.*

43.4 HIGH-SIDE GAGE

The high-side gage may be used in a similar manner. For example, suppose the high-side gage reads 278 psig and the outside ambient temperature is 80°F. (Ambient air is the air entering the condenser.) The head pressure seems too high. The gage manifold chart shows that the condenser is condensing at 125°F. However, a condenser should not condense at a temperature more than 35°F higher than the ambient temperature. The ambient temperature is 80°F + 35°F = 115°F, so the condensing temperature is actually 10°F too high. Probable causes are a dirty condenser or an overcharge of refrigerant.

The gage manifold is used whenever the pressures need to be known. Two types of pressure connections are used with air conditioning equipment: the Schrader valve and the service valve, Figure 43-4. The *Schrader valve* is a pressure connection only. The *service valve* can be used to isolate the system for service. It may have a Schrader connection for the gage port and a service valve for isolation, Figure 43-5.

43.5 WHEN TO CONNECT THE GAGES

When servicing small systems, a gage manifold should not be connected every time the system is serviced. A small amount of refrigerant escapes each time the gage is connected. Some residential and small commercial systems have a very critical refrigerant charge. When the high-side line is connected, high-pressure refrigerant will condense in the gage line. The refrigerant will escape when the gage line is disconnected from the Schrader connector. A gage line full of liquid refrigerant lost while checking pressures may be enough to affect the system charge. A short gage line connector for the high side will help prevent refrigerant loss, Figure 43-6. This can be used only to check pressure. You cannot use it to transfer refrigerant out of the system because it is not a manifold.

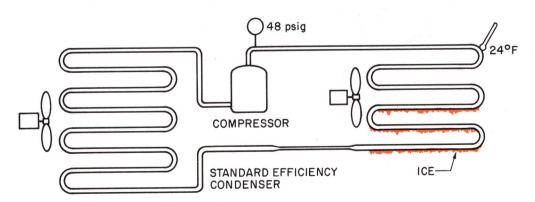

Figure 43-3. This coil was operated below freezing until the condensate on the coil froze.

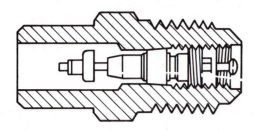

Figure 43–5. This service valve has a Schrader port instead of a back seat like a refrigeration service valve. *Photo by Bill Johnson*

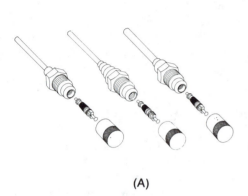

(A)

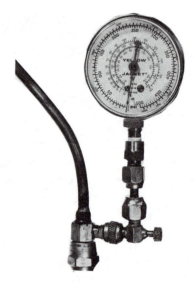

Figure 43–6. This short service connection is used on the high-side gage port to keep too much refrigerant from condensing in the gage line. *Photo by Bill Johnson*

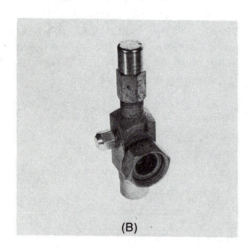

(B)

Figure 43–4. Two valves commonly used when taking pressure readings on modern air conditioning equipment. *Courtesy (A) J/B Industries (B) Photo by Bill Johnson*

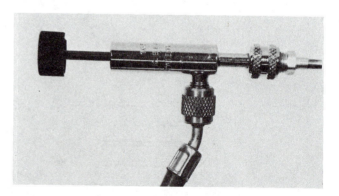

Figure 43–7. Hand valve used to press the Schrader valve stem and control the pressure. *Photo by Bill Johnson*

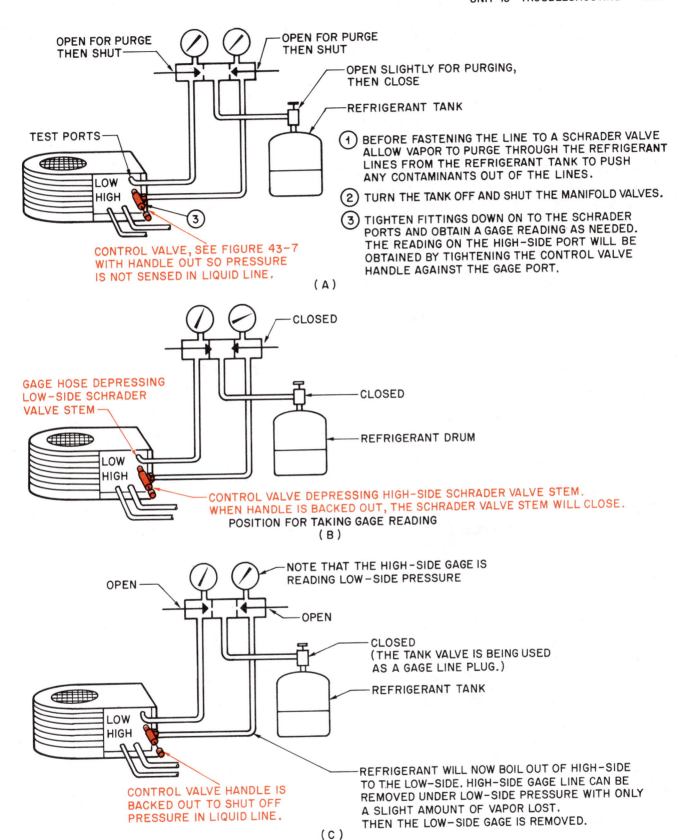

OPEN FOR PURGE THEN SHUT

OPEN FOR PURGE THEN SHUT

OPEN SLIGHTLY FOR PURGING, THEN CLOSE

REFRIGERANT TANK

TEST PORTS

LOW
HIGH

① BEFORE FASTENING THE LINE TO A SCHRADER VALVE ALLOW VAPOR TO PURGE THROUGH THE REFRIGERANT LINES FROM THE REFRIGERANT TANK TO PUSH ANY CONTAMINANTS OUT OF THE LINES.

② TURN THE TANK OFF AND SHUT THE MANIFOLD VALVES.

③ TIGHTEN FITTINGS DOWN ON TO THE SCHRADER PORTS AND OBTAIN A GAGE READING AS NEEDED. THE READING ON THE HIGH-SIDE PORT WILL BE OBTAINED BY TIGHTENING THE CONTROL VALVE HANDLE AGAINST THE GAGE PORT.

CONTROL VALVE, SEE FIGURE 43-7 WITH HANDLE OUT SO PRESSURE IS NOT SENSED IN LIQUID LINE.

(A)

CLOSED

GAGE HOSE DEPRESSING LOW-SIDE SCHRADER VALVE STEM

CLOSED

LOW
HIGH

REFRIGERANT DRUM

CONTROL VALVE DEPRESSING HIGH-SIDE SCHRADER VALVE STEM. WHEN HANDLE IS BACKED OUT, THE SCHRADER VALVE STEM WILL CLOSE.

POSITION FOR TAKING GAGE READING

(B)

NOTE THAT THE HIGH-SIDE GAGE IS READING LOW-SIDE PRESSURE

OPEN

OPEN

CLOSED
(THE TANK VALVE IS BEING USED AS A GAGE LINE PLUG.)

REFRIGERANT TANK

LOW
HIGH

CONTROL VALVE HANDLE IS BACKED OUT TO SHUT OFF PRESSURE IN LIQUID LINE.

REFRIGERANT WILL NOW BOIL OUT OF HIGH-SIDE TO THE LOW-SIDE. HIGH-SIDE GAGE LINE CAN BE REMOVED UNDER LOW-SIDE PRESSURE WITH ONLY A SLIGHT AMOUNT OF VAPOR LOST. THEN THE LOW-SIDE GAGE IS REMOVED.

(C)

Figure 43-8. Method of pumping refrigerant condensed in the high-pressure line over into the suction line. (A) Purging manifold. (B) Taking pressure readings. (C) Removing liquid refrigerant from high-side gage line.

43.6 GAGE PORT VALVES FOR SCHRADER VALVE

Another method of connecting the gage manifold to the Schrader valve service port is by a small hand valve, which has a depressor for depressing the stem in the Schrader valve, Figure 43–7. The hand valve can be used to keep the refrigerant in the system when disconnecting the gages; follow these steps: (1) Turn the stem on the valve out; this allows the Schrader valve to close. (2) Make sure that a plug is in the center line of the gage manifold. (3) Open the gage manifold handles. This will cause the gage lines to equalize to the low side of the system and pull the liquid refrigerant out of the high-pressure gage line to the low side of the system. The only refrigerant that will be lost is an insignificant amount of vapor in the three gage lines. This vapor is at the suction pressure while the system is running and should be about 70 psig. *This procedure should only be done with a clean and purged gage manifold, Figure 43–8.*

43.7 LOW-SIDE GAGE READINGS

We recommend that you use the gage manifold on the low side of the system to relate the evaporating pressure to the normal evaporating pressure. This verifies that the refrigerant is boiling at the correct temperature for the low side of the system at some load condition. It was pointed out in Unit 42 that there are high-efficiency systems and low-efficiency systems, and that high-efficiency systems often have oversized evaporators. This makes the suction pressure slightly higher than normal. A standard-efficiency system usually has a refrigerant boiling temperature of about 35°F cooler than the entering air temperature at the standard operating condition of 75°F return air with a 50% humidity.

If the space temperature is 85°F and the humidity is 70%, the evaporator has an oversized load. It is absorbing an extra heat load, both sensible heat and latent heat, from the moisture in the air. You need to wait a sufficient time for the system to reduce the load before you can determine if the equipment is functioning correctly. Gage readings at this time will not reveal the kind of information that will verify the system performance unless there is a manufacturer's performance chart available.

43.8 HIGH-SIDE GAGE READINGS

Gage readings obtained on the high-pressure side of the system are used to check the relationship of the condensing refrigerant to the ambient air temperature. Standard air conditioning equipment condenses the refrigerant at no more than 35°F higher than the ambient temperature. For a 95°F entering air temperature, the head pressure should correspond to 95° + 35° = 130°F or a corresponding head pressure of 296 psig

for R-22 and 181 psig for R-12. If the head pressure shows that the condensing temperature is higher than this, something is wrong.

When checking the condenser entering air temperature, be sure to check the actual temperature. Don't take the weather report as the ambient temperature. Air conditioning equipment located on a black roof has solar influence from the air being pulled across the roof, Figure 43–9. If the condenser is located close to any obstacles, such as below a sundeck, air may be circulated back through the condenser and cause the head pressure to be higher than normal, Figure 43–10.

43.9 HIGH-EFFICIENCY CONDENSER PRESSURES

High-efficiency condensers perform the same duties as standard-efficiency condensers except they operate at lower pressures and condensing temperatures. High-efficiency condensers normally condense the refrigerant at a temperature as low as 20°F higher than the ambient temperature. On a 95°F day the head pressure corresponds to a temperature of 95° + 20° = 115°F, which is a pressure of 243 psig for R-22.

43.10 TEMPERATURE READINGS

Temperature readings also can be useful. The four-lead electronic temperature tester is a very workable instrument, Figure 43–11. It has small temperature leads that respond quickly to temperature changes. The leads can be easily attached to the refrigerant piping with electrical tape and insulated from the ambient temperature with short pieces of foam line insulation, Figure 43–12.

It is important that the lead be insulated from the ambient if the ambient temperature is different from the line temperature. This temperature lead is a great deal better and easier to use than the glass thermometers used in the past. It was almost impossible to get a true line reading by strapping a glass thermometer to a copper line, Figure 43–13.

43.11 TYPES OF TEMPERATURE READINGS

Temperatures vary from system to system. The technician must be prepared to record accurately these temperatures in order to evaluate the various types of equipment. Some technicians record temperature readings of various equipment under different conditions for future reference.

43.12 INLET AIR TEMPERATURES

The inlet air temperature to the evaporator may be necessary for a complete analysis of a system. A wet bulb reading for determining the humidity may be necessary. Such a reading can be obtained by using one

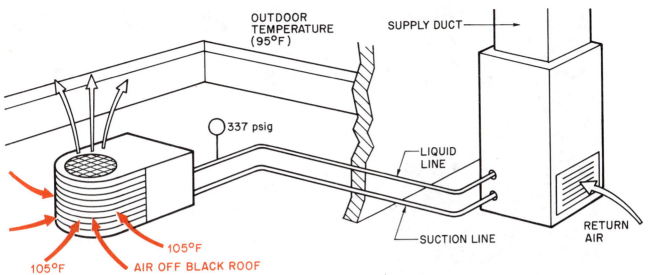

Figure 43-9. Condenser located low on a roof. Hot air from the roof enters it. A better installation would have the condenser mounted up about 20 in. so that the air could at least mix with the ambient air. Ambient air is 95°F but air off the roof is 105°F.

of the temperature leads with a cotton sock that has pure water on it. A wet bulb and a dry bulb reading may be obtained by placing a dry-bulb temperature lead next to a wet-bulb temperature lead in the return airstream, Figure 43-14. The velocity of the return air will be enough to accomplish the evaporation for the wet-bulb reading.

43.13 EVAPORATOR OUTLET TEMPERATURES

The evaporator outlet air temperature is seldom important. It may be obtained, however, in the same manner as the inlet air temperature. The outlet air temperature will normally be about 20°F less than the inlet air temperature. The temperature drop across an evaporator coil is about 20°F when it is operating at typical operating conditions of 75°F and 50% relative humidity return air. If the conditioned space temperature is high with a high humidity, the temperature drop across the same coil will be much less because of the latent-heat load of the moisture in the air.

If a WB reading is taken, there will be approximately a 10°F WB drop from the inlet to the outlet during standard operating conditions. The outlet humidity will

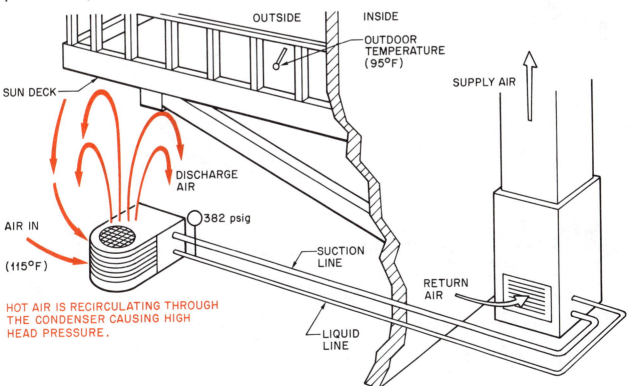

Figure 43-10. Condenser installed so that the outlet air is hitting a barrier in front of it.

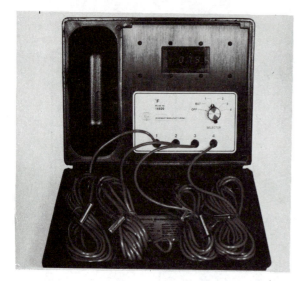

Figure 43-11. Four-lead temperature tester. *Photo by Bill Johnson*

Figure 43-13. Glass thermometer strapped to a refrigerant line. *Photos by Bill Johnson*

stopped up or if the evaporator coil is dirty, Figure 43-15. The suction gas may have a high superheat if the unit has a low charge or if there is a refrigerant restriction, Figure 43-16. The combination of suction-line temperature and pressure will help the service technician decide whether the system has a low charge or a stopped-up air filter in the air handler. For example, if the suction pressure is too low and the suction line is warm, the system has a starved evaporator, Figure 43-17. If the suction-line temperature is cold and the pressure indicates that the refrigerant is boiling at a low temperature, the coil is not absorbing heat as it should. The coil may be dirty, or there may not be enough airflow, Figure 43-18. The cold suction line indicates that the unit has enough charge, because the evaporator must be full for the refrigerant to get back to the suction line, Figure 43-19.

be almost 90%. Remember that this air is going to mix with the room air and will soon drop in humidity because it will expand to the room air temperature. It is very humid because it is contracted or shrunk.

SAFETY PRECAUTION: *The temperature lead must not be allowed to touch the moving fan while taking air temperature readings.*

43.14   SUCTION-LINE TEMPERATURES

The temperature of the suction line returning to the compressor and the suction pressure will tell the technician what the characteristics of the suction gas are. The suction gas may be part liquid if the filters are

43.15   DISCHARGE-LINE TEMPERATURES

The temperature of the discharge line may tell the technician that something is wrong inside the compres-

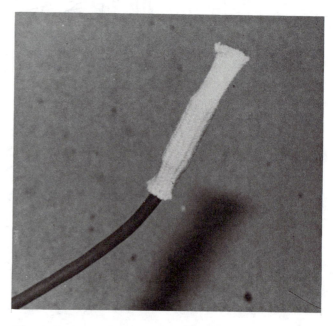

Figure 43-14. Electronic thermometer's temperature lead has a damp cotton sock wrapped around it to convert it to a wet-bulb lead. *Photo by Bill Johnson*

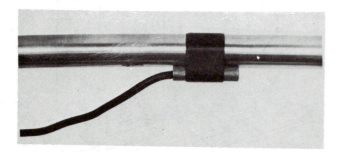

Figure 43-12. Temperature lead attached to a refrigerant line in the correct manner. It must also be insulated. *Photo by Bill Johnson*

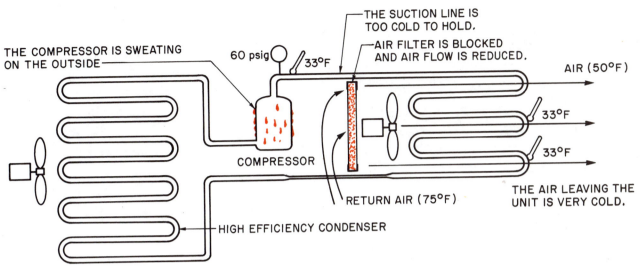

Figure 43–15. Evaporator flooded with refrigerant.

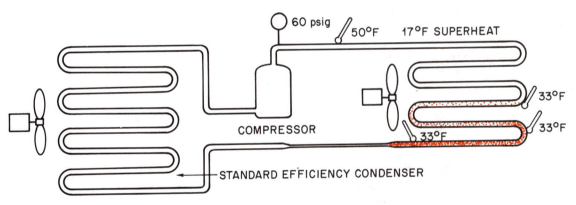

Figure 43–16. Evaporator is starved for refrigerant because of a low-refrigerant charge.

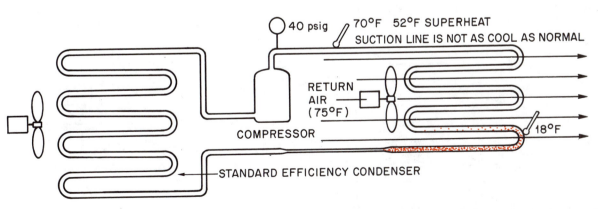

Figure 43–17. If the suction pressure is too low, and the suction line is not as cool as normal, the evaporator is starved for refrigerant. The unit may be low in refrigerant charge.

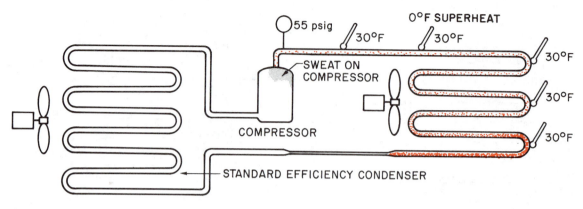

Figure 43–18. When the suction pressure is too low and the superheat is low, the unit is not boiling the refrigerant in the evaporator. The coil is flooded with liquid refrigerant.

sor. If there is an internal refrigerant leak from the high-pressure side to the low-pressure side, the discharge gas temperature will go up, Figure 43–20. Normally the discharge-line temperature at the compressor should not exceed 220°F for an air conditioning application even in very hot weather. When a high discharge-line temperature is discovered, the probable cause is an internal leak. The technician can prove this by building up the head pressure as high as 300 psig and then shutting off the unit. If there is an internal leak, this pressure difference between the high and the low sides can often be heard (as a whistle) equalizing through the compressor. If the suction line at the compressor shell starts to warm up immediately, the heat is coming from the discharge of the compressor.

SAFETY PRECAUTION: *The discharge line of a compressor may be as hot as 220°F under normal conditions, so be careful while attaching a temperature lead to this line.*

43.16 LIQUID-LINE TEMPERATURES

Liquid-line temperature may be used to check the subcooling efficiency of a condenser. Most condensers will subcool the refrigerant to between 10° and 20°F

below the condensing temperature of the refrigerant. If the condensing temperature is 130°F on a 95°F day, the liquid line leaving the condenser may be 110° to 120°F when the system is operating normally. If there is a slight low charge, there might not be as much subcooling and the system efficiency therefore will not be as good. Remember, the condenser performs three functions: (1) It removes the superheat from the discharge gas. (2) It condenses the liquid refrigerant. (3) It subcools the liquid refrigerant below the condensing temperature. All three of these must be successfully accomplished for the condenser to have its rated capacity.

It's a good idea for a new technician to take the time to completely check out a working system operating at the correct pressures. Apply the temperature probes and gages to all points to actually verify the readings. This will provide reference points to remember.

43.17 CHARGING PROCEDURES IN THE FIELD

A field charging procedure may be used to check the charge of some typical systems. The technician sometimes needs typical reference points to add small amounts of gas for adjusting the amount of refrigerant in equipment that has no charging directions.

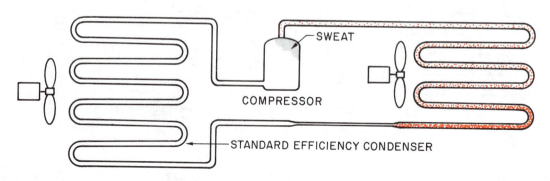

Figure 43–19. Evaporator full of refrigerant.

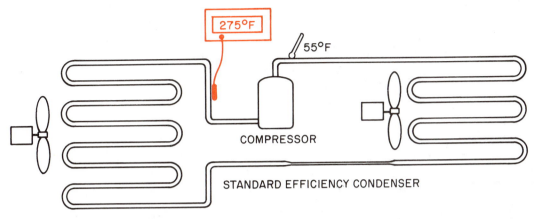

Figure 43–20. Temperature tester attached to the discharge line on this compressor. If the compressor has an internal leak the hot discharge gas will circulate back through the compressor. The discharge gas will be abnormally hot. When a compressor is cooled by suction gas, a high superheat will cause the compressor discharge gas to be extra hot.

43.18   FIXED-BORE METERING DEVICES—CAPILLARY TUBE AND ORIFICE TYPE

Fixed-bore metering devices like the capillary tube do not throttle the refrigerant as the thermostatic expansion valve does. They allow refrigerant flow based on the difference in the inlet and the outlet pressures. The one time when the system can be checked for the correct charge and everything will read normal is at the typical operating condition of 75°F, 50% humidity return air and 95°F outside ambient air. If other conditions exist, different pressures and different superheat readings will occur. The item that most affects the readings is the outside ambient temperature. When it is lower than normal, the condenser will become more efficient and will condense the refrigerant sooner in the coil. This will have the effect of partially starving the evaporator for refrigerant. Refrigerant that is in the condenser that is supposed to be in the evaporator starves the evaporator.

When you need to check the system for correct charge or to add refrigerant, the best method is to follow the manufacturer's instructions. If they are not available, the typical operating condition may be simulated by reducing the airflow across the condenser to cause the head pressure to rise. On a 95°F day the highest condenser head pressure is usually 296 psig (95°F ambient + 35°F added condensing temperature difference = 130°F condensing temperature or 296 psig for R-22). Since the high pressure pushes the refrigerant through the metering device, when the head pressure is up to the high normal end of the operating conditions there is no refrigerant held back in the condenser.

The evaporator side of this procedure requires some explanation. When the condenser is pushing the refrigerant through the metering device at the correct rate, the remainder of the charge must be in the evaporator. A superheat check at the evaporator is not always easy with a split air conditioning system, so a superheat check at the condensing unit for a split system will be described. The suction line from the evaporator to the condensing unit may be long or short. Let's use two different lengths: up to 30 ft, and from 30 to 50 ft for a test comparison. When the system is correctly charged, the superheat should be 10° to 15°F at the condensing unit with a line length of 10 to 30 ft. The superheat should be 15° to 18°F when the line is 30 to 50 ft long. Both of these conditions are with a head pressure of 296 psig ± 10 psig. At these conditions the actual superheat at the evaporator will be close to the correct superheat of 10°F. When using this method, be sure that you allow enough time for the system to settle down after adding refrigerant, before you draw any conclusions, Figure 43–21.

43.19   FIELD CHARGING THE THERMOSTATIC EXPANSION VALVE SYSTEM

The thermostatic expansion valve system can be charged in much the same way as the fixed-bore system, with some modifications. The condenser on a thermostatic expansion valve system will also hold refrigerant back in mild ambient conditions. This system always has a refrigerant reservoir or receiver to store refrigerant and will not be affected as much as the capillary tube by lower ambient conditions. To check the charge, restrict the airflow across the condenser until the head pressure stimulates a 95°F ambient, 296 psig head pressure for R-22. Using the superheat method will not work for this valve because if there is an overcharge, the superheat will remain the same. Superheats of 15° to 18°F are not unusual when measured at the condensing unit for thermostatic expansion valves. If the sight glass is full, the unit has at least enough refrigerant, but it may have an overcharge. If the unit

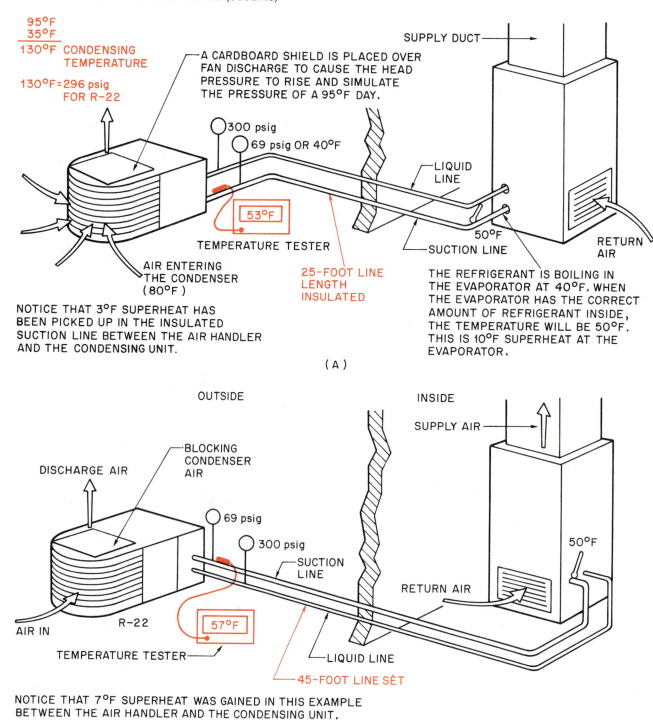

95°F
35°F
130°F CONDENSING
TEMPERATURE

130°F=296 psig
FOR R-22

A CARDBOARD SHIELD IS PLACED OVER
FAN DISCHARGE TO CAUSE THE HEAD
PRESSURE TO RISE AND SIMULATE
THE PRESSURE OF A 95°F DAY.

SUPPLY DUCT

300 psig
69 psig OR 40°F

LIQUID
LINE

53°F

TEMPERATURE TESTER

50°F

SUCTION LINE

RETURN
AIR

AIR ENTERING
THE CONDENSER
(80°F)

25-FOOT LINE
LENGTH
INSULATED

NOTICE THAT 3°F SUPERHEAT HAS
BEEN PICKED UP IN THE INSULATED
SUCTION LINE BETWEEN THE AIR HANDLER
AND THE CONDENSING UNIT.

THE REFRIGERANT IS BOILING IN
THE EVAPORATOR AT 40°F. WHEN
THE EVAPORATOR HAS THE CORRECT
AMOUNT OF REFRIGERANT INSIDE,
THE TEMPERATURE WILL BE 50°F.
THIS IS 10°F SUPERHEAT AT THE
EVAPORATOR.

(A)

OUTSIDE

INSIDE

SUPPLY AIR

BLOCKING
CONDENSER
AIR

DISCHARGE AIR

69 psig

300 psig

SUCTION
LINE

50°F

R-22

57°F

RETURN AIR

TEMPERATURE TESTER

LIQUID LINE

AIR IN

45-FOOT LINE SET

NOTICE THAT 7°F SUPERHEAT WAS GAINED IN THIS EXAMPLE
BETWEEN THE AIR HANDLER AND THE CONDENSING UNIT.

(B)

Figure 43-21. System charged by raising the discharge pressure to simulate a 95°F day.

does not have a sight glass, a measure of the subcooling of the condenser may tell you what you want to know. For example, a typical subcooling circuit will subcool the liquid refrigerant from 10° to 20°F cooler than the condensing temperature. A temperature lead attached to the liquid line should read 120° to 110°F, or 10° to 20°F cooler than the condensing temperature

of 130°F, Figure 43–22. If the subcooling temperature is 20° to 25°F cooler than the condensing temperature, the unit has an overcharge of refrigerant and the bottom of the condenser is acting as a large subcooling surface.

The charging procedures just described will also work for high-efficiency equipment. The head pressure does not need to be operated quite as high. A head

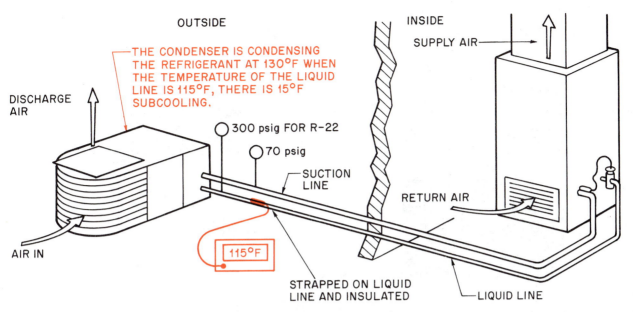

OUTSIDE

THE CONDENSER IS CONDENSING
THE REFRIGERANT AT 130°F WHEN
THE TEMPERATURE OF THE LIQUID
LINE IS 115°F, THERE IS 15°F
SUBCOOLING.

DISCHARGE
AIR

300 psig FOR R-22

70 psig

SUCTION
LINE

AIR IN

115°F

STRAPPED ON LIQUID
LINE AND INSULATED

INSIDE

SUPPLY AIR

RETURN AIR

LIQUID LINE

Figure 43-22. Unit with a thermostatic expansion valve cannot be charged using the superheat method. The head pressure is raised to simulate a 95°F day and the temperature of the liquid line is checked for the subcooling level. A typical system will have 10 to 20°F of subcooling when the condenser contains the correct charge.

pressure of 250 psig will be sufficient for an R-22 system when charging.

SAFETY PRECAUTION: *Refrigerant in drums and in the system is under great pressure (as high as 350 psi for R-22). Use proper safety precautions when transferring refrigerant, and be careful not to overfill refrigerant drums. Do not use disposable drums by refilling them. The high-side pressure may be reduced for attaching and removing gage lines by shutting off the unit and allowing the unit pressures to equalize.*

43.20 ELECTRICAL TROUBLESHOOTING

Electrical troubleshooting is often performed at the same time as mechanical troubleshooting. The VOM (volt-ohm-milliammeter) and the clamp-on ammeter are the main instruments used, Figure 43-23.

As we've said, you need to know what the readings should be in order to know whether the existing readings on a particular unit are correct. This is often not easy to determine because the desired reading may not be furnished. Figure 42-12 shows typical horsepower to amperage ratings. It is a very valuable tool for determining the correct amperage for a particular motor.

Troubleshooting electrical components and circuits begins with the three things that must be present to have current flow: (1) a power source, (2) a path for the power to flow on, (3) a load to consume the power.

For a residence or small commercial building one main power panel serves the building. This panel is divided into many circuits. For a split-system there are usually separate breakers (or fuses) in the main panel

for the air handler or furnace for the indoor unit and for the outdoor unit. For a package or self-contained system there is usually one breaker (or fuse) to serve the unit, Figure 43-24. The power supply voltage is stepped down by the control transformer to the control voltage of 24 V.

43.21 VERIFYING THE POWER SUPPLY

Begin any electrical troubleshooting by verifying that the power supply is energized and that the voltage is correct. One way to do this is to go to the room thermostat and see if the indoor fan will start with the FAN ON switch. See Figure 43-25 for a wiring diagram of a typical split system air conditioner.

The air handler or furnace, where the low-voltage transformer is located, is frequently under the house or in the attic. This quick check with the FAN ON switch can save you a trip under the house or to the attic. If the fan will start with the fan relay, several things are apparent: (1) the indoor fan works; (2) there is control voltage; and (3) there is line voltage to the unit because the fan will run. When taking a service call over the phone, ask the homeowner if the indoor fan will run. If it doesn't, take a transformer. This could save a trip to the supply house or the shop.

SAFETY PRECAUTION: *All safety practices must be observed while troubleshooting electrical systems. Many times the system must be inspected while power is on. Treat an energized power supply like a snake—keep your distance. Only let the insulated meter leads touch the hot terminals. Special care should be taken while troubleshooting the main power supply

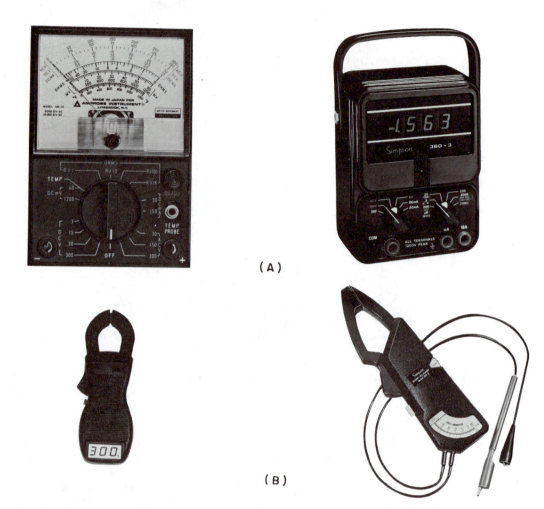

Figure 43–23. Instruments used to troubleshoot the electrical part of an air conditioning system. (A) VOM (volt-ohm-milliammeter) (B) Clamp-on ammeter. *Courtesy Amprobe Instrument Division of Core Industries Inc. and Simpson Electric Company*

because the fuses may be correctly sized large enough to allow great amounts of current to flow before they blow (for example, when a screwdriver slips in the panel and shorts across hot terminals). **Never** use a screwdriver in a hot panel.＊

43.22   COMPONENT SELECTION FOR TROUBLESHOOTING

If the power supply voltages are correct, move on to the various components. The path to the load may be the next item to check. If you're trying to get the compressor to run, remember that the compressor motor is operated by the compressor contactor. Is the contactor energized? Are the contacts closed? See Figure 43–26 for a diagram. Note that in the diagram the only thing that will keep the contactor coil from being energized is the thermostat, the path, or the low-pressure control. If the outdoor fan is operating and the com-

pressor is not, the contactor is energized because it also starts the compressor. If the fan is running and the compressor is not, either the path (wiring or terminals) are not making good contact or the compressor's internal overload is open.

43.23   COMPRESSOR OVERLOAD PROBLEMS

When the compressor's overload is open, touch the motor housing to see if it is hot. If you cannot hold your hand on the compressor shell, the motor is too hot. Ask yourself these questions: Can the charge be low (this compressor is suction-gas cooled)? or can the start-assist circuit not be working and the compressor not starting?

Allow the compressor to cool before restarting it. It is best that the unit be fixed so that it will not come back on for several hours. The best way to do this is to remove a low-voltage wire. If you pull the disconnect switch and come back the next day, the refrigerant charge may be in the crankcase because there was no crankcase heat. If you want to start the unit within the hour, pull the disconnect switch and run a small amount of water through a hose over the compressor. It will take about 30 min to cool. Have the gages on the unit and a drum of gas connected because when the compressor is started up by closing the disconnect, it may need refrigerant. *Remember the standing water poses a potential electrical hazard.* If the system has a low charge and you have to get set up to charge after starting the system, the compressor may cut off again from overheating before you have a chance to get the gages connected.

43.24 COMPRESSOR ELECTRICAL CHECKUP

You may need to do an electrical check of the compressor if the compressor will not start or if a circuit protector has opened. For example, suppose that the compressor can be heard trying to start. It will make a humming noise but will not turn. Check the compressor with an ohmmeter to see if all of the windings are correct. Remember, a load has to be present before current flows. The load must have the correct resistance. Let's say a compressor's specification calls for the run winding to have a resistance of 4 Ω and the start winding a resistance of 15 Ω. If the ohmmeter indicates that the start winding has only 10

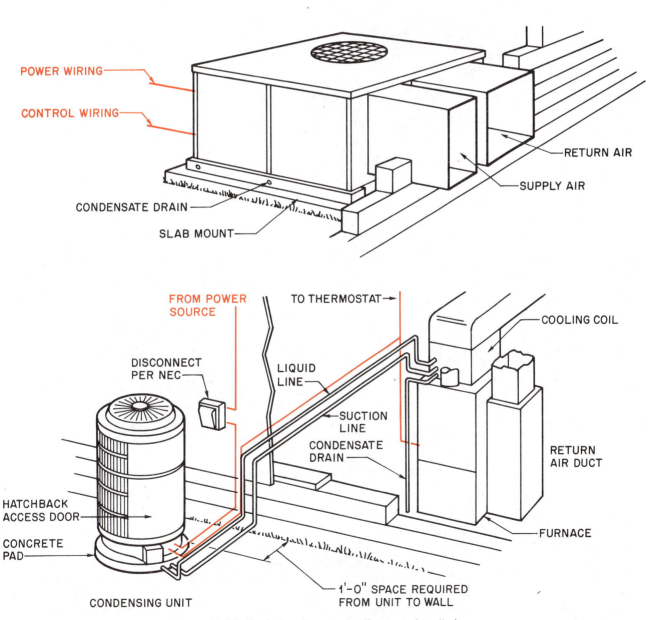

Figure 43-24. Typical package and split-system installation.

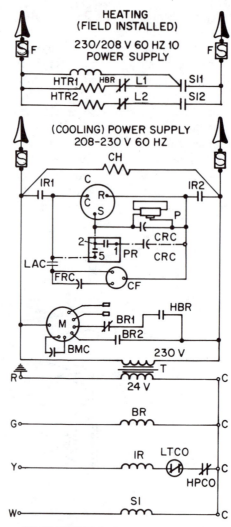

HEATING
(FIELD INSTALLED)
230/208 V 60 HZ 1Ø
POWER SUPPLY

(COOLING) POWER SUPPLY
208-230 V 60 HZ

COMPONENT PART IDENTIFICATION

BMC	BLOWER MOTOR CAPACITOR	HR	HEAT RELAY
BR	BLOWER RELAY	HTR	ELECTRIC HEATER
C	COMPRESSOR	IR	COMPRESSOR CONTACTOR
CF	CONDENSER FAN MOTOR	L	LIMIT
CH	CHRANKCASE HEATER	LAC	LOW AMBIENT CONTROL (0°F) (ALL
CRC	COMPRESSOR RUN CAPACITOR		RD/RG-D UNITS & ALL R-H RG-H3 PHASE
CS	EXHAUSTER CENTRIFUGAL SWITCH		UNITS)
CSC	COMPRESSOR START CAPACITOR	LTCO	LOW TEMPERATURE CUT-OUT (50°F)
EM	EXHAUSTER MOTOR		(ALL R-H/RG-H 1 PHASE UNITS)
F	FUSE	M	BLOWER MOTOR
FC	FAN CONTROL KLIXON	P	PTC STANDARD START ASSIST
FRC	FAN RUN CAPACITOR	PPK	POST PURGE KLIXON
FT	FAN TIMER	PR	POTENTIAL RELAY
GV	GAS VALVE	PR+	
HBR	BLOWER RELAY (HEATING)	CDC	OPTIONAL START ASSIST (REPLACES "P")
HPCO	HIGH-PRESSURE CUT-OUT	S1	SEQUENCER
		T	TRANSFORMER

Figure 43–25. Wiring diagrams of a split-system summer air conditioner. *Courtesy Climate Control*

Ω, then the winding has a short circuit (sometimes called a *shunt*). This will change the winding characteristics, and the compressor will not start. It is defective and must be changed. See Figure 43–27. It is **important** to have quality instruments for making these **checks.**

The ohmmeter check may show the compressor to **have** an open circuit in the start or run windings.

Suppose that the same symptom of a hot compressor is discovered. The compressor is allowed to cool and it still will not start. An ohm check shows that the start or run winding is open. This compressor is also defective and must be changed, Figure 43–28.

When the continuity check shows that the common circuit in the compressor is open, it could be that **the** internal overload is open because the motor may not

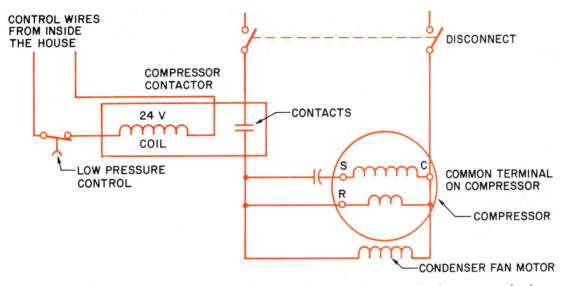

Figure 43–26. Wiring diagram of basic components that appear in a control and compressor circuit.

have cooled enough. Remember, the motor is suspended in a vapor space inside the shell, and it takes time to cool, Figure 43–29.

SAFETY PRECAUTION: *Never use an ohmmeter to check a live circuit.*

43.25 TROUBLESHOOTING THE CIRCUIT ELECTRICAL PROTECTORS—FUSES AND BREAKERS

One service call that must be treated cautiously is when a circuit protector such as a fuse or breaker opens the circuit. The compressor and fan motors have

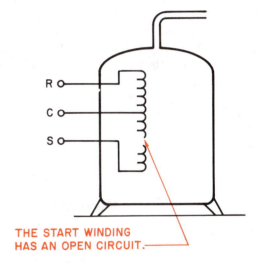

THE START WINDING HAS AN OPEN CIRCUIT.

Figure 43–28. Compressor with open start winding.

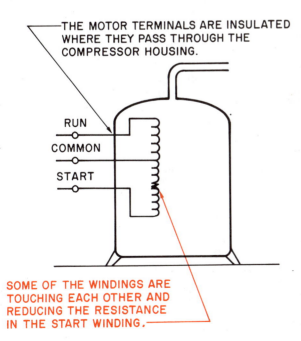

THE MOTOR TERMINALS ARE INSULATED WHERE THEY PASS THROUGH THE COMPRESSOR HOUSING.

SOME OF THE WINDINGS ARE TOUCHING EACH OTHER AND REDUCING THE RESISTANCE IN THE START WINDING.

Figure 43–27. Compressor with shunted winding.

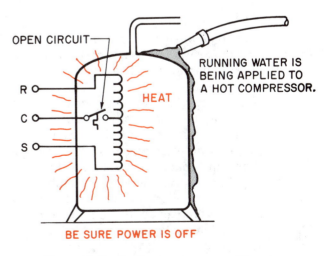

OPEN CIRCUIT

HEAT

RUNNING WATER IS BEING APPLIED TO A HOT COMPRESSOR.

BE SURE POWER IS OFF

Figure 43–29. Cooling a compressor with water.

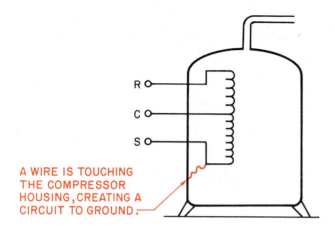

R

C

S

A WIRE IS TOUCHING THE COMPRESSOR HOUSING, CREATING A CIRCUIT TO GROUND

Figure 43-30. Compressor with winding shorted to the casing.

protection that will normally guard them from minor problems. The breaker or fuse is for large current surges in the circuit. When one is tripped, don't simply reset it. Perform a resistance check of the compressor section, including the fan motor. *The compressor might be grounded (has a circuit to the case of the compressor), and it will be harmful to try to start it.* **Be sure to isolate the compressor circuit before condemning the compressor.** Take the motor leads off the compressor to check for a ground circuit in the compressor, Figure 43-30.

43.26   SERVICE TECHNICIAN CALLS

Troubleshooting can take many forms and cover many situations. The following actual troubleshooting situations will help you understand what is going through the service technician's mind while solving actual problems.

Service Call 1

A residential customer calls. The central air conditioner at residence is not cooling enough and runs continuously. The problem is low refrigerant charge.

The technician arrives and finds the unit running. The temperature indicator on the room thermostat shows the thermostat is set at 72°F, and the thermometer on the thermostat shows the space temperature to be 80°F. The air feels very humid. The technician notices that the indoor fan motor is running. The velocity of the air coming out of the registers seems adequate, so the filters are not stopped up.

The technician goes to the condensing unit

and hears the fan running. The air coming out of the fan is not warm. This indicates that the compressor is not running. The door to the compressor compartment is removed and it is noticed that the compressor is hot to the touch. This is an indication that the compressor has been trying to run. The gages are installed on the service ports, and they both read the same because the system has been off. The gages read 144 psig, which corresponds to 80°F. The residence is 80°F inside, and the ambient is 85°F, so it can be assumed that the unit has some liquid refrigerant in the system because the pressure corresponds so closely to the chart. If the system pressure were 100 psig, it would be obvious that no liquid refrigerant is left in the system. A large leak should be suspected.

The technician decides that the unit must have a low charge and that the compressor is off because of an internal overload. *The technician pulls the electrical disconnect and takes a resistance reading across the compressor terminals with the motor leads removed.* The meter shows an open circuit from common to run and common to start. It shows a measurable resistance between run and start. This indicates that the motor winding thermostat must be open. This indication is verified by the hot compressor.

A water hose is connected, and a small amount of water is allowed to run over the top of the compressor shell to cool it. A tank of refrigerant is connected to the gage manifold so that when it is time to start the compressor, the technician will be ready to add gas and keep the compressor running. The gage lines are purged of any air that may be in them. While the compressor is cooling, the technician changes the air filters and lubricates the condenser and the indoor fan motors. After about 30 min the compressor seems cool. The water hose is removed, and the water around the unit is allowed a few minutes to run off. When the disconnect switch is closed, the compressor starts. The suction pressure drops to 40 psig. The normal suction pressure is 70 psig for the system because the refrigerant is R-22. Refrigerant is added to the system to bring the charge up to normal. Remember, the space temperature is 80°F, so the suction pressure will be higher than normal until the space temperature becomes normal. The technician must have a reference point to get the correct charge. The following reference points are used in this situation because there is no factory chart.

1. The outside temperature is 90°F; the normal operating head pressure should correspond to a temperature of 90° + 35° = 125°F or 260 psig. It should not exceed this when the space temperature is down to a normal 75°F. This is a standard-efficiency unit. The technician restricts the airflow to the condenser and causes the head pressure to rise to 295 psig as refrigerant is added.

2. The suction pressure should correspond to a temperature of about 35°F cooler than the space temperature of 80°F, 45°F, which corresponds to a pressure of 76 psig for R-22.

3. The system has a capillary tube metering device, so some conclusions can be drawn from the temperature of the refrigerant coming back to the compressor. A temperature tester lead is attached to the suction line at the condensing unit. As refrigerant is added, the technician notices that the refrigerant returning from the evaporator is getting cooler. The evaporator is about 30 ft from the condensing unit, and some heat will be absorbed into the suction line returning to the condensing unit. The technician uses a guideline of 15°F of superheat with a suction line this length. This assumes that the refrigerant leaving the evaporator has about 10°F of superheat and that another 5°F of superheat is absorbed along the line. When these conditions are reached, the charge is very close to correct and no more refrigerant is added.

A leak check is performed and a flare nut is found to be leaking. It is tightened and the leak is stopped. The technician loads the truck and leaves. A call later in the day shows that the system is working correctly.

Service Call 2

A residential customer calls. The unit has been cooling correctly until afternoon, then it just quit cooling. This is a residential unit that has been in operation for several years. The problem is a complete loss of charge.

The technician arrives at the residence and finds the thermostat to be set at 75°F and the space temperature to be 80°F. The air coming out of the registers feels the same as the return-air temperature. There is plenty of air velocity at the registers, so the filters appear to be clean.

The technician goes to the back of the house and finds that the fan and compressor in the condensing unit are not running. The breaker at the electrical box is in the ON position, so power must be available. A voltage check proves that the voltage is 235 V. A look at the wiring diagram shows that there should be 24 V between the C and Y terminals to energize the contactor. The voltage actually reads 25 V, slightly above normal, but so is the line voltage of 235 V. The conclusion is that the thermostat is calling for cooling, but the contactor is not energized. The only safety control in the contactor coil circuit is the no-charge protector, so the unit must be out of refrigerant.

Gages are fastened to the gage ports, and the pressure is 0 psig. The technician connects a drum of refrigerant and starts adding refrigerant. When the pressure is up to 30 psig, the refrigerant is stopped. Refrigerant can be heard leaking from the vicinity of the compressor suction line. A hole is found in the suction line where the cabinet had been rubbing against it. This accounts for the fact that the unit worked well up to a point and quit working almost immediately.

The refrigerant is allowed to escape from the system, and the hole is patched with silver solder. A liquid-line drier is installed in the liquid line, and a triple evacuation is performed to remove any contaminants that may have been pulled into the system. The system is charged and started. The technician follows the manufacturer's charging chart to assure the correct charge.

Service Call 3

A commercial customer calls. The unit is not cooling the small office building. The unit was operating and cooling yesterday afternoon when the office closed. The problem is someone turned the thermostat down to 55°F late yesterday afternoon, and the condensate on the evaporator froze solid overnight trying to pull the space temperature down to 55°F.

The technician arrives at the job and notices that the thermostat is set at 55°F. An inquiry shows that one of the employees was too warm in an office at the end of the building and turned down the thermostat to cool that office. The technician notices there is no air coming out of the registers. The air handler is located in a closet at the front of the building. Examination shows that the fan is running, and the suction line is frozen solid at the air handler.

The technician stops the compressor and leaves the evaporator fan operating by turning the heat-cool selector switch to OFF and the fan switch to ON. It is going to take a long time to thaw the evaporator, probably an hour. The technician leaves the following directions with the office manager.

1. Let the fan run until air comes out of the registers.
2. When air is felt at the registers, wait 30 min and turn the thermostat back to COOL to start the compressor.

The technician then looks in the ceiling to check the air damper on the duct run serving the back office of the employee who was not cool enough and turned down the thermostat. The damper is nearly closed; it probably was brushed against by a telephone technician who had been working in the ceiling. The damper is reopened.

A call later in the day proved the system was working again.

Service Call 4

A residential customer calls. The unit is not cooling correctly. The problem is a dirty condenser. The unit is in a residence and next to the side yard. The homeowner mows the grass, and the lawnmower throws grass on the condenser. This is the first hot day of summer, and the unit had been cooling the house until the weather became hot.

The technician arrives at the job and notices that the thermostat is set for 75°F but the space temperature is 80°F. There is plenty of slightly cool air coming out of the registers. The technician goes to the side of the house where the unit is located. The suction line feels cool, but the liquid line is very hot. An examination of the condenser coil shows that the coil is very clogged with grass and is dirty. *The technician shuts off the unit with the breaker at the unit and takes enough panels off to be able to spray coil cleaner on the coil.* A high detergent coil cleaner is applied to the coil and allowed to stand for about 15 min to soak into the coil dirt. While this is taking place, the motors are oiled and the filters are changed. *To keep the condenser fan motor from getting wet when the coil cleaner is washed off, it is covered with a plastic bag.*

A water hose with a nozzle to concentrate the water stream is used to wash the coil in the opposite direction of the airflow. One washing is not enough, so the coil cleaner is applied to the coil again and allowed to set for another 15 min. The coil is washed again and is now clean.

The unit is assembled and started. The suction line is cool and the liquid line is warm to the touch. The technician decides not to put gages on the system and leaves. A call back later in the day to the homeowner proves that the system is operating correctly.

Service Call 5

A commercial customer calls. The small office building unit is not cooling on the first warm day. The problem is the air filters are dirty, the unit evaporator has frozen condensate, and the airflow is blocked.

The technician first goes to the thermostat and finds that it is set at 75°F, and the space temperature is 82°F. The air feels very humid. There is no air coming from the registers. The air handler is in the attic. An examination of the condensing unit shows that the suction line has ice all the way back to the compressor. The compressor is still running. The thermostat is set to OFF and the fan to ON to keep air circulating for thawing the evaporator. When the coil is frozen solid, it will take at least an hour for air to start to flow through the coil. When it does start to flow through the coil, the ice will melt faster.

The thermostat was set correctly, so the technician looks for some problem that concerns airflow. The air filters are located in return-air filter grilles in the hall ceiling and are very dirty. The filters are all changed. The customer is told to run the fan until air is felt coming out of the grilles and to then let the fan run for another half hour before starting the compressor. A call back later in the day verified that the unit was cooling correctly.

Service Call 6

A residential customer calls. The customer reports the unit in their residence was worked on in the early spring under the service contract. The earlier work order shows that refrigerant was added by a technician new to the company. The unit is not running enough to keep the house cool. The problem is the new technician operated the unit and added refrigerant on a 70°F day. The unit has an overcharge of refrigerant. The outdoor temperature on this day is 85°F.

The technician finds that the thermostat is set at 73°F, and the house is 77°F. The unit is running, and cold air is coming out of the registers. Everything seems normal until the condensing unit is examined. The suction line is cold and sweating, but the liquid line is only warm. The unit has so much refrigerant that the condenser is acting as a large subcooler.

Gages are fastened to the gage ports, and the suction pressure is 85 psig, far from the correct pressure of 74 psig ($77° - 35° = 42°F$, which corresponds to about 74 psig). The head pressure is supposed to be 260 psig to correspond to a condensing temperature of 120°F ($85° + 35° = 120°F$ or 260 psig). The head pressure is 350 psig. The unit is cutting off because of high head-pressure after about 10 min running time. This unit has a time delay that keeps the compressor from restarting until it has been off for 8 min, so the unit is effectively running for 10 min then off for 8 min, not enough to keep the temperature down in the conditioned space.

The technician removes the compressor compartment door and sees the compressor sweating all over. The system has a capillary tube metering device that is flooding the compressor because of the overcharge. The charge is altered to the correct charge, and the unit performs correctly.

Service Call 7

A residential customer calls. The unit is not cooling enough to reduce the space temperature to the thermostat setting. It is running continuously on this first hot day. The problem is the compressor has a bad suction valve and is not pumping to capacity. It will cool the house on a mild day but has to run extra long periods of time.

The technician arrives and finds the thermostat set at 73°F; the house temperature is 78°F. Air is coming out of the registers, so the filters must be clean. The air is cool but not as cold as it should be.

At the condensing unit the technician finds that the suction line is cool but not cold. The liquid line seems extra cool. It is 90°F outside, and the condensing unit should be condensing at about 125°F. If the unit had 15°F of subcooling, the liquid line should be warmer than hand temperature, yet it feels cool.

Gages are fastened to the service ports to check the suction pressure. The suction pressure is 95 psig, and the discharge pressure is

225 psig. This has all the signs of a compressor that is not pumping to capacity. The airflow is restricted to the condenser, and the head pressure gradually climbs to 250 psig; the suction pressure goes up to 110 psig. The compressor is supposed to have a current draw of 27 A, but it only draws 15 A. The compressor is defective because it will not pump enough pressure difference from the low side to the high side and is not operating properly. This is determined by the lower current draw.

The compressor is changed the next day. When the correct charge is measured into the unit, the pressures are correct and the unit begins to cool. Late in the day, the technician calls the customer. The house is now cooled down and the unit is shutting off from time to time.

Service Call 8

A residential customer calls. The unit is not cooling. This is a residential high-efficiency unit. The problem is the thermostatic expansion valve is defective.

The service technician arrives and finds the house warm. The thermostat is set at 74°F, and the house temperature is 82°F. The indoor fan is running, and plenty of air is coming out of the registers.

The technician goes to the condensing unit and finds that the fan and compressor are running, but they quickly stop. The suction line is not cool. Gages are fastened to the gage ports. The low-side pressure is 25 psig, and the head pressure is 170 psig, this corresponds very closely to the ambient air temperature. The unit is off because of the low-pressure control. When the pressure rises, the compressor restarts but stops in about 15 sec. The liquid-line sight glass is full of refrigerant. The technician concludes that there is a restriction on the low side of the system. The expansion valve is a good place to start when there is an almost complete blockage. The system does not have service valves, so the charge has to be removed.

The expansion valve is soldered into the system. It takes an hour to complete the expansion valve change. When the valve is changed, the system is leak-checked and evacuated three times. A charge is measured into the system, and the system is started. It is evident from the beginning that the valve change has repaired the unit.

Service Call 9

A commercial customer calls. The unit in a small office building is not cooling. The unit was cooling correctly yesterday afternoon when the office closed. The problem is a night storm tripped a breaker on the air conditioner's air handler. The power supply is at the air handler.

The technician arrives and goes to the space thermostat. It is set at 73°F, and the temperature is 78°F. There is no air coming out of the registers. The fan switch is turned to ON, and the fan still does not start. It is decided to check the low-voltage power supply, which is in the attic at the air handler. The technician finds the tripped breaker. *Before resetting it, the technician decides to check the unit electrically.*

A resistance check of the fan circuit and the low-voltage control transformer proves there is a measurable resistance. The circuit seems to be safe. The circuit breaker is reset and stays in. The thermostat is set to COOL, and the system starts. A call later proves that the system is staying on.

Service Call 10

*A residential customer calls. The unit is not running. The homeowner found the breaker at the condensing unit tripped and reset it several times, but it did not stay set. The problem is the compressor is grounded and tripping the breaker. *The homeowner should be warned to reset a breaker no more than once.**

The technician arrives at the job and goes straight to the condensing unit with electrical test equipment. The breaker is in the tripped position and is moved to the OFF position. *The voltage is checked at the load side of the breaker to assure that it is not leaking voltage. The breaker has been reset several times and is not to be trusted.* When it is determined that the power is really off, the ohmmeter is connected to the load side of the breaker. The ohmmeter reads infinity, meaning no circuit. The compressor contactor is not energized, so the fan and compressor are not included in the reading. The technician pushes in the armature of the compressor contactor to make the contacts close. The ohmmeter now reads 0, or no resistance. Another term used is a *dead short*. It cannot be determined whether the fan or the compressor is the problem so the compressor wires are disconnected from the bottom of the contactor. The short does not exist when the compressor is disconnected. This verifies that the short is in the compressor circuit, possibly the wiring. The meter is moved to the compressor terminal box and the wiring is disconnected. The ohmmeter is attached to the motor terminals at the compressor. The short is still there, so the compressor motor is condemned.

The technician must return the next day to change the compressor. The technician disconnects the control wiring and insulates the disconnected compressor wiring so that power can be restored. This must be done to keep crankcase heat on the unit until the following day. If it is not done, most of the refrigerant charge will be in the compressor crankcase and oil will be blown out of the system when the system is vented to the atmosphere.

The technician performs one more task before leaving the job. The Schrader valve fitting at the compressor is slightly depressed to determine the smell of the refrigerant. The refrigerant has a strong acid odor. A suction-line filter-drier with high acid removal needs to be installed along with the normal liquid-line drier.

The technician returns on the following day and changes the compressor, adding the suction and liquid-line driers. The unit is leak-checked, evacuated, and the charge measured into the system with a graduated charging cylinder. The system is started, and the technician asks the customer to leave the air conditioner running, even though the weather is mild, to keep refrigerant circulating through the driers to clean it in case any acid is left in the system.

The technician returns on the fourth day and measures the pressure drop across the suction-line drier to make sure that it is not stopped up with acid from the burned-out compressor. It is well within the manufacturer's specification.

Service Call 11

A residential customer calls. The customer can hear the indoor fan motor running for a short time, then it cuts off. This happens repeatedly. The problem is the indoor fan motor is cycling on and off on its internal thermal overload because the fan capacitor is defective. The fan will start and run slowly then cut off.

The technician knows from the work order to go straight to the indoor fan section. It is in the crawl space under the house, so electrical instruments are carried on the first trip. The fan has been off long enough to cause the suction pressure to go so low that the evaporator coil is frozen. The breaker is turned off during the check. The technician suspects the fan capacitor or the bearings, so a fan capacitor is carried along. The fan capacitor is checked with the ohmmeter to see if it will charge and discharge. *It is important to short the two capacitor terminals together before touching the ohmmeter leads to the capacitor so that any charge in the capacitor will be drained off.* The capacitor will not charge and discharge, so it is changed.

The technician oils the fan motor and starts it from under the house with the breaker. The fan motor is drawing the correct amperage. The coil is still frozen, and the technician must set the space thermostat to operate only the fan until the homeowner feels air coming out of the registers. Then only the fan should be operated for one-half hour (to melt the rest of the ice off the coil) before the compressor is started.

Service Call 12

A commercial customer calls. The unit in a small office building was running but suddenly shut off. The problem is the control transformer is burned up. This stopped the indoor fan and the compressor. The compressor contactor coil is shorted and caused the low-voltage transformer to burn.

The technician arrives, goes to the space thermostat and finds it set at 74°F; and the space temperature is 77°F. The fan is not running. The fan switch is moved to ON, but the fan motor does not start. It is decided that the control voltage power supply should be checked first. The power supply is in the roof condensing unit on this particular job. *The ladder is propped against the building away from the power line entrance.* Electrical test instruments are taken to the roof along with tools to remove the panels. The breakers are checked and seem to be in the correct on position. The panel is removed where the low voltage terminal block is mounted. A voltmeter check shows there is no control voltage.

The technician turns off the breaker and checks the control transformer with an ohmmeter. Remember, the transformer must be isolated from the control circuit by removing one of the wires before it is checked with an ohmmeter. The secondary side of the transformer is found to be open. A new transformer is installed. The technician decides

to check the current draw in the secondary circuit (the 24-V circuit). The R wire from the transformer is removed and is wrapped around the ammeter 10 times with the aid of a length of wire. This is a 40-VA transformer and is capable of producing up to 1.66 A before overloading (40 VA ÷ 24 V = 1.66 A). The control amperage is only 0.12 A with just the fan relay energized. When the compressor contactor is energized, the amperage goes up to 5 A. The technician quickly turns off the system to keep from burning the new transformer. The compressor contactor coil is checked with an ohmmeter and found to only have a resistance of 5 Ω. This is an inductive load so it will not follow Ohm's law. A similar contactor is checked and found to have 30 Ω resistance. The contactor coil is shorted internally. It is changed and the system is started again. The current in the control circuit drops to 1.2 A which is considered correct.

SAFETY PRECAUTION: *All troubleshooting should be accomplished under the supervision of an experienced person. If you do not know whether a situation is safe, consider it unsafe.*

SUMMARY

- Troubleshooting air conditioning equipment involves both mechanical and electrical problems.
- Mechanical troubleshooting uses gages and thermometers for the instruments.
- The gage manifold is used along with the pressure–temperature chart to determine the boiling temperature for the low side and the condensing temperature for the high side of the system.
- The superheat for an operating coil is used to prove coil performance.
- Standard air conditioning conditions for rating equipment are 80°F return air with a humidity of 50% when the outside temperature is 95°F.
- The typical customer operates the equipment at 75°F return air with a humidity of 50%. This is the condition that the normal pressures and temperatures will be based on.
- There is a relationship between the evaporator and the return air at this typical condition. For a standard-efficiency unit, it is 35°F. The evaporator normally boils at 35°F lower than the 75°F return air or 40°F.
- A high-efficiency evaporator normally has a boiling temperature of 45°F.
- Gages fastened to the high side of the system are used to check the head pressure. The head pressure is a result of the condensing temperature.

- The condensing refrigerant temperature has a relationship to the medium to which it is giving up heat.
- A standard-efficiency unit normally condenses the refrigerant at no more than 35°F higher than the air to which the heat is rejected.
- A high-efficiency condensing unit normally condenses the refrigerant at a temperature as low as 20°F warmer than the air used as a condensing medium.
- High efficiency is obtained with more condenser surface area.
- Superheat is normally checked at the condensing unit with a temperature tester.
- The temperature tester may be used to check WB and DB temperatures (when using a wet wick).
- The condenser has three functions: to take the superheat out of the discharge gas, to condense the hot gas to a liquid, and to subcool the refrigerant.
- A typical condenser may subcool the refrigerant 10° to 20°F lower than the condensing temperature.
- Two types of metering devices are normally used on air conditioning equipment: the fixed bore (the capillary tube, for example) and the thermostatic expansion valve.
- The fixed-bore metering device uses the pressure difference between the inlet and outlet of the device for refrigerant flow. It does not vary in size.
- The thermostatic expansion valve modulates or throttles the refrigerant to maintain a constant superheat.
- The correct charging procedure is to follow the manufacturer's recommendations.
- The thermostatic expansion valve system normally has a sight glass in the liquid line to aid in charging. A subcooling temperature check may be used when there is no sight glass.
- The tools for electrical troubleshooting are the ammeter and the volt-ohmmeter.
- *Before checking a unit electrically, the proper voltage and the current draw of the unit should be known.*
- The main power panel may be divided into many circuits. Sometimes the air conditioning system is on two separate circuits. This occurs when the system is a split system. A package system is normally on one circuit.
- When a compressor is hot, such as when it has been running at low charge, the compressor's internal overload may stop the compressor.
- *When a hot compressor is started, assume that the system is low in refrigerant. A drum of refrigerant should be connected, so refrigerant may be added before the compressor shuts off again.*

- A compressor can be checked electrically with an ohmmeter. The run and start windings should have a known resistance, and there should be no circuit to ground.

REVIEW QUESTIONS

1. What are the two main troubleshooting areas in the air conditioning service field?
2. What information does the low-side pressure reading give the service technician?
3. What information does the high-side pressure reading give the service technician?
4. What is the correct operating superheat for an air conditioning coil?
5. What is the typical indoor air condition at which a homeowner operates a system?
6. What is the temperature at which the typical air conditioning coil operates?
7. What is the typical temperature difference between the entering air and the boiling refrigerant temperatures on a standard air conditioning evaporator?
8. What is the best method of charging an air conditioning system that is low in refrigerant?
9. When the outside ambient air temperature is 90°F, at what temperature should the refrigerant in the condenser be condensing, and what is the head pressure for an R-22 system?
10. What numbers are printed on a refrigerant gage besides pressure?
11. How can a service technician measure the superheat at the condenser if the system is a split-system?
12. When the ambient air temperature is 95°F and the unit is a high-efficiency unit, what is the lowest condensing temperature at which the refrigerant would normally operate, and what is the head pressure for R-22?
13. How is high efficiency accomplished in an air conditioning condensing unit?
14. Name two methods for fastening gage manifolds to an air conditioning system.
15. Which control in the compressor stops the compressor when the motor gets hot?
16. When the control in Question 15 has the compressor off, which terminals on the compressor will have continuity?
17. What instrument is used to measure the current draw of a compressor?
18. What instrument is used to measure continuity in a compressor winding?
19. What normally happens when a compressor that is electrically grounded tries to start?
20. What should a technician do if the circuit breaker is tripped?

ELECTRIC, GAS AND OIL HEAT
WITH ELECTRIC AIR CONDITIONING

OBJECTIVES

After studying this unit, you should be able to

- describe year-round air conditioning.
- discuss the three typical year-round air conditioning systems.
- list the five ways to condition air.
- state how much air is normally required to be distributed per ton for a cooling system.
- describe why a heating system normally uses less air than a cooling system does.
- explain two methods used to vary the airflow in the heating season from that in the cooling season.
- describe two types of control voltage power supplies used in add-on air conditioning.
- describe add-on air conditioning.
- explain package all-weather systems.

44.1 COMFORT ALL YEAR

Year-round air conditioning describes a system that conditions the living space for heating and cooling throughout the year. This is done in several ways: The most common are electric air conditioning with electric resistance heat, electric air conditioning with gas heat, and electric air conditioning with oil heat. This unit

describes how these systems work together. Each system has been covered individually in other units. Another common method is the heat pump, which is discussed in Unit 45. A less frequently used method is gas heat with gas air conditioning. This is a special system and will not be covered in this book.

44.2 FIVE PROCESSES FOR CONDITIONING AIR

Air is *conditioned* when it is heated, cooled, humidified, dehumidified, or cleaned. This unit discusses how air is heated and cooled with the same system. The systems discussed in this text are called *forced-air* systems. Air is distributed through duct work to the conditioned space. The fan is normally a component of the heating system and provides the force to move the air through the duct system. A typical system may have an electric, gas, or oil furnace with an evaporator in the airstream for cooling, Figure 44-1. The evaporator, located in the indoor air stream, is used in conjunction with a condensing unit outside the conditioned space. Interconnecting piping connects the evaporator to the condensing unit, Figure 44-2.

44.3 ADD-ON AIR CONDITIONING

Many systems are installed in stages. The furnace may be installed when the structure is built. The air

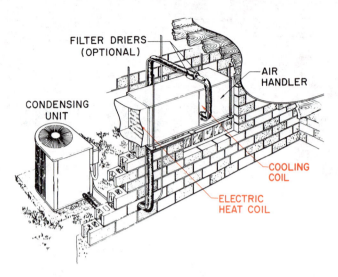

Figure 44-1. This is a typical electric furnace that has an air conditioning coil placed in the duct. The fan in the furnace is used to circulate air for both air conditioning and heating. *Courtesy Climate Control*

conditioning may be added at the same time or later (called *add-on*). When the heating system is installed first, some considerations must be made for air conditioning if it is to be added later.

Air conditioning systems must have the correct air circulation. Typically, they require an airflow of 400 cfm/ton, which is more than needed for an average forced-air heating system. A 3-ton cooling system, for example, requires 1200 cfm. The furnace on an existing system must be able to furnish the required amount of air, and the duct work must be sized for the airflow. If the duct work is installed outside the conditioned

space, it must be insulated or moisture ("sweat") from the ambient air will form on it during the cooling cycle. The insulation also helps prevent heat exchange between the air in the duct and the ambient air.

44.4 INSULATION FOR EXISTING DUCT WORK

It is popular in warm climates to install heating systems in the crawl space below the structure. The heat from the warm duct rises and warms the floor, Figure 44-3. Some people believe that heat lost under the house is not all lost because some of it rises through the floor. This is true only when the space under the house is sealed. Therefore many duct systems are not insulated. If air conditioning is added to this system, the ducts will sweat. The system would be more efficient if it were insulated from the beginning.

44.5 EVALUATION OF AN EXISTING DUCT SYSTEM

When the air conditioning system is added after the original furnace installation, several items must be considered: the airflow, the duct work, the air distribution system (registers and grilles), and the wiring are the most important.

44.6 THE AIR DISTRIBUTION SYSTEM

Air distribution was discussed in Unit 39. A heating system that is already installed needs to be evaluated to see if air distribution changes need to be made

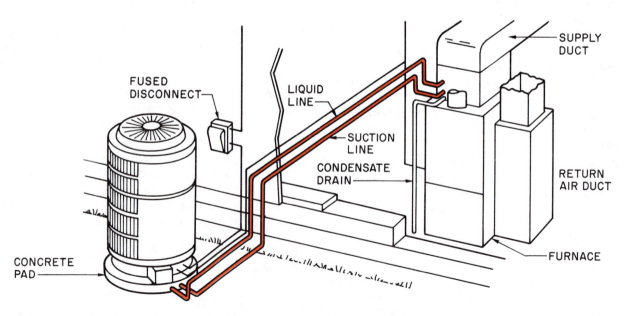

Figure 44-2. Complete installation with the piping shown. The evaporator is piped to the condensing unit on the outside of the structure.

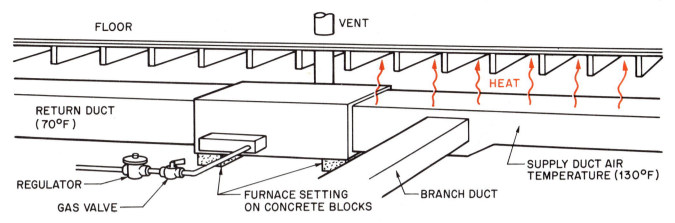

Figure 44–3. Heat rising off the duct in a crawl space under a house.

before air conditioning is installed. The blower size and the motor horsepower may be guides as to the amount of air the fan section is capable of moving. The manufacturer of the air handler or furnace is the best source from which to get this information. However, it is not always readily available. Figure 44–4 is a chart used by estimators and service technicians to help determine what the blower capabilities of a typical

QUICK-SIZING TABLE FOR HEATING AND COOLING DUCT SYSTEM

Air conditioning systems should never be sized on the basis of floor area only, but knowledge of the approximate floor area (sq. ft.) that can be cooled with a ton of air conditioning will be of invaluable assistance to you in avoiding serious mathematical errors.

Size of O.D. Unit	Normal Air Flow Req'd @ 400 cfm per Ton	Furnace		Supply Duct or Extended Plenum @ 800 fpm	Min. Number Supply Runs @ 600 fpm				Min. Return Duct Size at Furnace or Air Handler @ 800 fpm	Min. Return Air Grille Size (or equivalent) @ Face Velocity of 500 fpm
		Blower Motor hp	Blower Wheel Dia. X Width		5" Runs 80 cfm	6" Runs 115 cfm	7" Runs 155 cfm	$3\frac{1}{2}$ x 14" 170 cfm		
$1\frac{1}{2}$ ton 18,000 Btuh	600 cfm	$\frac{1}{4}$ hp	9" x 8" 10" x 8"	16" x 8" or 12" round	8	5	4	4	16" x 8" or 12" round	24" x 8"
2 ton 24,000 Btuh	800 cfm	$\frac{1}{4}$ hp	9" x 9" 10" x 8"	22" x 8" or 14" round	10	7	5	5	22" x 8" or 14" round	22" x 12"
$2\frac{1}{2}$ ton 30,000 Btuh	1000 cfm	$\frac{1}{3}$ hp	10" x 8" 10" x 10" 12" x 9"	20" x 10" or 18" round	13	9	7	6	20" x 10" or 16" round	30" x 12"
3 ton 36,000 Btuh	1200 cfm	$\frac{1}{3}$ hp	10" x 8" 10" x 10" 12" x 9"	24" x 10" or 18" round	—	11	8	7	24" x 10" or 18" round	30" x 12"
$3\frac{1}{2}$ ton 42,000 Btuh	1400 cfm	$\frac{1}{2}$ hp $\frac{3}{4}$ hp	10" x 8" 10" x 10" 12" x 9" 12" x 10"	24" x 12" or 18" round	—	12	9	8	24" x 12" or 18" round	30" x 14"
4 ton 48,000 Btuh	1600 cfm	$\frac{1}{2}$ hp $\frac{3}{4}$ hp	10" x 10" 12" x 9" 12" x 10" 12" x 12"	32" x 10" or 20" round	—	14	11	10	28" x 12" or 20" round	30" x 18"

Figure 44–4. This chart shows many system characteristics including the return-air characteristics of a typical system used for cooling and heating. *Copyright American Standard, Inc. 1985*

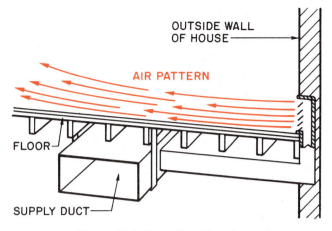

Figure 44-5. Low sidewall registers.

system may be. The blower wheel dimensions are an important factor.

The duct system may be evaluated with an evaluation chart also, Figure 44-4. *This chart should not be used when designing a system from the beginning.* It should only be used as a reference when estimating or troubleshooting. The estimator or service technician that suspects there is an airflow problem may consult the chart and compare the duct system to the chart. If the system does not meet the chart's minimum requirements, further investigation should be made.

The grilles and registers that are used for the final air distribution are important also, Figure 44-4. It is important that the air volume meet the minimum before the air conditioning system is added. The registers are responsible for distributing an air pattern into the conditioned space. If the original purpose of the installation was heating only, the registers may not provide the correct air pattern. This is particularly true for floor baseboard registers. They are designed to keep

the air down low on the floor if they are for heating only. Heated air will rise and warm the room. These registers will not work well in the air conditioning season because the cool air will stay on the floor, Figure 44-5.

When a distribution system does not meet the minimum standards of the charts discussed previously, it should be changed. If it is not changed, there will probably not be enough air for the air conditioning system, and the air might not be properly distributed. The changes may involve adding new duct or return-air runs or changing the blower or motor. If the furnace will not allow enough air to be circulated because the furnace heat exchanger is not large enough, a new furnace may be the only solution. The air distribution registers are usually easy to change to accommodate air conditioning.

44.7 SUMMER VERSUS WINTER AIR QUANTITY

Some designers and technicians like to have an airflow different in summer than in winter. It is sometimes desirable for less airflow in the winter, so the air that is entering the room through the registers will be hot. If the airflow of 400 cfm is used in the winter, the air may create drafts of slightly warm air instead of hot air. Changing the air volume is done by dampers or a multispeed fan.

Dampers are sometimes installed by the installing contractors. They have summer and winter positions. Someone must change the damper position for each change of season. Because this is often overlooked, some contractors install a low-voltage end switch that will not allow the cooling contactor to be energized unless the damper is in the summer position, Figure 44-6.

Multispeeds from the fan may be achieved by adjusting the fan motor pulley, or using multispeed motors.

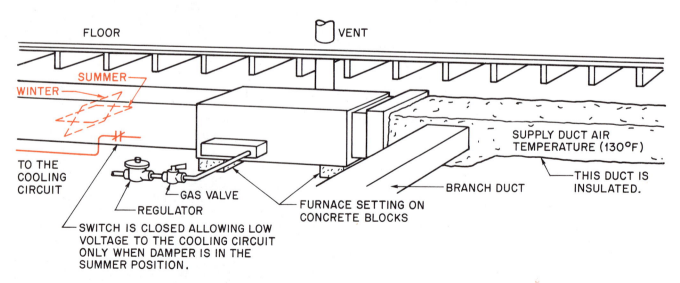

Figure 44-6. This system has a damper to allow more airflow in the summer than in the winter.

When the pulley has to be adjusted to change fan speeds for a new season, a service technician must visit the job twice a year. The pulley may be adjusted at the spring startup for air conditioning and at the fall furnace checkout. The filters may be changed and the motors oiled at the same time. This is a routine service for some installations each year. Changing the pulley is not a job that the homeowner should undertake unless they have experience with this type of service job.

Multispeed motors were discussed in another unit. They have a winding for each motor speed, Figure 44–7.

Figure 44-8. Typical HEAT–COOL thermostat. *Courtesy Robertshaw Controls Company*

44.8   CONTROL WIRING FOR SUMMER AND WINTER

The main consideration for year-round air conditioning is the control wiring. The control system must be capable of operating heating and air conditioning equipment at the proper times. *The heating must not be operated at the same time as the cooling.* The thermostat is the control that accomplishes this. See Figure 44–8 for an example of a typical HEAT-COOL thermostat with manual changeover from season to season. See Figure 44–9 for an example of a HEAT-COOL thermostat with automatic changeover.

44.9   TWO LOW-VOLTAGE POWER SUPPLIES

The control circuit may have more than one low-voltage power supply, which can cause much confusion for the technician trying to troubleshoot a problem. If the heating system was installed first and the air conditioning added later, there may be two power supplies. The furnace must have a low-voltage transformer to operate in the heating mode. When the air conditioning was added, a transformer may have been furnished with it. This is because the air conditioning manufacturer did not know if the low-voltage power supply

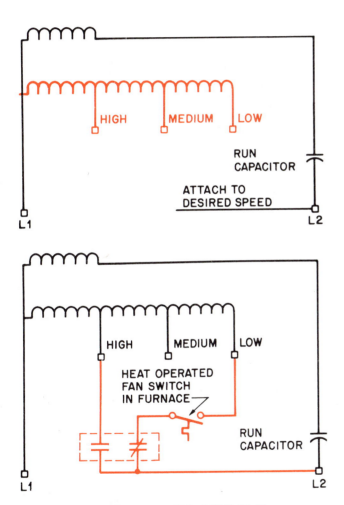

FAN RELAY: WHEN ENERGIZED, SUCH AS IN COOLING, THE FAN CANNOT RUN IN THE LOW–SPEED MODE. WHEN DEENERGIZED THE FAN CAN START IN THE LOW-SPEED MODE THROUGH THE CONTACTS IN THE HEAT OPERATED FAN SWITCH.

IF THE FAN SWITCH AT THE THERMOSTAT IS ENERGIZED WHILE THE FURNACE IS HEATING, THE FAN WILL MERELY SWITCH FROM LOW TO HIGH. THIS RELAY PROTECTS THE MOTOR FROM TRYING TO OPERATE AT 2 SPEEDS AT ONCE.

Figure 44-7. Three-speed motor.

Figure 44-9. Thermostat with automatic changeover from cooling to heating. *Photo by Bill Johnson*

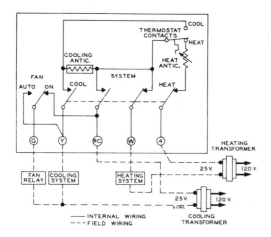

Figure 44–10. Wiring of an integrated subbase thermostat. *Courtesy White-Rodgers Division, Emerson Electric Co.*

Figure 44–12. Transformer and fan-relay package. *Photo by Bill Johnson*

44.10 PHASING TWO LOW-VOLTAGE TRANSFORMERS

＊When the two transformers are wired in parallel, they must be kept in phase or else the transformers will have opposing phases, Figure 44–11.＊

44.11 ADDING A FAN RELAY

When air conditioning is added to an existing system, the furnace may not have a fan relay to start the fan. Most fossil-fuel furnaces (oil or gas) will start the fan with a thermal type of fan switch in the heating mode and will have no provision to start the fan in the cooling mode. Some electric furnaces have a fan relay already built in. When there is none, a separate fan relay must be added when the air conditioning system is installed. This relay is sometimes furnished in a package with the control transformer called a *transformer relay package,* Figure 44–12.

furnished with the furnace was adequate. Air conditioning systems usually require a 40-VA transformer to supply enough current to energize the compressor contactor and the fan relay. Most basic furnaces can operate on a 25-VA transformer and may be furnished with one.

When two power supplies are used, a special arrangement must be made in the thermostat and subbase. In one arrangement the two circuits are separated; in the other the two transformers are wired in parallel by phasing them. ＊If the transformers are to be separated, two hot wires are wired into the subbase. The two circuits must be kept apart, or damage may occur, Figure 44–10. Care should be used around any power supply as electrical shock hazard is possible.＊

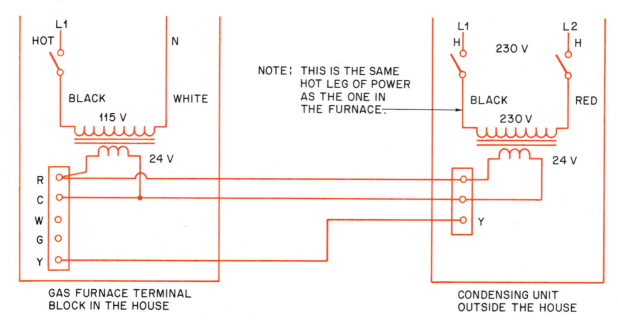

Figure 44–11. Transformers wired in parallel.

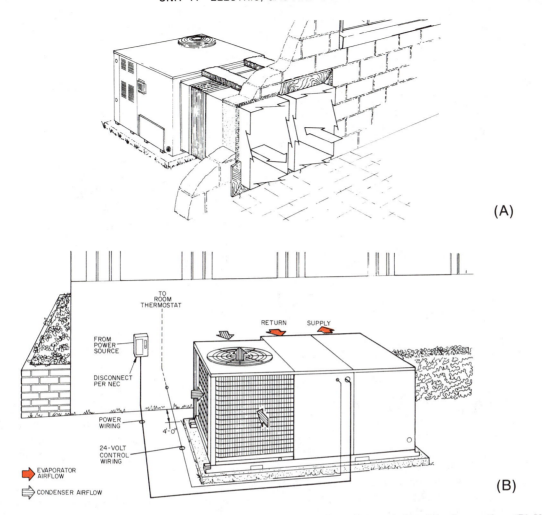

Figure 44-13. These two package units look alike. (A) Uses gas as the heat source and electricity for cooling. (B) Uses electric heat with electric air conditioning. *Courtesy (A) Climate Control (B) Reproduced courtesy of Carrier Corporation*

SAFETY PRECAUTION: *The fan relay is part of the low-voltage and high-voltage circuits. Use proper caution whenever working with electricity.*

44.12   NEW ALL-WEATHER SYSTEMS

When an all-weather system is installed from the beginning, the foregoing considerations can be correctly designed into the initial installation. The duct work is always designed around the cooling system because it requires more airflow.

New all-weather system installations are split systems or package systems. If split, a gas, oil, or electric furnace is used for heating, and electric air conditioning is used for cooling. A package system will be gas or electric heat with electric cooling. Heat pumps also come in package systems, and they are covered in the next unit.

44.13   ALL-WEATHER SPLIT SYSTEMS

Split systems are installed in the same way as a gas furnace with the duct sized to handle the air for the air conditioning. Remember, a summer air conditioning system must have more cubic feet per minute of air than a heating system alone. Some consideration may be given to the air distribution at the grilles for proper air distribution for cooling. Most furnace manufacturers have matching coil packages that will fit their furnaces. A review of the unit covering the particular furnace will describe how to install the furnace.

44.14   PACKAGE OR SELF-CONTAINED ALL-WEATHER SYSTEMS

Package systems are not normally made for oil with air conditioning. These systems are gas and electric or electric and electric, Figure 44-13, and are installed like electric air conditioning package systems. A review of the unit on air conditioning (cooling) will show how this is done. The units that use gas must have a gas line installed. The flue installed with a gas package unit is a fixed component and part of the unit. These units are sometimes called *gas packs,* Figure 44-14.

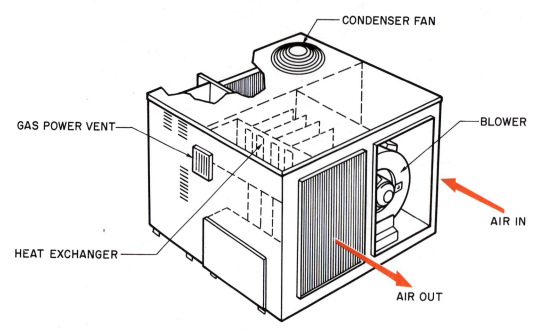

Figure 44-14. Flue vent on a gas package unit. *Courtesy Climate Control*

44.15   WIRING THE ALL-WEATHER SYSTEM

The wiring of all-weather systems is similar to that of air conditioning systems except for the extra power that may be required to operate the electric heat, Figure 44-15. The control wiring is much the same as a gas furnace and electric air conditioning except that the wiring is all done between the thermostat and the package unit. The wires that normally run to a remote furnace are not needed.

44.16   SERVICING THE ALL-WEATHER SYSTEM

Package equipment installations have the advantage that the whole system is outside the house. *Any gas hazard is virtually eliminated because gas leaks are dissipated outside.* A technician can service the unit without crawling to a furnace in an attic or under a house. All of the control wiring is at the unit and is easily accessible. See Figure 44-16 for an example of a typical installation.

When major repairs are needed on the compressor or expansion device, for example, the repair is simplified because they are all together and accessible. The proper panels may be removed to service the various components.

When troubleshooting an all-weather system, the technician usually is called to deal with a heating problem in winter or a cooling problem in summer. Seldom does a heating problem carry over to the cooling season, or a cooling problem to the heating season, because the controls are separated sufficiently. Dirty air filters are one type of problem that may not show up at the end of the cooling or heating season but will cause problems at the peak season. For example, when a homeowner does not change the air filters at the end of the heating season, the cooling will normally work fine until the first hot day that the cooling unit has long running times. When the running times are short, the unit may run below freezing due to the dirty filters but will cut off before the coil freezes up. When the running time is longer, the coil may freeze solid before the room thermostat shuts off the unit.

The reverse is true about filters when they are not changed at the end of the cooling season. The cooling unit may not run long enough to freeze solid before the thermostat shuts it off. When the heating season starts, the furnace will appear to be working fine when the weather is mild. *When the weather gets cold enough to cause the furnace to run for long periods, it may overheat with the high limit shutting it off.*

The other components of year-round air conditioning, such as the humidifier and the air cleaner, are discussed in other units in this text.

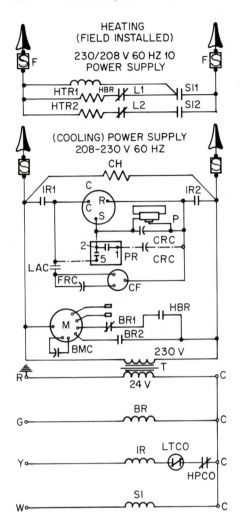

COMPONENT PART IDENTIFICATION

BMC	BLOWER MOTOR CAPACITOR	HR	HEAT RELAY
BR	BLOWER RELAY	HTR	ELECTRIC HEATER
C	COMPRESSOR	IR	COMPRESSOR CONTACTOR
CF	CONDENSER FAN MOTOR	L	LIMIT
CH	CHRANKCASE HEATER	LAC	LOW AMBIENT CONTROL (0°F) (ALL
CRC	COMPRESSOR RUN CAPACITOR		RD/RG-D UNITS & ALL R-H RG-H3 PHASE
CS	EXHAUSTER CENTRIFUGAL SWITCH		UNITS)
CSC	COMPRESSOR START CAPACITOR	LTCO	LOW TEMPERATURE CUT-OUT (50°F)
EM	EXHAUSTER MOTOR		(ALL R-H/RG-H 1 PHASE UNITS)
F	FUSE	M	BLOWER MOTOR
FC	FAN CONTROL KLIXON	P	PTC STANDARD START ASSIST
FRC	FAN RUN CAPACITOR	PPK	POST PURGE KLIXON
FT	FAN TIMER	PR	POTENTIAL RELAY
GV	GAS VALVE	PR+	
HBR	BLOWER RELAY (HEATING)	CDC	OPTIONAL START ASSIST (REPLACES "P")
HPCO	HIGH-PRESSURE CUT-OUT	S1	SEQUENCER
		T	TRANSFORMER

Figure 44–15. Wiring diagram for electric heat and electric air conditioning package unit. *Courtesy Climate Control*

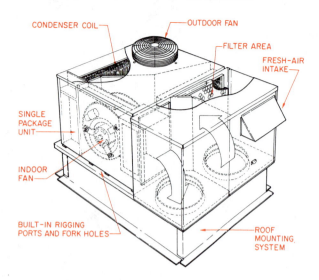

Figure 44–16. Installation of a package unit showing how the duct work is attached. *Courtesy Climate Control*

SUMMARY

- All-weather systems heat in the winter and cool in the summer.
- This is accomplished with combinations of equipment such as gas heat and electric cooling, oil heat and electric cooling, and electric heat and electric cooling.
- Air is conditioned by heating, cooling, humidifying, dehumidifying, and cleaning.
- Summer air conditioning is sometimes added to an existing heating system. This is called add-on cooling.
- When summer air conditioning is added to an existing heating system, the duct work, the terminal air distribution system, the fan on the existing furnace, and the control wiring must be considered.

- Different air volumes are sometimes desirable for the heating and cooling seasons. The air in the heating season is warmer at the terminal units when the air volume is reduced.
- Different air volumes are accomplished with dampers and variable fan speeds.
- The control circuit may have two transformers—one furnished with the furnace and one with the condensing unit.
- When there are two power supplies (transformers), the two may be kept separated in the thermostat or they may be wired in parallel.
- Package all-weather systems normally consist of gas heat and electric air conditioning or electric heat and electric air conditioning.
- An advantage of package systems is that the whole system is located outside the structure.

REVIEW QUESTIONS

1. Name the five things that are done to condition air.
2. What is an all-weather system?
3. What are the three common fuels used for heating in all-weather systems?
4. What is add-on summer air conditioning?
5. What are some of the things that must be considered with add-on air conditioning?
6. What are two methods to vary the airflow in an all-weather system?
7. Why is it desirable to have less airflow in the heating season?
8. What is the recommended airflow for summer air conditioning?
9. Name two ways to wire a system with two control transformers.
10. What is one advantage of a package all-weather system?

HEAT PUMPS

OBJECTIVES

After studying this unit, you should be able to

- **describe a reverse-cycle heat pump.**
- **list the components of a reverse-cycle heat pump.**
- **explain a four-way valve.**
- **state the various heat sources for heat pumps.**
- **compare electric heat to heat with a heat pump.**
- **state how heat pump efficiency is rated.**
- **determine by the line temperatures if a heat pump is in cooling or heating.**
- **discuss the terminology of heat pump components.**
- **list the special devices that must be added to the metering components if they are used on a heat pump.**
- **define COP.**
- **explain auxiliary heat.**
- **describe the control sequence on an air-to-air heat pump.**

45.1 REVERSE-CYCLE REFRIGERATION

Heat pumps are refrigeration machines. Refrigeration involves the removal of heat from a place where it is not wanted and depositing it in a place where it makes no difference. The heat can actually be deposited in a place where it will do some good. This is the difference between a heat pump and a summer air conditioner. The air conditioner can only pump heat one way. The heat pump is a refrigeration machine that can pump heat two ways. It is normally used for space conditioning, heating, and cooling.

All compression cycle refrigeration machines are heat pumps in that they pump heat-laden vapor. The evaporator of a heat pump absorbs heat into the refrigeration system, and the condenser rejects the heat from the refrigeration system. The compressor pumps the heat-laden vapor. The metering device controls the refrigerant flow. These four components—the evaporator, the condenser, the compressor, and the metering device—are essential to a compression cycle refrigeration machine. The same components are in a heat pump system along with the *four-way valve*.

45.2 HEAT SOURCES FOR WINTER

The cooling system in a typical residence absorbs heat into the refrigeration system through the evaporator and rejects this heat to the outside of the house through the condenser, where it makes no difference. The house might be 75°F inside while the outside temperature might be 95°F or higher. The summer air conditioner pumps heat from a low temperature inside the house to a higher temperature outside the house; that is, it pumps heat up the temperature scale. A freezer in a supermarket takes the heat out of ice cream at 0°F to cool it to −10°F so that it will be frozen hard. The heat removed can be felt at the condenser as hot air. This example shows that there is usable heat in air at 0°F, Figure 45–1. There is heat in

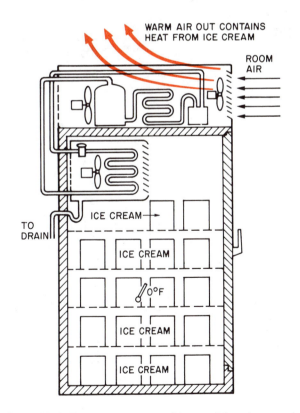

Figure 45–1. Low-temperature refrigerated box is removing heat from ice cream at 0°F. The heat coming out of the back of the box is coming out of the ice cream.

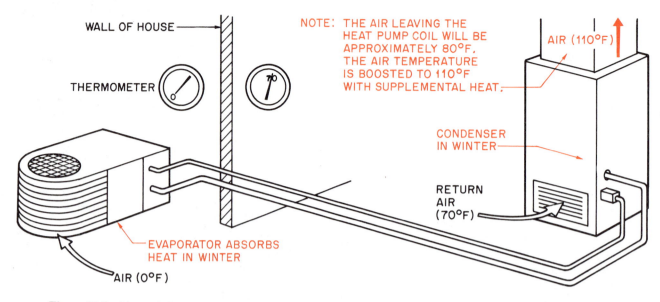

Figure 45–2. Air-to-air heat pump removing heat from 0°F air and depositing it in a structure for winter heat.

Figure 45–3. Outdoor portion of an air-to-air heat pump. It absorbs heat from the outside air for use inside the structure. *Reproduced courtesy of Carrier Corporation*

any substance until it is cooled down to −460°F. Review Unit 1 for examples of heat level.

If heat can be removed from 0°F ice cream, it can be removed from 0°F outside air. The typical heat pump does just that. It removes heat from the outside air in the winter and deposits it in the conditioned space to heat the house. (Actually, about 85% of the usable heat is still in the air at 0°F.) Hence, it is called an *air-to-air* heat pump, Figure 45–2. In summer the heat pump acts like a conventional summer air conditioner and removes heat from the house and deposits it outside. From the outside an air-to-air heat pump looks just like a central summer air conditioner, Figure 45–3.

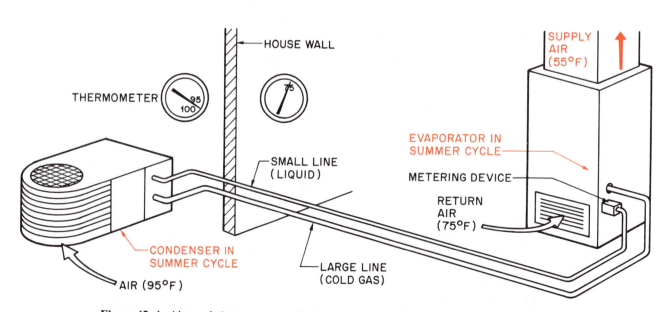

Figure 45–4. Air-to-air heat pump moving heat from the inside of a structure to the outside.

45.3   THE FOUR-WAY VALVE

The refrigeration principles that a heat pump uses are the same as those in refrigeration. However, a new component has to be added to allow the refrigeration machine to pump heat in either direction. The air-to-air heat pump in Figure 45–4 shows the heat pump moving heat from inside the conditioned space in summer to the outside. Then in winter the heat is moved from the outside to the inside, Figure 45–5. It is done with a special component called a *four-way valve*. This valve can best be described in the following way. The heat absorbed into the refrigeration system is pumped through the system with the compressor. The heat is contained in the discharge gas. The four-way valve diverts the discharge gas and the heat in the proper direction to either heat or cool the conditioned space. See Figure 45–6 for an example of a four-way valve. This valve is controlled by the space-temperature thermostat which positions it to either HEAT or COOL.

45.4   TYPES OF HEAT PUMPS, AIR-TO-AIR

Air is not the only source from which a heat pump can absorb heat, but it is the most popular. The most common sources of heat for a heat pump are air, water, and earth. For example, a structure located next to a large lake can remove heat from the lake and deposit it in the structure. The lake must be large enough so that the temperature will not drop appreciably in the lake. This system is a *water-to-air* heat pump, Figure 45–7. The water-to-air heat pump may also use a well

to supply the water. If so, some consideration must be given to where the water will be pumped after it is used. A typical water-to-air heat pump uses 3 gpm of water in the heating cycle and 1.5 gpm in the cooling cycle per ton of refrigeration (12,000 Btu/hr).

45.5   WATER-TO-AIR HEAT PUMPS

When a water-to-air heat pump uses a lake for the water source, the lake temperature may vary from the beginning of the season to the end of the season. This must be taken into consideration when choosing and sizing a system. The water may be pumped back into the lake after being used, so this is no problem. **However, local codes and authority must be followed for any installation.** Water-to-air heat pumps may get more attention from the local governing body than air-to-air heat pumps because of the handling of the water. For example, a 3-ton heat pump in the heating mode will use 9 gpm of water. This is a rate of 540 gph that must be drained off (9 gpm × 60 min/hr = 540 gph) or 12,960 gal/24 hr (540 gph × 24 hr/day = 12,960 gal/24 hr). Some installations have diversion valves to heat pools in the summer and to water the lawn, but at the rate of 12,960 gal/day the ground will become saturated very soon if there is not plenty of land. This amount of water cannot normally be just piped to the corner of the property to run off. It must be pumped to a planned location, such as a stream.

Industry and commercial buildings often use heat pumps to move heat from one part of the building or manufacturing process to another part. For example,

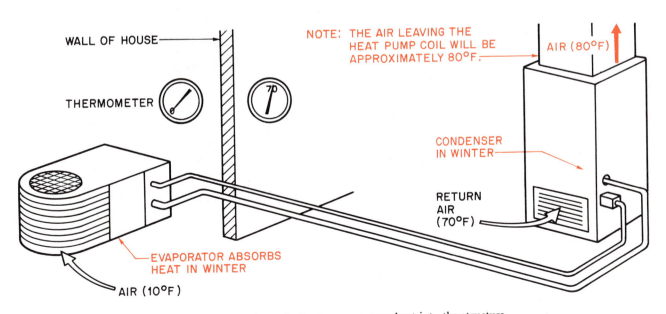

Figure 45–5. In winter the heat pump pumps heat into the structure.

Figure 45-6. Four-way valve. *Photo by Bill Johnson*

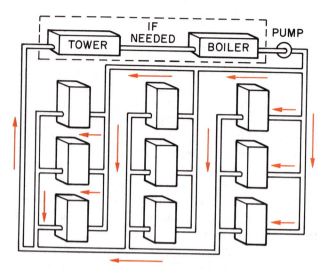

Figure 45-8. Water-to-air system that can absorb heat from one part of a building and deposit it in another. *Reproduced courtesy of Carrier Corporation*

a large office building may require cooling for some areas even in winter. The offices that are located in the interior of the building may have heat buildup due to lights, office machines, and people. It is economical in some office buildings to absorb this heat into a refrigeration system and to pump it to the outer perimeter of the building where heat is needed. This makes more sense than wasting this heat to the atmosphere and buying energy to heat the outside of the building. This moving of heat from one part of a commercial building to another is sometimes accomplished with many small residential-sized water-to-air heat pumps. The advantage is that it is economical

to operate and each zone has its own unit for zone control, Figure 45-8.

45.6 REMOVING HEAT FROM MANUFACTURING PROCESSES

Some manufacturing processes, such as metalworking, generate heat. Some of the energy that is used to form or cut metal in a machine shop is converted to heat. This energy was purchased from the local utility. The machines are cooled, and the excess heat removed

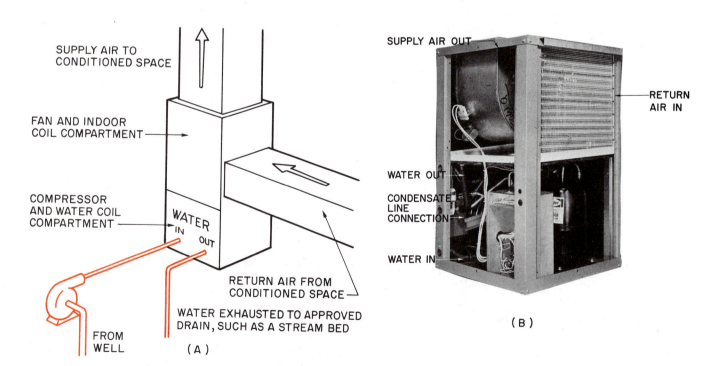

Figure 45-7. Water-to-air heat pump is absorbing heat from well water. *Photo reproduced courtesy of Carrier Corporation*

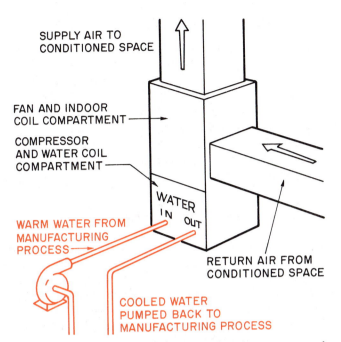

SUPPLY AIR TO
CONDITIONED SPACE

FAN AND INDOOR
COIL COMPARTMENT

COMPRESSOR
AND WATER COIL
COMPARTMENT

WATER IN OUT

WARM WATER FROM
MANUFACTURING
PROCESS

COOLED WATER
PUMPED BACK TO
MANUFACTURING PROCESS

RETURN AIR FROM
CONDITIONED SPACE

Figure 45-9. Heat from cooled manufacturing process used with heat pump to heat office.

from the machines. This heat may be absorbed into a refrigeration system and released into the office space to heat the offices, Figure 45-9.

45.7   REMOVING HEAT FROM THE GROUND

Groundwater heat pumps must have some method of removing heat from the ground. One way is to use a long ditch and buried pipe, Figure 45-10. Another method is to use well water as a heat source in the winter by removing the heat from it and pumping it to another well, Figure 45-11. A deep well with the pipe loop going to the bottom of the well may also be used. Be sure to check the local code for either type of installation. Technology with this type of heat pump is new, and the facts are still being gathered. When one of these systems is considered, someone with experience in this field should be consulted.

45.8   SOLAR-ASSISTED HEAT PUMPS

Many ways are being explored for using the reverse-cycle heat pump. Solar-assisted heat pumps are one example. These pumps capture heat from the sun and boost the level of the heat to usable levels for heating homes. Some manufacturers are designing systems especially for this purpose. This text will describe the basic heat pump cycle and the air-to-air heat pump in detail, because it is currently the most popular.

SUPPLY AIR TO
CONDITIONED SPACE

FAN AND INDOOR
COIL COMPARTMENT

COMPRESSOR
AND WATER COIL
COMPARTMENT

WATER IN OUT

RETURN AIR FROM
CONDITIONED SPACE

THE GROUND COIL IS BURIED IN THE GROUND TO MANUFACTURER'S SPECIFICATIONS FOR CORRECT HEAT TRANSFER.

Figure 45-10. Water-to-air heat pump using a long underground loop.

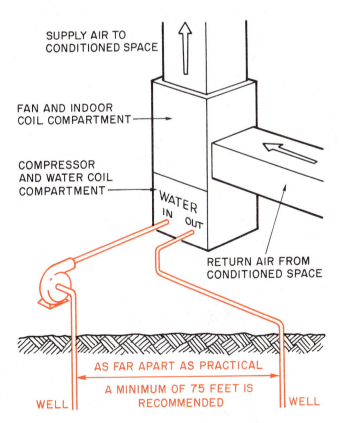

SUPPLY AIR TO
CONDITIONED SPACE

FAN AND INDOOR
COIL COMPARTMENT

COMPRESSOR
AND WATER COIL
COMPARTMENT

WATER IN OUT

RETURN AIR FROM
CONDITIONED SPACE

AS FAR APART AS PRACTICAL
A MINIMUM OF 75 FEET IS RECOMMENDED
WELL   WELL

Figure 45-11. Water-to-air heat pump using well water and pumping the water into another well.

HEATING CYCLE

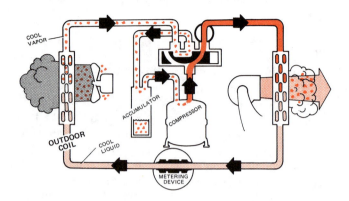

COOLING CYCLE

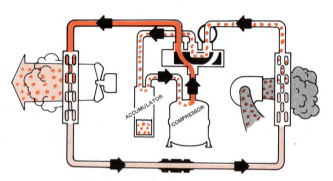

Figure 45–12. Heat pump refrigeration cycle shows the direction of the refrigerant gas flow. *Reproduced courtesy of Carrier Corporation*

45.9   THE AIR-TO-AIR HEAT PUMP

The air-to-air heat pump resembles the central air conditioning system. There are indoor and outdoor system components. When discussing summer air conditioning systems, these components are often called the *evaporator* (indoor unit) and the *condenser* (outdoor unit). This terminology will work for air conditioning but not for a heat pump. Like cooling equipment, heat pumps are manufactured in split systems and package systems.

45.10   SYSTEM IDENTIFICATION

The system's coils have new names when applied to a heat pump. The coil that serves the inside of the house is called the *indoor coil*. The unit outside the house is called the *outdoor unit* and contains the *outdoor coil*. The reason is that the indoor coil is a condenser in the heating mode and an evaporator in the cooling mode. The outdoor coil is a condenser in the cooling mode and an evaporator in the heating mode. This is all determined by which way the hot gas is flowing. In winter the hot gas is flowing toward the indoor unit and will give up heat to the conditioned space. The heat must come from the outdoor unit, which is the evaporator. See Figure 45–12 for an example of the direction of the gas flow. The system mode can easily be determined by gently touching the gas line to the indoor unit. *If it is hot (it can be 200° F, so be careful), the unit is in the heating mode.*

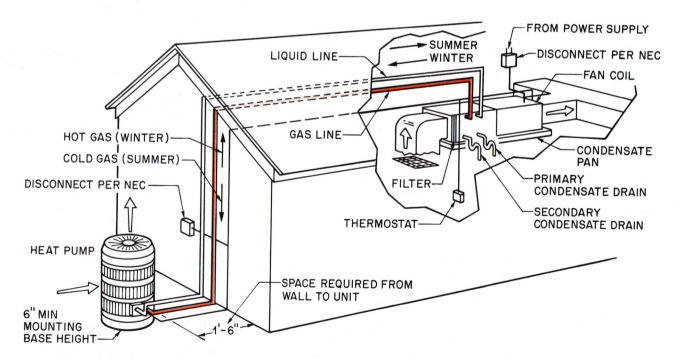

Figure 45–13. Split-system heat pump showing the interconnecting refrigerant lines. The large line is the gas line.

45.11   REFRIGERANT LINE IDENTIFICATION

When an air-to-air heat pump is a split system, the same lines are connected between the indoor unit and the outdoor unit except that they have a new name. The large line is called a *gas line* because it is always a gas line. In the unit on summer air conditioning it was called a cold gas line or a suction line. In a heat pump it is always a gas line because it is a suction line or cold gas line in summer and a hot gas line in winter, Figure 45–13.

The small line is a *liquid line* in summer and winter, so it keeps the same name. There are some changes that take place in the liquid line between summer operation and winter operation. The liquid flows toward the inside unit in summer and toward the outside unit in winter, Figure 45–14. The line is the same size as for cooling; the liquid direction is just reversed.

45.12   METERING DEVICES

Since the direction of the liquid flow is reversed from one season to another, some of the refrigeration components are slightly different. The metering devices used on heat pumps are different because there must be a metering device at the indoor unit as well as at the outdoor unit at the proper time. For example, when the unit is in the cooling mode, the metering device is at the indoor unit. When the system changes over to heating, a metering device must then meter refrigerant to the outdoor unit. This is accomplished in various ways with several combinations of metering devices.

45.13   THERMOSTATIC EXPANSION VALVES

The thermostatic expansion valve was the first metering device in common use with heat pumps. Since this device will only allow liquid to flow in one direction, it had to have a check valve piped in parallel

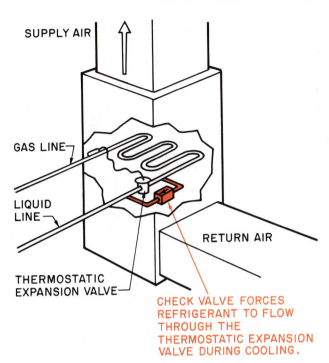

Figure 45–15. Piping diagram that must be used when a thermostatic expansion valve is used for a metering device on a heat pump. Notice the check valve.

with it to allow flow around it in the other mode of operation. For example, in the cooling mode the valve needs to meter the flow with the liquid refrigerant moving toward the indoor unit. When the system reverses to the heating mode, the indoor unit becomes a condenser and the liquid needs to be able to move freely toward the outdoor unit. The liquid flows through the check valve in winter and is metered through the thermostatic expansion valve in the summer cycle, Figure 45–15.

✱The thermostatic expansion valve applied to a heat pump must be chosen carefully. The sensing bulb for the valve is on the gas line, which may become too hot

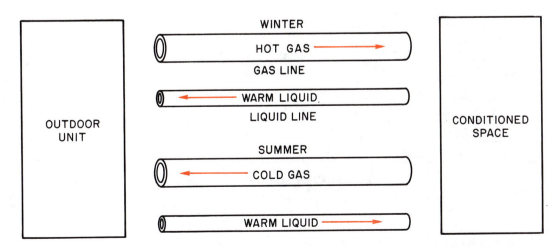

Figure 45–14. The small line is the liquid line in both the summer and winter operation.

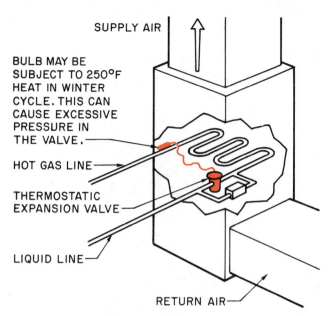

SUPPLY AIR

BULB MAY BE SUBJECT TO 250°F HEAT IN WINTER CYCLE. THIS CAN CAUSE EXCESSIVE PRESSURE IN THE VALVE.

HOT GAS LINE

THERMOSTATIC EXPANSION VALVE

LIQUID LINE

RETURN AIR

Figure 45-16. This illustration shows that care must be taken in the choice of thermostatic expansion valves. If an ordinary thermostatic expansion valve sensing bulb were to be mounted on this gas line, it would get hot enough to rupture the valve diaphragm.

for a typical expansion valve bulb.* It is not unusual for the hot-gas line to reach temperatures of 200°F. If the sensing bulb for a typical expansion valve is exposed to these temperatures, it is subject to rupture, Figure 45-16.

45.14   THE CAPILLARY TUBE

The capillary tube metering device is also used on heat pumps. This device allows refrigerant flow in either direction, but it is not normally used in this manner because capillary tubes of different sizes are required in summer and winter. There are several ways to use the capillary tube. One way is to use a check

valve to reverse the flow, as we have described. When this is done, two capillary tubes are usually used, Figure 45-17.

45.15   COMBINATIONS OF METERING DEVICES

Sometimes a combination of two metering devices is used. One popular combination is to use the capillary tube for the indoor metering device for summer operation, with a check valve piped parallel to allow flow in the other direction for winter operation. Then the outside unit may have a thermostatic expansion valve with a check valve. This system is very efficient because it uses the capillary tube in the summer mode only. It is an efficient metering device for summer operation because the load conditions are nearly constant. The thermostatic expansion valve is used in the winter on the outdoor coil. This is very efficient for the winter cycle because winter conditions are not constant. The thermostatic expansion device will reach maximum efficiency sooner than the capillary tube in the same application because it can open its metering port when needed. This allows the evaporator to become full of refrigerant earlier in the running cycle. This dual system uses each device at its best, Figure 45-18.

45.16   ELECTRONIC EXPANSION VALVES

The electronic expansion valve was discussed in the unit on refrigeration and is sometimes applied to heat pumps. If the heat pump is a close-coupled unit, such as a package unit (to be discussed later in this unit), a single valve can be used for the indoor and the outdoor unit. The reason the valve is best applied to a close-coupled unit is that the valve is metering in both directions, and if the liquid line were long it would need to be insulated. The valve will meter in either direction and maintain the correct superheat at the compressor's common suction line in the heating and cooling modes. The sensing element may be located on

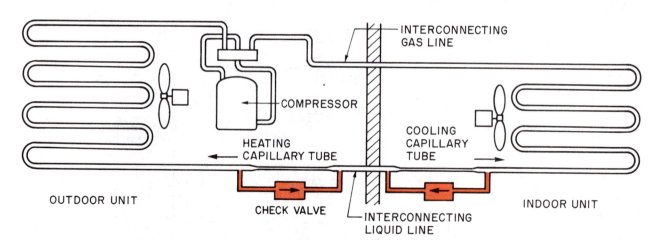

INTERCONNECTING GAS LINE

COMPRESSOR

HEATING CAPILLARY TUBE

COOLING CAPILLARY TUBE

OUTDOOR UNIT

CHECK VALVE

INTERCONNECTING LIQUID LINE

INDOOR UNIT

Figure 45-17. Capillary tube metering device used on a heat pump. Notice that check valves are used and there are two different capillary tubes.

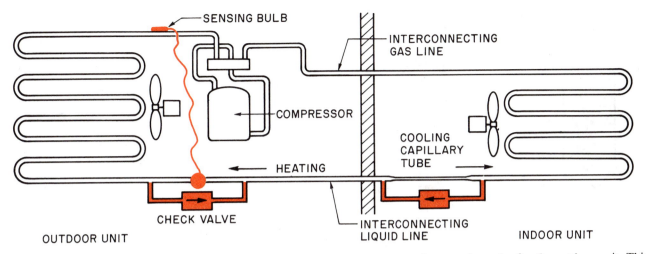

Figure 45–18. This unit uses a capillary tube for the indoor cycle and a thermostatic expansion valve for the outdoor unit. This thermostatic expansion valve has the sensing element mounted to the compressor permanent suction line.

the common suction line just before the compressor to maintain this correct refrigerant control.

45.17   LIQUID-LINE ACCESSORIES

Liquid-line filter-driers must be used in conjunction with all of the metering devices just described. When a standard liquid-line filter-drier is installed in a system that has check valves to control the flow through the metering devices, the drier is installed in series with the expansion device. The flow direction is the same as with the metering device. If the same filter-drier were installed in the common liquid line, it would filter in one direction and the particles would wash out in the other direction, Figure 45–19.

Special *biflow* filter-driers are manufactured to allow flow in either direction. They are actually two driers in one shell with check valves inside the drier shell to cause the liquid to flow in the proper direction at the proper time, Figure 45–20.

45.18   ORIFICE METERING DEVICES

Another common metering device used by some manufacturers is a combination flow device and check valve. This device allows full flow in one direction and restricted flow in the other direction, Figure 45–21. Two of these devices are necessary with a split system—one at the indoor coil and one at the outdoor coil. The metering device at the indoor coil has a larger bore than the one at the outdoor coil because the two coils have different flow characteristics. The summer cycle uses more refrigerant in normal operation. When this device is used, a biflow filter-drier is normally used in the liquid line when field repairs are done.

All of these components may be found on water-to-air heat pumps or air-to-air heat pumps. The components that control and operate the systems are much the same.

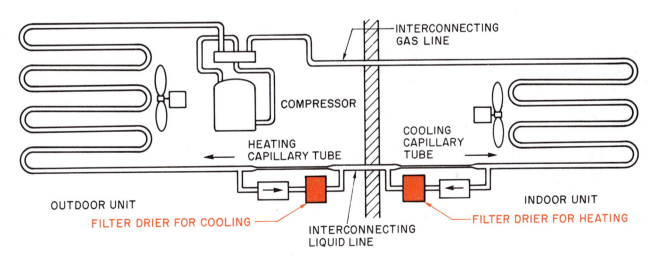

Figure 45–19. Placing a liquid-line filter-drier in the heat pump refrigerant piping.

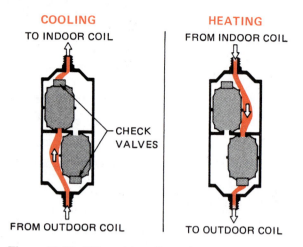

COOLING
TO INDOOR COIL

HEATING
FROM INDOOR COIL

CHECK VALVES

FROM OUTDOOR COIL

TO OUTDOOR COIL

Figure 45–20. Biflow drier. *Reproduced courtesy of Carrier Corporation*

45.19 APPLICATION OF THE AIR-TO-AIR HEAT PUMP

The air-to-air heat pump is the most popular and will be discussed in detail in this text. These units are normally installed in milder climates—in the "heat pump belt," which is basically those parts of the United States where winter temperatures can be as low as 10°F but seldom lower than 0°F.

The reason for this geographical line is the characteristic of the air-to-air heat pump. It absorbs heat from the outside air; as the outside air temperature drops, it is more difficult to absorb heat from it. For example, the evaporator must be cooler than the outside air for the air to transfer heat into the evaporator. Normally, in cold weather, the heat pump evaporator will be about 20° to 25°F cooler than the air from which it is absorbing heat. We will use 25°F temperature difference in this text as the example, Figure 45–22. On

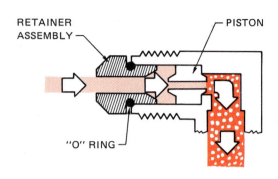

LIQUID SERVICE VALVE IN HEATING

RETAINER ASSEMBLY

PISTON

"O" RING

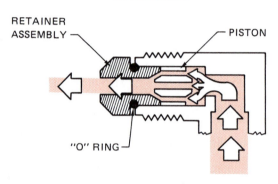

LIQUID SERVICE VALVE IN COOLING

RETAINER ASSEMBLY

PISTON

"O" RING

Figure 45–21. Combination fixed-bore (orifice) and check valve metering device. *Reproduced courtesy of Carrier Corporation*

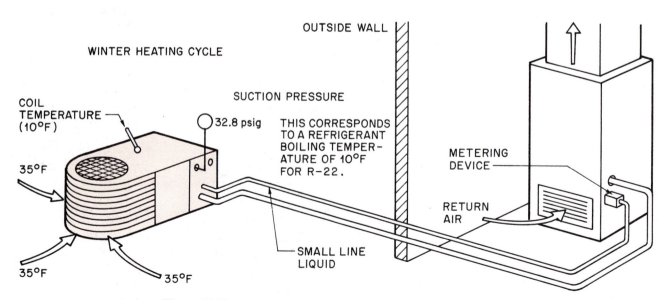

WINTER HEATING CYCLE

OUTSIDE WALL

COIL TEMPERATURE (10°F)

SUCTION PRESSURE

32.8 psig   THIS CORRESPONDS TO A REFRIGERANT BOILING TEMPERATURE OF 10°F FOR R-22.

35°F

35°F

35°F

SMALL LINE LIQUID

METERING DEVICE

RETURN AIR

Figure 45–22. Heat pump outdoor coil operating in the winter cycle.

a 10°F day the heat pump evaporator will be boiling (evaporating) the liquid refrigerant at 25°F lower than the 10°F air. This means that the boiling refrigerant temperature will be −15°F. Remember from Unit 21, as the evaporator temperature goes down, the compressor loses capacity. The compressor is a fixed-size pump. It will have more capacity on a 30°F day than on a 10°F day. **The heat pump loses capacity as the capacity need of the structure increases.** On a 10°F day the structure needs more capacity than on a 30°F day, but the heat pump's capacity is less. Thus, the heat pump must have help.

45.20 AUXILIARY HEAT

In an air-to-air heat pump system, the help that the heat pump gets is called *auxiliary heat*. The heat pump itself is the primary heat, and the auxiliary heat may be electric, oil, or gas. Electric auxiliary heat is the most popular because it is easier to adapt a heat pump to an electric system.

The structure that a heat pump is to heat has a different heat requirement for every outside temperature level. For example, a house might require 30,000 Btu of heat during the day when the outside temperature is 30°F. As the outside temperature drops, the structure requires more heat. It might need 60,000 Btu of heat in the middle of the night when the temperature has dropped to 0°F. The heat pump could have a capacity of 30,000 Btu at 30°F and 20,000 Btu/hr at 0°F. The difference of 40,000 Btu/hr must be made up with auxiliary heat.

45.21 BALANCE POINT

The *balance point* occurs when the heat pump can pump in exactly as much heat as the structure is leaking out. At this point the heat pump will completely heat the structure by running continuously. Above this point the heat pump will cycle off and on. Below this point the heat pump will run continuously but will not be able to heat the structure by itself.

45.22 COEFFICIENT OF PERFORMANCE

If the heat pump will not heat the structure all winter, why is it so popular? This can be answered in one word, *efficiency*. To understand the efficiency of an air-to-air heat pump, you need an understanding of electric heat. A customer receives 1 watt of usable heat for each watt of energy purchased from the power company while using electric resistance heat. This is called 100% efficient or a *coefficient of performance* (COP) of 1 to 1. The output is the same as the input. With a heat pump the efficiency may be improved as much as 3 to 1 for an air-to-air system with typical equipment. When the 1 W of electrical energy is used in the compression cycle to absorb heat from the out-

side air and pump this heat into the structure, the unit could furnish 3 W of usable heat. Thus its COP is 3 to 1, or it can be thought of as 300% efficient. See Figure 45–23 for a manufacturer's rating table for an air-to-air heat pump.

Such a COP only occurs during higher outdoor winter temperatures. As the temperature falls, the COP also falls. A typical air-to-air heat pump will have a COP of 1.5 to 1 at 0°F, so it is still economical to operate the compression system at these temperatures along with the auxiliary heating system. Some manufacturers have controls to shut off the compressor at low temperatures. When the temperature rises, the compressor will come back on. Some temperatures that manufacturers use to stop the compressor are 0° to 10°F. The long running times of the compressor will not hurt it or wear it out to any extent, so some manufacturers do not shut off the compressor at all.

The *water-to-air* heat pump might not need auxiliary heat because the heat source (the water) can be a constant temperature all winter. The heat loss (the heat requirement) and the heat gain (the cooling requirement) of the structure might almost be equal. Consequently, these heat pumps have COP ratings as high as 3.5 or even 4 to 1. Just remember that the higher COP ratings of water-to-air heat pumps are due not to the components of the refrigeration machine but to the temperatures of the heat source. Earth and lakes have a more constant temperature than air. Such pumps are very efficient for winter heating because of the COP and for summer cooling because they operate at water-cooled air conditioning temperatures and head pressures.

45.23 SPLIT-SYSTEM AIR-TO-AIR HEAT PUMP

Like summer air conditioning equipment, air-to-air heat pumps come in two styles: split systems and package (self-contained) systems.

The split-system air-to-air heat pump resembles the split-system summer air conditioning system. The components look exactly alike. An expert normally cannot tell if the equipment is air conditioning or heat pump equipment from the outside.

The indoor unit of an air-to-air system may be an electric furnace with a heat pump indoor coil where the summer air conditioning evaporator would be placed, Figure 45–24. The outdoor unit may resemble a cooling condensing unit.

45.24 THE INDOOR UNIT

The indoor unit is the portion of the unit that circulates the air for the structure. It contains the fan and coil. The airflow pattern may be upflow, downflow, or horizontal to serve different applications. Some

MODEL	541B034SHP					
SERIES	A					
ELECTRICAL						
Unit Volts—Hertz—Phase	208-230—60—1					
Operating Voltage Range	197-253					
Unit Ampacity for Wire Sizing	28.6					
Min Wire Size (60° Copper) (AWG)	10					
Max Branch Circuit Fuse Size (Amps)	45					
Total Unit Amps	23.1					
Compressor Rated Load Amps	21.8					
Locked Rotor Amps	88					
Fan Motor	1/6 HP, PSC					
Full Load Amps	1.3					
PERFORMANCE DATA						
ARI Sound Rating Number	19					
517B/HPFC (D)	030	036	—	—	—	—
519A/MCC	—	—	036	036	042	042
520B/BP	—	—	042	—	042	—
Rated Heating Capacity—47° F	33,000	34,500	34,500	34,000	35,000	34,500
Watts	4100	3950	3900	3850	3950	3850
COP	2.4	2.6	2.6	2.6	2.6	2.6
Rated Heating Capacity—17° F	18,000	19,000	19,000	18,500	19,000	18,500
Watts	3300	3200	3150	3100	3200	3100
COP	1.6	1.7	1.8	1.8	1.7	1.8
Rated Cooling Capacity (Btuh)	30,000	32,000	32,000	32,500	32,500	32,500
Watts	4500	4500	4450	4400	4600	4500
EER	6.7	7.1	7.2	7.4	7.1	7.2
Min Application Indoor Airflow (Ft³/Min)	1000	1050	1050	1050	1100	1100

541DJ030/ 517E030	Indoor Coil Airflow Ft³/Min* 1100	EDB* 70° F	OUTDOOR COIL ENTERING AIR TEMPERATURE °F																
			−13	−8	−3	2	7	12	17	22	27	32	37	42	47	52	57	62
Instantaneous Capacity (MBtuh)			10.50	12.00	13.50	15.00	16.50	18.10	19.60	21.30	22.90	24.70	26.80	28.00	31.00	33.60	36.50	39.50
Integrated Capacity (MBtuh)†			9.68	11.00	12.40	13.80	15.20	16.60	17.90	19.20	20.30	21.40	24.40	28.60	31.00	33.60	36.50	39.50
Total Power Input (KW)‡			2.04	2.13	2.22	2.31	2.41	2.51	2.60	2.71	2.80	2.91	3.04	3.14	3.28	3.44	3.61	3.80

Multipliers for Determining the Performance With Other Indoor Sections

Indoor Section	Size	Heating		Indoor Section	Size	Heating	
		Capacity	Power			Capacity	Power
506B	030	0.99	0.99	513C	030	1.01	1.00
	036	1.01	0.98	517E	030	1.00	1.00
507D & 518A	030	0.99	0.99		036	1.01	0.97
	036	1.01	1.00	518A036/ 520B042	—	1.00	0.96
507H & 518B	030	0.98	0.98				
	036	1.00	0.96				

*See the Heating Performance Correction Factors Table for Ft³/Min and indoor coil entering air temperature adjustments.

†The Btuh heating capacity values shown are net "integrated" values from which the defrost effect has been subtracted. The Btuh heating from supplement heaters should be added to those values to obtain total system capacity.

‡The KW values include the compressor, outdoor fan motor, and indoor blower motor. The KW from supplemental heaters should be added to these values to obtain total system KW.

Figure 45-23. Rating table of an air-to-air heat pump showing the capacities of the heat pump at various outdoor air temperatures. *Courtesy BDP Company*

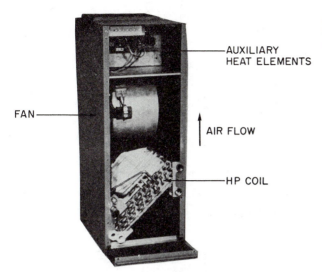

Figure 45–24. Air-to-air heat pump indoor unit. It is an electric furnace with a heat pump coil in it. Notice that the air flows through the heat pump coil and then through the heating elements. *Reproduced courtesy of Carrier Corporation*

manufacturers have very cleverly designed their units so that one unit may be adapted to all of these flow patterns. This is done by correctly placing the pan for catching and containing the summer condensate, Figure 45–25.

Coil placement in the airstream is very important in the indoor unit. The electric auxiliary heat and the primary heat (the heat pump) may need to operate at the same time in weather below the balance point of the house. **The refrigerant coil must be located in the airstream before the auxiliary heating coil.** Otherwise, heat from the auxiliary unit will pass through the refrigerant coil when both are operating in winter. If the auxiliary

HEATING & COOLING UNIT

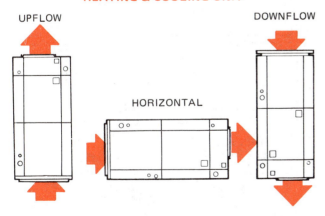

Figure 45–25. This heat pump indoor unit can be applied either in the vertical (upflow, downflow) or horizontal mode by placing the condensate pan under the coil in the respective position. *Reproduced courtesy of Carrier Corporation*

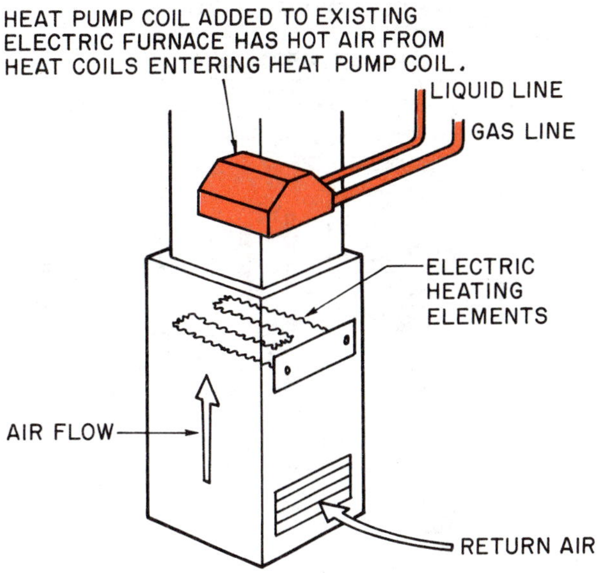

Figure 45–26. This illustration shows what will happen if a heat pump coil is applied to an electric furnace (oil or gas are the same) with the indoor coil after the electric heating coil.

heat is operating and is located before the heat pump coil, the head pressure will be too high and could rupture the coil or burn the compressor motor. Remember, the coil is operating as a *condenser* in the heating mode. It rejects heat from the refrigeration system, so any heat added to it will cause the head pressure to rise, Figure 45–26.

The indoor unit may be a gas or oil furnace. If this is the case, the indoor coil must be located in the outlet airstream of the furnace. When the auxiliary heat is gas or oil, the heat pump does not run while the auxiliary heat is operating. The coil must be located in the outlet of the furnace. If it isn't, the furnace heat exchanger would sweat in summer. The indoor coil then would be operating as an evaporator in summer, and the outlet air temperature could be lower than the dew-point temperature of the air surrounding the heat exchanger. Most local codes will not allow a gas or oil heat exchanger to be located in a cold airstream for this reason. When the auxiliary heat is gas or oil, special control arrangements must be made so that the heat pump will not operate while the gas or oil heat is operating. When a heat pump is added onto an electric furnace and the coil must be located after the heat strips, follow the same rules as for gas or oil. See Figure 45–27 for an example of this installation.

45.25   AIR TEMPERATURE OF THE CONDITIONED AIR

The heat pump indoor unit is installed in much the same way as a split-system cooling unit. The air distribution system must be designed more precisely due

THIS HEAT PUMP COIL IS INSTALLED IN THE HOT AIR STREAM OF THE GAS FURNACE. THIS SYSTEM HAS A CONTROL ARRANGEMENT THAT WILL NOT ALLOW THE FURNACE TO RUN AT THE SAME TIME AS THE HEAT PUMP. THIS CONTROL ARRANGEMENT CAN BE USED FOR THE ELECTRIC FURNACE IN FIGURE 45-26.

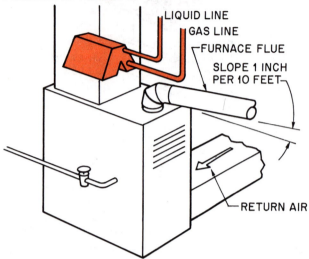

Figure 45-27. Gas furnace with a heat pump indoor coil instead of an air conditioner evaporator.

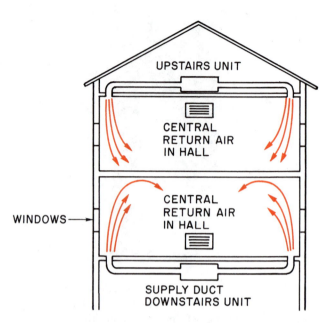

Figure 45-28. This heat pump installation shows a good method of distributing the warm air in the heating mode.

to the air temperatures leaving the air handler during the heating mode. The air temperature is not as hot as with gas and oil systems that are normally the heating system with electric cooling. The heat pump usually has leaving air temperatures of 100°F or lower when just the heat pump is operating. If the airflow is restricted, the air temperature will go up slightly but the unit COP will not be as good. The efficiency will be reduced. Most heat pumps require a minimum of 400 cfm/ton of capacity in the cooling mode. This will equate to approximately 400 cfm/ton in heating.

When 100°F air is distributed in the conditioned space, it must be distributed carefully or drafts will occur. Normally, the air distribution system is on the outside walls. The air registers are either in the ceiling or the floor, depending on the structure. It is common in two-story houses to have the first-story registers in the floor with the air handler in the crawl space below the house. The upstairs unit is commonly located in the attic crawl space with the registers in the ceiling near the outside walls. The outside walls are where the heat leaks out of a house in the winter and leaks into the house in the summer. If these walls are slightly heated in the winter with the airstream, less heat is taken out of the air in the conditioned space. The room air stays warmer, Figure 45-28. 100°F air mixed with room air does not feel like 130°F air mixed with room air from a gas or oil furnace. It may feel like a draft.

When 100°F air is distributed from the inside walls, such as high-side wall registers that were mentioned for

gas and oil systems in an earlier unit, the system may not heat satisfactorily. The air will mix with the room air and might feel drafty. The outside walls will also be lower in temperature than when the air is directed on them, and a "cold wall" effect could be noticed in cold weather, Figure 45-29.

45.26 THE OUTDOOR UNIT INSTALLATION

The outdoor unit installation for a heat pump is much like a central air conditioning system from an

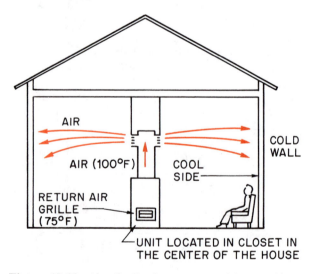

Figure 45-29. Air distribution system with the air being distributed from the inside using high-side wall registers shows the heat pump 100°F air mixing with the room air of 75°F.

airflow standpoint. The unit must have a good air circulation around it, and the discharge air must not be allowed to recirculate.

There are some more serious considerations that should be dealt with. The direction of the prevailing wind in the winter could lower the heat pump performance. If the unit is located in a prevailing north wind or a prevailing wind from a lake, the performance may not be up to standard. A prevailing north wind might cause the evaporator to operate at a lower than normal temperature. A wind blowing inland off a lake will be very humid and might cause freezing problems in the winter.

*The outdoor unit must *not* be located where roof water will pour into it.* The outdoor unit will be operating at below freezing much of the time, and any moisture or water that is not in the air itself should be kept from the unit's coil. If not, excess freezing will occur.

The outdoor unit is an evaporator in winter and will attract moisture from the outside air. If the coil is operating below freezing, the moisture will freeze on the coil. If the coil is above freezing, the moisture will run off the coil as it does in an air conditioning evaporator. This moisture must have a place to go. If the unit is in a yard, the moisture will soak into the ground. *If the unit is on a porch or walk, the moisture could freeze and create slippery conditions, Figure 45–30.*

The outdoor unit is designed with drain holes or pans in the bottom of the unit to allow free movement of the water away from the coil. If they are inadequate, the coil will become a solid block of ice in cold weather. When the coil is frozen solid, it is a poor heat exchanger with the outside air, and the COP will be reduced. Defrosting methods are discussed later.

Manufacturer's have recommended installation procedures for their particular units that will show the installer how to locate the unit for the best efficiency and water drainage.

The refrigerant lines that connect the indoor unit to the outdoor unit are much the same as for air conditioning. They come in line sets with the large line insulated. Quick-connect fittings with precharged lines or flare connectors with nitrogen-charged lines are typical. The only difference that should be considered is the large line, the gas line. In winter it may be 200°F and should be treated as a hot line in which heat must be contained. Therefore, many manufacturers use a thicker insulation on the gas line. The gas line should not be located next to any object that will be affected by its warm outside temperature.

SAFETY PRECAUTION: *Installing a heat pump involves observing the same safety precautions as installing a summer air conditioning system. The duct work must be insulated, which requires working with fiberglass. Be careful when working with metal duct and fasteners and especially when installing units on rooftops and other hazardous locations.*

45.27   PACKAGE AIR-TO-AIR HEAT PUMPS

Package air-to-air heat pumps are much like package air conditioners. They look alike and are installed in the same way. Therefore give the same considerations to the prevailing wind and water conditions. The pumps must have drainage in summer and winter. The package heat pump has all of the components in one housing and is easy to service, Figure 45–31. They have optional electric heat compartments that usually accept different electric heat sizes from 5000 W (5 kW) to 25 000 W (25 kW). When a system needs only 10 kW of auxiliary heat, the installing contractor need only install 10 kW of heat.

The metering devices used for package air-to-air heat pumps are much like those for split systems. Some manufacturers use a common metering device for the indoor and outdoor coils because the two are so close together. When one metering device is used, it must be

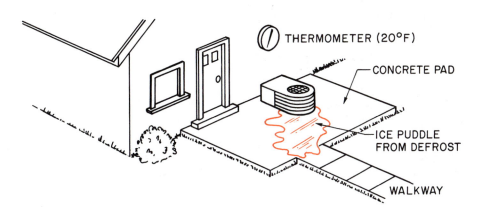

Figure 45–30. This heat pump was placed in the wrong place because the water that forms on the outside coil will run on to the walkway next to the coil and freeze.

Figure 45–31. Package heat pump. All components are contained in one housing. *Reproduced courtesy of Carrier Corporation*

able to meter both ways at a different rate in each direction because a different amount of refrigerant is used in the summer evaporator (indoor coil) than in the winter evaporator (outdoor coil).

Package air-to-air heat pumps must be applied correctly. For example, the duct must extend to the unit. Since the unit contains the outdoor coil also, the duct will need to extend all the way to an outside wall of the structure for a crawl space installation, Figure 45–32. Supply and return ducts must be insulated to prevent heat exchange between the duct and the ambient air. Insulating the duct is more important with a heat pump than with gas and oil installations because the cost of electricity to operate the heat pumps is much higher than the cost of gas or oil.

Package air-to-air heat pumps have the same advantage from a service standpoint as package summer air conditioners or package gas-burning equipment. All of the controls and components may be serviced from the outside. They also have the advantage of being factory assembled and charged. They are leak-checked and may even be operated before leaving the factory. This reduces the likelihood of having a noisy fan motor or other defect.

45.28   CONTROLS FOR THE AIR-TO-AIR HEAT PUMP

The air-to-air heat pump is different from any other combination of heating and cooling equipment. The following sequences of operation must be controlled at the same time for the heat pump to be efficient: space temperature, defrost cycle, indoor fan, the compressor, the outdoor fan, auxiliary heat, and emergency heat. Each manufacturer has its own method of controlling the heating and cooling sequence.

45.29   SPACE-TEMPERATURE CONTROL

The space temperature for an air-to-air heat pump is not controlled in the same way as a typical heating

and cooling system. There are actually two complete heating systems and one cooling system. The two heating systems are the refrigerated heating cycle from the heat pump and the auxiliary heat from the supplementary heating system—electric, oil, or gas. The auxiliary heating system may be operated as a system by itself if the heat pump fails. When the auxiliary heat becomes the primary heating system because of heat-pump failure, it is called *emergency heat* and is normally only operated long enough to get the heat pump repaired. The reason is that the COP of the auxiliary heating system is not as good as the COP of the heat pump. The electric system will be discussed in detail at this time. Oil and gas auxiliary heat systems will be discussed later.

The space-temperature thermostat is the key to controlling the system. It is normally a two-stage heating and two-stage cooling type of thermostat used exclusively for heat pump applications. Other variations in the thermostat concern the number of stages of heating or cooling. Figure 45–33 shows a typical heat-pump thermostat.

The following is the sequence of cooling and heating control for an automatic changeover thermostat—one that automatically changes from cooling to heating and back. The temperature-sensing element is a bimetal that controls mercury bulb contacts. The auxiliary heat is electric. The thermostat's fan switch is in AUTO.

45.30   COOLING CYCLE CONTROL

When the first-stage bulb of the thermostat closes its contacts (upon a rise in space temperature), the four-way valve magnetic coil is energized. When the compressor starts, it diverts the hot gas from the compressor to the outdoor coil, and the system is in the cooling cycle, Figure 45–34.

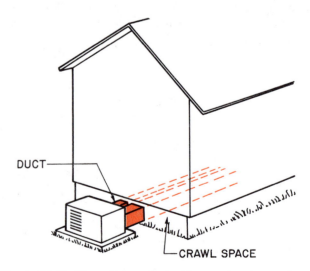

Figure 45–32. The self-contained unit contains the outdoor coil and the indoor coil. The duct work must be routed to the unit.

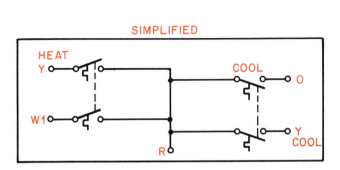

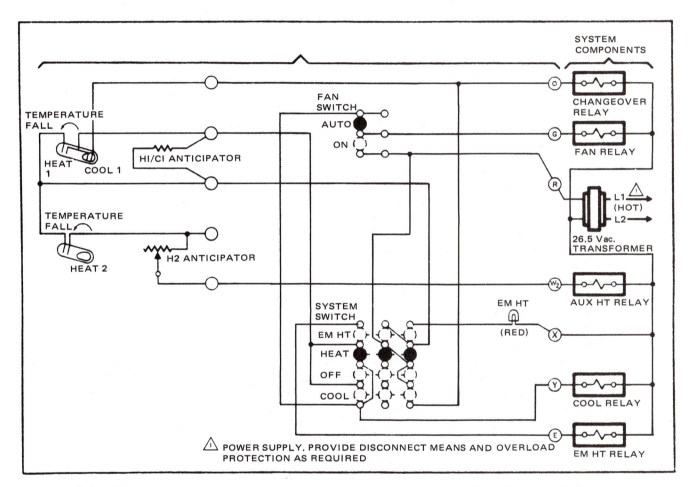

Figure 45–33. Heat pump thermostat with two-stage heating and two-stage cooling. *Photo by Bill Johnson,*

When the space temperature rises about 1°F, the second-stage contacts of the thermostat close. The second-stage contacts energize the compressor contactor and the indoor fan relay to start the compressor and outdoor fan and the indoor fan, Figure 45–35.

When the space temperature begins to fall (because the compressor is running and removing heat), the first contacts to open are the second-stage contacts. The compressor (and outdoor fan) and the indoor fan stop. The first-stage contacts remain closed, and the four-way valve remains energized.

When the space temperature rises again, the compressor (and outdoor fan) and indoor fan will start again. If the outdoor temperature is getting cooler (e.g.,

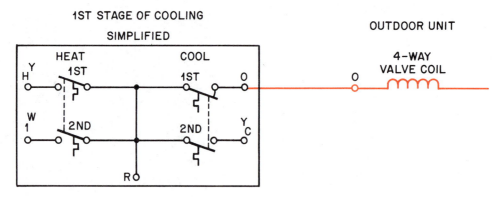

Figure 45–34. First stage of the cooling cycle where the four-way valve magnetic holding coil is energized.

in autumn), the space temperature will continue to fall, and the second-stage contacts will open. This deenergizes the four-way valve magnetic coil, and the unit can change over to heat when it starts up. **The four-way valve is a pilot-operated valve. It will not change over until the compressor starts and a pressure differential is built up.**

45.31   SPACE HEATING CONTROL

When the space temperature continues to fall due to outdoor conditions, the first-stage heating contacts close. This starts the compressor (and outdoor fan) and the indoor fan. The four-way valve is not energized, so the compressor's hot gas is directed to the indoor coil. The unit is now in the heating mode, Figure 45–36.

When the space temperature warms, the first-stage heating contacts open, and the compressor (and outdoor fan) and indoor fan stop.

If the outdoor temperature is cold, below the balance point of the structure, the space temperature will continue to fall because the heat pump alone will not heat the structure. The second-stage heating contacts will close and start the auxiliary electric heat to help the compressor. The second-stage heat was the last

component to be energized, so it will be the first component off. This is the key to how a heat pump can get the best efficiency from the refrigerated cycle. It will operate continuously because it was the first on, and the second-stage auxiliary heat will stop and start to assist the compressor, Figure 45–37. There are as many methods of controlling the foregoing of cooling and heating sequence as there are manufacturers. One common variation for controlling the heating cycle is to use outdoor thermostats to control the auxiliary heat, Figure 45–38. This keeps all of the auxiliary heat from coming on with the second stage of heating. It will come on based on outdoor temperature by the use of outdoor thermostats. These thermostats only allow the auxiliary heat to come on when needed, using the structure balance point as a guideline.

45.32   EMERGENCY HEAT CONTROL

The emergency heat feature is only used when the heat pump cannot function. For example, if a homeowner were to have a compressor problem, such as a noise that suddenly develops, on Sunday, the thermostat selector switch can be changed to the emergency heat setting. This will stop the compressor and ener-

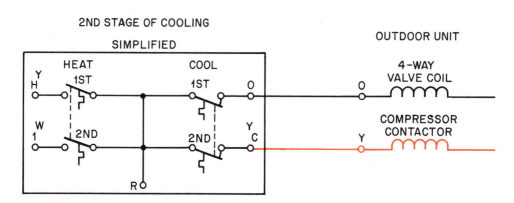

Figure 45–35. This diagram shows what happens with a rise in temperature. The first stage of the thermostat is already closed then the second stage closes and starts the compressor. The compressor was the last on and will be the first off.

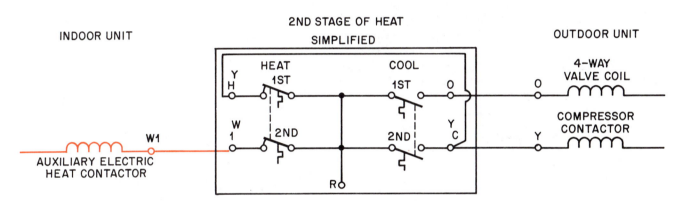

Figure 45-36. The space temperature drops to below the cooling set points for the first stage of heating, the compressor starts. This time, the "four-way valve" magnetic coil is de-energized and the unit will be in the heating mode.

Figure 45-37. When the outside temperature continues to fall, it will pass the balance point of the heat pump. When the space temperature drops approximately 1.5° F, the second-stage contacts of the thermostat will close and the auxiliary heat contactors will start the auxiliary heat.

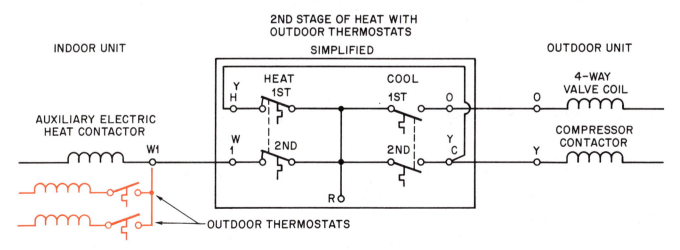

Figure 45-38. Wiring diagram with outdoor thermostats to control the electric auxiliary heat.

gize the auxiliary heat in the emergency mode. The outdoor thermostat contacts will be controlled out of the circuit for this mode so that maximum heat will be obtained from the auxiliary heat. The indoor space thermostat will control the auxiliary heat only, and it will be called emergency heat, Figure 45–39.

45.33   THE DEFROST CYCLE

The *defrost cycle* is used to defrost the ice from the outside coils during winter operation. The outside coils operate below freezing any time the outdoor temperature is below 45°F, because the outdoor coil (evaporator) operates 20° to 25°F colder than the outdoor air temperature. The outdoor coil is colder than the outdoor air so as to be able to absorb heat from the outdoor air, Figure 45–23. The need for defrost varies with the outdoor air conditions and the running time of the heat pump. For example, when the outdoor temperature is 45°F, the heat pump will satisfy and shut off from time to time. The frost will melt off the coil during this off time. As the temperature falls, more running time causes more frost to form on the coil. When the outside air contains more moisture, more frost forms. During cold rainy weather, when the air temperature is 35° to 45°F, frost will form so fast that the coil will be covered with ice between defrost cycles.

Defrost is accomplished with the air-to-air heat pump by reversing the system to the cooling mode and stopping the outdoor fan. This makes the outdoor coil the condenser with no fan, so the coil will get quite warm even in the coldest weather. Ice will melt from the coil and run down on the surface around the heat pump outdoor unit. The indoor coil is in the cooling mode and will blow cold air during defrost. One stage of the

auxiliary heat is normally energized during defrost to prevent the cool air from being noticed. The system is cooling and heating at the same time and is not efficient so the defrost time must be held to a minimum. It must not operate except when needed.

Demand defrost means defrosting only when needed. Combinations of time, temperature, and pressure drop across the outdoor coil are used to determine when defrost should be started and stopped.

45.34   INSTIGATING THE DEFROST CYCLE

Starting the defrost cycle with the correct frost build-up is desirable. Manufacturers design the systems to start defrost as close as possible to when the coil builds frost that affects performance. Some manufacturers use time and temperature to start defrost. This is called *time and temperature instigated* and is performed with a timer and temperature-sensing device. The timer closes a set of contacts for 10 to 20 sec for a trial defrost every 90 min. The timer contacts are in series with the temperature sensor contacts so that both must be made at the same time. This means that two conditions must be met before defrost can start, Figure 45–40. Normally the timer tries for a defrost every 90 min. The sensing device contacts will close if the coil temperature is as low as 26°F, and defrost will start. Figure 45–41 shows a typical timer and a sensor. Typically, the timer runs any time the compressor runs, even in the cooling mode.

Another common method of starting defrost uses an air-pressure switch that measures the air-pressure drop across the outdoor coil. When the unit begins to accumulate ice, a pressure drop occurs and the air switch contacts close. This can be wired in conjunction with the timer and temperature sensor to assure that there

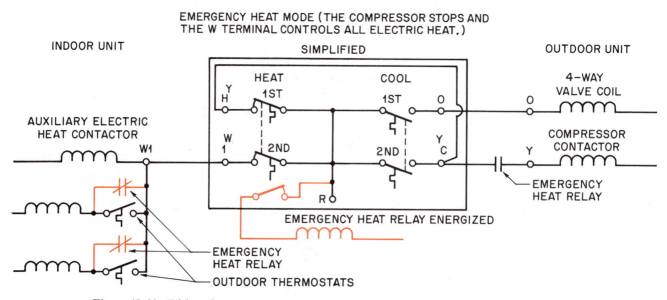

Figure 45–39. Wiring diagram of space-temperature thermostat set to the emergency heat mode.

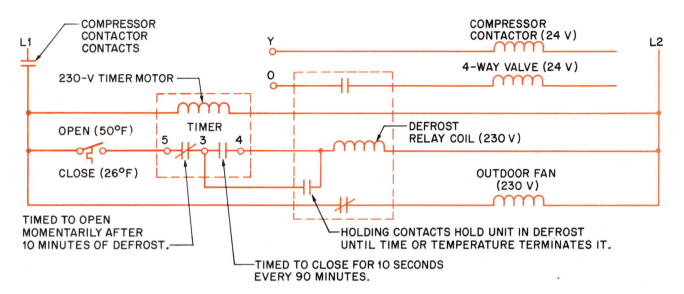

Figure 45–40. Wiring diagram of conditions to be met in this situation for defrost to start.

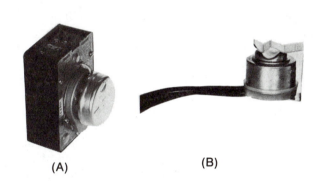

Figure 45–41. (A) Timer. (B) Sensor. *Photos by Bill Johnson*

is an actual ice buildup on the coil. The combination of time and temperature only assures that time has passed and the coil is cold enough to actually accumulate ice. It does not actually sense an ice thickness. It can be more efficient to defrost when there is an actual buildup of ice, Figure 45–42.

45.35   TERMINATING THE DEFROST CYCLE

Terminating the defrost cycle at the correct time is just as important as starting defrost only when needed. This is done in several ways. *Time and temperature* was the terminology used to start defrost. *Time or tempera-*

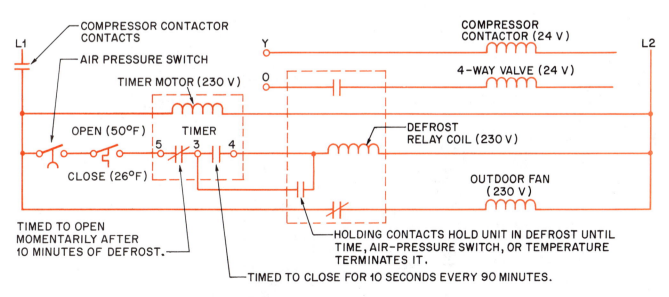

Figure 45–42. Wiring diagram taking ice buildup into consideration.

ture terminated is the terminology used to terminate one type of defrost cycle. The difference is that two conditions must be satisfied to start defrost, and either one of two conditions can terminate defrost. Once defrost has started, time or temperature can terminate it. For example, if the defrost temperature sensor warms up to the point that it is obvious that ice is no longer on the coil and its contacts break, defrost will stop. This temperature termination is normally 50°F at the location of the temperature sensor. If the contacts do not open because it is too cold outside, the timer will terminate defrost within a predetermined time period. Normally the maximum time the timer will allow for defrost is 10 min, Figure 45–43.

45.36 ELECTRONIC CONTROL OF DEFROST

Many heat pumps use electronic circuit boards for the timing portion of the defrost cycle. This electronic circuit board is used to keep the unit from cycling too often and to monitor voltage and amperage. Consult the manufacturer for service information if you need to work on one of these boards.

Starting the indoor fan motor of a heat pump differs from other heating systems. In other systems the indoor fan is started with a temperature-operated fan switch. With a heat pump the fan must start at the beginning of the cycle, which is controlled by the thermostat. The indoor fan is started with the fan terminal, sometimes labeled G. The compressor operates whenever the unit is calling for cooling or heating, so the G or fan terminal is energized whenever there is a call for cooling or heating.

45.37 AUXILIARY HEAT

During cold weather the air-to-air heat pump must have auxiliary heat. This is usually accomplished with an electric furnace with a heat pump coil for cooling and heating. Electric heat is often started with a sequencer that times to start and will time to stop. This means that the electric heat element is energized for a time period if the unit is shut off with the thermostat on–off switch. When the thermostat is allowed to operate the system normally, the heat pump will be the first component to come on and the last to shut off. The indoor fan will start with the compressor through the fan-relay circuit. The electric strip heat will operate as the last component to come on and the first to shut off. When the owner decides to shut off the heat and the strip heat is energized, the heating elements will continue to heat until the sequencers stop them. A temperature switch is often used to sense this heat and keep the fan running until the heaters cool. The electric furnaces used for heat pumps are like the ones described in Unit 30.

45.38 SERVICING THE AIR-TO-AIR HEAT PUMP

Servicing the air-to-air heat pump is much like servicing a refrigeration system. During the cooling season, the unit is operating as a high-temperature refrigeration system, and during the heating season it is operating as a low-temperature system with planned defrost. The servicing of the system is divided into electrical and mechanical servicing and occurs in the cooling and the heating seasons.

Servicing the electrical system is much like servicing any electrical system that is used to operate and control refrigeration equipment. The components that the manufacturer furnishes are built to last for many years of normal service. However, parts fail due to manufacturing faults and misuse. One significant point about a heat pump is that it has much more running time than a cooling only unit. A typical cooling unit in a

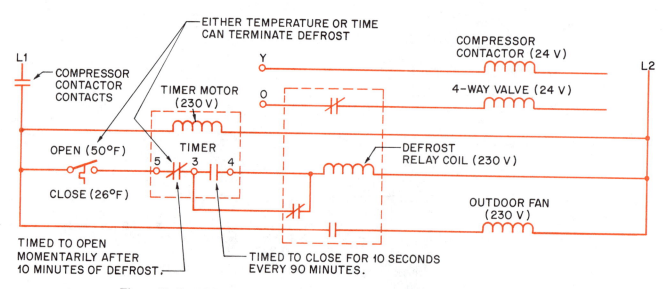

Figure 45–43. Wiring diagram for time and temperature instigated method of defrost.

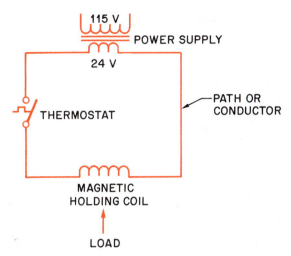

Figure 45–44. This is the basic example of a power supply, a path or conductor, and a load to consume power.

residence in Atlanta, Georgia, will run about 120 days per year. A heat pump may well run 250 days, more than twice the time. In winter the heat pump may run for days and not cut off when the temperature is below the balance point of the structure. This increase in running time shortens the life of the compressor and the electrical components. An average heat pump should last a few years less than a typical air conditioner.

45.39   TROUBLESHOOTING THE ELECTRICAL SYSTEM

Troubleshooting the electrical system of an air-to-air heat pump is similar to troubleshooting a summer air conditioning systems. Recall that there must be a power supply, a path for the electrical energy to flow on, and a load before electricity will flow. The power supply is the voltage to the unit, the path is the wiring, and the loads are the electrical components that perform the work. Every circuit in the heat pump's wiring can be reduced to these three items. It is important that you think in terms of power, conductor, and load, Figure 45–44.

It is usually easy to find electrical problems if some component will not function—for example, when a compressor contactor's coil is open. It is evident that the problem is in the contactor. A voltmeter reading can be taken on each terminal of the contactor coil. If there is voltage and it will not energize the contacts, a resistance check of the coil will show if there is a circuit

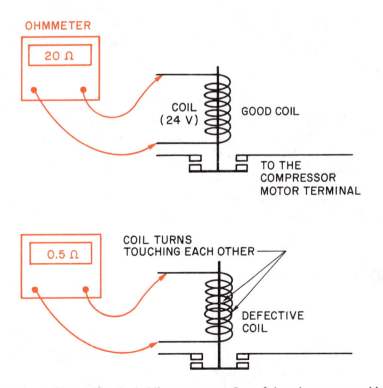

Figure 45–45. Magnetic holding coils used for the holding contactor. One of them has a measurable resistance of 0.5 ohm. The other one, the correct one, has a measurable resistance of 20 ohms. The one with the lower resistance is shorted and will draw too much current and place a load on the control transformer. If the load is too much, it will overload the transformer.

through it. Remember, a correctly operating circuit has the correct resistance. No resistance or too little resistance will create too much current flow. It sometimes helps to be able to compare a suspected bad component (for example, one with too little resistance) to a component of correct resistance, Figure 45-45.

Electrical malfunctions that cause the technician real problems are the intermittent ones, for example, a unit that does not always defrost and ice builds up on the outdoor coil. The heat pump can be very hard to troubleshoot in winter because of the weather. *If it is wet outside the emergency heat feature of the heat pump should be used until the weather improves. It is hard enough to find electrical problems in cold weather; the water makes it dangerous.*

45.40 TYPICAL ELECTRICAL PROBLEMS

1. Indoor fan motor or outdoor fan motor
   A. The electrical circuit is open; it will not start; draws no current.
   B. Burned; will not start; may draw high current.
   C. Bad capacitor; may run slowly and draw high current.
2. Compressor contactor, indoor fan relay or outdoor fan relay
   A. Coil winding is open; will not close contacts.
   B. Coil burned; might overload control transformer.
   C. Contacts burned; will cause local heating and burn wires.
3. Defrost relay
   A. Coil winding is open; will not close contacts.
   B. Coil burned; might overload control transformer.
   C. Contacts burned; cause defrost circuit problems; interfere with normal procedures of defrost.
4. Compressor
   A. Open winding; will not start; will draw high current if only one winding is open.
   B. Burned; high current; breaker tripped.
   C. Open internal overload; will not start; will show measureable resistance from R to S terminals.
5. Wiring problems
   A. Loose connections; local heat; insulation burned.
   B. Wires rubbing together; might cause electric short.

45.41 TROUBLESHOOTING MECHANICAL PROBLEMS

Mechanical problems can be very hard to find in a heat pump, particularly in winter operation. Summer operation of a heat pump is very similar to that of a conventional cooling unit. The pressures and temperatures are the same and the weather is not as difficult to work in. Mechanical problems are solved with gage manifolds, wet and dry bulb thermometers, and air-

measuring instruments. Refrigeration mechanical problems will be discussed in this text.

45.42 SOME TYPICAL MECHANICAL PROBLEMS

1. Indoor unit
   A. Air filter dirty (winter); high head pressure; low COP.
   B. Dirty coil (winter); high head pressure; low COP.
   C. Air filter dirty (summer); low suction pressure.
   D. Dirty coil (summer); low suction pressure.
   E. Refrigerant restriction (summer); low suction.
   F. Refrigerant restriction (winter); high head pressure.
   G. Defective thermostatic expansion valve (summer); low suction pressure.
   H. Defective thermostatic expansion valve (winter); no change.
   I. Leaking check valve (summer); flooded indoor coil.
2. Outdoor unit
   A. Dirty coil (winter); low suction pressure.
   B. Dirty coil (summer); high head pressure.
   C. Four-way valve will not shift; stays in same mode.
   D. Four-way valve halfway; pressures will equalize while running and appear as a compressor that will not pump.
   E. Compressor efficiency down; capacity down; suction and discharge pressures too close together. Compressor will not pump pressure differential.
   F. Defective thermostatic expansion valve (winter); low suction pressure.
   G. Leaking check valve (winter); flooded coil.

There are three mechanical problems to be aware of: (1) four-way valve leaking through; (2) compressor not pumping to capacity; and (3) charging the heat pump. These particular problems are different enough from commercial refrigeration that they need individual attention.

SAFETY PRECAUTION: *Be aware of the high-pressure refrigerant in the system while connecting gage lines. High-pressure refrigerant can pierce your skin and blow particles in your eyes; use eye protection. Liquid refrigerant can freeze your skin. The hot-gas line can cause serious burns and should be avoided.*

45.43 TROUBLESHOOTING THE FOUR-WAY VALVE

The four-way valve leaking through can very easily be confused with a compressor that is not pumping to capacity. The capacity of the system will not be up to normal in summer or winter cycles. This is the same symptom as small amounts of hot gas leaking from the high-pressure side of the system to the low-pressure side. When the gas is pumped around and around, work is accomplished in the compression process, but usable refrigeration is not available. When you suspect that a four-way valve is leaking from the high-pressure line to the low-pressure line, use a good-quality temperature tester to check the temperature of the low-side line, the suction line from the evaporator (the indoor coil in summer or the outdoor coil in winter), and the permanent suction line between the four-way valve and the compressor. The temperature difference should not be more than about 3°F. *Take these special precautions when recording temperatures: (1) Take the temperatures at least 5 in. from the valve body (to keep valve body temperature from affecting the reading); (2) insulate the temperature lead that is fastened tightly to the refrigerant line, Figure 45–46.*

45.44 TROUBLESHOOTING THE COMPRESSOR

Checking a compressor in a heat pump for pumping to capacity is much like checking a refrigeration compressor except there are normally no service valves to work with. Some manufacturers furnish a chart with the unit to show what the compressor characteristic should be under different operating conditions, Figure 45–47. If such a chart is available, use it.

The following is a reliable test, using field working conditions, that will tell if the compressor is pumping at near capacity. Any large inefficiencies will show up.

1. Whether summer or winter, operate the unit in the cooling mode. If winter the four-way valve may be switched by energizing it with a jumper or deenergizing it by disconnecting a wire to get the unit in the summer mode. This will allow the auxiliary heat to heat the structure and keep it from getting too cold.

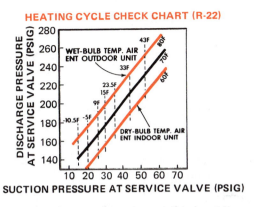

HEATING CYCLE CHECK CHART (R-22)

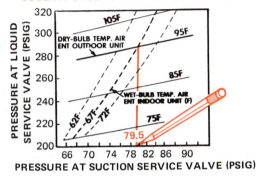

COOLING CYCLE CHARGING CHART (R-22)

Figure 45–47. Performance chart for a 36,000 Btu/h heat pump system. *Reproduced courtesy of Carrier Corporation*

2. Block the condenser airflow until the head pressure is 300 psig (this simulates a 95°F day) and the suction pressure is about 70 to 80 psig.

3. The compressor amperage should be at close to full load. The combination of discharge pressure, suction pressure, and amperage should reveal if the compressor is working at near full load. If the suction pressure is high and the discharge pressure is low, the amperage will be low. This indicates that the compressor is not pumping to capacity. Sometimes you can hear a whistle when the compressor is

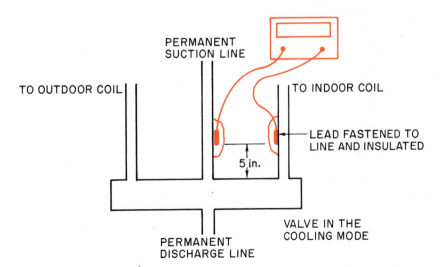

Figure 45–46. How to check performance of a four-way valve using temperature comparison.

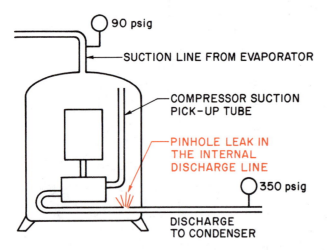

Figure 45–48. Compressor has vapor leaking from the high side to the low side.

shut off under these conditions. The suction line may also get warm immediately after shutting off the unit under these conditions. Such symptoms indicate that the compressor is leaking from the high side to the low side internally, Figure 45–48.

＊Remember, if there is any doubt as to where the leak through may be, perform a temperature check on the four-way valve.＊

45.45 CHECKING THE CHARGE

Most heat pumps have a critical refrigerant charge. The tolerance could be as close as $\pm \frac{1}{2}$ oz of refrigerant. Therefore don't install a standard gage manifold each time you suspect a problem. See Figure 45–49 for an example of a very short coupled system that may be used on the high side line that will not alter the operating charge. When a heat pump has a partial charge, it is obvious that the refrigerant leaked out of the system.

When a partial charge is found in a heat pump, some manufacturers recommend that the charge be vented to the atmosphere or captured in a spare drum. Then evacuate the system to a deep vacuum and recharge by measuring the charge into the system. Some manufacturers furnish a charging procedure to allow you to add a partial charge. It is always best to follow the manufacturer's recommendation. See Figure 45–50 for a sample performance chart for a heat pump.

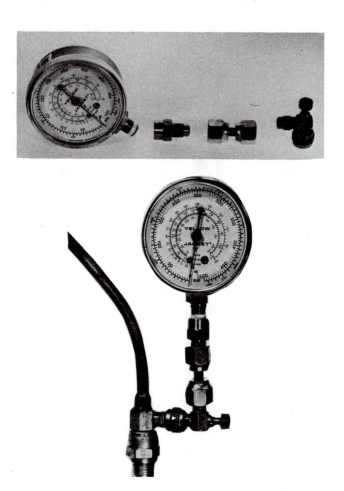

Figure 45–49. High-pressure gage for checking the head pressure on a heat pump with a critical charge. *Photos by Bill Johnson*

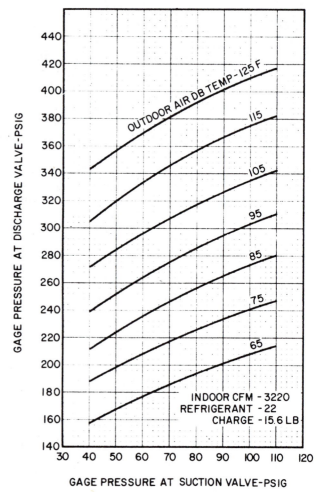

Figure 45–50. Heat pump performance chart.

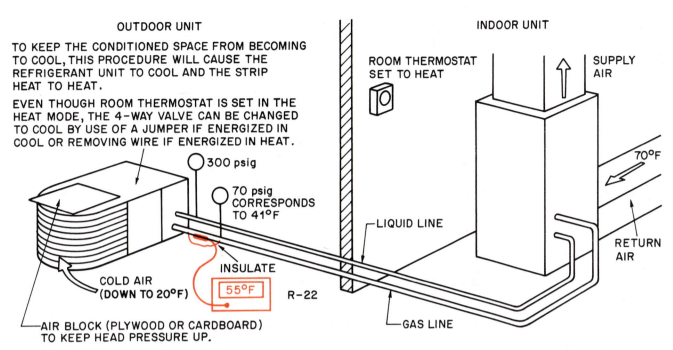

Figure 45–51. How a heat pump with a fixed-bore metering device may be charged when the manufacturer's chart or information is not available. See text for procedure.

If the manufacturer's recommendations are not available, the following method may be used to partially charge a system. This is specifically for systems with *fixed-bore metering devices*, not thermostatic expansion valves.

1. Start the unit in the cooling mode. If in the winter mode leave the electric heat where it can heat the structure. Block the condenser and cause the head pressure to rise to 300 psig. This will simulate a 95° F day operation. If the head pressure will not go up, refrigerant may have to be added.
2. Fasten a thermometer to the gas line—it will be cold while in the cooling mode. The suction pressure should be about 70 to 80 psig when fully charged, but may be lower if there is a reduced latent-heat load because of low humidity. If the system is split with a short gas line (10 to 30 ft), charge until the system has a 10° to 15° F superheat at the line entering the outdoor unit, Figure 45–51. If the line is long (30 to 50 ft), charge until the superheat is 12° to 18° F.

This charging procedure is very close to correct for a typical heat pump that has no liquid line to gas line heat exchange (some manufacturers have liquid to gas heat exchanges to improve their particular performance). This heat exchange is normally inside the outdoor unit and may not affect these procedures.

SAFETY PRECAUTION: * There is one particular precaution when working with the charge on any heat pump that has a suction-line accumulator, and this includes most heat pumps. Part of the charge can be stored in the accumulator and will boil out later. You can heat the accumulator by running water over it to drive the refrigerant out if you are in doubt. You may also give the liquid time to boil out on its own. Often, the accumulator will frost or sweat at a particular level if liquid refrigerant is contained in it, Figure 45–52.*

45.46 SPECIAL APPLICATIONS FOR HEAT PUMPS

The use of oil or gas furnaces for auxiliary heat is a special application. Several manufacturers have systems designed to accomplish this. There are several advantages in this type of system. Usually, natural gas is much cheaper to use as the auxiliary fuel than electricity is at a 1 to 1 COP. Fuel oil is usually not a good choice for auxiliary fuel because of cost. It may be considered where a fuel oil system is already installed or where fuel oil has a price advantage. Both systems have similar control arrangements. The heat pump indoor coil can be used in conjunction with an oil or gas furnace, but the coil must be downstream of the oil or gas heat exchanger. The air must flow through the oil or gas heat exchanger first, then through the heat pump indoor coil. This means that the gas or oil furnace must not operate at the same time that the heat pump is operating, Figure 45–53.

The control function can be accomplished with an outdoor thermostat set at the balance point of the structure to change the call for heating to the oil or

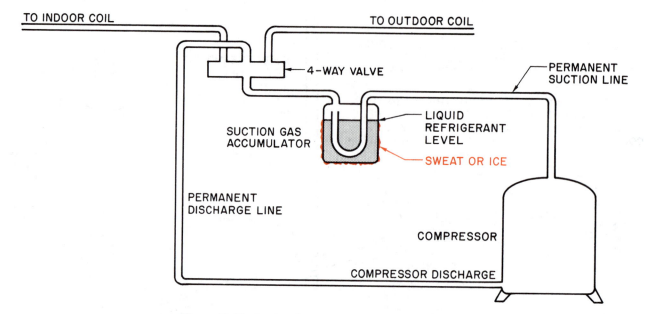

Figure 45–52. Suction-line accumulator that is sweating.

gas furnace and to shut off the heat pump. This allows the heat pump to operate down to the balance point when the oil or gas furnace will take over. When defrost takes place, the oil or gas furnace will come on during defrost, because the same controls are used as for a standard heat pump, and a call for auxiliary heat to warm the cool air during defrost is still part of the control sequence. *A high-pressure control can prevent the compressor from overloading if a defrost does not terminate due to a defective control.*

45.47 MAXIMUM HEAT PUMP RUNNING TIME

There are other control modifications that allow the heat pump to run below the balance point. Such arrangements have more relays and controls. The heat pump is designed to operate whenever it can. This is accomplished by using the second-stage contacts to start the oil or gas heat. This sequence stops the heat pump until the second-stage thermostat's contacts satisfy. The heat pump then starts again. This sequence

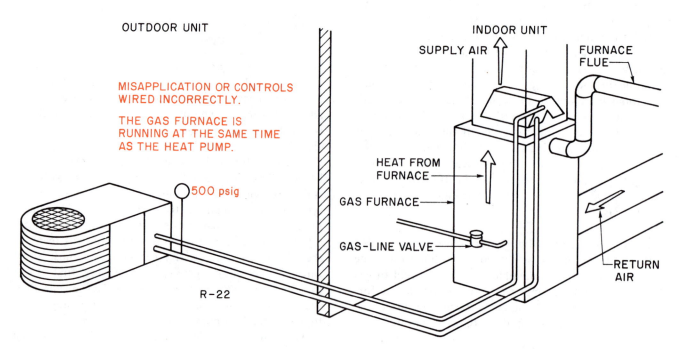

Figure 45–53. Operating conditions inside a heat pump coil while a gas furnace is operating at the same time the heat pump is operating.

repeats itself each cycle. Before considering this control sequence, a study of the system should be done to see if it is more economical to switch over from the heat pump to the auxiliary source or to restart the heat pump. The economic balance point of the structure may be determined using the cost comparison of fuels.

45.48 ADD-ON HEAT PUMP TO EXISTING ELECTRIC FURNACE

When a heat pump indoor coil is added to an existing electric furnace installation, the older furnace might not be equipped to have the heat pump coil located before the electric heating elements. If so, a wiring configuration similar to an oil or gas installation may be used, Figure 45-54.

Many manufacturers have different methods of building a heat pump for their own performance characteristics. These methods involve many different piping configurations with special heat exchangers. This text is intended to explain the typical heat pump. If a different type of system is encountered, the manufacturer should be consulted.

45.49 SERVICE TECHNICIAN CALLS

When reading through these service calls, keep in mind that three things are necessary for a circuit to be complete and current to flow: a power supply (line voltage), a path (wire), and a load (the measurable resistance motor or heat). The mechanical part of the system has four components that work together: the evaporator absorbs heat, the condenser rejects heat, the compressor pumps the heat-laden vapor, and the metering device meters the refrigerant.

Service Call 1

Customer calls. The air conditioning system is heating not cooling. This is the first time the unit has been operated in the cooling mode this season. The problem is the magnetic coil winding on the four-way valve is open and will not switch the valve over into the cooling mode.

The service technician turns the space thermostat to the cooling mode and starts the system. This is a split system with the air handler in a crawl space under the house. *The technician goes to the outdoor unit and carefully touches the gas line (the large line); it is hot, meaning the unit is in the heating mode.* The panel is removed and the voltage is checked at the four-way valve holding coil. 24 V are present; the coil should be energized. The unit is shut off and one side of the four-way valve coil is disconnected. A continuity check shows that the coil is open. The coil is replaced and the unit is started. It now operates correctly in the cooling mode.

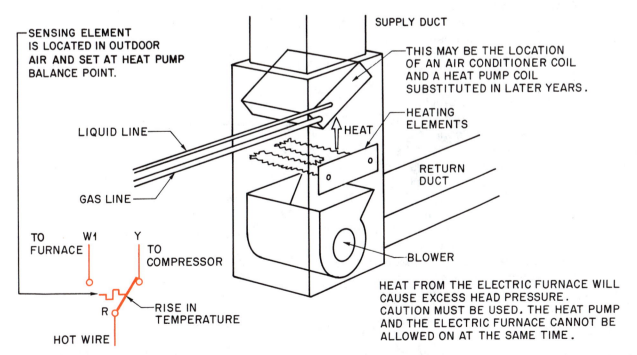

Figure 45-54. Electric furnace heat coils are below the heat pump coil and will cause high head pressure if used incorrectly with a heat pump.

Service Call 2

Customer calls. The air conditioning system in a small store will not start. This is a split-system heat pump with the outdoor coil on the roof. The problem is the four-way valve 24-V holding coil is burned and has overloaded the control transformer. This has blown the fuse in the 24-V control circuit.

The technician begins by trying to start the unit at the space-temperature thermostat. The unit will not start. When the FAN ON switch is turned to FAN ON, the fan will not start. A control voltage problem is suspected. The indoor unit is in a utility closet downstairs. *The power is turned off and the door to the indoor unit is removed.* The power is restored and it is found that there is no power at the low-voltage transformer secondary. The low voltage fuse is checked and found to be blown. A new fuse is in a box in the utility room. Before installing the new fuse, the technician turns off the power and then replaces the fuse. One lead is then removed from the control transformer. An ohmmeter is fastened to the two leads leaving the transformer and the resistance is 0.5 Ω. This seems very low and will likely blow another fuse if power is restored. Some 24-V component must be burned and have a shorted coil. (Instead of 20 Ω resistance, it has 0.5 Ω. This will increase the current flow enough to blow the fuse.) A look at the unit's diagram will show which components are 24 V. One of them must be burned. It is decided to check the components in the indoor unit first and save a trip on the roof if possible. Each component is checked, one at a time, by removing one lead and applying the ohmmeter. There are three sequencers for the electric heat, and one fan relay. There is no problem here. *The technician goes to the roof and turns off the power.* The compressor contactor is checked. It shows a resistance of 26 Ω. The four-way valve magnetic holding coil shows a resistance of 0.5 Ω. It is replaced with a new coil. The new coil has a resistance of 20 Ω. The problem has been found. The unit is put in operation with an ammeter attached to the lead leaving the control transformer to check for excessive current. An amperage multiplier made of thermostat wire wrapped 10 times around the ammeter lead is used. The current is 1 A which is normal.

Service Call 3

Customer calls. The heat pump cooling in the house has an ice bank on the outdoor coil. The customer is advised to turn the thermostat to the emergency heat mode until a technician can get there. The problem is the defrost relay coil winding is open, and the unit will not go into defrost.

The weather is below freezing, so when the technician arrives, there is still an ice bank on the outdoor coil. *With the unit off, the technician applies a jumper wire across terminals 3 and 4 on the defrost timer, Figure 45-55.* The unit is then started. This should force the unit into defrost if the timer contacts are not closing. This does not put the unit into defrost, so the jumper is left in place and the defrost temperature sensor is jumped. This satisfies the two conditions that must be satisfied to start defrost, but nothing happens. *The technician turns off the unit, turns off the power, removes the jumpers, and checks the defrost relay.* The coil is open. The relay is changed for a new one and the unit is started again. When the jumper is again applied to terminals 3 and 4, the unit goes into defrost. Remember that two conditions, time and temperature, must be satisfied before defrost can start. The temperature condition is satisfied because the coil is frozen. When the connection from terminals 3 to 4 is made, defrost is started. This defrost would have occurred normally when the time clock advanced to a trial defrost. By jumping from 3

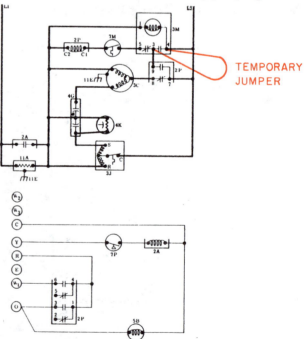

TEMPORARY JUMPER

Figure 45-55. *With the unit off, the technician applies a jumper wire from terminals 3 to 4 to cause a defrost. This is to keep from waiting for the timer. This is a high voltage circuit and this exercise should not be performed unless the technician is very experienced.* *Reproduced courtesy of Carrier Corporation*

to 4, the technician did not have to wait for the timer. The unit is allowed to go through defrost, and the coil is still iced. The technician decides to shut off the unit and use artificial means to defrost the coil. *The power is turned off. A water hose is pulled over to the unit and city water is used to melt the ice. City water is about 45°F. Care is taken that water is not directed on to the electrical components.* When the ice is melted, the unit is put back in operation. A call back the next morning to the customer verifies that the unit is not icing any more.

Service Call 4

Customer calls. The unit is running continuously and not heating as well as it used to. The problem is the outdoor fan is not running because the defrost relay has a set of burned contacts—the ones that furnish power to the outdoor fan. These contacts are used to stop the outdoor fan in the defrost cycle. They should also start the fan at the end of the cycle.

The technician finds that the outdoor fan is not running during the normal running cycle. The gas line is warm, and the coil is iced up, so the unit is in the heating cycle. The fan is not running. The unit is shut off and the panel to the fan control compartment is removed. A voltmeter is fastened to the fan motor leads and the unit is started. There is no voltage going to the fan motor. The wiring diagram shows that the power supply to the fan goes through the defrost relay. *The power is checked entering and leaving the defrost relay.* Power is going in but not coming out. The defrost relay is changed, and the unit starts and runs as it should.

Service Call 5

Customer calls. The heat pump serving the residence is blowing cold air for short periods of time. This is a split-system heat pump with the indoor unit in the crawl space under the house. The problem is the contacts in the defrost relay that bring on the auxiliary heat during defrost are open and will not allow the heat to come on during defrost.

The technician is familiar with the particular heat pump and believes that the problem is in the defrost relay or the first-stage heat sequencer. The first thing to do is to see if the sequencer is working correctly. The space-temperature thermostat is turned up until the second-stage contacts are closed. The technician goes under the house with an ammeter

and verifies that the auxiliary heat will operate with the space thermostat. Checking the defrost relay is the next step. This can be done by falsely energizing the relay, by waiting for a defrost cycle, or by simulating a defrost cycle. The technician decides to simulate a defrost cycle in the following manner. *The power is turned off. To prevent the outdoor fan from running, one of the motor leads is removed.* Then the unit is restarted. This causes the outdoor coil to form ice very quickly and make the defrost temperature thermostat contacts close. When the coil has ice on it for 5 min, the technician jumps the contacts on the defrost timer from 3 to 4, and the unit goes into defrost, Figure 45–55. The technician goes under the house and checks the strip heat; it is not operating. The problem must be the defrost relay. By the time the technician gets back to the outdoor unit, the unit is out of defrost. It is decided that the defrost cycle is functioning as it should, so the technician carefully energizes the defrost relay with a jumper wire. The 24-V power supply feeds terminal 4 and then should go out on terminal 6 to the auxiliary heat sequencer. These contacts are not closing. The defrost relay is changed, and the new relay is energized. The technician goes under the house and verifies that the strip heat is working. A call back later verifies all is well with the owner.

Service Call 6

Customer says that a large amount of ice is built up on the outdoor unit of the heat pump. The problem is the changeover contacts in the defrost relay are open. The system is not reversing in the defrost cycle. When this happens, the 10-min maximum time causes the unit to stay in heating without defrosting the outdoor coil because the timer is still functioning.

The technician goes to the outdoor unit first because it is obvious there is a defrost problem. The coil looks like a large bank of ice. The technician removes the control box panel and jumps the timer contacts from 3 to 4; the unit fan stops, but the cycle does not reverse as it should. See Figure 45–55 for the wiring diagram. A voltage check at the four-way valve coil shows that no voltage is getting to the valve. The technician then jumps from the R terminal to the O terminal, and the four-way valve changes over to cooling, so the jumper is removed. The valve must not be getting voltage through the defrost relay. The defrost relay contacts are jumped from 1 to 3, and the valve reverses. The defrost relay is changed, and another defrost is simulated by jumping the 3 to 4 contacts. The defrost thermostat contacts are still closed, and the unit goes into defrost. The unit has so much ice accumulation that the technician uses water to defrost the coil and get the unit back to a normal running condition. A call the next day proves that the unit is defrosting correctly.

Service Call 7

Customer calls and says the heat pump is running constantly. The unit is blowing cold air for short periods every now and then. The power bill was excessive last month, and it was not a cold month. This is a split system with 20-kW strip heat for quick recovery, so the customer can turn the thermostat back to 60° F on the weekend and have quick recovery on Monday morning. The outdoor unit is in the rear of the shop. The problem is the contacts in the defrost relay that change the unit to cooling in the defrost mode are stuck closed. The unit is running in the cooling mode, and the strip heat is heating the structure. It must also make up for the cooling effect of the cooling mode.

The technician sets the thermostat to the first stage of heating. The air at the registers is cool not warm. The unit should have warm air in the first stage of heat (85° to 100°F), depending on the outside air temperature. The indoor coil is in a closet in the back of the shop. The gas line is not hot but cold. This indicates that the unit is in the cooling mode. The technician thinks the four-way valve may be stuck in the cooling mode, but a voltage check should be performed. See Figure 45–56 for a wiring diagram. The technician removes the control voltage panel and finds 24 V at the C to O terminals. This means the four-way valve coil is energized and the unit should be in the cooling mode. Now, the technician must decide if the 24 V is coming from the space thermostat or the defrost relay. The field wire from the space thermostat is removed, and the valve stays energized. The voltage must be coming from the defrost relay. The wire is replaced on the O terminal and the wire on terminal 3 on the defrost relay is carefully removed. The unit changes to heating. The defrost relay is changed, and the unit operates correctly. A call back later proves the unit to be operating as expected.

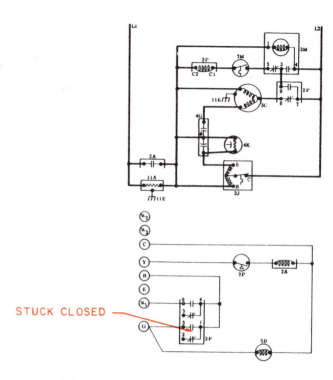

Figure 45–56. The contacts that change the unit to cooling for defrost are stuck closed. *Reproduced courtesy of Carrier Corporation*

STUCK CLOSED

Service Call 8

Customer calls. The heat pump heating a small retail store has large amounts of ice built up on the outdoor unit. The unit is located outside in back of the store. The problem is the defrost thermostat contacts are not closing when the timer advances to the defrost cycle. This is a combination control—the timer and the sensor built into the same control.

The technician goes to the outdoor unit and finds a large bank of ice on the coil. The control panel is removed. This is a unit that has a timer that can be advanced by hand. The timer and the defrost sensor are built into one control. It is obvious that the sensor is cold enough for the contacts to make, so the dial of the timer is advanced all the way around. The unit does not go into defrost, so it must be the contacts in the timer's sensor that are faulty. A new timer is installed and advanced by hand until the contacts make. The unit goes into defrost normally. One defrost cycle does not melt enough ice off the coil to leave the heat pump on its own. The technician forces the unit through three more defrosts before the ice is melted to a normal level. If a water hose had been available, it would have been used to melt the ice to save time.

Service Call 9

While on a routine service call on an apartment house complex, the technician notices a unit go into defrost when there is no ice on the coil. The unit stays in defrost for a long time, probably the full 10 min allowable. This is a split-system heat pump and should have defrosted for only 2 or 3 min with a minimum of ice build-up. The compressor was loud, as if it was pumping against a high head pressure. The problem is the defrost thermostat contacts are shut. The unit is going into defrost every 90 min and staying for the full 10 min. The timer is terminating the defrost cycle after 10 min. Toward the end of the cycle, the head pressure is very high.

The technician shuts off the unit, removes the panel to the control box, removes one lead of the defrost thermostat, and takes a continuity check across the thermostat. The circuit should be open if the coil temperature is above 50°F but the contacts are closed. (Remember, this control should close at about 25°F and open at about 50°F. Allow plenty of time for this control to function because it is large and it takes a while for the unit's line to exchange heat from and to the control.) The technician changes the defrost thermostat, and the unit is put back in operation.

Service Call 10

*Customer calls and says there is no heat in their home. It smells like something is hot or burning. *The customer is told to turn off the unit at the indoor breaker to stop all power to the unit.* The problem is the indoor fan motor has an open circuit, and the strip heat is coming on from time to time and then shutting off because of excessive temperature.*

The technician goes to the space thermostat and sets it to OFF. The electrical breakers are then energized. The thermostat is turned to FAN ON. The fan relay can be heard energizing, but the fan will not start. The indoor unit is under the house in the crawl space, so the technician takes tools and electrical test equipment to the unit. *The breaker under the house is turned off and the door to the fan compartment is opened. The fan motor is checked and found to have an open winding.* It is replaced with a new one and the unit is started. The technician uses an ammeter to check all three stages of electric heat for current draw. One stage draws no current. *The breaker is again turned off, and the electric

heat is checked with an ohmmeter.* The circuit in one unit is open because the internal fuse link is open. This happened because the fan was not running and the unit overheated. A new fuse link is installed and the heat is started again. This time it is drawing current. The technician puts the unit's panels back together and leaves the job.

Service Call 11

*The customer calls and says that the unit in his residence smells hot and not much air is coming out of the registers. *The customer is told to shut off the unit at the breaker for the indoor unit.* The problem is the indoor fan capacitor is bad, and the fan motor is not running up to speed. The fan motor is cutting off from internal overload periodically.*

The technician sets the thermostat to OFF and turns on the breaker. The FAN ON switch at the thermostat is set to ON. The fan relay can be heard energizing, and the fan starts. There does not seem to be enough air coming out of the registers. The fan motor is not running smoothly. The technician goes to the attic crawl space where the indoor unit is located, taking tools and electrical test instruments. *The breaker next to the indoor unit is shut off and the panel to the control box is removed. The breaker is then turned on. It is noticed that the fan is not turning up to speed. The breaker is turned off again. The fan's capacitor is removed and a new one is installed.* The breaker is energized again, and the fan comes up to speed. A current check of the fan motor shows the fan is operating correctly. *The power is shut off and the motor is oiled.* The technician then replaces all panels and turns on the power.

Service Call 12

Customer calls and says the unit in their residence is blowing cool air from time to time. One of the air registers is close to their television chair, and the cool air bothers them. The problem is the breaker serving the outdoor unit is tripped, apparently for no reason. The control voltage is supplied at the indoor unit so the electric auxiliary heat is still operable.

The technician turns the space temperature thermostat to the first stage of heat and notices that the outdoor unit does not start. When the thermostat is turned to a higher setting, the auxiliary heat comes on and heat comes out of the registers. With the thermostat in the first-stage setting, only the outdoor unit should be operating. The auxiliary heat should not be operating. In the second stage the heat pump and the auxiliary heat should be operating. The technician goes to the outdoor unit and finds that the breaker is tripped. *There is no way of telling how long the breaker has been off, but it is not wise to start the unit because there has been no crankcase heat on the compressor. The breaker is turned to the off position.* The technician uses an ohmmeter and checks the resistance across the contactor load-side circuit to make sure there is a measurable resistance. It is normal, about 2 Ω (remember, there is a compressor motor, fan motor, and defrost timer motor to read resistance through). The contactor load-side terminals are checked to ground to see if any of the power-consuming devices are grounded. The meter reads infinity, the highest resistance. It is assumed that someone may have turned the space-temperature thermostat off and back on before the system pressures were equalized and the compressor motor would not start. This means that the compressor tried to start in a locked-rotor condition, and the breaker tripped before the motor overload turned off the compressor. A sudden power outage then back on will cause the same thing.

The technician goes to the indoor thermostat, sets it to the emergency heat mode and tells the homeowner to switch back to normal heat mode in 10 hr. This gives the crankcase heater time to warm the compressor. The homeowner is instructed to see if the outdoor unit starts correctly or if the breaker trips. If the breaker trips, the thermostat should be set back to emergency heat, the breaker reset, and the technician should be called. If the technician must go back, the compressor can be started without fear of harming it. The compressor could have a starting problem, and it may not start the next time. The customer does not call back, so the unit must be operating correctly.

Service Call 13

Customer calls and says the unit serving the beauty shop is blowing cool air from time to time. The problem is the unit has a bad motor-winding thermostat. It has an open circuit and therefore will not close. The cool air the customer feels is actually recirculated air from the return air. The shop is heating by the second stage of the thermostat calling for auxiliary heat when the second-stage contacts are open

because the room temperature is satisfied. The first-stage contacts are still made, and the indoor fan is on without the compressor.

This is a split system with the outdoor unit on the roof. The technician turns the thermostat to the first stage of heat and cool air comes out of the air registers. The technician goes to the outdoor unit—the outdoor fan is running, but the compressor is not. *The power is turned off and the compressor access panel is removed.* The compressor is cool. A resistance reading of the compressor windings shows a measurable resistance between the R (run) and S (start) terminals. No reading can be obtained from R to C (common) or S to C. The winding-temperature thermostat is in this circuit and is open. Since the compressor is cool, it should be closed. This unit has only been installed for one year, so the compressor is in warranty. The technician obtains another compressor from the local supplier and proceeds to change the compressors. When the gages are connected and the refrigerant is exhausted from the unit, it is noticed that plenty of refrigerant seems to be in the unit. If the unit had only a small amount of refrigerant, the motor-winding open circuit may have been blamed on cycling because of a hot compressor due to the lack of suction gas cooling.

Service Call 14

Customer calls and says the heat pump outdoor unit is off because of a tripped breaker. The customer reset it three times, but it would not stay reset. The problem is the compressor motor is burned out and grounded. A new compressor will have to be installed.

The technician goes to the outdoor unit first and sees the breaker is in the tripped position. It is decided that before resetting it a motor test should be performed. *A volt reading is taken at the compressor contactor line side and it is found that 50 V is present.* This must be caused by the repeated resetting of the breaker. Voltage is leaking through the breaker. The main heating breaker is turned off, and the 50 V disappears. *If the technician had taken a continuity reading first, the meter would have been damaged.* After the voltage is reduced to 0 V, a resistance check on the load side of the compressor contactor shows a reading of 0 Ω to ground (the compressor suction line is a convenient ground). The circuits are isolated, and it is found that the compressor is defective. This unit is only two years old, and the compressor is still in warranty. The technician obtains a new compressor from the manufacturer and replaces the burned one. It is noticed that the smell of the refrigerant is very bad when the refrigerant is vented. This means that a bad motor burn has occurred, and therefore acid is present in the refrigerant. The compressor manufacturer insists that a suction-line filter-drier and a biflow (a drier that filters in both directions) liquid-line filter-drier be installed after a bad motor burn. The unit is pressured and leak-checked and a deep vacuum is pulled. A charge is measured into the unit, and the unit is started. While the unit is running, all amperages and voltages are checked. Everything seems normal. The motor burn is attributed to a random compressor failure. While the unit is running for its checkout, the technician fills out the necessary warranty paperwork for a factory compressor replacement.

Service Call 15

Customer calls and says the unit serving the upstairs to the house is blowing cold air sometimes and not heating like it should. The problem is the compressor has an open motor-winding thermostat due to a low charge of refrigerant. This compressor is a gas-cooled compressor, and even if the outdoor temperature is cold, the compressor will run hot without cool gas to cool the motor. The compressor's motor-winding-temperature thermostat contacts will close again when the compressor cools down. The compressor is not running continuously but the outdoor fan is. The auxiliary heat is heating the structure by using the second-stage contacts. When the space temperature heats to the set point of the second-stage contacts, they open and the first-stage contacts remain closed to keep the compressor on. It will not run, so the cool air the customer feels is the recirculated air from the return air. Remember, the outdoor fan is started with the compressor contactor. The compressor contactor's contacts are closed, but the compressor will not run because its internal-winding thermostat is open.

The technician turns the space temperature thermostat to the first stage of heating and notices that cool air is coming out of the air registers. The outdoor unit can be seen from a window and the fan is running. The technician goes to the outdoor unit and finds that the gas line leaving the unit is the same temperature as the cabinet of the unit—outside temperature. The compressor is either not run-

ning, or it is running and not working. The compressor access panel is removed and the compressor can be touched and seen. It is hot to the touch. *The technician turns off the power to the unit, installs a gage manifold and pulls a water hose around to the unit to cool the compressor.* The unit only has 20 psig of pressure, so it is evident that a low charge is the cause of the problem. While water is running slowly over the compressor housing (or can), the technician adds vapor refrigerant and boosts the pressure to 100 psig. A leak is found where the cabinet has rubbed a small hole in the liquid line. The charge is blown, and the leak is repaired. The unit never did operate in a vacuum, so no air will be in the unit. After repair and a leak check, the technician pulls a deep vacuum on the system and measures the correct charge into the unit, using a graduated cylinder. The deep vacuum is necessary to measure in the correct charge. The compressor is cool enough to start by the time all of this is complete, so the unit is started. All is normal, so all panels are replaced with all of the screws, and the unit is left running.

Service Call 16

Customer calls and says that the heat pump serving the dress shop will not cool. This is the first day of operation for the system this summer. The problem is the four-way valve is stuck in the heating mode. The compressor was changed about two weeks ago after a bad motor burn. The system has not operated much during the two-week period because the weather has been mild. The four-way valve is stuck because the technician who changed the compressor neglected to install a suction-line filter-drier, and sludge (burned oil) from the motor burn has migrated to the four-way valve and the piston is stuck.

The technician turns the thermostat to the cooling mode and starts the unit. The compressor is on the roof, and the indoor unit is in a closet. The technician goes to the indoor unit and feels the gas line; it is hot. The unit is in the heating mode with the thermostat set in the cooling mode. The technician goes to the roof and removes the low-voltage control panel. This unit has the four-way valve energized in the cooling mode, so it should be energized at this time. A voltmeter check shows that power is at the C to O terminals (the common terminal and the four-way valve coil terminal). The technician removes the wire from the O terminal and hears the pilot solenoid valve change (with a click). Every time the wire is touched to the terminal, the valve can be heard to position. This means that the pilot valve is moving, but the main slide valve is not changing position. The valve is stuck in the HEAT position. There is only one thing to do: change the valve. This is a difficult task, and the following procedure is used.

A. A new valve is acquired, an exact replacement because of the piping configuration.

B. The refrigerant is removed from the unit. This unit can be operated, so an empty drum is attached to the liquid-line gage port and the unit is operated in the heating mode (remember, the unit is stuck in the heating mode). The unit is started, and cold water is slowly allowed to run over the refrigerant drum to condense the refrigerant. The unit is allowed to operate until the low-side pressure drops to 5 psig, then the drum valve is shut off. This will save most of the refrigerant charge. The remainder of the refrigerant charge is vented to the atmosphere.

C. The unit is now at atmospheric pressure. The four-way valve is cut out of the unit as close as possible to the valve body. The valve is not unsoldered because it has three connections that would have to be melted at high temperature at nearly the same time (the evaporator suction line, the permanent suction line, and the hot-gas line to the condenser are all on the same side of the valve and very close to the valve body). These connections can be melted loose, but the next step would be to solder three connections at nearly the same time that have residue (solder left on the connection) still on the pipes. Reheating these three connections so close to the four-way valve body may create another stuck valve. Sawing the three pipes off and cleaning the filings from the pipe is probably the best procedure.

D. The three stubs of pipe are cleaned and readied for the new valve. The pipes are prepared for low-temperature 6% silver solder and inserted into the new valve. If the old pipes are not quite long enough, short stubs can be added. *The valve body is wrapped in a damp cloth, and a cup of water is positioned close by to pour on the cloth in case the valve heats up.* The wet cloth is wrapped around the valve body to keep it cool, Figure 45-57. Remember, as

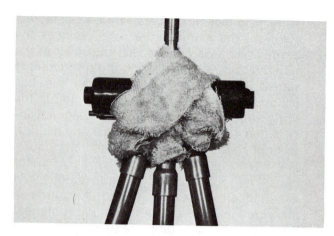

Figure 45-57. Wet cloth is wrapped around the valve body to keep it cool. *Photo by Bill Johnson*

long as the cloth is wet, the valve will be wrapped in a cloth that will be no higher than 212°F. The valve has Teflon parts inside that must not be overheated.

E. The valve is soldered into the system using low temperature 6% silver solder. When the connections are completed, water is slowly poured over the cloth to make sure it doesn't get hot.

Note: The technician is much better prepared to use low-temperature silver solder in the field. The factory is better prepared to successfully complete a high-temperature connection.

F. A suction-line drier is installed in the permanent suction-line (between the four-way valve and the compressor).

G. The system is pressured, leak-checked, and evacuated to a deep vacuum.

H. The correct charge is measured into the system with a charging cylinder.

I. The system is started and allowed to run in the cooling and the heating modes to ensure that the unit will change from cool to heat.

The technician cuts the old four-way valve open and finds that it is stuck due to burned oil and carbon from the motor burn.

Service Call 17

Customer says that the heat pump is running constantly in the heating mode in mild weather. It did not do this last year; it cut off and on in the same kind of weather. The problem is the check valve in the liquid line (parallel to the thermostatic expansion valve) is stuck open and not forcing liquid refrigerant through the thermostatic expansion valve. This is overfeeding the evaporator and causing the suction pressure to be too high, hurting the efficiency of the heat pump. For example, if it is 40°F outside, the evaporator should be operating at about 15°F for the outside air to give up heat to the evaporator. If the evaporator is operating at 30°F because it is overfeeding, it will not exchange as much heat as it should. The capacity will be down.

The technician turns the room thermostat to the first stage of heating and goes to the outdoor unit at the back of the house. The unit is running. The gas line temperature is warm but not as hot as it should be. The compressor compartment door is removed so the compressor can be observed. It is sweating all over from liquid refrigerant returning to the compressor. This means that the thermostatic expansion valve is not controlling the liquid refrigerant flow to the evaporator. A gage manifold is fastened to the high and low sides of the system. The suction pressure is higher than normal, and the head pressure is slightly lower than normal. The head pressure is down slightly because some of the charge that would normally be in the condenser is in the evaporator. A temperature tester lead is attached to the suction line just before the four-way valve and the superheat is found to be 0°F. The outdoor coil is flooded. Not enough refrigerant is flooding to the compressor to make a noise because the liquid is not reaching the cylinders. It is boiling away in the compressor crankcase. The technician tries adjusting the expansion valve and it makes no difference. The expansion valve is open and not closing or else the check valve is open and bypassing the expansion valve. The technician goes to the supply house and gets an expansion valve and a check valve. The unit charge is emptied into a refrigerant drum (see Service Call 16). The check valve is removed first. The technician can blow vapor refrigerant through the valve in either direction, so the check valve is defective. It is changed and the expansion valve is left alone. The unit is pressured, leak checked, evacuated, and charged. When the unit is started in the heating mode, it operates correctly. The check valve was the problem. The superheat is checked to be sure that it is normal (8° to 12°F) and the unit panels are assembled. A call later proves the unit is cycling as it should.

Service Call 18

Customer calls and says the heat pump serving the retail shop is blowing cold air from

time to time and not heating as it normally does. It runs constantly. The problem is the thermostatic expansion valve is not feeding refrigerant. The thermal element has lost its charge, and the valve is shut.

The technician turns the thermostat to the first stage of the heating mode. The air coming out of the air registers is cool; it feels like the return air. The outdoor unit is on the roof, and the indoor unit is in an attic crawl space. The technician goes to the roof; the thermostat is set with the first stage calling for heat. The outdoor unit should be on. The outdoor unit is running, but the gas line is cool. A gage manifold is installed on the high and low sides of the system. The low side is operating in a vacuum, and the high-side pressure is 122 psig, which corresponds to the inside temperature of 70°F. This indicates that liquid refrigerant is in the condenser (the indoor coil). If the condenser pressure were 60 psig, say, well below the 122 psig corresponding to 70°F, it would indicate that the unit is low in refrigerant charge. Because a refrigerant charge is in the unit and the suction pressure is low, the technician suspects that the thermostatic expansion valve is restricting the refrigerant flow. This unit has service valves on the gas and liquid lines at the outdoor unit. The liquid-line service valve is valved off, and the unit refrigerant charge is contained in the condenser. Remember that the condenser is the indoor coil. The four-way valve coil is not energized in the winter cycle. When the system is pumped down, the expansion valve is changed. The outdoor portion of the system is pressured, leak-checked, and evacuated. The system is started, and the charge is adjusted, using the manufacturer's charging chart furnished with the unit. The unit is now operating correctly.

Service Call 19

Customer calls. The power bill at the residence is much higher for the month of February than for the same month last year. The heating seasons for the two months were about the same. The problem is the air filters are stopped up. The homeowner is supposed to change these on a regular basis but has not. The unit is operating with a high head pressure and a low COP (coefficient of performance).

The technician turns the thermostat to the first heating stage. The airflow at the registers does not seem as strong as it should be. The air handler is in a hall closet. When the technician enters the closet, air can be heard sucking in around the fan compartment door because the restricted filters cause a pressure drop in the fan compartment. When the filters are removed, the problem is evident, they are dirty. When held up to a light, no light passes through the filters. New filters are installed, the motors are oiled and the system is left to operate. The homeowner is cautioned about keeping the air filters clean.

Service Call 20

Customer calls. The heat pump serving the home is not performing to capacity. The power bills are much higher than last year's bills, although the heating season is much the same. The customer would like a performance check on the heat pump. The problem is the heat pump has a two-cylinder compressor and one of the suction valves has been damaged in the compressor due to liquid floodback on startup. The owner had shut the disconnect off at the condensing unit last summer and left it off all summer. When the unit was started up in the fall, the compressor had enough liquid in the crankcase to damage one of the suction valves. If the crankcase heat had been allowed to operate for 8 to 10 hr before starting the unit, this would not have happened. Note that if one cylinder of a two-cylinder compressor is not pumping, the compressor will pump at half-capacity.

The technician fastens gages to the system and starts it. The manufacturer's performance chart is used to compare pressures. The unit is running a high suction pressure and a low discharge pressure. The technician switches the unit to cooling and blocks the outdoor-coil air flow to build up head pressure. The head pressure goes to 200 psi and the suction to 100 psi. The compressor current is below normal. The compressor is not pumping to capacity, it is defective. A temperature check of the gas lines at the four-way valve proves the valve is not bypassing gas. (See Service Call 22.)

The technician goes to a supply house and gets a new compressor. The old compressor is not in warranty, so there is no compressor exchange. The compressor is changed. A new liquid-line filter-drier is installed. The system is pressured, leak-checked, and evacuated before charging. The charge is measured into the system and the system is started. The system is operating correctly. A call to the customer later in the day proves the system is

functioning correctly. The outdoor unit is satisfying the room thermostat and shutting the unit off from time to time because the weather is above the balance point of the house.

Service Call 21

An apartment house complex with 100 package heat pumps is under contract with a local company for routine service. While on a routine visit, the technician noticed that a compressor was sweating all over in the heating cycle. The problem is one of the newer technicians serviced this unit during the last days of the cooling season and added refrigerant to the system. It is a capillary tube metering device system. The unit now has an overcharge.

The technician knows that a compressor should never sweat all over, so a gage manifold is installed to check the pressures. A charging chart is furnished with this unit to use when adjusting the charge. When the gage readings are compared to the chart readings, both the suction and the discharge pressures are too high. This symptom calls for reducing the charge. Refrigerant is allowed to flow into an empty drum until the pressures on the chart compare to the pressure in the machine. An observant technician has saved a compressor.

Service Call 22

Customer calls and says that the heat pump serving the residence must not be operating efficiently because the power bill has been abnormally high for the month of January compared to last year's bill. Both years have had about the same weather pattern. The problem is the four-way valve in the heat pump is stuck in mid-position and will not change all the way to heat. There is no apparent reason for the sticking valve, it must be a factory defect.

The technician turns the room thermostat to the first stage of heating and goes to the outdoor unit. The gas line leaving the unit is warm, not hot as it should be. Gages are installed on the system. The head pressure is low, and the suction pressure is high. The compressor amperage is low. All of these are signs of a compressor that is not pumping up to capacity. The outside temperature is 40°F, and the coil does not have an ice buildup, so the unit should be working at a high capacity. The technician suspects the compressor, so a compressor performance test is performed. A jumper wire is installed between the R and O terminals to energize the four-way valve coil and change the unit over to the cooling mode. The thermostat is left in the first-stage heat mode. If the house begins to cool off, the second-stage of heat will come on and keep the house from getting too cool. A plastic garbage bag is wrapped around the condenser coil to slow the air entering the coil. The head pressure begins to climb, but the suction pressure rises with it. The compressor will not build a sufficient differential between the high and low sides. Generally, the unit should be able to build 300 psig of head pressure with 70 to 80 psig suction. This unit has a 120 psig suction at 250 psig head pressure. Before the compressor is condemned, the technician decides to check the four-way valve operation. An electronic temperature tester is fastened with the two leads on the cold gas line entering and leaving the four-way valve. The difference in temperature should not be more than 3°F between these two lines, but it is 15°F. The four-way valve is leaking hot gas into the suction gas. The four-way valve is changed, using the procedures in Service Call 16. The unit is started after the pressure leak test, evacuation, and charging. The unit now performs to capacity with normal suction and head pressures.

SAFETY PRECAUTION: *Use common sense and caution while troubleshooting any piece of equipment. As a technician you are constantly exposed to *potential danger*. High-pressure refrigerant, electrical shock hazard, rotating equipment shafts, hot metal, and lifting heavy objects are among the most common hazards you will face. Experience and listening to experienced people will help to minimize the dangers.*

SUMMARY

- All compression cycle refrigeration machines are heat pumps; some have the capability of pumping heat either way.
- A new component, the four-way valve enables the heat pump to reverse the refrigeration cycle and reject heat in either direction.
- Air is not the only source of heat for a heat pump, but it is the most common. The ground and large bodies of water also contain heat that may be used.
- Air at 0°F still has about 85% of the available heat contained in it compared to no heat (−460°F).

- *Water to air* is the term used to describe heat pumps that absorb heat from water and transfer it to air.
- Water-to-air heat pumps use lakes, deep wells, and water from industrial cooling systems for a heat source.
- Commercial buildings may use water to absorb heat from one part of a building and reject the same heat to another part of the same building.
- Other methods, such as solar assistance, are being developed using the reverse-cycle heat pump.
- Air-to-air heat pumps are very similar to summer air conditioning equipment.
- There are two styles of equipment: split systems and package (self-contained) systems.
- New names are applied to the heat pump components because of the reverse-cycle operation.
- When applied to summer air conditioning equipment, the coil inside the house is called an evaporator. On a heat pump unit this is called an indoor coil. Sometimes it is operated as a condenser (in winter) and sometimes as an evaporator (in summer).
- When applied to summer air conditioning, the unit outside the house is called a condensing unit. With a heat pump this unit is a condenser in the summer and an evaporator in the winter. It is called the outdoor unit.
- The terms "indoor coil" and "outdoor coil" are used with package heat pumps also to avoid confusion.
- The refrigerant lines that connect the indoor coil with the outdoor coil are the gas line (hot gas in winter and cold gas in summer) and the liquid line.
- The liquid line is always the liquid line; the flow reverses from season to season.
- Several metering devices are used with heat pumps. The first was the thermostatic expansion valve. There must be two of them, with a check valve piped in parallel to the valve to force the liquid refrigerant to flow through the valve.
- Capillary tube metering devices are common. There must be two of them, one for summer and one for winter operations. They must have a check valve piped in parallel to force the liquid refrigerant to flow correctly.
- A fixed-bore metering device that will allow full flow in one direction and restricted flow in the other direction is used by many manufacturers.
- The electronic expansion valve is used by some manufacturers with close-coupled equipment because it will meter in both directions and maintain the correct superheat.
- When standard filter-driers are used with a heat pump installation, they must be placed in the circuit with the check valve to insure correct flow. *It must have a check valve piped in series with it, or else when the flow reverses, it will back wash and the particles will be pushed back into the system.*

- A special biflow drier may be used in the liquid line. The biflow drier is actually two driers in one shell with check valves to force the refrigerant through the correct circuit.
- The air-to-air heat pump loses capacity as the outside temperature goes down. The unit responds in reverse to the load. When the weather gets colder, the structure needs more heat and the air-to-air heat pump provides less.
- At the balance point the heat pump alone will run constantly and just heat the structure. If the outdoor temperature drops any lower, the heat pump must have help.
- Auxiliary heat is the heat that a heat pump uses as a supplement.
- Auxiliary heat is normally electric resistance heat. Oil and gas may be used in some installations.
- When the auxiliary heat is used as the only heat source, such as when the heat pump fails, it is called emergency heat.
- Emergency heat is controlled with a switch in the room thermostat. This switch turns off the heat pump and turns on the auxiliary heat.
- Coefficient of performance (COP) is determined from the heat pump's heating output divided by the input. A COP of 3 to 1 is common with air-to-air heat pumps at the 47°F outdoor temperature level. A COP of 1.5 to 1 is common at 0°F.
- The efficiency of a heat pump is a result of it being used to capture heat from the outdoors and pump that heat to the indoors.
- A heat pump is installed in much the same manner as a summer air conditioner. The air distribution requires at least 400 cfm of air per ton of cooling capacity.
- The terminal air must be distributed correctly because it is not as hot as the air in oil or gas installations. Heat pump air normally is not over 100°F with only the heat pump operating.
- *Provision for water drainage at the outdoor unit must be provided in winter.*
- Space-temperature control differs from controlling a combination cooling and heating system with electric air conditioning and gas or oil heat. There are two different heating systems: the heat pump and the auxiliary heat.
- The indoor fan must run when the compressor is operating, and the compressor operates in summer and winter modes.
- The four-way valve determines whether the unit is in the heating or cooling mode. In one position, the compressor discharge gas is directed toward the outdoor coil and the unit is in the cooling mode. In the other position, the hot discharge gas is directed toward the indoor coil and the unit is in the heating mode.
- The technician can tell which mode the unit is in by the temperature of the gas line. It can get very hot.
- The space-temperature thermostat controls the direc-

tion of the hot gas by controlling the position of the four-way valve. Many manufacturers energize the four-way valve in the summer cycle.
- Since the heat pump evaporator operates below freezing in the winter, frost and ice will build up on the outdoor coil. Defrost is accomplished by reversing the unit for a short time to remove this ice.
- When the system is in defrost, it is in the cooling mode with the outdoor fan off to aid in the buildup of heat. The indoor coil will blow cool air, so the auxiliary heat is usually energized during defrost.
- Defrost is normally instigated (started) by **time and temperature** and terminated by **time or temperature.**
- Servicing a heat pump involves both electrical and mechanical troubleshooting.
- The refrigerant charge is normally very critical with a heat pump. The recommended charging procedures are to follow the manufacturer's charging chart or to exhaust the charge, evacuate the system to a deep vacuum, and measure the charge back into the system.

REVIEW QUESTIONS

1. How does a heat pump resemble a refrigeration system?
2. What is the lowest temperature at which heat can be removed from a substance?
3. Name the three common sources of heat for heat pumps.
4. What component allows a heat pump to reverse its refrigerant cycle?
5. What large line connects the indoor unit to the outdoor unit?
6. What small line connects the indoor unit to the outdoor unit?
7. Why is the indoor unit not called an evaporator?
8. When is the outdoor unit an evaporator?
9. When is the indoor unit a condenser?
10. Why is it important to have drainage for the outdoor unit?
11. Name three common metering devices used with heat pumps.
12. Which metering device is most common?
13. Which metering device is most efficient?
14. Where is the only permanent suction line on a heat pump?
15. What type of drier may be used in the liquid line of a heat pump?
16. Where must a suction-line drier be placed in a heat pump after a motor burnout?
17. What controls the heat pump to determine whether it is in the heating cycle or the cooling cycle?
18. Can a heat pump switch from heating to cooling and from cooling to heating automatically?
19. What must be done when frost and ice builds up on the outdoor coil of a heat pump?
20. Are all heat pumps practical anywhere in the United States? Why?

Contents

APPLICATION GUIDE

Application Survey

APPLICATION GUIDE
The Trane Company, Dealer Products Group, Tyler, Texas 75711

Preliminary: An often overlooked preliminary to the actual technical survey is the necessity to talk to the person who will make or influence the actual "buy" decision. This person may be the owner, architect, consulting engineer, facilities engineer, buyer, finance manager, etc., but it may not be obvious who this individual is. Ask questions designed to determine who the decision maker is and then attempt to get his:

Objectives Does he want comfort air conditioning for his employees or a good environment for equipment or materials?

Requirements Must equipment be placed in a specific place, how much is he willing to spend, can duct work be visible?

Preferences Would he rather have several small systems or a few larger ones, split systems or single package, brand names?

The above knowledge will help you limit your survey to information applicable to the decision maker's idea of what the system should be and help in the actual equipment selection when the time comes.

Since this is the first "sales call" treat it as such by showing the proper interest and making an impression which will enhance your chances of getting the job.

The Survey

There is a practical, logical sequence of procedure which should be followed when engineering an air conditioning system, whether for a residence, office, factory or computer room. This procedure was established as a result of the accumulative experience of the best application engineers in the industry and has been accepted as standard practice throughout the industry.

There is no short-cut method for air conditioning equipment application. Each job requires individual attention to the details of application to assure a quality system providing satisfactory performance for the customer. Rules-of-thumb may be partially satisfactory for preliminary estimates but they cannot replace careful job engineering in the final stages.

The sequence of steps in application engineering consist of:

1. Survey
2. Load Calculations
3. Equipment Selection
4. Duct System Design
5. Register and Grille Selection
6. Installation
7. Start-up and Balancing

The purpose of the heating and air conditioning application survey is to gather into one source the information and data required to completely engineer a system. The accumulation of this material effectively eliminates chasing information from one source to another during application engineering and thus permits efficient engineering of the system without delay and lost motion. Therefore, a complete compilation of pertinent information at the very beginning will enhance the prospect of an efficiently engineered and profitable job.

It may be stated that to survey a proposed application is to collect all data and information necessary to: (1) calculate the cooling and heating loads, (2) size and select the equipment, (3) make a duct and system layout, (4) size ducts and select registers and grilles, (5) suggest any necessary modifications to utilities system, and (6) provide practical suggestions to the installation crew. From the description of the survey and the list of steps in application engineering procedure, it is evident that the entire procedure ties together into one important package aimed at providing the best possible installation at a reasonable cost with a maximum of satisfaction and a minimum of difficulty. This is a big order, but it can be accomplished by the practice of industry accepted methods and procedures.

The survey need not, however, consume an excessive amount of time and if planned properly, will not do so. The survey is simply a matter of gathering information, knowing where to look for it and what to look for, and why you need it, and then recording the information and data in such a manner that it may be effectively put to use at the right time. This is the reason for an "Application Survey" form. It should be completely filled in before the actual work begins and then referred to as often as necessary during the process of engineering and layout of the system. It is placed on permanent file as a record of conditions as they existed at the time the job was contracted for.

Types Of Surveys

Generally, there are two types of heating and air conditioning surveys. The first type is one made directly from blueprints, drawings, information supplied by the architect, builder or building owner, and the personal knowledge of the application engineer. The second type is an actual physical survey of an existing building.

In the first type you may be handed a print and told to engineer the heating and cooling system for the building, which does not yet exist. You must glean as much information as possible from the print and obtain other required information and data from any source possible. Since the facility has not been built, you must use your imagination to picture the finished structure as called for on the plans, and you must mentally picture the completed heating and air conditioning system within the building.

The second type of survey may be made wholly within the existing facility by visual observation and actual measurement of existing structural features. You can determine whether or not the ducts may be installed a certain way, or whether or not certain modification, cutting and patching, or changes in proposed design are necessary. You can measure actual lengths and dimensions in the existing building. This is different from the first type where you are compelled to depend on plans and specifications of a proposed structure which often change during construction, creating problems for the application engineer and installation crew.

The survey is equally important for both new and existing facilities. Good judgement and common sense are often necessary in filling out the survey form and adding pertinent notes. Good application engineers are developed through training, experience and hard work.

Information Required

In order to properly apply equipment in either a new or existing facility it is necessary to know the construction of the space to be conditioned as well as the use and occupancy. The time of day that the peak load occurs must be determined. The physical building characteristics and lighting layouts can normally be obtained from drawings or observed in the case of an existing building. A survey should include the following information:

1. Spaces to be air conditioned and heated

a) Orientation of building — direction building faces and shading from nearby structures.

b) Walls and partitions — construction, dimensions, type and thickness of insulation, temperature of adjoining unconditioned spaces, shaded or sunlit.

c) Ceilings, roofs and floors — same information as for walls.

d) Windows — size and location. Type of windows, single or multipane, shading devices — inside and exterior.

e) Doors — location, type, size, usage.

f) Ceiling height — floor to ceiling.

g) Sketch — if prints not available, sketch floor plan.

2. Location of Equipment and Air Distribution System

a) Possible location of supply and return outlets and duct work. Methods of introducing fresh air.

b) Equipment — single package or split system, can roof be used, space for air handler.

c) Condensate drains — location.

d) Power service — location, capacity, type and current limitations.

e) Delivery and rigging of equipment — access for delivery and placement. Any structural considerations.

f) Refrigerant lines — location and length.

3. Internal Loads

a) Lighting — total or average watts, locations if not uniform and usage. Type of lights.

b) Appliances — cooking, laboratory, gas or electric, hooded or unhooded.

c) Motors — horsepower and usage (if not constant). Location in conditioned space.

d) People — number on which design is to be based.

e) Peak occupancy — time of day or evening.

4. Ventilation or Infiltration

a) Fresh air design requirements — based on square feet of floor space or CFM per person

b) Local codes — requirements.

c) Method — how fresh air will be introduced.

d) Exhaust system — size, type, CFM

5. Future Plans.

a) Should future changes in loads be considered.

b) Will facilities be enlarged or should rearrangement of equipment at a later date be considered.

Obviously, the survey information for a restaurant will be different from a residence or a manufacturing concern. The sources of the pertinent information will also vary. The survey form used should be tailored to fit your personnel, your needs, and the job. Figure 1 is an example of a form which can be used for surveys. Figure 2 has been completed for illustrative purposes.

It will often be necessary to include explanatory information which may be written on the back of the form, or on a separate sheet and attached to it. Anyone familiar with the industry should be able to look at the sketch or drawing, read the survey form, and any explanatory details and understand the application.

An accurate survey may also be your protection at a later date. Always note source or sources of information. If assumptions are made, note them and get specific confirmation from a responsible individual that the assumptions are valid.

The survey is just the beginning of a job, but it is the foundation of everything that follows, and therefore, must be solid in fact and functional in service.

HVAC SURVEY INFORMATION FORM

Must be filled out completely and accompany architects prints.

PROJECT: _____ CITY: _____

Trane Rep.: _____ Address: _____ Phone: _____

Architect: _____ Address: _____ Phone: _____

Consulting Engr.: _____ Address: _____ Phone: _____

A/C Contractor: _____ Address: _____ Phone: _____

WINTER DESIGN CONDITIONS   **SUMMER DESIGN CONDITIONS**

Inside D.B. _____ °F.   Inside D.B. _____ °F.   Inside W.B. _____ °F.

Outside D.B. _____ °F.   Outside D.B. _____ °F.   Outside W.B. _____ °F.

Temp. Diff. _____ °F.   North Latitude _____ Degrees

CHECK PLANS TO BE SURE FOLLOWING ITEMS ARE COVERED

Floor Plan and Elevation View

Building Orientation ☐

	Front	**Rear**	**Sides**
Wall Construction	☐	☐	☐
Wall Color (Dark or Light)	☐	☐	☐
Wall Insulation	☐	☐	☐
Door Type	☐	☐	☐
Window Schedule	☐	☐	☐
Window Shading	☐	☐	☐

Roof   ☐ Construction   ☐ Insul.   ☐ Color (Dark or Light)

Floor   ☐ Type   ☐ Insulation

Exhaust Fans   ☐ Location   ☐ CFM

INTERNAL LOADS —

_____ People   _____ Watts Lighting   _____ Type

Heat Producing Equipment List   Show Usage Factor

Peak Hour, If Known _____ AM PM

Type of Unit Desired: _____ No Preference

Type of Air Distribution: ☐ Perimeter   ☐ High Wall   ☐ Other

Location of Equipment: _____ No Preference

Type of Fuel: Electricity [Volt. | Ph. | Hz.]   Gas [Nat. | LP]   Oil [   ]

Financing: FHA   Conventional   Other

Special Requirements Due To: Code   Lender Criteria   Customer Preference

_____

_____

_____

Customer's Name _____ Address _____

City _____ State _____ Zip _____ Telephone Number _____

WINTER: Inside Design Temp _____ °F — **Outside Design Temp** _____ °F = **Heating Temp Difference** _____ °F

SUMMER: Outside Design Temp _____ °F — **Inside Design Temp** _____ °F = **Cooling Temp Difference** _____ °F

HEATING		COMMON DATA SECTION			COOLING	
BTUH LOSS	HEATING FACTOR	SUBJECT		SQ. FT.	COOLING FACTOR	BTUH GAIN
		GROSS WALL				
		DOORS & WINDOWS (Table A or B)				
		NET WALL				
		CEILING				
		FLOORS				
		VOLUME OF BUILDING (cu. ft.)	$\times$ $\triangle$T $\times$ .007333		=	INFILTRATION BTU/HR.
			$\times$ $\times$		=	
		SUB-TOTAL BTUH LOSS (per 10°F)				
$\times$		ADJUSTMENT FACTOR (Table C)				
		TOTAL BTUH LOSS				
		① PEOPLE ____ $\times$ 300 BTUH GAIN				
		APPLIANCES BTUH				1200
		SUB-TOTAL BTUH GAIN (room sensible only)				
$\times$ 1.15 ④		DUCT LOSS/GAIN FACTOR ②				③ $\times$ 1.15
		SUB-TOTAL BTUH (Sensible Gain)				
		MOISTURE REMOVAL (sub total $\times$ 1.3)				$\times$ 1.3
		TOTAL BTUH LOSS/GAIN				

TABLE A—HEATING—DOORS & WOOD FRAME WINDOWS (PER 10° F)

WINDOW & DOOR TYPES	HEATING FACTORS CERTIFICATION		X AREA	= BTUH LOSS
Double Hung (cfm/lf)	**.75**	**.50**		
Single Glass	15.05	12.89		
Double Glass	11.28	9.12		
Single w/Storm	7.48	6.93		
Double w/Storm	5.82	5.27		
Fixed (cfm/lf)	—	**.20**		
Single	—	10.28		
Double	—	6.51		
Jalousie (cfm/lf)	**1.5**	—		
Without Storm	35.42	—		
w/Storm	17.21	—		
Sliding Doors (cfm/lf)	**1.0**	**.50**		
Single Glass	16.61	13.78		
Double Glass	12.43	9.60		
Door (cfm/lf)	**1.0**	**.50**		
Wood Only	10.21	7.13		
Wood w/Storm	5.90	4.99		
Insulated Core R-5	7.51	4.43		
Insulated Core R-5 w/Storm	4.41	3.50		
		TOTALS		

① Assume 2 persons per bedroom

② Calculate only if duct is located in an unconditioned space

③ For crawl space or basement use 1.05 multiplier in **COOLING**

④ Omit, if duct work is embedded in slab

TABLE B — COOLING — DOORS & WINDOWS
Factors assume windows have inside shading by draperies or venetian blinds and sliding glass doors are treated as windows.

	SINGLE GLASS			DOUBLE GLASS			TRIPLE GLASS			X Area	= BTUH GAIN
	TEMP. DIFF.			TEMP. DIFF.			TEMP. DIFF.				
Direction	15°	20°	25°	15°	20°	25°	15°	20°	25°		
N	20	25	30	15	20	20	12	14	16		
NE & NW	35	40	45	30	35	35	24	26	28		
E & W	55	55	60	45	50	50	36	38	40		
SE & SW	45	50	55	40	40	45	30	32	34		
S	30	30	35	25	25	30	18	20	22		
Skylights	146	150	154	139	143	147	132	136	140		
Doors	8.6	10.9	13.2	8.6	10.9	13.2	8.6	10.9	13.2		
									TOTALS		

TABLE C — ADJUSTMENT FACTORS — (HEATING)

°F. Temperature Diff.	30	40	50	60	70	80	90
Adjustment Factor	3	4	5	6	7	8	9

RESIDENTIAL SURVEY FORM

1. Direction House Faces _____

2. Windows: _____ Glazed _____ Storm

   _____ Certified _____ Non-certified

3. Insulation: Wall R= _____ Ceiling R= _____ Floor R= _____

4. Basement: Conditioned _____ Yes, _____ No

   Insulation R= _____ Above Grade _____ Below Grade

5. Crawl Space: Conditioned _____ Yes _____ No. Insulation R= _____

6. Ceiling Height _____

7. Cathedral Ceiling _____

8. Standard Doors: Type _____ Weatherstripped _____

   Storm _____

9. Sliding Glass Doors: _____ Glazed _____ Weatherstripped

   Storm _____ Certified _____ Non-certified _____

10. Age of House _____

11. Furnace or Air Handler: _____ Capacity _____ Fuel _____ Age

12. Blower Motor: _____ H.P. _____ Drive _____ Wheel Size

13. Fan Relay _____ Yes, _____ No.

14. Plenum Size: _____ High _____ Wide _____ Deep

15. Ductwork Size: _____ Return Air, _____ Supply Air

16. Duct Insulation: _____ Yes, _____ No.

    Where Needed? _____

17. Supply Registers: Type _____ High _____ Low

    Adequate for Cooling? _____

18. Outside Location of Condensing Unit _____

19. Refrigerant Line Length Needed _____

20. Drain Line Length Needed _____

21. Is Condensate Pump Necessary _____

22. Electrical Service Entrance Panel _____ Volts _____ Amps

23. Distance from Entrance Panel to Condenser Site _____

24. Is Attic Ventilation Adequate? _____ Yes _____ No

25. Has Heating and/or Cooling System been Satisfactory? ___ Yes ___ No

26. Does Customer Want: ___ Electronic Air Cleaner ___ Humidifier

27. Type of Thermostat _____

28. Location of Thermostat _____

HOUSE & DUCT LAYOUT (1 SQ. = _____ Feet)

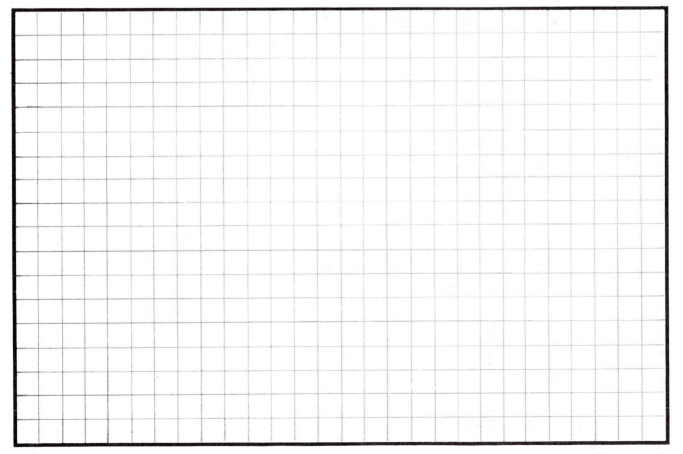

HEAT LOSS & GAIN FACTORS

TABLE D
CONSTRUCTION FACTORS HEATING & COOLING

Heating[1] Factor	TYPE OF CONSTRUCTION	Cooling Factor (°F. Temp. Diff.) 15°	20°	25°
	WALLS (Use Sq. Ft.)			
	Walls — wood frame w/sheeting & siding, veneer or other finish			
2.71	A) No insulation, 1/2" Gypsum Board	5.0	6.4	7.8
0.90	B) R-11 Cavity insulation + 1/2" Gypsum Board	1.7	2.1	2.6
0.80	C) R-13 Cavity insulation + 1/2" Gypsum Board	1.5	1.9	2.3
0.70	D) R-13 Cavity insulation + 3/4" Bead Board (R-2.7)	1.3	1.7	2.0
0.60	E) R-19 Cavity insulation + 1/2" Gypsum Board	1.1	1.4	1.7
0.50	F) R-19 Cavity insulation + 3/4" Extruded Poly	0.9	1.2	1.4
	Masonry Walls			
5.10	A) Above grade No insulation	5.8	8.3	10.9
1.44	B) Above grade + R-5	1.6	2.3	3.1
0.77	C) Above grade + R-11	0.9	1.3	1.6
1.47	D) Below grade No insulation	0.0	0.0	0.0
0.87	E) Below grade + R-5	0.0	0.0	0.0
0.60	F) Below grade + R-11	0.0	0.0	0.0
	CEILINGS (Use Sq. Ft.)			
5.99	A) No insulation	17.0	19.2	21.4
1.20	B) 2"-2½" insulation R-7	4.4	4.9	5.5
0.88	C) 3"-3½" insulation R-11	3.2	3.7	4.1
0.53	D) 5¼"-6½" insulation R-19	2.1	2.3	2.6
0.48	E) 6"-7" insulation R-22	1.9	2.1	2.4
0.33	F) 8"-9½" insulation R-30	1.3	1.5	1.6
0.26	G) 10"-12" insulation R-38	1.0	1.1	1.3
0.23	H) 12"-13" insulation R-44	0.9	1.0	1.1
3.08	I) Cathedral type No insulation (roof/ceiling combination)	11.2	12.6	14.1
0.72	J) Cathedral type R-11 (roof/ceiling combination)	2.8	3.2	3.5
0.49	K) Cathedral type R-19 (roof/ceiling combination)	1.9	2.2	2.4
0.45	L) Cathedral type R-22 (roof/ceiling combination)	1.8	2.0	2.2
0.40	M) Cathedral type R-26 (roof/ceiling combination)	1.6	1.8	2.0
	FLOORS (Use Sq. Ft. OR Linear Ft.)			
1.56	**Floors over unconditioned space (use sq. ft.)** A) Over basement or enclosed crawl space (not vented)	0.0	0.0	0.0
0.40	B) Same as "A" + R-11 insulation	0.0	0.0	0.0
0.26	C) Same as "A" + R-19 insulation	0.0	0.0	0.0
3.12	D) Over vented space or garage	3.9	5.8	7.7
0.80	E) Over vented space or garage + R-11 insulation	0.8	1.3	1.7
0.52	F) Over vented space or garage + R-19 insulation	0.5	0.8	1.1
0.28	**Basement Floors (use sq. ft.)**	0.0	0.0	0.0
8.10	**Concrete slab floor unheated (use linear ft.)** A) No edge insulation	0.0	0.0	0.0
4.10	B) 1" edge insulation R-5	0.0	0.0	0.0
2.10	C) 2" edge insulation R-9	0.0	0.0	0.0
19.00	**Concrete Slab floor duct in slab (use linear ft.)** A) No edge insulation	0.0	0.0	0.0
11.40	B) 1" edge insulation R-5	0.0	0.0	0.0
9.30	C) 2" edge insulation R-9	0.0	0.0	0.0

[1] Heating Factor for 10° Temperature Rise

ESTIMATING PROCEDURE

1. Fill in customer information.

2. Record inside and outside design temperatures; find temp difference.

3. Measure length of each outside wall multiply each by ceiling height. Record the total sq. ft. of exposed wall under "gross wall".

4. Using Tables A and B, determine the total area for windows & doors & enter in common data section.

5. Determine Net Wall by subtracting windows and doors from gross.

6. Measure & record total ceiling area.

7. Measure and record total floor area for floors over crawl space or basement. Total floor edge *length* (perimeter) if floor is a slab.

8. Using Table D select construction type and use the corresponding heat and cool factors on the form.

9. Determine BTUH Loss & Gain in Tables A and B by multiplying the area of glass and doors by the multiplier under the specified temperature difference. Enter total BTUH Loss/Gain on worksheet.

10. On worksheet, multiply the areas x the factors and total as instructed.

QUICK-SIZING TABLE for HEATING (FURNACE) ONLY DUCT SYSTEM
(Without a Cooling Coil in Place)

OUTPUT CAPACITY (See Notes)	Min. Air Flow Req'd	Supply Duct or Extended Plenum @800 FPM	Min. Sq. Inch Needed for Spec. CFM (Total Area of All Supply Duct)	Min. Number Supply Runs @600 FPM			MINIMUM SIZE	
				5" Runs 80 CFM	6" Runs 115 CFM	7" Runs 155 CFM	Return Duct Furnace or Air Handler @800 FPM	Return Air Grille (or equivalent) @ Face Velocity of 500 FPM
45,000 to 55,000	500 CFM	14" X 8" or 12" round	100	7	5	4	14" X 8" or 12" round	12" X 12"
60,000 to 70,000	700 CFM	18" X 8" or 14" round	140	10	6	5	18" X 8" or 14" round	24" X 10"
75,000 to 85,000	800 CFM	22" X 8" or 14" round	170	10	7	5	22" X 8" or 14" round	24" X 12"
95,000 to 105,000	900 CFM	24" X 8" or 15" round	190	12	8	6	24" X 8" or 14" round	24" X 12"
105,000 to 115,000	1100 CFM	22" X 10" or 16" round	220	—	10	7	22" X 10" or 16" round	30" X 12"
125,000 to 150,000	1400 CFM	24" X 12" or 18" round	280	—	12	9	24" X 12" or 18" round	30" X 14"
155,000 to 160,000	1600 CFM	1-35" X 10" or 20" round or 2-22" X 8"	360	—	14	10	32" X 10" or 20" round	30" X 18"

Notes: 1. BTUH with **maximum** temperature rise.
2. Gas furnaces are rated in input capacity. Rated output capacity is 80% of input.
3. Oil and electric furnace are rated in output capacity.

QUICK-SIZING TABLE for HEATING and COOLING DUCT SYSTEM

Air conditioning systems should never be sized on the basis of floor area only, but knowledge of the approximate floor area (sq. ft.) that can be cooled with a ton of air conditioning will be of invaluable assistance to you in avoiding serious mathematical errors.

Size of O.D. Unit	Normal Air Flow Req'd @ 400 CFM per Ton	Furnace		Supply Duct or Extended Plenum @800 FPM	Min. Number Supply Runs @600 FPM				Min. Return Duct Size at Furnace or Air Handler @800 FPM	Min. Return Air Grille Size (or equivalent) @Face Velocity of 500 FPM
		Blower Motor H.P.	Blower Wheel Dia. X Width		5" Runs 80 CFM	6" Runs 115 CFM	7" Runs 155 CFM	3½ X 14" 170 CFM		
1½ ton 18,000 BTUH	600 CFM	1/4 H.P.	9" X 8" 10" X 8"	16" X 8" or 12" round	8	5	4	4	16" X 8" or 12" round	24" X 8"
2 ton 24,000 BTUH	800 CFM	1/4 H.P.	9" X 9" 10" X 8"	22" X 8" or 14" round	10	7	5	5	22' X 8" or 14" round	22" X 12"
2½ ton 30,000 BTUH	1000 CFM	1/3 H.P.	10" X 8" 10" X 10" 12" X 9"	20" X 10" or 18" round	13	9	7	6	20" X 10" or 16" round	30" X 12"
3 ton 36,000 BTUH	1200 CFM	1/3 H.P.	10" X 8" 10" X 10" 12" X 9"	24" X 10" or 18" round	—	11	8	7	24" X 10" or 18" round	30" X 12"
3½ ton 42,000 BTUH	1400 CFM	1/2 H.P. 3/4 H.P.	10" X 8" 10" X 10" 12" X 9" 12" X 10"	24" X 12" or 18" round	—	12	9	8	24" X 12" or 18" round	30" X 14"
4 ton 48,000 BTUH	1600 CFM	1/2 H.P. 3/4 H.P.	10" X 10" 12" X 9" 12" X 10" 12" X 12"	32" X 10" or 20" round	—	14	11	10	28" X 12" or 20" round	30" X 18"

EQUIPMENT CAPACITY CALCULATIONS FROM WORKSHEET DATA

COOLING CAPACITY

TOTAL
GAIN _____ BTUH x ①_____ = _____ BTUH
ARI Rating

$$\frac{\text{ARI Rating} _____ \text{ BTUH}}{12,000 \text{ BTUH/TON}} = _____ \text{ TONS OF COOLING}$$

① **CAPACITY MULTIPLIER**
(BASED ON 75°F RETURN AIR TEMPERATURE)

90°F O.D.T. use 1.00	100°F O.D.T. use 1.10
95°F O.D.T. use 1.05	105°F O.D.T. use 1.15

HEATING CAPACITY

TOTAL
$$\frac{\text{LOSS} _____ \text{ BTUH}}{\text{(Select Factor Below)}} = _____ \text{ KW}$$

**ELECTRIC HEATER OR FURNACE DIVIDE USING 3413 BTUH/KW
NATURAL GAS OR PROPANE FURNACE . . . SELECT
OUTPUT TO BE AS CLOSE TO
CALCULATED HEAT LOSS AS POSSIBLE**

CALCULATION BY: _____

DATE: _____

EQUIPMENT SELECTION
Model Number(s)

HEAT PUMP................. _____

AIR HANDLER .... _____

SUPPLEMENTARY HEATER .. _____

CONDENSING UNIT......... _____

EVAPORATOR COIL .... _____

FURNACE, ELECTRIC........ _____

FURNACE, NATURAL GAS/
       PROPANE .... _____

REFRIGERANT LINES........ _____

THERMOSTAT.............. _____

ELECTRONIC AIR
CLEANER .... _____

HUMIDIFIER .... _____

ACCESSORIES .... _____

_____

_____

COST SCHEDULE(S)
MATERIAL/LABOR

Customer's Name _____ Address _____

City _____ State _____ Zip _____ Telephone Number _____

WINTER: Inside Design Temp _____ °F – Outside Design Temp _____ °F = Heating Temp Difference _____ °F

SUMMER: Outside Design Temp _____ °F – Inside Design Temp _____ °F = Cooling Temp Difference _____ °F

HEATING		COMMON DATA SECTION			COOLING	
BTUH LOSS	HEATING FACTOR	SUBJECT	SQ. FT.	COOLING FACTOR	BTUH GAIN	
		GROSS WALL				
		DOORS & WINDOWS (Table A or B)				
		NET WALL				
		CEILING				
		FLOORS				
		VOLUME OF BUILDING (cu. ft.)	× △T × .007333	= INFILTRATION BTU/HR.		
			× × =			
		SUB-TOTAL BTUH LOSS (per 10°F)				
×		ADJUSTMENT FACTOR (Table C)				
		TOTAL BTUH LOSS				
		① PEOPLE ____ × 300 BTUH GAIN				
		APPLIANCES BTUH			1200	
		SUB-TOTAL BTUH GAIN (room sensible only)				
× 1.15 ④		DUCT LOSS/GAIN FACTOR ②			③ × 1.15	
		SUB-TOTAL BTUH (Sensible Gain)				
		MOISTURE REMOVAL (sub total × 1.3)			× 1.3	
		TOTAL BTUH LOSS/GAIN				

TABLE A—HEATING—DOORS & WOOD FRAME WINDOWS (PER 10° F)

WINDOW & DOOR TYPES	HEATING FACTORS CERTIFICATION		X AREA	= BTUH LOSS
Double Hung (cfm/lf)	**.75**	**.50**		
Single Glass	15.05	12.89		
Double Glass	11.28	9.12		
Single w/Storm	7.48	6.93		
Double w/Storm	5.82	5.27		
Fixed (cfm/lf)	—	.20		
Single	—	10.28		
Double	—	6.51		
Jalousie (cfm/lf)	**1.5**	—		
Without Storm	35.42	—		
w/Storm	17.21	—		
Sliding Doors (cfm/lf)	**1.0**	.50		
Single Glass	16.61	13.78		
Double Glass	12.43	9.60		
Door (cfm/lf)	**1.0**	.50		
Wood Only	10.21	7.13		
Wood w/Storm	5.90	4.99		
Insulated Core R-5	7.51	4.43		
Insulated Core R-5 w/Storm	4.41	3.50		
		TOTALS		

① Assume 2 persons per bedroom

② Calculate only if duct is located in an unconditioned space

③ For crawl space or basement use 1.05 multiplier in **COOLING**

④ Omit, if duct work is embedded in slab

TABLE B — COOLING — DOORS & WINDOWS
Factors assume windows have inside shading by draperies or venetian blinds and sliding glass doors are treated as windows.

	SINGLE GLASS			DOUBLE GLASS			TRIPLE GLASS			X Area	= BTUH GAIN
	TEMP. DIFF.			TEMP. DIFF.			TEMP. DIFF.				
Direction	15°	20°	25°	15°	20°	25°	15°	20°	25°		
N	20	25	30	15	20	20	12	14	16		
NE & NW	35	40	45	30	35	35	24	26	28		
E & W	55	55	60	45	50	50	36	38	40		
SE & SW	45	50	55	40	40	45	30	32	34		
S	30	30	35	25	25	30	18	20	22		
Skylights	146	150	154	139	143	147	132	136	140		
Doors	8.6	10.9	13.2	8.6	10.9	13.2	8.6	10.9	13.2		
									TOTALS		

TABLE C — ADJUSTMENT FACTORS — (HEATING)

°F. Temperature Diff.	30	40	50	60	70	80	90
Adjustment Factor	3	4	5	6	7	8	9

HOUSE & DUCT LAYOUT (1 SQ. = _____ Feet)

HEAT LOSS & GAIN FACTORS

TABLE D
CONSTRUCTION FACTORS HEATING & COOLING

Heating① Factor	TYPE OF CONSTRUCTION	Cooling Factor (°F. Temp. Diff.)		
		15°	20°	25°
WALLS (Use Sq. Ft.)				
	Walls — wood frame w/sheeting & siding, veneer or other finish			
2.71	A) No insulation, 1/2" Gypsum Board	5.0	6.4	7.8
0.90	B) R-11 Cavity insulation + 1/2" Gypsum Board	1.7	2.1	2.6
0.80	C) R-13 Cavity insulation + 1/2" Gypsum Board	1.5	1.9	2.3
0.70	D) R-13 Cavity insulation + 3/4" Bead Board (R-2.7)	1.3	1.7	2.0
0.60	E) R-19 Cavity insulation + 1/2" Gypsum Board	1.1	1.4	1.7
0.50	F) R-19 Cavity insulation + 3/4" Extruded Poly	0.9	1.2	1.4
	Masonry Walls			
5.10	A) Above grade No insulation	5.8	8.3	10.9
1.44	B) Above grade + R-5	1.6	2.3	3.1
0.77	C) Above grade + R-11	0.9	1.3	1.6
1.47	D) Below grade No insulation	0.0	0.0	0.0
0.87	E) Below grade + R-5	0.0	0.0	0.0
0.60	F) Below grade + R-11	0.0	0.0	0.0
CEILINGS (Use Sq. Ft.)				
5.99	A) No insulation	17.0	19.2	21.4
1.20	B) 2"-2½" insulation R-7	4.4	4.9	5.5
0.88	C) 3"-3½" insulation R-11	3.2	3.7	4.1
0.53	D) 5¼"-6½" insulation R-19	2.1	2.3	2.6
0.48	E) 6"-7" insulation R-22	1.9	2.1	2.4
0.33	F) 8"-9½" insulation R-30	1.3	1.5	1.6
0.26	G) 10"-12" insulation R-38	1.0	1.1	1.3
0.23	H) 12"-13" insulation R-44	0.9	1.0	1.1
3.08	I) Cathedral type No insulation (roof/ceiling combination)	11.2	12.6	14.1
0.72	J) Cathedral type R-11 (roof/ceiling combination)	2.8	3.2	3.5
0.49	K) Cathedral type R-19 (roof/ceiling combination)	1.9	2.2	2.4
0.45	L) Cathedral type R-22 (roof/ceiling combination)	1.8	2.0	2.2
0.40	M) Cathedral type R-26 (roof/ceiling combination)	1.6	1.8	2.0
FLOORS (Use Sq. Ft. OR Linear Ft.)				
	Floors over unconditioned space (use sq. ft.)			
1.56	A) Over basement or enclosed crawl space (not vented)	0.0	0.0	0.0
0.40	B) Same as "A" + R-11 insulation	0.0	0.0	0.0
0.26	C) Same as "A" + R-19 insulation	0.0	0.0	0.0
3.12	D) Over vented space or garage	3.9	5.8	7.7
0.80	E) Over vented space or garage + R-11 insulation	0.8	1.3	1.7
0.52	F) Over vented space or garage + R-19 insulation	0.5	0.8	1.1
0.28	**Basement Floors (use sq. ft.)**	0.0	0.0	0.0
	Concrete slab floor unheated (use linear ft.)			
8.10	A) No edge insulation	0.0	0.0	0.0
4.10	B) 1" edge insulation R-5	0.0	0.0	0.0
2.10	C) 2" edge insulation R-9	0.0	0.0	0.0
	Concrete Slab floor duct in slab (use linear ft.)			
19.00	A) No edge insulation	0.0	0.0	0.0
11.40	B) 1" edge insulation R-5	0.0	0.0	0.0
9.30	C) 2" edge insulation R-9	0.0	0.0	0.0

① Heating Factor for 10° Temperature Rise

ESTIMATING PROCEDURE

1. Fill in customer information.

2. Record inside and outside design temperatures; find temp difference.

3. Measure length of each outside wall, multiply each by ceiling height. Record the total sq. ft. of exposed wall under "gross wall".

4. Using Tables A and B, determine the total area for windows & doors & enter in common data section.

5. Determine Net Wall by subtracting windows and doors from gross.

6. Measure & record total ceiling area.

7. Measure and record total floor area for floors over crawl space or basement. Total floor edge *length* (perimeter) if floor is a slab.

8. Using Table D select construction type and use the corresponding heat and cool factors on the form.

9. Determine BTUH Loss & Gain in Tables A and B by multiplying the area of glass and doors by the multiplier under the specified temperature difference. Enter total BTUH Loss/Gain on worksheet.

10. On worksheet, multiply the areas x the factors and total as instructed.

EQUIPMENT CAPACITY CALCULATIONS FROM WORKSHEET DATA

COOLING CAPACITY

TOTAL
GAIN _____ BTUH x _____ ① = _____ BTUH
ARI Rating

ARI Rating _____ BTUH
————————————————————— = _____ TONS OF
12,000 BTUH/TON   COOLING

① CAPACITY MULTIPLIER
(BASED ON 75°F RETURN AIR TEMPERATURE)

90°F O.D.T. use 1.00   100°F O.D.T. use 1.10
95°F O.D.T. use 1.05   105°F O.D.T. use 1.15

HEATING CAPACITY

TOTAL
LOSS _____ BTUH
——————————————————— = _____ KW
(Select Factor Below)

ELECTRIC HEATER OR FURNACE DIVIDE USING 3413 BTUH/KW
NATURAL GAS OR PROPANE FURNACE . . . SELECT
OUTPUT TO BE AS CLOSE TO
CALCULATED HEAT LOSS AS POSSIBLE

CALCULATION BY: _____

DATE: _____

EQUIPMENT SELECTION
Model Number(s)

HEAT PUMP. . . . . _____

AIR HANDLER . . . . _____

SUPPLEMENTARY HEATER . . _____

CONDENSING UNIT . . . . _____

EVAPORATOR COIL . . . . _____

FURNACE, ELECTRIC . . . . _____
FURNACE, NATURAL GAS/
        PROPANE . . . . _____

REFRIGERANT LINES . . . . _____

THERMOSTAT . . . . _____
ELECTRONIC AIR
CLEANER . . . . _____

HUMIDIFIER . . . . _____

ACCESSORIES . . . . _____

_____

_____

_____

COST SCHEDULE(S) MATERIAL/LABOR

_____

_____

_____

_____

_____

_____

_____

_____

_____

_____

_____

_____

_____

_____

_____

_____

_____

_____

_____

_____

_____

_____

_____

_____

_____

_____

_____

_____

_____

ROOM AIRFLOW DISTRIBUTION

1. Divide conditioned space into airflow zones.

2. Record linear feet of outside wall in each zone. Obtain total of all zones.

3. Distribute CFM for windows, doors, appliances, ceiling and people per instructions. Obtain total of all zones.

4. Obtain total CFM distributed in Step 6.

5. Find total airflow of selected system.

6. Use equation below and find CFM per linear foot of outside wall.

7. Using linear feet obtained in Step 2 find CFM for linear feet of wall in each zone. (Step 2 x Step 8).

8. Obtain total CFM in each zone (Step 9 plus 3, 4, and 5).

AIRFLOW ZONES	OUTSIDE WALLS			CFM				DOORS 15 CFM per STD. 3' DOOR / APPLIANCES 60 CFM TOTAL	TOTAL CFM PER ZONE
	LINEAR FEET	CFM PER FOOT	OUTSIDE WALLS	GLASS DOORS & WINDOWS		EXPOSED CEILING .1CFM/sq.ft.		PEOPLE 15 CFM per person	
				DIRECTION	CFM per sq. ft.				
				N	1				
				NE, NW, S	1½				
				E, SE, SW, W	2				
STEP	1	2	8	9		3	4		5	10
A										
B										
C										
D										
E										
F										
G										
H										
I										
J										
K										
L										

CONSTANT

STEP 6

STEP 7 [ ] − [ ] = [ ] ÷ [ ] = [ ]

DESIGN FRICTION LOSS LIMITS FOR LONGEST (EQUIVALENT) S & R RUN

1. Layout duct system on scaled drawing and show airflow volumes to each airflow zone.

2. Select apparent longest supply and longest return — consider actual length of ducts and equivalent length of duct fittings.

3. List the air distribution devices which make-up the longest supply and longest return runs: cooling coil, duct sections (main and branch), types of duct fittings, register, grille and filter as applicable.

4. Assign the airflow volumes through each component listed in Step 3.

5. Find static pressure losses of each of the fixed distribution devices: cooling coil, register, grille, filter as applicable. Total fixed pressure losses. Consult the manufacturer's engineering data for accurate information.

6. From the duct layout and equivalent length tables find the actual or equivalent length of each item listed in Step 3.

7. Select indoor air moving equipment from manufacturer's data. Find maximum external static pressure available at the required air volume.

8. Using equation below find allowable pressure loss per 100' of duct.

9. Size **ALL** ducts using duct calculator. Do not exceed friction loss found in Step 8 **or** recommend air velocities.

   Main Ducts: 900 FPM
   Branch Ducts: 600 FPM
   Branch Risers: 500 FPM

DUCT SECTIONS		PRESSURE LOSS	LENGTH	DUCT WORK SIZE
①IDENTIFICATION	CFM	①FIXED ITEMS	ACTUAL OR EQUIVALENT	MAINS, BRANCHES
3	4	5	6	9

NOTES

7 [ ] − [ ] = [ ] ÷ [ ] X [100] = [ ] 8

① Return Grille = .02 if Manufacturer's data is not available.

The Trane Company
Dealer Products Group

COMPUTER WORKSHEET FOR MJ6
LOAD CALCULATIONS FOR RESIDENTIAL AIR CONDITIONING

The MJ6 computer program uses the calculation procedure described in Manual J 6th edition published by the Air Conditioning Contractors of America (formerly NESCA) to perform residential heat loss and heat gain calculations. Both entire house and room by room load calculations are performed. There is provision for performing only an entire house calculation if desired. Data entry was designed to conform as closely as possible to that entered when performing a hand calculation. It is not necessary to fill in unused lines with zeros, leave them blank and enter a carriage return when the computer prints the line number. Pub. No. 22-3115-1 may be used if only a whole house calculation is desired.

For: **Name** _____

        Address _____

        City and State or Province _____

By: **Contractor** _____

        Address _____

        City _____

Winter Design Conditions

Outside_____°F   **Inside**_____°F   **Temperature Difference**_____ **Degrees**

(Insert data below only after all heat loss calculations have been completed)

Total Heat Loss (BTUH) _____   **Model No.** _____

Serial No. _____   **Manufactured by** _____

Rating Data: Input _____ **BTUH**   **Output at Bonnet** _____ **BTUH**

Description of Controls _____

Summer Design Conditions

Outside_____°F   **Inside** _____°F

North Latitude _____ **Degrees**   **Daily Range** _____

Summer Grain Difference _____.

(Insert data below only after all heat gain calculations have been completed)

Total Heat Gain (BTUH) _____

Equipment Capacity Multiplier _____   **Model No.** _____

Serial No. _____   **Manufactured by** _____

Rating Data: Cooling Capacity _____ **BTUH**   **Air Volume** _____ **Cfm**

Description of Controls _____

Winter Construction Data

Walls and Partitions _____

_____

_____

Windows and Doors _____

_____

_____

Ceilings _____

_____

_____

Floors _____

_____

Summer Construction Data

Direction House Faces _____

Windows and Doors _____

_____

_____

Walls and Partitions _____

_____

_____

Ceilings _____

_____

Floors _____

_____

FILE

759

CUSTOMER NAME	0010		
ADDRESS	0020		
CITY, STATE	0030		
PREPARED BY	0040		

		WINTER		SUMMER				
		OUTSIDE °F	INSIDE °F	OUTSIDE °F	INSIDE °F	GRAINS	LATITUDE	DAILY RANGE
DESIGN CONDITIONS	0050							

		* VENT CFM		NO. OF PEOPLE		NO. OF ROOMS		DUCT GAIN†		DUCT LOSS†	
ENTIRE HOUSE VALUES	0100										

			CONST. NO.	LENGTH	X	HEIGHT		LENGTH	X	HEIGHT
PRIMARY WALL CONSTRUCTION	GROSS EX. WALLS PARTITIONS	0110			X				X	
			CONST. NO.	GLASS NOTE C	SHADING NOTE A	DIRECTION NOTE B	NO. OF WINDOWS	LENGTH	X	HEIGHT
	WINDOWS AND GLASS DOORS IN WALL	0111							X	
		0112							X	
		0113							X	
		0114							X	
			CONST. NO.	WIDTH	X	HEIGHT		WIDTH	X	HEIGHT
	OTHER DOORS IN WALL	0119			X				X	

			CONST. NO.	LENGTH	X	HEIGHT		LENGTH	X	HEIGHT
SECONDARY WALL CONSTRUCTION	GROSS EX. WALL PARTITIONS	0120			X				X	
			CONST. NO.	GLASS NOTE C	SHADING NOTE A	DIRECTION NOTE B	NO. OF WINDOWS	LENGTH	X	HEIGHT
	WINDOWS AND GLASS DOORS IN WALL	0121							X	
		0122							X	
		0123							X	
		0124							X	
			CONST. NO.	WIDTH	X	HEIGHT		WIDTH	X	HEIGHT
	OTHER DOORS IN WALL	0129			X				X	

			CONST. NO.	LENGTH	X	HEIGHT		LENGTH	X	HEIGHT
SECONDARY WALL CONSTRUCTION	GROSS EX. WALLS PARTITIONS	0130			X				X	
			CONST. NO.	GLASS NOTE C	SHADING NOTE A	DIRECTION NOTE B	NO. OF WINDOWS	LENGTH	X	HEIGHT
	WINDOWS AND GLASS DOORS IN WALL	0131							X	
		0132							X	
		0133							X	
		0134							X	
			CONST. NO.	WIDTH	X	HEIGHT		WIDTH	X	HEIGHT
	OTHER DOORS IN WALL	0139			X				X	

			CONST. NO.	LENGTH	X	HEIGHT		LENGTH	X	HEIGHT
SECONDARY WALL CONSTRUCTION	GROSS EX. WALL PARTITIONS	0140			X				X	
			CONST. NO.	GLASS NOTE C	SHADING NOTE A	DIRECTION NOTE B	NO. OF WINDOWS	LENGTH	X	HEIGHT
	WINDOWS AND GLASS DOORS IN WALL	0141							X	
		0142							X	
		0143							X	
		0144							X	
			CONST. NO.	WIDTH	X	HEIGHT		WIDTH	X	HEIGHT
	OTHER DOORS IN WALL	0149			X				X	

		CONST. NO.	L—LIGHT D—DARK	LENGTH	X	WIDTH	LENGTH	X	WIDTH
CEILINGS AND ROOFS	0150				X			X	
	0151				X			X	

		CONST. NO.	*LENGTH	X	WIDTH	*LENGTH	X	WIDTH
FLOORS *ENTER ONLY LENGTH FOR CONST. NO. 20—21	0160			X			X	
	0161			X			X	

†SEE DUCT MULTIPLIER TABLES   *NOTE: FOR CONTINUOUS POSITIVE PRESSURE MECHANICAL VENTILATION ONLY.

				ROOM DIMENSIONS					
0200		**KITCHEN**							
				LENGTH	X	WIDTH	LENGTH	X	WIDTH
	0201				X			X	

				CONST. NO.	LENGTH	X	HEIGHT	LENGTH	X	HEIGHT
PRIMARY WALL CONSTRUCTION	**GROSS EX. WALLS PARTITIONS**		0210			X			X	

				CONST. NO.	GLASS NOTE C	SHADING NOTE A	DIRECTION NOTE B	NO. OF WINDOWS	LENGTH	X	HEIGHT
	WINDOWS AND GLASS DOORS IN WALL	0211							X	
		0212							X	
		0213							X	
		0214							X	

		CONST. NO.	WIDTH	X	HEIGHT	WIDTH	X	HEIGHT
OTHER DOORS IN WALL	0219			X			X	

			CONST. NO.	LENGTH	X	HEIGHT	LENGTH	X	HEIGHT
SECONDARY WALL CONSTRUCTION	**GROSS EX. WALL PARTITIONS**	0220			X			X	

			CONST. NO.	GLASS NOTE C	SHADING NOTE A	DIRECTION NOTE B	NO. OF WINDOWS	LENGTH	X	HEIGHT
WINDOWS AND GLASS DOORS IN WALL	0221							X	
	0222							X	
	0223							X	
	0224							X	

		CONST. NO.	WIDTH	X	HEIGHT	WIDTH	X	HEIGHT
OTHER DOORS IN WALL	0229			X			X	

		CONST. NO.	L–LIGHT D–DARK	LENGTH	X	WIDTH	LENGTH	X	WIDTH
CEILINGS AND ROOFS	0250				X			X	
	0251				X			X	

		CONST. NO.	*LENGTH	X	WIDTH	*LENGTH	X	WIDTH
FLOORS *ENTER ONLY LENGTH FOR CONST. NO. 20–23	0260			X			X	
	0261			X			X	

0300	**ROOM NAME**		ALL PEOPLE LOAD WILL BE INCLUDED IN THIS ROOM

		LENGTH	X	WIDTH	LENGTH	X	WIDTH
ROOM DIMENSIONS	0301		X			X	

			CONST. NO.	LENGTH	X	HEIGHT	LENGTH	X	HEIGHT
SECONDARY WALL CONSTRUCTION	**GROSS EX. WALLS PARTITIONS**	0310			X			X	

		CONST. NO.	GLASS NOTE C	SHADING NOTE A	DIRECTION NOTE B	NO. OF WINDOWS	LENGTH	X	HEIGHT
WINDOWS AND GLASS DOORS IN WALL	0311							X	
	0312							X	
	0313							X	
	0314							X	

		CONST. NO.	WIDTH	X	HEIGHT	WIDTH	X	HEIGHT
OTHER DOORS IN WALL	0319			X			X	

			CONST. NO.	LENGTH	X	HEIGHT	LENGTH	X	HEIGHT
SECONDARY WALL CONSTRUCTION	**GROSS EX. WALL PARTITIONS**	0320			X			X	

		CONST. NO.	GLASS NOTE C	SHADING NOTE A	DIRECTION NOTE B	NO. OF WINDOWS	LENGTH	X	HEIGHT
WINDOWS AND GLASS DOORS IN WALL	0321							X	
	0322							X	
	0323							X	
	0324							X	

		CONST. NO.	WIDTH	X	HEIGHT	WIDTH	X	HEIGHT
OTHER DOORS IN WALL	0329			X			X	

		CONST. NO.	L–LIGHT D–DARK	LENGTH	X	WIDTH	LENGTH	X	WIDTH
CEILINGS AND ROOFS	0350				X			X	
	0351				X			X	

		CONST. NO.	*LENGTH	X	WIDTH	*LENGTH	X	WIDTH
FLOORS *ENTER ONLY LENGTH FOR CONST. NO. 20–21	0360			X			X	
	0361			X			X	

0400 ROOM NAME:

		ROOM DIMENSIONS						MISCELLANEOUS HEAT GAINS (BTUH)	
		LENGTH	X	WIDTH	LENGTH	X	WIDTH		
	0401		X			X			

PRIMARY WALL CONSTRUCTION			CONST. NO.	LENGTH	X	HEIGHT			LENGTH	X	HEIGHT
	GROSS EX. WALLS PARTITIONS	0410			X					X	

			CONST. NO.	GLASS NOTE C	SHADING NOTE A	DIRECTION NOTE B	NO. OF WINDOWS	LENGTH	X	HEIGHT
WINDOWS AND GLASS DOORS IN WALL	0411								X	
	0412								X	
	0413								X	
	0414								X	

OTHER DOORS IN WALL	0419	CONST. NO.	WIDTH	X	HEIGHT	WIDTH	X	HEIGHT
				X			X	

SECONDARY WALL CONSTRUCTION

		CONST. NO.	LENGTH	X	HEIGHT	LENGTH	X	HEIGHT
GROSS EX. WALL PARTITIONS	0420			X			X	

		CONST. NO.	GLASS NOTE C	SHADING NOTE A	DIRECTION NOTE B	NO. OF WINDOWS	LENGTH	X	HEIGHT
WINDOWS AND GLASS DOORS IN WALL	0421							X	
	0422							X	
	0423							X	
	0424							X	

OTHER DOORS IN WALL	0429	CONST. NO.	WIDTH	X	HEIGHT	WIDTH	X	HEIGHT
				X			X	

CEILINGS AND ROOFS		CONST. NO.	L—LIGHT D—DARK	LENGTH	X	WIDTH	LENGTH	X	WIDTH
	0450				X			X	
	0451				X			X	

FLOORS *ENTER ONLY LENGTH FOR CONST. NO. 20—23		CONST. NO.	*LENGTH	X	WIDTH	*LENGTH	X	WIDTH
	0460			X			X	
	0461			X			X	

0500 ROOM NAME:

		ROOM DIMENSIONS						MISCELLANEOUS HEAT GAINS (BTUH)	
		LENGTH	X	WIDTH	LENGTH	X	WIDTH		
	0501		X			X			

SECONDARY WALL CONSTRUCTION

		CONST. NO.	LENGTH	X	HEIGHT	LENGTH	X	HEIGHT
GROSS EX. WALLS PARTITIONS	0510			X			X	

		CONST. NO.	GLASS NOTE C	SHADING NOTE A	DIRECTION NOTE B	NO. OF WINDOWS	LENGTH	X	HEIGHT
WINDOWS AND GLASS DOORS IN WALL	0511							X	
	0512							X	
	0513							X	
	0514							X	

OTHER DOORS IN WALL	0519	CONST. NO.	WIDTH	X	HEIGHT	WIDTH	X	HEIGHT
				X			X	

SECONDARY WALL CONSTRUCTION

		CONST. NO.	LENGTH	X	HEIGHT	LENGTH	X	HEIGHT
GROSS EX. WALL PARTITIONS	0520			X			X	

		CONST. NO.	GLASS NOTE C	SHADING NOTE A	DIRECTION NOTE B	NO. OF WINDOWS	LENGTH	X	HEIGHT
WINDOWS AND GLASS DOORS IN WALL	0521							X	
	0522							X	
	0523							X	
	0524							X	

OTHER DOORS IN WALL	0529	CONST. NO.	WIDTH	X	HEIGHT	WIDTH	X	HEIGHT
				X			X	

CEILINGS AND ROOFS		CONST. NO.	L—LIGHT D—DARK	LENGTH	X	WIDTH	LENGTH	X	WIDTH
	0550				X			X	
	0551				X			X	

FLOORS *ENTER ONLY LENGTH FOR CONST. NO. 20—23		CONST. NO.	*LENGTH	X	WIDTH	*LENGTH	X	WIDTH
	0560			X			X	
	0561			X			X	

1600 — ROOM NAME:

		ROOM DIMENSIONS				MISCELLANEOUS HEAT GAINS (BTUH)	
		LENGTH X	WIDTH	LENGTH X	WIDTH		
	1601	X		X			

PRIMARY WALL CONSTRUCTION

GROSS EX. WALLS PARTITIONS

		CONST. NO.	LENGTH X	HEIGHT	LENGTH X	HEIGHT
	1610		X		X	

WINDOWS AND GLASS DOORS IN WALL

	CONST. NO.	GLASS NOTE C	SHADING NOTE A	DIRECTION NOTE B	NO. OF WINDOWS	LENGTH X	HEIGHT
1611						X	
1612						X	
1613						X	
1614						X	

OTHER DOORS IN WALL

	CONST. NO.	WIDTH X	HEIGHT	WIDTH X	HEIGHT
1619		X		X	

SECONDARY WALL CONSTRUCTION

GROSS EX. WALL PARTITIONS

	CONST. NO.	LENGTH X	HEIGHT	LENGTH X	HEIGHT
1620		X		X	

WINDOWS AND GLASS DOORS IN WALL

	CONST. NO.	GLASS NOTE C	SHADING NOTE A	DIRECTION NOTE B	NO. OF WINDOWS	LENGTH X	HEIGHT
1621						X	
1622						X	
1623						X	
1624						X	

OTHER DOORS IN WALL

	CONST. NO.	WIDTH X	HEIGHT	WIDTH X	HEIGHT
1629		X		X	

CEILINGS AND ROOFS

	CONST. NO.	L—LIGHT D—DARK	LENGTH X	WIDTH	LENGTH X	WIDTH
1650			X		X	
1651			X		X	

FLOORS
*ENTER ONLY LENGTH FOR CONST. NO. 20—23

	CONST. NO.	*LENGTH X	WIDTH	*LENGTH X	WIDTH
1660		X		X	
1661		X		X	

DUCT HEAT GAIN MULTIPLIERS

Duct Location and Insulation	Duct Heat Gain Multipliers
Attic; 1 in. Flexible Blanket Insulation	0.15
Attic; 2 in. Flexible or 1 in. Rigid Duct Insulation	0.10
Unconditioned Above Grade Space; 1 in. Flexible Duct Insulation	0.05
Unconditioned Below Grade Space; No Duct Insulation	0.05
Ducts in Slab; No Duct Insulation	0.05

DUCT LOSS MULTIPLIERS

		Below 120°F		Above 120°F	
Duct Location and Insulation	Winter Design	Below 15°F	Above 15°F	Below 15°F	Above 15°F
Attic or Open Crawl Space - Insulation R value 3 to 5		.15	.10	.20	.15
Attic or Open Crawl Space - Insulation R value 7 to 9		.10	.05	.15	.10
Enclosed Unheated Crawl Space - Bare Duct		.15	.10	.20	.15
Enclosed Unheated Crawl Space - Insulated Duct		.10	.05	.15	.10
Duct in Slab with Perimeter insulation		.10	.05	.15	.10

Supply Temperatures

INCH TO DECIMAL FEET CONVERSION TABLE

Inches	Feet	Inches	Feet	Inches	Feet
1	.0	5	.4	9	.8
2	.2	6	.5	10	.8
3	.2	7	.6	11	.9
4	.3	8	.7	12	1.0

NOTE A

Type of Shading	Insert
No Shade	N
Drapes or Venetian Blinds	D or V
Roller Shades	R
Awnings, Porches or Other	S

NOTE B

Direction Glass Faces	Insert
North	N
Northeast and Northwest	NE or NW
East and West	E or W
Southeast and Southwest	SE or SW
South	S
Skylight	Sky

NOTE C

TYPE OF GLASS		INSERT
Single	Clear	S
	Tinted	ST
	Reflective	SR
Double	Clear	D
	Tinted	DT
	Reflective	DR
Triple	Clear	T
	Tinted	TT
	Reflective	TR

HEAT TRANSFER CONSTRUCTION NUMBERS

CONST. NO. 1-9 WINDOWS & GLASS DOORS, WOOD OR METAL FRAME. Factors include heat loss due to transmission and infiltration @ 15 MPH

No. 1 Fixed Glass Windows, Crack Ratio = 1.1:1.0, Infiltration Rate = 0.2 CFM/LF @ 25 MPH
- 1-A Wood Frame* - Single Glass
- 1-B Wood Frame* - Single Glass + Storm
- 1-C Wood Frame* - Double Glass
- 1-D Wood Frame* - Double Glass + Storm
- 1-E Metal Frame - Single Glass
- 1-F Metal Frame - Single Glass + Storm
- 1-G Metal Frame - Double Glass
- 1-H Metal Frame - Double Glass + Storm

No. 2 Movable Glass Windows, Crack Ratio = 1.4:1.0, Infiltration Rate = 0.2 CFM/LF @ 25 MPH
- 2-A Wood Frame* - Single Glass
- 2-B Wood Frame* - Single Glass + Storm
- 2-C Wood Frame* - Double Glass
- 2-D Wood Frame* - Double Glass + Storm
- 2-E Metal Frame - Single Glass
- 2-F Metal Frame - Single Glass + Storm
- 2-G Metal Frame - Double Glass
- 2-H Metal Frame - Double Glass + Storm

No. 3 Movable Glass Windows, Crack Ratio = 1.4:1.0, Infiltration Rate = 0.5 CFM/LF @ 25 MPH
- 3-A Wood Frame* - Single Glass
- 3-B Wood Frame* - Single Glass + Storm
- 3-C Wood Frame* - Double Glass
- 3-D Wood Frame* - Double Glass + Storm
- 3-E Metal Frame - Single Glass
- 3-F Metal Frame - Single Glass + Storm
- 3-G Metal Frame - Double Glass
- 3-H Metal Frame - Double Glass + Storm

No. 4 Movable Glass Windows, Crack Ratio = 1.4:1.0, Infiltration Rate = 0.75 CFM/LF @ 25 MPH
- 4-A Wood Frame - Single Glass
- 4-B Wood Frame - Single Glass + Storm
- 4-C Wood Frame - Double Glass
- 4-D Wood Frame - Double Glass + Storm
- 4-E Metal Frame - Single Glass
- 4-F Metal Frame - Single Glass + Storm
- 4-G Metal Frame - Double Glass
- 4-H Metal Frame - Double Glass + Storm

No. 5 Awning Type Windows, Crack Ratio = 1.7:1.0, Infiltration Rate = 0.5 CFM, LF @ 25 MPH
- 5-A Wood Frame* - Single Glass
- 5-B Wood Frame* - Single Glass + Storm
- 5-C Wood Frame* - Double Glass
- 5-D Wood Frame* - Double Glass + Storm
- 5-E Metal Frame - Single Glass
- 5-F Metal Frame - Single Glass + Storm
- 5-G Metal Frame - Double Glass
- 5-H Metal Frame - Double Glass + Storm

No. 6 Awning Type Windows, Crack Ratio = 1.7:1.0, Infiltration Rate + 0.75 CFM/LF @ 25 MPH
- 6-A Wood Frame - Single Glass
- 6-B Wood Frame - Single Glass + Storm
- 6-C Wood Frame - Double Glass
- 6-D Wood Frame - Double Glass + Storm
- 6-E Metal Frame - Single Glass
- 6-F Metal Frame - Single Glass + Storm
- 6-G Metal Frame - Double Glass
- 6-H Metal Frame - Double Glass + Storm

No. 7 Jalousie Windows, Crack Ratio = 3.0:1.0, Infiltration Rate = 1.50 CFM/LF @ 25 MPH
- 7-A Metal Frame - Single Glass
- 7-B Metal Frame - Single Glass + Storm

No. 8 Glass Doors, Crack Ratio = 1.1:1.0, Infiltration Rate = 0.5 CFM/LF @ 25 MPH
- 8-A Wood Frame* - Single Glass
- 8-B Wood Frame* - Single Glass + Storm
- 8-C Wood Frame* - Double Glass
- 8-D Wood Frame* - Double Glass + Storm
- 8-E Metal Frame - Single Glass
- 8-F Metal Frame - Single Glass + Storm
- 8-G Metal Frame - Double Glass
- 8-H Metal Frame - Double Glass + Storm

No. 9 Glass Doors, Crack Ratio = 1.1:1.0, Infiltration Rate = 1.0 CFM/LF @ 25 MPH
- 9-A Wood Frame - Single Glass
- 9-B Wood Frame - Single Glass + Storm
- 9-C Wood Frame - Double Glass
- 9-D Wood Frame - Double Glass + Storm
- 9-E Metal Frame - Single Glass
- 9-F Metal Frame - Single Glass + Storm
- 9-G Metal Frame - Double Glass
- 9-H Metal Frame - Double Glass + Storm

WOOD AND METAL DOORS. Factors include heat loss due to transmission and infiltration @ 15 MPH

No. 10 Doors, Crack Ratio = 1.0:1.0, Infiltration Rate = 0.5 CFM/LF @ 25 MPH
- 10-A Solid Wood
- 10-B Solid Wood + Storm Door
- 10-C Metal - Fiber Core
- 10-D Metal - Fiber Core + Storm Door
- 10-E Metal - Urethane Core
- 10-F Metal - Urethane Core + Storm Door
- 10-G Metal - Polystyrene Core
- 10-H Metal - Polystyrene Core + Storm

No. 11 Doors, Crack Ratio = 1.0:1.0, Infiltration Rate = 1.0 CFM/LF @ 25 MPH
- 11-A Solid Wood
- 11-B Solid Wood + Storm Door
- 11-C Metal - Fiber Core
- 11-D Metal - Fiber Core + Storm Door
- 11-E Metal - Urethane Core
- 11-F Metal - Urethane Core + Storm Door
- 11-G Metal - Polystyrene Core
- 11-H Metal - Polystyrene Core + Storm

WOOD FRAME AND MASONRY WALLS AND PARTITIONS. Factors include heat loss due to transmission only.

No. 12 Wood Frame Exterior Walls With Sheathing and Siding or Brick Veneer, or Other Exterior Finish.
- 12-A None ½" Gypsum Brd (R-0.5)
- 12-B None ½" Asphalt Brd (R-1.3)
- 12-C R-11 ½" Gypsum (R-0.5)
- 12-D R-11 ½" Asphalt Brd (R-1.3)
- 12-E R-11 ½" Bead Brd (R-1.8)
  - R-13 ½" Gypsum Brd (R-0.5)
  - R-11 ½" Extr Poly Brd (R-2.5)
  - R-11 ¾" Bead Brd (R-2.7)
  - R-13 ½" Asphalt Brd (R-1.3)
  - R-13 ¾" Bead Brd (R-1.8)
- 12-F R-11 1" Bead Brd (R-3.6)
  - R-11 ¾" Extr Poly Brd (R-3.8)
  - R-13 ½" Extr Poly Brd (R-2.5)
  - R-13 ¾" Bead Brd (R-2.7)
- 12-G R-13 ¾" Extr Poly Brd (R-3.8)
  - R-13 1" Bead Brd (R-3.6)
- 12-H R-11 1" Extr Poly Brd (R-5.0)
  - R-13 1" Extr Poly Brd (R-5.0)
- 12-I R-19 ½" Gypsum Brd (R-0.5)
  - R-19 ½" Asphalt Brd (R-1.3)
  - R-19 ½" Bead Brd (R-1.8)
- 12-J R-11 R-8 Sheathing
  - R-13 R-8 Sheathing
  - R-19 ½" or ¾" Extr Poly Brd
  - R-19 ¾" or 1" Bead Brd
- 12-K R-19 1" Extr Poly Brd (R-5.0)
- 12-L R-19 R-8 Sheathing

No. 13 Frame or Masonry Partitions Between a Conditioned and an Unconditioned Space. Use Htm from Construction No. 12 or 14. Select Htm for Actual Temperature Difference Expected Across the Partition.

No. 14 Masonry Walls, Block or Brick, Finished or Unfinished - Above Grade.
- 14-A 8" or 12" Block, No Insul., Unfinished
- 14-B 8" or 12" Block + R-5
- 14-C 8" or 12" Block + R-11
- 14-D 8" or 12" Block + R-19
- 14-E 4" Brick + 8" Block, No Insul.
- 14-F 4" Brick + 8" Block + R-5
- 14-G 4" Brick + 8" Block + R-11
- 14-H 4" Brick + 8" Block + R-19

No. 15 Masonry Walls, Block or Brick, Finished or Unfinished - Below Grade**
For Walls 2'-5' Below Grade
- 15-A 8" or 12" Block + No Insul.
- 15-B 8" or 12" Block + R-5
- 15-C 8" or 12" Block + R-11
- 15-D 8" or 12" Block + R-19

For Walls 5'-8' Below Grade
- 15-E 8" or 12" Block + No Insul.
- 15-F 8" or 12" Block + R-5
- 15-G 8" or 12" Block + R-11
- 15-H 8" or 12" Block + R-19

CEILINGS AND ROOFS. Factors include heat loss due to transmission only.

No. 16 Ceilings Under a Ventilated Attic Space or Unconditioned Room.
- 16-A No Insulation
- 16-B R-7 Insulation
- 16-C R-11 Insulation
- 16-D R-19 Insulation
- 16-E R-22 Insulation
- 16-F R-26 Insulation
- 16-G R-30 Insulation
- 16-H R-38 Insulation
- 16-I R-44 Insulation
- 16-J R-57 Insulation

No. 17 Roof on Exposed Beams or Rafters
- 17-A 1½" Wood Decking, No Insul.
- 17-B 1½" Wood Decking + R-4
- 17-C 1½" Wood Decking + R-5
- 17-D 1½" Wood Decking + R-6
- 17-E 1½" Wood Decking + R-8
- 17-F 2" Shredded Wood Planks
- 17-G 3" Shredded Wood Planks
- 17-H 1½" Fiber Board Insulation
- 17-I 2" Fiber Board Insulation
- 17-J 3" Fiber Board Insulation
- 17-K 1½" Wood Decking + R-13
- 17-L 1½" Wood Decking + R-19

No. 18 Roof-Ceiling Combination
- 18-A No Insulation
- 18-B R-11 Batts
- 18-C R-19 Batts (2" x 8" Rafters)
- 18-D R-22 Batts (2" x 8" Rafters)
- 18-E R-26 Batts (2" x 8" Rafters)
- 18-F R-30 Batts (2" x 10" Rafters)

WOOD FRAME OR CONCRETE SLAB FLOORS. (Btuh per sq. ft. or Btuh per lin. ft.) Factors include heat loss due to transmission only.

No. 19 Floors Over an Unheated Basement, Enclosed Crawl Space† or Crawl Space with Closable Vents.
- 19-A Hardwood Floor + No Insulation
- 19-B Hardwood Floor + R-11
- 19-C Hardwood Floor + R-13
- 19-D Hardwood Floor + R-19
- 19-E Hardwood Floor + R-30
- 19-F Carpeted Floor + No Insulation
- 19-G Carpeted Floor + R-11
- 19-H Carpeted Floor + R-13
- 19-I Carpeted Floor + R-19
- 19-J Carpeted Floor + R-30

No. 20 Floors Over an Open Crawl Space or Garage.
- 20-A Hardwood Floor + No Insulation
- 20-B Hardwood Floor + R-11
- 20-C Hardwood Floor + R-13
- 20-D Hardwood Floor + R-19
- 20-E Hardwood Floor + R-30
- 20-F Carpeted Floor + No Insulation
- 20-G Carpeted Floor + R-11
- 20-H Carpeted Floor + R-13
- 20-I Carpeted Floor + R-19
- 20-J Carpeted Floor + R-30

No. 21 Basement Floors

No. 22 Concrete Slab on Grade
- 22-A No Edge Insulation
- 22-B 1" Edge Insulation
- 22-C 1½" Edge Insulation
- 22-D 2" Edge Insulation

No. 23 Concrete Slab with Perimeter Warm Air Duct system
- 23-A No Edge Insulation
- 23-B 1" Edge Insulation
- 23-C 1½" Edge Insulation
- 23-D 2" Edge Insulation

---

* For metal frame with thermal break use wood frame data.
** Masonry wall less than two feet below grade should be considered as being above grade.
† Htm listed assume the temperature difference across the floor is equal to one half of the design temperature difference. For other temperature differences the Htm can be calculated by multiplying the UT valve by the expected temperature difference.

INSTRUCTIONS
COMPUTER WORKSHEET FOR MANUAL J

The instructions on this page are necessarily brief. If assistance is required when using this worksheet, contact your Trane sales representative. If only the entire house calculation is desired, enter "0" (zero) for the number of rooms, and enter only the entire house data. All dimensions should be in Feet or in Decimal Feet. It is not necessary to fill in unused lines with zeros, leave blank.

		WINTER		SUMMER				
		OUTSIDE °F	INSIDE °F	OUTSIDE °F	INSIDE °F	GRAINS	LATITUDE	DAILY RANGE
DESIGN CONDITIONS	0050	O	70	95	75	23	40	M

If not specified the design conditions can be found in Manual J.

		VENT CFM	NO. OF PEOPLE	NO. OF ROOMS	DUCT GAIN†	DUCT LOSS†
ENTIRE HOUSE VALUES	0100	O	6	7	.05	.10

The number of people and number of rooms must be filled out. The vent CFM and Duct Gain/Loss is optional and used only if applicable. The number of rooms should not include the entire house as a room. Duct Loss and Gain factors can be obtained from the Duct Heat Loss/Gain Multipliers Table. If Vent CFM is not used put "0" (zero).

		CONST. NO.		LENGTH	x	HEIGHT	LENGTH	x	HEIGHT
GROSS EX. WALLS PARTITIONS	0110	12	D	152	x	8		x	

For Gross Walls choose the proper construction number from the chart and enter the number first and the letter next as shown. If more than one type of wall construction is needed, use Secondary Wall Construction entries.

		CONST. NO.		GLASS NOTE C	SHADING NOTE A	DIRECTION NOTE B	NO. OF WINDOWS	LENGTH	x	HEIGHT
WINDOWS AND GLASS DOORS IN WALL	0111	2	B	D	D	W	7	4	x	3
	0112	8	C	D	D	E	1	6	x	7
	0113	4	B	D	D	S	1	3.5	x	2.8
	0114								x	

For Windows and Glass Doors choose the construction number from the chart. See Note "A" for shading factors. For the direction see Note "B". For the Entire House it will save space if you combine like directions. For example east and west, northeast and northwest, etc. Under Number of Windows indicate the number of similar windows that face the same direction and are the same size. If you run out of space for windows, you may combine several unlike sizes into one set of dimensions.

		CONST. NO.		WIDTH	x	HEIGHT	WIDTH	x	HEIGHT
OTHER DOORS IN WALL	0119	10	B	3	x	7	3	x	7

For Doors look up the construction number and enter the Width and Height. Again if you run out of space, you may combine several into one set of dimensions.

		CONST. NO.		L-LIGHT D-DARK	LENGTH	x	WIDTH	LENGTH	x	WIDTH
CEILINGS AND ROOFS	0150	16	D	D	28	x	48		x	
	0151					x			x	

For Ceilings choose the proper construction number and enter "L" for Light or "D" for Dark. There is space for two dimensions. If a different roof or ceiling is encountered, use the second line and repeat the process.

		CONST. NO.		*LENGTH	x	WIDTH	*LENGTH	x	WIDTH
FLOORS *ENTER ONLY LENGTH FOR CONST. NO. 20—23	0160	19	B	28	x	48		x	
	0161	22	D	12.5	x			x	

For Floors over unconditioned areas, enter the construction number and the Length and Width. For slab floors and conditioned crawl spaces enter only the length for the perimeter of exposed floor.

0200	KITCHEN					
	ROOM DIMENSIONS					
	LENGTH	x	WIDTH	LENGTH	x	WIDTH
0201	16	x	9		x	

The appliance load and number of people are already factored into the kitchen and living room respectively. Space is provided after each room fro the room dimensions. Additional dimensions are provided for "L" shaped rooms.

0400	ROOM NAME:					
	ROOM DIMENSIONS					MISCELLANEOUS HEAT GAINS (BTUH)
	LENGTH	x	WIDTH	LENGTH	x	WIDTH	
0401	12	x	17		x		600

For Additional Rooms enter the name of the room in line "00" and the dimensions in line "01". Additional space is provided for Miscellaneous BTUH Gains as required. For example, in rooms such as family and game rooms you may want to add additional BTUH for people since the living room is already calculated with the people load.

The Entire House and Room by Room is calculated separately. The total figures probably will not be exactly the same. It is only important if the figures are far off, since this would indicate that some inputs have been deleted. It is not important that Entire House entries be the same as Room by Room. For example, a room may have a wall in the Primary Wall Section which would be included in the Secondary Wall Section of the Entire House. Because you may have several wall constructions for the Entire House, four sections have been provided. For individual rooms two wall sections are provided, since very seldom would a single room involve more than two different kinds of walls.

The Trane Company
Dealer Products Group • Troup Highway • Tyler, TX 75711

APPENDIX II—REFRIGERATION PIPING

(Courtesy Copeland Corporation)

Probably the first skill that any refrigeration apprentice mechanic learns is to make a soldered joint, and running piping is so common a task that often its critical importance in the proper performance of a system is overlooked. It would seem elementary in any piping system that what goes in one end of a pipe must come out the other, but on a system with improper piping, it is not uncommon for a serviceman to add gallons of oil to a system, and it may seemingly disappear without a trace. It is of course lying on the bottom of the tubing in the system, usually in the evaporator or suction line. When the piping or operating condition is corrected, the oil will return and those same gallons of oil must be removed.

Refrigeration piping involves extremely complex relationships in the flow of refrigerant and oil. Fluid flow is the name given in mechanical engineering to the study of the flow of any fluid, whether it might be a gas or a liquid, and the inter-relationship of velocity, pressure, friction, density, viscosity, and the work required to cause the flow. These relationships evolve into long mathematical equations which form the basis for the fan laws which govern fan performance, and the pressure drop tables for flow through piping. But 99% of the theories in fluid flow textbooks deal with the flow of one homogenous fluid, and there is seldom even a mention of a combination flow of liquid, gas, and oil such as occurs in any refrigeration system. Because of its changing nature, such flow is just too complex to be governed by a simple mathematical equation, and practically the entire working knowledge of refrigeration piping is based on practical experience and test data. As a result, the general type of gas and liquid flow that must be maintained to avoid problems is known, but seldom is there one exact answer to any problem.

BASIC PRINCIPLES OF REFRIGERATION PIPING DESIGN

The design of refrigeration piping systems is a continuous series of compromises. It is desirable to have maximum capacity, minimum cost, proper oil return, minimum power consumption, minimum refrigerant charge, low noise level, proper liquid refrigerant control, and perfect flexibility of system operation from 0 to 100% of system capacity without lubrication problems. Obviously all of these goals cannot be satisfied, since some are in direct conflict. In order to make an intelligent decision as to just what type of compromise is desirable, it is essential that the piping designer clearly understand the basic effects on system performance of the piping design in the different parts of the system.

In general, pressure drop in refrigerant lines tends to decrease capacity and increase power requirements, and excessive pressure drops should be avoided. The magnitude of the pressure drop allowable varies depending on the particular segment of piping involved, and each part of the system must be considered separately. There are probably more tables and charts available covering line pressure drop and refrigerant line capacities at a given pressure drop than on any other single subject in the field of refrigeration.

It is most important, however, that the piping designer realize that pressure drop is not the only criteria that must be considered in sizing refrigerant lines, and that often refrigerant velocities rather than pressure drop must be the determining factor in system design. In addition to the critical nature of oil return, there is no better invitation to system difficulties than an excessive refrigerant charge. A reasonable pressure drop is far more preferable than oversized lines which can contain refrigerant far in excess of the system's needs. An excessive refrigerant charge can result in serious problems of liquid refrigerant control, and the flywheel effect of large quantities of liquid refrigerant in the low pressure side of the system can result in erratic operation of the refrigerant control devices.

The size of the service valve supplied on a compressor, or the size of the connection on a condenser, evaporator, accumulator, or other accessory does not determine the size of line to be used. Manufacturers select a valve size or connection fitting on the basis of its application to an average system, and such factors as the type of application, length of connecting lines, type of system control, variation in load, and other factors can be major factors in determining the proper line size. It is quite possible the required line size may be either smaller or larger than the fittings on various system components. In such cases, reducing fittings must be used.

Since oil must pass through the compressor cylinders to provide lubrication, a small amount of oil is always circulating with the refrigerant. Refrigeration oils are soluble in liquid refrigerant, and at normal room temperatures they will mix completely. Oil and refrigerant vapor, however, **do not** mix readily, and the oil can be properly circulated through the system only if the mass velocity of the refrigerant vapor is great enough to sweep the oil along. To assure proper oil circulation, adequate refrigerant velocities must be maintained not only in the suction and discharge lines, but in the evaporator circuits as well.

Several factors combine to make oil return most critical at low evaporating temperatures. As the suction pressure decreases and the refrigerant vapor becomes less dense, the more difficult it becomes to sweep the oil along. At the same time as the suction pressure falls, the compression ratio increases, and as a result compressor capacity is reduced, and the weight of refrigerant circulated decreases. Refrigeration oil alone becomes the consistency of molasses at temperatures below 0° F., but so long as it is mixed with sufficient liquid refrigerant, it flows freely. As the percentage of oil in the mixture increases, the viscosity increases.

At low temperature conditions all of these factors start to converge, and can create a critical condition. The density of the gas decreases, the mass velocity flow decreases, and as a result more oil starts accumulating in the evaporator. As the oil and refrigerant mixture becomes more viscous, at some point oil may start logging in the evaporator rather than returning to the compressor, resulting in wide variations in the compressor crankcase oil level in poorly designed systems.

Oil logging can be minimized with adequate velocities and properly designed evaporators even at extremely low evaporating temperatures, but normally oil separators are necessary for operation at evaporating temperatures below -50° F. in order to minimize the amount of oil in circulation.

COPPER TUBING FOR REFRIGERANT PIPING

For installations using R-12, R-22, and R-502, copper tubing is almost universally used for refrigerant piping. Commercial copper tubing dimensions have been standardized and classified as follows:

Type K   Heavy Wall
Type L   Medium Wall
Type M   Light Wall

Only types K or L should be used for refrigerant piping, since type M does not have sufficient strength for high pressure applications. Type L tubing is most commonly used, and all tables and data in this manual are based on type L dimensions.

It is highly recommended that only refrigeration grade copper tubing be used for refrigeration applications, since it is available cleaned, dehydrated, and capped to avoid contamination prior to installation. Copper tubing commonly used for plumbing usually has oils and grease or other contaminants on the interior wall, and these can cause serious operating problems if not removed prior to installation.

Table 22 lists the dimensions and properties of standard commercial copper tubing in the sizes commonly used in refrigeration systems, and Table 23 lists the weight of various refrigerants per 100 feet of piping in liquid, suction, and discharge lines.

FITTINGS FOR COPPER TUBING

For brazed or soldered joints, the required elbows, tees, couplings, reducers, or other miscellaneous fittings may be either forged brass or wrought copper. Cast fittings are not satisfactory since they may be porous and often lack sufficient strength.

Table 22

DIMENSIONS AND PROPERTIES OF COPPER TUBE
(Based on ASTM B-88)

Line Size O.D.	Type	Diameter		Wall Thickness In.	Surface Area Sq. Ft./Lin. Ft.		Inside Cross-section Area, Sq. In.	Lineal Feet Containing 1 Cu. Ft.	Weight Lb/Lin. Ft.	Working Pressure Psia
		OD In.	ID In.		OD	ID				
⅜	K	0.375	0.305	0.035	0.0982	0.0798	0.0730	1973.0	0.145	918
	L	0.375	0.315	0.030	0.0982	0.0825	0.0779	1848.0	0.126	764
½	K	0.500	0.402	0.049	0.131	0.105	0.127	1135.0	0.269	988
	L	0.500	0.430	0.035	0.131	0.113	0.145	1001.0	0.198	677
⅝	K	0.625	0.527	0.049	0.164	0.138	0.218	660.5	0.344	779
	L	0.625	0.545	0.040	0.164	0.143	0.233	621.0	0.285	625
¾	K	0.750	0.652	0.049	0.193	0.171	0.334	432.5	0.418	643
	L	0.750	0.666	0.042	0.193	0.174	0.348	422.0	0.362	547
⅞	K	0.875	0.745	0.065	0.229	0.195	0.436	331.0	0.641	747
	L	0.875	0.785	0.045	0.229	0.206	0.484	299.0	0.455	497
1⅛	K	1.125	0.995	0.065	0.295	0.260	0.778	186.0	0.839	574
	L	1.125	1.025	0.050	0.295	0.268	0.825	174.7	0.655	432
1⅜	K	1.375	1.245	0.065	0.360	0.326	1.22	118.9	1.04	466
	L	1.375	1.265	0.055	0.360	0.331	1.26	115.0	0.884	387
1⅝	K	1.625	1.481	0.072	0.425	0.388	1.72	83.5	1.36	421
	L	1.625	1.505	0.060	0.425	0.394	1.78	81.4	1.14	359
2⅛	K	2.125	1.959	0.083	0.556	0.513	3.01	48.0	2.06	376
	L	2.125	1.985	0.070	0.556	0.520	3.10	46.6	1.75	316
2⅝	K	2.625	2.435	0.095	0.687	0.638	4.66	31.2	2.93	352
	L	2.625	2.465	0.080	0.687	0.645	4.77	30.2	2.48	295
3⅛	K	3.125	2.907	0.109	0.818	0.761	6.64	21.8	4.00	343
	L	3.125	2.945	0.090	0.818	0.771	6.81	21.1	3.33	278
3⅝	K	3.625	3.385	0.120	0.949	0.886	9.00	16.1	5.12	324
	L	3.625	3.425	0.100	0.949	0.897	9.21	15.6	4.29	268
4⅛	K	4.125	3.857	0.134	1.08	1.01	11.7	12.4	6.51	315
	L	4.125	3.905	0.110	1.08	1.02	12.0	12.1	5.38	256

Table 23

WEIGHT OF REFRIGERANT IN COPPER LINES

Pounds per 100 feet of Type L Tubing

O.D. Line Size	Volume per 100 Ft. in Cu. Ft.	Weight of Refrigerant, Pounds					
		Liquid @ 100° F.	Hot Gas @ 120° F. Condensing	Suction Gas (Superheated to 65°)			
				−40° F.	−20° F.	20° F.	40° F.
R-12							
⅜	.054	4.25	.171	.011	.018	.044	.065
½	.100	7.88	.317	.021	.033	.081	.120
⅝	.162	12.72	.514	.033	.054	.131	.195
⅞	.336	26.4	1.065	.069	.112	.262	.405
1⅛	.573	45.0	1.82	.118	.191	.464	.690
1⅜	.872	68.6	2.76	.179	.291	.708	1.05
1⅝	1.237	97.0	3.92	.254	.412	1.01	1.49
2⅛	2.147	169.0	6.80	.441	.715	1.74	2.58
2⅝	3.312	260.0	10.5	.680	1.10	2.68	3.98
3⅛	4.728	371.0	15.0	.97	1.57	3.82	5.69
3⅝	6.398	503.0	20.3	1.32	2.13	5.18	7.70
4⅛	8.313	652.0	26.4	1.71	2.77	6.73	10.0
R-22							
⅜	.054	3.84	.202	.013	.021	.052	.077
½	.100	7.12	.374	.024	.04	.096	.143
⅝	.162	11.52	.605	.038	.064	.156	.232
⅞	.336	24.0	1.26	.079	.134	.323	.480
1⅛	.573	40.8	2.14	.136	.228	.550	.820
1⅜	.872	62.1	3.26	.207	.348	.839	1.25
1⅝	1.237	88.0	4.62	.294	.493	1.19	1.77
2⅛	2.147	153.0	8.04	.51	.858	2.06	3.06
2⅝	3.312	236.0	12.4	.78	1.32	3.18	4.72
3⅛	4.728	336.0	17.7	1.12	1.88	4.55	6.75
3⅝	6.398	456.0	24.0	1.51	2.55	6.15	9.14
4⅛	8.313	592.0	31.1	1.97	3.31	8.0	11.19
R-502							
⅜	.054	3.98	.284	.020	.033	.077	.112
½	.100	7.38	.525	.037	.061	.143	.208
⅝	.162	11.95	.852	.061	.098	.232	.337
⅞	.336	24.8	1.77	.126	.204	.481	.700
1⅛	.573	42.3	3.01	.215	.347	.820	1.19
1⅜	.872	64.4	4.60	.327	.527	1.25	1.81
1⅝	1.237	91.2	6.5	.465	.750	1.77	2.57
2⅛	2.147	159.0	11.3	.806	1.30	3.08	4.48
2⅝	3.312	244.0	17.4	1.24	2.0	4.74	6.90
3⅛	4.728	349.0	24.8	1.77	2.87	6.76	9.84
3⅝	6.398	471.0	33.6	2.40	3.87	9.15	13.32
4⅛	8.313	612.0	43.8	3.12	5.03	11.90	17.30

EQUIVALENT LENGTH OF PIPE

Each valve, fitting, and bend in a refrigerant line contributes to the friction pressure drop because of its interruption or restriction of smooth flow. Because of the detail and complexity of computing the pressure drop of each individual fitting, normal practice is to establish an equivalent length of straight tubing for each fitting. This allows the consideration of the entire length of line, including fittings, as an equivalent length of straight pipe. Pressure drop and line sizing tables and charts are normally set up on the basis of a pressure drop per 100 feet of straight pipe, so the use of equivalent lengths allows the data to be used directly.

The equivalent length of copper tubing for commonly used valves and fittings is shown in Table 24.

Table 24

EQUIVALENT LENGTH IN FEET OF STRAIGHT PIPE FOR VALVES AND FITTINGS

O.D., In. Line Size	Globe Valve	Angle Valve	90° Elbow	45° Elbow	Tee Line	Tee Branch
½	9	5	.9	.4	.6	2.0
⅝	12	6	1.0	.5	.8	2.5
⅞	15	8	1.5	.7	1.0	3.5
1⅛	22	12	1.8	.9	1.5	4.5
1⅜	28	15	2.4	1.2	1.8	6.0
1⅝	35	17	2.8	1.4	2.0	7.0
2⅛	45	22	3.9	1.8	3.0	10.0
2⅝	51	26	4.6	2.2	3.5	12.0
3⅛	65	34	5.5	2.7	4.5	15.0
3⅝	80	40	6.5	3.0	5.0	17.0

From Mueller Brass Co. Data

For accurate calculations of pressure drop, the equivalent length for each fitting should be calculated. As a practical matter, an experienced piping designer may be capable of making an accurate overall percentage allowance unless the piping is extremely complicated. For long runs of piping of 100 feet or greater, an allowance of 20% to 30% of the actual lineal length may be adequate, while for short runs of piping, an allowance as high as 50% to 75% or more of the lineal length may be necessary. Judgment and experience are necessary in making a good estimate, and estimates should be checked frequently with actual calculations to insure reasonable accuracy.

For items such as solenoid valves and pressure regulating valves, where the pressure drop through the valve is relatively large, data is normally available from the manufacturer's catalog so that items of this nature can be considered independently of lineal length calculations.

PRESSURE DROP TABLES

Figures 76, 77, and 78 are combined pressure drop charts for refrigerants R-12, R-22, and R-502. Pressure drops in the discharge line, suction line, and liquid line can be determined from these charts for condensing temperatures ranging from 80° F. to 120° F.

To use the chart, start in the upper right hand corner with the design capacity. Drop vertically downward on the line representing the desired capacity to the intersection with the diagonal line representing the operating condition desired. Then move horizontally to the left. A vertical line dropped from the intersection point with each size of copper tubing to the design condensing temperature line allows the pressure drop in psi per 100 feet of tubing to be read directly from the chart. The diagonal pressure drop lines at the bottom of the chart represent the change in pressure drop due to a change in condensing temperature.

For example, in Figure 78 for R-502, the dotted line represents a pressure drop determination for a suction line in a system having a design capacity of 5.5 tons or 66,000 BTU/hr operating with an evaporating temperature of -40° F. The 2⅝" O.D. suction line illustrated has a pressure drop of 0.22 psi per 100 feet at 85° F. condensing temperature, but the same line with the same capacity would have a pressure drop of 0.26 psi per 100 feet at 100° F. condensing, and 0.32 psi per 100 feet at 120° F. condensing.

In the same manner, the corresponding pressure drop for any line size and any set of operating conditions within the range of the chart can be determined.

"FREON" 12 REFRIGERANT
PRESSURE DROP IN LINES (65°F Evap. Outlet)

C-34 (65)

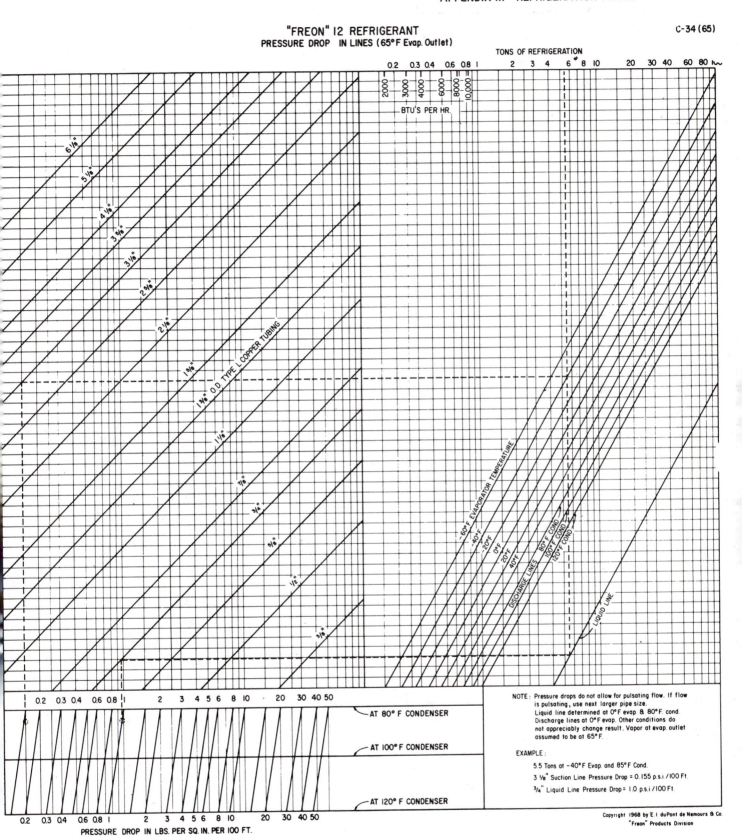

NOTE: Pressure drops do not allow for pulsating flow. If flow
is pulsating, use next larger pipe size.
Liquid line determined at 0°F evap. & 80°F. cond.
Discharge lines at 0°F evap. Other conditions do
not appreciably change result. Vapor at evap. outlet
assumed to be at 65°F.

EXAMPLE:

5.5 Tons at −40°F Evap. and 85°F Cond.

3 1/8" Suction Line Pressure Drop = 0.155 p.s.i./100 Ft.

3/4" Liquid Line Pressure Drop = 1.0 p.s.i./100 Ft.

TONS OF REFRIGERATION

BTU'S PER HR.

PRESSURE DROP IN LBS. PER SQ. IN. PER 100 FT.

AT 80°F CONDENSER

AT 100°F CONDENSER

AT 120°F CONDENSER

Figure 76

"FREON" 22 REFRIGERANT
PRESSURE DROP IN LINES (65°F Evap. Outlet)

C-35(6

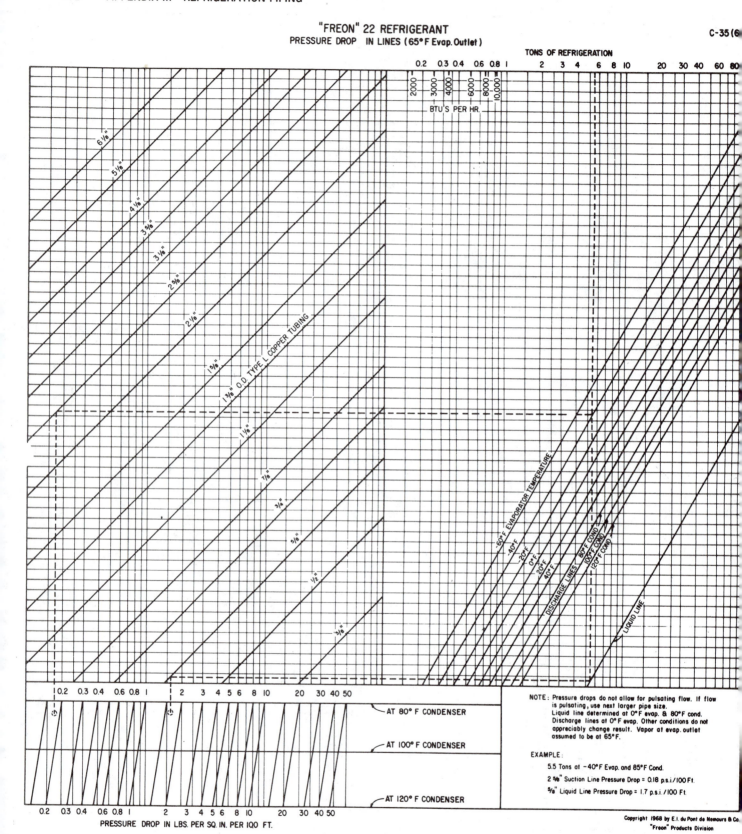

NOTE: Pressure drops do not allow for pulsating flow. If flow is pulsating, use next larger pipe size.
Liquid line determined at 0°F evap. & 80°F cond.
Discharge lines at 0°F evap. Other conditions do not appreciably change result. Vapor at evap. outlet assumed to be at 65°F.

EXAMPLE:

5.5 Tons at −40°F Evap. and 85°F Cond.

2 ⅝" Suction Line Pressure Drop = 0.18 p.s.i./100 Ft.

⅝" Liquid Line Pressure Drop = 1.7 p.s.i./100 Ft.

PRESSURE DROP IN LBS. PER SQ. IN. PER 100 FT.

Figure 77

"FREON" 502 REFRIGERANT
PRESSURE DROP IN LINES (65° F Evap. Outlet)

C-39(65)

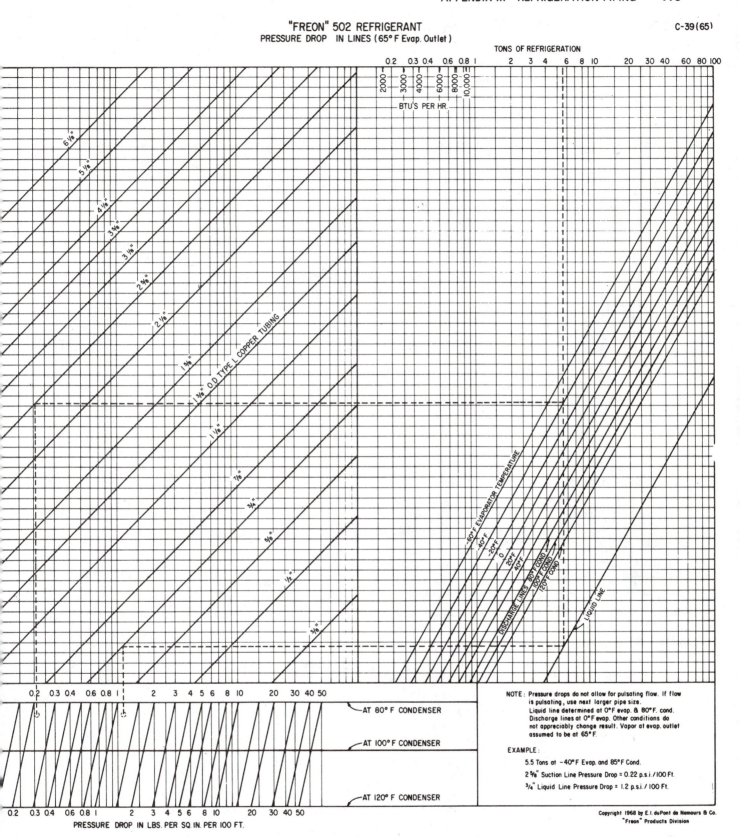

NOTE: Pressure drops do not allow for pulsating flow. If flow is pulsating, use next larger pipe size.
Liquid line determined at 0°F evap. & 80°F. cond.
Discharge lines at 0°F evap. Other conditions do not appreciably change result. Vapor at evap. outlet assumed to be at 65°F.

EXAMPLE:

5.5 Tons at −40° F Evap. and 85°F Cond.

2 5/8" Suction Line Pressure Drop = 0.22 p.s.i./100 Ft.

3/4" Liquid Line Pressure Drop = 1.2 p.s.i./100 Ft.

PRESSURE DROP IN LBS. PER SQ. IN. PER 100 FT.

Figure 78

SIZING HOT GAS DISCHARGE LINES

Pressure drop in discharge lines is probably less critical than in any other part of the system. Frequently the effect on capacity of discharge line pressure drop is over-estimated since it is assumed the compressor discharge pressure and the condensing pressure are the same. In fact, there are two different pressures, the compressor discharge pressure being greater than the condensing pressure by the amount of the discharge line pressure drop. An increase in pressure drop in the discharge line might increase the compressor discharge pressure materially, but have little effect on the condensing pressure. Although there is a slight increase in the heat of compression for an increase in head pressure, the volume of gas pumped is decreased slightly due to a decrease in volumetric efficiency of the compressor. Therefore the total heat to be dissipated through the condenser may be relatively unchanged, and the condensing temperature and pressure may be quite stable, even though the discharge line pressure drop and therefore the compressor discharge pressure might vary considerably.

The performance of a typical Copelametic compressor, operating at air conditioning conditions with R-22 and an air cooled condenser indicates that for each 5 psi pressure drop in the discharge line, the compressor capacity is reduced less than ½ of 1%, while the power required is increased about 1%. On a typical low temperature Copelametic compressor operating with R-502 and an air cooled condenser, approximately 1% of compressor capacity will be lost for each 5 psi pressure drop, but there will be little or no change in power consumption.

As a general guide, for discharge line pressure drops up to 5 psi, the effect on system performance would be so small as to be difficult to measure. Pressure drops up to 10 psi would not be greatly detrimental to system performance provided the condenser is sized to maintain reasonable condensing pressures.

Actually a reasonable pressure drop in the discharge line is often desirable to dampen compressor pulsation, and thereby reduce noise and vibration. Some discharge line mufflers actually derive much of their efficiency from pressure drop through the muffler.

Discharge lines on factory built condensing units usually are not a field problem, but on systems installed in the field with remote condensers, line sizes must be selected to provide proper system performance.

Because of the high temperatures existing in the discharge line, oil flows freely, and oil circulation through both horizontal and vertical lines can be maintained satisfactorily with reasonably low velocities. Since oil traveling up a riser usually creeps up the inner surface of the pipe, oil travel in vertical risers is dependent on the velocity of the gas at the tubing wall. The larger the pipe diameter, the greater will be the required velocity at the center of the pipe to maintain a given velocity at the wall surface. Figures 79 and 80 list the maximum recommended discharge line riser sizes for proper oil return for varying capacities. The variation at different condensing temperatures is not great, so the line sizes shown are acceptable on both water cooled and air cooled applications.

If horizontal lines are run with a pitch in the direction of flow of at least ½" in 10 feet, there is normally little problem with oil circulation at lower velocities in horizontal lines. However, because of the relatively low velocities required in vertical discharge lines, it is recommended wherever possible that both horizontal and vertical discharge lines be sized on the same basis.

To illustrate the use of the chart, assume a system operating with R-12 at 40° F. evaporating temperature has a capacity of 100,000 BTU/hr. The intersection of the capacity and evaporating temperature lines at point X on Figure 79 indicate the design condition. Since this is below the 2⅛" O.D. line, the maximum size that can be used to insure oil return up a vertical riser is 1⅝" O.D.

Oil circulation in discharge lines is normally a problem only on systems where large variations in system capacity are encountered. For example, an air conditioning system may have steps of capacity control allowing it to operate during periods of light load at capacities possibly as low as 25% or 33% of the design capacity. The same situation may exist on commercial refrigeration systems where compressors connected in parallel are cycled for capacity

MAXIMUM RECOMMENDED VERTICAL DISCHARGE LINE SIZES FOR PROPER OIL RETURN

R-12

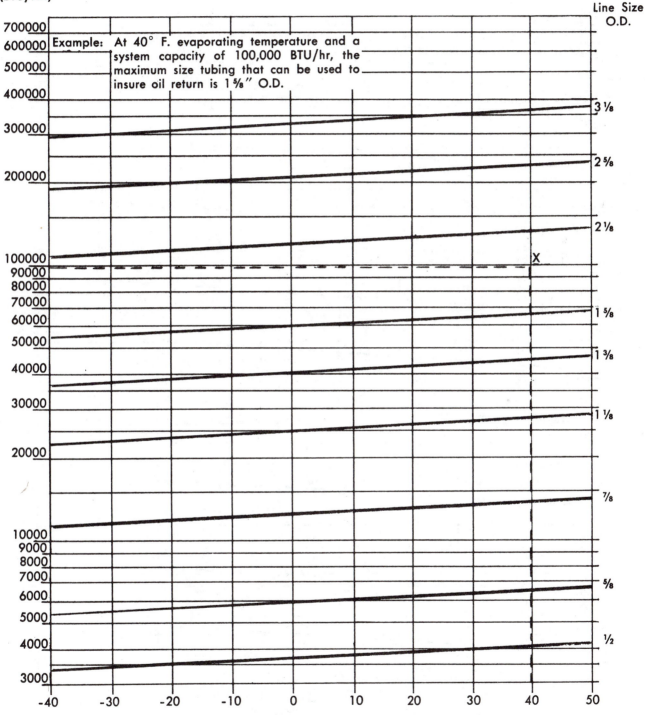

COMPRESSOR CAPACITY (BTU/Hr.)

Example: At 40° F. evaporating temperature and a system capacity of 100,000 BTU/hr, the maximum size tubing that can be used to insure oil return is 1⅝″ O.D.

Line Size O.D.

EVAPORATING TEMPERATURE (°F.)

Figure 79

MAXIMUM RECOMMENDED VERTICAL DISCHARGE LINE SIZES FOR PROPER OIL RETURN

R-22 and R-502

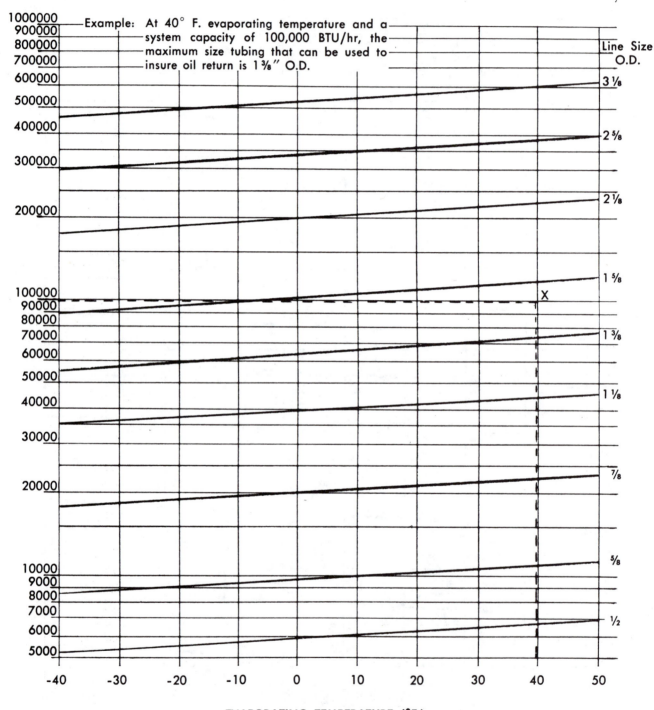

Example: At 40° F. evaporating temperature and a system capacity of 100,000 BTU/hr, the maximum size tubing that can be used to insure oil return is 1 ⅜″ O.D.

Figure 80

DISCHARGE LINE VELOCITIES FOR VARIOUS BTU/Hr. CAPACITIES

R-12

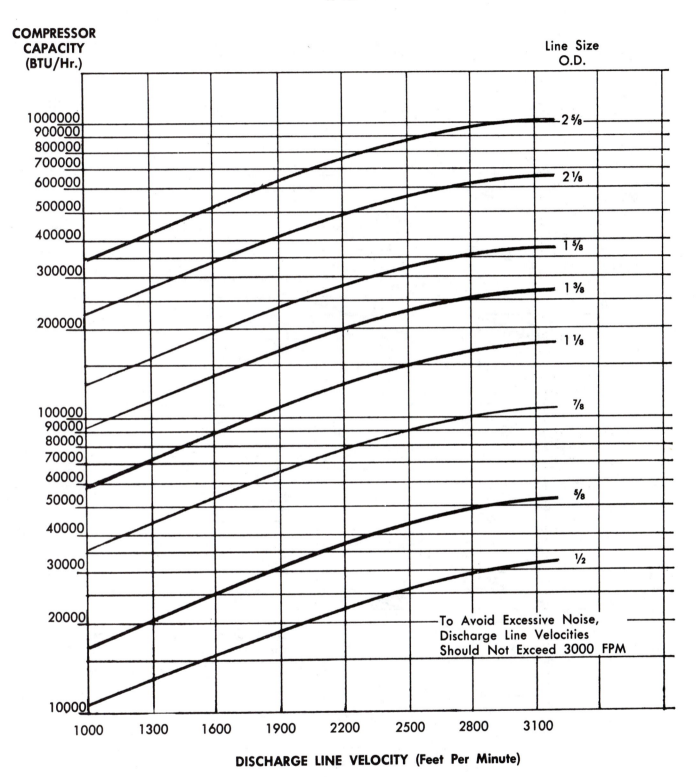

Figure 81

DISCHARGE LINE VELOCITIES FOR VARIOUS BTU/Hr. CAPACITIES

R-22 and R-502

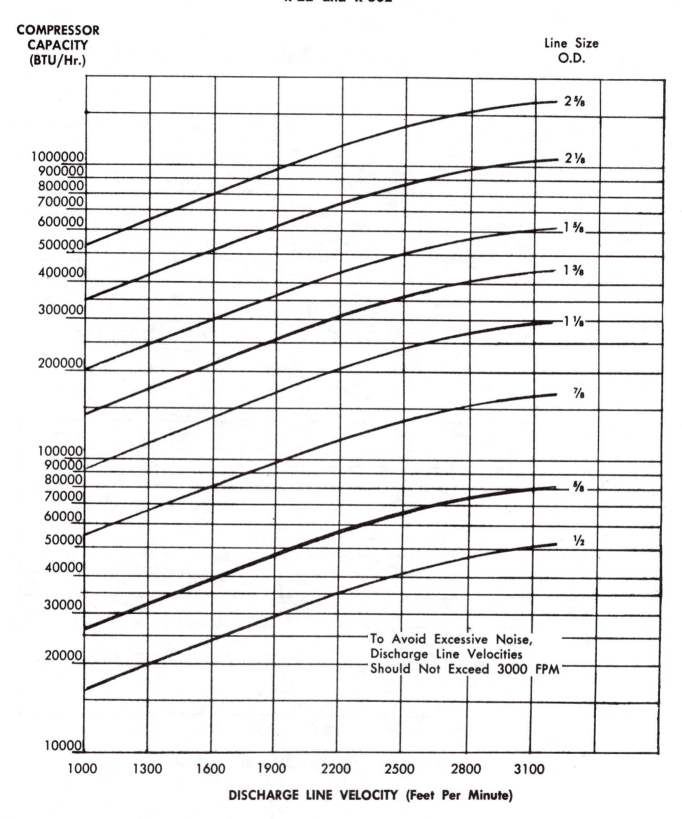

COMPRESSOR CAPACITY (BTU/Hr.)

Line Size O.D.

To Avoid Excessive Noise, Discharge Line Velocities Should Not Exceed 3000 FPM

DISCHARGE LINE VELOCITY (Feet Per Minute)

Figure 82

control. In such cases, vertical discharge lines **must** be sized to maintain velocities above the minimum necessary to properly circulate oil at the minimum load condition.

For example, consider an air conditioning system using R-12 having a maximum design capacity of 300,000 BTU/hr with steps of capacity reduction up to 66%. Although the 300,000 BTU/hr condition could return oil up a 2⅝″ O.D. riser, at light load conditions, the system would have only 100,000 BTU/hr capacity, so a 1⅝″ O.D. riser must be used. In checking the pressure drop chart, Figure 76, at maximum load conditions, a 1⅝″ O.D. pipe will have a pressure drop of approximately 4 psi per 100 feet at a condensing temperature of 120° F. If the total equivalent length of piping exceeds 150 feet, in order to keep the total pressure drop within reasonable limits, the horizontal lines should be the next larger size or 2⅛″ O.D., which would result in a pressure drop of only slightly over 1 psi per 100 feet.

Because of the flexibility in line sizing that the allowable pressure drop makes possible, discharge lines can almost always be sized satisfactorily without the necessity of double risers. If modifications are made to an existing system which result in the existing discharge line being oversized at light load conditions, the addition of an oil separator to minimize oil circulation will normally solve the problem.

One other limiting factor in discharge line sizing is excessive velocity which can cause noise problems. Velocities of 3,000 FPM or more may result in high noise levels, and it is recommended that maximum velocities be kept well below this level. Figures 81 and 82 give equivalent discharge line gas velocities for varying capacities and line sizes over the normal refrigeration and air conditioning range.

To summarize, in sizing discharge lines, it is recommended that a tentative selection of line size be made on the basis of a total pressure drop of approximately 5 psi plus or minus 50%, the actual design pressure drop to a considerable degree being a matter of the designer's judgment. Check Figure 79 or 80 to be sure that velocities at minimum load conditions are adequate to carry oil up vertical risers, and adjust vertical riser size if necessary. Check Figure 81 or 82 to be sure velocities at maximum load are not excessive.

Recommended discharge line sizes for varying capacities and equivalent lengths of line are given in Table 28, page 18-30.

SIZING LIQUID LINES

Since liquid refrigerant and oil mix completely, velocity is not essential for oil circulation in the liquid line. The primary concern in liquid line sizing is to insure a solid liquid head of refrigerant at the expansion valve. If the pressure of the liquid refrigerant falls below its saturation temperature, a portion of the liquid will flash into vapor to cool the liquid refrigerant to the new saturation temperature. This can occur in a liquid line if the pressure drops sufficiently due to friction or vertical lift.

Flash gas in the liquid line has a detrimental effect on system performance in several ways. It increases the pressure drop due to friction, reduces the capacity of the expansion device, may erode the expansion valve pin and seat, can cause excessive noise, and may cause erratic feeding of the liquid refrigerant to the evaporator.

For proper system performance, it is essential that liquid refrigerant reaching the expansion device be subcooled slightly below its saturation temperature. On most systems the liquid refrigerant is sufficiently subcooled as it leaves the condenser to provide for normal system pressure drops. The amount of subcooling necessary, however, is dependent on the individual system design.

On air cooled and most water cooled applications, the temperature of the liquid refrigerant is normally higher than the surrounding ambient temperature, so no heat is transferred into the liquid, and the only concern is the pressure drop in the liquid line. Besides the friction loss caused by flow through the piping, a pressure drop equivalent to the liquid head is involved in forcing liquid to flow up a vertical riser. A head of two feet of liquid refrigerant is approximately equivalent to 1 psi. For example, if a condenser or receiver in the basement of a building is to supply liquid refrigerant to an evaporator three floors above, or approximately 30 feet, then a pressure drop of approximately 15 psi must be provided for in system design for the liquid head alone.

On evaporative or water cooled condensers where the condensing temperature is below the ambient air temperature, or on any application where liquid lines must pass through hot areas such as boiler or furnace rooms, an additional complication may arise because of heat transfer into the liquid. Any subcooling in the condenser may be lost in the receiver or liquid line due to temperature rise alone unless the system is properly designed. On evaporative condensers where a receiver and subcooling coil are used, it is recommended that the refrigerant flow be piped from the condenser to the receiver and then to the subcooling coil. In critical applications it may be necessary to insulate both the receiver and the liquid line.

On the typical air cooled condensing unit with a conventional receiver, it is probable that very little subcooling of liquid is possible unless the receiver is almost completely filled with liquid. Vapor in the receiver in contact with the subcooled liquid will condense, and this effect will tend toward a saturated condition.

At normal condensing temperatures, the following relation between each 1° F. of subcooling and the corresponding change in saturation pressure applies.

Refrigerant	Subcooling	Equivalent Change in Saturation Pressure
R-12	1° F.	1.75 psi
R-22	1° F.	2.75 psi
R-502	1° F.	2.85 psi

To illustrate, 5° F. subcooling will allow a pressure drop of 8.75 psi with R-12, 13.75 psi with R-22, and 14.25 psi with R-502 without flashing in the liquid line. For the previous example of a condensing unit in a basement requiring a vertical lift of 30 feet or approximately 15 psi, the necessary subcooling for the liquid head alone would be 8.5° F. with R-12, 5.5° F. with R-22, and 5.25° F. with R-502.

The necessary subcooling may be provided by the condenser used, but for systems with abnormally high vertical risers, a suction to liquid heat exchanger may be required. Where long refrigerant lines are involved, and the temperature of the suction gas at the condensing unit is approaching room temperatures, a heat exchanger located near the condenser may not have sufficient temperature differential to adequately cool the liquid, and individual heat exchangers at each evaporator may be necessary.

In extreme cases, where a great deal of subcooling is required, there are several alternatives. A special heat exchanger with a separate subcooling expansion valve can provide maximum cooling with no penalty on system performance. It is also possible to reduce the capacity of the condenser so that a higher operating condensing temperature will make greater subcooling possible. Liquid refrigerant pumps may also be used to overcome large pressure drops.

Liquid line pressure drop causes no direct penalty in power consumption, and the decrease in system capacity due to friction losses in the liquid line is negligible. Because of this the only real restriction on the amount of liquid line pressure drop is the amount of subcooling available. Most references on pipe sizing recommend a conservative approach with friction pressure drops in the 3 to 5 psi range, but where adequate subcooling is available, many applications have successfully used much higher design pressure drops. The total friction includes line losses through such accessories as solenoid valves, filter-driers, and hand valves.

In order to minimize the refrigerant charge, liquid lines should be kept as small as practical, and excessively low pressure drops should be avoided. On most systems, a reasonable design criteria is to size liquid lines on the basis of a pressure drop equivalent to 2° F. subcooling.

LIQUID LINE VELOCITIES FOR VARIOUS PRESSURE DROPS

R-12, R-22, R-502

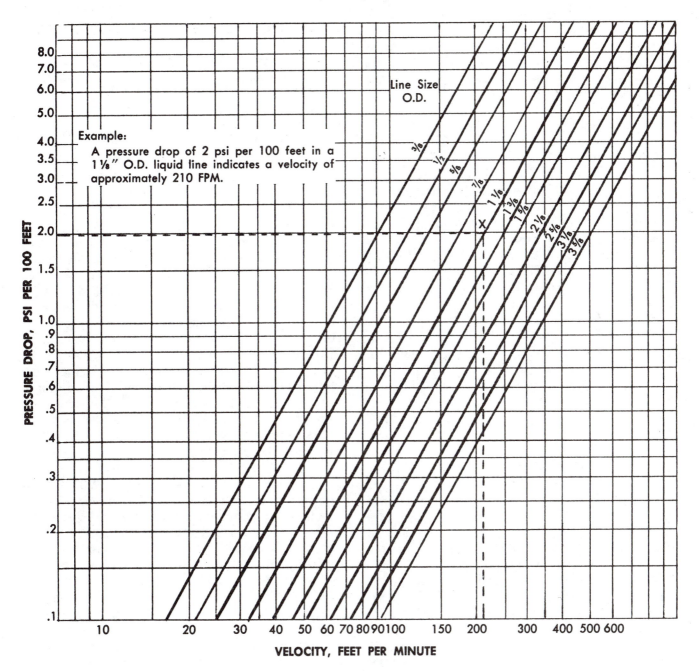

Example:
A pressure drop of 2 psi per 100 feet in a 1⅛″ O.D. liquid line indicates a velocity of approximately 210 FPM.

Figure 83

A limitation on liquid line velocity is possible damage to the piping from pressure surges or liquid hammer caused by the rapid closing of liquid line solenoid valves, and velocities above 300 FPM should be avoided when they are used. If liquid line solenoids are not used, then higher velocities can be employed. Figure 83 gives liquid line velocities corresponding to various pressure drops and line sizes.

To summarize, in sizing liquid lines, it is recommended that the selection of line size be made on the basis of a total friction pressure drop equivalent to 2° F. subcooling. If vertical lifts or valves with large pressure drops are involved, then the designer must make certain that sufficient subcooling is available to allow the necessary pressure drop without approaching a saturation condition at which gas flashing could occur. Check Figure 83 to be sure velocities do not exceed 300 FPM if a liquid line solenoid is used.

Recommended liquid line sizes for varying capacities and equivalent lengths of line are given in Table 27, page 18-29.

SIZING SUCTION LINES

Suction line sizing is the most critical from a design and system standpoint. Any pressure drop occurring due to frictional resistance to flow results in a decrease in the pressure at the compressor suction valve, compared with the pressure at the evaporator outlet. As the suction pressure is decreased, each pound of refrigerant returning to the compressor occupies a greater volume, and the weight of the refrigerant pumped by the compressor decreases. For example, a typical low temperature R-502 compressor at -40° F. evaporating temperature will lose almost 6% of its rated capacity for each 1 psi suction line pressure drop.

Normally accepted design practice is to use as a design criteria a suction line pressure drop equivalent to a 2° F. change in saturation temperature. Equivalent pressure drops for various operating conditions are shown in Table 25.

Table 25

PRESSURE DROP EQUIVALENT FOR 2° F. CHANGE IN SATURATION TEMPERATURE AT VARIOUS EVAPORATING TEMPERATURES

Evaporating Temperature	Pressure Drop, PSI		
	R-12	R-22	R-502
45° F.	2.0	3.0	3.3
20° F.	1.35	2.2	2.4
0° F.	1.0	1.65	1.85
-20° F.	.75	1.15	1.35
-40° F.	.5	.8	1.0

Of equal importance in sizing suction lines is the necessity of maintaining adequate velocities to properly return oil to the compressor. Studies have shown that oil is most viscous in a system after the suction vapor has warmed up a few degrees from the evaporating temperature, so that the oil is no longer saturated with refrigerant, and this condition occurs in the suction line after the refrigerant vapor has left the evaporator. Movement of oil through suction lines is dependent on both the mass and velocity of the suction vapor. As the mass or density decreases, higher velocities are required to force the oil along.

Nominal minimum velocities of 700 FPM in horizontal suction lines and 1500 FPM in vertical suction lines have been recommended and used successfully for many years as suction line sizing design standards. Use of the one nominal velocity provided a simple and convenient means of checking velocities. However, tests have shown that in vertical risers the oil tends to crawl up the inner surface of the tubing, and the larger the tubing, the greater velocity required in the center of the tubing to maintain tube surface velocities which will carry the oil. The exact velocity required in vertical lines is dependent on both the evaporating temperature and the line size, and under varying conditions, the specific velocity required might be either greater or less than 1500 FPM.

For better accuracy in line sizing, revised maximum recommended vertical suction line sizes based on the minimum gas velocities shown in the 1967 ASHRAE Guide and Data

MAXIMUM RECOMMENDED SUCTION LINE SIZES FOR PROPER OIL RETURN
VERTICAL RISERS R-12

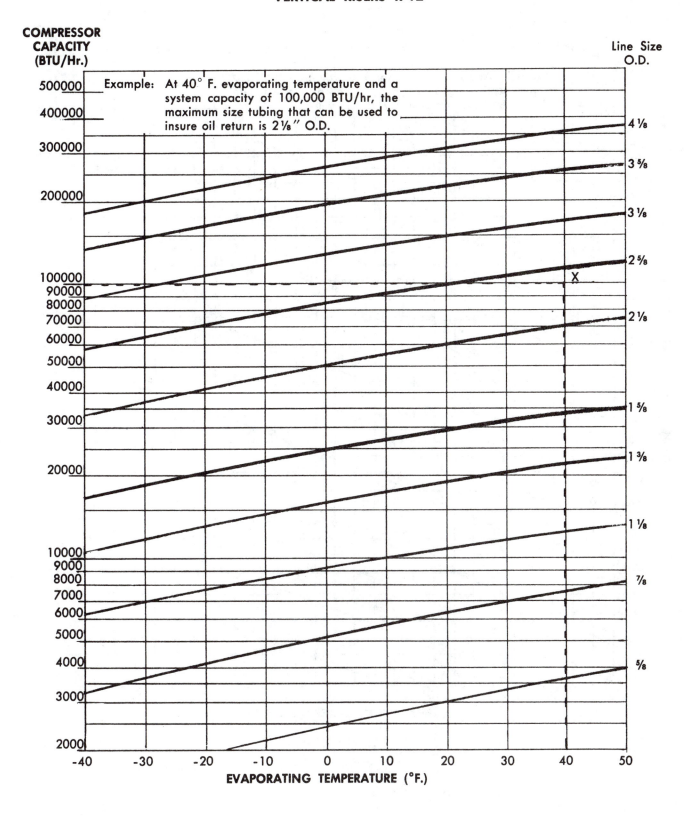

COMPRESSOR CAPACITY (BTU/Hr.)

Line Size O.D.

Example: At 40° F. evaporating temperature and a system capacity of 100,000 BTU/hr, the maximum size tubing that can be used to insure oil return is 2⅛" O.D.

EVAPORATING TEMPERATURE (°F.)

Figure 84

MAXIMUM RECOMMENDED HORIZONTAL SUCTION LINE SIZES FOR PROPER OIL RETURN

R-12

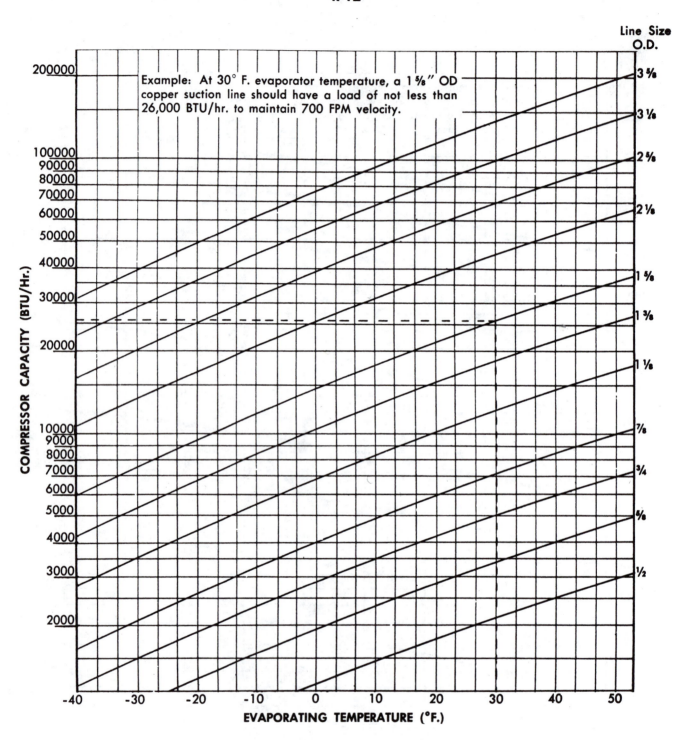

Example: At 30° F. evaporator temperature, a 1 5/8" OD copper suction line should have a load of not less than 26,000 BTU/hr. to maintain 700 FPM velocity.

Figure 85

MAXIMUM RECOMMENDED SUCTION LINE SIZES FOR PROPER OIL RETURN
VERTICAL RISERS R-22 and R-502

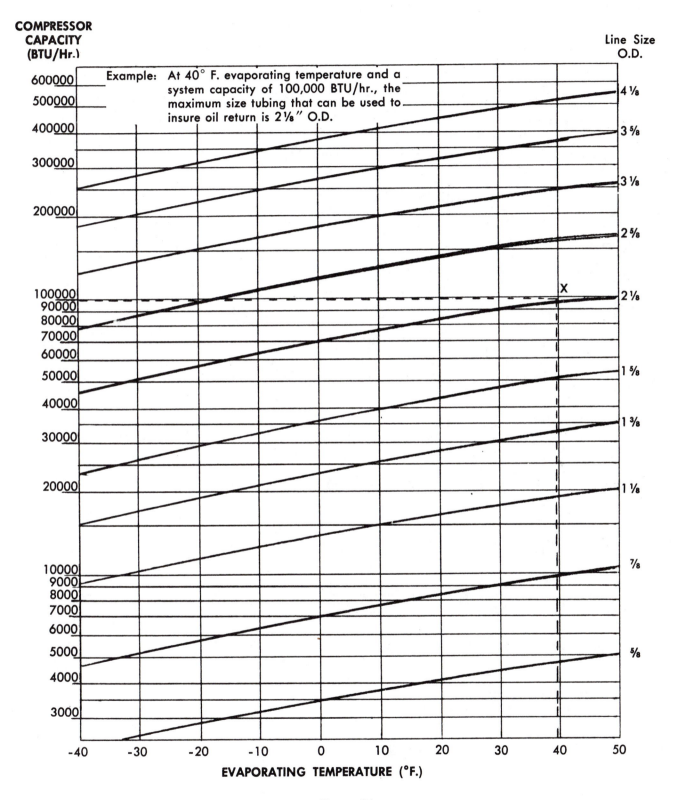

Figure 86

MAXIMUM RECOMMENDED HORIZONTAL SUCTION LINE SIZES FOR PROPER OIL RETURN

R-22 and R-502

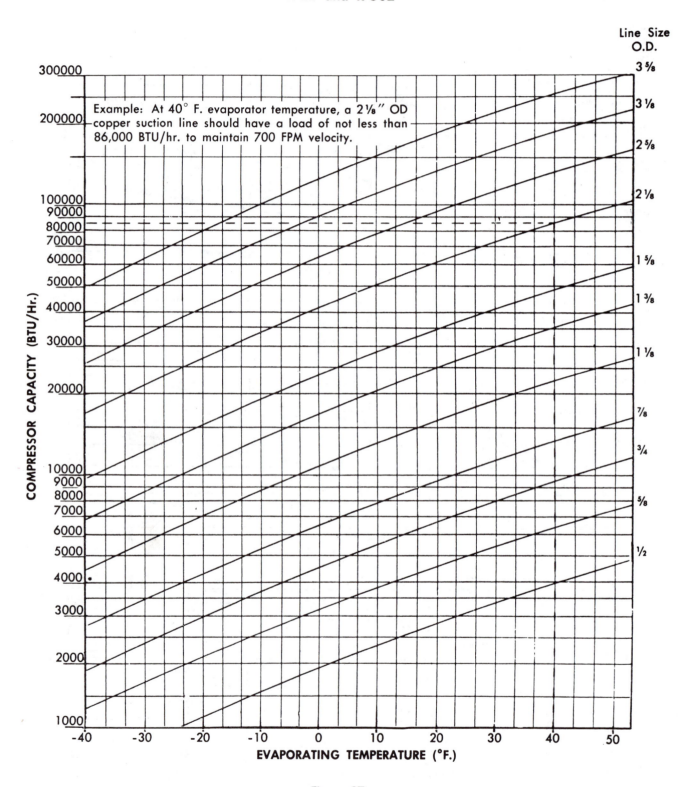

Figure 87

Book have been calculated and are plotted in chart form for easy usage in Figures 84 and 86. These revised recommendations supersede previous Copeland vertical suction riser recommendations. No change has been made in the 700 FPM minimum velocity recommendation for horizontal suction lines, and Figures 85 and 87 cover maximum recommended horizontal line sizes for proper oil return.

To illustrate, again assume a system operating with R-12 at 40° F. evaporating temperature has a capacity of 100,000 BTU/hr. On Figure 84, the intersection of the evaporating temperature and capacity lines indicate that a 2⅛″ O.D. line will be required for oil return in the vertical suction risers.

Even though the system might have a much larger design capacity, the suction line sizing must be based on the minimum capacity anticipated in operation under light load conditions after allowing for the maximum reduction in capacity from capacity control if provided.

Since the dual goals of low pressure drop and high velocities are in direct conflict, obviously compromises must be made in both areas. As a general approach, in suction line design, velocities should be kept as high as possible by sizing lines on the basis of the maximum pressure drop that can be tolerated, but in no case should gas velocity be allowed to fall below the minimum levels necessary to return oil. It is recommended that a tentative selection of suction line sizes be made on the basis of a total pressure drop equivalent to a 2° F. change in the saturated evaporating temperature. Check Figures 84 or 86 to be sure that velocities in vertical risers are satisfactory. Where refrigerant lines are lengthy, it may be desirable to use as large tubing as practical to minimize pressure drop, and Figure 85 or 87 should be checked to determine the maximum permissible horizontal line size. The final consideration must always be to maintain velocities adequate to return oil to the compressor, even if this results in a higher pressure drop than is normally desirable.

Recommended suction line sizes for varying capacities and equivalent lengths of line are given in Tables 29 to 41, starting on page 18-31.

DOUBLE RISERS

On systems equipped with capacity control compressors, or where tandem or multiple compressors are used with one or more compressors cycled off for capacity control, single suction line risers may result in either unacceptably high or low gas velocities. A line properly sized for light load conditions may have too high a pressure drop at maximum load, and if the line is sized on the basis of full load conditions, then velocities may not be adequate at light load conditions to move oil through the tubing. On air conditioning applications where somewhat higher pressure drops at maximum load conditions can be tolerated without any major penalty in overall system performance, it is usually preferable to accept the additional pressure drop imposed by a single vertical riser. But on medium or low temperature applications where pressure drop is more critical and where separate risers from individual evaporators are not desirable or possible, a double riser may be necessary to avoid an excessive loss of capacity.

A typical double riser configuration is shown in Figure 88. The two lines should be sized so that the total cross-sectional area is equivalent to the cross-section area of a single riser that would have both satisfactory gas velocity and acceptable pressure drop at maximum load conditions. The two lines normally are different in

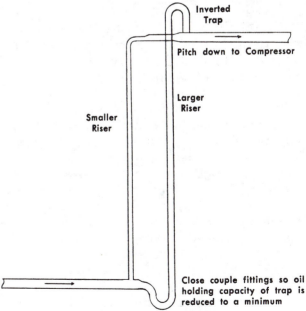

SUCTION LINE DOUBLE RISER

Figure 88

size, with the larger line trapped as shown, and the smaller line must be sized to provide adequate velocities and acceptable pressure drop when the entire minimum load is carried in the smaller riser.

In operation, at maximum load conditions gas and entrained oil will be flowing through both risers. At minimum load conditions, the gas velocity will not be high enough to carry oil up both risers. The entrained oil will drop out of the refrigerant gas flow, and accumulate in the "P" trap, forming a liquid seal. This will force all of the flow up the smaller riser, thereby raising the velocity and assuring oil circulation through the system.

For example, assume a low temperature system as follows:

Maximum capacity	150,000 BTU/hr.
Minimum capacity	50,000 BTU/hr.
Refrigerant	R-502
Evaporating Temperature	-40° F.
Equivalent length of piping, horizontal	125 ft.
Vertical Riser	25 ft.
Desired design pressure drop (equivalent to 2° F.)	1 psi

A preliminary check of the R-502 pressure drop chart, Figure 78, indicates for a 150 foot run with 150,000 BTU/hr capacity and a total pressure drop of approximately 1 psi, a 3⅛" O.D. line is indicated. At the minimum capacity of 50,000 BTU/hr, Figure 87 shows a 3⅝" O.D. horizontal suction line is acceptable, but Figure 86 indicates that the maximum vertical riser size is 2⅛" O.D. Referring again to the pressure drop chart, Figure 78, the pressure drop for 150,000 BTU/hr through 2⅛" O.D. tubing is 4 psi per 100 feet, or 1.0 psi for the 25 foot suction riser. Obviously, either a compromise must be made in accepting a greater pressure drop at maximum load conditions, or a double riser must be used.

If the pressure drop must be held to a minimum, then the size of the double riser must be determined. At maximum load conditions, a 3⅛" O.D. riser would maintain adequate veloc-

ities, so a combination of line sizes approximating the 3⅛" O.D. line can be selected for the double riser. The cross sectional area of the line sizes to be considered are:

3⅛" O.D.	6.64	sq. in.
2⅝" O.D.	4.77	sq. in.
2⅛" O.D.	3.10	sq. in.
1⅝" O.D.	1.78	sq. in.

At the minimum load condition of 50,000 BTU/hr., the 1⅝" O.D. line will have a pressure drop of approximately .5 psi, and will have acceptable velocities, so a combination of 2⅝" O.D. and 1⅝" O.D. tubing should be used for the double riser.

In a similar fashion, double risers can be calculated for any set of maximum and minimum capacities where single risers may not be satisfactory.

SUCTION PIPING FOR MULTIPLEX SYSTEMS

It is common practice in supermarket applications to operate several fixtures, each with liquid line solenoid valve and expansion valve control, from a single compressor. Temperature control of individual fixtures is normally achieved by means of a thermostat opening and closing the liquid line solenoid valve as necessary. This type of system, commonly called multiplexing, requires careful attention to design to avoid oil return problems and compressor overheating.

Since the fixtures fed by each liquid line solenoid valve may be controlled individually, and since the load on each fixture is relatively constant during operation, individual suction lines and risers are normally run from each fixture or group of fixtures controlled by a liquid line solenoid valve for minimum pressure drop and maximum efficiency in oil return. This provides excellent control so long as the compressor is operating at its design suction pressure, but there may be periods of light load when most or all of the liquid line solenoids are closed. Unless some means of controlling compressor capacity is provided, this can result in compressor short cycling or operation at excessively low suction pressures, which can result not only in overheating the compressor, but in reducing the suction pressure to a level where the gas becomes so rarified it can no longer return oil properly in lines sized for much greater gas density.

Because of the fluctuations in refrigeration load caused by closing of the individual liquid line solenoid valves, some means of compressor capacity control must be provided. In addition, the means of capacity control must be such that it will not allow extreme variations in the compressor suction pressure.

Where multiple compressors are used, cycling of individual compressors provides satisfactory control. Where multiplexing is done with a single compressor, a hot gas bypass system has proven to be the most satisfactory means of capacity reduction, since this allows the compressor to operate continuously at a reasonably constant suction pressure while compressor cooling can be safely controlled by means of a desuperheating expansion valve.

In all cases, the operation of the system under all possible combinations of heavy load, light load, defrost, and compressor capacity must be studied carefully to be certain that operating conditions will be satisfactory.

Close attention must be paid to piping design on multiplex systems to avoid oil return problems. Lines must be properly sized so that the minimum velocities necessary to return oil are maintained in both horizontal and vertical suction lines under minimum load conditions. Bear in mind that although a hot gas bypass maintains the suction pressure at a proper level, the refrigerant vapor being bypassed is not available in the system to aid in returning oil.

PIPING DESIGN FOR HORIZONTAL AND VERTICAL LINES

Horizontal suction and discharge lines should be pitched downward in the direction of flow to aid in oil drainage, with a downward pitch of at least ½ inch in 10 feet. Refrigerant lines should always be as short and should run as directly as possible.

Piping should be located so that access to system components is not hindered, and so that any components which could possibly require future maintenance are easily accessible. If piping must be run through boiler rooms or other areas where they will be exposed to abnormally high temperatures, it may be necessary to insulate both the suction and liquid lines to prevent excessive heat transfer into the lines.

Every vertical suction riser greater than 3 to 4 feet in height should have a "P" trap at the base to facilitate oil return up the riser as shown in Figure 89. To avoid the accumulation of large quantities of oil, the trap should be of minimum depth and the horizontal section should be as short as possible. Prefabricated wrought copper traps are available, or a trap can be made by using two street ells and one regular ell. Traps at the foot of hot gas risers are normally not required because of the easier movement of oil at higher temperatures. However it is recommnded that the discharge line from the compressor be looped to the floor prior to being run vertically upwards to prevent the drainage of oil back to the compressor head during shut down periods. See Figure 90.

For long vertical risers in both suction and discharge lines, additional traps are recommended for each full length of pipe (approximately 20 feet) to insure proper oil movement.

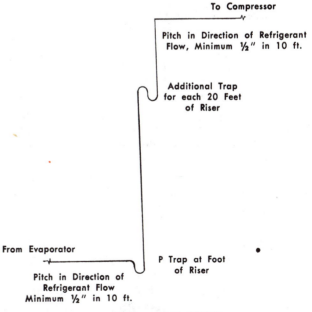

To Compressor

Pitch in Direction of Refrigerant Flow, Minimum ½" in 10 ft.

Additional Trap for each 20 Feet of Riser

From Evaporator

Pitch in Direction of Refrigerant Flow Minimum ½" in 10 ft.

P Trap at Foot of Riser

SUCTION LINE RISERS

Figure 89

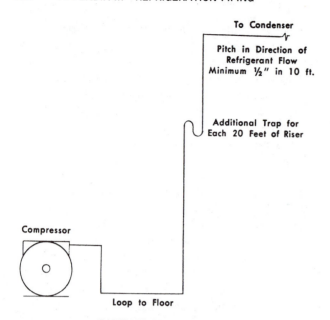

To Condenser

Pitch in Direction of
Refrigerant Flow
Minimum ½" in 10 ft.

Additional Trap for
Each 20 Feet of Riser

Compressor

Loop to Floor

DISCHARGE LINE RISERS

Figure 90

In general, trapped sections of the suction line should be avoided except where necessary for oil return. Oil or liquid refrigerant accumulating in the suction line during the off cycle can return to the compressor at high velocity as liquid slugs on start up, and can break compressor valves or cause other damage.

SUCTION LINE PIPING DESIGN AT THE EVAPORATOR

If a pumpdown control system is not used, each evaporator must be trapped to prevent liquid refrigerant from draining back to the compressor by gravity during the off cycle. Where multiple evaporators are connected to a common suction line, the connections to the common suction line must be made with inverted traps to prevent drainage from one evaporator from affecting the expansion valve bulb control of another evaporator.

Where a suction riser is taken directly upward from an evaporator, a short horizontal section of tubing and a trap should be provided ahead of the riser so that a suitable mounting for the thermal expansion valve bulb is available. The trap serves as a drain area, and helps to prevent the accumulation of liquid under the bulb which

could cause erratic expansion valve operation. If the suction line leaving the evaporator is free draining or if a reasonable length of horizontal piping precedes the vertical riser, no trap is required unless necessary for oil return.

Typical evaporator connections are illustrated in Figure 91.

RECEIVER LOCATION

Gas binding at the receiver can occur when the receiver is exposed to an ambient temperature higher than the condensing temperature. Heat transfer through the receiver shell causes some of the liquid in the receiver to evaporate, creating a pressure in the receiver higher than in the condenser. This forces liquid refrigerant to back up into the condenser until its efficiency is reduced to the point where the condensing pressure again exceeds the pressure in the receiver.

The best remedy for this problem is to make sure the receiver is always exposed to ambient temperatures lower than the condensing temperature. If this is not possible, the receiver should be insulated to minimize heat transfer. Various types of venting arrangements for receivers have been proposed, but these require extreme care in circuiting to avoid flow problems. When the receiver is vented back to the condenser, the only force causing flow from the condenser to the receiver is gravity. Vented piping arrangements are complicated at best, and should be avoided if possible.

Even though there may not be sufficient heat transfer into a receiver to cause gas binding, on systems where the condensing temperature is lower than the ambient (for example water cooled or evaporative condensers with remote receivers) the liquid refrigerant may be warmed sufficiently in the receiver to lose most and possibly all of its subcooling. As mentioned previously, special subcooling coils or insulation may be required for proper operation under these conditions.

If a difference in temperature exists between two parts of an idle refrigeration system with

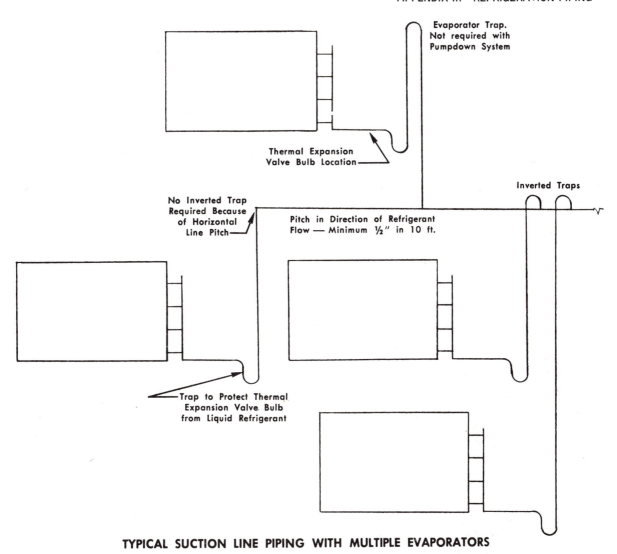

Evaporator Trap.
Not required with
Pumpdown System

Thermal Expansion
Valve Bulb Location

Inverted Traps

No Inverted Trap
Required Because
of Horizontal
Line Pitch

Pitch in Direction of Refrigerant
Flow — Minimum ½" in 10 ft.

Trap to Protect Thermal
Expansion Valve Bulb
from Liquid Refrigerant

TYPICAL SUCTION LINE PIPING WITH MULTIPLE EVAPORATORS

Figure 91

interconnecting piping, this actually creates a little built-in static refrigeration system. The liquid refrigerant at the high temperature point will slowly vaporize, travel through the system as vapor, and recondense at the lowest temperature point. This most often is a matter of concern with a roof mounted remote condenser when the compressor is located in an inside machine room. If the system is idle, the sun is shining on the condenser, and the machine room is cool, then liquid is going to move out of the condenser and back down the discharge line to the machine room. Occasionally inverted traps are made in the discharge line at the condenser in the belief they will prevent this type of reverse

flow. Actually with even a few degrees temperature difference, an inverted loop 20 feet high would be of no value.

However, if the receiver is located either in the machine room, or at some other point where it will not be exposed to the roof heat, reverse flow from the condenser seldom is a source of operating difficulty. The amount of refrigerant actually returning down the discharge line will be minimized and rarely if ever will this cause compressor damage if good piping practice is followed. It is possible to mount a check valve in the discharge line near the condenser as a means of preventing refrigerant backflow of this nature, but check valves in this location are noisy, expensive, and subject to damage, and should be employed only if absolutely essential.

VIBRATION AND NOISE

No matter how well the compressor is isolated, some noise and vibration will be transmitted through the piping, but both can be minimized by proper design and support of the piping.

On small units a coil of tubing at the compressor may provide adequate protection against vibration. On larger units, flexible metallic hose is frequently used. When the compressor is supported by vibration absorbing mounts allowing compressor movement, refrigerant lines should not be anchored solidly at the unit, but at a point beyond the vibration absorber, so the vibration can be isolated and not transmitted into the piping system.

The noise characteristics of a large refrigeration or air conditioning system, particularly when installed with long refrigerant lines and remote condensers, are not predictable. Variations in piping configuration, the pattern of gas flow, line sizes, operating pressures, the compressor and unit mounting, all can affect the noise generated by the system. Occasionally a particular combination of gas flow and piping will result in a resonant frequency which may amplify the sound and vibration to an undesirable level. Gas pulsation from the compressor may also be amplified in a similar manner.

If gas pulsation or resonant frequencies are encountered on a particular application, a discharge line muffler may be helpful in correcting the problem. The purpose of a muffler is to dampen the pulses of gas in the discharge line and to change the frequency to a level which is not objectionable. A muffler normally depends on multiple internal baffles and/or pressure drop to obtain an even flow of gas. In general, the application range of a muffler depends on the mass flow of gas through the muffler, so the volume and density of the refrigerant gas discharged from the compressor are both factors in muffler performance.

A given muffler may work satisfactorily on a fairly wide range of compressor sizes, but it is also quite possible that a given system may require a muffler with a particular pressure drop to effectively dampen pulsations. On problem applications, trial and error may be the only final guide. While larger mufflers are often more efficient in reducing the overall level of compressor discharge noise, in order to satisfactorily dampen pulsations, smaller mufflers with a greater pressure drop are usually more effective. Adjustable mufflers are often helpful since they allow tuning of the muffler pressure characteristics to the exact system requirement.

Occasionally, a combination of operating conditions, mounting and piping arrangement may result in a resonant condition, which tends to magnify compressor pulsation and cause a sharp vibration, although noise may not be a problem. For larger Copelametic compressors, discharge muffler plates have been developed for use when necessary to dampen excessive pulsation. The muffler plate fits between the discharge valve and the compressor body and has a number of muffling holes to provide the proper characteristics for the particular compressor displacement. The muffling holes break up the pattern of gas flow and create sufficient restriction to reduce the gas pulsation to a minimum.

When piping passes through walls or floors, precautions should be taken to see that the piping does not touch any structural members and is properly supported by hangers in order o prevent the transmission of vibration into the building. Failure to do so may result in the building structure becoming a sounding board.

Table 26 gives the maximum recommended spacing for pipe supports.

Table 26
MAXIMUM RECOMMENDED SPACING BETWEEN PIPE SUPPORTS FOR COPPER TUBING

Line Size, O.D., In.	Maximum Span, Feet
5/8	5
1 1/8	7
1 5/8	9
2 1/8	10
3 1/8	12
3 5/8	13
4 1/8	14

From 1967 ASHRAE Guide and Data Book
Reprinted by Permission

RECOMMENDED LINE SIZING TABLES

Tables 27 to 41 give recommended line sizes for single stage applications at various capacities and for equivalent lengths of pipe based on the design criteria discussed previously. (For piping recommendations on two stage systems, refer to Section 19).

Vertical suction line sizes have been selected on the basis of a total vertical rise up to 30 feet. For longer risers, individual calculations should be made since the increased pressure drop may require different line sizes and possibly the use of double risers in place of the single riser shown.

Discharge line sizes have been calculated on the basis of a nominal pressure drop of 5 psi.

Vertical line sizes have been selected so that minimum velocities necessary to carry oil up the riser will be maintained under the reduced load conditions shown, and velocities have been checked to see that they do not exceed 2,700 FPM at maximum load conditions. Because of the relatively small variation in discharge line velocity over the normal refrigeration and air conditioning range, the line sizes shown may be safely used for evaporating temperatures from -40° F. to 45° F., and condensing temperatures from 80° F. to 130° F.

Liquid line sizes have been calculated on the basis of a nominal pressure drop equivalent to 2½° F. subcooling, and velocities have been checked to see that they do not exceed 250 FPM. Liquid lines from the condenser to receiver have been selected on the basis of 100 FPM velocity in accordance with standard industry practice in order to allow a free draining line with gas equalization where piping allows. As in the case with discharge lines, the relatively small variation in liquid line velocities over the normal refrigeration and air conditioning range allows use of the recommended line sizing for evaporating temperatures from -40° F. to 45° F., and condensing temperatures from 80° F. to 130° F.

Suction line sizes have been calculated on the basis of a nominal pressure drop equivalent to a 2° F. change in the saturated evaporating temperature. Both horizontal and vertical line sizes have been checked to see that the necessary minimum velocities are maintained under the reduced load conditions shown. Line sizes have been calculated for various evaporating temperatures, and may be safely applied for condensing temperatures from 80° F. to 130° F.

Table 27

RECOMMENDED LIQUID LINE SIZES

Capacity BTU/hr	R-12					R-22					R-502				
	Condenser to Receiver	Receiver to Evaporator Equivalent Length, Ft.				Condenser to Receiver	Receiver to Evaporator Equivalent Length, Ft.				Condenser to Receiver	Receiver to Evaporator Equivalent Length, Ft.			
		50	100	150	200		50	100	150	200		50	100	150	200
6,000	3/8	3/8	3/8	3/8	3/8	3/8	1/4	3/8	3/8	3/8	3/8	1/4	3/8	3/8	3/8
12,000	1/2	3/8	3/8	1/2	1/2	1/2	3/8	3/8	3/8	3/8	1/2	3/8	1/2	1/2	1/2
18,000	1/2	1/2	1/2	1/2	1/2	1/2	3/8	3/8	1/2	1/2	5/8	1/2	1/2	1/2	1/2
24,000	5/8	1/2	1/2	1/2	5/8	5/8	3/8	1/2	1/2	1/2	5/8	1/2	5/8	5/8	5/8
36,000	5/8	1/2	5/8	5/8	5/8	5/8	1/2	1/2	1/2	1/2	7/8	1/2	5/8	5/8	5/8
48,000	7/8	1/2	5/8	5/8	7/8	7/8	1/2	5/8	5/8	5/8	7/8	5/8	5/8	5/8	7/8
60,000	7/8	5/8	5/8	7/8	7/8	7/8	1/2	5/8	5/8	5/8	7/8	5/8	7/8	7/8	7/8
75,000	7/8	5/8	7/8	7/8	7/8	7/8	1/2	5/8	5/8	5/8	7/8	5/8	7/8	7/8	7/8
100,000	1 1/8	7/8	7/8	7/8	7/8	7/8	5/8	7/8	7/8	7/8	1 1/8	7/8	7/8	7/8	7/8
150,000	1 1/8	7/8	7/8	1 1/8	1 1/8	1 1/8	7/8	7/8	7/8	7/8	1 3/8	7/8	7/8	1 1/8	1 1/8
200,000	1 3/8	7/8	1 1/8	1 1/8	1 1/8	1 1/8	7/8	7/8	1 1/8	1 1/8	1 3/8	1 1/8	1 1/8	1 1/8	1 1/8
300,000	1 5/8	1 1/8	1 1/8	1 3/8	1 3/8	1 3/8	1 1/8	1 1/8	1 1/8	1 1/8	1 5/8	1 3/8	1 3/8	1 3/8	1 3/8
400,000	1 5/8	1 3/8	1 3/8	1 3/8	1 3/8	1 5/8	1 1/8	1 1/8	1 3/8	1 3/8	1 5/8	1 3/8	1 3/8	1 3/8	1 5/8
500,000	1 5/8	1 3/8	1 3/8	1 5/8	1 5/8	1 5/8	1 1/8	1 3/8	1 3/8	1 3/8	2 1/8	1 3/8	1 3/8	1 5/8	1 5/8
600,000	2 1/8	1 5/8	1 5/8	1 5/8	1 5/8	1 5/8	1 3/8	1 3/8	1 3/8	1 5/8	2 1/8	1 5/8	1 5/8	1 5/8	1 5/8
750,000	2 1/8	1 5/8	1 5/8	1 5/8	2 1/8	2 1/8	1 5/8	1 5/8	1 5/8	1 5/8	2 1/8	2 1/8	2 1/8	2 1/8	2 1/8

Recommended sizes are applicable with evaporating temperatures from -40° F. to 45° F. and condensing temperatures from 80° F. to 130° F.

Table 28
RECOMMENDED DISCHARGE LINE SIZES

Capacity BTU/hr	Light Load Capacity Reduction	R-12 Equivalent Length, Ft.				R-22 Equivalent Length, Ft.				R-502 Equivalent Length, Ft.			
		50	100	150	200	50	100	150	200	50	100	150	200
6,000	0	½	½	½	⅝*	⅜	½	½	½	½	½	½	⅝*
12,000	0	⅝	⅝	⅝	⅞*	½	½	⅝	⅝	⅝	⅝	⅝	⅞*
18,000	0	⅝	⅞	⅞	⅞	⅝	⅝	⅝	⅞	⅝	⅞*	⅞*	⅞*
24,000	0	⅞	⅞	⅞	⅞	⅝	⅞	⅞	⅞	⅞	⅞	⅞	⅞
36,000	0	⅞	⅞	⅞	1⅛	⅞	⅞	⅞	⅞	⅞	⅞	1⅛*	1⅛*
48,000	0	⅞	1⅛	1⅛	1⅛	⅞	⅞	⅞	1⅛*	⅞	1⅛	1⅛	1⅛
60,000	0	1⅛	1⅛	1⅛	1⅜	⅞	1⅛	1⅛	1⅛	1⅛	1⅛	1⅛	1⅜*
	33%	1⅛	1⅛	1⅛	1⅜*	⅞	1⅛	1⅛	1⅛	1⅛	1⅛	1⅛	1⅜**
75,000	0	1⅛	1⅛	1⅛	1⅜	⅞	1⅛	1⅛	1⅛	1⅛	1⅛	1⅜	1⅜
	33%	1⅛	1⅛	1⅛	1⅜	⅞	1⅛	1⅛	1⅛	1⅛	1⅛	1⅜*	1⅜*
100,000	0	1⅛	1⅜	1⅜	1⅝	1⅛	1⅛	1⅜	1⅜	1⅛	1⅜	1⅜	1⅝*
	33% to 50%	1⅛	1⅜	1⅜	1⅝*	1⅛	1⅛	1⅜*	1⅜*	1⅛	1⅜*	1⅜*	1⅝**
150,000	0	1⅜	1⅝	1⅝	2⅛	1⅛	1⅜	1⅜	1⅜	1⅜	1⅜	1⅝	1⅝
	33% to 50%	1⅜	1⅝	1⅝	2⅛*	1⅛	1⅜*	1⅜*	1⅜*	1⅜	1⅜	1⅝*	1⅝*
	66%	1⅜	1⅝*	1⅝*	2⅛**	1⅛	1⅜*	1⅜*	1⅜*	1⅜*	1⅜*	1⅝**	1⅝**
200,000	0	1⅝	1⅝	2⅛	2⅛	1⅜	1⅜	1⅝	1⅝	1⅜	1⅝	1⅝	2⅛*
	33% to 50%	1⅝	1⅝	2⅛*	2⅛*	1⅜	1⅜	1⅝*	1⅝*	1⅜	1⅝*	1⅝*	2⅛**
	66%	1⅝	1⅝	2⅛*	2⅛*	1⅜*	1⅜*	1⅝**	1⅝**	1⅜	1⅝*	1⅝*	2⅛**
300,000	0	2⅛	2⅛	2⅛	2⅛	1⅜	1⅝	1⅝	2⅛	1⅝	2⅛	2⅛	2⅛
	33% to 50%	2⅛	2⅛	2⅛	2⅛	1⅜	1⅝	1⅝	2⅛*	1⅝	2⅛*	2⅛*	2⅛*
	66%	2⅛*	2⅛*	2⅛*	2⅛*	1⅜	1⅝*	2⅛**	2⅛**	1⅝*	2⅛**	2⅛**	2⅛**
400,000	0	2⅛	2⅛	2⅛	2⅝	1⅝	2⅛	2⅛	2⅛	2⅛	2⅛	2⅛	2⅝
	33% to 66%	2⅛	2⅛	2⅛	2⅝*	1⅝	2⅛*	2⅛*	2⅛*	2⅛*	2⅛*	2⅛*	2⅝**
500,000	0	2⅝	2⅝	2⅝	2⅝	2⅛	2⅛	2⅛	2⅛	2⅛	2⅛	2⅝	2⅝
	33% to 50%	2⅝	2⅝	2⅝	2⅝	2⅛	2⅛	2⅛	2⅛	2⅛	2⅛	2⅝*	2⅝*
	66%	2⅝*	2⅝*	2⅝*	2⅝*	2⅛*	2⅛*	2⅛*	2⅛*	Horizontal 2⅝ Double Riser 1⅜ - 2⅛			
600,000	0	2⅝	2⅝	2⅝	3⅛	2⅛	2⅛	2⅛	2⅝	2⅛	2⅝	2⅝	3⅛
	33% to 50%	2⅝	2⅝	2⅝	3⅛*	2⅛	2⅛	2⅛	2⅝*	2⅛	2⅝*	2⅝*	3⅛**
	66%	2⅝*	2⅝*	3⅛**	3⅛**	2⅛*	2⅛*	2⅛*	2⅝**	2⅛	2⅝*	2⅝*	3⅛**
750,000	0	3⅛	3⅛	3⅛	3⅛	2⅛	2⅝	2⅝	2⅝	2⅝	2⅝	2⅝	3⅛
	33% to 50%	3⅛	3⅛	3⅛	3⅛	2⅛	2⅝*	2⅝*	2⅝*	2⅝	2⅝	2⅝	3⅛*
	66%	3⅛*	3⅛*	3⅛*	3⅛*	2⅛	2⅝*	2⅝*	2⅝*	2⅝*	2⅝*	2⅝*	3⅛**

* Use one line size smaller for vertical riser
** Use two line sizes smaller for vertical riser

Recommended sizes are applicable for applications with evaporating temperatures from -40° F. to 45° F. and condensing temperatures from 80° F. to 130° F.

Table 29
RECOMMENDED SUCTION LINE SIZES

R-12   40° F. Evaporating Temperature

Capacity BTU/hr.	Light Load Capacity Reduction	Equivalent Length, Ft.							
		50		100		150		200	
		H	V	H	V	H	V	H	V
6,000	0	⅝	⅝	⅝	⅝	⅝	⅝	⅝	⅝
12,000	0	⅞	⅞	⅞	⅞	⅞	⅞	⅞	⅞
18,000	0	⅞	⅞	⅞	⅞	1⅛	⅞	1⅛	1⅛
24,000	0	⅞	⅞	1⅛	1⅛	1⅛	1⅛	1⅛	1⅛
36,000	0	1⅛	1⅛	1⅛	1⅛	1⅜	1⅛	1⅜	1⅜
48,000	0	1⅛	1⅛	1⅜	1⅜	1⅜	1⅜	1⅝	1⅝
60,000	0 to 33%	1⅛	1⅛	1⅜	1⅜	1⅝	1⅜	1⅝	1⅝
75,000	0 to 33%	1⅜	1⅜	1⅝	1⅜	1⅝	1⅜	1⅝	1⅝
100,000	0 to 50%	1⅜	1⅜	1⅝	1⅝	2⅛	1⅝	2⅛	1⅝
150,000	0 to 33%	1⅝	1⅝	2⅛	1⅝	2⅛	1⅝	2⅝	2⅛
	50% to 66%	1⅝	1⅝	2⅛	1⅝	2⅛	1⅝	2⅛	1⅝
200,000	0	2⅛	2⅛	2⅛	2⅛	2⅝	2⅛	2⅝	2⅝
	33% to 50%	2⅛	2⅛	2⅛	2⅛	2⅝	2⅛	2⅝	2⅛
	66%	2⅛	2⅛	2⅛	2⅛	2⅛	2⅛	2⅛	2⅛
300,000	0 to 50%	2⅛	2⅛	2⅝	2⅛	2⅝	2⅛	3⅛	2⅝
	66%	2⅛	2⅛	2⅝	2⅛	2⅝	2⅛	2⅝	2⅛
400,000	0 to 50%	2⅝	2⅝	3⅛	2⅝	3⅛	2⅝	3⅛	3⅛
	66%	2⅝	2⅝	3⅛	2⅝	3⅛	2⅝	3⅛	2⅝
500,000	0 to 50%	2⅝	2⅝	3⅛	2⅝	3⅛	2⅝	3⅝	3⅛
	66%	2⅝	2⅝	3⅛	2⅝	3⅛	2⅝	3⅝	2⅝
600,000	0 to 66%	3⅛	2⅝	3⅛	3⅛	3⅝	3⅛	3⅝	3⅛
750,000	0 to 66%	3⅛	3⅛	3⅝	3⅛	3⅝	3⅛	4⅛	3⅝

Recommended sizes are applicable for applications with
condensing temperatures from 80° F. to 130° F.

H - Horizontal
V - Vertical

Table 30
RECOMMENDED SUCTION LINE SIZES

R-12   25° F. Evaporating Temperature

Capacity BTU/hr.	Light Load Capacity Reduction	Equivalent Length, Ft.							
		50		100		150		200	
		H	V	H	V	H	V	H	V
6,000	0	5/8	5/8	7/8	5/8	7/8	5/8	7/8	5/8
12,000	0	7/8	7/8	7/8	7/8	1 1/8	7/8	1 1/8	7/8
18,000	0	1 1/8	1 1/8	1 1/8	1 1/8	1 1/8	1 1/8	1 1/8	1 1/8
24,000	0	1 1/8	1 1/8	1 1/8	1 1/8	1 3/8	1 1/8	1 3/8	1 1/8
36,000	0	1 3/8	1 1/8	1 3/8	1 3/8	1 3/8	1 3/8	1 5/8	1 3/8
48,000	0	1 3/8	1 3/8	1 5/8	1 3/8	1 5/8	1 5/8	1 5/8	1 5/8
60,000	0 to 33%	1 5/8	1 3/8	1 5/8	1 5/8	2 1/8	1 5/8	2 1/8	1 5/8
75,000	0 to 33%	1 5/8	1 5/8	2 1/8	1 5/8	2 1/8	1 5/8	2 1/8	1 5/8
100,000	0 to 33%	1 5/8	1 5/8	2 1/8	1 5/8	2 1/8	2 1/8	2 1/8	2 1/8
	50%	1 5/8	1 5/8	2 1/8	1 5/8	2 1/8	1 5/8	2 1/8	1 5/8
150,000	0 to 33%	2 1/8	2 1/8	2 5/8	2 1/8	2 5/8	2 5/8	2 5/8	2 5/8
	50% to 66%	2 1/8	2 1/8	2 5/8	2 1/8	2 5/8	1 5/8*2 1/8	2 5/8	1 5/8*2 1/8
200,000	0 to 50%	2 5/8	2 5/8	2 5/8	2 5/8	3 1/8	2 5/8	3 1/8	2 5/8
	66%	2 5/8	2 1/8	2 5/8	1 5/8*2 1/8	2 5/8	1 5/8*2 1/8	2 5/8	1 5/8*2 1/8
300,000	0 to 50%	2 5/8	2 5/8	3 1/8	2 5/8	3 5/8	2 5/8	3 5/8	2 5/8
	66%	2 5/8	2 5/8	3 1/8	2 5/8	3 1/8	1 5/8*2 5/8	3 1/8	1 5/8*2 5/8
400,000	0 to 50%	3 1/8	3 1/8	3 1/8	3 1/8	3 5/8	3 1/8	3 5/8	3 1/8
	66%	3 1/8	2 5/8	3 1/8	1 5/8*2 5/8	3 1/8	1 5/8*2 5/8	3 1/8	1 5/8*2 5/8

Recommended sizes are applicable for applications with condensing temperatures from 80° F. to 130° F.

* Double Riser

H - Horizontal
V - Vertical

Table 31
RECOMMENDED SUCTION LINE SIZES
R-12   15° F. Evaporating Temperature

Capacity BTU/hr.	Light Load Capacity Reduction	Equivalent Length, Ft.							
		50		100		150		200	
		H	V	H	V	H	V	H	V
6,000	0	⅞	⅞	⅞	⅞	⅞	⅞	⅞	⅞
12,000	0	⅞	⅞	1⅛	⅞	1⅛	⅞	1⅛	⅞
18,000	0	1⅛	1⅛	1⅛	1⅛	1⅜	1⅛	1⅜	1⅛
24,000	0	1⅛	1⅛	1⅜	1⅛	1⅜	1⅛	1⅜	1⅜
36,000	0	1⅜	1⅜	1⅜	1⅜	1⅝	1⅜	2⅛	1⅝
48,000	0	1⅜	1⅜	1⅝	1⅜	1⅝	1⅜	2⅛	1⅝
60,000	0 to 33%	1⅝	1⅝	1⅝	1⅝	2⅛	1⅝	2⅛	1⅝
75,000	0 to 33%	1⅝	1⅝	2⅛	1⅝	2⅛	1⅝	2⅝	1⅝
100,000	0 to 33%	2⅛	1⅝	2⅛	1⅝	2⅝	1⅝	2⅝	2⅛
	50%	2⅛	1⅝	2⅛	1⅝	2⅝	1⅝	2⅝	1⅝
150,000	0 to 33%	2⅛	2⅛	2⅝	2⅛	2⅝	2⅛	3⅛	2⅛
	50% to 66%	2⅛	2⅛	2⅝	2⅛	2⅝	1⅝*2⅛	2⅝	1⅝*2⅛
200,000	0 to 50%	2⅝	2⅛	2⅝	2⅝	2⅝	2⅝	3⅛	2⅝
	66%	2⅝	2⅛	3⅛	2⅛	2⅝	1⅝*2⅛	2⅝	1⅝*2⅛
300,000	0 to 50%	3⅛	2⅝	3⅛	2⅝	3⅛	3⅛	3⅝	3⅛
	66%	3⅛	2⅝	3⅛	2⅝	3⅛	1⅝*2⅝	3⅛	1⅝*2⅝
400,000	0 to 50%	3⅛	2⅝	3⅛	3⅛	3⅝	3⅛	3⅝	3⅛
	66%	3⅛	2⅝	3⅛	1⅝*2⅝	3⅝	1⅝*2⅝	3⅝	1⅝*2⅝

Recommended sizes are applicable for applications with condensing temperatures from 80° F. to 130° F.

H - Horizontal
V - Vertical

* Double Riser

Table 32
RECOMMENDED SUCTION LINE SIZES
R-12   -20° F. Evaporating Temperatures

Capacity BTU/hr.	Light Load Capacity Reduction	Equivalent Length, Ft.							
		50		100		150		200	
		H	V	H	V	H	V	H	V
6,000	0	⅞	⅞	1⅛	⅞	1⅛	⅞	1⅛	⅞
12,000	0	1⅛	1⅛	1⅜	1⅛	1⅜	1⅛	1⅜	1⅛
18,000	0	1⅜	1⅜	1⅜	1⅜	1⅝	1⅜	1⅝	1⅜
24,000	0	1⅜	1⅜	1⅝	1⅝	1⅝	1⅝	2⅛	1⅝
36,000	0	1⅝	1⅝	2⅛	1⅝	2⅛	1⅝	2⅛	1⅝
48,000	0	2⅛	1⅝	2⅛	1⅝	2⅛	2⅛	2⅝	2⅛
60,000	0 to 33%	2⅛	2⅛	2⅝	2⅛	2⅝	2⅛	2⅝	2⅛
75,000	0 to 33%	2⅛	2⅛	2⅝	2⅛	2⅝	2⅛	3⅛	2⅛
100,000	0 to 50%	2⅝	2⅛	2⅝	2⅛	2⅝	2⅛	3⅛	2⅛
150,000	0 to 50%	2⅝	2⅝	3⅛	2⅝	3⅛	2⅝	3⅝	2⅝
	66%	2⅝	1⅝*2⅛	3⅛	1⅝*2⅛	3⅛	1⅝*2⅛	3⅝	1⅝*2⅝

Recommended sizes are applicable for applications with condensing temperatures from 80° F. to 130° F.

H - Horizontal
V - Vertical

* Double Riser

Table 33
RECOMMENDED SUCTION LINE SIZES
R-12   -40° F. Evaporating Temperature

Capacity BTU/hr.	Light Load Capacity Reduction	Equivalent Length, Ft.							
		50		100		150		200	
		H	V	H	V	H	V	H	V
6,000	0	1 1/8	1 1/8	1 1/8	1 1/8	1 3/8	1 1/8	1 3/8	1 1/8
12,000	0	1 3/8	1 1/8	1 5/8	1 3/8	1 5/8	1 3/8	1 5/8	1 3/8
18,000	0	1 3/8	1 3/8	1 5/8	1 3/8	1 5/8	1 3/8	2 1/8	1 5/8
24,000	0	1 5/8	1 5/8	2 1/8	1 5/8	2 1/8	1 5/8	2 5/8	1 5/8
36,000	0	2 1/8	2 1/8	2 1/8	2 1/8	2 5/8	2 1/8	2 5/8	2 1/8
48,000	0	2 1/8	2 1/8	2 5/8	2 1/8	2 5/8	2 1/8	2 5/8	2 1/8
60,000	0 to 33%	2 5/8	2 1/8	3 1/8	2 1/8	3 1/8	2 1/8	3 1/8	2 1/8
75,000	0	2 5/8	2 5/8	3 1/8	2 5/8	3 1/8	2 5/8	3 1/8	2 5/8
	33%	3 1/8	2 1/8	3 1/8	2 1/8	3 1/8	2 1/8	3 1/8	2 1/8
100,000	0 to 33%	3 1/8	2 5/8	3 5/8	2 5/8	3 5/8	2 5/8	3 5/8	2 5/8
	50%	3 1/8	1 5/8 *2 5/8	3 1/8	1 5/8 *2 5/8	3 5/8	1 5/8 *2 5/8	3 5/8	1 5/8 *2 5/8

Recommended sizes are applicable for applications with condensing temperatures from 80° F. to 130° F.

H - Horizontal
V - Vertical

* Double Riser

Table 34
RECOMMENDED SUCTION LINE SIZES
R-22   40° F. Evaporating Temperature

Capacity BTU/hr.	Light Load Capacity Reduction	Equivalent Length, Ft.							
		50		100		150		200	
		H	V	H	V	H	V	H	V
6,000	0	1/2	1/2	1/2	1/2	5/8	1/2	5/8	1/2
12,000	0	5/8	5/8	5/8	5/8	7/8	5/8	7/8	5/8
18,000	0	7/8	7/8	7/8	7/8	7/8	7/8	7/8	7/8
24,000	0	7/8	7/8	7/8	7/8	7/8	7/8	1 1/8	7/8
36,000	0	7/8	7/8	1 1/8	7/8	1 1/8	7/8	1 1/8	1 1/8
48,000	0	1 1/8	1 1/8	1 1/8	1 1/8	1 1/8	1 1/8	1 3/8	1 1/8
60,000	0 to 33%	1 1/8	1 1/8	1 1/8	1 1/8	1 3/8	1 1/8	1 3/8	1 1/8
75,000	0 to 33%	1 1/8	1 1/8	1 3/8	1 1/8	1 3/8	1 1/8	1 5/8	1 3/8
100,000	0 to 50%	1 3/8	1 3/8	1 3/8	1 3/8	1 3/8	1 3/8	1 5/8	1 3/8
150,000	0 to 66%	1 3/8	1 3/8	1 5/8	1 5/8	1 5/8	1 5/8	2 1/8	1 5/8
200,000	0 to 66%	1 5/8	1 5/8	2 1/8	1 5/8	2 1/8	1 5/8	2 1/8	1 5/8
300,000	0 to 50%	2 1/8	2 1/8	2 1/8	2 1/8	2 1/8	2 1/8	2 5/8	2 1/8
	66%	2 1/8	2 1/8	2 1/8	2 1/8	2 1/8	2 1/8	2 1/8	2 1/8
400,000	0 to 66%	2 1/8	2 1/8	2 1/8	2 1/8	2 5/8	2 1/8	2 5/8	2 1/8
500,000	0 to 66%	2 1/8	2 1/8	2 5/8	2 1/8	2 5/8	2 1/8	2 5/8	2 5/8
600,000	0 to 66%	2 5/8	2 5/8	2 5/8	2 5/8	2 5/8	2 5/8	3 1/8	2 5/8
750,000	0 to 66%	2 5/8	2 5/8	3 1/8	2 5/8	3 1/8	2 5/8	3 1/8	2 5/8

Recommended sizes are applicable with condensing temperatures from 80° F. to 130° F.

H - Horizontal
V - Vertical

Table 35
RECOMMENDED SUCTION LINE SIZES

R-22   25° F. Evaporating Temperature

Capacity BTU/hr.	Light Load Capacity Reduction	Equivalent Length, Ft.							
		50		100		150		200	
		H	V	H	V	H	V	H	V
6,000	0	½	½	⅝	⅝	⅝	⅝	⅝	⅝
12,000	0	⅝	⅝	⅞	⅝	⅞	⅝	⅞	⅞
18,000	0	⅞	⅞	⅞	⅞	⅞	⅞	1⅛	⅞
24,000	0	⅞	⅞	⅞	⅞	1⅛	⅞	1⅛	⅞
36,000	0	1⅛	1⅛	1⅛	1⅛	1⅛	1⅛	1⅜	1⅛
48,000	0	1⅛	1⅛	1⅛	1⅛	1⅜	1⅛	1⅜	1⅛
60,000	0 to 33%	1⅛	1⅛	1⅜	1⅜	1⅜	1⅜	1⅜	1⅜
75,000	0 to 33%	1⅜	1⅜	1⅜	1⅜	1⅝	1⅜	1⅝	1⅜
100,000	0 to 50%	1⅜	1⅜	1⅝	1⅜	1⅝	1⅜	1⅝	1⅜
150,000	0 to 50%	1⅝	1⅝	2⅛	1⅝	2⅛	1⅝	2⅛	1⅝
	66%	1⅝	1⅝	1⅝	1⅝	1⅝	1⅝	1⅝	1⅝
200,000	0 to 50%	2⅛	2⅛	2⅛	2⅛	2⅛	2⅛	2⅛	2⅛
	66%	2⅛	1⅜*1⅝	2⅛	1⅜*1⅝	2⅛	1⅜*1⅝	2⅛	1⅜*1⅝
300,000	0 to 50%	2⅛	2⅛	2⅝	2⅛	2⅝	2⅛	2⅝	2⅝
	66%	2⅛	2⅛	2⅝	2⅛	2⅝	2⅛	2⅝	2⅛
400,000	0 to 50%	2⅝	2⅛	2⅝	2⅛	3⅛	2⅛	3⅛	2⅛
	66%	2⅝	2⅛	2⅝	2⅛	2⅝	1⅝*2⅛	2⅝	1⅝*2⅛
500,000	0 to 66%	2⅝	2⅝	2⅝	2⅝	3⅛	2⅝	3⅛	2⅝
600,000	0 to 66%	2⅝	2⅝	3⅛	2⅝	3⅝	2⅝	3⅝	2⅝
750,000	0 to 66%	3⅛	3⅛	3⅛	3⅛	3⅝	3⅛	3⅝	3⅛

Recommended sizes are applicable with condensing temperatures from 80° F. to 130° F.

* Double Riser

H - Horizontal
V - Vertical

Table 36

RECOMMENDED SUCTION LINE SIZES

R-22   **15° F. Evaporating Temperature**

Capacity BTU/hr.	Light Load Capacity Reduction	Equivalent Length, Ft.							
		50		100		150		200	
		H	V	H	V	H	V	H	V
6,000	0	5/8	5/8	5/8	5/8	5/8	5/8	5/8	5/8
12,000	0	5/8	5/8	7/8	5/8	7/8	5/8	7/8	7/8
18,000	0	7/8	7/8	7/8	7/8	1 1/8	7/8	1 1/8	7/8
24,000	0	7/8	7/8	1 1/8	7/8	1 1/8	7/8	1 1/8	7/8
36,000	0	1 1/8	1 1/8	1 1/8	1 1/8	1 1/8	1 1/8	1 3/8	1 1/8
48,000	0	1 1/8	1 1/8	1 3/8	1 1/8	1 3/8	1 1/8	1 3/8	1 1/8
60,000	0 to 33%	1 3/8	1 3/8	1 3/8	1 3/8	1 5/8	1 3/8	1 5/8	1 3/8
75,000	0 to 33%	1 3/8	1 3/8	1 5/8	1 3/8	1 5/8	1 3/8	1 5/8	1 3/8
100,000	0 to 50%	1 3/8	1 3/8	1 5/8	1 5/8	1 5/8	1 5/8	1 5/8	1 5/8
150,000	0 to 50%	1 5/8	1 5/8	2 1/8	1 5/8	2 1/8	1 5/8	2 1/8	1 5/8
	66%	1 5/8	1 5/8	1 5/8	1 5/8	1 5/8	1 5/8	1 5/8	1 5/8
200,000	0 to 50%	2 1/8	2 1/8	2 1/8	2 1/8	2 5/8	2 1/8	2 5/8	2 1/8
	66%	2 1/8	1 3/8 *1 5/8	2 1/8	1 3/8 *1 5/8	2 1/8	1 3/8 *1 5/8	2 1/8	1 3/8 *1 5/8
300,000	0 to 50%	2 1/8	2 1/8	2 5/8	2 1/8	3 1/8	2 1/8	3 1/8	2 1/8
	66%	2 1/8	2 1/8	2 5/8	2 1/8	2 5/8	2 1/8	2 5/8	1 5/8 *2 1/8
400,000	0 to 50%	2 5/8	2 5/8	2 5/8	2 5/8	3 1/8	2 5/8	3 1/8	2 5/8
	66%	2 5/8	2 5/8	2 5/8	2 5/8	3 1/8	1 5/8 *2 1/8	3 1/8	1 5/8 *2 1/8
500,000	0 to 50%	2 5/8	2 5/8	3 1/8	2 5/8	3 1/8	2 5/8	3 5/8	2 5/8
	66%	2 5/8	2 5/8	3 1/8	2 5/8	3 1/8	2 5/8	3 1/8	2 5/8
600,000	0 to 66%	3 1/8	2 5/8	3 1/8	3 1/8	3 5/8	3 1/8	3 5/8	3 1/8
750,000	0 to 66%	3 1/8	3 1/8	3 5/8	3 1/8	3 5/8	3 1/8	3 5/8	3 1/8

Recommended sizes are applicable with condensing temperatures from 80° F. to 130° F.

* Double Riser

H - Horizontal
V - Vertical

Table 37
RECOMMENDED SUCTION LINE SIZES
R-22   -20° F. Evaporating Temperature

Capacity BTU/hr.	Equivalent Length, Ft.							
	50		100		150		200	
	H	V	H	V	H	V	H	V
6,000	⅞	⅞	⅞	⅞	⅞	⅞	⅞	⅞
12,000	1⅛	1⅛	1⅛	1⅛	1⅛	1⅛	1⅛	1⅛
18,000	1⅛	1⅛	1⅜	1⅛	1⅜	1⅛	1⅜	1⅛
24,000	1⅜	1⅜	1⅜	1⅜	1⅝	1⅜	1⅝	1⅜
36,000	1⅝	1⅝	1⅝	1⅝	1⅝	1⅝	1⅝	1⅝
48,000	1⅝	1⅝	2⅛	1⅝	2⅛	1⅝	2⅛	1⅝

Recommended sizes are applicable with condensing temperatures from 80° F. to 130° F.

H - Horizontal
V - Vertical

Table 38
RECOMMENDED SUCTION LINE SIZES
R-502   25° F. Evaporating Temperature

Capacity BTU/hr.	Light Load Capacity Reduction	Equivalent Length, Ft.							
		50		100		150		200	
		H	V	H	V	H	V	H	V
6,000	0	⅝	⅝	⅝	⅝	⅝	⅝	⅝	⅝
12,000	0	⅞	⅞	⅞	⅞	⅞	⅞	⅞	⅞
18,000	0	⅞	⅞	1⅛	⅞	1⅛	⅞	1⅛	⅞
24,000	0	⅞	⅞	1⅛	⅞	1⅛	⅞	1⅛	1⅛
36,000	0	1⅛	1⅛	1⅛	1⅛	1⅜	1⅛	1⅜	1⅛
48,000	0	1⅛	1⅛	1⅜	1⅛	1⅜	1⅛	1⅝	1⅜
60,000	0 to 33%	1⅜	1⅛	1⅜	1⅛	1⅜	1⅜	1⅝	1⅜
75,000	0 to 33%	1⅜	1⅜	1⅝	1⅜	1⅝	1⅜	1⅝	1⅝
100,000	0 to 33%	1⅜	1⅜	1⅝	1⅝	1⅝	1⅝	2⅛	1⅝
	50%	1⅜	1⅜	1⅝	1⅝	1⅝	1⅝	1⅝	1⅝
150,000	0 to 50%	1⅝	1⅝	2⅛	1⅝	2⅛	1⅝	2⅛	1⅝
	66%	1⅝	1⅝	1⅝	1⅝	1⅝	1⅝	1⅝	1⅝
200,000	0 to 50%	2⅛	2⅛	2⅛	2⅛	2⅝	2⅛	2⅝	2⅛
	66%	2⅛	1⅜*1⅝	2⅛	1⅜*1⅝	2⅛	1⅜*1⅝	2⅛	1⅜*1⅝
300,000	0 to 50%	2⅛	2⅛	2⅝	2⅝	2⅝	2⅝	2⅝	2⅝
	66%	2⅛	2⅛	2⅝	1⅝*2⅛	2⅝	1⅝*2⅛	2⅝	1⅝*2⅛
400,000	0 to 50%	2⅝	2⅝	2⅝	2⅝	3⅛	2⅝	3⅛	2⅝
	66%	2⅝	1⅝*2⅛	2⅝	1⅝*2⅛	2⅝	1⅝*2⅛	2⅝	1⅝*2⅛
500,000	0 to 50%	2⅝	2⅝	3⅛	2⅝	3⅛	2⅝	3⅛	3⅛
	66%	2⅝	2⅝	3⅛	1⅝*2⅛	3⅛	1⅝*2⅛	3⅛	1⅝*2⅛
600,000	0 to 50%	2⅝	2⅝	3⅛	2⅝	3⅝	2⅝	3⅝	3⅛
	66%	2⅝	2⅝	3⅛	2⅝	3⅝	2⅝	3⅝	1⅝*2⅝
750,000	0 to 50%	3⅛	3⅛	3⅝	3⅛	3⅝	3⅛	4⅛	3⅛
	66%	3⅛	3⅛	3⅝	3⅛	3⅝	3⅛	3⅝	3⅛

Recommended sizes are applicable with condensing temperatures from 80° F. to 130° F.
* Double Riser

H - Horizontal
V - Vertical

Table 39
RECOMMENDED SUCTION LINE SIZES

R-502   **15° F. Evaporating Temperature**

Capacity BTU/hr.	Light Load Capacity Reduction	Equivalent Length, Ft.							
		50		100		150		200	
		H	V	H	V	H	V	H	V
6,000	0	5/8	5/8	5/8	5/8	5/8	5/8	5/8	5/8
12,000	0	7/8	7/8	7/8	7/8	7/8	7/8	7/8	7/8
18,000	0	7/8	7/8	7/8	7/8	1 1/8	7/8	1 1/8	7/8
24,000	0	1 1/8	7/8	1 1/8	7/8	1 1/8	7/8	1 1/8	1 1/8
36,000	0	1 1/8	1 1/8	1 3/8	1 1/8	1 3/8	1 1/8	1 3/8	1 1/8
48,000	0	1 3/8	1 3/8	1 3/8	1 3/8	1 3/8	1 3/8	1 5/8	1 3/8
60,000	0 to 33%	1 3/8	1 3/8	1 5/8	1 3/8	1 5/8	1 3/8	1 5/8	1 3/8
75,000	0 to 33%	1 3/8	1 3/8	1 5/8	1 3/8	1 5/8	1 3/8	1 5/8	1 5/8
100,000	0 to 33%	1 5/8	1 5/8	1 5/8	1 5/8	2 1/8	1 5/8	2 1/8	1 5/8
	50%	1 5/8	1 5/8	1 5/8	1 5/8	1 5/8	1 5/8	1 5/8	1 5/8
150,000	0 to 50%	2 1/8	1 5/8	2 1/8	1 5/8	2 1/8	1 5/8	2 1/8	1 5/8
	66%	1 5/8	1 5/8	1 5/8	1 5/8	1 5/8	1 5/8	1 5/8	1 5/8
200,000	0 to 50%	2 1/8	2 1/8	2 5/8	2 1/8	2 5/8	2 1/8	2 5/8	2 1/8
	66%	2 1/8	1 3/8 *1 5/8	2 1/8	1 3/8 *1 5/8	2 1/8	1 3/8 *1 5/8	2 1/8	1 3/8 *1 5/8
300,000	0 to 50%	2 1/8	2 1/8	2 5/8	2 1/8	3 1/8	2 1/8	3 1/8	2 5/8
	66%	2 1/8	2 1/8	2 5/8	2 1/8	2 5/8	1 5/8 *2 1/8	2 5/8	1 5/8 *2 1/8
400,000	0 to 50%	2 5/8	2 5/8	3 1/8	2 5/8	3 1/8	2 5/8	3 1/8	2 5/8
	66%	2 5/8	2 5/8	3 1/8	2 5/8	3 1/8	2 5/8	2 5/8	1 5/8 *2 1/8
500,000	0 to 50%	2 5/8	2 5/8	3 1/8	2 5/8	3 5/8	2 5/8	3 5/8	3 1/8
	66%	2 5/8	2 5/8	3 1/8	2 5/8	3 1/8	1 5/8 *2 5/8	3 1/8	1 5/8 *2 5/8
600,000	0 to 50%	3 1/8	3 1/8	3 5/8	3 1/8	3 5/8	3 1/8	3 5/8	3 5/8
	66%	3 1/8	3 1/8	3 5/8	3 1/8	3 5/8	3 1/8	3 5/8	2 1/8 *3 1/8
750,000	0 to 50%	3 1/8	3 1/8	3 5/8	3 1/8	3 5/8	3 1/8	4 1/8	3 5/8
	66%	3 1/8	3 1/8	3 5/8	3 1/8	3 5/8	3 1/8	3 5/8	2 1/8 *3 1/8

Recommended sizes are applicable with condensing temperatures from 80° F. to 130° F.

*Double Riser

H - Horizontal
V - Vertical

Table 40

RECOMMENDED SUCTION LINE SIZES

R-502   -20° F. Evaporating Temperature

Capacity BTU/hr.	Light Load Capacity Reduction	Equivalent Length, Ft.							
		50		100		150		200	
		H	V	H	V	H	V	H	V
6,000	0	⅞	⅞	⅞	⅞	⅞	⅞	⅞	⅞
12,000	0	1⅛	1⅛	1⅛	1⅛	1⅛	1⅛	1⅛	1⅛
18,000	0	1⅛	1⅛	1⅜	1⅛	1⅜	1⅛	1⅝	1⅛
24,000	0	1⅛	1⅛	1⅜	1⅜	1⅜	1⅜	1⅝	1⅜
36,000	0	1⅜	1⅜	1⅝	1⅜	1⅝	1⅝	2⅛	1⅝
48,000	0	1⅝	1⅝	2⅛	1⅝	2⅛	1⅝	2⅛	1⅝
60,000	0 to 33%	1⅝	1⅝	2⅛	1⅝	2⅝	1⅝	2⅝	1⅝
75,000	0 to 33%	2⅛	1⅝	2⅝	1⅝	2⅝	1⅝	2⅝	1⅝
100,000	0 to 33%	2⅛	2⅛	2⅝	2⅛	2⅝	2⅛	2⅝	2⅛
	50%	2⅛	1⅜*1⅝	2⅝	1⅜*1⅝	2⅝	1⅜*1⅝	2⅝	1⅜*1⅝
150,000	0 to 50%	2⅝	2⅛	2⅝	2⅛	3⅛	2⅛	3⅛	2⅛
	66%	2⅝	2⅛	2⅝	2⅛	2⅝	1⅝*2⅛	2⅝	1⅝*2⅛
200,000	0 to 50%	2⅝	2⅝	3⅛	2⅝	3⅛	2⅝	3⅝	2⅝
	66%	2⅝	1⅝*2⅛	3⅛	1⅝*2⅛	3⅛	1⅝*2⅛	3⅛	1⅝*2⅛
300,000	0 to 50%	3⅛	2⅝	3⅛	3⅛	3⅝	3⅛	4⅛	3⅛
	66%	3⅛	2⅝	3⅛	1⅝*2⅝	3⅝	1⅝*2⅝	3⅝	1⅝*2⅝

Recommended sizes are applicable with condensing temperatures from 80° F. to 130° F.

* Double Riser

H - Horizontal
V - Vertical

Table 41

RECOMMENDED SUCTION LINE SIZES

R-502   -40° F. Evaporating Temperature

Capacity BTU/hr.	Light Load Capacity Reduction	Equivalent Length, Ft.							
		50		100		150		200	
		H	V	H	V	H	V	H	V
6,000	0	⅞	⅞	1⅛	⅞	1⅛	⅞	1⅛	⅞
12,000	0	1⅛	1⅛	1⅜	1⅛	1⅜	1⅛	1⅜	1⅛
18,000	0	1⅜	1⅜	1⅝	1⅜	1⅝	1⅜	1⅝	1⅜
24,000	0	1⅜	1⅜	1⅝	1⅜	1⅝	1⅜	2⅛	1⅜
36,000	0	1⅝	1⅝	2⅛	1⅝	2⅛	1⅝	2⅛	1⅝
48,000	0	2⅛	1⅝	2⅛	2⅛	2⅝	2⅛	2⅝	2⅛
60,000	0	2⅝	1⅝	2⅝	2⅛	2⅝	2⅛	2⅝	2⅛
	33%	2⅝	1⅝	2⅝	1⅜*1⅝	2⅝	1⅜*1⅝	2⅝	1⅜*1⅝
75,000	0 to 33%	2⅛	2⅛	2⅝	2⅛	2⅝	2⅛	2⅝	2⅛
100,000	0 to 50%	2⅛	2⅛	2⅝	2⅛	3⅛	2⅛	3⅛	2⅛
150,000	0 to 33%	2⅝	2⅝	3⅛	2⅝	3⅝	2⅝	3⅝	2⅝
	50% to 66%	3⅛	2⅛	3⅛	1⅝*2⅛	3⅝	1⅝*2⅛	3⅝	1⅝*2⅛
200,000	0 to 33%	3⅛	3⅛	3⅝	3⅛	3⅝	3⅛	4⅛	3⅛
	50% to 66%	3⅛	1⅝*2⅝	3⅝	1⅝*2⅝	3⅝	1⅝*2⅝	3⅝	1⅝*2⅝

Recommended sizes are applicable with condensing temperatures from 80° F. to 130° F.

* Double Riser

H - Horizontal
V - Vertical

APPENDIX III—TEMPERATURE CONVERSION TABLE

TEMPERATURE CONVERSION TABLE

°F	Temperature to be Converted	°C	°F	Temperature to be Converted	°C
− 76.0	− 60	−51.1	10.4	− 12	−24.4
− 74.2	− 59	−50.6	12.2	− 11	−23.9
− 72.4	− 58	−50.0	14.0	− 10	−23.3
− 70.6	− 57	−49.4	15.8	− 9	−22.8
− 68.8	− 56	−48.9	17.6	− 8	−22.2
− 67.0	− 55	−48.3	19.4	− 7	−21.7
− 65.2	− 54	−47.8	21.2	− 6	−21.1
− 63.4	− 53	−47.2	23.0	− 5	−20.6
− 61.6	− 52	−46.7	24.8	− 4	−20.0
− 59.8	− 51	−46.1	26.6	− 3	−19.4
− 58.0	− 50	−45.6	28.4	− 2	−18.9
− 56.2	− 49	−45.0	30.2	− 1	−18.3
− 54.4	− 48	−44.4	32.0	0	−17.8
− 52.6	− 47	−43.9	33.8	1	−17.2
− 50.8	− 46	−43.3	35.6	2	−16.7
− 49.0	− 45	−42.8	37.4	3	−16.1
− 47.2	− 44	−42.2	39.2	4	−15.6
− 45.4	− 43	−41.7	41.0	5	−15.0
− 43.6	− 42	−41.1	42.8	6	−14.4
− 41.8	− 41	−40.6	44.6	7	−13.9
− 40.0	− 40	−40.0	46.4	8	−13.3
− 38.2	− 39	−39.4	48.2	9	−12.8
− 36.4	− 38	−38.9	50.0	10	−12.2
− 34.6	− 37	−38.3	51.8	11	−11.7
− 32.8	− 36	−37.8	53.6	12	−11.1
− 31.0	− 35	−37.2	55.4	13	−10.6
− 29.2	− 34	−36.7	57.2	14	−10.0
− 27.4	− 33	−36.1	59.0	15	− 9.4
− 25.6	− 32	−35.6	60.8	16	− 8.9
− 23.8	− 31	−35.0	62.6	17	− 8.3
− 22.0	− 30	−34.4	64.4	18	− 7.8
− 20.2	− 29	−33.9	66.2	19	− 7.2
− 18.4	− 28	−33.3	68.0	20	− 6.7
− 16.6	− 27	−32.8	69.8	21	− 6.1
− 14.8	− 26	−32.2	71.6	22	− 5.6
− 13.0	− 25	−31.7	73.4	23	− 5.0
− 11.2	− 24	−31.1	75.2	24	− 4.4
− 9.4	− 23	−30.6	77.0	25	− 3.9
− 7.6	− 22	−30.0	78.8	26	− 3.3
− 5.8	− 21	−29.4	80.6	27	− 2.8
− 4.0	− 20	−28.9	82.4	28	− 2.2
− 2.2	− 19	−28.3	84.2	29	− 1.7
− 0.4	− 18	−27.8	86.0	30	− 1.1
1.4	− 17	−27.2	87.8	31	− 0.6
3.2	− 16	−26.7	89.6	32	0.0
5.0	− 15	−26.1	91.4	33	0.6
6.8	− 14	−25.6	93.2	34	1.1
8.6	− 13	−25.0	95.0	35	1.7

°F	Temperature to be Converted	°C	°F	Temperature to be Converted	°C
96.8	36	2.2	174.2	79	26.1
98.6	37	2.8	176.0	80	26.7
100.4	38	3.3	177.8	81	27.2
102.2	39	3.9	179.6	82	27.8
104.0	40	4.4	181.4	83	28.3
105.8	41	5.0	183.2	84	28.9
107.6	42	5.6	185.0	85	29.4
109.4	43	6.1	186.8	86	30.0
111.2	44	6.7	188.6	87	30.6
113.0	45	7.2	190.4	88	31.1
114.8	46	7.8	192.2	89	31.7
116.6	47	8.3	194.0	90	32.2
118.4	48	8.9	195.8	91	32.8
120.2	49	9.4	197.6	92	33.3
122.0	50	10.0	199.4	93	33.9
123.8	51	10.6	201.2	94	34.4
125.6	52	11.1	203.0	95	35.0
127.4	53	11.7	204.8	96	35.6
129.2	54	12.2	206.6	97	36.1
131.0	55	12.8	208.4	98	36.7
132.8	56	13.3	210.2	99	37.2
134.6	57	13.9	212.0	100	37.8
136.4	58	14.4	213.8	101	38.3
138.2	59	15.0	215.6	102	38.9
140.0	60	15.6	217.4	103	39.4
141.8	61	16.1	219.2	104	40.0
143.6	62	16.7	221.0	105	40.6
145.4	63	17.2	222.8	106	41.1
147.2	64	17.8	224.6	107	41.7
149.0	65	18.3	226.4	108	42.2
150.8	66	18.9	228.2	109	42.8
152.6	67	19.4	230.0	110	43.3
154.4	68	20.0	231.8	111	43.9
156.2	69	20.6	233.6	112	44.4
158.0	70	21.1	235.4	113	45.0
159.8	71	21.7	237.2	114	45.6
161.8	72	22.2	239.0	115	46.1
163.4	73	22.8	240.8	116	46.6
165.2	74	23.3	242.6	117	47.2
167.0	75	23.9	244.4	118	47.7
168.8	76	24.4	246.2	119	48.3
170.6	77	25.0	248.0	120	48.8
172.4	78	25.6			

Example 1. To find 37°F as a Celsius equivalent, find 37 in the Temperature to be Converted column and read the value in the °C column which is 2.8°C.

Example 2. To find 75°C as a Fahrenheit equivalent, find 75 in the Temperature to be Converted column and read the value in the °F column which is 167.0°F.

APPENDIX IV—ELECTRICAL SYMBOLS CHART

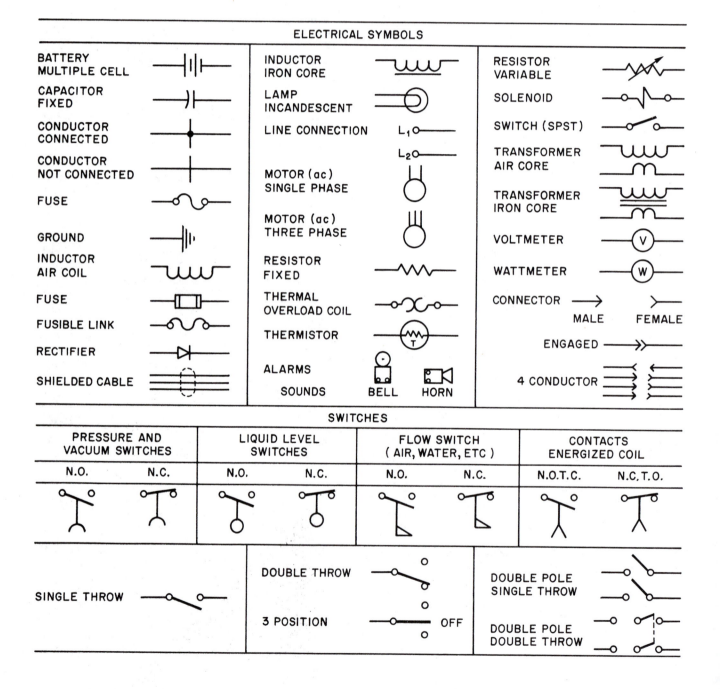

ELECTRICAL SYMBOLS

BATTERY MULTIPLE CELL		INDUCTOR IRON CORE		RESISTOR VARIABLE	
CAPACITOR FIXED		LAMP INCANDESCENT		SOLENOID	
CONDUCTOR CONNECTED		LINE CONNECTION	L_1 / L_2	SWITCH (SPST)	
CONDUCTOR NOT CONNECTED				TRANSFORMER AIR CORE	
FUSE		MOTOR (ac) SINGLE PHASE		TRANSFORMER IRON CORE	
GROUND		MOTOR (ac) THREE PHASE		VOLTMETER	V
INDUCTOR AIR COIL		RESISTOR FIXED		WATTMETER	W
FUSE		THERMAL OVERLOAD COIL		CONNECTOR	MALE FEMALE
FUSIBLE LINK		THERMISTOR		ENGAGED	
RECTIFIER		ALARMS		4 CONDUCTOR	
SHIELDED CABLE		SOUNDS	BELL HORN		

SWITCHES

PRESSURE AND VACUUM SWITCHES		LIQUID LEVEL SWITCHES		FLOW SWITCH (AIR, WATER, ETC)		CONTACTS ENERGIZED COIL	
N.O.	N.C.	N.O.	N.C.	N.O.	N.C.	N.O.T.C.	N.C.T.O.

SINGLE THROW	DOUBLE THROW	DOUBLE POLE SINGLE THROW
	3 POSITION OFF	DOUBLE POLE DOUBLE THROW

Glossary

ABSOLUTE PRESSURE. Gage pressure plus the pressure of the atmosphere, normally 14.696 at sea level at 68°F.

ABSOLUTE ZERO TEMPERATURE. The lowest obtainable temperature where molecular motion stops, −460°F and −273°C.

ACCUMULATOR. A storage tank located in the suction line of a compressor. It allows small amounts of liquid refrigerant to boil away before entering the compressor. Sometimes used to store excess refrigerant in heat pump systems during the winter cycle.

ACID-CONTAMINATED SYSTEM. A refrigeration system that contains acid due to contamination.

ACR TUBING. *A*ir *C*onditioning and *R*efrigeration tubing that is very clean, dry, and normally charged with dry nitrogen. The tubing is sealed at the ends to contain the nitrogen.

ACTIVATED ALUMINA. A chemical desiccant used in refrigerant driers.

AIR HEAT EXCHANGER. A device used to exchange heat between air and another medium at different temperature levels, such as air-to-air, air-to-water, or air-to-refrigerant.

AIR CONDITIONER. Equipment that conditions air by cleaning, cooling, heating, humidifying or dehumidifying it. A term often applied to comfort cooling equipment.

AIR CONDITIONING. A process that maintains comfort conditions in a defined area.

AIR-COOLED CONDENSER. One of the four main components of an air-cooled refrigeration system. It receives hot gas from the compressor and rejects it to a place where it makes no difference.

AIR GAP. The clearance between the rotating rotor and the stationary winding on an open motor. Known as a vapor gap in a hermetically sealed compressor motor.

AIR HANDLER. The device that moves the air across the heat exchanger in a forced-air system—normally considered to be the fan and its housing.

AIR SENSOR. A device that registers changes in air conditions such as pressure, velocity, temperature, or moisture content.

AIR, STANDARD. Dry air at 70°F and 14.696 psia at which it has a mass density of 0.075 lb/ft³ and a specific volume of 13.33 ft³/lb., ASHRAE 1986.

AIR VENT. A fitting used to vent air manually or automatically from a system.

ALGAE. A form of green or black, slimy plant life that grows in water systems.

ALLEN HEAD. A recessed hex head in a fastener.

ALTERNATING CURRENT. An electric current that reverses its direction at regular intervals.

AMBIENT TEMPERATURE. The surrounding air temperature.

AMERICAN STANDARD PIPE THREAD. Standard thread used on pipe to prevent leaks.

AMMETER. A meter used to measure current flow in an electrical circuit.

AMPERAGE. Amount of electron or current flow (the number of electrons passing a point in a given time) in an electrical circuit.

AMPERE. Unit of current flow.

ANEMOMETER. An instrument used to measure the velocity of air.

ANGLE VALVE. Valve with one opening at a 90° angle from the other opening.

A.S.A. Abbreviation for the American Standards Association [now known as American National Standards Institute (ANSI)].

ASHRAE. Abbreviation for the American Society of Heating, Refrigerating, and Air Conditioning Engineers.

ASME. Abbreviation for the American Society of Mechanical Engineers.

ASPECT RATIO. The ratio of the length to width of a component.

ATMOSPHERIC PRESSURE. The weight of the atmosphere's gases pressing down on the earth. Equal to 14.696 psi at sea level and 70°F.

ATOM. The smallest particle of an element.

ATOMIZE. Using pressure to change liquid to small particles of vapor.

AUTOMATIC CONTROL. Controls that react to a change in conditions to cause the condition to stabilize.

AUTOMATIC DEFROST. Using automatic means to remove ice from a refrigeration coil.

AUTOMATIC EXPANSION VALVE. A refrigerant control valve that maintains a constant pressure in an evaporator.

BACK PRESSURE. The pressure on the low-pressure side of a refrigeration system (also known as *suction pressure*).

BACK SEAT. The position of a refrigeration service valve when the stem is turned away from the valve body and seated.

BAFFLE. A plate used to keep fluids from moving back and forth at will in a container.

BALL CHECK VALVE. A valve with a ball-shaped internal assembly that only allows fluid flow in one direction.

BAROMETER. A device used to measure atmospheric pressure that is commonly calibrated in inches or millimeters of mercury. There are two types: mercury column and aneroid.

BATTERY. A device that produces electricity from the interaction of metals and acid.

BEARING. A device that surrounds a rotating shaft and provides a low-friction contact surface to reduce wear from the rotating shaft.

BELLOWS. An accordion-like device that expands and contracts when internal pressure changes.

BELLOWS SEAL. A method of sealing a rotating shaft or valve stem that allows rotary movement of the shaft or stem without leaking.

BENDING SPRING. A coil spring that can be fitted inside or outside a piece of tubing to prevent its walls from collapsing when being formed.

BIMETAL. Two dissimilar metals fastened together to create a distortion of the assembly with temperature changes.

BIMETAL STRIP. Two dissimilar metal strips fastened back to back.

BLEEDING. Allowing pressure to move from one pressure level to another very slowly.

BLEED VALVE. A valve with a small port usually used to bleed pressure from a vessel to the atmosphere.

BOILER. A container in which a liquid may be heated using any heat source. When the liquid is heated to the point that vapor forms and is used as the circulating medium, it is called a steam boiler.

BOILING POINT. The temperature level of a liquid at which it begins to change to a vapor. The boiling temperature is controlled by the vapor pressure above the liquid.

BORE. The inside diameter of a cylinder.

BOURDON TUBE. C-shaped tube manufactured of thin metal and closed on one end. When pressure is increased inside, it tends to straighten. It is used in a gage to indicate pressure.

BRAZING. High-temperature (above 800°F) soldering of two metals.

BREAKER. A heat-activated electrical device used to open an electrical circuit to protect it from excessive current flow.

BRITISH THERMAL UNIT. The amount (quantity) of heat required to raise the temperature of 1 lb of water 1°F.

Btu. British thermal unit.

BULB, SENSOR. The part of a sealed automatic control used to sense temperature.

BURNER. A device used to prepare and burn fuel.

BUTANE GAS. A liquified petroleum gas burned for heat.

CALIBRATE. To adjust instruments or gages to the correct setting for conditions.

CAPACITANCE. The term used to describe the electrical storage ability of a capacitor.

CAPACITOR. An electrical storage device used to start motors (start capacitor) and to improve the efficiency of motors (run capacitor).

CAPACITY. The rating system of equipment used to heat or cool substances.

CARBON MONOXIDE. A poisonous, colorless, odorless tasteless gas generated by incomplete combustion.

CAVITATION. A vapor formed due to a drop in pressure in a pumping system.

CELSIUS SCALE. A temperature scale with 100 graduations between water freezing (0°C) and water boiling (100°C).

CENTIGRADE SCALE. See Celsius.

CENTRIFUGAL COMPRESSOR. A compressor used for large refrigeration systems. It is not positive displacement but it is similar to a blower.

CHANGE OF STATE. The condition that occurs when a substance changes from one physical state to another, such as ice to water and water to steam.

CHARGE. The quantity of refrigerant in a system.

CHECK VALVE. A device that permits fluid flow in one direction only.

CHILL FACTOR. A factor or number that is a combination of temperature, humidity, and wind velocity that is used to compare a relative condition to a known condition.

CHIMNEY. A vertical shaft used to convey flue gases above the rooftop.

CHIMNEY EFFECT. A term used to describe air or gas when it expands and rises when heated.

CIRCUIT. An electron or fluid-flow path that makes a complete loop.

CIRCUIT BREAKER. A device that opens an electric circuit when an overload occurs.

CLEARANCE VOLUME. The volume at the top of the stroke in a compressor cylinder between the top of the piston and the valve plate.

CLOSED CIRCUIT. A complete path for electrons to flow on.

CLOSED LOOP. Piping circuit that is complete and not open to the atmosphere.

CODE. The local, state, or national rules that govern safe installation and service of systems and equipment for the purpose of safety of the public and trade personnel.

COEFFICIENT OF PERFORMANCE. (COP): The ratio of usable output energy divided by input energy.

CO$_2$ INDICATOR. An instrument used to detect the quantity of carbon dioxide in flue gas for efficiency purposes.

COLD. The word used to describe heat at lower levels of intensity.

COLD JUNCTION. The opposite junction to the hot junction in a thermocouple.

COLD WALL. The term used in comfort heating to describe a cold outside wall and its effect on human comfort.

COMFORT CHART. A chart used to compare the relative comfort of one temperature and humidity condition to another condition.

COMPOUND GAGE. A gage used to measure the pressure above and below the atmosphere's standard pressure. It is a Bourdon tube sensing device and can be found on all gage manifolds used for air conditioning and refrigeration service work.

COMPRESSION. A term used to describe a vapor when pressure is applied and the molecules are compacted closer together.

COMPRESSION RATIO. A term used with compressors to describe the actual difference in the low- and high-pressure sides of the compression cycle. It is absolute discharge pressure divided by absolute suction pressure.

COMPRESSOR. A vapor pump that pumps vapor (refrigerant or air) from one pressure level to a higher pressure level.

COMPRESSOR DISPLACEMENT. The internal volume of a compressor, used to calculate the pumping capacity of the compressor.

COMPRESSOR SHAFT SEAL. The seal that prevents refrigerant inside the compressor from leaking around the rotating shaft.

CONDENSATE. The moisture collected on an evaporator coil.

CONDENSATE PUMP. A small pump used to pump condensate to a higher level.

CONDENSATION. Liquid formed when a vapor condenses.

CONDENSE. Changing a vapor to a liquid at a particular pressure.

CONDENSER. The component in a refrigeration system that transfers heat from the system by condensing refrigerant.

CONDENSING PRESSURE. The pressure that corresponds to the condensing temperature in a refrigeration system.

CONDENSING UNIT. A complete unit that includes the compressor and the condensing coil.

CONDUCTIVITY. The ability of a substance to conduct electricity or heat.

CONDUCTOR. A path for electrical energy to flow on.

CONTAMINANT. Any substance in a refrigeration system that is foreign to the system, particularly if it causes damage.

CONTROL. A device to stop, start, or modulate flow of electricity or fluid to maintain a preset condition.

CONTROL SYSTEM. A network of controls to maintain desired conditions in a system or space.

CONVECTION. Heat transfer from one place to another using a fluid.

CONVERSION FACTOR. A number used to convert from one equivalent value to another.

COOLER. A walk-in or reach-in refrigerated box.

COOLING TOWER. The final device in many water-cooled systems, which rejects heat from the system into the atmosphere by evaporation of water.

COPPER PLATING. Small amounts of copper are removed by electrolysis and deposited on the ferrous metal parts in a compressor.

CORROSION. A chemical action that eats into or wears away material from a substance.

COUNTER EMF. Voltage generated or induced above the applied voltage in a single-phase motor.

COUNTERFLOW. Two fluids flowing in opposite directions.

COUPLING. A device for joining two fluid-flow lines. Also the device connecting a motor drive shaft to the driven shaft in a direct-drive system.

CRACKAGE. Small spaces in a structure that allow air to infiltrate the structure.

CRANKSHAFT SEAL. Same as the compressor shaft seal.

CRANKSHAFT THROW. The off-center portion of a crankshaft that changes rotating motion to recipro-cating motion.

CRISPER. A refrigerated compartment that maintains a high humidity and a low temperature.

CROSS CHARGE. A control with a sealed bulb that con-tains two different fluids that work together for a common specific condition.

CURRENT, ELECTRICAL. Electrons flowing along a con-ductor.

CURRENT RELAY. An electrical device activated by a change in current flow.

CUT-IN AND CUT-OUT. The two points at which a con-trol opens or closes its contacts based on the condi-tion it is supposed to maintain.

CYCLE. A complete sequence of events (from start to finish) in a system.

CYLINDER. A circular container with straight sides used to contain fluids or to contain the compression process (the piston movement) in a compressor.

CYLINDER, COMPRESSOR. The part of the compressor that contains the piston and its travel.

CYLINDER HEAD, COMPRESSOR. The top to the cylinder on the high-pressure side of the compressor.

CYLINDER, REFRIGERANT. The container that holds refrigerant.

DAMPER. A component in an air distribution system that restricts airflow for the purpose of air balance.

DEFROST. Melting of ice.

DEFROST CYCLE. The portion of the refrigeration cycle that melts the ice off the evaporator.

DEFROST TIMER. A timer used to start and stop the defrost cycle.

DEGREASER. A cleaning solution used to remove grease from parts and coils.

DEHUMIDIFY. To remove moisture from air.

DEHYDRATE. To remove moisture from a sealed sys-tem or a product.

DENSITY. The weight per unit of volume of a sub stance.

DESICCANT. Substance in a refrigeration system drier that collects moisture.

DESIGN PRESSURE. The pressure at which the system is designed to operate under normal conditions.

DETECTOR. A device to search and find.

DEW. Moisture droplets that form on a cool surface.

DEW POINT. The exact temperature at which moisture begins to form.

DIAPHRAGM. A thin flexible material (metal, rubber, or plastic) that separates two pressure differences.

DIFFERENTIAL. The difference in the cut-in and cut-out points of a control, pressure, time, temperature, or level.

DIFFUSER. The terminal or end device in an air dis-tribution system that directs air in a specific direc-tion using louvers.

DIRECT CURRENT. Electricity in which all electron flow is continuously in one direction.

DIRECT EXPANSION. The term used to describe an evaporator with an expansion device other than a low-side float type.

DOUBLE FLARE. A connection used on copper, alumi-num, or steel tubing that folds tubing wall to a dou-ble thickness.

DOWEL PIN. A pin, that may or may not be tapered, used to align and fasten two parts.

DRAFT GAGE. A gage used to measure very small pres-sures (above and below atmospheric) and compare them to the atmosphere's pressure. Used to deter-mine the flow of flue gas in a chimney or vent.

DRIER. A device used in a refrigerant line to remove moisture.

DRIP PAN. A pan shaped to collect moisture condensing on an evaporator coil in an air conditioning or refrigeration system.

DRY-BULB TEMPERATURE. The temperature measured using a plain thermometer.

DUCT. A sealed channel used to convey air from the system to and from the point of utilization.

ECCENTRIC. An off-center device that rotates in a circle around a shaft.

EFFECTIVE TEMPERATURE. Different combinations of temperature and humidity that provide the same comfort level.

ELECTRIC HEAT. The process of converting electrical energy, using resistance, into heat.

ELECTROMAGNET. A coil of wire wrapped around a soft iron core that creates a magnet.

ELECTRON. The smallest portion of an atom that carries a negative charge.

ELECTRONIC AIR FILTER. A filter that charges dust particles using high-voltage direct current and then collects these particles on a plate of an opposite charge.

ELECTRONIC LEAK DETECTOR. An instrument used to detect gases in very small portions by using electronic sensors and circuits.

ELECTRONICS. The use of electron flow in conductors, semiconductors, and other devices.

END BELL. The end structure of an electric motor that normally contains the bearings and lubrication system.

END PLAY. The amount of lateral travel in a motor or pump shaft.

ENERGY. The capacity for doing work.

ENVIRONMENT. Our surroundings, including the atmosphere.

ETHANE GAS. The fossil fuel, natural gas, used for heat.

EVACUATION. The removal of any gases not characteristic to a system or vessel.

EVAPORATION. The condition that occurs when heat is absorbed by liquid and it changes to vapor.

EVAPORATOR. The component in a refrigeration system that absorbs heat into the system and evaporates the liquid refrigerant.

EVAPORATOR FAN. A forced convector used to improve the efficiency of an evaporator by air movement over the coil.

EVAPORATOR TYPES. Flooded—an evaporator where the liquid refrigerant level is maintained to the top of the heat exchange coil. Dry type—an evaporator coil that achieves the heat exchange process with a minimum of refrigerant charge.

EXHAUST VALVE. The movable component in a refrigeration compressor that allows hot gas to flow to the condenser and prevents it from refilling the cylinder on the downstroke.

EXPANSION JOINT. A flexible portion of a piping system or building structure that allows for expansion of the materials due to temperature changes.

EXTERNAL DRIVE. An external type of compressor motor drive, as opposed to a hermetic compressor.

EXTERNAL EQUALIZER. The connection from the evaporator outlet to the bottom of the diaphragm on a thermostatic expansion valve.

FAHRENHEIT SCALE. The temperature scale that places the boiling point of water at 212°F and the freezing point at 32°F.

FAN. A device that produces a pressure difference in air in order to move it.

FARAD. The unit of capacity of a capacitor. Capacitors in our industry are rated in microfarads.

FEMALE THREAD. The internal thread in a fitting.

FILTER. A fine mesh or porous material that removes particles from passing fluids.

FIN COMB. A hand tool used to straighten the fins on an air-cooled condenser.

FLAPPER VALVE. See reed valve.

FLARE. The angle that may be fashioned at the end of a piece of tubing to match a fitting and create a leak-free connection.

FLARE NUT. A connector used in a flare assembly for tubing.

FLASH GAS. The term used to describe the pressure drop in an expansion device when some of the liquid passing through the valve is changed quickly to a gas and cools the remaining liquid to the corresponding temperature.

FLOAT, VALVE OR SWITCH. An assembly used to maintain or monitor a liquid level.

FLOODED SYSTEM. A refrigeration system operated with the liquid refrigerant level very close to the outlet of the coil for improved heat exchange.

FLOODING.   The term applied to a refrigeration system when the liquid refrigerant reaches the compressor.

FLUE.   The duct that carries the products of combustion out of a structure for a fossil- or a solid-fuel system.

FLUID.   The state of matter of liquids and gases.

FLUSH.   The process of using a fluid to push contaminants from a system.

FLUX.   A substance applied to soldered and brazed connections to prevent oxidation during the heating process.

FOAMING.   A term used to describe oil when it has liquid refrigerant boiling out of it.

FOOT-POUND.   The amout of work accomplished by lifting 1 lb of weight 1 ft; a unit of energy.

FORCE.   Energy exerted.

FORCED CONVECTION.   The movement of fluid by mechanical means.

FREEZER BURN.   The term applied to frozen food when it becomes dry and hard from dehydration due to poor packaging.

FREEZE UP.   Excess ice or frost accumulation on an evaporator to the point that airflow may be affected.

FREEZING.   The change of state of water from a liquid to a solid.

FREON.   The trade name for refrigerants manufactured by E. I. du Pont de Nemours & Co., Inc.

FROST BACK.   A condition of frost on the suction line and even the compressor body usually due to liquid refrigerant in the suction line.

FROZEN.   The term used to describe water in the solid state, also used to describe a rotating shaft that will not turn.

FUEL OIL.   The fossil fuel used for heating; a petroleum distillate.

FURNACE.   Equipment used to convert heating energy, such as fuel oil, gas, or electricity, to usable heat. It usually contains a heat exchanger, a blower, and the controls to operate the system.

FUSE.   A safety device used in electrical circuits for the protection of the circuit conductor and components.

FUSIBLE LINK.   An electrical safety device normally located in a furnace that burns and opens the circuit during an overheat situation.

FUSIBLE PLUG.   A device (made of low-melting-temperature metal) used in pressure vessels that is sensitive to low temperatures and relieves the vessel contents in an overheating situation.

GAGE.   An instrument used to detect pressure.

GAGE MANIFOLD.   A tool that may have more than one gage with a valve arrangement to control fluid flow.

GAGE PORT.   The service port used to attach a gage for service procedures.

GAS.   The vapor state of matter.

GAS VALVE.   A valve used to stop, start, or modulate the flow of natural gas.

GASKET.   A thin piece of flexible material used between two metal plates to prevent leakage.

GRAIN.   Unit of measure. One pound = 7000 grains.

GRILLE.   A louvered, often decorative, component in an air system at the inlet or the outlet of the airflow.

GROMMET.   A rubber, plastic, or metal protector usually used where wire or pipe goes through a metal panel.

GROUND, ELECTRICAL.   A circuit or path for electron flow to the earth ground.

GROUND WIRE.   A wire from the frame of an electrical device to be wired to the earth ground.

HALIDE REFRIGERANTS.   Refrigerants that contain halogen chemicals, R-12, R-22, R-500, and R-502 are among them.

HALIDE TORCH.   A torch-type leak detector used to detect the halogen refrigerants.

HALOGENS.   Chemical substances found in many refrigerants containing chlorine, bromine, iodine, and fluorine.

HANGER.   A device used to support tubing, pipe, duct, or other components of a system.

HEAD.   Another term for pressure, usually referring to gas or liquid.

HEAD PRESSURE CONTROL.   A control that regulates the head pressure in a refrigeration or air conditioning system.

HEADER.   A pipe or containment to which other pipe lines are connected.

HEAT.   Energy that causes molecules to be in motion and to raise the temperature of a substance.

HEAT EXCHANGER.   A device that transfers heat from one substance to another.

HEAT OF COMPRESSION.   That part of the energy from the pressurization of a gas or a liquid converted to heat.

HEAT OF FUSION.   The heat released when a substance is changing from a liquid to a solid.

HEAT OF RESPIRATION. When oxygen and carbon hydrates are taken in by a substance or when carbon dioxide and water are given off. Associated with fresh fruits and vegetables during their aging process while stored.

HEAT PUMP. A refrigeration system used to supply heat or cooling using valves to reverse the refrigerant gas flow.

HEAT SINK. A low-temperature surface to which heat can transfer.

HEAT TRANSFER. The transfer of heat from a warmer to a colder substance.

HEATING COIL. A device made of tubing or pipe designed to transfer heat to a cooler substance by using fluids.

HERMETIC SYSTEM. A totally enclosed refrigeration system where the motor and compressor are sealed within the same system with the refrigerant.

HERTZ. Cycles per second.

Hg. Abbreviation for the element mercury.

HIGH-PRESSURE CUT-OUT. A control that stops a boiler heating device or a compressor when the pressure becomes too high.

HIGH SIDE. A term used to indicate the high-pressure or condensing side of the refrigeration system.

HIGH-VACUUM PUMP. A pump that can produce a vacuum in the low micron range.

HORSEPOWER. A unit equal to 33,000 ft-lb of work per minute.

HOT-GAS BYPASS. Piping that allows hot refrigerant gas into the cooler low-pressure side of a refrigeration system usually for system capacity control.

HOT-GAS DEFROST. A system where the hot refrigerant gases are passed through the evaporator to defrost it.

HOT JUNCTION. That part of a thermocouple or thermopile where heat is applied.

HOT-WATER HEAT. A heating system using hot water to distribute the heat.

HOT WIRE. The wire in an electrical circuit that has a voltage potential between it and another electrical source or between it and ground.

HUMIDIFIER. A device used to add moisture to the air.

HUMIDISTAT. A control operated by a change in humidity.

HUMIDITY. Moisture in the air.

HYDRAULICS. Producing mechanical motion by using liquids under pressure.

HYDROCARBONS. Organic compounds containing hydrogen and carbon found in many heating fuels.

HYDROMETER. An instrument used to measure the specific gravity of a liquid.

HYDRONIC. Usually refers to a hot-water heating system.

HYGROMETER. An instrument used to measure the amount of moisture in the air.

IDLER. A pulley on which a belt rides. It does not transfer power but is used to provide tension or reduce vibration.

IGNITION TRANSFORMER. Provides a high-voltage current, usually to produce a spark to ignite a furnace fuel, either gas or oil.

IMPEDANCE. A form of resistance in an alternating current circuit.

IMPELLER. The rotating part of a pump that causes the centrifugal force to develop fluid flow and pressure difference.

INDUCED MAGNETISM. Magnetism produced, usually in a metal, from another magnetic field.

INDUCTION MOTOR. An alternating current motor where the rotor turns from induced magnetism from the field windings.

INDUCTIVE REACTANCE. A resistance to the flow of an alternating current produced by an electromagnetic induction.

INFILTRATION. Air that leaks into a structure through cracks, windows, doors, or other openings due to less pressure inside the structure than outside the structure.

INSULATION, ELECTRIC. A substance that is a poor conductor of electricity.

INSULATION, THERMAL. A substance that is a poor conductor of the flow of heat.

JUNCTION BOX. A metal or plastic box within which electrical connections are made.

KELVIN. A temperature scale where absolute 0 equals 0 or where molecular motion stops at 0. It has the same graduations per degree of change as the Celsius scale.

KILOWATT. A unit of electrical power equal to 1000 watts.

KILOWATT-HOUR. 1 kilowatt (1000 watts) of energy used for 1 hour.

KING VALVE. A service valve at the liquid receiver.

LATENT HEAT. Heat energy absorbed or rejected when a substance is changing state and there is no change in temperature.

LEAK DETECTOR. Any device used to detect leaks in a pressurized system.

LIMIT CONTROL. A control used to make a change in a system, usually to stop it when predetermined limits of pressure or temperature are reached.

LIQUID. A substance where molecules push outward and downward and seek a uniform level.

LIQUID LINE. A term applied in the industry to refer to the tubing or piping from the condenser to the expansion device.

LIQUID NITROGEN. Nitrogen in liquid form.

LIQUID RECEIVER. A container in the refrigeration system where liquid refrigerant is stored.

LOW SIDE. A term used to refer to that part of the refrigeration system that operates at the lowest pressure, between the expansion device and the compressor.

LP FUEL. Liquified petroleum. A substance used as a gas for fuel. It is transported and stored in the liquid state.

MAGNETIC FIELD. A field or space where magnetic lines of force exist.

MAGNETISM. A force causing a magnetic field to attract ferrous metals, or where like poles of a magnet repel and unlike poles attract each other.

MALE THREAD. A thread on the outside of a pipe, fitting, or cylinder; an external thread.

MANOMETER. An instrument used to check low vapor pressures. The pressures may be checked against a column of mercury or water.

MASS. Matter held together to the extent that it is considered one body.

MEGOHM. A measure of electrical resistance equal to 1,000,000 ohms.

MELTING POINT. The temperature at which a substance will change from a solid to a liquid.

MICRO. A prefix meaning 1/1,000,000.

MICROFARAD. Capacitor capacity equal to 1/1,000,000 of a farad.

MICROMETER. A precision measuring instrument.

MICRON. A unit of length equal to 1/1000 of a millimeter, 1/1,000,000 of a meter.

MICRON GAGE. A gage used when it is necessary to measure pressures close to a perfect vacuum.

MILLI. A prefix meaning 1/1000.

MODULATOR. A device that adjusts by small increments or changes.

MOISTURE INDICATER. A device for determining moisture in a refrigerant.

MOLECULE. The smallest particle that a substance can be broken into and still retain its chemical identity.

MONOCHLORODIFLUOROMETHANE. The refrigerant R-22.

MOTOR STARTER. Electromagnetic contactors that contain motor protection and are used for switching electric motors on and off.

MUFFLER, COMPRESSOR. Sound absorber at the compressor.

MULLION. Stationary frame between two doors.

MULLION HEATER. Heating element mounted in mullion of a refrigerator to keep moisture from forming on it.

NATURAL CONVECTION. The natural movement of a gas or fluid caused by differences in temperature.

NEEDLEPOINT VALVE. A device having a needle and a very small orifice for controlling the flow of a fluid.

NEOPRENE. Synthetic flexible material used for gaskets and seals.

NEUTRALIZER. A substance used to counteract acids.

NOMINAL. A rounded off stated size. The nominal size is the closest rounded off size.

NONCONDENSABLE GAS. A gas that does not change into a liquid under normal operating conditions.

NONFERROUS. Metals containing no iron.

NORTH POLE, MAGNETIC. One end of a magnet.

OFF CYCLE. A period when a system is not operating.

OHM. A unit of measurement of electrical resistance.

OHMMETER. A meter that measures electrical resistance.

OHM'S LAW. A law involving electrical relationships discovered by Georg Ohm: $E = I \times R$.

OIL, REFRIGERATION. Oil used in refrigeration systems.

OIL SEPARATOR. Apparatus that removes oil from a gaseous refrigerant.

OPEN COMPRESSOR. A compressor with an external drive.

OPERATING PRESSURE. The actual pressure under operating conditions.

ORGANIC. Materials formed from living organisms.

ORIFICE. A small opening through which fluid flows.

OVERLOAD PROTECTION. A system or device that will shut down a system if an overcurrent condition exists.

PACKAGE UNIT. A refrigerating system where all major components are located in one cabinet.

PACKING. A soft material that can be shaped and compressed to provide a seal. It is commonly applied around valve stems.

PARALLEL CIRCUIT. An electrical or fluid circuit where the current or fluid takes more than one path at a junction.

PERMANENT MAGNET. An object that has its own permanent magnetic field.

PHASE. One distinct part of a cycle.

PISTON. The part that moves up and down in a cylinder.

PISTON DISPLACEMENT. The volume within the cylinder that is displaced with the movement of the piston from top to bottom.

PITOT TUBE. Part of an instrument for measuring air velocities.

PLENUM. A sealed chamber at the inlet or outlet of an air handler. The duct attaches to the plenum.

POLYPHASE. Three or more phases.

PORCELAIN. A ceramic material.

POTENTIOMETER. An instrument that controls electrical current.

POWER. The rate at which work is done.

PRESSURE DROP. The difference in pressure between two points.

PRESSURE LIMITER. A device that opens when a certain pressure is reached.

PRESSURE REGULATOR. A valve capable of maintaining a constant outlet pressure when a variable inlet pressure occurs. Used for regulating fluid flow such as natural gas, refrigerant, and water.

PRESSURE SWITCH. A switch operated by a change in pressure.

PRIMARY CONTROL. Controlling device for an oil burner to insure ignition within a specific time span, usually 90 sec.

PROPANE. An LP gas used for heat.

PROTON. That part of an atom having a positive charge.

psi. Abbreviation for pounds per square inch.

psia. Abbreviation for pounds per square inch absolute.

psig. Abbreviation for pounds per square inch gage.

PSYCHROMETER. An instrument for determining relative humidity.

PSYCHROMETRIC CHART. A chart that shows the relationship of temperature, pressure, and humidity in the air.

PUMP. A device that forces fluids through a system.

PUMP DOWN. To use a compressor to pump the refrigerant charge into the condenser and/or receiver.

PURGE. To remove or release fluid from a system.

QUENCH. To submerge a hot object in a fluid for cooling.

QUICK-CONNECT COUPLING. A device designed for easy connecting or disconnecting of fluid lines.

R-12. Dichlorodifluoromethane, a popular refrigerant for refrigeration systems.

R-22. Monochlorodifluoromethane, a popular refrigerant for air conditioning systems.

R-502. An azeotropic mixture of R-22 and R-115, a popular refrigerant for low-temperature refrigeration systems.

RADIANT HEAT. Heat that passes through air heating solid objects that in turn heat the surrounding area.

RADIATION. Heat transfer. See radiant heat.

RANKINE. The absolute Fahrenheit scale with 0 at the point where all molecular motion stops.

REACTANCE. A type of resistance in an alternating current circuit.

RECEIVER-DRIER. A component in a refrigeration system for storing and drying refrigerant.

RECIPROCATING. Back-and-forth motion.

RECTIFIER. A device for changing alternating current to direct current.

REED VALVE. A thin steel plate used as a valve in a compressor.

REFRIGERANT. The fluid in a refrigeration system that changes from a liquid to a vapor and back to a liquid at practical pressures.

REGISTER. A terminal device on an air distribution system that directs air but also has a damper to adjust airflow.

RELATIVE HUMIDITY. The amount of moisture contained in the air as compared to the amount the air could hold at that temperature.

RELAY. A small electromagnetic device to control a switch, motor, or valve.

RELIEF VALVE. A valve designed to open and release liquids at a certain pressure.

REMOTE SYSTEM. Often called a split system where the condenser is located away from the evaporator and/or other parts of the system.

RESISTANCE. The opposition to the flow of an electrical current or a fluid.

RESISTOR. An electrical or electronic component with a specific opposition to electron flow. It is used to create voltage drop or heat.

RESTRICTOR. A device used to create a planned resistance to fluid flow.

REVERSE CYCLE. The ability to direct the hot-gas flow into the indoor or the outdoor coil in a heat pump to control the system for heating or cooling purposes.

ROTARY COMPRESSOR. A compressor that uses rotary motion to pump fluids. It is a positive displacement pump.

ROTOR. The rotating or moving component of a motor, including the shaft.

RUNNING TIME. The time a unit operates, also called the on time.

RUN WINDING. The electrical winding in a motor that draws current during the entire running cycle.

SADDLE VALVE. A valve that straddles a fluid line and is fastened by solder or screws. It normally contains a device to puncture the line for pressure readings.

SAFETY CONTROL. An electrical, mechanical, or electromechanical control to protect the equipment or public from harm.

SAFETY PLUG. A fusible plug.

SATURATION. A term used to describe a substance when it contains all of another substance it can hold.

SCAVENGER PUMP. A pump used to remove the fluid from a sump.

SCHRADER VALVE. A valve similar to the valve on an auto tire that allows refrigerant to be charged or discharged from the system.

SCOTCH YOKE. A mechanism used to create reciprocating motion from the electric motor drive in very small compressors.

SCREW COMPRESSOR. A form of positive displacement compressor that squeezes fluid from a low-pressure area to a high-pressure area, using screw-type mechanisms.

SEALED UNIT. The term used to describe a refrigeration system, including the compressor, that is completely welded closed. The pressures can be accessed by saddle valves.

SEAT. The stationary part of a valve that the moving part of the valve presses against for shutoff.

SEMICONDUCTOR. A component in an electronic system that is considered neither an insulator nor a conductor but a partial conductor.

SEMIHERMETIC COMPRESSOR. A motor compressor that can be opened or disassembled by removing bolts and flanges; also known as a serviceable hermetic.

SENSIBLE HEAT. Heat that causes a change in the level of a thermometer.

SENSOR. A component for detection that changes shape, form or resistance when a condition changes.

SEQUENCER. A control that causes a staging of events, such as a sequencer between stages of electric heat.

SERIES CIRCUIT. An electrical or piping circuit where all of the current or fluid flows through the entire circuit.

SERVICE VALVE. A manually operated valve in a refrigeration system used for various service procedures.

SERVICEABLE HERMETIC. See semihermetic compressor.

SHADED-POLE MOTOR. An AC motor used for very light loads.

SHELL AND COIL. A vessel with a coil of tubing inside that is used as a heat exchanger.

SHELL AND TUBE. A heat exchanger with straight tubes in a shell that can normally be mechanically cleaned.

SHORT CIRCUIT. A circuit that does not have the correct measurable resistance; too much current flows and will overload the conductors.

SHORT CYCLE. The term used to describe the running time (on time) of a unit when it is not running long enough.

SHROUD. A fan housing that insures maximum airflow through the coil.

SIGHT GLASS. A clear window in a fluid line.

SILICA GEL. A chemical compound often used in refrigerant driers to remove moisture from the refrigerant.

SILVER BRAZING A high-temperature (above 800°F) brazing process for bonding metals.

SINE WAVE. The graph or curve used to describe the characteristics of AC voltage.

SINGLE PHASE. The electrical power supplied to equipment or small motors, normally under $7\frac{1}{2}$ hp.

SLING PSYCHROMETER. A device with two thermometers, one a wet bulb and one a dry bulb, used for checking air conditions, temperature and humidity.

SLUGGING. A term used to describe the condition when large amounts of liquid enter a pumping compressor cylinder.

SMOKE TEST. A test performed to determine the amount of unburned fuel in an oil-burner flue-gas sample.

SOLAR HEAT. Heat from the sun's rays.

SOLDERING. Fastening two base metals together by using a third, filler metal that melts at a temperature below 800°F.

SPECIFIC GRAVITY. The weight of a substance compared to the weight of an equal volume of water.

SPECIFIC HEAT. The amount of heat required to raise the temperature of 1 lb of a substance 1°F.

SPECIFIC VOLUME. The volume occupied by one pound of a fluid.

SPLASH LUBRICATION SYSTEM. A system of furnishing lubrication to a compressor by agitating the oil.

SPLIT-PHASE MOTOR. A motor with run and start windings.

SPLIT SYSTEM. A refrigeration or air conditioning system that has the condensing unit remote from the indoor (evaporator) coil.

SQUIRREL CAGE FAN. A fan assembly used to move air.

STANDARD ATMOSPHERE, or STANDARD CONDITIONS. Air at sea level at 70°F when the atmosphere's pressure is 14.696 psia (29.92 in. Hg). Air at this condition has a volume of 13.33 ft^3/lb.

STARTING RELAY. An electrical relay used to disconnect the start winding in a hermetic compressor.

STARTING WINDING. The winding in a motor used primarily to give the motor extra starting torque.

STATOR. The component in a motor that contains the windings, it does not turn.

STEAM. The vapor state of water.

STRAINER. A fine-mesh device that allows fluid flow and holds back solid particles.

STRATIFICATION. The condition where a fluid appears in layers.

SUBCOOLING. The temperature of a liquid when it is cooled below its condensing temperature.

SUBLIMATION. When a substance changes from the solid state to the vapor state without going through the liquid state.

SUCTION LINE. The pipe that carries the heat-laden refrigerant gas from the evaporator to the compressor.

SUCTION SERVICE VALVE. A manually operated valve with front and back seats located at the compressor.

SUPERHEAT. The temperature of vapor refrigerant above its saturation change of state temperature.

SWAGING TOOL A tool used to enlarge a piece of tubing for a solder or braze connection.

SWAMP COOLER. A slang term used to describe an evaporative cooler.

SWEATING. A word used to describe moisture collection on a line or coil that is operating below the dew-point temperature of the air.

TANK. A closed vessel used to contain a fluid.

TAP. A tool used to cut internal threads in a fastener or fitting.

TEMPERATURE. A word used to describe the level of heat or molecular activity expressed in Fahrenheit, Rankine, Celsius, or Kelvin.

TEST LIGHT. A light-bulb arrangement used to prove the presence of electrical power in a circuit.

THERM. Quantity of heat, 100,000 Btu.

THERMISTOR. A semiconductor electronic device that changes resistance with a change in temperature.

THERMOCOUPLE. A device made of two unlike metals that generates electricity when there is a difference in temperature from one end to the other. Thermocouples have a hot and cold junction.

THERMOMETER. An instrument used to detect differences in the level of heat.

THERMOPILE. A group of thermocouples connected in series to increase voltage output.

THERMOSTAT. A device that senses temperature change and changes some dimension or condition within to control an operating device.

THERMOSTATIC EXPANSION VALVE. A valve used in refrigeration systems to control the superheat in an evaporator by metering the correct refrigerant flow to the evaporator.

THREE-PHASE POWER. A type of power supply usually used for operating heavy loads. It consists of 3 sine waves that are out of phase with each other.

THROTTLING. Creating a restriction in a fluid line.

TIMERS. Clock-operated devices used to time various sequences of events in circuits.

TON OF REFRIGERATION. The amount of heat required to melt a ton (2000 lb) of ice at 32°F, 288,000 Btu/24 hr, 12,000 Btu/hr, or 200 Btu/min.

TORQUE. The twisting force often applied to the starting power of a motor.

TORQUE WRENCH. A wrench used to apply a prescribed amount of torque or tightening to a connector.

TOTAL HEAT. The total amount of sensible heat and latent heat contained in a substance from a reference point.

TRANSFORMER. A coil of wire wrapped around an iron core that induces a current to another coil of wire wrapped around the same iron core. Note: A transformer can have an air coil.

TUBE WITHIN A TUBE COIL. A coil used for heat transfer that has a pipe in a pipe and is fastened together so that the outer tube becomes one circuit and the inner tube another.

TUBING. Pipe with a thin wall used to carry fluids.

TWO-TEMPERATURE VALVE. A valve used in systems with multiple evaporators to control the evaporator pressures and maintain different temperatures in each evaporator. Sometimes called a hold-back valve.

URETHANE FOAM. A foam that can be applied between two walls for insulation.

VACUUM. The pressure range between the earth's atmosphere and no pressure, normally expressed in inches of mercury (in. Hg) vacuum.

VACUUM PUMP. A pump used to remove some fluids such as air and moisture from a system at a pressure below the earth's atmosphere.

VALVE. A device used to control fluid flow.

VALVE PLATE. A plate of steel bolted between the head and the body of a compressor that contains the suction and discharge reed or flapper valves.

VAPOR. The gaseous state of a substance.

VAPOR BARRIER. A thin film used in construction to keep moisture from migrating through building materials.

VAPOR LOCK. A condition where vapor is trapped in a liquid line and impedes liquid flow.

VAPORIZATION. The changing of a liquid to a gas or vapor.

VARIABLE PITCH PULLEY. A pulley whose diameter can be adjusted.

V-BELT. A belt that has a V-shaped contact surface and is used to drive compressors, fans, or pumps.

VELOCITY METER. A meter used to detect the velocity of fluids, air, or water.

VELOCITY. The speed at which a substance passes a point.

VOLTAGE. The potential electrical difference for electron flow from one line to another in an electrical circuit.

VOLTMETER. An instrument used for checking electrical potential.

VOLUMETRIC EFFICIENCY. The pumping efficiency of a compressor or vacuum pump that describes the pumping capacity in relationship to the actual volume of the pump.

WALK-IN COOLER. A large refrigerated space used for storage of refrigerated products.

WATER-COOLED CONDENSER. A condenser used to reject heat from a refrigeration system into water.

WATT. A unit of power applied to electron flow. One watt equals 3.414 Btu.

WATT-HOUR. The unit of power that takes into consideration the time of consumption. It is the equivalent of a 1-W bulb burning for 1 hr.

WET-BULB TEMPERATURE. The wet-bulb temperature of air is used to evaluate the humidity in the air. It is obtained with a wet thermometer bulb to record the evaporation rate with an airstream passing over the bulb to help in evaporation.

WET HEAT. A heating system using steam or hot water as the heating medium.

WINDOW UNIT. An air conditioner installed in a window that rejects the heat outside the structure.